HANDBOOK OF ANTENNAS IN WIRELESS COMMUNICATIONS

THE ELECTRICAL ENGINEERING AND APPLIED SIGNAL PROCESSING SERIES

Edited by Alexander Poularikas

The Advanced Signal Processing Handbook: Theory and Implementation for Radar, Sonar, and Medical Imaging Real-Time Systems
Stergios Stergiopoulos

The Transform and Data Compression Handbook
K.R. Rao and P.C. Yip

Handbook of Multisensor Data Fusion
David Hall and James Llinas

Handbook of Antennas in Wireless Communications
Lal Chand Godara

Forthcoming Titles

Propagation Data Handbook for Wireless Communications
Robert Crane

The Digital Color Imaging Handbook
Guarav Sharma

Handbook of Neural Network Signal Processing
Yu Hen Hu and Jeng-Neng Hwang

Applications in Time Frequency Signal Processing
Antonia Papandreou-Suppappola

Noise Reduction in Speech Applications
Gillian Davis

Signal Processing in Noise
Vyacheslav Tuzlukov

Electromagnetic Radiation and the Human Body: Effects, Diagnosis and Therapeutic Technologies
Nikolaos Uzunoglu and Konstantina S. Nikita

Digital Signal Processing with Examples in MATLAB®
Samuel Stearns

HANDBOOK OF ANTENNAS IN WIRELESS COMMUNICATIONS

Edited by

LAL CHAND GODARA

CRC Press
Boca Raton London New York Washington, D.C.

Library of Congress Cataloging-in-Publication Data

Catalog record is available from the Library of Congress

Direct all inquiries to CRC Press LLC, 2000 N.W. Corporate Blvd., Boca Raton, Florida 33431.

Visit the CRC Press Web site at www.crcpress.com

International Standard Book Number 0-8493-0124-6
Printed in the United States of America 1 2 3 4 5 6 7 8 9 0
Printed on acid-free paper

Preface

I authored a two-part article for the Proceedings of the Institute of Electrical and Electronics Engineers (IEEE) on the application of antenna arrays to mobile communications in 1997. It provided the current state of antenna array research and described how an array of antennas may be used to help meet the ever-growing demand of increased channel capacity for wireless mobile communications services. The amount and the kind of feedback I received on the subject, particularly from graduate students and practicing engineers, indicated to me that there is a need for a more comprehensive source of this material than a journal article.

One day in late 1998, I received an e-mail from Dr. Alexander D. Poularikas, who coordinates the *Electrical Engineering and Signal Processing* series for CRC Press, inviting me to be the editor of a handbook covering the fundamental developments of this field so that the engineers in practice or the ones who want to start in this area have a good source to guide them.

I accepted his invitation and prepared a list of topics to be covered by the handbook. Because the handbook was meant to be a major reference source on this subject, I invited the leading experts in the field to contribute material on topics of their special interest.

I am very excited about the final outcome and trust that you share my enthusiasm as I briefly describe what the handbook has to offer.

The handbook has successfully brought together every aspect of antennas in wireless communications with 26 chapters filled with the latest research and development results compiled by leading researchers in a manner that is easy to follow. The material has been developed logically, requiring no prerequisite and thus making it extremely useful not only for researchers and practicing engineers as a reference book but also for newcomers as a great source of learning.

It is a unique book covering all facets of antennas for wireless communications providing detailed treatment of cellular systems, antenna design techniques, practical antennas, phased-array technology, theory and implementation of smart antennas, and interaction of EM radiation with the human body. It contains more than 1200 references for the readers to probe further.

The handbook would be useful for

- Practicing electrical engineers, in general, and communication engineers, in particular, as a reference book
- Academics in the area of mobile communications, signal processing, antenna theory, and smart antennas
- Graduate students and researchers in this area
- Antenna designers in general
- Those who are fascinated by the field of mobile communications and smart antennas

The chapters in the book have been selected to provide coverage of different topics. However, some overlap between various chapters has been allowed to provide discussion from a different point of view.

The handbook has been organized into six parts outlined as follows:

A Wireless communication systems and channel characteristics
B Antenna technology and numerical methods
C Antenna developments and practical antennas
D Smart antennas and array theory
E Implementation of smart antenna systems
F Electromagnetic radiation and the human body

Chapters 1 through 4 are devoted to wireless communications systems and channel characterization. Chapter 1, "Cellular Systems," presents cellular fundamentals by describing the working of mobile communications systems and discussing concepts of multiple access schemes, channel reuse, channel allocation and assignments, and handoff and power control. It then briefly describes various popular standards. Chapter 2, "Satellite-Based Mobile Communications," discusses satellite orbital fundamentals and the satellite radiopath, and describes various mobile satellite communications systems. Chapter 3, "Propagation Prediction for Urban Systems," treats the prediction of the average signal strength for a variety of physical parameters and conditions such as range, antenna height, presence of foliage, and terrain; and discusses site specific predictions using ray models. Chapter 4, "Fading Channels," emphasizes fundamental fading manifestations, types of degradation, and methods for mitigating the degradation. It presents examples to mitigate the effects of frequency-selective fading in time division multiple access (TDMA) and code division multiple access (CDMA) systems.

Chapters 5 through 10 provide coverage of antenna technology and numerical methods. Chapter 5 introduces basic antenna parameters and terminology; and discusses commonly used antenna types, impedance matching, feeding arrangements, and available software for antenna analysis and design. Chapter 6 introduces microstrip patch antennas by discussing their general characteristics. It describes various feed techniques and methods to enhance bandwidth of patch antenna and to reduce the size of conductors. Examples of active patch antennas are also included in this chapter. Chapter 7 introduces the finite difference time domain (FDTD) method with emphasis on its applications to printed antenna and antenna arrays. The chapter discusses fundamentals of FDTD, absorbing boundary conditions, and radiation patterns; and presents examples of various microstrip antenna analyses. Chapter 8, "Method of Moments Applied to Antennas," concentrates on the application of integral equations to antenna problems and their solution using the method of moments (MOM). It presents the basic philosophy of MOM and its application to wire antennas, arbitrary metallic structures, and combined metallic and dielectric structures. Chapter 9 introduces genetic algorithms and shows how these may be applied to find good solutions to wireless antenna problems. Chapter 10, "High-Frequency Techniques," presents high-frequency applications for antennas by discussing modern geometric optics, geometric theory of diffraction, physical optics, and physical theory of diffraction.

Chapters 11 through 15 constitute Part C of the handbook and are devoted to antenna developments and practical antennas. Chapter 11 presents development in outdoor and indoor base station antennas in Japan by describing various base station antennas for cellular systems, diversity antennas for macrocellular systems, antennas for micro- and picocellular systems, and personal handy phone system (PHS) base station antennas. Chapter 12, "Handheld Antennas," describes various antennas used for handheld phones and presents a detailed study of meander line antennas for personal wireless communications. Chapters 13 and 14 provide coverage on antenna development for satellite communications; Chapter 13 concentrates on aeronautical and maritime antennas whereas Chapter 14 focuses on fixed and mobile antennas. Chapter 13 presents antennas and tracking systems for International Maritime Satellite (INMARSAT)-A, -B, -C, -F, -M, and -AERO; and antennas for land mobile earth stations and hand-carried terminals. Chapter 14 presents space segment antennas, earth-segment antennas, and gateway antennas for satellite communications; microstrip antennas for fixed and mobile satellite communications; and mobile antennas for receiving direct-broadcast satellite service television (DBS TV) and SATPHONE antenna systems. Chapter 15, "Shaped-Beam Antennas," focuses on shaped dielectric lens antennas and presents design guidelines for these antennas along with the discussion of some practical aspects, focusing on mobile applications.

Part D of this handbook on smart antennas and array theory contains Chapters 16 to 21. Chapter 16 presents basic array theory and pattern synthesis techniques by discussing basic theory of antenna arrays, array weight synthesis techniques, and array geometry consideration for pattern adjustment. Many examples are included in the chapter to emphasize the concepts. Chapter 17, "Electromagnetic Vector Sensors with Beamforming Applications," describes advantages and developments of vector sensors, solves a beamforming problem using these sensors, and compares the results with that of scalar sensors. Chapter 18, "Optimum and Suboptimum Transmit Beamforming," discusses channel characterization and presents beamforming strategies for transmit arrays including beamforming algorithms and robust beamforming methods.

Chapter 19, "Spatial Diversity for Wireless Communications," treats the basic principles of spatial diversity combining and discusses the performance improvement that can be accomplished by a diversity array using various combined techniques. The chapter also presents the results on the effect of branch correlation and mutual coupling. Chapter 20, "Direction-of-Arrival Estimation in Mobile Communication Environments," presents various methods for estimating direction of arrival (DOA) of point sources and tracking of moving sources. A detailed treatment of estimation for the wireless channel is also included in the chapter. Chapter 21, "Blind Channel Identification and Source Separation in Space Division Multiple Access Systems," addresses the problem of discriminating radio sources in the context of cellular mobile wireless digital communications systems. The chapter describes several deterministic as well as stochastic maximum likelihood methods to solve the blind sources separation and channel identification problem.

Chapter 22 through Chapter 24 are devoted to implementation of smart antenna systems. Chapter 22, "Smart Antenna System Architecture and Hardware Implementation," presents an overview of system architecture and implementation and discusses various important design issues. The chapter describes some real-time implemented systems using digital signal processor (DSP) modules. Chapter 23 presents phased-array technology for wireless systems by discussing phased-array antennas for land mobile communications systems, stratospheric communications systems, and satellite communications systems. Chapter 24, "Adaptive Antennas for Global System for Mobile Communications and Time Division Multiple Access (Interim Standard-136) Systems," starts with an overview of these systems and then outlines some of the most important issues to consider when applying adaptive antenna techniques to existing cellular systems. A discussion of some possible system architectures suitable for implementation is presented and issues related to signal-processing algorithms are considered. The chapter presents a detailed simulation of the system and compares the results with those obtained from field trials.

Chapters 25 and 26 are devoted to the final part on electromagnetic radiation and the human body. Chapter 25 mainly deals with the effect on the human body of the radiation characteristics of handheld antennas whereas Chapter 26 concentrates on health hazards of electromagnetic (EM) radiation. Chapter 25, "Electromagnetic Interactions of Handheld Wireless Communication Antennas with the

Human Body," reviews exposure standards for radio-frequency (RF) fields and different types of handheld wireless devices, and describes numerical techniques and experimental methods used to quantify and characterize the interactions of the radiated field with humans. Examples showing the effect of these interactions on the radiation and input impedance characteristics of antennas in handheld devices are presented. Chapter 26, "Safety Aspects of Radio-Frequency Effects in Humans from Communication Devices," considers how guidelines for human exposures to RF are derived, known interactions with human tissues and their measurements, and the evidence for the existence of health effects.

Contributors

Sören Andersson
Ericsson Radio Systems
Stockholm, Sweden

Hiroyuki Arai
Division of Electric and Computer Engineering
Yokohama National University
Yokohama, Japan

Victor Barroso
Instituto Superior Tecnico
Instituto de Sistemas e Robotica
Lisboa, Portugal

Mats Bengtsson
Department of Signals, Sensors and Systems
Royal Institute of Technology
Stockholm, Sweden

Magnus Berg
Ericsson Radio Systems
Stockholm, Sweden

Jennifer T. Bernhard
Department of Electrical and Computer Engineering
University of Illinois at Urbana-Champaign
Urbana, Illinois

Henry L. Bertoni
Department of Electrical & Computer Engineering
Polytechnic University
Brooklyn, New York

Marek E. Bialkowski
School of Computer Science & Electrical Engineering
University of Queensland
Brisbane, Queensland, Australia

Christos Christodoulou
University of New Mexico
Albuquerque, New Mexico

Henrik Dam
Ericsson LMD
Copenhagen, Denmark

Paul W. Davis
School of Computer Science and Electrical Engineering
University of Queensland
St. Lucia, Queensland, Australia

Antonije R. Djordjevic
School of Electrical Engineering
University of Belgrade
Belgrade, Yugoslavia

Atef Z. Elsherbeni
Electrical Engineering Department
University of Mississippi
University, Mississippi

Meng Hwa Er
Nanyang Technological University
School of Electrical and Electronic Engineering
Singapore, Republic of Singapore

Carlos A. Cardoso Fernandes
Instituto Superior Técnico
Instituto de Telecomunicações
Lisboa, Portugal

Ulf Forssén
Ericsson Radio Systems
Stockholm, Sweden

Lal C. Godara
School of Electrical Engineering
University College, University of New South Wales
Australian Defence Force Academy
Canberra, Australia

Javier Gómez-Tagle
Electrical Engineering Department
ITESM
Guadalajara, Mexico

Bo Hagerman
Ericsson Radio Systems
S-164 80 Stockholm, Sweden

Kwok Chiang Ho
Addest Technovation Pte. Ltd.
Singapore, Republic of Singapore

Chun-Wen Paul Huang
Electrical Engineering Department
University of Mississippi
University, Mississippi

Magdy F. Iskander
Electrical Engineering Department
University of Utah
Salt Lake City, Utah

Ramakrishna Janaswamy
Code EC/Js, Naval Postgraduate School
Monterey, California

Ami Kanazawa
Yokosuka Radio Communications Research Center
Communication Research Laboratory
Ministry of Posts and Telecommunications
Yokosuka, Japan

Jonas Karlsson
Ericsson Radio Systems
Stockholm, Sweden

Nemai C. Karmakar
School of Electrical and Electronic Engineering
Nanyang Technological University
Singapore, Republic of Singapore

Branko M. Kolundzija
School of Electrical Engineering
University of Belgrade
Belgrade, Yugoslavia

Fredric Kronestedt
Ericsson Radio Systems
Stockholm, Sweden

Te-Hong Lee
Department of Electrical Engineering/ESL
The Ohio State University
Columbus, Ohio

Sara Mazur
Ericsson Radio Systems
Stockholm, Sweden

Eric Michielssen
Center for Computational Electromagnetics
Department of Electrical and Computer Engineering
University of Illinois at Urbana-Champaign
Urbana, Illinois

Ryu Miura
Yokosuka Radio Communications Research Center
Communications Research Laboratory
Ministry of Posts and Telecommunications
Yokosuka, Kanagawa, Japan

Karl J. Molnar
Ericsson Inc.
Research Triangle Park, North Carolina

José M. F. Moura
Department of Electrical and Computer Engineering
Carnegie Mellon University
Pittsburgh, Pennsylvania

Arye Nehorai
Department of EECs (M/C 154)
University of Illinois at Chicago
Chicago, Illinois

Boon Poh Ng
School of Electrical and Electronic Engineering
Nanyang Technological University
Singapore, Republic of Singapore

H. Ogawa
Communications Research Laboratory
Ministry of Posts and Telecommunications
Yokosuka, Kanagawa, Japan

Shingo Ohmori
Communication Systems Division
Communications Research Laboratory
Tokyo, Japan

Björn Ottersten
Department of Signals, Sensors and Systems
Royal Institute of Technology
Stockholm, Sweden

A. W. Preece
Medical Physics University Research Centre
Bristol Oncology Centre
Bristol, United Kingdom

Sembiam R. Rengarajan
Department of Electrical and Computer Engineering
California State University-Northridge
Northridge, California

Roberto G. Rojas
Department of Electrical Engineering/ESL
The Ohio State University
Columbus, Ohio

Michael J. Ryan
School of Electrical Engineering
Australian Defence Force Academy
Canberra, Australia

Tapan K. Sarkar
Department of Electrical and Computer Engineering
Syracuse University
Syracuse, New York

Bernard Sklar
Communications Engineering Services
Tarzana, California

Charles E. Smith
Electrical Engineering Department
University of Mississippi
University, Mississippi

Hyok J. Song
HRL Laboratories, LLC
Malibu, California

Thomas Svantesson
Department of Signals and Systems
Chalmers University of Technology
Göteborg, Sweden

B. T. G. Tan
Faculty of Science
National University of Singapore
Singapore, Republic of Singapore

Masato Tanaka
Kashima Space Research Center
Communications Research Laboratory
Ministry of Posts and Telecommunications
Kashima, Ibaraki, Japan

Saúl A. Torrico
Comsearch
Reston, Virginia

Hiroyuki Tsuji
Yokosuka Radio Communications Research Center
Communication Research Laboratory
Ministry of Posts and Telecommunications
Yokosuka, Japan

Mats Viberg
Department of Signals and Systems
Chalmers University of Technology
Göteborg, Sweden

T. Bao Vu
Department of Electronic Engineering
City University of Hong Kong
Kowloon, Hong Kong

Rod Waterhouse
Department of Communication and Electronic Engineering
RMIT University
Melbourne, Victoria, Australia

Wesley O. Williamson
TRW, Inc.
Redondo Beach, California

J. Xavier
Instituto Superior Tecnico,
Instituto de Sistemas e Robotica
Lisboa, Portugal

Zhengqing Yun
Electrical Engineering Department
University of Utah
Salt Lake City, Utah

Contents

PART A Wireless Communication Systems and Channel Characteristics

PART B Antenna Technology and Numerical Methods

PART C Antenna Developments and Practical Antennas

PART D Smart Antennas and Array Theory

PART E Implementation of Smart Antenna Systems

PART F Electromagnetic Radiation and the Human Body

PART A

Wireless Communication Systems and Channel Characteristics

1

Cellular Systems

Lal C. Godara
University of New South Wales

1.1 Introduction

The cellular concept was invented by Bell Laboratories and the first commercial analog voice system was introduced in Chicago in October 1983 [1, 2]. The first generation analog cordless phone and cellular systems became popular using the design based on a standard known as Advanced Mobile Phone Services (AMPS). Similar standards were developed around the world including Total Access Communication System (TACS), Nordic Mobile Telephone (NMT) 450, and NMT 900 in Europe; European Total Access Communication System (ETACS) in the United Kingdom; C-450 in Germany; and Nippon Telephone and Telegraph (NTT), JTACS, and NTACS in Japan.

In contrast to the first-generation analog systems, second-generation systems are designed to use digital transmission. These systems include the Pan-European Global System for Mobile Communications (GSM) and DCS 1800 systems, North American dual-mode cellular system Interim Standard (IS)-54, North American IS-95 system, and Japanese personal digital cellular (PDC) system [1, 3].

The third-generation mobile communication systems are being studied worldwide, under the names of Universal Mobile Telecommunications System (UMTS) and International Mobile Telecommunications (IMT)-2000 [4, 5]. The aim of these systems is to provide users advance communication services, having wideband capabilities, using a single standard. Details on various systems could be found in References [1, 6–9]. In third-generation communication systems, satellites are going to play a major

0-8493-0124-6/02/$0.00+$1.50

role in providing global coverage [10–16]. Chapter 2 of this book provides more details on satellite communications.

The aim of this chapter is to present fundamental concepts of cellular systems by explaining various terminology used to understand the working of these systems. The chapter also provides details on some popular standards. More details on cellular fundamentals may be found in References [17–20]. The chapter is organized as follows.

In Section 1.2 fundamentals of cellular systems are presented for understanding how these systems work. Sections 1.3 and 1.4 are devoted to first-generation and second-generation systems, respectively, where a brief description of some popular standards is presented. A discussion on third-generation systems is included in Section 1.5.

1.2 Cellular Fundamentals

The area served by mobile phone systems is divided into small areas known as cells. Each cell contains a base station that communicates with mobiles in the cell by transmitting and receiving signals on radio links. The transmission from the base station to a mobile is typically referred to as downstream, forward-link, or downlink. The corresponding terms for the transmission from a mobile to the base are upstream, reverse-link, and uplink. Each base station is associated with a mobile switching center (MSC) that connects calls to and from the base to mobiles in other cells and the public switched telephone network. A typical setup depicting a group of base stations to a switching center is shown in Fig. 1.1. In this section terminology associated with cellular systems is introduced with a brief description to understand how these systems work [21].

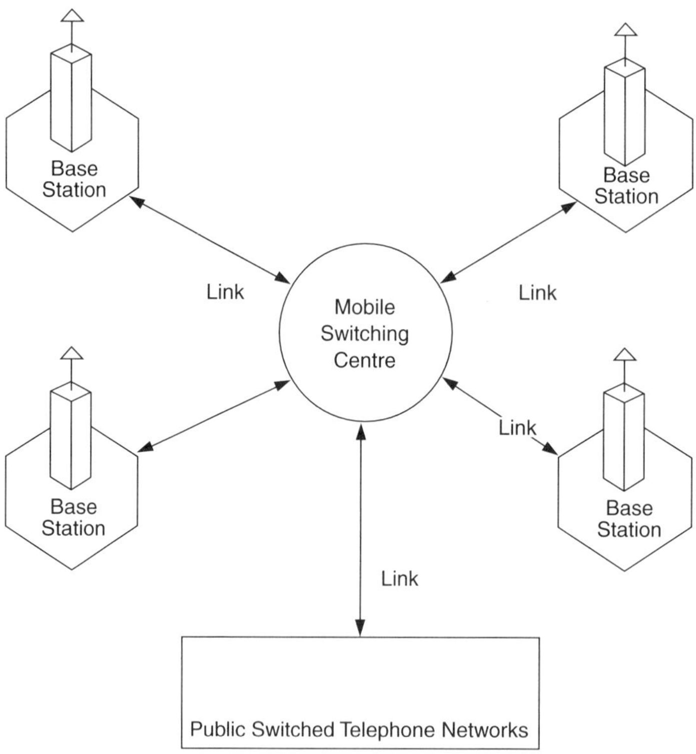

FIGURE 1.1 A typical cellular system setup.

1.2.1 Communication Using Base Stations

A base station communicates with mobiles using two types of radio channels, control channels to carry control information and traffic channels to carry messages. Each base station continuously transmits control information on its control channels. When a mobile is switched on, it scans the control channels and tunes to a channel with the strongest signal. This normally would come from the base station located in the cell in which the mobile is also located. The mobile exchanges identification information with the base station and establishes the authorization to use the network. At this stage, the mobile is ready to initiate and receive a call.

1.2.1.1 A Call from a Mobile

When a mobile wants to initiate a call, it sends the required number to the base. The base station sends this information to the switching center that assigns a traffic channel to this call because the control channels are only used for control information. Once the traffic channel is assigned, this information is relayed to the mobile via the base station. The mobile switches itself to this channel. The switching center then completes the rest of the call.

1.2.1.2 A Call to a Mobile

When someone calls a mobile, the call arrives at the mobile switching center. It then sends a paging message through several base stations. A mobile tuned to a control channel detects its number in the paging message and responds by sending a response signal to the nearby base station. The base station informs the switching center about the location of the desired mobile. The switching center assigns a traffic channel to this call and relays this information to the mobile via the base. The mobile switches itself to the traffic channel and the call is complete.

1.2.1.3 Registration

A mobile is normally located by transmitting a paging message from various base stations. When a large number of base stations are involved in the paging process, it becomes impractical and costly. It is avoided by a registration procedure where a roaming phone registers with an MSC closer to itself. This information may be stored with the switching center of the area as well as the home switching center of the phone. The home base of the phone is the one where it is permanently registered. Once a call is received for this phone, its home switching center contacts the switching center where the phone is currently roaming. Paging in the vicinity of the previous known location helps to locate the phone. Once it responds, the call may be connected as discussed previously.

1.2.2 Channel Characteristics

An understanding of propagation conditions and channel characteristics is important for an efficient use of a transmission medium. Attention is being given to understanding the propagation conditions where a mobile is to operate and many experiments have been conducted to model the channel characteristics. Many of these results could be found in review articles [22–24] and references therein. Two chapters of this book are devoted to propagation prediction and channel characterization.

1.2.2.1 Fading Channels

The signal arriving at a receiver is a combination of many components arriving from various directions as a result of multipath propagation. This depends on terrain conditions and local buildings and structures, causing the received signal power to fluctuate randomly as a function of distance. Fluctuations on the order of 20 dB are common within the distance of one wavelength (I λ). This phenomenon is called fading. One may think this signal as a product of two variables.

The first component, also referred to as the short-term fading component, changes faster than the second one and has a Rayleigh distribution. The second component is a long-term or slow-varying quantity and has lognormal distribution [17, 25]. In other words, the local mean varies slowly with lognormal distribution and the fast variation around the local mean has Rayleigh distribution.

A movement in a mobile receiver causes it to encounter fluctuations in the received power level. The rate at which this happens is referred to as the fading rate in mobile communication literature [26] and it depends on the frequency of transmission and the speed of the mobile. For example, a mobile on foot operating at 900 MHz would cause a fading rate of about 4.5 Hz whereas a typical vehicle mobile would produce the fading rate of about 70 Hz.

1.2.2.2 Doppler Spread

The movement in a mobile causes the received frequency to differ from the transmitted frequency because of the Doppler shift resulting from its relative motion. As the received signals arrive along many paths, the relative velocity of the mobile with respect to various components of the signal differs, causing the different components to yield a different Doppler shift. This can be viewed as spreading of the transmitted frequency and is referred to as the Doppler spread. The width of the Doppler spread in frequency domain is closely related to the rate of fluctuations in the observed signal [22].

1.2.2.3 Delay Spread

Because of the multipath nature of propagation in the area where a mobile is being used, it receives multiple and delayed copies of the same transmission, resulting in spreading of the signal in time. The root-mean-square (rms) delay spread may range from a fraction of a microsecond in urban areas to on the order of 100 μsec in a hilly area, and this restricts the maximum signal bandwidth between 40 and 250 kHz. This bandwidth is known as coherence bandwidth. The coherence bandwidth is inversely proportional to the rms delay spread. This is the bandwidth over which the channel is flat; that is, it has a constant gain and linear phase.

For a signal bandwidth above the coherence bandwidth the channel loses its constant gain and linear phase characteristic and becomes frequency selective. Roughly speaking, a channel becomes frequency selective when the rms delay spread is larger than the symbol duration and causes intersymbol interference (ISI) in digital communications. Frequency-selective channels are also known as dispersive channels whereas the nondispersive channels are referred to as flat-fading channels.

1.2.2.4 Link Budget and Path Loss

Link budget is a name given to the process of estimating the power at the receiver site for a microwave link taking into account the attenuation caused by the distance between the transmitter and the receiver. This reduction is referred to as the path loss. In free space the path loss is proportional to the second power of the distance; that is, the distance power gradient is two. In other words, by doubling the distance between the transmitter and the receiver, the received power at the receiver reduces to one fourth of the original amount.

For a mobile communication environment utilizing fading channels the distance power gradient varies and depends on the propagation conditions. Experimental results show that it ranges from a value lower than two in indoor areas with large corridors to as high as six in metal buildings. For urban areas the path loss between the base and the cell site is often taken to vary as the fourth power of the distance between the two [22].

Normal calculation of link budget is done by calculating carrier to noise ratio (CNR), where noise consists of background and thermal noise, and the system utility is limited by the amount of this noise. However, in mobile communication systems the interference resulting from other mobile units is a dominant noise compared with the background and man-made noise. For this reason these systems are limited by the amount of total interference present instead of the background noise as in the other case. In other words, the signal to interference ratio (SIR) is the limiting factor for a mobile communication system instead of the signal to noise ratio (SNR) as is the case for other communication systems. The calculation of link budget for such interference-limited systems involves calculating the carrier level, above the interference-level contributed by all sources [27].

1.2.3 Multiple Access Schemes

The available spectrum bandwidth is shared in a number of ways by various wireless radio links. The way in which this is done is referred to as a multiple access scheme. There are basically four principle schemes. These are frequency division multiple access (FDMA), time division multiple access (TDMA), code division multiple access (CDMA), and space division multiple access (SDMA) [29–40].

1.2.3.1 Frequency Division Multiple Access Scheme

In an FDMA scheme the available spectrum is divided into a number of frequency channels of certain bandwidth and individual calls use different frequency channels. All first-generation cellular systems use this scheme.

1.2.3.2 Time Division Multiple Access Scheme

In a TDMA scheme several calls share a frequency channel [29]. The scheme is useful for digitized speech or other digital data. Each call is allocated a number of time slots based on its data rate within a frame for upstream as well as downstream. Apart from the user data, each time slot also carries other data for synchronization, guard times, and control information.

The transmission from base station to mobile is done in time division multiplex (TDM) mode whereas in the upstream direction each mobile transmits in its own time slot. The overlap between different slots resulting from different propagation delay is prevented by using guard times and precise slot synchronization schemes.

The TDMA scheme is used along with the FDMA scheme because there are several frequency channels used in a cell. The traffic in two directions is separated either by using two separate frequency channels or by alternating in time. The two schemes are referred to as frequency division duplex (FDD) and time division duplex (TDD), respectively. The FDD scheme uses less bandwidth than TDD schemes use and does not require as precise synchronization of data flowing in two directions as that in the TDD method. The latter, however, is useful when flexible bandwidth allocation is required for upstream and downstream traffic [29].

1.2.3.3 Code Division Multiple Access Scheme

The CDMA scheme is a direct sequence (DS), spread-spectrum method. It uses linear modulation with wideband pseudonoise (PN) sequences to generate signals. These sequences, also known as codes, spread the spectrum of the modulating signal over a large bandwidth, simultaneously reducing the spectral density of the signal. Thus, various CDMA signals occupy the same bandwidth and appear as noise to each other. More details on DS spread-spectrum may be found in Reference [36].

In the CDMA scheme, each user is assigned an individual code at the time of call initiation. This code is used both for spreading the signal at the time of transmission and despreading the signal at the time of reception. Cellular systems using CDMA schemes use FDD, thus employing two frequency channels for forward and reverse links.

On forward-link a mobile transmits to all users synchronously and this preserves the orthogonality of various codes assigned to different users. The orthogonality, however, is not preserved between different components arriving from different paths in multipath situations [34]. On reverse links each user transmits independently from other users because of their individual locations. Thus, the transmission on reverse link is asynchronous and the various signals are not necessarily orthogonal.

It should be noted that these PN sequences are designed to be orthogonal to each other. In other words, the cross correlation between different code sequences is zero and thus the signal modulated with one code appears to be orthogonal to a receiver using a different code if the orthogonality is preserved during the transmission. This is the case on forward-link and in the absence of multipath the signal received by a mobile is not affected by signals transmitted by the base station to other mobiles.

On reverse link the situation is different. Signals arriving from different mobiles are not orthogonalized because of the asynchronous nature of transmission. This may cause a serious problem when the base station is trying to receive a weak signal from a distant mobile in the presence of a strong signal from a nearly mobile. This situation where a strong DS signal from a nearby mobile swamps a weak DS signal from a distant mobile and makes its detection difficult is known as the "near–far" problem. It is prevented by controlling the power transmitted from various mobiles such that the received signals at the base station are almost of equal strength. The power control is discussed in a later section.

The term *wideband CDMA* (WCDMA) is used when the spread bandwidth is more than the coherence bandwidth of the channel [37]. Thus, over the spread bandwidth of DS-CDMA, the channel is frequency selective. On the other hand, the term *narrowband CDMA* is used when the channel encounters flat fading over the spread bandwidth. When a channel encounters frequency-selective fading, over the spread bandwidth, a RAKE receiver may be employed to resolve the multipath component and combine them coherently to combat fading.

A WCDMA signal may be generated using multicarrier (MC) narrowband CDMA signals, each using different frequency channels. This composite MC-WCDMA scheme has a number of advantages over the single-carrier WCDMA scheme. It not only is able to provide diversity enhancement over multipath fading channels but also does not require a contiguous spectrum as is the case for the single-carrier WCDMA scheme. This helps to avoid frequency channels occupied by narrowband CDMA, by not transmitting MC-WCDMA signals over these channels. More details on these and other issues may be found in Reference [37] and references therein.

1.2.3.4 Comparison of Different Multiple Access Schemes

Each scheme has its advantages and disadvantages such as complexities of equipment design, robustness of system parameter variation, and so on. For example, a TDMA scheme not only requires complex time synchronization of different user data but also presents a challenge to design portable RF units that overcome the problem of a periodically pulsating power envelope caused by short duty cycles of each user terminal. It should be noted that when a TDMA frame consists of N users transmitting equal bit rates, the duty cycles of each user is 1/N. TDMA also has a number of advantages [29].

1. A base station communicating with a number of users sharing a frequency channel only requires one set of common radio equipment.
2. The data rate, to and from each user, can easily be varied by changing the number of time slots allocated to the user as per the requirements.
3. It does not require as stringent power control as that of CDMA because its interuser interference is controlled by time slot and frequency-channel allocations.
4. Its time slot structure is helpful in measuring the quality of alternative slots and frequency channels that could be used for mobile-assisted handoffs. Handoff is discussed in a later section.

It is argued in Reference [34] that, though there does not appear to be a single scheme that is the best for all situations, CDMA possesses characteristics that give it distinct advantages over others.

1. It is able to reject delayed multipath arrivals that fall outside the correlation interval of the PN sequence in use and thus reduces the multipath fading.
2. It has the ability to reduce the multipath fading by coherently combing different multipath components using a RAKE receiver.
3. In TDMA and FDMA systems a frequency channel used in a cell is not used in adjacent cells to prevent co-channel interference. In a CDMA system it is possible to use the same frequency channel in adjacent cells and thus increase the system capacity.
4. The speech signal is inherently bursty because of the natural gaps during conversation. In FDMD and TDMA systems once a channel (frequency and/or time slot) is allocated to a user, that channel cannot be used during nonactivity periods. However, in CDMA systems the background noise is roughly the average of transmitted signals from all other users and thus a nonactive period in speech reduces the background noise. Hence, extra users may be accommodated without the loss of signal quality. This in turn increases the system capacity.

1.2.3.5 Space Division Multiple Access

The SDMA scheme also referred to as space diversity uses an array of antennas to provide control of space by providing virtual channels in angle domain [38]. This scheme exploits the directivity and beam-shaping capability of an array of antennas to reduce co-channel interference. Thus, it is possible that by using this scheme simultaneous calls in a cell could be established at the same carrier frequency. This helps to increase the capacity of a cellular system.

The scheme is based on the fact that a signal arriving from a distant source reaches different antennas in an array at different times as a result of their spatial distribution, and this delay is utilized to differentiate one or more users in one area from those in another area. The scheme allows an effective transmission to take place between a base station and a mobile without disturbing the transmission to other mobiles. Thus, it has the potential such that the shape of a cell may be changed dynamically to reflect the user movement instead of currently used fixed size cells. This arrangement then is able to create an extra dimension by providing dynamic control in space [39, 40]. A number of chapters in this book deal with various aspects of antenna array processing.

1.2.4 Channel Reuse

The generic term *channel* is normally used to denote a frequency in FDMA system, a time slot in TDMA system, and a code in CDMA system or a combination of these in a mixed system. Two channels are different if they use different combinations of these at the same place. For example, two channels in a FDMA system use two different frequencies. Similarly, in TDMA system two separate time slots using the same frequency channel is considered two different channels. In that sense, for an allocated spectrum the number of channels in a system is limited. This limits the capacity of the system to sustain simultaneous calls and may only be increased by using each traffic channel to carry many calls simultaneously. Using the same channel again and again is one way of doing it. This is the concept of channel reuse.

The concept of channel reuse can be understood from Fig. 1.2. Figure 1.2a shows a cluster of three cells. These cells use three separate sets of channels. This set is indicated by a letter. Thus, one cell uses set A, the other uses set B, and so on. In Fig. 1.2b this cluster of three cells is being repeated to indicate that three sets of channels are being reused in different cells. Figure 1.3 shows a similar arrangement with cluster size of seven cells. Now let us see how this helps to increase the system capacity.

Assume there are a total of F channels in a system to be used over a given geographic area. Also assume that there are N cells in a cluster that use all the available channels. In the absence of channel reuse this cluster covers the whole area and the capacity of the system to sustain simultaneous calls is F. Now if the cluster of N cells is repeated M times over the same area, then the system capacity increases to MF as each channel is used M times.

The number of cells in a cluster is referred to as the cluster size, the parameter $1/N$ is referred to as the frequency reuse factor, and a system using a cluster size of N sometimes is also referred to as a system using N frequency reuse plan. The cluster size is an important parameter. For a given cell size, as the cluster size is decreased, more clusters are required to cover the given area leading to more reuse of channels and hence the system capacity increases. Theoretically, the maximum capacity is attained when cluster size is one, that is, when all the available channels are reused in each cell. For hexagonal cell geometry, the cluster size can only have certain values. These are given by $N = i^2 + j^2 + ij$, where i and j are nonnegative integers.

The cells using the same set of channels are known as co-channel cells. For example, in Fig. 1.2, the cells using channels A are co-channel cells. The distance between co-channel cells is known as co-channel distance and the interference caused by the radiation from these cells is referred to as co-channel interference. For proper functioning of the system, this needs to be minimized by decreasing the power transmitted by mobiles and base stations in co-channel cells and increasing the co-channel distance. Because the transmitted power normally depends on the cell size, the minimization of co-channel interference requires a minimum co-channel distance; that is, the distance cannot be smaller than this minimum distance.

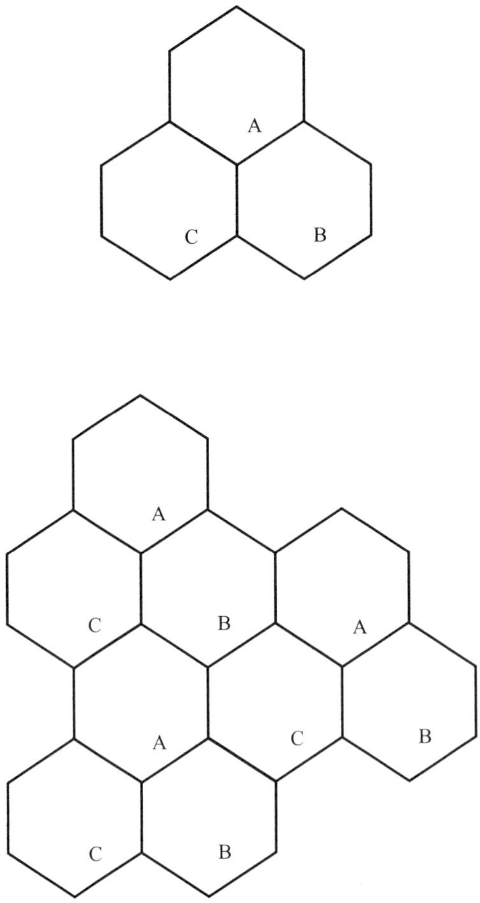

FIGURE 1.2 (a) A cluster of three cells. (b) Channel reuse concept using a three-cell cluster.

In a cellular system of equal cell size, the co-channel interference is a function of a dimensionless parameter known as co-channel reuse ratio Q. This is a ratio of the co-channel distance D and the cell radius R, that is,

$$Q = \frac{D}{R}$$

For hexagonal geometry,

$$Q = \sqrt{3N}$$

It follows from these equations that an increase in Q increases the co-channel distance and thus minimizes the co-channel interference. On the other hand, a decrease in Q decreases the cluster size N and hence maximizes the system capacity. Thus, the selection of Q is a trade-off between the two parameters, namely, the system capacity and co-channel interferences. It should be noted that for proper functioning of the system, the signal to co-channel interference ratio should be above a certain minimum value [19].

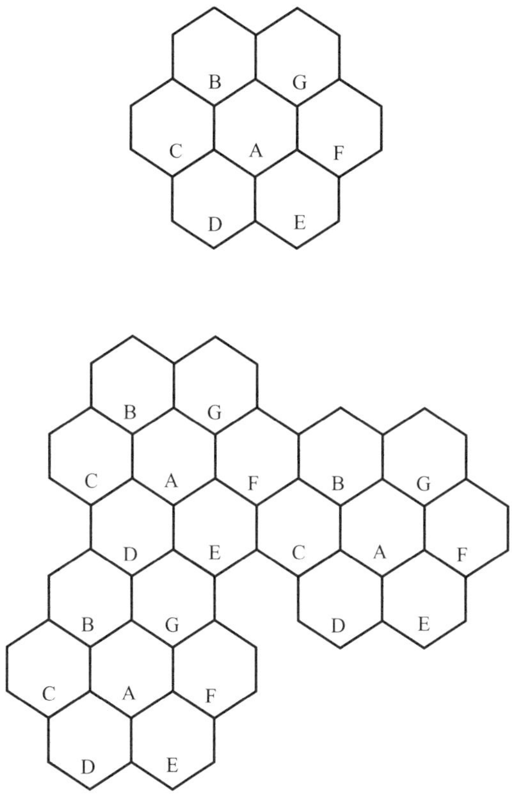

FIGURE 1.3 (a) A cluster of seven cells. (b) Channel reuse concept using a seven-cell cluster.

1.2.5 Cellular Configuration

A cellular system may be referred to as a macrocell, a microcell, or a picocell system depending on the size of cells. Some characteristics of these cellular structures are now described.

1.2.5.1 Macrocell System

A cellular system with its cell size of several kilometers is referred to as macrocell systems. Base stations of these systems transmit several watts of power from antennas mounted on high towers. Normally there is no line of sight (LOS) between the base station and mobiles and thus a typical received signal is a combination of various signals arriving from different directions. The received signal in these systems experience spreading of several microseconds because of the nature of propagation conditions.

1.2.5.2 Microcell Systems

As cells are split and their boundaries are redefined, their size becomes very small. At a radius less than about a kilometer, the system is referred to as a microcell system. In these systems a typical base station transmits less than 1 W of power from an antenna mounted at a few meters above the ground and normally an LOS exists between the base and a mobile. Cell radius in microcell systems is less than a kilometer giving rms delay spread on the order of few tens of nanoseconds compared with a few micoseconds for macrocell systems. This impacts on the maximum data rate a channel could sustain. For microcell systems maximum bit rate is about 1 Mbps compared with that of about 300 kbps for macocell systems [27].

Microcell systems are also useful in providing coverage along roads and highways. Because the antenna height is normally lower than the surrounding buildings the propagation is along the streets and an LOS exists between the base and a mobile. When a mobile turns a corner, sometimes a sudden drop in received signal strength is experienced because of loss of LOS. Depending on how antennas are mounted on intersections and corners, various cell plans are possible. More details on these aspects may be found in Reference [41] and references therein.

1.2.5.3 Picocell Systems

When cell sizes are reduced below about 100 m covering areas such as large rooms, corridors, underground stations, large shopping centers, and so on, cellular systems are sometimes referred to as picocell systems with antennas mounted below rooftop levels or in buildings. These in-building areas have different propagation conditions than those covered by macrocell and microcell systems, and thus require different considerations for developing channel models. Details on various models to predict propagation conditions may be found in Reference [24]. Sometimes the picocell and microcell systems are also referred to as cordless communication systems with the term *cellular* identifying a macrocell system. Mobiles within these smaller cell systems are called cordless terminals or cordless phones [1, 6, 42].

Providing in-building communication services using wireless technology, based on cell shapes dictated by floors and walls, is a feasible alternative and offers many advantages. It is argued in Reference [43] that radio frequencies in 18-GHz band are ideal for such services because these do not penetrate concrete and steel structures, eliminating the problem of co-channel interferences. These frequencies offer huge bandwidth and require millimeter size antennas that are easy to manufacture and install.

1.2.5.4 Overlayed System

Small cell systems make very efficient use of the spectrum, allowing large frequency reuse resulting in an increased capacity of a system. However, these are not suitable for all conditions because of their large handoff requirement. A system of mixed cells with the concept of overlaying is discussed in References [41, 44–46]. In this system a hierarchy of cells is assumed to exist. A macrocell system is assumed at the top of the hierarchy with smaller cells systems at its bottom. A mobile with high mobility is assigned to a macrocell system whereas the one with a low mobility, to smaller cell systems. A design incorporating various combinations of different multiple access schemes reflects the ease of handoff and other traffic management strategies. A space division multiple access scheme has an important role to play in this concept, with various beams placed at the bottom of the hierarchy.

1.2.6 Channel Allocation and Assignment

Various multiple access schemes discussed in a previous section are used to divide a given spectrum into a set of disjoint channels. These channels are then allocated to various cells for their use. Channel allocation may be carried out using one of the three basic schemes, namely, fixed channel allocation, dynamic channel allocation, and hybrid channel allocation [47].

1.2.6.1 Fixed Channel Allocation Schemes

In fixed channel allocation schemes a number of channels are allocated to a cell permanently for its use such that these channels satisfy certain channel reuse constraints as discussed in the previous section. In its simplest form the same number of channels are allocated to each cell. For a system with uniform traffic distribution across all cells, this uniform channel allocation scheme is efficient in the sense that the average call blocking probability in each cell is the same as that of the overall system. For systems where the distribution is not uniform, the call blocking probability differs from cell to cell, resulting in the call being blocked in some cells when there are spare channels available in other cells.

This situation could be improved by allocating channels nonuniformally as per the expected traffic in each cell or employing one of many prevailing channel borrowing schemes. One of these is referred to as a static borrowing scheme where some channels are borrowed from cells with light traffic and allocated to those with heavy traffic. Rearrangements of channels between cells are performed periodically to meet the variation in traffic load. In this scheme the borrowed channels stay with the new cell until reallocated. There are other temporary borrowing schemes where a cell that has used all its channels is allowed to borrow a free channel from a neighbor provided it does not interfere with existing calls. The borrowed channel is returned to the original cell once the call is complete. Some temporary borrowing schemes allow any channel from a cell to be borrowed, whereas in others only nominated channels are allowed to be borrowed. Many borrowing strategies are available for selecting a channel, ranging from a simple scheme to pick the first available channel that satisfies the co-channel distance constraints to the one that performs an exhaustive search to select a channel that yields maximum signal to interference ratio and minimizes the future probability of call blocking.

1.2.6.2 Dynamic Channel Allocation Schemes

Fixed channel allocation schemes discussed thus far are simple to implement and are generally useful for relatively stable traffic conditions. These schemes are not very efficient for fast changing user distribution because they are not designed to adapt to short-term variations in traffic conditions. Dynamic channel allocation schemes are most suited for such situations. In these schemes channels are not allocated to various cells but are kept in a central pool and are assigned to calls as they arrive. At the completion of a call the assigned channel is released and goes back to the pool. The process of channel assignment involves working out a cost of assigning a channel to a call and a channel with the minimum cost is chosen for the purpose. The various channel assignment schemes differ in the way the cost function is selected using various parameters of interest such as reuse distance, SIR ratio, probability of call blocking, and so on. Some schemes base their assignment only on the current traffic conditions in the service area whereas the others take the past and the present conditions into account.

Dynamic channel assignment schemes may be implemented centrally where a central controller assigns the channels to calls from the pool. The central controller is able to achieve very efficient channel assignment but requires high overhead. The channel assignment may also be implemented in a distributed manner by base stations where calls are originated. The channel implementation by base stations requires less overhead than that required by a central controller and is more suitable for microcell systems. The distributed channel assignment schemes can be divided into two categories. In one case each base station keeps detailed status information about current available channels in its neighborhood by exchanging status information with other base stations. The schemes in this category may provide near optimum allocation but pay a heavy price in terms of increased communication with other base stations, particularly in heavy traffic. The other category of distributed channel assignment schemes uses simple algorithms that rely on mobiles to measure signal strength to decide the suitability of a channel.

1.2.6.3 Hybrid Channel Allocation Schemes

The fixed channel allocation schemes are efficient under uniformly distributed heavy traffic. On the other hand, the dynamic channel allocation schemes perform better under low traffic conditions with varying and nonuniformly distributed loads. The hybrid channel allocation schemes maximize advantages of both these schemes by dividing channels into fixed and dynamic sets. The channels in fixed sets are allocated as per fixed channel allocation strategies and those in the other set are free to be assigned to calls in a cell that has used all its allocated channels. The channels in this set are assigned as per the dynamic channel allocation procedures. Apparently, no optimum ratio of channels are assigned to two sets and the design parameter is dependent on local traffic conditions. More details on these and related issues may be found in Reference [47] and references therein.

1.2.7 Handoff

It is common for a mobile to move away from its servicing base station while a call is in progress. As the mobile approaches the cell boundary, the strength and quality of the signal it receives starts to deteriorate. At some stage, near the cell boundary, it receives a stronger signal from a neighboring base station than it does from its serving base station. At this point the control of the mobile is handed over to the new base station by assigning a channel belonging to the new cell. This process where a radio channel used by a mobile is changed, is referred to as handoff or handover [41, 44, 48–50]. When handoff is between two base stations as described earlier, it is referred to as intercell handoff. On the other hand, when handoff is between two channels belonging to the same base stations, it is referred to as intracell handoff. The situation arises when the network, while monitoring its channels, finds a free channel of better quality than that used by a mobile and decides to move the mobile to this new channel to improve the quality of channels in use. Sometimes, the network rearranges channels to avoid congestion and initiates intracell handoff.

Handoff is also necessary between different layers of overlayed systems consisting of microcells and macrocells. In these systems, the channels are divided into microcell channels and macrocell channels. When a mobile moves from one microcell to another and there is no available channel for handoff, a macrocell channel is used to meet the handoff request. This avoids the forced termination of a call. Later if a channel becomes available at an underlayed microcell, then the macrocell channel may be released and a microcell channel is assigned to the call by initiating a new handoff.

Forced termination of a call in progress is undesirable and to minimize it a number of strategies are employed. These include reserving channels for handoff, using channel assignment schemes that give priority to a handoff request over new calls, and queuing the handoff request. The channel reservation and handoff priority scheme reduce the probability of forced termination by increasing the probability of blocking new calls. The queuing schemes are effective when handoff requests arrive in groups and there is a reasonable likelihood of channel availability in the near future.

The handoff is initiated when the quality of current channels deteriorates below an acceptable threshold or a better channel is available. The channel quality is measured in terms of bit-error rate (BER), received signal strength, or some other signal quality such as eye opening of radio signal that indicates signal to interference plus noise ratio.

For handoff initiation the signal strength is used as an indication of the distance between the base and the mobile. For this reason, a drop in signal strength resulting from Rayleigh fading is normally not used to initiate handoff and some kind of averaging is used to avoid the problem. In some systems the round-trip delay between mobile and base is also used as an indication of the distance.

The measurement of various parameters may be carried out either at the mobile or at the base. Depending on where the measurements are made and who initiates the handoff, various handoff implementation schemes are possible including network-controlled handoff, mobile-controlled handoff, and mobile-assisted handoff.

1.2.7.1 Network-Controlled Handoff

In network-controlled handoff, each base station monitors the signal strength received from mobiles in their cells and makes periodic measurements of the received signal from mobiles in their neighboring cells. The MSC then initiates and completes the handoff of a mobile as and when it decides. The decision is based on the received signal strength at the base station serving the mobiles and base stations in neighboring cells. Because of its centralized nature, the collection of these measurements generates a large network traffic. This could be reduced to an extent by making measurements less frequently and by not requiring the neighboring base station to send the measurements continually. However, this reduces the accuracy. The execution of handoff by this method takes a few seconds and for this reason the method is not preferred by microcellular systems where a quick handoff is desirable.

1.2.7.2 Mobile-Controlled Handoff

Mobile-controlled handoff, is a highly decentralized method and does not need any assistance from the MSC. In this scheme a mobile monitors signal strength on its current channel and measures signals received from the neighboring base stations. It receives BER and signal strength information from its serving base stations about uplink channels. Based on all this information, it initiates the handoff process by requesting the neighboring base for allocation of a low interference channel. The method has a handoff execution time on the order of 100 ms and is suitable for microcell systems.

1.2.7.3 Mobile-Assisted Handoff

In mobile-assisted handoff methods, as the name suggests, a mobile helps the network in the handoff decision making by monitoring the signal strength of its neighboring base stations and passing them to

MSC via its serving base station. The handoff is initiated and completed by the network. The execution time is on the order of 1 s.

1.2.7.4 Hard Handoff and Soft Handoff

Handoff may be classified into hard handoff and soft handoff. During hard handoff the mobile can communicate only with one base station. The communication link gets broken with the current base station before the new one is established and there is normally a small gap in communication during the transition. In the process of soft handoff, the mobile is able to communicate with more than one base station. It receives signals from more than one base station and the received signals are combined after appropriate delay adjustment. Similarly, more than one station receives signals from mobiles and the network combines different signals. This scheme is also known as macroscopic diversity and is mostly employed by CDMA systems.

Hard handoff, on the other hand, is more appropriate for TDMA and FDMA systems. It is also simple to implement compared with soft handoff. However, it may lead to unnecessary handoff back and forth between two base stations when the signals from two base stations fluctuate. The situation may arise when a mobile, currently being served for example by Base 1 receives a stronger signal, from say Base 2 and is handed over to Base 2. Immediately after that it receives a stronger signal from Base 1 compared to that it receives from Base 2, causing a handoff. This phenomenon, known as the ping-pong effect, may continue for some time and is undesirable because every handoff has a cost associated with it requiring network signaling of varying amount for authentication, database updates, circuit switching, and so on. This is avoided by using a hysteresis margin such that the handoff is not initiated until the difference between the signal received from the two base stations is more than the margin. For example, if the margin is Δ dB then the handoff is initiated when the signal received by the mobile from base 2 is Δ dB more than that from Base 1. More details on various handoff implementation issues may be found in References [41, 48, 49] and references therein.

1.2.8 Cell Splitting and Cell Sectorization

Each cell has a limited channel capacity and thus could only serve so many mobiles at a given time. Once the demand in that cell exceeds this limit the cell is further subdivided into smaller cells, each new cell with its own base station and its frequency allocation. The power of the base station transmitters is adjusted to reflect the new boundaries. The power transmitted by new base stations is less than that of the old one.

The consequence of the cell splitting is that the frequency assignment has to be done again, which affects the neighboring cells. It also increases the handoff rate because the cells are now smaller and a mobile is likely to cross cell boundaries more often compared with the case when the cells are big. Because of altered signaling conditions, this also affects the traffic in control channels.

Cell sectorization is referred to the case when a given cell is subdivided into several sectors and all sectors are served by the same base station. This is normally done by employing directional antennas such that the energy in each sector is directed by separate antennas. This has the effect of increased channel capacity similar to cell splitting. However, it uses the same base station and thus does not incur the cost of establishing new base stations associated with the cell splitting. This helps in reducing the co-channel interference because the energy is directed in the direction of the sector that does not cause interference in the co-channel cells, particularly in co-channel cells in the opposite direction to the sector. As in the case of cell splitting, this also affects the handoff rate.

1.2.9 Power Control

It is important that a radio receiver receives a power level that is enough for its proper function but not high enough for this level to disturb other receivers. This is achieved with maintaining constant power level at the receiver by transmitter power control. The receiver controls the power of the transmitter at the other end. For example, a base would control the power transmitted by mobile phones and vice versa.

It is done by a receiver monitoring its received power and sending a control signal to the transmitter to control its power transmission as required. Sometimes a separate pilot signal is used for this purpose.

Power control reduces the near–far problem in CDMA systems and helps to minimize the interference near the cell boundaries when used in forward-link [32, 33].

1.3 First-Generation Systems

These systems use analog frequency modulation for speech transmission and frequency shift keying (FSK) for signaling, and employ FDMA to share the allocated spectrum. Some of the popular standards developed around the world include Advanced Mobile Phone Service (AMPS), Total Access Communication System (TACS), Nordic Mobile Telephone (NMT), Nippon Telephone and Telegraph (NTT) and C450. These systems use two separate frequency channels, one for base to mobile and the other for mobile to base for full duplex transmission [1].

1.3.1 Characteristics of Advanced Mobile Phone Service

AMPS system uses bands of 824 to 849 MHz for uplink and 869 to 894 MHz for downlink transmission. This spectrum is divided into channels of 30-kHz bandwidth. In a two-way connection two of these channels are used. A pair of channels in a connection is selected such that channels used for uplink and downlink transmission are separated by 45 MHz. This separation was chosen so that inexpensive but highly selective duplexers could be utilized. A typical frequency-reuse plan in this system either uses clusters of 12 cells with omnidirectional antennas or 7-cell clusters with three sectors per cell.

There are a total of 832 duplex channels. Of these, 42 are used as control channels and the remaining 790 channels are used as voice channels. The control channels used for downlink and uplink transmission are referred to as forward control channels (FCC) and reverse control channels (RCC), respectively. Similarly, voice channels are referred to as forward voice channels (FVC) and reverse voice channels (RVC).

Each base continuously broadcasts FSK data on FCC and receives on RCC. A mobile scans all FCCs and locks on an FCC with the strongest signal. Each mobile needs to be locked on an FCC signal to receive and send a call. Base stations monitor their RCCs for transmission from mobiles that are locked on the matching FCCs.

1.3.2 Call Processing

When a mobile places a call, it transmits a message on RCC consisting of destination phone number, its mobile identification number, and other authorization information. The base station monitoring an RCC receives this information and sends it to the MSC. The MSC in turn checks the authentication of the mobile; assigns a pair of FVC and RVC, a supervisory audio tone (SAT), and a voice mobile attenuation code (VMAC); and connects the call to a public switched telephone network (PSTN). The mobile switches itself to the assigned channels. The SAT is used to ensure the reliable voice communication and the VMAC is used for power control.

The SAT is an analog tone of 5970, 6000, or 6030 Hz. It is transmitted during a call on both FVC and RVC. It is superimposed on voice signal and is barely audible to the user. It helps the mobile and the base to distinguish each other from co-channel users located in nearby cells. It also serves as a handshake between the base station and the mobile. The base transmits it on FVC and the mobile retransmits it on RVC after detection. If the SAT is not detected within 1 s, both the mobile and the base stop transmission.

A call to a mobile originating at a PSTN is processed by the MSC in a similar fashion. When a call arrives at an MSC, a paging message with mobile identification number (MIN) is sent out on FCCs of every base station controlled by the MSC. A mobile terminal recognizes its MIN and responds on RCC. The MSC assigns a pair of FVC and RVC, SAT and VMAC. The mobile switches itself to the assigned channel.

TABLE 1.1 Parameters of Some First-Generation Cellular Standards

Parameters	AMPS	C450	NMT 450	NTT	TACS
Tx Frequency (MHz)					
Mobile	824–849	450–455.74	453–457.5	925–940	890–915
Base Station	869–894	460–465.74	463–467.5	870–885	935–960
Channel bandwidth (kHz)	30	20	25	25	25
Spacing between forward and reverse channels (MHz)	45	10	10	55	45
Speech signal FM deviation	±12	±5	±5	±5	±9.5
Control signal data rate (kbps)	10	5.28	1.2	0.3	8
Handoff decision is based on	Power received at base	Round-trip delay	Power received at base	Power received at base	Power received at base

While a call is in progress on voice channels, the MSC issues several blank-and-burst commands to transmit signaling information using binary FSK at a rate of 10 kbps. In this mode, voice and SAT are temporarily replaced with this wideband FSK data. The signaling information is used to initiate handoff, to change mobile power level, and to provide other data as required.

The handoff decision is taken by the MSC when the signal strength on RVC falls below threshold or when the SAT experiences interference level above a predetermined value. The MSC uses scanning receivers in nearby base stations to determine the signal level of a mobile requiring handoff.

The termination of a call by a mobile is initiated using a signaling tone (ST). ST is a 10-kbps data burst of 1 and 0 s. It is sent at the end of a message for 200 ms indicating "end-of-call." It is sent along with SAT and indicates to the base station that the mobile has terminated the call instead of the call dropping out or prematurely terminating. It is sent automatically when a mobile is switched off.

1.3.3 Narrowband Advanced Mobile Phone Service, European Total Access Communication System, and Other Systems

A narrowband AMPS (N-AMPS) was developed by Motorola to provide three 10-kHz channels using FDMA in a 30-kHz AMPS channel. By replacing one AMPS channel by three N-AMPS channels at a time, the service providers are able to increase the system capacity by three times. It uses SAT, ST, and blank-and-burst similar to AMPS. Because it uses 10-kHz channels, FM deviation is smaller compared with AMPS and hence it has a lower signal to noise plus interference ratio (SNIR) resulting in degradation of audio quality. It has taken measures to compensate this degradation.

The European Total Access Communication Systems (ETACS) is identical to AMPS except that it uses 25-kHz wide channels compared with 30-kHz channels used by AMPS. It also formats its MIN differently to AMPS to accommodate different country codes in Europe. Parameters for some other popular analog systems are shown in Table 1.1.

1.4 Second-Generation Systems

In contrast to the first-generation analog systems, second-generation systems are designed to use digital transmission and to employ TDMA or CDMA as a multiple access scheme. These systems include North American dual-mode cellular system IS-54, North American IS-95 systems, Japanese Personal Digital Cellular (PDC) systems, and European GSM and DCS 1800 systems. GSM, DCS 1800, IS-54, and PDC systems use TDMA and FDD whereas IS-95 is a CDMA system and also uses FDD for a duplexing technique. Other parameters for these systems are shown in Table 1.2. In this section a brief description of these systems is presented [1].

TABLE 1.2 Parameters of Some Second-Generation Cellular Standards

Parameters	IS-54	GSM	IS-95	PDC
TX frequencies (MHz)				
Mobile	824–849	890–915	824–849	940–956 and 1429–1453
Base station	869–894	935–960	869–894	810–826 and 1477–1501
Channel bandwidth (kHz)	30 kHz	200 kHz	1250 kHz	25 kHz
Spacing between forward and reverse channels (MHz)	45	45	45	30/48
Modulation	π/4 DQPSK	GMSK	BPSK/QPSK	π/4 DQPSK
Frame duration (ms)	40	4.615	20	20

1.4.1 United States Digital Cellular (Interim Standard-54)

United States Digital Cellular (IS-54) is a digital system and uses TDMA as a multiple access technique compared with AMPS, which is an analog system and uses FDMA. It is referred to as a dual-mode system because it was designed to share the same frequency, frequency-reuse plan, and base stations with AMPS. It was done so that the mobile and base stations can be equipped with AMPS and IS-54 channels within the same equipment to help migrate from an analog to a digital system and simultaneously to increase system capacity. In this system each frequency channel of 30 kHz is divided into six time slots in each direction. For full-rate speech, three users equally share six slots where two slots are allocated per user. For half-rate speech, each user only uses one slot. Thus, the system capacity is three times more than that of the AMPS for full-rate speech and double that for the half-rate speech. This system also uses the same signaling (FSK) technique as that of AMPS for control whereas it uses π/4 DQPSK for the voice.

It was standardized as IS-54 by the Electronic Industries Association and Telecommunication Industry Association (EIA/TIA) and was later revised as IS-136. The revised version has digital control channels (DCCs) that provide an increased signaling rate as well as additional features such as transmission of point-to-point short messages, broadcast messages, group addressing, and so on.

As was discussed in the previous section, AMPS has 42 control channels. The IS-54 standard specifies these as primary channels and additional 42 channels as secondary channels. Thus, it has twice the control channels as that of AMPS and is able to carry twice the control traffic in a given area. The secondary channels are not monitored by AMPS users and are for the exclusive use of IS-54 users.

Each time slot in each voice channel has one digital traffic channel (DTC) for user data and digitized speech and three supervisory channels to carry control information.

A full duplex DTC consists of forward DTC to carry data from the base station to the mobile and reverse DTC to carry data from the mobile to the base station. The three supervisory channels are coded digital verification color code (CDVCC), slow associated control channel (SACCH), and fast associated control channel (FACCH).

The CDVCC is a 12-b message transmitted every slot containing 8-b color code number between 1 and 255. The 12-b message is generated using shortened Hamming code. It has a similar function to SAT in AMPS. A station transmits this number on CDVCC channels and expects a handshake from each mobile that must retransmit this value on a reverse voice channel. If the number is not returned within a specified time, the time slot is relinquished.

The SACCH is a signaling channel and carries control information between base and mobile while a call is in progress. It is sent with every slot carrying information about power level change, handoff, and so on. Mobiles use this channel to send signal strength measurement of neighboring base stations so that the base may implement mobile-assisted handoff (MAHO).

The FACCH is a second signaling channel to carry control information when the call is in progress. It does not have a dedicated time during each slot as is the case for CDVCC and SACCH. It is similar to blank-and-burst in AMPS and replaces speech data when used. It carries call release instructions, MAHO, and mobile status requests.

1.4.2 Personal Digital Cellular System

The PDC system, established in Japan, employs TDMA technique. It uses three time slots per frequency channel and has a frame duration of 20 ms. It can support three users at full-rate speech and six half-rate speech users similar to IS-54. It has a channel spacing of 25 kHz and uses $\pi/4$ DQPSK modulation. It supports a frequency-reuse plan with cluster size four and uses MAHO.

1.4.3 Code Division Multiple Access Digital Cellular System (Interim Standard-95)

This CDMA digital system uses CDMA as a multiple access technique and occupies the same frequency band as that occupied by AMPS; that is, the forward-link frequency band is from 869 to 894 MHz and the reverse-link band is from 824 to 849 MHz. Forward-link and reverse-link carrier frequencies are separated by 45 MHz.

Each channel in IS-95 occupies a 1.25-MHz bandwidth and this is shared by many users. The users are separated from each other by allocating 1 of 64 orthogonal spreading sequences (Walsh functions).

The user data are grouped into 20-ms frames and are transmitted at a basic user rate of 9600 bps. This is spread to a channel chip rate of 1.2288 Mchip/s giving a spreading factor of 128.

RAKE receivers are used at both base station and mobiles to resolve and combine multipath components. During handoff the standard allows for base station diversity whereby a mobile keeps link with both the base stations and combines signals from both the stations to improve signal quality as it would combine multipath signals.

In forward-link a base station transmits simultaneously to all users using 1 of 64 spreading sequences for each user once the user data are encoded using a half-rate convolution code and are interleaved. All signals in a cell are also scrambled using a PN sequence of length 2^{15} to reduce the co-channel interference. During the scrambling process the orthogonality between different users is preserved.

The forward channel consists of 1 pilot channel, 1 synchronization channel, up to 7 paging channels, and up to 63 traffic channels. The pilot channel transmits higher power than other channels and is used by mobiles to acquire timing for forward channel and to compare signal strength of different base stations. It also provides phase reference for coherent detection.

The synchronization channel operates at 1200 bps and broadcasts a synchronization message to mobiles. The paging channels are used to transmit paging messages from the base station to mobiles and to operate at 9600, 4800, or 2400 bps. The traffic channels support variable data rate operating at 9600, 4800, 2400, and 1200 bps.

On reverse channels, mobiles transmit asynchronously to the base, and orthogonally between different users in a cell is not guaranteed. A strict control is applied to the power of each mobile so that a base station receives constant power from each user, thus eliminating the near–far problem. Power control command is sent by the base to mobiles at a rate of 800 bps. The reverse channels are made up of access channels and reverse traffic channels.

The reverse channels contain up to 32 access channels per paging channel, operate at 4800 bps, and are used by mobiles to initiate communication with base and to respond to paging messages. The reverse traffic channel is a variable data rate channel and operates similar to the forward channels at 9600, 4800, 2400, and 1200 bps.

1.4.4 Pan European Global System for Mobile Communications

The "Groupe Special Mobile" was established in 1982 to work toward the evolution of digital system in Europe, and its work now has become the Global System for Mobile (GSM) Communications system. Two frequency bands have been allocated for this system. The primary band is at 900 MHz and the

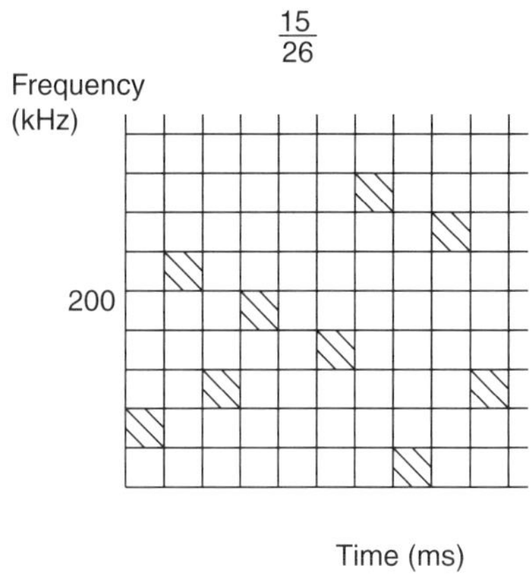

FIGURE 1.4 GSM multiple access.

secondary band is at 1800 MHz. The description here mainly concerns the primary band. It has been divided into two subbands of 25 MHz each, separated by 20 MHz. The lower band is used for uplink and the upper band is used for downlink. Operators are assigned a portion of the spectrum for their use.

The carrier frequencies are separated by 200 kHz. This gives the total number of frequency channels over the 25-MHz band as 124. The first carrier is at 890.2 MHz, the second one is at 890.4 MHz, and so on. These carriers are numbered as 0, 1, 2 and so on, respectively. Similarly, 374 different carriers are allocated in the secondary band, which is 75 MHz wide.

1.4.4.1 Multiple Access Scheme

GSM employs a combination of TDMA and FDMA schemes with slow frequency hopping. GSM transmission takes place by modulating a bundle of about 100 b known as a burst. A burst occupies a finite duration in time and frequency plane. The center frequency of these bursts is 200 kHz apart and these are 15/26 ms in duration.

The duration of these bursts is the time unit and is referred to as burst period (BP). Thus, time is measured in BP. When this burst is combined with slow frequency hopping, a typical transmission appears as shown in Fig. 1.4. The hopping sequence is selected randomly using a PN sequence.

A channel is defined by specifying which time slot it may use for transmit burst. That means specifying time instant and specific frequency. Time slots of a channel are not contiguous in time. All time slots are numbered and the description of a channel sent to the mobile by the base refers to this numbering scheme. The numbering is cyclic and each time slot is uniquely identified in this cycle, which is about 3.5 h (3 h 28 min 53 s, and 760 ms).

Many types of channels are defined in GSM, and each are cyclic. The simplest cycle is of 8 BP. This cycle of eight time slots is also called a *slot*, which is 60/13 ms in duration. The duration of the BP is chosen such that 26 slots equal to 120 ms, which is a multiple of 20 ms to obtain synchronization with other networks such as the Integrated Services Digital Network (ISDN).

A full dedicated channel is thus cyclic in 120 ms and uses 26 slots. Note that each slot is made up of eight time slots of 15/26 ms known as BP. Out of these 26 slots, 24 slots are used for traffic burst, 1 slot is used for control burst, and 1 slot is not used.

The transmission between the uplink and the downlink is not independent. Transmission in uplink follows the downlink reception by 3 BP later. When the mobile is far from the base, the mobile advances its transmission from the reception to compensate the propagation delay. Hopping sequences in the uplink and the downlink are also related. The hopping sequence in the uplink direction is derived from the one in the downlink direction in by adding 45 MHz.

1.4.4.2 Common Channels

These channels do not carry traffic and are organized based on a cycle of 51 slots (51×8 BP). This cycle is deliberately chosen differently from 26 slots of traffic channels so as not to have a common divider between the two. This allows mobiles in a dedicated channel to listen to synchronization channels (SCHs) and frequency correction channels (FCCHs) of the surrounding cells, which helps mobiles to stay in synchronization. Each SCH and FCCH uses 5 slots in a 51-slot cycle with SCH following FCCH 8 BP later. This helps mobiles to find SCH once it has located FCCH. The other downlink common channels defined include BCCH for broadcasting and PAGCH for paging. For uplink, a channel RACH is defined.

1.4.4.3 Burst Format

The quantum of transmission in GSM is 1 BP, which is 7500/13 s in duration and is occupied by about $(156 + 1/4)$ b. In GSM several burst formats have been defined and these are used for different purposes.

Access burst is used in uplink direction from the mobile to the base during the initial phase. This is the first access of the mobile to the base. The burst has constant amplitude for the period of 87 b. The structure of the burst consists of 7 b of tail followed by 41 b of training sequence, 36 b of information, and 3 b of tail on the other side. A single training sequence is specified for this burst. The access burst is the first burst from the mobile to the base and contains required demodulation information for the base.

The S burst is similar to the access burst but it is transmitted from the base to the mobile. It is the first burst from the base and has 64 training sequence bits surrounded by 39 information bits and 3 tail bits. The training sequence is unique and chosen so that the mobile knows which sequence the base has chosen.

The F burst enables the mobile to find and demodulate the S burst. All of its 148 b are set to zero, resulting in a pure sine wave of 1625/24 kHz.

The normal burst is used for all other purposes. Its amplitude stays constant covering 147 b. It has 26 training sequence bits surrounded by 58 information bits and 3 tail bits. Eight different training sequences have been specified to distinguish co-channel signals. For more details on the GSM system, see, for example, Reference [51].

1.4.5 Cordless Mobiles

The first generation analog cordless phones were designed to communicate with a single base station, effectively replacing telephone cord with a wireless link to provide terminal mobility in a small coverage area such as a house or an office. The aim of the second-generation digital cordless system is to use the same terminal in residential as well as public access areas such as offices, shopping malls, train stations, and so on to receive and to originate calls. The cordless systems differ from cellular systems in a number of ways. Their cell size is small, typically less than half a kilometer, and their antenna elevation is low. These are designed for low-speed mobiles, typically on foot, and provide coverage in specific zones instead of continuous wide-area coverage provided by cellular systems. Cordless handsets transmit very low power. A typical average transmitted power is about 5 to 10 mW compared with a few hundred milliwatt for cellular handsets [1].

Most of the cordless systems use TDD as duplexing techniques compared with FDD employed by cellular systems.

Some of the popular digital cordless standard include CT2, a British standard; digital European cordless telecommunication (DECT) standard; personal handyphone system (PHS) of Japan, and personal access communication service (PACS) of the United States. Some of the parameters of these systems are compared in Table 1.3. More details on these may be found in Reference [1] and references therein.

1.5 Third-Generation Systems

The third-generation systems aim to provide a seamless network that can provide users voice, data, multimedia, and video services regardless of their location on the network: fixed, cordless, cellular,

TABLE 1.3 Digital Cordless System Parameters

Parameter	CT2	DECT	PACS	PHS
Frequency (MHz)	864–868	1880–1900	1850–1910 and 1930–1990	1895–1918
Channel spacing (kHz)	100	1728	300	300
Duplexing	TDD	TDD	FDD	TDD
Channel rate (kbps)	72	1152	384	384
Transmitted power (mW)				
Average	5	10	25	10
Peak	10	250	200	80
Frame duration (ms)	2	10	2.5	5
Channels per carrier	1	12	8	4

satellite, and so on. The networks support global roaming while providing high-speed data and multimedia applications of up to 144 kbps on the move and up to 2 Mbps in a local area.

Third-generation systems are currently being defined by both the International Telecommunications Union (ITU) and regional standardization bodies. Globally, the ITU has been defining third-generation systems since the late 1980s through work on the IMT-2000 system, formerly called the Future Public Land Mobile Telecommunications Service (FPLMTS) [52]. The ITU is now in the process of seeking candidate technologies to be evaluated in accordance with agreed guidelines [53]. The European proposal for IMT-2000 is known as the Universal Mobile Telecommunications System (UMTS) — see discussions in References [54–57] — and is being defined by the European Telecommunications Standards Institute (ETSI), which has been responsible for UMTS standardization since the 1980s.

Although UMTS will provide significant changes for customers and technologies, systems will be deployed within a short time frame. Japan plans to launch its UMTS network in the year 2001; and the United Kingdom wants its UMTS radio interface working alongside enhanced GSM networks by the year 2002, with a fully working UMTS network operating by 2005. Third-generation networks are planned for the United States sometime between 2003 and 2005 [58]. Additional updates on IMT-2000 developments in the Asia Pacific Region can be found in References [59–62].

IMT-2000 defines systems capable of providing continuous mobile telecommunications coverage for any point on the earth's surface. Access to IMT-2000 is via either a fixed terminal or a small, light, portable mobile terminal (MT) [63].

Several different radio environments are utilized to provide the required layers of coverage. These range from very small indoor picocells with high capacity, through to terrestrial micro- and macrocells, to satellite megacells. IMT-2000 recommendations aim to maximize commonality between the various radio interfaces involved, to simplify the task of developing multimode terminals for the various operating environments [64]. In this section some salient features of IMT-2000 are discussed [4, 5].

1.5.1 Key Features and Objectives of International Mobile Telecommunications-2000

The key features and objectives of IMT-2000 include [64]

1. Integration of current first- and second-generation terrestrial and satellite-based communications systems into a third-generation system
2. Ensuring a high degree of commonality of design at a global layer
3. Compatibility of services within IMT-2000 and with fixed networks
4. Ensuring high quality and integrity of communications, comparable to the fixed network
5. Accommodation of a variety of types of terminals including pocket-size terminals
6. Use of terminals worldwide
7. Provision for connection of mobile users to other mobile users or fixed users

8. Provision of services by more than one network in any coverage area
9. Availability to mobile users of a range of voice and nonvoice services
10. Provision of services over a wide range of user densities and coverage areas
11. Efficient use of the radio spectrum consistent with providing service at acceptable cost
12. Provision of a framework for the continuing expansion of mobile network services and for the access to services and facilities of the fixed network
13. Number portability independent of service provider
14. Open architecture that accommodates advances in technology and different applications
15. Modular structure that allows the system to grow as needed

1.5.2 International Mobile Telecommunications-2000 Services

IMT-2000 supports a wide range of services, including those based on the fixed telecommunication network and those that are specific to mobile users. Services are available in a variety of environments ranging from dense urban situations, including high intensity office use, through to suburban and rural areas [64]. The actual services obtained by a user depend on the capabilities of their terminal, their subscribed set of services, and the services offered by the relevant network operator and service provider [65].

Global roaming users have access to at least a minimum set of services comprising voice telephony, selection of data services, and indication of other services available. IMT-2000 also provides services to fixed users and if required, can provide rapid and economical implementation of wide-area communications, which is particularly relevant to developing countries [64].

The general service objectives of IMT-2000 are to [66]

1. Provide a wide range of telecommunication services to mobile or stationary users by means of one or more radio links
2. Make these services available for mobile terminals located anywhere (subject to economic constraints)
3. Provide for flexibility of service provision
4. Promote flexible introduction of services
5. Ensure that a user is provided with an indication of service availability
6. Provide access to voice telephony
7. Provide access to a selection of data services
8. Provide services that depend on terminal type, location, and availability from the network operator
9. Provide a temporary or permanent substitute to fixed networks in rural or urban areas under conditions approved by the appropriate national or regional regulation authority

The general service requirements of IMT-2000 are to provide

1. Validation and authentication procedures to facilitate billing and accounting based on ITU-T X509
2. Additional layer of security for mobile telecommunications services
3. Privacy of location of a roaming user when desired by the called or calling party
4. Quality of service comparable with that of fixed networks

The general access requirements are as follows:

1. For access to fixed networks, IMT-2000 may be either an adjunct to, or an integral part of, the PSTN/ISDN.
2. For global use, IMT-2000 should allow international operation and automatic roaming of terminals.
3. For maritime and aeronautical use, operation should be facilitated to the extent permitted by the relevant regulatory body.
4. For satellite operation, IMT-2000 should facilitate direct and indirect satellite operation.

The first phase of IMT-2000 provides several telecommunications services, most of which are based on ITU-T E and F series recommendations.

Network services — These services are provided by IMT-2000 [67]:

1. Voice telephony (ITU-T E105)
2. Program sound
3. Message handling (ITU-T F400)
4. Teletex (ITU-T F200)
5. Paging (open loop, closed loop, user acknowledged)
6. Telefax (ITU-T F160 and F180)
7. Point to multipoint
8. Data
9. Video services using MPEG derived algorithms [68]
10. Short messages
11. Location monitoring
12. Multimedia

Supplementary services — The following services may be provided by IMT-2000:

1. Separation of answering from alerting: Currently, the alerting function resides in the same device used for answering. In IMT-2000, the alerting device may be a pager and the answering device may be a terminal of the user's choice.
2. Advice of charging: Parties to a call should be able to receive charging information before, during, and after the call.

1.5.3 Planning Considerations

In defining IMT-2000, several factors required consideration: radio access, spectrum requirements, security, network issues, and regulatory environments.

1.5.3.1 Radio Access

IMT-2000 provides access, by means of one or more radio links, to a wide range of services in a wide variety of operating environments. High data rates are required to provide users with the necessary quality of service for multimedia communications, ranging from a few tens of kilobits per second for image transfer, to a couple of hundreds of bits per second for peak Internet transfers, to 2 Mbps for video. The bearers for IMT-2000 are therefore defined as 384 kbps for full area coverage and 2 Mbps for local area coverage.

It is essential to optimize third-generation techniques to cater for variable bit rate and packet capabilities because many multimedia applications are packet oriented. Similarly, multimedia support implies flexibility to handle services with different bit rates and E_b/N_0 requirements [69].

The mode of delivery is via either terrestrial or satellite-based radio links, with the possibility of incorporating two or more radio links in tandem. Although it would be desirable for a common radio interface to be provided for the terrestrial and satellite components, this is unlikely to be practical because of spectral and power efficiency design constraints. Therefore, terminals will most likely be required to operate over more than one type of interface, with adaptation controlled by software using digital signal processing technology. Dual-mode handsets already exist to combine GSM at different frequencies, GSM/DECT, and GSM/satellite. The IMT-2000 design allows for the provision of competitive services to the user in each of these operating conditions [65].

The UMTS radio interface, called UMTS terrestrial radio access (UTRA) will consist of a number of hierarchical layers. The higher layer will use W-CDMA, where each user will be given a special CDMA code and full access to the bandwidth allocated. The macrolayer will provide basic data rates to 144 kbps. The lower layers will provide higher data rates of 384 kbps and 2 Mbps, through the use of a frequency division duplex. It may also be possible to use TDD through time division CDMA (TD-CDMA) for higher data rates by dividing the frequency allocation into time slots for the lower layers [69, 70].

This compromise between the two competing standards of W-CDMA and TD-CDMA means that Europe will have a "family of standards." TD-CDMA provides greater efficiency than GSM and offers reuse of the existing GSM network structure as well as efficient interworking with GSM. TD-CDMA has the same basic frame structure as GSM, each having eight time slots per frame length, but provides higher data rates, up to 2 Mbps indoors. The combination of different access methods is intended to provide flexibility and network efficiency, with the UTMS terminal adopting the access method that best seeks its environment.

1.5.3.2 Spectrum Requirements

The work of ITU on the IMT-2000 is aimed at the establishment of advanced global communication services within the frequency bands, 1885 to 2025 MHz and 2110 to 2200 MHz, identified by the World Administrative Radio Convention (WARC-92). Within this bandwidth, the bands 1980 to 2010 MHz and 2170 to 2200 MHz will be utilized by the satellite component [64]. It is important to note that although the WARC-92 frequencies were intended for IMT-2000, their use by other systems such as personal communication services (PCS) and UMTS is not precluded [71]. WARC-92 resolved that administrations implementing IMT-2000 should make spectrum available in the identified bands for system development and implementation and should use the relevant international technical characteristics that will be developed to facilitate worldwide use and roaming.

Although the intention was to reserve this bank of the spectrum on a worldwide basis for IMT-2000, the Federal Communications Commission (FCC) in the United States engaged in a spectrum auction in late 1994, which resulted in the allocation of large portions of bandwidth in North America to operators providing PCS. The European DECT service and the Japanese PHS service also have spectrum overlaps with the IMT-2000/WARC-92 allocation. The use of this spectrum for other than IMT 2000 services indicates that the allocated spectrum is not enough to meet the growing demand for additional spectrum to provide services such as mobile data services, mobile e-commerce, wireless Internet access, and mobile video services. It should be noted that this spectrum identified in 1992 for global communication services was based on a model where voice services were assumed to be the major source of traffic and the services indicated above were not foreseen. In the current climate where the number of users worldwide is expected to reach 2 billion by 2010 and there is a need to provide common spectrum for global roaming, the World Radio Communication Conference in June 2000 (WRC-2000) decided to increase the available spectrum for IMT-2000 use on a global basis.

This additional spectrum has been identified in three bands: one below 1 GHz (806–960 MHz), another at 1.7 GHz (1710–1885 MHz), and the third band at 2.5 GHz (2500–2690 MHz).

Even though these bands are made available on a global basis for countries to implement IMT-2000, a good degree of flexibility has been provided for operators to evolve towards IMT-2000 as per market and other national considerations. The flexibility allows the use of these bands by services other than those for which the spectrum has been made available. Furthermore, it not only enables each country to decide on timing of availability based on national needs, but also permits countries to select those parts of bands where sharing with existing services is most suitable.

1.5.3.3 Security

Because of the radiating nature of wireless communications, IMT-2000 needs to incorporate security measures to prevent easy reception by parties other than the intended recipient. In addition, because of the nature of mobile communications, security measures are also required to prevent fraudulent use of services [72]. The security provisions for IMT-2000 are defined with the objective of ensuring interoperability with roaming across international and national network boundaries. Virtually all security requirements and features are related to the radio interface. IMT-2000 security features are categorized as user-related or service provider related. Within these categories, they are further categorized as essential or optional [73].

1.5.3.4 Intelligent Networks

The standardization of IMT-2000 is considering new and evolving technologies on the telecommunications network side, such as intelligent networks (INs). IMT-2000 network issues are studied in close cooperation by ITU-R/ITU-T and to a great extent as an integral part of ITU-T work on IN concepts and capabilities. It is anticipated that future versions of IN switching and signaling standards will include the management of mobile and radio access as a natural part of the protocols. This includes location registration/updating and paging as well as the various types of handover between radio cells [68].

1.5.3.5 Regulatory Environments

The regulatory considerations for the introduction and use of IMT-2000 include determining the conditions for regulated and nonregulated systems, spectrum sharing, identifying the number of operators and service providers, licensing procedures, and call charging. The provision and establishment of IMT-2000 is subject to the regulatory process in each country's telecommunications authority. It may be necessary to develop new regulatory environments for IMT-2000, which will enable the provision of new services in a variety of ways not anticipated by existing regulations.

1.5.4 Satellite Operation

The satellite component of IMT-2000 enhances the overall coverage and attractiveness of the service and facilitates the development of telecommunications services in developing countries [74]. Satellites are particularly useful in mobile communications because they are able to achieve coverage of very large areas of the earth's surface [75]. To provide service at an acceptable cost, the catchment area must include as many users as possible. In this situation a globally unique standard formulated by IMT-2000 is preferable to adopting regional solutions [58]. The current version of the recommendation pertaining to the satellite component of IMT-2000 is very generic and does not provide specific details in relation to service, equipment, architecture, or interfaces and protocols [76].

Currently many satellite PCSs have been proposed based on constellations of orbiting satellites offering continental and worldwide communications, data, tracking, and paging services. The experience gained from these networks in the next few years will provide valuable input into the satellite component of third-generation systems. Depending on the lessons learned, the three possible levels of integration of the satellite component into the terrestrial network include

1. Network integration at the call level
2. Equipment integration, requiring common service standards and dual-mode terminals
3. System integration, where the satellite is an integral part of the network and handoff can be supported between terrestrial and satellite megacells [75]

It is anticipated that IMT-2000 will use several satellite constellations, each comprising a number of satellites, radio (*service*) links from the satellite to the IMT-2000 terminal, and radio (*feeder*) links from the satellites to the land earth stations (LESs) [77]. Because the satellite component will have a limited number of LESs, the operation of the network will inherently involve international terrestrial connections, and access to the network may therefore also involve an international connection. A number of non-geostationary earth orbit (GEO) satellites based on low/mid-earth orbit (LEO/MEO) constellations have been or are being deployed to deliver mobile voice and broadband data services. For the first time at WRC-1997, spectrum was made available for the operation of these satellites and a provisional power limit was imposed so that they could share the spectrum with GEO satellites. Studies conducted since 1997 on spectrum sharing have found in favor of the concept and the WRC-2000 has decided to limit the power of non-GEO satellites to enable their co-existence with GEO satellites, which aim to provide high-speed local access to global broadband services without unacceptable interference [78].

Continuity of coverage will be provided by contiguous footprints of spot beams from one or more satellites in a constellation. For nongeostationary satellites, these footprints will be in motion and continuity of calls in progress will be achieved by handover between beams, using functionality in both the mobile and satellite components [77].

Key features — The following list identifies the key features of the satellite component [77, 79, 80]:

1. Coverage of any one satellite will be much larger than that of any cluster of terrestrial base stations.
2. Coverage is likely to be by means of a number of spot beams, which will form megacells, with each spot beam larger than any terrestrial macrocell.
3. Satellite coverage can be regional, multiregional, or global.
4. A range of orbit constellations may be used.
5. The number of LES will be limited.
6. The terrestrial and satellite components should be optimized with respect to each other.
7. The LESs will connect to the satellites using feeder links that operate in frequency bands outside those identified for IMT-2000 operation; the feeder link frequencies may be used by other satellite systems and terrestrial systems, with appropriate sharing criteria.
8. Inter-satellite links (ISLs), if used, will operate outside the IMT-2000 band.
9. Provisions must exist to allow multiple service providers to compete in the satellite component.

References

1. J. E. Padgett, C. G. Gunther and T. Hattori, Overview of wireless personal communications, *IEEE Commun. Mag.*, vol. 33, pp. 28–41, January 1995.
2. W. W. Erdman, Wireless communications: A decade of progress, *IEEE Commun. Mag.*, vol. 31, pp. 48–51, December 1993.
3. D. J. Goodman, Second generation wireless information networks, *IEEE Trans. Veh. Technol.*, vol. 40, pp. 366–374, 1991.
4. L. C. Godara, M.J. Ryan and N. Padovan, Third generation mobile communication systems: overview and modelling considerations, *Annals of Telecommunications*, vol. 54, No.1–2, pp. 114–136, 1999.
5. N. Padovan, M. Ryan and L. Godara, An Overview of Third Generation Mobile Communications Systems, IEEE Tencon '98: IEEE Region 10 Annual Conference, New Delhi, December 17–19, 1998.
6. R. Pandya, Emerging mobile and personal communication systems, *IEEE Commun. Mag.*, vol. 33, pp. 44–52, June 1995.
7. P. W. Baier, P. Jung and A. Klein, Taking the challange of multiple access for third generation cellular mobile radio systems — A European view, *IEEE Commun. Mag.*, vol. 34, pp. 82–89, February 1996.
8. J. S. Dasilva, B. Arroyo, B. Barni and D. Ikonomou, European third-generation mobile systems, *IEEE Commun. Mag.*, vol. 34, pp. 68–83, October 1996.
9. E. D. Re, A coordinated European effort for the definition of a satellite integrated environment for future mobile communications, *IEEE Commun. Mag.*, vol. 34, pp. 98–104, February 1996.
10. W.W. Wu, E. F. Miller, W. L. Pritchard and R. L. Pickholtz, Mobile satellite communications, *IEEE Proc.*, vol. 82, pp. 1431–1448, 1994.
11. F. Abrishamkar and Z. Siveski, PCS global mobile satellites, *IEEE Commun. Mag.*, vol. 34, pp. 132–136, September 1996.
12. J. R. Norbury, Satellite land mobile communication systems, *IEE Electron. Commun. Eng. J.*, pp. 245–253, November/December 1989.
13. F. Ananasso and F. D. Priscoli, The role of satellite in personal communication services, *IEEE J. Selected Areas Commun.*, vol. 13, pp. 180–196, 1995.
14. R. D. Gaudenzi, F. Giannetti and M. Luise, Advances in satellite CDMA transmission for mobile and personal communications, *IEEE Proc.*, vol. 84, pp. 18–39, 1996.
15. E. D. Re, C. L. Devieux, Jr., S. Kato, S. Raghavan, D. Taylor and R. Ziemer, Eds., Special issue on mobile satellite communications for seamless PCS, *IEEE Trans. Selected Areas Commun.*, vol. 13, February 1995.
16. R. Laane, Ed.-in-Chief, Special issue on satellite and terrestrial systems and services for travelers, *IEEE Commun. Mag.*, vol. 29, November 1991.
17. W. C. Y. Lee, *Mobile Communication Design Fundamentals*, John Wiley & Sons, New York, 1993.

18. W. C. Y. Lee, Mobile Cellular Telecommunications, McGraw-Hill, New York, 1995.
19. T. S. Rappaport, *Wireless Communications: Principles and Practice,* Prentice-Hall, Englewood Cliffs, NJ, 1996.
20. V. K. Garg and J. E. Wilks, *Wireless and Personal Communications Systems,* Prentice-Hall, Englewood Cliffs, NJ, 1996.
21. L. C. Godara, Application of antenna arrays to mobile communications — Part I: Performance improvement, feasibility and system considerations, *Proc. IEEE,* vol. 85, No. 7, pp. 1031–1062, July 1997.
22. K. Pahlavan and A. H. Levesque, Wireless data communications, *IEEE Proc.,* vol. 82, pp. 1398–1430, 1994.
23. H. L. Bertoni, W. Honcharenko, L. R. Maceil and H. H. Xia, UHF propagation prediction for wireless personal communications, *IEEE Proc.,* vol. 82, pp. 1333–1359, 1994.
24. B. H. Fleury and P. E. Leuthold, Radiowave propagation in mobile communications: An overview of European research, *IEEE Commun. Mag.,* vol. 34, pp. 70–81, February 1996.
25. R. C. French, The effect of fading and shadowing on channel reuse in mobile radio, *IEEE Trans. Veh. Technol.,* vol. 28, pp. 171–181, 1979.
26. J. H. Winters, Optimum combining for indoor radio systems with multiple users, *IEEE Trans. Commun.,* vol. COM-35, pp. 1222–1230, 1987.
27. J. B. Andersen, T. S. Rappaport and S. Yoshida, Propagation measurements and models for wireless communications channels, *IEEE Commun. Mag.,* vol. 33, pp. 42–49, January 1995.
28. F. Adachi, M. Sawahashi and H. Suda, Wideband DS-CDMA for next-generation mobile communications systems, *IEEE Commun. Mag.,* vol. 36, pp. 56–69, September 1998.
29. D. D. Falconer, F. Adachi and B. Gudmundson, Time division multiple access methods for wireless personal communications, *IEEE Commun. Mag.,* vol. 33, pp. 50–57, January 1995.
30. K. Raith and J. Uddenfeldt, Capacity of digital cellular TDMA systems, *IEEE Trans. Veh. Technol.,* vol. 40, pp. 323–332, 1991.
31. W. C. Y. Lee, Overview of cellular CDMA, *IEEE Trans. Veh. Technol.,* vol. 40, pp. 291–302, 1991.
32. K. S. Gilhousen et al., On the capacity of cellular CDMA system, *IEEE Trans. Veh. Technol.,* vol. 40, pp. 303–312, 1991.
33. R. L. Pickholtz, L.W. Milstein and D. L. Schilling, Spread spectrum for mobile communications, *IEEE Trans. Veh. Technol.,* vol. 40, pp. 313–322, 1991.
34. R. Kohno, R. Meidan and L. B. Milstein, Spread spectrum access methods for wireless communications, *IEEE Commun. Mag.,* vol. 33, pp. 58–67, January 1995.
35. N. Abramson, Multiple access in wireless digital networks, *IEEE Proc.,* vol. 82, pp. 1360–1370, 1994.
36. L. Pickholtz, D.L. Schilling and L.B. Milstein, Theory of spread spectrum communications — A tutorial, *IEEE Trans. Commun.,* pp. 855–884, May 1982.
37. L.B. Milstein, Wideband Code Division Multiple Access, *IEEE J. Sel. Areas Commun.,* vol. 18, No. 8, pp. 1344–1354, August 2000.
38. J. H. Winters, J. Salz, and R. D Gitlin, The impact of antenna diversity on the capacity of wireless communication systems, *IEEE Trans. Commun.,* vol. 42, pp. 1740–1751, 1994.
39. L. C. Godara, Application of antenna arrays to mobile communications- Part II: Beamforming and DOA considerations, *Proc. IEEE,* vol. 85, No. 8, pp. 1195–1247, August 1997.
40. M. Mizuno and T Ohgane, Application of adaptive array antennas to radio communications, Electronics & Communications in Japan, Part I: Communications, vol. 77, pp. 48–59, 1994.
41. N. D. Tripathi, J. H. Reed, and H. F. Van Landingham, Handoff in Cellular Systems, *IEEE Personal Commun.,* pp. 26–37, December 1998.
42. W. Tuttlebee, Cordless personal telecommunications, *IEEE Commun. Mag.,* vol. 30, pp. 42–53, December 1992.
43. T. A. Freeburg, Enabling technologies for wireless in-building network communications — Four technical challenges, Four solutions, *IEEE Commun. Mag.,* vol. 29, pp. 58–64, April 1991.
44. G. P. Pollini, Trends in handover design, *IEEE Commun. Mag.,* vol. 34, pp. 82–90, March 1996.

45. S.S. Rappaport and L. R. Hu, Microcellular communication systems with hierarchical macrocell overlays: Traffic performance models and analysis, *IEEE Proc.*, vol. 82, pp. 1383–1397, 1994.
46. R. Steel, J. Whitehead and W. C. Wong, System aspects of cellular radio, *IEEE Commun. Mag.*, vol. 33, pp. 80–86, January 1995.
47. I. Katzola and M. Naghsineh, Channel assignment schemes for cellular mobile telecommunication systems: A comparative survey, *IEEE Personal Commun.*, pp. 10–31, June 1996.
48. A. Noerpel and Y. B. Lin, Handover management for a PC network, *IEEE Personal Commun.*, pp. 18–24, December 1997.
49. D. Wong and T. J. Lim, Soft handoff in CDMA mobile systems, *IEEE Personal Commun.*, pp. 6–17, December 1997.
50. S. Tekinay and B. Jabbary, Handover and channel assignment in mobile cellular networks, *IEEE Commun. Mag.*, vol. 29, pp. 42–46, November 1991.
51. M. Mouli and M. B. Pautet, The GSM system for mobile communications. Cell & Sys, Paris 1992.
52. W. Van Den Broek et al., Functional models of UMTS and integration into future networks, *Electron. Commun. Eng. J.*, vol. 5, No. 3, pp. 165–172, June 1993.
53. Special Issue on IMT-2000, *IEEE Personal Commun.*, vol. 4, No. 4, August 1997.
54. M. Shafi, Wireless Communications in the Twenty-First Century: A Perspective, *Proc. IEEE*, vol. 85, No. 10, pp. 1622–1638, October 1997.
55. I. Kriaris et al., Third-generation mobile network architectures for the universal mobile telecommunications system (UMTS), *Bell Labs Technical Journal*, pp. 99–117, Summer 1997.
56. Y. Hu and R. Sheriff, The potential demand for the satellite component of the universal mobile telecommunications system, *Electron. Commun. Eng. J.*, pp. 59–67, April 1997.
57. W. Johnston, Europe's future mobile telephony system, *IEEE Spectrum*, pp. 49–53, October 1998.
58. D. O'Mahony, UMTS: The fusion of fixed and mobile networking, *IEEE Internet Computing*, pp. 49–55, January–February 1998.
59. A. Sasaki and M. Yabusaki, The Current Situation of IMT-2000 Standardization Activities in Japan, *IEEE Commun. Mag.*, vol. 36, No. 9, pp. 145–153, September 1998.
60. T.A. Rahman et al., The cellular phone industry in Malaysia: Toward IMT-2000, *IEEE Commun. Mag.*, vol. 36, No. 9, pp. 154–156, September 1998.
61. C. Shumin, Current Development of IMT-2000 in China, *IEEE Commun. Mag.*, vol. 36, No. 9, pp. 157–159, September 1998.
62. K. J. Wee and Y. S. Shin, Current IMT-2000 R&D status and views in Korea, *IEEE Commun. Mag.*, vol. 36, No. 9, pp. 160–164, September 1998.
63. S. Chia, The Universal Mobile Telecommunications System, *IEEE Commun. Mag.*, vol. 30, No. 12, pp. 54–62, December 1992.
64. P. W. Baier et al., Taking the challenge of multiple access for third generation cellular mobile radio systems — A European view, *IEEE Commun. Mag.*, vol. 34, No. 2, pp. 82–89, February 1996.
65. International Telecommunications Union, IMT-2000 Home Page, www.itu.int, June 1997.
66. ITU-M Recommendation M.816, Framework for Services Supported on Future Public Land Mobile Telecommunications Systems, pp. 22–30, 1994.
67. ITU-M Recommendation M.817, Future Public Land Mobile Telecommunications Systems Network Architectures, pp. 1–18, 1994.
68. J. S. Da Silva et al., European Mobile Communications on the Move, *IEEE Commun. Mag.*, vol. 34, No. 2, February 1996.
69. E. Dahlman et al., UMTS/IMT-2000 Based on Wideband CDMA, *IEEE Commun. Mag.*, pp. 70–80, September 1998.
70. T. Ojanperä, An overview of air interface multiple access for IMT-2000/UMTS, *IEEE Commun. Mag.*, pp. 82–95, September 1998.
71. P. G. Andermo, System Flexibility and its Requirements on a Third Generation Mobile System, Third IEEE International Symposium on Personal, Indoor and Mobile Radio Communications, Ottawa, 1992, pp. 397–401.

72. M. J. Beller, L. Chang and Y. Yacobi, Security for Personal Communications Services: Public Key vs Private Key Approaches, Third IEEE International Symposium on Personal, Indoor and Mobile Radio Communications, Boston, 1992, pp. 26–31.
73. ITU-M Recommendation M.1078, Speech and Voiceband Data Performance Requirements for Future Public Land Mobile Telecommunications Systems, pp. 110–135, 1994.
74. F. Ananasso et al., Issues on the Evolution Towards Satellite Personal Communications Networks, IEEE Global Telecommunications Conference, Singapore, November 1995, pp. 541–545.
75. P. Dondl, Standardization of the Satellite Component of the UMTS, *IEEE Personal Commun.*, vol. 2, No. 5, pp. 68–74, Ocober 1995.
76. ITU-M Recommendation M.818-1, Satellite Operation within the Future Public Land Mobile Telecommunications Systems, pp. 49–51, 1994.
77. ITU-M Recommendation M.1167, Framework for the Satellite Component of Future Public Land Mobile Telecommunications Systems, pp. 2–10, 1995.
78. WRC-2000, http://www.itu.int./brconf/wrc-2000/docs/index.html.
79. E. Del Re, A coordinated European effort for the definition of a satellite integrated environment for future mobile communications, *IEEE Commun. Mag.*, vol. 34, No. 2, pp. 98–104, February 1996.
80. ITU-M Recommendation M.1182, Integration of Terrestrial and Satellite Mobile Communications Systems, pp. 9–15, 1995.

2

Satellite-Based Mobile Communications

Michael John Ryan
University of New South Wales

2.1 Introduction

Mobile satellite communications began in 1976 with the launch by COMSAT of the MARISAT satellites to provide communications to ships at sea. The International Maritime Satellite Organisation (INMARSAT) was subsequently formed in 1979, and that organization now provides mobile satellite communications services to aircraft and land-based terminals. A number of national mobile satellite communications systems also serve the United States, Canada, Australia, and Japan with many more planned.

The spectacular growth of terrestrial mobile communications systems has provided a catalyst for efforts to provide global mobile communications through the use of mobile satellite communications systems in low-, medium-, and geostationary-Earth orbit.

Until now, second-generation terrestrial and satellite mobile communications systems have existed as two independent environments. However, these environments are beginning to combine to form a third-generation global mobile communications system in which terrestrial and satellite systems have complementary instead of independent roles and form a single universal integrated system.

This chapter addresses satellite mobile communications systems, that is, satellite systems providing telecommunication services directly to mobile end users.

2.1.1 A Brief History of Satellite Communications

In an article in *Wireless World* in 1945, Arthur C. Clarke proposed the idea of placing satellites in geostationary orbit around Earth such that three equally spaced satellites could provide worldwide coverage. However, it was not until 1957 that the Soviet Union launched the first satellite *Sputnik 1*, which was followed in early 1958 by the U.S. Army's *Explorer 1*. Both *Sputnik* and *Explorer* transmitted telemetry information.

0-8493-0124-6/02/$0.00+$1.50

The first communications satellite, the *Signal Communicating Orbit Repeater Experiment (SCORE)*, was launched in 1958 by the U.S. Air Force. *SCORE* was a delayed-repeater satellite, which received signals from Earth at 150 MHz and stored them on tape for later retransmission. A further experimental communication satellite, *Echo 1*, was launched on August 12, 1960 and placed into inclined orbit at about 1500 km above Earth. *Echo 1* was an aluminized plastic balloon with a diameter of 30 m and a weight of 75.3 kg. *Echo 1* successfully demonstrated the first two-way voice communications by satellite.

On October 4, 1960, the U.S. Department of Defense launched *Courier* into an elliptical orbit between 956 and 1240 km, with a period of 107 min. Although *Courier* lasted only 17 days, it was used for real-time voice, data, and facsimile transmission. The satellite also had five tape recorders onboard; four were used for delayed repetition of digital information, and the other for delayed repetition of analog messages.

Direct-repeated satellite transmission began with the launch of *Telstar I* on July 10, 1962. *Telstar I* was an 87-cm, 80-kg sphere placed in low-Earth orbit between 960 and 6140 km, with an orbital period of 158 min. *Telstar I* was the first satellite to be able to transmit and receive simultaneously and was used for experimental telephone, image, and television transmission. However, on February 21, 1963, *Telstar I* suffered damage caused by the newly discovered Van Allen belts. *Telstar II* was made more radiation resistant and was launched on May 7, 1963. *Telstar II* was a straight repeater with a 6.5-GHz uplink and a 4.1-GHz downlink. The satellite power amplifier used a specially developed 2-W traveling wave tube. Along with its other capabilities, the broadband amplifier was able to relay color TV transmissions. The first successful trans-Atlantic transmission of video was accomplished with *Telstar II*, which also incorporated radiation measurements and experiments that exposed semiconductor components to space radiation.

The first satellites placed in geostationary orbit were the *synchronous communication* (SYNCOM) satellites launched by NASA in 1963. *SYNCOM I* failed on injection into orbit. However, *SYNCOM II* was successfully launched on July 26, 1964 and provided telephone, teletype, and facsimile transmission. *SYNCOM III* was launched on August 19, 1964 and transmitted TV pictures from the Tokyo Olympics.

The International Telecommunications by Satellite (INTELSAT) consortium was founded in July 1964 with the charter to design, construct, establish, and maintain the operation of a global commercial communications system on a nondiscriminatory basis. The INTELSAT network started with the launch on April 6, 1965, of *INTELSAT I*, also called *Early Bird*. On June 28, 1965, *INTELSAT I* began providing 240 commercial international telephone channels as well as TV transmission between the United States and Europe.

A second global satellite system was established by Intersputnik, founded in 1971, to serve the 14 socialist countries that had not joined the INTELSAT consortium. In 1979, INMARSAT established a third global system. In 1995, the INMARSAT name was changed to the International Mobile Satellite Organisation to reflect the fact that the organization had evolved to become the only provider of global mobile satellite communications at sea, in the air, and on the land.

Early telecommunication satellites were mainly used for long-distance continental and intercontinental broadband, narrowband, and TV transmission. With the advent of broadband optical fiber transmission, satellite services shifted focus to TV distribution, and to point-to-multipoint and very small aperture terminal (VSAT) applications. Satellite transmission is currently undergoing further significant growth with the introduction of mobile satellite systems for personal communications and fixed satellite systems for broadband data transmission.

2.1.2 Types of Telecommunications Satellite Services

Because satellite communications cover the whole range of voice, data, and video transmission, telecommunication satellite services are normally classified into three types:

- *Fixed satellite service (FSS)* networks are mainly intended for long-distance operation of telecommunications networks. FSS satellites are employed to relay signals between large, complex, and expensive Earth stations, which are connected to the terrestrial telecommunications network.

- *Direct-broadcast satellite service (DBS)* networks transmit broadcast and TV signals from a large central Earth station, via a satellite to receive-only Earth stations. DBS receive stations either are distribution heads for cable TV or are located in homes for direct-to-home transmission.
- *Mobile satellite services (MSS)* networks are relayed via satellite between large fixed Earth stations and small mobile terminals fitted to a ship, an aircraft, or a vehicle. Increasingly, MSS networks are formed to relay communications to portable handheld terminals.

In 1996, the ITU defined the *Global Mobile Personal Communications by Satellite (GMPCS)* as comprising the following systems:

- *Geostationary Earth Orbit (GEO) MSS* are for voice and low-speed data mobile personal communications services.
- *Non-GEO (NGEO) MSS* are for narrowband mobile personal communications services excluding voice — because these are invariably based on low-Earth orbit (LEO) satellites, they are also called *Little-LEO*.
- *NGEO MSS* for narrowband mobile personal communications include voice, operating in LEO, medium-Earth orbit (MEO), or highly elliptical orbit (HEO) — also called *Big-LEO*.
- *GEO* and *NGEO FSS* offer fixed and transportable multimedia broadband services — also called *Super-LEO*.

This chapter focuses on MSS for personal communications and therefore considers the first three of these systems: GEO MSS, Little-LEO MSS, and Big-LEO MSS.

2.2 Satellite Orbit Fundamentals

2.2.1 Orbital Mechanics

An understanding of how orbits are chosen and what limitations they have for satellite performance is important in the design and use of a satellite system. Although the field of orbital mechanics is complex, a limited understanding is sufficient to consider the utility of various satellite orbits for MSS applications.

2.2.1.1 Kepler's Laws

Johann Kepler's three laws apply to the motion of satellites in elliptical orbits. Kepler developed these laws empirically, based on conclusions drawn from the extensive observations of Mars by Tycho Brahe (taken around the year 1600). They were originally defined in terms of the motion of the planets about the sun, but apply equally to the motion of natural or artificial satellites about Earth.

First Law

Kepler's first law states that the satellite will follow an elliptical path in its orbit around the primary body, in this case Earth, which is at one of the foci of the ellipse. As illustrated in Fig. 2.1, the point at which the satellite is closest to Earth is called the *perigee*, and the point that it is farthest away is called the *apogee*. In many cases, the orbits selected for use by communications satellites are circular, which is the special case of an ellipse when both foci are coincident.

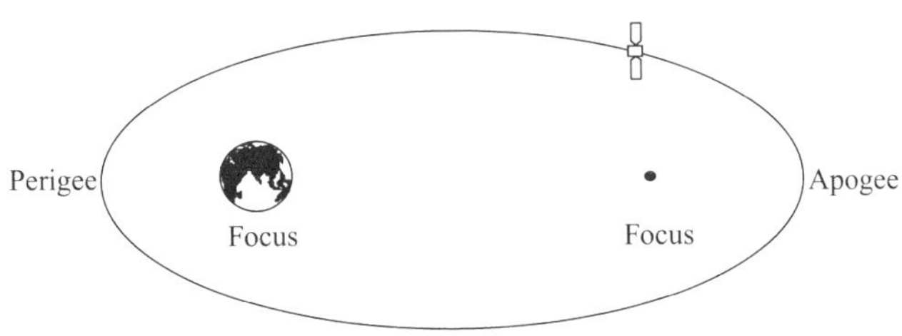

FIGURE 2.1 Illustration of Kepler's first law.

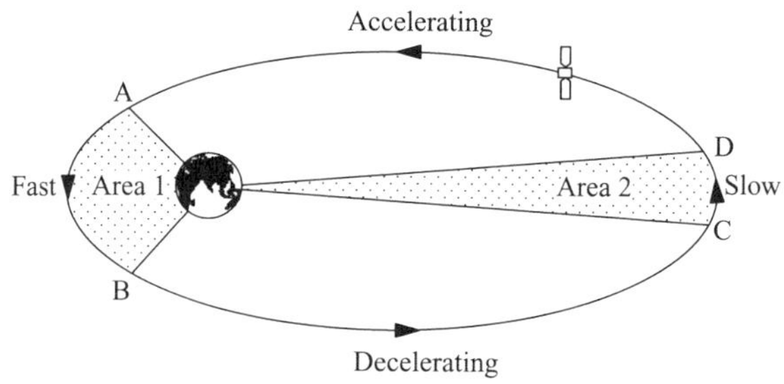

FIGURE 2.2 Illustration of Kepler's second law.

Second Law

Kepler's second law states that the line joining the satellite with the center of Earth sweeps out equal areas in equal times, as illustrated in Fig. 2.2. Area 1 is equal to area 2; the satellite sweeps out both areas in the same time. It follows that the satellite must move more quickly between points A and B than between points C and D.

Third Law

In his third law, Kepler stated that the cube of the mean distance of the satellite from Earth is proportional to the square of its period. Specifically, the orbit period of a satellite, T, at a height h above Earth's surface is

$$T = 2\pi\sqrt{\frac{(R+h)^3}{GM_e}} \text{ hours}$$

where Earth's radius $R = 6{,}378{,}155$ km, the gravitational constant $G = 6.67 \times 10^{-11}$ Nm^2/kg^2, and Earth's mass $M_e = 5.95 \times 10^{24}$ kg (neglecting the satellite mass).

2.2.2 Orbital Variations

Kepler's laws are sufficient to study simple satellite motion, but are not accurate enough to describe real satellites whose orbits are perturbed by Earth's gravitational anomalies and the effects of other bodies such as the sun and the moon. Earth is not a perfect sphere and the sun and the moon exert a gravitational pull on an orbiting satellite. These effects give rise to orbital perturbations and discrepancies between the motion predicted above and the actual motion of the satellite. Additionally, Earth's orbit and rotation must be taken into account to orient the satellite's orbit with respect to the stars.

2.2.2.1 Oblateness and Equatorial Ellipticity

Kepler's laws apply to bodies that are perfectly spherical. Earth is not a true sphere, with a flattening at the poles, and the equatorial circumference not quite circular. The distance from the center of Earth to the north or south poles is less than at the equator by about 20 km. This oblateness of Earth, together with the fact that the equator is not a circle, means that the direction of the force of gravity acting on an orbiting satellite is not toward the center of Earth but is slightly displaced. These effects cause satellites to drift and adjustments are typically required every few months to return them to a nominally geostationary position.

2.2.2.2 Lunisolar Perturbations

Although the effects of the gravitational forces from the moon, and to a lesser extent the sun, are small compared with Earth, there are noticeable perturbations to a geostationary satellite through an inclination of the orbit to the equator. This is evident whether or not the orbit is initially inclined. The energy required to correct for lunar–solar perturbations (perform *station-keeping*) is prohibitive for most satellite applications, although some satellites carry sufficient fuel for such corrections. The more usual approach for small geostationary satellites is to set the orbital inclination initially at 2° to 3°. The lunar–solar perturbation then causes the inclination to move through 0° and back to the initial angle in a period of 4 to 5 years, such that the angle of inclination remains small over the expected life of the satellite.

2.2.2.3 Solar Radiation Pressure

For geostationary satellites, the effects of solar radiation pressure must also be considered. The effect increases with an increase in the size of the surface area of the satellite that is projected in the direction of the sun. This is the case for large powerful satellites that use large solar arrays. The net effect of the solar radiation pressure on a geostationary satellite is to increase the orbital eccentricity and to introduce a disturbing torque affecting the north–south axis of the satellite. Such perturbations are corrected periodically.

2.2.2.4 Atmospheric Drag

Low-orbit satellites suffer atmospheric drag from the friction caused by collision with atoms and ions. The effect of drag is to reduce the ellipticity of an elliptical orbit, making it more circular, and to cause a loss of altitude of a circular orbit. At very low orbital altitudes the friction causes excessive heat on a satellite that finally results in its loss by burning. The orbital lifetime of a satellite (limited by drag) is a complex function of initial orbit height, geometry and mass of the satellite, and ionospheric conditions. However, in predicting a satellite's life, the orbital life must be distinguished from the operational life. The latter is the period during which a satellite performs the planned mission successfully. Moreover, a satellite can continue to orbit for some time after ceasing to function. Conversely, a satellite on low orbit may well reach its orbital life well before its operational life would expire.

For example, the orbital life of a small satellite in a 400-km circular Earth orbit is typically a few months, whereas the orbital life of a similar satellite in an 800-km circular orbit could be several decades. In the case of the 400-km orbit, the functional life of the satellite is mainly governed by the orbital lifetime (except for the less likely situation where the satellite equipment fails earlier), whereas in the latter the functional life depends on the lifetime of the satellite equipment.

2.2.3 Types of Orbit

The height of a satellite above Earth is a major factor in its utility for use within a communications system. Satellite height determines the orbit period, the time that the satellite is visible to a ground station, the footprint (coverage area on Earth's surface), the propagation delay of signals to and from the satellite, and the path attenuation. Other physical influences include the two Van Allen radiation belts (Fig. 2.3);

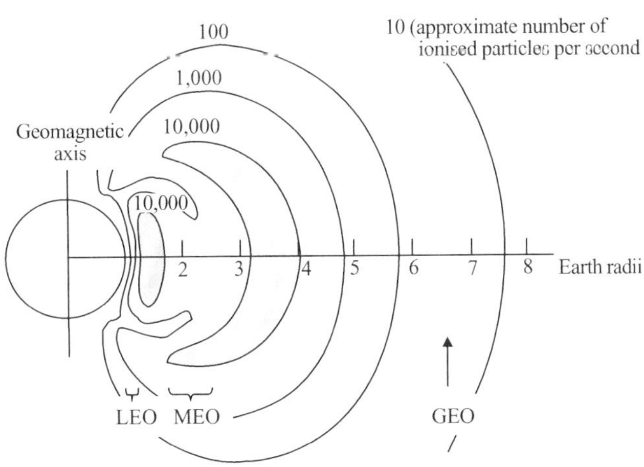

FIGURE 2.3 Satellite orbits and the Van Allen radiation belts.

peaks of radiation occur at altitudes of around 3,000 to 7,000 km and around 13,000 to 20,000 km where prolonged exposure can seriously shorten satellite lifetime. Additionally, altitudes lower than 400 km are uneconomical because of Earth's gravity and the drag of the atmosphere that reduces satellite lifetimes.

Four types of orbit are generally used for telecommunications satellites:

- *Low-earth orbit (LEO)* at heights of between 500 and 2,000 km
- *Medium-earth orbit (MEO)* between 5,000 and 15,000 km — also called *intermediate circular orbit (ICO)*
- *Geostationary earth orbit (GEO)* at 35,786 km
- *Highly elliptical orbit (HEO)* with an apogee that may be beyond GEO

2.2.3.1 Geostationary Earth Orbit

At geostationary altitudes, satellites have an orbital period equal to the period of rotation of Earth. Consequently, they remain in a fixed position in respect to a given point on Earth. An obvious advantage is they are available to all Earth stations within their shadow 100% of the time. The shadow of a satellite includes all Earth stations that have a line-of-sight (LOS) path to it and lie within the radiation pattern of the satellite antenna. At geostationary altitude, the satellite has coverage of 17.3° cone angle, or about 40% of Earth's surface. However, the polar areas are not covered.

Early launch vehicles did not have sufficient capacity to lift useful payloads into geostationary orbit so lower orbits were employed. As launch vehicles improved, GEO satellites became the most popular for telecommunications satellites because they have the following advantages:

- Only three satellites are needed to provide global coverage.
- Each satellite has a large footprint.
- Each satellite remains stationary with respect to Earth, minimizing the need for the terminal to track the satellite.
- The transmitted signal has only a small Doppler shift (affecting synchronous digital systems) caused by satellite movement.
- The technology required is well known after almost 40 years of operating geostationary satellites.

However, for mobile communications, GEO systems have the following major problems:

- High latency — round-trip delay of 250 ms leading to a total of 500 ms for a two-way conversation
- Poor coverage beyond 75° to 80° north and south of the equator
- High path attenuation (mostly from free-space loss) requiring high-power transmitters and large antennas at both the satellite and the mobile terminal
- Large losses resulting from shadowing by buildings in the urban environment
- Limited number of orbital slots available above each country

2.2.3.2 Low-Earth Orbit

LEOs have an altitude of 500 to 1000 km and move over Earth's surface at velocities ranging from 6 to 8 km/s. In contrast to terrestrial mobile communications systems, LEO networks have a highly dynamic network topology. Handover occurs, not only because of movement of mobiles but also because of the satellites passing over the terminals. Because a single LEO cannot provide a communications link for more than a short period, constellations of LEOs are required to provide continuous coverage. The LEO concept was first considered in the 1960s for transoceanic communications using passive reflectors. However, cost and technological constraints restricted large-scale use of LEO systems until the 1990s, when interest in their use grew because of advances in technology coupled with the growing cost of placing GEO satellites into orbit, the congestion of the geostationary orbit, and the development of low-cost launchers.

Reasons why LEO systems are more suited to personal communications services include

- They can make use of small handheld terminals because they require less power and use small omnidirectional antennas.
- They can make more efficient reuse of frequencies.
- They have minimum propagation delay — even at a maximum LEO height of 1000 km, the latency is less than 3.3 ms for each uplink and downlink leading to a total of 13 ms for a two-way conversation.
- They are below the Van Allen belts.
- They are less subject to shadowing than GEO systems.
- They can provide communications coverage of the entire globe.
- They have low satellite launching costs because satellites can be injected directly into orbit, with the ability to launch several satellites in one launch.

However, the use of LEO systems does have some difficulties, such as

- A large number of satellites are required to provide the same coverage as GEO systems.
- To provide uninterrupted communications, frequent handover is required between satellites because of the speed at which the satellite is traveling across the ground.
- Satellites spend considerable time covering empty space (such as oceans and deserts).
- Satellites have a shorter lifetime because of orbital decay.
- Satellite lifetimes are also limited to 5 to 7.5 years because the satellite spends time in eclipse by Earth, providing a significant demand on battery power.

2.2.3.3 Medium-Earth Orbit

MEO — also called intermediate circular orbits — orbit between 5,000 and 15,000 km above Earth's surface. At those altitudes, a terminal would only see the satellite for slightly longer than 1 h, requiring approximately ten satellites in two planes (each plane inclined at 45° to the equator) for complete global coverage. MEOs have larger footprints than LEOs and require fewer satellites to provide the same coverage. However, they require larger, more capable payloads (larger antennas and higher power transmitters) to cater to increased transmission losses.

MEO systems are generally better suited to personal communications services than GEO and have the following advantages when compared with other orbits:

- They are between the Van Allen belts.
- Latency of 17 to 50 ms for each uplink and downlink leads to a total of 67 to 200 ms for a two-way conversation.
- Only a few satellites are required to cover the whole Earth, and each satellite has a relatively large footprint.
- Intersatellite handover is not as frequent as for LEO systems.
- Fewer eclipse cycles than LEO allow longer battery lifetimes of more than 7 years.
- Lower cosmic radiation leads to longer expected lifetimes.
- Higher average elevation angle from terminal to satellite reduces shadowing of LOS.
- Shorter slant ranges require less power than GEO systems, resulting in smaller satellites and mobile terminals.

However, the MEO systems also have some disadvantages, including

- Doppler frequency offsets are larger than for GEO because of higher relative satellite motion, but lower than for LEO.
- Although fewer satellites are required than for LEO, the trade-off between number of satellites and latency is generally considered suboptimum.
- Satellites spend considerable time covering empty space.

TABLE 2.1 Comparison of Satellite Systems as a Function of Orbit

Characteristic	LEO	MEO	GEO
Satellite height (km)	600–1,500	9,000–11,000	35,800
Orbital period (hr)	1–2	6-8	24
Number of satellites	40–80	8–20	2–4
Two-way propagation delay (ms)	10–15	150–250	480–540
Satellite life (years)	3–7	10-15	10–15
Elevation angle	Medium	Best	Good
Visibility of satellite	Short	Medium	Permanent
Handheld terminal	Possible	Possible	Restricted
Handover	Frequent	Infrequent	None
Cost of satellite	Maximum	Minimum	Medium
Gateway cost	Highest	Medium	Lowest
Network complexity	Complex	Medium	Simplest
Radio frequency output power	Low	Medium	High
Propagation loss	Low	Medium	High

2.2.3.4 Highly Elliptical Orbit

HEO satellites have an apogee that may be beyond GEO. Unlike GEO, HEO systems also cover the polar regions, which is why a HEO orbit was chosen for the Molniya system to provide coverage of the former Union of Soviet Socialist Republics. Molniya has a period of 12 h, 8 of which provides coverage of the operational region — for continuous coverage three satellites are required. HEO systems have the advantages of

- Coverage of the polar regions with a small number of satellites
- Lower launching costs
- Higher elevation angle for ground stations
- Lower atmospheric loss

However, HEO systems have significant disadvantages such as

- Requirement for continual tracking of the satellite by Earth station
- Extensive eclipse periods
- Signal fading
- More complex control of the satellites and Earth stations

2.2.4 Orbit Selection

The selection of orbit is effectively dictated by the specifications of the ground terminal. If the terminal is to be handheld, satellites need to be in LEO — or at least MEO — so that the terminal can use low powers and a small omnidirectional antenna. However, at these lower altitudes, the satellite footprint is significantly reduced and a number of satellites are required depending on the altitude.

Table 2.1 summarizes the satellite systems as a function of orbit. Further design trade-off issues are discussed in later sections when specific systems are described.

2.3 Satellite Radio Path

In the radio path between the terminal and the satellite, the major attenuation tends to be caused by free-space loss. There is a multipath component, although this tends to be somewhat different from terrestrial paths. Doppler frequency shift is a significant factor in all types of orbit, although it is most significant in low-Earth orbits where the satellite has the highest speed across the ground. Propagation between the satellite and the ground is also influenced by a number of factors that depend on the polarization of the wave.

2.3.1 Path Loss in a Satelllite Link

2.3.1.1 Free-Space Loss

The major loss in an Earth-satellite path is free-space loss, L_{FS}, which is given by

$$L_{FS} = -92.44 - 20\log d_s f \ (\text{dB})$$

where d_s is the slant range in kilometers and f is in gigahertz.

Figure 2.4 shows the significant increase in free-space loss as the altitude is increased from LEO to GEO altitudes (for a zenith path at 2 GHz).

2.3.1.2 Atmospheric Absorption

Losses occur in Earth's atmosphere as a result of energy absorbed by atmospheric gases. Absorption at any frequency is a function of temperature, pressure, humidity of the atmosphere, and elevation angle of the satellite. Absorption increases with frequency, as elevation angle is reduced and as propagation path is increased. Specific frequency bands have high absorption; the first absorption band, caused by water vapor, is centered around 22.2 GHz, whereas the second band, caused by oxygen, is centered around 60 GHz.

2.3.1.3 Ionospheric Effects

Radio waves traveling between satellites and Earth stations must pass through the ionosphere, which is the upper region of Earth's atmosphere that has been ionized by solar radiation. The free electrons of this layer are not uniformly distributed but form layers. Furthermore, clouds of electrons (known as *traveling ionospheric disturbances*) may travel through the ionosphere and give rise to fluctuations in the signal that can only be determined on a statistical basis. The effects include *scintillation, absorption, variation in the direction of arrival, propagation delay, dispersion, frequency change*, and *polarization rotation*. All these effects decrease as frequency increases, most in inverse proportion to the frequency squared, and only the polarization rotation and the scintillation effects are of major concern for satellite communications. Other effects are negligible at the frequencies of main interest for satellite communication, except for a small fraction of time under events such as solar flares.

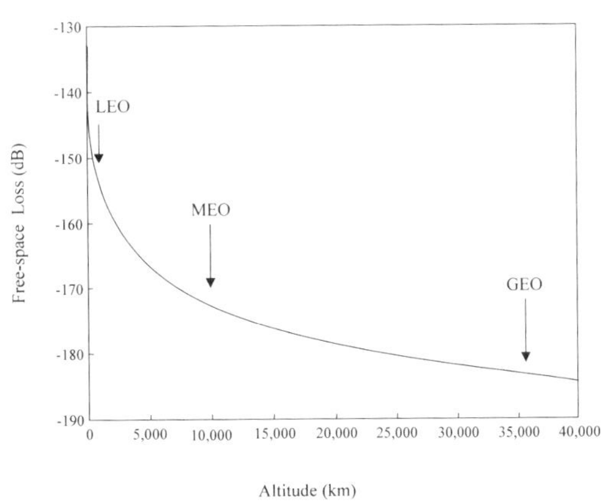

FIGURE 2.4 Free-space loss vs. altitude for a zenith path for a 2-GHz signal.

Polarization Rotation

The radio path between the satellite and the ground is affected by a number of factors that depend on the polarization of the wave. Because the relative geometry of the path is time varying, linear polarization may be difficult to orient, particularly for handheld terminals that may be operated in random orientation. Additionally, the signal is subjected to varying degrees of Faraday rotation, particularly at low frequencies. Both these effects can be overcome by the use of circular polarization in personal communications systems. However, circular polarization is still sensitive to diffraction and reflection of the signal resulting from obstacles in the path, because both reflection and diffraction are polarization sensitive.

Ionospheric Scintillation

Ionospheric scintillations are rapid variations in the amplitude, phase, polarization, or angle of arrival of radio waves. They are caused by small-scale refractive index variations in the F region of the ionosphere caused by local concentrations of ionization. The main effect of scintillations is fading of the signal. The fades can be quite severe, and they may last up to several minutes.

Ionospheric scintillation decreases in proportion to the inverse square of the frequency. Major scintillation is confined to frequencies below about 4 GHz. However, in extreme conditions such as magnetic storms, scintillation can cause problems up to 7 GHz.

Maximum levels of scintillation are observed in a region around the equator. Scintillation also increases in conditions of high solar activity and has a diurnal variation with high sunspot numbers; the peak levels occur approximately 1 to 2 h after sunset. It is interesting to note that, unlike tropospheric scintillation, ionospheric scintillation is independent of the elevation angle of the radio path.

When scintillation is expected at locations of interest, a link margin is provided to mitigate the effect. Link margins of several decibels could be required to achieve high link reliabilities for poorly sited Earth stations and operations below 3 to 4 GHz. At 4 GHz, attenuation resulting from ionospheric scintillation may vary from 0.5 to ~10 dB for 0.1% of the time, depending on the location and the level of solar activity.

2.3.1.4 Attenuation from Hydrometers

Hydrometer is a general term referring to condensed water vapors in the atmosphere, which includes rain, hail, ice, fog, cloud, and snow. All types of hydrometers produce transmission impairments. However, raindrops produce, by far, the maximum attenuation by absorbing and scattering radio waves.

Rain attenuation is a function of *rain rate* (measured in millimeters per hour) in the region of an Earth station. The *specific attenuation*, α, can then be calculated:

$$\alpha = aR_p^b \quad \left(\text{dB km}^{-1}\right)$$

where R_p is the rain rate that would be exceeded p percentage time of the year, and a and b are coefficients that depend on frequency and polarization.

Clouds and fog are suspended water droplets, usually less than 0.1 mm in diameter. Attenuation in the radio path depends on the liquid water content of the atmosphere along the propagation path. The liquid content of fog is generally low and the attenuation from fog is negligible for satellite communications. Hail, ice, and snow have little effect on attenuation because of the low water content. However, ice can cause depolarization and subsequent polarization loss.

Only a small link margin is required below 2 GHz to accommodate the additional loss introduced by rainfall. However, at higher frequencies such as 20 to 30 GHz, the rainfall loss is very significant, especially at low elevation angles where an additional link margin of tens of decibels may be required.

2.3.1.5 Multipath Propagation

Unlike terrestrial communications, there are few multiple paths in the earth–satellite path. However, depending on the elevation of the satellite, the path can be obstructed by buildings and trees. From 0° to 20° elevation, the signal propagation is similar to that of terrestrial mobile communications systems in that it suffers frequent multipath effects and significant blockage from obstacles. Above 20°, few multiple paths exist and the signal can generally be considered to be in free space. Unfortunately, most radio paths in personal communications are generally to LEO and MEO heights so that the typical maximum elevation of the satellite is not far above the lowest design elevation angle. The signals at the mobile terminal consist of a direct ray and a Rayleigh distribution of reflected rays with various amplitudes and phases. The resulting distribution may be modeled by the Rice distribution.

Mitigation effects for multipath include the selection of coding and modulation techniques that are able to provide robust operation in a multipath fading environment. A common time-domain strategy is to use convolutional coding and interleaving. Alternatively, a frequency-domain approach is to use convolutional coding and a spread-spectrum technique. ARQ schemes are used to ensure that lost data is retransmitted. The use of high-gain antennas can also significantly reduce multipath.

2.3.1.6 Doppler Shift

The signal transmitted from the satellite is subject to a Doppler shift that results from motion of the satellite as well as the movement of the ground station because of the rotation of Earth. The received signal therefore has the frequency:

$$f_r = f_t \pm f_d = f_t \pm \frac{v_r f_t}{c}$$

where f_r is the received frequency, f_d is the Doppler shift, and v_r is the relative velocity of the satellite relative to the receiver. Note that the Doppler shift is affected by whether the satellite is approaching or receding, which results in the ambiguous ± sign.

2.3.2 Frequency Selection

Table 2.2 lists the frequency bands allocated to satellite transmission. Commercial MSS systems predominantly operate in L-band, although systems will use C-band as well as Ka-band for satellite control and connection to gateways. As discussed earlier, lower frequencies experience lower attenuation. To minimize the power that must be generated onboard the satellite, the downlink frequency is chosen to be lower than the frequency for the uplink.

L-band is very useful for mobile communications because long wavelengths can penetrate many structures and less powerful transmitters are required. L-band is also the least affected by rain because rain attenuation is negligible in this band. However, L-band does suffer from ionospheric scintillation, which results in a split of the signal into direct and refracted paths. L-band also uses circular polarization to reduce the effect of Faraday rotation. S-band suffers less than L-band from ionospheric effects. However, it suffers slightly higher atmospheric attenuation from rain.

Ku-band has medium wavelengths that can penetrate many obstacles and provide high bandwidths. The band requires smaller antennas, but is more affected by rain attenuation.

Ka-band frequencies provide large amounts of available spectrum and high bandwidths. However, powerful transmitters are required and the short wavelengths are subject to strong rain fade — large rain attenuation creates a major technical challenge.

TABLE 2.2 Frequency Bands Allocated to Satellite Transmission

Frequency Band	Frequency (Uplink/Downlink)	User	Service
UHF	400/225 MHz	Military	Mobile
L-band	1.6/1.5 GHz	Commercial	Mobile
S-band	3/2 GHz	Commercial	Satellite control
C-band	6/4 GHz	Commercial	Fixed
X-band	8/7 GHz	Military	Fixed/mobile
Ku-band	14/12 GHz	Commercial	Fixed
Ka-band	30/20 GHz	Commercial	Fixed
Ka-band	44/20 GHz	Military	Fixed/mobile

2.4 Multiple Access Schemes

Whenever access to a satellite is required for a number of Earth stations, it is necessary to employ a multiple-access scheme to allow for a distinct separation between the uplink and downlink transmissions to and from each Earth station. Three *multiple access* schemes are commonly used: *frequency division multiple accessing (FDMA)*, *time division multiple accessing (TDMA)*, and *code division multiple accessing (CDMA)*. Each scheme has its own characteristics, advantages, and disadvantages.

With FDMA, each Earth station is assigned specific uplink and downlink frequency bands within an allotted satellite channel bandwidth. That is, Earth station transmissions are separated in the frequency domain. Each terminal may operate independently with its own carrier, bandwidth, modulation, coding, and data rate. Antenna size varies for each terminal, based on individual link budget calculations. The disadvantages of FDMA include the requirement to reduce the transponder power by approximately one half to minimize intermodulation products. Precise control is also required over uplink power.

With TDMA, all Earth stations use the same carrier frequency and each station transmits a short burst of information during a specific time slot (epoch) within a TDMA frame. The bursts must be synchronized so that each station's burst arrives at the satellite at a different time. That is, transmissions from different Earth stations are separated in the time domain. TDMA is generally superior to FDMA because intermodulation distortion is lower and the full downlink power is available to all users. The main disadvantage is the requirement of accurate timing and synchronization.

With CDMA (sometimes called spread-spectrum multiple access), all Earth stations transmit within the same frequency band and, for all practical purposes, have no limitation on when they may transmit. Signal separation is accomplished by spread-spectrum techniques where each user has a unique code that is used to generate a pseudorandom sequence that is used to spread out the transmission to occupy the entire bandwidth. That is, users are separated by a unique code. At the receiver, the spread signal is despread by applying the same code to the received data. Although CDMA provides high immunity to jamming and has a low probability of interception, it is expensive and can cater to only a limited number of users.

2.5 Mobile Satellite Communications Systems

As discussed earlier, the ITU GMPCS defines three systems for the provision of MSS for personal communications as follows:

- GEO MSS — satellites operating in geostationary orbit providing voice and low-speed data mobile personal communications services
- Little-LEO — NGEO MSS for narrowband mobile personal communications services excluding voice, generally operating in LEO
- Big-LEO — NGEO MSS for narrowband mobile personal communications including voice, operating in LEO, MEO, or HEO

The next section briefly describes the major MSS systems for personal communications.

2.5.1 Geostationary Earth Orbit Mobile Satellite Services Systems

This section briefly gives a short outline of major GEO MSS systems, which use satellites in geostationary orbit to provide voice and low-speed data mobile personal communications.

INMARSAT was created in 1979 with approximately 40 members and has grown to more than 81 member countries. In 1990, INMARSAT launched four satellites into geostationary orbit and launched a fifth satellite in 1998. The new satellites are capable of providing spot beams to increase capacity by more than 20 times over earlier generations. The main advantage is that the increased power and sensitivity of the newer satellites has allowed a significant reduction in the size of the earth terminals, which are now little bigger than a laptop computer. Although INMARSAT provides a number of services, the most mobile is the INMARSAT-M terminal, which is briefcase sized and can provide global digital 6.8-kbps voice and 2.4-kbps fax and data.

Agrani is Afro-Asian Satellite Communications (ASC). Agrani is expected to cover 54 countries — Turkey to Singapore and Byelorussia to Somalia including the Middle East, India, and much of China. Two GEO satellites provide mainly voice, but also provide fax, data, and messaging for fixed and mobile users. The first satellite (expected in 2002) will mainly cover the Indian region, with the second extending coverage across Africa.

The American Mobile Satellite Corporation (AMSC) satellite provides coverage of the United States and regions of U.S. coastal waters, providing voice, messaging, and data services.

The Asian Cellular Satellite System (ACes) *Garuda-1* satellite was successfully launched in February 2000. ACes covers most of Asia from Pakistan to Japan, including southern China and all Association of Southeast Asian Nation (ASEAN) countries and provides mobile and fixed voice, data, fax, and paging.

Euro-African Satellite Telecommunications (EAST) is planned to provide a range of services including narrowband communications to handheld user terminals and higher bandwidth services to VSAT antennas. *Optus MobileSat II* comprises the Australian Optus B1 and B3 satellites and ground stations in Perth and Sydney. Over 1500 mobile terminals operate in the D-band with uplinks and downlinks at 1.6 and 1.5 GHz, respectively. These enable users to communicate from any location on the Australian mainland and 200 km out to sea. Both voice and low-rate data communications are provided.

Thuraya plans to provide high-powered GEO satellites that will cover a footprint of 99 countries throughout Europe, North and Central Africa, the Middle East, Central Asia, and the Indian Subcontinent. The first satellite was launched in October 2000. *Thuraya* will provide voice, fax, data, messaging, and global positioning system (GPS) within handheld and vehicular telephones that will integrate both satellite and cellular communications. A second satellite is planned to provide backup and expansion; a third satellite may follow if required. Other planned GEO MSS systems include: *Africom*, *APMT*, and *Cyprus GEM*.

2.5.2 Little-Low-Earth Orbit Systems

Little-LEO systems are non-GEO (NGO) MSS for narrowband mobile personal communications services excluding voice. Advances in antenna design, signal reception, and miniaturization have allowed the deployment of small LEO systems that can support data transmission at 100 to 300 bps. These systems operate at a frequency below 1 GHz and are only appropriate for nonvoice, store-and-forward messaging.

2.5.2.1 OrbComm

OrbComm was granted a license by the U.S. Federal Communications Commission (FCC) in 1994 and the first two satellites were launched in April 1995. OrbComm provides mobile tracking, remote monitoring, and commercial and personal messaging services. The final OrbComm constellation will consist of 48 small (40 kg) satellites in LEO with a planned life of approximately 4 years. Space-to-ground communications is via very high frequency (VHF): 148 to 150.05 MHz uplink and 137 to 138 MHz downlink. Terminals will transmit at a burst rate of 2.4 kbps and receive at a burst rate of 4.8 kbps giving an effective throughput of 300 bps.

Messages will be sent via OrbComm satellites to gateway stations on Earth, which will either forward the message directly to the destination via terrestrial networks, or store it for access on demand. Although the effective throughput is low, global coverage is provided.

2.5.2.2 Other Systems

VITAsat is managed by Volunteers in Technical Assistance (VITA), a U.S. nonprofit organization dedicated to bringing technical assistance to the developing world. It has entered into agreements permitting usage of two already-orbiting satellites (*HealthSat-2* and *UoSAT-12*) to bring low-cost e-mail services to rural and isolated areas of developing countries.

Leo One Worldwide has announced a little LEO data-messaging system based on 48 satellites at 950 km in eight orbital planes inclined at 50°. The system will provide data rates of 2.4 to 9.6 kbps for the subscriber uplink and 2.4 kbps for the subscriber downlink. Gateway uplinks and downlinks will operate at 50 kbps. Other systems proposed include *E-Sat*, and *GEMnet*.

2.5.3 Big-Low-Earth Orbit Mobile Satellite Services Systems

Big-LEO systems are NGEO MSS for narrowband mobile personal communications including voice, operating in LEO, MEO, or HEO. These systems operate between 1 and 3 GHz and provide the full range of mobile services including voice and data. They are generally more complex than little LEOs and more satellites are deployed.

2.5.3.1 Iridium

The first fully functioning big-LEO system was Motorola's Iridium, which was probably the most complex of the MSS personal communications systems, providing worldwide voice, data, facsimile, and paging. Iridium satellites perform onboard processing, not only to translate a waveform frequency and retransmit it, but also to process the waveform down to baseband. This allows the satellites to support a variety of packet-oriented services, including routing, flow control, and error detection and correction.

The Iridium space segment consists of 72 satellites arranged in six orbital planes with 11 active and 1 spare satellite per plane at an altitude of 780 km. Satellites were expected to have a life of 5 to 7 years. Each satellite had 48 spot beams, each of which could be shared four ways using time division — the L-band, TDMA-TDD system provided 230 simultaneous conversations per satellite.

Iridium is the only big LEO to use intersatellite links to obviate the need to downlink traffic to hub stations on Earth. Each satellite has four cross-links; one forward within a plane, one backward within a plane, and two across planes. Cross-links operate at 25 Mbps at frequencies between 22.55 and 23.55 GHz. Although onboard processing and satellite cross-links increase flexibility, the advantages are offset by an increase in complexity and satellite weight (500 kg) leading to a satellite cost of U.S. $62 million.

Iridium phone calls are made through the shortest route including the satellite constellation to the terrestrial gateway closest to the destination. Intersatellite links and onboard satellite switching are used to route calls back to a satellite that can "see" a ground terminal. To provide global coverage, 12 ground stations are employed.

Despite meeting most planning deadlines, Iridium failed to attract sufficient subscribers to avoid bankruptcy and, in March 2000, the service was closed. Plans to bring the constellation of satellites down in a controlled reentry were aborted in late 2000 when the assets of Iridium LLC were acquired by Iridium Satellite LLC. The new owner does not see Iridium as a mass consumer service and intends to concentrate on industrial markets and specialized segments.

2.5.3.2 Globalstar

Globalstar has adopted a much simpler approach than Iridium and has employed satellites as a bent pipe with the same technology as used in GEO-based systems. The Globalstar system has 48 satellites twice as high as Iridium at an altitude of 1414 km inclined at 45° and 135° to the equator. Satellites have a lifetime of 5 to 15 years, weigh 704 lb, and have a capacity of 2800 full-duplex circuits. The constellation has eight orbital planes and covers from 70° south to 70° north. Globalstar signals go up to the nearest satellite and down from there to a terrestrial gateway that the satellite can see and are then passed on to existing fixed and cellular telephone networks; the satellite must be in sight of a ground station to complete a call. Complex intersatellite routing is avoided by having 38 ground stations worldwide.

Globalstar offers voice, SMS, roaming, positioning, facsimile, and data. Satellites have six spot beams using L-band CDMA; the full constellation can provide 28,000 simultaneous voice and data channels at 4.8 kbps.

2.5.3.3 Ellipso

Ellipso provides UHF CDMA voice, messaging, and positioning services through 17 satellites. Two inclined, elliptical orbital planes (with an apogee of 2903 km and a perigee of 425 km), each with five satellites effectively providing a LEO system in the Southern Hemisphere and a MEO system in the

Northern Hemisphere. Ellipso will also launch seven of its satellites into two slightly inclined circular planes around the equator to supplement services in the tropics.

The Ellipso ground segment employs 12 ground control stations (GCSs), deploying them near regional fiber hubs for networking efficiency. Each GCS typically tracks and uses two satellites while acquiring a third. The GCS also determines the position of subscriber terminals for administrative, billing, and application purposes.

2.5.3.4 Intermediate Circular Orbit

The Intermediate Circular Orbit (ICO) system was developed as a spin-off from INMARSAT (originally called INMARSAT-P, or Project 21). ICO provides full Earth coverage with ten satellites at 10,355 km arranged in two planes of five satellites in each plane. Satellites have an orbital period of 6 h with each satellite being visible from one point on Earth's surface for typically about 20 min. Twelve ground stations, called *satellite access nodes (SANs)*, are linked by high-speed terrestrial networks. ICO terminals are dual-mode (satellite and terrestrial) handsets and all operations are through existing cellular networks. ICO signals go from ground station to satellite to ground station with users at each end connecting through cellular and land networks. Each ICO satellite has a capacity of around 4500 voice channels using S-band CDMA.

The original ICO system suffered a number of early problems including launch failures during which one satellite was lost. ICO has emerged from bankruptcy with investment from ICO-Teledesic. NewICO has a planned service start date of 2003.

2.5.3.5 Constellation ECCO

The Constellation ECCO system will consist initially of one plane of 12 satellites (11 operational and 1 spare) in circular orbit around the equator, at an altitude of 2000 km. The satellites will transmit directly to mobile and fixed-site users in the 2483.5 to 2500 MHz band and receive directly from these users in the 1610 to 1626.5 MHz band. The system will provide phone and data services to all areas between the tropics.

Each satellite will have 24 antenna beams that collectively cover one eleventh of the equatorial belt. Each of the first-generation satellites will be able to support up to 192,000 subscribers and the whole system will be able to support as many as 1,392,000 subscribers. The design lifetime of these satellites is a minimum of 5 to 7 years.

2.5.3.6 Other Systems

Other planned big-LEO systems include *Movisat*, *MSAT*, and *Satphone*.

2.6 Summary

Over the next few years, a large number of consortia have launched, or are preparing to launch, constellations of satellites that will provide continuous, global phone and data services to mobile terminals. Most systems will be launched into LEO where there are considerable advantages to be gained over higher orbits. Because of the lower altitude, mobile terminals can be small personal handsets with low antenna gain and low-output power. These systems hold the promise of achieving third-generation mobile communications systems of integrating terrestrial and satellite networks providing a similar interface to a small handheld terminal.

References

Comparetto, G., and Ramirez, R., Trends in mobile satellite technology, *IEEE Computer*, 44, 1997.

Elbert, B.R., *The Satellite Communications Application Handbook*, Artech House, Norwood MA, 1997.

Evans, J.V., Satellite systems for personal communications, *IEEE Antennas and Propagation Magazine*, 39, 1997.

Jamalipour, A., *Low Earth Orbital Satellites for Personal Communication Networks*, Artech House, Norwood MA, 1998.

Kiesling, J.D., Land mobile satellite systems, *Proceedings of the IEEE,* 78, 1107, 1990.
Miller B., Satellites free the mobile phone, *IEEE Spectrum*, 26, 1998.
Ohmori, S., Wakana, H., and Kawase, S., *Mobile Satellite Communications*, Artech House, Norwood MA, 1998.
Pattan, B., *Satellite-Based Global Cellular Communications*, McGraw-Hill, New York, 1998.
Silk, R., and Bath, M., Future trends in satellite communications, *British Telecommunications Engineering*, 17, 73, 1998.
Wu, W.W., Miller, E.F., Pritchard, W.L., and Pickholtz, R.L., Mobile satellite communications, *Proceedings of the IEEE,* 9, 1994.

3

Propagation Prediction for Urban Systems

Henry L. Bertoni
Polytechnic University

Saúl A. Torrico
Comsearch

3.1 Introduction

To provide mobile communication services to subscribers, the radio links in modern wireless systems cover relatively small distances. The short links allow for reuse of the spectrum in different regions, and work with lower radiated power to save battery life. Because the customers for modern wireless applications are predominately among buildings, the buildings have the greatest influence on the propagation over short distances. For the frequencies used, the wavelength is small compared with building dimensions. As one result of the interaction with the buildings, the received signal exhibits a dependence on the link distance (range dependence) that is different than that of free space. The small wavelength also results in significant variations of the signal over the scale of the building dimensions (shadow fading). Finally, strong scattering by the buildings causes the signals to arrive at the antennas via many different paths.

In all types of systems the received power level is of fundamental importance, so that knowledge of the range dependence and shadow fading are critical in system design. However, the way in which

0-8493-0124-6/02/$0.00+$1.50

multipath is perceived depends on the type of system. Narrowband systems, such as first-generation cellular, are affected by the rapid spatial variation because of interference among the multipath arrivals. Systems that make use of digital signaling require a wider bandwidth, and are affected by the time response of the channel resulting from the presence of many delayed versions of the transmitted signal (echo). For these systems the time-delay characteristics of the channel are of importance, in addition to the spatial amplitude variation. Smart antennas and other multiple antenna systems, which are being developed to improve system capacity, require knowledge of the distribution in angle of arrival of the scattered signals.

3.1.1 Need for Predicting Channel Characteristics

The channel characteristics are found to vary from one mobile location to another, so that they are usually described by means of statistical measures and distribution functions. For example, narrowband signals exhibit fast fading over the scale of one half wavelength because of the interference of waves arriving from all directions. For non-line-of-sight paths (non-LOS), the fast fading causes the magnitude of the received voltage to have statistical properties that are approximately those of a Rayleigh distribution [1, 2]. When the fast-fading variations are averaged out over a distance of 20 or so wavelengths, the resulting average shows random variations resulting from shadowing by the buildings. This middle scale of variation has the properties of a lognormal distribution [1, 3]. On the largest scale, ranging from hundreds of meters to kilometers, the received signal is found to have range dependence of the form A/R^n, where the factors A and n are found after suitable signal averaging [4].

To experimentally obtain reliable statistical measures of these and other statistical properties of interest requires measurements at many locations. Considering the number of frequencies, base station locations, building environments, statistical channel parameters, and so forth that are of interest to the designer, it is unrealistic to find the characteristic solely through experiments. Theoretical models, when calibrated against measurements for one set of parameters, can be used to find the statistical measures for other frequencies, for other link geometry, and in other building environments. Theoretical models also help to identify when two different environments will give rise to the same statistics. In general, predicting the channel statistics requires matching the theoretical approach to the building environment and the particular characteristic being sought. For example, we may replace a relatively homogeneous building environment by an idealized "average environment" to compute the range dependence. We make extensive use of this approach in Sections 3.2 to 3.5 to predict the range dependence of the spatial average signal over flat earth, and to account for trees and terrain. Alternatively, we may account for variations in the buildings, or use a detailed representation of the buildings, to compute the signal received at many subscriber locations along the streets. In this way we can simulate an actual measurement campaign, and extract from the results the statistical properties of the desired channel characteristic. This approach is used in Section 3.6.

3.1.2 Matching Prediction Methods to the Building Environment

Outside of the high-rise core, cities are composed of rows of buildings having nearly uniform height, with an occasional high building or cluster of high buildings. The buildings are taller and closer together near the core, but become lower and more spread farther from the core and into the suburbs. The buildings are located in rows along the streets that provide access. The cost of land results in a close side-to-side spacing between neighboring houses, except in the distant suburbs. In some neighborhoods, the buildings are attached to each other with no gap between them. In cities, the spacing between neighboring houses is generally less than the width of the houses. The presence of the access streets imposes a significant spacing between the fronts of the houses. Likewise, there is ample back-to-back spacing to accommodate a yard. As seen from the air, the separation of houses across the streets is about the same as the back-to-back separation across the yards.

The street grid, or portions of it, forms a rectangular mesh that divides the land into blocks. The narrow width of these blocks ranges from 80 m or less, up to more than 120 m. A row of buildings lines each side of the block, and sometimes the ends of the blocks as well. The length of a block is two or three times its width. Outside the core area, the base station antennas are near to, or above, the rooftops, and propagation takes place primarily over the buildings. When viewed from above, the Fresnel zone about the ray crossing the rows of buildings is only the width of several buildings at its widest point. As a result, modeling the range dependence can be carried out by replacing the individual buildings in a row by a continuous smooth prism of uniform height, as done in Sections 3.2 and 3.3. The spacing between rows, as measured along the ray, are roughly equal even when the street grid is not rectangular. Random variations in building height contribute to the shadow loss for propagation over low buildings. The statistical properties of the shadow loss can be found from Monte Carlo simulation of the propagation by assigning random heights to the buildings [5, 6]. Vegetation can be modeled as a partially absorbing medium either by itself, or in conjunction with buildings, as is discussed in Section 3.4. Terrain effects can be examined by giving the buildings a systematic height variation with respect to sea level, as in Section 3.6.

For short links, the received signal may depend on the exact shape of the buildings and the placement of the antennas in relation to them. If the antennas are well below buildings, such as in Manhattan, propagation will be around the sides of the buildings rather than over the tops, and the footprints of the buildings will be important for the predictions. However, if the buildings are of mixed heights, propagation is over, as well as around the buildings, so that the building height is also important. For links such as these, site-specific prediction codes have been developed that use ray tracing in conjunction with a building database, as discussed in Section 3.6. Codes working with two-dimensional (2D) building databases are useful for predicting propagation in building environments like Manhattan, whereas codes working with a three-dimensional (3D) building database have wider application. Site-specific predictions can also be used for Monte Carlo simulation of various statistical channel parameters aside from received power. This new area of prediction is discussed in Section 3.6.

When describing the effect of the buildings on the received signal, it is convenient to introduce the concept of path gain for isotropic antennas that are match terminated. The path gain PG in decibels is the received power in decibels measured, less the sum of the transmitted power in decibels measured and the antenna gains. With this definition, the path gain is the negative of the path loss L for isotropic antennas. Because of reciprocity, the path gain is the same no matter if the base station antenna is transmitting and the mobile receiving, or vice versa, provided that the transmission is at the same frequency. In many cases it is convenient to derive the path gain by thinking of the base station as the transmitter and the mobile as the receiver.

3.2 Range Dependence for Macrocellular Applications

Macrocellular networks make use of high base station antennas to achieve coverage out to distances R from the base station that is in the range from 1 to 20 km. For most cities, such distances lie outside the high rise core, in regions where the buildings are of more or less uniform height, and the propagation takes place past many rows of buildings. If we restrict the prediction to that of the range dependence, we need only use average descriptors of the buildings. However, over such distances vegetation and terrain effects can be significant, as discussed in Sections 3.4 and 3.5. Except on a few streets aligned with the base station, the LOS path from the base station antenna to mobiles at street level is blocked by the intervening buildings. Thus, propagation must take place through the buildings, between them, or over the rooftops. Signals propagating through the buildings are highly attenuated by the exterior and interior walls. Except in the distant suburbs, the gaps between buildings are small and not generally aligned with the base station, or aligned from row to row. As a result, the signals cannot readily propagate through the gaps. Therefore, propagation must primarily take place over the rooftops.

To predict the range dependence of the spatial average path gain for macrocells, the individual buildings in a row are replaced by a continuous smooth prism, as seen in the end view in Fig. 3.1. All rows are assumed to have the same height, and each row of buildings is separated by the same distance d, where d is the average of the separation from back-to-back across the yards and front-to-front across the streets.

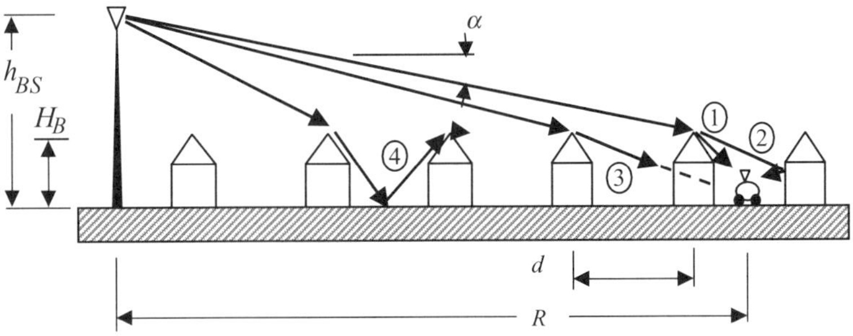

FIGURE 3.1 Propagation paths of UHF signals over the rows of buildings to a mobile located in a low-rise building environment.

By using this model of the buildings, the mean path gain PG in decibels is given by the sum of three factors [7, 8]

$$PG_{dB} = 10\log PG_0 + 10\log PG_1 + 10\log PG_2 \tag{3.1}$$

Here, PG_0 is the free space path gain, which is the ratio of received to radiated power for isotropic antennas, and is given by

$$PG_0 = \left(\frac{\lambda}{4\pi R}\right)^2 \tag{3.2}$$

where λ is the wavelength. The horizontal separation R is used in Eq. (3.2) to approximate the distance from the base station to the mobile. The term PG_1 in Eq. (3.1) is the reduction of the fields arriving at the buildings near to the mobile as a result of diffraction past the previous rows of buildings. The term PG_2 is the reduction of the fields resulting from diffraction from the rooftops down to ground level. These two terms are discussed in more detail in the following sections.

3.2.1 Diffraction from Rooftops Down to Street Level

Parsons [9] was the first to suggest that the radio waves propagate over the buildings and are diffracted down to street level at the rooftops next to the mobile. Ikegami et al. [10] analyzed the diffraction process by which the rays reach the street level receiver from different directions after diffraction at the rooftop. This process is suggested in the side view in Fig. 3.2, where the two rays giving the major contribution are shown. The first of these is diffracted from the rooftop of the building nearest the mobile in the direction of the base station, while the second is reflected from the face of the building across the street. Additional rays contribute to the process by which the field goes from the rooftops down to street level.

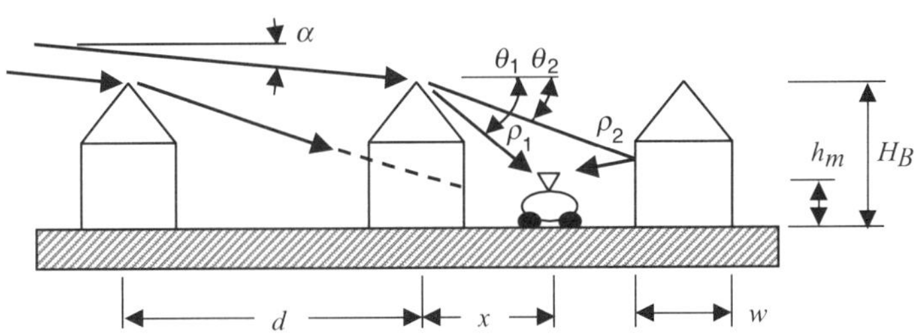

FIGURE 3.2 Some of the many paths by which the rooftop fields are diffracted down to street level.

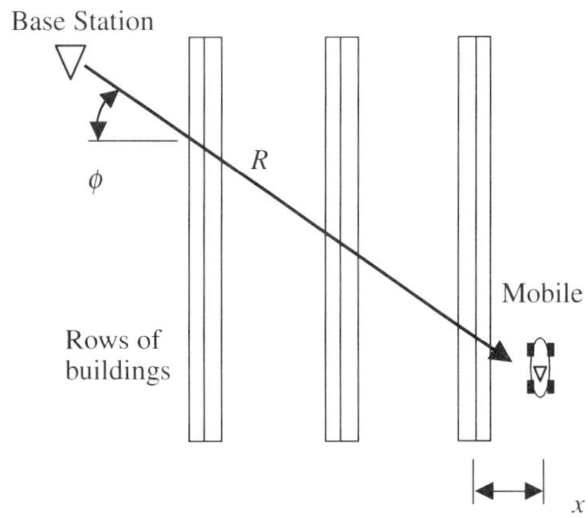

FIGURE 3.3 Top view of propagation across the rows of buildings to a mobile located in a low-building environment.

Fields diffracted at a previous rows of building, such as ray 3 in Fig. 3.1, suffer additional loss in passing through the buildings before the mobile. Fields diffracted at the rooftop of the building after the mobile will be weaker as a result of diffraction through a large angle. These and other contributions are neglected for simplicity.

To evaluate PG_2, the field incident on the last rooftop before the receiver is assumed to be a plane wave propagating at an angle α to the horizontal that is small, and at angle ϕ to the direction perpendicular to the rows, as see from above in Fig. 3.3. The fields resulting from diffraction at the building edge is in the form of a cylindrical wave, with the edge acting as an equivalent line source. Because of the rapid spatial variation resulting from the interference of the two waves, the spatial average power will be the sum of the individual ray powers. With the foregoing assumptions, the spatial average received power is given by

$$PG_2 = \left[\frac{1}{\rho_1} \left| D(\theta_1, \phi) \right|^2 + \left| \Gamma \right|^2 \frac{1}{\rho_2} \left| D(\theta_2, \phi) \right|^2 \right] \tag{3.3}$$

Here, Γ is the reflection coefficient of the building opposite to the mobile, $k = 2\pi/\lambda$ and $D(\theta_i)$ is the diffraction coefficient.

For a receiver in the middle of the street, the distances ρ_1 and ρ_2 in Fig. 3.2 from the diffracting edge are given by

$$\begin{aligned} \rho_1 &= \sqrt{\left(H_B - h_m\right)^2 + x^2} \\ \rho_2 &= \sqrt{\left(H_B - h_m\right)^2 + \left(2d - w - x\right)^2} \end{aligned} \tag{3.4}$$

whereas the angles θ_i for $i = 1, 2$ are

$$\theta_i = \arcsin\left[\left(H_B - h_m\right)/\rho_i\right] \tag{3.5}$$

In these expressions, H_B is the building height, h_m is the mobile height, w is the front-to-back dimension of the building, and x is the distance of the receiver from the diffracting edge of the building just before the mobile. The variation of Eq. (3.3) with x has been validated by measurements in Japan [10].

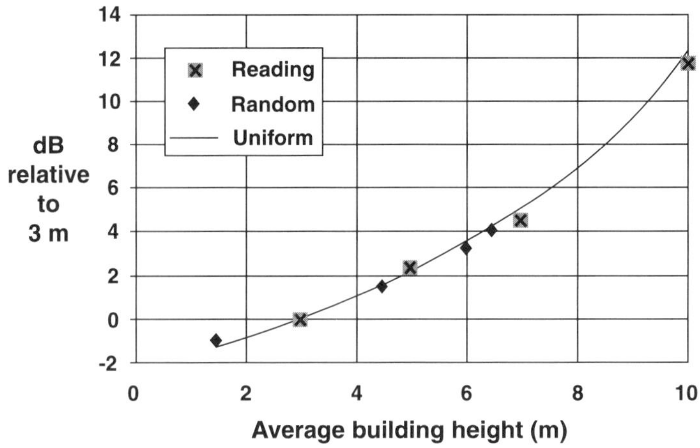

FIGURE 3.4 Measured and computed average mobile antenna height gain at 191.25 MHz for Reading, United Kingdom. Solid curve computed from theory and squares are from a numerical simulation accounting for random building heights. (From Bertoni, H. L. and Maciel, L., *Proc. IEEE ICUPC Conf.*, pp. 1–4, 1992. With permission.)

The diffraction coefficient $D(\theta_i)$ in Eq. (3.3) depends on the boundary condition at the rooftop edge of the building, which is not usually known. However, for small diffraction angles θ_i the diffraction coefficient is not sensitive to the boundary conditions. Thus, we may use the diffraction coefficient for an absorbing edge, which is given by [11]

$$D(\theta_i) = \frac{1}{\sqrt{2\pi k \cos\phi}} \left(\frac{1}{\theta_i - \alpha} + \frac{1}{2\pi + \theta_i - \alpha} \right) \approx \frac{1}{\sqrt{2\pi k \cos\phi}} \frac{1}{\theta_i} \tag{3.6}$$

The term $\cos\phi$ in Eq. (3.6) reflects the fact that the effective wave number in the plane perpendicular to the rows of buildings is $k\cos\phi$, whereas the effective wavelength is $\lambda/\cos\phi$. The diffraction coefficient is only valid in the shadow region outside of the transition region. The transition region is centered on the shadow boundary of the diffracting edge, and at the center of the street has a half width given by $\sqrt{\lambda d/2\cos\phi}$ [12]. Inside the transition region, $D(\theta_i)$ is given by the uniform asymptotic theory of diffraction [11, 13].

We can simplify Eq. (3.3) by accounting for Γ, which for common building materials is $\Gamma \approx 0.3$. The value of Γ compensates for the differences in sizes of $D(\theta_1, \phi)/\sqrt{\rho_1}$ and $D(\theta_2, \phi)/\sqrt{\rho_2}$, so that the second term is close to the first term. The near equality of the two terms is seen from the deep fades observed in the fast-fading pattern. Thus, the path gain for diffraction down to street level can be rewritten as

$$PG_2 \approx \frac{2}{\rho_1}\left[D(\theta_1, \phi)\right]^2 \approx \frac{1}{\pi k \cos\phi}\frac{1}{\rho_1 \theta_1^2} \tag{3.7}$$

The antenna height gain predicted by Eq. (3.7) has been compared with measurements made in a fairly homogeneous building environment (Reading, United Kingdom) using a signal from a distant TV transmitter broadcasting at 191.25 MHz [14]. In Fig. 3.4, the measured height gain, normalized to a receiver at 3 m, is plotted by Xs for antenna heights of 5, 7, and 10 m [6]. The solid curve in Fig. 3.4 is the normalized height gain predicted by Eq. (3.7). The close agreement serves as a further validation of diffraction as the mechanism by which the rooftop signals reach street level. As a final simplification of Eq. (3.7), we assume that the mobile is in the center of the street where $x = d/2$. We further approximate ρ_1 by $d/2$ and replace θ_1 by its tangent, which is $2(H_B - h_m)/d$, to achieve the following simplified expression:

$$PG_2 = \frac{1}{\pi k \cos\phi} \frac{d/2}{\left(H_B - h_m\right)^2} \tag{3.8}$$

3.2.2 Reduction of the Rooftop Fields

In macrocellular networks the base station antenna is located well above the average building height. However, because the range R is greater than about 1 km, the glancing angle α shown in Fig. 3.1 is still small, typically less than 2°, so that for a flat earth

$$\alpha = \tan^{-1}\left(\frac{h_{BS} - H_B}{R}\right) \approx \frac{h_{BS} - H_B}{R} \tag{3.9}$$

Also, the number of rows of buildings crossed by the radio wave is large. For example, if row separation is $d = 50$ m and the range is $R = 2$ km, then the number of rows is 40. As a result of diffraction past the rows up to the last row of buildings, the fields will be smaller than for free space. When computing this reduction, we make use of approximations based on the large number of rows that are crossed by the radio wave, and on the small size of α.

For low angles α, the forward diffraction from one row to the next is not sensitive to irregularities in the row spacing between buildings, or the lack of parallelism between the rows. Also because of the low-glancing angle α, the diffraction is not sensitive to the shape of the rooftops or to their boundary conditions, so that the rows of buildings can be represented as thin absorbing screens. Rays diffracted through large angles down to the ground and reflected back up to the rooftops, as depicted by ray 4 in Fig. 3.1, have small amplitudes and are neglected. Thus, the canonical problem that must be solved consists of forward diffraction past a series of parallel absorbing screens. When a spherical wave is radiated by an antenna in this geometry, the variation of the fields in the direction parallel to the screens is nearly that of a spherical wave, so that the multiple diffraction can be reduced to a 2D problem. In the case of low base station antennas for microcells, where a limited number of screens are crossed, the 2D problem involves line source excitation. For macrocell applications, where propagation takes place past many rows, the reduction of the rooftop fields for an incident spherical wave is the same as for an incident plane wave [8]. This use of the plane wave reduction factor is similar to employing the plane wave reflection coefficient when a spherical wave is reflected by a surface.

Diffraction past a few screens may be evaluated using the uniform theory of diffraction (UTD) including the slope diffraction term [15]. For many screens and for small glancing angles α, researchers have used repeated application of the physical optics approach, where the fields above one screen are found from the fields above the previous screen. The repeated integration can be carried out in a semianalytic way for screens of uniform height [16, 17]. Alternatively, the parabolic equation method can be used even when the screens are not of uniform height [18]. The most general approach makes use of direct numerical integration of the physical optics integrals [7, 8], and has been used when the screens are not of uniform height [5, 6, 19]. It can also be used when the screens are only partially absorbing [20, 21], and even when ground reflection is taken into account.

When a unit amplitude plane wave is diffracted past a series of absorbing screens for low glancing angles α, the field incident on successive edges is found to decrease initially to a value Q, called the settled field, and then to oscillate with decreasing amplitude about this value. This settling process is associated with the end effects of the starting screens. The settled value gives the reduction of the rooftop fields resulting from diffraction by previous rows of buildings, so that PG_1 is given by

$$PG_1 = 10\log\left[Q^2\left(g_p\right)\right] \tag{3.10}$$

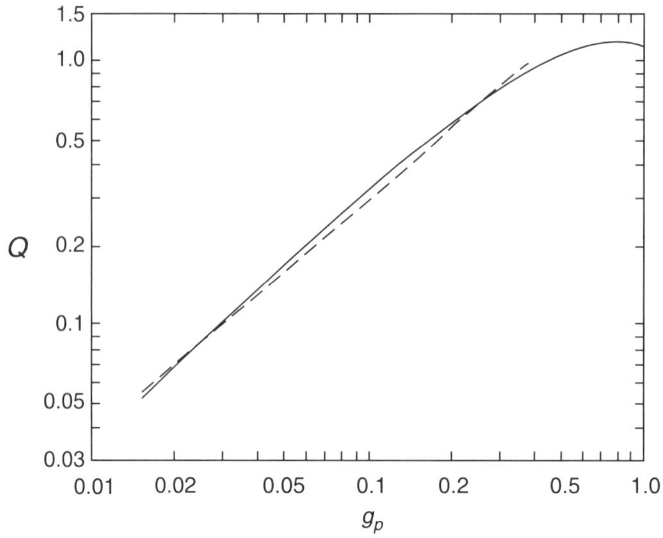

FIGURE 3.5 Dependence on the parameter g_p of the factor Q giving the reduction of the rooftop fields resulting from propagation past previous rows of buildings. (From Walfisch, J. and Bertoni, H. L., *IEEE Trans. Antennas Propag.*, 36, 1778, 1988. With permission.)

Here we have indicated that the settled value $Q(g_p)$ depends on the angles α and ϕ, row spacing d, and wavelength λ through the dimensionless parameter g_p, which is defined by [7, 16]

$$g_p = \alpha\sqrt{\frac{d}{\lambda\cos\phi}} \tag{3.11}$$

As discussed after Eq. (3.6), the term $\cos\phi$ is from the fact that along the oblique path the distance between rows of buildings is $d/\cos\phi$. The variation of $Q(g_p)$ obtained from the physical optics integrals is plotted in Fig. 3.5.

To apply the numerical results for $Q(g_p)$ to path loss prediction, a third-order polynomial was fit to the numerical results. This polynomial is given by

$$Q(g_p) = 3.502g_p - 3.327g_p^2 + 0.962g_p^3 \tag{3.12}$$

and is accurate to within 0.5 dB over the range $0.01 < g_p < 1$ [8]. For large angles α, such as on satellite links, $g_p > 1$ and $Q(g_p) \approx 1$ so that only the last row of buildings before the mobile affects the received signal. An even simpler approximation to $Q(g_p)$ is given by the straight line shown dashed in Fig. 3.5. This approximation is given by

$$Q(g_p) = 2.35g_p^{0.9} \tag{3.13}$$

and is accurate to within 0.8 dB over the range $0.01 < g_p < 0.4$ [8].

3.2.3 Summary of Macrocell Path Gain

Typical values for macrocellular systems are $h_{BS} - H_B = 12$ m and 1 km $< R <$ 10 km. In cities, $d \approx 50$ m, so that g_p falls in the range of 0.015 to 0.15 at 900 MHz and 0.021 to 0.21 at 1800 MHz, and we may use Eq. (3.13) for $Q(g_p)$. By combining Eqs. (3.9) to (3.11) into Eq. (3.13), and with the help of Eq. (3.2) and Eq. (3.8), the range dependence of the path gain of Eq. (3.1) becomes

$$PG_{dB} = 10\log\left(\frac{\lambda}{4\pi R}\right)^2 + 10\log\left[(2.35)^2\left(\frac{h_{BS}-H_B}{R}\right)^{1.8}\left(\frac{d}{\lambda\cos\phi}\right)^{0.9}\right] + 10\log\left(\frac{1}{\pi k\cos\phi}\frac{d/2}{\left(H_B-h_m\right)^2}\right) \quad (3.14)$$

By substituting $k = 2\pi/\lambda$ and $\lambda = c/f$, combining the various constant terms in Eq. (3.13), expressing the frequency f_{MH} in megahertz and the range R_k in kilometers, and leaving d and the various heights in meters, the path gain can be written as

$$PG_{dB} = -92.5 - 38\log R_k - 21\log f_{MH} + 18\log\left(h_{BS}-H_B\right) + 10\log\left(\frac{d^{1.9}/(\cos\phi)^{1.9}}{\left(H_B-h_m\right)^2}\right) \quad (3.15)$$

Eq. (3.15) shows that the R dependence of Q combines with the free-space path to give the overall range dependence of $1/R^{3.8}$ or 38 log R_k. The exponent of R is referred to as the slope index, and the theoretical value of $n = 3.8$ is close to that reported in North American cities [4]. As a result of the near cancellation of the frequency dependence in PG_1 and PG_2 in Eqs. (3.7) or (3.8) and (3.13), the path gain is seen to vary inversely with frequency to the 2.1 power, which is nearly that of the free-space path gain.

The predictions given by Eq. (3.15) with $\phi = 0^0$ for the received signal are shown in Fig. 3.6 superimposed on the small area average received power measured in Philadelphia [4]. The horizontal range is plotted on a logarithmic scale, for which Eq. (3.15) plots as a straight line. Excellent agreement is seen with the slope index of propagation and the average signal level. Equation (3.15) shows agreement with the well-known Hata model [22], although the Hata formulas do not have building height or row separation as input variables. The dependency on cos ϕ in Eq. (3.15) does not apply when the propagation path is aligned with the streets so that $\phi = 90°$, and the propagation takes place down the street, not over the building. The variation of received signal with the angle ϕ has been observed over short paths in some cities [23]. However, the variation observed is not as great as suggested by Eq. (3.15).

3.3 Range Dependence for Microcells in Low-Rise Environments

Microcellular systems make use of base station antennas located at about the height of three-story buildings, or on lampposts, to cover cells of radius 1 km or less. Even in low-rise building environments, the base station antenna is near to, or below, the rooftops, so that its location relative to the buildings needs to be taken into account. For such small cells, the street grid is more likely to be rectangular, and LOS streets are a more significant fraction of all the streets in a cell, calling for their separate treatment. To find path loss in a low-rise area, we again approximate the rows of buildings by parallel absorbing screens and compute the forward diffraction past the rows. For a high-rise environment, propagation takes place around the sides of the buildings, and may be evaluated using ray methods [8], as discussed in Section 3.6.

3.3.1 Line-of-Sight Propagation Along Streets

Measurements have shown that a simple two-ray model consisting of the direct and the ground-reflected ray, as shown in Fig. 3.7, is sufficient to predict the path gain (loss) for propagation over a flat earth [24, 25]. For isotropic antennas the path gain of the two ray model is given by [25]

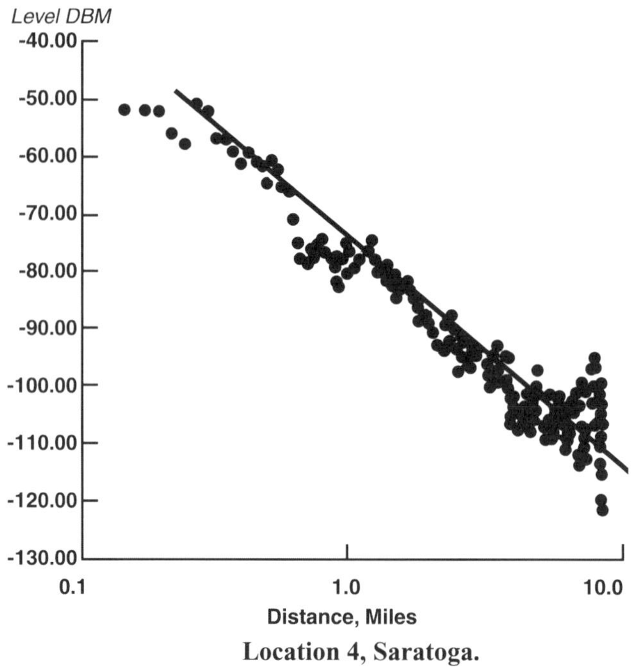

FIGURE 3.6 Measured sector averages signal (dots) plotted vs. distance R on a logarithmic scale, and showing clustering about an average dependence of the form A/R^n. (From Walfisch, J. and Bertoni, H. L., *IEEE Trans. Antennas Propag.*, 36, 1778, 1988. With permission.)

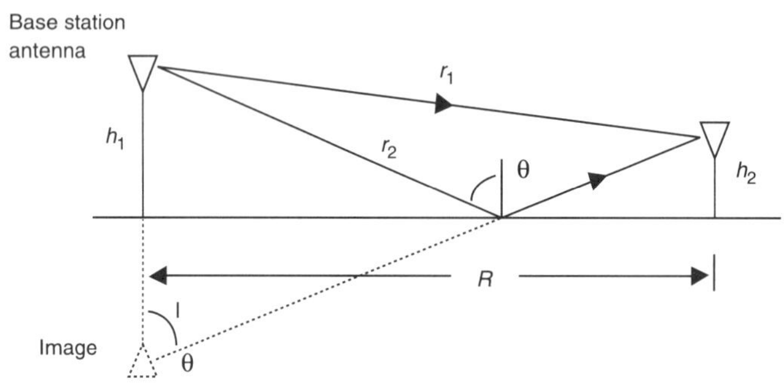

FIGURE 3.7 Two-ray model of propagation over a flat earth.

$$PG = \left(\frac{\lambda}{4\pi}\right)^2 \left| \frac{e^{-jkr_1}}{r_1} + \Gamma(\theta)\frac{e^{-jkr_2}}{r_2} \right|^2 \tag{3.16}$$

where r_1 is the direct distance from the transmitter to the receiver, r_2 is the distance through reflection point, and $\Gamma(\theta)$ is the reflection coefficient. The reflection coefficient, which depends on the angle of incidence θ and the polarization, is given by

$$\Gamma(\theta) = \frac{\cos\theta - a\sqrt{\varepsilon_r - \sin^2(\theta)}}{\cos\theta + a\sqrt{\varepsilon_r - \sin^2(\theta)}} \tag{3.17}$$

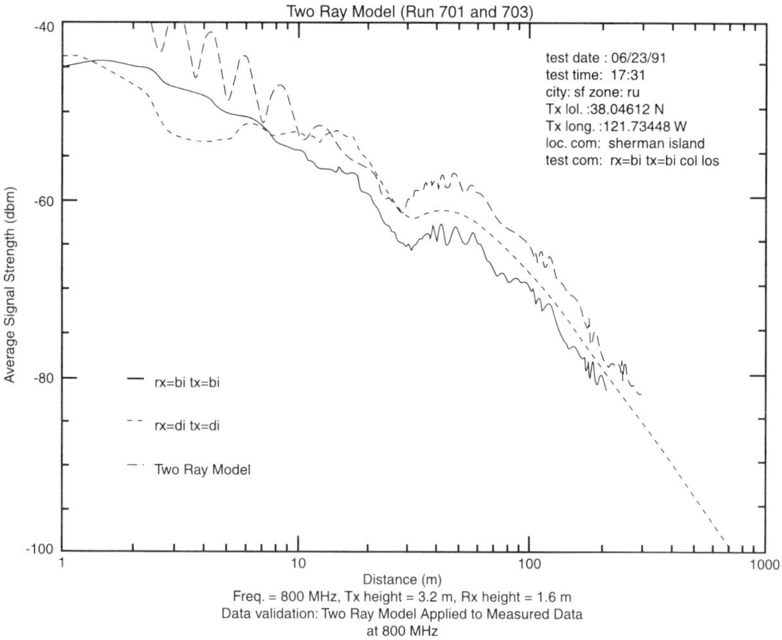

FIGURE 3.8 Comparison of the predictions of the two-ray model with measurements made at 800 MHz in a rural environment using bicone and dipole antennas. (From Xia, H. H. et al., *IEEE Trans. Antennas Propag.*, 41, 1439, 1993. With permission.)

Here $a = 1/\varepsilon_r$ for vertical polarization and $a = 1$ for horizontal polarization, where ε_r is the relative dielectric constant of the ground. For average ground, the relative dielectric constant is $\varepsilon_r = 15 - j\sigma\lambda$, and the conductivity σ is around 0.005 mho/m [25]. We note that as the distance between the transmitter and receiver increases, the angle θ approaches 90° and Γ approaches –1.

Figure 3.8 shows a plot of the path loss measured with a pair of vertically polarized dipole antennas and a pair of vertically polarized bi-cone antennas along a road in a flat area whose only features were low vegetation and wooden telephone poles [25]. In the measurements the frequency was 800 MHz and the antenna heights were 3.2 and 1.8 m. For comparison we have plotted the signal predicted by Eq. (3.16) for vertical polarization as the dashed curve. For horizontal separation $R < 10$ m, the antenna patterns have an influence on the measurements. When $R > 10$ m, only the antenna gains are important and they result in a vertical offset of the curves (the signal for dipoles is a few decibels greater than for isotropic antennas, and for bi-cones it is a few decibels smaller).

By using the logarithmic scale for the horizontal separation R in Fig. 3.8, it is clearly seen that the variation of the received signal has two distinct slopes separated by a break point R_B that lies near the last peak in the two ray model. Before the R_B, the radio signal oscillates severely because of alternating regions of destructive and constructive combination of the two rays, whereas after the R_B it decreases more rapidly with distance due. The break point distance can be understood in terms of the first Fresnel zone clearance. The first Fresnel zone is an ellipsoid of revolution having the two antennas as foci, and is such that the distance from one antenna to a point on the ellipsoid and back to the other antenna is $\lambda/2$ greater than the direct path distance between the antennas.

The break point is defined here as the distance R_B between the antennas such that the ground just touches the first Fresnel zone, so that $r_2 = r_1 + \lambda/2$ in Fig. 3.7. Since $\Gamma \approx -1$, the two rays add in phase at R_B. For $R > R_B$, the distance r_2 approaches r_1 and the two rays interfere destructively, leading to a power dependence of $1/R^4$, instead of the $1/R^2$ of free space. For high frequencies, R_B can be expressed as a simple function of wavelength and antenna heights as follows:

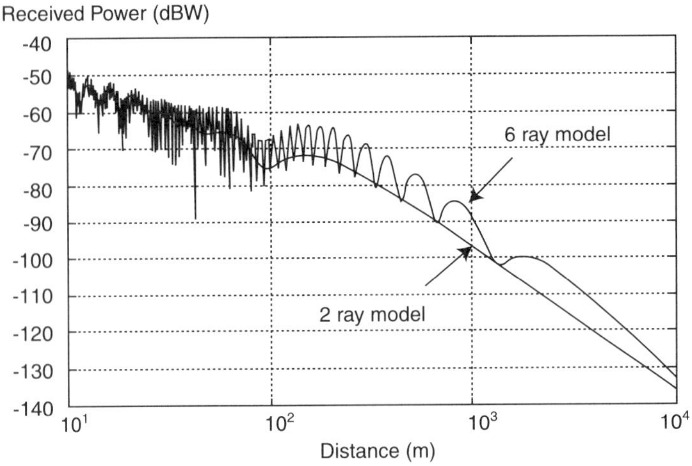

FIGURE 3.9 Path gain at 900 MHz computed using the six-ray model for LOS propagation in urban canyons, and the two-ray model for a flat earth.

$$R_B = 4\frac{h_1 h_2}{\lambda} \tag{3.18}$$

In an urban environment with buildings lining both sides of the streets, reflections can also take place at the faces of the buildings. Rays may be multiply reflected between the buildings lining the streets. Building-reflected rays result in additional rapid variations about the simple two-ray model, but do not change the overall variation. Accounting for a single reflection in the building walls and ground reflection leads to the six-ray model, which has been used to obtain the plot of Fig. 3.9. Similar results are obtained from measurements in urban environments [24, 25]. Statistical models that account for the gaps between buildings suggest an even more rapid decrease with R for large distances [26].

3.3.2 Propagation Over Buildings for Low Antennas

For application to microcellular environments, the European study group COST-231 has developed a semiempirical model for predicting the range dependence of the path loss in an environment of low buildings. Their approach is based on using Eq. (3.12) for the factor Q at small ranges, and incorporating measurements to extrapolate the results to base station antennas somewhat above, at, or slightly below the average building heights [23]. An alternative theoretical approach to predict the range dependence evaluates the 2D diffraction process in the plane perpendicular to the rows of buildings for low base station antennas. This approach requires the study of fields radiated by a source that is localized in the vertical plane, instead of an incident plane wave. A semianalytical treatment has been carried out for the special case when the horizontal separation between the base station and first row of buildings is d and all the buildings are of the same height [16, 17].

The analytic approach again replaces the rows of buildings by absorbing screens that are assumed to have the same height, and the average row separation d as shown in Fig. 3.1. The path gain is computed as was done for the macrocellular case by using Eqs. (3.1), (3.2), and (3.7) or (3.8), except that PG_1 is now computed from

$$PG_1 = 10\log\left[Q_M^2\left(g_c\right)\right] \tag{3.19}$$

Here Q_M is the reduction of the rooftop field at the Mth row past the base station and the dimensionless parameter g_c is given by [16, 17]

$$g_c = \Delta h \sqrt{\frac{\cos\phi}{\lambda d}} \tag{3.20}$$

where $\Delta h = h_{BS} - H_B$ is the height of the base station relative to the buildings and ϕ is the angle shown in Fig. 3.3.

When the base station antenna is located at a distance d from the first row, and all the rows (absorbing screens) are the same height, $Q_M(g_c)$ has been shown to be [16]

$$Q_M(g_c) = \sqrt{M}\left|\sum_{q=o}^{\infty} \frac{1}{q!}\left(2g_c\sqrt{j\pi}\right)^q I_{M-1,q}\right| \tag{3.21}$$

where $I_{M-1,q}$ are the Boersma functions [27]. The Boersma functions are found from the recursion relation

$$I_{M-1,q} = \frac{(M-1)(q-1)}{2M} I_{M-1,q-2} + \frac{1}{2\sqrt{\pi M}} \sum_{n=1}^{M-2} \frac{I_{n,q-1}}{\sqrt{M-1-n}} \tag{3.22}$$

The initial terms for the recursion relation are

$$I_{M-1,0} = \frac{1}{M^{3/2}} \quad \text{and} \quad I_{M-1,1} = \frac{1}{4\sqrt{\pi}} \sum_{n=1}^{M-1} \frac{1}{\left[n(M-n)\right]^{3/2}} \tag{3.23}$$

When $\Delta h = 0$, it is seen from Eq. (3.20) that $g_c = 0$ and so that Eq. (3.21) reduces to a single term, which with the help of Eq. (3.23) is seen to be

$$Q_M(0) = \frac{1}{M} = \frac{d}{R\cos\phi} \tag{3.24}$$

The second equality in Eq. (3.24) is obtained by recognizing that $R \cos \phi$ is the projection onto the plane perpendicular to the rows of buildings of the distance from the base station to the rooftop before the mobile. Dividing this distance by the row separation d gives the number of rows M. When Eq. (3.24) is used in Eq. (3.19) for PG_1, the path gain is seen to vary as $1/R^4$, just as was found beyond the break point for a flat earth.

For $\Delta h \neq 0$, the variation of $Q_M(g_c)$ with M is different than that given by Eq. (3.24) for small values of M, but eventually has the same slope. To make the slope index quantitative for $\Delta h \neq 0$, we use the logarithmic derivative of $Q_M(g_c)$ defined by

$$s = -\frac{\log Q_{M+1} - \log Q_M}{\log(M+1) - \log M} \tag{3.25}$$

in which case the slope index is $n = 2(1 + s)$. The variation of n with Δh is plotted as the continuous curves in Fig. 3.10 for $f = 900$ MHz and $d = 50$ m. The curve with the smaller variation is for the case when $M = 10$ and $\phi = 0^0$, whereas the other is for $M = 5$ and $\phi = 60^0$. Both cases correspond to a distance $R = Md/\cos\phi$ of 0.5 km. It is seen from Fig. 3.10 that $n > 4$ when the base station is below the buildings ($\Delta h < 0$), while $n < 4$ when the base station is above the buildings ($\Delta h > 0$).

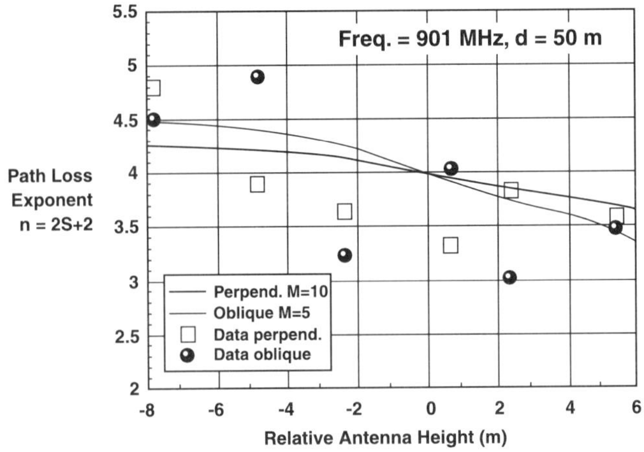

FIGURE 3.10 Comparison of the path loss index n at 901 MHz obtained from measurements in San Francisco and from the theory for low base station antennas for propagation at right angles to the street grid ($\phi = 0$) and oblique to the street grid ($\phi \approx 60°$). (From Maciel, L. R., Bertoni, H. L., and Xia, H. H., *Proc. IEEE Int. Symp. Personal, Indoor and Mobile Radio Commun.*, Boston, MA, p. 75, 1992. With permission.)

Measurements at 901 and 1937 MHz have been made in the Mission and Sunset districts of San Francisco, California, at three antenna curb heights, and for propagation perpendicular to the rows, and at an angle $\phi \approx 60°$ [17, 28]. In these areas the building heights are approximately 11.5 and 8 m, respectively, giving six different antenna heights relative to the buildings. The measurement range was from 100 m to 1.2 km. When expressed in terms of $\log R$, the measurement range is approximately centered on a distance of 500 m, which for the row separation $d = 50$ m corresponds to $M = 10$ for $\phi = 0$, and $M = 5$ for $\phi = 60°$. The slopes taken from the regression fits to the measured data at 901 MHz are shown in Fig. 3.10. It is difficult to accurately measure slope index over short distance, so that the measurements show considerable variation. However, the trend in the measurements mimics the computed slope index.

The height gain of the base station antenna is given by the variation of Q_M with $\Delta h = h_{BS} - H_B$. The variation of Q_M in decibels with antenna height is plotted in Fig. 3.11 for 901 MHz propagation and $R = 1$ km for the case when $\phi = 0$ ($M = 20$) and the case when $\phi = 60°$ ($M = 10$). The values of Q_M obtained

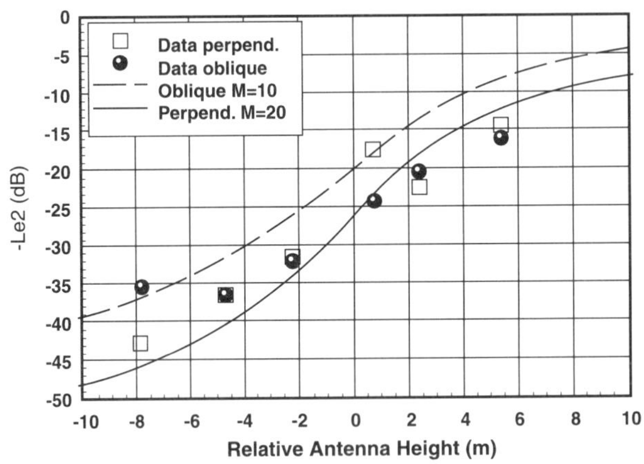

FIGURE 3.11 Comparison of base station antenna height gain at 901 MHz obtained from measurements in San Francisco and from the theory for low base station antennas for propagation at right angles to the street grid ($\phi = 0$) and oblique to the street grid ($\phi \approx 60°$). (From Maciel, L. R., Bertoni, H. L., and Xia, H. H., *Proc. IEEE Int. Symp. Personal, Indoor and Mobile Radio Commun.*, Boston, MA, p. 75, 1992. With permission.)

from the San Francisco measurements are also plotted in Fig. 3.11. Relatively good agreement is seen between the theory and measurements for perpendicular propagation. However, for oblique propagation there is more scatter in the measurements, and for low base station antennas the theory is significantly more pessimistic than the measurements. Similar results were obtained at 1937 MHz, with the theory again giving lower values than the measurements for low base station antennas, both for perpendicular and oblique propagation [28].

When the base station antenna is sufficiently above the rooftops, so that the first building lies outside the first Fresnel zone between the base station and rooftop before the mobile, the path gain PG_1 found from $Q_M(g_c)$ is equal to the path gain found from $Q(g_P)$ with $g_p = g_c/M = g_c d/R \cos \phi$ [8]. For base station antennas well below the rooftop, as in mobile-to-mobile communications, the multiple diffraction process can be separated into two simpler processes. First, the wave excited by the source below the average rooftop is diffracted by the first row of buildings. The first row of buildings then acts approximately as a line source for the multiple diffraction past the remaining rows of buildings [17]. The first process can be treated by geometric theory of diffraction (GTD) as in the case of diffraction from rooftop to street level. The multiple diffraction process for the cylindrical wave generated at the first edge is as described above for $\Delta h = 0$, taking the number of edges to be $M - 1$. Let θ_s be the angle from the horizontal to the rooftop of the first row of buildings and ρ_s be the perpendicular distance from the mobile to the diffracting edge of the buildings. Then the field reduction resulting from the combination diffraction processes leads to the expression [17]

$$Q_M = \frac{1}{M-1} \frac{\left|D(\theta_S)\right|}{\sqrt{\rho_s}} \tag{3.26}$$

3.4 Effects of Vegetation

Personal communications service (PCS) measurements made in environments with trees have shown the importance of considering the effect of the trees on the propagation loss [3, 30, 31]. Measured transmission loss at 900 MHz through large free-standing trees has found attenuation on the order of 10 dB [31]. Propagation past a forest of trees into a clearing has been shown to involve diffraction at the treetops down to ground level [32]. Simple path loss models can be devised accounting for this diffraction [8]. Additional attenuation has been observed for propagation to mobiles located inside a forested area [33]. However, the available theory for glancing incidence over a forest is based on mean field approximations [34] that apply at very high frequencies (VHF) but not at ultra high frequencies (UHF). In suburban areas it is common to see one or more trees planted in front of, or behind, each house to form nearly continuous rows along the streets. In this section we examine the effect of trees in such a suburban setting.

3.4.1 Propagation Through the Tree Canopy

The tree canopy can be viewed as a random collection of leaves and branches. The leaves are modeled by randomly oriented flat, circular disks with radius a_d, thickness t, and complex relative dielectric constant ε_{rd}. The branches are modeled as randomly oriented finitely long, circular cylinders, which are assumed to have a radius a_c, length l, and complex relative dielectric constant ε_{rc}. The mean field in the canopy can then be calculated using the discrete scattering theory of Foldy and Lax — see References [20, 35–38]. The solution for a plane wave propagating at an angle θ to the vertical (z-axis) incident on a layer of the random media ($x > 0$) is given by

$$\mathbf{E}(x,z;\mathbf{a}) = \mathbf{a}\exp\left[-j\left(k\sin\theta + \kappa\right)x - jk\left(\cos\theta\right)z\right] \tag{3.27}$$

Here $\mathbf{a}$ is the polarization vector of the mean field, k is the free-space wave number, and $\kappa = \kappa' - j\kappa''$ gives the effect of the random medium on the propagation constant.

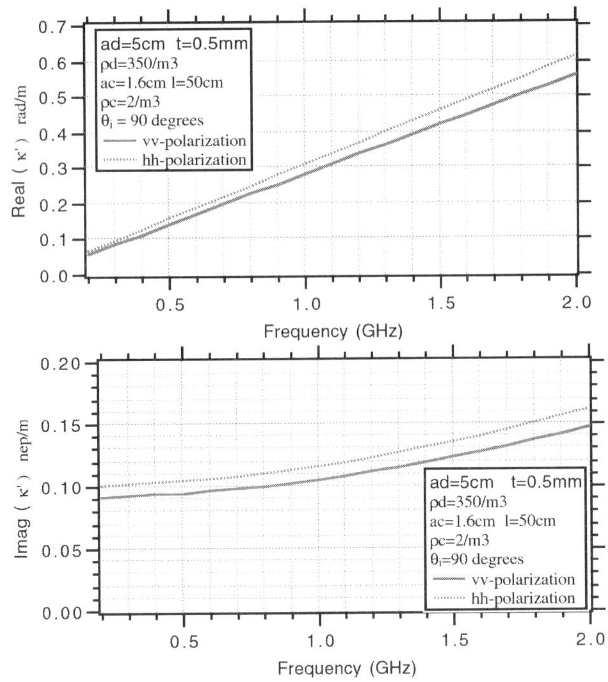

FIGURE 3.12 Frequency dependence of the real part κ' and imaginary part κ'' of the additional propagation constant through a tree canopy for horizontal propagation ($\theta = 90°$). (From Torrico, S. A., *Theoretical Modeling of Foliage Effects on Path Loss for Residential Environments*, D.Sc. dissertation, George Washington University, Washington, D.C., 1998. With permission.)

Computation of $\kappa = \kappa' - j\kappa''$ has been carried out for a set of leaf parameters ($a_d = 5$ cm, $t = 0.5$ mm, $\varepsilon_{rd} = 25 + i7$ and density $\rho_d = 350/\text{m}^3$) and branch parameters ($a_c = 1.6$ cm, $l = 50$ cm, $\varepsilon_{rc} = 19 + i7$, and density $\rho_c = 2/\text{m}^3$) [39]. The orientation of the leaves and the branches is assumed to be uniform in the horizontal plane. The probability density in the polar angle is taken to be uniform over a range of angles: for the leaves the range is from 0° to 180° and for the branches the range is from 0° to 60°. In Fig. 3.12 we have plotted κ' and κ'' vs. frequency for both polarizations of the incident field when $\theta = 90°$. It is seen that both κ' and κ'' for vertical polarization are higher than for horizontal polarization because of the statistical distribution of leaves and branches. The attenuation in decibels per meter is given by $8.69\,\kappa''$. A tree with a canopy 10 m in diameter, therefore has an attenuation of about 8 dB at 800 MHz and increases to 13 dB at 2 GHz. These values of attenuation are consistent with measurements.

3.4.2 Rows of Trees in a Suburban Setting

In suburban environments, it is common to see rows of trees planted along the streets, often one in front of each house, and with the trees taller than the houses. One possible arrangement of rows of trees next to rows of houses is depicted in Fig. 3.13. To study the influence of the trees in this arrangement, each row of houses or buildings is represented by an absorbing screen, while the canopy of the adjacent row of trees is represented by a partially absorbing phase screen, as shown in Fig. 3.13. With this approximation, the field passing any row is equal to the field incident on the row multiplied by the transmission coefficient past the tree canopy. The phase and amplitude of the transmission coefficient are found from the width of the canopy and the additional propagation constant κ of the canopy. Numerical evaluation of the physical optics integral is then used to find the field incident on the next screen [20].

The attenuation and phase of the partially absorbing phase screen is found by taking the tree canopy to have an elliptical cross-section, with semiminor axis equal to a and semimajor axis equal to b. The center of the canopy is assumed to be at the same height as the buildings, so that the canopy extends a

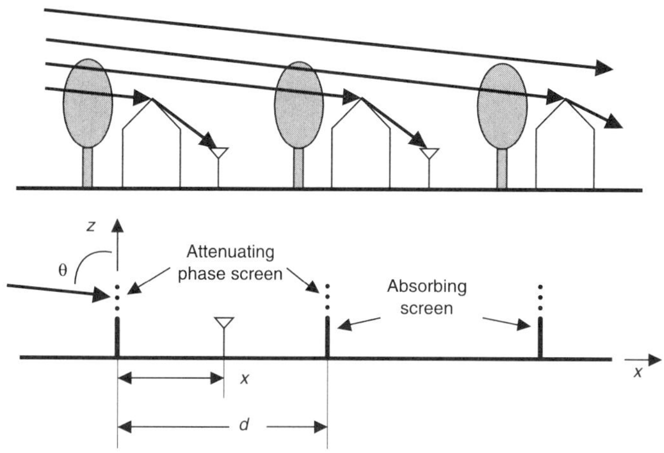

FIGURE 3.13 Rows of trees next to rows of buildings are modeled as a partially transmitting phase screen, whose properties vary with height.

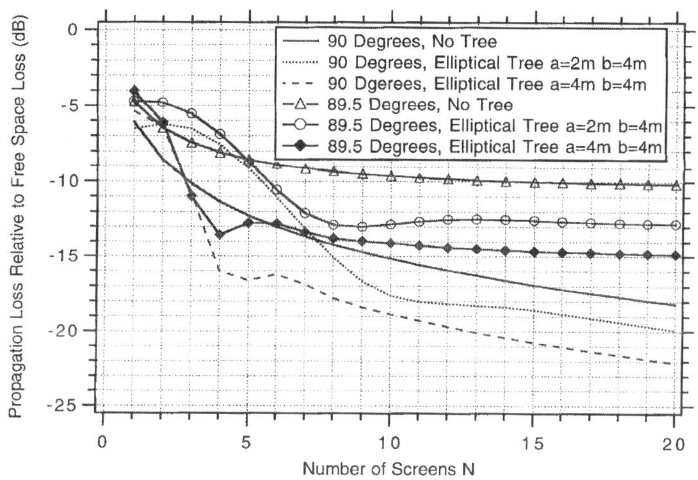

FIGURE 3.14 Path gain at the tops of the buildings in the Nth row for a 900-MHz plane wave incident at angles of 0° and 0.5° to the horizontal (90° and 89.5° to the vertical) for row separation of $d = 50$ m. (From Torrico, S. A., Bertoni, H. L., and Lang, R. H. *IEEE Trans. Antennas Propag.*, 46, 872, 1998. With permission.)

distance b above the buildings. Figure 3.14 shows the field amplitude incident at the top of the $N + 1$ row of buildings when a plane wave of unit amplitude is incident on the first row of buildings. The calculations assume $d = 50$ m, $f = 900$ MHz, $\theta = 90°$, 89.5° (corresponding to $\alpha = 0°$, 0.5°), and tree width $2a = 4$, 8 m for trees that are $b = 4$ m higher than the buildings. It is seen from the results that for $\alpha = 0.5°$, the field amplitude at the top of the buildings, with and without trees, decreases to a settled value for N greater than about 15. However, for $\alpha = 0$, the field amplitude continues to decrease with increasing N. This behavior is similar to that found previously for buildings alone. After ten rows the wider trees are seen to have 4 to 5 dB more path loss than the buildings by themselves.

3.5 Accounting for Terrain

Cities are frequently built on undulating terrain or on rolling hills so that propagation may be simultaneously affected by both buildings and terrain. Statistical and deterministic approaches have been used

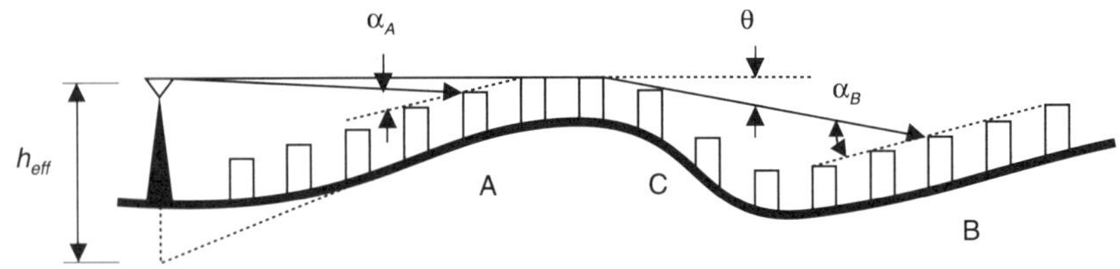

FIGURE 3.15 Propagation to the rooftops of buildings at different locations in rolling terrain. (From Bertoni, H. L., *Radio Propagation for Modern Wireless Systems,* Prentice-Hall PTR, Upper Saddle River, NJ, 2000. With permission.)

to account for terrain effects on radio wave propagation. One of the statistical models, which is based on measurements taken in Japan using very high base station antennas, accounts for terrain variability through the use of field strength correction factors [1]. However, this approach does not allow the evaluation of the signal at specific locations even when the terrain is known. In regard to deterministic models, an extensive literature deals with deterministic propagation over terrain in the absence of buildings. Approaches based on the geometric diffraction theory account for the tops of hills that block the LOS path — see, for example, References [40–42]. Full wave approaches, such as the parabolic equation method [43, 44] and the method of moments [45], take the entire terrain profile into account. In the presence of buildings, Lee proposed using an effective base station antenna height when LOS conditions exist to rooftops in the vicinity of the mobile [46]. Later, diffraction over terrain obstacles was treated in connection with buildings [19, 47].

Transmission paths from an elevated base station are depicted in Fig. 3.15 for three classes of subscriber locations in a metropolitan area built on rolling terrain. To model the path gain for these cases, the rows of buildings are assumed to be oriented perpendicular to the plane of curvature of the hills, as shown in the side view in Fig. 3.15, which is also assumed to be the plane of propagation [8, 19]. At location *A* in Fig. 3.15, the rooftops are within LOS of the rooftops near the mobile. In this case the path loss is determined using the flat terrain approach discussed in Section 3.3, with the difference that the angle α of Eq. (3.9) is replace by α_A when computing PG_1. This approach is equivalent to replacing the base station antenna height by the effective value h_{eff}, as proposed by Lee [46]. At location *B* in Fig. 3.15, the path gain is calculated by adding to the three terms in Eq. (3.1) a term that accounts for diffraction over the intervening hill, and computing PG_1 using the glancing angle α_B. When the hill is rounded, it may be necessary to model the diffraction by a creeping ray, as discussed later. At location *C* in Fig. 3.15, reduction of the rooftop fields near the mobile cannot be separated from diffraction loss at the houses on top of the hill. Instead, propagation involves multiple diffraction past the buildings at the top of the hill down to the buildings near the mobile, and may be modeled by a creeping ray that goes over the hill.

3.5.1 Creeping Ray Representation

Multiple diffraction past rows of buildings located on hills has been studied by Piazzi and Bertoni [19] using numerical integration of the physical optics integrals relating the field in the plane above one row to the fields above the previous row. Because these numerical techniques are too cumbersome to be incorporated into a cellular planning tool, the results have been approximated in terms of creeping ray optics to obtain a compact approximation [19]. Creeping rays were originally developed to describe the fields behind smooth metal cylinders. As suggested in Fig. 3.16, the ray from the source that is tangent to the cylinder launches a creeping ray that follows around the cylinder. The creeping ray and subsequently sheds rays that are tangent to the cylinder illuminate points in the geometric shadow region. As a result of the shedding, the field strength along the creeping ray decreases exponentially with arc length traveled on the surface of the cylinder. Excitation coefficients give the amplitude of the fields launched onto the cylinder, and the amplitudes of the fields launched along the rays that are shed. From the numerical evaluation of the diffracted fields for rows of buildings on cylindrical hills, Piazzi and Bertoni [19] extracted the attenuation coefficient and the excitation coefficients.

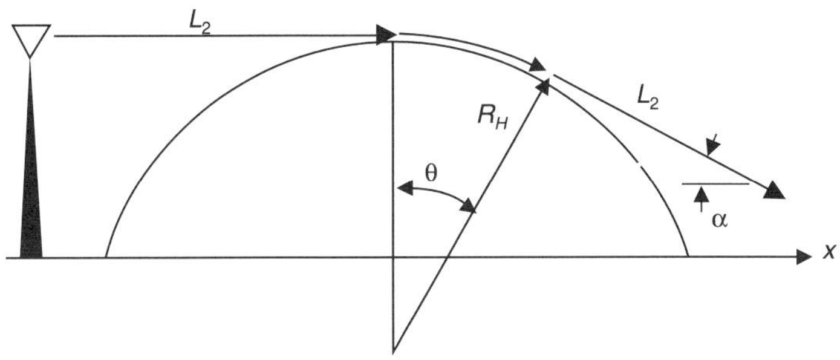

FIGURE 3.16 Geometry for describing diffraction by a circular cylinder in terms of the creeping ray.

On the back side of the hill, such as location *C* in Fig. 3.15, the path gain at the subscriber is given by

$$PG_{dB} = 10\log\left[\left(\frac{\lambda}{4\pi}\right)^2 \frac{\exp(-2\psi\theta)}{RL_1} D_H^2\right] + 10\log PG_2 \tag{3.28}$$

As shown in Fig. 3.16, L_1 is the distance from the base station antenna to the point of ray tangency on the hill, and θ is the angle from the tangent point to the building just before the mobile, as measured from the center of curvature of the hill. The total length of the ray path is approximated by the horizontal distance *R* from the base station to the mobile. As before, PG_2 is the path gain [Eq. (3.3)] from the rooftop down to the mobile. In Eq. (3.28) ψ is the attenuation coefficient and D_H is the excitation coefficient, both of which depend on the hill radius R_h, the row separation *d* and the frequency.

For locations such as *B* in Fig. 3.15 that are shadowed by a previous hill, the path gain at the subscriber is given by:

$$PG_{dB} = 10\log\left[\left(\frac{\lambda}{4\pi}\right)^2 \frac{\exp(-2\psi\theta)}{RL_1L_2} D_1^2 Q^2(g_B)\right] + 10\log PG_2 \tag{3.29}$$

As shown in Fig. 3.16, L_1 is the distance from the base station to the point of ray tangency on the hill, and L_2 is the distance from the launch point on the hill to the building just before the mobile. Again, the total length of the ray path is approximated by the horizontal distance *R* between the base station and the mobile. In Eq. (3.29) $Q(g_B)$ is the multiple diffraction loss resulting from the rows of houses before the mobile, and is given by Eqs. (3.12) or (3.13) for the angle α_B shown in Figs. 3.15 and 3.16.

3.5.2 Coefficients for Hills with Houses

Piazzi and Bertoni [19] constructed simple expressions for ψ, D_H, and D_1 by fitting Eqs. (3.28) and (3.29) to the numerically generated diffraction calculations. The attenuation constant ψ is given by:

$$\Psi = 2.02\left(\frac{\pi R_h}{\lambda}\right)^{1/3} - 1.04\sqrt{\frac{d}{\lambda}} \tag{3.30}$$

where the first term is the same as found for metal cylinder with electric field polarized parallel to the surface. The variation of D_H with the various parameters has been approximated by

$$\ln(D_H) = 3.75 + \left[-0.648 + 0.072\ln\left(\frac{d}{\lambda}\right)\right]\ln\left(\frac{R_h}{\lambda}\right) - 0.259\ln\left(\frac{d}{\lambda}\right) \tag{3.31}$$

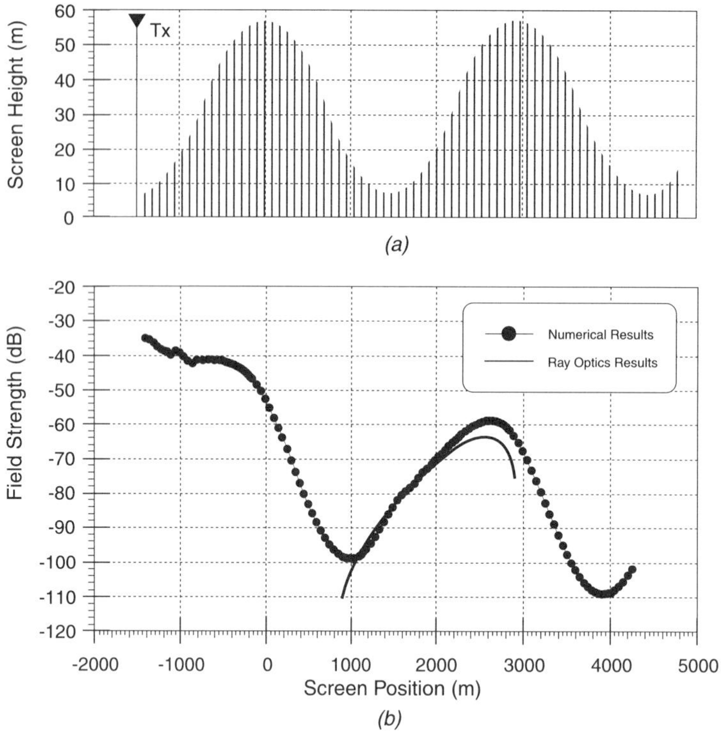

FIGURE 3.17 Profile of the screens representing rows of buildings on sinusoidal hills (a) and the amplitude of the field incident on the tops of the buildings (b) obtained from numerical integration (dots) and the creeping ray approximation (solid curve). (From Piazzi, L. and Bertoni, H. L., *Proc. IEEE Veh. Technol. Conf.*, Ottawa, Canada, p. 159, 1998. With permission.)

The excitation coefficient D_1 has been approximated by:

$$\ln(D_1) = 2.22 + \left[0.19 + 0.031\ln\left(\frac{d}{\lambda}\right)\right]\ln\left(\frac{R_h}{\lambda}\right) - 0.79\ln\left(\frac{d}{\lambda}\right) \tag{3.32}$$

The creeping ray models have not been tested against measurements. However, Eq. (3.29) has been compared against numerical evaluations for buildings on rolling hills [19, 47]. Figure 3.17a shows the screen heights representing buildings on 50-m high sinusoidal hills. The transmitter is at the height of the rooftops at the tops of the hills. The field strength at the top of the screens for the line source case, when the $1/R$ dependence in Eq. (3.29) is omitted, is shown in Fig. 3.17b for $f = 900$ MHz. The dots are obtained from the numerical integration, and the curve is found from Eq. (3.29). The starting point of the curve is near the inflection point, where α_B is very small and the plane wave reduction factor $Q(g_B)$ is overly pessimistic because it applies to propagation past many rows of buildings. The end of the curve is near the top of the second hill, where again the creeping ray model is pessimistic. Between these extremes, the ray formulation is seen to give a good approximation.

3.6 Site-Specific Predictions

When the buildings are of variable height and irregular shape, as in the high-rise core of a city, or in an office park, propagation takes place around the sides of tall buildings, as well as over the tops of lower buildings. If all the buildings are much taller than the base station antenna, propagation takes place around the buildings through the "urban canyons" formed by the buildings. In some regions, such as parts of Manhattan, the tall buildings completely fill the blocks, which are arranged in a rectangular grid. For these regions it is possible to characterize the path loss without reference to the shapes of individuals buildings [8, 48, 49]. However, in most urban cores, or office parks, the buildings have a more random appearance when viewed over an area of 1 km^2 or greater. Some buildings may be low, plazas and other open spaces break up the canyon walls, and building shapes and orientations may not conform to a rectangular grid. In these environments accurate predictions of the path loss requires that the actual building shapes be accounted for. Site-specific predictions in these environments have been made using ray optical methods. In addition to received power, site-specific predictions may be used to predict other statistical parameters of the channel, such as rms time-delay spread and angle of arrival spread [50, 51].

3.6.1 Ray Methods

At UHF frequencies the wavelength λ is small compared with the building size. The electromagnetic fields can therefore be viewed as propagating along rays that are reflected by the building walls and diffracted at the corners of buildings. In most prediction codes, meter scale variations in the building faces are ignored and the faces are treated as specular reflectors. Irregularities in the faces produce diffuse scattering, which some codes try to incorporate. When smoothed over a meter scale, most building walls are flat. The few buildings that have curved faces are typically approximated by a series of flat segments. Another simplifying feature is that almost all walls are vertical. Some prediction codes further simplify the buildings by assuming the roofs to be flat, although other codes can handle peaked roofs and wedding cake shapes. The simplifying assumptions are important because the codes must treat as many as seven reflections and two diffractions at vertical corners to describe the signal reaching behind buildings [52].

The Fresnel zone width provides a second scale requirement that must be satisfied to include these reflections in a straightforward way. The width of the Fresnel zone about the reflected ray must be smaller than the width of the reflecting walls. If L is the unfolded path length of a multiply reflected ray, then the maximum width of the first Fresnel zone about the ray is $\sqrt{\lambda L}$. For terrestrial links that are entirely among the buildings, $\sqrt{\lambda L}$ should be less than the typical building width W. This condition implies that $L < W^2/\lambda$. Taking 20 m for W, at 900 MHz the inequality is $L < 1.2$ km, whereas at 1,800 MHz the limit on L is 2.4 km. Thus, the ray methods are limited to predictions over microcells having a radius up to about 1 km. When the Fresnel width is larger than the building width, the buildings act as scatter [53, 54]. Because the buildings are in the near-field regions of the neighbors, the scattering description is difficult to use, especially because rays undergoing as many as seven reflections must be taken into account [52].

One application of the ray methods is to predict the received signal level for coverage and interference evaluation. Ray methods might, in principle, be able to predict the multipath interference pattern for narrowband signals, which is experienced as fast fading. In practice they can only give its statistical properties, but not the actual pattern. This is because the limited accuracy of the building database and the building simplifications limit the accuracy with which the ray phases and amplitudes can be computed. Also, the codes do not include moving objects that cause local scattering. Of greatest interest is the small area average received power, which is found by spatially averaging the received power. Because this average is equal to the sum of the ray powers, it is usual to add ray powers instead of adding the ray fields. Because the ray codes can output the direction of the rays at both ends of the link, as well as ray path length and ray powers, they can also be used to find time delay and angle of arrival and other statistical information.

The ray procedure starts by tracing multiply reflected rays from the source to the mobile location, and to the edges of buildings. The building corners are treated as equivalent sources of diffracted rays as suggested in top view in Fig. 3.18. Multiply reflected rays are then traced from these equivalent sources to the mobile locations, and again to the building edges, which act as equivalent sources of the multiply diffracted rays. Finding the multiply reflected ray paths connecting the actual or equivalent sources and receiver points is the most time-consuming aspect of the ray code. The two well-known approaches are the image method and the pincushion method. Because of the large fraction of the computation devoted to finding ray intersections with walls, ray codes are efficient at finding the paths from a single base station to many subscriber locations at the same time.

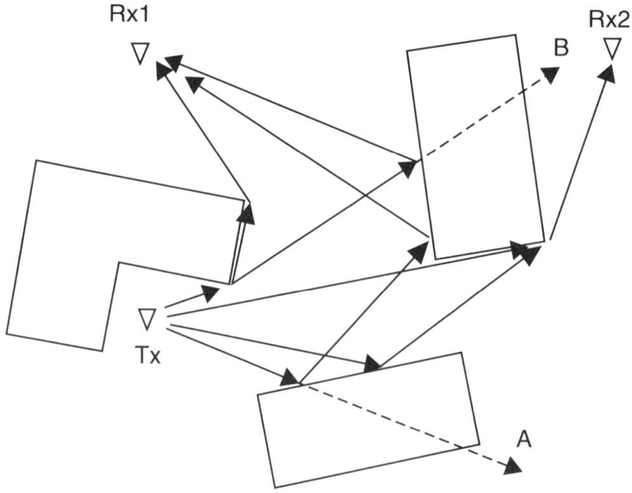

FIGURE 3.18 Rays from a base station can reach subscribers by a combination of multiple reflection and diffraction events, including paths that go in part or entirely over buildings.

The image method starts by constructing the image of the source in all the building surfaces that are visible to it. The image is itself imaged in all the surfaces visible to it, and the process is repeated up to the number of multiple reflections that are to be accounted for. The lines between each image and the receiver points are then constructed. At this point in the process it is necessary to determine whether the lines cross the image plane at the location of the building wall, or are just its analytic extension. This process of checking that the ray path intersects the building wall, instead of its analytic extension, must be continued back all the way to the source. At any step in the process the reflection point may not lie on the building surface and the path is not a legitimate ray. When the database of buildings is large, and several reflections are allowed, the process of checking to see that all the reflection points lie on building surfaces is the most time-consuming aspect of the program. After identifying the legitimate multiple reflections, for each ray the angles of incidence at all surfaces are found to compute the reflection coefficient, and the unfolded path length s of the ray is determined.

The pincushion method starts rays from the source (or secondary source) with some angular separation δ. Each ray is tested to see if it illuminates any of the receiver locations, and then the intersection of each ray with all the walls in the database is computed. The intersection closest to the source is chosen as the reflecting wall, because the other intersections are shadowed by the first wall. Computing the angle of incidence, gives the direction of the reflected ray. The reflected ray is tested to see if it illuminates any of the receiver locations, and then the next wall intersection is found. This process is repeated for any desired number of reflections. Finding the intersections of the rays with the walls is the most time-consuming aspect of the pincushion codes. Because there is vanishing probability that a ray intersects a point, some procedure is required to capture one and only one ray from each family of rays that passed by the receiver. One such approach is to give the receiver a finite width perpendicular to the ray that is the product of angular separation δ between the rays and the unfolded distance along the ray. The value δ must be such that the spacing between the rays at the edge of the computational area is less than the building size. For example, if the ray length $L = 1$ km and the building width $W = 10$ m, then δ must be less than 0.01 radians, or 0.6^0.

3.6.2 Codes Using a Two-Dimensional Building Database

When the buildings are tall compared with the antenna heights, rays that are diffracted over the buildings, such as A and B in Fig. 3.18, experience large path loss and are neglected. The rays that are reflected and diffracted around tall buildings are found using only the two-dimensional (2D) footprint of the buildings.

For example, receiver R1 is illuminated by one ray that undergoes two reflections, by another ray that undergoes diffraction at a corner followed by a reflection, and by a ray that undergoes diffraction at two corners. Rays reach R2 only after diffraction at a corner. The path loss on diffraction through a large angle, as in turning a corner, is much greater that at a reflection, so that the twice reflected ray reaching R1 is likely to give the dominant contribution. However, in some cases no ray is reflected, such as at R2 in Fig. 3.18; or in other cases only diffracted rayes that have undergone many reflections give the dominant contribution. For some locations rays must undergo diffraction at two corners [52]. Rays that undergo diffraction at more than two corners are typically neglected because most areas can be reached by rays that are reflected and/or diffracted at two corners, and because significant addition to the computation time would be required.

Each ray seen in the top view in Fig. 3.18 may represent two or more rays that have different components of displacement in the vertical direction. In the absence of terrain blockage, one such ray has a vertical component of travel that goes directly from transmitter to receiver. If the ground is flat, then a second ray exists that is reflected in the ground, and appears to come from the image of the transmitter in the ground plane, as previously discussed for LOS paths in Section 3.5. When viewed from above, the direct and ground-reflected rays appear as one. If the terrain is not flat, then there may be more than one ground-reflected ray, or the rays may undergo diffraction over terrain obstacles. After the ray paths have been constructed, the reflection coefficients and UTD diffraction coefficients are computed to find the ray field amplitude. Although the treatment of reflection and diffraction can account for the polarization coupling, the observed coupling is more likely to be a result of scattering from irregular objects near the transmitter and receiver.

Several groups have written computer codes that work with a 2D database of buildings to find the multiply reflected and diffracted rays [55–61]. One such set of predictions made using a 2D code have been compared with measurements in the business section of Rosslyn, Virginia, where there are many tall buildings but also some as low as two stories. The footprints of a building covering an area of 600 × 500 m^2 are shown in Fig. 3.19, together with a sequence of subscriber locations indicated by triangles, and several base station locations. Computations were made allowing for rays that experience two diffractions at building corners and up to six reflections on each portion of the link between the base station, corners, and subscriber locations.

FIGURE 3.19 Footprints of the buildings in the high-rise section of Rosslyn, Virginia. Various subscriber locations are indicated by x's and various base station sites by the Tx's. (From Liang, G. and Bertoni, H. L., *IEEE Trans. Antennas Propag.*, AP-46, p. 853, 1998. With permission.)

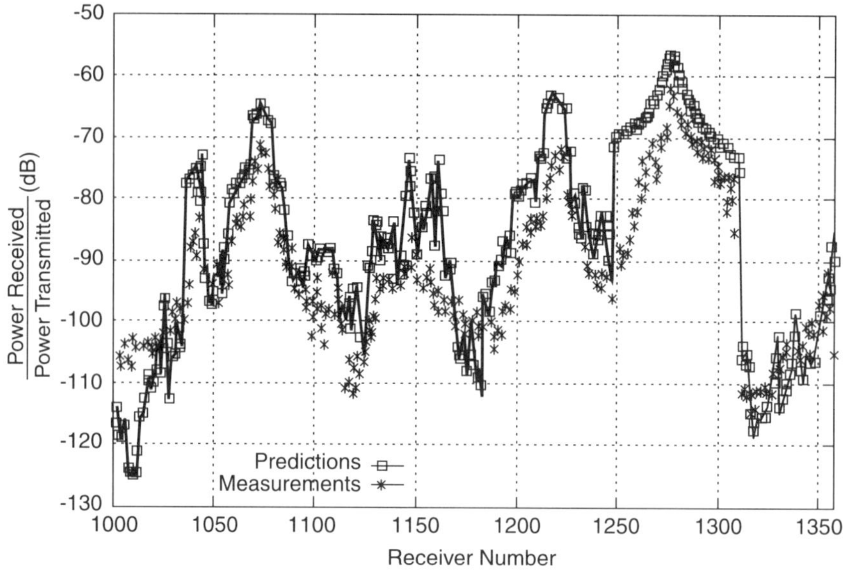

FIGURE 3.20 Comparison between measurements at 1.9 GHz and predictions made using a 2D building database for the footprints of Fig. 3.19 and a base station antenna height of 5 m at site Tx4b. (From Bertoni, H. L., *Radio Propagation for Modern Wireless Systems,* Prentice-Hall, Upper Saddle River, NJ, 2000. With permission.)

Figure 3.20 shows a comparison of the predictions and measurements of the small averages for a mobile antenna height of 2.5 m and for an omnidirectional base station antenna at a height of 5 m located at site Tx4b in Fig. 3.19. The comparison in Fig. 3.20 indicates the accuracy that can be obtained by site-specific predictions. To the left of the base station several buildings only two stories high were located. Propagation over these buildings may explain why the signals measured at locations 1001 to 1020 are significantly higher than the predictions. Although the predictions follow the measurements, they show greater variations than the measurements, which is typical of ray methods.

3.6.3 Codes Using a Three-Dimensional Building Database

For high base station antennas, or when the buildings are of mixed height, it is necessary to account for propagation paths such as A and B in Fig. 3.18 that go over some buildings, and possibly around other buildings. Various approaches have been taken to make use of a 3D database of buildings to find such paths. One such code involves numerical integration, and although it is very accurate, it is limited to a few buildings [60]. Another uses UTD approximations, but limits the buildings to rectangular parallelepipeds on a rectangular street grid, with individual streets having the same width for their entire length, in order to find the ray paths [61].

A ray-tracing code developed at Bell Laboratories [62, 63] assumes that the building walls and roofs can be described by a collection of vertical and horizontal planes, but otherwise does not restrict building shape. This code makes use of the pincushion method in 3D for finding the rays. The rays are launched from the source at small incremental angles over the surface of a unit sphere, and traced in much the same manner as is done in 2D. Predictions obtained using the 3D code have been compared with measurements made in Rosslyn, Virginia [63]. Figure 3.21 shows such a comparison for subscriber locations 1001 through 1092 in the building footprint of Fig. 3.19, and for base station site Tx6, which is situated on top of a building at a height $h_{BS} = 44$ m. The mobile height is $h_m = 2.5\ m$, the frequency is 908 MHz, and directional antenna of 30° beam width and 6° down tilt was used at the base station. The predictions are seen to follow the measurements fairly closely along this one street, and are much more accurate than those made using a 2D building database, whose predictions are also shown in Fig. 3.21. Because the 2D database assumes the buildings to be infinitely tall, it can give high predictions

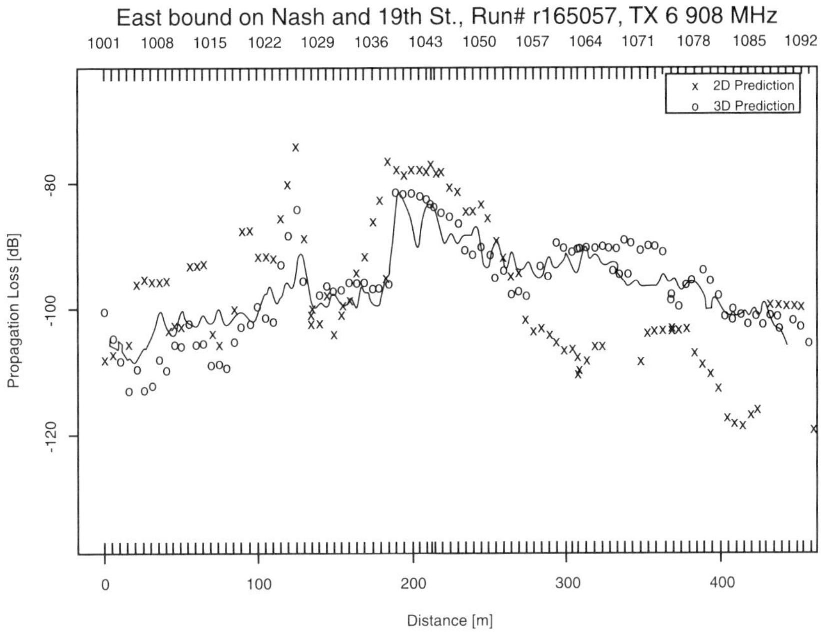

FIGURE 3.21 Comparison between 908-MHz measurements and prediction made using 3D and 2D building database for Rosslyn, Virginia. Measurements were made using a directive base station antenna pointing east and located at Tx6 at a height of 44m. (From Kim, S. C. et al., *IEEE Trans. Veh. Technol.*, 48, 931, 1999. With permission.)

in some locations by including rays that are reflected at height above the actual building, and give low predictions at other locations by blocking rays that actually pass over low buildings.

The pincushion approach in 3D can allow for an adequate number of reflections at building surfaces, but running time restricts the total number of diffractions at vertical or horizontal building corners to two. The long running time for diffraction results from the fact that each ray incident on a corner generates a different diffracted ray cone. Each ray in each cone must then be traced to find its subsequent path. Because the cones all have different angles with respect to the corner, for the purpose of finding the subsequent paths, the corner must be divided into discrete segments, each of which is treated as a secondary source of rays. For a large number of buildings, the number of computations needed to do all the indicated ray traces from all the segments of all the corners sets is enormous.

3.6.4 Vertical Plane Launch Approximation for Three-Dimensional Ray Tracing

The vertical plane launch (VPL) method reduces running time and increases functionality by assuming that the building walls are vertical. This assumption simplifies the determination of the vertical component of travel for the rays reflected from the walls and diffracted at vertical corners. For rays that do not undergo diffraction at horizontal corners, all segments of the ray between the antennas, or between antennas and the ground reflection point, have the same slope in the vertical plane. The sections of vertical planes containing the segments of such rays can be found from the 2D pincushion method, and the actual ray segments within the planes are then found by analytic methods.

Although the method can include rays that travel over buildings, as indicated by rays A and B in Fig. 3.18, it requires a further approximation to include rays diffracted at horizontal corners. Because the diffracted rays lie on a cone, when viewed from above, as in Fig. 3.18, the rays diffracted at the horizontal corners appear to turn through an angle dependent on the angle of diffraction in the vertical plane. To include these rays in the simplified method, it is necessary to assume that the ray segments leaving the horizontal corner lie in the vertical plane of the incident ray or in the vertical plane of the reflected rays.

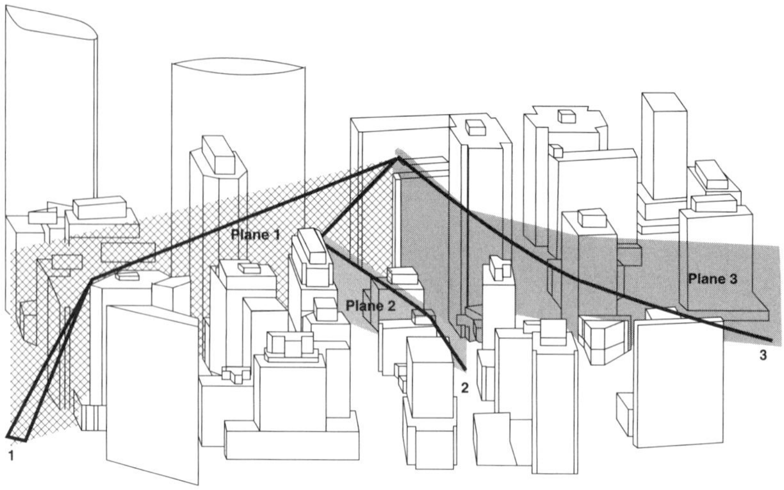

FIGURE 3.22 A perspective view of the buildings in Rosslyn, Virginia, showing the vertical planes launched from a base station at site Tx5 of Fig. 3.19, and the ray paths in those planes. (From Liang, G. and Bertoni, H. L., *IEEE Trans. Antennas Propag.*, AP-46, 853, 1998. With permission.)

In other words, the ray cone generated at the horizontal corner is broken at the top and bottom, and unrolled into two vertical planes. The resulting distortion is small when the horizontal corner is close to being perpendicular to the vertical plane containing the incident ray, or when the slope of the incident and diffracted rays in the vertical plane is small.

The VPL method is suggested in Fig. 3.22, where the base station is treated as launching a series of vertical planes, which are allowed to go over buildings, as well as being reflected at vertical walls, and diffracted at vertical corners [64, 65]. By unfolding the vertical planes, the vertical trajectory of the rays can be found analytically, thereby giving the ray path in 3D, as shown for several rays in Fig. 3.22. Because of the branching of the rays to account for reflection and transmission over buildings, this method is not as fast as the simple 2D case, and various algorithms are employed to prune the branches [65]. However, the VPL method can easily account for multiple diffraction over buildings, which can be very important when the buildings are of roughly uniform height, in addition to double diffraction at vertical edges.

Predictions made with the VPL method have been compared with 900-MHz measurements made in Rosslyn, Virginia, for a base station at site Tx5 in Fig. 3.19, and for all the subscriber locations shown [65]. Figure 3.22 is a perspective view of the area and the base station location, with the bottom of Fig. 3.22 corresponding to the top of Fig. 3.19. The base station employed a directional antenna having a 30° beam width and 6° down tilt, and is located at a height $h_{BS} = 42$ m. Figure 3.23 shows a comparison of predictions with the measurements. Again the predictions follow the measurements, but give greater variations of the received signal. For all base station locations, the average prediction error is typically a few decibels, and a standard deviation of the prediction error is 7 to 10 dB [65].

Other methods further restrict and approximate the actual rays in 3D to gain computational speed. The vertical plane–slant plane method [66] erects a vertical plane and a slant plane that contain the base station and subscriber antennas, and ray tracing is limited to these two planes. For low base station antennas, tracing is done in the horizontal plane, instead of slant planes, so that all subscriber locations can be treated at the same time. Rays diffracted over buildings are accounted for by the vertical plane. However, for the vertical plane trace, reflected rays are forced to lie in the vertical plane, which can be a significant distortion if the wall of the reflecting building is not perpendicular to the vertical plane. This and other forms of out of plane reflection and diffraction at vertical corners are sources of error, which are found to be significant for a high base station antenna [65]. Notwithstanding, this method offers improvement in predictions over statistical methods, such as the Hata model [66].

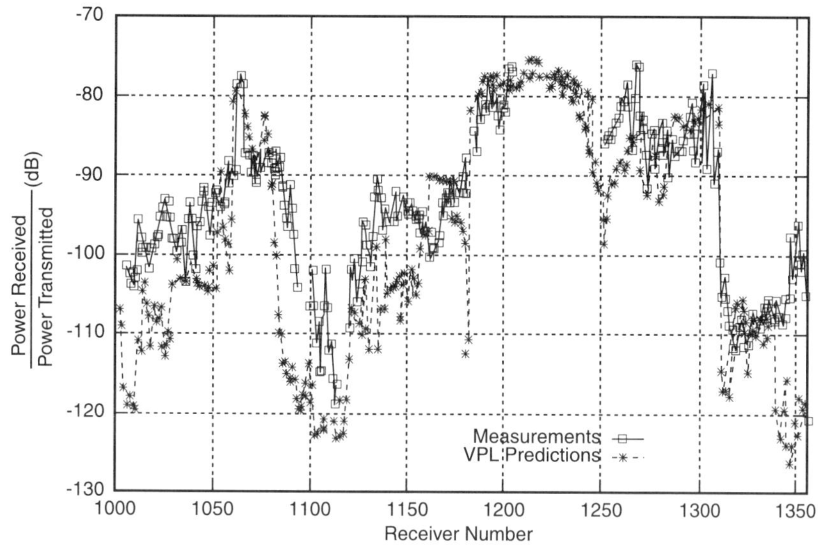

FIGURE 3.23 Comparison between 908-MHz measurements and prediction made using the VPL method for the 3D building database of Rosslyn, Virginia. Measurements were made using a directive base station antenna pointing north and located at Tx5 at a height of 42 m. (From Liang, G. and Bertoni, H. L., *IEEE Trans. Antennas Propag.*, AP-46, 853, 1998. With permission.)

3.6.5 Predicting Delay and Angle Spread Statistics

In wireless systems using digital transmission, multipath echo leads to a spreading of the received signal in time. Much of the signal energy can be recovered in the detection process through the use of equalizers. In a similar way, the signals arriving from different directions can be coherently combined using multiple antennas at the receiver. The most advanced system concepts yield even higher capacity through space–time signal processing that employs multiple antennas at the receiver (smart antennas), or multiple input and multiple output (MIMO) antennas at both ends of the radio link. Design of such systems requires an understanding of the spreading of the signal in time and in angle of arrival (space).

Although time-delay spread has been measured by many groups — see [67] for a summary — measurement of the angle spread has only recently been addressed. Because of the time and expense of making measurements, they are only made for a few base station sites in convenient cities. Because the link geometry and the building environment are different for each set of measurements, it is not clear how to compare the measured values of delay and angle spread, or whether they are applicable to other building environments, antenna heights, etc. To clarify these issues, the ray-tracing codes can be used for Monte Carlo simulations of the channel characteristics. The simulations allow for comparison of channel properties in different cities, and allow the operator to modify the distribution of building properties, such as building height, to see its affect on the properties. Ray-tracing simulations time-delay spread have been shown to give good accuracy compared with measurements [50], and have been used to compare delay and angle spread in different cities [51].

The root-mean-square (rms) delay spread (DS) and the rms angle spread (AS) are commonly used measures of the spread in time of the echo and the angle of arrival spread. Let A_m be the voltage amplitude of the mth ray, let $\tau_m = L_m/c$ be its propagation delay for the ray path length L_m, and let ϕ_m be the angle of arrival as measured from the geometric direction to the transmitter. Then we may find DS using

$$DS = \sqrt{\frac{\sum_m |A_m|^2 (\tau_m - \bar{\tau})^2}{\sum_m |A_m|^2}} \tag{3.33}$$

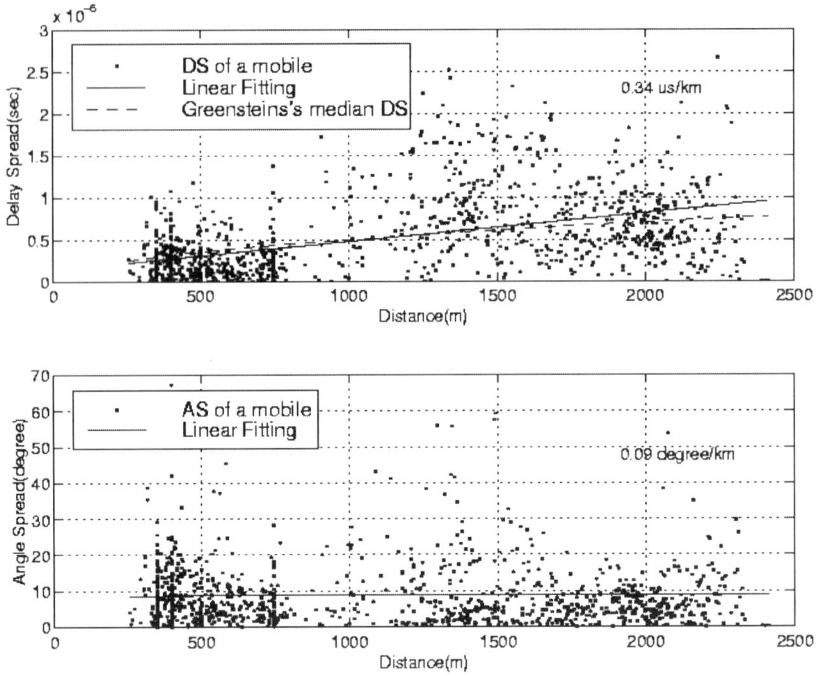

FIGURE 3.24 Scatter plots of *DS* and *AS* at a 74-m-high base station vs. distance to street level mobiles in Munich for a frequency of 900 MHz. (From Cheon, C. and Bertoni, H. L., *Proc. IEEE Veh. Technol. Conf.*, Boston, MA, September 2000. With permission.)

where $\bar{\tau}$ is the average delay of all the rays. Angle of arrival spread is properly defined in terms of the vector direction of arrival over the unit sphere [68, 69]. When the arrivals are grouped around the geometric direction to the transmitter, as in the case of rays arriving at an elevated base station antenna, the rms angle spread *AS* can be defined in terms of ϕ_m [69]. This definition is

$$AS = \sqrt{\frac{\sum_m |A_m|^2 \left(\phi_m - \bar{\phi}\right)^2}{\sum_m |A_m|^2}} \tag{3.34}$$

where $-180° < \phi_m \leq +180°$ and $\bar{\phi}$ is the average of ϕ_m over all the rays.

One set of simulations computed *DS* and *AS* for rays arriving at an elevated base station from 1200 street level mobiles in a building environment constructed from a building database for a section of Munich, Germany [51]. The building database was copied and shifted over to create a 3.8 × 1.5 km area having the same statistical distribution of building heights. The values of *DS* and *AS* for each mobile vs. the horizontal distance from the base station are plotted as dots in Fig. 3.24 for a 74-m-high base station antenna and a frequency of 900 MHz. Regression lines, which represent the distance dependence of the median values of *DS* and *AS*, for these simulated points are shown by the solid straight line. The *DS* and *AS* for the various mobile locations appear to be randomly distributed about the fit lines. The regression line for *DS* increases with distance, whereas the regression line for *AS* indicates that the angle spread does not depend on distance.

Greenstein et al. [67] have proposed a model for the distance dependence of the delay spread based on the various measurements reported in the literature. In their model, *DS* is a random variable that is approximately lognormal in distribution about a median value whose distance dependence is given by $\langle DS \rangle = T\sqrt{R_k}$. Here T is the median value at a distance of 1 km, which is in the range of 0.4 to 1.0 μsec,

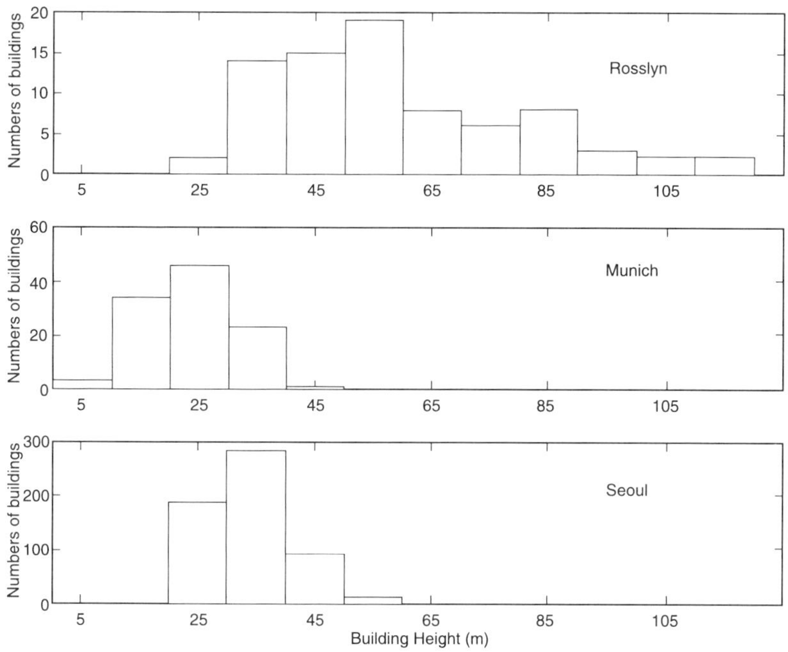

FIGURE 3.25 Histograms of building heights in three cities. (From Bertoni, H. L., *Radio Propagation for Modern Wireless Systems,* Prentice-Hall, Upper Saddle River, NJ, 2000. With permission.)

and R_k is the distance in kilometers. The Greenstein model is plotted at the dashed curve in Fig. 3.24 for $T = 0.5$ μsec. This curve is seen to be close to the regression line. The Greenstein model indicates that the delay spread is not strongly dependent on the city, or on the antenna height. Simulation support this interpretation, although some increase in delay spread was found as the base station antenna was lowered below the rooftops [51].

In contrast to the angle spread, the delay spread was found to be dependent on the distribution of building heights, and to increase significantly as the base station is lowered below the rooftops. As a basis for comparing the angle spread in different cities, Fig. 3.25 shows histograms of building heights in sections of Rosslyn (Virginia), Munich (Germany), and Seoul (Korea) [8]. Rosslyn is a city with many tall buildings, whereas Munich and Seoul are cities with lower and less variation in building height. Simulations in the three cities were carried out for mobiles located 300 m to 1.2 km from the base station, with antennas 5 m above the tallest building, including steeples. The cumulative distribution functions of *AS* for the three cities are plotted in Fig. 3.26, and are seen to be significantly different [70]. For example, the median delay spread is about 4° in Rosslyn, 7° in Seoul, and 10° in Munich. By contrast, the distribution functions of the delay spread *DS* in the three cities are almost the same [70].

3.7 Conclusions

Theoretical models of radio propagation in cities can give useful predictions of signal characteristic in the UHF band (300 MHz to 3 GHz). The theoretical models have the advantage over measurement-based models of showing which of the physical parameters, such as frequency, building height, and street width, are of importance; and how the signal characteristics depend on the parameters. The theoretical model enhances the value of a set of measurements, in that experimentally validated model can be used to translate the measurements to a different set of physical parameters.

To treat the wide range of environments and characteristics, the theoretical models must be selected to match the building environment as well as the property to be predicted. There is no one model that will do everything for all ranges of parameters. Building environments of interest include suburban,

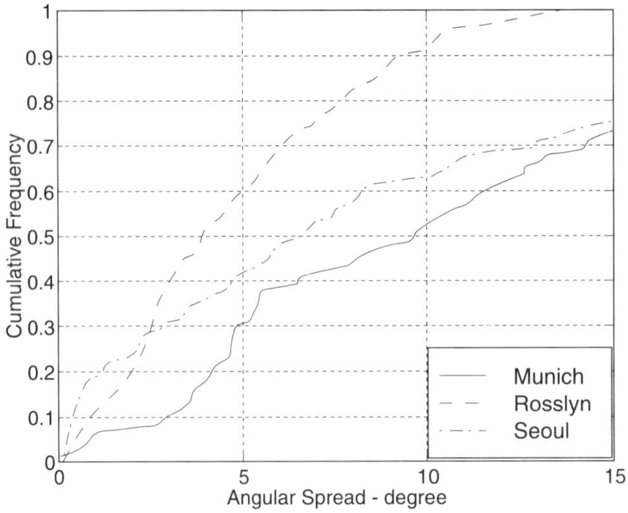

FIGURE 3.26 Comparison of the cumulative distribution functions of *AS* at 900 MHz in the three cities for high base station antennas. (From Bertoni, H. L. et al., *Proc. IEEE Veh. Technol. Conf.*, Houston, TX, May 1999. With permission.)

urban, and the high-rise urban core. Examples of the propagation characteristics of interest include range dependence of the received signal, shadow fading, time-delay spread, and angle of arrival spread. Most of the characteristics have a random nature that is described by a distribution function, or by mean values and standard deviations. Correlation properties of the random variations are also of interest. The discussion of Sections 3.2 to 3.5 has treated the prediction of the average signal strength for a variety of physical parameters and conditions, such as range, antenna height, and presence of foliage and terrain. The approached discussed in these sections can be adapted to other parameter ranges and other propagation characteristics, such a shadow fading. In Section 3.6 it was shown that ray methods may be applied to site-specific predictions that make use of a database of the building. This powerful approach not only predicts the signal strength but also allows for the prediction of many other channel characteristics. The ray methods allow time-consuming and expensive measurements to be replaced by computer simulations for many different link and building geometries.

References

1. Y. Okumura, E. Ohmori, T. Kawano, and K. Fukuda, "Field strength and its variability in VHF and UHF land-mobile radio service," *Rec. Elec. Comm. Lab.*, vol. 16, pp. 825–873, 1968.
2. M. Lecours, I. Y. Chouinard, G. Y. Delisle, and J. Roy, "Statistical Modeling of the Received Signal Envelope in a Mobile Radio Channel," *IEEE Trans. Veh. Technol.*, vol. VT-37, pp. 204–212, 1988.
3. P. E. Mogensen, P. Eggers, C. Jensen, and J. B. Andersen, "Urban area radio propagation measurements at 955 and 1845 MHz for small and micro cells," in *Proc. GLOBECOM'91* (Phoenix, AZ, 1991), pp. 1297–1302.
4. G. D. Ott and A. Plitkins, "Urban path-loss characteristics at 820 MHz," *IEEE Trans. Veh. Technol.*, vol. VT-27, pp. 189–197, 1978.
5. C. Chrysanthou and H. L. Bertoni, "Variability of sector averaged signals for UHF propagation in cities," *IEEE Trans. Veh. Technol.*, vol. 39, pp. 352–358, 1990.
6. H. L. Bertoni and L. Maciel, "Theoretical Predictions of Slow Fading Statistics in Urban Environments," *Proc. IEEE ICUPC Conf.*, pp. 1–4, 1992.
7. J. Walfisch and H. L. Bertoni, "A theoretical model of UHF propagation in urban environments," *IEEE Trans. Antennas Propag.*, vol. 36, pp. 1788–1796, 1988.

8. H. L. Bertoni, *Radio Propagation for Modern Wireless Systems*. Prentice-Hall PTR, Upper Saddle River, NJ, 2000.
9. J. D. Parsons, *The Mobile Radio Channel*, John Wiley & Sons, New York, 1992.
10. F. Ikegami, S. Yoshida, T. Takeuchi, and M. Umehira, "Propagation factors controlling mean field strength on urban streets," *IEEE Trans. Antennas Propag.*, vol. 32, pp. 822–829, 1984.
11. L. B. Felsen and N. Marcuvitz, *Radiation and Scattering of Waves*, Prentice-Hall, Englewood Cliffs, NJ, 1973.
12. M. Born and E. Wolf, *Principles of Optics*, 6th ed. Pergamon Press, Oxford, 1980.
13. D. A. McNamara, C. W. I. Pistorius, and J. A. G. Malherbe, *Introduction to the Uniform Geometrical Theory of Diffraction*, Artech House, Norwood, MA, 1990.
14. M. J. Brooking and R. Larsen, "Results of height gain measurements taken in different environments," Tech. Rep. MTR 84/42, GEC Res, Lab., Marconi Res. Cen., Chelmsford, England, 1985.
15. J. B. Andersen, "UTD multiple-edge transition zone diffraction," *IEEE Trans. Antennas Propag.*, vol. 45, pp. 1093–1097, 1997.
16. H. H. Xia and H. L. Bertoni, "Diffraction of cylindrical and plane waves by an array of absorbing half screens," *IEEE Trans. Antennas Propag.*, vol. 40, pp. 170–177, 1992.
17. L. R. Maciel, H. L. Bertoni, and H. H. Xia, "Unified approach to prediction of propagation over buildings for all ranges of base station antenna height," *IEEE Trans. Veh. Technol.*, vol. 42, pp. 41–45, 1993.
18. R. Janaswamy, J. B. Andersen, "Path loss predictions in urban areas with irregular terrain topography," *Wireless Personal Communications*, 1998.
19. L. Piazzi and H. L. Bertoni, "Effect of terrain on path loss in urban environments for wireless applications," *IEEE Trans. Antennas Propag.*, vol. 46, pp. 1138–1147, 1998.
20. S. A. Torrico, H. L. Bertoni, and R. H. Lang, "Modeling tree effects on path loss in a residential environment," *IEEE Trans. Antennas Propag.*, vol. 46, pp. 872–880, 1998.
21. S. A. Torrico, *Theoretical Modeling of Foliage Effects on Path Loss for Residential Environments*, D.Sc. dissertation, George Washington University, Washington, D.C., 1998.
22. M. Hata, "Empirical formula for propagation loss in land mobile radio services," *IEEE Trans. Veh. Technol.*, vol. VT-29, pp. 317–325, 1980.
23. E. Damosso, Ed., *COST Action 231: Digital Mobile Radio Towards Future Generation Systems*, European Commission, Directorate G, Brussels, Belgium, Ch. 4, 1999.
24. A. J. Rustako, Jr., N. Amitay, G. J. Owens, and R. S. Roman, "Radio propagation at microwave frequencies for line-of-sight microcellular mobile and personal communications," *IEEE Trans. Veh. Technol.*, vol. 40, pp. 203–210, 1991.
25. H. H. Xia, H. L. Bertoni, L. R. Maciel, A. Lindsay-Stewart, and R. Rowe, "Radio propagation characteristics for line-of-sight microcellular and personal communications," *IEEE Trans. Antennas Propag.*, vol. 41, pp. 1439–1447, 1993.
26. N. Blaunstein, R. Giladi, and M. Levin, "Characteristics prediction in urban and suburban environments," *IEEE Trans. Veh. Technol.*, vol. 47, pp. 225–234, 1998.
27. J. Boersma, "On certain multiple integrals occurring in waveguide scattering problem," *SIAM J. Math. Anal.*, vol. 9, pp. 377–393, 1978.
28. L. R. Maciel, H. L. Bertoni, and H. H. Xia, "Propagation over buildings for paths oblique to the street grid," in *Proc. Int. Symp. Personal, Indoor and Mobile Radio Commun.*, (Boston, MA, 1992), pp. 75–79.
29. H. H. Xia, H. L. Bertoni, L. R. Maciel, A. Lindsay-Stewart, and R. Rowe, "Microcellular propagation characteristics for personal communications in urban and suburban environments," *IEEE Trans. Veh. Technol.*, vol. 43, pp. 743–752, 1994.
30. K. Rizk, A. Mawira, J. F. Wagen, and F. Gardiol, "Propagation in urban microcells with high rise buildings," *Proc. IEEE Veh. Technol. Conf.* (Atlanta, GA, 1996), pp. 859–863.
31. W. J. Vogel and J. Goldhirsh, "Tree attenuation at 869 MHz derived from remotely piloted aircraft measurements," *IEEE Trans. Antennas Propag.*, vol. AP-34, pp. 1460–1464, 1986.

32. A. H. LaGrone, "Propagation of VHF and UHF electromagnetic waves ofer a grove of trees in full leaf," *IEEE Trans. Antennas Propag.*, vol. 25, pp. 866–869, 1977.
33. J. C. R. Dal Bello, G. L. Siqueira, and H. L. Bertoni, "Theoretical analysis and measurement results of vegetation effects on path loss for mobile cellular communication systems," *IEEE Trans. Veh. Technol.*, vol. 49, pp. 1285–1293, 2000.
34. T. Tamir, "On radio-wave propagation in forest environments," *IEEE Trans. Antennas Propag.*, vol. 15, pp. 806–817, 1967.
35. L. L. Foldy, "The multiple scattering of waves," *Phys. Rev.*, vol. 67, pp. 107–119, 1945.
36. M. X. Lax, "Multiple scattering of waves," *Rev. Mod. Phys.*, vol. 23, pp. 287–310, 1951.
37. V. Twersky, "Multiple scattering of electromagnetic waves by arbitrary configurations," *J. Math. Phys.*, vol. 8, No. 3, pp. 589–610, 1967.
38. R. H. Lang, "Electromagnetic backscattering from a sparse distribution of lossy dielectric scatterers," *Radio Sci.*, vol. 16, pp. 15–30, 1981.
39. S. A Torrico and R. H. Lang, "Simple engineering expressions to predict the specific attenuation of a tree," Rep. COST 259 TD (98) 5, Thessaloniki, Greece, 1998.
40. K. Bullington, "Radio propagation for vehicular communications," *IEEE Trans. on Veh. Technol.*, pp. 295–308, 1977.
41. R. J. Luebbers, "Finite conductivity uniform GTD versus knife edge diffraction in prediction of propagation path loss," *IEEE Trans. Antennas Propag.*, vol. AP-32, pp. 70–76, 1984.
42. G. Lampard and T. Vu-Dinh, "The effect of terrain on radio propagation in urban microcells," *IEEE Trans. Veh. Technol.*, vol. VT-42, pp. 314–317, 1993.
43. M. F. Levy, "Parabolic equation modeling of propagation over irregular terrain," *Electron. Lett.*, vol. 26, pp. 1153–1155, 1990.
44. R. Janaswamy and J. B. Andersen, "A curvilinear coordinate-based split-step parabolic equation method for propagation prediction over terrain," *IEEE Trans. Antennas Propag.*, vol. AP-46, pp. 1089–1097, 1998.
45. J. T. Hviid, J. B. Andersen, J. Toftgard, and J. Bojer, "Terrain-based propagation model for rural area-an integral equation approach," *IEEE Trans. Antennas Propag.*, vol. AP-43, pp. 41–46, 1995.
46. W. C. Y. Lee, "Studies of base-station antenna height effect on mobile radio," *IEEE Trans. Veh. Technol.*, vol. VT-29, pp. 252–260, 1980.
47. L. Piazzi and H.L. Bertoni, "A path loss formulation for wireless applications considering terrain effects for urban environments," *Proc. IEEE-VTC'98 Conf.*, pp.159–163, 1998.
48. V. Erceg, A. J. Rustako, Jr., and R. S. Roman, "Diffraction around corners and its effects on the microcell coverage area in urban and suburban environments at 900 MHz, 2 GHz, and 6 GHz," *IEEE Trans. Veh. Technol.*, vol. 43, pp. 762–766, 1994.
49. F. Nui and H.L. Bertoni, "Path loss and cell coverage of urban microcells in high-rise building environment," *Proc. IEEE GLOBECOM*, pp. 266–270, 1993.
50. G.E. Athanasiadou, A.R. Nix and J.P. McGeehan, "A microcellular ray-tracing propagation model and evaluation of its narro-band and wide-band predictions," *IEEE J. Selected Areas Commun.*, vol. 18, pp. 322–335, 2000.
51. C. Cheon and H. L. Bertoni, "Monte Carlo simulation of delay and angle spread in different building environments," *Proc. IEEE Veh. Technol. Conf.* (Boston, MA, September 2000).
52. G. E. Athanasiadou and A. R. Nix, "Investigation into the sensitivity of the power predictions of a microcellular ray tracing propagation model," *IEEE Trans. Veh. Technol.*, vol. 49, pp. 1140–1151, 2000.
53. M. O. Al-Nuaimi and M. S. Ding, "Prediction models and measurements of microwave signal scattered from buildings," *IEEE Trans. Antennas Propag.*, vol. AP-42, pp. 1126–1137, 1994.
54. U. Dersch and E. Zollinger, "Propagation mechanisms in microcell and indoor environments," *IEEE Trans. Veh. Technol.*, vol. 43, pp. 1058–1066, 1994.
55. S. T. S. Chia, R. Steele, E. Green, and A. Baran, "Propagation and bit error ratio measurements for macrocellular systems," *J. IRE*, vol. 57, pp. S255–S266, 1987.

56. M. C. Lawton and J. P. McGeehan, "The application of a deterministic ray launching algorithm for the prediction of radio channel characteristics in small-cell environments," *IEEE Trans. Veh. Technol.*, vol. 43, pp. 955–969, 1994.
57. S. Y. Tan and H. S. Tan, "Propagation model for microcellular communications applied to path loss measurements in Ottawa city streets," *IEEE Trans. Veh. Technol.*, vol. 44, pp. 313–317, 1995.
58. S. Y. Tan and H. S. Tan, "A microcellular communications propagation model based on the uniform theory of diffraction and multiple image theory," *IEEE Trans. Antennas Propag.*, vol. AP-44, pp. 1317–1325, 1996.
59. K. Rizik, J-F. Wagen, and F. Gardiol, "Two-dimensional ray-tracing modeling for propagation prediction in microcellular environments," *IEEE Trans. Veh. Technol.*, vol. 46, pp. 508–517, 1997.
60. C. C. Constantinou and L. C. Ong, "Urban radiowave propagation: A 3-D path-integral wave analysis," *IEEE Trans. Antennas Propag.*, vol. AP-46, pp. 211–217, 1998.
61. A. G. Kanatas, I. D. Kountouris, G. B. Kostraras, and P. Constantinou, "A UTD propagation model in urban microcellular environments," *IEEE Trans. Veh. Technol.*, vol. 46, pp. 185–193, 1997.
62. V. Erceg, S. J. Fortune, J. Ling, A. J. Rustako, and R. A. Valenzuela, "Comparisons of a computer-based propagation prediction tool with experimental data collected in urban microcellular environments," *IEEE J. Selected Areas Commun.*, vol. SAC-15, pp. 677–684, 1997.
63. S. C. Kim, B. J. Guarino, Jr., T. M. Willis, III, V. Erceg, S. J. Fortune, R. Valenzuela, L. W. Thomas, J. Ling, and J. D. Moore, "Radio propagation measurements and prediction using three dimensional ray tracing in urban environments at 908 MHz and 1.9 GHz," *IEEE Trans. Veh. Technol.*, vol. 48, pp. 931–946, 1999.
64. J. P. Rossi, J. C. Bie, A. J. Levy, Y. Gabillet, and M. Rosen, "A ray-launching method for radio-mobile propagation in urban area," *Digest IEEE APS Symp.* (London, Ontario, June 1991), pp. 1540–1543.
65. G. Liang and H. L. Bertoni, "A new approach to 3-D ray tracing for propagation prediction in cities," *IEEE Trans. Antennas Propag.*, vol. AP-46, pp. 853–863, 1998.
66. T. Kurner, D. J. Cichon, and W. Wiesbeck, "Evaluation and verification of the VHF/UHF propagation channel based on a 3-D wave propagation model," *IEEE Trans. Antennas Propag.*, vol. 44, pp. 393–404, 1996.
67. L. J. Greenstein, V. Erceg, Y. S. Yeh, and M. V. Clark, "A new path-gain/delay-spread propagation model for digital cellular channels," *IEEE Trans. Veh. Technol.*, vol. 46, pp. 477–485, 1997.
68. G. Durgin and T. Rappaport, "Basic relationship between multipath angular spread and narrow-band fading in wireless channels," *Electron. Lett.*, vol. 34, November 25, 1998.
69. B. H. Fleury, "Direction dispersion and space selectivity in the mobile radio channel," *Proc. IEEE Veh. Technol. Conf.* (Boston, MA, September 2000).
70. H. L. Bertoni, P. Pongsilamanee, C. Cheon and G. Liang, "Sources and Statistics of Multipath Arrival at Eevated Base Station Antennas," *Proc. IEEE Veh. Technol. Conf.* (Houston, TX, May 1999).

4

Fading Channels*

Bernard Sklar
Communications Engineering Services

When the mechanisms that cause fading in communication channels were first modeled in the 1950s and 1960s, the principles developed were primarily applied to over-the-horizon communications covering a wide range of frequency bands. The 3 to 30 MHz high-frequency (HF) band used for ionospheric propagation, and the 300 MHz to 3 GHz ultra-high-frequency (UHF) as well as the 3 to 30 GHz super-high-frequency (SHF) bands used for tropospheric scatter, are examples of channels that are affected by fading phenomena. Although the fading effects in mobile radio channels are somewhat different than those encountered in ionospheric and tropospheric channels, the early models are still quite useful in helping to characterize the fading effects in mobile digital communication systems. This chapter emphasizes

*A version of this chapter appears in the book, *Digital Communications: Fundamentals and Applications*, 2nd ed., by Bernard Sklar, Prentice-Hall, Englewood Cliffs, NJ, 2001.

0-8493-0124-6/02/$0.00+$1.50

so-called *Rayleigh fading*, primarily in the UHF band, that affects mobile systems such as cellular and personal communication systems (PCS). Emphasis is also placed on the fundamental fading manifestations, types of degradation, and methods for mitigating the degradation. Two examples of specific mitigation techniques are examined, namely, the Viterbi equalizer implemented in the Global System for Mobile (GSM) Communications, and the RAKE receiver used in CDMA systems built to meet Interim Standard-95 (IS-95).

4.1 The Challenge of Communicating over Fading Channels

In the analysis of communication system performance, the classical (ideal) additive-white-Gaussian-noise (AWGN) channel, with statistically independent Gaussian noise samples corrupting data samples free of intersymbol interference (ISI), is the usual starting point for developing basic performance results. An important source of performance degradation is thermal noise generated in the receiver. Another source of degradation stems from both natural and man-made sources of noise and interference that enter the receiving antenna, and can be quantified by a parameter called *antenna temperature*. Thermal noise typically has a flat power spectral density over the signal band and a zero-mean Gaussian voltage probability density function (pdf). In mobile communication systems, the external noise and interference is often more significant than the receiver thermal noise. When modeling practical systems, the next step is the introduction of band-limiting filters. Filtering in the transmitter usually serves to satisfy some regulatory requirement on spectral containment. Filtering in the receiver is often the result of implementing a matched filter. Because of the band-limiting and phase-distortion properties of filters, special signal design and equalization techniques may be required to mitigate the filter-induced ISI.

If propagating characteristics of a radio channel are not specified, one usually infers that the signal attenuation vs. distance behaves as if propagation takes place over ideal free space. The model of free space treats the region between the transmit and receive antennas as being free of all objects that might absorb or reflect radio frequency (RF) energy. It also assumes that, within this region, the atmosphere behaves as a perfectly uniform and nonabsorbing medium. Furthermore, the earth is treated as being infinitely far away from the propagating signal (or, equivalently, as having a reflection coefficient that is negligible). Basically, in this idealized free-space model, the attenuation of RF energy between the transmitter and receiver behaves according to an inverse square law. The received power expressed in terms of transmitted power is attenuated by a factor, $L_s(d)$, where this factor is called *path loss* or *free-space loss*. When the receiving antenna is isotropic, this factor is expressed as

$$L_s(d) = \left(\frac{4\pi d}{\lambda} \right)^2 \tag{4.1}$$

In Eq. (4.1), d is the distance between the transmitter and the receiver, and λ is the wavelength of the propagating signal. For this case of idealized propagation, received signal power is very predictable. For most practical channels, where signal propagation takes place in the atmosphere and near the ground, the free-space propagation model is inadequate to describe the channel behavior and predict system performance. In a wireless mobile communication system, a signal can travel from transmitter to receiver over multiple reflective paths. This phenomenon, referred to as *multipath propagation*, can cause fluctuations in the received signal amplitude, phase, and angle of arrival, giving rise to the terminology *multipath fading*. Another name, *scintillation*, having originated in radio astronomy, is used to describe the fading caused by physical changes in the propagating medium, such as variations in the electron density of the ionospheric layers that reflect HF radio signals. Both names, fading and scintillation, refer to the random fluctuations of a signal; the main difference is that scintillation involves mechanisms (e.g., electrons) that are much smaller than a wavelength. The end-to-end modeling and design of systems that incorporate techniques to mitigate the effects of fading are usually more challenging than those whose sole source of performance degradation is AWGN.

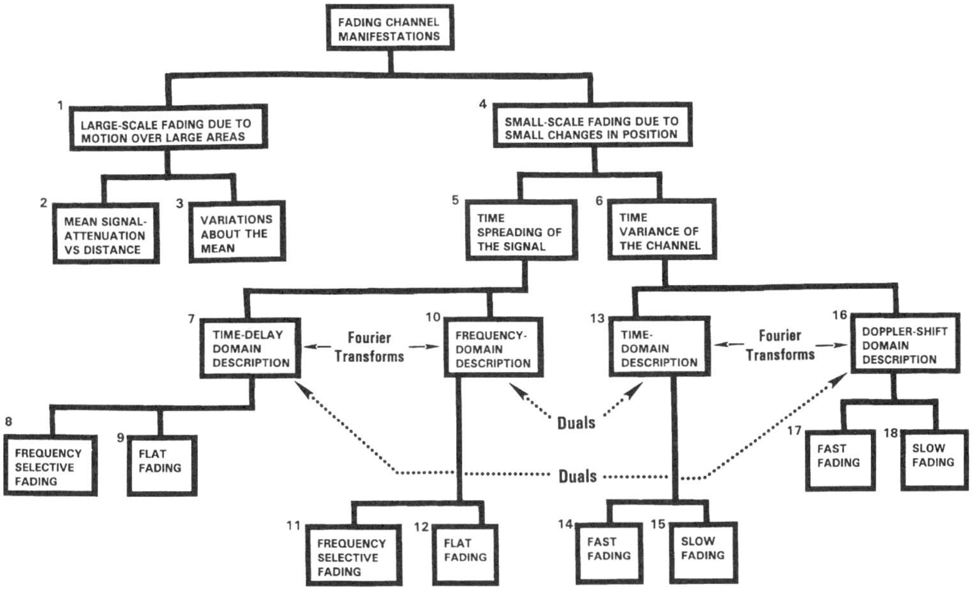

FIGURE 4.1 Fading-channel manifestations.

4.2 Characterizing Mobile-Radio Propagation

Figure 4.1 represents an overview of fading-channel manifestations. It starts with two types of fading effects that characterize mobile communications: large-scale fading and small-scale fading. Large-scale fading represents the average signal power attenuation or the path loss resulting from motion over large areas. In Fig. 4.1, the large-scale fading manifestation is shown in blocks 1, 2, and 3. This phenomenon is affected by prominent terrain contours (e.g., hills, forests, billboards, or clumps of buildings) between the transmitter and receiver. The receiver is often said to be "shadowed" by such prominences. The statistics of large-scale fading provide a way of computing an estimate of path loss as a function of distance. This is often described in terms of a mean-path loss (nth-power law) and a log normally distributed variation about the mean. Small-scale fading refers to the dramatic changes in signal amplitude and phase that can be experienced as a result of small changes (as small as a half wavelength) in the spatial positioning between a receiver and a transmitter. As indicated in Fig. 4.1 (blocks 4, 5, and 6), small-scale fading manifests itself in two mechanisms, namely, time spreading of the signal (or signal dispersion) and time variant behavior of the channel. For mobile-radio applications, the channel is time variant because motion between the transmitter and receiver results in propagation path changes. The rate of change of these propagation conditions accounts for the fading rapidity (rate of change of the fading impairments). Small-scale fading is called Rayleigh fading if there are multiple reflective paths that are large in number, and there is no line-of-sight (LOS) signal component; the envelope of such a received signal is statistically described by a Rayleigh pdf. When there is a dominant nonfading signal component present, such as LOS propagation path, the small-scale fading envelope is described by a Rician pdf [1]. In other words, the small-scale fading statistics are said to be Rayleigh whenever the LOS path is blocked, and Rician otherwise. A mobile radio roaming over a large area must process signals that experience both types of fading: small-scale fading superimposed on large-scale fading.

Large-scale fading (attenuation or path loss) can be considered to be a spatial average over the small-scale fluctuations of the signal. It is generally evaluated by averaging the received signal over 10 to 30 wavelengths to decouple the small-scale (mostly Rayleigh) fluctuations from the large-scale shadowing effects (typically log normal). Three basic mechanisms impact signal propagation in a mobile communication system. They are reflection, diffraction, and scattering [1].

- Reflection occurs when a propagating electromagnetic wave impinges on a smooth surface with very large dimensions compared with the RF signal wavelength (λ).
- Diffraction occurs when the propagation path between the transmitter and receiver is obstructed by a dense body with dimensions that are large when compared with λ, causing secondary waves to be formed behind the obstructing body. Diffraction is a phenomenon that accounts for RF energy traveling from transmitter to receiver without an LOS path between the two. It is often termed *shadowing* because the diffracted field can reach the receiver even when shadowed by an impenetrable obstruction.
- Scattering occurs when a radio wave impinges on either a large rough surface or any surface with dimensions on the order of λ or less, causing the energy to be spread out (scattered) or reflected in all directions. In an urban environment, typical signal obstructions, that yield scattering are lampposts, street signs, and foliage. The name *scatterer* applies to any obstruction in the propagation path that causes a signal to be reflected or scattered.

Figure 4.1 may serve as a table of contents for the sections that follow. The two manifestations of small-scale fading, signal time spreading (signal dispersion) and the time-variant nature of the channel, are examined in two domains: time and frequency, as indicated in Fig. 4.1 (blocks 7, 10, 13, and 16). For signal dispersion, the fading degradation types are categorized as being frequency selective or frequency nonselective (flat), as listed in blocks 8, 9, 11, and 12. For the time-variant manifestation, the fading degradation types are categorized as fast fading or slow fading, as listed in blocks 14, 15, 17, and 18. The labels indicating Fourier transforms and duals are explained later.

Figure 4.2 is a convenient pictorial (not a precise graphic representation) showing the various contributions that must be considered when estimating path loss for link budget analysis in a mobile-radio application [2]. These contributions are (1) mean path loss as a function of distance, as a result of large-scale fading, (2) near-worst-case variations about the mean path loss or large-scale fading margin (typically 6 to 10 dB), (3) near-worst-case Rayleigh or small-scale fading margin (typically 20 to 30 dB). In Fig. 4.2, the annotations "≈1 to 2%" indicate a suggested area (probability) under the tail of each pdf as

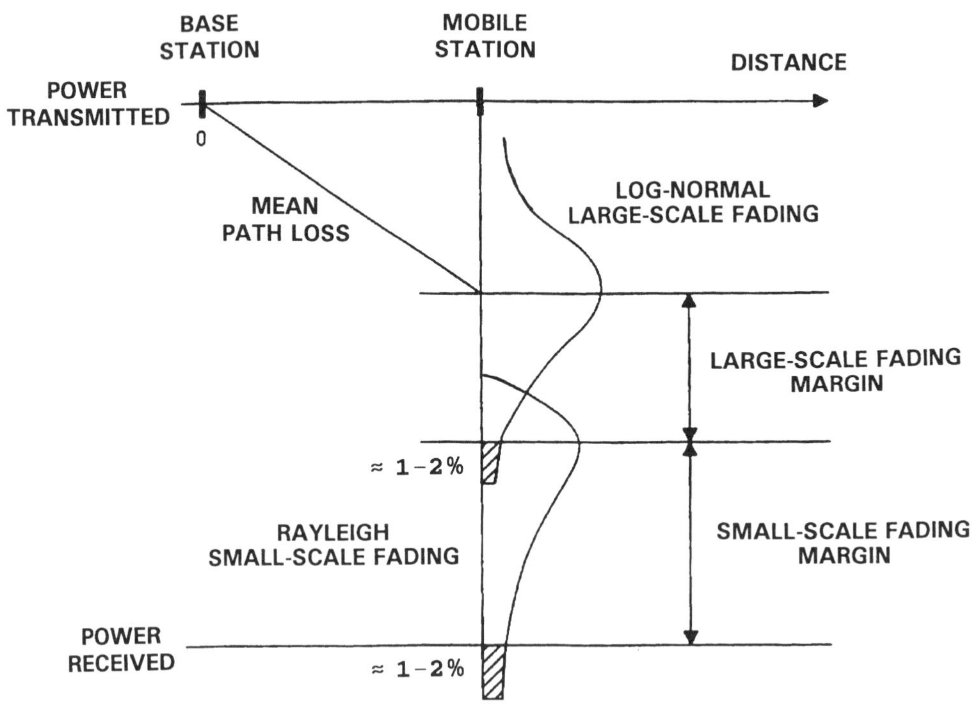

FIGURE 4.2 Link budget considerations for a fading channel.

a design goal. Hence, the amount of margin indicated is intended to provide adequate received signal power for approximately 98 to 99% of each type of fading variation (large and small scale).

By using complex notation, a transmitted signal is written as

$$s(t) = \mathrm{Re}\left\{g(t)e^{j2\pi f_c t}\right\} \tag{4.2}$$

where Re{·} denotes the real part of {·}, and f_c is the carrier frequency. The baseband waveform $g(t)$ is called the complex envelope of $s(t)$, and can be expressed as

$$g(t) = |g(t)|e^{j\phi(t)} = R(t)e^{j\phi(t)} \tag{4.3}$$

where $R(t) = |g(t)|$ is the envelope magnitude, and $\phi(t)$ is its phase. For a purely phase- or frequency-modulated signal, $R(t)$ is constant, and, in general, varies slowly compared with $t = 1/f_c$.

In a fading environment, $g(t)$ is modified by a complex dimensionless multiplicative factor $\alpha(t)e^{-j\theta(t)}$. (Later, we show this derivation.) The *modified* baseband waveform can be written as $\alpha(t)e^{-j\theta(t)}g(t)$; but for now, let us examine the magnitude, $\alpha(t)R(t)$ of this envelope, which can be expressed in terms of three positive terms, as follows [3]:

$$\alpha(t)R(t) = m(t) \times r_0(t) \times R(t) \tag{4.4}$$

where $m(t)$ is called the large-scale-fading component of the envelope, and $r_0(t)$ is called the small-scale-fading component. Sometimes, $m(t)$ is referred to as the *local mean* or *lognormal fading* because generally its measured values can be statistically described by a lognormal pdf; or equivalently, when measured in decibels, $m(t)$ has a Gaussian pdf. Furthermore, $r_0(t)$ is sometimes referred to as multipath or Rayleigh fading. For the case of a mobile radio, Fig. 4.3 illustrates the relationship between $\alpha(t)$ and $m(t)$. In Fig. 4.3, we consider that an *unmodulated* carrier wave is being transmitted, which in the context of Eq. (4.4) means that for all time, $R(t) = 1$. Figure 4.3a is a representative plot of signal power received vs. antenna displacement (typically in units of wavelength). The signal power received is, of course, a function of the multiplicative factor $\alpha(t)$. Small-scale fading superimposed on large-scale fading can be readily identified. The typical antenna displacement between adjacent signal-strength nulls, resulting from small-scale fading, is approximately a half wavelength. In Fig. 4.3b, the large-scale fading or local mean, $m(t)$, has been removed to view the small-scale fading, $r_0(t)$, referred to some average constant power. Recall that $m(t)$ can generally be evaluated by averaging the received envelope over 10 to 30 wavelengths. The lognormal fading is a relatively slowly varying function of position, whereas the Rayleigh fading is a relatively fast varying function of position. Note that for an application involving motion, such as the case of a radio in a moving vehicle, a function of position is tantamount to a function of time. In the sections that follow, some of the details concerning the statistics and mechanisms of large-scale and small-scale fading are enumerated.

4.2.1 Large-Scale Fading

For mobile radio applications, Okumura et al. [4] made some of the earlier comprehensive path-loss measurements for a wide range of antenna heights and coverage distances. Hata [5] transformed Okumura's data into parametric formulas. In general, propagation models, for both indoor and outdoor radio channels, indicate that the mean path loss, $\overline{L}_p(d)$, as a function of distance, d, between a transmitter and receiver is proportional to an nth-power of d relative to a reference distance d_0 [1].

$$\overline{L}_p(d) \propto \left(\frac{d}{d_0}\right)^n \tag{4.5}$$

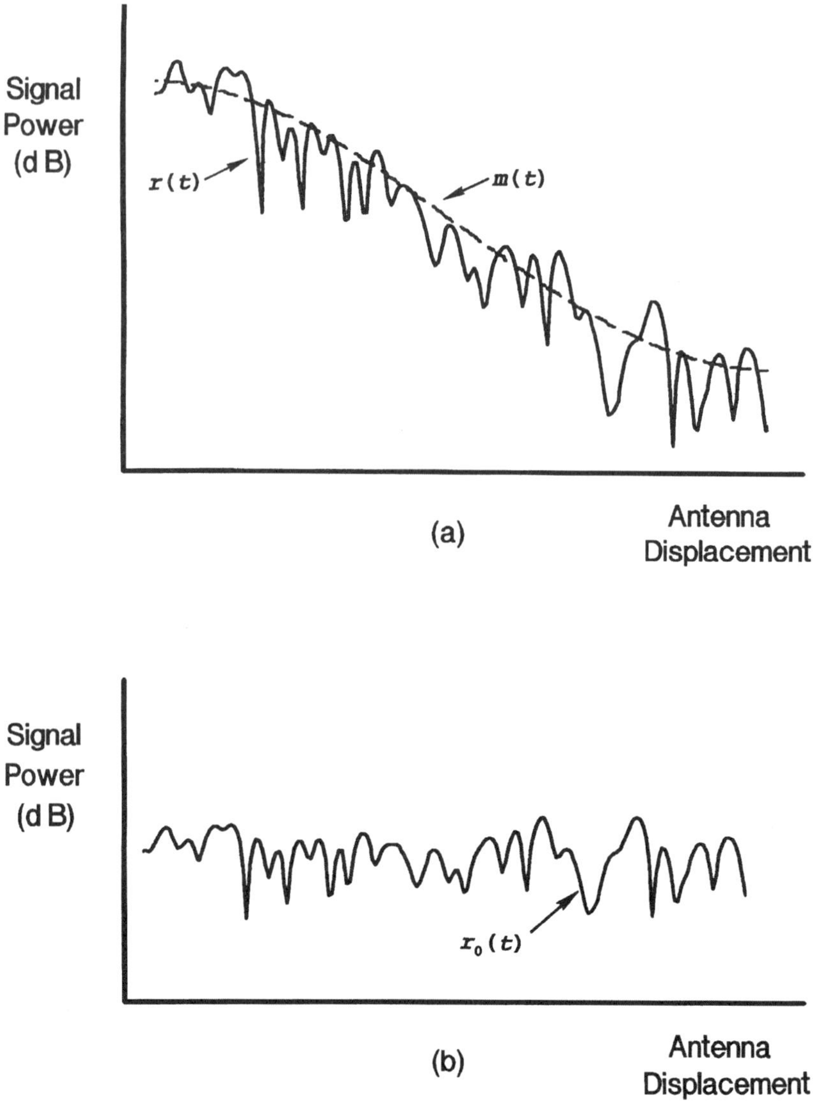

FIGURE 4.3 Large-scale fading and small-scale fading.

$\overline{L_p}(d)$ is often stated in decibels, shown as follows:

$$\overline{L_p}\left(d\right)\left(dB\right)=L_s\left(d_0\right)\left(dB\right)+10n\ \log\left(\frac{d}{d_0}\right) \tag{4.6}$$

The reference distance d_0, corresponds to a point located in the far-field of the transmit antenna. Typically, the value of d_0 is taken to be 1 km for large cells, 100 m for microcells, and 1 m for indoor channels. Moreover, $L_s(d_0)$ is evaluated by using Eq. (4.1) or by conducting measurements. $\overline{L_p}(d)$ is the average path loss (over a multitude of different sites) for a given value of d. When plotted on a log–log scale, $\overline{L_p}(d)$ vs. d (for distances greater than d_0) yields a straight line with a slope equal to $10n$ dB/decade. The value of the *path–loss exponent n* depends on the frequency, antenna heights, and propagation

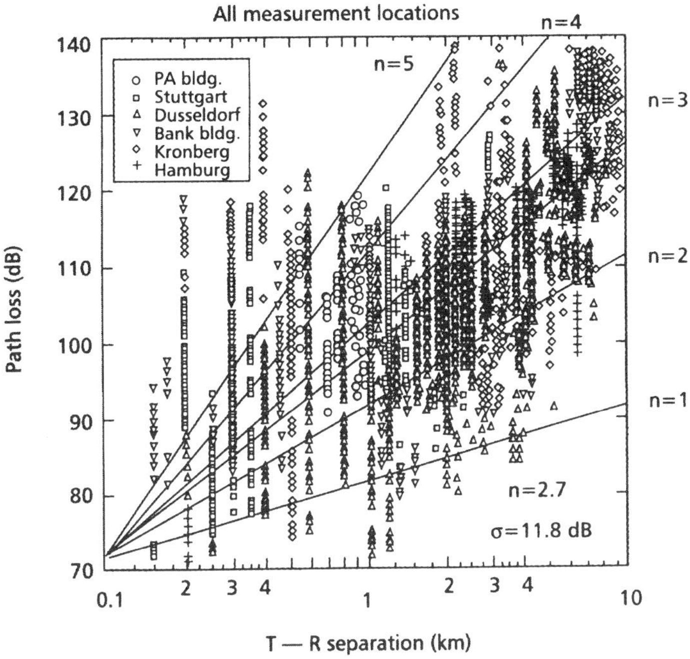

FIGURE 4.4 Path loss vs. distance measured in several German cities.

environment. In free space where signal propagation follows an inverse square law, n is equal to 2, as seen in Eq. (4.1). In the presence of very strong guided wave phenomena (like urban streets), n can be lower than 2. When obstructions are present, n is larger. Figure 4.4 shows a scatter plot of path loss vs. distance for measurements made at several sites in Germany [6]. Here, the path loss has been measured relative to a reference distance $d_0 = 100$ m. Also shown are straight-line fits to various exponent values.

The path loss vs. distance expressed in Eq. (4.6) is an average, and therefore not adequate to describe any particular setting or signal path. It is necessary to provide for variations about the mean because the environment of different sites may be quite different for similar transmitter–receiver (T-R) separations. Figure 4.4 illustrates that path-loss variations can be quite large. Measurements have shown that for any value of d, the path loss $L_p(d)$ is a random variable having a lognormal distribution about the mean distant-dependent value $\overline{L}_p(d)$ [7]. Thus, path loss $L_p(d)$ can be expressed in terms of $\overline{L}_p(d)$ as expressed in Eq. (4.6), plus a random variable X_σ, as follows [1]:

$$L_p(d)\,(dB) = L_s(d_0)\,(dB) + 10n\log_{10}(d/d_0) + X_\sigma\,(dB) \tag{4.7}$$

where X_σ denotes a zero-mean, Gaussian random variable (in decibels) with standard deviation σ (also in decibels). X_σ is site and distance dependent. Because X_σ and $L_p(d)$ are random variables, if Eq. (4.7) is used as the basis for computing an estimate of path loss or link margin, some value for X_σ must first be chosen. The choice of the value is often based on measurements (made over a wide range of locations and T-R separations). It is not unusual for X_σ to take on values as high as 6 to 10 dB or greater. Thus, the parameters needed to statistically describe path loss resulting from large-scale fading, for an arbitrary location with a specific transmitter–receiver separation are (1) the reference distance, (2) the path–loss exponent, (3) the standard deviation σ of X_σ. There are several good references dealing with the measurement and estimation of propagation path loss for many different applications and configurations [1, 5–9].

4.2.2 Small-Scale Fading

In this section we develop the small-scale fading component, $r_0(t)$. Analysis proceeds on the assumption that the antenna remains within a limited trajectory, so that the effect of large-scale fading, $m(t)$, is a constant (assumed unity). Assume that the antenna is traveling, and that there are multiple scatterer paths, each associated with a time-variant propagation delay $\tau_n(t)$, and a time-variant multiplicative factor $\alpha_n(t)$. By neglecting noise, the received band-pass signal, $r(t)$, can be written as

$$r(t)=\sum_n \alpha_n(t)\, s\left[t-\tau_n(t)\right] \tag{4.8}$$

By substituting Eq. (4.2) into Eq. (4.8), we write the received band-pass signal as follows:

$$\begin{aligned} r(t) &= \mathrm{Re}\left(\left\{\sum_n \alpha_n(t) g\left[t-\tau_n(t)\right]\right\} e^{j2\pi f_c\left[t-\tau_n(t)\right]}\right) \\ &= \mathrm{Re}\left(\left\{\sum_n \alpha_n(t) e^{-j2\pi f_c \tau_n(t)} g\left[t-\tau_n(t)\right]\right\} e^{j2\pi f_c t}\right) \end{aligned} \tag{4.9}$$

From Eq. (4.9), it follows that the equivalent received baseband signal is

$$z(t)=\sum_n \alpha_n(t) e^{-j2\pi f_c \tau_n(t)} g\left[t-\tau_n(t)\right] \tag{4.10}$$

Consider the transmission of an unmodulated carrier at frequency f_c. In other words, for all time, $g(t) = 1$. Then, the received baseband signal, for this case of an unmodulated carrier and discrete multipath components given by Eq. (4.10), reduces to

$$z(t)=\sum_n \alpha_n(t) e^{-j2\pi f_c \tau_n(t)} = \sum_n \alpha_n(t) e^{-j\theta_n(t)} \tag{4.11}$$

where $\theta_n(t) = 2\pi f_c \tau_n(t)$. The baseband signal $z(t)$ consists of a sum of time-variant phasors having amplitudes $\alpha_n(t)$ and phases $\theta_n(t)$. Notice that $\theta_n(t)$ changes by 2π radians whenever τ_n changes by $1/f_c$ (typically, a very small delay). For a cellular radio operating at $f_c = 900$ MHz, the delay $1/f_c = 1.1$ nsec. In free space, this corresponds to a change in propagation distance of 33 cm. Thus, in Eq. (4.11), $\theta_n(t)$ can change significantly with relatively small propagation delay changes. In this case, when two multipath components of a signal differ in path length by 16.5 cm, one signal will arrive 180° out of phase with respect to the other signal. Sometimes the phasors add constructively and sometimes they add destructively, resulting in amplitude variations, namely, fading of $z(t)$. Equation (4.11) can be expressed more compactly as the net received envelope, which is the summation over all the scatterers, as follows:

$$z(t)=\alpha(\tau) e^{-j\theta(t)} \tag{4.12}$$

where $\alpha(t)$ is the resultant magnitude, and $\theta(t)$ is the resultant phase. Equation (4.12) represents the same complex dimensionless multiplicative factor that was described earlier. It is an important result because the expression tells us that even though a *bandpass* signal $s(t)$, as expressed in Eq. (4.2), is the signal that experienced the fading effects and gave rise to the received signal $r(t)$, these effects can be described by analyzing $r(t)$ at the *baseband* level.

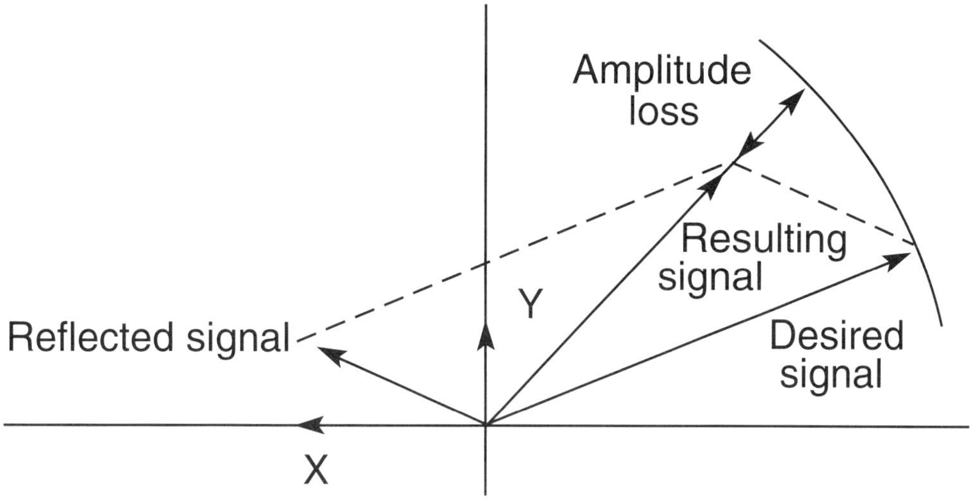

FIGURE 4.5 Effect of a multipath reflected signal on a desired signal.

Figure 4.5 illustrates the primary mechanism that causes fading in multipath channels, as described by Eqs. (4.11) and (4.12). In Fig. 4.5, a reflected signal has a phase delay (a function of additional path length) with respect to a desired signal. The reflected signal also has reduced amplitude (a function of the reflection coefficient of the obstruction).

Reflected signals can be described in terms of orthogonal components, $x_n(t)$ and $y_n(t)$, where $x_n(t) + j\,y_n(t) = \alpha_n(t)e^{-j\theta_n(t)}$. If the number of such stochastic components is large and none is dominant, then at a fixed time the variables $x_r(t)$ and $y_r(t)$ resulting from their addition have a Gaussian pdf. These orthogonal components yield the small-scale random variable, $r_0(t)$, the envelope of the received signal. Thus, $r_0(t)$, the factor that was defined in Eq. (4.4), can be expressed in terms of $x_r(t)$ and $y_r(t)$. For the case of an unmodulated carrier wave as shown in Eq. (4.12), $r_0(t)$ can be designated the small-scale component of the envelope of $z(t)$, as follows:

$$r_0(t) = \sqrt{x_r^2(t) + y_r^2(t)} \tag{4.13}$$

When the received signal is made up of multiple reflective rays plus a significant LOS (nonfaded) component, the received envelope amplitude has a Rician pdf, shown as follows, and the fading is referred to as *Rician fading* [2].

$$p(r_0) = \begin{cases} \dfrac{r_0}{\sigma^2}\exp\left[-\dfrac{(r_0^2 + A^2)}{2\sigma^2}\right] I_0\left(\dfrac{r_0 A}{\sigma^2}\right) & \text{for } r_0 \geq 0,\ A \geq 0 \\ 0 & \text{otherwise} \end{cases} \tag{4.14}$$

Although $r_0(t)$ varies dynamically with motion (time), at any fixed time it is a random variable, with a value that stems from the ensemble of real positive numbers. Hence, in describing probability density functions, it is appropriate to drop the functional dependence on time. The parameter $2\sigma^2$ is the predetection mean power of the multipath signal, A denotes the peak magnitude of the *nonfaded* signal component, called the *specular component*, and $I_0(\cdot)$ is the modified Bessel function of the first kind and zero order [11]. The Rician distribution is often described in terms of a parameter K, which is defined as the ratio of the power in the specular component to the power in the multipath signal. It is given by

$K = A^2/(2\sigma^2)$. As the magnitude of the specular component approaches zero, the Rician pdf approaches a Rayleigh pdf, expressed as

$$p(r_0) = \begin{cases} \dfrac{r_0}{\sigma^2} \exp\left[-\dfrac{r_0^2}{2\sigma^2}\right] & \text{for } r_0 \geq 0 \\ 0 & \text{otherwise} \end{cases} \tag{4.15}$$

The Rayleigh faded component is sometimes called the *random* or *scatter* or *diffuse* component. The Rayleigh pdf results from having no specular signal component; thus, for a single link (no diversity) it represents the pdf associated with the worst case of fading per mean received signal power. For the remainder of this chapter, it is assumed unless stated otherwise that loss of signal to noise ratio (SNR) resulting from fading follows the Rayleigh model described. It is also assumed that the propagating signal is in the UHF band, encompassing cellular and PCSs with nominal frequency allocations of 1 and 2 GHz, respectively. As indicated in Fig. 4.1 (blocks 4, 5, and 6), small-scale fading manifests itself in two mechanisms: (1) time spreading of the underlying digital pulses within the signal, and (2) time-variant behavior of the channel resulting from motion (e.g., a receive antenna on a moving platform).

Figure 4.6 illustrates the consequences of both these manifestations by showing the response of a multipath channel to a narrow pulse vs. delay, as a function of antenna position (or time, assuming a mobile traveling at a constant velocity). In Fig. 4.6, it is important to distinguish between two different

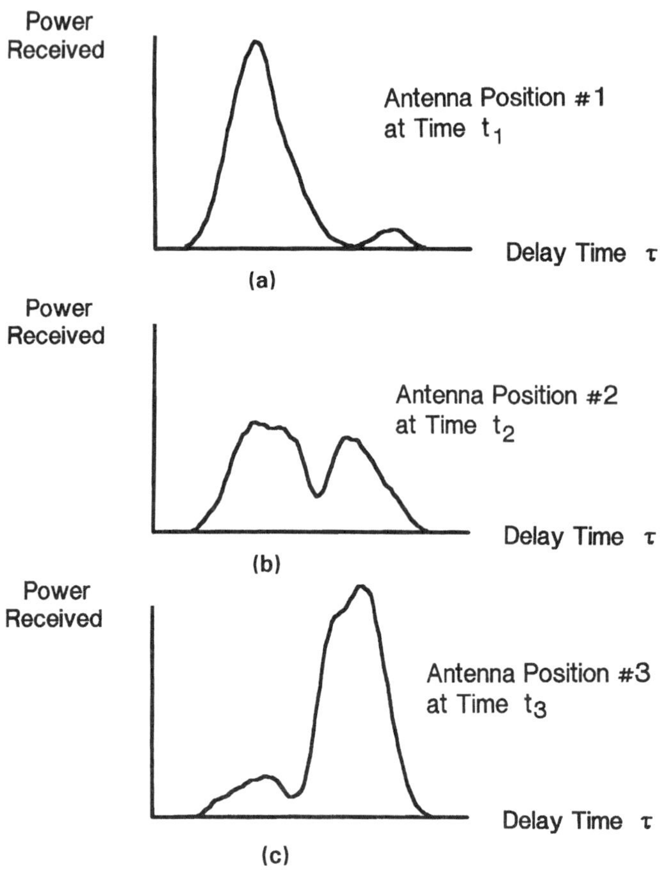

FIGURE 4.6 Response of a multipath channel to a narrow pulse vs. delay, as a function of antenna position.

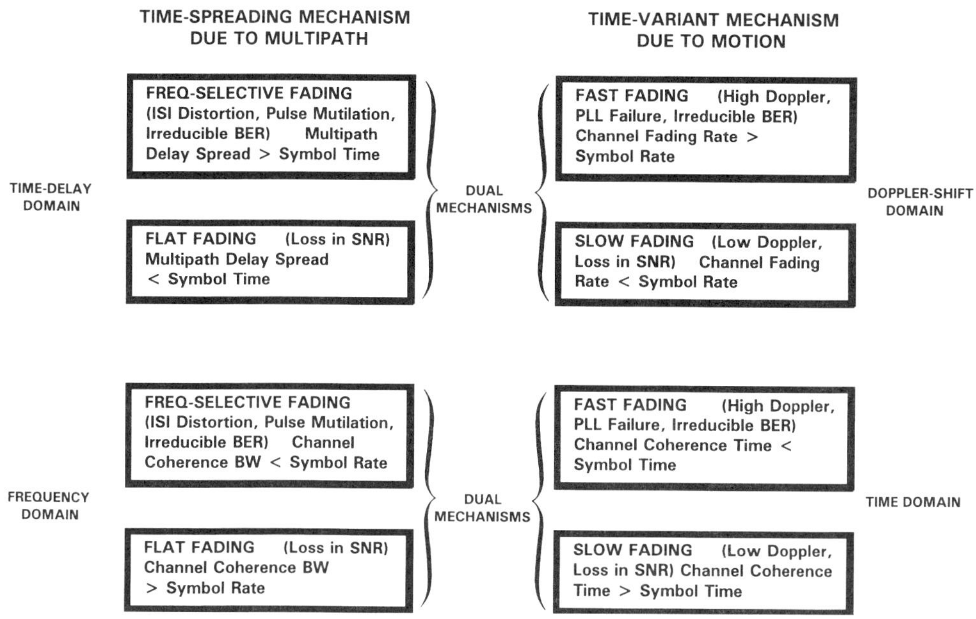

FIGURE 4.7 Small-scale fading: mechanisms, degradation categories, and effects.

time references — delay time τ and transmission or observation time t. Delay time refers to the time-spreading effect that results from the nonoptimum impulse response of the fading channel. The transmission time, however, is related to the antenna motion or spatial changes, accounting for propagation path changes that are perceived as the channel time-variant behavior. Note that for constant velocity, as is assumed in Fig. 4.6, either antenna position or transmission time can be used to illustrate this time-variant behavior. Figures 4.6a to c show the sequence of received pulse–power profiles as the antenna moves through a succession of equally spaced positions. In this case, the interval between antenna positions is 0.4 λ [12], where λ is the wavelength of the carrier frequency. For each of the three cases shown, the response pattern differs significantly in the delay time of the largest signal component, the number of signal copies, their magnitudes, and the total received power (area in each received power profile).

Figure 4.7 summarizes these two small-scale fading mechanisms, the two domains (time or time delay and frequency or Doppler shift) for viewing each mechanism, and the degradation categories each mechanism can exhibit. Note that any mechanism characterized in the time domain can be characterized equally well in the frequency domain. Hence, as outlined in Fig. 4.7, the time-spreading mechanism is characterized in the time-delay domain as a multipath delay spread, and in the frequency domain as a channel coherence bandwidth. Similarly, the time-variant mechanism is characterized in the time domain as a channel coherence time, and in the Doppler-shift (frequency) domain as a channel fading rate or Doppler spread. These mechanisms and their associated degradation categories are examined in the sections that follow.

4.3 Signal Time Spreading

4.3.1 Signal Time Spreading Viewed in the Time-Delay Domain

A simple way to model the fading phenomenon was introduced by Bello [13] in 1963; he proposed the notion of wide-sense stationary uncorrelated scattering (WSSUS). The model treats signals arriving at a receive antenna with different delays as uncorrelated. It can be shown [2, 13] that such a channel is effectively WSS in both the time and frequency domains. With such a model of a fading channel, Bello

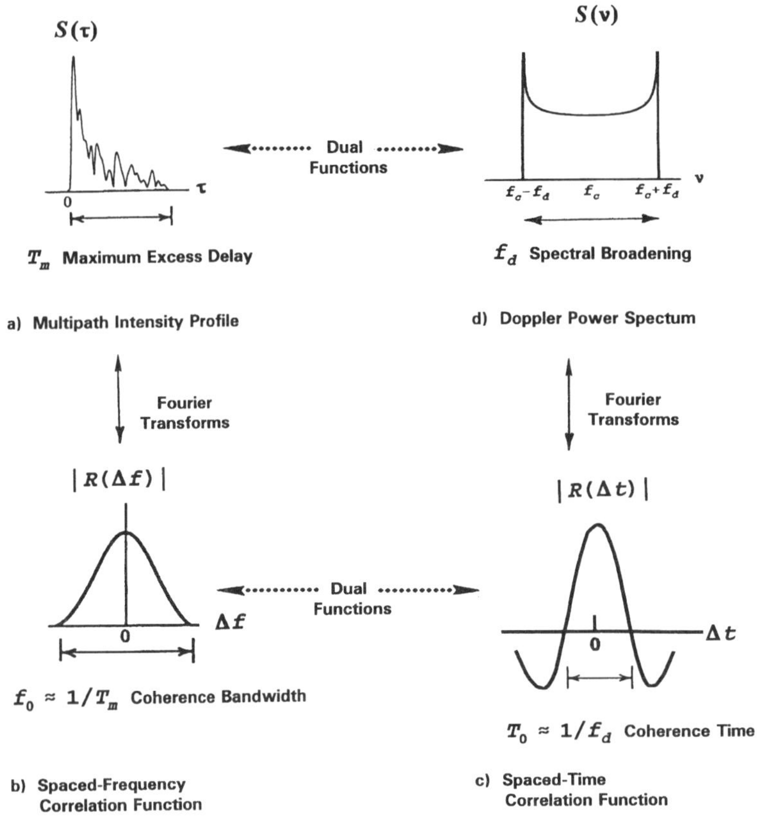

FIGURE 4.8 Relationships among the channel correlation functions and power density functions.

was able to define functions that apply for all time and all frequencies. For the mobile channel, Fig. 4.8 contains four functions that make up this model [2, 10, 13–15]. These functions are now examined, starting with Fig. 4.8a and proceeding counterclockwise toward Fig. 4.8d.

In Fig. 4.8a, a multipath-intensity profile, $S(\tau)$ vs. time delay τ is plotted. Knowledge of $S(\tau)$ helps answer the question, "For a transmitted impulse, how does the average received power vary as a function of time delay, τ?" The term *time delay* is used to refer to the excess delay. It represents the signal propagation delay that exceeds the delay of the first signal arrival at the receiver. For a typical wireless channel, the received signal usually consists of several discrete multipath components, causing $S(\tau)$ to exhibit multiple isolated peaks, sometimes referred to as *fingers* or *returns*. For some channels, such as the tropospheric scatter channel, received signals are often seen as a continuum of multipath components [10, 15]. In such cases, $S(\tau)$ is a relatively smooth (continuous) function of τ. For making measurements of the multipath intensity profile, wideband signals (impulses or spread spectrum) need to be used [15]. For a single transmitted impulse, the time, T_m, between the first and last received component represents the maximum excess delay, after which the multipath signal power falls below some threshold level relative to the strongest component. The threshold level might be chosen at 10 or 20 dB below the level of the strongest component. Note that for an ideal system (zero excess delay), the function $S(\tau)$ would consist of an ideal impulse with weight equal to the total average received signal power.

4.3.1.1 Degradation Categories Resulting from Signal Time Spreading Viewed in the Time-Delay Domain

In a fading channel, the relationship between maximum excess delay time, T_m, and symbol time, T_s, can be viewed in terms of two different degradation categories, frequency-selective fading and frequency nonselective or flat fading, as indicated in Fig. 4.1 (blocks 8 and 9), and Fig. 4.7. A channel is said to exhibit frequency-selective fading if $T_m > T_s$. This condition occurs whenever the received multipath components of a symbol extend beyond the symbol time duration. Such multipath dispersion of the signal yields the same kind of intersymbol interference (ISI) distortion that is caused by an electronic filter. In fact, another name for this category of fading degradation is channel-induced ISI. In the case of frequency-selective fading, mitigating the distortion is possible because many of the multipath components are resolvable by the receiver. Later, several such mitigation techniques are described.

A channel is said to exhibit frequency nonselective or flat fading if $T_m < T_s$. In this case, all the received multipath components of a symbol arrive within the symbol time duration; hence, the components are not resolvable. Here, there is no channel-induced ISI distortion, because the signal time spreading does not result in significant overlap among neighboring received symbols. There is still performance degradation because the unresolvable phasor components can add up destructively to yield a substantial reduction in SNR. Also, signals that are classified as exhibiting flat fading can sometimes experience the distortion effects of frequency-selective fading. This is explained later when viewing degradation in the frequency domain, where the phenomenon is more easily described. For loss in SNR resulting from flat fading, the mitigation technique called for is to improve the received SNR (or reduce the required SNR). For digital systems, introducing some form of signal diversity and using error-correction coding is the most efficient way to accomplish this.

4.3.2 Signal Time Spreading Viewed in the Frequency Domain

A completely analogous characterization of signal dispersion can be specified in the frequency domain. In Fig. 4.8b, the function $|R(\Delta f)|$, designated as a *spaced-frequency* correlation function, can be seen; it is the Fourier transform of $S(\tau)$. The function $R(\Delta f)$ represents the correlation between the channel response to two signals as a function of the frequency difference between the two signals. It can be thought of as the channel frequency transfer function. Therefore, the time-spreading manifestation can be viewed as if it were the result of a filtering process. Knowledge of $R(\Delta f)$ helps answer the question, "What is the correlation between received signals that are spaced in frequency $\Delta f = f_1 - f_2$?" The function $R(\Delta f)$ can be measured by transmitting a pair of sinusoids separated in frequency by Δf, cross correlating the two separately received signals, and repeating the process many times with ever larger separation Δf. Therefore, the measurement of $R(\Delta f)$ can be made with a sinusoid that is swept in frequency across the band of interest (a wideband signal). The *coherence bandwidth*, f_0, is a statistical measure of the range of frequencies over which the channel passes all spectral components with approximately equal gain and linear phase. Thus, the coherence bandwidth represents a frequency range over which frequency components have a strong potential for amplitude correlation. That is, signal spectral components in that range are affected by the channel in a similar manner, for example, exhibiting fading or no fading. Note that f_0 and T_m are reciprocally related (within a multiplicative constant). As an approximation, it is possible to say that

$$f_0 \approx 1/T_m \tag{4.16}$$

The maximum excess delay, T_m, is not necessarily the best indicator of how any given system can perform when signals propagate on a channel, because different channels with the same value of T_m can exhibit very different signal-intensity profiles over the delay span. A more useful parameter is the delay spread. It is most often characterized in terms of its root-mean-square (rms) value, called the rms delay spread, σ_τ, where

$$\sigma_\tau = \sqrt{\overline{\tau^2} - (\bar{\tau})^2} \tag{4.17}$$

$\bar{\tau}$ is the mean excess delay, $(\bar{\tau})^2$ is the mean squared, $\overline{\tau^2}$ is the second moment, and σ_τ is the square root of the second central moment of $S(\tau)$ [1].

A universal relationship between coherence bandwidth and delay spread that would be useful for all applications does not exist. An approximation can be derived from signal analysis (usually using Fourier transform techniques) of actual signal dispersion measurements in particular channels. Several approximate relationships have been developed. If coherence bandwidth is defined as the frequency interval over which the channel complex frequency transfer function has a correlation of at least 0.9, the coherence bandwidth is approximately [16]

$$f_0 \approx \frac{1}{50\,\sigma_\tau} \tag{4.18}$$

For the case of a mobile radio, an array of radially uniformly spaced scatterers, all with equal-magnitude reflection coefficients but independent, randomly occurring reflection phase angles [17, 18] is generally accepted as a useful model for an urban propagation environment. This model is referred to as the *dense-scatterer* channel model. With the use of such a model, coherence bandwidth has similarly been defined [17] for a bandwidth interval over which the channel complex frequency transfer function has a correlation of at least 0.5, to be

$$f_0 = \frac{0.276}{\sigma_\tau} \tag{4.19}$$

Studies involving ionospheric effects often employ the following definition [19]:

$$f_0 = \frac{1}{2\pi\sigma_\tau} \tag{4.20}$$

A more popular approximation of f_0 corresponding to a bandwidth interval having a correlation of at least 0.5 is [1]

$$f_0 = \frac{1}{5\sigma_\tau} \tag{4.21}$$

The delay spread and coherence bandwidth are related to channel multipath characteristics, differing for different propagation paths (such as, metropolitan areas, suburbs, hilly terrain, and indoors). It is important to note that the parameters in Eq. (4.21) do not depend on signaling speed. System signaling speed only influences its transmission bandwidth, W.

4.3.2.1 Degradation Categories Resulting from Signal Time Spreading Viewed in the Frequency Domain

A channel is referred to as frequency selective if $f_0 < 1/T_s \approx W$, where the symbol rate, $1/T_s$ is nominally taken to be equal to the signaling rate or signal bandwidth W. In practice, W may differ from $1/T_s$ because of system filtering or data modulation type (e.g., QPSK, MSK, or spread spectrum) [20]. Frequency-selective fading distortion occurs whenever signal spectral components are not all affected equally by the channel. Some of the signal spectral components, falling outside the coherence bandwidth, are affected differently (independently) compared with those components contained within the coherence bandwidth. Figure 4.9 contains three examples. Each one illustrates the spectral density vs. frequency of a transmitted signal having a bandwidth of W Hz. Superimposed on the plot in Fig. 4.9a is the frequency transfer function of a frequency-selective channel ($f_0 < W$). Figure 4.9a shows that various spectral components of the transmitted signal are affected differently.

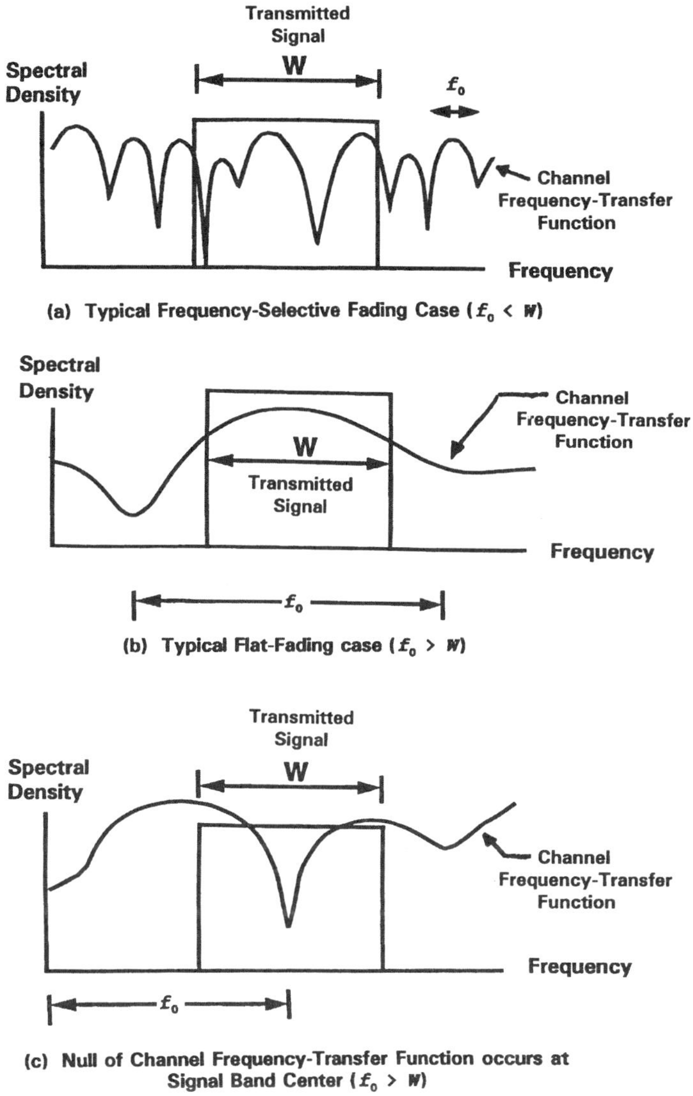

FIGURE 4.9 Relationships between the channel frequency transfer function and a transmitted signal with band width *W*.

Frequency-nonselective or flat-fading degradation occurs whenever $f_0 > W$. Hence, all the signal spectral components are affected by the channel in a similar manner (e.g., fading or no fading). This is illustrated in Fig. 4.9b, which features the spectral density of the same transmitted signal having a bandwidth of W Hz. However, superimposed on this plot is the frequency transfer function of a flat-fading channel ($f_0 > W$). Figure 4.9b illustrates that all the spectral components of the transmitted signal are affected in approximately the same way. Flat fading does not introduce channel-induced ISI distortion, but performance degradation can still be expected because of the loss in SNR whenever the signal is fading. To avoid channel-induced ISI distortion, the channel is required to exhibit flat fading. This occurs provided that

$$f_0 > W \approx \frac{1}{T_s} \tag{4.22}$$

Hence, the channel coherence bandwidth f_0 sets an upper limit on the transmission rate that can be used without incorporating an equalizer in the receiver.

For the flat-fading case, where $f_0 > W$ (or $T_m < T_s$), Fig. 4.9b shows the usual flat-fading pictorial representation. However, as a mobile radio changes its position, there are times when the received signal experiences frequency-selective distortion even though $f_0 > W$. This is seen in Fig. 4.9c, where the null of the channel frequency transfer function occurs near the band center of the transmitted signal spectral density. When this occurs, the baseband pulse can be especially mutilated by deprivation of its low-frequency components. One consequence of such loss is the absence of a reliable pulse peak on which to establish the timing synchronization, or from which to sample the carrier phase carried by the pulse [17]. Thus, even though a channel is categorized as flat fading (based on rms relationships), it can still manifest frequency-selective fading on occasions. It is fair to say that a mobile-radio channel, classified as exhibiting flat-fading degradation, cannot exhibit flat fading all the time. As f_0 becomes much larger than W (or T_m becomes much smaller than T_s), less time is spent exhibiting the type of condition shown in Fig. 4.9c. By comparison, it should be clear that in Fig. 4.9a the fading is independent of the position of the signal band, and frequency-selective fading occurs all the time, not just occasionally.

4.3.3 Examples of Flat Fading and Frequency-Selective Fading

Figure 4.10 shows some examples of flat fading and frequency-selective fading for a direct-sequence–spread-spectrum (DS/SS) system [19, 20]. In Fig. 4.10, there are three plots of the output of a pseudonoise (PN) code correlator vs. delay as a function of time (transmission or observation time). Each amplitude vs. delay plot is akin to $S(\tau)$ vs. τ shown in Fig. 4.8a. The key difference is that the

a) $f_0 T_{ch} \approx 1$

b) $f_0 T_{ch} = 0.25$

c) $f_0 T_{ch} = 0.1$

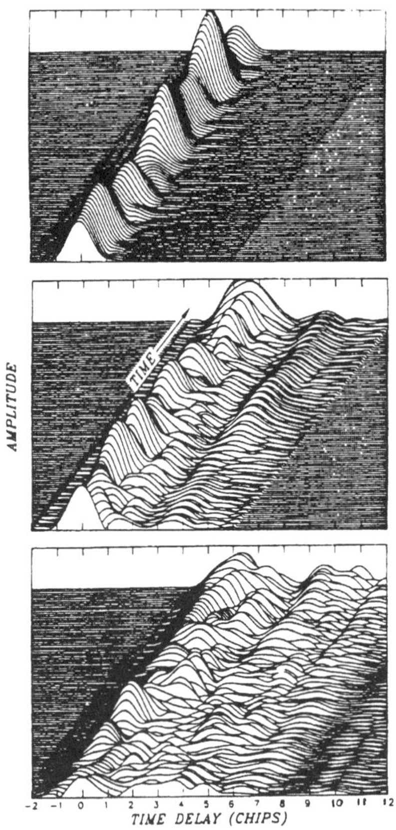

FIGURE 4.10 DS/SS matched-filter output time-history examples for three levels of channel conditions, where T_{ch} is the time duration of a chip.

amplitudes shown in Fig. 4.10 represent the output of a correlator; hence, the waveshapes are a function not only of the impulse response of the channel but also of the impulse response of the correlator. The delay time is expressed in units of chip durations (chips), where the chip is defined as the spread-spectrum–minimum-duration keying element. For each plot, the observation time is shown on an axis perpendicular to the amplitude vs. time-delay plane. Figure 4.10 is drawn from a satellite-to-ground communications link exhibiting scintillation because of atmospheric disturbances. However, Fig. 4.10 is still a useful illustration of three different channel conditions that might apply to a mobile-radio situation. A mobile radio that moves along the observation-time axis is affected by changing multipath profiles along the route, as seen in Fig. 4.10.

The scale along the observation-time axis is also in units of chips. In Fig. 4.10a, the signal dispersion (one "finger" of return) is on the order of a chip time duration, T_{ch}. In a typical DS/SS system, the spread-spectrum signal bandwidth is approximately equal to $1/T_{ch}$; hence, the normalized coherence bandwidth $f_0 T_{ch}$ of approximately unity in Fig. 4.10a implies that the coherence bandwidth is about equal to the spread-spectrum bandwidth. This describes a channel that can be called frequency nonselective or slightly frequency selective. In Fig. 4.10b, where $f_0 T_{ch} = 0.25$, the signal dispersion is more pronounced. There is definite interchip interference, resulting from the coherence bandwidth being approximately 25% of the spread-spectrum bandwidth. In Fig. 4.10c, where $f_0 T_{ch} = 0.1$, the signal dispersion is even more pronounced, with greater interchip–interference effects, because of the coherence bandwidth being approximately 10% of the spread-spectrum bandwidth. The coherence bandwidths (relative to the spread-spectrum signaling speed) shown in Figs. 4.10b and c depict channels that can be categorized as moderately and highly frequency selective, respectively. Later, it is shown that a DS/SS system operating over a frequency-selective channel at the chip level, does not necessarily experience frequency-selective distortion at the symbol level.

The signal dispersion manifestation of a fading channel is analogous to the signal spreading that characterizes an electronic filter. Figure 4.11a depicts a wideband filter (narrow impulse response) and its effect on a signal in both the time domain and the frequency domain. This filter resembles a flat-fading channel yielding an output that is relatively free of distortion. Figure 4.11b shows a narrowband filter (wide impulse response). The output signal suffers much distortion, as shown in both time and frequency. Here, the process resembles a frequency-selective channel.

4.4 Time Variance of the Channel Caused by Motion

4.4.1 Time Variance Viewed in the Time Domain

Signal dispersion and coherence bandwidth, described earlier, characterize the channel time-spreading properties in a local area. However, they do not offer information about the time-varying nature of the channel caused by relative motion between a transmitter and receiver, or by movement of objects within the channel. For mobile-radio applications, the channel is time variant because motion between the transmitter and receiver results in propagation-path changes. For a transmitted continuous wave (CW) signal, such changes cause variations in the signal amplitude and phase at the receiver. If all scatterers making up the channel are stationary, whenever motion ceases, the amplitude and phase of the received signal remains constant; that is, the channel appears to be time invariant. Whenever motion begins again, the channel appears time variant. Because the channel characteristics are dependent on the positions of the transmitter and receiver, time variance in this case is equivalent to spatial variance.

Figure 4.8c shows the function $R(\Delta t)$, designated the *spaced-time* correlation function; it is the autocorrelation function of the channel response to a sinusoid. This function specifies the extent to which there is correlation between the channel response to a sinusoid sent at time t_1 and the response to a similar sinusoid sent at time t_2, where $\Delta t = t_2 - t_1$. The *coherence time*, T_0, is a measure of the expected time duration over which the channel response is essentially invariant. Earlier, measurements of signal dispersion and coherence bandwidth were made by using wideband signals. Now, to measure the time-variant nature of the channel, a narrowband signal is used [15]. To measure $R(\Delta t)$, a single sinusoid

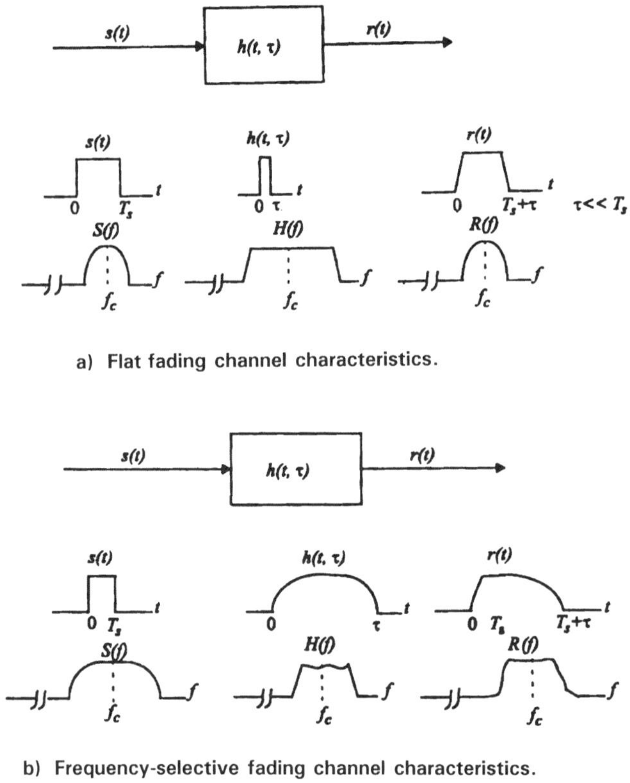

FIGURE 4.11 Flat-fading and frequency-selective-fading characteristics.

($\Delta f = 0$) can be transmitted at times t_1 and t_2, and the cross correlation function of the received signals is determined. The function $R(\Delta t)$ and the parameter T_0 provide knowledge about the fading rapidity of the channel. Note that for an ideal time-invariant channel (e.g., transmitter and receiver exhibiting no motion at all), the channel response would be highly correlated for all values of Δt; thus, $R(\Delta t)$ as a function of Δt would be a constant. For example, if a stationary user's location is characterized by a multipath null, then that null remains unchanged until there is some movement (either by the transmitter or receiver or by objects within the propagation path). When using the dense-scatterer channel model described earlier, with constant mobile velocity, V, and an unmodulated CW signal having wavelength λ, the normalized $R(\Delta t)$ is described as [18]

$$R\left(\Delta t\right) = J_0\left(kV\Delta t\right) \tag{4.23}$$

where $J_0(\cdot)$ is the zero-order Bessel function of the first kind [11], $V\,\Delta t$ is distance traversed, and $k = 2\pi/\lambda$ is the free-space phase constant (transforming distance to radians of phase). Coherence time can be measured in terms of either time or distance traversed (assuming some fixed velocity). Amoroso [17] described such a measurement using a CW signal and a dense-scatterer channel model. He measured the statistical correlation between the combination of received magnitude and phase sampled at a particular antenna location x_0, and the corresponding combination sampled at some displaced location $x_0 + \zeta$, with displacement measured in units of wavelength λ. For a displacement ζ of 0.4 λ between two antenna locations, the combined magnitudes and phases of the received CW are statistically uncorrelated. In other words, the signal observation at x_0 provides no information about the signal at $x_0 + \zeta$. For a given velocity, this displacement is readily transformed into units of time (coherence time).

4.4.1.1 Basic Fading Manifestations are Independent of One Another

For a moving antenna, the fading of a transmitted carrier wave is usually regarded as a random process, even though the fading record may be completely predetermined from the disposition of scatterers and the propagation geometry from the transmitter to the receiving antenna. This is because the same waveform received by two antennas that are displaced by at least 0.4 λ are statistically uncorrelated [17, 18]. Because such a small distance (about 13 cm for a carrier wave at 900 MHz) corresponds to statistical decorrelation in received signals, the basic fading manifestations of signal dispersion and fading rapidity can be considered to be independent of each other. Any of the cases in Fig. 4.10 can provide some insight. At each instant of time (corresponding to a spatial location) we see a multipath intensity profile $S(\tau)$ as a function of delay, τ. The multipath profiles are primarily determined by the surrounding terrain (buildings, vegetation, etc.). Consider Fig. 4.10b, where the direction of motion through regions of differing multipath profiles is indicated by an arrow labeled *time* (it might also be labeled *antenna displacement*). As the mobile moves to a new spatial location characterized by a different profile, there are changes in the fading state of the channel as characterized by the profile at the new location. However, because one profile is decorrelated with another profile at a distance as short as 13 cm (for a carrier at 900 MHz), the rapidity of such changes only depends on the speed of movement, not on the underlying geometry of the terrain.

4.4.1.2 Concept of Duality

The mathematical concept of duality can be defined as follows: two processes (functions, elements, or systems) are *dual* to each other if their mathematical relationships are the same even though they are described in terms of different parameters. In this chapter, it is interesting to note duality when examining time-domain vs. frequency-domain relationships.

In Fig. 4.8, we can identify functions that exhibit similar behavior across domains. For the purpose of understanding the fading channel model, it is useful to refer to such functions as duals. For example, the phenomenon of signal dispersion can be characterized in the frequency domain by $R(\Delta f)$, as shown in Fig. 4.8b. It yields knowledge about the range of frequencies over which two spectral components of a received signal have a strong potential for amplitude and phase correlation. Fading rapidity is characterized in the time domain by $R(\Delta t)$, as shown in Fig. 4.8c. It yields knowledge about the span of time over which two received signals have a strong potential for amplitude and phase correlation. These two correlation functions, $R(\Delta f)$ and $R(\Delta t)$, have been labeled as duals. This is also noted in Fig. 4.1 as the duality between blocks 10 and 13, and in Fig. 4.7 as the duality between the time-spreading mechanism in the frequency domain and the time-variant mechanism in the time domain.

4.4.1.3 Degradation Categories Resulting from Time Variance, Viewed in the Time Domain

The time-variant nature or fading rapidity mechanism of the channel can be viewed in terms of two degradation categories as listed in Fig. 4.7: fast fading and slow fading. The terminology *fast fading* is used for describing channels in which $T_0 < T_s$, where T_0 is the channel coherence time and T_s is the time duration of a transmission symbol. Fast fading describes a condition where the time duration in which the channel behaves in a correlated manner is short compared with the time duration of a symbol. Therefore, it can be expected that the fading character of the channel will change several times during the time span of a symbol, leading to distortion of the baseband pulse shape. Analogous to the distortion previously described as channel-induced ISI, distortion takes place because the received signal components are not all highly correlated throughout time. Hence, fast fading can cause the baseband pulse to be distorted, often resulting in an irreducible error rate. Such distorted pulses cause synchronization problems (failure of phase-locked-loop receivers), in addition to difficulties in adequately designing a matched filter.

A channel is generally referred to as introducing slow fading if $T_0 > T_s$. Here, the time duration that the channel behaves in a correlated manner is long compared with the time duration of a transmission symbol. Thus, one can expect the channel state to virtually remain unchanged during the time in which a symbol is transmitted. The propagating symbols likely do not suffer from the pulse distortion described earlier. The primary degradation in a slow-fading channel, as with flat fading, is loss in SNR.

4.4.2 Time Variance Viewed in the Doppler-Shift Domain

A completely analogous characterization of the time-variant nature of the channel can be presented in the Doppler-shift (frequency) domain. Figure 4.8d shows a *Doppler power spectral density* (or Doppler spectrum), $S(\nu)$, plotted as a function of Doppler-frequency shift. For the case of the dense-scatterer model, a vertical receive antenna with constant azimuthal gain, a uniform distribution of signals arriving at all arrival angles throughout the range $(0, 2\pi)$, and an unmodulated CW signal, the signal spectrum at the antenna terminals is [18]

$$S(\nu) = \frac{1}{\pi f_d \sqrt{1 - \left(\frac{\nu - f_c}{f_d}\right)^2}} \tag{4.24}$$

The equality holds for frequency shifts of ν that are in the range $\pm f_d$ about the carrier frequency f_c, and would be zero outside that range. The shape of the RF Doppler spectrum described by Eq. (4.24) is classically bowl shaped, as seen in Fig. 4.8d. Note that the spectral shape is a result of the dense-scatterer channel model. Equation (4.24) has been shown to match experimental data gathered for mobile-radio channels [22]; however, different applications yield different spectral shapes. For example, the dense-scatterer model does not hold for the indoor radio channel; the channel model for an indoor area assumes $S(\nu)$ to be a flat spectrum [23].

In Fig. 4.8d, the sharpness and steepness of the boundaries of the Doppler spectrum are a result of the sharp upper limit on the Doppler shift produced by a vehicular antenna traveling among the stationary scatterers of the dense-scatterer model. The largest magnitude (infinite) of $S(\nu)$ occurs when the scatterer is directly ahead of the moving antenna platform or directly behind it. In that case, the magnitude of the frequency shift is given by

$$f_d = \frac{V}{\lambda} \tag{4.25}$$

where V is relative velocity, and λ is the signal wavelength. When the transmitter and receiver move toward each other, f_d is positive, whereas f_d is negative when they move away from each other. For scatterers directly broadside of the moving platform, the magnitude of the frequency shift is zero. The fact that Doppler components arriving at exactly 0° and 180° have an infinite power spectral density is not a problem, because the angle of arrival is continuously distributed and the probability of components arriving at exactly these angles is zero [1, 18].

$S(\nu)$ is the Fourier transform of $R(\Delta t)$. The Fourier transform of the autocorrelation function of a time series equals the magnitude squared of the Fourier transform of the original time series. Therefore, measurements can be made by simply transmitting a sinusoid (narrowband signal) and using Fourier analysis to generate the power spectrum of the received amplitude [15]. This Doppler power spectrum of the channel yields knowledge about the spectral spreading of a transmitted sinusoid (impulse in frequency) in the Doppler-shift domain. As indicated in Fig. 4.8, $S(\nu)$ can be regarded as the dual of the multipath intensity profile, $S(\tau)$, because the latter yields knowledge about the time spreading of a transmitted impulse in the time-delay domain. This is also noted in Fig. 4.1 as the duality between blocks 7 and 16, and in Fig. 4.7 as the duality between the time-spreading mechanism in the time-delay domain and the time-variant mechanism in the Doppler-shift domain.

Knowledge of $S(\nu)$ allows estimating how much spectral broadening is imposed on the signal as a function of the rate of change in the channel state. The width of the Doppler power spectrum, denoted f_d, is referred to in the literature by several different names: *Doppler spread*, or *fading rate*, or *fading bandwidth*, or *spectral broadening*. Equation (4.24) describes the Doppler frequency shift. In a typical multipath environment, the received signal travels over several reflected paths, each with a different distance and a different angle of arrival. The Doppler shift of each arriving path is generally different from that of other paths. The effect on the received signal manifests itself as a Doppler spreading of the transmitted signal frequency, instead of a shift. Note that the Doppler spread, f_d, and the coherence time, T_0, are reciprocally related (within a multiplicative constant), resulting in an approximate relationship between the two parameters given by

$$T_0 \approx \frac{1}{f_d} \tag{4.26}$$

Hence, the Doppler spread f_d (or $1/T_0$) is regarded as the typical fading rate of the channel. Earlier, T_0 was described as the expected time duration over which the channel response to a sinusoid is essentially invariant. When T_0 is defined more precisely as the time duration over which the channel response to sinusoids yields a correlation of at least 0.5 between them, the relationship between T_0 and f_d is approximately [2]

$$T_0 \approx \frac{9}{16\pi f_d} \tag{4.27}$$

A popular rule of thumb is to define T_0 as the geometric mean of Eq. (4.26) and (4.27). This yields

$$T_0 = \sqrt{\frac{9}{16\pi f_d^2}} = \frac{0.423}{f_d} \tag{4.28}$$

For the case of a 900-MHz mobile radio, Fig. 4.12 illustrates the typical effect of Rayleigh fading on signal envelope amplitude vs. time [1]. The figure shows that the distance traveled by the mobile in a time interval corresponding to two adjacent nulls (small-scale fades) is on the order of a half wavelength (λ/2).

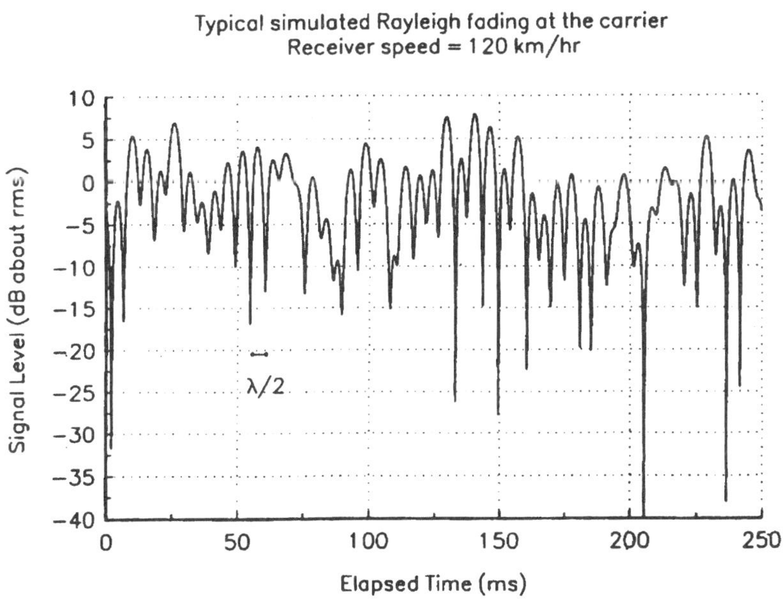

FIGURE 4.12 A typical Rayleigh-fading envelope at 900 MHz.

Thus, from Fig. 4.12 and Eq. (4.25), the time required to traverse a distance λ/2 (approximately the coherence time) when traveling at a constant velocity, V, is

$$T_0 \approx \frac{\lambda/2}{V} = \frac{0.5}{f_d} \tag{4.29}$$

Thus, when the interval between fades is approximately λ/2, as in Fig. 4.12, the resulting expression for T_0 in Eq. (4.29) is quite close to the geometric mean shown in Eq. (4.28). From Eq. (4.29), and by using the parameters shown in Fig. 4.12 (velocity = 120 km/h, and carrier frequency = 900 MHz), it is straightforward to determine that the channel coherence time is approximately 5 ms and the Doppler spread (channel-fading rate) is approximately 100 Hz. Therefore, if this example represents a channel over which digitized speech signals are transmitted with a typical rate of 10^4 symbols per second, the fading rate is considerably less than the symbol rate. Under such conditions, the channel would manifest slow-fading effects. Note that if the abscissa of Fig. 4.12 were labeled in units of wavelength instead of time, the plotted fading characteristics would look the same for any radio frequency and any antenna speed.

4.4.2.1 Analogy for Spectral Broadening in Fading Channels

Let us discuss the reason why a signal experiences spectral broadening as it propagates from, or is received by, a moving platform, and why this spectral broadening (also called the fading rate of the channel) is a function of the speed of motion. An analogy can be used to explain this phenomenon. Figure 4.13 shows the keying of a digital signal (such as amplitude shift keying or frequency shift keying), where a single tone $\cos 2\pi f_c t$ defined for $-\infty < t < \infty$ is characterized in the frequency domain in terms of impulses (at $\pm f_c$). This frequency domain representation is ideal (i.e., zero bandwidth), because the tone is a single frequency with infinite time duration. In practical applications, digital signaling involves switching (keying) signals on and off at a required rate. The keying operation can be viewed as multiplying the infinite-duration tone in Fig. 4.13a by an ideal rectangular on-off (switching) function in Fig. 4.13b. The frequency-domain description of this switching function is of the form sinc fT.

In Fig. 4.13c, the result of the multiplication yields a tone, $\cos 2\pi f_c t$, that is time-duration limited. The resulting spectrum is obtained by convolving the spectral impulses shown in part (a) of Fig. 4.13 with

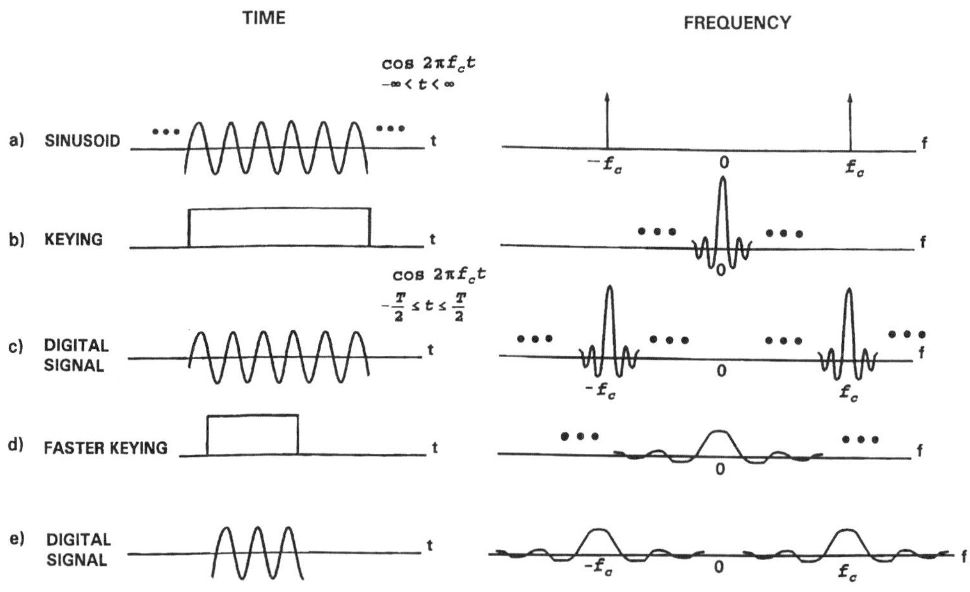

FIGURE 4.13 Analogy between spectral broadening in fading and spectral broadening in keying a digital signal.

the sinc fT function of part (b), yielding the broadened spectrum depicted in part (c). It is further seen that if the signaling occurs at a faster rate characterized by the rectangle of shorter duration in part (d), the resulting signal spectrum in part (e) exhibits greater spectral broadening. The changing state of a fading channel is somewhat analogous to the on-off keying of digital signals. The channel behaves like a switch, turning the signal on and off. The greater the rapidity of the change in the channel state, the greater is the spectral broadening experienced by signals propagating over such a channel. The analogy is not exact because the on and off switching of signals may result in phase discontinuities, whereas the typical multipath-scatterer environment induces phase-continuous effects.

4.4.2.2 Degradation Categories Resulting from Time Variance, Viewed in the Doppler-Shift Domain

A channel is said to be fast fading if the symbol rate, $1/T_s$ (approximately equal to the signaling rate or bandwidth W) is less than the fading rate, $1/T_0$ (approximately equal to f_d); that is, fast fading is characterized by

$$W < f_d \tag{4.30}$$

or

$$T_s > T_0 \tag{4.31}$$

Conversely, the channel is referred to as slow fading if the signaling rate is greater than the fading rate. Thus, to avoid signal distortion caused by fast fading, the channel must be made to exhibit slow fading characteristics by ensuring that the signaling rate exceeds the channel fading rate. That is,

$$W > f_d \tag{4.32}$$

or

$$T_s < T_0 \tag{4.33}$$

In Eq. (4.22), it was shown that because of signal dispersion, the coherence bandwidth, f_0, sets an *upper limit* on the signaling rate that can be used without suffering frequency-selective distortion. Similarly, Eq. (4.32) shows that because of Doppler spreading, the channel fading rate, f_d, sets a *lower limit* on the signaling rate that can be used without suffering fast-fading distortion. For HF communication systems, when teletype or Morse-coded messages were transmitted at low-data rates, the channels often exhibited fast-fading characteristics. However, most present-day terrestrial mobile-radio channels can generally be characterized as slow fading.

Equations (4.32) and (4.33) do not go far enough in describing the desirable behavior of the channel. A better way to state the requirement for mitigating the effects of fast fading is that W must be *much greater* than f_d (or for T_s to be *much less* than T_0). If this condition is not satisfied, the random frequency modulation (FM) resulting from varying Doppler shifts degrades system performance significantly. The Doppler effect yields an irreducible error rate that cannot be overcome by simply increasing E_b/N_0 [24]. This irreducible error rate is most pronounced for any transmission scheme that involves modulating the carrier phase. A single specular Doppler path, without scatterers, registers an instantaneous frequency shift, classically calculated as $f_d = V/\lambda$. However, a combination of specular and multipath components yields a rather complex time dependence of instantaneous frequency that can cause frequency swings much larger than $\pm V/\lambda$ when the information is recovered by an instantaneous frequency detector (a nonlinear device) [25]. Figure 4.14 illustrates how this can happen. At time t_1, because of vehicle motion, the specular phasor has rotated through an angle θ, whereas the net phasor has rotated through an angle ϕ, which is about four times greater than θ. The rate of change of phase at a time near this particular

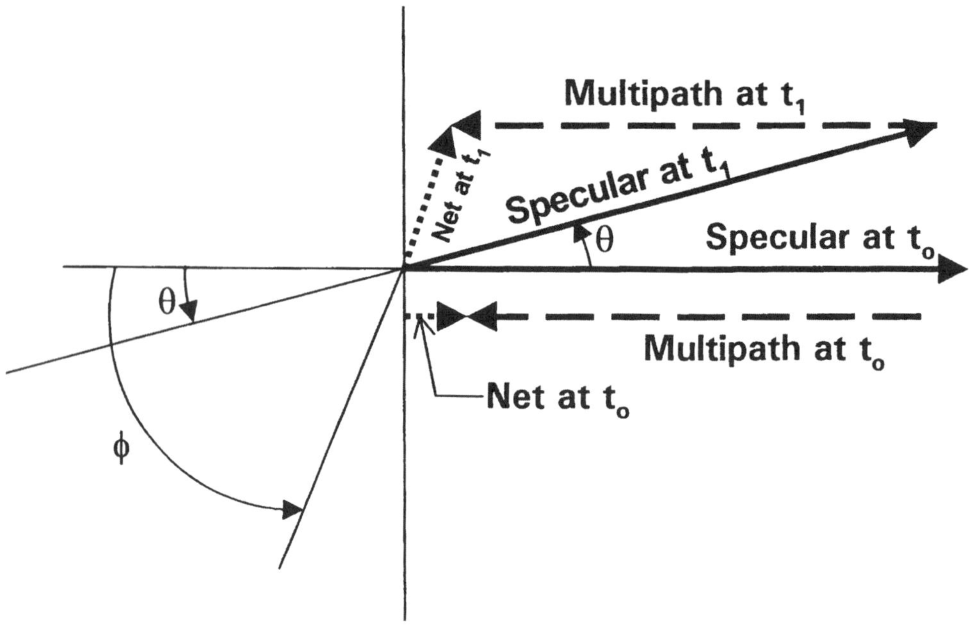

FIGURE 4.14 A combination of specular and multipath components can register much larger frequency swings than $\pm V/\lambda$.

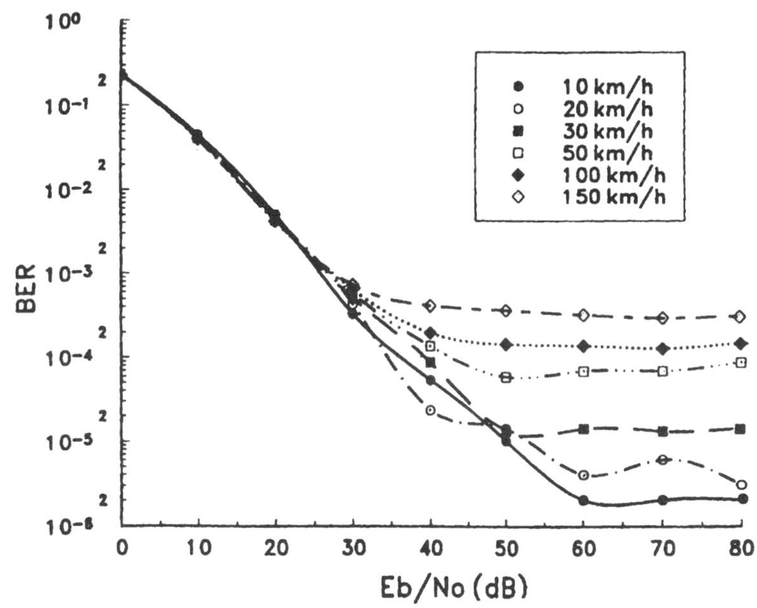

FIGURE 4.15 Error performance vs. E_b/N_0 for $\pi/4$ DQPSK for various mobile speeds: f_c = 850 MHz, R_s = 24 ksymbols per second.

fade is about four times that of the specular Doppler alone. Therefore, the instantaneous frequency shift, $d\phi/dt$, would be about four times that of the specular Doppler shift.

The peaking of instantaneous frequency shifts at a time near deep fades is akin to the phenomenon of FM "clicks" or "spikes." Figure 4.15 illustrates the seriousness of this problem. This figure shows bit-error rate vs. E_b/N_0 performance plots for $\pi/4$ DQPSK signaling at f_0 = 850 MHz for various simulated mobile speeds [26]. It should be clear that at high speeds, the performance curve bottoms out at an error-rate level that may be unacceptably high. Ideally, coherent demodulators that lock onto and track the information signal should suppress the effect of this FM noise and thus cancel the impact of Doppler shift. However, for large values of f_d, carrier recovery is difficult to implement because very wideband (relative to the data rate) phase-lock loops (PLLs) need to be designed. For voice-grade applications with bit-error rates of 10^{-3} to 10^{-4}, a large value of Doppler shift is considered to be on the order of $0.01 \times W$. Therefore, to avoid fast-fading distortion and the Doppler-induced irreducible error rate, the signaling rate should exceed the fading rate by a factor of 100 to 200 [27]. The exact factor depends on the signal modulation, receiver design, and required error rate [1, 25–29]. Davarian et al. [29] showed that a frequency-tracking loop can help lower, but not completely remove, the irreducible error rate in a mobile system by using differential minimum shift keyed (DMSK) modulation.

4.4.3 Performance over a Slow- and Flat-Fading Rayleigh Channel

For the case of a discrete multipath channel, with a complex envelope $g(t)$ described by Eq. (4.3), a demodulated signal (neglecting noise) is described by Eq. (4.10), which is rewritten as follows:

$$z(t) = \sum_n \alpha_n(t)\, e^{-j2\pi f_c \tau_n(t)}\, R\left[t - \tau_n(t)\right] e^{j\phi(t-\tau_n)} \tag{4.34a}$$

where $R(t) = |g(t)|$ is the envelope magnitude, and $\phi(t)$ is its phase. Assume that the channel exhibits flat fading so that the multipath components are not resolvable. Then the $\{\alpha_n(t)\}$ terms in Eq. (4.34a), in one signaling interval, T, need to be expressed as a resultant amplitude $\alpha(T)$ of all the n phasors received in that interval. Similarly, the preceding phase terms, in one signaling interval, need to be expressed as the resultant phase $\theta(T)$ of all the n-fading phasors plus the information phase received in that interval. Assume also that the channel exhibits slow fading so that the phase can be estimated from the received signal without significant error using PLL circuitry or some other appropriate techniques. Therefore, for a slow- and flat-fading channel, we can express a received test statistic $z(T)$ out of the demodulator in each signaling interval, including the noise $n_0(T)$, as

$$z(T) = \alpha(T)\, R(T)\, e^{-j[\theta(T) - \phi(T)]} + n_0(T) \tag{4.34b}$$

For simplicity, we now replace $\alpha(T)$ with α. For binary signaling over an AWGN channel with a fixed attenuation of $\alpha = 1$, the bit-error probabilities vs. E_b/N_0 for the basic coherent and noncoherent phase shift keying (PSK) and orthogonal frequency shift keying (FSK) each manifest a classical exponential relationship (a waterfall shape associated with AWGN performance). However, for multipath conditions, if there is no specular signal component, α is a Rayleigh-distributed random variable, or equivalently, α^2 is described by a chi-square pdf. Under these Rayleigh-fading conditions, Fig. 4.16 depicts the performance curves. When $(E_b/N_0)\,\mathbf{E}(\alpha^2)$ is *much greater* than 1, where $\mathbf{E}(\cdot)$ represents statistical expectation, then the bit-error probability expressions for the basic binary signaling schemes shown in Fig. 4.16 are given in Table 4.1. Each of the signaling schemes that had manifested a waterfall-shaped performance plot under AWGN, now exhibits performance that takes the form of an inverse linear function, as a result of the Rayleigh fading.

4.5 Mitigating the Degradation Effects of Fading

Figure 4.17 subtitled "the good, the bad, and the awful," highlights three major performance categories in terms of bit-error probability, P_B, vs. E_b/N_0. The leftmost exponentially shaped curve highlights the performance that can be expected when using any nominal modulation scheme in AWGN interference. Observe that at a reasonable E_b/N_0 level, good performance can be expected. The middle curve, referred

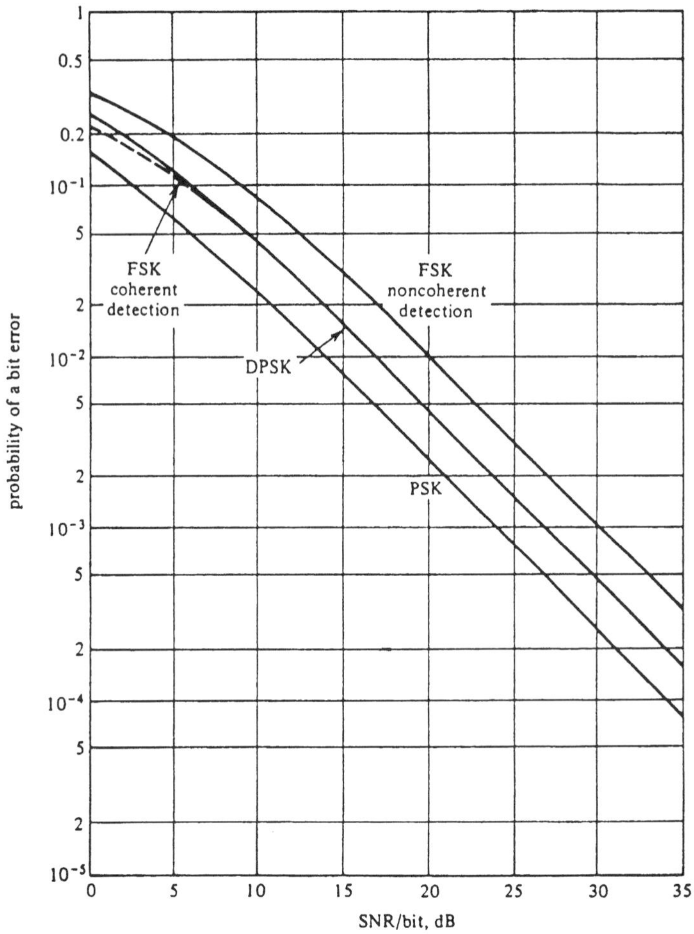

FIGURE 4.16 Performance of binary signaling over a slow Rayleigh-fading channel.

TABLE 4.1 Rayleigh-Limit Bit-Error Performance Where (E_b/N_0) E(α^2) Is Much Greater than 1

Modulation	P_B
PSK (coherent)	$\frac{1}{4(E_b/N_0)\mathbf{E}(\alpha^2)}$
DPSK (differentially coherent)	$\frac{1}{2(E_b/N_0)\mathbf{E}(\alpha^2)}$
Orthogonal FSK (coherent)	$\frac{1}{2(E_b/N_0)\mathbf{E}(\alpha^2)}$
Orthogonal FSK (noncoherent)	$\frac{1}{(E_b/N_0)\mathbf{E}(\alpha^2)}$

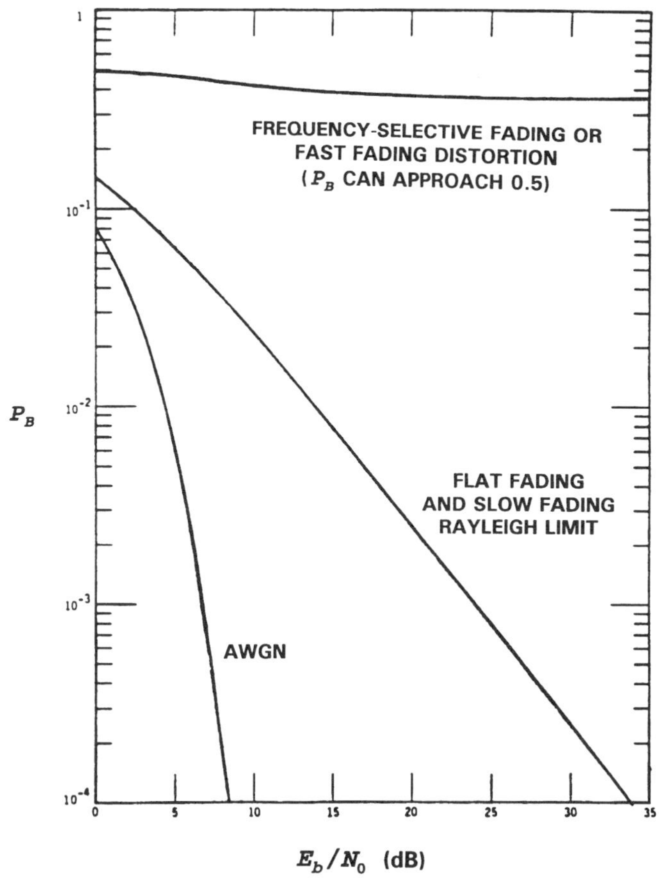

FIGURE 4.17 Error performance: the good, the bad, and the awful.

to as the *Rayleigh limit,* shows the performance degradation resulting from a loss in E_b/N_0 that is characteristic of flat fading or slow fading when there is no LOS signal component present. The curve is a function of the reciprocal of E_b/N_0 (an inverse-linear function), so for practical values of E_b/N_0, performance is generally bad. In the case of Rayleigh fading, parameters with overbars are often introduced to indicate that an average is being taken over the "ups" and "downs" of the fading experience. Therefore, one often sees such bit-error probability plots with averaged parameters denoted by $\overline{P_B}$ and $\overline{E_b}/N_0$. This notation emphasizes the fact that the fading channel has memory; thus, received samples of the signal are correlated to one another in time. Therefore, when producing such error probability plots for a fading channel, one needs to examine the process over a window of time that is much larger than the channel coherence time. The curve that reaches an irreducible error-rate level, sometimes called an *error floor* represents awful performance, where the bit-error probability can level off at values nearly equal to 0.5. This shows the severe performance degrading effects that are possible with frequency-selective fading or fast fading.

If the channel introduces signal distortion as a result of fading, the system performance can exhibit an irreducible error rate at a level higher than the desired error rate. In such cases, no amount of E_b/N_0 can help achieve the desired level of performance, and the only approach available for improving performance is to use some form of mitigation to remove or reduce the signal distortion. The mitigation method depends on whether the distortion is caused by frequency-selective fading or fast fading. Once the signal distortion has been mitigated, the P_b vs. E_b/N_0 performance can transition from the awful category to the merely bad Rayleigh-limit curve. Next, it is possible to further ameliorate the effects of fading and strive to approach AWGN system performance by using some form of diversity to provide

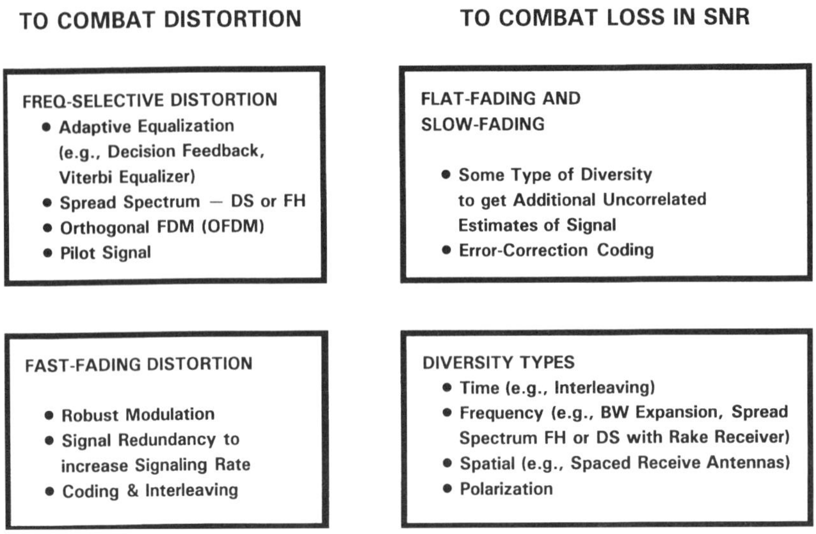

FIGURE 4.18 Basic mitigation types.

the receiver with a collection of uncorrelated replicas of the signal, and by using a powerful error-correction code.

In Fig. 4.18, several mitigation techniques for combating the effects of both signal distortion and loss in SNR are listed. Just as Figs. 4.1 and 4.7 serve as a guide for characterizing fading phenomena and their effects, Fig. 4.18 can similarly serve to describe mitigation methods that can be used to ameliorate the effects of fading. The mitigation approaches to be used when designing a system should be considered in two basic steps: first, choose the type of mitigation to reduce or remove any distortion degradation; next, choose a diversity type that can best approach AWGN system performance.

4.5.1 Mitigation to Combat Frequency-Selective Distortion

Equalization can mitigate the effects of channel-induced ISI that is brought on by frequency-selective fading. That is, it can help modify system performance described by the curve that is awful in Fig. 4.17 to the one that is merely bad. The process of equalizing for mitigating ISI effects involves using methods to gather the dispersed symbol energy back into its original time interval. In effect, an equalizer is an inverse filter of the channel. If the channel is frequency selective, the equalizer enhances the frequency components with small amplitudes and attenuates those with large amplitudes. The goal is for the combination of channel and equalizer filter to provide a flat composite received frequency response and linear phase [30]. Because in a mobile system the channel response varies with time, the equalizer filter must also change or adapt to the time-varying channel characteristics. Such equalizer filters are therefore adaptive devices that accomplish more than distortion mitigation; they also provide diversity. Because distortion mitigation is achieved by gathering the dispersed symbol energy back into the symbol original time interval so that it does not hamper the detection of other symbols, the equalizer is simultaneously providing the receiver with symbol energy that would otherwise be lost.

4.5.1.1 Decision Feedback Equalizer

The decision feedback equalizer (DFE) has a feed-forward section that is a linear transversal filter [30] with stage length and tap weights that are selected to coherently combine virtually all the current symbol energy. The DFE also has a feedback section that removes energy remaining from previously detected symbols [10, 30–32]. The basic idea behind the DFE is that once an information symbol has been detected, the ISI that it induces on future symbols can be estimated and subtracted before the detection of subsequent symbols.

4.5.1.2 Maximum Likelihood Sequence Estimation Equalizer

A maximum likelihood sequence estimation (MLSE) equalizer tests all possible data sequences (instead of detecting each received symbol by itself) and chooses the data sequence that is the most probable of all the candidates. The MLSE equalizer was first proposed by Forney [33] and implemented by using the Viterbi decoding algorithm [34]. The MLSE is optimum in the sense that it minimizes the probability of a sequence error. Because the Viterbi decoding algorithm is typically used in the implementation of the MLSE equalizer, this device is often referred to as the *Viterbi equalizer*. Later in this chapter, we illustrate the adaptive equalization performed in the Global System for Mobile (GSM) Communications using the Viterbi equalizer.

4.5.1.3 Direct-Sequence/Spread-Spectrum Techniques

Direct-sequence/spread-spectrum (DS/SS) techniques can be used to mitigate frequency-selective ISI distortion because the hallmark of spread-spectrum systems is their capability to reject interference, and ISI is a type of interference. Consider a DS/SS binary PSK communication channel comprising one direct path and one reflected path. Assume that the propagation from transmitter to receiver results in a multipath wave that is delayed by τ compared with the direct wave. The received signal, $r(t)$, neglecting noise, can be expressed as

$$r(t) = Ax(t)p(t)\cos(2\pi f_c t) + \alpha Ax(t-\tau)p(t-\tau)\cos(2\pi f_c t + \theta) \tag{4.35}$$

where $x(t)$ is the data signal, $p(t)$ is the PN spreading code, and τ is the differential time delay between the two paths. The angle θ is a random phase, assumed to be uniformly distributed in the range $(0, 2\pi)$, and α is the attenuation of the multipath signal relative to the direct path signal. The receiver multiplies the incoming $r(t)$ by the code $p(t)$. If the receiver is synchronized to the direct path signal, multiplication by the code signal yields

$$r(t)p(t) = Ax(t)p^2(t)\cos(2\pi f_c t) + \alpha Ax(t-\tau)p(t)p(t-\tau)\cos(2\pi f_c t + \theta) \tag{4.36}$$

where $p^2(t) = 1$. If τ is greater than the chip duration, then,

$$\left|\int p(t)p(t-\tau)dt\right| = \left|\int p^2(t)dt\right| \tag{4.37}$$

over some appropriate interval of integration (correlation), where τ is equal to, or larger than, the PN chip duration. Thus, the spread-spectrum system effectively eliminates the multipath interference by virtue of its code-correlation receiver. Even though channel-induced ISI is typically transparent to DS/SS systems, such systems suffer from the loss in energy contained in the multipath components rejected by the receiver. The need to gather up this lost energy belonging to a received chip was the motivation for developing the RAKE receiver [35–37]. The RAKE receiver dedicates a separate correlator to each multipath component (finger), and coherently adds the energy from each finger by selectively delaying each (the earliest component gets the longest delay) so that they can all be coherently combined.

Earlier, we described a channel that could be classified as flat fading, but occasionally exhibits frequency-selective distortion when the null of the channel frequency transfer function occurs at the center of the signal band. The use of DS/SS is a practical way of mitigating such distortion because the wideband SS signal can span many lobes of the selectively faded channel frequency response. Hence, a great deal of pulse energy is passed by the scatterer medium, in contrast to the channel nulling effect on a relatively narrowband signal (see Fig. 4.9c) [17]. The ability of the signal spectrum to span over many lobes of the frequency-selective channel transfer function is the key to how DS/SS signaling can overcome the degrading effects of a multipath environment. This requires that the spread-spectrum bandwidth, W_{ss} (or the chip rate, R_{ch}), be greater than the coherence bandwidth, f_0. The larger the ratio of W_{ss} to f_0, the more effective is the mitigation. A time-domain view of such mitigation can be similarly described. That is, to resolve multipath components requires that the spread-spectrum signal dispersion be greater than a chip time.

4.5.1.3 Frequency Hopping/Spread Spectrum

Frequency hopping/spread spectrum (FH/SS) can be used as a technique to mitigate the distortion caused by frequency-selective fading, provided the hopping rate is at least equal to the symbol rate. Compared with DS/SS, mitigation takes place through a different mechanism. FH systems avoid the degradation effects resulting from multipath by rapidly changing in the receiver the carrier-frequency band. Interference is avoided by similarly changing the band position in the receiver before the arrival of the multipath signal.

4.5.1.4 Orthogonal Frequency Division Multiplexing

Orthogonal frequency division multiplexing (OFDM) can be used for signal transmission in frequency-selective fading channels to avoid the use of an equalizer by lengthening the symbol duration. The approach is to partition (demultiplex) a high symbol-rate sequence into N symbol groups, so that each group contains a sequence of a lower symbol rate (by the factor $1/N$) than the original sequence. The signal band is made up of N orthogonal carrier waves, and each one is modulated by a different symbol group. The goal is to reduce the symbol rate (signaling rate), $W \approx 1/T_s$, on each carrier to be less than the channel coherence bandwidth f_0. OFDM, originally referred to as *Kineplex*, is a technique that has been implemented in the United States in mobile-radio systems [38], and has been chosen by the European community, under the name coded OFDM (COFDM), for high-definition television (HDTV) broadcasting [39].

4.5.1.5 Pilot Signal

Pilot signal is the name given to a signal intended to facilitate the coherent detection of waveforms. Pilot signals can be implemented in the frequency domain as in-band tones [40], or in the time domain as digital sequences that can also provide information about the channel state and thus improve performance in fading conditions [41].

4.5.2 Mitigation to Combat Fast-Fading Distortion

Fast-fading distortion calls for the use of a robust modulation (noncoherent or differentially coherent) scheme that does not require phase tracking, and reduces the detector integration time [19]. Another technique is to increase the symbol rate, $W \approx 1/T_s$, to be greater than the fading rate, $f_d \approx 1/T_0$, by adding signal redundancy. Error-correction coding can also provide mitigation; instead of providing more signal energy, a code reduces the required E_b/N_0 for a desired error performance. For a given E_b/N_0, with coding present, the error floor out of the demodulator is not lowered, but a lower error rate out of the decoder can be achieved [19]. Thus, with coding, one can get acceptable error performance and in effect withstand a large error floor from the demodulator that might have otherwise been unacceptable. To realize these coding benefits, errors out of the demodulator should be uncorrelated (which generally is the case in a fast-fading environment) or an interleaver must be incorporated into the system design.

An interesting filtering technique can provide mitigation when fast-fading distortion and frequency-selective distortion occur simultaneously. The frequency-selective distortion can be mitigated by the use of an OFDM signal set. Fast fading, however, typically degrades conventional OFDM because the Doppler spreading corrupts the orthogonality of the OFDM subcarriers. A polyphase filtering technique [42] is used to provide time-domain shaping and partial-response coding to reduce the spectral sidelobes of the signal set, and thus helps preserve its orthogonality. The process introduces known ISI and adjacent channel interference (ACI), which are then removed by a postprocessing equalizer and canceling filter [43].

4.5.3 Mitigation to Combat Loss in Signal to Noise Ratio

After implementing some mitigation technique to combat signal distortion resulting from frequency-selective or fast fading, the next step is to use diversity methods to move the system-operating point from the error-performance curve labeled as bad in Fig. 4.17 to a curve that approaches AWGN performance. The term *diversity* is used to denote the various methods available for providing the receiver with uncorrelated renditions of the signal of interest. Uncorrelated is the important feature here, because it would not help the receiver to have additional copies of a signal if the copies were all equally poor. Listed next are some of the ways in which diversity methods can be implemented.

- Time diversity is transmitting the signal on L different time slots with time separation of at least T_0. Interleaving, when used along with error-correction coding, is a form of time diversity.
- Frequency diversity is transmitting the signal on L different carriers with frequency separation of at least f_0. Bandwidth expansion is a form of frequency diversity. The signal bandwidth, W, is expanded so as to be greater than f_0, thus providing the receiver with several independently fading signal replicas. This achieves frequency diversity on the order of $L = W/f_0$. Whenever W is made larger than f_0, there is the potential for frequency-selective distortion unless mitigation in the form of equalization is provided. Thus, an expanded bandwidth can improve system performance (via diversity) only if the frequency-selective distortion that the diversity may have introduced is mitigated.
- Spread-spectrum systems, in which the signal bandwidth is much narrower than the coherence bandwidth of the channel, have no means to resolve the different multipath contributions. Such contributions interfere to create the fading conditions. In spread-spectrum systems, the delayed signals do not contribute to the fading, but to interchip interference. Spread spectrum is a bandwidth expansion technique that excels at rejecting interfering signals. In the case of DS/SS, it was demonstrated earlier that multipath components are rejected if they are time delayed by more than the duration of one chip. However, to approach AWGN performance, it is necessary to compensate for the loss in energy contained in those rejected components. The RAKE receiver (described later) makes it possible to coherently combine the energy from several of the multipath components arriving along different paths (with sufficient differential delay). Thus, used with a RAKE receiver, DS/SS modulation can be said to achieve path diversity. The RAKE receiver is needed in phase-coherent reception, but in differentially coherent bit detection, a simple delay, equivalent to the duration of 1 b, with complex conjugation, can be implemented [44].
- FH/SS is sometimes used as a diversity mechanism. The GSM system uses slow FH (217 hops per second) to compensate for those cases where the mobile unit is moving very slowly (or not at all) and experiences deep fading as a result of a spectral null.
- Spatial diversity is usually accomplished through the use of multiple receive antennas, separated by a distance of at least ten wavelengths when located at a base station (and less when located at a mobile unit). Signal-processing techniques must be employed to choose the best antenna output or to coherently combine all the outputs. Systems have also been implemented with multiple transmitters, each at a different location, as in the Global Positioning System (GPS).
- Polarization diversity [45] is yet another way to achieve additional uncorrelated samples of the signal.
- Any diversity scheme may be viewed as a trivial form of repetition coding in space or time. However, techniques exist for improving the loss in SNR in a fading channel that are more efficient and more powerful than repetition coding. Error-correction coding represents a unique mitigation technique, because instead of providing more signal energy, it reduces the required E_b/N_0 needed to achieve a desired performance level. Error-correction coding coupled with interleaving [19, 46–51] is probably the most prevalent of the mitigation schemes used to provide improved system performance in a fading environment. Note that the time diversity mechanism obtained through interleaving relies on the vehicle motion to spread the errors during the fading. The faster the speed of the mobile unit, the more effective is the interleaver. The interleaver is less effective at slow speeds. This speed vs. interleaver performance is demonstrated in Section 4.5.6.

4.5.4 Diversity Techniques

The goal in implementing diversity techniques is to utilize additional independent (or at least uncorrelated) signal paths to improve the received SNR. Diversity can provide improved system performance at relatively low cost; unlike equalization, diversity requires no training overhead. In this section, we show the error-performance improvements that can be obtained with the use of diversity techniques. The bit-error probability, $\bar{P}_B$, averaged through all the ups and downs of the fading experience in a slow-fading channel can be computed as follows:

$$\bar{P}_B = \int_0^{\infty} P_B(x)\, p(x)\, dx \tag{4.38}$$

where $P_B(x)$ is the bit-error probability for a given modulation scheme at a specific value of SNR = x, where $x = \alpha^2 E_b/N_0$, and $p(x)$ is the pdf of x resulting from the fading conditions. With E_b and N_0 constant, α is used to represent the amplitude variations resulting from fading.

For Rayleigh fading, α has a Rayleigh distribution so α^2, and consequently x, have a chi-square distribution. Thus, following the form of Eq. (4.15):

$$p(x) = \frac{1}{\Gamma}\exp\left(-\frac{x}{\Gamma}\right) \qquad x \geq 0 \tag{4.39}$$

where $\Gamma = \overline{\alpha^2}\, E_b/N_0$ is the SNR averaged through the ups and downs of fading. If each diversity (signal) branch, $i = 1, \ldots, M$ has an instantaneous SNR = γ_i, and we assume that each branch has the same average SNR given by Γ, then

$$p(\gamma_i) = \frac{1}{\Gamma}\exp\left(-\frac{\gamma_i}{\Gamma}\right) \qquad \gamma_i \geq 0 \tag{4.40}$$

The probability that a single branch has SNR less than some threshold γ, is

$$\begin{aligned} P(\gamma_i \leq \gamma) &= \int_0^{\gamma} p(\gamma_i)\, d\gamma_i = \int_0^{\gamma} \frac{1}{\Gamma}\exp\left(-\frac{\gamma_i}{\Gamma}\right) d\gamma_i \\ &= 1 - \exp\left(-\frac{\gamma}{\Gamma}\right) \end{aligned} \tag{4.41}$$

The probability that all M independent signal diversity branches are received simultaneously with an SNR less than some threshold value γ, is

$$P(\gamma_1, \cdots, \gamma_M \leq \gamma) = \left[1 - \exp\left(-\frac{\gamma}{\Gamma}\right)\right]^M \tag{4.42}$$

The probability that any single branch achieves SNR > γ,

$$P(\gamma_i > \gamma) = 1 - \left[1 - \exp\left(-\frac{\gamma}{\Gamma}\right)\right]^M \tag{4.43}$$

This is the probability of exceeding a threshold when selection diversity is used.

Example 1

Assume that four-branch diversity is used, and that each branch receives an independently Rayleigh-fading signal. If the average SNR is $\Gamma = 20$ dB, determine the probability that all four branches are received simultaneously with an SNR less than 10 dB (and also, the probability that this threshold will be exceeded). Compare the results to the case when no diversity is used.

Solution.
By using Eq. (4.42) with $\gamma = 10$ dB, and $\gamma/\Gamma = 10\text{ dB} - 20\text{ dB} = -10\text{ dB} = 0.1$, we solve for the probability that the SNR will drop below 10 dB, as follows:

$$P(\gamma_1, \gamma_2, \gamma_3, \gamma_4 \leq 10\text{ dB}) = \left[1 - \exp(-0.1)\right]^4 = 8.2 \times 10^{-5}$$

or, by using selection diversity, we can say that

$$P(\gamma_i > 10\text{ dB}) = 1 - 8.2 \times 10^{-5} = 0.9999$$

Without diversity

$$P(\gamma_1 \leq 10\text{ dB}) = \left[1 - \exp(-0.1)\right]^1 = 0.095$$

$$P(\gamma_1 > 10\text{ dB}) = 1 - 0.095 = 0.905$$

4.5.4.1 Diversity-Combining Techniques

The most common techniques for combining diversity signals are *selection*, *feedback*, *maximum ratio*, and *equal gain*. For systems using spatial diversity, selection involves the sampling of M antenna signals, and sending the largest one to the demodulator. Selection diversity combining is relatively easy to implement; however, it is not optimum because it does not make use of all the received signals simultaneously.

With feedback or scanning diversity, instead of using the largest of M signals, the M signals are scanned in a fixed sequence until one that exceeds a given threshold is found. This one becomes the chosen signal until it falls below the established threshold and the scanning process starts again. The error performance of this technique is somewhat inferior to the other methods, but feedback diversity is quite simple to implement.

In the case of maximum-ratio combining, the signals from all the M branches are weighted according to their individual SNRs and then summed. The individual signals must be cophased before being summed. The control algorithms for setting gains and delays are similar to those used in equalizers and in RAKE receivers. Maximum-ratio combining produces an average SNR, $\overline{\gamma_M}$, equal to the sum of the individual average SNRs, shown as follows [30]:

$$\overline{\gamma_M} = \sum_{i=1}^{M} \overline{\gamma_i} = \sum_{i=1}^{M} \Gamma = M\Gamma \tag{4.44}$$

where we assume that each branch has the same average SNR given by $\overline{\gamma_i} = \Gamma$. Thus, maximum-ratio combining can produce an acceptable average SNR, even when none of the individual $\overline{\gamma_i}$ is acceptable. It uses each of the M branches in a cophased and weighted manner such that the largest possible SNR is available at the receiver. Equal-gain combining is similar to maximum-ratio combining, except that the weights are all set to unity. The possibility of achieving an acceptable output SNR from a number of unacceptable inputs is still retained. The performance is marginally inferior to maximum-ratio combining. See Reference [52] for a detailed treatment of diversity combining.

TABLE 4.2 Doppler Spread vs. Vehicle Speed

Velocity		Doppler (Hz) 900 MHz (λ = 33 cm)	Doppler (Hz) 1800 MHz (λ = 16.6 cm)
mi/h	km/h		
3	5	4	8
20	32	27	54
50	60	66	132
80	108	106	212
120	192	160	320

4.5.5 Modulation Types for Fading Channels

It should be apparent that an amplitude-based signaling scheme such as amplitude shift keying (ASK) or quadrature amplitude modulation (QAM) is inherently vulnerable to performance degradation in a fading environment. Thus, for fading channels, the preferred choice for a signaling scheme is a frequency or phase-based modulation type.

In considering orthogonal FSK modualtion for fading channels, the use of MFSK with $M = 8$ or larger, is useful because its error performance is better than binary signaling. In slow Rayleigh-fading channels, binary DPSK and 8-FSK perform within 0.1 dB of each other [19]. At first glance, one might argue that a higher order orthogonal alphabet expands the transmission bandwidth, which at some point may cause the coherence bandwidth of the channel to be exceeded, leading to frequency-selective fading. However, for MFSK, the transmission bandwidth that must be available is much larger than the bandwidth of the propagating signal. For example, consider the case of 8-FSK and a symbol rate of 10,000 symbols per second. The transmission bandwidth is MR_s = 80,000 Hz. This is the bandwidth that must be available for the use of the system. However, each time that a symbol is transmitted, only one single-sideband tone (having a spectral occupancy of 10,000 Hz) is sent — not the whole alphabet. In considering PSK modulation for fading channels, higher order modulation alphabets perform poorly. MPSK with $M = 8$ or larger should be avoided [19]. Example 2 that follows examines a mobile communication system to substantiate such avoidance.

Example 2 Phase Variations in a Mobile Communication System

The Doppler spread $f_d = V/\lambda$ shows that the fading rate is a direct function of velocity. Table 4.2 shows the Doppler spread vs. vehicle speed at carrier frequencies of 900 and 1800 MHz. Calculate the phase variation per symbol for the case of signaling with QPSK modulation at the rate of 24.3 kilosymbols per second. Assume that the carrier frequency is 1800 MHz and that the velocity of the vehicle is 50 mi/h (80 km/h). Repeat for a vehicle speed of 100 mi/h.

Solution.

$$\Delta\theta/\text{symbol} = \frac{f_d \text{ Hz}}{R_s \text{ symbols/s}} \times 360^\circ$$

$$= \frac{132 \text{ Hz}}{24.3 \times 10^3 \text{ symbols/s}} \times 360^\circ$$

$$= 2^\circ/\text{symbol}$$

At a velocity of 100 mi/h: $\Delta\theta$/symbol = 4°/symbol. Thus, it should be clear why MPSK with a value of $M > 4$ is not generally used to transmit information in a multipath environment.

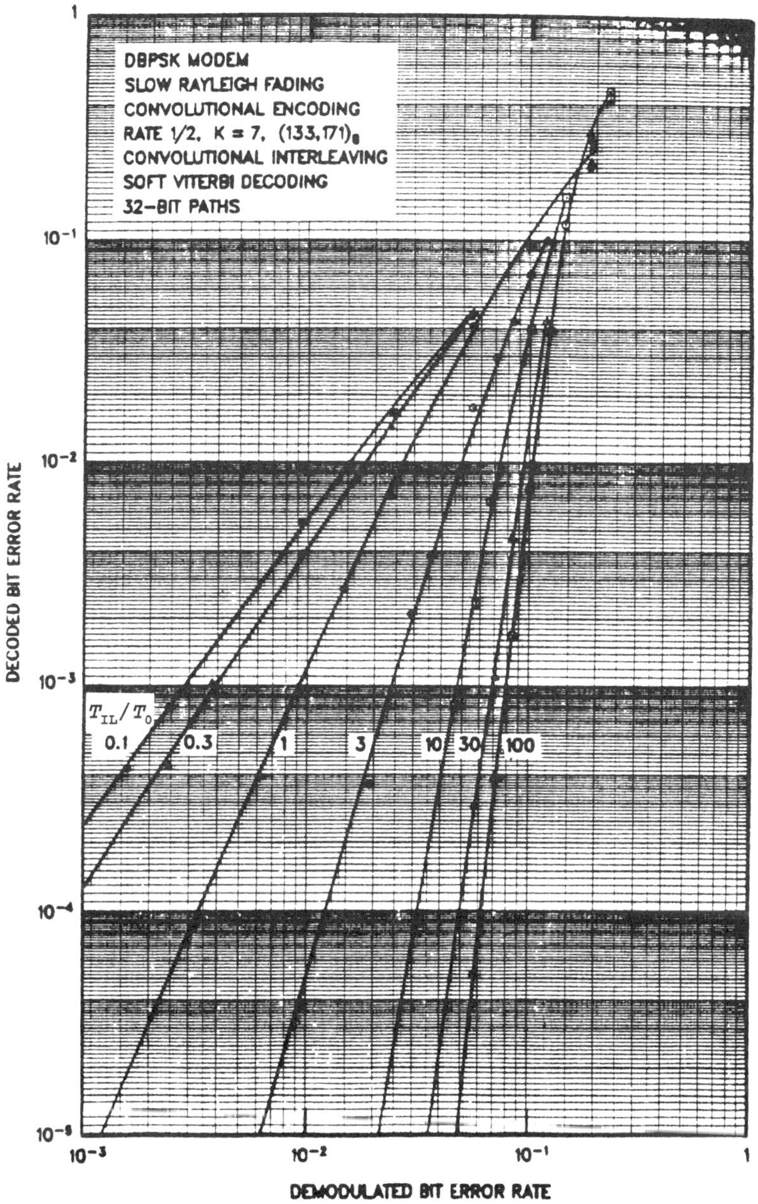

FIGURE 4.19 Error performance for various ratios of interleaver span to coherence time.

4.5.6 Role of an Interleaver

For transmission in a multipath environment, the primary benefit of an interleaver is to provide time diversity (when used along with error-correction coding). The larger the time span over which the channel symbols are separated, the greater chance there is that contiguous bits (after deinterleaving) have been subjected to uncorrelated fading manifestations; thus, the greater the chance there is to achieve effective diversity. Figure 4.19 illustrates the benefits of providing an interleaver time span, T_{IL}, which is large compared with the channel coherence time, T_0, for the case of DBPSK modulation with soft-decision decoding of a rate ½, K = 7 convolutional code, over a slow Rayleigh-fading channel. It should be apparent that an interleaver having the largest ratio of T_{IL}/T_0 is the best performing — large demodulated bit-error

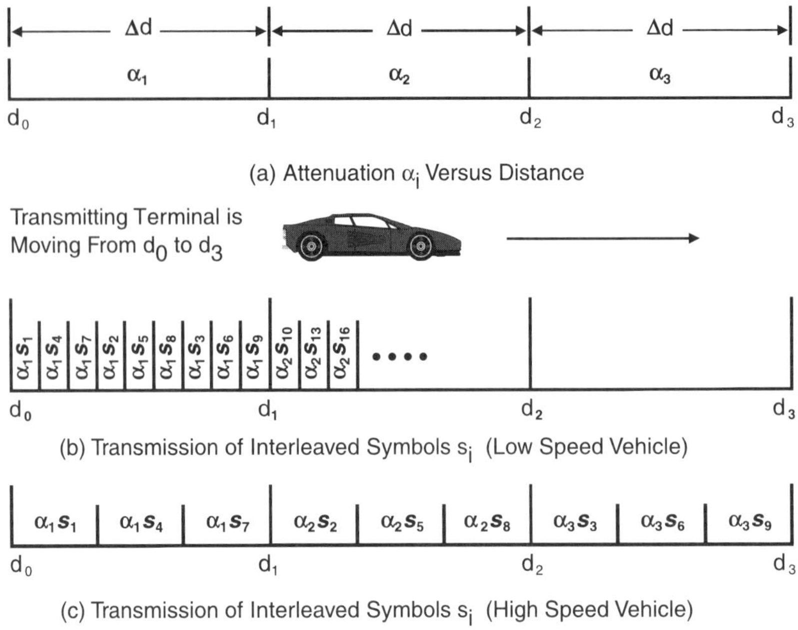

FIGURE 4.20 The benefits of interleaving improve with increased vehicle speed.

rate (BER) leading to small decoded BER. This leads to the conclusion that T_{IL}/T_0 should be some large number, for example, 1,000 or 10,000. However, in a real-time communication system this is not possible because the inherent time delay associated with an interleaver would be excessive. For the case of a block interleaver, before the first row of an array can be transmitted, virtually the entire array must be loaded. Similarly at the receiver, before the array can be deinterleaved, virtually the entire array must be stored. This leads to a delay of one block of data at each the transmitter and the receiver. Example 3 that follows shows that for a cellular telephone system with a carrier frequency of 900 MHz, a T_{IL}/T_0 ratio of 10 is about as large as one can implement without suffering excessive delay.

It is interesting to note that the interleaver provides no benefit against multipath unless there is motion between the transmitter and receiver (or motion of objects within the signal-propagating paths). As the motion increases in velocity, so does the benefit of a given interleaver to the error-performance of the system. (Do not use this as an excuse for exceeding a highway speed limit). This is shown in Fig. 4.20 where part (a) of the figure shows a terrain that is mapped out with attenuation factors, $\{\alpha_i\}$, for a particular mobile communications link over a particular terrain. In the region between the points d_0 and d_1 the attenuation factor is α_1. Between the points d_1 and d_2 the attenuation factor is α_2, and so forth. Assume that the points d_i are equally separated by a distance Δd. Part (b) of the figure shows an automobile that is traveling at a slow speed; as the vehicle traverses a distance Δd, nine symbols are emitted from its transmitter. Assume that the interleaver has a span of three-symbol intervals, so that symbols s_1 through s_9 appear in the permuted order shown in part (b) of the figure. Notice that all nine of the symbols experience the same attenuation α_1, so that after deinterleaving there is no benefit obtained by using an interleaver with this small a span.

Next consider part (c) of Fig. 4.20 where the vehicle is moving three times faster than in part (b); thus, as the vehicle traverses a distance Δd, only three symbols are emitted from its transmitter. As before, the symbols are affected by the regional attenuation yielding the nine-symbol sequence as shown in part (c) of the figure. After deinterleaving of the sequence shown in part (c), we have the following attenuation factor–symbol pairs result: $\alpha_1 s_1$, $\alpha_2 s_2$, $\alpha_3 s_3$, $\alpha_1 s_4$, $\alpha_2 s_5$, $\alpha_3 s_6$, $\alpha_1 s_7$, $\alpha_2 s_8$, $\alpha_3 s_9$. It can be seen that adjacent symbols are affected by different attenuation factors. Thus, the interleaver with too small of a span to yield any benefit at low speeds does provide benefit at faster speeds.

Figure 4.21 also provides evidence that, although communications degrade with increased speed of the mobile unit (the fading rate increases), the benefit of an interleaver is enhanced with increased speed.

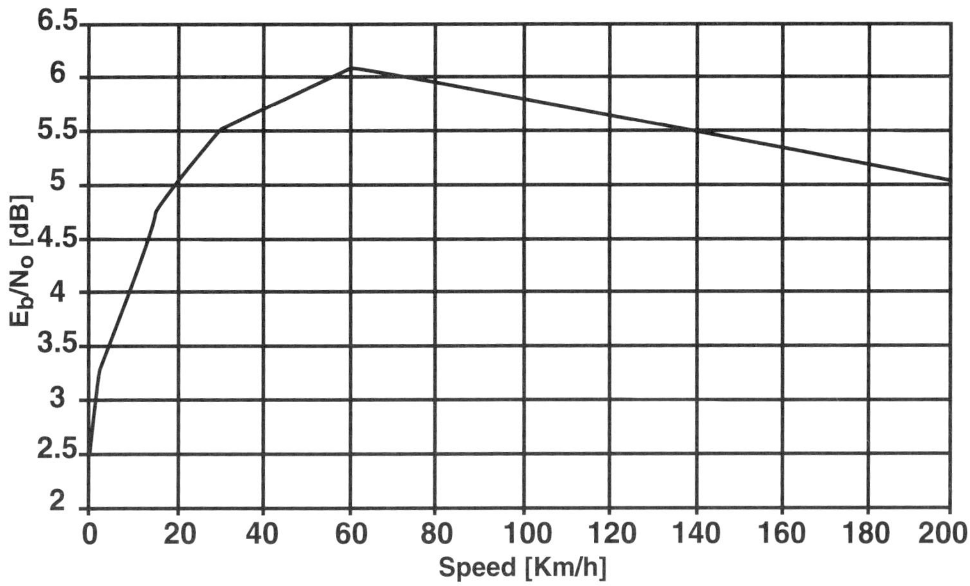

FIGURE 4.21 Typical E_b/N_0 performance vs. vehicle speed for 850-MHz links to achieve an FER = 1%. Rayleigh channel with two independent paths.

Figure 4.21 shows the results of field testing performed on a code division multiple access (CDMA) system meeting the IS-95 over a link comprising a moving vehicle and a base station [53]. This figure shows a plot of required E_b/N_0 vs. vehicle speed to maintain a frame (20 ms of data) error rate of 1%. The best performance (smallest E_b/N_0 requirement) is achieved at low speeds from 0 to 20 km/h. This slow-speed region is where the system power-control methods can most effectively compensate for the effects of slow fading; at these slow speeds the interleaver cannot provide any benefit, and the plot shows a steep degradation as a function of speed. For velocity in the range of 20 to 60 km/h the steepness of this degradation is reduced. This is the range where the dynamics of the system power control cannot quite keep up with the increase in fading rapidity; and at the same time, the interleaver does not yet provide sufficient benefit. The speed of 60 km/h represents the worst error-performance case for this system. As the vehicle goes faster than 60 km/h, the power control no longer provides any benefits against fading, but the interleaver provides a steadily increasing (with speed) performance improvement. The interleaver task of transforming the effects of a deep fade (time-correlated degradation events) into random events, becomes easier with increased speed.

In summary, the system error performance over a fading channel typically degrades with increased speed because of the increase in Doppler spread or fading rapidity. However, the action of an interleaver in the system provides mitigation, which becomes more effective at higher speeds. This trend toward improved error performance does not continue indefinitely. Eventually, the irreducible error floor seen in Fig. 4.15 dominates. Therefore, if the type of measurements shown in Fig. 4.21 were made at speeds beyond 200 km/h, there would be a point where the curve would turn around and steadily show the degradation effects resulting from the increased Doppler.

4.6 Summary of the Key Parameters Characterizing Fading Channels

We summarize the conditions that must be met so that the channel does not introduce frequency-selective distortion and fast-fading distortion. By combining the expressions given by Eqs. (4.22), (4.32), and (4.33), we obtain

$$f_0 > W > f_d \tag{4.45}$$

or

$$T_m < T_s < T_0 \tag{4.46}$$

In other words, it is desired that the channel coherence bandwidth should exceed the signaling rate, which in turn should exceed the fading rate of the channel. Recall, that without distortion mitigation, f_0 sets an upper limit and f_d sets a lower limit on the signaling rate.

4.6.1 Fast-Fading Distortion: Case One

If the conditions of Eqs. (4.45) and (4.46) are not met, distortion results, unless appropriate mitigation is provided. Consider the fast-fading case where the signaling rate is less than the channel fading rate. That is,

$$f_0 > W < f_d \tag{4.47}$$

Mitigation consists of using one or more of the following methods (see Fig. 4.18).

- Choose a modulation/demodulation technique that is most robust under fast-fading conditions. That means, for example, avoiding schemes that require PLLs for carrier recovery, because the fast fading could keep a PLL from achieving lock conditions.
- Incorporate sufficient redundancy so that the transmission symbol rate exceeds the channel fading rate, but at the same time does not exceed the coherence bandwidth. The channel can then be classified as flat fading. However, as pointed out in Section 4.3.3, even flat-fading channels experience frequency-selective distortion whenever a channel transfer function exhibits a spectral null near the signal band center. Because this happens only occasionally, mitigation can be accomplished by adequate error-correction coding and interleaving.
- The preceding two mitigation approaches should result in the demodulator operating at the Rayleigh limit [19] (see Fig. 4.17). However, the probability of error vs. E_b/N_0 curve may exhibit flattening as seen in Fig. 4.15, because of the FM noise that results from the random Doppler spreading. The use of an in-band pilot tone and a frequency-control loop can decrease the level at which the performance curve exhibits the flattening effect.
- To avoid the error floor described earlier because of random Doppler spreading, the signaling rate should be increased to about 100 to 200 times the fading rate [27]. This is one motivation for designing mobile communication systems to operate in a time division multiple access (TDMA) mode.
- Incorporate error-correction coding and interleaving to further improve system performance.

4.6.2 Frequency-Selective Fading Distortion: Case Two

Consider the frequency-selective case where the coherence bandwidth is less than the symbol rate, while the symbol rate is greater than the Doppler spread. That is,

$$f_0 < W > f_d \tag{4.48}$$

Because the transmission symbol rate exceeds the channel fading rate, there is no fast-fading distortion. However, mitigation of frequency-selective effects is necessary. One or more of the following techniques may be considered (see Fig. 4.18):

- Included are adaptive equalization, spread spectrum (DS or FH), OFDM, and pilot signal. The European GSM system uses a midamble training sequence in each transmission time slot so that the receiver can estimate the impulse response of the channel. A Viterbi equalizer (explained later) is implemented for mitigating the frequency-selective distortion.
- Once the distortion effects have been reduced, diversity techniques as well as error-correction coding and interleaving should be introduced to approach AWGN performance. For DS/SS signaling, the use of a RAKE receiver (explained later) can be used for providing diversity by coherently combining multipath components that would otherwise be lost.

4.6.3 Fast-Fading and Frequency-Selective Fading Distortion: Case Three

Consider the case where the channel coherence bandwidth is less than the signaling rate, which in turn is less than the fading rate. This condition is mathematically described by

$$f_0 < W < f_d \tag{4.49}$$

or

$$f_0 < f_d \tag{4.50}$$

Clearly, the channel exhibits both fast fading and frequency-selective fading. Recall from Eq. (4.45) and (4.46) that f_0 sets an upper limit and f_d sets a lower limit on the signaling rate. Thus, the condition described by Eq. (4.50) presents a difficult design problem, because, unless distortion mitigation is provided, the *maximum* allowable signaling rate is, strictly speaking, *less* than the *minimum* allowable signaling rate. Mitigation in this case is similar to the initial approach outlined in case one.

- Choose a modulation/demodulation technique that is most robust under fast-fading conditions.
- Use transmission redundancy to increase the transmitted symbol rate.
- Provide some form of frequency-selective fading mitigation in a manner similar to that outlined in case two.
- Once the distortion effects have been reduced, introduce some form of diversity (as well as error-correction coding and interleaving), to approach AWGN performance.

Example 3 Equalizers and Interleavers for Mobile Communications

Consider a cellular telephone that is located in a vehicle traveling at 60 mi/h (96 km/h). The carrier frequency is 900 MHz. Use the GSM equalizer test profile shown in Fig. 4.22 to determine the following: (1) the rms delay spread, σ_τ, and (2) the maximum allowable signal bandwidth $W \approx 1/T_s$ that does not require the use of an equalizer. (3) Also determine when operating over a channel with the delay spread found in part (a), which of the following systems requires an equalizer — the United States Digital Cellular Standard (USDC) known as IS-54 (updated to IS-136), the Global System for Mobile (GSM) Communications, or CDMA systems designed to meet IS-95. The bandwidths and symbol rates of these systems are USDC: W = 30 kHz, $1/T_s$ = 24.3 kilosymbols per second; GSM: W = 200 kHz, $1/T_s$ = 271 kilosymbols per second; and IS-95: W = 1.25 MHz, $1/T_s$ = 9.6 kilosymbols per second. (4) Then ascertain the total (transmitter plus receiver) time delay caused by the interleaver, when the ratio of interleaver span to coherence time T_{IL}/T_0 is equal to 10. If the total tolerable time delay (transmitter plus receiver) for speech is 100 ms, can such an interleaver be implemented for speech? (5) Finally, repeat parts (1) through (4) for a carrier frequency of 1900 MHz.

Solution.

1. In Fig. 4.22, the GSM test profile shows an idealized multipath component (finger) located at each of six delay times $\{\tau_k\}$ from 0 to 16 μsec. Each finger can be designated by $S(\tau_k)$, its average relative power, which in this profile, is unity (0 dB). The profile represents a *fictitious* multipath environment,

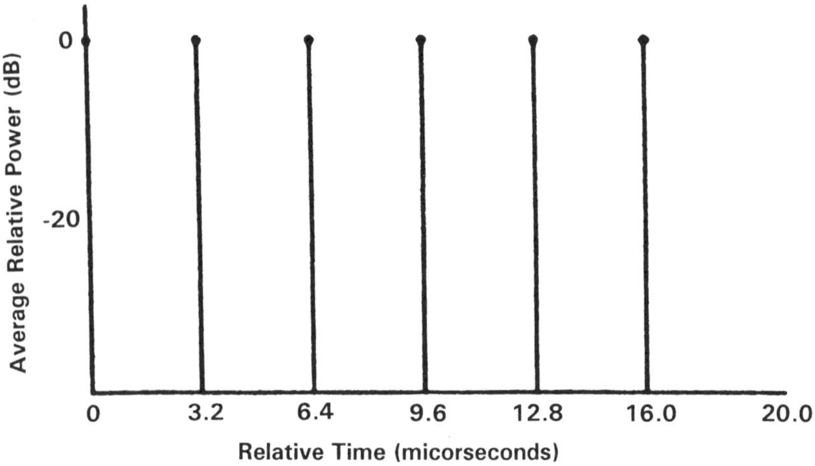

FIGURE 4.22 GSM equalizer test profile.

used for equalization testing [15]. With the finger locations shown on Fig. 4.22, the mean delay spread, $\bar{\tau}$, is computed as follows:

$$\bar{\tau} = \frac{\sum_k S(\tau_k)\,\tau_k}{\sum_k S(\tau_k)} = \frac{0+3.2+6.4+9.6+12.8+16.0}{6} = 8\ \mu\text{sec}$$

The second moment of delay spread, $\overline{\tau^2}$, and the rms delay spread, σ_τ, are computed as follows:

$$\overline{\tau^2} = \frac{\sum_k S(\tau_k)\,\tau_k^2}{\sum_k S(\tau_k)} = \frac{0+3.2^2+6.4^2+9.6^2+12.8^2+16.0^2}{6} = 93.87\ \mu\text{sec squared}$$

$$\sigma_\tau = \sqrt{\overline{\tau^2} - \left(\bar{\tau}\right)^2} = \sqrt{93.87-8^2} = 5.5\ \mu\text{sec}$$

2. By using Eq. (4.21), the channel coherence bandwidth is determined to be

$$f_0 = \frac{1}{5\sigma_\tau} = \frac{1}{5\times 5.5\ \mu\text{sec}} = 36.4\ \text{kHz}$$

Thus, the maximum allowable signal bandwidth that does not require the implementation of an equalizer is $W = 36.4$ kHz.

3. For the various system bandwidths given in this example, it is apparent that the need for an equalizer in USDC is marginal, whereas in GSM it is definitely required. With regard to systems that are designed to meet IS-95, because the signaling rate or transmission bandwidth W of 1.25 MHz is much larger than the coherence bandwidth f_0 of 36.4 kHz, the system exhibits frequency-selective fading. However, in such DS/SS systems, W is purposely spread with the intent

of exceeding f_0 and thus mitigating the effects of frequency-selective fading. An equalizer is only required if ISI poses a problem, and ISI is not a problem if the symbol rate is smaller than the coherence bandwidth (or the symbol duration is larger than the multipath spread). Hence, in the IS-95 example, because the symbol rate of 9.6 ksymbols per second is considerably smaller than the coherence bandwidth, an equalizer is not needed. A RAKE receiver, which is described in Section 4.7.2, is used for exploiting path diversity; at the chip level, its implementation resembles that of an equalizer.

4. To determine the interleaver delay, we compute the Doppler spread and coherence time using Eq. (4.25) and (4.29), as follows:

$$f_d = \frac{V}{\lambda} = \frac{\dfrac{96{,}000 \text{ m/h}}{3600 \text{ s/h}}}{\dfrac{3\times10^8 \text{ m/h}}{9\times10^8 \text{ Hz}}} = 80 \text{ Hz}$$

Thus

$$T_0 \approx \frac{0.5}{f_d} = 6.3 \text{ ms}$$

Based on the requirement that T_{IL}/T_0, the interleaver span is T_{IL} = 63 ms, making the total transmitter plus receiver delay time equal to 126 ms. For speech, this may be in the marginally acceptable range. Mobile systems often use interleavers with shorter spans that produce one-way delays in the range of 20 to 40 ms.

5. Repeating the preceding for a carrier frequency of 1900 MHz, the coherence bandwidth calculations are unaffected by the change in carrier frequency, but the Doppler spread, coherence time, and interleaver delay must be computed again. The results are

$$f_d = \frac{V}{\lambda} = 169 \text{ Hz}$$

Thus

$$T_0 \approx \frac{0.5}{f_d} = 3 \text{ ms}$$

Thus, the interleaver span is T_{IL} = 30 ms, making the total transmitter plus receiver delay equal to 60 ms, which is acceptable for speech signals.

4.7 Applications: Mitigating the Effects of Frequency-Selective Fading

4.7.1 Viterbi Equalizer as Applied to Global System for Mobile Communications

Figure 4.23 shows the GSM TDMA frame, having a duration of 4.615 ms and comprising eight slots, one assigned to each active mobile user. A normal transmission burst, occupying one time slot contains 57 message bits on each side of a 26-b midamble called a *training* or *sounding sequence.* The slot-time

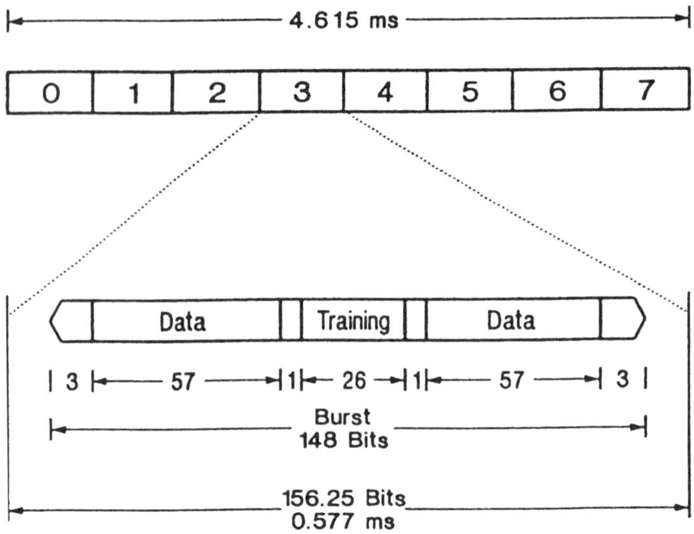

FIGURE 4.23 The GSM TDMA frame and time slot containing a normal burst.

duration is 0.577 ms (or the slot rate is 1733 slots per second). The purpose of the midamble is to assist the receiver in estimating the impulse response of the channel adaptively (during the time duration of each 0.577-ms slot). For the technique to be effective, the fading characteristics of the channel must not change appreciably during the time interval of one slot. In other words, there cannot be any fast-fading degradation during a time slot when the receiver analyzes the midamble distortion; otherwise, efforts to compensate for the channel fading characteristics are not effective. Consider, for example, a GSM receiver used aboard a high-speed train, traveling at a constant velocity of 200 km/h (55.56 m/s). Assume the carrier frequency to be 900 MHz (the wavelength is $\lambda = 0.33$ m). From Eq. (4.29), the distance corresponding to a half wavelength is traversed in

$$T_0 \approx \frac{\lambda/2}{V} \approx 3 \text{ ms} \tag{4.51}$$

which as indicated in Eq. (4.51) corresponds approximately to the coherence time.

Therefore, the channel coherence time is more than 5 times greater than the slot time of 0.577 ms. The time needed for a significant change in channel-fading characteristics is relatively long compared with the time duration of one slot. Note, that the choices made for GSM in the design of its TDMA slot time and midamble were undoubtedly influenced by the need to preclude fast-fading effects that could cause the equalizer to be ineffective. The GSM symbol rate (or bit rate, because the modulation is binary) is 271 kilosymbols per second and the bandwidth, W, is 200 kHz. Because the typical rms delay spread, σ_τ, in an urban environment is on the order of 2 µsec, then by using Eq. (4.21) it can be seen that the resulting coherence bandwidth, f_0, is approximately 100 kHz. It should therefore be apparent that because $f_0 < W$, the GSM receiver must utilize some form of mitigation to combat frequency-selective distortion. To accomplish this goal, the Viterbi equalizer is typically implemented.

Figure 4.24 illustrates the basic functional blocks used in a GSM receiver for estimating the channel impulse response. This estimate is used to provide the detector with channel-corrected reference waveforms [54] as explained later. In the final step, the Viterbi algorithm is used to compute the MLSE of the message bits. A received signal can be described in terms of the transmitted signal convolved with the impulse response of the channel. Let $s_{tr}(t)$ denote the transmitted midamble training sequence, and $r_{tr}(t)$ denote the corresponding received midamble training sequence. Thus,

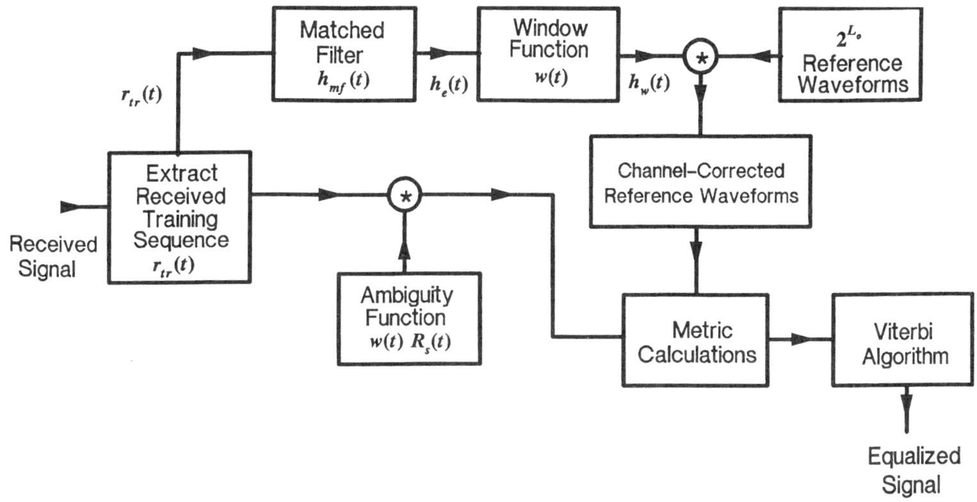

FIGURE 4.24 The Viterbi equalizer as applied to GSM.

$$r_{\text{tr}}(t) = s_{\text{tr}}(t) * h_c(t) \tag{4.52}$$

where * denotes convolution, and noise has been neglected. At the receiver, because $r_{\text{tr}}(t)$ is part of the received normal burst, it is extracted and sent to a filter having impulse response, $h_{\text{mf}}(t)$, that is matched to $s_{\text{tr}}(t)$. This matched filter yields at its output, an estimate of $h_c(t)$, denoted $h_e(t)$, and developed from Eq. (4.52) as follows:

$$\begin{aligned} h_e(t) &= r_{\text{tr}}(t) * h_{\text{mf}}(t) \\ &= s_{\text{tr}}(t) * h_c(t) * h_{\text{mf}}(t) \\ &= R_s(t) * h_c(t) \end{aligned} \tag{4.53}$$

where $R_s(t) = s_{\text{tr}}(t) * h_{\text{mf}}(t)$ is the autocorrelation function of $s_{\text{tr}}(t)$. If $s_{\text{tr}}(t)$ is designed to have a highly peaked (impulse-like) autocorrelation function $R_s(t)$, then $h_e(t) \approx h_c(t)$. Next, by using a windowing function, $w(t)$, we truncate $h_e(t)$ to form a computationally affordable function, $h_w(t)$. The time duration of $w(t)$, denoted L_o, must be large enough to compensate for the effect of typical channel-induced ISI. The term L_o consists of the sum of two contributions, namely, L_{CISI}, corresponding to the controlled ISI caused by Gaussian filtering of the baseband waveform (which then modulates the carrier using MSK), and L_C, corresponding to the channel-induced ISI caused by multipath propagation. Therefore, L_o can be written as

$$L_o = L_{\text{CISI}} + L_C \tag{4.54}$$

The GSM system is required to provide distortion mitigation caused by signal dispersion having delay spreads of approximately 15 to 20 μsec. Because in GSM the bit duration is 3.69 μsec, we can express L_o in units of bit intervals. Thus, the Viterbi equalizer used in GSM has a memory of 4- to 6-b intervals. For each L_o-bit interval in the message, the function of the Viterbi equalizer is to find the most likely L_o-bit sequence out of the 2^{L_o} possible sequences that might have been transmitted. Determining the most likely transmitted L_o-bit sequence requires that 2^{L_o} meaningful reference waveforms be created by modifying (or disturbing) the 2^{L_o} ideal waveforms (generated at the receiver) in the same way that the channel has

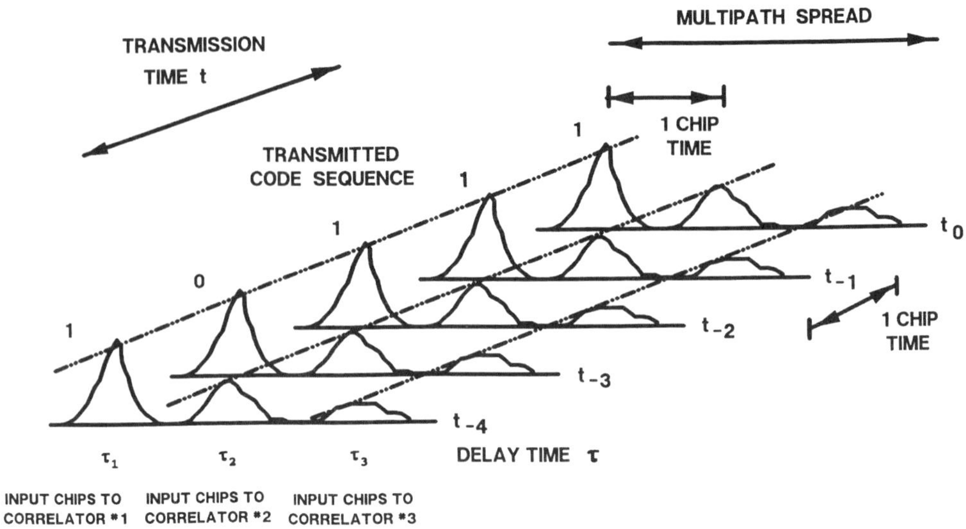

FIGURE 4.25 Example of received chips seen by a three-finger RAKE receiver.

disturbed the transmitted slot. Therefore, the 2^{L_o} reference waveforms are convolved with the windowed estimate of the channel impulse response, $h_w(t)$ to generate the disturbed or so-called channel-corrected reference waveforms. Next, the channel-corrected reference waveforms are compared against the received data waveforms to yield metric calculations. However, before the comparison takes place, the received data waveforms are convolved with the known windowed autocorrelation function $w(t)R_s(t)$, transforming them in a manner comparable to the transformation applied to the reference waveforms. This filtered message signal is compared with all possible 2^{L_o} channel-corrected reference signals, and metrics are computed in a manner similar to that used in the Viterbi decoding algorithm (VDA). The VDA yields the maximum likelihood estimate of the transmitted data sequence [34].

4.7.2 RAKE Receiver Applied to Direct-Sequence/Spread-Spectrum Systems

IS-95 describes a DS/SS cellular system that uses a RAKE receiver [35–37] to provide path diversity. The RAKE receiver searches through the different multipath delays for code correlation and thus recovers delayed signals that are then optimally combined with the output of other independent correlators. In Fig. 4.25, the power profiles associated with the five-chip transmissions of the code sequence 1 0 1 1 1 are shown, where the observation times are labeled t_{-4} for the earliest transmission and t_0 for the latest. Each abscissa shows three "fingers" or components arriving with delays τ_1, τ_2, and τ_3. Assume that the intervals between the transmission times, t_i, and the intervals between the delay times, τ_i, are each one chip in duration. From this, one can conclude that the finger arriving at the receiver at time t_{-4}, with delay τ_3, is time coincident with two other fingers, namely, the fingers arriving at times t_{-3} and t_{-2} with delays τ_2 and τ_1, respectively. In this example, because the delayed components are separated by at least one chip time, they can be resolved. At the receiver, there must be a sounding device that is dedicated to estimating the τ_i delay times. Note that for a terrestrial mobile-radio system, the fading rate is relatively slow (on the order of milliseconds) or the channel coherence time large compared with the chip time duration ($T_0 > T_{ch}$). Hence, the changes in τ_i occur slowly enough so that the receiver can readily adapt to them.

Once the τ_i delays are estimated, a separate correlator is dedicated to recovering each resolvable multipath finger. In this example, there would be three such dedicated correlators, each one processing a delayed version of the same chip sequence 1 0 1 1 1. In Fig. 4.25, each correlator receives chips with power profiles represented by the sequence of fingers shown along a diagonal line. For simplicity, the

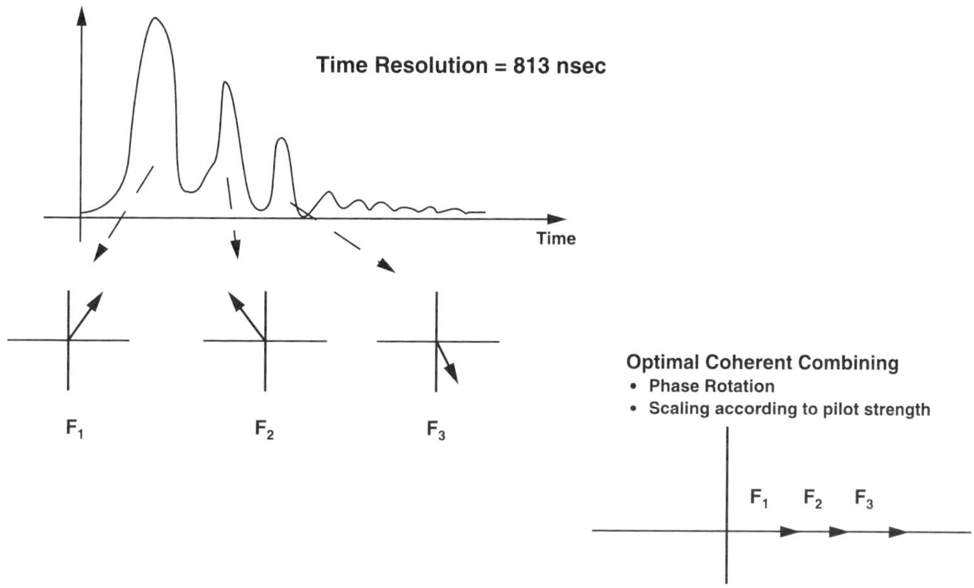

FIGURE 4.26 Coherent combining of multipath returns in the RAKE receiver.

chips are all shown as positive signaling elements. In reality, these chips form a PN sequence, which of course, contains both positive and negative pulses. Each correlator attempts to correlate these arriving chips with the same appropriately synchronized PN code. At the end of a symbol interval (typically there may be hundreds or thousands of chips per symbol), the outputs of the correlators are coherently combined, and a symbol detection is made. Figure 4.26 illustrates the phase rotation (provided by the RAKE receiver) of three fingers labeled F_1, F_2, and F_3, to facilitate their coherent combining. At the chip level, the RAKE receiver resembles an equalizer, but its real function is to exploit the path diversity.

The interference–suppression capability of DS/SS systems stems from the fact that a code sequence arriving at the receiver time-shifted by merely one chip, has very low correlation to the particular PN code with which the sequence is correlated. Therefore, any code chips that are delayed by one or more chip times are suppressed by the correlator. The delayed chips only contribute to raising the interference level (correlation sidelobes). The mitigation provided by the RAKE receiver can be termed path diversity, because it allows the energy of a chip that arrives via multiple paths to be combined coherently. Without the RAKE receiver, this energy would be transparent and therefore lost to the DS/SS receiver. In Fig. 4.25, looking vertically above point t_3, it is clear that there is interchip interference resulting from different fingers arriving simultaneously. The spread-spectrum processing gain allows the system to endure such interference at the chip level. No other equalization is deemed necessary in IS-95.

4.8 Conclusion

In this chapter, the major elements that contribute to fading in certain communication channels have been characterized. Figure 4.1 was presented as a guide for characterizing the fading phenomena. Two types of fading, large-scale and small-scale, were described. Two manifestations of small-scale fading (signal dispersion and fading rapidity) were examined. Each examination involved two views, one in time and the other in frequency. Two degradation categories were defined for dispersion: frequency-selective fading and flat fading. Two degradation categories were defined for fading rapidity: fast and slow fading. The small-scale fading degradation categories were summarized in Fig. 4.7. A mathematical model using correlation and power density functions was presented in Fig. 4.8. This model yields a useful symmetry to help us visualize the Fourier transform and duality relationships that describe the fading phenomena. Further, mitigation techniques for ameliorating the effects of each degradation category were presented; these were summarized in Fig. 4.18. Finally, mitigation methods that have been applied in two different systems, GSM and CDMA meeting IS-95, were described.

References

1. Rappaport, T.S., *Wireless Communications,* Chapters 3 and 4, Prentice-Hall, Upper Saddle River, NJ, 1996.
2. Greenwood, D. and Hanzo, L., "Characterisation of Mobile Radio Channels," *Mobile Radio Communications,* edited by R. Steele, Chapter 2, Pentech Press, London, 1994.
3. Lee, W.C.Y., "Elements of Cellular Mobile Radio Systems," *IEEE Trans. Veh. Technol.,* vol. V-35, no. 2, May 1986, pp. 48–56.
4. Okumura, Y. et al., "Field Strength and Its Variability in VHF and UHF Land Mobile Radio Service," *Rev. Elec. Commun. Lab.,* vol. 16, nos. 9 & 10, 1968, pp. 825–873.
5. Hata, M., "Empirical Formulae for Propagation Loss in Land Mobile Radio Services," *IEEE Trans. Veh. Technol.,* vol. VT-29, no. 3, 1980, pp. 317–325.
6. Seidel, S.Y. et al., "Path Loss, Scattering and Multipath Delay Statistics in Four European Cities for Digital Cellular and Microcellular Radiotelephone," *IEEE Trans. Veh. Technol.,* vol. 40, no. 4, November 1991, pp. 721–730.
7. Cox, D.C., Murray, R., and Norris, A., "800 MHz Attenuation Measured in and around Suburban Houses," *AT&T Bell Lab. Tech. J.,* vol. 673, no. 6, July–August 1984, pp. 921–954.
8. Schilling, D.L. et al., "Broadband CDMA for Personal Communications Systems," *IEEE Commun. Mag.,* vol. 29, no. 11, November 1991, pp. 86–93.
9. Andersen, J.B., Rappaport, T.S., and Yoshida, S., "Propagation Measurements and Models for Wireless Communications Channels, *IEEE Commun. Mag.,* vol. 33, no. 1, January 1995, pp. 42–49.
10. Proakis, J.G., *Digital Communications,* Chapter 7, McGraw-Hill, New York, 1983.
11. Schwartz, M., *Information, Transmission, Modulation, and Noise,* Second Edition, McGraw-Hill, New York, 1970.
12. Amoroso, F., "Investigation of Signal Variance, Bit Error Rates and Pulse Dispersion for DSPN Signalling in a Mobile Dense Scatterer Ray Tracing Model," *Intl. J. Satellite Commun.,* vol. 12, 1994, pp. 579–588.
13. Bello, P.A., "Characterization of Randomly Time-Variant Linear Channels," *IEEE Trans. Commun. Syst.,* December 1963, pp. 360–393.
14. Green, P.E., Jr., "Radar Astronomy Measurement Techniques," *MIT Lincoln Laboratory,* Lexington, MA, Tech. Report No. 282, December 1962.
15. Pahlavan, K. and Levesque, A.H., *Wireless Information Networks,* Chapters 3 and 4, John Wiley & Sons, New York, 1995.
16. Lee, W.Y.C., *Mobile Cellular Communications,* McGraw-Hill, New York, 1989.
17. Amoroso, F. "Use of DS/SS Signaling to Mitigate Rayleigh Fading in a Dense Scatterer Environment," *IEEE Personal Commun.,* vol. 3, no. 2, April 1996, pp. 52–61.
18. Clarke, R.H., "A Statistical Theory of Mobile Radio Reception," *Bell System Tech. J.,* vol. 47, no. 6, July–August 1968, pp. 957–1000.
19. Bogusch, R.L., *Digital Communications in Fading Channels: Modulation and Coding,* Mission Research Corp., Santa Barbara, CA, Report no. MRC-R-1043, March 11, 1987.
20. Amoroso, F., "The Bandwidth of Digital Data Signals," *IEEE Commun. Mag.,* vol. 18, no. 6, November 1980, pp. 13–24.
21. Bogusch, R.L. et al., "Frequency Selective Propagation Effects on Spread-Spectrum Receiver Tracking," *Proc. IEEE,* vol. 69, no. 7, July 1981, pp. 787–796.
22. Jakes, W.C. (Ed.), *Microwave Mobile Communications,* John Wiley & Sons, New York, 1974.

23. *Joint Technical Committee of Committee T1 R1P1.4 and TIA TR46.3.3/TR45.4.4 on Wireless Access,* "Draft Final Report on RF Channel Characterization,"Paper No. JTC(AIR)/94.01.17-238R4, January 17, 1994.
24. Bello, P.A. and Nelin, B.D., "The Influence of Fading Spectrum on the Binary Error Probabilities of Incoherent and Differentially Coherent Matched Filter Receivers, "*IRE Trans. Commun. Syst.,* vol. CS-10, June 1962, pp. 160–168.
25. Amoroso, F., "Instantaneous Frequency Effects in a Doppler Scattering Environment," *IEEE Int. Conf. Commun.,* June 7–10, 1987, pp. 1458–1466.
26. Fung, V., Rappaport, T.S., and Thoma, B., "Bit-Error Simulation for π/4 DQPSK Mobile Radio Communication Using Two-Ray and Measurement-Based Impulse Response Models," *IEEE J. Sel. Areas Commun.,* vol. 11, no. 3, April 1993, pp. 393–394.
27. Bateman, A.J. and McGeehan, J.P., "Data Transmission over UHF Fading Mobile Radio Channels," *IEEE Proc.,* vol. 131, Pt. F, No. 4, July 1984, pp. 364–374.
28. Feher, K., *Wireless Digital Communications,* Prentice-Hall, Upper Saddle River, NJ, 1995.
29. Davarian, F., Simon, M., and Sumida, J., "DMSK: A Practical 2400-bps Receiver for the Mobile Satellite Service," *Jet Propulsion Laboratory* Publication 85-51 (MSAT-X Report No. 111), June 15, 1985.
30. Rappaport, T.S., *Wireless Communications,* Chapter 6, Prentice-Hall, Upper Saddle River, NJ, 1996.
31. Bogusch, R.L., Guigliano, F.W., and Knepp, D.L., "Frequency- Selective Scintillation Effects and Decision Feedback Equalization in High Data-Rate Satellite Links," *Proc. IEEE,* vol. 71, no. 6, June 1983, pp. 754–767.
32. Qureshi, S.U.H., "Adaptive Equalization," *Proc. IEEE,* vol. 73, no. 9, September 1985, pp. 1340–1387.
33. Forney, G.D., "The Viterbi Algorithm," *Proc. IEEE,* vol. 61, no. 3, March 1978, pp. 268–278.
34. Viterbi, A.J. and Omura, J.K., *Principles of Digital Communication and Coding,* McGraw-Hill, New York, 1979.
35. Price, R. and Green P.E. Jr., "A Communication Technique for Multipath Channels," *Proceedings of the IRE,* March 1958, pp. 555–70.
36. Turin, G.L., "Introduction to Spread-Spectrum Antimultipath Techniques and their Application to Urban Digital Radio," *Proc. IEEE,* vol. 68, no. 3, March 1980, pp. 328–353.
37. Simon, M.K., Omura, J.K., Scholtz, R.A., and Levitt, B.K., *Spread Spectrum Communications Handbook,* McGraw-Hill, New York, 1994.
38. Birchler, M.A. and Jasper, S.C., "A 64 kbps Digital Land Mobile Radio System Employing M-16QAM," *Proc. 1992 IEEE Intl. Conf. Selected Topics Wireless Communications,* Vancouver, British Columbia, June 25–26, 1992, pp. 158–162.
39. Sari, H., Karam, G., and Jeanclaude, I., "Transmission Techniques for Digital Terrestrial TV Broadcasting," *IEEE Commun. Mag.,* vol. 33, no. 2, February 1995, pp. 100–109.
40. Cavers, J.K., The Performance of Phase Locked Transparent Tone-in-Band with Symmetric Phase Detection, *IEEE Trans. Commun.,* vol. 39, no. 9, September 1991, pp. 1389–1399.
41. Moher, M.L. and Lodge, J.H., "TCMP — A Modulation and Coding Strategy for Rician Fading Channel," *IEEE J. Selected Areas in Commun.,* vol. 7, no. 9, December 1989, pp. 1347–1355.
42. Harris, F., "On the Relationship Between Multirate Polyphase FIR Filters and Windowed, Overlapped FFT Processing," *Proceedings of the Twenty-Third Annual Asilomar Conference on Signals, Systems, and Computers,* Pacific Grove, CA, Oct. 30–Nov. 1, 1989, pp. 485–488.
43. Lowdermilk, R.W. and Harris, F., "Design and Performance of Fading Insensitive Orthogonal Frequency Division Multiplexing (OFDM) using Polyphase Filtering Techniques," *Proceedings of the Thirtieth Annual Asilomar Conference on Signals, Systems, and Computers,* Pacific Grove, CA, November 3–6, 1996.
44. Kavehrad, M. and Bodeep, G.E., "Design and Experimental Results for a Direct-Sequence Spread-Spectrum Radio Using Differential Phase-Shift Keying Modulation for Indoor Wireless Communications," *IEEE JSAC,* vol. SAC-5, no. 5, June 1987, pp. 815–823.
45. Hess, G.C., *Land-Mobile Radio System Engineering,* Artech House, Boston, 1993.

46. Hagenauer, J. and Lutz, E., "Forward Error Correction Coding for Fading Compensation in Mobile Satellite Channels," *IEEE JSAC,* vol. SAC-5, no. 2, February 1987, pp. 215–225.
47. McLane, P.I. et al., "PSK and DPSK Trellis Codes for Fast Fading, Shadowed Mobile Satellite Communication Channels," *IEEE Trans. Commun.*, vol. 36, no. 11, November 1988, pp. 1242–1246.
48. Schlegel, C. and Costello, D.J., Jr., "Bandwidth Efficient Coding for Fading Channels: Code Construction and Performance Analysis," *IEEE JSAC,* vol. 7, no. 9, December 1989, pp. 1356–1368.
49. Edbauer, F., "Performance of Interleaved Trellis-Coded Differential 8-PSK Modulation over Fading Channels," *IEEE J. Selected Areas Commun.*, vol. 7, no. 9, December 1989, pp. 1340–1346.
50. Soliman, S. and Mokrani, K., "Performance of Coded Systems over Fading Dispersive Channels," *IEEE Trans. Commun.*, vol. 40, no. 1, January 1992, pp. 51–59.
51. Divsalar, D. and Pollara, F. "Turbo Codes for PCS Applications," *Proc. ICC '95,* Seattle, WA, June 18–22, 1995, pp. 54–59.
52. Simon, M. and Alouini, M-S., *Digital Communications over Fading Channels: A Unified Approach to Performance Analysis*, John Wiley & Sons, New York, 2000.
53. Padovani, R., "Reverse Link Performance of IS-95 Based Cellular Systems," *IEEE Personal Commun.*, Third Quarter 1994, pp. 28–34.
54. Hanzo, L. and Stefanov, J., "The Pan-European Digital Cellular Mobile Radio System — Known as GSM," *Mobile Radio Communications,* edited by R. Steele, Chapter 8, Pentech Press, London, 1992.

PART B

Antenna Technology and Numerical Methods

5

Antenna Parameters, Various Generic Antennas and Feed Systems, and Available Software

Jennifer Bernhard
University of Illinois at Urbana-Champaign

Eric Michielssen
University of Illinois at Urbana-Champaign

5.1 Introduction

Antennas serve as transducers between electromagnetic waves traveling in free space and guided electromagnetic signals in circuits. As such, they play a critical role in the performance of wireless communication systems. With the proliferation of mobile wireless services that deliver voice and/or data in smaller and smaller devices, the task to design an antenna for a portable unit that meets not only operational requirements but also aesthetic and packaging restrictions is becoming more and more challenging. As a result, engineers rely on a combination of theory, simulation, and experimental investigation to arrive at a design that meets all the demands of a particular application.

This chapter provides an overview of basic antenna operational parameters, including input impedance, efficiency, radiation pattern, directivity, gain, and noise temperature. It then describes several basic types of antennas that are implemented in today's wireless systems. These include linear antennas, loops, planar inverted-F antennas, and dielectric resonator antennas. Although these antennas serve as the basis for many designs, the real-world electromagnetic environments of portable devices can affect significantly antenna performance. This chapter discusses some of these nonideal design factors, such as finite and imperfect grounds and dielectric packaging effects. Additionally, common feeding methods, impedance matching techniques, and balance-to-unbalance transformers (baluns) are presented. Finally, an overview of available software for antenna analysis and design is provided that illustrates a broad range of capabilities that can help engineers develop antennas for specialized applications and deployment scenarios.

Phasor notation is used throughout this chapter, specifically, an $\exp(j\omega t)$ dependence is assumed and suppressed.

0-8493-0124-6/02/$0.00+$1.50

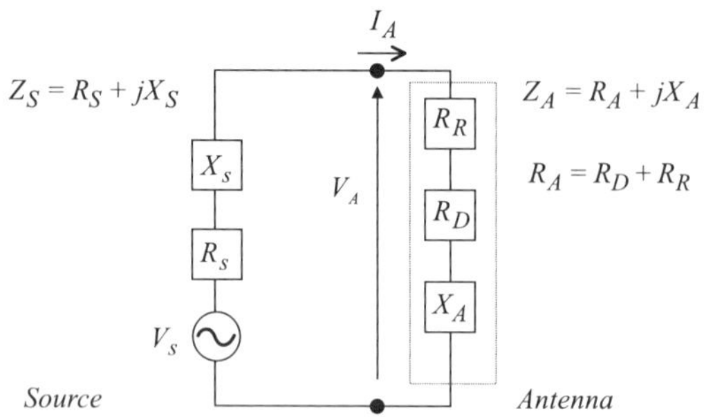

FIGURE 5.1 Equivalent circuit of the transmitting antenna.

5.2 Basic Antenna Parameters and Antenna Links

This section reviews basic antenna theory and introduces antenna parameters such as antenna impedance, mismatch and ohmic efficiency, radiation pattern and polarization, directivity, gain and equivalent isotropically radiated power, and effective height and aperture. In addition, the celebrated Friis equation is derived, and equations for the signal to noise ratio (SNR) of an antenna link presented. A subsection on source-field relationships concludes this section. For a more detailed treatment of the material presented in this section, the reader is referred to References 1 through 4.

5.2.1 Antenna Input Impedance and Efficiency

When used in transmitting mode, an antenna is a device for radiating electromagnetic energy. The input impedance of the antenna, Z_A, equals the ratio of the voltage V_A to the current I_A at its input terminals, and is a key parameter in the study of power transfer in an antenna link.

Consider a source with open circuit signal amplitude V_S and internal impedance $Z_S = R_S + jX_s$, which drives an antenna with input impedance $Z_A = R_A + jX_A$ where R_A and X_A are the antenna input resistance and reactance, respectively (Fig. 5.1). The voltage and current at the antenna terminals are

$$\begin{aligned} V_A &= \frac{V_S Z_A}{Z_S + Z_A} \\ I_A &= \frac{V_S}{Z_S + Z_A} \end{aligned} \tag{5.1}$$

The time-average power absorbed by the antenna terminals, P_A, is

$$\begin{aligned} P_A &= \frac{1}{2}\mathrm{Re}\left[V_A I_A^*\right] \\ &= \frac{|V_S|^2 R_A}{2|Z_S + Z_A|^2} \end{aligned} \tag{5.2}$$

where * and $|\cdot|$ denote a phasor's complex conjugate and magnitude, respectively.

It is well known from circuit theory that power deposited into a load is maximized when the source and load impedances are conjugate matched. This implies that P_A is maximum when $Z_A = Z_S^*$, or, equivalently, when $R_A = R_S$ and $X_A = -X_S$. Under these conditions, the power delivered to the antenna, also termed the available source power P_S, is

$$P_S = P_A \Big|_{[Z_A = Z_S^*]} = \frac{|V_S|^2}{8R_S} \tag{5.3}$$

When conjugate matching conditions are not satisfied, P_A is only a fraction of P_S, or

$$P_A = qP_S \tag{5.4}$$

where q is termed the antenna reflection or mismatch efficiency. From Eq. (5.2) to (5.4), it follows that

$$q = \frac{4R_A R_S}{\left|Z_S + Z_A\right|^2} \tag{5.5}$$

When Z_S or Z_A are purely real (i.e., when $X_S = 0$ or $X_A = 0$), the mismatch efficiency can be expressed as

$$q = 1 - \left|\Gamma\right|^2 \tag{5.6}$$

with Γ, the reflection coefficient, given by

$$\Gamma = \frac{Z_A - Z_S}{Z_A + Z_S} \tag{5.7}$$

Ideally, a transmitting antenna radiates all power absorbed by its terminals. In practice, however, ohmic and dielectric losses in the antenna prevent this from happening. The total power radiated by the antenna, further denoted P_R, is expressed as a fraction of P_A as

$$P_R = eP_A \tag{5.8}$$

where e is the antenna efficiency. Obviously, the difference between P_A and P_R must equal the power P_D dissipated by the antenna, or

$$\begin{aligned} P_D &= P_A - P_R \\ &= (1-e)P_A \end{aligned} \tag{5.9}$$

It is often convenient to decompose the antenna resistance as

$$R_A = R_R + R_D \tag{5.10}$$

where R_R and R_D are the antenna radiation and loss resistances, respectively. By using

$$\begin{aligned} P_A &= \frac{1}{2} R_A \left|I_A\right|^2 \\ P_R &= \frac{1}{2} R_R \left|I_A\right|^2 \\ P_D &= \frac{1}{2} R_D \left|I_A\right|^2 \end{aligned} \tag{5.11}$$

it follows that

$$e = \frac{R_R}{R_A} = \frac{R_R}{R_R + R_D} \tag{5.12}$$

5.2.2 Electromagnetic Fields Produced by an Antenna

The power P_R is carried by the electromagnetic field radiated by the antenna. Far away from the antenna, the antenna electric field can always be expressed as

$$\mathbf{E}(\mathbf{r}) = \frac{e^{-jkr}}{r}\left[FF_\phi\left(\hat{\mathbf{k}}\right)\hat{\phi} + FF_\theta\left(\hat{\mathbf{k}}\right)\hat{\theta}\right] \tag{5.13}$$

where $k = 2\pi/\lambda$ is the free-space wavenumber; $\lambda = c/f$ represents the free-space wavelength; $c \cong 3.10^8$ m/s is the free-space speed of light; f is the frequency of operation; $\hat{\mathbf{k}} = \hat{\mathbf{x}} \sin\theta \cos\phi + \hat{\mathbf{y}} \sin\theta \sin\phi + \hat{\mathbf{z}} \cos\theta$ is a unit vector that points from the antenna toward the observer $\mathbf{r}$; $\hat{\mathbf{x}}$, $\hat{\mathbf{y}}$, and $\hat{\mathbf{z}}$ are unit vectors in the x, y, and z directions; and $r = |\mathbf{r}|$. In other words, far away from the antenna (Fig. 5.2) the following occur:

- The electric field propagates radially away from the antenna (i.e., the field propagates in the $\hat{\mathbf{r}} = \mathbf{r}/r$ direction).
- The electric field attenuates as $1/r$ because of the spherical spreading.
- The electric field has no $\hat{\mathbf{r}}$-directed component (i.e., no component along its direction of propagation); the field can be fully characterized in terms of its $\phi = (\hat{\mathbf{z}} \times \hat{\mathbf{r}})/|\hat{\mathbf{z}} \times \hat{\mathbf{r}}|$ and $\hat{\theta} = \hat{\phi} \times \hat{\mathbf{r}}$ components. The amplitudes of the field along $\hat{\phi}$ and $\hat{\theta}$ depend on the direction of radiation, and are specified by the far-field functions $FF_\phi(\hat{\mathbf{k}})$ and $FF_\theta(\hat{\mathbf{k}})$.

Note that Eq. (5.13) only describes accurately the antenna electric far-field, and fails to capture the behavior of the field near the antenna. If d measures the diameter of the smallest sphere that contains the antenna, it can be shown that Eq. (5.13) only holds true at distances r from the antenna satisfying

$$\begin{aligned} &r \gg d \\ &r \gg \lambda \\ &r \gg \frac{2d^2}{\lambda} \end{aligned} \tag{5.14}$$

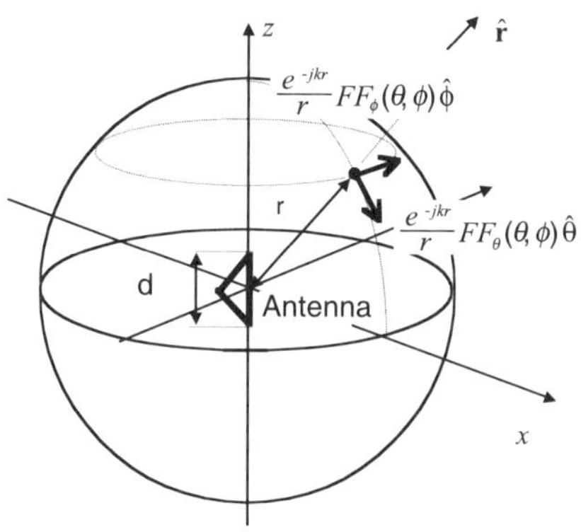

FIGURE 5.2 Electric far-field of the transmitting antenna.

For a "far-field observer," the electric field radiated by the antenna locally behaves as a plane wave. Hence, the antenna's magnetic far-field can be expressed as

$$
\begin{aligned}
\mathbf{H}(\mathbf{r}) &= \frac{1}{\eta}\left[\hat{\mathbf{r}} \times \mathbf{E}(\mathbf{r})\right] \\
&= \frac{1}{\eta}\frac{e^{-jkr}}{r}\left[FF_{\theta}(\hat{\mathbf{k}})\hat{\phi} - FF_{\phi}(\hat{\mathbf{k}})\hat{\theta}\right]
\end{aligned}
\tag{5.15}
$$

where $\eta = \sqrt{\mu_0/\varepsilon_0}$ is the free-space wave impedance, and μ_0 and ε_0 the free-space permeability and permittivity, respectively.

The Poynting vector $\mathbf{S}(\mathbf{r}) = [\mathbf{E}(\mathbf{r}) \times \mathbf{H}(\mathbf{r})^*]/2$ characterizes the power flow associated with the electromagnetic field. It follows from Eq. (5.13) and Eq. (5.15) that in the far-field $\mathbf{S}(\mathbf{r})$ is purely real and given by

$$
\mathbf{S}(\mathbf{r}) = \frac{1}{2\eta r^2}\left[\left|FF_{\phi}(\hat{\mathbf{k}})\right|^2 + \left|FF_{\theta}(\hat{\mathbf{k}})\right|^2\right]\hat{\mathbf{r}}
\tag{5.16}
$$

The radiation intensity $U(\hat{\mathbf{k}})$ of the antenna is

$$
\begin{aligned}
U(\hat{\mathbf{k}}) &= r^2\left|\mathbf{S}(\mathbf{r})\right| \\
&= \frac{1}{2\eta}\left[\left|FF_{\phi}(\hat{\mathbf{k}})\right|^2 + \left|FF_{\theta}(\hat{\mathbf{k}})\right|^2\right]
\end{aligned}
\tag{5.17}
$$

The normalized radiation intensity is denoted $U_{nor}(\hat{\mathbf{k}}) = U(\hat{\mathbf{k}})/U_{\max}$, where $U_{\max}$ is the maximum of $U(\hat{\mathbf{k}})$ over all angles $0 \le \theta \le \pi$ and $0 \le \phi < 2\pi$.

The total power radiated by the antenna is obtained by computing the flux of the real part of the Poynting vector through any closed surface S that contains the antenna. If S is a sphere of radius r, large enough such that conditions in Eq. (5.14) are satisfied and therefore $\mathbf{S}(\mathbf{r})$ is real on S, then

$$
P_R = \oint\!\!\oint_S \mathbf{S}(\mathbf{r}) \cdot d\mathbf{s}
\tag{5.18}
$$

where $d\mathbf{s} = \hat{\mathbf{r}}r^2 \sin\theta \, d\theta \, d\phi$ is an elemental surface vector. It follows from Equations Eq. (5.17) and Eq. (5.18) that

$$
\begin{aligned}
P_R &= \int_0^{\pi} \sin\theta d\theta \int_0^{2\pi} d\phi U(\hat{\mathbf{k}}) \\
&= \oint\!\!\oint U(\hat{\mathbf{k}}) d\Omega
\end{aligned}
\tag{5.19}
$$

with $d\Omega = \sin\theta \, d\theta \, d\phi$. In conclusion, P_R, previously expressed as the efficiency of the antenna multiplied by the power accepted by its terminals, can also be obtained by integrating the antenna radiation intensity over a unit sphere. When $U(\hat{\mathbf{k}}) \equiv 1$, Eq. (5.19) yields $P_R = 4\pi$. Hence, the average radiation intensity is $U_{ave} = P_R/4\pi$.

The electric and magnetic fields generated by an antenna behave much like plane waves when observed *locally* in the far-field: the electric field is perpendicular to the magnetic field, and both the electric and magnetic fields are perpendicular to the direction of propagation. It is convenient to define the polarization of a transmitting antenna in a given direction as the polarization of the far-field radiated wave. The polarization of an electromagnetic wave is defined as the figure traced by the extremity of the electric field vector, and the sense in which it is traced, as observed looking into the direction of propagation. Antenna polarization can be classified as linear, circular, or elliptical. For circular and elliptically polarized waves, the sense of rotation is either clockwise (right-hand polarized) or counterclockwise (left-hand polarized). Note that the polarization of an antenna is a direction-dependent characteristic. In what follows, antenna polarization is characterized in terms of the polarization vector $\hat{\mathbf{p}}(\hat{\mathbf{k}})$, defined as

$$\hat{\mathbf{p}}\left(\hat{\mathbf{k}}\right)=\frac{FF_{\phi}\left(\hat{\mathbf{k}}\right)\hat{\phi}+FF_{\theta}\left(\hat{\mathbf{k}}\right)\hat{\theta}}{FF\left(\hat{\mathbf{k}}\right)} \tag{5.20}$$

where

$$FF\left(\hat{\mathbf{k}}\right)=\sqrt{\left|FF_{\phi}\left(\hat{\mathbf{k}}\right)\right|^{2}+\left|FF_{\theta}\left(\hat{\mathbf{k}}\right)\right|^{2}} \tag{5.21}$$

The quantity $FF(\hat{\mathbf{k}})$ is termed the antenna field strength in the $\hat{\mathbf{k}}$ direction. Obviously, $\hat{\mathbf{p}}(\hat{\mathbf{k}})\cdot\hat{\mathbf{p}}(\hat{\mathbf{k}})^{*} = 1$ and $\hat{\mathbf{p}}(\hat{\mathbf{k}})\cdot\hat{\mathbf{k}}=0$. The normalized field strength is defined as $FF_{nor}(\hat{\mathbf{k}}) = FF(\hat{\mathbf{k}})/FF_{\max}$, where $FF_{\max}$ is the maximum of $FF(\hat{\mathbf{k}})$ over all angles $0 \le \theta \le \pi$ and $0 \le \phi < 2\pi$. It is easily shown that $U_{nor}(\hat{\mathbf{k}}) = [FF_{nor}(\hat{\mathbf{k}})]^2$. As an example, consider an antenna that radiates a field characterized by $FF_{\phi}(\hat{\mathbf{k}}) = A \sin\theta$ and $FF_{\theta}(\hat{\mathbf{k}}) = B\sin\theta\cos\phi$. If $A = 0$ or $B = 0$, then the antenna field is linearly polarized along $\hat{\theta}$ or $\hat{\phi}$, respectively. If A and B are nonzero but real, then the polarization remains linear with a direction-dependent mixture of $\hat{\theta}$ and $\hat{\phi}$. If $A = 1$ and $B = j$, then radiation into the $(\phi, \theta) = (0, \pi/2)$ direction (i.e., along the x-axis) is right-hand circularly polarized, whereas radiation into the $(\phi, \theta) = (\pi, \pi/2)$ (i.e., along the $-x$-axis) is left-hand circularly polarized; fields radiated along the $\pm y$-axes are linearly polarized.

5.2.3 Radiation Pattern

The radiation field of an antenna is fully specified once we characterize its electric and magnetic fields by Eq. (5.13) and (5.15) (i.e., once we specify $FF_{\phi}(\hat{\mathbf{k}})$ and $FF_{\theta}(\hat{\mathbf{k}})$. The radiation pattern of an antenna is a "graphic representation of the radiation properties of an antenna..." and could include information on the energy distribution, phase, and polarization of the radiated field. Often, we are most interested in plots of the relative energy distribution $U_{nor}(\hat{\mathbf{k}})$ on spheres surrounding the antenna (or specific cuts through these spheres), and such graphs are referred to as power patterns. Similar plots of field magnitudes $FF_{nor}(\hat{\mathbf{k}})$ are referred to as field patterns. We may, of course, also plot absolute (i.e., unnormalized patterns). When plotting an antenna pattern, we refer to the region of strong radiation as the antenna's "main beam." Radiation in other directions occurs in the form of sidelobes. These concepts, as well as some that follow, are illustrated in Fig. 5.3.

The half power beam width (*HPBW*) and beamwidth between first nulls (*BWFN*) are two figures of merit that characterize two-dimensional cross-sections of an antenna main beam. The *HPBW* of the mainlobe of an antenna (in a given plane) measures the angle surrounding the direction of maximum radiation across which the antenna normalized radiation intensity is larger than one-half. Defining θ^{right}_{HPBW} and $\tilde{\theta}^{left}_{HPBW}$ as the angles measured from the direction of the mainlobe maximum toward its left and right for which

$$U_{nor}\left(\hat{\mathbf{k}}\left(\tilde{\theta}^{left}_{HPBW}\right)\right)=U_{nor}\left(\hat{\mathbf{k}}\left(\tilde{\theta}^{right}_{HPBW}\right)\right)=\frac{1}{2} \tag{5.22}$$

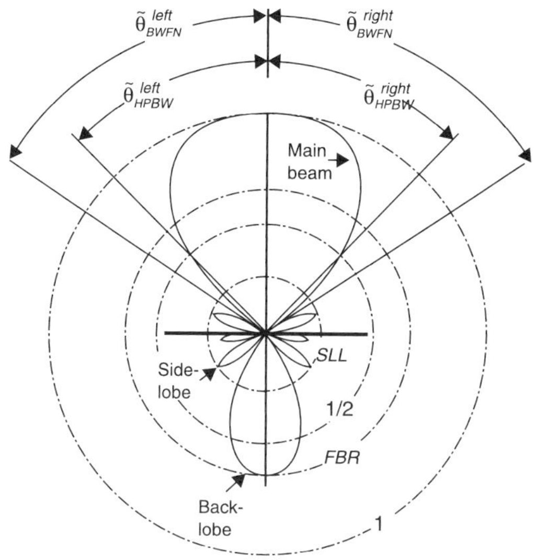

FIGURE 5.3 Normalized (power) radiation pattern. Numbers on circles indicate their radii.

the *HPBW* is expressed as

$$HPBW = \tilde{\theta}^{left}_{HPBW} + \tilde{\theta}^{right}_{HPBW} \tag{5.23}$$

The *BWFN* (in a given plane) is the angle in between the first pattern nulls adjacent to the mainlobe. Defining $\tilde{\theta}^{right}_{BWFN}$ and $\tilde{\theta}^{left}_{BWFN}$ as the angles measured from the direction of the mainlobe maximum toward its left and right for which

$$U_{nor}\left(\hat{\mathbf{k}}\left(\tilde{\theta}^{left}_{BWFN}\right)\right) = U_{nor}\left(\hat{\mathbf{k}}\left(\tilde{\theta}^{right}_{BWFN}\right)\right) = 0 \tag{5.24}$$

the *BWFN* is

$$BWFN = \tilde{\theta}^{left}_{BWFN} + \tilde{\theta}^{right}_{BWFN} \tag{5.25}$$

The sidelobe level (*SLL*) of an antenna is defined as the ratio of the radiation intensity in the direction of the largest sidelobe (which is often, though not always, the first sidelobe adjacent to the main antenna beam) to the maximum radiation intensity, that is,

$$\begin{aligned} SLL &= \frac{U\left(\hat{\mathbf{k}}\right)\Big|_{\text{largest side lobe}}}{U_{\max}} = \frac{U_{SLL}}{U_{\max}} \\ &= \frac{FF^2\left(\hat{\mathbf{k}}\right)\Big|_{\text{largest side lobe}}}{FF^2_{\max}} = \frac{FF^2_{SLL}}{FF^2_{\max}} \end{aligned} \tag{5.26}$$

The front to back ratio (*FBR*) of an antenna is defined as the ratio of the radiation intensity in the direction of maximum radiation to the radiation intensity in the opposite direction, that is,

$$FBR = \frac{U_{\max}}{U(\hat{\mathbf{k}})\Big|_{\substack{\text{in the direction opposite}\\ \text{to that of maximum}\\ \text{radiation}}}} = \frac{U_{\max}}{U_{FBR}}$$
$$= \frac{FF^2_{\max}}{FF(\hat{\mathbf{k}})^2\Big|_{\substack{\text{in the direction opposite}\\ \text{to that of maximum}\\ \text{radiation}}}} = \frac{FF^2_{\max}}{FF^2_{FBR}} \tag{5.27}$$

The antenna beam solid angle is the solid angle of the main beam of a hypothetical antenna that radiates the same total power as the antenna being characterized, but with a constant radiation intensity equal to the maximum radiation intensity $U_{\max}$ of the latter. The antenna beam solid angle, often denoted Ω_A, can be expressed as

$$\Omega_A = \frac{\oiint_S U(\theta,\phi)\, d\Omega}{U_{\max}} \tag{5.28}$$

Equivalent expressions for Ω_A are

$$\Omega_A = \int_0^{\pi} \sin\theta\, d\theta \int_0^{2\pi} d\phi\, U_{nor}(\hat{\mathbf{k}})$$
$$= \int_0^{\pi} \sin\theta\, d\theta \int_0^{2\pi} d\phi \left[FF_{nor}(\hat{\mathbf{k}})\right]^2 \tag{5.29}$$

It is easily verified that for an antenna with constant radiation intensity $U(\hat{\mathbf{k}}) = U_{\max}$, $\Omega_A = 4\pi$. Such an antenna is termed an isotropic radiator. For all other antennas, $\Omega_A < 4\pi$. The antenna beam solid angle concept is illustrated in Fig. 5.4.

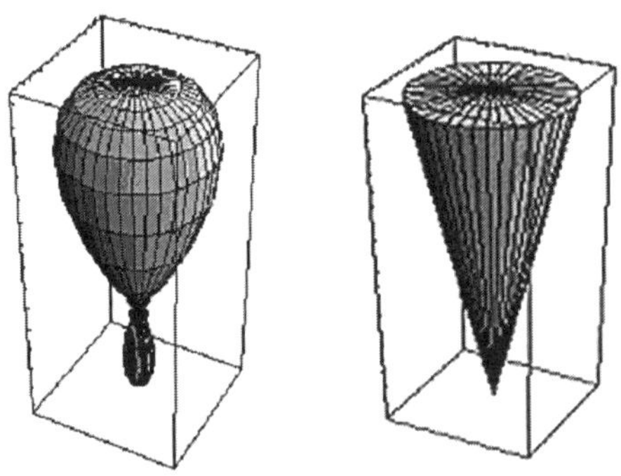

FIGURE 5.4 Antenna beam solid angle: The Ω_A corresponds to the area, as measured on a unit sphere, of the normalized power radiation pattern of a hypothetical antenna that radiates the same amount of energy as the antenna under study, but in a single cone with constant radiation intensity, (left) normalized power pattern of antenna being characterized, (right) normalized power pattern of hypothetical antenna being considered.

5.2.4 Antenna Directivity, Gain, and Equivalent Isotropically Radiated Power

The directivity of an antenna, $D(\hat{\mathbf{k}})$, is a figure of merit that quantifies the antenna directive properties by comparing them with those of a hypothetical isotropic antenna that radiates the same total power as the antenna being characterized. Therefore, by definition, the radiation intensity of this hypothetical radiator equals U_{ave}, the average radiation intensity of the antenna being characterized. Antenna directivity $D(\hat{\mathbf{k}})$ is defined as the ratio of the antenna radiation intensity to the antenna average radiation intensity, which is also the radiation intensity of the previously mentioned hypothetical isotropic antenna, or

$$D(\hat{\mathbf{k}}) = \frac{U(\hat{\mathbf{k}})}{U_{ave}} = \frac{4\pi U(\hat{\mathbf{k}})}{P_R} \tag{5.30}$$

The quantity D refers to the maximum directivity over all radiation angles, or

$$D = D(\hat{\mathbf{k}})\Big|_{\max} = \frac{U_{\max}}{U_{ave}} \tag{5.31}$$

It is easily shown that $D = 4\pi/\Omega_A$.

The gain of the antenna is defined as

$$G(\hat{\mathbf{k}}) = \frac{4\pi U(\hat{\mathbf{k}})}{P_A} \tag{5.32}$$

Obviously, because $P_R = eP_A$, antenna gain and directivity are related as

$$G(\hat{\mathbf{k}}) = e\,D(\hat{\mathbf{k}}) \tag{5.33}$$

Finally, the maximum of the gain is often denoted G

$$G = G(\hat{\mathbf{k}})\Big|_{\max} = \frac{4\pi U_{\max}}{P_A} \tag{5.34}$$

The quantity $qG(\hat{\mathbf{k}})$ is sometimes termed realized antenna gain.

The equivalent isotropically radiated power (*EIRP*) is the total power that would be radiated by a hypothetical isotropic antenna with radiation intensity equal to the maximum radiation intensity of the antenna being characterized. Because the total radiated power equals 4π times the radiation intensity for an isotropic antenna, we have

$$EIRP = 4\pi U_{\max} \tag{5.35}$$

By using Eqs. (5.8), (5.34), and (5.35), the *EIRP* can also be expressed as

$$\begin{aligned} EIRP &= G\,P_A \\ &= e\,D\,P_A \\ &= D\,P_R \end{aligned} \tag{5.36}$$

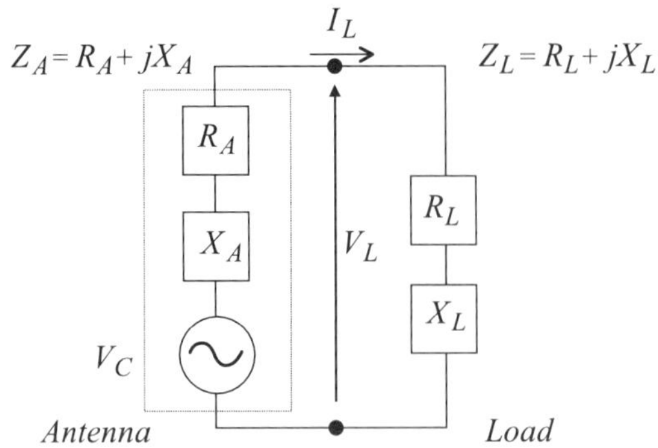

FIGURE 5.5 Receiving antenna and its equivalent circuit.

5.2.5 Receiving Antennas

This section focuses on antennas that operate in receiving mode. Assume that an antenna is illuminated by an electric field $\mathbf{E}_{inc}(\mathbf{r}) = \mathbf{E}^0_{inc}\, e^{-j\mathbf{k}_{inc}\cdot\mathbf{r}}$ (i.e., a plane wave with wave vector $\mathbf{k}_{inc} = k\hat{\mathbf{k}}_{inc}$ traveling in the $\hat{\mathbf{k}}_{inc}$ direction with field $\mathbf{E}^0_{inc}$ at the origin). The polarization vector of this wave is denoted $\hat{\mathbf{p}}_{inc} = \mathbf{E}^0_{inc}/|\mathbf{E}^0_{inc}|$. Assume that a load $Z_L = R_L + jX_L$ is connected to the antenna terminals. How much power can be deposited into R_L?

To answer this question, we consider the equivalent circuit of the receiving antenna, shown in Fig. 5.5. Excitation by a plane wave results in an equivalent voltage source $V_C(|\mathbf{E}^0_{inc}|, \hat{\mathbf{p}}_{inc}, \hat{\mathbf{k}}_{inc})$ appearing in series with the antenna impedance Z_A; as implied by its arguments, V_C depends on the magnitude, polarization, and direction of propagation of the incident field. Note that Z_A is independent of the mode of operation of the antenna; in other words, if this receiving antenna would be used in transmitting mode, it would present an impedance Z_A to the signal source. Of course, once $V_C(|\mathbf{E}^0_{inc}|, \hat{\mathbf{p}}_{inc}, \hat{\mathbf{k}}_{inc})$ is known, computing total received power is a trivial matter. Our discussion parallels that of the transmitting antenna. Under conjugate matching conditions (i.e., when $Z_A = Z_L^*$) the power P_C delivered to Z_L equals

$$P_C = \frac{\left|V_C\left(\left|\mathbf{E}^0_{inc}\right|, \hat{\mathbf{p}}_{inc}, \hat{\mathbf{k}}_{inc}\right)\right|^2}{8R_A} \tag{5.37}$$

P_C is referred to as the power available at the receiver. When conjugate matching conditions do not hold (i.e., when $Z_A \neq Z_L^*$), the power actually delivered to the load is smaller than P_C and expressed as

$$P_L = qP_C \tag{5.38}$$

where q is the antenna mismatch loss factor, given by

$$q = \frac{4R_A R_L}{\left|Z_L + Z_A\right|^2} \tag{5.39}$$

When either the load or the antenna impedance is purely real, i.e., when $X_L = 0$ or $X_A = 0$, q can also be expressed as

$$q = 1 - \left|\Gamma\right|^2 \tag{5.40}$$

where

$$\Gamma = \frac{Z_A - Z_L}{Z_A + Z_L} \tag{5.41}$$

Typically, the power P_C available at the receiver is not computed directly using Eq. (5.37), but indirectly, by using the concept of the "antenna effective aperture" (or area), $A_e(\hat{\mathbf{p}}_{inc}, \hat{\mathbf{k}}_{inc})$. The antenna effective aperture is defined such that P_C equals the power density $S_{inc} = |\mathbf{E}^0_{inc}|^2/(2\eta)$ of the incident wave, multiplied by $A_e(\hat{\mathbf{p}}_{inc}, \hat{\mathbf{k}}_{inc})$:

$$P_C = A_e\left(\hat{\mathbf{p}}_{inc}, \hat{\mathbf{k}}_{inc}\right) S_{inc} \tag{5.42}$$

From P_C we can compute P_L using Eq. (5.38) and $V_C(|\mathbf{E}^0_{inc}|, \hat{\mathbf{p}}_{inc}, \hat{\mathbf{k}}_{inc})$ using Eq. (5.37), if so desired. Equation (5.42) has a simple physical interpretation: when a plane with polarization $\hat{\mathbf{p}}_{inc}$ travels in the $\hat{\mathbf{k}}_{inc}$ direction and impinges on a conjugate matched receiving antenna, the latter "intercepts and captures" all power carried by the incident wave over an area of $A_e(\hat{\mathbf{p}}_{inc}, \hat{\mathbf{k}}_{inc})$ square meters. From the reciprocity theorem, it follows that $A_e(\hat{\mathbf{p}}_{inc}, \hat{\mathbf{k}}_{inc})$ can be expressed as

$$A_e\left(\hat{\mathbf{p}}_{inc}, \hat{\mathbf{k}}_{inc}\right) = \frac{\lambda^2}{4\pi} G\left(-\hat{\mathbf{k}}_{inc}\right) \left|\hat{\mathbf{p}}\left(-\hat{\mathbf{k}}_{inc}\right) \cdot \hat{\mathbf{p}}_{inc}\right|^2 \tag{5.43}$$

where $G(-\hat{\mathbf{k}}_{inc})$ and $\hat{\mathbf{p}}(-\hat{\mathbf{k}}_{inc})$ are the gain and polarization vector of the receiving antenna when operated in transmitting mode, looking into the direction of arrival of the incoming wave. In other words, the antenna effective aperture for a given direction of arrival and polarization of the impinging plane wave is proportional to

- The gain of the receiving antenna in the direction of arrival of the plane wave — When an antenna is an ineffective transmitter in a given direction, it is also an ineffective receiver for waves impinging on the antenna from that direction.
- The square of the norm of the dot product of the state of polarization of the receiving antenna and the polarization vector of the incident wave — When an antenna cannot radiate a certain polarization in a given direction, it also cannot receive that polarization from that direction.
- The wavelength squared of the incident signal — Note that the λ^2 factor gives $A_e(\hat{\mathbf{p}}_{inc}, \hat{\mathbf{k}}_{inc})$ its desired dimensions of square meters.

The maximum of $A_e(\hat{\mathbf{p}}_{inc}, \hat{\mathbf{k}}_{inc})$ overall directions and incident polarizations is referred to as the maximum effective aperture and denoted $A_{e,\max}$:

$$A_{e,\max} = \frac{\lambda^2}{4\pi} G \tag{5.44}$$

Another often-quoted measure is that of antenna effective height. Consider a receiving antenna that is open circuited, i.e., not connected to any load. The voltage at its terminals is $V_c(|\mathbf{E}^0_{inc}|, \hat{\mathbf{p}}_{inc}, \hat{\mathbf{k}}_{inc})$. This voltage equals the dot product of the vector antenna effective height and the incident field vector $\mathbf{E}^0_{inc}$

$$V_C\left(\left|\mathbf{E}^0_{inc}\right|, \hat{\mathbf{p}}_{inc}, \hat{\mathbf{k}}_{inc}\right) = \mathbf{h}\left(\left|\mathbf{E}^0_{inc}\right|, \hat{\mathbf{p}}_{inc}, \hat{\mathbf{k}}_{inc}\right) \cdot \mathbf{E}^0_{inc} \tag{5.45}$$

The maximum of $|\mathbf{h}(|\mathbf{E}^0_{inc}|, \hat{\mathbf{p}}_{inc}, \hat{\mathbf{k}}_{inc})|$ over all angles of incidence and polarizations is the scalar effective height h. It can be shown that

$$A_{e,\max} = \frac{h^2 \eta}{4 R_A} \tag{5.46}$$

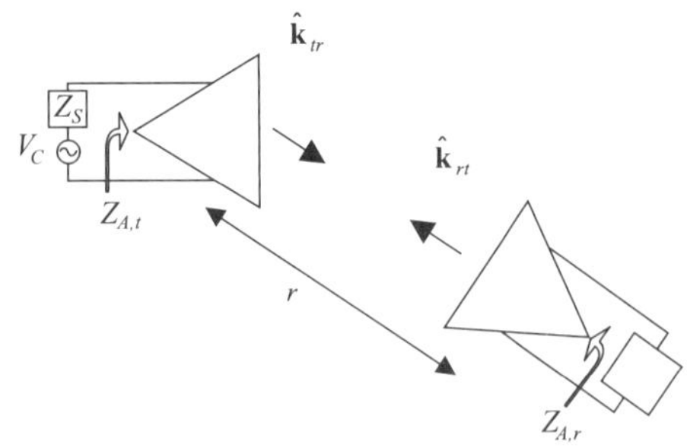

FIGURE 5.6 Antenna link.

5.2.6 Antenna Links

In this section, the terminology developed in the previous sections is used to describe a complete antenna link. Consider the two-antenna system depicted in Fig. 5.6; these two antennas are located in each other's far field. The transmitting antenna is fed by a voltage source with voltage V_S and internal impedance Z_S. The impedance of the transmitting antenna is denoted by $Z_{A,t}$. (Subscripts "t" and "r" stand for "transmitting" and "receiving" and are added to symbols whenever a notational conflict arises). As seen from the transmitting antenna, the receiving antenna is located in direction $\hat{\mathbf{k}}_{tr}$; similarly, from the perspective of the receiving antenna, the transmitting antenna resides in direction $\hat{\mathbf{k}}_{rt} = -\hat{\mathbf{k}}_{tr}$. The receiving antenna has input impedance $Z_{A,r}$ and is terminated by a load Z_L. The transmitting and receiving antennas are separated by a distance r, hence, the vector connecting the terminals of the transmitting and receiving antennas is $\mathbf{r}_{tr} = r\hat{\mathbf{k}}_{tr}$. How much power is deposited into Z_L?

The power developed in Z_L equals

$$P_L = q_r P_C \tag{5.47}$$

where q_r is the antenna mismatch loss factor of the receiving antenna, and P_C is

$$P_C = A_e\left(\hat{\mathbf{p}}_t\left(\hat{\mathbf{k}}_{tr}\right), \hat{\mathbf{k}}_{tr}\right) S_{inc,tr} \tag{5.48}$$

where $A_e(\hat{\mathbf{p}}_t(\hat{\mathbf{k}}_{tr}), \hat{\mathbf{k}}_{tr})$ is the effective aperture of the receiving antenna for waves of polarization $\hat{\mathbf{p}}_t(\hat{\mathbf{k}}_{tr})$ traveling in the $\hat{\mathbf{k}}_{tr}$ direction, and $S_{inc,tr}$ is the power density of the field radiated by the transmitter, measured at the receiver. From Eq. (5.16) we have

$$\begin{aligned} S_{inc,tr} &= \left|\mathbf{S}_t\left(\mathbf{r}_{tr}\right)\right| \\ &= \frac{1}{2\eta r_{tr}^2}\left[\left|FF_{\phi,t}\left(\hat{\mathbf{k}}_{tr}\right)\right|^2 + \left|FF_{\theta,t}\left(\hat{\mathbf{k}}_{tr}\right)\right|^2\right] \\ &= \frac{1}{r_{tr}^2} U_t\left(\hat{\mathbf{k}}_{tr}\right) \end{aligned} \tag{5.49}$$

However, by using Eq. (5.32), $U_t(\hat{\mathbf{k}}_{tr})$ can also be expressed as

$$U_t\left(\hat{\mathbf{k}}_{tr}\right) = \frac{G_t\left(\hat{\mathbf{k}}_{tr}\right) P_A}{4\pi} \tag{5.50}$$

Combining Eq. (5.49) and (5.50) yields

$$S_{inc,tr} = \frac{G_t(\hat{\mathbf{k}}_{tr})P_A}{4\pi r_{tr}^2} \tag{5.51}$$

This equation implies that the power density of the field radiated by the transmitter, measured at the receiver, equals the power density that would result from an equally powerful isotropic transmitter, multiplied by the gain of the transmitting antenna in the direction of the receiving antenna.

Use of Eq. (5.43) in conjunction with Eq. (5.48) to (5.51) yields

$$P_C = \frac{\lambda^2 G_r(\hat{\mathbf{k}}_{rt})\left|\hat{\mathbf{p}}_r(\hat{\mathbf{k}}_{rt})\cdot\hat{\mathbf{p}}_t(\hat{\mathbf{k}}_{tr})\right|^2 G_t(\hat{\mathbf{k}}_{rt})P_A}{(4\pi r_{tr})^2} \tag{5.52}$$

Equation (5.52) is the celebrated Friis formula for characterizing antenna transmission links. Equivalent forms of this equation result from substituting $P_L = q_r P_C$ and $P_A = q_t P_S$, as well as $G_r(\hat{\mathbf{k}}_{rt}) = e_r D_r(\hat{\mathbf{k}}_{rt})$ and $G_t(\hat{\mathbf{k}}_{tr}) = e_t D_t(\hat{\mathbf{k}}_{tr})$ as follows:

$$P_L = \frac{\lambda^2 q_r G_r(\hat{\mathbf{k}}_{rt})\left|\hat{\mathbf{p}}_r(\hat{\mathbf{k}}_{rt})\cdot\hat{\mathbf{p}}_t(\hat{\mathbf{k}}_{tr})\right|^2 G_t(\hat{\mathbf{k}}_{rt})q_t P_S}{(4\pi r_{tr})^2} \tag{5.53}$$

$$P_L = \frac{\lambda^2 q_r e_r D_r(\hat{\mathbf{k}}_{rt})\left|\hat{\mathbf{p}}_r(\hat{\mathbf{k}}_{rt})\cdot\hat{\mathbf{p}}_t(\hat{\mathbf{k}}_{tr})\right|^2 D_t(\hat{\mathbf{k}}_{rt})e_t q_t P_S}{(4\pi r_{tr})^2} \tag{5.54}$$

The factors $e_t q_t$ and $q_r e_r$ are sometimes referred to as the total loss factors of the transmitting and receiving antennas, respectively. The factors $|\hat{\mathbf{p}}_r(\hat{\mathbf{k}}_{rt})\cdot\hat{\mathbf{p}}_t(\hat{\mathbf{k}}_{tr})|^2$ and $[\lambda/(4\pi r)]^2$ are termed the polarization mismatch and space loss factors of the link. By using this terminology, Eq. (5.54) says that the received power equals the power available at the transmitter, multiplied by the free-space, transmitter, receiver, and polarization mismatch loss factors and multiplied by the directivity of both antennas along the link axis.

5.2.7 Antenna Noise Temperature and Signal to Noise Ratio

A crucial factor limiting the ability of an antenna to receive small signals is noise. Noise in an antenna is characterized by the antenna noise temperature. Knowledge of the antenna noise temperature allows for the computation of the SNR.

Principal noise sources important to antenna analysis include

- Man-made (power grid, appliances, etc.)
- Cosmic (sun, galactic, extragalactic, etc.)
- Absorption (water vapor)
- Atmospheric

To construct an equation for the noise power picked up by a receiving antenna, it is assumed that the antenna is lossless and is in thermodynamic equilibrium with its surroundings. Both restrictions will be relaxed later on. If the antenna is in thermodynamic equilibrium with its surroundings, then the total noise powers radiated and received by the antenna are equal. The load resistance acts as a source of noise power, provided that the antenna (and the load) are at nonzero temperature. In what follows, it is assumed that this load is matched to the antenna.

The amount of noise power available for radiation from the antenna load impedance equals

$$P_{S,n} = KT_0B \tag{5.55}$$

where $K = 1.3801\ 10^{-23}$ J/K (joule/Kelvin) is the Boltzmann constant, T_0 is the antenna physical temperature (approximately 300 K if the antenna resides on Earth and is not cooled), and B is the bandwidth of the system. Under matched conditions, the noise power that is radiated by the antenna is

$$P_{R,n} = KT_0B \tag{5.56}$$

On radiation, this power is not distributed evenly over all directions. The noise radiation intensity $U_{R,n}(\hat{\mathbf{k}})$ and the total radiated noise power are given by

$$\begin{aligned} U_{R,n}\left(\hat{\mathbf{k}}\right) &= KT_0B\frac{D\left(\hat{\mathbf{k}}\right)}{4\pi} \\ P_{R,n} &= \oiint U_{R,n}\left(\hat{\mathbf{k}}\right)d\Omega \end{aligned} \tag{5.57}$$

A blackbody radiator is an object that re-radiates all the energy it absorbs. Assume that the antenna surroundings comprise blackbody radiators that are in thermal equilibrium with the antenna. To maintain this equilibrium, the noise power radiated by the antenna through a given solid angle must equal the noise power received by the antenna through the same solid angle. Therefore, by denoting the noise radiation intensity and total noise power absorbed by the antenna by $U_{L,n}(\hat{\mathbf{k}})$ and $P_{L,n}$, we have

$$\begin{aligned} U_{L,n}\left(\hat{\mathbf{k}}\right) &= KT_0B\frac{D\left(\hat{\mathbf{k}}\right)}{4\pi} \\ P_{L,n} &= \oiint U_{L,n}\left(\hat{\mathbf{k}}\right)d\Omega \end{aligned} \tag{5.58}$$

In reality, the antenna surroundings comprise imperfect blackbody radiators that only re-radiate a fraction $\alpha(\hat{\mathbf{k}})$ of the energy they absorb; $\alpha(\hat{\mathbf{k}})$ is often referred to as the emmissivity of the imperfect blackbody radiator. Let $T(\hat{\mathbf{k}})$ denote the temperature of the imperfect blackbody radiator. In thermodynamic equilibrium $T_0 = \alpha(\hat{\mathbf{k}})T(\hat{\mathbf{k}})$ holds. If there is no thermal equilibrium, then $T_0 \neq \alpha(\hat{\mathbf{k}})T(\hat{\mathbf{k}})$; and, under those conditions, noise radiation intensities and power received by the antenna are

$$\begin{aligned} U_{L,n}\left(\hat{\mathbf{k}}\right) &= K\alpha\left(\hat{\mathbf{k}}\right)T\left(\hat{\mathbf{k}}\right)B\frac{D\left(\hat{\mathbf{k}}\right)}{4\pi} \\ P_{L,n} &= \oiint U_{L,n}\left(\hat{\mathbf{k}}\right)d\Omega \end{aligned} \tag{5.59}$$

If the antenna is lossy, the preceding derivation does not hold true. Indeed, not all noise power picked up by the antenna is delivered to the load. A fraction $\alpha_0 = 1 - e_r$ of the noise power picked up by the antenna is absorbed by the antenna and then transferred to the load at its own temperature; the remaining fraction e_r makes it into the load unaffected as follows:

$$P_{L,n} = (1 - e_r) K T_0 B + \int e_r K\alpha(\hat{\mathbf{k}}) T(\hat{\mathbf{k}}) B \frac{D(\hat{\mathbf{k}})}{4\pi} d\Omega$$
$$= K\alpha_0 T_0 B + \int e_r K\alpha(\hat{\mathbf{k}}) T(\hat{\mathbf{k}}) B \frac{D(\hat{\mathbf{k}})}{4\pi} d\Omega \qquad (5.60)$$

Because $G(\hat{\mathbf{k}}) = e_r D(\hat{\mathbf{k}})$, this equation can also be cast into the following form:

$$P_{L,n} = K\alpha_0 T_0 B + \int K\alpha(\hat{\mathbf{k}}) T(\hat{\mathbf{k}}) B \frac{G(\hat{\mathbf{k}})}{4\pi} d\Omega \qquad (5.61)$$

The antenna noise temperature T_a is defined as

$$T_a = \alpha_0 T_0 + \int \alpha(\hat{\mathbf{k}}) T(\hat{\mathbf{k}}) \frac{G(\hat{\mathbf{k}})}{4\pi} d\Omega \qquad (5.62)$$

To effectively use this equation, knowledge of temperatures of different noise sources is required. Figure 5.7 depicts the variation of noise temperatures of sources ranging from human-made to galactic to atmospheric vs. frequency. By using this definition, the total noise power picked up by the receiving antenna is

$$P_{L,n} = K T_a B \qquad (5.63)$$

Finally. the SNR of an antenna is given by

$$S/N = \frac{P_L}{P_{L,n}} = \frac{P_L}{K T_a B} \qquad (5.64)$$

In practice, noise sources in the receiver also contribute to the SNR; to this end, T_a in Eq. (5.64) is replaced by $T_s = T_a + (F_s - 1)\, T_o$, where F_s and T_o are the system noise figure and ambient temperature, respectively.

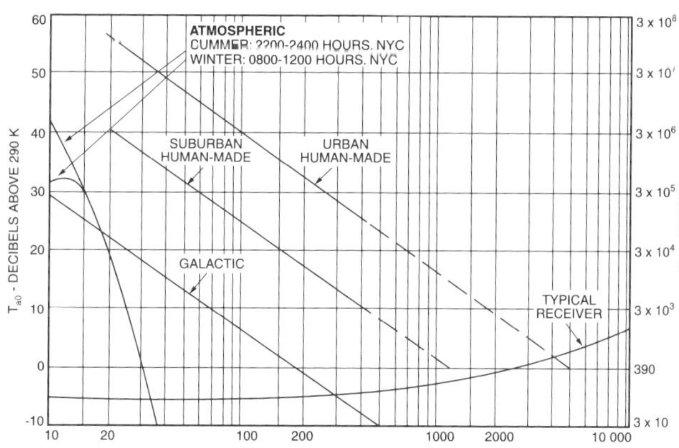

FIGURE 5.7 Median values of antenna noise temperature for an isotropic antenna near the Earth's surface. (From Lo, Y. T. and Lee, S. W. *Antenna Handbook*, Van Nostrand Reinhold, 1988. With permission.)

The preceding formulas indicate that to increase the SNR figure, one should try minimize the antenna noise temperature. Assume that the receiving antenna points toward the sky. Given that the temperature of an ambient sky can be as low as a few Kelvin (this is, of course, not true if the antenna points toward the sun) and that the temperature of an uncooled antenna on Earth is around 300 K, one should

- Minimize antenna losses
- Minimize sidelobes that point toward Earth
- Cool the antenna/receiver

By using the equations developed in the previous section for the power absorbed in the load, P_L, Eq. (5.64) can also be expressed as

$$S/N = \frac{\lambda^2 G_r(\hat{\mathbf{k}}_{rt}) \left|\hat{\mathbf{p}}_r(\hat{\mathbf{k}}_{rt}) \cdot \hat{\mathbf{p}}_t(\hat{\mathbf{k}}_{rt})\right|^2 G_t(\hat{\mathbf{k}}_{rt}) P_A}{KT_s B (4\pi r_{tr})^2} \tag{5.65}$$

The normalized signal to noise figure S/N_0 refers to the SNR assuming a unit bandwidth signal:

$$S/N_0 = \frac{\lambda^2 G_r(\hat{\mathbf{k}}_{rt}) \left|\hat{\mathbf{p}}_r(\hat{\mathbf{k}}_{rt}) \cdot \hat{\mathbf{p}}_t(\hat{\mathbf{k}}_{rt})\right|^2 G_t(\hat{\mathbf{k}}_{rt}) P_A}{KT_s (4\pi r_{tr})^2} \tag{5.66}$$

Provided that the transmitting antenna points its principal beam toward the receiver, $EIRP = G_t(\hat{\mathbf{k}}_{tr})P_A$; hence, Eq. (5.66) can also be expressed as

$$S/N_0 = \frac{\lambda^2}{K(4\pi r_{tr})^2} \left|\hat{\mathbf{p}}_r(\hat{\mathbf{k}}_{rt}) \cdot \hat{\mathbf{p}}_t(\hat{\mathbf{k}}_{tr})\right|^2 \frac{G_r(\hat{\mathbf{k}}_{rt})}{T_a} EIRP \tag{5.67}$$

5.2.8 Source-Field Relationships

To characterize a given antenna, the functions $FF_\phi(\hat{\mathbf{k}})$ and $FF_\theta(\hat{\mathbf{k}})$ are either measured or computed from knowledge of the electric currents $\mathbf{J}(\mathbf{r})$ that reside on the antenna and are known. These currents are often computed themselves (see Section 5.4). For certain simple antennas (linear antennas and loops; see Section 5.3), they can be estimated using transmission line arguments.

By assuming that the electric currents $\mathbf{J}(\mathbf{r})$ that reside on the antenna are known, the radiation integral $\mathbf{RI}(\hat{\mathbf{k}})$ is defined as

$$\mathbf{RI}(\hat{\mathbf{k}}) = \frac{1}{4\pi} \iiint \mathbf{J}(\mathbf{r}) e^{jk\mathbf{r}\cdot\hat{\mathbf{k}}} d\mathbf{r} \tag{5.68}$$

Aside from a constant, the transverse components of $\mathbf{RI}(\hat{\mathbf{k}})$ yield the $FF_\phi(\hat{\mathbf{k}})$ and $FF_\theta(\hat{\mathbf{k}})$ functions:

$$\begin{aligned} FF_\phi(\hat{\mathbf{k}}) &= -j\omega\mu_o \, \mathbf{RI}(\hat{\mathbf{k}}) \cdot \hat{\phi} \\ FF_\theta(\hat{\mathbf{k}}) &= -j\omega\mu_o \, \mathbf{RI}(\hat{\mathbf{k}}) \cdot \hat{\theta} \end{aligned} \tag{5.69}$$

Equations (5.68) and (5.69) are referred to as source-field relationships.

5.3 Commonly Used Antenna Types

This section briefly reviews four types of antennas commonly used in wireless links: linear antennas (dipoles and monopoles), loops, planar antennas, and dielectric resonators. Other types are reviewed elsewhere in this volume.

5.3.1 Linear Wire Antennas

Linear wire antennas are built from solid or tubular wire and are very popular in many applications because of their ease of construction and useful properties. The basic properties of dipole and monopole antennas are reviewed in this section.

5.3.1.1 Dipoles

A short dipole is a center-fed wire of length $L \ll \lambda$ that is assumed z directed and centered at the origin and that supports a current density $\mathbf{J}(\mathbf{r}) = \mathbf{I}_{sd}(z)\delta(x)\delta(y)$ with $\mathbf{I}_{sd}(z)$ given by

$$\begin{aligned} \mathbf{I}_{sd}(z) &= I_0\left(1-\frac{2|z|}{L}\right)\hat{\mathbf{z}} \qquad |z|<\frac{L}{2} \\ &= \mathbf{0} \qquad\qquad\qquad\quad \text{elsewhere} \end{aligned} \tag{5.70}$$

Clearly, the current on the wire is triangularly shaped and peaks at $I_0 = I_A$ (Fig. 5.8.a). The preceding model assumes that the current flows on the wire axis although in reality it flows on its outer skin; this model, however, suffices to determine radiation resistances and patterns. Other quantities (e.g., the ohmic

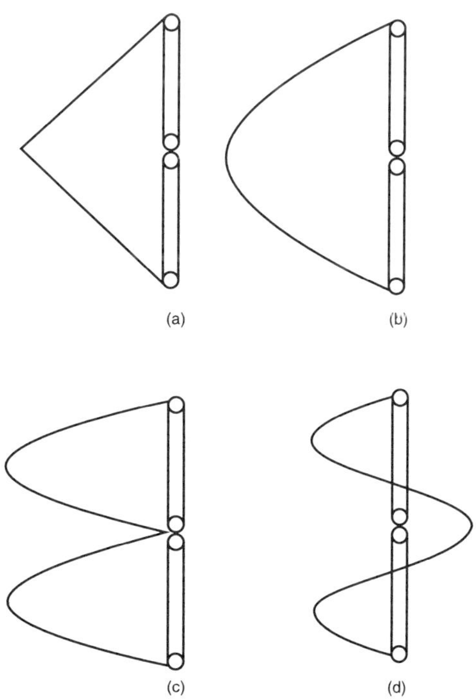

FIGURE 5.8 Current on finite length dipole (a) $L = 0.1\ \lambda$, (b) $L = 0.5\ \lambda$, (c) $L = 1.0\ \lambda$, (d) $L = 1.5\ \lambda$. The current for case (a) approaches the triangular shape specified by Eq. (5.70).

resistance and input reactance of the short dipole require knowledge of the wire radius r_w). The radiation integral of the short dipole is

$$\mathbf{RI}(\hat{\mathbf{k}}) = \frac{1}{4\pi}\int_{-L/2}^{L/2} I_o\left(1-\frac{2|z|}{L}\right)\hat{\mathbf{z}}e^{jk\hat{\mathbf{k}}}dz \\ \cong \frac{1}{8\pi}I_o L\hat{\mathbf{z}} \tag{5.71}$$

It was assumed that the wire was short enough to approximate the exponential in the integrand by one. It follows from Eqs. (5.69) and (5.71) that $FF_\phi(\hat{\mathbf{k}}) = 0$ and that

$$FF_\theta(\hat{\mathbf{k}}) = \frac{jk\eta I_o L\sin\theta}{8\pi} \tag{5.72}$$

The field of the short dipole antenna therefore is linearly polarized along the $\hat{\theta}$ direction. By using Eqs. (5.11), (5.17), (5.19), and (5.30) it follows that

$$U(\hat{\mathbf{k}}) = \frac{k^2\eta}{128\pi^2}(I_o L)^2\sin^2\theta \tag{5.73}$$

$$P_R = \frac{k^2\eta(I_o L)^2}{48\pi} \tag{5.74}$$

$$R_R = 20\pi^2\left(\frac{L}{\lambda}\right)^2 \tag{5.75}$$

$$D(\hat{\mathbf{k}}) = \frac{3}{2}\sin^2\theta \tag{5.76}$$

To estimate the efficiency of a short dipole, P_D is computed as

$$P_D = \frac{1}{2}\int_{-\frac{L}{2}}^{\frac{L}{2}} \frac{R_s}{2\pi r_w} I_o^2\left(1-\frac{2|z|}{L}\right)^2 dz \\ = \frac{R_s}{12\pi r_w} I_o^2 L \tag{5.77}$$

In the preceding equation, R_s is the so-called "resistance per square" of the wire surface, and is given by $R_s = \sqrt{\omega\mu_o/2\sigma}$ where σ is the conductivity of the wire material. It is clear that in the preceding equation, $R_S/2\pi r_w$ represents the wire resistance per unit length. By using Eq. (5.11), the short dipole ohmic resistance equals

$$R_D = \frac{L}{2\pi a}\frac{R_S}{3} \tag{5.78}$$

For truly short dipole and typical wire conductivities, the ohmic resistance of a short dipole is often of the same order of magnitude as its radiation resistance. Short dipoles therefore often have low efficiencies. The preceding theory fails to predict the reactance of the short dipole; this reactance can be determined using variational methods[1,5,7] and approximately equals

$$X_A = \frac{-\eta\lambda}{\pi^2 L}\left(\ln\left(L/2r_w - 1\right)\right) \tag{5.79}$$

which shows that the short dipole is highly capacitive.

A finite length dipole is a z-directed wire of length L, not necessarily short compared with the wavelength, that supports a current density $\mathbf{J}(\mathbf{r}) = \mathbf{I}_{fl}(z)\delta(x)\delta(y)$ with

$$\mathbf{I}_{fl}(z) = I_o \sin\left[k\left(\frac{L}{2} - |z|\right)\right]\hat{\mathbf{z}} \tag{5.80}$$

This current is plotted in Fig. 5.8 for various L. In what follows, it is assumed that the dipole is center fed. Comments pertaining to the exact location of current flow proffered for the short dipole also apply here. The terminal current is $I_A = I_o \sin(kL/2)$. The radiation integral associated with this current distribution is given by

$$\begin{aligned}\mathbf{RI}(\hat{\mathbf{k}}) &= \frac{1}{4\pi}\int_{-L/2}^{L/2} I_o \sin\left[k\left(\frac{L}{2} - |z|\right)\right]\hat{\mathbf{z}}e^{jk\mathbf{r}\cdot\hat{\mathbf{k}}}dz \\ &= \frac{1}{4\pi} I_o \hat{\mathbf{z}}\frac{2}{k}\frac{\cos[(kL/2)\cos\theta] - \cos(kL/2)}{\sin^2\theta}\end{aligned} \tag{5.81}$$

It follows from Eqs. (5.69) that $FF_\phi(\hat{\mathbf{k}}) = 0$ and that

$$FF_\theta(\hat{\mathbf{k}}) = \frac{jk\eta}{4\pi} I_o \sin\theta \frac{2}{k}\frac{\cos[(kL/2)\cos\theta]\ \ \cos(kL/2)}{\sin^2\theta} \tag{5.82}$$

Therefore, the electric field resulting from a finite length dipole is linearly polarized along the $\hat{\theta}$ direction. The field pattern of the finite length dipole is plotted in Fig. 5.9 for various ratios of L/λ. By using Eq. (5.17), the radiation intensity of the finite length dipole is

$$U(\hat{\mathbf{k}}) = I_o^2 \frac{\eta}{8\pi^2}\sin^2\theta\left[\frac{\cos[(kL/2)\cos\theta] - \cos(kL/2)}{\sin^2\theta}\right]^2 \tag{5.83}$$

The total power radiated by the dipole is obtained by integrating the radiation density over a unit sphere. From an expression for the total radiated power, the radiation resistance is computed using Eq. (5.11). One obtains, following extensive mathematical manipulations, that[6]

FIGURE 5.9 Patterns of various finite length dipoles (vertically oriented).

$$R_R = \frac{\eta}{2\pi\sin^2(kL/2)}\Big\{\gamma + \ln(kL) - Ci(kL) + 0.5\sin(kL)\big[Si(2kL) - 2Si(kL)\big] + 0.5\cos(kL)\big[\gamma + \ln(kL/2) + Ci(2kL) - 2Ci(kL)\big]\Big\} \tag{5.84}$$

where $\gamma = 0.5772$ (Euler's constant), and $Ci(x)$ and $Si(x)$ are the cosine and sine integrals, defined as

$$Ci(x) = -\int_x^\infty \frac{\cos y}{y}dy \qquad Si(x) = \int_0^x \frac{\sin y}{y}dy \tag{5.85}$$

A procedure similar to that adopted to calculate the short dipole ohmic resistance leads to the following equation for the finite length dipole resistance:

$$R_R = \frac{R_s}{\pi r_w \sin^2(kL/2)}\left(\frac{L}{4} - \frac{\sin(kL)}{4k}\right) \tag{5.86}$$

This equation is not valid when L approaches an even number of half wavelengths. For resonant structures (e.g., the half-wavelength dipole discussed later) dipole ohmic resistance is often small compared with radiation resistance; hence, the antenna efficiency is high. The antenna reactance again cannot be deduced based on the preceding simplifying theory. By using variational arguments, it is found that[6]

$$X_A = \frac{\eta}{4\pi\sin^2(kL/2)}\left\{\begin{array}{l} 2Si(kL) + \cos(kL)\big[2Si(kL) - Si(2kL)\big] \\ \quad -\sin(kL)\big[2Ci(kL) - Ci(2kL) - Ci(2kr_w^2/L)\big] \end{array}\right\} \tag{5.87}$$

Figure 5.10 depicts input (radiation) resistances and reactances of dipoles vs. their length. The input resistance of the dipole is (quasi) independent of its radius. This is not true for its reactance. The thinner the wire is, the faster the variation of the reactance with dipole length (or, alternatively, for a fixed dipole length, with frequency). This implies that thick dipoles are more broadband than thin ones. In addition, although all dipoles have $X_A = 42.5\ \Omega$ when $L = 0.5\ \lambda$ — this again holds true quasi independent of the wire radius – the dipole actual resonance length ($X_A = 0$) does depend on the radius. Generally speaking, resonant dipoles are shorter than $L = 0.5\ \lambda$, and the amount of shortening required to achieve resonance increases with wire radius. Combining the preceding equations also permits the computation of the finite length dipole directivity.

The half-wavelength dipole is of great practical importance. For $L = \lambda/2$ the expressions for $FF_\theta(\hat{\mathbf{k}})$ and $U(\hat{\mathbf{k}})$ reduce to

$$FF_\theta(\hat{\mathbf{k}}) = \frac{j\eta}{2\pi} I_o \frac{\cos\left[\frac{\pi}{2}\cos\theta\right]}{\sin\theta} \qquad U(\hat{\mathbf{k}}) = I_o^2 \frac{\eta}{8\pi^2}\left[\frac{\cos\left[\frac{\pi}{2}\cos\theta\right]}{\sin\theta}\right]^2 \tag{5.88}$$

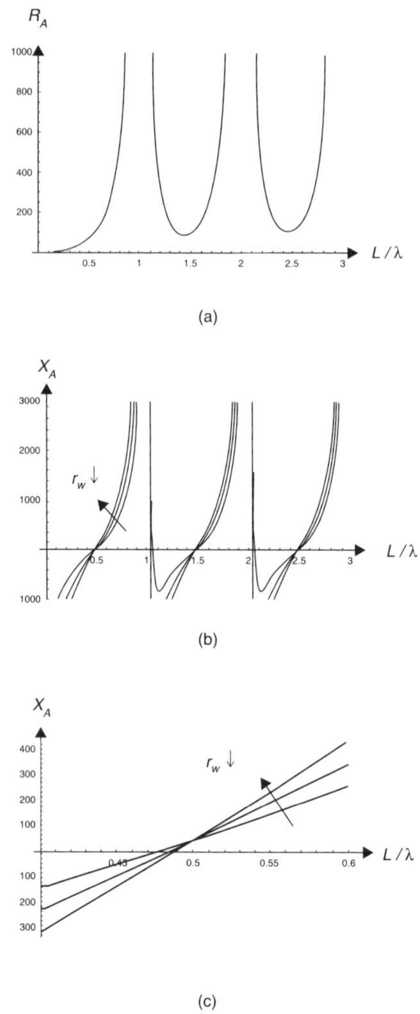

FIGURE 5.10 Dipole input impedances (a) input (radiation) resistance of dipole, (b) input reactance of dipole for three different wire radii, namely, $r_w = 0.001\ \lambda$, $0.0001\ \lambda$, and $0.00001\ \lambda$, (c) close-up of (b) near the first resonance.

The total radiated power can be obtained by simplifying Eq. (5.84) to

$$\begin{aligned} P_R &= I_o^2 \frac{\eta}{8\pi}\left\{\gamma + \ln(2\pi) - Ci(2\pi)\right\} \\ &= 2.435\, I_o^2 \frac{\eta}{8\pi} \end{aligned} \tag{5.89}$$

The directivity is obtained by combining Eqs. (5.88) and (5.89):

$$D(\hat{\mathbf{k}}) = \frac{4}{2.435}\left[\frac{\cos\left[\dfrac{\pi}{2}\cos\theta\right]}{\sin\theta}\right]^2 \tag{5.90}$$

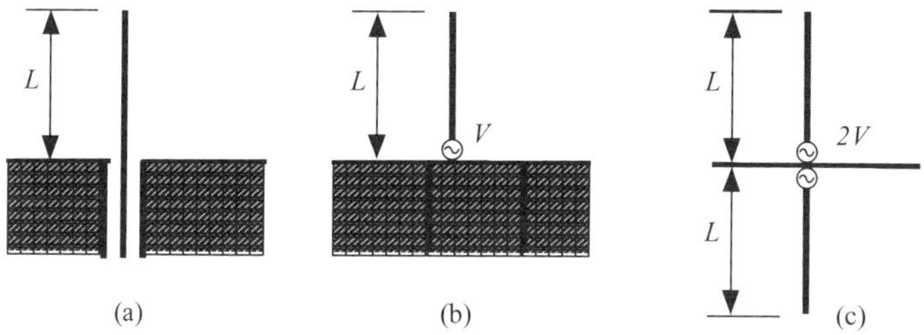

FIGURE 5.11 Monopole antennas. (a) Coax fed, (b) single half-space equivalent model, (c) image equivalent model.

The maximum directivity is attained for $\theta = \pi/2$ and equals $D = 1.64$. The radiation resistance of the half-wavelength dipole equals $R_R = 2P_R/I_o^2 = 73\ \Omega$. The input reactance of the half-wavelength dipole is $42.5\ \Omega$.

5.3.1.2 Monopoles

A monopole is a wire connected to a ground plane, and excited at (or close to) its base. Such an antenna is easily constructed by connecting the outer conductor of a coaxial line to a perfectly conducting ground plane and by using the extended inner conductor as the radiating element (Fig. 5.11a). The following discussion pertains to monopoles that reside on infinite, perfectly conducting ground planes. An equivalent model for the monopole antenna of length L is shown in Fig. 5.11b. By invoking image theory, it can be shown that in the half space where the monopole resides, its fields are identical to those of the structure shown in Fig. 5.11c: a dipole of length $2L$ excited by a source $2V$. Image theory also predicts that the current on the upper half of the wire in Fig. 5.11c is identical to that on the original monopole. The input impedance of the monopole can be computed as

$$\begin{aligned} Z_A^m &= \frac{V}{I} \\ &= \frac{1}{2}\frac{(2V)}{I} \\ &= \frac{1}{2} Z_A^d \end{aligned} \tag{5.91}$$

where superscripts m and d refer to monopole and dipole related quantities, respectively. Hence, the input impedance of a quarter-wavelength monopole equals half the input impedance of a half-wavelength dipole, or, ignoring losses,

$$\begin{aligned} Z_A &= \frac{1}{2}(73 + j\,42.5)\Omega \\ &= 36.5 + j\,21.25\ \Omega \end{aligned} \tag{5.92}$$

Because the fields radiated by the monopole and dipole of Fig. 5.11b and c are identical in for $z > 0$, the radiation intensities of both systems are also identical in the $z > 0$ half space. However, the total power radiated by the dipole is twice that of the monopole, because the dipole also radiates power in the $z < 0$ half plane, whereas the monopole does not; therefore,

$$D^m(\hat{\mathbf{k}}) = \frac{4\pi U^m(\hat{\mathbf{k}})}{P_R^m} = \frac{4\pi U^d(\hat{\mathbf{k}})}{\frac{1}{2} P_R^d} = 2\frac{4\pi U^d(\hat{\mathbf{k}})}{P_R^d} = 2\, D^d(\hat{\mathbf{k}}) \tag{5.93}$$

It is concluded that the directivity of a short monopole equals 3, and that the directivity of a quarter-wavelength monopole equals $2 \times 1.64 = 3.28$.

5.3.1.3 Practical Aspects

The preceding discussion reveals that monopole and dipole properties are highly related. In practice, monopole configurations are preferred over dipole arrangements because of the (relative) simplicity of monopole feed systems when compared with those for dipoles. Dipole feeds typically require a matched transition from an unbalanced (coax) to a balanced (twin wire/dipole) field configuration. Many so-called baluns for accomplishing this transition have been developed; some are discussed in Section 5.3.6.1.

Short dipoles are to be avoided whenever available real estate permits because of their undesirable impedance characteristics. As discussed earlier, (near) half-wavelength dipoles/quarter-wavelength monopoles can be tuned to resonate and their impedances are easily matched to 50 Ω feed lines. Their directivities are just above those of small dipoles. Longer dipoles/monopoles are seldom used because they are typically hard to match and/or have undesirable patterns. There is one important exception: the 5/8 λ monopole (or 1.25 λ dipole). This antenna has a broadside main beam and its sidelobes are small. The directivity of the 5/8 λ monopole is 6.56; that is virtually double that of the λ/4 monopole. Its input impedance is near 50 Ω after a matching coil is added to the base of the antenna; the effective length of this coil is around λ/8 (Fig. 5.12.a).[4] The 5/8 λ monopole has one more advantage over its λ/4 cousin: pattern stability when mounted on a finite ground plane. Typically, ground plane truncation effects cause the main beam of a λ/4 monopole to tilt upward, reducing the directivity of the antenna in the horizontal plane, often below that of a λ/2 dipole. This effect is less outspoken for a 5/8 λ monopole, which further accentuates its directive advantages.[4,8]

The 5/8 λ monopole is only one example of an antenna with lumped coils. Indeed, coils are often integrated into linear antennas, not only for matching purposes (as is the case for the 5/8 λ monopole) but also for enhancing their directive properties. Shown in Fig. 5.12b is a high-gain 3/4 λ whip loaded with an air-wound coil λ/4 from its base. The coil ensures that currents in the upper λ/2 element (and its image) are in phase with those in the center λ/2 element formed by the λ/4 wire extending from the base to the coil and its image. Similar principles drive the design of Franklin and coaxial collinear antennas (Figures 5.12c to d).

Another family of monopole enhancements, not entirely unrelated to the loaded variants discussed earlier, constitute sleeve antennas. Sleeve structures come in many variants. Sleeves are tubular conductors that typically are mounted on wire antennas to improve their bandwidth characteristics; the interior of the sleeve serves as the outer conductor of the coaxial antenna feed and its exterior and the extension of the coax inner conductor become the radiating elements. The most commonly used sleeve structure is shown in Fig. 5.13. The parameters that feature into the design of the sleeve antenna are the sleeve and monopole height d and H, the source and gap height L and Δ, and the monopole and sleeve radii a and b. Careful optimization of these parameters renders the sleeve monopole a wideband radiator (e.g., 4:1).[9,10] Sleeve dipole structures are also popular.[11]

One final note relating to the construction of linear wire antennas is mentioned. Although wires of circular cross section are used most often in the design of the preceding systems, other cross sections can also be used. For many typical cross sections, effective radii of equivalent systems with circular cross sections have been computed. Some of these data are summarized in Fig. 5.14.[6]

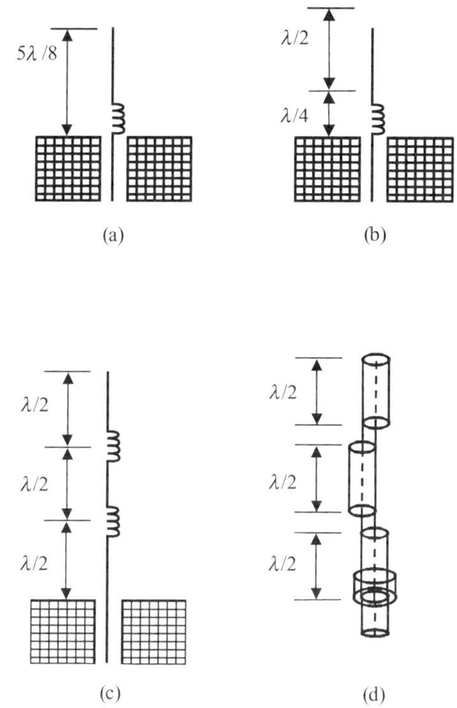

FIGURE 5.12 Various coil-loaded structures (a) typical 5/8 λ monopole, (b) high-gain whip, (c) Franklin antenna, (d) coaxial collinear antenna.

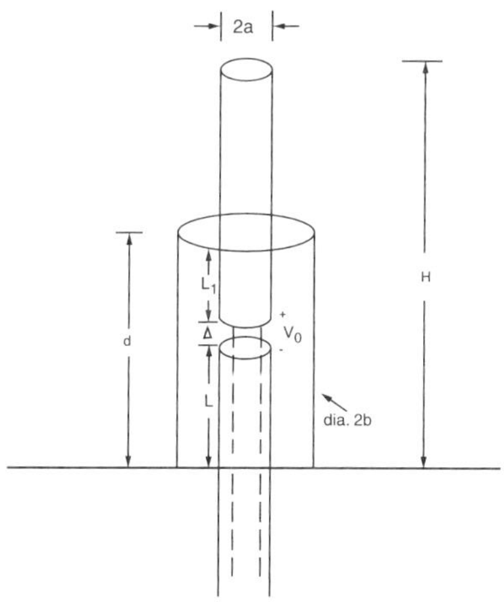

FIGURE 5.13 A common sleeve monopole structure. (From Bailey, M. C., *IEEE Trans. Antennas Propagat.*, 32, 4, 1984. ©IEEE. With permission.)

5.3.2 Loops

Loop antennas are, just like linear wire antennas, often used in practice. This section rigorously derives the properties of small circular loop antennas and discusses, in more general terms, those of large loops.

A loop antenna is considered small if its circumference is small compared with the wavelength. In what follows, it is assumed that the loop is of radius a, is centered at the origin, and that the plane of

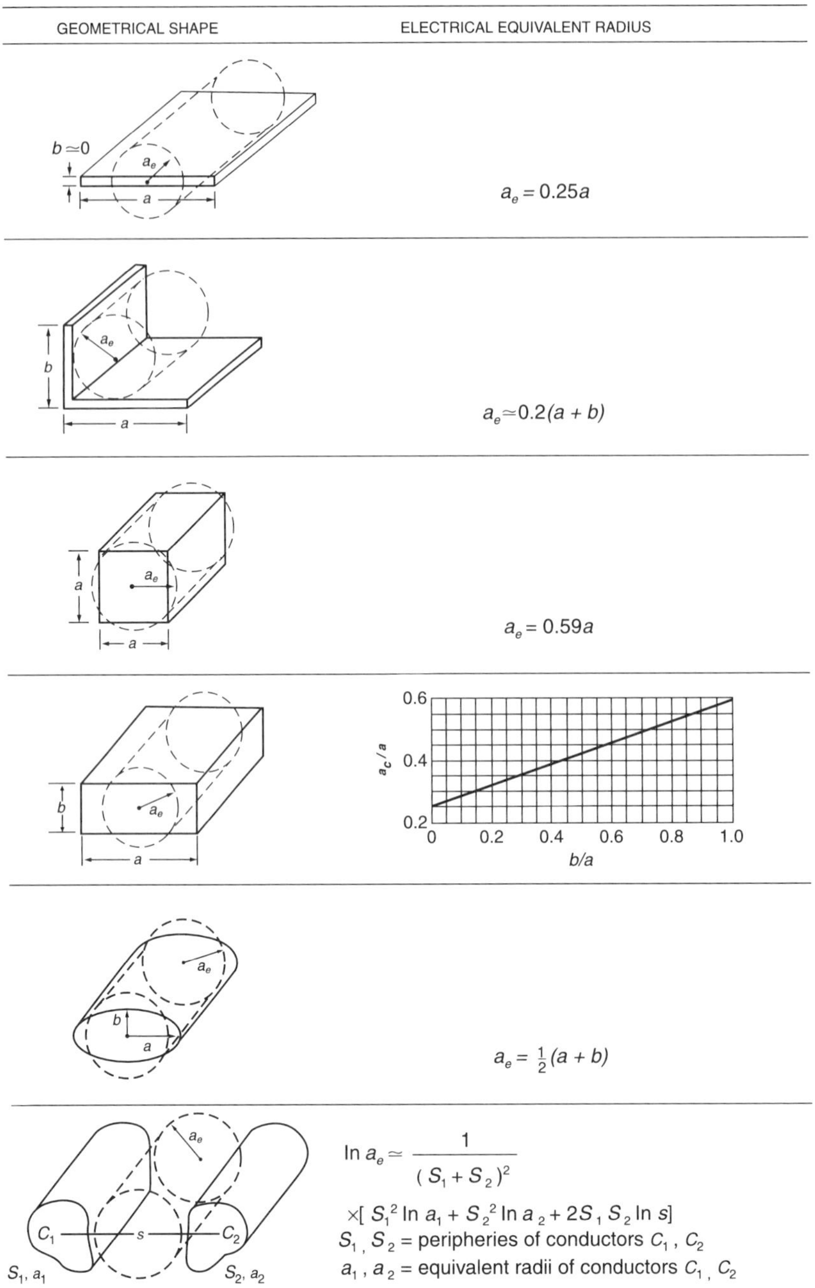

FIGURE 5.14 Effective radii for noncircular wire cross sections. (From Balanis, C., *Antenna Theory,* John Wiley & Sons, New York, 1982. With permission.)

the loop coincides with the *xy* plane (Fig. 5.15a). The current I_o is assumed constant and its flow is counterclockwise. The radiation integral of such a loop is

$$\mathbf{RI}\left(\hat{\mathbf{k}}\right) = \frac{1}{4\pi}\int_0^{2\pi} a I_o \hat{\phi}' e^{jka\left(\hat{\mathbf{x}}\cos\phi' + \hat{\mathbf{y}}\sin\phi'\right)\cdot\hat{\mathbf{k}}}\, d\phi' \tag{5.94}$$

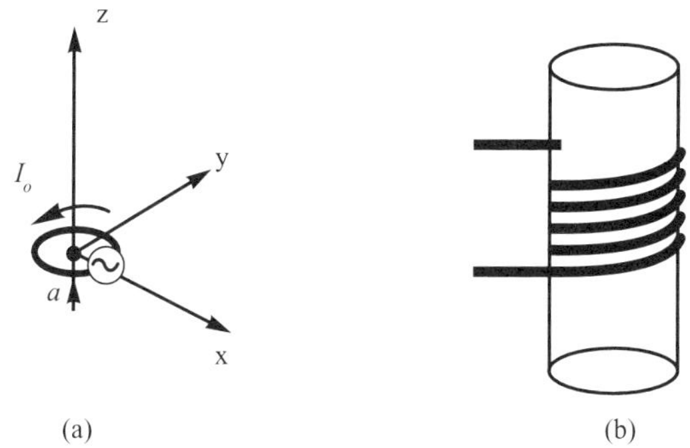

FIGURE 5.15 Small loop antenna, (a) geometry, (b) multiturn loop mounted on a ferrite core.

$$= \frac{1}{4\pi}\int_0^{2\pi} aI_o\hat{\phi}' e^{jka\sin\theta\cos(\phi'-\phi)} d\phi'$$

$$\cong \frac{1}{4\pi}\int_0^{2\pi} aI_o\hat{\phi}'\left(1 + jka\sin\theta\cos(\phi'-\phi)\right)d\phi'$$

$$= \frac{jka^2\sin\theta I_o}{4\pi}\int_0^{2\pi} \hat{\phi}'\cos(\phi'-\phi)d\phi'$$

$$= \frac{jkS_l\sin\theta I_o}{4\pi}\hat{\phi}$$

where $S_l = \pi a^2$ is the area of the loop. It immediately follows that $FF_\theta(\hat{\mathbf{k}})$ is zero, and that

$$FF_\phi(\hat{\mathbf{k}}) = \frac{\eta k^2 S_l \sin\theta I_o}{4\pi}\hat{\phi} \tag{5.95}$$

Small loops and dipoles are dual to one another: comparing Eqs. (5.72) and (5.95), it appears that, scale factors aside, the roles of $\mathbf{E}(\mathbf{r})$ and $\mathbf{H}(\mathbf{r})$ are reversed. It can be shown that the fields resulting from a small loop of electric current are identical to those of a magnetic Hertzian dipole.[6] By using Eqs. (5.11), (5.17), (5.19), and (5.30), the radiation intensity, total radiated power, and radiation resistance and directivity of the small loop are

$$U(\hat{\mathbf{k}}) = \frac{k^4\eta(S_l I_o)^2}{32\pi^2}\sin^2\theta \tag{5.96}$$

$$P_R = 10(k^2 S I_o)^2 \tag{5.97}$$

$$R_R \approx 31{,}170\left(\frac{S_l}{\lambda^2}\right)^2 \tag{5.98}$$

$$D(\hat{\mathbf{k}}) = \frac{3}{2}\sin^2\theta \tag{5.99}$$

Because all field components of the small loop antenna are proportional to sin θ, the directivity of the small loop antenna equals that of the short dipole. The ohmic resistance of the small loop antenna may be approximated as $R_D = a/r_w R_s$. Multiturn loops and loops loaded with a ferrite core (Fig. 5.15b) are often used to improve on the impedance characteristics of single turn small loops; for such loops,[1] the radiation resistance is given by

$$R_R \approx 31{,}200\left(n\mu_{core}\frac{C}{\lambda^2}\right)^2 \Omega \tag{5.100}$$

where n denotes the number of turns on the loop and μ_{core} is the effective relative permeability of the core. For an n-turn loop, $R_D = a/r_w R_s(R_p/R_0 + 1)$, where the ratio R_p/R_0 is a result of the proximity effect of various turns; for reasonable turn spacings, $0 \leq R_p/R_0 \leq 2$.[6] Most of the preceding equations are valid not only for circular loops but also for loops of other shapes.

Electrically large loops offer significant benefits over small loop antennas in terms of input impedance, directivity, and radiation pattern; in addition, the shape of the loop can be used to fine-tune its properties for certain applications. Unfortunately, the complete analysis of electrically large loop antennas is complex and best carried out using numerical techniques (see Section 5.4). The input impedance of circular loop antennas is plotted in Fig. 5.16 for varying loop sizes. For small loop circumferences, the loop behaves inductively. As the loop circumference approaches half a wavelength, the loop starts acting capacitively. Resonance occurs for loops constructed out of relatively thin wire when the loop circumference is just over a wavelength. The patterns of a small circular loop and a loop of circumference equal to 1 λ are plotted in Fig. 5.17. The pattern of the loop antenna changes dramatically as the loop radius is increased: for small loops the main antenna beam is in the plane of the loop, but the main beam tilts to the direction perpendicular to the loop plane for a circumference equal to 1 λ.

Loop shapes other than circular ones are often preferred because their mechanical construction is facilitated by the presence of long straight wire segments (Fig. 5.18). A circular loop typically has higher directivity than a triangular or rectangular one. In practice, loops with a circumference of half a wavelength are not often used because they exhibit very large reactances. Nonetheless, when space is at a premium, one sometimes does not have a choice. In such cases, the loop is loaded (inductively) to reduce its reactance. Alternatively, for base-fed square loop antennas, the loop is sometimes opened in the center of the side of the loop that opposes the source. This in essence reduces the loop to a half wavelength dipole antenna that is bent into a loop shape. Loops with a circumference of one wavelength are more often used though. One wavelength loop antennas typically come in four flavors, represented by the shapes shown in Figs. 5.18a, c to e. For square geometries, each side of the loop measures approximately a quarter wavelength and for triangular loops each side measures approximately a third of a wavelength. The input impedances of these configurations differ from one another, and a variety of matching networks has been devised for them. The half delta sloper shown in Fig. 5.18f is essentially half of a 1 λ triangular loop antenna; its properties can be deduced from those of the triangular loop antenna by image theory. In practice, if the antenna is mounted on a ground that does not conduct well, a buried ground wire is often added to the system.

5.3.3 Planar Antennas

Planar antennas possess a number of advantages, such as low-profile and durable form factors, light weight, and low cost, which make them ideal for implementation in wireless communication applications.

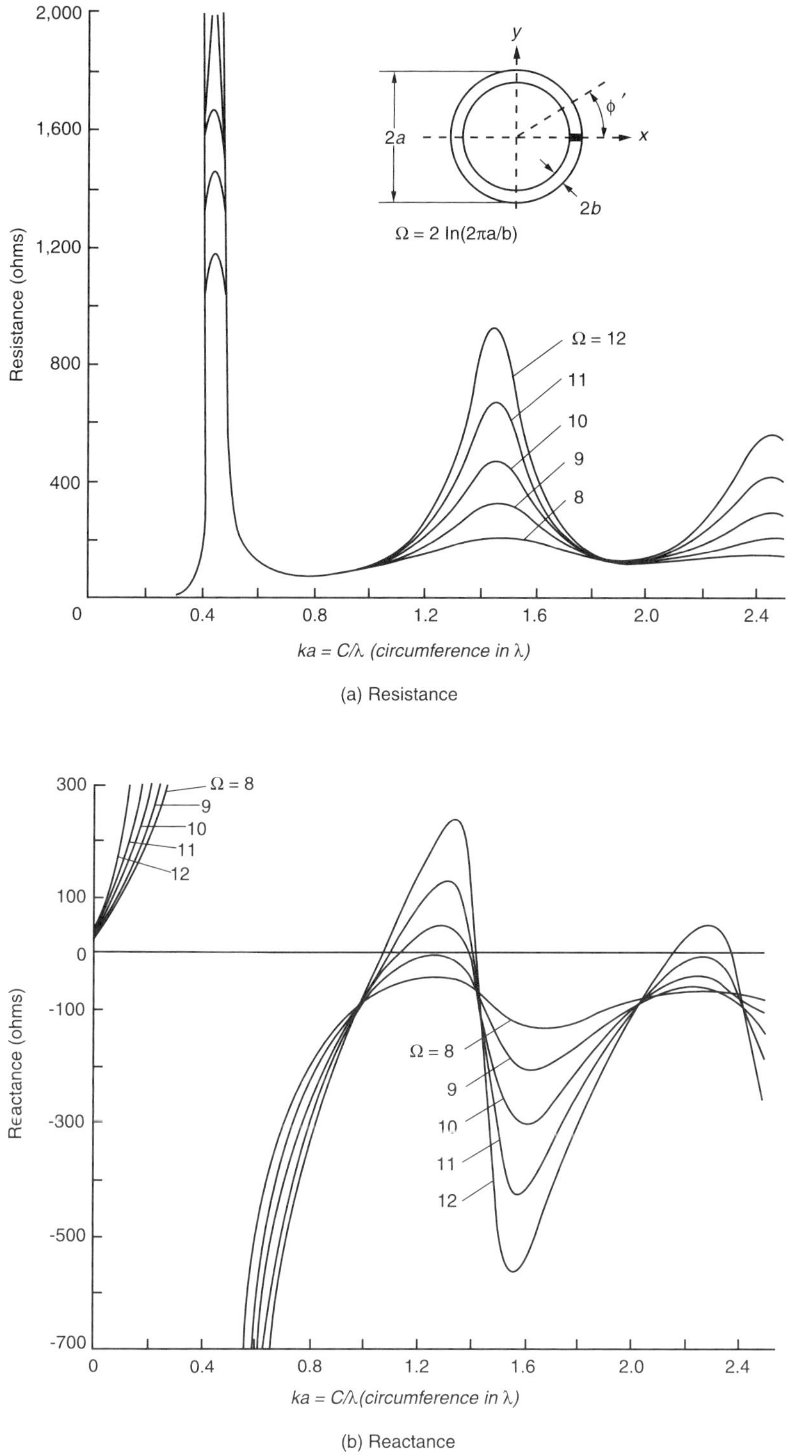

FIGURE 5.16 Input reactance of loop antennas; $\Omega = 2\ \ln(2\pi a/r_w)$ measures wire thickness. (From Stover, J. E., *AIEE Trans.*, 75, 1956. ©IEEE. With permission.)

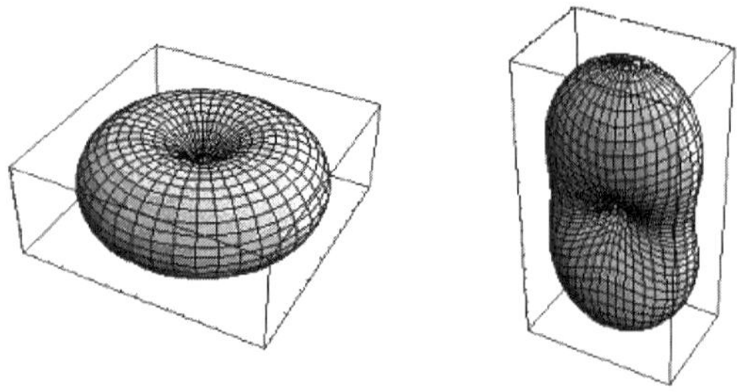

FIGURE 5.17 Radiation patterns of small loop and 1 λ circumference loop (source location, see Fig. 5.15.a).

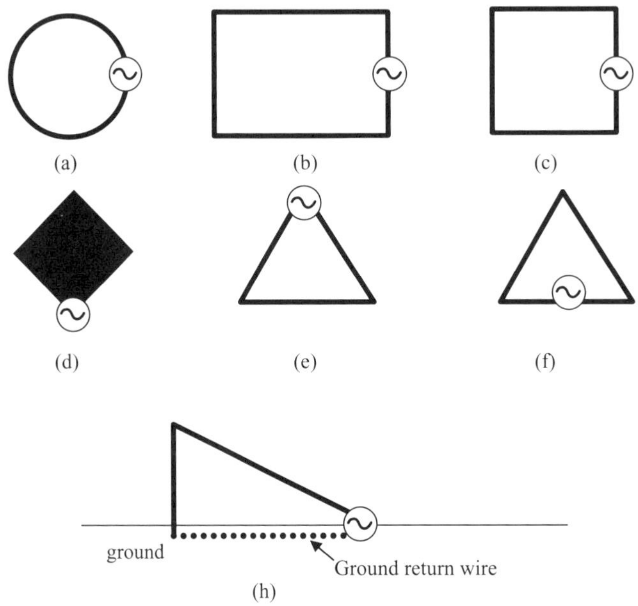

FIGURE 5.18 Various loop shapes (a) circular loop antenna, (b) rectangular loop antenna, (c) quad loop antenna, (d) bottom-fed quad, (e) top-fed triangular loop (delta loop antenna), (f) base-fed triangular loop, (g) half delta sloper.

This class of antennas includes microstrip antennas and planar inverted-F antennas (PIFAs). Microstrip antennas are discussed in detail in the next chapter. PIFAs have received much attention because their size makes them suitable for integration into portable communication devices. For instance, Ogawa et al. designed a PIFA suitable for shoulder mounting,[12] Virga and Rahmat-Samii[13] included a PIFA on a handset, and Ng et al. developed a PIFA for inclusion on a laptop computer chassis.[14] The following section discusses the basic structure of a PIFA antenna as well as some published design approaches that widen its bandwidth and improve its radiation characteristics in wireless communication scenarios.

A PIFA consists of a top plate element, a ground plane, a feed wire attached to the top element through a hole in the ground plane, and a shorting wire or strip that is directly connected between the top element and the ground plane. Related to both the linear inverted-F antenna[15] and the short-circuited microstrip antenna,[16] a typical design is shown in Fig. 5.19.[17] The design of the antenna involves selection of the top plate width, length, and height above the ground plane as well as the location and width of the feed wire and the shorting strip. For PIFA antennas on an infinite ground plane, the resonant frequency varies with the width of the shorting strip and the dimensional ratio of the top element.[13,17] On finite ground planes, these relationships are not as straightforward and design often requires the trial-and-error approach.[17] In general, the impedance bandwidth of the basic antenna is determined by the height of

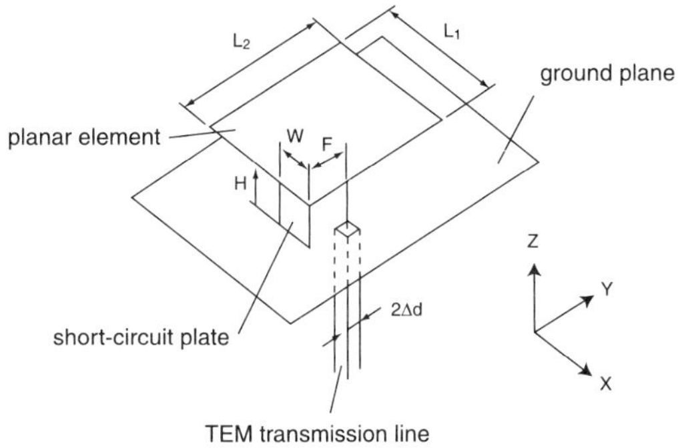

FIGURE 5.19 Structure of the planar inverted-F antenna. (From Hirasawa, K. and Haneishi, M., *Analysis, Design, and Measurement of Small Low Profile Antennas,* Artech House. Norwood, MA, 1992. With permission.)

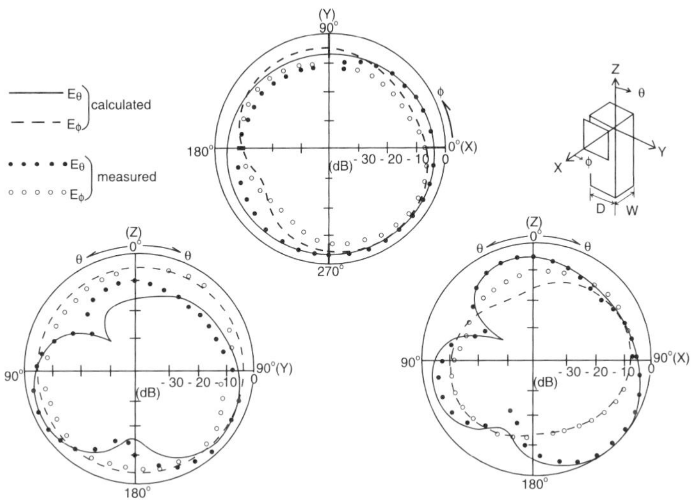

FIGURE 5.20 Radiation patterns of PIFA mounted on the side of the case (D = 0.17 λ, W = 0.11 λ, L = 0.55 λ, f = 920 MHz). (From Taga, T. and Tsunekawa, K., *IEEE J. Selected Area Commun.*, 5, 924, 1983. ©IEEE. With permission.)

the top element above the ground plane with some additional bandwidth resulting from surface currents on a noninfinite ground plane or conductive mounting structure.[17] Figure 5.20 illustrates the radiation patterns of a PIFA antenna mounted on the side of a radio housing.[18] These patterns demonstrate the PIFA's ability to send and receive both vertically and horizontally polarized fields, making it an appropriate choice for portable devices where the antenna orientation is not fixed.

Many approaches have been used to design PIFA smaller sizes and larger impedance bandwidths. For example, Wong and Yang[19] developed a modified PIFA with a total length of less than an 1/8 of a wavelength, a height less than 1/100 of a wavelength, and an impedance bandwidth ten times greater than that of a standard PIFA. The reduction in antenna length was achieved by meandering the radiating patch and the enhanced bandwidth despite the small antenna height resulted from use of a chip-resistor load in place of the shorting post. Rowell and Murch[20,21] achieved 7% impedance bandwidth at 900 MHz with a PIFA by loading the structure with a slot in the top plate and a capacitive load to ground. Several researchers (e.g., References 13 and 22) have used the positions of shorting pins and tuning capacitors or diodes to actively tune PIFAs for different frequency bands. Dual- and triple-band operation of PIFAs has also been achieved by including multiple resonating structures and feed points within the footprint of the lowest frequency PIFA.[20,23]

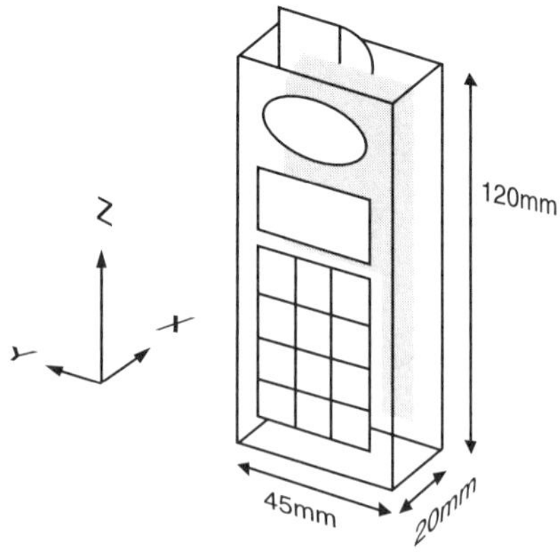

FIGURE 5.21 The 180° circular sector DRA design with a = 12 mm, d = 22 mm on a 100 × 40 mm² ground plane. (From Tam, M. T. K. and Murch, R. D., *IEEE Trans. Antennas Propag.*, 47, 841, 1999. ©IEEE. With permission.)

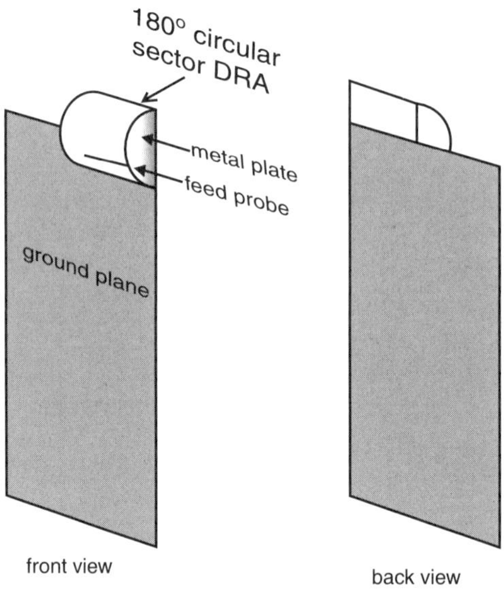

FIGURE 5.22 Illustration of the proposed circular sector DRA integrated into a mobile telephone handset. (From Tam, M. T. K. and Murch, R. D., *IEEE Trans. Antennas Propag.*, 47, 841, 1999. ©IEEE. With permission.)

5.3.4 Dielectric Resonators

The dielectric resonator antenna (DRA) has emerged in the last decade as alternative to wire and planar antennas for many wireless applications. As shown in Figs. 5.21 and 5.22, they can be easily integrated into portable communication devices.[24] The antenna is made of low-loss dielectric material; its resonant frequency, impedance bandwidth, and radiation pattern are functions of the dielectric structure shape, size, and permittivity. Because the dimensions of a DRA are on the order of a wavelength in the dielectric material, choosing a high value of permittivity can produce a physically small antenna. A DRA also has high radiation efficiency because of the absence of surface wave or conductor losses. The resonator permittivity and aspect ratio determine the impedance bandwidth.[25] Low-permittivity resonators have larger impedance bandwidths than those with higher permittivity, whereas selection of a high-permittivity resonator results in an antenna with a very compact size.[25] The ability of the resonator to support several modes of excitation produces a variety of radiation patterns for different coverage requirements. Patterns

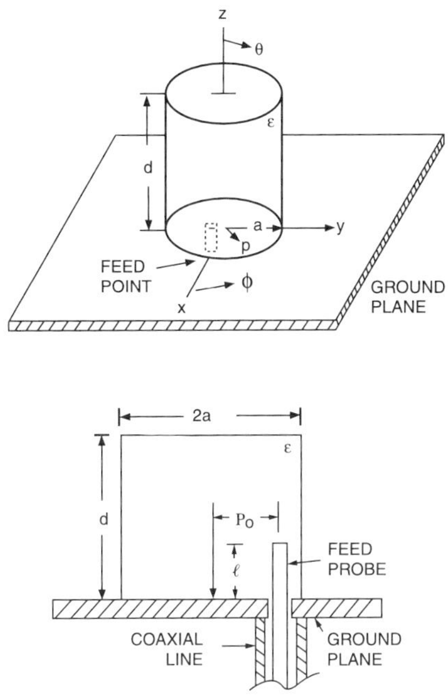

FIGURE 5.23 Cylindrical DRA geometry and feed configuration. (From Long, S. A., McAllister, M. W., and Shen, L. C., *IEEE Trans. Antennas Propag.*, 31, 407, 1983. ©IEEE. With permission.)

can range between broadside[26] to nearly omnidirectional,[27] but may be significantly affected by the size and shape of the ground plane on which they reside.[25] Mongia and Bhartia[28] provide a good overview of the fundamental work on DRAs, including equations for resonant frequency and impedance bandwidth for a number of common resonator configurations.

Dielectric resonator antennas of various shapes can offer options in frequency and radiation characteristics as well as packaging. For example, Long et al. used a cylindrical dielectric resonator with probe feeding as illustrated in Fig. 5.23.[29] The first-order approximation to the lowest resonant frequency of DRAs assumes that the surfaces of the dielectric shape, in this case, a cylinder, are perfect magnetic conductors.[29] Although this approximation neglects the effects of the probe feed, it does provide a relatively accurate first-order model of the antenna behavior.

The structure of a spherical dielectric resonator antenna supports an analytic solution for its modes of operation. Determination of the resonant frequency and impedance bandwidth of a spherical resonator requires solution of transcendental equations involving Bessel and Hankel functions of fractional orders.[28] Gastine et al.[30] have computed and plotted values of normalized wave number and radiation Q for a number of modes of the spherical DRA with permittivities between 10 and 100. Because of spherical symmetry, degenerate modes exist that may result in undesired or distorted radiation patterns. The modes of a more practical antenna configuration, the hemispheric dielectric resonator, can be derived on the basis of the isolated spherical resonator.[28] A related structure, the spherical cap dielectric resonator antenna,[31] is depicted in Fig. 5.24. This design provides a wider bandwidth than that of a hemispheric DRA and has a broadside radiation pattern.

The rectangular DRA, shown in Fig. 5.25, radiates like a short horizontal magnetic dipole in its first resonant mode.[25] A rectangular DRA offers an advantage over cylindrical and spherical dielectric resonators in that the resonant frequencies of different modes can be chosen to be different from each other by properly choosing the three dimensions of the resonator. Rectangular DRAs with relative permittivities less than ten have been implemented that provide 2:1 VSWR impedance bandwidths up to 10%.[25] Other resonator shapes that have been proposed include circular sectors,[24,26] annular sectors,[24] and equilateral triangles (Fig. 5.26).[3]

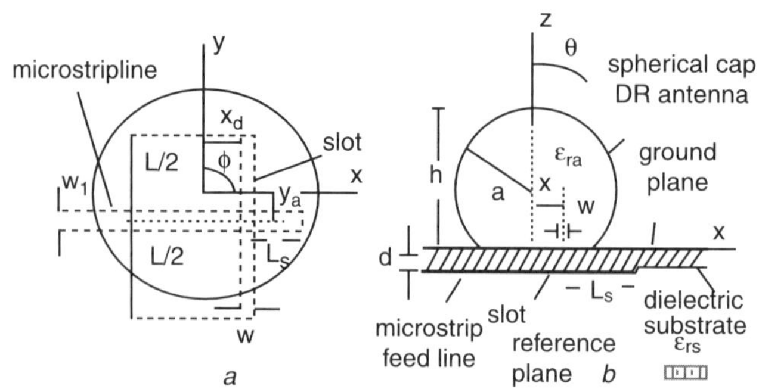

FIGURE 5.24 Geometry of aperture-coupled spherical cap dielectric resonator antenna. (From Leung, K. W. Luk, K. M., and Yung, E. K. N., *Electron Lett.*, 30, 1366, 1994. With permission.)

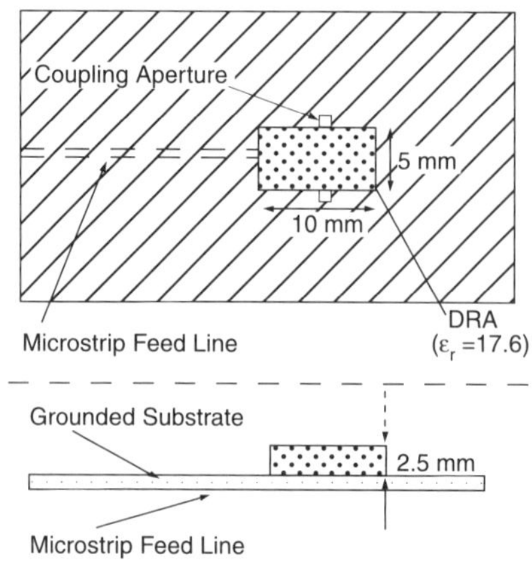

FIGURE 5.25 A basic rectangular DRA fed by a slot-coupled microstrip line. (From Petosa, A., Ittipiboon, A., Antar, Y. M. M., Roscoe, D., and Cuhaci, M., *IEEE Antennas Propag. Mag.*, 40, 35, 1998. ©IEEE. With permission.)

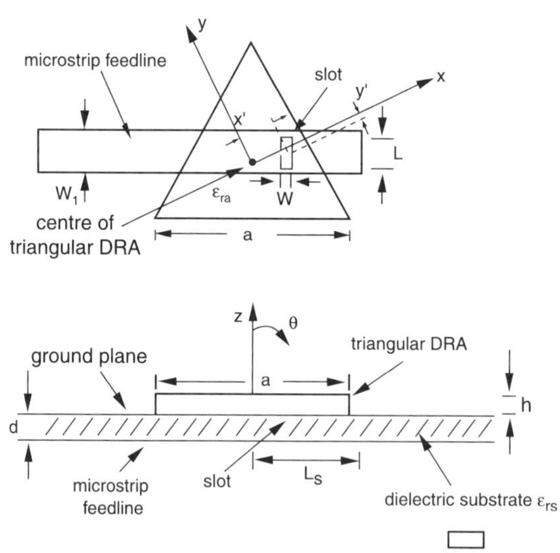

FIGURE 5.26 Configuration of aperture-coupled equilateral-triangular DRA. (From Lo, H. Y., Leung, K. W., Luk, K. M., and Yung, E. K. N., *Electron. Lett.*, 35, 2164, 1999. With permission.)

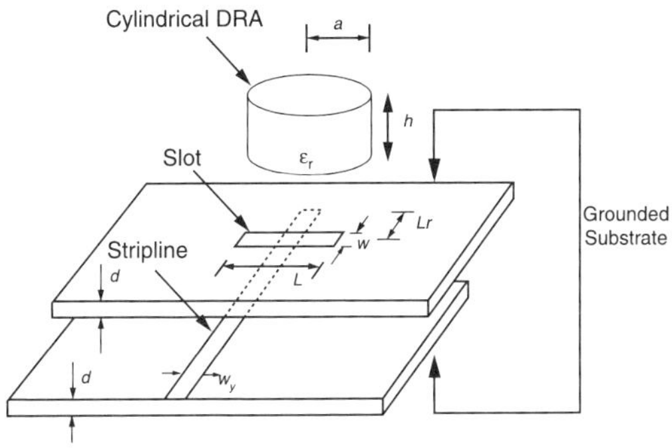

FIGURE 5.27 Configuration of the aperture-coupled cylindrical DRA using a strip-line feed. (From Leung, K. W., Poon, M. L., Wong, W. C., Luk, K. M., and Yung, E. K. N., *Microwave Optical Technol. Lett.*, 24, 120, 2000. With permission.)

Another advantage of DRAs is that they can be used with a myriad of feeding methods, including probes (Fig. 5.23), slots, microstrip lines (Fig. 5.24), striplines (Fig. 5.27),[33] dielectric image guides, and co-planar lines. This flexibility provides compatibility with a range of systems and packaging schemes.

5.3.5 Nonideal Design Factors

When an antenna is designed and implemented for wireless communication applications, nonideal design factors can change the antenna's operating characteristics, creating significant deviations from theoretical performance. The following sections describe several of these factors, including imperfect and finite ground planes and packaging effects.

5.3.5.1 Finite Ground Planes

In practical applications, antennas seldom have infinite ground planes. For instance, in the case of an antenna for a portable communication device, the ground plane is usually provided by the chassis ground of the device, which is seldom symmetrical or even completely planar.

In general, the radiation behavior of a vertically polarized antenna centered in a finite ground plane differs from that of the ideal case in that the pattern maximum is shifted upward away from the ground plane as shown in Fig. 5.28.[34] Radiation also occurs in the plane below the ground plane.[34] If the antenna is not centered on the finite ground plane, the gain of the antenna is usually increased in the direction of the larger portion of the ground plane, again with significant radiation in the plane beneath that of the ground. This behavior also emerges as a packaging effect for many kinds of antennas, because the chassis of many portable communication devices serve as asymmetrical ground planes for antennas.

Many researchers have examined the effects of finite ground plane size on microstrip antennas. These include References 35 through 45. Microstrip antennas with finite ground planes usually experience shifts

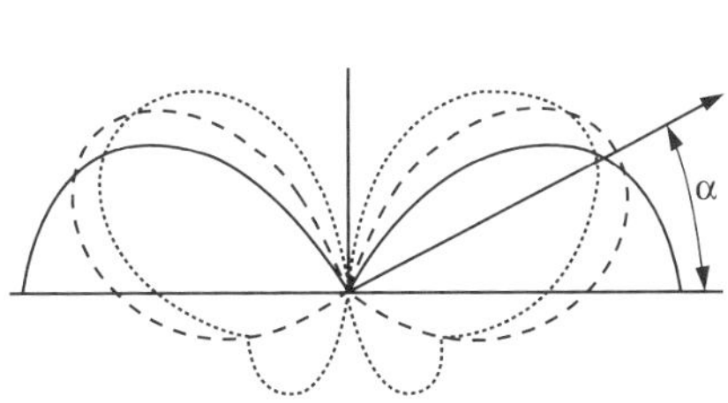

FIGURE 5.28 Vertical-plane patterns of $\lambda/4$ stub antenna on infinite ground plane (solid), and on finite ground planes several wavelengths in diameter (dashed) and about 1 λ in diameter (dotted). (From Kraus, J. D., *Antennas*. second edition. New York, McGraw-Hill, 1988. With permission.)

in their resonant frequencies, slightly increased directivity in the broadside direction, and more backplane radiation depending on the size of the ground plane and the thickness of the dielectric substrate.

Several others have analyzed microstrip antennas with cavity-backing to mitigate finite ground plane effects as well as the generation of surface waves.[46-54] Cavity-backing results in decreased bandwidth, increased structure complexity, and a more directive radiation pattern. Layering of dielectrics and/or use of stacked antennas inside the cavity can combat the small bandwidth of the antenna.[50,53,54]

5.3.5.2 Imperfect Ground Planes

Just as noninfinite ground planes can affect antenna performance, ground planes with noninfinite conductivity can move antenna operation away from ideal behavior. The presence of a lossy ground plane can substantially influence the input resistance of an antenna.[55-58] A related increase in frequency bandwidth also occurs, but at the expense of the radiating efficiency of the antenna, because some power is deposited in the lossy ground or is lost in surface waves. The placement of surrogate conducting planes in the form of radial wires or metallic disks on the ground beneath the antenna often ameliorates these effects.[56,58] Surface waves caused by the lossy ground plane tend to reduce overall antenna gain and may cause some null filling compared with ideal radiation patterns.[2,59-61]

5.3.5.3 Packaging Effects

As consumer expectations of wireless communication equipment continue to rise, antenna design becomes more and more critical to ensure reliable system performance. To arrive at a complete system design, therefore, all components of the electromagnetic system need to be considered in the design process. These components consist of the antenna and the surrounding structures, including the chassis to which the antenna is attached, and the user if the equipment is portable. User effects on antenna characteristics are described in Chapter 25.

As discussed in the previous sections, a conductive chassis can present a nonuniform and imperfect ground plane for an antenna, resulting in detuning and radiation pattern distortion. Several groups have studied the behavior of monopole antennas on handset-sized conducting boxes.[62-65] In Reference 64, the authors use a simple conducting box to model the handset chassis and investigate the changes on the radiation pattern caused by changes in the box dimensions. The geometry and the associated dimensions are given in Figs. 5.29 and 5.30.[64] With the antenna centered in the middle of the top of the box, the radiation patterns from the monopole changed significantly with increases in the vertical length of the box as shown in Figs. 5.31 to 5.33. Longer boxes tended to produce patterns with deeper nulls and more downward-directed radiation, which is undesirable for cellular telephone applications that prefer most power be radiated in the horizontal direction. Moreover, Yamaguchi et al.[63] demonstrated that movement of the monopole away from the center of top of the box causes asymmetries in the radiation patterns as shown in Fig. 5.34.

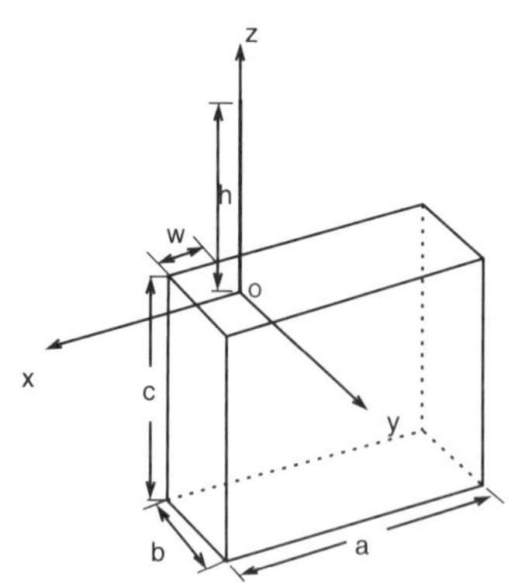

FIGURE 5.29 Geometry of monopole antenna on conducting box. (From Luebbers, R., Chen, L., Uno, T., and Adachi, S., *IEEE Trans. Antennas Propag.*, 40, 1578, 1992. ©IEEE. With permission.)

$a = 60$ mm	$b = 10$ mm
$h = 50$ mm	$r = 0.5$ mm
$c = 50$, 130, or 200 mm	
$w = 20$ or 30 mm	

FIGURE 5.30 Dimensions of monopole on metal box shown in Fig. 5.29. The monopole radius is r. (From Luebbers, Chen, R. L., Uno, T., and Adachi, S., *IEEE Trans. Antennas Propag.*, 40, 1579, 1992. ©IEEE. With permission.)

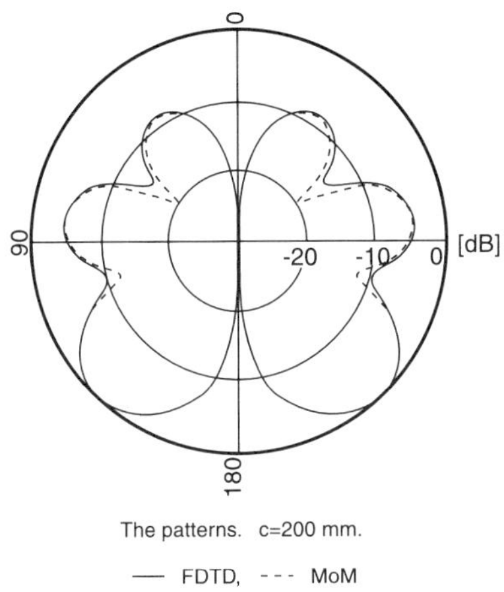

FIGURE 5.31 Calculated and measured xz plane radiation pattern at 1.5 GHz for monopole antenna centered ($w = 30$ mm) on smallest box ($c = 50$ mm). (From Luebbers, R., Chen, L., Uno, T., and Adachi, S., *IEEE Trans. Antennas Propag.*, 40, 1581, 1992. ©IEEE. With permission.)

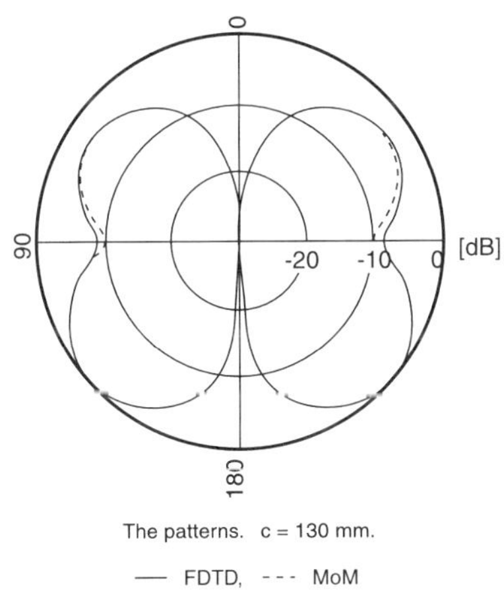

FIGURE 5.32 Calculated and measured xz plane radiation pattern at 1.5 GHz for monopole antenna centered ($w = 30$ mm) on intermediate size box ($c = 130$ mm). (From Luebbers, R., Chen, L., Uno, T., and Adachi, S., *IEEE Trans. Antennas Propag.*, 40, 1581, 1992. ©IEEE. With permission.)

As another example, consider the case of a quarter-wave monopole mounted on a small ground plane that protrudes from the side of a keyboard of a laptop computer.[66] The radiation pattern of this antenna, although intended to be omnidirectional, is instead skewed significantly in the direction across the keyboard because of the asymmetrical ground plane presented by the laptop keyboard as well as reflections from the computer display panel as shown in Fig. 5.35. Figure 5.36 illustrates that these pattern effects also change as the deployment angle of the laptop display changes from 90° to 135°.[66]

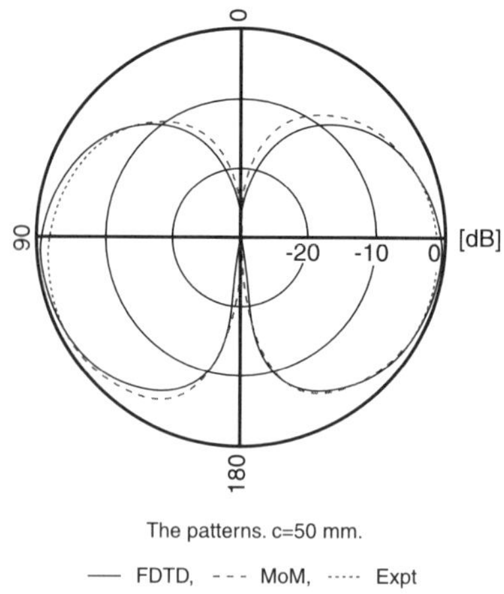

FIGURE 5.33 Calculated and measured *xz* plane radiation pattern at 1.5 GHz for monopole antenna centered ($w = 30$ mm) on largest box (c = 200 mm). (From Luebbers, R., Chen, L., Uno, T., and Adachi, S., *IEEE Trans. Antennas Propag.*, 40, 1581, 1992. ©IEEE. With permission.)

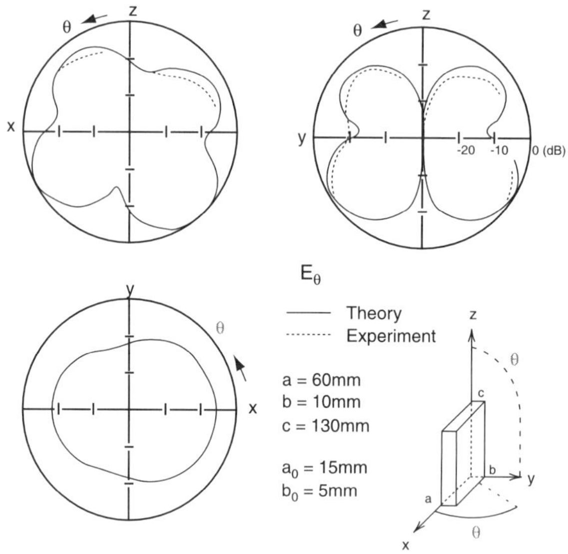

FIGURE 5.34 Radiation pattern of monopole antenna offset on conducting box. (From Yamaguchi, R., Sawaya, K., Fujino, Y., and Adachi, S., *IEICE Trans.*, E76-B, 1529, 1993. With permission.)

In addition to the effects of a conductive chassis on antenna characteristics, the coupling of an antenna to the internal circuitry of a communication device is also a concern. This undesired coupling could cause device malfunctions, deterioration in modulation accuracy, or interference with internal digital clocks. Fukasawa et al.[67] have analyzed this coupling with a monopole antenna and a conductive handset chassis using the finite difference time domain (FDTD) and electromotive force (EMF) methods. Coupling between an antenna and device circuitry can occur if the device is not shielded at all or if the shielding structure has unavoidable gaps resulting from fabrication constraints. The authors concluded that the coupling can be lowered with changes in three design factors: (1) increases in the number of internal connections between the circuit ground and the shielding; (2) decreases in the substrate height of the stripline/microstrip circuitry; and (3) selection of an external antenna length (~$\lambda/2$ instead of $\lambda/4$) that minimizes the induced current on the outside of the conductive chassis shielding. This third factor

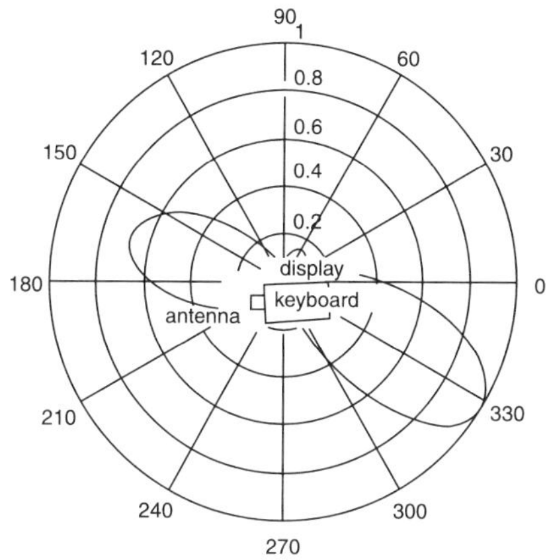

FIGURE 5.35 Normalized power for 800-MHz quarter-wave whip on laptop keyboard side (display 90°). (From Ponnapalli, S., *IEEE Trans. Components Packaging Manuf. Technol. Part B*, 19, 494, 1996. ©IEEE. With permission.)

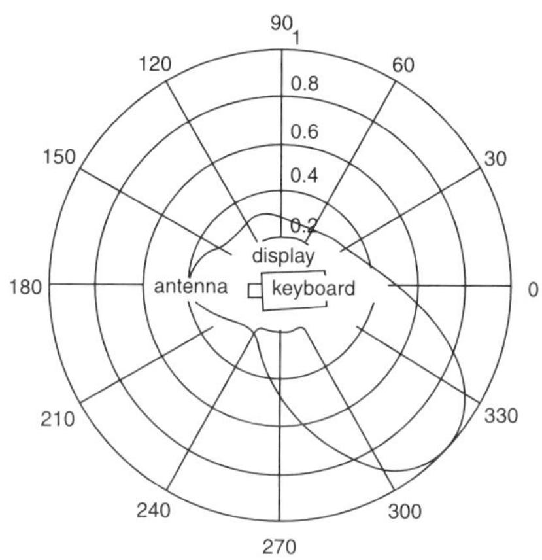

FIGURE 5.36 Normalized power for 800-MHz quarter-wave whip on laptop keyboard side (display 135°). (From Ponnapalli, S., *IEEE Trans. Components Packaging Manuf. Technol.*, Part B, 19, 494, 1996. ©IEEE. With permission.)

implies that use of antennas that rely less on the presence of a ground plane to operate properly may improve system performance by reducing coupling to internal circuits as well as minimizing chassis effects.

Chassis and coupling effects such as these can be accentuated in the case of embedded antennas, because their positions in a chassis make them more susceptible to shadowing and blockage caused by the chassis and also put them closer to internal device circuitry. Most proposed embedded antenna designs (e.g., References 13, 68, and 69) do rely on the presence of a ground plane, so chassis currents can greatly affect the radiating characteristics of the complete system. Work in the area of embedded antenna placement in packages[70] suggests that preferred antenna positions in a chassis can be acquired by considering the chassis as part of the radiating system.

The current trend toward implementation of embedded antennas presents other challenges, including reduction of antenna size to fit into a specified package and the effects of covers over antennas. Reduction in overall antenna size can be accomplished using a number of techniques, including meander lines,[71-73] reactive loading,[74,75] and inclusion of dielectrics in the antenna structure. Each of these approaches reduces antenna size at the expense of another design factor, such as fabrication complexity, efficiency, weight, or cost.

Planar (e.g., References 13 and 18) and microstrip antennas are popular choices for embedded applications that require the antenna(s) to be placed out of the sight of the consumer, beneath plastic covers or protective dielectric superstrates. Embedding a resonant antenna beneath a dielectric cover shifts the antenna operating frequency by changing the effective permittivity of the structure. Several researchers have analyzed the effect of spaced or flush dielectric covers on microstrip antenna characteristics, particularly resonant frequency. Examples of these studies include References 76 through 84. Bernhard and Tousignant[84] provide closed-form design equations to predict changes in resonant frequency with spaced or flush dielectric covers as well as an analysis that takes uncertainty in material parameters and dimensions into account.

5.3.6 Impedance Matching and Feeding Arrangements

5.3.6.1 Feeding Arrangements

Most antennas are connected to radio modules using one of three methods. The first of these, and perhaps the most common, is the coaxial feed. In this configuration, the antenna structure is connected to the inner conductor and the outer conductor is connected to the antenna ground plane. Figures 5.37 and 5.38 depict coaxial feeding of a monopole antenna[85] and a circular microstrip antenna,[6] respectively. Typically, quarter-wave monopoles, helical antennas, and some kinds of planar antennas use coaxial feeds in wireless communication equipment. A removable coaxial connection in this application, either through a snap-in or screw-in connector, permits simple replacement of the antenna in the communication system if necessary. Coaxial connections can be made over a range of characteristic impedance to simplify matching between the antenna and the radio, although most standard coaxial connectors use a 50-Ω

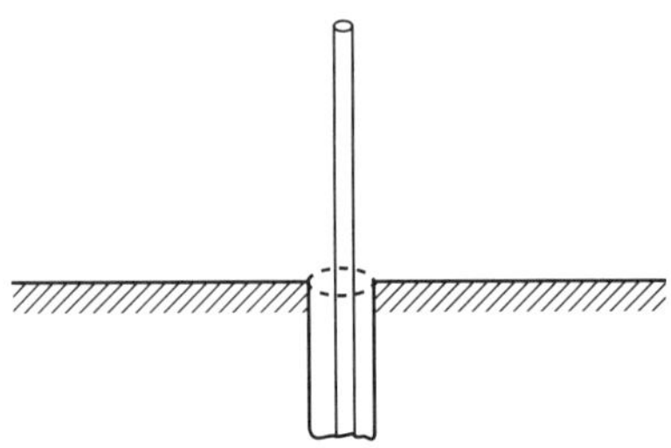

FIGURE 5.37 Coaxial line feeding a monopole through a ground plane. (From Ramo, S., Whinnery, J. R., and Van Duzer, T., *Fields and Waves in Communication Electronics.* third edition. John Wiley & Sons, New York, 1994. With permission.)

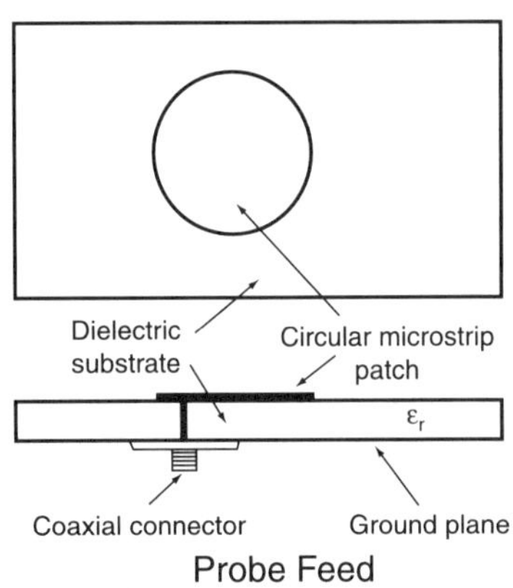

FIGURE 5.38 Probe feeding of a microstrip antenna. (From Balanis, C. A., *Antenna Theory Analysis and Design,* second edition, John Wiley & Sons, New York, 1997. With permission.)

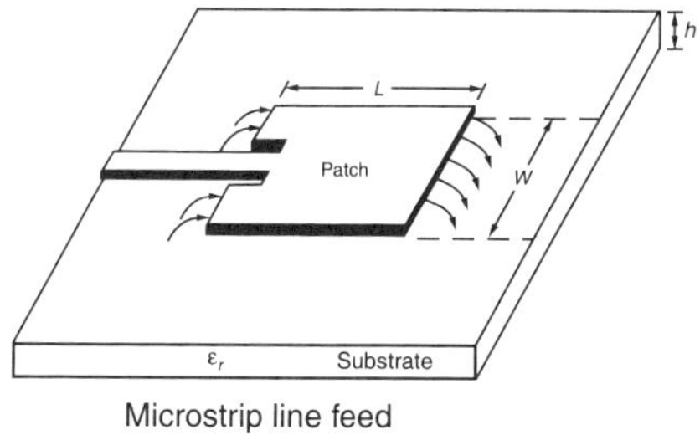

FIGURE 5.39 Microstrip line feed of microstrip antenna. (From Balanis, C. A., *Antenna Theory Analysis and Design*, second edition, John Wiley & Sons, New York, 1997. With permission.)

characteristic impedance. Coaxial connections are unbalanced and require a balun to feed a balanced antenna properly.

The second connection option is use of a microstrip feed between the radio and antenna as shown in Fig. 5.39.[6] This choice is more suitable for cases where the antenna itself is a microstrip structure and/or the radio module and the antenna are on the same circuit board. This type of connection can directly incorporate impedance matching circuitry and be fabricated along with the antenna using common printed circuit techniques.

Electromagnetic coupling, a third connection option that is becoming more popular for portable wireless equipment, relies on the transmission of signals between the radio module and the antenna without a direct connection between the two. Much of its popularity results from its decreased cost relative to other fixed connection methods that require molded, machined, or complex feed structures. This kind of connection can take a variety of forms depending on the antenna and the space available in the package to establish the coupling. Electromagnetic coupling of a microstrip or planar antenna via a small slot in the ground plane (aperture coupling)[86,87] or via a buried microstrip line (proximity coupling)[88,89] often eliminates the need for matching networks and can broaden the bandwidth of the antenna.[90] Aperture and proximity coupling to a microstrip antenna are depicted in Fig. 5.40.[6] Monopole or helix/monopole combination antennas can use either capacitive or inductive coupling that eliminates the need for a contact point between the antenna and the radio module.[91,92]

5.3.6.2 Impedance Matching

Usually the radio module and the antenna for a wireless communication device are designed for a standard characteristic impedance, such as 50 Ω, so that no impedance matching network is necessary. However, if an antenna impedance does not match with the input impedance of the radio module, an impedance-matching network is required. In the case of resonant antennas, an impedance match can be established using quarter-wavelength transformers, transmission line stubs, or lumped elements. These solutions provide matches over relatively narrow bandwidths, but usually meet system requirements given the narrow bandwidths of resonant antennas themselves. Cascading multiple quarter-wavelength transformers or using tapered transmission lines can provide more broad impedance bandwidths with prescribed frequency responses.[6] Baluns can also be used to achieve impedance transitions from unbalanced transmission lines to balanced antennas.

5.3.6.3 Baluns

When a balanced antenna is fed with an unbalanced transmission line (such as a coaxial cable), currents on the transmission line can radiate and distort the antenna radiation pattern. Baluns solve this problem by eliminating the unwanted portion of the transmission line current and providing a balanced state at the input terminals of the antenna. A schematic diagram of a two-terminal (a and b) balanced antenna fed using a balun is shown in Fig. 5.41.[93] The balance is characterized by nearly zero current on the outer surface of the coaxial cable ($I_s = 0$), equal impedances looking into each terminal of the antenna ($Z_{ag} = Z_{bg}$), equal driving voltages that are opposite in phase at each antenna terminal ($V_a = -V_b$), and terminal currents that are equal in amplitude and opposite in direction ($I_a = -I_b$).[93]

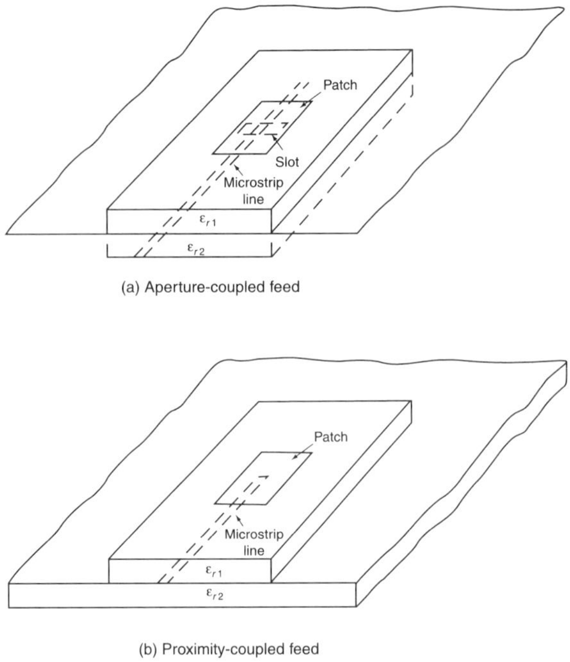

FIGURE 5.40 Aperture-coupled feed (a) and proximity-coupled feed (b) for a microstrip antenna. (From Balanis, C. A., *Antenna Theory Analysis and Design*, 2nd edition, John Wiley & Sons, New York, 1997. With permission.)

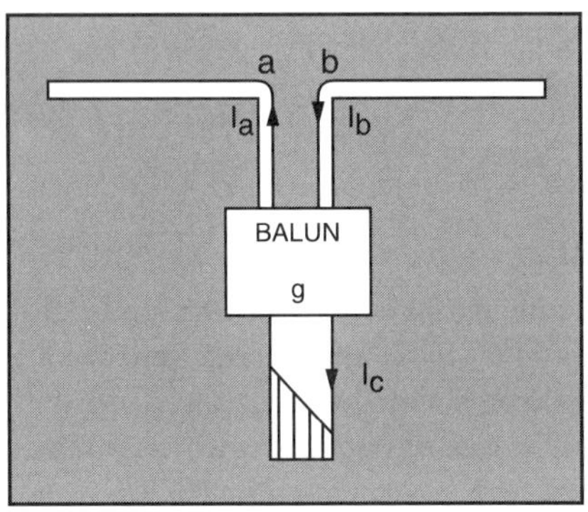

FIGURE 5.41 A simple antenna system illustrating the effect of a balun. (From Shuhao, H., *Microwave J.*, 30, 227, 1987. With permission.)

There are many kinds of baluns from which to choose, depending on system requirements.[93,94] A simple narrowband balun is constructed by attaching a dummy quarter-wavelength cable in parallel with the active coaxial line as shown in Fig. 5.42.[95] A more wideband balun consists of the coaxial cable winding around a ferrite core placed at the center of the balanced antenna.[2,96] Depending on the coiling configuration, this type of balun can also be used as an impedance transformer.[2,96,97]

5.4 Available Software for Antenna Analysis and Design

Closed-form design equations, whether developed with theory, simulations, or measurements, can provide clear relationships between antenna dimensions, material properties, and antenna resonance, creating the ability to make acceptable first-pass designs and analyze antenna behavior in terms of intuitive

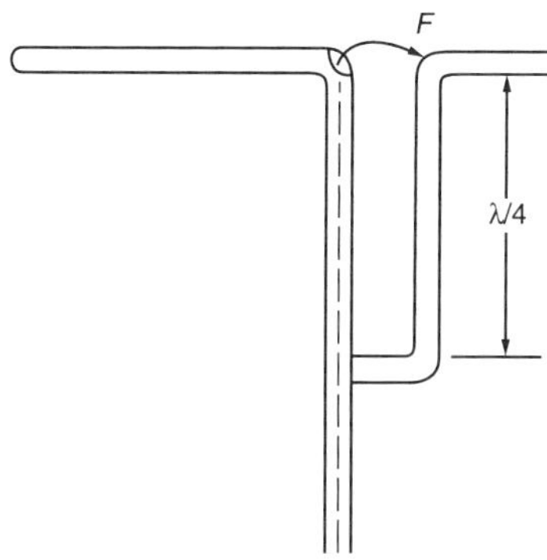

FIGURE 5.42 One type of balun used for feeding a balanced antenna from an unbalanced transmission line (coaxial in this case). (From Jordan, E. C. and Balmain, K. G., *Electromagnetic Waves and Radiating Systems*. 2nd edition, Prentice-Hall, Englewood Cliffs, NJ, 1968. With permission.)

models. For instance, models for quarter-wave monopoles,[2,6] half-wave dipoles,[2,6] helical antennas,[2] log-periodic antennas,[98] and microstrip antennas[6] have helped to make these antennas ubiquitous in the wireless communications industry. Several texts, for example, References 99 and 100, even provide software that implements closed-form design equations through simple interfaces to speed computer-aided antenna design. Development of other antennas that do not have established design rules must rely on a combination of computational modeling and experimental tuning to arrive at designs that meet operational specifications. Of course, when the antenna is included on or inside a chassis with a user present, even a very common antenna may require adjustments to operate properly.

Many commercial software packages are capable of modeling a variety of antennas. Selection of a particular package should be governed by the kinds of antennas to be modeled, available computer resources, software cost, and learning curve of the individual user. In this section we note several packages used in both industry and academia. This list of packages is not complete, and mention or omission of any particular package does not constitute a positive or negative appraisal of its capabilities. Of course, many groups in industry and academia choose to develop their own computational codes that perform more specialized and often more accurate simulations of particular antennas of interest. However, advances in computer capabilities and high-speed computation have resulted in commercial packages that are flexible and powerful enough for a wide range of antenna research, development, and analysis. Antenna researchers have compared several commercial and academic software packages for their abilities to model planar antenna structures.[101]

The first group of packages implements the full-wave moment method.[102,103] These packages include Accufield 2000©,[104] Antenna Solver©,[105] AWAS©,[106] CONCEPT©,[107] EMPLAN©,[108] EMSight©,[109] Ensemble©,[110] Expert MININEC©,[111] IE3D©,[112] Momentum©,[113] Sonnet©,[114] WinNEC©,[115] and WIPL-D©.[116] These packages are capable of simulating 2.5-dimensional planar structures (defined by an infinite ground plane and infinite lateral extents) as well as 3-dimensional (3D) structures in some cases. Some of the codes developed around the Numerical Electromagnetics Code core (NEC2) do not include capabilities for modeling dielectric materials. All these packages can provide impedance and frequency behavior as well as far-field radiation characteristics. As with all method of moment (MOM) codes, each simulation is done one frequency at a time. Depending on the package, finite ground planes, antenna gain, axial ratio, radar cross-section analyses, and optimizers are also available.

A second group of packages uses the finite element method to calculate antenna properties.[117,118] These packages include HFSS© from Ansoft,[110] High-Frequency Structure Simulator© from Agilent EEsof,[113] JMAG-Studio©,[119] and PAM-CEM©.[120] Although these packages are not designed specifically for antenna analysis, they can model full 3D structures as well as planar antennas. Each code can provide far-field pattern cuts and far-field 3D plots as well as antenna input impedance and frequency characteristics. As with the MOM codes, simulations are performed one frequency at a time with the results of multiple simulations compiled to produce swept frequency characteristics. Some of these codes also include optimizers that may be useful for some antenna designs. In general, codes based on the finite element method take longer that those using the MOM for antenna structures that require fine detail modeling.

A third group of commercial codes implement the FDTD method.[121,122] These include EMA3D© and EMA3D_CYL©,[123] EMLAB-SE©,[124] Empire©,[125] EZ-FDTD©,[126] Fidelity©,[112] and XFDTD©.[127] These packages are capable of modeling antenna systems that incorporate lossy dielectric and magnetic materials, finite ground planes, and perfect and nonideal conductors in three dimensions. Many packages support 2D and 3D displays of electromagnetic fields as well as transient fields. Because the simulations take place in the time domain, one complete simulation run provides impedance and radiation results over a wide frequency range. Several packages provide localized meshing for detailed structures to minimize the total simulation run time. Complete simulation time depends on the number of cells contained within the problem space, which in turn is a function of the maximum frequency where data are desired and the electrical properties of the materials included in the antenna system.

Although many packages use one of the three computational methods discussed previously (the MOM, the finite element method (FEM), and the FDTD method), there are also several computational tools that use other methods. For instance, Antenna Software, Ltd. has a number of special-purpose numerical codes for antenna analysis, such as ALDAS©, a geometric theory of diffraction/uniform theory of diffraction (GTD/UTD) code that computes the radiation patterns of low-gain antennas installed on conducting structures; PLANAR©, a code that analyses planar arrays; and REFLECT©, a physical optics (PO) code that addresses single- and dual-reflector antennas.[128] CST Microwave Studio© uses the finite integration method (FIM) to model 3D structures.[129] MMP© simulates electrodynamic fields based on the generalized multipole technique.[130] The boundary element method (BEM) is implemented for 3D field calculations in SINGULA©.[131] Additionally, a number of texts are accompanied by software for antenna analysis and design. Some of these include Reference 32, which uses physical optics, and References 133 and 134 which implement the finite element method.

To exploit the capabilities of several computational methods, several commercial packages provide linked or hybrid codes that combine two or more techniques. Some of the programs that can handle 3D radiating structures include EMC2000©, a code based on MOM linked with a hybrid IPO asymptotic technique;[135] EMIT©, a combination of MOM and FDTD kernels in one package;[136] FEKO©, a hybrid MOM/PO/UTD code;[137] GEMACS©, a hybrid MOM/UTD/FD (finite difference) code;[138] and SuperNEC©, a hybrid MOM/UTD code.[139]

References

1. Stutzman, W. L. and Thiele, A. T. *Antenna Theory and Design*, second edition. New York, John Wiley & Sons, 1998.
2. Kraus, J. D. *Antennas*, second edition. New York, McGraw-Hill, 1988.
3. Lo, Y. T. and Lee, S. W., Eds. *Antenna Handbook*. New York, Van Nostrand Reinhold, 1988.
4. Fujimoto, K. and James, J. R., Eds. *Mobile Antenna Systems Handbook*, Norwood, MA, Artech House. 1994.
5. Schelnukoff, S. A. *Advanced Antenna Theory*. New York, John Wiley & Sons, 1952.
6. Balanis, C. *Antenna Theory*. New York, John Wiley & Sons, 1982.
7. King, R. W. P. *The Theory of Linear Antennas*. Cambridge, MA, Harvard University Press, 1956.
8. Hahn, R. F. and Fikioris, J. G. "Impedance and radiation pattern of antennas above flat disks," *IEEE Trans. Antennas Propagat.*, vol. AP-21, no. 1, Jan. 1973, pp. 97–100.
9. Poggio, A. J. and Mayes, P. E. "Pattern bandwidth optimization of the sleeve monopole antenna," *IEEE Trans. Antennas Propagat.*, vol. AP-14, Sept. 1966, pp. 643–645.
10. Wunsch, A. D. "Fourier series treatment of the sleeve monopole antenna," *IEE Proc. Pt. H*, vol. 135, no. 4, Aug. 1988, pp. 217–225.
11. Weeks, W. L. *Antenna Engineering*. New York, McGraw-Hill, 1968.

12. Ogawa, K., Uwano, T., and Takahashi, M. "A shoulder-mounted planar antenna for mobile radio applications," *IEEE Trans. Veh. Technol.*, vol. 49, no. 3, May 2000, pp. 1041–1044.
13. Virga, K. L. and Rahmat-Samii, Y. "Low-profile enhanced-bandwidth PIFA antennas for wireless communications packaging," *IEEE Trans. Microwave Theory Tech.*, vol. 45, no. 10, Oct. 1997, pp. 1879–1888.
14. Ng, A. K. P. Lau, J., and Murch, R. D. "2.4 GHz ISM band antenna for PC cards," *IEEE Antennas and Propagation Society International Symposium. 1999 Digest.* vol.3, 1999, pp. 2066–2069.
15. King, R. W. P., Harrison, C. W., Jr., and Denton, D. H., Jr. "Transmission-line missile antennas," *IRE Trans. Antennas Propagat.*, vol. 8, no. 1, Jan. 1960, pp. 88–90.
16. Sanford, G. and Klein, L. "Increasing the beamwidth of a microstrip radiating element," *IEEE AP-S Int. Symp. Digest*, June 1979, pp. 126–129.
17. Hirasawa, K. and Haneishi, M. *Analysis, Design, and Measurement of Small Low Profile Antennas.* Norwood, MA, Artech House. 1992.
18. Taga, T. and Tsunekawa, K. "Performance analysis of built-in planar inverted F antenna for 800 MHz band portable radios," *IEEE J. Selected Area Comm.*, vol. 5, May 1983, pp. 921–929.
19. Wong, K.-L. and Yang, K.-P. "Modified planar inverted F antenna," *Electron. Lett.*, vol. 34, no. 1, 8 Jan. 1998, pp. 7–8.
20. Rowell, C. R. and Murch, R. D. "A compact PIFA suitable for dual-frequency 900/1800-MHz operation," *IEEE Trans. Antennas Propagat.*, vol. 46, no. 4, April 1998, pp. 596–598.
21. Rowell, C. R. and Murch, R. D. "A capacitively loaded PIFA for compact mobile telephone handsets," *IEEE Trans. Antennas Propagat.*, vol. 45, no. 5, May 1997, pp. 837–842.
22. Panayi, P. K. Al-Nuaimi, M. and Ivrissimtzis, L. P. "Tuning techniques for the planar inverted-F antenna," *IEE National Conference on Antennas and Propagation*, 1999, pp. 259–262.
23. Song, C. T. P., Hall, P. S., Ghafouri-Shiraz, H., and Wake, D. "Triple band planar inverted F antennas for handheld devices," *Electron. Lett.*, vol. 36, no. 2, Jan. 20 2000, pp. 112–114.
24. Tam, M. T. K. and Murch, R. D. "Compact circular sector and annular sector dielectric resonator antennas," *IEEE Trans. Antennas Propagat.*, vol. 47, no. 5, 1999, pp. 837–842.
25. Petosa, A., Ittipiboon, A., Antar, Y. M. M., Roscoe, D., and Cuhaci, M., "Recent advances in dielectric-resonator antenna technology," *IEEE Antennas Propagat. Mag.*, vol. 40, no. 3, June 1998, pp. 35–48.
26. Mongia, R. K. Ittipiboon, A., Antar, Y. M. M., Bhartia, P., and Cuhaci, M. "A half-split dielectric resonator using slot coupling," *IEEE Microwave Guided Wave Lett.*, vol. 2, no. 2, Feb. 1993, pp. 38–39.
27. Mongia, R. K., Ittipiboon, A., Bhartia, P., and Cuhaci, M. "Electric monopole antenna using a dielectric ring resonator," *Electron. Lett.*, vol. 29, Aug. 1993, pp. 1530–1531.
28. Mongia, R. K. and Bhartia, P. "Dielectric resonator antennas — a review and general design relations for resonant frequency and bandwidth," *Int. Microwave Millimeter-Wave Computer-Aided Eng.*, vol. 4, no. 3, 1994, 230–247.
29. Long, S. A., McAllister, M. W., and Shen, L. C. "The resonant cylindrical dielectric cavity antenna," *IEEE Trans. Antennas Propagat.*, vol. 31, no. 3, 1983, pp. 406–412.
30. Gastine, M., Courtois, L., and Dormann, J. J. "Electromagnetic resonances of free dielectric spheres," *IEEE Trans. Microwave Theory Tech.*, vol. 15, no. 12, Dec. 1967, pp. 694–700.
31. Leung, K. W., Luk, K. M., and Yung, E. K. N. "Spherical cap dielectric resonator antenna using aperture coupling," *Electron Lett.*, vol. 30, no. 17, Aug. 18, 1994, pp. 1366–1367.
32. Lo, H. Y., Leung, K. W., Luk, K. M., and Yung, E. K. N. "Low profile equilateral-triangular dielectric resonator antenna of very high permittivity," *Electron. Lett.*, vol. 35, no. 25, 1999, pp. 2164–2166.
33. Leung, K. W., Poon, M. L., Wong, W. C., Luk, K. M., and Yung, E. K. N. "Aperture coupled dielectric resonator antenna using a strip-line feed," *Microwave Optical Technol. Lett.*, vol. 24, no. 2, January 20, 2000, pp. 120–121.
34. Kraus, J. D. *Antennas.* second edition. New York, McGraw-Hill, 1988.

35. Bhattacharyya, A. K. "Effects of ground plane and dielectric truncations on the efficiency of a printed structure," *IEEE Trans. Antennas Propagat.*, vol. 39, Mar. 1991, pp. 303–308.
36. Bokhari, S. A., Mosig, J. R., and Gardiol, F. E. "Radiation pattern computation of microstrip antennas on finite size ground planes," *IEE Proc.-Microw. Antennas Propag.*, vol. 139, Jun. 1992, pp. 278–286.
37. Huang, J. "The finite ground plane effect on the microstrip antenna radiation patterns," *IEEE Trans. Antennas Propagat.*, vol. 31, Jul. 1983, pp. 649–653.
38. Lier, E. and Jakobsen, K. R. "Rectangular microstrip patch antennas with infinite and finite ground plane dimensions," *IEEE Trans. Antennas Propagat.*, vol. 31, Nov. 1983, pp. 978–984.
39. Noghanian, S. and Shafai, L. "Control of microstrip antenna radiation characteristics by ground plane size and shape," *IEE Proc. — Microw. Antennas Propag*, vol. 145, no. 3, June 1998, pp. 207–212.
40. Otero, M. F. and Rojas, R. G. "Analysis and treatment of edge effects on the radiation pattern of a microstrip patch antenna," *IEEE AP-S Int. Symp. Digest*, 1995, pp. 1050–1053.
41. Sanad, M. "Microstrip antennas on very small ground planes for portable communication systems," *IEEE AP-S Int. Symp. Digest*, 1994, pp. 810–813.
42. Schaubert, D. H. and Yngvesson, K. S. "Experimental study of a microstrip array on high permittivity substrate," *IEEE Trans. Antennas Propagat.*, vol. 34, Jan. 1986, pp. 92–97.
43. Vaudon, P., Reineix, A., and Jecko, B. "Asymptotic method in prediction of E plane radiation pattern of rectangular microstrip patch antennas on finite ground," *Elect. Lett.*, vol. 26, June 21, 1990, pp. 956–957.
44. Wang, S.-M. and Wong, K.-L. "Effects of a finite ground plane on the mutual coupling between rectangular microstrip antennas," *Microwave Opt. Technol. Lett.*, vol. 11, Mar. 1996, pp. 201–202.
45. Zhou, W. and Wahid, P. F. "Analysis of microstrip antennas on finite ground planes," *Microwave Opt. Technol. Lett.*, vol. 15, July 1997, pp. 204–207.
46. Aberle, J. T. and Zavosh, F. "Analysis of probe-fed circular microstrip patches backed by circular cavities," *Electromagnetics*, vol. 14, 1994, pp. 239–258.
47. Cheng, J.-C., Dib, N. I., and Katehi, L. P. B. "Theoretical modeling of cavity-backed patch antennas using a hybrid technique," *IEEE Trans. Antennas Propagat.*, vol. 43, Sept. 1995, pp. 1003–1013.
48. Duffy, S. M. and Gouker, M. A. "Experimental comparison of the radiation efficiency for conventional and cavity backed microstrip antennas," *IEEE AP-S Int. Symp. Digest*, 1996, pp. 196–199.
49. Flynt, R. A., Fan, L., Navarro, J. A., and Chang, K. "Low cost and compact active integrated antenna transceiver for system applications," *IEEE Trans. Microwave Theory Tech.*, vol. 44, Oct. 1996, pp. 1642–1649.
50. Gentili, G. G., Garcia-Castillo, L. E., Salazar-Palma, M., and Perez-Martinez, F. "Green's function analysis of single and stacked rectangular microstrip patch antennas enclosed in a cavity," *IEEE Trans. Antennas Propagat.*, vol. 45, Apr. 1997, pp. 573–579.
51. Jellett, S. T. and Bialkowski, M. E. "A circularly polarized MSAT antenna element with increased operational bandwidth," *IEEE AP-S Int. Symp. Digest*, 1994, pp. 1044–1047.
52. K. Kaneta, T. Kondo, M. Ando and N. Goto, "A flush-mounted antenna for mobile communications," *IEEE AP-S Int. Symp. Digest*, 1988, pp. 1323–1326.
53. Navarro, J. A., Chang, K., Tolleson, J., Sanzgiri, S., and Lee, R. Q. "A 29.3 GHz cavity-enclosed aperture-coupled circular-patch antenna for microwave circuit integration," *IEEE Microw. Guided Wave Lett.*, vol. 1, July 1991, pp. 170–171.
54. Zavosh, F. and Aberle, J. T. "Improving the performance of microstrip-patch antennas," *IEEE Antennas Propagat. Mag.*, vol. 38, Aug. 1996, pp. 7–12.
55. Cui, T. J. and Chew, W. C. "Accurate model of arbitrary wire antennas in free space, above or inside lossy ground," *IEEE Antennas Propagat. Int. Symp. Digest*, vol. 3, 1999, pp. 1696–1699.
56. Weiner, M. M., Zamoscianyk, S., and Burke, G. J. "Radiation efficiency and input impedance of monopole elements with radial-wire ground planes in proximity to earth," *Elect. Lett.*, vol. 28, no. 16, Jul. 30, 1992, pp. 1550–1551.

57. Stalzer, H. J. Jr., Fathy, A., Hessel, A., and Shmoys, J. "Effect of lossy ground on performance of planar and cylindrical arrays," *Radio Science*, vol. 25, no. 2, Mar.–Apr. 1990, pp. 133–147.
58. Balanis, C. A. *Antenna Theory: Analysis and Design*, second edition, New York, John Wiley & Sons, 1997.
59. Yang, J. J., Chow, Y. L., and Fang, D. G. "Discrete complex images of a three-dimensional dipole above and within a lossy ground," *IEE Proc.-H Microw. Antennas Propagat.*, vol. 138, no. 4, Aug. 1991, pp. 319–326.
60. Helaly, A. "Radiation from a horizontal electric dipole in the presence of a lossy ground," *Canadian J. Physics*, vol. 72, no. 9–10, Sept.–Oct. 1994, pp. 672–680.
61. Cai, Z. and Bornemann, J. "Rigorous analysis of radiation properties of lossy patch resonators on complex anisotropic media and lossy ground metallization," *IEEE Trans. Antennas Propagat.*, vol. 42, no. 10, Oct. 1994, pp. 1443–1446.
62. Hirasawa, K. and Fujimoto, K. "Characteristics of wire antennas on a rectangular conducting body," *IEICE Trans.*, vol. J65-B, no. 9, Sept. 1982, pp. 1133–1139.
63. Yamaguchi, R., Sawaya, K., Fujino, Y., and Adachi, S. "Effect of dimension of conducting box on radiation pattern of a monopole antenna for portable telephone," *IEICE Trans.*, vol. E76-B, no. 12, Dec. 1993, pp. 1526–1531.
64. Luebbers, R., Chen, L., Uno, T., and Adachi, S. "FDTD calculation of radiation patterns, impedance, and gain for a monopole antenna on a conducting box," *IEEE Trans. Antennas Propagat.*, vol. 40, no. 12, Dec. 1992, pp. 1577–1583.
65. Jensen, M. A. and Rahmat-Samii, Y. "Performance analysis of antennas for handheld transceivers using FDTD," *IEEE Trans. Antennas Propagat.*, vol. 42, no. 8, Aug. 1994, pp. 1106–1113.
66. Ponnapalli, S. "Modeling and design of antennas for RF wireless systems," *IEEE Trans. Components Packaging & Manufact. Technol. Part B-Advanced Packaging*, vol. 19, no. 3, Aug. 1996, pp. 487–502.
67. Fukasawa, T., Miyazaki, C., Makino, S., and Urasaki, S. "Calculation of coupling between a monopole antenna and a strip line in a shield case for a portable telephone," *IEICE Trans. Commun.*, vol. E83-B, no. 3, March 2000, pp. 505–510.
68. Chiba, N., Amano, T., and Iwasaki, H. "Dual-frequency planar antenna for handsets," *Electron. Lett.*, vol. 34, no. 25, Dec. 10, 1998, pp. 2362–2363.
69. Fuhl, J., Nowak, P., and Bonek, E. "Improved internal antenna for handheld terminals," *Electron. Lett.*, vol. 30, no. 22, Oct. 27, 1994, pp. 1816–1818.
70. Bernhard, J. T. "Analysis of integrated antenna positions on a laptop computer for mobile data communication," *IEEE AP-S Int. Symp. Digest*, vol. 4, 1997, pp. 2210–2213.
71. Wang, H. Y., Simkin, J., Emson, C., and Lancaster, M. J. "Compact meander slot antennas," *Microw. Opt. Technol. Lett.*, vol. 24, no. 6, March 20, 2000, pp. 377–380.
72. Williams, T., Rahman, M., and Stuchly, M. A. "Dual-band meander antenna for wireless telephones," *Microw. Opt. Technol. Lett.*, vol. 24, no. 2, Jan. 20, 2000, pp. 81–85.
73. Chien-Jen, W. and Jou, C. F. "Compact microstrip meander antenna," *Microw. Opt. Technol. Lett.*, vol. 22, no. 6, Sept. 20, 1999, pp. 413–414.
74. Chen, H.-M. "Dual-frequency microstrip antenna with embedded reactive loading," *Microw. Opt. Technol. Lett.*, vol. 23, no. 3, Nov. 5, 1999, pp. 186–188.
75. Fang, S.-T., Chiou, T.-W., and Wong, K.-L. "Broadband equilateral-triangular microstrip antenna with asymmetric bent slots and integrated reactive loading," *Microw. Opt. Technol. Lett.*, vol. 23, no. 3, Nov. 5, 1999, pp. 149–151.
76. Afzalzadeh, R. and Karekar, R. N. "Characteristics of a rectangular microstrip patch antenna with protecting spaced dielectric superstrate," *Microw. Opt. Technol. Lett.*, vol. 7, no. 2, Feb. 1994, pp. 62–66.
77. Bahl, I. J., Bhartia, P., and Stuchly, S. "Design of microstrip antennas covered with a dielectric layer," *IEEE Trans. Antennas Propagat.*, vol. 30, no. 2, Mar. 1982, pp. 314–318.
78. Shavit, R. "Dielectric cover effect on rectangular microstrip antenna array," *IEEE Trans. Antennas Propagat.*, vol. 42, no. 8, Aug. 1994, pp. 1180–1184.

79. Fan, Z. and Lee, K. F. "Input impedance of annular-ring microstrip antenna with a dielectric cover," *IEEE Trans. Antennas Propagat.*, vol. 40, no. 8, Aug. 1992, pp. 992–995.
80. Ramahi, O. M. and Lo, Y. T. "Superstrate effect on the resonant frequency of microstrip antennas," *Microw. Opt. Technol. Lett.*, vol. 5, no. 6, Jun. 1992, pp. 254–257.
81. Bhattacharyya, A. and Tralman, T. "Effects of dielectric superstrate on patch antennas," *Electron. Lett.*, vol. 24, no. 6, Mar. 1988, pp. 356–358.
82. Al-Charchafchi, S. H. and Ibrahim, M. R. "An experimental investigation of flat radome-loaded microstrip patch antennas," *Microwave J.*, vol. 40, no. 2, Feb. 1997, pp. 100–109.
83. Zhong, S.-S., Liu, G., and Qasim, G. "Closed form expressions for resonant frequency of rectangular patch antennas with multidielectric layers," *IEEE Trans. Antennas Propagat.*, vol. 42, no. 9, Sept. 1994, pp. 1360–1363.
84. Bernhard, J. T. and Tousignant, C. "Resonant frequencies of rectangular microstrip antennas with flush and spaced dielectric superstrates," *IEEE Trans. Antennas Propagat.*, vol. 47, no. 2, February 1999, pp. 302–308.
85. Ramo, S., Whinnery, J. R., and Van Duzer, T. *Fields and Waves in Communication Electronics.* third edition. New York, John Wiley & Sons, 1994.
86. Oltman, H. G. and Huebner, D. A "Electromagnetically coupled microstrip dipoles," *IEEE Trans. Antennas Propagat.*, vol. 29, no.1, Jan. 1981, pp. 151–157.
87. Pozar, D. M. "A microstrip antenna aperture coupled to a microstrip line," *Electron. Lett.*, vol. 21, Jan. 1985, pp. 49–50.
88. Tsai, M.-J. and Alexopoulos, N. G. "Via hole and parasitically coupled microstrip antennas of arbitrary shape in multi-layered substrates," *Electromagnetics,* vol. 16, no. 3, May–Jun. 1996, pp. 229–252.
89. Pozar, D. M. and Kaufman, B. F. "Increasing the bandwidth of a microstrip antenna by proximity coupling," *Electron. Lett.*, vol. 23, Apr. 1987, pp. 368–369.
90. Duffy, S. M. "An enhanced bandwidth design technique for electromagnetically coupled microstrip antennas," *IEEE Trans. Antennas Propagat.*, vol. 48, no. 2, Feb. 2000, pp. 161–164.
91. Fukasawa, T., Endo, T., Makino, S., and Urasaki, S. "Analysis of whip antenna fed by a slot antenna with electromagnetic coupling," *IEEE AP-S Int. Symp. Digest,* vol. 3, 1996, pp. 1574–1577.
92. Egashira, S., Sakitani, A., and Harada, T. "Mobile telephone antenna fed with electromagnetic coupling," *IEICE Trans. Commun.*, Part B, vol. J71-B, no. 12, Dec. 1988, pp. 1719–1721.
93. Shuhao, H. "The balun family," *Microwave J.*, vol. 30, Sept. 1987, pp. 227–229.
94. Johnson, R. C. and Jasik, H. *Antenna Engineering Handbook.* second edition. New York, McGraw-Hill. 1984.
95. Jordan, E. C. and Balmain, K. G. *Electromagnetic Waves and Radiating Systems.* second edition. Englewood Cliffs, NJ, Prentice-Hall. 1968.
96. Ruthroff, C. L. "Some broad-band transformers," *Proc. IRE,* vol. 47, Aug. 1959, pp. 1337–1342.
97. Weeks, W. L. *Antenna Engineering.* New York, McGraw-Hill. 1968.
98. DuHamel, R. H. and Isbell, D. E. "Broadband Logarithmically Periodic Antenna Structures," *1957 IRE National Convention Record,* pt. 1, pp. 119–128.
99. Sainati, R. A. *CAD of Microstrip Antennas for Wireless Applications.* Cambridge, MA, Artech House. 1996.
100. Siwiak, K. *Radiowave Propagation and Antennas for Personal Communications.* second edition. Boston, Artech House. 1998.
101. Pozar, D., Duffy, S., Targonski, S., and Herscovici, N. "A comparison of commercial software packages for microstrip antenna analysis," *IEEE AP-S Int. Symp. Digest,* vol. 1, 2000, pp. 152–155.
102. Harrington, R. F. *Field Computation by Moment Methods.* second edition. New York, IEEE Press. 1993.
103. Miller, E. K., Medgyesi-Mitschang, L., and Newman, E. H., eds. *Computational Electromagnetics: Frequency-Domain Method of Moments.* New York, IEEE Press. 1992.

104. Fujitsu Limited, Simulation Development Dept., 1-1-4 Kamikodanaka, Nkahara, Kawasaki, 211-8588 Japan.
105. Grating Solver Development Company, P.O. Box 353, Allen, TX 75013.
106. Djordjevic, A. R., Bazdar, M. B., Sarkar, T. K., and Harrington. R. F. *AWAS for Windows: Analysis of Wire Antennas and Scatterers.* Boston, Artech House. 1995.
107. Singer, H. and Brüns, H.-D. Technical University of Hamburg-Harburg, Hamburg, Germany.
108. Kinayman, N. and Aksun, M. I. *EMPLAN: Electromagnetic Analysis of Printed Structures in Planarly Layered Media.* Boston, Artech House. 2000.
109. Applied Wave Research, Inc., El Segundo, CA.
110. Ansoft Corporate Headquarters, Four Station Square, Suite 200, Pittsburgh, PA, 15129.
111. EM Scientific, Inc., 2533 N. Carson Street, Suite 2107, Carson City, Nevada 89706.
112. Zeland Software, Inc., 48890 Milmont Drive, Suite 105D, Fremont, CA 94538.
113. Agilent Technologies, P.O. Box #10395, Palo Alto, CA 94303.
114. Sonnet Software, Inc., 1020 Seventh North Street, Suite 210, Liverpool, NY 13088.
115. Nittany Scientific, 1733 West 12600 South, Suite 420, Riverton, UT 84065.
116. Kolundzija, B., Ognjanovic, J. S., and Sarkar, T. K. WIPL-D: Electromagnetic Modeling of Composite Metallic and Dielectric Structures — Software and User's Manual. Boston, Artech House. 2000.
117. Jin, J. *The Finite Element Method in Electromagnetics.* New York, John Wiley & Sons. 1993.
118. Volakis, J. L., Chatterjee, A., and Kempel, L. C. *Finite Element Method For Electromagnetics: Antennas, Microwave Circuits, and Scattering Applications.* New York, IEEE Press. 1998.
119. The Japan Research Institute, Ltd., 16 Ichibancho, Chiyoda-ku, Tokyo, 102-0082, Japan.
120. ESI North America, 13399 West Star, Shelby Township, MI 48315-2701.
121. Kunz, K. S. and Luebbers, R. J. *The Finite Difference Time Domain Method for Electromagnetics.* New York, CRC Press. 1993.
122. Taflove, A. and Hagness, S. C. *Computational Electrodynamics: The Finite-Difference Time-Domain Method.* second edition. Cambridge, MA, Artech House. 2000.
123. ElectroMagnetic Applications, Inc., 7655 W. Mississippi Avenue, Suite 300, Lakewood, CO 80226.
124. SE Integrated Systems Engineering AG, Balgriststrasse 102, CH 8008 Zurich, Switzerland.
125. IMST GmbH, Carl-Friedrich-Gauss Str. 2, D-47475 Kamp-Lintfort, Germany.
126. EMS-PLUS, P.O. Box 71884, Durham, NC 27722-1884.
127. Remcom Inc., 315 South Allen Street, Suite 222, State College, PA 16801.
128. Antenna Software, Ltd., 16 Peachfield Road, Malvern, WORCS WR14 4AP, Great Britain.
129. CST GmbH, Computer Simulation Technology, Büdinger Str. 2a, D-64289 Darmstadt, Germany.
130. John Wiley & Sons, Inc., 605 Third Avenue, New York, NY 10158-0012.
131. Integrated Engineering Inc., 1999–2000, 300 Cree Cresent, Winnipeg, MB, Canada, R3J 3W9.
132. Diaz, L. and Milligan, T. *Antenna Engineering Using Physical Optics: Practical CAD Techniques and Software.* Cambridge, MA, Artech House. 1996.
133. Pelosi, G., Coccioli, R., and Selleri, S. *Quick Finite Element Method for Electromagnetic Waves.* Boston, Artech House. 1998.
134. Herscovici, N. *CAD of Aperture-Fed Microstrip Transmission Lines and Antennas.* Boston, Artech House. 1996.
135. Matra Systeme Information, Toulouse Cedex, France.
136. SETH Corporation, Johnstown, PA.
137. EM Software and Systems, P.O. Box 1354, Stellenbosch, 7599, South Africa.
138. Advanced Electromagnetics, 4516 Stockbridge Ave. NW, Albuquerque, NM 87120–5406.
139. Poynting Innovations (Pty) Ltd., P.O. Box 318, Wits, 2050, Republic of South Africa.

6

Microstrip Patch Antennas

Rod Waterhouse
RMIT University

6.1 Introduction

Microstrip patch antennas were first proposed in the early 1970s and since then a plethora of activity in this area of antenna engineering has occurred, probably more than in any other field of antenna research and development. Microstrip patch antennas have several well-known advantages over other antenna structures, including their low profile and hence conformal nature, light weight, low cost of production, robust nature, and compatibility with microwave monolithic integrated circuits (MMICs) and optoelectronic integrated circuits (OEICs) technologies. Because of these merits, forms of the microstrip patch antenna have been utilized in many applications such as in mobile communication base stations, space-borne satellite communication systems, and even mobile communication handset terminals.

Unfortunately, the expression, "there is no such thing as a free lunch," also applies to microstrip patch technology. Despite the previously mentioned features, microstrip patch antennas suffer from several inherent disadvantages of this technology in its pure form, namely, they have small bandwidth and relatively poor radiation efficiency resulting from surface wave excitation and conductor and dielectric losses. Also, to accurately predict the performance of this form of radiator, in particular, its input impedance nature, typically a full-wave computationally intensive numerical analysis is required.

0-8493-0124-6/02/$0.00+$1.50

Probably the saving grace of the microstrip patch antenna that prevented this technology from becoming one of the numerous white elephant technologies was its ease of fabrication or development. Because of this characteristic, universities and other research institutions (with somewhat limited budgets) throughout the world could make serious impacts on the shortcomings of microstrip patch technology without incurring huge expenses. Fortunately, because of the efforts of many research and development teams throughout the world, most of these issues have been addressed and solved so that the area of microstrip patch antennas is a thriving technology and will continue to be for many years to come.

The outline of this chapter is as follows: Section 6.2 gives a brief review of the general characteristics of a single-layer microstrip patch antenna, including performance trends as a function of the dielectric material used to form the radiating element. From this some of the shortcomings of this technology are evident. Also in this section, the different conductor shapes used for metallic patch conductor are examined, highlighting the pros and cons of each shape. Section 6.3 presents the different excitation methods for a microstrip patch antenna, including edge fed, probe fed, proximity coupled, and aperture coupled. Advantages and disadvantages of each technique are presented. In Section 6.4, methods to enhance the bandwidth of a microstrip patch antenna are summarized, including using parasitic elements, such as a stacked-patch configuration. In this section a brief comparison to other broadband printed antenna technologies are given, highlighting the advantages of disadvantages of each. Section 6.5 gives a brief summary on techniques for generating circular and dual polarization using microstrip patch technology. Section 6.6 looks into means of reducing the conductor size of the microstrip patch antenna and also the issues related to doing this, including the effect on bandwidth and gain. Section 6.7 summarizes some of the problems associated with integrating printed antennas with active circuits and how high-performance, fully monolithic solutions can be obtained. Finally, examples of integrated active patches are presented.

6.2 General Characteristics

Figure 6.1 shows a schematic diagram of a microstrip patch antenna. Here an arbitrarily shaped metallic conductor is etched on a grounded dielectric laminates. A microstrip patch antenna is a resonant-style radiator so one of its dimensions must be approximately $\lambda_g/2$, where λ_g is a guided wavelength taking into consideration the surrounding environment of the printed antenna. It is apparent that the properties of the substrate, namely, its dielectric constant, ε_r and its height play a fundamental role in the performance of the printed antenna. For a detailed explanation of how a microstrip patch antenna conceptually operates in terms of equivalent slots, etc., please consult one of the many articles on this subject.[1]

Figure 6.2 shows some very important performance trends of a single-layer microstrip patch antenna performance as a function of the laminate properties used to fabricate the antenna. These performance trends represent the properties of microstrip patch antennas in their pure form with a simple, ideal excitation method. Although the trends are for rectangular patches, other conductor shapes have similar responses. In Fig. 6.2a, the impedance bandwidth (defined as 10-dB return loss) is given for various

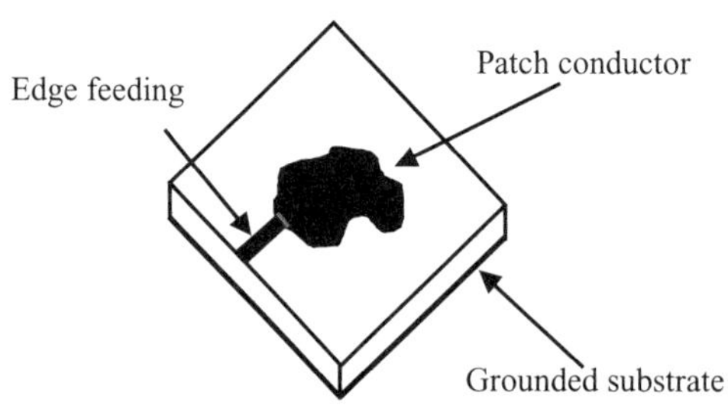

FIGURE 6.1 Schematic diagram of arbitrarily shaped microstrip patch antenna.

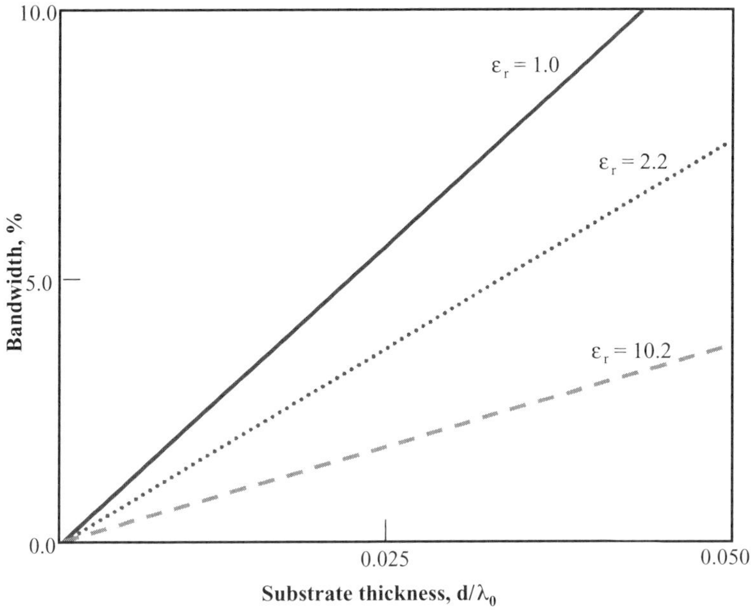

(a)

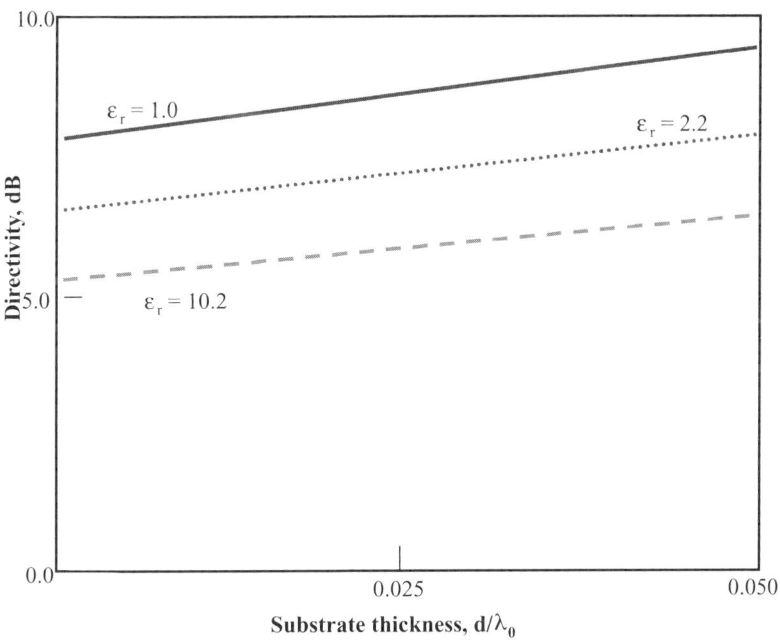

(b)

FIGURE 6.2 Performance trends of single-layered microstrip patch antenna: (a) impedance bandwidth; (b) directivity; (c) surface wave efficiency.

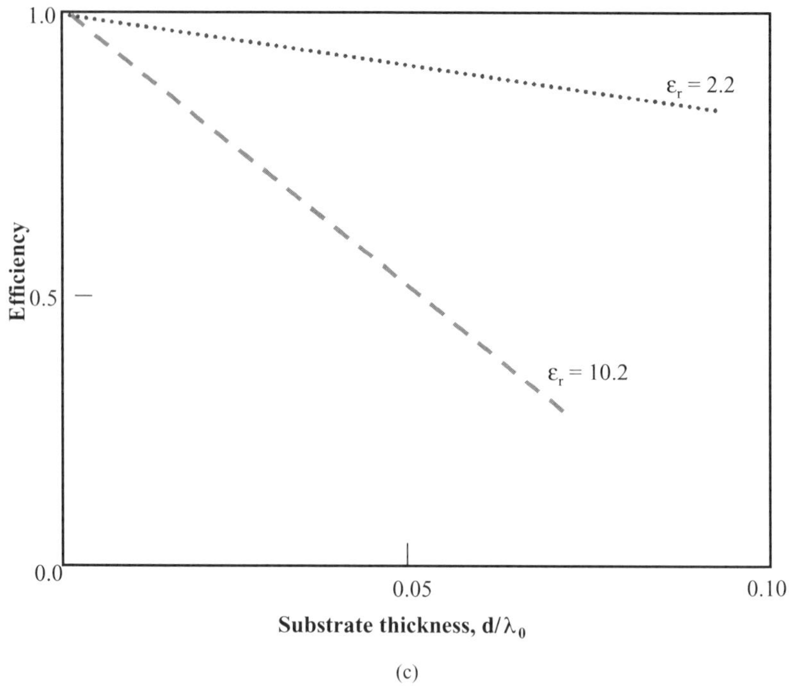

FIGURE 6.2 (continued).

dielectric constant values as a function of the electrical thickness of the laminate. As can be seen from this plot, the thicker the material is, the greater the bandwidth. Please note that these values are somewhat lower than what is generally required for present-day communication systems. Methods to significantly enhance these are given in Section 6.4. The other important observation is that the lower the dielectric constant, the greater the bandwidth that can be obtained.

Figure 6.2b shows the directivity of a patch antenna once again for different dielectric constants as a function of the electrical thickness. Simplistically, what is observed here is that because the microstrip patch antenna mounted on the low dielectric constant material is physically bigger than the antenna on the high dielectric constant laminate, it has a larger collecting area and therefore greater directivity. The directivity slightly increases as the thickness increases because of the increasing volume of the antenna. Please note that efficiency is not included in this plot.

Figure 6.2c shows the surface wave efficiency of a microstrip patch antenna for several dielectric constants as a function of thickness of the substrate. As can be seen from this figure, the higher the dielectric constant, the more power is lost to the surface wave and therefore the antenna is less efficient. Please note there are no surface waves excited for the case when $\varepsilon_r = 1.0$.

From Fig. 6.2 there appears to be a fundamental problem concerning the integration of microstrip patch antennas with MMIC and OEIC technology. MMICs and OEICs are typically developed on high dielectric constant, thin material (note that the dielectric constant for GaAs and AlGasInP, common materials for MMICs and OEICs, is approximately 13). Trying to develop a microstrip patch antenna in this environment would result in an antenna with poor bandwidth and radiation performance. Even trying to directly integrate microstrip patch technology with passive microwave circuits such as filters and couplers presents a problem, because to make these circuits compact, once again high dielectric constant and thin laminates are typically utilized, such as alumina materials ($\varepsilon_r = 10.2$). Fortunately, the ways in which to overcome this problem are addressed later in Section 6.7.2.

Over the years there have been many conductor shapes proposed and investigated for a microstrip patch antenna. Schematic diagrams of these are shown in Fig. 6.3, and a brief summary of the advantages and disadvantages follow:

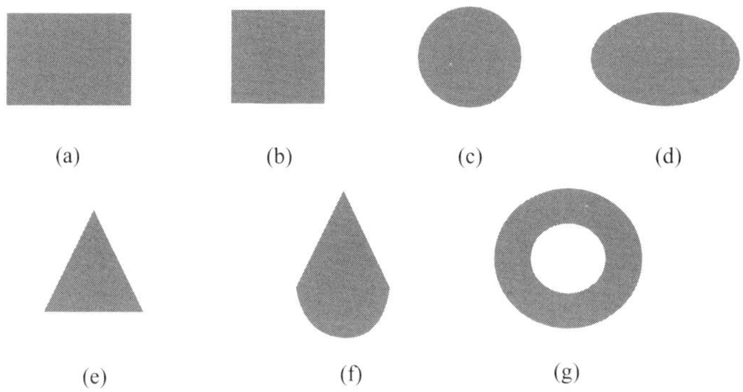

FIGURE 6.3 Common patch conductor shapes.

1. Rectangular and square patches (Figs. 6.3a and b) are the first and probably the most utilized patch conductor geometries. Rectangular patches tend to have the largest impedance bandwidth, simply because they are larger than the other shapes. Square patches can be used to generate circular polarization.
2. Circular and elliptical patches (Figs. 6.3c and d) are probably the second most common shape. These patches are slightly smaller than their rectangular counterpart and as a result have slightly lower gain and bandwidth. One of the primary reasons the circular geometry was quite expensively investigated in the past was because of its inherent symmetry. This allowed full-wave analysis tools utilizing a spectral domain technique to be written that were computationally more efficient than their rectangular counterpart. This was important in the early stages of patch design and development; however, with the advent of several rigorous, computationally fast full-wave design tools, such as Ensemble© and IE3D© systems, incorporating circular patch antennas are becoming increasingly rare.
3. Triangular and disk sector patch (Figs. 6.3e and f) geometries are smaller than their rectangular and circular counterparts, although at the expense of further reduction in bandwidth and gain. Triangular patches also tend to generate higher cross-polarization levels, because of their lack of symmetry in the configuration. Dual-polarized patches can be developed using these conductor shapes; however, the bandwidth is typically very narrow.
4. Annular ring (Fig. 6.3g) geometries are the smallest conductor shape, once again at the expense of bandwidth and gain. One problem associated with an annular ring is that it is not a simple process to excite the lowest order mode and obtain a good impedance match at resonance. Noncontact forms of excitation are typically required. The symmetry issues mentioned for the circular patch cases also apply here.

6.3 Feeding Techniques

Four fundamental techniques to feed or excite a microstrip patch antenna include edge fed, probe fed, aperture coupled, and proximity coupled. These can be further simplified into direct (edge and probe) and noncontact (aperture and proximity-coupled) methods. Some new excitation techniques are being developed, such as the L-shape probe;[2] however, this is really a hybrid representation of the probe and proximity-coupled versions. The properties of each feeding method are summarized below.

6.3.1 Edge-Fed Patches

One of the original excitation methods for a microstrip patch antenna is the edge-fed, or microstrip-line fed technique.[3] A schematic diagram representing this method is shown in Fig. 6.4. Here a microstrip feed line of width w_f is in direct contact with a rectangular patch conductor of length L and width W.

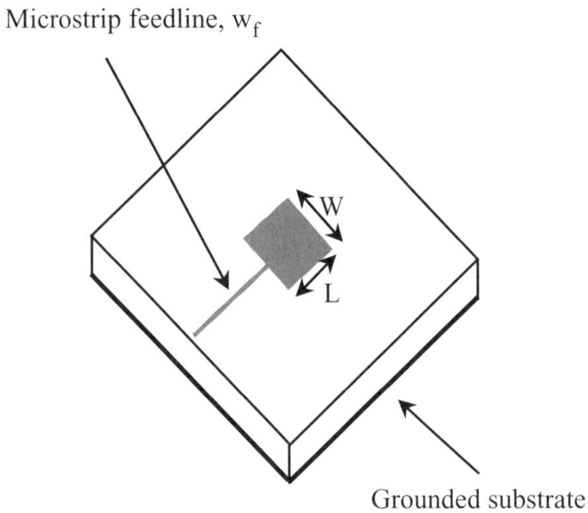

FIGURE 6.4 Schematic diagram of edge-fed microstrip patch antenna.

Typically the microstrip feed line comes in contact with one of the radiating edges of the patch, as shown in Fig. 6.4, although cases where the contact is located along the width of the patch have also been examined.

Edge-fed patches have several advantages over other feeding techniques. One of the key features of this technology is its ease of fabrication, because the feed layout and patches can be etched on one board. For this reason many large planar arrays have been developed using edge-fed patches. It is also very easy to control the level of the input impedance of an edge-fed patch. Simply by inserting the feed into the patch conductor the impedance at resonance can be adjusted from very high 150 to 250 Ω when the contact point of the feed line and the patch at the radiating edge of the patch, down to a couple of ohms if the contact point is near the center of the patch.

Edge-fed patches in their simplest form are relatively easy to model, if electrically thin material is used. Simple transmission line models can be utilized to give estimations of the input impedance performance of the antenna. For cases when thicker materials are used, the modeling of the performance is not too straightforward. This is because of the current distribution of the discontinuities associated with the contact point between the microstrip line and the patch antenna.

Edge-fed microstrip patches have bandwidth and gain characteristics consistent with Fig. 6.2, that is, these patches are relatively narrow bandwidth antennas. As is evident from Fig. 6.2, if high dielectric constant material is used as the substrate, the surface wave efficiency is poor. Also this form of feeding technique suffers from relatively high spurious feed radiation. This is simply because the feed network is not separated from the antenna and thus material suitable for efficient radiation for the antenna also causes the feed network to radiate, too.

6.3.2 Probe-Fed Patches

Probe feeding a microstrip patch antenna is another form of the original excitation methods proposed in the mid-1970s.[3] A schematic diagram representing this configuration is shown in Fig. 6.5, in which a probe of radius r_0 extends through the ground plane and is connected to the patch conductor, typically soldered to it. The probe or feeding pin is usually the inner conductor of a coaxial line; hence, probe feeding is often referred to as a coaxial feed. The probe position provides the impedance control in a similar manner to inserting the feed for an edge-fed patch. Because of the direct contact between the feed transmission line and the patch antenna, probe feeding is referred to as a direct contact excitation mechanism.

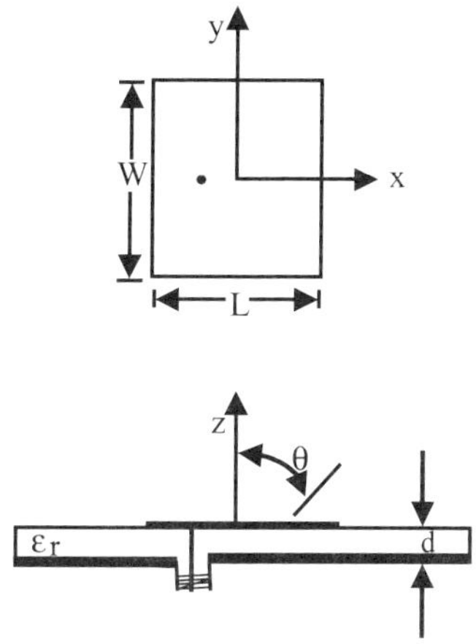

FIGURE 6.5 Schematic diagram of probe-fed microstrip patch antenna.

The probe-fed patch has several key advantages. First, the feed network, where phase shifters and filters may be located, is isolated from the radiating elements via a ground plane. This feature allows independent optimization of each layer. Second, of all the excitation methods, probe feeding is probably the most efficient because the feed mechanism is in direct contact with the antenna and most of the feed network is isolated from the patch, minimizing spurious radiation. The high efficiency of this printed antenna has seen a renaissance of the probe-fed-styled patch, despite the added complexity of developing a connection.

Probe-fed microstrip patches have similar issues to edge-fed patches; namely, their bandwidth is somewhat small and these printed antennas are somewhat difficult to accurately analyze. The probe used to couple power to the patch can generate somewhat high cross-polarized fields if electrically thick substrates are used. Also because this antenna is no longer a single-layer geometry, as a result of the location of the feed network, it is more complicated to manufacture.

6.3.3 Aperture-Coupled Patches

Because of the shortcomings of the direct contract feeding techniques, namely, the small inherent bandwidth and the detrimental effect of surface waves, noncontact excitation mechanisms were introduced. The first of these is the aperture-coupled patch.[4] A schematic diagram of this printed antenna, as given in Fig. 6.6, shows how separate laminates are used for the feed network and the patch antenna.

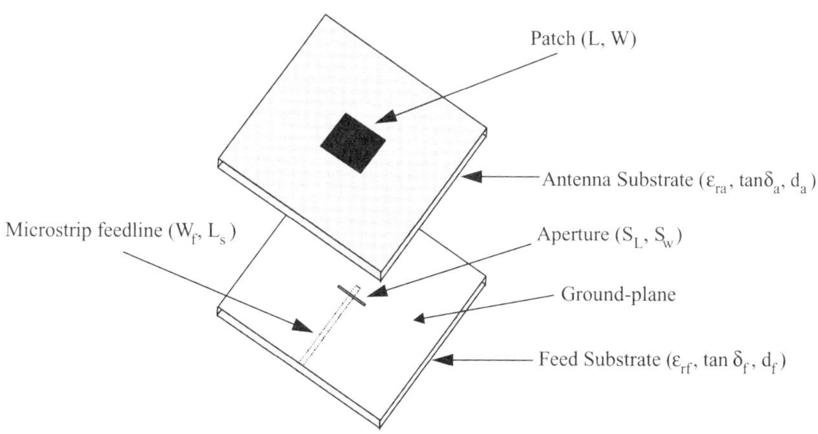

FIGURE 6.6 Schematic diagram of aperture-coupled microstrip patch antenna.

The laminates are separated by a ground plane and coupling between the feed, in this case a microstrip line, and the patch antenna is achieved via a small slot in the ground plane.

The configuration shown in Fig. 6.6 has advantages over its direct contact counterparts. Unlike the edge-fed patch antenna, independent optimization of the feed and antenna substrates can be achieved. Unlike the probe-fed configuration, no vertical interconnects are required, simplifying the fabrication processes and also adhering to the conformal nature of printed circuit technology. However, alignment issues can be important and also multilevel fabrication processes are typically required. A multilayered antenna can create other problems. The presence of small gaps between the layers of dielectric can significantly alter the input impedance nature of the antenna, especially at high frequencies, where these gaps appear larger electrically. Also, the material required to bond the layers could play a significant role in the performance of the antenna. If the bonding material is lossy and is located near, for example, the slot, the efficiency of the antenna is reduced.

The aperture-coupled patch has more design parameters than the direct contact fed patches and therefore has more flexibility or degrees of freedom for the antenna designer. Despite its somewhat complex appearance, the aperture-coupled microstrip patch antenna is relatively easy to accurately model, even when using full-wave analyses. The reason for this is that unlike for the direct contact fed patches, there are no abrupt current discontinuities. Thus, relatively simple, accurate, computationally fast, full-wave analyses are easy to develop.

In its original form the aperture-coupled patch has similar bandwidth and gain responses as the direct fed patches; however, it is very easy to significantly enhance the impedance bandwidth of this antenna. This is discussed in the next section. Aperture-coupled microstrip patch antennas are probably the most utilized microstrip patches in today's global market.

6.3.4 Proximity-Coupled Patches

The second form of noncontact fed patches created to overcome the shortcomings of the direct contact fed patches is the proximity-coupled patch.[5] A schematic diagram of this printed antenna is shown in Fig. 6.7. The microstrip antenna consists of a grounded substrate where a microstrip feed line is located. Above this material is another dielectric laminate with a microstrip patch etched on its top surface. Please note there is no ground plane separating the two dielectric layers. The power from the feed network is coupled to the patch electromagnetically, as opposed to a direct contact. This is why this form of microstrip patch is sometimes referred to as an electromagnetically coupled patch antenna.

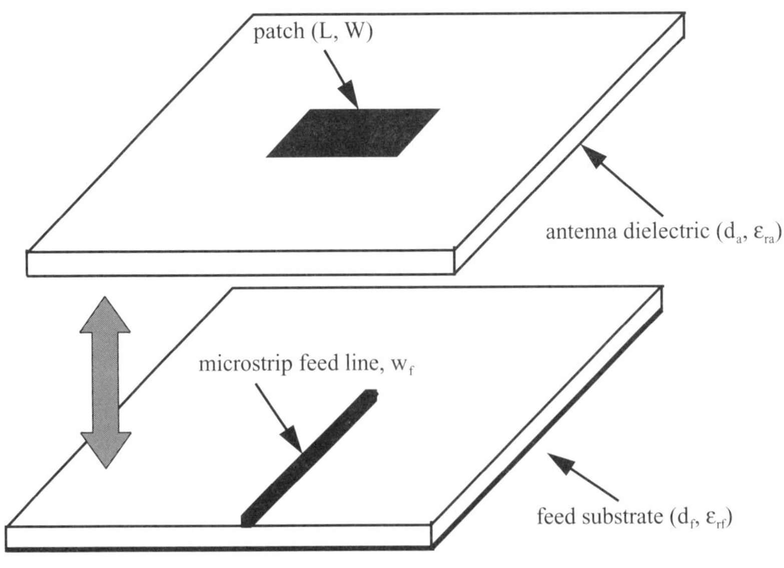

FIGURE 6.7 Schematic diagram of proximity-coupled microstrip patch antenna.

A key attribute of the proximity-coupled patch is that its coupling mechanism is capacitative in nature. This is in contrast to the direct contact methods, which are predominantly inductive. The difference in coupling significantly affects the obtainable impedance bandwidth, because the inductive coupling of the edge- and probe-fed geometries limits the thickness of the material useable. Thus, bandwidth of a proximity-coupled patch is inherently greater than the direct contact feed patches.

As with the aperture-coupled patch, full-wave analyses are not too difficult to develop for the proximity-coupled patch because of the lack of a current discontinuity between the feed network and the radiating element. The proximity-coupled microstrip patch has some shortcomings. The feed and antenna layers are not fully independent, because power must be coupled efficiently to the antenna. Therefore, these printed antennas can have relatively high spurious feed radiation, although not as high as for an edge-fed case. The antenna is a multilevel structure and thus alignment procedures are important. Small air gaps between the feed substrate and the laminate for the antenna can affect the coupling to the patch and, therefore, care must be taken when fabricating these antennas.

6.4 Enhancing Bandwidth

As mentioned previously, the microstrip patch in its pure form cannot satisfy the bandwidth requirements for most wireless communication systems. Typically, a modification to the structure of the printed antenna must be undertaken to meet the impedance bandwidth specifications. Over the years there have been numerous bandwidth enhancements investigated, all with varying degrees of success and complexity. The general philosophy of most of these techniques is to add one or more resonant antennas to the microstrip patch configuration. These additional elements may be in the form of slots or other patches. Once the additional radiator or radiators have been chosen, the objective of the antenna designer is to ensure there is the right degree of interaction between these elements so that the performance of the printed antenna is enhanced. In this section a summary of some of the more recognized methods are presented.

6.4.1 Parasitically Coupled (or Gap-Coupled) Patches

In the early to mid-1980s, parasitically coupling patches in a horizontal manner to the driven patch were proposed and investigated.[6] The philosophy behind this technique is that if the resonant frequency of the coupled element or elements is slightly different to that of the driven patch, then the bandwidth of the entire antenna may be increased. Figure 6.8 shows an example of a driven probe-fed rectangular patch with two parasitic patches positioned on either side of the excited patch in the *y*-axis direction. The critical parameters are the lengths and widths of each patch for the control of resonant frequency and bandwidth, as well as the gap between the elements. The gaps tend to control the coupling between the patches and therefore the tightness of the *resonant loop* (or loops) in the impedance locus of the antenna.

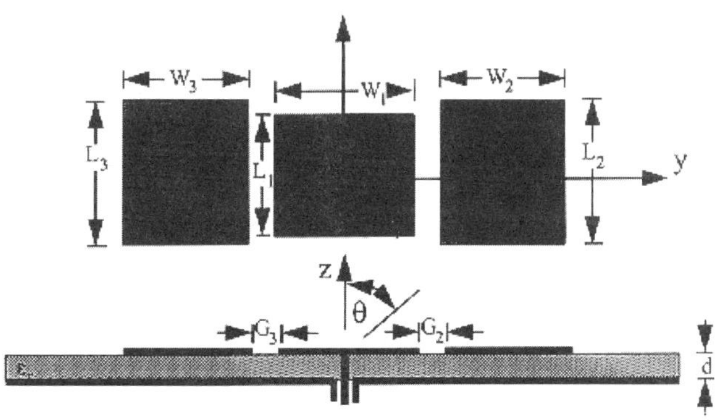

FIGURE 6.8 Schematic diagram of parasitically coupled (horizontal direction) microstrip patch antennas.

Bandwidths on the order of 20% have been achieved using this enhancement technique,[6,7] although there are several shortcomings of using parasitically coupled patches. First, to achieve these reasonable bandwidths, wide parasitic elements are required to make the overall size of the printed antenna configuration electrically large and, therefore, it is difficult to develop an array without incurring grating lobe problems.[7] Second, the radiation patterns of parasitically coupled patches tend to be somewhat distorted across the useful impedance bandwidth, because of the lack of symmetry of the generated currents with respect to the center of the printed antenna.

6.4.2 Stacked Microstrip Patches

Stacking patches on top of each other is probably the most common procedure utilized to enhance the bandwidth of a microstrip antenna. Figure 6.9 shows a schematic diagram of an edge-fed stacked patch configuration, where an arbitrarily shaped patch is etched on a grounded substrate and is fed by a microstrip transmission line. Another patch antenna is mounted on a second laminate (with no ground plane) and is placed directly above the driven patch.

Interestingly, when stacking was first proposed in the late 1970s to increase the bandwidth of direct contact fed patches, only moderate improvements were achieved.[8] One possible reason as to why such minor improvements were observed can be attributed to the relative complexity nature of these printed antennas. By looking at Fig. 6.9, there are many variables in this configuration and thus a rigorous full-wave analysis to accurately model the performance of the antenna is required. Importantly, the analysis needs to be not only accurate but also computationally fast so that trends in the impedance nature can be accurately and rapidly observed and then later optimized. Such accurate, fast codes are available nowadays, thanks mainly to the order of magnitude increases in the computational speeds of computers. A thorough investigation into how to design broadband direct contact stacked patches was undertaken, in particular, focusing on what optimum parameters are needed to achieve good bandwidth characteristics.[9] From this study, direct contact feed-stacked patches with bandwidths approaching 30% have been achieved. This order of bandwidth can also be achieved using noncontact, feed-stacked patches, such as aperture-coupled stacked patches,[10] of which a schematic diagram is shown in Fig. 6.10. Advantages of utilizing direct contact feed-stacked patches over aperture-coupled stacked patches include ease of fabrication and a minimal backward-directed radiation. As mentioned previously, aperture-coupled patches do have more degrees of freedom than direct contact fed patches and therefore an aperture-coupled stacked patch is somewhat easier to design.

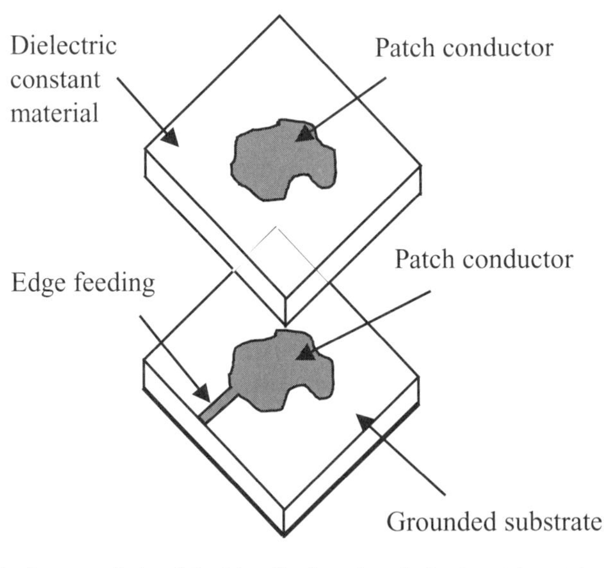

FIGURE 6.9 Schematic diagram of edge-fed arbitrarily shaped stacked microstrip patches.

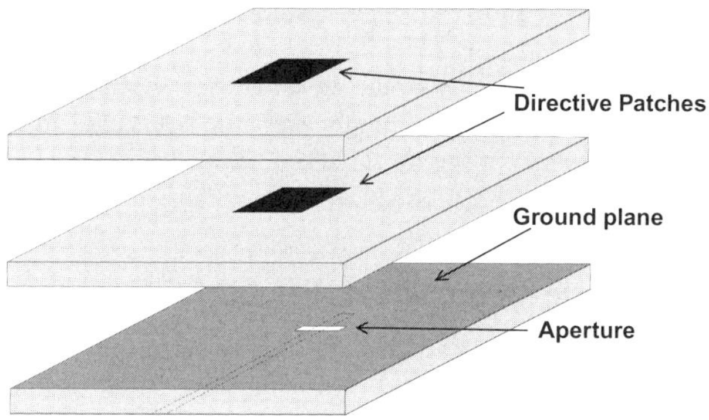

FIGURE 6.10 Schematic diagram of aperture-coupled rectangular stacked microstrp patches.

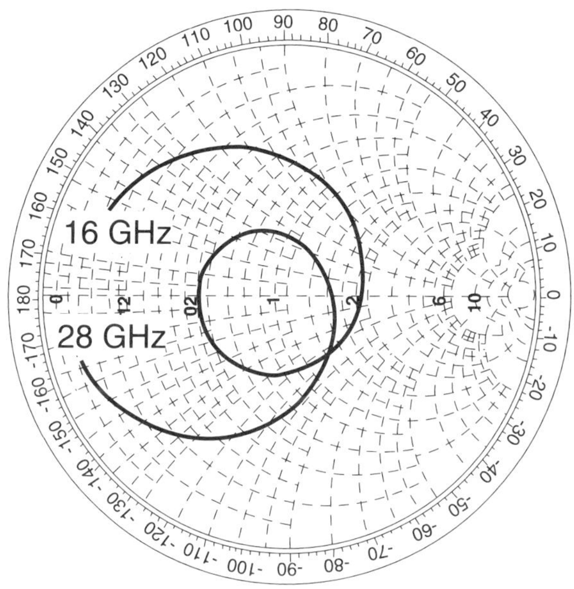

FIGURE 6.11 Typical input impedance locus of stacked microstrip patches.

Figure 6.11 shows a typical impedance locus of a stacked patch antenna configuration. The important characteristic to observe is the *resonant loop* created by the interaction between the two patches. This interaction governs the achievable bandwidth. In any stacked patch configuration there are many degrees of freedom and to say that one parameter controls the resonant loop over the other parameters is not necessarily true. However, the gap between the two patches (d_2) does primarily control the tightness of the resonant loop. The closer the patches are, the bigger the loop. It can be postulated simplistically that the closer the patches are together, the more coupling or interaction. In the extreme, with the second patch located directly on top of the driven patch, the loop becomes very large because only one patch can be seen. As the patches are moved farther apart, the coupling becomes less and the loop contracts. Eventually the loop closes entirely until the locus looks identical to a single-element configuration, because the driven patch no longer sees the parasitic antenna. Readers interested in the design procedures of stacked microstrip patches may refer to the articles within the literature[9,10] for details.

A stacked patch geometry over other bandwidth enhancement techniques has several advantages: they are relatively easy to design — once the design trends have been established; their radiation pattern remains reasonably constant over the 10-dB return loss bandwidth; and they can be easily accommodated into an array environment.

It is possible to stack more patches on top of the driven element, such as a triple-stacked patch. The success of such a configuration is very susceptible to the dielectric layers used and in fact has been shown to give minimum improvement over a conventional stacked patch when low dielectric constant laminates are utilized.[11] Triple-stacked patches are discussed further in Section 6.7.3.

6.4.3 Large Slot Aperture-Coupled Patches

A simple way of enhancing the bandwidth of an aperture-coupled patch without increasing the complexity of the antenna by means of stacking is to increase the size of the slot.[12] By having the slot close to resonance allows an impedance response similar to that presented in Fig. 6.11, because there are now two strongly coupled radiating structures, namely, the slot and the patch. In fact, having a relatively large slot is a natural progression to achieve large bandwidth. As shown in Fig. 6.2, to increase the bandwidth of a microstrip patch antenna, thick dielectric material is required. In an aperture-coupled patch environment, to ensure power is coupled to the patch located on a thick dielectric layer, the size of the slot must be increased. Bandwidths in excess of 40% have been achieved using this relatively simple technique. There are two problems with using a large slot aperture-coupled patch:

1. The front to back ratio can become poor. Because a slot in a ground plane radiates equally in both the upper and lower half spaces (ignoring the effect of the dielectric materials surrounding the ground plane), the level of backward radiation can become somewhat high. This leads to less power available in the power budget of a link and more importantly for sectored wireless communication systems, it can lead to increased levels of interference. This latter factor is very important for mobile communications systems. Typical base stations/radio hubs provide a sectoral coverage area to increase the system capacity; for example, for local multipoint distribution services (LMDS) the 360° azimuth plane is usually split into four 90° sectors, whereas for a mobile base station it is usually three 120° sectors. Imperative for these sectoral systems is to minimize the back radiation from each antenna to ensure interference from adjoining subcells that are also minimized. Several ways to overcome the inherent back radiation problem of a large slot aperture-coupled patch antenna are available, although these are usually accompanied by added structural complexity. These include using a cavity-backed configuration[13] or a reflector element[14] to minimize the backward directed radiation.
2. Scalping of the radiation pattern, particularly in the E-plane, can occur. This results from using a large slot in conjunction with a small ground plane. Because a reasonable portion of the power radiated from a slot is directed toward endfire, this power can be easily diffracted off a truncated ground plane, resulting in a deformation of the radiation pattern. Figure 6.12 shows an example of this phenomenon associated with the large slot of an aperture-coupled patch antenna. Possible ways to alleviate this problem are to ensure the ground plane extends a relatively large distance with respect to the center of the patch and slot, on the order of a couple of wavelengths. Also, absorbing material can be used to coat the edges of the ground plane.

Despite these problems, large slot aperture-coupled patches, or modifications of this microstrip patch antenna, are currently utilized as the antenna for several mobile communications base stations throughout the world.

6.4.4 Aperture-Stacked Patches

From the previous two bandwidth enhancement cases, a natural progression would be to use a large slot aperture-coupled stacked patch configuration to further improve the bandwidth. This printed antenna, referred to as an aperture-stacked patch (ASP) has been proposed, designed, and developed.[15,16] A schematic diagram of the antenna is shown in Fig. 6.13 and a photograph of the original ASP[15] is shown

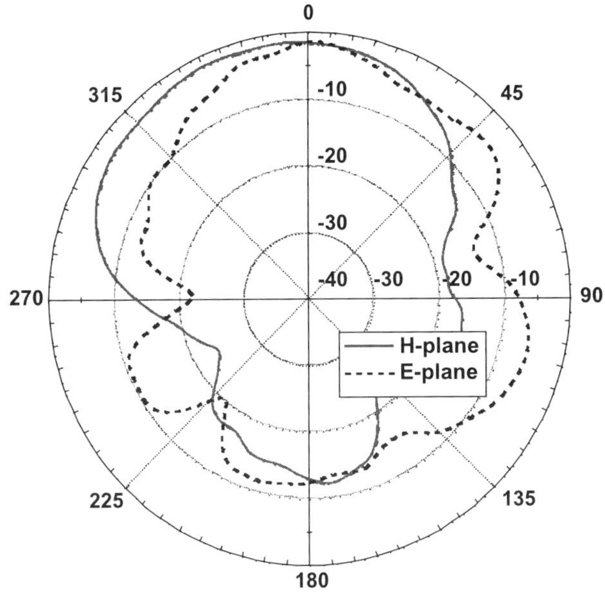

FIGURE 6.12 Example of radiation pattern scalping for large slot aperture-coupled microstrip patch antenna.

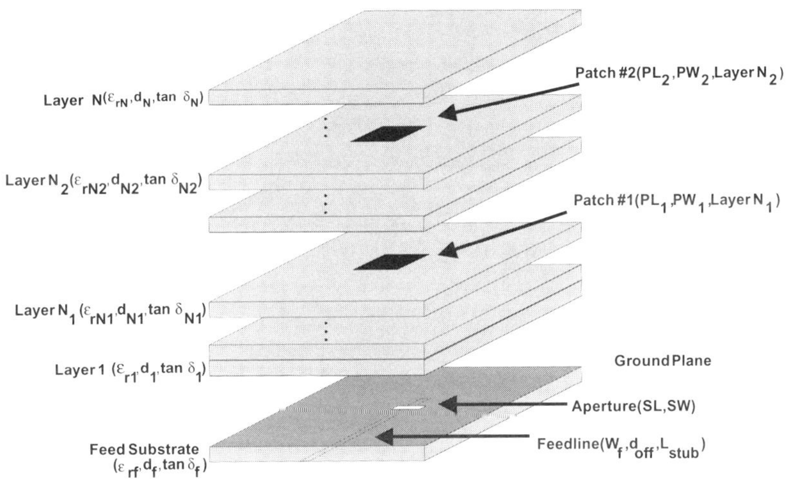

FIGURE 6.13 Schematic diagram of aperture-stacked patch (ASP) antenna.

in Fig. 6.14. The printed antenna consists of a large slot and two directive patches. Impedance bandwidths in excess of an octave have been achieved using this printed antenna configuration. A typical impedance locus of an ASP is shown in Fig. 6.15. As can be seen in this plot, there are two *resonant loops* resulting from the mutual resonances set up between with the three radiators, namely, the slot and the two directive patches. The gain of the ASP is also relatively constant over the octave bandwidth. The front to back ratio is not as poor as for that of a large slot aperture-coupled patch because of the additional directive patch, typically with values no less than 12 dB across the entire octave impedance bandwidth. This has been improved by utilizing cavity-backed solutions[17] as well as reflector patch solutions to values greater than 30 dB.[18] The design procedure for these wideband printed antennas is relatively straightforward.[16]

Presently, ASPs are probably the ultimate wideband printed antennas based on microstrip patch technology. These antennas have all the characteristics deemed important for wideband operation,

FIGURE 6.14 Photograph of ASP.

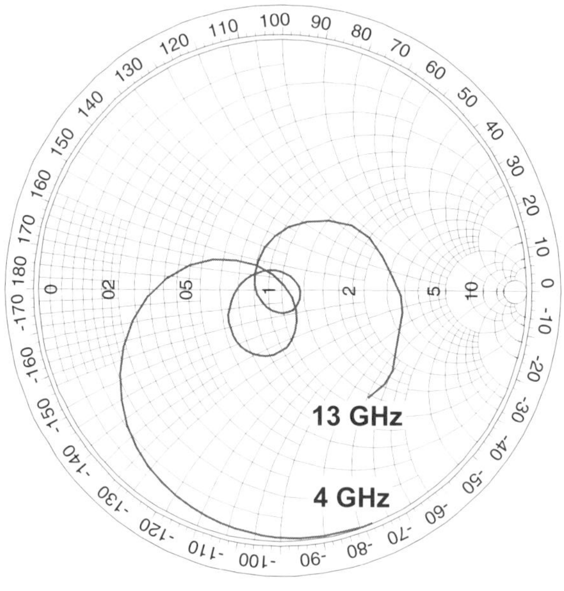

FIGURE 6.15 Typical input impedance locus of ASP.

namely, good impedance and gain bandwidth, good polarization control, compactness, and relatively simple development. In essence, the ASP can be thought of as a vertical log-periodic printed antenna; however, importantly it does not suffer from surface wave problems despite its electrical thickness. This is because the surface wave power is coupled to the adjoining patches and radiated into free space. ASP antennas, when designed appropriately have surface wave efficiencies greater than 85% even though the material used is greater than 0.1 λ_0, where λ_0 is the free-space wavelength at the center of the operation band. The large bandwidth encountered using ASPs potentially solves one of the problems that have dogged the universal deployment of microstrip patch technology into communication systems. The development of a printed antenna that is fundamentally broadband should enable various radio service applications to share some equipment hardware, thereby reducing the number of hub sites required to deliver these services and the overall cost of the radio network. Having a wideband printed antenna also potentially reduces the design stages required for the antenna development, because it tends to be more tolerant of fabrication discrepancies.

6.4.5 Alternative Printed Antenna Solutions

Of course, other printed antenna solutions can provide very wide bandwidths: printed spirals,[19] tapered slots,[20] and printed bow-tie antennas.[21] However, unless the bandwidth required is well in excess of an octave, these solutions are not as simple or as elegant as the ASP configuration. Tapered slots as well as printed spirals can yield multioctave bandwidth performance; however, the structural complexity and overall size of these antennas is a deterrent, especially if such bandwidths are not required.

Other printed antennas that have recently shown reasonable promise at providing broad bandwidth responses (10-dB return loss bandwidths between 30 and 45%). These include the L-shaped excited stacked patch,[2] incorporating slots into the patch conductors,[22] and also the printed quasi yagi antenna.[23]

6.5 Circular Polarization Techniques

One of the stated advantages of microstrip patch technology is the relative ease to generate dual and hence circular polarization (CP). This feature is one reason why microstrip patch antennas have been utilized for space-borne communication antennas on satellites. Three methods are used for generating CP: using a single feed and perturbation approach, a dual-feed excitation technique, or a synchronous subarray approach. Each method has its merits and disadvantages, which are summarized in this section.

6.5.1 Single-Feed Circular Polarization Patch Antenna

This approach requires no external circuitry for the microstrip patch antenna to generate CP. The principle relies on the fact that CP is obtained when two orthogonal but otherwise identical modes are excited with a 90° phase difference between them. To do so on a single patch with only one feed is not too difficult to implement. Figure 6.16 shows a schematic diagram of how this can be done. The patch is fed on one of its diagonal planes; therefore, it excites two orthogonal modes, one resonant in the x direction and the other in the y direction, satisfying one of the requirements for CP. To achieve 90° phase shift between the two modes, one mode (or field) is slightly perturbed with respect to the other. This can be done by making one resonant dimension slightly longer than the other, truncating a set of the corners of the patch or even putting a slot within the patch. The philosophy here is that by slightly

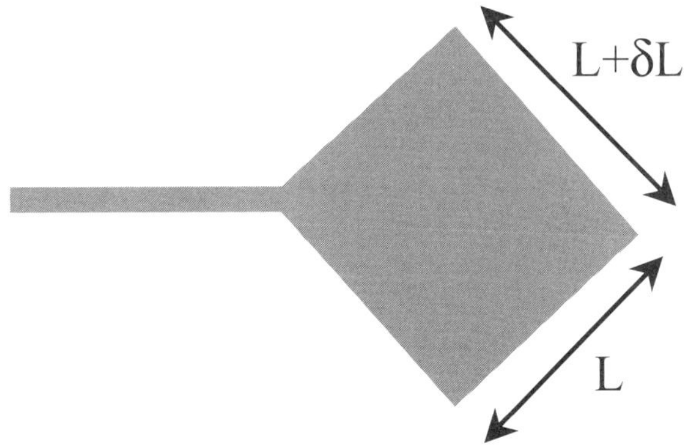

FIGURE 6.16 Schematic diagram of single-feed CP edge-fed microstrip patch antenna.

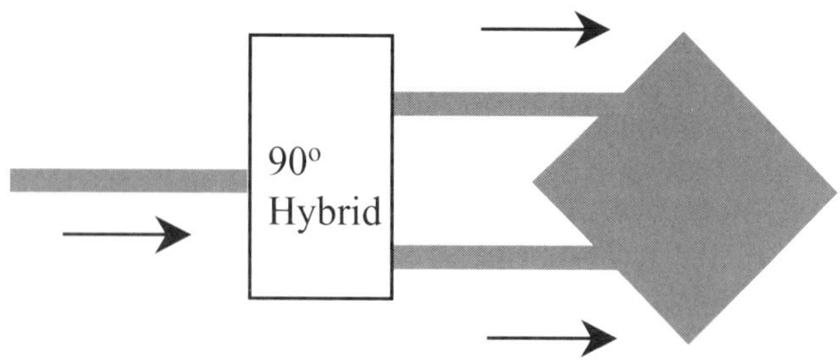

FIGURE 6.17 Schematic diagram of dual-feed CP edge-fed microstrip patch antenna.

adjusting the resonant frequency of one mode with respect to the other, if properly designed, then the overall phase difference between the self-impedances of the modes and therefore the fields will be 90°, satisfying the second requirement for CP generation.

Although relatively simple to implement, there is a critical flaw in this design procedure: the CP bandwidth (axial ratio less than 3 dB) is extremely narrow, typically a fraction of the impedance (10-dB return loss) bandwidth. Effectively the problem here is to ensure the phase relationship (90° phase difference) holds across a range of frequencies. This is quite difficult to do, especially when considering the rapid impedance variation of a typical microstrip patch antenna near resonance. A case has been presented where the CP bandwidth was greater than 10%, but still a fraction of the impedance bandwidth of the antenna.[24] The key to successfully broadening the CP bandwidth of this antenna is reducing the impedance variation of each mode as a function of frequency.

6.5.2 Dual-Feed Circular Polarization Patch Antenna

Probably the simplest and most common means of producing CP on a microstrip patch antenna is to excite two orthogonal modes using separate feeds and ensuring there is 90° phase difference between the modes by incorporating a 90° phase shift within one of the feed lines, as shown in Fig. 6.17. CP bandwidths comparable to the impedance bandwidth of the patch antenna are usually observed for this configuration. It can be somewhat difficult to obtain true CP (0-dB axial ratio) for a direct contact dual-feed CP patch antenna because of the inherent asymmetry of the configuration as well as the asymmetrical cross-polarization levels generated by the patch (note that the H-plane levels are higher than the E-plane levels off broadside, $\theta = 0°$).

6.5.3 Synchronous Subarrays

By far the best CP bandwidths and lowest axial ratios can be achieved using a synchronous subarray.[25] A four-element synchronous subarray in Fig. 6.18 shows each patch spatially rotated 90° with respect to the previous patch. In addition to the spatial rotation, the feeds are also phase rotated sequentially by 90°. Thus, port 2 leads port 1 by 90°, port 3 leads port 1 by 180°, and port 4 leads port 1 by 270°. Having such a feed arrangement cancels the generated cross-polarization fields of the individual patches. It has been shown in the past that very good axial ratio bandwidths can be achieved using a synchronous subarray of elements that individually do not radiate CP.[26] Also the impedance bandwidth of this CP antenna is typically larger than that of the single element because of the feed network. Synchronous subarrays do not necessarily consist of four elements; cases of three (0°, 120°, and 240°) and higher order cases have been investigated. As long as there is a spatial symmetry with respect to the center of the configuration as well as the appropriate phase rotation, this approach should provide good axial ratio performance. A typical axial ratio plot of a four-element synchronous subarray is shown in Fig. 6.19.

One significant disadvantage of the synchronous subarray is its electrical size. Because of this a synchronous subarray can have severe grating lobe problems when implemented in an array. It is potentially possible to utilize the synchronous subarray concept on a single patch; however, the limited

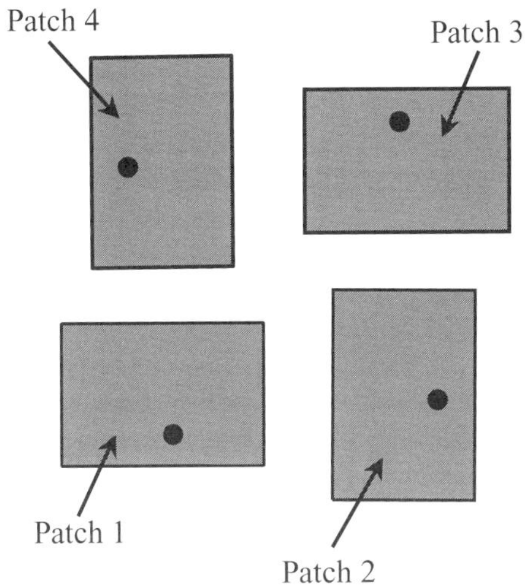

FIGURE 6.18 Schematic diagram of synchronous subarray of four probe-fed microstrip patch antennas.

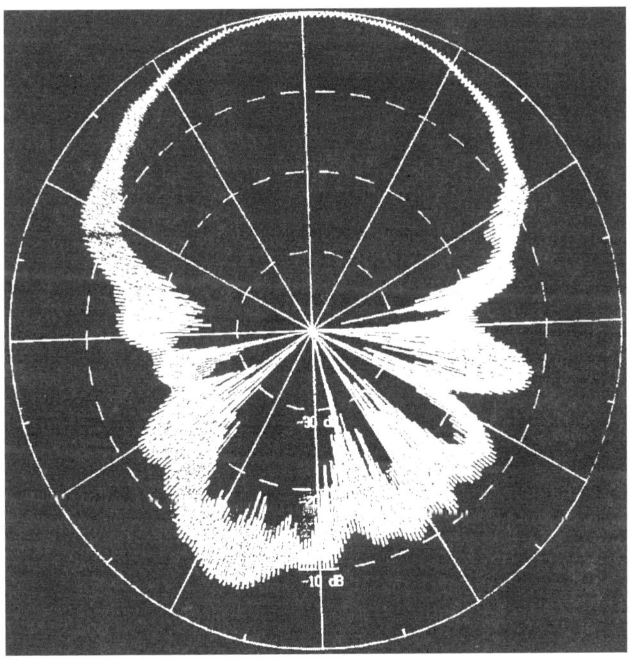

FIGURE 6.19 Typical axial ratio plot of synchronous subarray of microstrip patch antennas.

patch real estate as well as the effect of the coupling of each feed on the input impedance of the printed antenna could make this difficult to implement.

6.6 Reducing Conductor Size

With the advent and popularity of many wireless services, a quandary has arisen in the antenna community about how to develop small antennas that can satisfy the performance requirements for these systems as well as be aesthetically pleasing to the user. It is interesting to note that the latter point is not insignificant. As with the computer industry, mobile communications is very much a customer-driven market and thus the user requests, although technically with little merit, must be addressed. Ideally, an antenna that is unobtrusive and low cost, and can be located within the casing of the handset would ensure the compactness of the handset terminal and therefore please the users.

Microstrip antennas would appear to be possible candidates because of several attractive features outlined in Section 6.1, including their low profile, light weight, and ease of fabrication. Unfortunately, most present-day mobile communication systems are in the lower microwave region of the spectrum (less than 3 GHz) where these antennas in their conventional form are too large for wireless communication handsets. Several approaches have been reported in the literature to effectively reduce the size of the printed conductor of a microstrip patch antenna and are summarized as follows:

1. The quarter-wave patch antenna[27] is probably the first technique used to reduce the size of a patch antenna. Here one of the radiating edges is terminated in a short-circuit plane connecting the patch antenna to the ground plane. By doing so, the short-circuit plane acts as a mirror and the size of the patch conductor can be effectively halved, similar to the relation between a wire dipole and a wire monopole. Several cases of successfully developed quarter-wave patch antennas have been reported in the literature. Unfortunately, two drawbacks associated with this size reduction technique make it unsuitable for present-day mobile communication handset terminals. First, the size reduction is not great enough. Second, impedance bandwidth is reduced. Another issue related to this technology, although not that critical for mobile communication handset terminals, is the increased level of cross-polarized radiation.
2. The patch antenna on high dielectric constant material[28] is probably the most obvious means of reducing the patch conductor size. The main issues are the associated reduction in bandwidth and efficiency when using such a technique (refer to Fig. 6.2 for details). Having such narrowband radiators makes the performance of these antennas very susceptible to manufacturing tolerance errors. Also the cost associated with high dielectric constant material can be somewhat prohibitive, although the recent availability of some low-cost, high dielectric constant ceramic materials may alleviate the cost issue to an extent. Care, however, must be taken when using these materials because the high loss tangents can considerably degrade the radiation performance of the printed antenna.
3. Shorted patch antennas[29] have shown that a significant reduction in physical size of the patch conductor can be achieved if a single shorting post is used, instead of the conventional short-circuit plane of a quarter-wave patch. A schematic diagram of a probe-fed shorted patch is shown in Fig. 6.20. A rectangular patch is loaded with a shorting pin located in close proximity to the probe feed. Reductions in size of greater than a further halving have been achieved using this procedure and the bandwidth and efficiency issues appear not to be as severe as for the case of using very high dielectric constant substrates. However, the fundamental problem associated with this form of printed antenna is that it is still difficult to achieve the necessary bandwidth, and also the close proximity between the feed and the shorting post, especially when high dielectric constant or thick materials are used, makes the antenna difficult to fabricate. A summary of the design procedure and performance of shorted patch antennas is available in the literature.[30] Since then, there have been several variations of shorted patch that have alleviated some of the encountered problems to an extent, although typically at the expense of increased volume (e.g., a stacked shorted patch[31]), or increased overall complexity and component count. These factors are very important when considering high-volume manufacturing runs. It should be noted that with respect to size reduction, in the limit, shorted patches are very similar to planar inverted-F antennas (PIFAs).[32] The difference really is associated with how the problem is formulated. Shorted patches were derived from the patch antenna concept and therefore are electrically thin, whereas PIFAs originated from folded wire antennas and thus the thickness between the radiating conductor and the ground plane is somewhat large. One means of increasing the bandwidth of a shorted patch is to

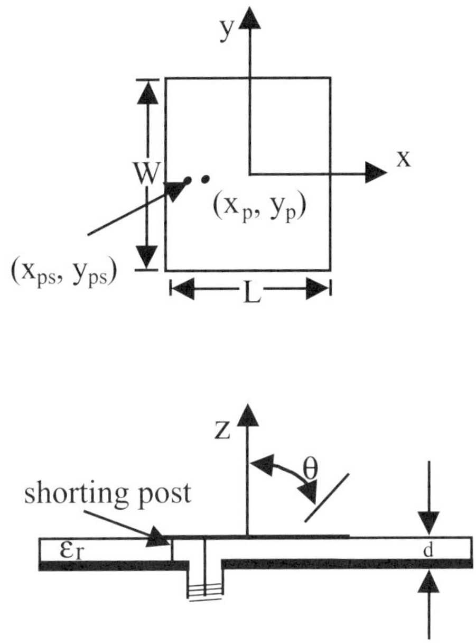

FIGURE 6.20 Schematic diagram of probe-fed shorted microstrip patch antenna.

increase the thickness of the substrate. As this is done, the characteristics, both electrical and mechanical, of the shorted patch antenna become more and more similar to that of a PIFA.

4. Resistive loading patch antennas[33] using resistive terminations as opposed to a shorting pin have been proposed to give size reduction of the conductor of a microstrip patch without reducing the associated impedance bandwidth. A very important disadvantage of using resistive terminations to reduce the size of an antenna is the associated reduction in radiation efficiency. Thus although an antenna using a resistor as a load can yield small size and reasonable bandwidth, the associated decrease in efficiency has a more dramatic effect on the system performance. This can be simply proved by applying the Friis transmission equation[1] to a link utilizing the antenna.

6.7 Integration with Active Devices and Examples of Active Patches

As is evident in Fig. 6.2 and was previously discussed, the microstrip patch antenna in its pure form despite the initial enthusiasm is not suited to monolithic integration with active devices, because of the low achievable bandwidths and poor radiation efficiency. Over the years there have been several techniques developed to allow such integration and these are summarized next.

6.7.1 Noncontact Feed Mechanisms

As shown in Section 6.3, noncontact feed mechanisms can alleviate these problems to an extent. Proximity coupled patches, although allowing some degree of independent optimization of the feed substrate and antenna laminate, are not well suited to integration with MMIC and OEIC technology because of the relatively poor surface wave efficiencies and also the difficulty in coupling power to the patch antenna. The first problem results from the patch antenna still seeing some of the high dielectric constant material used for the active devices and therefore affecting the efficiency. The latter problem can be simply explained by the fact that the microstrip line on the high dielectric constant material has most of its field contained between the conductor and the ground plane, therefore making it difficult to couple power to

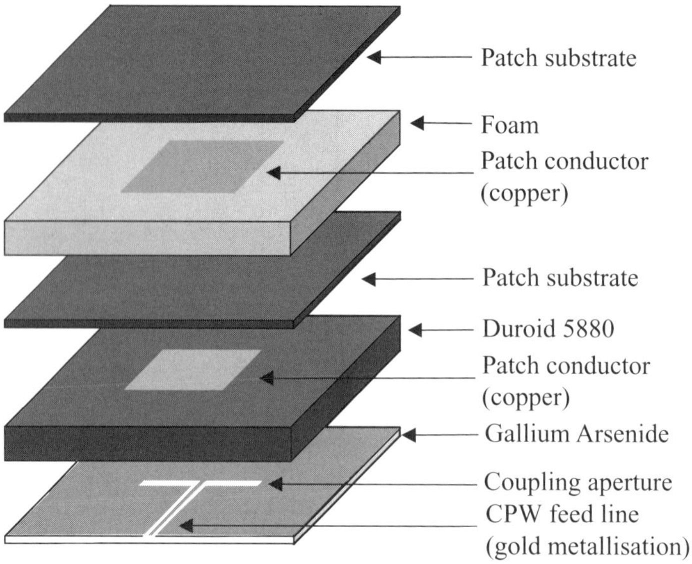

FIGURE 6.21 Schematic diagram of CPW-fed ASP.

the antenna. To do so, the high dielectric constant substrate must be increased in thickness, or the antenna laminate must be reduced in thickness. Both of these measures result in poor bandwidth and surface wave efficiency.

Aperture-coupled solutions are better suited to integration with active devices, although care must be taken to ensure surface wave efficiencies remain high. When the aperture or slot is small and a high dielectric constant substrate is used for the feed, the surface wave losses associated with the feed can be high, significantly reducing the radiation efficiency of the antenna.[34] However, as the aperture size is increased, the reduction in surface wave efficiency because of the high dielectric constant feed material seems to diminish. Simply, the slot tends to draw power from the surface wave thereby increasing the efficiency of the antenna. Versions of co-planar waveguide (CPW)-fed ASP antennas (refer to Section 6.4 for a description of the ASP antenna), using high dielectric constant substrates for the feed, have been designed and developed with impedance bandwidths greater than 40% and surface wave efficiencies greater than 85% across this band.[35] A schematic diagram of this antenna is shown in Fig. 6.21. CPW feed configurations were considered because of the popularity of this transmission line for integration with active devices. Similar results for microstrip line fed ASP antennas using a high dielectric constant feed substrate can also be achieved. As pointed out in Section 6.4, for large aperture or slot-coupled printed antennas care must be taken with backward radiation. Although the ASP can be easily integrated with millimeter-wave and photonic devices in its present form, one potential difficulty is that these active devices need some form of heat sinking and support structure. To implement this for the ASP is not typically straightforward because the performance of the antenna may be compromised if a metal plate for mechanical support is located in close proximity to the electrically large slot or if the ground plane in which the slot resides is electrically thick.[36] One possible solution to the design problem is to incorporate a cavity-backed structure as outlined in Section 6.4, where a cavity is located under the slot of the antenna. This not only provides a good support structure for the antenna and the active devices but also improves the front to back ratio of the antenna itself.

6.7.2 Antennas with Reduced Surface Wave Excitation

Another technique to overcome the problem of inefficient patch antennas resulting from surface wave excitation has been proposed.[37] In this case, it was shown that if the size of the antenna were beyond a critical dimension related to the propagation constant of the TM_0 surface wave, the TM_0 surface wave

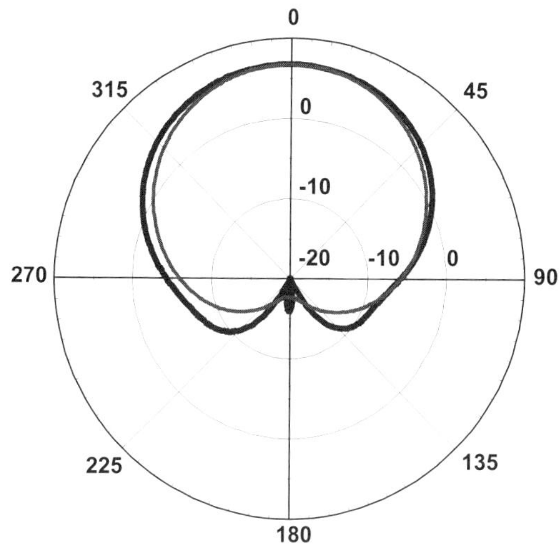

FIGURE 6.22 Measured radiation pattern of circular microstrip patch with concentric annular ring.

mode would not propagate. Several printed antennas were demonstrated, although these microstrip patch antennas are relatively difficult to construct and are extremely narrowband. Simpler microstrip patch antennas, consisting of a circular patch and a concentric annular coupled ring have been developed using the same principal and showing similar improved efficiency and also slightly enhanced bandwidths.[38] The measured co-polar radiation patterns of a circular patch and a concentric annular coupled ring[38] mounted on an electrically thick, high dielectric constant material is shown in Fig. 6.22 at the resonant frequency of the antenna. Note that despite a small ground plane used, there is no detectable ripple in either the E-plane or H-plane patterns. This can be attributed to the minimum surface wave excitation and hence the minimum diffraction off the ground plane of the patch. There are two relatively fundamental drawbacks of the oversized patch technique: only moderate bandwidths can be achieved; and also the large size of the patch conductor is required. This latter problem is indeed an issue when only limited space on a wafer is available.

6.7.3 Stacked Patch Structures

It is interesting to note that throughout the history of the development of stacked patch technology, a thorough investigation into the most appropriate materials/laminates to achieve the optimum performance had not been undertaken. This can be explained by the results presented in Fig. 6.2, namely, that single-layer patch antennas do not perform well when high dielectric constant materials are used as a substrate; thus, it was thought that similar results would result when a stacked configuration was utilized. However, this has been shown to be incorrect. It has been reported that a combination of high dielectric constant and low dielectric constant laminates, referred to as hi–lo configurations,[24] can yield similar impedance bandwidths and radiation performance to that of a conventional stacked patch.[9] Unlike cases where a single-layer case using a high dielectric constant substrate where the surface wave efficiency is poor, the hi–lo stacked patch has very high surface wave efficiency across the impedance bandwidth. This somewhat startling revelation opens up a new avenue in the area of integration research. Hi–lo stacked patches have several salient features that are useful for integration with MMICs and OEICs: (1) good bandwidth can be achieved; (2) minimum back radiation is present because of the ground plane; (3) they are relatively simple to design;[24] (4) low cross-polarization levels are radiated, indicating that good quality CP can easily be generated.[24]

Maximum bandwidths of approximately 30% have been achieved with a hi–lo stacked patch configuration. This has been further extended utilizing a third patch, or a triple-stacked patch configuration.[11] A schematic diagram of the triple patch is shown in Fig. 6.23. A triple-stacked patch arrangement only

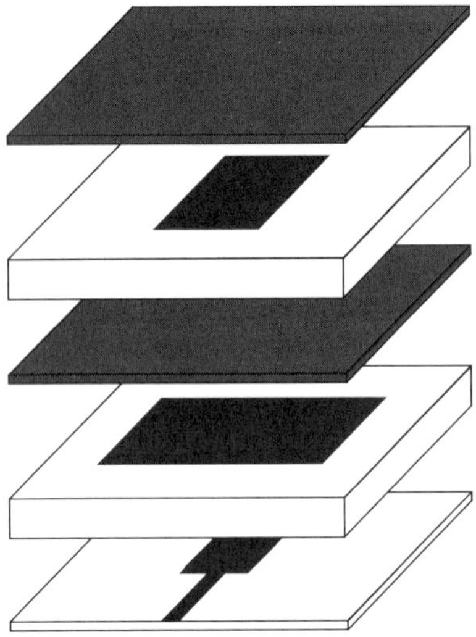

FIGURE 6.23 Schematic diagram of edge-fed, triple-stacked patch antenna.

shows improved characteristics, beyond that achievable for a conventional stacked patch antenna, when the lowest layer material has a relatively high dielectric constant. Bandwidth performance approaching that of a large slot aperture-coupled patch has been achieved without the problems of backward-directed radiation. A comparison of the radiation patterns of a hi–lo patch structure and a CPW-fed ASP antenna is shown in Fig. 6.24.[39] As can be seen from these plots, the hi–lo patch configuration has a significantly better front to back ratio compared with the CPW-fed ASP. The ultimate bandwidth that can be achieved as well as the best combination of materials is ongoing research and hopefully these questions will be answered soon.

6.7.4 Patch Antennas on Photonic Bandgap Structures

Considerable attention has been directed toward the development of metallic electromagnetic structures that are characterized by having high-surface impedances. These structures, unlike conventional surfaces, do not support propagating surface waves. Several versions of these frequency-dependent impedance surfaces have been reported, commonly referred to photonic bandgap (PBG) structures or substrates.[40,41] The material is periodically loaded to create an "electromagnetic crystal" whose surface-wave dispersion diagram presents a forbidden range of frequencies hopefully near the operation frequency of the antenna. Because surface waves cannot propagate along the substrate, an increased amount of radiated power couples to space waves, enhancing the radiation efficiency. Some promising results have been achieved for microstrip antenna structures mounted on PBG materials. However, there are still several questions that need to be resolved before this technology is universally accepted, namely: (1) how large does the periodic structure need to be before it works effectively; and (2) how broadband can the structure be?

6.7.5 Examples of Integrated Antennas

Although from the previous subsections it is evident that there are several techniques to overcome the problem of direct integration of MMICs and OEICs with patch technology, very few successful cases are reported in the literature on direct integration. This is simply because this technology is quite immature at present. On the other hand, the area of hybrid integration, commonly referred to as active integrated antennas, is somewhat more established and there are many cases of such printed antennas. An excellent review article can be found in the literature.[42]

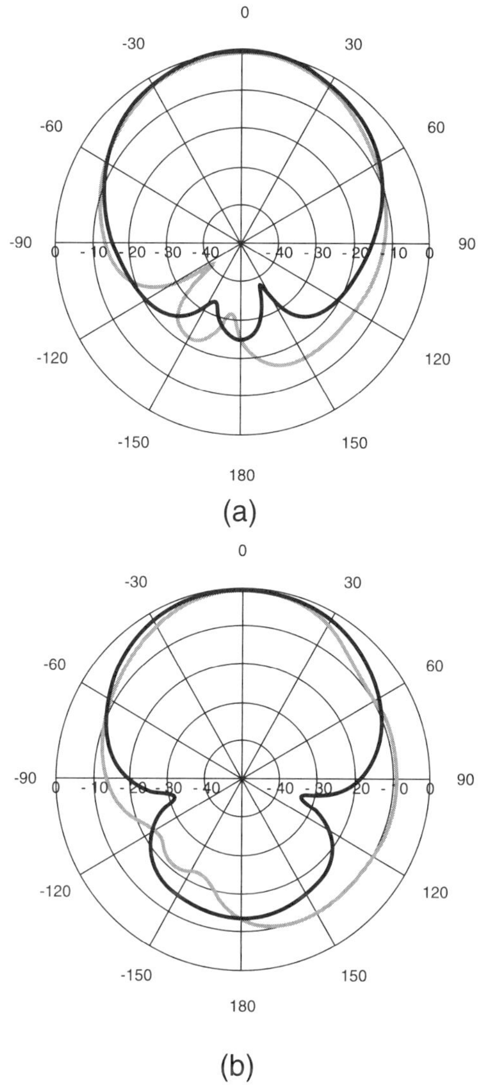

FIGURE 6.24 Comparison of radiation patterns of hi–lo stacked patch antenna and CPW-fed ASP antenna.

The purpose of an active integrated antenna is to have the active circuitry within the antenna structure itself, thereby removing the lossy and sometimes bulky interconnects between the devices and antenna. This concept, if successfully mastered, can yield systems that are compact, low cost, and low profile, with minimum power consumption and a high degree of flexibility. The degree of flexibility is apparent because the designer is no longer restricted to a 50-Ω system and thus the impedance response of the antenna can be tailored to that which gives, for example, the minimum noise response for the amplifier. Over the years there have been several reported cases of printed active integrated antennas including high efficiency power amplifiers,[43,44] quasi-optical power combiners,[45,46] beam-steering arrays,[47,48] optically assisted active integrated antennas,[49,50] retrodirective arrays,[51,52] and transceivers and transponders.[53,54]

It should be noted at this stage that active integrated antennas cannot provide full duplex operation (transmit and receive at the same time) for present-day styled mobile communications systems. This can be attributed to one fundamental reason. Typically, some form of frequency diversity is required, namely, there is an upper band and lower band for transmitting and receiving. The natural filtering response of a patch antenna, whether it be frequency or polarization based, is not pure enough to satisfy both the

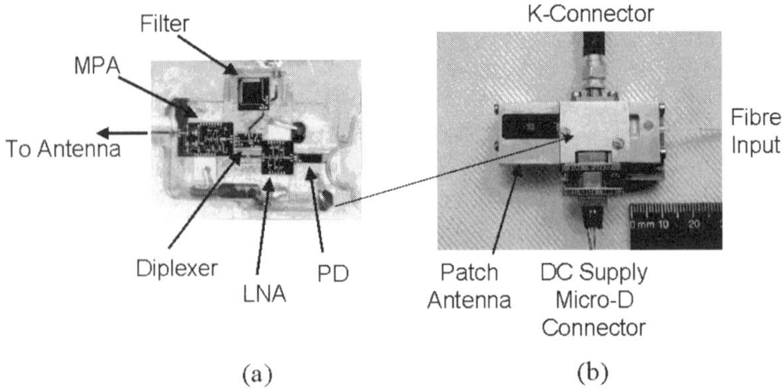

FIGURE 6.25 Photograph of integrated photonic/antenna module.

bandwidth requirements for transmitting and receiving only and to give the isolation required for the small guard band typically used in today's mobile communication systems.

Cases of hybrid integrated antenna photonic modules have also been reported.[55-57] The objectives here are to have modules that can be readily connected to fiber-backbone networks for such applications as picocellular mobile communications. Using fiber technology for these wireless access networks allows most of the expensive switching and signal processing functions to be located at a central office, thereby reducing the overall cost of the system. Imperative for these systems are small, efficient base station units, or photonic/antenna modules. Figure 6.25 shows a photograph of a hybrid integrated antenna photonic module for operation between 28 and 43 GHz.[57] Once again, probably because of the immaturity of the field, successful directly integrated cases have yet to be developed at this stage.

6.8 Conclusions

In this chapter, several concepts related to microstrip patch antenna technology have been summarized. The fundamental characteristics of microstrip patch antennas have been presented; methods to couple power to and from the printed antenna have been summarized; techniques to enhance the bandwidth and efficiency have been reported; and how to reduce the conductor size of a patch antenna and methods to generate circular polarization have been discussed. It is hoped that this chapter has provided the reader with the essentials required to understand microstrip patch technology and how to optimize the performance of this form of antenna for the application at hand. For further information on the topics presented the reader should consult the given reference material.

References

1. Balanis, C.A., *Antenna Theory: Analysis and Design* (second edition), John Wiley & Sons, New York, 1997.
2. Kuo, J.-S. and Wong, K.-L., A dual-frequency L-shaped patch antenna, *Microwave & Optical Technology Letters*, vol. 27, pp. 177–179, Nov. 2000.
3. Munson, R.E., Conformal microstrip antennas and microstrip phased arrays, *IEEE Transactions Antennas & Propagation*, vol. 22, pp. 74–78, Jan. 1974.
4. Pozar, D.M., Microstrip antenna aperture-coupled to a microstrip-line, *Electron. Lett.*, vol. 21, pp. 49–50, Jan. 1985.
5. Pozar, D.M., Increasing the bandwidth pf a microstrip antenna by proximity coupling, *Electron. Lett.*, vol. 23, pp. 368–369, 1987.

6. Kumar, G. and Gupta, K.C., Nonradiating edges and four edges gap-coupled multiple resonator broad-band microstrip antennas, *IEEE Transactions Antennas & Propagation*, vol. 33, pp. 173–178, Feb. 1985.
7. Waterhouse, R.B., Rigorous analysis of probe-fed microstrip antennas incorporating parasitic elements to enhance the bandwidth, (*invited*) *Progress in Electromagnetics Research Symposium*, Seattle, WA, p. 596, July 1995.
8. Long, S.A. and Walton, D.M., A dual-frequency stacked circular-disc antenna, *IEEE Transactions Antennas & Propagation*, vol. 27, pp. 270–273, Mar. 1979.
9. Waterhouse, R.B., Design of probe-fed stacked patches, *IEEE Transactions Antennas & Propagation*, vol. 47, pp. 1780–1784, Dec. 1999.
10. Croq, F. and Pozar, D.M., Millimeter wave design of wide-band aperture-coupled stacked microstrip antennas, *IEEE Transactions Antennas & Propagation*, vol. 39, pp. 1770–1776, Dec. 1991.
11. Rowe, W.S.T. and Waterhouse, R.B., Broadband microstrip patch antenna for MMICs, *Electron. Lett.*, vol. 36, pp. 597–599, Apr. 2000.
12. Zurcher, J.-F. and Gardiol, F.E., *Broadband Patch Antennas*, Artech House, Boston, 1995.
13. Pozar, D.M. and Schaubert, D.H., *Microstrip Antennas*, IEEE Press, Piscataway, NJ, 1995.
14. Targonski, S.D., Waterhouse, R.B., and Pozar, D.M., Wideband aperture coupled microstrip patch array with backlobe reduction, *Electron. Lett.*, vol. 33, pp. 2005–2006, Nov. 1997.
15. Targonski, S.D., Waterhouse, R.B., and Pozar, D.M., A wideband aperture coupled stacked patch antenna using thick substrates, *Electron. Lett.*, vol. 32, pp. 1941–1942, Oct. 1996.
16. Targonski, S.D., Waterhouse, R.B., and Pozar, D.M., Design of wideband aperture-stacked patch microstrip antennas, *IEEE Transactions Antennas & Propagation*, vol. 46, pp. 1246–1251, Sept. 1998.
17. Waterhouse, R.B., Novak, D., Nirmalathas, A., and Lim, C., Broadband printed antennas for point-to-point and point-to-multipoint wireless millimetre-wave applications, *IEEE Antennas & Propagation Symposium*, Salt Lake City, Utah, pp. 1390–1393, July 2000.
18. Waterhouse, R.B., Novak, D., Nirmalathas, A., and Lim, C., Broadband printed antennas with reflector elements for millimetre-wave wireless applications, *2000 IEEE-APS Conference on Antennas & Propagation for Wireless Communications*, pp. 47–50, Nov. 2000.
19. Wang, J.J.H. and Tripp, V.K. Design of multioctave spiral-mode microstrip antennas, *IEEE Transactions Antennas & Propagation*, vol. 39, pp. 332–335, March 1991.
20. Chio, T. and Schaubert, D.H., Effects of slotline cavity on dual-polarized tapered slot antenna arrays, *1999 Antennas & Propagation International Symposium*, Orlando, FL, pp. 130–133, July 1999.
21. Serrano-Vaello, A. and Sanchez-Hernandez, D., Printed antennas for dual-band GSM/DCS 1800 mobile handsets, *Electron. Lett.*, vol. 34, pp. 140–141, Jan. 22, 1998.
22. Huynh, T. and Lee, K.-F., Single-layer single-patch wideband microstrip antenna, *Electron. Lett.*, vol. 31, pp. 1310–1312, Aug. 3, 1995.
23. Qian, Y., Deal, W.R., Kaneda, N. and Itoh, T., A uniplanar quasi-yagi antenna with wide bandwidth and low mutual coupling characteristics, *1999 IEEE AP-S International Symposium*, Orlando, FL, pp. 924–927, July 1999.
24. Waterhouse, R.B., Stacked patches using high and low dielectric constant material combination, *IEEE Transactions Antennas & Propagation*, vol. 47, pp. 1767–1771, Dec. 1999.
25. Huang, J., A technique for an array to generate circular polarization with linearly polarized elements, *IEEE Trans. Antennas & Propagat.*, vol. 34, pp. 1113–1124, Sept. 1986.
26. Pozar, D.M., Scanning characteristics of infinite arrays of printed antenna subarrays, *IEEE Trans. Antennas & Propagat.*, vol. 40, pp. 666–674, June 1992.
27. Zaid, L., Kossiavas, G., Dauvignac, J.-Y., Cazajous, J. and Papiernik, A., Dual-frequency and broadband antennas with stacked quarter wavelength elements, *IEEE Transactions Antennas & Propagation*, vol. 47, pp. 654–659, April 1999.

28. Lo, T.K., Ho, C.-O., Hwang, Y., Lam, E.K.W. and Lee, B., Miniature aperture-coupled microstrip antenna of very high permittivity, *Electron. Lett.*, vol. 33, pp. 9–10, Jan. 1997.
29. Waterhouse, R.B., Small microstrip patch antenna, *Electron. Lett.*, vol. 31, pp. 604–605, Apr. 1995.
30. Waterhouse, R.B., Targonski, S.D. and Kokotoff, D.M., Design and performance of small printed antennas, *IEEE Transactions on Antennas and Propagation*, vol. 46, pp. 1629–1633, Nov. 1998.
31. Waterhouse, R.B., A broadband stacked shorted patch, *Electron. Lett.*, vol. 35, pp. 98–100, Jan. 1999.
32. Taga, T. and Tsunekawa, K., Performance analysis of a built-in planar inverted F antenna for 800 MHz band portable radio units, *IEEE Journal on Selected Areas in Communications*, vol. 5, pp. 921–929, June 1987.
33. Wong, K.–L. and Lin, Y.F., Small broadband rectangular microstrip antenna with chip-resistor loading, *Electron. Lett.*, vol. 33, pp. 1593–1594, Sept. 1997.
34. Pozar, D.M., Analysis of an infinite phased array of aperture coupled microstrip patches, *IEEE Trans. Antennas & Propagat.*, vol. 37, pp. 418–424, April 1989.
35. Rowe, W.S.T. and Waterhouse, R.B., Broadband CPW fed stacked patch antenna, *Electron. Lett.*, vol. 35, pp. 681–682, April 1999.
36. Haddad, P.R. and Pozar, D.M., Analysis of two aperture-coupled cavity-backed antennas, *IEEE Trans. Antennas & Propagat.*, vol. 45, pp. 1717–1726, Dec. 1997.
37. Jackson, D.R., Williams, T.J., Bhattacharyya, A.K., Smith, R.L., Buchheit, S.J. and Long, S.A., Microstrip patch designs that do not excite surface waves, *IEEE Trans. Antennas & Propagat.*, vol. 41, pp. 1026–1037, Aug. 1993.
38. Kokotoff, D.M., Waterhouse, R.B., Birtcher, C.R. and Aberle, J.T., Annular ring coupled circular patch with enhanced performance, *Electron. Lett.*, vol. 33, pp. 2000–2001, Nov. 1997.
39. Rowe, W.S.T. and Waterhouse, R.B., Comparison of broadband millimetre-wave antenna structures for MMIC and optical device integration, *IEEE Antennas & Propagation Symposium*, Salt Lake City, Utah, pp. 1406–1409, July 2000.
40. Sievenpiper, D., Zhang, L., Broas, R.F.J., Alexopolous, N.G. and Yablonovitch, E., High impedance electromagnetic surfaces with a forbidden frequency band, *IEEE Transactions on Microwave Theory & Techniques*, vol. 47, pp. 2059–2074, Nov. 1999.
41. Coccioli, R., Yang, F.-R., Ma, K.-P. and Itoh, T., Aperture-coupled patch antenna on UC-PBG substrate, *IEEE Transactions on Microwave Theory & Techniques*, vol. 47, pp. 2123–2130, Nov. 1999.
42. Qian, Y. and Itoh, T., Progress in active integrated antennas and their applications, *IEEE Transactions on Microwave Theory & Techniques*, vol. 46, pp. 1891–1900, Nov. 1998.
43. Schultz, V., Chew, S.T., Qian, Y. and Itoh, T., High efficiency power amplifier integrated with antenna, *IEEE Microwave & Guided Wave Letters*, vol. 7, pp. 39–41, Feb. 1997.
44. Radisic, V., Qian, Y. and Itoh, T., Class-F power amplifier integrated with circular sector microstrip antenna, *IEEE MTT-S International Microwave Symposium*, Denver, CO, pp. 687–690, June 1997.
45. York, R. and Compton, R.C., Quasi-optical power combining using mutually synchronized oscillator arrays, *IEEE Transactions on Microwave Theory & Techniques*, vol. 39, pp. 1000–1009, June 1991.
46. Mortazawi, A., Foltz, H.D. and Itoh, I., A periodic spatial power combining oscillator, *IEEE Transactions on Microwave Theory & Techniques*, vol. 40, pp. 851–856, May 1992.
47. Liao, P. and York, R.A., A 1-W X-band power combining array using coupled VCO's, *IEEE MTT-S International Microwave Symposium*, San Diego, CA, pp. 1235–1238, June 1994.
48. Nogi, S., Sanagi, M., Sono, M. and Miyake, F., Injection signal controlled active phased array with a frequency shifted end element, *Asia Pacific Microwave Conference*, Hong Kong, pp. 953–956, Dec. 1997.
49. Deal, W.R. and Itoh, T., An active phased array with optical control and beam-scanning capability, *Microwave Photonics '97*, Duisberg, Germany, pp. 175–178, Sept. 1997.
50. Chew, S.T., Tong, D.T.K., Wu, M.C. and Itoh, T., Use of direct-modulated/gain-switched optical links in monopulse-type active phased arrays, *IEEE Transactions on Microwave Theory & Techniques*, vol. 44, pp. 326–330, Feb. 1996.

51. Pobanz, C.W. and Itoh, T., A conformal retrodirective array for radar applications using a heterodyne phased scattering element, *IEEE MTT-S International Microwave Symposium*, Orlando, FL, pp. 905–908, May 1995.
52. Karode, S.L. and Fusco, V.F., Novel retrodirective beam forming techniques, *27th European Microwave Conference*, Jerusalem, Israel, pp. 81–86, Sept. 1997.
53. Singer, M., Strohm, K.M., Luy, J.-F. and Biebl, E.M., Active SIMMWIC antenna for automotive applications, *IEEE MTT-S International Microwave Symposium*, Denver, CO, pp. 1265–1268, June 1997.
54. Singh, D., Gardner P., and Hall, P.S., Frequency doubling integrated push-pull active microstrip transponder, *27th European Microwave Conference*, Jerusalem, Israel, pp. 1181–1185, Sept. 1997.
55. Nagatsuma, T., Hirata, A., Royter, Y., Shinagawa, M., Furuta, T. and Ito, H., A 120 GHz integrated photonic transmitter, *IEEE International Topical Meeting on Microwave Photonics (MWP 2000)*, Oxford, UK, pp. 225–228, Sept. 2000.
56. Paolella, A.C., Joshi, A.M. and Bauerle, A., Photonic modules for millimeter wave communication systems, *IEEE International Topical Meeting on Microwave Photonics (MWP 2000)*, Oxford, UK, pp. 233–236, Sept. 2000.
57. Novak, D., Nirmalathas, A., Lim, C., Waterhouse, R.B. and Smith G.H., Low cost fiber radio antenna modules, *presented at the IMS 2001 Workshop on Microwave Photonic Component, Integration and System Techniques for Broadband Fiber-Fed Wireless*, Phoenix, AZ, May 2001.

7

The Finite Difference Time Domain Technique for Microstrip Antenna Applications

Atef Z. Elsherbeni
University of Mississippi

Christos G. Christodoulou
University of New Mexico

Javier Gómez-Tagle
ITESM

7.1 Introduction

The finite difference time domain (FDTD) method has gained tremendous popularity in the past decade as a tool for solving Maxwell's equations. FDTD has been used very successfully in the design of antennas for several applications ranging from simple microstrip antennas to complex phased-array antennas. Some of these antennas are currently being used in mobile communications (cellular, personal communication systems and networks), satellite communications, global positioning system (GPS), and aeronautical and radar systems.

Before such antennas are fabricated, it is necessary to perform the analysis at several stages so that various candidate antennas can be considered viable. The designer proceeds to select a few of such candidate antennas based on both computer-aided performance analysis and cost of fabrication. The final system may be selected based on a trade-off between cost and performance parameters.

In determining which technique to use in modeling any antenna, it is useful to consider the different methodologies available. These methods are divided into the following categories: empirical models, semiempirical models, and full-wave models.

The empirical models are generally based on some fundamental simplifying assumptions concerning the radiation mechanism of the antenna. The assumptions are extracted from experimental experience. Phenomena such as surface wave propagation in a microstrip antenna and dispersion are generally not included in these models. The importance of empirical models in providing a qualitative idea about the antenna behavior and at least a first-order solution to a design problem cannot be denied.

0-8493-0124-6/02/$0.00+$1.50

Semiempirical models are a hybrid of empirical and full-wave analyses. The analytic and computational complexity involved is more than that of the empirical models and less than that of the full-wave models, and the effects of surface wave modes are taken into account in many of these models.

The full-wave analyses include formulations that are electromagnetically rigorous (no empirical or semiempirical assumptions are made) as well as computationally extensive. The most common numerical techniques are the method of moments (MOM), the finite difference technique, and the finite element method (FEM). Each one of these approaches can be implemented in the time domain and the frequency domain, and can be used to obtain the radiation pattern, gain and input impedance (for radiation problems), scattering pattern, and gain and radar cross section (for scattering problems).

In this chapter, the FDTD is introduced with emphasis on its applications to printed antennas and antenna arrays. The main advantages of the FDTD method are

1. It can accommodate very general and intricate modeling requirements.
2. When the geometric and material configuration becomes very complex, the algorithm complexity remains low. This is particularly true when modeling highly inhomogeneous dielectric objects.
3. The method requires O(N) storage for N unknowns. This is in contrast to finite element and moment method techniques that require storing a matrix consisting of $O(N^2)$ entries.
4. This technique does not require a linear system solution (matrix inversion).
5. Because it is a time-domain technique, it can predict the transient response of an electromagnetic system. When these transient data are transformed to the frequency domain, they translate into a wideband system response.

The advent of high-speed, large-memory personal and supercomputer systems has made possible the development of accurate electromagnetic algorithms such as the FDTD method. This method solves Maxwell's time-dependent equations directly in the time domain by converting them into finite difference equations. The finite difference equations are then solved in a time marching sequence by alternately calculating the electric and magnetic field components on an interlaced spatial grid.

Yee introduced the first FDTD algorithm to the electromagnetic community in 1966 [1]. Since then, the algorithm has been used in electromagnetic wave scattering both in the time and frequency domains, electromagnetic interactions with biological tissues, antenna radiation, and microwave circuit design, among several other applications. Research efforts in the area of the FDTD method are currently directed toward advancing its capabilities with parallel algorithms that can be used in parallel and distributed computers.

7.2 Finite Difference Time Domain Fundamentals

7.2.1 Basic Formulation

FDTD is a discrete representation of Maxwell's equations using, in its common form, a central difference scheme in both time and space. To simulate the wave propagation in three dimensions, Yee's original algorithm (also called leapfrog algorithm) is usually adopted. Maxwell's equations are replaced by a system of finite difference equations. The differential time domain Maxwell equations needed to specify the field behavior over time of linear, isotropic, and nondispersive materials are [2, 3]

$$\nabla \times \mathbf{E} = -\frac{\partial \mathbf{B}}{\partial t} - \mathbf{J}_m, \quad \nabla \times \mathbf{H} = \frac{\partial \mathbf{D}}{\partial t} + \mathbf{J}_e$$
$$\nabla \cdot \mathbf{D} = \rho, \quad \nabla \cdot \mathbf{B} = 0, \quad \mathbf{D} = \varepsilon \mathbf{E}, \mathbf{B} = \mu \mathbf{H} \tag{7.1}$$

E is the electric field vector in volts per meter. **D** is the electric flux density vector in coulombs per square meter. **H** is the magnetic field vector in amperes per meter. **B** is the magnetic flux density vector in Weber per square meter. $\mathbf{J}_e$ is the electric conduction current density in ampere per square meter. $\mathbf{J}_m$ is the equivalent magnetic conduction current density in volt per square meter.

The FDTD formulation is based on two coupled-curl equations in time domain derived from Maxwell's equations [2, 3]

$$\frac{\partial \mathbf{H}}{\partial t} = -\frac{1}{\mu}\left(\nabla \times \mathbf{E} + \sigma^* \mathbf{H}\right)$$
$$\frac{\partial \mathbf{E}}{\partial t} = \frac{1}{\varepsilon}\left(\nabla \times \mathbf{H} - \sigma \mathbf{E}\right) \qquad (7.2)$$

Here, μ is the magnetic permeability in henrys per meter, ε is the electric permittivity in farads per meter, and σ is the electric conductivity in siemens per meter. The electric current $\mathbf{J}_e = \sigma\mathbf{E}$ is used to allow for lossy dielectric materials, and the magnetic current $\mathbf{J}_m = \sigma^*\mathbf{H}$ has been included to have the possibility of magnetic loss by adding an equivalent magnetic conductivity term σ^* in ohms per meter.

The formulation only treats the electromagnetic fields **E** and **H**, and not the fluxes **D** and **B**. All four constitutive parameters ε, μ, σ, and σ^* are present so that any linear isotropic material property can be specified. By taking the divergence of the curl equations we only need to consider the curl equations because the divergence equations are contained in them [3]. Although the divergence equations are not part of the FDTD formalism, they can be used as a test on the predicted field response, so that after forming $\mathbf{D} = \varepsilon\mathbf{E}$ and $\mathbf{B} = \mu\mathbf{H}$ from the predicted fields, the resulting **D** and **B** must satisfy the divergence equations.

Writing the vector components of the curl operators yields the following system of six coupled scalar equations equivalent to Maxwell's curl equations in three-dimensional rectangular coordinate system (x, y, z), that form the basis of the FDTD numerical algorithm for electromagnetic wave interactions with general three-dimensional objects:

$$\frac{\partial H_x}{\partial t} = \frac{1}{\mu}\left(\frac{\partial E_y}{\partial z} - \frac{\partial E_z}{\partial y} - \sigma^* H_x\right)$$
$$\frac{\partial H_y}{\partial t} = \frac{1}{\mu}\left(\frac{\partial E_z}{\partial x} - \frac{\partial E_x}{\partial z} - \sigma^* H_y\right) \qquad (7.3a)$$
$$\frac{\partial H_z}{\partial t} = \frac{1}{\mu}\left(\frac{\partial E_x}{\partial y} - \frac{\partial E_y}{\partial x} - \sigma^* H_z\right)$$

$$\frac{\partial E_x}{\partial t} = \frac{1}{\varepsilon}\left(\frac{\partial H_z}{\partial y} - \frac{\partial H_y}{\partial z} - \sigma E_x\right)$$
$$\frac{\partial E_y}{\partial t} = \frac{1}{\varepsilon}\left(\frac{\partial H_x}{\partial z} - \frac{\partial H_z}{\partial x} - \sigma E_y\right) \qquad (7.3b)$$
$$\frac{\partial E_z}{\partial t} = \frac{1}{\varepsilon}\left(\frac{\partial H_y}{\partial x} - \frac{\partial H_x}{\partial y} - \sigma E_z\right)$$

In 1966, Kane Yee [1] originated a set of finite difference equations for the time-dependent Maxwell's curl equations system as given by Eqs. (7.3a) and (7.3b). These equations can be represented in a discrete

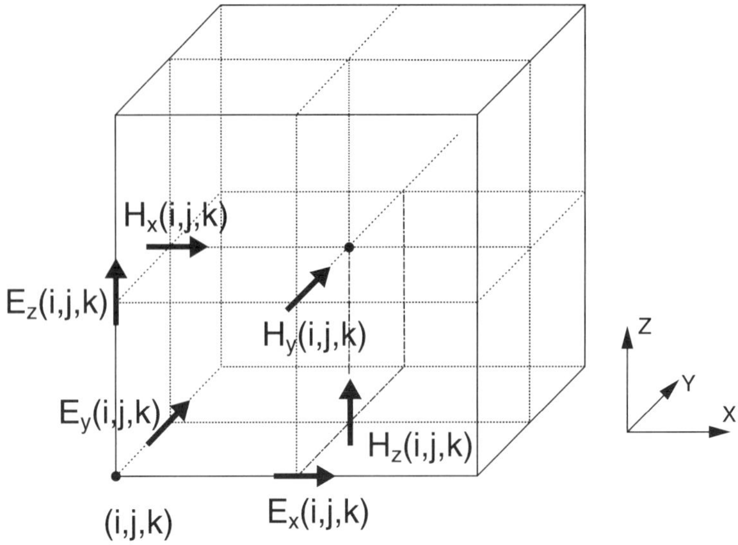

FIGURE 7.1 Yee's FDTD cell.

form, both in space and time, employing a second-order approximation resulting from the Taylor series expansion as

$$\frac{\partial f}{\partial x}\left(x_o\right)=\frac{f\left(x_o+\frac{\Delta}{2}\right)-f\left(x_o-\frac{\Delta}{2}\right)}{\Delta}+O\left(\Delta^2\right) \tag{7.4}$$

The Yee algorithm centers the **E** and **H** components in three-dimensional space such that every **E** component is surrounded by four circulating **H** components and every **H** component is surrounded by four circulating **E** components as shown in Fig. 7.1. This provides a simple picture of three-dimensional space being filled by an interlinked array of Faraday's law and Ampere's law contours. The computational domain is thus obtained by stacking these unit cells into a large problem space. The resulting finite difference expressions for the time and space derivatives used in the curl operators are central in nature (central difference) and second-order accurate. For the rectangular Yee mesh, a stepped or "staircase" approximation of the surface and internal geometry of the structure of interest is made, with a space resolution set by the size of the lattice unit cell. The Yee algorithm also centers its **E** and **H** components in time in what is termed as a leapfrog arrangement. All the **E** computations in the three-dimensional space are completed and stored in memory for a particular time point using **H** data previously stored in the computer memory. Then all the **H** computations in the modeled space are completed and stored in memory using the **E** data just computed. The cycle can begin again with the recomputation of the **E** components based on the newly obtained **H**. This leapfrog time-stepping process is fully explicit, thereby completely avoiding the problems involved with simultaneous equations and matrix inversion. This procedure is illustrated by the flowchart in Fig. 7.2.

By using Yee's notation, a space point in a uniform, rectangular lattice is denoted as

$$\left(i,j,k\right)=\left(i\Delta x, j\Delta y, k\Delta z\right) \tag{7.5}$$

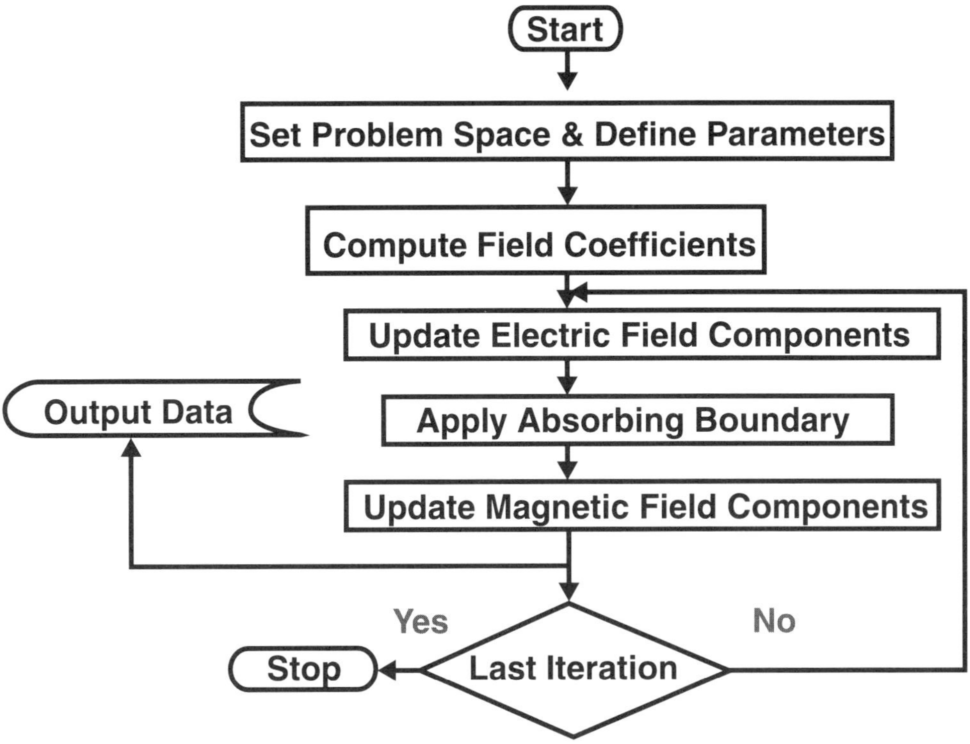

FIGURE 7.2 The computational flowchart based on Yee's FDTD procedure.

where Δx, Δy, Δz are the lattice space increments in the x, y, and z coordinate directions and i, j, k are integers assuming the values of 0 to nx, 0 to ny, and 0 to nz, respectively.

Further, let us denote any function u of space and time evaluated at a discrete point in the grid and at a discrete point in time as

$$u\left(i\Delta x, j\Delta y, k\Delta z, n\Delta t\right) = u^{n}_{i,j,k} \tag{7.6}$$

Using the centered finite difference expressions for the space and time derivatives yields the desired explicit time-stepping relations for **E** and **H** fields in a source free-lossy media as follows

$$\begin{aligned}
E_x\Big|^{n+1}_{i,j,k} &= C_a\Big|_{i,j,k} E_x\Big|^{n}_{i,j,k} + C_b\Big|_{i,j,k}\left[\left(\frac{H_z\Big|^{n+1/2}_{i,j+1/2,k} - H_z\Big|^{n+1/2}_{i,j-1/2,k}}{\Delta y}\right) - \left(\frac{H_y\Big|^{n+1/2}_{i,j,k+1/2} - H_y\Big|^{n+1/2}_{i,j,k-1/2}}{\Delta z}\right)\right] \\
E_y\Big|^{n+1}_{i,j,k} &= C_a\Big|_{i,j,k} E_y\Big|^{n}_{i,j,k} + C_b\Big|_{i,j,k}\left[\left(\frac{H_x\Big|^{n+1/2}_{i,j,k+1/2} - H_x\Big|^{n+1/2}_{i,j,k-1/2}}{\Delta z}\right) - \left(\frac{H_z\Big|^{n+1/2}_{i+1/2,j,k} - H_z\Big|^{n+1/2}_{i-1/2,j,k}}{\Delta x}\right)\right] \\
E_z\Big|^{n+1}_{i,j,k} &= C_a\Big|_{i,j,k} E_z\Big|^{n}_{i,j,k} + C_b\Big|_{i,j,k}\left[\left(\frac{H_y\Big|^{n+1/2}_{i+1/2,j,k} - H_y\Big|^{n+1/2}_{i-1/2,j,k}}{\Delta x}\right) - \left(\frac{H_x\Big|^{n+1/2}_{i,j+1/2,k} - H_x\Big|^{n+1/2}_{i,j-1/2,k}}{\Delta y}\right)\right]
\end{aligned} \tag{7.7a}$$

$$C_a\Big|_{i,j,k} = \frac{1-\dfrac{\sigma_{i,j,k}\Delta t}{2\varepsilon_{i,j,k}}}{1+\dfrac{\sigma_{i,j,k}\Delta t}{2\varepsilon_{i,j,k}}}, C_b\Big|_{i,j,k} = \frac{\dfrac{\Delta t}{\varepsilon_{i,j,k}}}{1+\dfrac{\sigma_{i,j,k}\Delta t}{2\varepsilon_{i,j,k}}} \tag{7.7b}$$

$$H_x\Big|_{i,j,k}^{n+1/2} = D_a\Big|_{i,j,k} H_x\Big|_{i,j,k}^{n-1/2} + D_b\Big|_{i,j,k}\left[\left(\frac{E_y\Big|_{i,j,k+1/2}^{n} - E_y\Big|_{i,j,k-1/2}^{n}}{\Delta z}\right) - \left(\frac{E_z\Big|_{i,j+1/2,k}^{n} - E_z\Big|_{i,j-1/2,k}^{n}}{\Delta y}\right)\right]$$

$$H_y\Big|_{i,j,k}^{n+1/2} = D_a\Big|_{i,j,k} H_y\Big|_{i,j,k}^{n-1/2} + D_b\Big|_{i,j,k}\left[\left(\frac{E_z\Big|_{i+1/2,j,k}^{n} - E_z\Big|_{i-1/2,j,k}^{n}}{\Delta x}\right) - \left(\frac{E_x\Big|_{i,j,k+1/2}^{n} - E_x\Big|_{i,j,k-1/2}^{n}}{\Delta z}\right)\right] \tag{7.8a}$$

$$H_z\Big|_{i,j,k}^{n+1/2} = D_a\Big|_{i,j,k} H_z\Big|_{i,j,k}^{n-1/2} + D_b\Big|_{i,j,k}\left[\left(\frac{E_x\Big|_{i,j+1/2,k}^{n} - E_x\Big|_{i,j-1/2,k}^{n}}{\Delta y}\right) - \left(\frac{E_y\Big|_{i+1/2,j,k}^{n} - E_y\Big|_{i-1/2,j,k}^{n}}{\Delta x}\right)\right]$$

$$D_a\Big|_{i,j,k} = \frac{1-\dfrac{\sigma^*_{i,j,k}\Delta t}{2\mu_{i,j,k}}}{1+\dfrac{\sigma^*_{i,j,k}\Delta t}{2\mu_{i,j,k}}}, D_b\Big|_{i,j,k} = \frac{\dfrac{\Delta t}{\mu_{i,j,k}}}{1+\dfrac{\sigma^*_{i,j,k}\Delta t}{2\mu_{i,j,k}}} \tag{7.8b}$$

The numerical algorithm requires that the time increment Δt have a specific bound relative to the lattice space increments Δx, Δy, and Δz. The upper bound on Δt to guarantee numerical stability (known as Courant's stability criterion) is [1–3]

$$\Delta t \le \frac{1}{c\sqrt{\left(\dfrac{1}{\Delta x}\right)^2 + \left(\dfrac{1}{\Delta y}\right)^2 + \left(\dfrac{1}{\Delta z}\right)^2}} \tag{7.9}$$

where c is the maximum velocity of the wave propagating in the media (i.e., speed of light in free space).

7.2.2 Source Waveform

In addition to the basic aspects of terminating the computational domain by an absorbing boundary condition (ABC), discrete representation and stability of Maxwell's vector-field equations when numerically approximated in a Cartesian FDTD space lattice, it is important to consider the introduction of electromagnetic wave excitations into the FDTD mesh. These excitations are associated with time domain signals that take different waveform shapes. An example of an excitation is a "hard source" that can be specified by assigning a desired time function to electric or magnetic field components in the FDTD space lattice. Because the field is specified at a certain region i_s without regard to any possible reflected waves in the grid (hence the terminology *hard source*), the hard source causes a spurious, nonphysical reflection at i_s back toward the material structure of interest. It prevents the movement of reflected wave energy through the source position, and thereby fails to properly simulate a physical incident wave. A simple way to mitigate the reflective nature of a pulsed hard source is to remove it from the algorithm after the pulse has decayed essentially to zero and apply instead the normal Yee field update [4].

The following source waveform on E_z could be established at the grid source location to generate a continuous sinusoidal wave of frequency f_o that is switched on at time $t = 0$ or the time index $n = 0$

$$E_z\Big|_{i,j,k}^{n} = E_o \sin\left(2\pi f_o n\Delta t\right) \qquad (7.10)$$

$$t = n\Delta t$$

A wideband Gaussian pulse source waveform provides a smooth roll off in frequency content and a finite direct current (DC) content. The pulse is centered at time t_0 or time step n_o and has a $1/e$ characteristic of time decay τ or of n_{decay} time steps

$$E_z\Big|_{i,j,k}^{n} = E_o e^{-\left(\frac{t-t_o}{\tau}\right)^2} = E_o e^{-\left(\frac{n-n_o}{n_{decay}}\right)^2} \qquad (7.11)$$

$$t_o = n_o \Delta t$$

A band-pass Gaussian pulse source waveform provides zero DC component, that has a Fourier spectrum symmetrical about f_0. The pulse is again centered at time t_0 or time step n_o and has a $1/e$ characteristic of time decay τ or of n_{decay} time steps:

$$E_z\Big|_{i,j,k}^{n} = E_o e^{-\left(\frac{n-n_o}{n_{decay}}\right)^2} \sin\left[2\pi f_o\left(n-n_o\right)\Delta t\right] \qquad (7.12)$$

Although an ideal Gaussian pulse extends infinitely in time, one must truncate the pulse in the calculations and the effects of this truncation must be considered, as well as the time duration of the pulse so that it has a suitable bandwidth. To examine the effects of these two parameters, consider the Fourier transform of the Gaussian pulse

$$\Im\left\{e^{-\left(\frac{n-n_o}{n_{decay}}\right)^2}\right\} = \sqrt{\pi}\, n_{decay}\Delta t e^{-\left(\pi f n_{decay}\Delta t\right)^2} e^{-j\omega \cdot t_o} \qquad (7.13)$$

By normalizing the magnitude of the preceding expression, it is possible to choose its minimum value at the maximum frequency of interest. For example, if the desired minimum magnitude is –20 dB, the value of n_{decay} can be obtained from

$$20 \log\left[e^{-\left(\pi f_{\max} n_{decay}\Delta t\right)^2}\right] = -20 \text{ dB} \qquad (7.14)$$

Based on the assumption just stated, the following expressions yield a desired pulse within a dynamic range characterized by τ or n_{decay} as

$$\tau = n_{decay}\,\Delta t = \frac{1}{2f_{\max}} \tag{7.15}$$

where $f_{\max}$ is the maximum frequency of interest.

Although an ideal Gaussian pulse extends infinitely in time, it must be truncated in the calculations. Consider that the Gaussian pulse exists from 0 to $2n_o$ time steps with peak value at n_o and approximately zero values outside this range. At the truncation points, the Gaussian pulse has a value of $e^{-(\pm n_o/n_{decay})^2}$. Thus, by choosing $n_o = 4n_{decay}$ the Gaussian pulse is down by e^{-16} or almost –140 dB.

7.2.3 Conductor Treatment

The conductor treatment in the FDTD formulation is not difficult. As an example, assume the geometry of a microstrip-type antenna with perfect electric conductors. The tangential components of the electric field and the normal components of the magnetic field should be set to zero on the ground plane and on the surface of the conducting patch of the microstrip antenna and the feed line conductors. To take into account the singularity near the metallic edge, the known asymptotic field behavior can be employed to modify the finite difference approximations of the magnetic field components near the metallic edge. By using this procedure, it is possible to show that the standard finite difference equations can still represent the accurate field behavior near the metallic edge.

7.2.4 Dielectric Interface Treatment

One procedure for modeling objects in the FDTD computational procedure is by assigning material parameters to each unit cell; alternatively, one can assign the material properties to each related field component directly. Regardless of which procedure is to be used, at the interface between two different materials, the material parameters are not well defined. Thus, it is necessary to find the proper material parameters at the interface to correctly apply the boundary conditions.

Consider a dielectric interface, in the xy plane, separating two media of permittivity ε_1 and ε_2. The tangential electric field component E_x can be carried out via Maxwell's equations in the two media as [5–7]

$$\begin{aligned} \varepsilon_1 \frac{\partial E_x^1}{\partial t} &= \frac{\partial H_z^1}{\partial y} - \frac{\partial H_y^1}{\partial z} \\ \varepsilon_2 \frac{\partial E_x^2}{\partial t} &= \frac{\partial H_z^2}{\partial y} - \frac{\partial H_y^2}{\partial z} \end{aligned} \tag{7.16}$$

Based on the boundary conditions, the tangential electric field and the normal derivative of the magnetic field components $\partial H_y/\partial z$ must maintain continuity. Thus, adding the preceding equations and applying the continuity property yields

$$\left(\frac{\varepsilon_1 + \varepsilon_2}{2}\right)\frac{\partial E_x}{\partial t} = \frac{1}{2}\left(\frac{\partial H_z^1}{\partial y} + \frac{\partial H_z^2}{\partial y}\right) - \frac{\partial H_y}{\partial z} \tag{7.17}$$

Numerical experience shows that the average of $\partial H_z/\partial y$ in the preceding equation is not necessary because there is only very little difference between these two H_z derivatives. Therefore, the fields behave just as if the permittivity were

$$\varepsilon = \frac{\varepsilon_1 + \varepsilon_2}{2} \tag{7.18}$$

The dielectric constant on the interface of two different materials is thus modeled as an average of these two dielectric constants as given by Eq. (7.18) for uniformly discretized domain in the direction normal to the interface.

7.3 Absorbing Boundary Conditions

For many applications that require to model scattering from an object or a radiating antenna situated in free space, it is desired that the scattered or radiated fields propagate into boundless space. Unfortunately, the FDTD computational space is bounded, and when the scattered or radiated fields arrive at the boundary, they are reflected back into the computation space. Therefore, it is necessary to have an ABC that absorbs these fields when they arrive at the limits of the FDTD space such that scattering or radiation into boundless free space is at least approximately simulated.

7.3.1 Mur's Absorbing Boundary Condition

A simple, yet very useful, ABC was proposed by Mur [8]. A first-order Mur condition looks back one step in time and into the space one cell location, while a second-order Mur condition looks back two steps in time and inward two cell locations.

Mur's first-order ABC assumes that the waves are normal by incident on the outer mesh walls. This assumption leads to a simple approximate continuous ABC: the tangential fields at the outer boundaries obey the one-dimensional wave equation in the direction normal to the mesh wall.

For the wave normal to the x direction

$$\left(\frac{\partial}{\partial x}-\frac{1}{c}\frac{\partial}{\partial t}\right)E_{\tan}=0 \tag{7.19a}$$

For the wave normal to the y direction

$$\left(\frac{\partial}{\partial y}-\frac{1}{c}\frac{\partial}{\partial t}\right)E_{\tan}=0 \tag{7.19b}$$

For the wave normal to the z direction

$$\left(\frac{\partial}{\partial z}-\frac{1}{c}\frac{\partial}{\partial t}\right)E_{\tan}=0 \tag{7.19c}$$

By considering the E_x and E_y components located at $x = i\Delta x$, $j\Delta y$, and $z = nz\,\Delta z$, the first-order Mur estimate at the boundary walls is

$$\begin{aligned}
E_x\Big|_{i,j,nz}^{n+1} &= E_x\Big|_{i,j,nz-1}^{n} + \frac{c\Delta t-\Delta z}{c\Delta t+\Delta z}\left(E_x\Big|_{i,j,nz-1}^{n+1} - E_x\Big|_{i,j,nz}^{n}\right)\\
E_y\Big|_{i,j,nz}^{n+1} &= E_y\Big|_{i,j,nz-1}^{n} + \frac{c\Delta t-\Delta z}{c\Delta t+\Delta z}\left(E_y\Big|_{i,j,nz-1}^{n+1} - E_y\Big|_{i,j,nz}^{n}\right)
\end{aligned} \tag{7.19d}$$

By considering the E_x and E_y components located at $x = i\Delta x$, $y = j\Delta y$, and $z = 0$, the first-order Mur estimate at the boundary walls is

$$E_x\Big|_{i,j,0}^{n+1} = E_x\Big|_{i,j,1}^{n} + \frac{c\Delta t - \Delta z}{c\Delta t + \Delta z}\left(E_x\Big|_{i,j,1}^{n+1} - E_x\Big|_{i,j,0}^{n}\right)$$
$$E_y\Big|_{i,j,0}^{n+1} = E_y\Big|_{i,j,1}^{n} + \frac{c\Delta t - \Delta z}{c\Delta t + \Delta z}\left(E_y\Big|_{i,j,1}^{n+1} - E_y\Big|_{i,j,0}^{n}\right) \quad (7.19e)$$

By considering the E_y and E_z components located at $x = nx\,\Delta x$, $y = j\Delta y$, and $z = k\Delta z$, the first-order Mur estimate at the boundary walls is

$$E_y\Big|_{nx,j,k}^{n+1} = E_y\Big|_{nx-1,j,k}^{n} + \frac{c\Delta t - \Delta x}{c\Delta t + \Delta x}\left(E_y\Big|_{nx-1,j,k}^{n+1} - E_y\Big|_{nx,j,k}^{n}\right)$$
$$E_z\Big|_{nx,j,k}^{n+1} = E_z\Big|_{nx-1,j,k}^{n} + \frac{c\Delta t - \Delta x}{c\Delta t + \Delta x}\left(E_z\Big|_{nx-1,j,k}^{n+1} - E_z\Big|_{nx,j,k}^{n}\right) \quad (7.19f)$$

By considering the E_y and E_z components located at $x = 0$, $y = j\Delta y$, and $z = k\Delta z$, the first-order Mur estimate at the boundary walls is

$$E_y\Big|_{0,j,k}^{n+1} = E_y\Big|_{1,j,k}^{n} + \frac{c\Delta t - \Delta x}{c\Delta t + \Delta x}\left(E_y\Big|_{1,j,k}^{n+1} - E_y\Big|_{0,j,k}^{n}\right)$$
$$E_z\Big|_{0,j,k}^{n+1} = E_z\Big|_{1,j,k}^{n} + \frac{c\Delta t - \Delta x}{c\Delta t + \Delta x}\left(E_z\Big|_{1,j,k}^{n+1} - E_z\Big|_{0,j,k}^{n}\right) \quad (7.19g)$$

By considering the E_x and E_z components located at $x = i\Delta x$, $y = ny\,\Delta y$, and $z = k\Delta z$, the first-order Mur estimate at the boundary walls is

$$E_x\Big|_{i,ny,k}^{n+1} = E_x\Big|_{i,ny-1,k}^{n} + \frac{c\Delta t - \Delta y}{c\Delta t + \Delta y}\left(E_x\Big|_{i,ny-1,k}^{n+1} - E_x\Big|_{i,ny,k}^{n}\right)$$
$$E_z\Big|_{i,ny,k}^{n+1} = E_z\Big|_{i,ny-1,k}^{n} + \frac{c\Delta t - \Delta y}{c\Delta t + \Delta y}\left(E_z\Big|_{i,ny-1,k}^{n+1} - E_z\Big|_{i,ny,k}^{n}\right) \quad (7.19h)$$

By considering the E_x and E_z components located at $x = i\Delta x$, $y = 0$, and $z = k\Delta z$, the first-order Mur estimate at the boundary walls is

$$E_x\Big|_{i,0,k}^{n+1} = E_x\Big|_{i,1,k}^{n} + \frac{c\Delta t - \Delta y}{c\Delta t + \Delta y}\left(E_x\Big|_{i,1,k}^{n+1} - E_x\Big|_{i,0,k}^{n}\right)$$
$$E_z\Big|_{i,0,k}^{n+1} = E_z\Big|_{i,1,k}^{n} + \frac{c\Delta t - \Delta y}{c\Delta t + \Delta y}\left(E_z\Big|_{i,1,k}^{n+1} - E_z\Big|_{i,0,k}^{n}\right) \quad (7.19i)$$

In other words, the tangential electric field on the artificial boundary wall can be obtained from the previous value of that field, and the field components one node inside of the mesh wall in the current and next time step.

To implement the first-order Mur ABC, the normal components of the electric field at the boundary walls are obtained using the regular Yee algorithm. The tangential components of the electric field at the intersection of two of the terminating planes (boundary walls) are obtained by taking the average of the values of that component around the specific point.

The second-order absorbing boundary condition for a wave normal to the x-constant plane is

$$\left[\frac{1}{c}\frac{\partial^2}{\partial x \partial t}+\frac{1}{c^2}\frac{\partial^2}{\partial t^2}-\frac{1}{2}\left(\frac{\partial^2}{\partial x^2}-\frac{\partial^2}{\partial y^2}\right)\right]E_{\tan}=0 \tag{7.20}$$

Hence, the second-order estimate for E_z at the boundary $x = 0$ is

$$E_z\Big|_{0,j,k}^{n+1}=-E_z\Big|_{1,j,k}^{n-1}+EQ_1+EQ_2+EQ_3+EQ_4 \tag{7.21a}$$

$$EQ_1=\frac{c\Delta t-\Delta x}{c\Delta t+\Delta x}\left(E_z\Big|_{1,j,k}^{n+1}+E_z\Big|_{0,j,k}^{n-1}\right)$$

$$EQ_2=\frac{2\Delta x}{c\Delta t+\Delta x}\left(E_z\Big|_{0,j,k}^{n}+E_z\Big|_{1,j,k}^{n}\right)$$

$$EQ_3=\frac{\Delta x(c\Delta t)^2}{2(\Delta y)^2(c\Delta t+\Delta x)}(C_a+C_b) \tag{7.21b}$$

$$EQ_4=\frac{\Delta x(c\Delta t)^2}{2(\Delta z)^2(c\Delta t+\Delta x)}(C_c+C_d)$$

$$C_a=E_z\Big|_{0,j+1,k}^{n}-2E_z\Big|_{0,j,k}^{n}+E_z\Big|_{0,j-1,k}^{n}$$

$$C_b=E_z\Big|_{1,j+1,k}^{n}-2E_z\Big|_{1,j,k}^{n}+E_z\Big|_{1,j-1,k}^{n}$$

$$C_c=E_z\Big|_{0,j,k+1}^{n}-2E_z\Big|_{0,j,k}^{n}+E_z\Big|_{0,j,k-1}^{n} \tag{7.21c}$$

$$C_d=E_z\Big|_{1,j,k+1}^{n}-2E_z\Big|_{1,j,k}^{n}+E_z\Big|_{1,j,k-1}^{n}$$

While the second-order estimate for E_z at the boundary $x = n\,\Delta x$ is

$$E_z\Big|_{nx,j,k}^{n+1}=-E_z\Big|_{nx-1,j,k}^{n-1}+EQ_1+EQ_2+EQ_3+EQ_4 \tag{7.21d}$$

$$EQ_1=\frac{c\Delta t-\Delta x}{c\Delta t+\Delta x}\left(E_z\Big|_{nx-1,j,k}^{n+1}+E_z\Big|_{nx,j,k}^{n-1}\right)$$

$$EQ_2=\frac{2\Delta x}{c\Delta t+\Delta x}\left(E_z\Big|_{nx,j,k}^{n}+E_z\Big|_{nx-1,j,k}^{n}\right) \tag{7.21e}$$

$$EQ_3=\frac{\Delta x(c\Delta t)^2}{2(\Delta y)^2(c\Delta t+\Delta x)}(C_a+C_b)$$

$$EQ_4 = \frac{\Delta x (c\Delta t)^2}{2(\Delta z)^2 (c\Delta t + \Delta x)} (C_c + C_d)$$

$$\begin{aligned}
C_a &= E_z\Big|^n_{nx,\, j+1,k} - 2E_z\Big|^n_{nx,\, j,k} + E_z\Big|^n_{nx,\, j-1,k} \\
C_b &= E_z\Big|^n_{nx-1,\, j+1,k} - 2E_z\Big|^n_{nx-1,\, j,k} + E_z\Big|^n_{nx-1,\, j-1,k} \\
C_c &= E_z\Big|^n_{nx,\, j,k+1} - 2E_z\Big|^n_{nx,\, j,k} + E_z\Big|^n_{nx,\, j,k-1} \\
C_d &= E_z\Big|^n_{nx-1,\, j,k+1} - 2E_z\Big|^n_{nx-1,\, j,k} + E_z\Big|^n_{nx-1,\, j,k-1}
\end{aligned} \tag{7.21f}$$

The equations needed to determine other field components with second-order Mur estimate at other limiting surfaces of the FDTD space are determined by modification of the preceding expressions.

One important consideration for implementing the second-order Mur ABC is that because the second-order estimate requires field values from adjacent Yee cells, it cannot be used for determining electric field values that are adjacent to the intersection of two of the terminating planes (boundary walls). Therefore, even if second-order Mur is being applied, first-order Mur must be used for field components located at the edges of the problem space.

For Mur's ABCs, the farther from the object the outer boundary is located, the better the absorption of the outward traveling waves. This is because these waves become more like plane waves as they travel farther from the structure that radiates them. However, the number of cells that can be placed between the object and the outer boundary is limited by computer memory. Moving the outer boundary too close to the object may cause instabilities in the absorbing boundary implementation. Also, some fields that are required for an accurate solution may be absorbed if the outer boundary is too close to the object.

7.3.2 Perfectly Matched Layer

In several applications Mur's second-order ABCs have been shown to be accurate enough. However, in many other applications, these conditions produce reflection above the acceptable levels; furthermore, they create instabilities, especially if the outer boundary is close to the simulated object. To avoid these instabilities and to reduce the reflection from Mur's ABC, the computational domain must be made large enough, and the computation must be carried out over a certain limited period of time over which useful information can been obtained [8]. These conditions are not feasible for practical applications that involves high permittivities, antenna arrays, array of stacked microstrip antennas, and many others. It is, therefore, necessary to have a boundary condition that can absorb all the waves coming in every direction toward the boundary walls while keeping the computational domain within bounds.

In 1994, Berenger introduced the perfectly matched layer (PML), a technique which is based on the use of an absorbing layer especially designed to absorb the electromagnetic waves without noticeable reflection [9–11].

PML splits all six Cartesian field vector components and introduces the electric and magnetic conductivity (σ and σ^*) to cause the decay of propagating fields, yielding 12 equations as follows [12]:

$$\frac{\partial H_{xy}}{\partial t} = -\frac{1}{\mu}\left[\frac{\partial (E_{zx} + E_{zy})}{\partial y} - \sigma_y^* H_{xy}\right], \quad \frac{\partial H_{xz}}{\partial t} = \frac{1}{\mu}\left[\frac{\partial (E_{yx} + E_{yz})}{\partial z} - \sigma_z^* H_{xz}\right] \tag{7.22a}$$

$$\frac{\partial H_{yz}}{\partial t} = -\frac{1}{\mu}\left[\frac{\partial\left(E_{xy}+E_{xz}\right)}{\partial z} - \sigma_z^* H_{yz}\right], \frac{\partial H_{yx}}{\partial t} = \frac{1}{\mu}\left[\frac{\partial\left(E_{zx}+E_{zy}\right)}{\partial x} - \sigma_x^* H_{yx}\right] \tag{7.22b}$$

$$\frac{\partial H_{zx}}{\partial t} = -\frac{1}{\mu}\left[\frac{\partial\left(E_{yx}+E_{yz}\right)}{\partial x} - \sigma_x^* H_{zx}\right], \frac{\partial H_{zy}}{\partial t} = \frac{1}{\mu}\left[\frac{\partial\left(E_{xy}+E_{xz}\right)}{\partial y} - \sigma_y^* H_{zy}\right] \tag{7.22c}$$

$$\frac{\partial E_{xy}}{\partial t} = \frac{1}{\varepsilon}\left[\frac{\partial\left(H_{zx}+H_{zy}\right)}{\partial y} - \sigma_y E_{xy}\right], \frac{\partial E_{xz}}{\partial t} = -\frac{1}{\varepsilon}\left[\frac{\partial\left(H_{yx}+H_{yz}\right)}{\partial z} - \sigma_z H_{xz}\right] \tag{7.22d}$$

$$\frac{\partial E_{yz}}{\partial t} = \frac{1}{\varepsilon}\left[\frac{\partial\left(H_{xy}+H_{xz}\right)}{\partial z} - \sigma_z E_{yz}\right], \frac{\partial E_{yx}}{\partial t} = -\frac{1}{\varepsilon}\left[\frac{\partial\left(H_{zx}+H_{zy}\right)}{\partial x} - \sigma_x E_{yx}\right] \tag{7.22e}$$

$$\frac{\partial E_{zx}}{\partial t} = \frac{1}{\varepsilon}\left[\frac{\partial\left(H_{yx}+H_{yz}\right)}{\partial x} - \sigma_x E_{zx}\right], \frac{\partial E_{zy}}{\partial t} = -\frac{1}{\varepsilon}\left[\frac{\partial\left(H_{xy}+H_{xz}\right)}{\partial y} - \sigma_y E_{zy}\right] \tag{7.22f}$$

By using the centered finite difference expressions for the space and time derivatives, the desired time-stepping relations for electric and magnetic fields are as follows:

$$\begin{aligned} H_{xy}\Big|_{i,j,k}^{n+1/2} &= D_{y1}\Big|_{i,j,k} H_{xy}\Big|_{i,j,k}^{n-1/2} - D_{y2}\Big|_{i,j,k}\left(\frac{E_z\Big|_{i,j+1/2,k}^{n} - E_z\Big|_{i,j-1/2,k}^{n}}{\Delta y}\right) \\ H_{xz}\Big|_{i,j,k}^{n+1/2} &= D_{z1}\Big|_{i,j,k} H_{xz}\Big|_{i,j,k}^{n-1/2} - D_{z2}\Big|_{i,j,k}\left(\frac{E_y\Big|_{i,j,k+1/2}^{n} - E_y\Big|_{i,j,k-1/2}^{n}}{\Delta z}\right) \end{aligned} \tag{7.23a}$$

$$\begin{aligned} H_{yz}\Big|_{i,j,k}^{n+1/2} &= D_{z1}\Big|_{i,j,k} H_{yz}\Big|_{i,j,k}^{n-1/2} - D_{z2}\Big|_{i,j,k}\left(\frac{E_x\Big|_{i,j,k+1/2}^{n} - E_x\Big|_{i,j,k-1/2}^{n}}{\Delta z}\right) \\ H_{yx}\Big|_{i,j,k}^{n+1/2} &= D_{x1}\Big|_{i,j,k} H_{yx}\Big|_{i,j,k}^{n-1/2} - D_{x2}\Big|_{i,j,k}\left(\frac{E_z\Big|_{i+1/2,j,k}^{n} - E_z\Big|_{i-1/2,j,k}^{n}}{\Delta x}\right) \end{aligned} \tag{7.23b}$$

$$H_{zx}\Big|_{i,j,k}^{n+1/2} = D_{x1}\Big|_{i,j,k} H_{zx}\Big|_{i,j,k}^{n-1/2} - D_{x2}\Big|_{i,j,k}\left(\frac{E_y\Big|_{i+1/2,j,k}^{n} - E_y\Big|_{i-1/2,j,k}^{n}}{\Delta x}\right) \tag{7.23c}$$

$$H_{zy}\Big|_{i,j,k}^{n+1/2} = D_{y1}\Big|_{i,j,k} H_{zy}\Big|_{i,j,k}^{n-1/2} - D_{y2}\Big|_{i,j,k}\left(\frac{E_x\Big|_{i,j+1/2,k}^{n} - E_x\Big|_{i,j-1/2,k}^{n}}{\Delta y}\right)$$

$$E_{xy}\Big|_{i,j,k}^{n+1} = C_{xy1}\Big|_{i,j,k} E_{xy}\Big|_{i,j,k}^{n} - C_{xy2}\Big|_{i,j,k}\left(\frac{H_z\Big|_{i,j+1/2,k}^{n+1/2} - E_z\Big|_{i,j-1/2,k}^{n+1/2}}{\Delta y}\right)$$

$$E_{xz}\Big|_{i,j,k}^{n+1} = C_{xz1}\Big|_{i,j,k} E_{xz}\Big|_{i,j,k}^{n} - C_{xz2}\Big|_{i,j,k}\left(\frac{H_y\Big|_{i,j,k+1/2}^{n+1/2} - H_y\Big|_{i,j,k-1/2}^{n+1/2}}{\Delta z}\right) \tag{7.23d}$$

$$E_{yz}\Big|_{i,j,k}^{n+1} = C_{yz1}\Big|_{i,j,k} E_{yz}\Big|_{i,j,k}^{n} - C_{yz2}\Big|_{i,j,k}\left(\frac{H_x\Big|_{i,j,k+1/2}^{n+1/2} - H_x\Big|_{i,j,k-1/2}^{n+1/2}}{\Delta z}\right)$$

$$E_{yx}\Big|_{i,j,k}^{n+1} = C_{yx1}\Big|_{i,j,k} E_{yx}\Big|_{i,j,k}^{n} - C_{yx2}\Big|_{i,j,k}\left(\frac{H_z\Big|_{i+1/2,j,k}^{n+1/2} - H_z\Big|_{i-1/2,j,k}^{n+1/2}}{\Delta x}\right) \tag{7.23e}$$

$$E_{zx}\Big|_{i,j,k}^{n+1} = C_{zx1}\Big|_{i,j,k} E_{zx}\Big|_{i,j,k}^{n} - C_{zx2}\Big|_{i,j,k}\left(\frac{H_y\Big|_{i+1/2,j,k}^{n+1/2} - H_y\Big|_{i-1/2,j,k}^{n+1/2}}{\Delta x}\right)$$

$$E_{zy}\Big|_{i,j,k}^{n+1} = C_{zy1}\Big|_{i,j,k} E_{zy}\Big|_{i,j,k}^{n} - C_{zy2}\Big|_{i,j,k}\left(\frac{H_x\Big|_{i,j+1/2,k}^{n+1/2} - H_x\Big|_{i,j-1/2,k}^{n+1/2}}{\Delta y}\right) \tag{7.23f}$$

$$E_x\Big|_{i,j,k}^{n} = E_{xy}\Big|_{i,j,k}^{n} + E_{xz}\Big|_{i,j,k}^{n},\ E_y\Big|_{i,j,k}^{n} = E_{yx}\Big|_{i,j,k}^{n} + E_{yz}\Big|_{i,j,k}^{n},\ E_z\Big|_{i,j,k}^{n} = E_{zx}\Big|_{i,j,k}^{n} + E_{zy}\Big|_{i,j,k}^{n} \tag{7.24a}$$

$$H_x\Big|_{i,j,k}^{n+1/2} = H_{xy}\Big|_{i,j,k}^{n+1/2} + H_{xz}\Big|_{i,j,k}^{n+1/2},\ H_y\Big|_{i,j,k}^{n+1/2} = H_{yx}\Big|_{i,j,k}^{n+1/2} + H_{yz}\Big|_{i,j,k}^{n+1/2},\ H_z\Big|_{i,j,k}^{n+1/2} = H_{zx}\Big|_{i,j,k}^{n+1/2} + H_{zy}\Big|_{i,j,k}^{n+1/2} \tag{7.24b}$$

The coefficients in Eqs. (7.23a to 7.23f) for a media characterized by ε, μ, σ, and σ^* are

$$C_{xy1}\Big|_{i,j,k} = \frac{\dfrac{\varepsilon_x\big|_{i,j,k}}{\Delta t} - \dfrac{\sigma_y\big|_{i,j,k}}{2}}{\dfrac{\varepsilon_x\big|_{i,j,k}}{\Delta t} + \dfrac{\sigma_y\big|_{i,j,k}}{2}},\ C_{xy2}\Big|_{i,j,k} = \frac{1}{\dfrac{\varepsilon_x\big|_{i,j,k}}{\Delta t} + \dfrac{\sigma_y\big|_{i,j,k}}{2}} \tag{7.25a}$$

$$C_{xz1}\Big|_{i,j,k} = \frac{\dfrac{\varepsilon_x|_{i,j,k}}{\Delta t} - \dfrac{\sigma_z|_{i,j,k}}{2}}{\dfrac{\varepsilon_x|_{i,j,k}}{\Delta t} + \dfrac{\sigma_z|_{i,j,k}}{2}}, C_{xz2}\Big|_{i,j,k} = \frac{1}{\dfrac{\varepsilon_x|_{i,j,k}}{\Delta t} + \dfrac{\sigma_z|_{i,j,k}}{2}} \tag{7.25b}$$

$$C_{yz1}\Big|_{i,j,k} = \frac{\dfrac{\varepsilon_y|_{i,j,k}}{\Delta t} - \dfrac{\sigma_z|_{i,j,k}}{2}}{\dfrac{\varepsilon_y|_{i,j,k}}{\Delta t} + \dfrac{\sigma_z|_{i,j,k}}{2}}, C_{yz2}\Big|_{i,j,k} = \frac{1}{\dfrac{\varepsilon_y|_{i,j,k}}{\Delta t} + \dfrac{\sigma_z|_{i,j,k}}{2}} \tag{7.25c}$$

$$C_{yx1}\Big|_{i,j,k} = \frac{\dfrac{\varepsilon_y|_{i,j,k}}{\Delta t} - \dfrac{\sigma_x|_{i,j,k}}{2}}{\dfrac{\varepsilon_y|_{i,j,k}}{\Delta t} + \dfrac{\sigma_x|_{i,j,k}}{2}}, C_{yx2}\Big|_{i,j,k} = \frac{1}{\dfrac{\varepsilon_y|_{i,j,k}}{\Delta t} + \dfrac{\sigma_x|_{i,j,k}}{2}} \tag{7.25d}$$

$$C_{zx1}\Big|_{i,j,k} = \frac{\dfrac{\varepsilon_z|_{i,j,k}}{\Delta t} - \dfrac{\sigma_x|_{i,j,k}}{2}}{\dfrac{\varepsilon_z|_{i,j,k}}{\Delta t} + \dfrac{\sigma_x|_{i,j,k}}{2}}, C_{zx2}\Big|_{i,j,k} = \frac{1}{\dfrac{\varepsilon_z|_{i,j,k}}{\Delta t} + \dfrac{\sigma_x|_{i,j,k}}{2}} \tag{7.25e}$$

$$C_{zy1}\Big|_{i,j,k} = \frac{\dfrac{\varepsilon_z|_{i,j,k}}{\Delta t} - \dfrac{\sigma_y|_{i,j,k}}{2}}{\dfrac{\varepsilon_z|_{i,j,k}}{\Delta t} + \dfrac{\sigma_y|_{i,j,k}}{2}}, C_{zy2}\Big|_{i,j,k} = \frac{1}{\dfrac{\varepsilon_z|_{i,j,k}}{\Delta t} + \dfrac{\sigma_y|_{i,j,k}}{2}} \tag{7.25f}$$

$$D_{x1}\Big|_{i,j,k} = \frac{\dfrac{\mu_o}{\Delta t} - \dfrac{\sigma^*_x|_{i,j,k}}{2}}{\dfrac{\mu_o}{\Delta t} + \dfrac{\sigma^*_x|_{i,j,k}}{2}}, D_{x2}\Big|_{i,j,k} = \frac{1}{\dfrac{\mu_o}{\Delta t} + \dfrac{\sigma^*_x|_{i,j,k}}{2}} \tag{7.25g}$$

$$D_{y1}\Big|_{i,j,k} = \frac{\dfrac{\mu_o}{\Delta t} - \dfrac{\sigma^*_y|_{i,j,k}}{2}}{\dfrac{\mu_o}{\Delta t} + \dfrac{\sigma^*_y|_{i,j,k}}{2}}, D_{y2}\Big|_{i,j,k} = \frac{1}{\dfrac{\mu_o}{\Delta t} + \dfrac{\sigma^*_y|_{i,j,k}}{2}} \tag{7.25h}$$

$$D_{z1}\Big|_{i,j,k} = \frac{\dfrac{\mu_o}{\Delta t} - \dfrac{\sigma_z^*\big|_{i,j,k}}{2}}{\dfrac{\mu_o}{\Delta t} + \dfrac{\sigma_z^*\big|_{i,j,k}}{2}}, \quad D_{z2}\Big|_{i,j,k} = \frac{1}{\dfrac{\mu_o}{\Delta t} + \dfrac{\sigma_z^*\big|_{i,j,k}}{2}} \tag{7.25i}$$

At the interfaces between normal FDTD and PML regions, it is desired that waves can propagate through without any reflections. Once waves enter the PML region, they get attenuated and absorbed. If η_1 and η_2 are the wave impedances in the FDTD and PML region, respectively, they can be expressed by the following equations, assuming normal incidence [9]:

$$\eta_1 = \eta_o \sqrt{\frac{\mu_r}{\varepsilon_r}}$$

$$\eta_2 = \sqrt{\frac{\mu_o\mu_r + \dfrac{\sigma^*}{j\omega}}{\varepsilon_o\varepsilon_r + \dfrac{\sigma}{j\omega}}} = \eta_o \sqrt{\frac{\mu_r}{\varepsilon_r}} \sqrt{\frac{1 + \dfrac{\sigma^*}{j\omega\mu_o}}{1 + \dfrac{\sigma}{j\omega\varepsilon_o}}} \tag{7.26}$$

where η_0 is the characteristic impedance of free space and ω is the frequency of the incident wave. The necessary condition for no reflection as waves propagate through the interface is the impedance matching between these two regions, $\eta_1 = \eta_2$, which yields

$$\frac{\sigma}{\varepsilon_o} = \frac{\sigma^*}{\mu_o} \tag{7.27}$$

If ψ represents any component of a plane wave propagating in a two-dimensional PML medium, TE case as shown in Fig. 7.3, Berenger shows that with the matching condition given earlier, ψ can be expressed as

$$\psi = \psi_o e^{j\omega\left(t - \frac{x\cos\phi + y\sin\phi}{v}\right)} e^{-\frac{\sigma_x \cos\phi}{\varepsilon_o\varepsilon_r v}x} e^{-\frac{\sigma_y \sin\phi}{\varepsilon_o\varepsilon_r v}y} \tag{7.28}$$

where ψ_0 is the magnitude of that component, ϕ is the incident angle with the y-axis, and v is the wave velocity in the medium. This equation shows that the wave in the PML region travels at exactly the same speed as that in the normal FDTD region, but decays exponentially along the x and y directions. In many antenna applications a set of air buffer layers exists between the geometry of the antenna and the PML region. In such cases, ε_r and μ_r in Eq. (7.26) to (7.28) reduce to 1; hence, the wave impedance of the PML region is the same as that of free space. These results can be extended to the three-dimensional case [11–14].

When a wave propagates in the x direction only, σ_x and σ_x^* are matched according to the matching condition, $\sigma_y = \sigma_y^* = 0$ and $\sigma_z = \sigma_z^* = 0$. Similarly, when a wave propagates in the y or z direction only, the conductivity values that are not in the direction of propagation are set equal to zero. At the corner regions of a two-dimensional domain (as shown in Fig. 7.4) where there is an overlap of two PML lattices, four losses are nonzero (σ_x, σ_x^*, and σ_y, σ_y^*). At the corner regions of a three-dimensional domain (as shown in Fig. 7.5), where there is an overlap of three PML lattices, six losses are nonzero (σ_x, σ_x^*, σ_y, σ_y^*, σ_z, and σ_z^*).

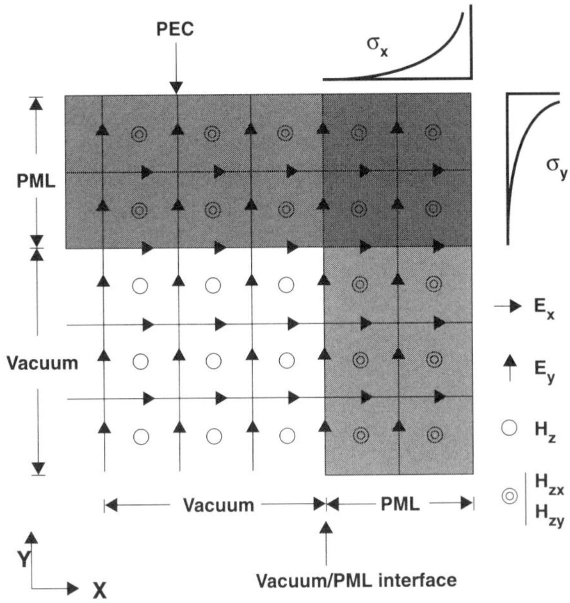

FIGURE 7.3 The arrangement of field components of a two-dimensional TE problem with PML absorbing boundary.

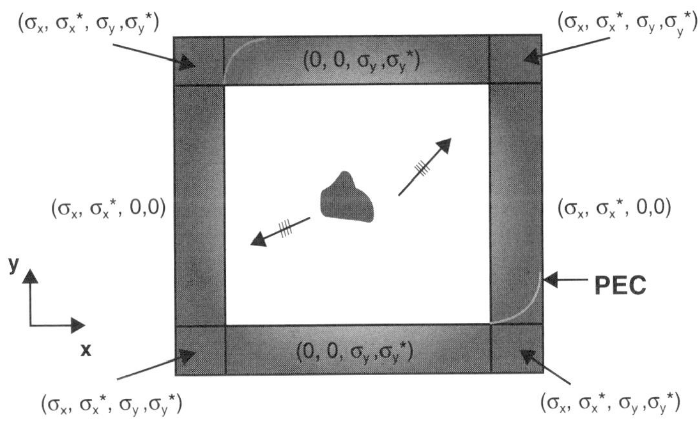

FIGURE 7.4 The assignment of losses for a two-dimensional FDTD domain with PML absorbing boundary.

A PML layer is defined by three parameters: the number of cells (thickness), the kind of conductivity scaling, and the theoretical reflection coefficient at normal incidence, $R(0)$.

7.3.2.1 Conductivity Scaling

In the preceding equations, the electric and magnetic conductivity values satisfy the matching impedance condition. With that treatment, theoretically there is no reflection at the vacuum-layer interface. Nevertheless, the PML layer does have a reflection because the outgoing waves are reflected by the perfectly conducting conditions set on its outer boundary and they can return into the vacuum. Thus, for a plane wave, an apparent reflection is defined as a function of the PML layer with thickness δ and a conductivity profile $\sigma(\rho)$. This conductivity is either σ_x in a layer normal to x, σ_y in a layer normal to y, or σ_z in a layer normal to z [9–11].

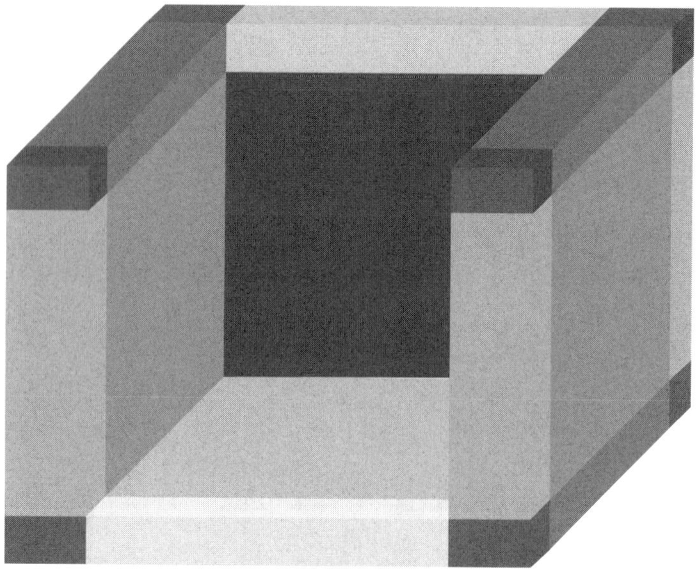

FIGURE 7.5 Classification of regions of PML boundary for a three-dimensional computational domain.

The reflection factor at an angle of incidence θ is

$$R(\theta) = \left[R(0)\right]^{\cos\theta}$$
$$R(0) = e^{-\frac{2\delta\sqrt{\varepsilon_r}}{\varepsilon_o c}\langle\sigma_i\rangle} \tag{7.29}$$
$$\langle\sigma_i\rangle = \frac{1}{\delta}\int_0^{\delta}\sigma(\rho)d\rho$$

The average conductivity in terms of the reflection at normal incidence is given by

$$\langle\sigma_i\rangle = -\frac{\varepsilon_o c \ln\left(R(0)\right)}{2\sqrt{\varepsilon_r}\delta} \tag{7.30}$$

Theoretically, one can obtain reflection factors as small as required by increasing the thickness δ or the conductivity σ(ρ), or both. In practice, a small amount of numerical reflection occurs, which depends on the choice of δ and σ(ρ).

Generally for a microstrip antenna type geometry, ε_r is inhomogeneous, causing an ambiguity in the preceding equation. This problem can be solved by applying quasi-static theory to determine the effective dielectric constant of the structure. Therefore, ε_r represents the effective dielectric constant of a multilayer space.

Polynomial Scaling

For a polynomial scaling of the conductivity [10, 11]

$$\sigma(\rho) = \sigma_o(\rho/\delta)^n \tag{7.31}$$

Typically, n in the range between 2 and 4 has been found to be suitable for the analysis of microstrip-type antennas. For a PML layer of thickness δ, the reflection coefficient factor at normal incidence is

$$R(0) = e^{-\frac{2\sigma_o \sqrt{\varepsilon_r}\delta}{\varepsilon_o c(n+1)}} \tag{7.32}$$

The conductivity σ_o can be obtained from the preceding equation as follows:

$$\sigma_o = -\frac{\varepsilon_o c(n+1)}{2\sqrt{\varepsilon_r}\delta} \ln R(0) \tag{7.33}$$

The conductivity at the mesh points is implemented as the average value in the cell around the index location. At index i, the

$$\sigma_n(i) = \frac{1}{\Delta}\int_{\rho(i)-\Delta/2}^{\rho(i)+\Delta/2} \sigma(u)du \tag{7.34}$$

Therefore [10]

$$\begin{aligned} \sigma_n(0) &= -\frac{\varepsilon_o c \ln(R(0))}{2^{n+2}\Delta N^{n+1}} \\ \sigma_n(i>0) &= \sigma_n(0)\left[(2i+1)^{n+1} - (2i-1)^{n+1}\right] \end{aligned} \tag{7.35}$$

where Δ is the spatial increment of the FDTD and N is the number of PML layers such that $\delta = N\Delta$. Berenger showed that for solving wave-structure interaction problems, the PML layer must satisfy two constraints. First, the FDTD conductivity $\sigma_n(0)$ is bounded by

$$\sigma_n(0) < \frac{2\pi\varepsilon_o}{D_c} \tag{7.36}$$

where D_c is the duration of the computation (number of time steps times Δt); second, the ratio of successive conductivity values must be lower than a value S, that is,

$$\frac{\sigma_n(i+1/2)}{\sigma_n(i)} \leq S \tag{7.37}$$

For polynomial scaling, the preceding ratio is not constant and depends on the position in the PML layer. Therefore, polynomial scaling is not an optimum profile of conductivity because it could result in a needlessly thickness of the PML layer.

Nevertheless, because all its parameters can be determined easily, the FDTD simulation can be set to run at least for a time

$$D_c = \frac{2\pi\varepsilon_o}{\Theta\sigma_n(0)} \tag{7.38}$$

The value Θ has been determined empirically to be about ten for a parabolic conductivity profile (polynomial scaling with $n = 2$).

Geometric Scaling

To reduce the numerical reflections from the PML interface, layers with conductivity increasing geometrically can be used. Denoting by Δ the spatial increment of the FDTD mesh and σ_o the conductivity in the vacuum-layer interface, such conductivity profile is [9–11]

$$\sigma(\rho) = \sigma_o \left(g^{\frac{1}{\Delta}} \right)^{\rho} \tag{7.39}$$

such that the conductivity increases by the factor g from one cell to the next. For an N-cell layer, the reflection coefficient factor at normal incidence is

$$R(0) = e^{-\frac{2}{\varepsilon_o c}\frac{g^N - 1}{\ln(g)}\sigma_o \sqrt{\varepsilon_r}\Delta} \tag{7.40}$$

The conductivity σ_o can be obtained from N, g, and $R(0)$ as follows:

$$\sigma_o = -\frac{\varepsilon_o c}{2\sqrt{\varepsilon_r}\Delta} \frac{\ln(g)}{g^N - 1} \ln R(0) \tag{7.41}$$

The conductivity at the mesh points is implemented as the average value in the cells around the index location. Therefore

$$\begin{aligned} \sigma_n(0) &= \sigma_o \frac{\sqrt{g} - 1}{\ln(g)} \\ \sigma_n(i > 0) &= \sigma_o \frac{g - 1}{\sqrt{g}\ln(g)} g^i \end{aligned} \tag{7.42}$$

The ratio S of Eq. (7.37) is constant and equal to $\sqrt{g}$. Therefore, for solving wave-structure interaction problems, geometric scaling is the optimum profile of conductivity. Once g is known, an approximate number for the duration of the simulation can be obtained from Eq. (7.38).

7.3.3 Perfectly Matched Layer Parameter Selection

From Eq. (7.29), it is clear that inside the perfectly matched layer (PML) region the field decays as $e^{-\delta/d}$ where $d = \varepsilon_o c/2 < \sigma_i > \sqrt{\varepsilon_r}$. The cell size in the PML region is chosen to be a fraction m of this decay length d. Thus, $d = m\,\Delta$.

Therefore, the average conductivity becomes

$$\langle \sigma_i \rangle = \frac{\varepsilon_o c}{2\sqrt{\varepsilon_r}\, m\,\Delta} \tag{7.43}$$

Using the preceding equation and Eq. (7.30) yields

$$m = -\frac{N}{\ln R(0)} \tag{7.44}$$

A reasonable value for m has been obtained through extensive experimental study. In a broad range of applications, an optimum choice for a 10-cell thick PML is $R(0) \approx e^{-16}$. Therefore, $m \approx 10/16$. The larger the value of m, the thicker the PML will be. Typically, m takes a value of 1.

Based on the desired reflectivity $R(0)$ and the value of m, $\langle \sigma_i \rangle$ is calculated in Eq. (7.43) and it is straightforward to show from Eq. (7.29) that the length of the PML in meters is determined to be

$$\delta = -\frac{\varepsilon_o c \ln\left(R(0)\right)}{2\sqrt{\varepsilon_r}\left\langle \sigma_i \right\rangle} \tag{7.45}$$

Therefore, the PML has a thickness $N = \delta/\Delta$ cells. Typically, PML has a thickness of 7, 9, or 16 layers.

By using Eq. (7.29), we show that for polynomial scaling, the maximum conductivity obtained is

$$\sigma_o = (n+1)\left\langle \sigma_i \right\rangle \tag{7.46}$$

For geometric scaling, the maximum conductivity is

$$\sigma_o = \frac{N \ln(g)}{g^N - 1}\left\langle \sigma_i \right\rangle \tag{7.47}$$

The value of g can be calculated by solving simultaneously the preceding equation with Eqs. (7.38) and (7.42), yielding

$$\frac{g^N - 1}{\sqrt{g} - 1} = N\left\langle \sigma_i \right\rangle \frac{\Theta D_c}{2\pi\varepsilon_o} \tag{7.48}$$

Although at this point the duration of the computation is unknown, a first-order estimate can be obtained assuming that it is equal to that of the parabolic profile.

7.4 Radiation Patterns

7.4.1 Background

The fields calculated with FDTD are only within the computational space and therefore the FDTD calculations only produce information of the electromagnetic fields that are adjacent to the radiating object. However, FDTD can also be applied to analyze radiation from antennas, with the desired results being the far zone scattered or radiated fields that lie outside the FDTD space. This can be accomplished through the use of the equivalence principle, where the near-field information is used to obtain the equivalent tangential electric and magnetic currents. These currents are then transformed into the far-field region to obtain the radiated field [15–17].

The four near-zone to far-zone transformation procedures that can be used to obtain the radiation pattern of an antenna are discussed next.

7.4.1.1 Single-Frequency Far-Zone Calculations

With a sinusoidal time-harmonic source being specified, the FDTD calculations are stepped through time until steady-state conditions are reached. The complex time-harmonic electric and magnetic currents flowing on a closed surface surrounding the object are then obtained. This involves very little computer storage, being four complex tangential fields (two electric fields and two magnetic fields) or surface currents for each Yee cell face on the closed surface. If these complex fields or currents are written to disk, then in postprocessing the far-zone radiated fields can be calculated in any direction. This is a good method to apply when far-zone radiation or scattering patterns are desired at only one frequency.

7.4.1.2 Multiple-Frequencies Hybrid Approach

Although this approach uses pulsed excitation for the FDTD calculations, it supplies frequency domain far-zone fields. For each frequency of interest a running discrete Fourier transform (DFT) of the time-harmonic tangential fields (surface currents) on a closed surface surrounding the FDTD geometry is updated at each time step. The running DFT provides the complex frequency domain currents for any number of frequencies using pulse excitation for the FDTD calculation. This is more efficient than using time-harmonic excitation, which requires a separate FDTD calculation for each frequency of interest. The running DFT requires no more computer storage (per frequency) for the complex surface currents than the frequency domain far-zone transform described earlier, and like it, provides frequency domain far-zone fields at any far-zone angle. If far-zone results are desired at a few frequencies, then the running DFT approach seems to be the optimum choice.

7.4.1.3 Full Time-Domain Approach

This is the most straightforward approach; it requires saving the time-domain tangential fields (surface currents) over a closed surface containing the FDTD geometry for all time steps. Transient far-zone fields can be computed in postprocessing for all angles, and frequency domain results are computed using fast Fourier transforms (FFTs) for all frequencies at all angles. This method is extremely versatile; it provides all possible far-zone results available from the FDTD calculation; however, except for a very small geometry (thin plates or wires), a large amount of computer storage is required to save all the surface current components at all the time steps. Therefore, it does not seem to be a practical approach.

7.4.1.4 Transient Far-Zone Fields for Discrete Angles

In many situations, transient and/or broadband frequency domain results are required at a limited number of angles, for example, transient antenna radiation for a limited number of pattern cuts. For these situations it is better to directly compute the transient far-zone fields at each angle of interest as a running summation. This approach requires storing transient results for each time step for six far-zone vector potentials per far-zone angle instead of four tangential field components (surface currents) per cell face on the transformation surface. If frequency domain results are desired, they can be efficiently obtained from the far-zone transient results with application of the FFT. In this way, one FDTD computation using pulse excitation along with an FFT produces wideband far-zone scattering or radiation results at one specific angle [15, 16].

7.4.1.5 Conclusions

Out of all-frequency-domain methods described previously the most efficient one is the running DFT especially if results at more than one frequency (at all angles) are desired. The running DFT method requires more computational effort than the all-time-domain approach if far-zone results are desired at more than a few frequencies and at all angles. The transient approach at discrete angles is much more efficient in its use of memory than the full-transient approach, if a broadband response is necessary for only a few discrete angles.

7.4.2 Near-Field to Far-Field Transformation

In this study, the frequency domain approach based on the multiple frequencies hybrid approach is adopted as described in Reference [3], and is briefly outlined here. The field values computed by the FDTD updating equations are the near-zone transient field. The far fields can be obtained through a near-field to far-field transformation procedure by using a pair of vector potentials defined as follows:

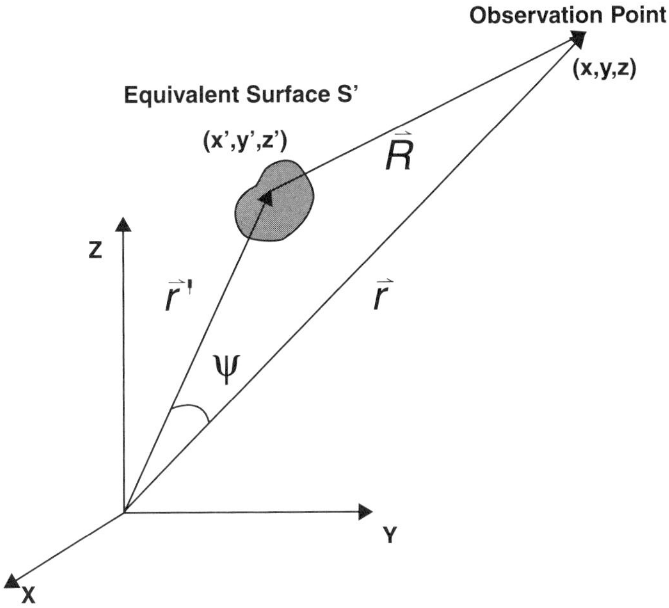

FIGURE 7.6 The equivalent surface current source and a far-field observation point.

$$\mathbf{A} = \frac{\mu_0 \, e^{-jkR}}{4\pi R}\mathbf{N}$$
$$\mathbf{F} = \frac{\varepsilon_0 \, e^{-jkR}}{4\pi R}\mathbf{L} \tag{7.49}$$

where

$$\mathbf{N} = \iint_s \mathbf{J}_s e^{jkr'\cos\psi} ds$$

$$\mathbf{L} = \iint_s \mathbf{M}_s e^{jkr'\cos\psi} ds$$

$\bar{\mathbf{r}} = r\hat{r} \equiv$ position of observation point (x, y, z), $\bar{\mathbf{r}}' = r'\hat{r}' \equiv$ position of source point (x', y', z') on S, $\bar{R} = R\hat{R} \equiv \bar{r} - \bar{r}'$, and $\psi \equiv$ angle between $\bar{r}$ and $\bar{r}'$.

The distance R is defined in Fig. 7.6 and is approximated as

$$R = \left[r^2 + (r')^2 - 2rr'\cos\psi \right]^{\frac{1}{2}}$$

$$= \begin{cases} r - r'\cos\psi & \text{for the phase terms} \\ r & \text{for the amplitude terms} \end{cases}$$

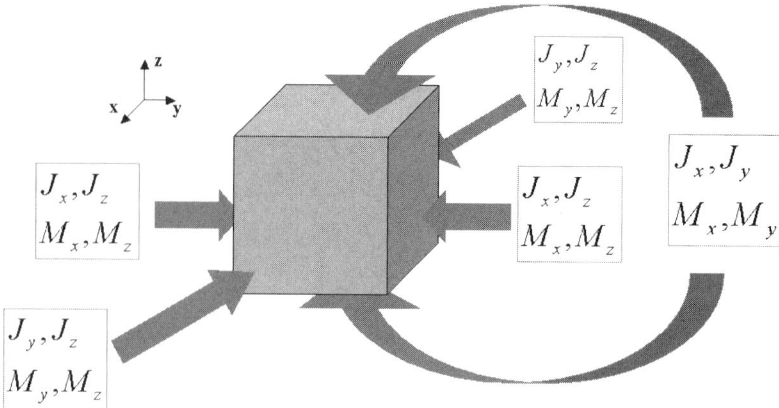

FIGURE 7.7 The electric and magnetic surface current components on a closed surface *S*.

The computation of the electric and magnetic field components in the far fields can be obtained using the vector potentials, such that

$$E_r = 0 \tag{7.50a}$$

$$E_\theta = -\frac{jke^{-jkr}}{4\pi r}\left(L_\phi + \eta_0 N_\theta\right) \tag{7.50b}$$

$$E_\phi = +\frac{jke^{-jkr}}{4\pi r}\left(L_\theta - \eta_0 N_\phi\right) \tag{7.50c}$$

$$H_r = 0 \tag{7.50d}$$

$$H_\theta = +\frac{jke^{-jkr}}{4\pi r}\left(N_\phi - \frac{L_\theta}{\eta_0}\right) \tag{7.50e}$$

$$H_\phi = -\frac{jke^{-jkr}}{4\pi r}\left(N_\theta + \frac{L_\phi}{\eta_0}\right) \tag{7.50f}$$

where η_0 is the intrinsic impedance of free space. When the Cartesian near-to-far-field transformation surface S′ is chosen, the surface currents $\mathbf{J}_s$ and $\mathbf{M}_s$ in frequency domain can be obtained by transferring the time domain surface currents using DFT. The components of the surface currents needed for such analysis are shown in Fig. 7.7. Once the currents are known in the frequency domain, the parameters N_θ, N_ϕ, L_θ, and L_ϕ can be computed using the following expressions [18]:

$$N_\theta = \iint_s \left(J_x \cos\theta\cos\phi + J_y \cos\theta\sin\phi - J_z \sin\theta\right) e^{jkr'\cos\psi} ds \tag{7.51a}$$

$$N_\phi = \iint_s \left(-J_x \sin\phi + J_y \cos\phi\right) e^{jkr'\cos\psi} ds \tag{7.51b}$$

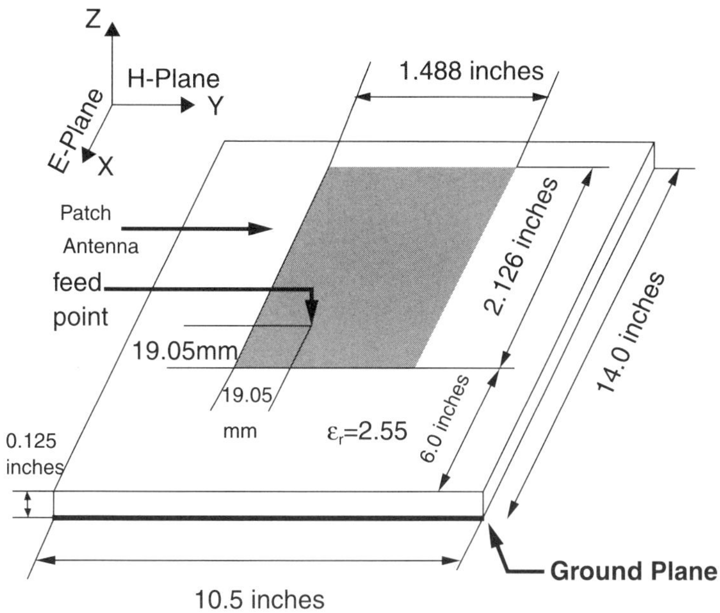

FIGURE 7.8 Geometry of a microstrip patch antenna.

$$L_\theta = \iint_s \left(M_x \cos\theta\cos\phi + M_y \cos\theta\sin\phi - M_z \sin\theta\right) e^{jkr'\cos\psi} ds \tag{7.51c}$$

$$L_\phi = \iint_s \left(-M_x \sin\phi + M_y \cos\phi\right) e^{jkr'\cos\psi} ds \tag{7.51d}$$

Once N_θ, N_ϕ, L_θ, and L_ϕ are determined, the far-field components can be obtained at any observation point. With this approach, the far-field patterns at various frequencies can be obtained. As an example, the FDTD computed results of the E- and H-plane radiation patterns for a microstrip patch antenna (shown in Fig. 7.8) are compared in Fig. 7.9 with the measured data reported in Reference [19]. The

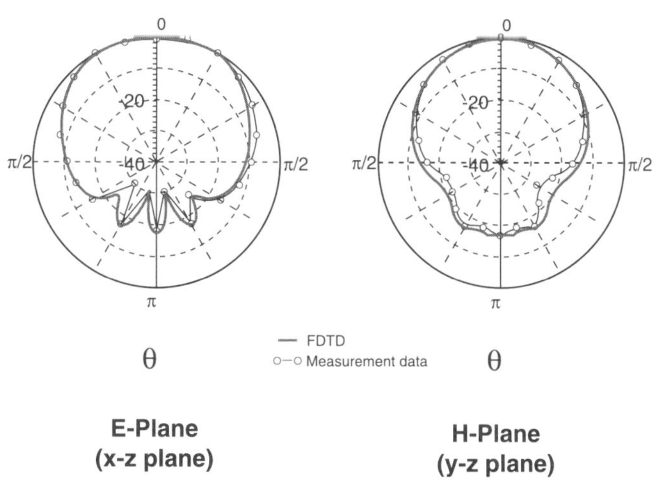

FIGURE 7.9 The radiation pattern of a microstrip patch antenna.

radiation efficiency is a very important indication for the effectiveness of an antenna. The radiation efficiency can be obtained using the FDTD technique. First, the radiated power, P_{rad}, of an antenna is obtained by applying the surface equivalence theorem, such that

$$P_{rad} = \frac{1}{2}\mathrm{Re}\left\{\iint_s \mathbf{E}\times\mathbf{H}^* \cdot \hat{n}ds\right\} = \frac{1}{2}\mathrm{Re}\left\{\iint_s \mathbf{J}^* \times \mathbf{M}\cdot\hat{n}ds\right\} \tag{7.52}$$

Then, the delivered power, P_{del}, of an antenna is determined by the product of the voltage and current provided by the source, which can be expressed as

$$P_{del} = \frac{1}{2}R_e\left[V_s(w)I_s^*(w)\right] \tag{7.53}$$

where $V_s(w)$ and $I_s^*(w)$ represent Fourier transformed value of the source voltage and the complex conjugate current. The radiation efficiency η_a is then defined as

$$\eta_a = \frac{P_{rad}}{P_{del}} \tag{7.54}$$

Another important quantity in the far fields, for circularly polarized antennas, is the axial ratio, which indicates the polarization of the far fields. The axial ratio presented in Reference [18] is defined as

$$AR = \frac{\left[\frac{1}{2}\left\{E_\theta^2 + E_\phi^2 + \left[E_\theta^4 + E_\phi^4 + 2E_\theta^2 E_\phi^2 \cos(2\delta)\right]^{1/2}\right\}\right]^{1/2}}{\left[\frac{1}{2}\left\{E_\theta^2 + E_\phi^2 + \left[E_\theta^4 + E_\phi^4 - 2E_\theta^2 E_\phi^2 \cos(2\delta)\right]^{1/2}\right\}\right]^{1/2}} \tag{7.55}$$

where δ is the phase difference between E_θ and E_ϕ.

By using the formulation presented previously, both the near- and far-field characteristics of antennas can be obtained. A few examples of such analysis are presented in the next sections.

7.5 Example: Stacked Microstrip Antenna Analysis

7.5.1 Coaxial Line and Microstrip Antenna Analysis

To make an accurate characterization and modeling of a stacked microstrip antenna fed with a coaxial line, shown in Fig. 7.10, a full-wave analysis is required to deliver accurate results in a wideband frequency region. By using full-wave analysis, all the parasitic effects are included in the model, as well as fringing fields, dielectric discontinuities, and characteristic impedance of the coaxial line. FDTD meets these characteristics, because it is capable of solving Maxwell's equations directly [20–28].

The FDTD space grid size must be chosen such that over one increment the electromagnetic field does not change significantly. The minimum dimension of the grid must be only a fraction of the minimum wavelength expected in the model; through extensive experimental study it has been determined to be $\lambda/20$ or less for rectangular patches, and $\lambda/40$ or less for circular patches. For computational stability, the time stepping is chosen to satisfy Courant's stability criterion.

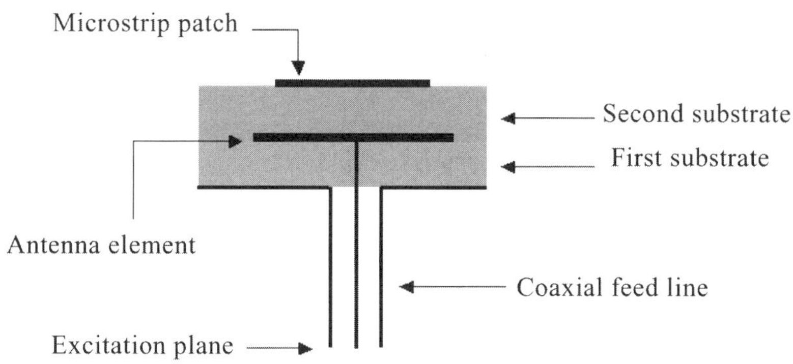

FIGURE 7.10 Coaxial feed model. The computational domain for the feed used for the simulations is 7 × 7 × 70 and $\varepsilon_r = 2.5$.

At $t = 0$, all the field components are made equal to zero over the entire computational region. Because the frequency spectrum of a Gaussian pulse is also Gaussian and gives a frequency domain information from DC to the desired frequency range of interest by adjusting its width, a Gaussian pulse is used as the excitation of the fields in this example.

The common focuses on the FDTD numerical implementation of the microstrip antenna geometry are the modeling of substrates and the modeling of the discontinuities. Microstrip patch antennas have relatively small substrate thickness as compared with the patch resonant length width. Because the substrate thickness has a major effect on the antenna characteristics, it is necessary for this substrate thickness to be modeled by at least three cells to take into account the effect of substrate material [6, 25, 26]. In this analysis the circular boundaries are approximated using "staircasing," but making sure that all the grid cells representing an electric conductor are entirely within the circular area covered by the patch.

7.5.1.1 Coaxial Feed Models

At least three different coaxial feed models can be used in combination with the FDTD method. The first model, proposed by Reineix and Jecko [5], is a simple gap feed model in terms of total-field voltage and current; it was later modified by Luebbers and Langdon [29]. The second one, proposed by Jensen and Rahmat-Samii [30], involves simulating a coaxial feeding cable. The third one, proposed by Wu et al. [31], involves the explicit insertion of the coaxial feed in the FDTD lattice.

Gap Feed Model

The gap feed model, which turns out to be the simplest one, consists of a feeding structure realized as a voltage generator with an internal resistor placed at a small gap between the patch and ground plane. This model works well assuming that the coaxial probe is very thin. An added internal resistor can provide an additional loss mechanism to decrease the current amplitude rapidly and then reduce the required simulation steps. A larger resistance can provide more loss and further reduce the time steps required for convergence. However, it has been shown through numerical experiments that the resistance cannot be too large; otherwise the accuracy level is not satisfied, and instabilities might occur because of neglecting the displacement current through the FDTD cell containing the source. The value of the internal source resistance is generally chosen as the system characteristic impedance to physically mimic the system. Because all the microwave equipment is set to 50 Ω, the source resistance is thus chosen to be 50 Ω [5, 29].

The FDTD implementation of this voltage source with an internal resistor can be illustrated by its equivalent circuit representation. A voltage source V_s^n in series with the internal resistor $R_s = 50\ \Omega$ is located between nodes (i, j, k) and $(i, j, k + 1)$ is shown in Fig. 7.11. The voltage source waveform can be any of the source waveform types presented in Section 7.2.2. The voltage at the feed point node (i, j, k),

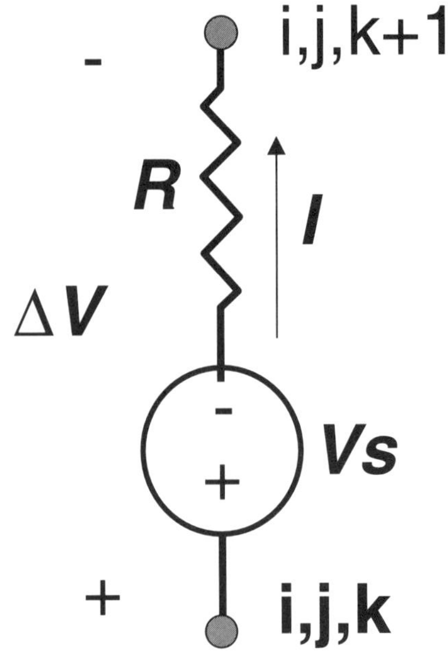

FIGURE 7.11 A resistive voltage source between nodes (i, j, k) and $(i, j, k + 1)$.

V_z^n is represented in the FDTD simulation by an imposed E_z field at the feed point. The E_z field component is updated based on Ampère's Maxwell's equation, that is,

$$\frac{\partial E_z}{\partial t} = \frac{1}{\varepsilon}\left(\frac{\partial H_y}{\partial x} - \frac{\partial H_x}{\partial y} - \frac{I_{sz}}{\Delta x \Delta y}\right) \tag{7.56}$$

which transforms to the FDTD updating form as

$$\frac{E_z^{n+1}(i,j,k) - E_z^n(i,j,k)}{\Delta t} = \frac{1}{\varepsilon(i,j,k)}\left[\frac{H_y^{n+1/2}(i,j,k) - H_y^{n+1/2}(i-1,j,k)}{\Delta x} - \frac{H_x^{n+1/2}(i,j,k) - H_x^{n+1/2}(i,j-1,k)}{\Delta y} - \frac{I_{sz}^{n+1/2}(i,j,k)}{\Delta x \Delta y}\right] \tag{7.57}$$

Then, because the voltage between the nodes (i, j, k) and $(i, j, k + 1)$ is the sum of the voltage drop across the resistance and the voltage of the source, we get

$$I_{sz}^{n+1/2} = \frac{\Delta V_z^{n+1/2} - V_s^{n+1/2}}{R} = \frac{\Delta z\left(\frac{E_z^{n+1}(i,j,k) + E_z^n(i,j,k)}{2}\right) - V_s^{n+1/2}}{R} \tag{7.58}$$

From Eq. (7.57) and (7.58), the expression for the z component of the electric field at (i, j, k) can be obtained.

It is worth noting that the connecting wire with infinitesimal diameter in this model of the feeding structure is actually modeled in the FDTD problem space by setting to zero the tangential component of the electric field (i.e., E_z).

This simple model can also be used to simulate a microstrip feeding structure. However, because the wire is a nonphysical structure for the microstrip feed, a series inductance needs to be removed from the final results during the postprocessing. The inductance introduced by the connecting wire can be found by a calibration run, which is done by comparing a known result with the FDTD simulation result. It has been determined empirically that the wire, for this example, has an approximate inductance of 0.01 μH/cm.

Wire Model

This model involves simulating a coaxial wire feed along the z direction. The inner conductor radius r_a, outer conductor radius r_b, and permittivity ε within the coaxial line are chosen to represent the desired characteristic impedance. A gap voltage is introduced in the coaxial center conductor, and the standard FDTD relations are used within the coaxial line to propagate the fields toward the antenna [30].

At the end of the coaxial line, special interfacing relations are required to update the radial electric and circumferential magnetic fields. The circumferential magnetic field is located a half cell over the ground plane in the FDTD lattice. Its expression is derived using the integral forms of Maxwell's equations and assuming a $1/\rho$ dependence of the fields. By using this approach, the modified time-stepping equation assumes the following form:

$$H_x\Big|_{i,j,k}^{n+1/2} = H_x\Big|_{i,j,k}^{n-1/2} + \frac{\Delta t}{\mu_{i,j,k}}\left[\frac{E_y\Big|_{i,j,k}^{n} - \psi E_y\Big|_{i,j,k-1}^{n}}{\Delta z} - \frac{E_z\Big|_{i,j,k}^{n} - E_z\Big|_{i,j-1,k}^{n}}{(\Delta y/2)\ln(\Delta y/r_a)}\right]$$

$$\psi = \frac{r_b \ln(r_b/r_a)}{\Delta y \ln(\Delta y/r_a)} \tag{7.59}$$

The tangential electric field $E_y\Big|_{i,j,k-1}^{n+1}$ at the coaxial–antenna interface can be calculated using the standard FDTD relations with the modification that $H_x\Big|_{i,j,k}^{n+1/2}$ is substituted by $(\Delta y/r_b)H_x\Big|_{i,j,k}^{n+1/2}$ to align $H_x\Big|_{i,j,k}^{n+1/2}$ (antenna region) with $H_x\Big|_{i,j,k-1}^{n+1/2}$ (coaxial region).

Similar relations can be derived for the other circumferential magnetic and radial electric fields at the interface. The feed point voltage is obtained from the computed radial electric field using

$$V_s^n = \frac{r_b}{2}\ln\left(\frac{r_b}{r_a}\right)E_y\Big|_{i,j,k-1}^{n} \tag{7.60}$$

The current in the wire can be obtained from the discrete form of Ampère's law. Because reverse propagating modes in the coaxial line are excited by the source voltage and by reflections from the coaxial–antenna transition, it is essential to terminate the coaxial line with an absorbing boundary condition. A first-order accuracy boundary condition works well assuming that the propagating mode in the coaxial line is transverse electromagnetic.

In the antenna region, the circumferential magnetic fields in the cells immediately adjacent to the wire, over the coaxial–antenna transition and toward the microstrip patch, can be calculated using an expression similar to Eq. (7.59) but with $\psi = 1$ The term $E_z\Big|_{i,j,k-1}^{n+1}$ is set to zero for a perfectly conducting wire (it could also be forced to some functional form if it is to be a source point). The computation of the other field components and for the fields not adjacent to the wire can be accurately performed with the standard FDTD equations.

Explicit Coaxial Feed Model

This is a rigorous, explicit feed model. We need a simple feed model that does not introduce a very large computational overhead, that can be used to simulate thick coaxial probes, and that can be used in a microstrip array environment. In this explicit coaxial feed model, the computational space is divided into two regions. One is the coaxial line region and the other is the antenna region. If the coaxial structure is defined with a small number of FDTD cells, thus the processing time spent in the coaxial line region is very small compared with that spent in the antenna region [8, 31].

The space increments Δx, Δy, Δz of the FDTD lattice in the coaxial region are chosen to be equal to those of the antenna region. Because the characteristic impedance of a SMA connector and its coaxial line is about 50 Ω, a calibration run is needed to make sure that the numerical characteristic impedance of the coaxial line is equal to 50 Ω over the frequency range of interest. Figure 7.10 illustrates the coaxial line feed and how it is merged into the antenna.

Given that a coaxial line can excite TEM modes up to a predictable cutoff frequency, it can be assumed that the electric field is traveling perpendicular to the coaxial–antenna transition and to the excitation plane. Thus, the model is valid below this cutoff frequency and Mur's first-order ABC can be used in the coaxial region, as long as the coaxial–line length is kept long enough to account for the decay of any higher order modes that could be present in the numerical model. The non-TEM modes that are excited by the nonphysical excitation decay after propagating, at most, a few lattices. The only mode that is able to propagate down the coaxial line is the TEM mode.

A radial field distribution is specified at the excitation plane, projected to the Cartesian components of the FDTD lattice,

$$\begin{aligned} V_s^n &= -E_{feed}\Big|_{i,j,k}^n \Delta x \\ E_{x-feed}\Big|_{i,j,k}^n &= E_\rho\Big|_{i,j,k}^n \cos\phi \\ E_{y-feed}\Big|_{i,j,k}^n &= E_\rho\Big|_{i,j,k}^n \sin\phi \\ \phi &= \tan^{-1}\left(\frac{\Delta\rho_y}{\Delta\rho_x}\right) \end{aligned} \tag{7.61}$$

The increments $\Delta\rho_x$, $\Delta\rho_y$ represent the distance from the point being evaluated to the center of the coaxial line. The electric field $E_\rho\big|_{i,j,k}^n$ takes on the values of the Gaussian excitation. Mur's first-order ABC is applied at the excitation plane once the Gaussian pulse has completely departed from that plane. It is also used at the end of the coaxial line for the calibration run. The curved boundary of the inner and outer conductor of a coaxial line can be approximated with a staircase, although a square coaxial line works well as long as it preserves the desired electrical characteristics (TEM excitation and characteristic impedance).

The radial electric and circumferential magnetic fields are merged at the coaxial–antenna transition. The radial electric field components are located at the coaxial–antenna interface, and therefore it is enough to set the continuity condition of the tangential electric fields at this interface:

$$E_x\Big|_{ic,jc,kc}^n = E_x\Big|_{i,j,0}^n, \quad E_y\Big|_{ic,jc,kc}^n = E_y\Big|_{i,j,0}^n \tag{7.62}$$

The index (i_c, j_c, k_c) is used in the coaxial–region whereas the index (i, j, k) is used in the antenna region. The standard FDTD equations can be used for the circumferential magnetic fields and for the rest of the fields in the computational space.

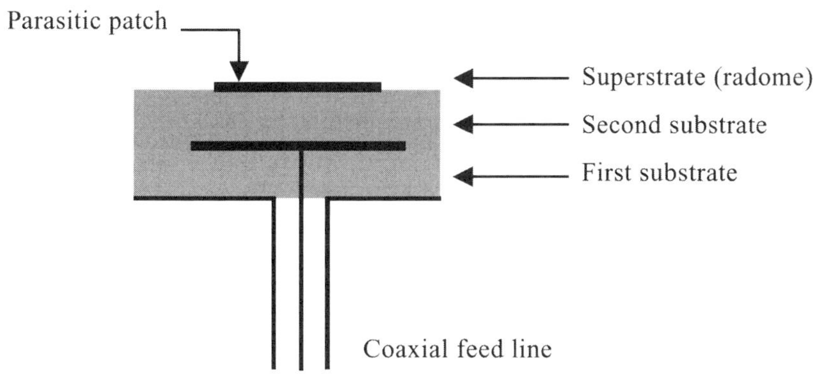

FIGURE 7.12 Single microstrip stacked element with a dielectric superstrate.

7.5.2 Radome Coverings

A radar dome, or radome, is a protective dielectric housing for a microwave or millimeter wave antenna. The function of the radome is to protect the antenna from adverse environments in ground-based, shipboard, airborne, and aerospace applications while having insignificant effect on the electrical performance of the enclosed antenna or antennas. The frequency band of application for radomes is approximately from 1 to 1000 GHz.

Radomes are generally composed of low-loss dielectrics of thickness comparable to a wavelength that are shaped to cover the antenna and, if necessary, to conform to aerodynamic streamlining. A multilayered antenna can be modeled using the FDTD method (Fig. 7.12). It is important to study the radome covering to (1) determine the effects of the superstrate on the resonance and input impedance on a single stacked element and (2) build an accurate model for a single element before the array environment is tackled.

7.5.3 Phased-Array Antennas

Figure 7.13 depicts an eight-element linear array of stacked microsctrip patches that was analyzed using the FDTD method. The arrays consist of coaxial-fed stacked microstrip antennas. Figure 7.14 depicts the layout of the antenna element used for these arrays. The width and thickness of the ground plane were 5 and 0.125 mm, respectively; the side length (width) of the dielectric and the foam was 1.25 mm, whereas the dielectric thickness was 0.127 mm; and the foam thickness was 0.215 mm. The width of the microstrip

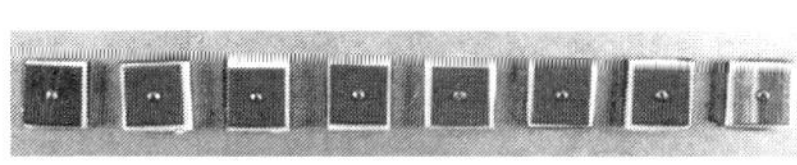

FIGURE 7.13 An eight-element linear array of stacked microstrip patches.

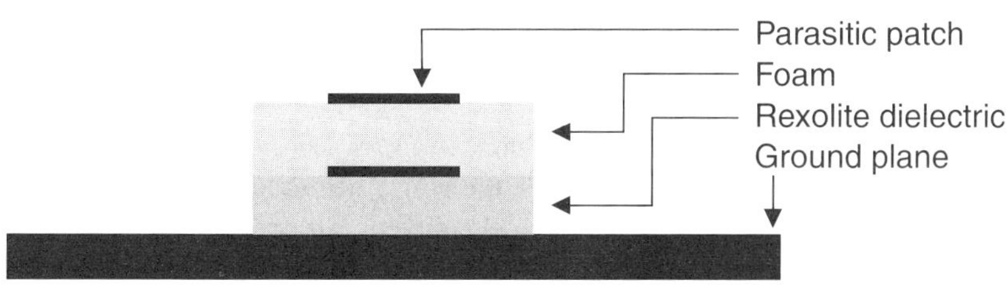

FIGURE 7.14 Layout of the antenna element used for the linear and planar array.

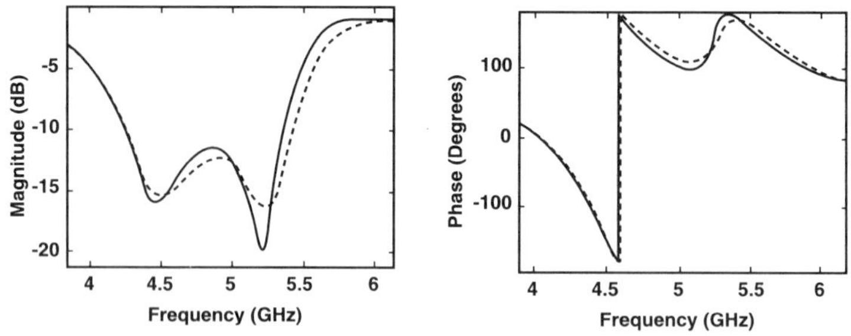

FIGURE 7.15 S44 of the microstrip patches array configuration. The solid line represents the calculations and the dashed line represents the measurements.

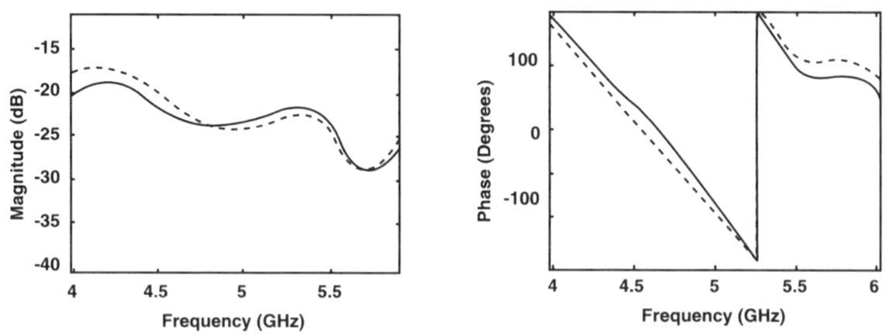

FIGURE 7.16 S_{45} of the microstrip patches array configuration. The solid line represents the calculations and the dashed line represents the measurements.

patch and the parasitic patch were 0.8 and 0.84 mm, respectively. Measurements were performed on a stacked antenna consisting of two square patches and a foam layer ($\varepsilon_r = 1.1$) on top of a Rexolite substrate ($\varepsilon_r = 2.53$, $\tan\delta = 0.00066$), with truncated dielectric layers. The coaxial probe is connected to the lower patch and the upper patch (parasitic) is excited through coupling from the main radiating patch. For these calculations, a nine-layer PML was used. The conductivity was varied using geometric scaling with $g = 4.0$, and a reflection coefficient at normal incidence $R(0) = 10^{-4}$. The computational grid size was $\Delta x = \Delta y = 1.016$ mm, $\Delta z = 0.80645$ mm. The coaxial line length was 70 cells long and the voltage sampling point was located 10 cells below the ground plane, inside the coaxial line. The S-parameters and input impedance are calculated and compared with measurements.

The results of the microstrip patches array are shown in Figs. 7.15 through 7.18. For the shown results, an incident Gaussian pulse is sent through element no. 4, whereas all other elements are terminated with an ABC. The voltage at element no. 5 is sampled. Therefore, S_{45} (the coupling between element no. 5 and element no. 4) is obtained.

7.6 Conclusions

This chapter presents the basics of the FDTD technique for the analysis of microstrip antennas and antenna arrays. From the presented sample results, the FDTD method proves to be a very flexible and powerful simulation tool not only for the analysis and design of this class of antennas but also for many other practical applications. This chapter does not cover all the details of the FDTD technique, but concentrates on the main topics related to the design of antennas.

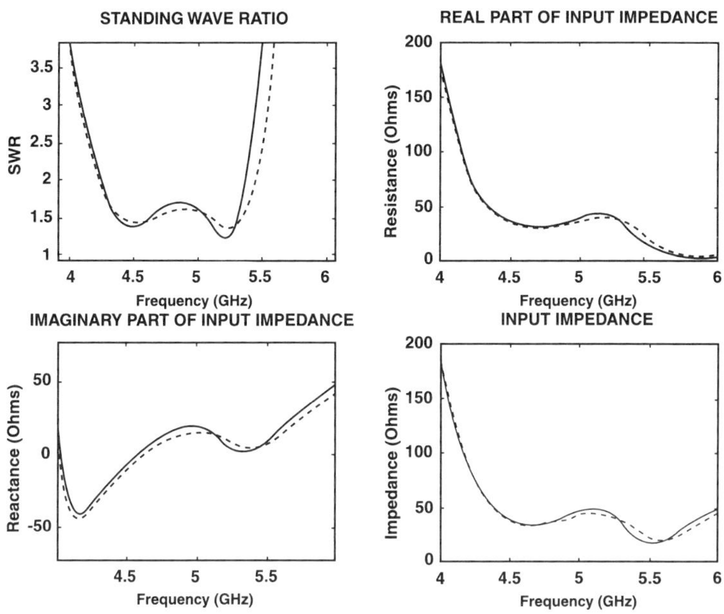

FIGURE 7.17 Impedance analysis for element no. 4 of the microstrip patches array configuration. The solid line represents the calculations and the dashed line represents the measurements.

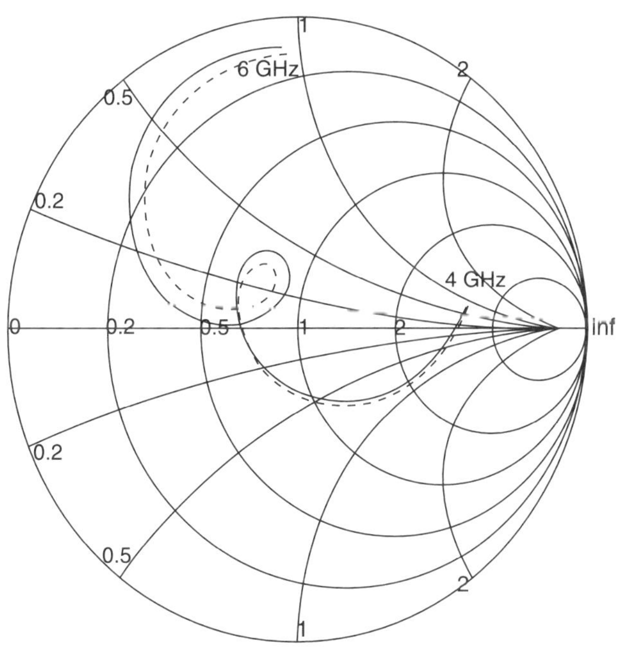

FIGURE 7.18 Input impedance for element no. 4 of the microstrip patches array configuration. The solid line represents the calculations and the dashed line represents the measurements.

References

1. K. S. Yee, "Numerical Solution of Initial Boundary Value Problems involving Maxwell's Equations in Isotropic Media," *IEEE Trans. Antennas Propagat.*, vol. 14, no. 5, pp. 302–307, May 1966.
2. K. S. Kunz and R. J. Luebbers, "The Finite Difference Time Domain Method for Electromagnetics," first edition, CRC Press, Boca Raton, FL, 1993.
3. A. Taflove and S. C. Hagness, "Computational Electrodynamics: The Finite Difference Time Domain Method," second edition, Artech House, Norwood, MA, 2000.
4. A. Taflove, "Advances in Computational Electrodynamics: The Finite Difference Time Domain Method," Artech House, Norwood, MA, 1998.
5. A. Reineix and B. Jecko, "Analysis of Microstrip Patch Antennas Using Finite Difference Time Domain Method," *IEEE Trans. Antennas Propagat.*, vol. 37, no. 11, pp. 1361–1369, Nov. 1989.
6. D. M. Sheen, S. M. Ali, M. D. Abouzahra, and J. A. Kong, "Application of the Three Dimensional Finite Difference Time Domain Method to the Analysis of Planar Microstrip Circuits," *IEEE Trans. Microwave Theory Tech.*, vol. 38, no. 7, pp. 849–857, Jul. 1990.
7. X. Zhang and K. K. Mei, "Time-Domain Finite Difference Approach to the Calculation of the Frequency Dependent Characteristics of Microstrip Discontinuities," *IEEE Trans. Microwave Theory Tech.*, vol. 36, no. 12, pp. 1775–1787, Dec. 1988.
8. G. Mur, "Absorbing Boundary Conditions for the Finite Difference Approximation of the Time Domain Electromagnetic Field Equations," *IEEE Trans. Electromagnetic Compat.*, vol. 23, no. 4, pp. 377–382, Nov. 1981.
9. J. P. Berenger, "A Perfectly Matched Layer for the Absorption of Electromagnetic Waves," *J. Comp. Phys.*, vol. 114, pp. 185–200, Oct. 1994.
10. J. P. Berenger, "Perfectly Matched Layer for the FDTD Solution of Wave Structure Interaction Problems," *IEEE Trans. Antennas Propagat.*, vol. 44, no. 1, pp. 110–117, Jan. 1996.
11. J. P. Berenger, "Improved PML for the FDTD Solution of Wave Structure Interaction Problems," *IEEE Trans. Antennas Propagat.*, vol. 45, no. 3, pp. 466–473, Mar. 1997.
12. D. S. Katz, E. T. Thiele, and A. Taflove, "Validation and Extension to Three Dimensions of the Berenger PML Absorbing Boundary Condition for FDTD Meshes," *IEEE Microwave Guided Wave Lett.*, vol. 4, no. 8, pp. 268–270, Aug. 1994.
13. D. T. Prescott, "Reflection Analysis of FDTD Boundary Conditions — Part I: Time Space Absorbing Boundaries," *IEEE Trans. Microwave Theory Tech.*, vol. 45, pp. 1162–1170, Aug. 1997.
14. D. T. Prescott, "Reflection Analysis of FDTD Boundary Conditions — Part II: Berenger's PML Absorbing Layers," *IEEE Trans. Microwave Theory Tech.*, pp. 1171–1178, vol. 45, Aug. 1997.
15. R. J. Luebbers, K. S. Kunz, M. Schnider, and F. Hunsberger, "A Finite-Difference Time-Domain Near Zone to Far Zone Transformation," *IEEE Trans. Antennas Propagat.*, vol. 39, no. 4, pp. 429–433, Apr. 1991.
16. R. Luebbers, D. Ryan, and J. Beggs, "A Two Dimensional Time Domain Near Zone to Far Zone Transformation," *IEEE Trans. Antennas Propagat.*, vol. 40, no. 7, pp. 848–851, Jul. 1992.
17. K. L. Shlager and G. S. Smith, "Near-field to near-field transformation for use with FDTD method and its application to pulsed antenna problems," *Electron. Lett.*, vol. 30, pp. 1262–1264, Aug. 1994.
18. C. A. Balanis, *Antenna Theory Analysis and Design*, second edition, John Wiley & Sons, New York, 1997.
19. J. Huang, "The Finite Ground Plane Effect on the microstrip Antenna Radiation Patterns," *IEEE Trans. Antennas Propagat.*, vol. AP-31, No. 4, pp. 645-653, Jul. 1983.
20. J. G. Maloney, G. S. Smith, and W. R. Scott, Jr., "Accurate Computation of the Radiation from Simple Antennas Using the Finite Difference Time Domain Method," *IEEE Trans. Antennas Propagat.*, vol. 38, no. 7, pp. 1059–1068, Jul. 1990.
21. P. A. Tirkas and C. A. Balanis, "Finite Difference Time Domain Method for Antenna Radiation," *IEEE Trans. Antennas Propagat.*, vol. 40, no. 3, pp. 334–340, Mar. 1992.

22. R. J. Luebbers, "FDTD Calculation of Wide Band Antenna Gain and Efficiency," *IEEE Trans. Antennas Propagat.*, vol. 40, no. 11, pp. 1403–1407, Nov. 1992.
23. R. Luebbers, L. Chen, T. Uno, and S. Adachi, "FDTD Calculation of Radiation Patterns, Impedance, and Gain for a Monopole Antenna on a Conducting Box," *IEEE Trans. Antennas Propagat.*, vol. 40, no. 12, pp. 1577–1583, Dec. 1992.
24. G. M. Turner and C. G. Christodoulou, "FDTD Analysis of Periodic Phased Array Antennas," *IEEE Tran. Antennas Propagat.*, vol. 47, no. 4, pp. 661–667, April 1999.
25. C. G. Christodoulou, J. Gómez-Tagle and J. Zalewsky, "Applying MPI to Electromagnetic Field Calculations," *Informatica Int. J. Computing,* vol. 22, no. 4, pp. 485–490, December 1998.
26. J. Gómez-Tagle and C. G. Christodoulou, "FDTD Analysis of Stacked Microstrip Ring Antennas," IEEE AP/S International Symposium, Montreal, Canada, July 1997.
27. S. Dai, A. Z. Elsherbeni, and C. E. Smith,"Nonuniform FDTD Formulation for the Analysis and Reduction of Crosstalk on Coupled Microstrip Lines," *J. Electromag. Waves Appl.* (JEWA), vol. 10, pp. 1663–1682, 1996.
28. C. P. Huang, J. B. Chen, A. Z. Elsherbeni and C. E. Smith, "FDTD Characterization of Meander Line Antennas for RF and Wireless Communications," Electromagnetic Wave Monograph Series, Progress in Electromagnetic Research (PIER 24), Chief Editor: J. A. Kong, vol. 24, pp. 185–200, 1999.
29. R. J. Luebbers and H. S. Langdon, "A Simple Feed Model that Reduces Time Steps Needed for FDTD Antenna and Microstrip Calculations," *IEEE Trans. Antennas Propagat.*, vol. 44, no. 7, pp. 1000–1005, Jul. 1996.
30. M. A. Jensen and Y. Rahmat-Samii, "Performance Analysis of Antennas for Hand-Held Transceivers Using FDTD," *IEEE Trans. Antennas Propagat.*, vol. 42, no. 8, pp. 1106–1113, Aug. 1994.
31. C. Wu, K.-L. Wu, Z.-Q. Bi, and J. Litva, "Accurate Characterization of Planar Printed Antennas Using Finite Difference Time Domain Method," *IEEE Trans. Antennas Propagat.*, vol. 40, no. 5, pp. 526–534, May 1992.

8

Method of Moments Applied to Antennas

Tapan K. Sarkar
Syracuse University

Antonije R. Djordjevic
University of Belgrade

Branko M. Kolundzija
University of Belgrade

8.1 Introduction

The art of the electrical engineering design partly relies on the ability to properly model the physical structure under consideration. Good models enable an efficient and accurate analysis, so that designers can reach their goals with a few iterations on the models and, usually, a few steps of experimental verification.

Most electrical and electronic engineers use circuit-theory models to analyze various passive and active circuits. Such models are simple and straightforward to implement, they do not require bulky theoretical background, and they are easy to visualize. However, they may fail to predict circuit behavior even at power frequencies, let alone analyze radiation phenomena. Let us not forget that the circuit-theory models need a link to the physical structure they represent to provide meaningful results. For example, we need to know how to evaluate the resistance of a wire in order to represent it by a resistor.

Electromagnetic field models are predominantly used by antenna and microwave engineers. The analysis starts from the physical structure (i.e., the geometry and electrical properties of materials involved), and it gives a full insight into the properties of devices and circuits (including propagation, radiation, parasitic effects, etc.). Most electromagnetic field problems do not have an analytic solution and a numerical approach is required. However, writing a computer code for the solution of a class of problems is a hard task. Even to properly use codes for the electromagnetic field analysis, a lot of background and experience is required. This software is usually very sophisticated, it covers only a narrow region of applications, and it may sometimes require a long central processor unit (CPU) time to produce results.

An efficient and accurate computer simulation of various electromagnetic field problems, including antennas, is made possible by modern fast computers and well-developed numerical techniques. This simulation enables an antenna designer to visualize the targeted antenna on the desktop, providing in

0-8493-0124-6/02/$0.00+$1.50

many cases more information than can ever be measured in the laboratory or *in situ*, at a lower cost and higher efficiency. A good personal computer and appropriate software may cost significantly less than antenna measurement instrumentation required to equip an antenna laboratory. The turnaround time required to obtain antenna properties after changing antenna shape or dimensions is usually measured by minutes or hours for a computer simulation, but it may require days to build a new antenna prototype and perform measurements. Designers can tune the antenna by modifying certain parameters of the simulation model (e.g., antenna dimensions or material properties), and thus faithfully reflect results they would be getting in the laboratory by trimming the antenna structure. The accuracy of available numerical models is often such that only a small degree of adjustment is required, if any, on the laboratory prototype or on the final product. However, proper interpretation of computed results is necessary, bearing in mind inherent limitations of the technique applied. Hence, a proper selection and evaluation of the computer code is a prerequisite for obtaining reliable results. In many cases, users strive for user-friendly programs, which have ample graphics input and output capabilities, and even include movies. However, in code evaluation, it is more important to be sure that the implemented models can be applied to the actual problem to be solved, and that results can be obtained with a sufficient speed and accuracy.

A variety of numerical methods exist for the analysis of electromagnetic fields. They are based on the solution of Maxwell's equations or certain equations derived from them. Maxwell's equations are fundamental equations for electromagnetic fields [1] and they can be in integral or differential form. Maxwell's equations are presented in Section 8.2 of this chapter.

The numerical methods for field analysis can be classified in a variety of ways. Most numerical techniques deal with linear systems, as are most antenna structures. Such systems can always be described in terms of linear operator equations. An operator is a mapping of a function space to a function space [2]. Hence, the unknown in an operator equation is a function. Some techniques deal with nonlinear systems, but they are not within our scope here.

Another classification is based on the quantity that is solved for in the numerical technique (further referred as the unknown quantity). One group of methods directly solves for the electric or magnetic field vectors, or for quantities tightly related with them (e.g., the Lorentz potentials). The starting equations are Maxwell's equations in differential form or their derivatives (e.g., the wave equation). The unknowns are, hence, spread throughout the volume occupied by the fields. For linear media, as we assume in this chapter, the resulting equations are linear partial differential equations in terms of the unknowns. To this group belong the finite element method (FEM) and the finite difference (FD) method. The latter method includes, for example, the technique for solving the Laplace equation in electrostatics and the finite difference time domain technique described in Chapter 7 of this book. Both the FEM and FD are relatively straightforward to program, and they can handle highly inhomogeneous and even nonlinear media. However, they usually require a lot of spatial and temporal samples to provide a satisfactory accuracy, and, consequently, they demand large computer resources.

The second group of methods solves for the field sources (currents and charges). These sources can be either physical sources, or mathematical (equivalent) sources introduced through various electromagnetic field theorems [3, 4]. In the numerical analysis, the electromagnetic fields, or the related potentials, are expressed in terms of these sources, usually through the Lorentz potentials. The expressions are integral forms, where the sources appear under some integrals, multiplied by appropriate functions, which are referred to as kernels. For example, for fields in a vacuum, the kernel for the Lorentz potentials is the free-space Green's function. On the other hand, certain equations are imposed based on the boundary conditions or constitutive relations. The boundary conditions relate tangential and normal components of the field vectors at a surface of discontinuity. For example, on the surface of a perfectly conducing body, the tangential component of the electric field vanishes. The constitutive equations reflect material properties: dielectric polarization, current conduction, and magnetization. Finally in the derivation, the quantities involved in the boundary conditions and constitutive relations are expressed in terms of the field sources. As the result, an integral equation (or a set of integral equations) is obtained for the unknown sources. For linear media, as assumed here, these integral equations are linear. In some cases, the unknowns are distributed through a volume, like direct current (DC) and associated charges in a conducting medium. In many other cases, the sources are distributed only over surfaces, thus depending on two local coordinates (e.g., scattering from a thin metallic plate in a vacuum), or along lines, thus depending on one local coordinate (e.g., a wire antenna). The resulting equations are integral equations in terms of the unknowns, though, in some cases, derivatives of the unknowns may appear somewhere in the equation. The techniques of this group are most often based on the method of moments (MOM), which is the main topic of this chapter. As a rule, techniques of this group require a lot of analytic preparation and implementation of sophisticated numerical procedures. They are usually inefficient when applied to highly inhomogeneous media, and they are not applicable to nonlinear media.

Combinations of these two groups of methods are also possible. They are referred to as hybrid methods, and they can combine the respective advantages of each group. Thereby, the differential equation formulation is applied to highly inhomogeneous (and possibly anisotropic and nonlinear) regions, and the integral equation formulation for the remaining space.

At this point, a remark should be made on the dimensionality of the electromagnetic fields and unknowns. An electromagnetic field is always a three-dimensional (3D) spatial phenomenon, meaning that it exists within a finite or infinite region (volume). In most cases, the field vectors are functions of three spatial coordinates (e.g., the Cartesian x, y, and z coordinates), and such problems are referred to as 3D electromagnetic field problems. In some problems, the fields are functions of only two coordinates. For example, the electrostatic field of an infinitely long two-wire line depends only on the transverse coordinates. The related problems are referred to as two-dimensional (2D) problems. Even simpler cases are when the fields depend only on one spatial coordinate. For example, the electric and magnetic fields of a uniform plane wave depend only on the longitudinal coordinate. In such cases we speak about one-dimensional (1D) problems.

The dimensionality of an electromagnetic field problem should not be confused with the mathematical dimensionality of the unknowns. They may or may not coincide. For example, when the unknowns are fields in a 3D electromagnetic problem, the unknowns are also functions of three spatial coordinates, and we have a 3D mathematical problem. However, if we solve for the field sources, the situation may be different. For example, if we analyze scattering from a rectangular metallic plate in a vacuum, the unknowns are currents induced on the plate, which depend on two local coordinates associated with the plate. Hence, the unknowns constitute a 2D mathematical problem. If we consider scattering from a thin wire in a vacuum, the unknown is the current distribution along the wire, and we have a 1D mathematical problem.

For the analysis in the time domain, the temporal variable increases the mathematical dimensionality of the problem by one. In this chapter, however, we deal exclusively with the frequency domain analysis. Efficiency of a numerical solution significantly depends on the mathematical dimensionality of the unknowns. In most cases, faster and more accurate solutions are obtained when the dimensionality is smaller.

In this chapter, the emphasis is on the application of integral equations to antenna problems, and their solution using the MOM. In Section 8.3, the basic philosophy of the MOM is presented, without going into details and omitting rigorous proofs. An interested reader should refer to several excellent books [2, 5–15] for an in-depth coverage of the MOM. In Section 8.4, specifics of the MOM application to antennas are presented. This section is further divided into three parts, according to the increased complexity of structures analyzed. Section 8.4.2 deals with wire antennas, Section 8.4.3 deals with arbitrarily shaped metallic structures, whereas Section 8.4.4 is devoted to the most general case – combined metallic and dielectric structures. In Sections 8.3 and 8.4, illustrative examples are given showing various possibilities of the MOM.

8.2 Maxwell's Equations

8.2.1 Basic Equations, Constitutive Relations, and Boundary Conditions

Maxwell's equations are general equations that govern macroscopic electromagnetic fields. In the time domain, in differential form, the four basic Maxwell's equations read [1] as follows:

$$\mathrm{curl}\,\mathbf{E} = -\frac{\partial \mathbf{B}}{\partial t}$$

$$\mathrm{curl}\,\mathbf{H} = \mathbf{J} + \frac{\partial \mathbf{D}}{\partial t} \tag{8.1}$$

$$\mathrm{div}\,\mathbf{D} = \rho$$

$$\mathrm{div}\,\mathbf{B} = 0$$

where **E** is the electric field intensity; **H**, the magnetic field intensity; **D**, the electric flux density (also referred to as the electric displacement or the electric induction); **B**, the magnetic flux density (also referred to as the magnetic induction); **J**, the electric current density; and ρ, the volume charge density. All quantities in Eq. (8.1) depend on the position vector (**r**) and time (t). To obtain a complete system, the four basic equations should be complemented by the constitutive relations, which read in the general form

$$\mathbf{D} = \mathbf{D}(\mathbf{E})$$

$$\mathbf{J} = \mathbf{J}(\mathbf{E}) \tag{8.2}$$

$$\mathbf{B} = \mathbf{B}(\mathbf{H})$$

In particular, for linear media,

$$\mathbf{D} = \varepsilon\mathbf{E}$$

$$\mathbf{J} = \sigma\mathbf{E} + \mathbf{J}_i \tag{8.3}$$

$$\mathbf{B} = \mu\mathbf{H}$$

where ε is the permittivity, σ is the conductivity, and μ is the permeability of the medium, whereas $\mathbf{J}_i$ is the density of an impressed electric currents, which model the excitation. The excitation in Eq. (8.3) corresponds to a current generator in the circuit theory. The impressed currents create an electromagnetic field, just like ordinary electric currents. The excitation can alternatively be modeled by the impressed electric field, $\mathbf{E}_i$, using the relation $\mathbf{J} = \sigma(\mathbf{E} + \mathbf{E}_i)$, which corresponds to a voltage generator in the circuit theory.

In practical electromagnetic field problems, the geometry and constitutive parameters of the structure are usually given along with the excitation, and the objective is to evaluate other quantities of interest.

From the second and third equation in (1), the continuity equation can be derived,

$$\mathrm{div}\,\mathbf{J} = -\frac{\partial \rho}{\partial t} \tag{8.4}$$

In the circuit theory, the continuity equation corresponds to Kirchhoff's current law.

Equations (8.1) and (8.4) are valid provided the vectors **E**, **H**, **D**, **B**, and **J** are differentiable functions of the position vector. These vectors may not be differentiable at an interface surface between two media (which differ in parameters ε, σ, or μ), shown in Fig. 8.1. At such an interface, instead of Maxwell's equations in differential form, fields satisfy boundary conditions. These conditions are relations between tangential and normal components of the field vectors. They are expressed in vector form as

$$\mathbf{n}\times\mathbf{E}_1 - \mathbf{n}\times\mathbf{E}_2 = 0$$

$$\mathbf{n}\times\mathbf{H}_1 - \mathbf{n}\times\mathbf{H}_2 = \mathbf{J}_s \tag{8.5}$$

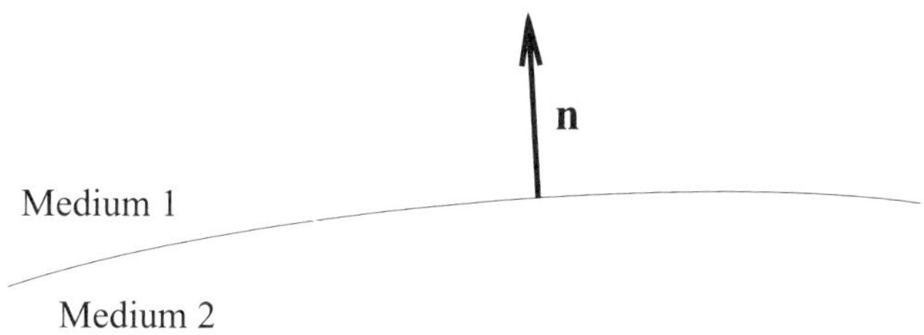

FIGURE 8.1 Interface between two media.

$$\mathbf{n}\cdot\mathbf{D}_1 - \mathbf{n}\cdot\mathbf{D}_2 = \rho_s$$
$$\mathbf{n}\cdot\mathbf{B}_1 - \mathbf{n}\cdot\mathbf{B}_2 = 0$$

where **n** is the unit normal directed from medium 1 toward medium 2, $\mathbf{J}_s$ is the density of surface currents, and ρ_s the density of surface charges on the interface.

Note that the integral form of Maxwell's equations is more general than the differential form, and Eqs. (8.1) and (8.5) are directly derivable from the integral form. However, differential form is more convenient for our present needs.

A perfect electric conductor (PEC) is a fictitious conductor whose conductivity (σ) is infinitely large. In such a medium, time-dependent electromagnetic fields cannot exist. Hence, if medium 2 is a PEC, Eqs. (8.5) reduce to

$$\begin{aligned}\mathbf{n}\times\mathbf{E}_1 &= 0\\ \mathbf{n}\times\mathbf{H}_1 &= \mathbf{J}_s\\ \mathbf{n}\cdot\mathbf{D}_1 &= \rho_s\\ \mathbf{n}\cdot\mathbf{B}_1 &= 0\end{aligned} \qquad (8.6)$$

To analyze an electromagnetic problem, we essentially have to solve the complete system of Maxwell's equations, with appropriate boundary conditions, for a given excitation. Some solution techniques directly solve the differential equations, whereas others first relate the fields to the field sources (currents and charges), leading to integral equations. For the latter case, solution is facilitated if the electric and magnetic fields are expressed in terms of the electric scalar potential (V) and the magnetic vector potential (**A**),

$$\begin{aligned}\mathbf{E} &= -\frac{\partial\mathbf{A}}{\partial t} - \operatorname{grad} V\\ \mathbf{B} &= \operatorname{curl}\mathbf{A}\end{aligned} \qquad (8.7)$$

These potentials are related to the field sources in a simpler form than the fields themselves. Various definitions for the potentials exist. For the numerical analysis of antennas by the method of moments, the Lorentz potentials are predominantly used. They are related by the Lorentz gauge,

$$\operatorname{div}\mathbf{A} = -\frac{\partial V}{\partial t} \qquad (8.8)$$

The Lorentz potentials are elaborated in Section 8.2.3.

For completeness, we note that the density of the power flow in an electromagnetic field (i.e., the Poynting vector) is given by

$$\mathrm{P} = \mathbf{E} \times \mathbf{H} \tag{8.9}$$

8.2.2 Phasor Representation and Equations in Complex Domain

In principle, the field vectors can be arbitrary functions of time. For engineering applications (e.g., narrowband signals), it is often sufficient to assume a steady-state (sinusoidal) regime. In this chapter, we consider only such a regime.

Before going on, we define complex vectors, because they are essential for the analysis. We shall reveal the canonical form of a sinusoidal scalar quantity on the example of a current that is a sinusoidal function of time. This form reads

$$i(t) = I_{\mathrm{m}} \cos(\omega t + \psi) \tag{8.10}$$

where $i(t)$ is the instantaneous current, $I_m = i_{\mathrm{rms}}\sqrt{2}$ is its amplitude (peak value), I_{rms} is the root-mean-square (rms, or effective) value, ω is the angular frequency ($\omega = 2\pi f$, where f is the frequency), and ψ is the initial phase. The standard procedure in the analysis of sinusoidal regimes is to switch to the domain of the complex frequency, because differential equations in the time domain are converted to ordinary algebraic equations. More precisely, the derivative with respect to time is replaced in the complex domain by the multiplication by $j\omega$, which significantly facilitates the analysis.

The complex-domain counterpart of the current $i(t)$ (i.e., the phasor current, I, is introduced in two ways). The first one (commonly used, e.g., in Europe) is by the equation

$$i(t) = \mathrm{Re}\left(\sqrt{2} I e^{j\omega t}\right) \tag{8.11}$$

where Re denotes the real part, and j is the imaginary unit ($j = \sqrt{-1}$). The phasor I is referred to as the complex rms or effective value, because $|I|$ equals the rms value of the current $i(t)$. Another possibility (commonly used, e.g., in the United States) is

$$i(t) = \mathrm{Re}\left(I e^{j\omega t}\right) \tag{8.12}$$

in which case I is referred to as the complex amplitude, because $|I|$ now equals the amplitude of the current $i(t)$. The choice of one of the preceding definitions does not affect any of the equations in the following sections that are linear relations between complex representatives. However, it does affect relations for power, as well as relations between the complex numbers and the quantities in the time domain that these complex numbers represent. We assume the definition Eq. (8.11), but we shall point out equations in this chapter that differ depending on the choice of Eqs. (8.11) or (8.12).

A sinusoidal time-domain vector, like, for example, the electric-field vector, $\mathbf{E}(t)$, is defined in the following way. It is a vector separable into three orthogonal (e.g., Cartesian) components:

$$\mathbf{E}(t) = E_x(t)\mathbf{u}_x + E_y(t)\mathbf{u}_y + E_z(t)\mathbf{u}_z \tag{8.13}$$

where each component is a sinusoidal function of time,

$$\left.\begin{aligned} E_x(t) &= E_{xm}\cos(\omega t+\theta_x)\\ E_y(t) &= E_{ym}\cos(\omega t+\theta_y)\\ E_z(t) &= E_{zm}\cos(\omega t+\theta_z)\end{aligned}\right\} \tag{8.14}$$

having arbitrary amplitudes (E_{xm}, E_{ym}, E_{zm}) and initial phases (θ_x, θ_y, θ_z), but the same angular frequency (ω). The complex (phasor) electric-field vector, **E**, is obtained by finding complex representatives of $E_x(t)$, $E_y(t)$, and $E_z(t)$, according to Eqs. (8.11) or (8.12), denoted by E_x, E_y, and E_z, respectively. These phasors are then used as components of the resulting phasor vector,

$$\mathbf{E} = E_x\mathbf{u}_x + E_y\mathbf{u}_y + E_z\mathbf{u}_z \tag{8.15}$$

We shall not introduce separate notations for field vectors in the time domain and in the frequency domain. This should not cause confusion, because in this chapter we do not deal with the vectors in the time domain.

A sinusoidal vector in the time domain is, generally, elliptically polarized. Hence, both its magnitude and direction vary as a function of time. The tip of the vector describes an ellipse. As special cases, the vector can be linearly polarized, when it has a constant direction, but changes the magnitude and sense, or circularly polarized, when it has a constant magnitude, but rotates at a uniform speed. The complex vector, however, does not have a physically defined direction, except for linearly polarized fields. If the definition in Eq. (8.11) is used, the magnitude of the phasor electric field, $|\mathbf{E}|$, has a clear meaning: it is the rms of $|\mathbf{E}(t)|$.

Maxwell's equations in the complex domain can be written only for linear media, because the sinusoidal regime cannot exist in nonlinear media. Equations (8.1), (8.4), (8.7), and (8.8) become, respectively,

$$\begin{aligned} \operatorname{curl}\mathbf{E} &= -j\omega\mathbf{B}\\ \operatorname{curl}\mathbf{H} &= \mathbf{J} + j\omega\mathbf{D}\\ \operatorname{div}\mathbf{D} &= \rho\\ \operatorname{div}\mathbf{B} &= 0\end{aligned} \tag{8.16}$$

$$\operatorname{div}\mathbf{J} = -j\omega\rho \tag{8.17}$$

$$\begin{aligned} \mathbf{E} &= -j\omega\mathbf{A} - \operatorname{grad} V\\ \mathbf{B} &= \operatorname{curl}\mathbf{A}\end{aligned} \tag{8.18}$$

$$\operatorname{div}\mathbf{A} = -j\omega V \tag{8.19}$$

where all quantities depend only on the position vector, **r**. Equations (8.3) are still formally valid, but all quantities should now be interpreted as being phasors (i.e., in the frequency domain).

If the definition in Eq. (8.11) is used, the complex Poynting vector is

$$\mathrm{P} = \mathbf{E}\times\mathbf{H}^* \tag{8.20}$$

where the asterisk denotes complex conjugate. If Eq. (8.12) is used, the complex Poynting vector is

$$\mathrm{P} = \frac{1}{2}\mathbf{E}\times\mathbf{H}^* \tag{8.21}$$

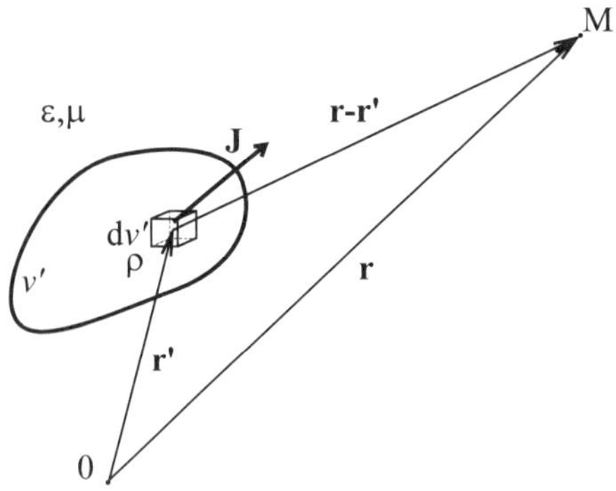

FIGURE 8.2 Coordinate system for evaluation of potentials.

8.2.3 Lorentz Potentials and Green's Function

In a linear homogeneous lossless medium, the electric and magnetic fields can be expressed in terms of the field sources (currents and charges) through the Lorentz potentials, starting from Eq. (8.18). Referring to Fig. 8.2, the potentials are related to the field sources as

$$\mathbf{A}(\mathbf{r}) = \mu \int_{v'} \mathbf{J}(\mathbf{r}')g(\mathbf{r},\mathbf{r}')dv'$$
$$V(\mathbf{r}) = \frac{1}{\varepsilon}\int_{v'} \rho(\mathbf{r}')g(\mathbf{r},\mathbf{r}')dv' \quad (8.22)$$

where $\mathbf{r}$ is the coordinate of the field point M (i.e., the point at which the potentials and fields are evaluated), v' the volume occupied by the sources, $\mathbf{r}'$ the coordinate of the source point (i.e., the point at which the field source element dv' is located),

$$g(\mathbf{r},\mathbf{r}') = \frac{\exp(-jk|\mathbf{r}-\mathbf{r}'|)}{4\pi|\mathbf{r}-\mathbf{r}'|} \quad (8.23)$$

is the Green's function, and $k = \omega\sqrt{\varepsilon\mu}$ is the phase coefficient. Losses in media can be incorporated in the preceding equations by taking the permittivity and permeability to be complex. Note that Green's function is, generally, the response to an impulse function (Dirac's delta function). It gives the potential resulting from a point source, which can be regarded as a spatial delta function.

The expressions in Eq. (8.22) are written assuming currents and charges distributed throughout the source volume, v'. In many cases the currents and charges can be assumed distributed over surfaces, like charges on conducting bodies in electrostatics, or currents and charges on metallic bodies when the skin effect is fully pronounced. It is also possible to have the sources practically distributed along lines (filaments), like currents and charges on thin-wire conductors. For surface sources Eq. (8.22) is to be modified appropriately by taking the densities of the surface currents ($\mathbf{J}_s$) and charges (ρ_s), and integrating over the source surface (S'), that is,

$$\mathbf{A}(\mathbf{r}) = \mu \int_{S'} \mathbf{J}_s(\mathbf{r}')g(\mathbf{r},\mathbf{r}')dS'$$
$$V(\mathbf{r}) = \frac{1}{\varepsilon}\int_{S'} \rho_s(\mathbf{r}')g(\mathbf{r},\mathbf{r}')dS' \quad (8.24)$$

For filamental currents and charges the current intensity (I) and the per-unit-length charge density (ρ_l) should be used, and integrated along the source line (L'), yielding

$$\mathbf{A}(\mathbf{r}) = \mu \int_{L'} \mathbf{u}(\mathbf{r}') I(\mathbf{r}') g(\mathbf{r},\mathbf{r}') dl'$$
$$V(\mathbf{r}) = \frac{1}{\varepsilon} \int_{L'} \rho_l(\mathbf{r}') g(\mathbf{r},\mathbf{r}') dl' \qquad (8.25)$$

where **u** is the unit vector tangential to the line. For surface and line sources the continuity Eq. (8.17) is replaced by

$$\mathrm{div}_s\,\mathbf{J}_s = -j\omega\rho_s \qquad (8.26)$$

$$\frac{dI}{ds} = -j\omega\rho_l \qquad (8.27)$$

respectively, where the surface divergence (div_s) in Eq. (8.26) implies differentiation only with respect to two local coordinates on the surface, and s in Eq. (8.27) is a local coordinate along the line.

To simplify the analysis, it is convenient to relate the fields only to the currents, thus avoiding dealing with the charges. Two basic possibilities can express the fields in terms of only the current density. The first way is to combine the expressions in Eq. (8.18) with the Lorentz gauge in the complex domain, Eq. (8.19). As the result, the electric field is expressed only in terms of the magnetic vector potential as

$$\mathbf{E} = -j\omega\left(\mathbf{A} + \frac{1}{k^2}\,\mathrm{grad}\,\mathrm{div}\,\mathbf{A}\right) \qquad (8.28)$$

With the use of Eq. (8.28) and the first expression in Eq. (8.22), the electric field is related only to the currents. The second way is to express the charge density from the continuity Eq. (8.17) in terms of the current density, substituting into the second expression in Eq. (8.22), and then using Eq. (8.18).

For 3D static problems (including electrostatics), Green's function [Eq. (8.23)] reduces to

$$g(\mathbf{r},\mathbf{r}') = \frac{1}{4\pi|\mathbf{r}-\mathbf{r}'|} \qquad (8.29)$$

For 2D high-frequency problems, Green's function is

$$g(\mathbf{r},\mathbf{r}') = -\frac{j}{4} H_0^{(2)}\left(k|\mathbf{r}-\mathbf{r}'|\right) \qquad (8.30)$$

where $H_0^{(2)}(x)$ is Hankel's function of the second kind and order 0. In this case, Green's function gives the potential of a uniform, infinitely long line source, which is the elemental source in 2D problems. For low frequencies, Green's function [Eq. (8.30)] can be approximated by

$$H_0^{(2)}\left(k|\mathbf{r}-\mathbf{r}'|\right) \approx 1 - j\frac{2}{\pi}\log\frac{\gamma k|\mathbf{r}-\mathbf{r}'|}{2} \qquad (8.31)$$

where $\gamma = 1.781 \ldots$ is Euler's constant, yielding

$$g(\mathbf{r},\mathbf{r}') \approx -\frac{j}{4} - \frac{1}{2\pi}\log\frac{\gamma k|\mathbf{r}-\mathbf{r}'|}{2} \tag{8.32}$$

As frequency diminishes, tending toward the static case, Green's function [Eq. (8.32)] can be substituted by

$$g(\mathbf{r},\mathbf{r}') = -\frac{1}{2\pi}\log|\mathbf{r}-\mathbf{r}'| \tag{8.33}$$

under the condition that the integral of the field sources (e.g., the total charge of the system) is zero. If this condition is violated, the static potentials resulting from Eq. (8.32) become infinitely large.

Only very few electromagnetic field problems have analytic solutions. Most such solutions can be found in Reference [16]. Examples of analytically solvable problems in electrostatics are a conducting sphere and an infinite conducting circular cylinder. Among high-frequency problems, analytic solutions exist for the propagation of uniform plane waves, and for the wave propagation along certain transmission lines (e.g., coaxial lines) and waveguides (rectangular and circular waveguides), but there are no analytic solutions for antennas. Note that the well-known sinusoidal current distribution along a thin wire [1] is only an approximation; the thinner the wire is the better the antenna conductor. Most problems that have closed-form solutions are impractical because realistic structures often have complicated shapes, but they can serve as an estimate of properties of the realistic structures. For example, the capacitance of an arbitrarily shaped conductor is larger than the capacitance of the largest inscribed sphere, but smaller than the capacitance of the smallest circumscribed sphere.

The only available way to precisely analyze practical structures is to implement numerical techniques. The MOM is one of them, particularly suitable for structures that are not too large in terms of the wavelength. The limits depend on the complexity of the structure analyzed, numerical implementation, and computer resources. As estimation on the order of magnitude, the MOM can commonly handle wire structures that are 1000 wavelengths long, and surfaces whose area is 100 square wavelengths and solve them efficiently on personal computers.

8.3 Method of Moments

8.3.1 Linear Operator Equations

As stated in Section 8.1, numerical solutions of electromagnetic field problems are usually classified into two groups. The first one attacks directly electromagnetic fields, and the second one attacks the field sources. In both cases, the equations that are to be solved are linear operator equations in terms of the unknowns (the fields, namely, the sources). However, in the first case the equations are differential, whereas in the second case they are integral. Both classes of equations belong to the general class of linear operator equations, which have the common form

$$L(f) = g \tag{8.34}$$

where L is the operator; g is the source or excitation, which is assumed to be a known function; and f is the field or response, which is the unknown function to be determined. The linearity of the operator follows from the linearity of Maxwell's equations and the constitutive equations, as we consider only linear media. We assume that a unique solution to Eq. (8.34) exists.

For the first group of numerical methods L is a differential operator. It generally involves derivatives with respect to three spatial coordinates. For the time-domain analysis, derivatives with respect to time are also involved. Further, f is a field vector or potential (depending on the formulation), whereas g is a known quantity (e.g., the field or potential resulting from an incident wave). For the second group of numerical methods L is an integral operator, f represents the field sources, and g is, again, a known quantity that models the excitation.

Irrespective of the approach, the operator Eq. (8.34) can be solved following the numerical procedure known under the generic name of the MOMs, which is a general technique for solving linear operator equations.

8.3.2 Basic Steps of the Method of Moments

The basic idea of the MOM is as follows. The unknown quantity (f) is expanded in terms of a set of linearly independent known functions, f_n (referred to as basis or expansion functions); that is, it is approximated by the following finite series:

$$f \approx \sum_{n=1}^{N} \alpha_n f_n \tag{8.35}$$

where α_n are unknown coefficients yet to be determined. The expansion functions should be chosen, usually based on experience, so that reasonable approximation of f is obtained with a small number of terms, N.

When Eq. (8.35) is substituted into Eq. (8.34), one obtains the approximate equation

$$L\left(\sum_{n=1}^{N} \alpha_n f_n\right) \approx g \tag{8.36}$$

Because of the linearity of the operator, we can rewrite Eq. (8.36) as

$$\sum_{n=1}^{N} \alpha_n L(f_n) \approx g \tag{8.37}$$

Note that Eq. (8.37) cannot be exactly satisfied at all points, because we have a finite number of terms in the series. Exceptions are rare examples that do have analytic solutions, but are not of our interest here. The unknown coefficients (α_n) should now be determined such that Eq. (8.37) is satisfied in some sense. Hence, a measure is needed describing the degree of accuracy to which the left side and the right side of Eq. (8.37) match.

In the MOM, this measure is obtained in the following way. Both sides of Eq. (8.37) are multiplied by a known, properly selected function, referred to as the weighting function, w_m, and the results integrated over a spatial region. This integration is a special, but very frequent case of an inner product of two functions, f and g, which is denoted by $\langle f, g\rangle$. Generally, the inner product of elements f and g of a given space is a scalar, which satisfies the following conditions: $\langle f, g\rangle = \langle g, f\rangle$, $\langle \alpha f + \beta g, h\rangle = \alpha\langle f, h\rangle + \beta\langle g, h\rangle$, $\langle f, f^*\rangle > 0$ if $f \neq 0$, and $\langle f, f^*\rangle = 0$ if $f = 0$, where α and β are arbitrary scalars, and h is another element of the same space.

The choice of the weighting functions and the inner product is, again, based on experience. Now we have

$$\sum_{n=1}^{N} \alpha_n \langle w_m, L(f_n)\rangle = \langle w_m, g\rangle \tag{8.38}$$

The inner products in Eq. (8.38) are definite numbers, because they can be evaluated analytically or, more frequently, numerically. Hence, Eq. (8.38) represents a linear equation in coefficients α_n. To obtain a determined system of linear equations for these coefficients, the weighting procedure is done for a linearly independent set of N functions, yielding

$$\sum_{n=1}^{N} \alpha_n \langle w_m, L(f_n) \rangle = \langle w_m, g \rangle \qquad m = 1, \ldots, N \tag{8.39}$$

Equation (8.39) represents a system of N ordinary linear equations in N unknowns, and it can be solved using various techniques. As a rule, the methods based on differential equations result in huge, but sparse systems of linear equations, which are solved using specific techniques. The methods based on integral equations result in more compact, but full systems, which are usually solved using the Gaussian elimination or similar techniques [17], such as the LU decomposition. Note that the classical matrix inversion is an inefficient approach, because it requires about three times more operations, and thus three times longer CPU time, than the Gaussian elimination. Large, full MOM systems of linear equations have also been successfully solved using other techniques, such as the conjugate gradients [18] alone or in combination with the fast Fourier transform [19].

To prepare a computer code that uses the MOM to solve a complex electromagnetic field problem, usually requires a lot of work and experience. Often, codes are specialized for certain classes of problems. There is no guarantee of convergence, and in most cases a useful measure of accuracy does not exist for the solution obtained. In spite of all these deficiencies, the MOM is the most powerful tool available nowadays for analysis of fairly general electromagnetic field problems that involve linear media.

The expansion and testing functions can be arbitrary. However, to provide an efficient solution, the expansion functions should be selected such that the solution can be well approximated by a relatively small number of functions. Similarly, the weighting functions should provide a reliable measure of discrepancy between the two sides of Eq. (8.37). On the other hand, all these functions should be selected bearing in mind complexity and speed of computations, and flexibility to accommodate to a wide range of problems [20].

Expansion and testing functions may coincide (i.e., we can take $f_n = w_n$, $n = 1, \ldots, N$). In this case we have a Galerkin solution, which is equivalent to the Rayleigh–Ritz variational method, often used in the finite element approach.

In the literature there is a certain confusion between the terms *method of moments* (MOM) and *finite element method* (FEM), emerging from the existence of two distinct groups of practitioners. One group usually deals with integral equations and solves them using the MOM, thus identifying the MOM with the solution of integral equations. The other group usually deals with differential equations and solves them also using the MOM, but with subsectional basis functions referred to as finite elements, thus identifying the FEM with the solution of differential equations. This second group also claims that solving integral equations with subsectional basis functions is an application of the FEM. To add to the confusion, in the FEM, the starting differential equation that is to be solved is often formulated from the variational (energy) principle, thus obscuring the fact that the same result can be obtained if the Galerkin procedure is directly applied to a differential equation derivable from Maxwell's equation. The truth seems to be that both groups essentially do similar things, but they speak somewhat different languages. In this chapter we predominantly solve integral equations using the MOM, so there should be no confusion about the terms.

Both expansion and testing functions can be divided into two categories. The first category is subdomain functions. The domain, where the unknown function (f) is defined, is divided into a number of small subdomains. Each basis function is defined only on one subdomain (i.e., it is assumed zero elsewhere), and it is a very simple function. Such a choice simplifies evaluation of matrix elements, and it can relatively easily accommodate an arbitrary geometry. However, it may result in instabilities as the approximation of the unknown function is discontinuous or has discontinuous derivatives, and it may require a large number of basis functions for an accurate solution.

The simplest subdomain approximation is using samples (impulses, Dirac's delta functions). This procedure is seldom used for expansion in the MOM, except with the FD method. If used, it may require modifying the original operator to better suit the expansion. For example, instead of the derivative, an FD scheme is used. However, impulses are often used as testing functions (i.e., $w_m = \delta_m(P_m)$, where $\delta_m(P_m)$) denotes an impulse centered at a point P_m, amounting to the point-matching (collocation) technique. In this approach, the integration of the product of a function with the impulse, involved in the inner product, yields simply the value of the function at the center of the impulse, P_m (i.e., $\langle w_m, g\rangle = g(P_m)$) and Eq. (8.38) can be interpreted as equating (matching) the values of the left and right sides at this point. Thus, Eq. (8.39) is interpreted as requiring Eq. (8.37) to be simultaneously satisfied at N discrete points, P_m, $m = 1, \ldots, N$, referred to as matching points. The point-matching method simplifies evaluation of the matrix elements as the integration involved in the inner product is avoided. It annihilates the error in the operator equation at matching points, but there is no guarantee about the behavior of the error elsewhere, between adjacent matching points.

Slightly more complicated are pulse functions. When used for expansion, they yield a staircase (piecewise-constant) approximation of f. A pulse is defined analytically as

$$f_n = \begin{cases} 1 \text{ in subdomain } n \\ 0 \text{ elsewhere} \end{cases} \tag{8.40}$$

Figure 8.3a shows a set of pulse expansion functions in one dimension and the resulting staircase approximation.

The piecewise-constant approximation is discontinuous. A better approximation is the piecewise-linear (triangular, rooftop) approximation, which is continuous, but has a discontinuous first derivative. Analytically, this approximation can be constructed in two ways. For simplicity, we consider 1D expansion. The first way is assuming a linear function on a subdomain, and then matching the approximations on adjacent subdomains to obtain continuity. Alternatively, a subdomain function can be assumed a triangle, each triangle defined on two adjacent subdomains. Hence, the triangles partially overlap, as shown in Fig. 8.3b.

More sophisticated functions can be designed using more complicated subdomain functions and introducing additional constraints. Examples are spline approximations and functions that include edge

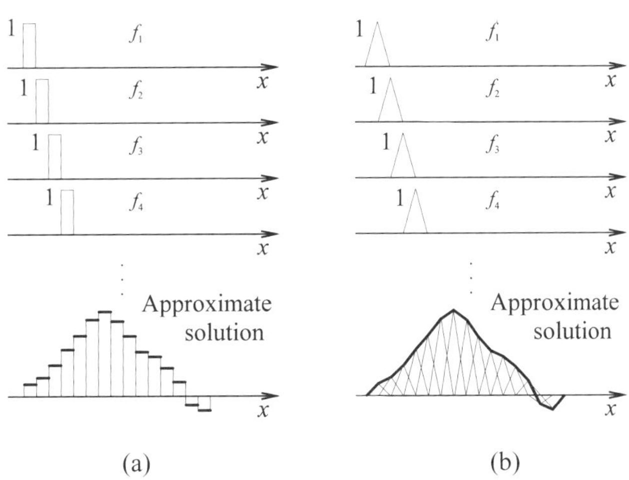

FIGURE 8.3 Subdomain approximations: (a) piecewise constant, and (b) piecewise linear.

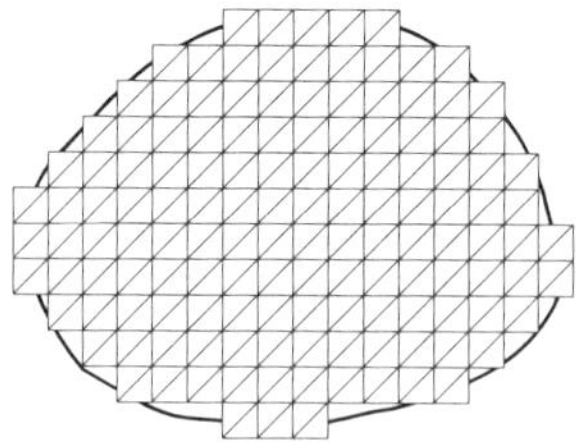

FIGURE 8.4 Surface patches associated with the pulse approximation of the surface charges of a conducting body in electrostatics.

effects. The edge effect is pronounced, for example, along sharp edges and wedges of perfectly conducting bodies, where the current and charge densities tend to infinity, but are integrable. An expansion function that closely resembles such source distributions may expedite the numerical solution.

The approximation by expansion functions involved in the MOM means an approximation not only of the unknown function but also of the geometry of the problem analyzed. The approximation of the geometry means a modification of the shape of the domain where the unknown function is defined, because the subdomains may not exactly match the shape of the domain. As an example, let us consider a conducting body in electrostatics, which is analyzed using an integral equation for its surface charges (Fig. 8.4). We assume the pulse approximation to be implemented. A pulse can be defined on a simple surface (usually a triangle or a quadrilateral) that is often referred to as a patch. In this case, pulses are 2D functions. Hence, the original surface is approximated by a set of patches. Obviously, the approximate charges are distributed over a different surface than the original surface of the conducting body. To minimize the error introduced by the geometry approximation, it is usually advisable to make the new (approximating) surface "oscillate" around the original surface.

The second category of expansion and testing functions is entire domain functions. Each function is defined on the entire domain of interest, so that all functions are nonzero on the whole domain. An example is the power series ($1, x, x^2, x^3, \ldots$) which, when combined into Eq. (8.35), yield a polynomial approximation [21, 22]. Another example is a set of trigonometric functions, amounting to the Fourier expansion. Sometimes rational functions are used, or functions that involve special effects, like the asymptotic charge and current distribution behavior near edges or wedges.

In practice, however, the entire domain is divided into a small number of relatively large subdomains. For example, a wire yagi-uda antenna is divided into its physical segments (i.e., dipoles). The expansion and testing functions are then defined on these large subdomains. This procedure is referred to as the almost entire domain approximation.

In the numerical implementation of the entire domain or almost entire domain approximations, a complicated evaluation of matrix elements is often encountered, requiring high-precision computations. These kinds of functions may well accommodate complex geometries and yield good results with a smaller number of unknowns and in a substantially shorter CPU time than with the subdomain functions. However, the technique is prone to instabilities with increasing the order of approximation because of an ill-conditioned system of linear Eqs. (8.39).

The more complicated the basis functions, the more analytic preparation is usually required before starting to write the computer code. A set of basis functions is usually suitable for a certain class of problems, but not for a general structure. Hence, a code customized for a class of problems is usually more efficient than a general code.

Convergence of the MOM solution cannot be guaranteed in most cases. At first, results usually improve with increasing the number of unknowns, but then they suddenly diverge. This is caused by various problems: approximations involved in the starting equation that is solved, inadequacy of the basis functions, insufficient accuracy of computing the basic integrals, propagation of numerical errors when solving the system of linear equations, etc.

8.3.3 Formulation of Integral Equations

We restrict our attention here to the MOM applications to solving integral equations, where the unknowns are field sources (currents and charges). These integral equations are, generally, formulated in the following three steps.

The first step is to enforce a boundary condition from Eq. (8.5) or Eq. (8.6) for the electric or magnetic field, or to utilize a constitutive relation from Eq. (8.3). For example, if we analyze a body made of a perfect conductor (a PEC body), the tangential component of the electric field on its surface is zero.

The second step is to express the fields in terms of the potentials, according to Eq. (8.18) or Eq. (8.28), and plug into the boundary condition or constitutive relation, as appropriate.

The third step is to express the potentials in terms of the sources, according to the expressions in Eq. (8.22). Instead of those in Eq. (8.22), Eqs. (8.24) should be taken for surface sources, and Eq. (8.25), for filamental sources. On a substitution into the equation derived in the second step, we finally obtain the integral equation for the unknown field sources.

A variety of equations can be formulated in this way. A given, particular problem can usually be solved using several equations. Depending on the field involved in the first step (the electric field or the magnetic field), the integral equations are, generally, categorized as electric-field integral equations (EFIEs) and magnetic-field integral equations (MFIE). There are some cases when the two fields are involved simultaneously, resulting in combined-field integral equations.

Example

To illustrate the basic MOM concepts, we consider an example of a conducting body in an electrostatic field in a vacuum (Fig. 8.5). The body is equipotential, its potential is a constant, V_o, and the tangential component of the electric field at its surface (S) is zero. Hence, there are two approaches to start with: imposing the boundary condition for the potential, V_o = constant on the body surface, and imposing the condition for the electric field (i.e., the first expression of Eq. (8.6). Theoretically, both approaches should have the same answer, but there are differences in the numerical implementation in the two cases. We adopt the first approach, because the kernel of the resulting integral equation is simpler and easier for evaluation.

We take a field point $M(\mathbf{r})$ at the conductor surface (see Fig. 8.5). The boundary condition is simply $V = V_o$ for any such point. The unknown is the distribution of conductor surface charges (ρ_s), and the potential is expressed in terms of these charges using Eqs. (8.24) and (8.29). The resulting integral equation reads

$$\frac{1}{\varepsilon_0}\int_{S'} \rho_s(\mathbf{r}')g(\mathbf{r},\mathbf{r}')dS' = V_o \text{ for arbitrary } \mathbf{r} \text{ on } S \tag{8.41}$$

Note that the source surface and the field surface coincide in this case (i.e., the body surface is both S and S').

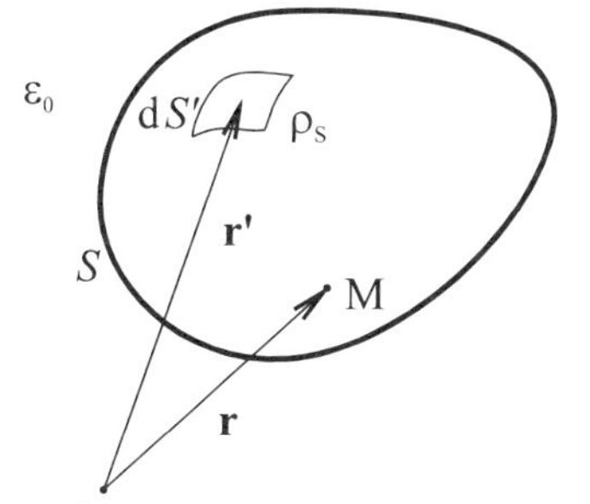

FIGURE 8.5 Coordinate system for setting up an integral equation for a charged conducting body in electrostatics.

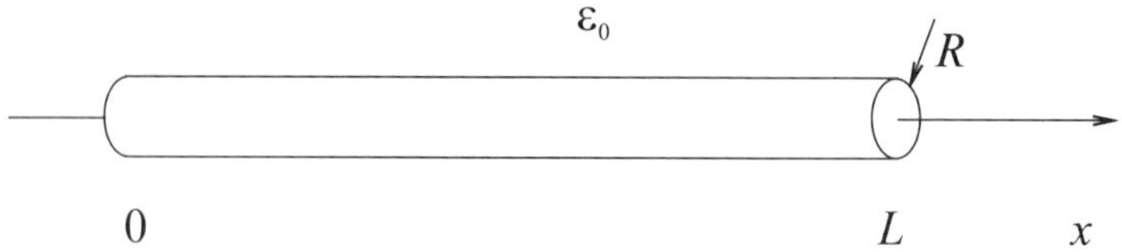

FIGURE 8.6 A conducting rod in electrostatics.

Then assume the conducting body is a cylindrical rod, as shown in Fig. 8.6. Although we consider an electrostatic example, the cylindrical rod leads us to certain conclusions important in the analysis of wire antennas. Let the rod length be L and radius R. Let us compute the capacitance of the rod.

The integration over S' in Eq. (8.41) means a double integral: one integration along the x-coordinate, and another along the circumferential coordinate (e.g., the azimuthal angle ϕ) around the rod. Because of symmetry, ρ_s is a function of x alone (i.e., the unknown charge distribution does not depend on ϕ). However, the resulting double integral is still hard for evaluation (regardless of the basis functions used) because it has a singularity when $\mathbf{r}'$ approaches $\mathbf{r}$. One integration can be carried out explicitly, but the second integration can be carried out only numerically, still with significant difficulties associated with singularities.

A simplification of Eq. (8.41) for this case can be made using the concept of the extended boundary conditions [23]. The rod is assumed to be a solid conducting body. Hence, in electrostatics, its potential V_o is not only on the surface but also at any point of its interior. In particular, we have $V = V_o$ at any point on the x-axis for $0 \leq x \leq L$. Alternatively speaking, $E_x = 0$ for $0 < x < L$. According to this concept, instead of imposing a boundary condition on the surface of the body, we impose a condition at some points at the body interior. Certain restrictions apply regarding where the extended boundary conditions are to be imposed to obtain a numerically stable solution, but we cannot discuss this question here. For simplicity, we impose that along the x-axis $V = V_o$, which reduces Eq. (8.41) to

$$\frac{1}{\varepsilon_0} \int_{x'=0}^{L} \rho_l(x') g(x, x') dx' = V_o \text{ for } 0 \leq x \leq L \tag{8.42}$$

where $\rho_l(x') = 2\pi R \rho_x(x')$ is the per-unit-length charge density and $g(x, x') = 1/(4\pi\sqrt{(x-x')^2 + R^2})$ is the kernel of the integral equation. Note that x and x' are measured along the same coordinate line. In Eq. (8.42) we have only a single integral (over x'). An identical result would be obtained if the surface charges of the rod were located along a filament on the surface generatrix (i.e., if they constituted a line charge).

Note, however, that the integral in Eq. (8.42) has a trouble spot. At $x = 0$ and $x = L$, in reality, there are two charged circular surfaces (disks, i.e., the end caps), which are not encompassed by Eq. (8.42). In other words, we have neglected the charges on the two disks closing the rod. We impose the potential to be constant along the axis of the rod. However, this condition cannot be satisfied exactly without taking into account the effect of the caps [21]. As a consequence, Eq. (8.42) gives diverging results when a very high order approximation for the charge distribution is taken.

An alternative interpretation of Eq. (8.42) can be made in terms of the equivalent sources [24]. We observe the boundary conditions on the surface of the rod. However, instead of considering the original sources (i.e., the charges located on the surface of the rod) we consider some equivalent sources encapsulated by S. Certain rules apply regarding where the equivalent sources should be located to obtain a numerically stable solution, but we can not discuss this question here. Equation (8.42) amounts to assuming the equivalent sources to be a nonuniform line charge, of per-unit length density $\rho_l(x')$, located along the x-axis for $0 \leq x' \leq L$.

Regardless of interpretation, Eq. (8.42) is simpler than Eq. (8.41) because the dimensionality of the mathematical problem is reduced by one. Such an approach is not only used in electrostatics; it is almost always implemented in the analysis of wire antennas and scatterers [21, 25], with the dynamic Green's function [Eq. (8.23)], when it is referred to as the thin-wire approximation.

Once we have formulated the integral expression in Eq. (8.42), we shall solve it by the MOM. We adopt a simple procedure: the pulse approximation for the unknown charge distribution as a function of x' and the point-matching testing. For the approximation, we take N uniform pulses along the x-axis. The choice of the approximation functions is arbitrary, and the selection here is targeted for simplicity. The choice of the uniform pulse distribution is not the most efficient one. For example, taking nonuniform pulses, shorter toward the ends of the rod, would yield a more efficient solution.

For the uniform pulse distribution the length of each pulse is $\Delta x = L/N$. The nth pulse, belonging to the nth subdomain, is located on $x' \in [n-1)\Delta x, n\Delta x]$, where $n = 1, \ldots, N$. The matching points are assumed to be located at the subdomain midpoints (i.e., at $x_m = (m - 0.5)\Delta x$, $m = 1, \ldots, N$, which is, from experience, a good policy, although not the only possibility. This choice of the expansion and testing functions reduces Eq. (8.42) to the following system of linear equations:

$$\frac{1}{4\pi\varepsilon_0}\sum_{n=1}^{N}\alpha_n \int_{x'=(n-1)\Delta x}^{n\Delta x} \frac{1}{\sqrt{\left(x_m - x'\right)^2 + R^2}}\,dx' = V_o,\ m = 1, \ldots, N \tag{8.43}$$

We can arbitrarily adopt V_o = 1V (because this choice does not affect the capacitance). The integral in Eq. (8.43) can be evaluated analytically using

$$\int_{x'=x_1}^{x_2} \frac{1}{\sqrt{\left(x_m - x'\right)^2 + R^2}}\,dx' = \log\frac{x_2 - x_m + \sqrt{\left(x_m - x_2\right)^2 + R^2}}{x_1 - x_m + \sqrt{\left(x_m - x_1\right)^2 + R^2}} \tag{8.44}$$

Note that Eq. (8.44) may lead to numerical difficulties when R is small compared with $|x_m - x_1|$ for $x_1 - x_m < 0$, or with $|x_m - x_2|$ for $x_2 - x_m < 0$. The remedy is to rationalize the denominator, or the numerator, as deemed appropriate.

Once the system of linear Eq. (8.43) is solved, we obtain the approximate charge distribution. The capacitance of the rod can then be evaluated as

$$C = \frac{\sum_{n=1}^{N}\alpha_n \Delta x}{V_o} \tag{8.45}$$

As a numerical example, we take L = 1 m and three different rod radii, R = 1 mm, R = 10 mm, and R = 100 mm. By linking these data to wire antennas, the first radius corresponds to a thin wire, and the third radius, to a thick wire. The classification is based on the ratio of the cylinder length to its diameter.

Table 8.1 shows the rod capacitance as a function of the number of pulses (N). For all three rods, the results initially converge with increasing N. However, the capacitance of the thickest rod starts oscillating already for N = 64. The capacitances of the other two rods also start oscillating, but for much larger N

TABLE 8.1 Capacitance (C), in pF, of the Rod Shown in Fig. 8.6, for L = 1 m and Three Different Radii (R), vs. the Number of Pulses

N	2	4	8	16	32	64	128	256	512
R = 1 mm	8.225	8.331	8.394	8.432	8.456	8.470	8.480	8.487	8.492
R = 10 mm	12.469	12.731	12.905	13.026	13.114	13.182	13.237	13.286	13.331
R = 100 mm	25.521	26.778	27.764	28.579	29.314	30.017	29.798	30.328	30.273

than shown in Table 8.1. This breakdown is a consequence of neglecting the end effects. The effect is more pronounced if the charge distribution is observed, since it has an erratic behavior in the vicinity of the ends even for low values of N (e.g., for $N = 64$ for $R = 10$ mm, and $N = 8$ for $R = 100$ mm) when Δx becomes on the order of magnitude of R.

8.4 Antenna Analysis

8.4.1 Introduction

The MOM is applicable to many antenna types. The analysis can also involve, to a certain extent, the environment where the antenna is located, like a mounting mast, or a stratified ground. The MOM can handle antennas with dimensions that are very small, a fraction of the wavelength, up to about 1000 wavelengths for wire antennas. To have an antenna that radiates efficiently, its dimensions must not be too small: the order of magnitude of 1/10 of the operating frequency is considered as a practical minimum. Well-written MOM codes, however, can analyze structures with dimensions that are many orders of magnitude smaller.

The applicability of MOM is limited by the complexity of the antenna, which requires a precise modeling of various antenna parts, and the antenna overall dimensions, because both factors influence the total number of unknowns required to obtain an accurate solution. Depending on computer resources, the number of unknowns is nowadays usually limited to a few tens of thousands, but this limit is pushed higher with the increase of available CPU power and fast memory. For higher frequencies, when the dimensions of the antenna and nearby relevant objects are many wavelengths, other high-frequency (HF) techniques are used, as described in Chapter 10 of this book.

For the present purpose, antennas are classified according to the complexity of their analysis into the following three groups:

- Wire antennas
- Surface (metallic) antennas
- Metallodielectric antennas

The basics of the analysis of each group are presented next.

8.4.2 Wire Antennas

8.4.2.1 Definition of Wire Antennas

Wire antennas are structures made of wirelike conductors: conductor radii are much smaller than their lengths and the wavelength at the operating frequency (Fig. 8.7). Conductors can be perfect (PEC) or the wires can be loaded (e.g., resistively or inductively). Our primary interest here is PEC structures.

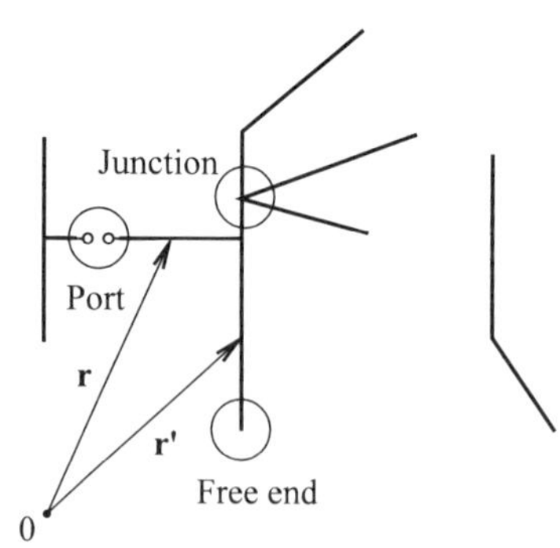

FIGURE 8.7 A wire antenna.

Examples of wire antennas are simple wire dipoles; V-antennas; loops and rhombic antennas used for HF communications; tower broadcast antennas for medium-frequency (MF) and low-frequency (LF) bands; yagi-uda antennas and log-periodic dipole arrays used in the HF, very high frequency (VHF), and ultra-high-frequency (UHF) bands, etc. However, the analysis of such structures can be extended to some other antennas and scatterers that can be approximated by wire structures, like aircraft at lower frequencies and some printed-circuit antennas, or whose surfaces can be approximated by wire-grid models. Some structures in the analysis of the electromagnetic compatibility (EMC) and electromagnetic interference (EMI) can be modeled by wires and wire grids. Examples are cages, shields with openings, power lines, etc.

We consider wire structures assembled from one or more straight PEC wires, referred to as segments, each having a circular cross section of a constant radius, arbitrarily oriented and interconnected. A generalization toward curved wire segments and wires with varying radii is straightforward, but not always easy for implementation. Alternatively, a curved segment or a segment with a varying radius can be approximated by a chain of straight segments with uniform cross-sections.

The segments can also have concentrated or distributed loadings, but we shall not present the corresponding analysis because of the lack of space. A further possibility is to approximate a conductor of an arbitrary cross section by an equivalent wire of a circular cross section, by using the concept of equivalent radius [26], and to approximate a printed-circuit trace on a substrate (usually without a ground plane) or a dielectric-coated wire by an equivalent wire of a circular cross section and a series distributed inductive loading.

The wire segments can be isolated in space or placed near an object, such as above a perfectly conducting ground plane. In the presence of certain objects of well-defined shapes, the antenna analysis using the MOM can be carried out by modifying Green's functions, instead of treating the object itself by the MOM approach. For example, the influence of a PEC ground plane is substituted by the antenna image and Green's function contains two terms of the form of Eq. (8.23) — one for the original and another for the image. Another example is an antenna placed above or in a stratified medium, in which case Sommerfeld's theory is applicable [27] (beyond our scope here).

The wire structure can be driven at one or more ports or excited by a plane wave of an arbitrary polarization. We are interested in evaluating the current distribution along antenna conductors, near and far fields, port impedance, admittance, scattering parameters, etc. The primary goal is to evaluate the distribution of the currents and charges along the wires. Other quantities of interest can thereafter be found by postprocessing. The current distribution can be evaluated only numerically and the MOM is the key tool that has been used for decades for this purpose. Generally, the analysis can be carried out in the frequency domain (steady state), or in the time domain (transients). We shall limit our attention here to the frequency-domain analysis.

8.4.2.2 Integral Equations and Their Solutions

For the frequency-domain analysis, various integral and integro-differential equations have been used: Pocklington's equation, two-potential equation, Schelkunoff's equation, and Hallén's equation [21, 25, 28]. The first three equations are formulated starting from the boundary condition for the electric field, which is the first expression in Eq. (8.6). They differ in the way the electric field is related to the wire currents. In Pocklington's equation, the electric field is expressed only in terms of the magnetic vector potential, using Eq. (8.28). The resulting integral equation involves only the antenna currents, but the kernel of the equation is hard for integration, because it involves the first-order and the second-order derivatives of Green's function. The two-potential equation uses the first expression in Eq. (8.18), and the result is an equation where the unknowns are both the current and its first derivative with respect to a local coordinate along the wire axis. The kernel of this equation is easier to handle than that in Pocklington's equation. This is the most widely used equation for the analysis of wire antennas and scatterers, and its extension is straightforward to more complicated antenna structures, like surface and metallodielectric antennas. Schelkunoff's equation is convenient for parallel wires. It has a mild kernel, but it involves the current and its first two derivatives. Hallén's equation is most complicated to set up in the general case, because it is formulated in terms of the magnetic vector potential, and not for the electric field. The magnetic vector potential is solved from this equation, and then it is expressed in terms of the currents. Hallén's equation yields most stable and accurate results, but it is not available for generalization to other antenna structures. Hence, in Section 8.4.2.3 we focus our attention to the two-potential equation.

In the analysis of wire antennas, the thin-wire approximation is almost always used. As the consequence, we deal with filamental currents (in the direction of the wire axis), and the unknown quantity is the distribution of the current along the axes of the wire segments.

Various approximations (basis functions) are used for the current distribution. Examples of subdomain approximations are the pulse (piecewise-constant) approximation, as, for example, used in Reference [29]; and triangular (piecewise-linear), and piecewise-sinusoidal approximation [30]. Among almost entire domain approximations, polynomials [21, 22] have been used predominantly, either alone, or in combination with trigonometric functions. The subdomain approximations are easier for computer programming. In particular, the basic integrals encountered in the sinusoidal approximation can be evaluated explicitly. However, the most efficient codes are claimed to be those based on the polynomial approximation (e.g., References [31, 32]). This may result from the fact that the subdomain approximations applied to long, smooth wire segments artificially introduce significant discontinuities, which deteriorate the quality of the solution. For example, the pulse approximation for the current (see Fig. 8.3a) is discontinuous at subsegment ends, and the associated charge distribution is singular. The electric field produced by such an approximation has large peaks at subsegment boundaries. The piecewise-linear and piecewise-sinusoidal approximations have a continuous current, but a discontinuous charge distribution, also leading to artificial peaks in the electric field, though milder than for the pulse approximation. On the other hand, the almost entire domain approximations produce a smooth electric field along a wire segment, except in the vicinity of the segment ends. However, in regions where the current distribution suffers rapid variations, such as in the excitation region, it is often necessary to split a physical wire segment into a number of shorter segments to provide a more flexible approximation of the current and charge distributions. Hence, a skillful blend of subdomain and entire domain functions gives the best results in the general case.

For weighting, the most frequent choices are the point-matching procedure [33], pulse-weighting functions [29, 31], and the Galerkin procedure [32, 34]. The point-matching procedure is the simplest one. However, it does not properly take care of large fields in the vicinity of antenna discontinuities, like junctions, bends, and excitation regions, except with Hallén's equation; and other measures may be necessary for these regions to provide an accurate solution [21, 31]. The pulse-weighting functions associated with the two-potential equation enable an explicit integration of the $-\text{grad}\, V$ term in Eq. (8.18), which leads to numerical simplification [29]. The Galerkin procedure requires the most analytic preparation, but it is reported to yield most accurate and stable results [32].

8.4.2.3 Two-Potential Equation

As an example, we shall outline the solution of the two-potential equation with the polynomial-testing and pulse-weighting functions. Details can be found in Reference [31].

We have assumed wires to be perfectly conducting. On the wire surface, the tangential component of the electric field must vanish, according to the first expression in Eq. (8.6). We separate the electric field into two components, $\mathbf{E} = \mathbf{E}_w + \mathbf{E}_i$. The first component ($\mathbf{E}_w$) is produced by the currents and charges of the wire structure. This component is related to the potentials and field sources (currents and charges), following the principles explained in connection with Eqs. (8.18) and (8.25). The second component ($\mathbf{E}_i$) is the impressed electric field. It models the excitation of the antenna, and is assumed to be a known function. This component can be given directly or evaluated as a field produced by known impressed currents, $\mathbf{J}_i$.

Hence, we rewrite the boundary condition for the electric field as

$$\left(\mathbf{E}_w + \mathbf{E}_i\right)_{\tan} = 0 \tag{8.46}$$

A transmitting antenna is driven by a lumped generator, with dimensions that are always assumed much smaller than the wavelength at the operating frequency. The input impedance (or admittance) can be defined only if we have two closely spaced terminals. If the separation between the terminals is a significant fraction of the wavelength, then there is no way to uniquely define the input parameter.

For a lumped generator, the impressed electric field is localized in a small region of a wire segment, referred to as the excitation region. Treatment of excitation regions is a delicate problem when the size of the region is above about 1/100 of the wavelength, and details can be found elsewhere [21].

A receiving antenna is excited by an incident electromagnetic wave, which may arrive at the antenna after reflections from nearby objects, like a perfectly conducting ground plane. The impressed electric field exists at all points of the receiving antenna structure.

By implementing the thin-wire approximation described in Section 8.3.4, we can avoid dealing with the surface integrals in Eq. (8.24). Namely, by using Eq. (8.25), the two potentials can be expressed in terms of the wire current and the per-unit-length charge density, which are filamental and located on the wire surface. Then Eq. (8.46) should be interpreted in terms of the extended boundary conditions as postulating the axial component of the total electric field to be zero on the wire axis. In the thin-wire approximation, the current is only axially directed. By changing the notation in Eq. (8.25), assuming a vacuum everywhere, the two potentials are evaluated as

$$\mathbf{A}(\mathbf{r}) = \mu_0 \int_{L'} \mathbf{u}(s') I(s') g_w(\mathbf{r},\mathbf{r}')\, ds'$$
$$V(\mathbf{r}) = \frac{1}{\varepsilon_0} \int_{L'} \rho_l(s') g_w(\mathbf{r},\mathbf{r}')\, ds' \tag{8.47}$$

where $\mathbf{r}$ is the position vector of the field point, $I(s) = 2\pi R|\mathbf{J}_s(s)|$ is the wire current, $\rho_l = 2\pi R\rho_s$ is the per-unit-length charge density, R is the wire radius, $\mathbf{J}_s(s)$ is the surface-current density, ρ_s is the surface-charge density, s is the local coordinate along the wire axis (L'), and $\mathbf{u}(s')$ is the unit vector of the axis,

$$g_w(\mathbf{r},\mathbf{r}') = \frac{\exp\left(-jk\sqrt{|\mathbf{r}-\mathbf{r}'|^2 + R^2}\right)}{4\pi\sqrt{|\mathbf{r}-\mathbf{r}'|^2 + R^2}} \tag{8.48}$$

is known as the thin-wire (reduced) kernel, and $\mathbf{r}'$ is the position vector of the element ds' of the wire axis.

Equations (8.47) and (8.48) produce exact results for points along the axis of a cylindrical wire segment. Otherwise, they yield a good approximation except in the immediate vicinity of discontinuities (junctions and ends).

The wire current and charge are related by the equation of continuity Eq. (8.27). Hence, the electric field can be expressed only in terms of the wire current and its first derivative as

$$\mathbf{E}_w(\mathbf{r}) = -j\omega\mu_0 \int_s \left(\mathbf{u}(s') I(s') g_w(\mathbf{r},\mathbf{r}') + \frac{1}{k^2}\frac{dI(s')}{ds'} \operatorname{grad} g_w(\mathbf{r},\mathbf{r}') \right) ds' \tag{8.49}$$

where the gradient is evaluated by differentiating the kernel with respect to $\mathbf{r}$.

The wire structure is divided into N straight segments. Each segment has its local axis (s_m, $m = 1, \ldots, N$), which starts at one segment end, where we assume $s_m = 0$, and is directed toward the other segment end, where $s_m = h_m$, and h_m is the segment length. The reference direction for the current coincides with the orientation of the s-axis. After substituting Eq. (8.49) into the boundary condition of Eq. (8.46), the two-potential equation (also referred to as the vector-scalar-potential equation) is finally obtained as

$$\sum_{m=1}^{N}\int_{0}^{h_m}\mathbf{u}_p\cdot\left(\mathbf{u}_m I_m(s'_m)g_w(\mathbf{r},\mathbf{r}')+\frac{1}{k^2}\frac{dI_m(s'_m)}{ds'_m}\text{grad }g_w(\mathbf{r},\mathbf{r}')\right)ds'_m=\frac{\mathbf{u}_p\cdot\mathbf{E}_i(\mathbf{r})}{j\omega\mu_0}\qquad(8.50)$$

$$p=1,\ldots,N$$

where p is the index of the wire segment where the boundary condition is imposed.

We omit further details here. We only note that Eq. (8.50) can be enhanced to incorporate skin effect losses and distributed loadings by modifying the boundary condition in Eq. (8.46), and include lumped loadings by controlled generator models [31]. Various loadings are deliberately inserted into antennas [21]. For example, resistors are used to dampen resonances and thus increase the operating bandwidth, inductors can apparently electrically lengthen the antenna or provide an increased gain (both at the expense of reducing bandwidth), and capacitors can improve broadband properties. Also, often the matching and filtering circuit of an antenna is analyzed simultaneously with the antenna, which extends applications of the loadings.

The presence of a perfectly conducting ground plane is replaced by the taking the image of the wire structure. Other kinds of symmetries that exist in an antenna structure may also be incorporated to expedite the analysis.

We solve Eq. (8.50) using the polynomials for expansion and pulses for testing. In Reference [31] we see that the polynomial expansion is superior both in accuracy and speed compared with the pulse expansion. The current distribution, $I_m(s_m)$, is approximated along each wire segment by a polynomial (power series) with unknown coefficients, which amounts to an almost entire domain approximation,

$$I_m(s_m)=\sum_{i=0}^{n_m}I_{mi}\left(\frac{s_m}{h_m}\right)^i$$

$$0\le s_m\le h_m\qquad(8.51)$$

$$m=1,\ldots,N$$

where n_m is a chosen degree of the polynomial, I_{mi} are unknown complex coefficients, and s_m/h_m is the normalized local coordinate along the segment. The total number of the unknown coefficients for a segment is (n_m+1). Numerical experiments have indicated that $n_m=4,\ldots,8$ per wavelength is sufficient to yield accurate results for the antenna characteristics in most practical cases.

Expansion (8.51) is substituted into Eq. (8.50). A set of pulses is selected for testing. Pulses are distributed along wire segments, but some pulses lie partly on pairs of wire segments at junctions. Details of the scheme can be found in Reference [31]. An integration over a pulse located on wire segment p $(s_{p1}<s_p<s_{p2})$ annihilates the gradient in Eq. (8.50), reducing this equation to

$$\sum_{m=1}^{N}\sum_{i=0}^{n_m}I_{mi}\left(\int_{s_{p1}}^{s_{p2}}\int_{0}^{h_m}\mathbf{u}_p\cdot\mathbf{u}_m\left(\frac{s_m}{h_m}\right)^i g_w(\mathbf{r},\mathbf{r}')ds'_m ds_p\right.$$

$$\left.+\frac{1}{k^2}\frac{i}{h_m}\int_{0}^{h_m}\left(\frac{s_m}{h_m}\right)^{i-1}\left(g_w(\mathbf{r},\mathbf{r}')_{s_{p2}}-g_w(\mathbf{r},\mathbf{r}')_{s_{p1}}\right)ds'_m\right)\qquad(8.52)$$

$$=\int_{s_{p1}}^{s_{p2}}\frac{\mathbf{u}_p\cdot\mathbf{E}_i(\mathbf{r})}{j\omega\mu_0}ds_p,\quad p=1,\ldots,N$$

Equations of the form Eq. (8.52) are augmented with equations expressing Kirchhoff's current law for each junction and free wire end.

The integrals appearing in Eq. (8.52) are solved numerically. Generally, the numerical integration is the only possibility, because there is no analytic solution in most cases. The integrals that appear in antenna problems are often hard for evaluation, because the integrands have singularities or pseudo-singularities when $\mathbf{r}$ and $\mathbf{r}'$ become close or coincide. The singularity is such that, for example, the Green's function (kernel) in Eq. (8.25) goes to infinity when $\mathbf{r} = \mathbf{r}'$. The kernel in Eq. (8.48) and in Eq. (8.52) is finite, but it has a very sharp peak with an amplitude on the order of $1/R$, centered at $\mathbf{r} = \mathbf{r}'$. This peak is referred to as the pseudosingularity.

A useful strategy is to subtract the static term, which dominates near the pseudosingularity, from the kernel in Eq. (8.48), or even extract several terms that can be integrated analytically. The remainder is a reasonably well-behaved function, small in magnitude, so it can be integrated numerically with a satisfactory accuracy.

The resulting system of linear equations is solved for the coefficients I_{mi}, using Gaussian elimination or LU decomposition, thus yielding the approximate current distribution.

8.4.2.4 Evaluation of Antenna Characteristics

Once the current distribution is known, one can relatively easily evaluate various antenna characteristics. The current distribution along the wires is readily available, because the solution has determined the expansion polynomials in Eq. (8.51). If the electric field in the antenna vicinity is required, which is referred to as the near field, it can be evaluated from Eq. (8.49). This field is needed, for example, to establish the safety region for humans in the vicinity of transmitting antennas (e.g., radio and TV broadcast antennas, or mobile phones), analyze corona problems associated with high-power antennas, and in EMC/EMI considerations.

However, for most practical cases, the key characteristics of an antenna are its input impedance, or, equivalently, reflection coefficient with respect to the given characteristic impedance of the feeder, and the radiation pattern. Because of reciprocity [3, 4], these characteristics are identical when the antenna is in the transmitting mode and when the same antenna is in the receiving mode, although the current distributions in the two cases are different. The numerical analysis is somewhat simpler for the transmitting mode, and we consider this mode in what follows.

We consider an antenna that has only one port. We assume the antenna is driven by one lumped ideal voltage generator. The driving voltage (i.e., the generator electromotive force, V) equals the integral of the impressed field ($\mathbf{E}_i$) in the excitation region, along the wire axis. It is, hence, a known quantity. The numerical analysis yields the current distribution and, consequently, the current at the generator (I_0). The antenna input admittance is simply $Y = I_0/V$. It is now a straightforward matter to evaluate the input impedance and the reflection coefficient with respect to a given reference impedance.

A multiport antenna is characterized by an admittance, impedance, or scattering matrix. The simplest procedure is to evaluate the admittance matrix, [$\mathbf{y}$], first, by driving the antenna one port at a time, following a similar procedure as for a single-port antenna. The other two matrices can be evaluated by matrix manipulations as explained in Reference [31]. If the multiport antenna is actually an antenna array, then one of these three matrices could be needed to solve for the feeding voltages by analyzing the feeding network terminated with the antenna matrix. Thereafter, the array is analyzed with all ports simultaneously driven by these voltages to evaluate the radiation pattern.

The far (radiated) electric field of an antenna is related to the magnetic vector potential (as in Fig. 8.8),

$$\mathbf{E} = j\omega \mathbf{u}_r \times \left(\mathbf{u}_r \times \mathbf{A}\right) \tag{8.53}$$

where $\mathbf{u}_r$ is the unit vector directed from the coordinate origin (located in the antenna vicinity) toward the field point. We suppress indices "w" and "i" with the vector $\mathbf{E}$, because in the far-field zone of a transmitting antenna the impressed currents usually radiate negligibly and the impressed electric field

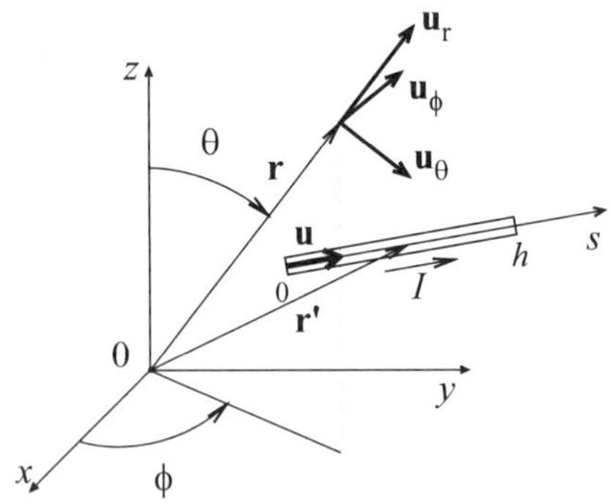

FIGURE 8.8 Coordinate system for evaluation of far fields.

does not exist. Hence, only the antenna currents and charges produce the radiated fields. The radiated electric field has only the transverse components with respect to the radius vector (**r**). In spherical coordinates,

$$\mathbf{E} = E_\theta \mathbf{u}_\theta + E_\phi \mathbf{u}_\phi \tag{8.54}$$

where $\mathbf{u}_\theta$ and $\mathbf{u}_\phi$ are the unit vectors of the spherical coordinate system. In the far-field zone, at a point with spherical coordinates (r, θ, ϕ), instead of using the first expression in Eq. (8.47), the magnetic vector potential can be evaluated in a simpler way by neglecting variations of $|\mathbf{r} - \mathbf{r}'|$ in the denominator of Green's function, leading to the following expression for the radiated electric field:

$$\mathbf{E}(\mathbf{r}) = -j\omega \frac{\mu_0}{4\pi} \frac{\exp(-jkr)}{r} \int_0^h \left(\mathbf{u}_\theta \left(\mathbf{u}_\theta \cdot \mathbf{u}(s') \right) + \mathbf{u}_\phi \left(\mathbf{u}_\phi \cdot \mathbf{u}(s') \right) \right) I(s') \exp\left(jk\mathbf{r}' \cdot \mathbf{u}_r \right) ds' \tag{8.55}$$

where $r = |\mathbf{r}|$. The radiated magnetic field is related to the electric field by

$$\mathbf{H} = \frac{\mathbf{u}_r \times \mathbf{E}}{\zeta_0} \tag{8.56}$$

where $\zeta_0 = \sqrt{\mu_0/\varepsilon_0}$ is the wave impedance (intrinsic impedance) of a vacuum.

The Poynting vector can be evaluated from Eq. (8.20) or Eq. (8.21), as appropriate. The power gain (with respect to an isotropic radiator) is then given by

$$G_p = \frac{|\mathrm{P}|}{P_{\mathrm{fed}}} 4\pi r^2 \tag{8.57}$$

where P_{fed} is the average power fed to the antenna, which can be evaluated from the voltages and currents at the antenna ports. The power fed to the antenna in the transmitting mode is $P_{\mathrm{fed}} = P_{\mathrm{rad}} + P_{\mathrm{loss}}$, where P_{rad} is the radiated power and P_{loss} is the lost power. The antenna efficiency is $\eta = P_{\mathrm{rad}}/P_{\mathrm{fed}}$, and the directive gain is $G_d = G_p/\eta$. In decibels, the gain (power or directive) is evaluated as $g = 10 \log_{10} G$ dB*i*.

Examples

Two examples of the analysis of wire antennas follow to illustrate the capabilities of the MOM solution.

The first example is a log-periodic dipole array for UHF television (TV) reception, with 16 elements, shown in Fig. 8.9a. The antenna has a feeding line made of two booms (rods) of a square cross-section,

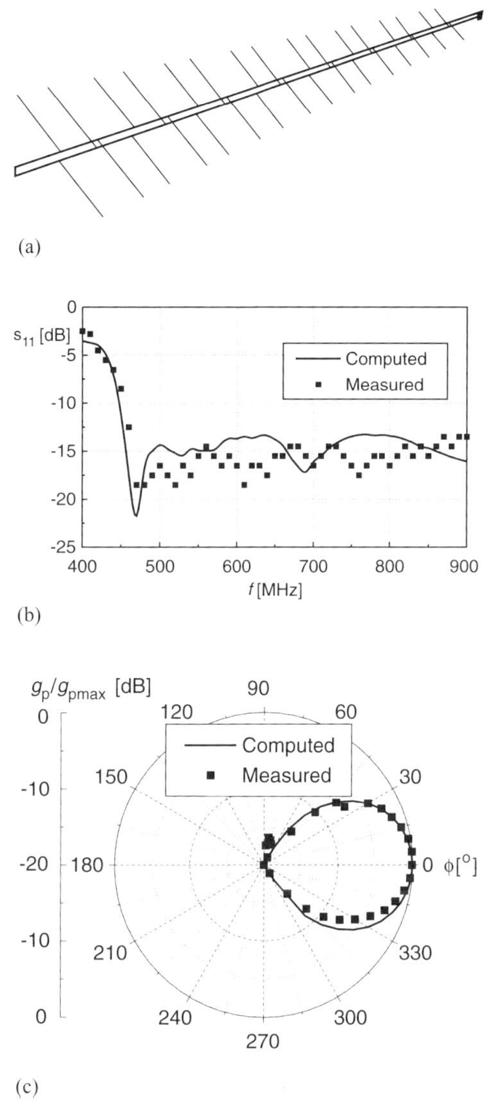

FIGURE 8.9 A 16-element log-periodic dipole array: (a) sketch, (b) input reflection coefficient with respect to 75 Ω (s_{11}) as a function of frequency (f), and (c) E-plane radiation pattern at 760 MHz: relative power gain (g_p/g_{pmax}) as a function of azimuth angle (ϕ).

which form a two-wire line. The dipoles are attached to this line, with alternating orientations to provide the required phasing of the dipole excitations. The dipoles are made of wires that have a circular cross section. The input port to the antenna is at the "nose," where a 75-Ω coaxial cable is attached. The cable passes through one of the booms, but the cable is not included in the computer simulation. In the wire antenna model the booms were replaced by equivalent conductors of a circular cross section. The equivalence is such as to keep intact the characteristic impedance of the feeder. The dipoles and the feeder were then analyzed using the program cited in Reference [31], as a unique wire structure.

Figure 8.9b shows the input reflection coefficient of the antenna, computed and measured on a laboratory prototype. In measurements, there were two major difficulties that affected the quality of the results. First, the network analyzer was a 50-Ω system; thus, minimum-loss pads were inserted to convert it to a 75-Ω system. The second problem was the calibration. A commercial 75-Ω coaxial cable (1 m long) was used to check the antenna performance under realistic practical conditions. The cable was attached to the antenna two-wire feeder by pigtails, and by a connector on the other side. A precise calibration of the network analyzer was performed at the reference plane of this connector, because the calibration kit could not be connected to the pigtails.

Figure 8.9c shows the radiation pattern of the antenna, measured in outdoor conditions. Some small reflections can be noted in the measured pattern, causing an asymmetry. In spite of all these problems, the agreement between the theory and experiment can be qualified as satisfactory for most practical purposes.

The second example is a global positioning system (GPS) ring-resonator antenna, designed for the L2 band, shown in Fig. 8.10a [35]. The antenna consists of a ring, placed parallel to a ground plane, and two capacitive probes. The ring and the plane play the role of a reentrant resonator. The objective is to excite a traveling wave on the ring, of a proper orientation. The ring current corresponding to this wave

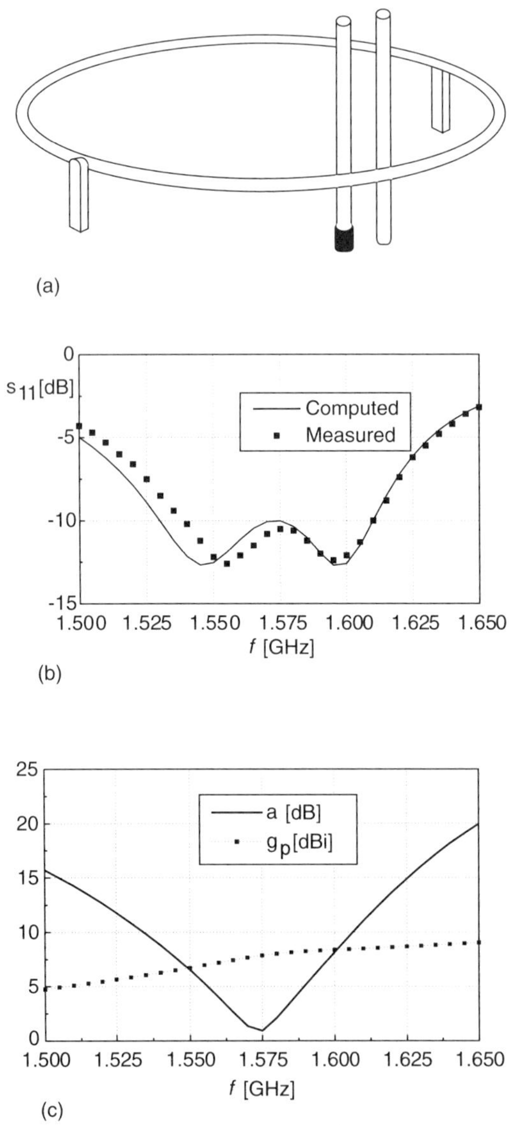

FIGURE 8.10 A ring antenna for right-hand circular polarization: (a) sketch, (b) input reflection coefficient with respect to 50 Ω (s_{11}) as a function of frequency (f), and (c) axial ratio (a) and power gain (g_p) in zenith direction as a function of frequency (f).

then radiates a circularly polarized wave in the zenith direction. The ring resonator is excited by one vertical probe (a piece of wire), fed by a coaxial line of a 50-Ω characteristic impedance. The probe is capacitively coupled to the ring. However, if this probe were alone, it would excite two waves traveling in opposite senses, which would correspond to a standing wave. The antenna would then radiate a linearly polarized wave. One of the two traveling waves can be suppressed by using another, grounded probe, which is capacitively coupled to the ring at an optimum location. The ring is supported by two plastic poles, which add small parasitic capacitances between the ring and the ground.

Figure 8.10b shows the computed and measured reflection coefficient of the antenna, demonstrating a good agreement. Figure 8.10c shows the computed antenna power gain in the zenith direction and the axial ratio of the polarization ellipse. The axial ratio is the ratio of the major to the minor axis of the ellipse. If the axial ratio is 1 (i.e., 0 dB), a perfect circular polarization is obtained. For GPS applications, an RHC (right-hand circular) polarization is required, which is provided by the disposition of the feeding and passive capacitive probes, as shown in Fig. 8.10a. Reversing the roles of the probes would yield an LHC (left-hand circular) polarization.

8.4.3 Metallic (Surface) Antennas

8.4.3.1 Definition of Metallic Antennas

In this section we analyze PEC bodies and surfaces. They are assumed to be placed in a vacuum. This is an approximation of real metallic bodies and surfaces that always have some losses.

The currents and charges on a PEC body (Fig. 8.11a), because of the skin effect, are localized on the surface of the body (S). This surface is, of course, closed. Because there are no electromagnetic fields in the interior region of the body (i.e., within the space enclosed by S), we can assume this region filled by a vacuum instead of the PEC. Hence, we can reduce the body to an empty, zero-thickness shell (Fig. 8.11b), that is, a closed PEC surface (S), without affecting the fields in the exterior space. The analysis of closed PEC surfaces, in turn, can follow the same steps as the analysis of open PEC surfaces. This is the reason why we treat PEC bodies and surfaces in a unique way.

Basically, radiation and scattering by PEC surfaces are analyzed following similar guidelines as the analysis of PEC wire structures. One starts from an appropriate boundary condition, expresses fields in terms of sources using potentials, and formulates an integral equation for the current distribution. This equation is then solved following the general MOM steps.

The analysis of surfaces is a more complex problem than the analysis of wire structures for three major reasons. The first one is that no approximation similar to the thin-wire approximation can be made for arbitrary surfaces. Hence, the potentials in Eq. (8.18) must be evaluated in terms of the field sources (surface currents and charges) using surface integrals as in Eq. (8.24), except that ε and μ should be substituted by ε_0 and μ_0, respectively. Integration in Eq. (8.24) is a much more complicated task than the evaluation of single integrals in Eq. (8.47). An exception is PEC bodies of revolution [36], for which the starting equation for the analysis can be fully reduced to the thin-wire integral expression in Eq. (8.50).

The second reason is the dimensionality of the unknowns. For wires the unknown current distribution is a function of only one coordinate (i.e., the local coordinate along the wire axis). For surfaces, the current distribution is a function of two coordinates (e.g., two local coordinates of a system bound to

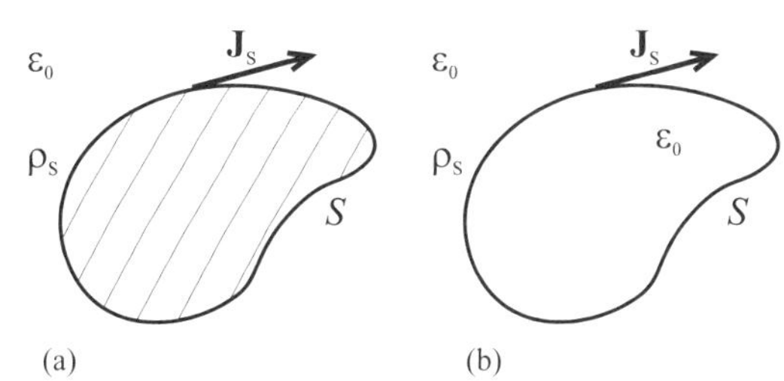

FIGURE 8.11 A PEC body (a) can be replaced by a zero-thickness PEC shell (b).

the surface). Hence, the basis functions must be more complicated: first, if a segmentation of the surface is used, which is needed both for subdomain and almost entire domain approximations, it must be two dimensional. For example, the surface is divided into a set of triangles [37] or quadrilaterals [32, 38]. Second, the basis function defined on a subdomain is most often such that it depends on two coordinates. (If it is taken to depend only on one local coordinate, it amounts to a piecewise-constant approximation in terms of the other coordinate.) As a result, the number of unknowns required to analyze a "simple" surface, such as a rectangular plate, is approximately proportional to the frequency squared. Alternatively speaking, it is proportional to the surface area divided by the wavelength squared. The number of unknowns for a simple wire is linearly proportional to the frequency, which is substantially less than for a rectangular plate of similar linear dimensions.

The third reason is that the surface–current density is a vector quantity that has two components. In a numerical solution, we need to approximate both components, which further doubles the number of unknowns.

8.4.3.2 Integral Equations and Their Solutions

We have seen that electrostatic problems and wire antennas are analyzed starting from the boundary condition for the electric field, or, equivalently, for the scalar-potential in electrostatics. An analog approach can be applied to the analysis of perfectly conducting surfaces, resulting in an EFIE (Electric Field Integral Equation). This equation has a form analogous to that for wire antennas, except for having surface integrals, and the currents and charges being functions of two local coordinates in the surface. The EFIE for PEC surfaces thus reads

$$\int_{S'} \mathbf{u}(\mathbf{r}) \cdot \left(\mathbf{J}_s(\mathbf{r}') g(\mathbf{r},\mathbf{r}') + \frac{1}{k^2} \operatorname{div}'_s \mathbf{J}_s(\mathbf{r}') \operatorname{grad} g(\mathbf{r},\mathbf{r}') \right) dS' = \frac{\mathbf{u}(\mathbf{r}) \cdot \mathbf{E}_i(\mathbf{r})}{j\omega\mu_0} \tag{8.58}$$

where S' coincides with the surface S of the PEC body, Green's function $g(\mathbf{r}, \mathbf{r}')$ is given by Eq. (8.23), div′ denotes differentiation with respect to $\mathbf{r}'$, and $\mathbf{u}(\mathbf{r})$ is the unit vector tangential to S at the point defined by $\mathbf{r}$. The differentiation implied by the operator is performed with respect to $\mathbf{r}$. There are, generally, two local components of the vector $\mathbf{J}_s$. Consequently, to provide a sufficient number of conditions, two orthogonal vectors $\mathbf{u}(\mathbf{r})$ are used at any point in Eq. (8.58); resulting in a pair of scalar equations for any $\mathbf{r}$. Equation (8.58) can be applied to both open and closed PEC surfaces.

Based on the boundary condition for the magnetic field, that is, the second expression in Eq. (8.6), one can formulate an MFIE (Magnetic Field Integral Equation) for closed PEC surfaces. For open surfaces, an MFIE cannot be set because of difficulties encountered with distinguishing between the fields and surface currents on the two faces of the zero-thickness surface. Hence, the EFIE is usually the preferred choice.

For closed PEC surfaces, there exists, however, a problem of spurious solutions associated with both the EFIE and MFIE. The spatial region encompassed by the surface is an ideal electromagnetic resonator, which, theoretically, can support free oscillations at a set of discrete resonant frequencies. The resonant field is confined to this region and it is not coupled with the surrounding space. Hence, such a field cannot be induced by an external field. However, a solution of the EFIE (or MFIE) cannot distinguish between currents induced on the inner and outer faces of the PEC shell. Because of various approximations and errors involved in the numerical procedure, in the vicinity of resonant frequencies the solution for the current distribution contains spurious components, which are similar to the theoretical resonant modes. These spurious components create a nonzero field in the spatial region outside S because they are not identical to the resonant modes. Hence, they modify the field in the exterior of the PEC surface and cause errors in the numerical results for the input impedance and radiation pattern of the analyzed antenna.

The problem of spurious solutions can be bypassed in several ways, which include an artificial insertion of losses in the system [39] and modifications of the integral equation, like a combination of the EFIE and the MFIE [40, 41].

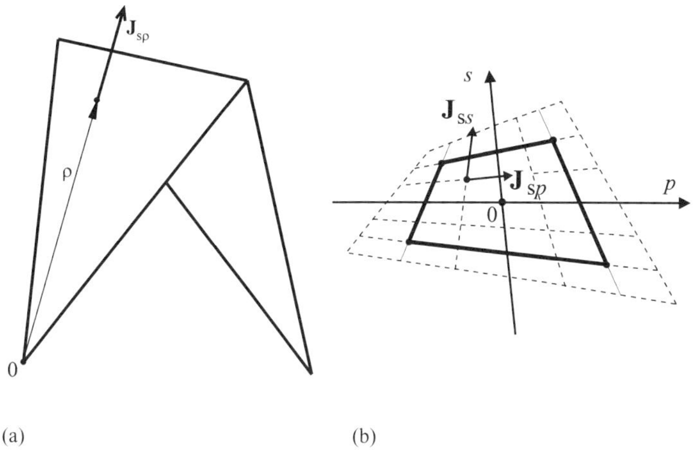

FIGURE 8.12 Two basis functions for the analysis of surface antennas: (a) triangular patches, and (b) quadrilateral patches.

Among a variety of useful and efficient approximations for the current distribution on PEC bodies, we stress two techniques. The first one [37] is a subdomain approximation. The surface of the PEC body is divided into a set of triangular patches. A basis function is defined on a pair of adjacent triangles for one component of the surface-current density vector ($\mathbf{J}_{\text{sp}}$), as shown in Fig. 8.12a. The basis function is such that there is no current flow outside these two triangles. On the common edge of the two triangles, the normal components of $\mathbf{J}_{\text{sp}}$ are equal on both triangles. Hence, the continuity equation is automatically satisfied for the triangle edges without creating line charges on the edges. The direction of the vector $\mathbf{J}_{\text{sp}}$ on a triangle is radial with respect to the vertex (i.e., it has the direction of the corresponding radius vector ρ in Fig. 8.12a). The intensity of the vector $\mathbf{J}_{\text{sp}}$ is a linear function of $|\rho|$. On each triangle, except for triangles that are at the boundary of the surface, three such basis functions are defined, each of them tangential to the triangle surface. When added, these three vectors yield the resulting current density vector. This approximation of the current density over a triangle results in a constant charge density on that triangle, which amounts to a pulse approximation for the surface charges.

The second technique [32, 38] divides the surface into a set of so-called bilinear surfaces (quadrilateral patches). On each patch, shown in Fig. 8.12b, the current density is split into two components ($\mathbf{J}_{ss}$ and $\mathbf{J}_{sp}$) with respect to a local *ps* coordinate system. Each component is approximated by a power series (polynomial). The polynomials are constructed such that the continuity is satisfied for the normal component of the current over each edge, as for the triangular patches. Individual quadrilateral patches can be relatively large, on the order of a wavelength in linear dimensions, which helps reduce the number of unknowns at the expense of much harder basic integrals that are to be evaluated numerically. The charge density associated with the approximate current distribution is also a polynomial function, and it is thus smooth over a patch. This results in a better behaved local field than for the pulse approximation, which further adds to the accuracy of the method.

Regardless of the set of expansion functions used, we approximate a given surface by a set of relatively simple, special surfaces, like triangles and quadrilaterals for the preceding quoted techniques. Hence, we not only approximate the variations of the vector $\mathbf{J}_s$ but also approximate the original surface S by a composite surface consisting of a number of smaller (subdomain) surfaces, as depicted in Fig. 8.4. This approximation of the geometry should be performed with care, to minimize the associated error. For example, if we approximate a sphere by a set of triangles, the original sphere should be between the inscribed and circumscribed sphere with respect to the triangularized surface. In particular, we may postulate the original sphere and the triangularized surface to have the same surface area or to encompass the same volume.

Losses in real conductors can be accounted for by the perturbation technique [2] if the skin effect is fully pronounced. Instead of having a zero tangential component of the electric field, as postulated by

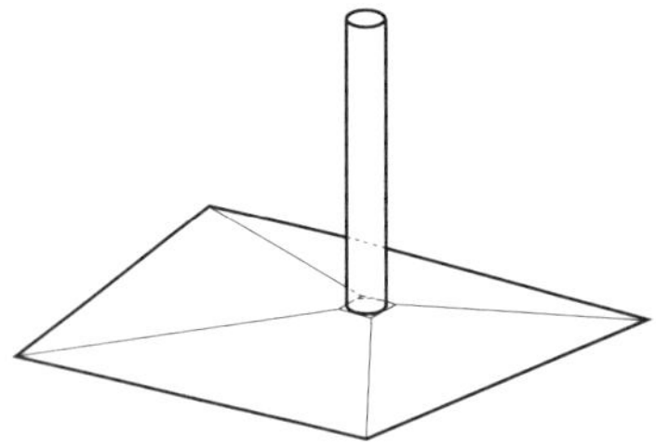

FIGURE 8.13 Modeling of attachment of a wire to a metallic plate, by subdividing the plate into four quadrilaterals.

the first expression in Eq. (8.6), we have a modified boundary condition that this field component is proportional to the density of the surface currents, where the proportionality coefficient is the intrinsic (wave) impedance of the metal, Z_m. Thus, instead of the first condition in Eq. (8.6), we have

$$\mathbf{n} \times \mathbf{E}_1 = Z_m \mathbf{n} \times \mathbf{J}_s \tag{8.59}$$

The intrinsic impedance is given by $Z_m = \sqrt{j\omega\mu/\sigma}$, where μ is the permeability and σ is the conductivity of the metal. This impedance is complex, and its real part is referred to as the surface resistance. The modified boundary condition does not impose a particular complication in the numerical solution, but one should take care about the conditions under which Eq. (8.59) is valid.

A special problem is the combined analysis of wires and PEC surfaces. If one wants to preserve the simplicity of the analysis of wires, but to include surfaces as well, there is a need to carefully model attachments (junctions) of wires to PEC surfaces. One technique is to define attachment modes [42]. Such a mode is a special current distribution that exists on the metallic surface in the vicinity of the junction. This distribution continues into the wire current at the center of the junction area, and it vanishes at a certain distance from the junction. Another technique is to subdivide the metallic plate (Fig. 8.13) so that the surface current on the plate is concentrated in the junction area and has a continuous transition to the wire current without introducing any special current distribution [43].

Once the current distribution on the surfaces and wires is known, antenna characteristics can be evaluated following similar guidelines as described in Section 8.4.2.4.

Examples

As the first example of analysis of metallic antennas we consider the UHF TV panel antenna shown in Fig. 8.14a [44]. The antenna consists of two flat dipoles, placed parallel to a finite-size conducting reflector. The dipoles are supported by posts. The dipoles are fed by a coaxial line that passes through one of the posts, and continues into the horizontal two-wire lines. The supporting posts act like a balance-to-unbalance transformer (balun). The antenna was analyzed numerically using the program of Reference [32]. Figure 8.14b shows the computed and measured antenna input impedance. An excellent agreement between the two sets of data can be observed.

As the second example we consider the rectangular horn antenna shown in Fig. 8.15a. The horn is fed by a rectangular waveguide, and the waveguide is excited by a small dipole placed inside. Figure 8.15b shows results for the radiation pattern of the antenna, computed using the program of Reference [32] along with experimental results from Reference [45], again demonstrating an excellent agreement.

8.4.4 Metallodielectric Antennas

8.4.4.1 Definition of Metallodielectric Antennas

Most practical antennas are made of various materials, which include good conductors (metallic parts) and dielectrics (insulators). In some cases, the dielectric parts do not play a vital role in the antenna electrical performance. For example, these parts can serve only as a mechanical support, similar to insulators in guy ropes of LF and MF tower antennas, or they can cover the antenna to protect it, such as

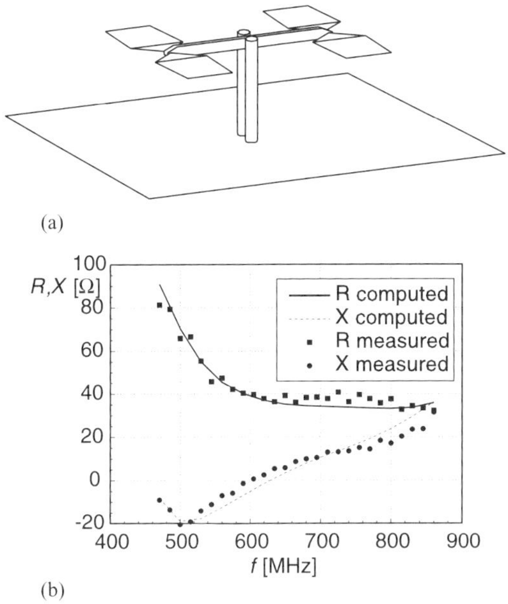

FIGURE 8.14 A panel UHF TV antenna: (a) sketch and (b) input resistance (R) and reactance (X) as a function of frequency (f).

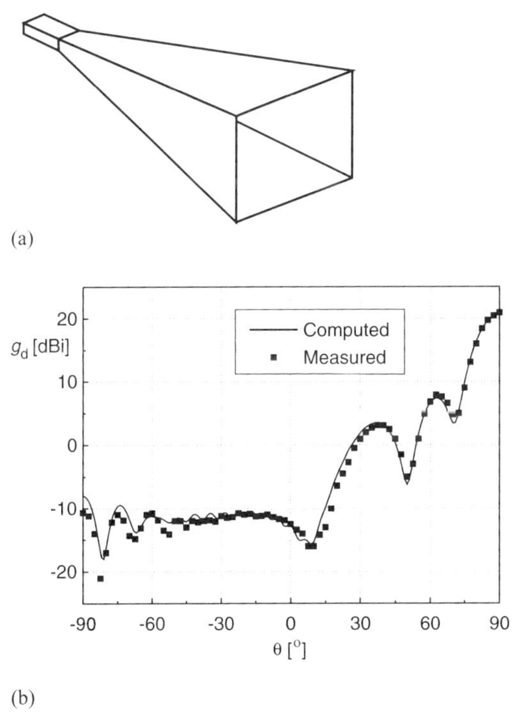

FIGURE 8.15 A rectangular horn antenna: (a) sketch and (b) E-plane radiation pattern: directive gain (g_d) as a function of zenith angle (θ).

a radome. In other cases, the dielectrics are deliberately placed to modify some properties of the antenna. For example, a high-permittivity dielectric substrate is used to reduce the size of a printed-circuit (patch) antenna. Materials can also have magnetic properties, but this is beyond our scope here. To add to the complexity, the conductors are never perfect and dielectrics also have certain losses. Even superconductors have a finite surface resistance at high frequencies, which increases as frequency squared. Electromagnetic analysis of structures that consist of a variety of materials is one of the hardest numerical problems.

In contrast to the situation described in Sections 8.4.2 and 8.4.3, where PEC wires and surfaces were embedded in a homogeneous medium (a vacuum), we deal here with an inhomogeneous medium. Generally, the medium properties, characterized by the permittivity, conductivity, and permeability, are arbitrary functions of the spatial position. In many practical cases, however, the medium is homogeneous within certain regions, and material properties change abruptly at their boundary surfaces. For example, for an ordinary printed-circuit antenna, one region with a homogeneous dielectric is the substrate, and the surrounding medium is a vacuum, which is another homogeneous region. Such cases are referred to as piecewise-homogeneous media.

Several basic approaches exist for the analysis of electromagnetic systems that consist of a variety of materials. We shall divide them into three groups. The first two groups are based on integral equations. One of them is founded on the volume equivalence theorem and results in a volume integral equation. The other stems from the surface equivalence theorem and results in surface integral equations. The third group is based on differential equations, and it includes the FEM and FD methods. In this chapter, we briefly describe only the FEM, whereas Chapter 7 of this book is devoted to the FD approach.

8.4.4.2 Volume Integral Equation and Its Solution

For antennas that consist of conductors and dielectrics, the field sources are electric currents and charges induced in conductors, as well as polarization charges and polarization currents throughout dielectric volumes. The conductor currents and charges are pure surface sources for a perfect conductor, but they are distributed throughout the conductor volume in the case of an imperfect, real conductor. In a leaky dielectric, volume conduction currents and associated free charges also exist. However, mathematically, they can automatically be taken into account through the imaginary part of the dielectric complex permittivity. All these sources produce the electromagnetic fields as if they were located in a vacuum. This is because these sources completely replace the conducting and dielectric bodies with respect to the electric field (**E**) and the magnetic flux density (**B**) they create. This fact is basically the statement of the volume equivalence theorem [3, 4].

The approach to analyzing antennas that consist of conductors and dielectrics is to explicitly find all the field sources, by solving an appropriate integral equation [46–50].

For simplicity, we first consider scattering from the dielectric body shown in Fig. 8.16a. The body is located in a vacuum. We assume to know the impressed (illuminating) electric field ($\mathbf{E}_i$) (e.g., the electric field of an incident plane wave). As the consequence of the illumination, polarization currents and charges are induced in the body, which create electromagnetic fields at any point within the body and outside the body. We also assume the permittivity (ε) of the material to be a differentiable function of spatial coordinates, whereas the permeability is μ_0 everywhere. Hence, the only surface of discontinuity is the

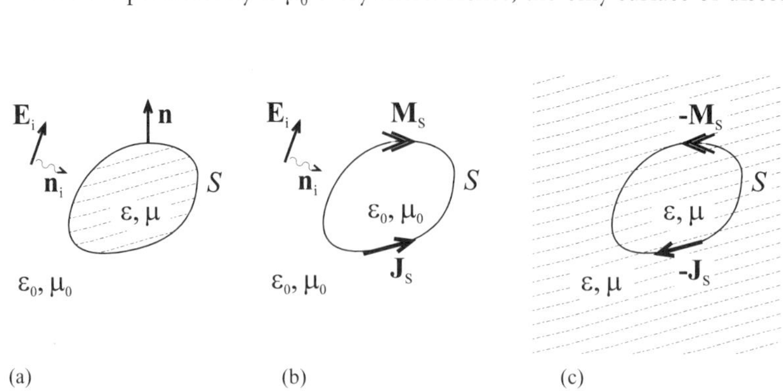

FIGURE 8.16 Illustration of the equivalence theorems: (a) a dielectric body illuminated by an incident plane wave, and equivalent systems for (b) the exterior region and (c) the interior region.

surface S bounding the body, where the material parameters have an abrupt change. In this case the field sources are volume polarization currents ($\mathbf{J}_p$) and charges (ρ_p), which are spread within the volume bounded by the surface S, and surface polarization charges (ρ_{sp}) spread over the surface. All these sources are assumed located in a vacuum. They can be determined starting from a properly adopted constitutive equation.

We start with the first constitutive expression in Eq. (8.3). It can be written in a more general form, which includes the polarization vector ($\mathbf{P}$), as

$$\mathbf{D} = \varepsilon_0 \mathbf{E} + \mathbf{P} \tag{8.60}$$

By combining Eqs. (8.3) and (8.60), the polarization vector is related to the electric field as $\mathbf{P} = \varepsilon - \varepsilon_0/\varepsilon\ \mathbf{E}$. Because the density of the volume polarization currents is related to the polarization vector as $\mathbf{J}_p = j\omega\mathbf{P}$, the polarization currents are related to the electric field as

$$\mathbf{J}_p = j\omega \frac{\varepsilon - \varepsilon_0}{\varepsilon} \mathbf{E} \tag{8.61}$$

This equation actually represents another form of the constitutive relation for the dielectric.

Let us express the electric field vector on the right side of Eq. (8.61) in terms of the known excitation and unknown sources. First, we separate the electric field into two components, $\mathbf{E} = \mathbf{E}_p + \mathbf{E}_i$, where $\mathbf{E}_p$ is the electric field produced by the polarization currents and charges, and $\mathbf{E}_i$ is the impressed electric field. The electric field $\mathbf{E}_p$ at an arbitrary point can now be evaluated using the first expression in Eq. (8.18), that is, as $\mathbf{E}_p = -j\omega\mathbf{A} - \text{grad}V$, where $\mathbf{A}$ and V are potentials produced by the polarization currents and charges placed in a vacuum. By having in mind expressions in Eq. (8.22) and (8.24), these potentials can be expressed as

$$\begin{aligned} \mathbf{A}(\mathbf{r}) &= \mu_0 \int_{v'} \mathbf{J}_p(\mathbf{r}') g(\mathbf{r},\mathbf{r}') dv' \\ V(\mathbf{r}) &= \frac{1}{\varepsilon_0} \left(\int_{v'} \rho_p(\mathbf{r}') g(\mathbf{r},\mathbf{r}') dv' + \int_{S'} \rho_{sp}(\mathbf{r}') g(\mathbf{r},\mathbf{r}') dS' \right) \end{aligned} \tag{8.62}$$

where v' denotes the spatial region enclosed by S, whereas the surfaces S' and S coincide. Finally, the volume polarization charges are related to the polarization currents by the continuity equation,

$$\text{div}\,\mathbf{J}_p = -j\omega\rho_p \tag{8.63}$$

and the surface polarization charges are related to the polarization currents by the corresponding boundary condition,

$$\rho_{sp} = \mathbf{n} \cdot \mathbf{P} = \frac{\mathbf{n} \cdot \mathbf{J}_p}{j\omega} \tag{8.64}$$

where $\mathbf{n}$ is the unit outward normal on S and $\mathbf{P}$ is the polarization vector in the dielectric at the inner face of S.

By substituting the resulting expression for the electric field into Eq. (8.61), a volume integral equation (VIE) for the volume polarization currents is obtained. In this equation the unknown vector $\mathbf{J}_p$ appears both under the integral and as a free term.

To solve this integral equation using the MOM, we have to expand the volume polarization currents. This is a more complicated task than for the PEC surfaces in Section 8.4.3.2, because there are three components of the vector $\mathbf{J}_p$, which depend on three spatial coordinates. Several notable techniques have been published. For example, in Reference [48] the basis functions are tetrahedrons, which are a 3D generalization of the triangular patches from Reference [37]. In Reference [50] the basis functions are almost entire domain polynomials in terms of three spatial coordinates defined in large hexahedrons.

For an antenna problem, we need to include some metallic wires and plates. If the conductors are assumed perfect, on their surfaces there exist surface currents ($\mathbf{J}_s$) and charges (ρ_s). In addition, there are surface polarization charges at the conductor surfaces, which add to the conductor charges to yield the total charges, $\rho_{st} = \rho_s + \rho_{sp}$. All these sources should be included in Eqs. (8.62), resulting in

$$\mathbf{A}(\mathbf{r}) = \mu_0\left(\int_{v'} \mathbf{J}_p(\mathbf{r}')g(\mathbf{r},\mathbf{r}')dv' + \int_{S'} \mathbf{J}_s(\mathbf{r}')g(\mathbf{r},\mathbf{r}')dS'\right)$$
$$V(\mathbf{r}) = \frac{1}{\varepsilon_0}\left(\int_{v'} \rho_p(\mathbf{r}')g(\mathbf{r},\mathbf{r}')dv' + \int_{S'} \rho_{st}(\mathbf{r}')g(\mathbf{r},\mathbf{r}')dS'\right) \quad (8.65)$$

where S' now denotes the union of all surfaces of discontinuity (i.e., the dielectric boundary surface and the conductor surfaces). The densities of the conductor surface currents and free charges are related by Eq. (8.26).

For the PEC surfaces the boundary condition for the tangential component of the electric field in Eq. (8.6) still holds. Hence, for a composite structure consisting of dielectric and PEC objects a set of two integral equations is formulated. One equation is obtained by imposing the boundary condition in Eq. (8.6) for the PEC surfaces and expressing the electric field in terms of the current distributions $\mathbf{J}_s$ and $\mathbf{J}_p$. The other equation is obtained from the constitutive relation in Eq. (8.61) in an analog way as explained earlier for the dielectric scatterer. In the numerical solution of these equations one has to expand the conductor currents as in Section 8.4.3.2, and the polarization currents as explained earlier.

The approach based on the volume integral equation inherently requires a large number of unknowns, because one has to approximate field sources within volumes. However, it can efficiently handle arbitrarily inhomogeneous media, unlike the approach described in the following section.

8.4.4.3 Surface Integral Equations and Their Solutions

This approach is tailored for piecewise-homogeneous media. It is based on the surface equivalence theorem [3, 4]. This theorem, basically, states the following. If we consider a spatial region, encompassed by a closed surface S, all field sources outside this region can be substituted by fictitious (equivalent) surface electric and magnetic currents ($\mathbf{J}_s$ and $\mathbf{M}_s$), placed on S, without affecting the field in the region considered. The region under consideration can also be the space external to S. Such a region is assumed bounded by S and by a closed surface at infinity.

Note that magnetic currents are a mathematically introduced quantity as being dual to the electric currents. They enable a symmetrization of Maxwell's equations and boundary conditions. More details about these currents can be found, for example, in Reference [3].

As an example of the application of this theorem, we consider a body made of a homogeneous dielectric (Fig. 8.16a), of parameters ε and μ, located in a vacuum, in an incident electromagnetic field. The polarization currents and charges are induced in the body to create electromagnetic fields inside and outside the body, as in the corresponding example in Section 8.4.4.2.

If we now consider the outer region (i.e., the region exterior to S), the fields resulting from the induced polarization currents and charges can be, according to the surface equivalence theorem, substituted by the fields produced by the equivalent surface currents on S (Fig. 8.16b). The substitution means that the equivalent surface currents are placed on S, and the original field sources within S are removed. The densities of the equivalent currents are related to the electric and magnetic field (**E**, namely, **H**) on the outer face of *S* by

$$\mathbf{J}_s = \mathbf{n} \times \mathbf{H}$$
$$\mathbf{M}_s = -\mathbf{n} \times \mathbf{E} \tag{8.66}$$

where **n** is the outside normal to *S*. By starting from the boundary conditions and using the uniqueness theorem [3, 4], it can be proved that the electromagnetic fields inside *S* are zero in the system of Fig. 8.16b. If the fields within a region are zero, the material parameters of that region are irrelevant because nothing is induced in the region. Hence, we can assume the medium within the region inside *S* to be a vacuum, as in the outside region. Then by having a homogeneous medium everywhere, we can use the standard expression for the potentials. In the presence of the magnetic currents, a total of four potentials are needed. The magnetic vector potential and the electric scalar potential are still given by Eq. (8.24), with ε and μ replaced by ε_0 and μ_0, respectively. The new potentials are the electric vector potential (**F**) and the magnetic scalar potential (V_m). In a homogeneous medium of parameters ε and μ these two potentials for surface sources are given by

$$\mathbf{F}(\mathbf{r}) = \varepsilon \int_{S'} \mathbf{M}_s(\mathbf{r}') g(\mathbf{r}, \mathbf{r}') dS'$$
$$V_m(\mathbf{r}) = \frac{1}{\mu} \int_{S'} \tau_s(\mathbf{r}') g(\mathbf{r}, \mathbf{r}') dS' \tag{8.67}$$

where τ_s are fictitious surface magnetic charges (dual to the electric charges). Equations (8.67) are dual to Eq. (8.24), and the surface magnetic currents are related to the magnetic charges by the continuity equation dual to Eq. (8.26),

$$\operatorname{div}_s \mathbf{M}_s = -j\omega\tau_s \tag{8.68}$$

The electric and magnetic field, in a homogeneous medium whose parameters are ε and μ, are expressed in terms of the four potentials as

$$\mathbf{E} = -j\omega\mathbf{A} - \operatorname{grad} V - \frac{1}{\varepsilon} \operatorname{curl} \mathbf{F}$$
$$\mathbf{H} = -j\omega\mathbf{F} - \operatorname{grad} V_m + \frac{1}{\mu} \operatorname{curl} \mathbf{A} \tag{8.69}$$

Of course, for a vacuum, we have to use ε_0 and μ_0 instead of ε and μ, respectively. In contrast to Eq. (8.18), we now deal with the vector **H** instead of **B**, to stress the symmetry (duality) in relations for the vectors **E** and **H**. From the preceding equations, we can express the vectors **E** and **H** in terms of $\mathbf{J}_s$ and $\mathbf{M}_s$.

The fields within *S* in Fig. 8.16b are zero because the equivalent sources, $\mathbf{J}_s$ and $\mathbf{M}_s$, along with the associated charges, annihilate the actual fields within *S*. Consequently, the fields produced by $\mathbf{J}_s$ and $\mathbf{M}_s$ within *S* are the negatives of the actual fields. If we consider the system of Fig. 8.16c, where we have the negatives of $\mathbf{J}_s$ and $\mathbf{M}_s$ distributed over *S*, with all other field sources outside *S* removed, the resulting fields within *S* are the same as the original fields in Fig. 8.16a. It can be proved that in Fig. 8.16c the fields outside *S* are zero. Hence, the outside region can be assumed to be filled with the same medium as is inside *S*, and expressions for potentials in Eqs. (8.24) and (8.67) can be used again with the actual parameters of the dielectric body under consideration, ε and μ.

Following the preceding example, we shall outline the procedure for solving the scattering problem of the dielectric body shown in Fig. 8.16a. We assume to know the impressed (illuminating) electric field ($\mathbf{E}_i$). However, we do not know the actual induced polarization currents and charges, and the equivalent surface electric and magnetic currents. The objective is to find the equivalent surface sources, because then we can evaluate the fields produced (scattered) by the dielectric body. We simultaneously consider the systems of Figures 8.16b and c. In the system of Fig. 8.16b the fields on the inner face of S are zero, as the fields at any point of the region encapsulated by S are zero. Hence, we can impose the condition that the tangential component of the electric field be zero on the inner face of S. This field is the sum of the illuminating field and the field produced by $\mathbf{J}_s$ and $\mathbf{M}_s$. By using Eq. (8.69), the expressions for the potentials, and this boundary condition, we obtain one integral equation for $\mathbf{J}_s$ and $\mathbf{M}_s$. Similarly, in the system of Fig. 8.16c the tangential component of the electric field on the outer face of S can be imposed to be zero. This field is produced by the negatives of $\mathbf{J}_s$ and $\mathbf{M}_s$, and this boundary condition yields another integral equation for $\mathbf{J}_s$ and $\mathbf{M}_s$. Note that, compared with the case of a PEC body, we now have doubled the unknowns (i.e., we have two surface currents instead of one, but we have also doubled the number of boundary conditions). The resulting equations are EFIEs. In an analogous way, MFIEs can be derived. However, another approach, referred to as the PMCHW formulation [51], has been found to have certain advantages and it has been used more frequently than the EFIE and MFIE.

The resulting system of simultaneous integral equations can be solved using the MOM, following a similar procedure as for PEC bodies. Now we have to approximate both $\mathbf{J}_s$ and $\mathbf{M}_s$. We can use similar basis functions for both currents and similar testing procedures for both integral equations. This approach can be generalized to the case of an arbitrary number of homogeneous media. For an antenna problem, we need to include some metallic parts into the structure, like wires and plates. This can be done in a straightforward manner, but further details are omitted here.

Based on the preceding outline, several techniques have been developed for solving antennas and scatterers with piecewise-homogeneous media. Two prominent techniques exist that are extensions of the PEC-body solutions. They are based on the triangular patches [52] and the bilinear surfaces with polynomial basis functions [32, 53], respectively. The techniques based on the surface integral equations are considered to be the most efficient methods for the analysis of practical systems that consist of metallic and dielectric bodies. Generally, these methods are applicable not only to radiating structures but also to virtually any electromagnetic field problem, including, for example, microwave components and circuits. However, they cannot treat problems that involve highly inhomogeneous and anisotropic materials.

8.4.4.4 Finite Element Method

The FEM is a technique that can be used to efficiently analyze electromagnetic structures that include inhomogeneous and anisotropic materials. We shall only outline this technique here. An extensive survey of the FEM can be found, for example, in Reference [54].

As stated in Section 8.1, the FEM is based on solving for the field distribution, or, equivalently, solving for the potentials. The equation that is to be solved numerically is usually derived from differential form of Maxwell's equations following the so-called variational approach, which is equivalent to applying the Galerkin method. The region where the fields exist is divided into a large number of subdomains, which are of a finite size. Within each subdomain, the field or potential distribution is approximated by a basis function, which is most often a linear or a quadratic function. For 3D electromagnetic problems, the basis functions depend on three spatial coordinates.

Because of the finite size of the subdomains, the basic FEM is most suitable for the analysis of fields within an electromagnetically shielded region, such as a microwave cavity or a shielded microwave circuit. In antenna problems, however, the space occupied by the fields is infinite. Hence, to analyze antennas, it is necessary to bypass the limitation of the finite-size subdomains. This can be done following two distinct approaches.

The first approach is to construct special basis functions on subdomains that extend to infinity. This technique has not been found suitable enough for antenna applications. The second approach is to assume the region where the finite elements are distributed to be bounded by a finite closed surface, S. This surface encompasses all material inhomogeneities, to make an efficient use of the FEM. In radiation (antenna) problems, this surface must also simulate an infinite open space into which the antenna radiates. This simulation is performed in two ways.

The first way is to impose a local boundary condition on *S*. The simplest approach is to assume *S* to be a sphere located in the far-field region (referred to as the radiation sphere). In this case, the electric and magnetic field vectors are practically tangential to the sphere, they are mutually orthogonal, and they are related by the so-called radiation boundary condition, that is, Eq. (8.56). In this case, the unit vector $\mathbf{u}_r$ is the unit outward normal on *S*. The basic problem is to have *S* really far away, in the far-field zone, to be sure Eq. (8.56) is valid with a sufficient accuracy. This request may overly extend the size of *S*. Modifications of the radiation boundary condition have been constructed, collectively referred to as the absorbing boundary conditions, which yield good results for smaller sized *S*, also possibly having an arbitrary shape. Nevertheless, even for these modified conditions, there is a limitation of the size of *S*, and this surface may still need to be impractically large.

The second way is to use the so-called nonlocal boundary conditions. This is hybridization between the FEM and the surface integral equation approach described in Section 8.4.4.3. Thereby, *S* can have an arbitrary shape and it can be shrunk to minimum dimensions sufficient to encompass all media inhomogeneities. We separately consider two problems, one for the region exterior to the surface *S*, and another for the region encapsulated by *S*. On the surface, *S* fictitious electric and magnetic currents are placed following the same equivalence principles described in connection with Eq. (8.66) and shown in Fig. 8.16b. For the exterior region these currents replace all field sources within *S*. An identical set of integral equations is formulated based on the boundary conditions as described in Section 8.4.4.3. For the interior region the classical FEM is applied, with finite-size elements. However, on *S* the negatives of the equivalent currents are placed, as shown in Fig. 8.16c, which replace the influence of the exterior region. The approximation for the equivalent surface currents is deducted from the finite elements. Appropriate boundary conditions are imposed on *S* requiring that the field outside *S* be zero. The whole procedure results in a system of simultaneous equations, which ultimately yields the distribution of equivalent surface currents and the field distribution within *S*.

The FEM hybridized with the surface integral equations can handle antenna and scatterer problems that involve highly inhomogeneous and anisotropic media more efficiently than the volume integral equation approach described in Section 8.4.4.2. However, such systems are very rare in practical antenna designs. For piecewise-homogeneous and isotropic media, the surface integral equation approach described in Section 8.4.4.3 is more efficient than the FEM because it involves a significantly smaller number of unknowns. Finally, the finite elements cannot easily be accommodated to thin wires and plates. As the result, the FEM has not found a wider practical use for antenna applications.

Example

As an example of the analysis of composite metallic and dielectric structures, we consider the microstrip patch antenna shown in Fig. 8.17a. The patch is almost a square: its dimensions are 57.6 by 58.8 mm, and it is designed for the GPS L1 band applications. It is printed on an FR-4 substrate, 1.59 mm thick. Measurements of the parameters of FR-4 in a wide frequency range have shown that its relative permittivity steadily decreases with frequency, whereas the loss tangent is fairly constant (about 0.02). At the GPS band the relative permittivity of the substrate is $\varepsilon_r = 43$. The antenna is fed by a coaxial line, connected near one diagonal. Figure 8.17b shows measured results for the antenna reflection coefficient as a function of frequency along with results computed using the program of Reference [32]. The agreement is excellent, bearing in mind the very narrow frequency band. Figure 8.17c shows the computed radiation pattern in a vertical plane. The numerical model takes into account the finite size of the ground plane, and it also includes effects of the surface waves in the dielectric, unlike many simplified techniques for the analysis of patch antennas.

8.5 Conclusion

This chapter presents the basic principles of the method of moments (MOM), emphasizing its applications to the analysis of antennas. This technique enables a computer simulation of practically arbitrary

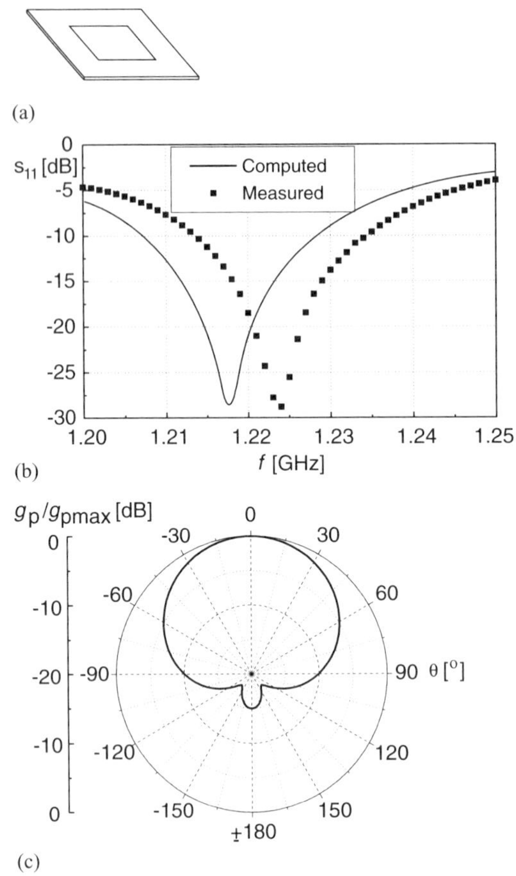

FIGURE 8.17 A microstrip patch antenna: (a) sketch, (b) input reflection coefficient with respect to 50 Ω (s_{11}) as a function of frequency (f), and (c) radiation pattern in a vertical plane at 1225 MHz: relative power gain (g_p/g_{pmax}) as a function of zenith angle (θ).

antennas whose dimensions range from a small fraction of the wavelength up to several tens or even hundreds of wavelengths. Details of the MOM can be found summarized in several excellent monographs, including References [2, 5–15]. Examples are given of various antennas illustrating the power of this technique and showing the degree of agreement between the theory and experiment that can usually be expected in practice.

The analysis presented in this chapter is for the steady-state sinusoidal regime. To complete the overview of numerical techniques, the reader should refer to Chapter 7 devoted to FD methods, which deals with the time-domain analysis; and to Chapter 10 devoted to HF techniques, which deals with methods that are more efficient than MOM for electrically large structures.

Acknowledgment

The authors wish to thank the following staff and students of the Electromagnetic Group of the Faculty of Electrical Engineering, University of Belgrade, as well as the staff of Imtel Microwaves, Belgrade, for their help in preparing and performing experiments presented in this chapter (in alphabetical order): Dr. Radivoje Biljic; Professor Emeritus Dr. Momcilo Dragovic; Ms. Andjelija Ilic, B.Sc.; Dr. Vladana Likar-Smiljanic; Mr. Miodrag Mikavica, M.Sc.; Ms. Aleksandra Stekovic, B.Sc.; Mr. Ivica Stevanovic; Professor Emeritus Dr. Jovan Surutka; Mr. Miodrag Tasic, B.Sc.; and Ms. Alenka Zajic. The authors also wish to thank Dr. Luca Niccolai, managing director of Dynaflex, S.R.L., Osimo, Italy, for permission to publish Fig. 8.10.

References

1. R. Plonsey and R.E. Collin, *Principles and Applications of Electromagnetic Fields*, New York: McGraw-Hill, 1961.
2. R.F. Harrington, *Field Computation by Moment Methods*, Piscataway, NJ: IEEE Press, 1993.
3. R.F. Harrington, *Time-Harmonic Electromagnetic Fields*, New York: McGraw-Hill, 1961.
4. B.D. Popovic, "Electromagnetic field theorems (a review)," *Proc. IEE, Pt. A*, pp. 47–63, 1981.
5. R. Mittra (Ed.), *Computer Techniques for Electromagnetics*, Elmsford, NY: Pergamon Press, 1973.
6. R. Mittra (Ed.), *Numerical and Asymptotic Techniques in Electromagnetics*, New York: Springer-Verlag, 1975.
7. B.J. Strait (Ed.), *Application of the Method of Moments to Electromagnetic Fields*, St. Cloud, FL: SCEEE Press, 1980.
8. J. Moore and R. Pizer, *Moment Methods in Electromagnetics: Techniques and Applications*, New York: John Wiley & Sons, 1984.
9. T. Itoh (Ed.), *Numerical Techniques for Microwave and Millimeter-Wave Passive Structures*, New York: John Wiley & Sons, 1990.
10. R.C. Hansen (Ed.), *Moment Methods in Antennas and Scattering*, Boston: Artech House, 1990.
11. J.H.H. Wang, *Generalized Moment Methods in Electromagnetics*, New York: John Wiley & Sons, 1991.
12. E.K. Miller, L. Medgyesi-Mitschang, and E.H. Newman (Eds.), *Computational Electromagnetics: Frequency-Domain Method of Moments*, Piscataway, NJ: IEEE Press, 1992.
13. N. Morita, N. Kumagai, and J. Mautz, *Integral Equation Methods for Electromagnetics*, Boston: Artech House, 1992.
14. K. Umashankar, *Computational Electromagnetics*, Boston: Artech House, 1993.
15. A. Peterson, S.C. Ray, and R. Mittra, *Computational Methods for Electromagnetics*, New York: IEEE Press and Oxford: Oxford University Press, 1998.
16. W. R. Smythe, *Static and Dynamic Electricity*, New York: McGraw-Hill, 1968.
17. T.K. Sarkar, K.R. Siarkiewicz, and R.F. Stratton, "Survey of numerical methods for solution of large systems of linear equations for electromagnetic field problems," *IEEE Trans. Antennas Propag.*, vol. AP-29, pp. 847–856, Nov. 1981.
18. T.K. Sarkar, *Application of the Conjugate Gradient Method in Electromagnetics and Signal Processing*, New York: Elsevier Science Publications, 1991.
19. T.K. Sarkar and E. Arvas, "Application of FFT and the conjugate gradient method for the solution of electromagnetic radiation from electrically large and small conducting bodies," *IEEE Trans. Antennas Propag.*, vol. 34, pp. 635–640, May 1986.
20. T.K. Sarkar, A.R. Djordjevic, and E. Arvas, "On the choice of expansion and weighting functions in the numerical solution of operator equations," *IEEE Trans. Antennas Propag.*, vol. AP-33, pp. 988–996, Sept. 1985.
21. B.D. Popovic, M.B. Dragovic, and A.R. Djordjevic, *Analysis and Synthesis of Wire Antennas*, Chichester, United Kingdom: Research Studies Press (a division of John Wiley & Sons), 1982.
22. B.M. Kolundzija, "Comparison of a class of sub-domain and entire-domain basis functions automatically satisfying KCL," *IEEE Trans. Antennas Propag.*, vol. AP-44, pp. 1362–1366, Oct. 1996.
23. P.C. Waterman, "Matrix formulation of electromagnetic scattering," *Proc. IEEE*, vol. 53, pp. 805–812, 1965.
24. Y. Leviatan, A. Boag, and A. Boag, "Generalized formulations for electromagnetic scattering from perfectly conducting and homogeneous material bodies-theory and numerical solution," *IEEE Trans. Antennas Propag.*, vol. AP-36, pp. 1722–1734, Dec. 1988.
25. E.K. Miller and F.J. Deadrick, "Some computational aspects of thin-wire modeling," in R. Mittra (Ed.), *Numerical and Asymptotic Techniques in Electromagnetics*, Berlin: Springer-Verlag, 1975.
26. B.D. Popovic and A. Nesic, "Generalization of the concept of equivalent radius of thin cylindrical antennas," *Proc. IEE, Pt. H*, pp. 153–159, 1984.

27. A. Baños, *Dipole Radiation in the Presence of a Conducting Half-Space*, New York: Pergamon Press, 1966.
28. W.A. Imbraile, "Applications of the method of moments to thin-wire elements and arrays," in R. Mittra (Ed.), *Numerical and Asymptotic Techniques in Electromagnetics*, Berlin: Springer-Verlag, 1975.
29. J. Rockway, J. Logan, D. Tam, and S. Li, *The MININEC SYSTEM: Microcomputer Analysis of Wire Antennas*, Boston: Artech House, 1988.
30. J.H. Richmond, *Radiation and Scattering by Thin-Wire Structures in the Complex Frequency Domain*, OSU Research Foundation Report RF 2902-10, 1973.
31. A.R. Djordjevic, M.B. Bazdar, T.K. Sarkar, and R.F. Harrington, *AWAS for Windows: Analysis of Wire Antennas and Scatterers, Software and User's Manual*, Boston: Artech House, 1995.
32. B.M. Kolundzija, J.S. Ognjanovic, and T.K. Sarkar, WIPL-D: *Electromagnetic Modeling of Composite Metallic and Dielectric Structures, Software and User's Manual*, Boston: Artech House, 2000.
33. A.R. Djordjevic, B.D. Popovic, and M.B Dragovic, "A method for rapid analysis of wire-antenna structures," *Archiv. Elektrotechnik*, vol. 61, pp. 17–23, Jan. 1979.
34. B. M. Kolundzija and B. D. Popovic, "Entire-domain Galerkin method for analysis of generalized wire antennas and scatterers," *Proc. IEE, Pt. H,* vol. 139, pp. 17–24, Feb. 1992.
35. A.R. Djordjevic, L.N. Niccolai, and T.K. Sarkar, "Vehicular antenna for AM/FM radio, mobile phone, and GPS," *Proc. Millennium Conf. on Antennas Propagat.* (*AP2000*), Davos, April 2000.
36. C.D. Taylor and D.R. Wilton, "The extended boundary condition solution of the dipole antenna of revolution," *IEEE Trans. Antennas Propag.*, vol. AP-22, pp. 407–413, May 1974.
37. S.M. Rao, D.R. Wilton, and A.W. Glisson, "Electromagnetic scattering by surfaces of arbitrary shape," *IEEE Trans. Antennas Propag.*, vol. AP-30, pp. 409–418, May 1982.
38. B.M. Kolundzija and B. D. Popovic, "Entire-domain Galerkin method for analysis of metallic antennas and scatterers," *Proc. IEE, Pt. H,* vol. 140, pp. 1–10, Feb. 1993.
39. J.C. Monson and N.J. Damaskos, "A scheme for eliminating internal resonances: the parasitic body technique," *IEEE Trans. Antennas Propag.*, vol. AP-42, pp. 1089–1096, Nov. 1982.
40. A.J. Poggio and E.K. Miller, "Integral equation solutions of three-dimensional scattering problems," in R. Mittra (Ed.), *Computer Techniques for Electromagnetics*, Oxford: Pergamon Press, 1973.
41. J.R. Mautz and R.F. Harrington, "H-field, E-field and combined-field solutions for conducting bodies of revolution," *A.E.Ü.*, vol. 32, pp. 157–164, 1978.
42. E. H. Newman and D. M. Pozar, "Electromagnetic modeling of composite wire and surface geometries," *IEEE Trans. Antennas Propag.*, vol. AP-26, pp. 784–789, Nov. 1978.
43. B.M. Kolundzija and B.D. Popovic, "General localized junction model in the analysis of wire-to-plate junctions," *Proc. IEE, Pt. H,* vol. 141, pp. 1–7, Feb. 1994.
44. B. Kolundzija, J. Surutka, and B. Miletic, "New approach to the design of TV panel antenna for UHF frequency range," *Proc. XLIII Yugoslav ETRAN Conf.*, Zlatibor, pp. 25–32, Sept. 1999.
45. K. Liu, C. A. Balanis, C. R. Birtcher, and G. C. Barber, "Analysis of pyramidal horn antennas using moment methods," *IEEE Trans. Antennas Propag.*, vol. 41, pp. 1379–1388, Oct. 1993.
46. J.H. Richmond, "Scattering by a dielectric cylinder of arbitrary cross section shape," *IEEE Trans. Antennas Propag.*, vol. AP-13, pp. 334–341, Mar. 1965.
47. D. E. Livesay and K. M. Chen, "Electromagnetic fields induced inside arbitrarily shaped biological bodies," *IEEE Trans. Microwave Theory Technol.*, vol. MTT-22, pp. 1273–1280, Dec. 1974.
48. H. Schaubert, D.R. Wilton, and A.W. Glisson, "A tetrahedral modeling method for electromagnetic scattering by arbitrarily shaped inhomogeneous dielectric bodies," *IEEE Trans. Antennas Propag.*, vol. AP-32, pp. 77–85, Jan. 1984.
49. T.K. Sarkar, E. Arvas, and S. Ponnapalli, "Electromagnetic scattering from dielectric bodies," *IEEE Trans. Antennas Propag.*, vol. AP-37, pp. 673–676, May 1989.
50. B.M. Notaros and B.D. Popovic, "Large-domain integral-equation method for analysis of general 3-D electromagnetic structures," *IEE Proc. Microw. Antennas Propag.*, vol. 145, pp. 491–495, Dec. 1998.

51. J.R. Mautz and R.F. Harrington, "Electromagnetic scattering from a homogeneous material body of revolution," *A.E.Ü.*, vol. 33, pp. 71–80, 1979.
52. S.M. Rao, T.K.Sarkar, P.Midya, and A.R.Djordjevic, "Electromagnetic radiation and scattering from finite conducting and dielectric structures: surface/surface formulation," *IEEE Trans. Antennas Propag.*, vol. AP-39, pp. 1034–1037, July 1991.
53. B.M. Kolundzija, "Electromagnetic modeling of composite metallic and dielectric structures," *IEEE Trans. Microwave Theory Technol.*, vol. MTT-47, pp. 1021–1032, July 1999.
54. M. Salazar-Palma, T.K. Sarkar, L.E. Garcia-Castillo, T. Roy, and A.R. Djordjevic, *Iterative and Self-Adaptive Finite-Elements in Electromagnetic Modeling*, Boston: Artech House, 1998.

9

Genetic Algorithms

Wesley O. Williamson
TRW, Inc.

Sembiam R. Rengarajan
California State University-Northridge

9.1 Introduction

Genetic algorithms (GAs) are becoming increasingly popular in electromagnetic applications. Originally developed by Holland[1,2] and popularized by Goldberg,[3] they have been successfully applied to a number of electromagnetic design problems such as array antennas, wire and patch antennas, antenna pattern synthesis, and electromagnetic filters.[4] They are capable of solving design problems with a myriad of design parameters and multiple design goals.

The field of wireless antenna design presents many challenges. Today's cell-phone antennas must be inexpensive, small, and insensitive to polarization misalignment and multipath, with minimum radiation into the human head.[5] Medium to high-gain antennas for satellite-to-vehicle communication must be small and insensitive to multipath, and able to track the satellite signal in any direction. Although many complicated analysis tools exist to aid the antenna designer, a simple software driver, such as a GA, can reduce the time required to evaluate the many possible design solutions for a typical problem.

This chapter is an introduction to GAs and some of the methods whereby they may be applied to optimizing or finding good solutions to wireless antenna problems. For most of the problems one finds in engineering, straightforward solutions based on calculus are not applicable. Rarely does one find a structure whose performance can be represented by a simple model such as a parabola or a straight line.

0-8493-0124-6/02/$0.00+$1.50

Instead, the structures found in nature are typically represented by a myriad of parameters, all interrelated. Consider an antenna optimization problem, for example. Very seldom does the solution result in a simple structure such as a monopole on an infinite ground plane. Instead, complicated interactions between a multitude of antenna elements and the surrounding structures must often be considered.

An example of an easily optimized mathematical function is a simple cosine function, constrained between two adjacent nulls and exhibiting one peak. Suppose one were trying to find the argument for which this function has a maximum value. By using the derivative information, one could easily obtain the peak value.

An illustration of a function that might elude a simple calculus-based optimization is a Bessel function. Many traditional optimization techniques require one to start off with a best "guess" for a solution. For the Bessel function, depending on which argument value one used as a guess, one could easily wind up "stuck" on a local maximum, when what one really wants is the global maximum. One way to be assured that one has found the global maximum, besides a visual inspection, is to perform an incremental search over a large argument domain. Another way is to randomly sample the domain, performing a calculus-based optimization for each value.

Obviously, these techniques are the method of choice for simple functions such as the Bessel function. However, what if the function is a method-of-moments (MOMs) simulation that requires a substantial amount of computer time? Finding the values of the input parameters that optimize certain output parameters can become quite time consuming, especially if the number of variables is increased from 1 to, for example, 25. For avoiding such exhaustive simulations, global optimization techniques such as the GA and simulated annealing have found substantial applications. This chapter is devoted to GAs.

9.2 What Are Genetic Algorithms?

GAs are sophisticated ways of guessing the best answer to an optimization problem. They are particularly well suited to problems involving a number of mutually interacting design parameters and a poorly behaved solution space.

Although traditional algorithms either perform an exhaustive search of all possible solutions or, based on some "best guess," follow a simulated performance curve to the closest optimum value, the GA begins with a random sampling of the entire solution space. Samples with the best performance are combined with other high-performance samples to produce new sample locations, using the principles of biological genetics. The process of generating new sample locations based on the performance of previous samples is continued, iteratively, until some convergence criterion is met. Occasionally, new sample locations are generated, at random, using the principles of genetic mutation.

In a genetic algorithm, sample locations are represented by an important data structure called a *chromosome*. Chromosomes are similar to artificial DNA structures for a particular species of design. They come in two basic varieties. First, there are binary chromosomes, wherein all the design parameters are encoded into a long binary string, such as 0110101101001101101. Next, there are real-valued or continuous-parameter chromosomes. For these, all the design parameters remain as floating-point numbers, and, instead of being strung together, they are merely stored together in an array.

The binary string is probably a closer representation of a chromosome found in nature (e.g., where people have either blue eyes or brown eyes, but not much in between). The binary string can reserve certain bit locations for making decisions, such as whether an antenna element is turned off or on. However, the real-valued chromosomes are sometimes easier to work with. They are discussed in detail in Section 9.3.

Consider the fictitious antenna shown in Fig. 9.1. This is a simple Yagi-Uda-type antenna with a feed element and a reflector. In this example, the reflector length is 0.50 λ, the feed element length is 0.48 λ, and the element spacing is 0.25 λ. This design may or may not have any practical value. Its dimensions can be input into a wire antenna simulator, for example, the numerical electromagnetic code (NEC),[6] from which will come a predicted gain, voltage standing wave ratio (VSWR), and sidelobe performance. As shown in Fig. 9.1, the design can be represented by either a binary or a real-valued chromosome.

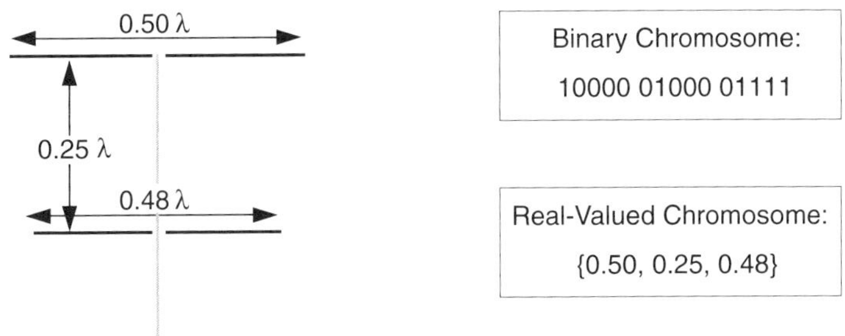

FIGURE 9.1 A fictitious antenna design. For a real-valued or continuous parameter algorithm, the chromosome for this particular design could be {0.50, 0.25, 0.48}. For a binary algorithm, each of these values would first be converted to binary. If the lowest bit is worth 2^{-5} in a 5-bit encoding, 0.50 would become 10000, 0.25 would be 01000, and 0.48 would be 01111. The chromosome would then be 100000100001111.

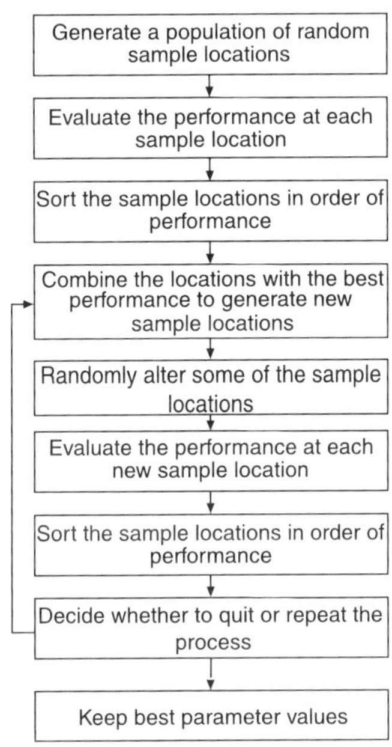

FIGURE 9.2 General flow diagram of a genetic algorithm.

For the binary chromosome, the number of bits per parameter has been arbitrarily set to 5 (i.e., numerical values between 0 and 1 are encoded into 32 discrete values). Then 0.50 would become 10000, 0.25 would be 01000, and 0.48 would be 01111. The binary chromosome for this case would then be 100000100001111. The real-valued chromosome is quite simply represented as {0.50, 0.25, 0.48}.

The general flow diagram of a GA is shown in Fig. 9.2. First, a population of sample locations is generated by a pseudorandom number generator. Each of these sample locations corresponds to some combination of antenna design parameters, such as reflector length, element spacing, amplitude weight, or wire diameter.

Next, the performance of the antenna is evaluated for each combination of parameter values. Then, the various antenna designs are sorted by some performance criteria, such as antenna gain, beamwidth, relative sidelobe level, or radiation pattern fitness. Some of the parameter values from the best-performing antenna configurations are mixed and matched to obtain new antenna designs.

To illustrate the concept of mixing and matching antenna designs, consider the example shown in Fig. 9.3, where two antenna designs A and B are blended to produce designs C and D. The reflector element length from A becomes the reflector length for C, whereas the driver element from A becomes

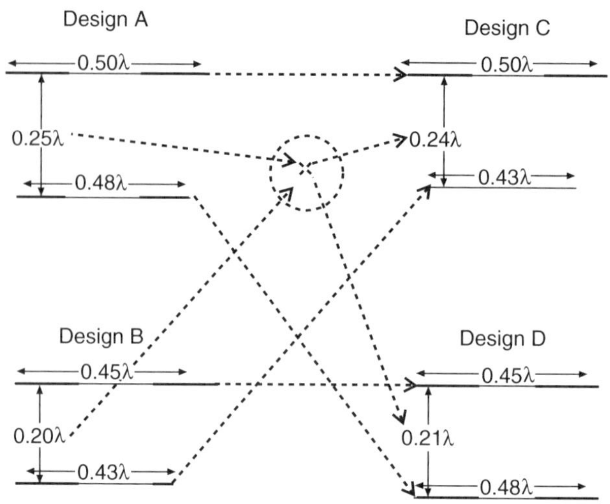

FIGURE 9.3 Combining two 2-element yagi antenna designs. Designs A and B are combined to produce designs C and D. In this example, the reflector length from design A is transferred to design C, and so on. The element spacings from A and B are "mixed" together to produce the spacings for C and D.

the driver element for D. Similarly, the reflector length from B becomes that of design D, whereas the driver from B is copied to C. The element spacings from designs A and B are combined to produce the spacings for C and D, using the following formulas:

$$0.24\,\lambda = 0.25\,\lambda - 0.187 * (0.25\,\lambda - 0.20\,\lambda) \tag{9.1}$$

$$0.21\,\lambda = 0.20\,\lambda - 0.187 * (0.20\,\lambda - 0.25\,\lambda) \tag{9.2}$$

where the factor 0.187 is a pseudorandom number.[7] (These equations are just one example of the various ways in which two parameter values may be combined to produce new values.)

After the combinatorial process, randomly selected parameter values are replaced with new pseudorandom numbers. This perturbation helps to prevent the GA from converging too soon. Next, the new and altered antenna designs are reevaluated and sorted together with the best performers of the original population. At this point, either the algorithm is terminated or the process of combining and perturbing designs is repeated.

9.2.1 A Simple Example — Optimizing a Linear Array with Nonuniform Spacing

Consider the nonuniformly spaced, linear antenna array depicted in Fig. 9.4. This array is symmetrical about its center, and each element has the same amplitude and phase. The number of elements, not shown in Fig. 9.4, is 48. The design goal is to find the 47 element-to-element spacings that yield the lowest maximum sidelobe levels relative to the amplitude of the main beam. Because the array is symmetrical, only 24 spacings need to be found.[8]

The equation for the array factor of this antenna array, assuming a sin(ϕ) element pattern, is

$$AF(\phi) = 2\sin(\phi)\sum_{n=1}^{\frac{N}{2}} \cos\left[k\left(\sum_{m=1}^{n} d_m - \frac{d_1}{2}\right)\cos(\phi)\right] \tag{9.3}$$

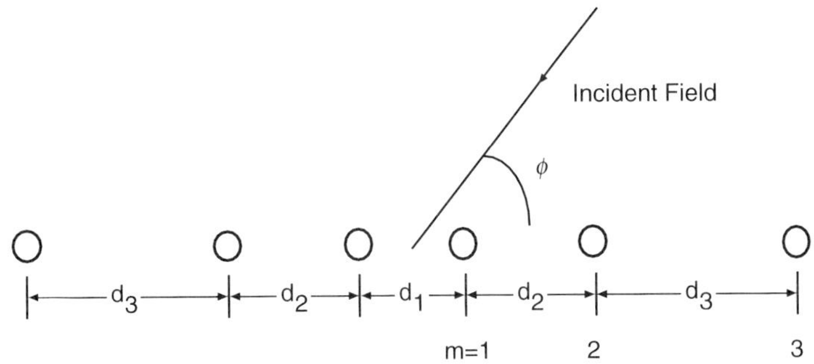

FIGURE 9.4 A symmetrical linear array with nonuniform spacing. (From Haupt, R. L., *IEEE Antennas Propagat. Mag.*, 37, 7, 1995. ©IEEE. With permission.)

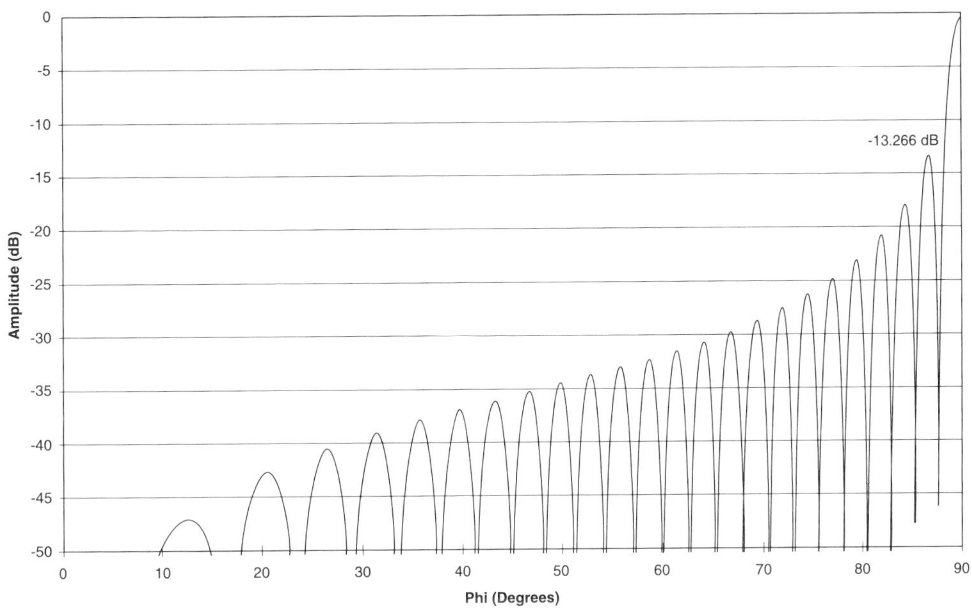

FIGURE 9.5 Radiation pattern for a 48-element array with uniform spacing.

where N is the number of elements and is assumed to be even, d_1 is the spacing between the center two elements, and d_m is the spacing between the mth and the $(m\text{-}1)$th elements, as shown in Fig. 9.4. $k = 2\pi/\lambda$ is the free-space wave number, where λ is the free-space wavelength, and ϕ is the antenna look angle. The spacings are constrained to lie between 0.25 λ and 1 λ. All elements are excited, uniformly.

The antenna pattern resulting from uniform spacing is shown in Fig. 9.5. Note that the maximum relative sidelobe level is –13.266 dB. An example of an array with random, nonuniform spacings is shown in Fig. 9.6.

An example of an optimized array pattern using GA is shown in Fig. 9.7. The element-to-element spacings for this configuration in wavelengths are 0.345, 0.558, 0.341, 0.450, 0.318, 0.493, 0.501, 0.447, 0.455, 0.353, 0.638, 0.364, 0.579, 0.515, 0.568, 0.616, 0.542, 0.632, 0.712, 0.880, 0.921, 0.999, 0.999, and 0.998. The maximum relative sidelobe level is –26.04 dB. Better performances have been achieved; Haupt has reported a maximum relative sidelobe level of –27.2 dB. In other words, the GA does not guarantee the *most optimum* solution to a problem; instead, it is a tool for providing a *good* solution to a problem.

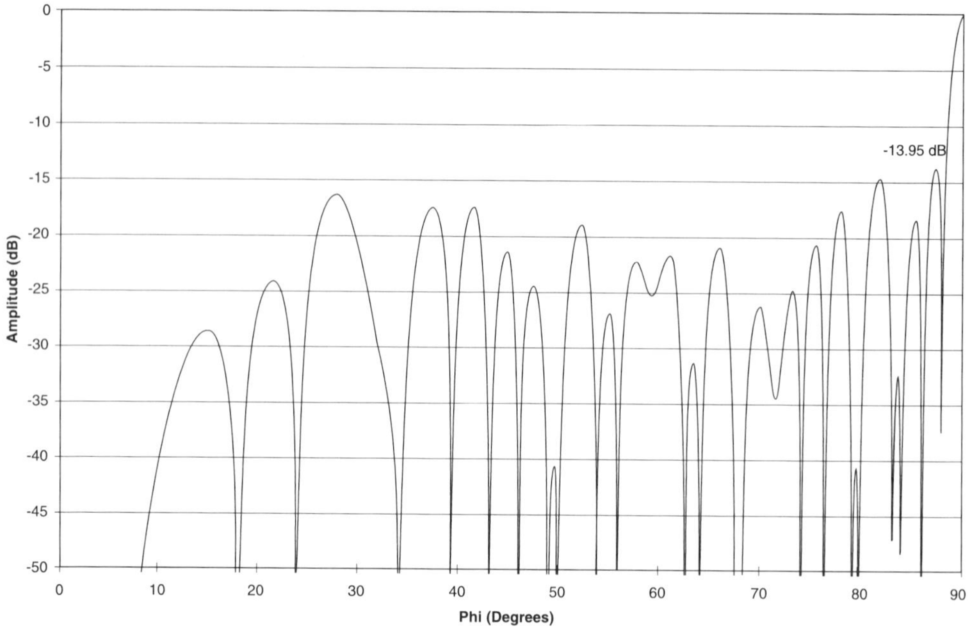

FIGURE 9.6 Radiation pattern for a 48-element array with random spacings.

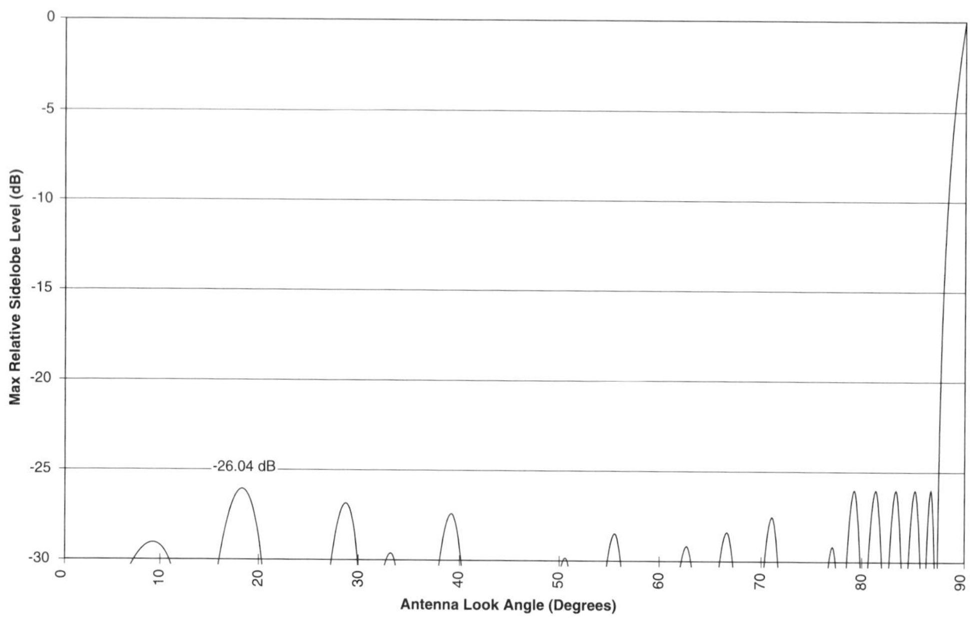

FIGURE 9.7 Radiation pattern for an optimized linear array with nonuniform spacings. The maximum relative sidelobe level is –26.04 dB.

9.2.2 Genetic Algorithm Terminology

Many of the terms associated with GAs are derived from the principles of biology, evolution, or natural selection.[3] For example, an antenna design may be referred to as a chromosome or an *individual,* whereas the combining of two or more antenna designs may be referred to as *mating* or *breeding.* Antenna performance may be referred to as *fitness,* whereas the process of picking the best antenna designs may be called *selection.*

Chromosomes are made up of *genes*, where a gene may be considered as a single antenna parameter, such as a wire diameter or the excitation coefficient of an element. The numerical encoding of a gene may be called a *genotype*, whereas the physical manifestation of the genotype is called a *phenotype*. A specific attribute of a gene, such as whether an array element is turned off or on, is synonymous with an *allele*. When two or more *parent* chromosomes or designs are combined to form new ones, the resultant designs are called *children* or *offspring*. The process of randomly altering or perturbing a chromosome or gene is called *mutation*. Often, it is advantageous to protect the fittest individual from mutation, a concept also known as *elitism*.

A term not derived from genetics is *cost*, the inverse of *fitness* and performance, any of which may be used for deciding which antenna designs are superior to others. Thus, a design with the lowest cost also has the best performance.

9.3 Kinds of Genetic Algorithms

Genetic algorithms may be classified into two categories, *binary* and *real-valued* or *continuous-parameter*, depending on the particular methods of encoding the various design parameters. In a binary GA, each of the parameters one is attempting to optimize is first encoded as a binary number. Then, all the binary numbers for a particular design are placed together in a single string. On the other hand, in a real-valued GA, all the parameters are represented as floating-point numbers. Because of these differences in numerical representation between the two types of algorithms, separate methods exist for combining and perturbing designs.

9.3.1 Continuous Parameter or Real-Valued Genetic Algorithms

Continuous parameter or real-valued GAs are easier to comprehend than their binary counterparts. However, this does not mean that they are the best variety for a particular application.

9.3.1.1 Initialization

Before initializing a population of continuous-parameter individuals, it is helpful to know the range of allowed values for each parameter. In the example of the linear antenna array with nonuniform spacings, each spacing must lie between 0.25 λ and 1 λ, where λ is the free-space wavelength. Because pseudo-random number generators in most computers generate a number between 0 and 1, the formula for generating an initial value for a specific parameter is

$$\text{Initial\_Val} = \text{Min\_Val} + \text{Random\_Num} \times \left(\text{Max\_Val} - \text{Min\_Val}\right) \tag{9.4}$$

For the 48-element symmetrical antenna array, there are 24 unique element-to-element spacings. In a typical problem with this many design parameters, there may be several thousand individual designs in a population. In addition, the initial population may be larger than the normal or working population, thereby helping the algorithm to get a better start. For example, an optimization may start out by keeping the best 500 individuals out of an initial population of, for example, 2000.

9.3.1.2 Evaluation and Sorting

Computing the cost or fitness values for each design is simply a matter of plugging the numbers into the simulation, which can be anything from an involved computation such as the use of a MOM code to a simple array factor equation. Once the fitness values have been computed, the individuals are ranked or sorted, according to fitness.

9.3.1.3 Selection

Combining the best performers from a population of antenna designs is a two-step process. First, there are many ways of choosing which performers to combine, and second, there are various ways of combining them.

TABLE 9.1 Fitness Values for Four Individuals of a Population

Individual No.	Fitness	Probability of Survival	Cumulative Probability
1	50	50/109 = 0.459	0.459
2	39	39/109 = 0.358	0.459 + 0.358 = 0.817
3	15	15/109 = 0.138	0.817 + 0.138 = 0.955
4	5	5/109 = 0.046	0.955 + 0.046 ≈ 1.000
Total	109	1.000	

By far the simplest method of choosing which individuals to combine is to go down the list of sorted fitness values and group the individuals by pairs. Thus, 1 and 2 would form a pair, 3 and 4, and so on. This method of selection is called top-to-bottom pairing.

Generally, only the best half of the population is chosen for mating, whereas the individuals in the worst-performing half are replaced by the new offspring. For example, suppose there are 16 individuals in the working population. (There may have been a lot more individuals in the *initial* population.) The best eight of these are chosen for mating. The offspring from the combination of 1 and 2 are used to replace 9 and 10, those from 3 and 4 replace 11 and 12, and so on.

A more complicated method of selection for mating, called *performance-weighted random* (roulette-wheel) *selection*, uses the fitness or cost of the individual to generate a *probability of survival.* These probabilities are added up to give a cumulative probability for each individual. A random number generator is then used to decide which individual is to be selected for mating.

For example, suppose the population is comprised of eight individuals, four of whom are chosen to be in the *mating pool.* The fitnesses of these four are shown in column 2 of Table 9.1. The total fitness for these 4 individuals is 109. The probability of survival for individual 1 is 50/109 = 0.459, for 2, 39/109 = 0.358, and so forth, as shown in column 3. The cumulative probability for an individual is its probability of survival plus all the probabilities of its superiors. Thus, the cumulative probability for individual 1 is the same as its probability of survival, the cumulative probability for 2 is the sum of its probability and that of 1, and so forth. This is demonstrated in column 4.

A random number generator is used to decide which individuals are to become parents. The first individual whose cumulative probability is greater than the random number becomes a parent. For example, suppose a random number generator generates the following pair of random numbers: (0.228, 0.535). Going down the list of cumulative probabilities, 0.459 is greater than 0.228, whereas 0.817 is greater than 0.535. Hence, the parents are 1 and 2.

The method of using probabilities for selection assures that the best performers have a greater chance of reproducing or combining than the poorer performers. However, there is a chance that some individuals or designs are not chosen at all, or that an individual might be chosen to mate with itself, thereby producing two clones of itself.

9.3.1.4 Blending

Continuous-parameter algorithms have several different ways for combining chromosomes or parameter sets, the simplest of which is the method illustrated in Fig. 9.3. A pseudorandom number generator picks a crossover point among the parameters in the parent chromosomes, as shown in Fig. 9.8, where an ellipse has been drawn around the crossover parameter. Parameters to the left of the crossover point are copied to one offspring, whereas parameters to the right of the crossover point are copied to the other offspring, as illustrated in Fig. 9.9.

Next, the crossover parameters from the parent designs are blended together using the following equations:[7,9]

$$Offspring\ 1_{xover} = Parent\ 1_{xover} - \beta\left(Parent\ 1_{xover} - Parent\ 2_{xover}\right) \tag{9.5}$$

$$Offspring\ 2_{xover} = Parent\ 2_{xover} - \beta\left(Parent\ 2_{xover} - Parent\ 1_{xover}\right) \tag{9.6}$$

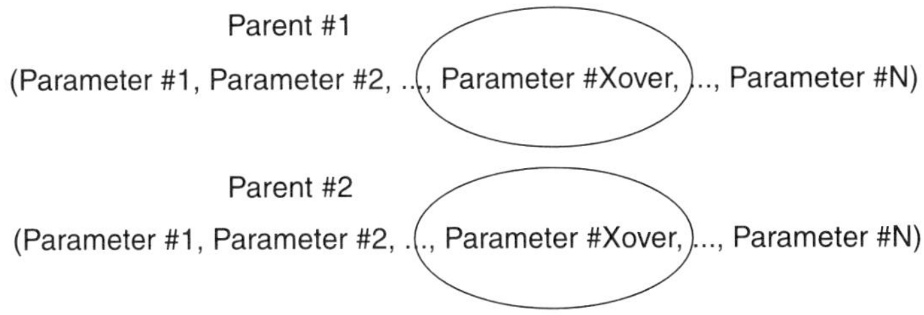

FIGURE 9.8 Selection of crossover parameter. A random number generator selects the crossover point, shown encircled, for mating or blending two design configurations.

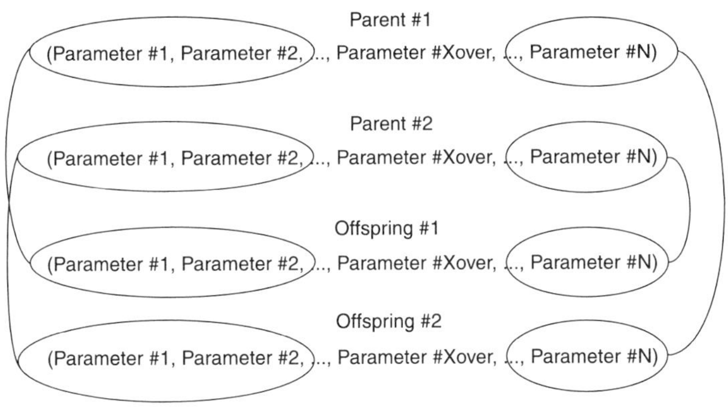

FIGURE 9.9 Transfer of parent parameters to offspring. For each parent design, parameters to the left of the crossover point are copied to one offspring, whereas parameters to the right are copied to the other offspring.

where β is a pseudorandom number between 0 and 1. This type of blending, known as *single-point complementary heuristic blending*, does not result in crossover-parameter values outside of the range between the parent crossover parameters.

Table 9.2 shows the blending of designs 1 and 2 to produce new offspring designs 5 and 6. The crossover parameter is spacing 11. For offspring design 5, parameter values for parameters before or above the crossover parameter are copied from design 1, whereas values for parameters after or below the crossover parameter are copied from design 2. The crossover-parameter values are a blend of the crossover-parameter values from the parent designs. A β value of 0.840 has been used in this exercise.

A more severe variation of complementary heuristic blending is to mix each of the parent design parameters with those of its mate. This has the effect of bringing each of the offspring parameters into the region between the parent parameters, often resulting in premature convergence of the algorithm.

A more complicated method of blending designs, which provides some local optimization of parameters, uses three parents to form parabolic curves for each parameter number.[10] Suppose the parent designs have performances of $Perf_1$, $Perf_2$, and $Perf_3$, and kth parameter values of P_{k1}, P_{k2}, and P_{k3}, respectively. For each parameter number, if the parameter values are noncollinear, a parabola may exist, as shown in Fig. 9.10. For the kth parameter, the equations for the parabola are[9]

$$Performance\left(P_k\right) = a_k P_k^2 + b_k P_k + c_k \tag{9.7}$$

where

$$a_k = \frac{1}{P_{k3} - P_{k1}} \left(\frac{Perf_3 - Perf_2}{P_{k3} - P_{k2}} - \frac{Perf_2 - Perf_1}{P_{k2} - P_{k1}} \right) \tag{9.8}$$

TABLE 9.2 Continuous Parameter Design Offspring Produced from Single-Point Complementary Heuristic Blending ($\beta = 0.840$)

Design No. / Spacing	1	2	5	6
1	0.7353	0.9124	0.7353	0.9124
2	0.7073	0.4552	0.7073	0.4552
3	0.3593	0.4461	0.3593	0.4461
4	0.4326	0.4162	0.4326	0.4162
5	0.3995	0.3185	0.3995	0.3185
6	0.6930	0.8292	0.6930	0.8292
7	0.6244	0.3225	0.6244	0.3225
8	0.5311	0.5206	0.5311	0.5206
9	0.4830	0.5088	0.4830	0.5088
10	0.4243	0.7762	0.4243	0.7762
11	0.5561	0.3642	0.3949	0.5254
12	0.3460	0.9628	0.9628	0.3460
13	0.5059	0.6688	0.6688	0.5059
14	0.4513	0.7709	0.7709	0.4513
15	0.4040	0.8248	0.8248	0.4040
16	0.4362	0.4396	0.4396	0.4362
17	0.8165	0.7422	0.7422	0.8165
18	0.7911	0.9579	0.9579	0.7911
19	0.9282	0.5272	0.5272	0.9282
20	0.5544	0.4383	0.4383	0.5544
21	0.3043	0.7703	0.7703	0.3043
22	0.5655	0.3874	0.3874	0.5655
23	0.7434	0.5455	0.5455	0.7434
24	0.5505	0.4044	0.4044	0.5505

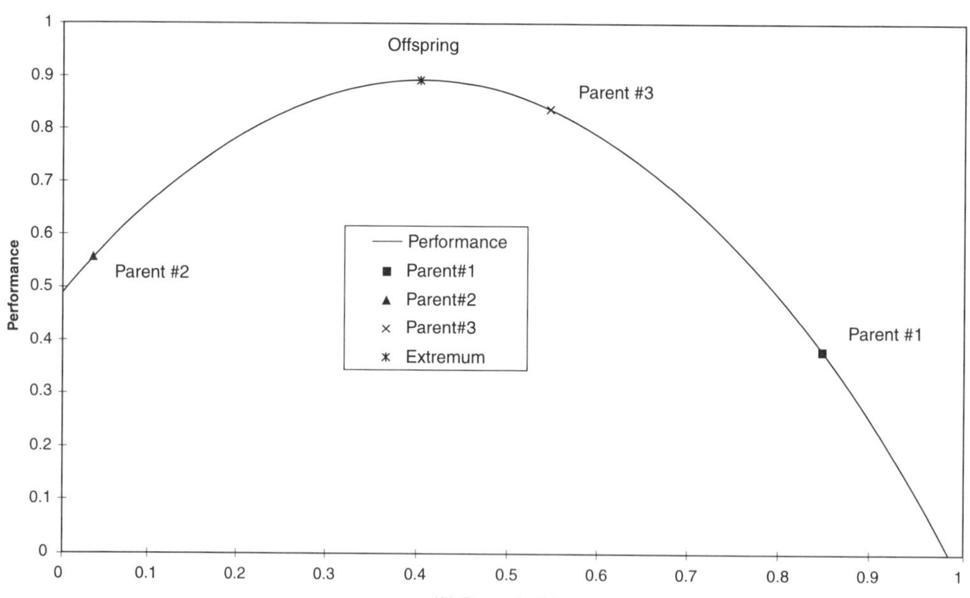

FIGURE 9.10 Quadratic crossover for *k*th parameter value. In this case, the extremum is a maximum. If a minimum value of performance or cost were desired for these particular values, then an alternative blending method would be desired.

$$b_k = \frac{Perf_2 - Perf_1}{P_{k2} - P_{k1}} - a_k\left(P_{k2} + P_{k1}\right) \tag{9.9}$$

$$c_k = Perf_1 - a_k\left(P_{k1}\right)^2 - b_k P_{k1} \tag{9.10}$$

and P_k represents the domain of the kth parameter. The extremum of the parabola is at

$$P_{k\_extremum} = -\frac{b_k}{2a_k} \tag{9.11}$$

Depending on whether a maximum performance or minimum cost is desired, the extremum of the parabola may not be the optimum value for the offspring. In Fig. 9.10, where a maximum performance is desired, the extremum of the parabola is the optimum value for the kth parameter of this particular offspring. If the extremum is the wrong type or outside the range of acceptable parameter values, or if it cannot be determined because the three parent values are collinear, then the first two parent values are used in a linear interpolation. Assuming the performance of parent 2 is superior to that of parent 1, the equation for determining the offspring parameter value is

$$P_{k\_offspring} = \beta\left(P_{k2} - P_{k1}\right) + P_{k2} \tag{9.12}$$

where β is a pseudorandom number such that $0 \le \beta \le 1$. This method of blending is illustrated in Fig. 9.11. Note that, if the performance of the first parent is greater than that of the second parent, Eq. (9.12) becomes

$$P_{k\_offspring} = \beta\left(P_{k1} - P_{k2}\right) + P_{k1} \tag{9.13}$$

If $P_{k\_offspring}$ is outside the range of acceptable parameter values, then it is given the value of one of the three initial parents, chosen at random.

This optimization scheme may not work if the various optimization parameters are not independent of one another (i.e., if adjusting the value of one parameter affects the contributions of the other parameter values). Of course, if the parameters are truly noninteracting, then perhaps the problem could be broken down into separate optimizations for each independent parameter.

9.3.1.5 Mutation

After the various design parameters have been blended together to make new ones, a number of parameters are selected from the population, at random, and replaced with randomly generated values. The best design in the population is often kept intact, a practice known as *elitism*.

The simplest method of mutating a design parameter is to randomly generate another value, anywhere within the range of allowable values for that parameter. A slightly more complicated method is to pick a random number from a Gaussian or uniform distribution centered about zero, with a standard deviation equal to 10% of the allowable range of the parameter.[9] This number is then added to the original parameter value to obtain the new value for the parameter. If the new value is outside of acceptable limits, then either the new parameter can be given the value of the nearest design limit or a new mutation can be performed until an acceptable value is found. Typical values for mutation rate are less than 10%.

9.3.2 Binary Genetic Algorithms

Binary GAs are a closer representation of natural processes than are continuous-parameter algorithms. In a binary GA, all design parameters are encoded into strings of 0's and 1's. For example, an antenna with a 3-m reflector, a 2-m feed, and 1-m element spacing could be encoded as: 11*10*01, where the first

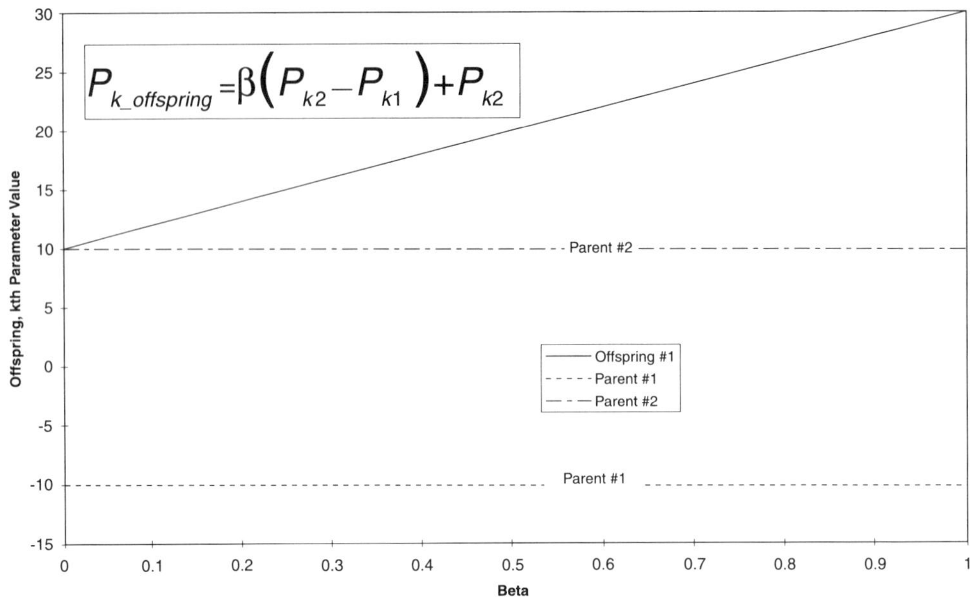

FIGURE 9.11 Linear interpolation of continuous parameters. This method of blending allows the *k*th parameter of the offspring to venture outside of the region bounded by the two parent values, provided the resultant parameter stays within the range of acceptable values.

2 b represent the reflector length, the next 2 b represent the feed length, and the last 2 b represent the element spacing. Because we are using 2 b for each parameter, only four discrete values are permitted for each. With a sufficient number of bits, it is possible to have a resolution for each parameter on the order of manufacturing tolerance. Other bits can be involved in decision making, such as whether to use a dipole or a loop, or whether an array element is turned off or on.

With binary parameter encodings, the processes of mating and mutation can be made much simpler than with continuous-parameter representations. Mating or blending can consist of selecting a point of crossover and copying bits from parent designs to offspring, while mutation is just the toggling of a bit from 0 to 1 or vice versa.

The general flow of a binary GA is the same as that of the continuous-parameter algorithm. The primary difference is the added overhead of binary encoding and decoding associated with the binary algorithm. Once the number of bits per parameter has been decided on, a population of bits is generated using a pseudorandom number generator, which may be seeded from the computer's internal clock. Each of these randomly generated strings of bits is parsed and decoded into a format from which the parameters can be evaluated by the design simulator, fitness, or cost function.

Following evaluation, the various designs are ranked by performance and selected for mating using some scheme such as rank, roulette wheel, or tournament selection. Offspring designs are generated and used to replace a certain percentage of the population. Next, a random number of bits in the population are toggled to simulate mutation. Then, the new or altered designs are reevaluated, the population is sorted again, and the process of mating and mutation is repeated. Subsequent generations of designs are created and evaluated until the algorithm converges, gets stuck, or reaches some limitation such as performance, number of simulations, or number of generations. This process is illustrated in Fig. 9.12 and is described next.

9.3.2.1 Population Initialization and Evaluation

As in the case of the continuous-parameter genetic algorithm, the size of the initial population can be quite large, to give the optimization a good start. The binary strings can be initialized using the computer random number generator, which typically generates a real number between 0 and 1. A rounding function is used to convert the real number into an integer or Boolean value.

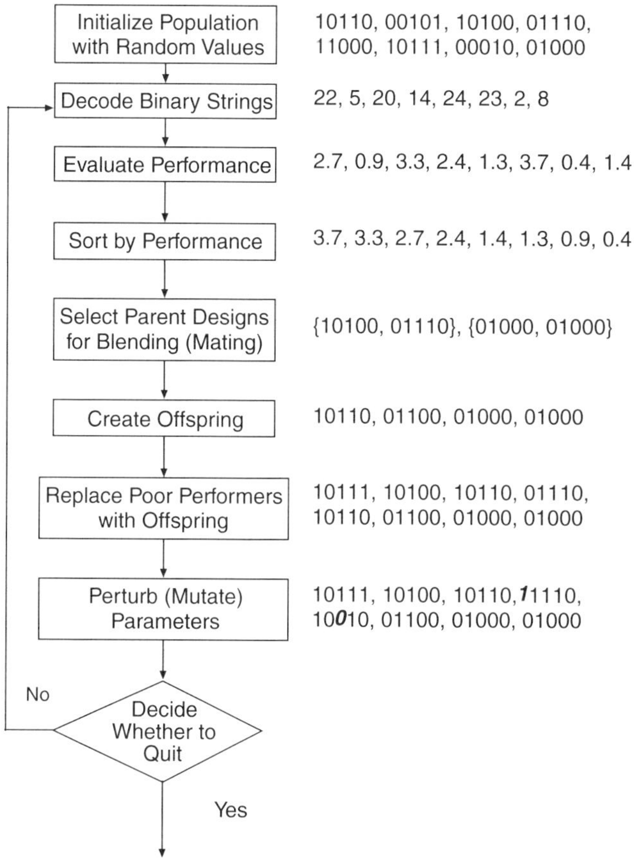

FIGURE 9.12 Flowchart of a binary genetic algorithm.

If there is a constraint on some of the parameters, then either a new set of random bits must be generated or the bits are set to some prescribed value. For example, suppose one of the parameters has a 3-b binary encoding and a maximum allowed value of 6, or in representation, 110. If, during initialization, a binary value of 111 (or integer value of 7) is encountered, these 3 b can either be re-randomized or set to 000, 110, or some other value less than 110.

Actually, it is often better to handle any parameter constraints in the encoding/decoding process. For example, if the maximum allowed value for a 3-b representation is 6, then binary 111 ought to be scaled to integer 6. An equation for encoding a real parameter value into an integer is

$$x_{integer} = round\left[\frac{(x_{real} - x_{min})}{(x_{max} - x_{min})} \cdot \left(2^{Number\_of\_bits} - 1\right)\right] \tag{9.14}$$

where $x_{\max}$ and $x_{\min}$ are maximum and minimum values of x_{real}.

Similarly, to decode an integer into a real parameter value,

$$x_{real} = x_{min} + \frac{x_{integer}}{\left(2^{Number\_of\_bits} - 1\right)} \cdot (x_{max} - x_{min}) \tag{9.15}$$

Of course, there must also be a conversion between integer and binary values.

9.3.2.2 Parent Selection

Parent selection is essentially the same as that of continuous-parameter algorithms. The basic idea is to give better performers a higher probability of being selected as parents.

9.3.2.3 Mating or Blending of Designs

As indicated earlier, the blending of two binary-encoded designs can be quite simple indeed. First, a crossover point between two adjacent bits in the parent designs is determined by a random number generator as follows:

Parent Designs
00101 | *10001011010*
10101 | 10110110001

Next, a pair of offspring designs are created by copying the bit values to the right and left of the crossover point to opposing strings:

Offspring Designs
10101*10001011010*
*00101*10110110001

More complicated variations exist. For example, a control called *probability of crossover* determines whether the crossover actually takes place. If the crossover probability is 0.5, then there is a 50% chance that the parent designs will merely create copies of themselves.

Another variation, called *two-point crossover*, copies all the bits between two randomly generated crossover points to opposing offspring designs:

Parent Designs
0010110 | *0010* | *11010*
1010110 | 1101 | 10001

Offspring Designs
*0010110*1101*11010*
1010110*0010*10001

The advantage of this technique is that so-called *building blocks* (or important substrings of bits) containing bits at the beginning and end of design strings are allowed to propagate into a new generation.

9.3.2.4 Mutation

There are at least two ways of performing bit mutation. The first involves a control called *probability of mutation*, which is usually quite small. For each bit position in a population of designs, a random number is generated. If the random number is less than the probability of mutation, the bit for that position is toggled from a 0 to a 1 or vice versa. The second method is to generate a certain number of random numbers corresponding to various design strings and bit positions within those strings. Bits that lie in these positions are then toggled.

As in the continuous-parameter algorithms, the best-performing (elite) design for a particular design is often left out of the mutation process.

9.3.2.5 Gray Coding

Although perhaps not very useful, some words must be said about a variation of binary encoding known as *gray coding*. The reason for gray coding is to prevent large jumps in the design search space resulting from small binary mutations. For example, suppose the first bit of the 3-b combination *000* is mutated so that the new 3-b word is *100*. In phenotypic (or value) space, the decoded parameter would change from *0* to *4*, which could be construed as a large jump in the search space.

Gray coding attempts to circumvent this problem by scrambling the bit values so that toggling a single bit does not create so much of a change in the decoded value. For example, integer 4 is encoded as *110*, instead of *100*, and integer 3 is encoded as *010* instead of *011*. Hence, toggling the most significant bit of integer 4 (gray code *110*) changes the value of the parameter to integer 3. However, toggling the second bit of integer 4 creates integer 7 (gray code *100*), which is still a considerable jump in the search space. In addition, unless one uses some clever encoding technique, the process of translating straight binary codes to gray codes and vice versa can cause an increase in computer execution time.

9.3.3 Genetic Algorithm Design Considerations

This section discusses the trade-offs between binary and continuous-parameter encodings. In addition, alternative selection schemes, population sizes, and mutation rates are considered.

9.3.3.1 Binary vs. Continuous

Many authors have expressed an opinion as to which type of GA is best, binary or continuous parameter. The answer appears to be one of personal preference. Often, the continuous-parameter algorithm can achieve slightly better performance than can the binary algorithm, although run times can be considerably longer. However, for an antenna design, one should be aware of engineering tolerances. A 3-b binary-parameter representation may be well within manufacturing tolerances. In addition, where decision-making variables are concerned, a binary encoding can be much easier to implement.

A comparison of the two techniques is shown in Fig. 9.13. The optimization problem is the 48-element, linear antenna array with nonuniform spacings, discussed earlier. Data have been averaged over ten computer runs. In both cases, there are 2000 individuals in the initial population and 1000 individuals in the normal population. Mutation rates are 500 per generation. Because there are 24 element-to-element spacings to be optimized, this equates to a mutation rate of $500/(24 \times 1000) = 2.1\%$ for the continuous-parameter and $500/(24 \times 3 \times 1000) = 0.7\%$ for the binary algorithm.

Data can often be interpreted in contradictory ways. In this case, the data could be used to show superiority of the continuous-parameter algorithm over the binary algorithm. However, adjustments in

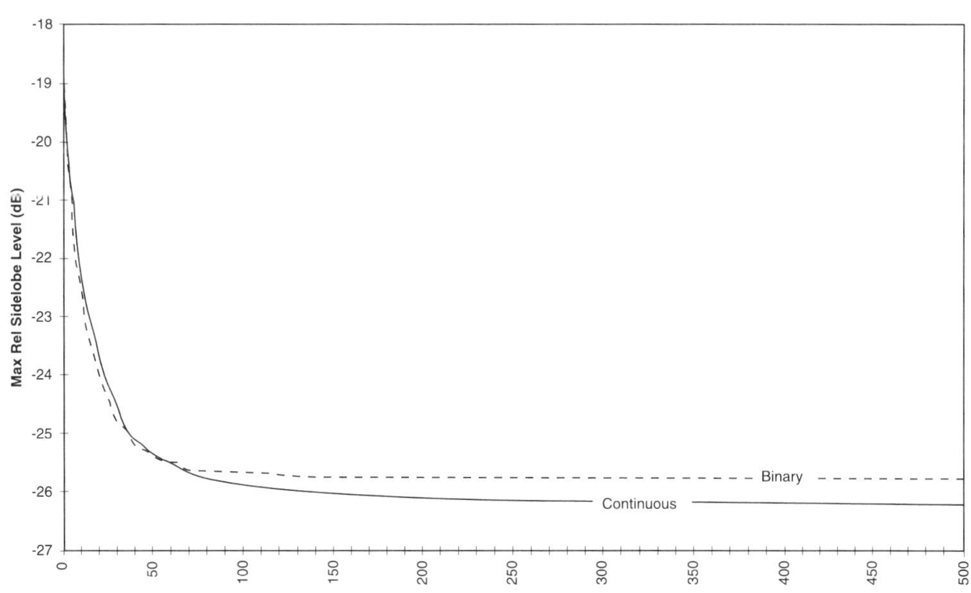

FIGURE 9.13 Binary vs. continuous parameter comparison.

mutation rate, selection criterion, population size, etc. could result in slight improvements for either algorithm. What is interesting to note is that both algorithms converge to a good solution. Hence, genetic algorithms of either type could be applied to this kind of problem.

As far as programming is concerned, binary algorithms require a significant amount of computer time for binary decoding during performance evaluations of the individual designs; however, blending and perturbing designs is relatively simple. Also, binary designs readily lend themselves to decision making. Conversely, continuous-parameter algorithms provide easily accessible parameters, although blending and perturbation are usually more complicated than for their binary cousins.

Continuous-parameter algorithms, by virtue of their floating-point parameter representations, have a much larger search space, and are therefore more likely to come up with a better solution, given enough time. The designer should consider whether the improvement in parameter accuracy is realizable in construction. In general, the binary GA is preferred.

9.3.3.2 Alternate Selection Schemes

Many methods are available for choosing parent designs from which to generate offspring. Thus far, only top-to-bottom pairing and performance-weighted random (or roulette wheel) selection have been discussed at length. Some additional methods are random selection, rank-weighted random selection, and tournament selection.[7] In each case, it is assumed that the poorest performers have already been eliminated from the population. For example, suppose there are 1000 individuals in a population, and only the top 50% of them are allowed to survive from generation to generation. Then, parent designs are only selected from the top 500 performers of the previous generation.

Random Selection

Random selection uses a pseudorandom number generator with a uniform distribution to select the identification (ID) numbers of parent designs. For example, suppose there are 16 designs in the selection pool, ranked in order of performance. The ranking or ID number of a parent design would be

$$Parent_n = \text{roundup}\left(\beta_n \times 16\right) \tag{9.16}$$

where β_n is a random number between 0 and 1 and *roundup* is a function that rounds up to the next integer. Some example parent pairs are (8,14), (12,4), (3,14), (3,3), (12,12), (7,5), (16,11), (3,12).

Rank-Weighted Random Selection

This method of selection is essentially the same as cost-weighted random selection, except the parent designs are now weighted by rank. An example weighting function is

$$P_{rank} = \frac{Number\_of\_Ranks - rank + 1}{\sum_{rank=1}^{Number\_of\_Ranks} rank} \tag{9.17}$$

where P_{rank} is the probability of selection, *Number_of_Ranks* is the number of individuals in the selection pool, and *rank* is an integer between 1 and *Number_of_Ranks*. For example, suppose there are eight individuals in the selection pool. Then

$$P_{rank} = \frac{9 - rank}{36} \tag{9.18}$$

These probabilities all add up to a cumulative probability of one.

Tournament Selection

In tournament selection, the best designs out of groups of randomly chosen designs are selected for blending. For example, 6 designs may be chosen, at random, out of a population of 100. Out of these six designs, the best design is selected to become a parent design. Then, another group of six designs is chosen, at random — these may include some of the previous six competitors — out of which the next parent is chosen, and so on. One advantage of tournament selection is that the population may never have to be sorted. Instead, by using a technique called *thresholding* or *survival-of-the-fittest*, only the designs whose performance exceeds a certain threshold are kept in the population, whereas the remaining designs are discarded. The tournament competition groups are then picked, at random, from the surviving designs. This technique is thought to closely mimic reproduction processes found in nature.

9.3.3.3 Simple vs. Steady-State Genetic Algorithms

It is difficult to deduce much meaning from the terms *simple* and *steady state* as applied to GAs. For one thing, simple GAs have an additional operation called a crossover test, while both simple *and* steady-state algorithms can have a certain percentage of new designs vs. old designs from generation to generation. The difference is, the worst performers in a given generation are typically discarded (through population decimation) in a steady-state algorithm, whereas they are given some small chance of survival in a simple algorithm.

For the case of the simple GA, no design is guaranteed to exist in the succeeding generation. Instead, each blending operation is accompanied by the generation of a new random number between 0 and 1. If this number is greater than the probability of crossover (typically between 0.6 and 0.8), then the parent designs are blended; if it is not, then the parent designs are simply copied (or replicated or cloned) into the offspring designs. Of course, designs with superior performances are still more likely to be selected for blending, depending on the selection process, and are, therefore, given a greater chance of survival than their poorer performing siblings.

In a steady-state GA, a certain percentage (typically between 30 and 50%) of higher performing designs in a given generation is copied to the next generation, whereas the rest are replaced by blended (offspring) designs. Although some potentially useful genetic material may be lost from the discarded designs, steady-state algorithms may converge faster than simple algorithms.[11]

The easiest way to decide which designs to throw out is to sort each generation of designs in order of performance; however, another method is to throw out any design that fails to meet some performance criteria or threshold. Other options for the steady-state algorithm include giving each design a lifetime of several generations depending on its performance.[12] Thus, poorer performers die off sooner than better performing designs, resulting in a population size that varies from generation to generation.

9.3.3.4 Convergence

There are many ways of determining when a GA run should stop, the simplest of which is to set up a loop counter to run a certain number of generations or iterations. A variation is to stop after a certain number of cost function or performance-simulator evaluations.

It is generally helpful to monitor the progress of a run while it is executing. An easy parameter to display on a computer monitor is the performance of the highest ranking design for each generation. Another parameter of interest is the average performance for each generation. Averaging the performance of the top 10% of the population for each generation is yet another option. A possible indication of convergence is when the top 10% of the population has stopped improving over the last ten generations. Another simple test of when to stop is when the performance is deemed good enough. Why waste more time when the design goal has been achieved?

9.4 Tips for Implementing or Using GA

In this section, tips for implementing or modifying existing GA software codes are discussed, including suggestions for enhancing algorithm performance and troubleshooting.

9.4.1 Using a Generic Genetic Algorithm Driver

A simple way to implement a GA is to use an existing driver. Remember, a GA is essentially a shell program for feeding design parameters to a performance-simulator function. One such driver has been written by David Carroll.[13]

The trick to using a generic driver is to locate the existing cost function or performance simulator. Some scrutiny reveals the array variables used for storing the individual designs and design performances. Then the desired cost function can be implemented in place of the existing function. For the optimization of the nonuniformly spaced antenna array, a cost function for the David Carroll driver is found in the Appendix. (Comments have been removed from the code for brevity.) Note that all the parameter values are stored in the array-variable *parent*(*Gene_Number, j*), where *j* is the number of the individual in the population.

Next, one must learn how to adjust the control variables such as mutation rate, number of bits per parameter, and population size. In David Carroll's software code, an auxiliary file contains this information.

Another approach is to start with the pseudocode written by Randy Haupt.[8] This steady-state binary implementation uses top-to-bottom pairing for parent selection and simple crossover for blending. Once one has a basic algorithm up and running, it is a simple matter to introduce more complex selection and blending techniques. A top-level code segment, based on Haupt's pseudocode, is shown next. The argument lists have been left out for the sake of clarity.

```
CALL Initialize_Population()
CALL Evaluate_Cost()
CALL Sort_Population_by_Cost()
DO Generation_Number = 1, Number_of_Iterations
  CALL Mate_Best_Half_of_Population()
  CALL Mutate_Random_Bit()
  CALL Evaluate_Cost()
  CALL Sort_Population_by_Cost()
END DO
CALL Write_Best_Genes_to_File()
```

The first subroutine, *Initialize_Population,* uses the pseudorandom number generator to fill all the chromosome strings in the initial population. For each bit in each chromosome string, the number generator provides a real number between 0 and 1. Then the nearest integer (NINT) function is used to generate either a 0 or a 1. Next, *Evaluate_Cost* computes the performance of each of the randomly generated strings in the population. Then, *Sort_Population_by_Cost* ranks each of the strings in ascending order.

The DO loop controls how many iterations (or generations) the population will go through (typically between 50 and 500, depending on how big the problem is and how long you are willing to wait for the answer).

Inside the loop, *Mate_Best_Half_of_Population* goes down the best half of the list of strings and mates each pair of strings to produce a pair of offspring. The offspring replace the strings in the "worst" half of the population. Then *Mutate_Random_Bit* inverts a prescribed number of bits out of all the members of the population (except for the best one) at random bit locations for individuals, also picked at random. Then *Evaluate_Cost* evaluates the performance of all the new or modified members of the population. Finally, *Sort_Population_by_Cost* re-sorts the new population in ascending order. The loop repeats for a prescribed number of iterations.

After the loop is done executing, *Write_Best_Genes_to_File* saves the parameters of the best-performing string to a file, from where they can be retrieved as a final result or for restarting additional optimizations.

9.4.2 Avoiding Repetitious Cost Calculations

For optimizations involving time-consuming performance evaluations, it is a good idea to limit the number of evaluations to only those designs that have been changed for a given generation. If a design chromosome is copied, intact, there is no need to reevaluate its performance. If computer memory is not a concern, an extra Boolean variable can be used to indicate whether a design has already been evaluated. Otherwise, the performance variable itself can be used to indicate whether it needs to be evaluated, by placing its value outside the range of possible values. For example, the performance variable *beam width* can be set to 50,000°, an impossible value, but one that can be used to trigger the software to evaluate this particular design, while skipping over the designs that have already been evaluated in a previous generation.

9.4.3 Keeping Solutions from Previous Runs

One may sometimes wish to use the design from a previous optimization run to initialize some of the designs in a subsequent problem. This may result in a significant time savings, provided the previous best-case design is close to an optimum solution. On the other hand, the previous design could be a local maximum or minimum, requiring some time for the algorithm to break away from it.

9.4.4 Twins

For binary algorithms involving time-consuming performance simulations, it could be more efficient to do a bit-by-bit comparison with other competing designs, to search for a twin. Once the twin is found, its performance values are copied, thereby avoiding a lengthy computation.

9.4.5 Population Sizes

For binary algorithms, a population size which is 5 to 10 times the number of bits per design string is a good starting point; whereas, for continuous algorithms, the size should be 5 to 10 times the number of real-valued parameters per design. Population size can be the most important parameter in an optimization problem, but it appears to be dependent on the type of problem. As noted, the initial population can be more than twice as large as the running population, to give the algorithm a good start.

9.4.6 Percentage of Overlap

For steady-state algorithms, a 50% population overlap is a good starting place. This means that the top 50% of the designs from a particular generation is passed on to the next generation.

9.4.7 Mutation Rates

Mutation rate is generally regarded as a secondary genetic operator, much the same as it appears in nature. If it is adjusted too high, the algorithm can never converge, because the population does not settle down. On the other hand, some mutation is necessary to encourage the design parameters to explore new numerical ranges. A mutation rate of 5% can be a good place to start, although the optimum value could prove to be anywhere from 4 to 10%. This means that each binary bit or real-valued parameter has a 4 to 10% chance of being toggled or perturbed.

9.4.8 Random Number Seeding

For large design problems, some consideration should be given to reseeding the computer random number generator using the system clock. How often this occurs depends on the problem. If the reseeding occurs too often, the random numbers lose their stochastic properties.

9.4.9 Troubleshooting

GAs often appear to work even though there is an error in the software. Rigorous debugging of new cost-function implementations is required. It may be helpful to have the new cost function implemented in an external program to be used as a check.

If a bug-free GA is converging too soon, before an adequate solution has been found, increasing the population or mutation rate may help.[14] Conversely, if the algorithm does not converge, decreasing the population or mutation rate may help. Other parameters to vary include the percentage of overlap for the population (for steady-state algorithms) or crossover rate (for simple algorithms). One may also try changing the scale of the cost function. If one is trying to obtain a performance value equal to 1.00 for a population whose starting average is 0.95, obtaining or identifying a meaningful design improvement may be difficult.

9.5 Applications of Genetic Algorithms

GAs have been applied to many types of engineering problems. In the area of wireless antenna design, GAs have been used to optimize array thinning, pattern synthesis, wire antenna loading, and patch antenna bandwidth, to name a few. In addition, exotic, three-dimensional wire antennas have been designed. These topics are discussed in the following sections.

9.5.1 Antenna Array Thinning

Haupt[15] has demonstrated using GAs for thinning 200-element linear and two-dimensional antenna arrays. Because uniformly spaced, uniformly fed arrays are easier to design and analyze than arrays with aperiodic spacing or weighted-feed networks, it is desirable to simulate amplitude tapers by switching off individual array elements. The goal of the amplitude taper is to generate an antenna radiation pattern with low relative sidelobe levels or perhaps with a pattern null in the direction of an interfering signal. Because each antenna element has an *on* and an *off* state, the number of available switching patterns is 2^{200}. Hence, finding the optimum switching pattern for a desired antenna pattern can be a formidable task.

Traditionally, statistical density tapering has been used to determine which elements are turned off or on. A desired amplitude taper, normalized between 0 and 1, is used in conjunction with a random number generator to compute the probability of an element being turned on. This method generates switching patterns, which, in turn, create antenna patterns with a certain probability of having the desired sidelobe level.

Another method for computing switching patterns is called *simulated annealing*.[16,17] This algorithm models the antenna array as a liquid undergoing crystallization. An optimum solution is obtained when the "liquid" is in its crystalline or lowest energy state.

Array thinning is well suited to GA optimization because the switching pattern is readily encoded into a binary chromosome. *Off* is 0, and *On* is 1. Whereas an unthinned array has a maximum relative sidelobe level of –13 dB, the GA has achieved levels of –22.09 dB for an array with isotropic element patterns and –23.69 dB for an array with sin(ϕ) element patterns. The percentage of *on* elements for each case is 77 and 75%, respectively.

Cases of a steered antenna beam have been analyzed. For a beam steered 30° off of normal, the maximum relative sidelobe level achieved is –18.8 dB, using directional elements.

The GA has been used for nulling out an interfering signal over a small angular extent, while minimizing overall sidelobe levels. Nulling levels of –26.8 dB have been achieved for an angular extent of 2.9° and a 10% bandwidth, while maintaining overall maximum sidelobe levels of –16.4 dB.

Sidelobe levels have been optimized for a 20- × 10-element planar array of isotropic elements in a square lattice. For the principal plane cut along the broad dimension of the array, a –22.17 dB maximum relative sidelobe level has been achieved with 54% of the elements turned on. In the orthogonal plane, a sidelobe level of –12.93 dB has been achieved with 58% of the elements turned on.

9.5.2 Pattern Synthesis

Most antenna designs have a desired radiation pattern, for which techniques exist to determine the requisite antenna element excitations. However, optimum element excitations may be difficult to realize because of mutual-coupling effects and network constraints. For example, an antenna element-current may be set to ten times the current of its nearest neighbor. GAs have been used to search among possible element excitation sets for the excitation that yields the lowest maximum-current ratio, $|I_n/I_{n\pm1}|_{max}$, where the currents $I_{n\pm1}$ are adjacent to current I_n.

To elaborate, the array factor of a linear array may be expressed as a polynomial

$$f(w) = \sum_{n=0}^{N} \left(\frac{I_n}{I_N} \right) w^n = \prod_{n=1}^{N} (w - w_n) \tag{9.19}$$

where $N + 1$ is the number of array elements, $w = e^{jkd\cos\theta}$, k is the wave number, and d is the interelement distance.[18] The roots of the polynomial, w_n, may be further broken down as

$$w_n = e^{a_n + jb_n} \tag{9.20}$$

where b_n corresponds to the locations of radiation pattern nulls on the Schelkunoff unit circle and a_n controls the null depth. Setting $a_n = 0$ creates a null as opposed to a local minimum. Replacing w_n by

$$w'_n = e^{-a_n + jb_n} \tag{9.21}$$

does not alter the shape of the power pattern $|f(w)|^2$, but it does affect the individual element currents, when $a_n \neq 0$.

Thus, an antenna array comprised of $N + 1$ elements creates an antenna pattern with N nulls, the depth of which can be controlled by adjusting a_n. If there are m nonzero pattern nulls, then there are 2^m array factors to choose from, all of which create the same power pattern. The GA has been used to search this vast solution space for the array factor that produces the most realizable current excitations. The nonzero a_n's are encoded into a binary chromosome by using a 1 to represent a positive a_n and a 0 to represent a negative a_n.

Ares-Pena et al.[18] have used GAs to obtain array factors yielding substantial improvements in $|I_n/I_{n\pm1}|_{max}$ for a 100-element linear array, based on an initial Orchard solution.

9.5.3 Loaded Antennas

GAs have been used in conjunction with various MOM codes to optimize wire antennas interspersed with lumped-element resistance-inductance-capacitance (RLC) circuits.[19] These loads serve to alter the current distribution along the radiating element in a frequency-dependent manner, thereby increasing the bandwidth and other performance characteristics of the antenna structure. The load locations and impedance values are combined with matching-network parameters into a binary chromosome.

A method has been devised whereby the admittance matrix of the unloaded antenna is computed, only once. For each prospective loaded design, the admittances of the loads are combined with the admittances of the unloaded antenna at each load location to yield the loaded current distribution, which is, in turn, used to evaluate the performance of the antenna. Load locations, impedance values, and circuit parameters for the antenna matching network are all combined into a binary chromosome for each individual design.

An example of an optimized antenna design using this technique is the biconical antenna operating in the frequency range from 30 to 300 MHz. The unloaded design has a pattern null of –35 dBi at the horizon at 200 MHz. The optimized, loaded design eliminates the pattern null and achieves a horizon gain of –2 to 0 dBi across the band. Similarly, a loaded monopole design over the same bandwidth eliminates pattern nulls at 140 and 275 MHz, to achieve a gain of –1 to +4 dBi across the band at the horizon. Optimized loading has also been used on a 6 λ dipole to shift the pattern maximum toward the horizon.

9.5.4 Crooked-Wire Genetic Antenna

A remarkable antenna design is the crooked-wire genetic antenna designed by Altshuler and Linden.[20] This antenna, which resembles a seven-segment, three-dimensional, randomly twisted paper clip, has achieved hemispheric, right-hand circular polarization coverage with less than 4 dB of gain variation for elevation angles greater than 10° above the horizon. The bandwidth of the antenna is 1300 to 1900 MHz. Other crooked-wire antennas have been optimized for 1225 to 1625 MHz.

The design parameters for the crooked-wire antenna are the three-dimensional starting and ending points for each wire segment, constrained within a certain volume. Both binary and real-valued GAs have been used with success.

9.5.5 Wideband/Dual-Band Patch Antenna Design

The genetic algorithm has been used in conjunction with the direct matrix manipulation (DMM) method to optimize the bandwidth of patch antennas.[21] To begin with, a patch antenna is subdivided into a checkerboard grid of rectangular elements. Each of these rectangular elements is further subdivided into a pair of triangular patches. The current density function for each pair of triangular patches constitutes a Rao-Wilton-Glisson (RWG) basis set.[22] The basis functions for all the antenna subelements are used in a MOM electric field integral equation to compute an N-by-N impedance matrix. This impedance matrix is subsequently inverted and used to compute the current density for the patch antenna for a given voltage excitation. Removal of a rectangular subelement from the patch antenna is simulated by removing the corresponding row and column from the impedance matrix.

The array of subelements is represented in the GA by a binary chromosome, where a bit value of *0* indicates the absence of metalization for the corresponding subelement. In other words, a chromosome string of *1110 1111 1111 1111*, represents a patch antenna with a small rectangular piece of metal removed from the upper right corner. In evaluating the performance of this particular chromosome, the row and column corresponding to the removed piece of metal would be removed from the impedance matrix. The resulting submatrix would be inverted and used to compute the current density of the antenna design. The current density would be used to derive input VSWR or antenna radiation pattern performance. GAs have been used to increase the bandwidth (or to provide a dual bandwidth) over which patch antennas have a good VSWR.

9.6 Advanced Topics

This section introduces two advanced topics for GAs, meta-GA and multiobjective optimization.

9.6.1 Meta-Genetic Algorithms

A meta-GA optimizes the operational parameters of a slave GA. For example, a meta-GA could be used to determine the optimum population size, percentage overlap, and mutation rate for another genetic algorithm that is, in turn, used to optimize a single design problem.

9.6.2 Multiobjective Optimization

The term *multiobjective optimization* refers to a problem where several design objectives are to be met at the same time. For example, an antenna system may be required to have high gain and a large bandwidth, or a narrow beamwidth and low sidelobes. In general, these problems are solved by integrating all performance indicators into a single expression. High gain and large bandwidth can be combined into a single-gain bandwidth product. However, more sophisticated methods exist.[3,9,23]

Suppose there is a population of eight antenna designs, each with a calculated beam width and maximum relative sidelobe level, as shown in Figure 9.14. A narrow beam width and small relative sidelobes are desired. Design 6 has the narrowest beam width, but design 3 has the lowest maximum relative sidelobe levels. Design 7 has lower sidelobes than design 6 and a narrower beam width than

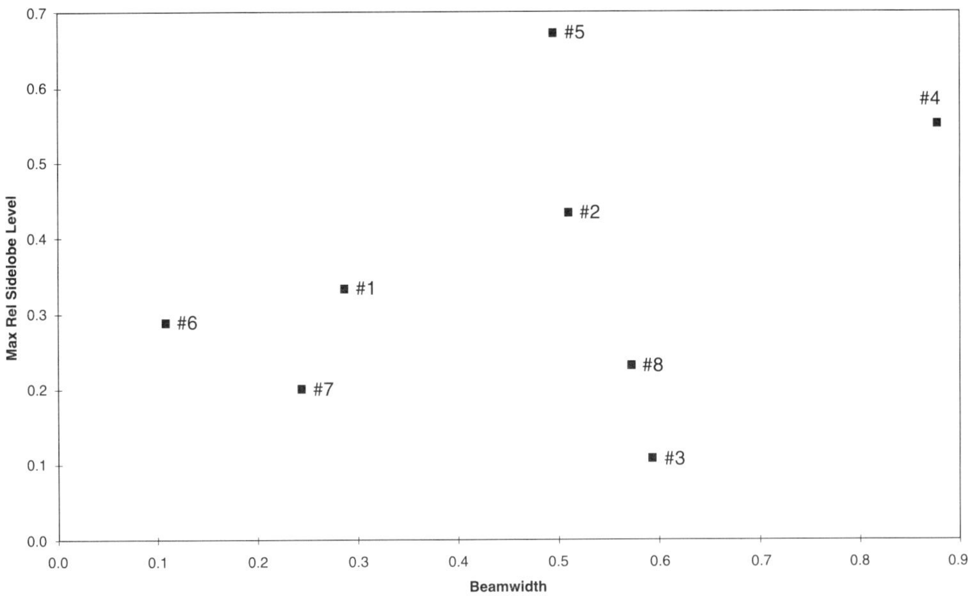

FIGURE 9.14 Multiobjective performance of eight arbitrary antenna designs.

design 3. The performances of the remaining five designs are inferior to those of designs 3, 6, or 7. For example, design 1 has a wider beam width and higher sidelobes than designs 6 and 7. Therefore, design 1 is said to be *dominated* by designs 6 and 7. Designs 3, 6, and 7 are said to be *nondominated*, because there are no designs with performances superior on all design fronts. The set of nondominated designs is called the *Pareto front.*

By using the *Pareto dominance principle,*[23] competing designs are ranked according to the *Pareto rank* in which they lie. This is illustrated in Fig. 9.15. First, the designs in the Pareto front are given the rank of 1 and then temporarily removed from the population. A new Pareto front is formed and given a rank

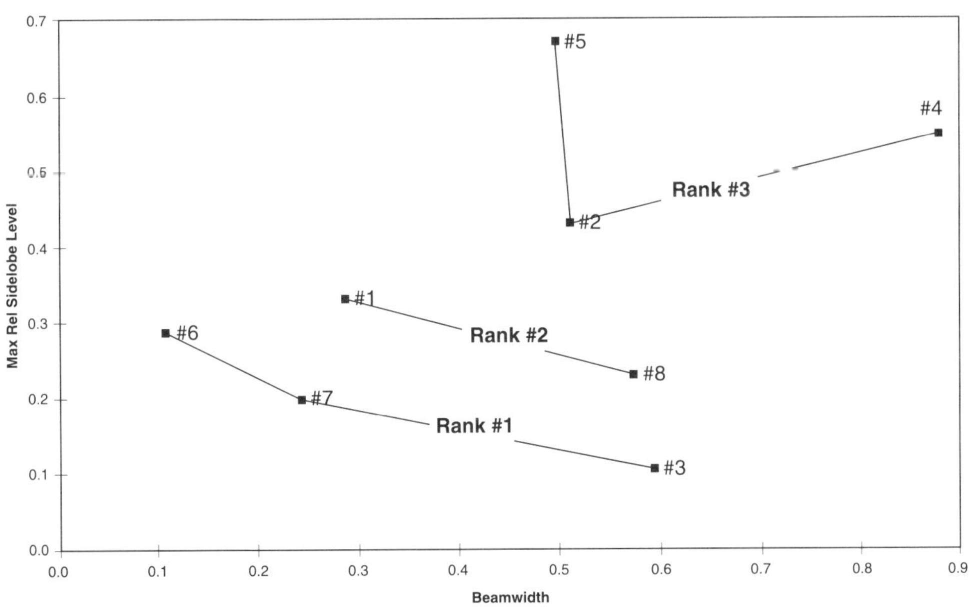

FIGURE 9.15 Nondominance ranking.

TABLE 9.3 Pareto Parameters

	Design			
Rank	m	n	d	$s_{m,n}(d)$
1	6	7	0.16207	0.59482
	6	3	0.51796	0
	7	3	0.36166	0.09585
2	1	8	0.30336	0.24160
3	5	2	0.23823	0.40442
	5	4	0.40096	0
	2	4	0.38610	0.03476

of 2. These designs are removed and the process is repeated until all the designs have been ranked. These rankings could be used to determine which designs are to be blended to form offspring designs for the next generation; however, the algorithm would then be prone to prematurely converge on a single point. A method of promoting population diversity to allow the algorithm to continue exploring new designs is called *sharing* or *niching*.

Niching and sharing occur in nature when two or more designs are competing for the same niche. For example, suppose there are a number of rodents competing for a limited food supply. As more and more rodents are able to access the food, the amount of food available to each rodent is reduced because of sharing.

The impact of sharing on a design performance is taken into account by first computing the proximity of each design to each of the other current designs in the population. In general, this computation can be performed in either *genotypic* or *phenotypic* space. (Genotypic refers to the design parameters, whereas phenotypic refers to the performance of the design.) However, phenotypic or performance differences are most often used for multiobjective optimization, because the Pareto ranking, which occurs before sharing, is usually based on performance.

Once the relative proximities of each design have been computed, the performance values for each design are scaled down, accordingly. Design performances that are similar to one another are scaled down more than isolated performances. In the multiobjective-optimization problem, sharing only occurs among the designs in a common rank (i.e., designs with a Pareto ranking of 1 are only required to share their performance values with other rank 1 designs).

Consider the population of eight antenna designs represented in Fig. 9.14. As shown in Fig. 9.15, the population may be subdivided into three Pareto ranks or niches. For each rank, a series of performance differences may be computed using the following equation:

$$d(Design_m, Design_n) = \sqrt{w_{SLL}\left(SLL_m - SLL_n\right)^2 + w_{BW}\left(BW_m - BW_n\right)^2} \tag{9.22}$$

where $d(Design_m, Design_n)$ is the difference between the performance of $Design_m$ and $Design_n$, SLL_m is the maximum relative sidelobe level of $Design_m$, BW_m is the antenna beam width of $Design_m$, and w_{SLL} and w_{BW} are positive weighting coefficients. By assuming w_{SLL} and w_{BW} are equal to 1, the performance differences for the designs in Fig. 9.14 are tabulated in Table 9.3. Based on these differences, a *niching function* $s_{m,n}(d)$ may be defined as

$$s_{m,n}(d) = \begin{cases} 1 - \left(\dfrac{d}{\sigma}\right)^p & d \le \sigma \\ 0 & d > \sigma \end{cases} \tag{9.23}$$

where σ is a so-called niche radius and p is an exponent greater than 1. By assuming $\sigma = 0.4$ and $p = 1$, the niching function values are tabulated in Table 9.3.

TABLE 9.4 Example Niche Counts and Performance

Rank	Design	Niche Count	Performance
1	6	1.5948	0.6270
	7	1.6907	0.5915
	3	1.0959	0.9125
2	1	1.2416	0.4764
	8	1.2416	0.4764
3	5	1.4044	0.3392
	2	1.4392	0.3310
	4	1.0348	0.4604

Next, all the niching function values for each design must be summed up into a *niche count*:

$$niche\_count_m = \sum_{n=1}^{Num\_Designs(rank)} s_{m,n} \tag{9.24}$$

Note that, because

$$s_{m,n}(d) = 1\Big|_{m=n} \tag{9.25}$$

$niche\_count_m \geq 1$ for all m. Example niche counts are shown in Table 9.4.

The performance values for each of the designs of Pareto rank 1 are simply the reciprocal of the niche counts for each design. However, the values for rank 2 are set to the quotient of the worst adjusted performance value from rank 1 divided by the respective niche counts of the rank 2 designs. Otherwise, the performance values for rank 2 designs would exceed those of rank 1. Similarly, rank 3 performances are computed using the worst value from rank 2, and so on. Example performances are tabulated in Table 9.4. Note that, whereas design 7 appears to have the best performance (see Fig. 9.14), it is given the lowest performance value of all the rank 1 designs. In this way, the algorithm is encouraged to continue seeking designs that are away from the superior performer.

Finally, the performance values are used to determine survival probabilities in a roulette wheel selection scheme. Ranking or tournament selection schemes are *not* recommended for multiobjective optimization.

9.7 Conclusion

Various types of GA have been discussed in this chapter. In solving a problem, one must first decide what the cost function is. How is each design chromosome to be evaluated? For the example of the nonuniformly spaced, linear antenna array depicted in Fig. 9.4, Eq. (9.3) is used to compute the antenna pattern array factor for each prospective design. Then, a search algorithm is used to find the maximum relative sidelobe level for each array factor.

Next, one must decide whether to use a binary or a continuous-parameter chromosomes. Binary chromosomes are easy to work with, once the binary encoding and decoding subroutines have been worked out. If the binary route has been chosen, one must decide how many bits to allocate to each design parameter. For the antenna array, mentioned earlier, each spacing can be encoded into a 3-b binary value. There are a total of 24 unique spacings, so that, when all the 3-b words are concatenated into a string, there are a total of 72 b per chromosome.

If designers want to use a canned GA driver, such as the code written by David Carroll,[13] they should replace the generic cost function with the cost function for their designs. A flexible GA driver should allow the antenna designer to arbitrarily pick the number of bits to allocate for each design parameter. Otherwise, a design-specific GA can be written.

The first step in designing one's own GA is to decide how the design parameters are to be stored. From a software design standpoint, a rigorous implementation is to create a *record* for each design chromosome. Each record contains variables for performance, binary and floating-point representations of the various design parameters, and perhaps a Boolean variable indicating whether the design performance needs to be evaluated. In the following code segment, a record of type *Gene* has been defined to contain binary and floating point representations of a design parameter. At the next level, a record of type *Chromosome* contains all the genes for a chromosome, plus an integer string to contain the binary representation of the chromosome. Finally, type *Individual_Type* contains a chromosome and a performance evaluation or cost for that chromosome.

```
TYPE Gene
  INTEGER Bits(1:Max_Number_of_Bits_per_Gene)
  REAL(KIND = 8) Value
END TYPE
!
TYPE Chromosome
  TYPE (Gene) Genes(1:Max_Number_of_Genes_per_Chromosome)
  INTEGER String(1:Max_Number_of_Bits_per_Gene * &
      Max_Number_of_Genes_per_Chromosome)
END TYPE
!
TYPE Individual_Type
  TYPE (Chromosome) Chrom
  REAL Cost
END TYPE
```

The previously mentioned software structure can be simplified into a single record type or avoided altogether, by using separate array variables for binary, floating-point, and Boolean constituents of each design chromosome. Using array variables is certainly more efficient, but is also more confusing.

The next step in software coding one's own GA is to write a routine for initializing new chromosomes using the pseudorandom number generator. One should be careful in seeding the random number generator. If the seeding occurs too often, the output will no longer be random. A good idea is to seed the random number generator, using the system clock, at the beginning of each program execution.

A top level design, repeated from Section 9.4, is shown next. Once the data types have been defined and the population initialized, the subroutines *Evaluate_Cost* and *Sort_Population_by_Cost* should be rather straightforward. The subroutine, *Mate_Best_Half_of_Population* assumes that a steady-state algorithm is being used (see Section 9.3.3.3.) The mating process, discussed in Section 9.3.2.3, consists of pairing off each of the chromosomes in the best-performing half of the population, using a random number to determine the crossover point for each pair of parents, and exchanging bits on either side of the crossover point to form new offspring designs.

```
CALL Initialize_Population()
CALL Evaluate_Cost()
CALL Sort_Population_by_Cost()
DO Generation_Number = 1, Number_of_Iterations
     CALL Mate_Best_Half_of_Population()
     CALL Mutate_Random_Bit()
     CALL Evaluate_Cost()
     CALL Sort_Population_by_Cost()
END DO
CALL Write_Best_Genes_to_File()
```

The *Number_of_Iterations* should be nominally set to 100, to start. The population size should be about ten times the number of bits per chromosome. For the 72-b chromosome of the nonuniform antenna array example, the population size would be about 700. Note that the initial population size can be set to 1400, to give the algorithm a good start, and reduced to 700 after the first generation. Figure 9.12 shows the typical exponential performance curve of a GA. Once the GA has been debugged, the various control parameters, such as number of bits per parameter and population size, may be adjusted for better convergence.

For the interested reader, an excellent bibliography is provided by Johnson and Rahmat-Samii.[24] Haupt and Haupt's[7] *Practical Genetic Algorithms* is a good introduction to GAs for engineering, while Goldberg's[3] *Genetic Algorithms Search, Optimization and Machine Learning* is a classical presentation of the subject. Finally, Rahmat-Samii and Michielssen's[19] *Electromagnetic Optimization by Genetic Algorithms* is a compilation of various GA applications to antenna engineering problems.

9.8 Appendix: A Cost Function Code for a David Carroll Genetic Algorithm Driver

```
subroutine func(j,funcval)
implicit real*8 (a-h,o-z)
INTEGER Index, Gene_Number, m, Index_1, Index_2, Index_of_Minimum
INTEGER, PARAMETER:: Number_of_Antenna_Pattern_Points = 901
REAL(KIND = 8) Array_Factors(0:Number_of_Antenna_Pattern_Points - 1),
      Sorted_Array_Factors(0:Number_of_Antenna_Pattern_Points - 1),
      Angle, Pi, k, Sum, Sum_dm, d1, Peak_Array_Factor,
      Maximum_Relative_Sidelobe_Level, Fitness, Temp
include 'params.f'
dimension parent(nparmax,indmax)
dimension iparent(nchrmax,indmax)
common/ga2/nparam,nchrome
common/ga3/parent,iparent
Pi = DATAN2 (0.0d0, -1.0d0)
k = 2 * Pi
d1 = parent(1,j)
Sum = 0.0d0
DO Gene_Number = 1, nparam
  Sum_dm = 0.0d0
  DO m = 1, Gene_Number
      Sum_dm = Sum_dm + parent(Gene_Number, j)
  END DO
  Sum = Sum + DCOS(k*(Sum_dm - d1/2) * DCOSD(90.0d0))
END DO
Peak_Array_Factor = 2 * DSIND(90.0d0) * Sum
DO Index = 0, Number_of_Antenna_Pattern_Points - 1
  Angle = 90.0d0 * Index/(Number_of_Antenna_Pattern_Points - 1)
  Sum = 0.0d0
  DO Gene_Number = 1, nparam
      Sum_dm = 0.0d0
      DO m = 1, Gene_Number
          Sum_dm = Sum_dm + parent(m, j)
      END DO
      Sum = Sum + DCOS(k*(Sum_dm - d1/2) * DCOSD(Angle))
  END DO
  Array_Factors(Index) = DABS(2 * DSIND(Angle) * Sum/Peak_Array_Factor)
END DO
```

```
Sorted_Array_Factors = Array_Factors
DO Index_1 = 0, Number_of_Antenna_Pattern_Points - 2
  Index_of_Minimum = Index_1
  DO Index_2 = Index_1, Number_of_Antenna_Pattern_Points - 1
      IF (Sorted_Array_Factors(Index_2) .LT.
      Sorted_Array_Factors(Index_of_Minimum)) THEN
      Index_of_Minimum = Index_2
      END IF
  END DO
  Temp = Sorted_Array_Factors(Index_of_Minimum)
  Sorted_Array_Factors(Index_of_Minimum) = Sorted_Array_Factors(Index_1)
  Sorted_Array_Factors(Index_1) = Temp
END DO
DO Index = Number_of_Antenna_Pattern_Points - 1, 0, -1
  IF (DABS(Array_Factors(Index) - Sorted_Array_Factors(Index)) .GT. 1.0d0
    * 10**(-10)) THEN
      Maximum_Relative_Sidelobe_Level = Sorted_Array_Factors(Index)
      EXIT
  END IF
END DO
Fitness = 1.0d0 - Maximum_Relative_Sidelobe_Level
funcval = Fitness
RETURN
```

References

1. Holland, J.H., Adaptation in Natural and Artificial Systems, University of Michigan Press, Ann Arbor, MI, 1975.
2. Holland, J.H., Genetic Algorithms, *Scientific American*, 7, 66, 1992.
3. Goldberg, D.E., *Genetic Algorithms in Search, Optimization and Machine Learning,* New York, Addison Wesley Longman, 1989.
4. *Electromagnetic Optimization by Genetic Algorithms* (Y. Rahmat-Samii and E. Michielssen, eds.), John Wiley & Sons, New York, 1999.
5. Mittra, Raj, "Challenging Issues in Antenna Design for Personal Communication Systems," Proceedings of 10emes Journees Internationales de Nice sur les Antennes International Symposium on Antennas, November 17-19, 1998, Nice, France, pp. 609–618.
6. Burke, G.J. and Poggio, A.J., *Numerical Electromagnetics Code (NEC)-Method of Moments.* Lawrence Livermore Laboratory, Livermore, CA, 1981.
7. Haupt, R.L. and Haupt, S.E., *Practical Genetic Algorithms,* New York, John Wiley & Sons, 1998.
8. Haupt, R.L., "An Introduction to Genetic Algorithms for Electromagnetics," *IEEE Antennas and Propagation Magazine,* 37, 7, 1995.
9. Weile, D.S. and Michielssen, E., Genetic Algorithms: Theory and Advanced Techniques. In *Electromagnetic Optimization by Genetic Algorithms* (Y. Rahmat-Samii and E. Michielssen, eds.), John Wiley & Sons, New York, 1999.
10. Adewuya, A., New methods in genetic search with real-valued chromosomes. M. S. Thesis, Massachusetts Institute of Technology, Cambridge, MA, 1997.
11. Johnson, J. M. and Rahmat-Samii, Y., An Introduction to Genetic Algorithms. In *Electromagnetic Optimization by Genetic Algorithms* (Yahya Rahmat-Samii and Eric Michielssen, eds.), John Wiley & Sons, New York, 1999.
12. Michalewicz, Z., *Genetic Algorithms + Data Structures = Evolution Programs,* New York: Springer-Verlag, 1992.
13. Carroll, D.L., *FORTRAN Genetic Algorithm Driver, Version 1.7,* Download from: <http://www.staff.uiuc.edu/~carroll/ga.html>, 12/11/98.

14. Linden, D.S., Rules of Thumb for GA Parameters, GA Monitoring, GA Parameter Optimization, and Enhancement of GA Efficiency. In *Electromagnetic Optimization by Genetic Algorithms* (Y. Rahmat-Samii and E. Michielssen, eds.), John Wiley & Sons, New York, 1999.
15. Haupt, R.L., Thinned Arrays Using Genetic Algorithms. In *Electromagnetic Optimization by Genetic Algorithms* (Y. Rahmat-Samii and E. Michielssen, eds.), John Wiley & Sons, New York, 1999.
16. Haupt, R.L. and Haupt, S.E., *Practical Genetic Algorithms,* New York, John Wiley & Sons, 1998, 16.
17. Ares-Pena, F., Application of Genetic Algorithms and Simulated Annealing to Some Antenna Problems. In *Electromagnetic Optimization by Genetic Algorithms* (Y. Rahmat-Samii and E. Michielssen, eds.), John Wiley & Sons, New York, 1999.
18. Ares-Pena, F., et al., Application of Genetic Algorithms in Optimising the Aperture Distributions of Antenna Array Patterns, *Electronics Letters,* 32(3), 148, 1996.
19. Altman, Z. et al., Application of the Genetic Algorithm to Broadband Antenna Design. In *Electromagnetic Optimization by Genetic Algorithms* (Y. Rahmat-Samii and E. Michielssen, eds.), John Wiley & Sons, New York, 1999.
20. Altshuler, E.E. and Linden, D.S., Design of Wire Antennas Using Genetic Algorithms. In *Electromagnetic Optimization by Genetic Algorithms* (Y. Rahmat-Samii and E. Michielssen, eds.), John Wiley & Sons, New York, 1999.
21. Johnson, J.M. and Rahmat-Samii, Y., Evolutionary Designs of Integrated Antennas Using Genetic Algorithms and Method of Moments (GA/MoM). In *Electromagnetic Optimization by Genetic Algorithms* (Y. Rahmat-Samii and E. Michielssen, eds.), John Wiley & Sons, New York, 1999.
22. Rao, S.M., Wilton, D.R., and Glisson, A.W., "Electromagnetic Scattering by Surfaces of Arbitrary Shape," *IEEE Transactions Antennas and Propagation,* 30, 3, 1982.
23. Bristeau, Marie-Odile et al., Genetic Algorithms for Electromagnetic Backscattering: Multiobjective Optimization. In *Electromagnetic Optimization by Genetic Algorithms* (Y. Rahmat-Samii and E. Michielssen, eds.), John Wiley & Sons, New York, 1999.
24. Johnson, J.M. and Rahmat-Samii, Y., "Genetic Algorithms in Engineering Electromagnetics," *IEEE Antennas and Propagation Magazine,* 39, 4, 1997.

10

High-Frequency Techniques

Roberto G. Rojas
The Ohio State University

Teh-Hong Lee
The Ohio State University

10.1 Introduction

A variety of numerical, asymptotic, and hybrid techniques are available for analyzing complex electromagnetic radiation, propagation, and scattering problems. Although numerical techniques such as the method of moments (MOM), finite difference time domain (FDTD) methods, finite element method (FEM), and others can be used to analyze complex geometrics, they become less efficient as the electrical size of the computational domain increases. On the other hand, high-frequency (HF) techniques, which are based on the asymptotic solution of Maxwell's equations, become more efficient and accurate as the electrical size of the structure under study increases. An even more important feature of HF techniques is that they provide physical insight into the dominant radiation and scattering mechanisms. This makes them a very powerful diagnostic and design tool for antenna engineers as well engineers who design wireless communications systems. Note that HF electromagnetic (EM) techniques in conjunction with ray tracing tools are particularly useful in the wireless communications industry where the understanding of EM propagation in urban environments is important. Although this chapter deals with HF techniques, it is important to emphasize that the design and analysis of very complex EM systems usually requires a combination of various techniques. Thus, it is imperative that engineers become familiar with all the tools available to them so that they are aware of the most appropriate technique for each particular problem.

This chapter summarizes several HF solutions currently available. Unfortunately, not all the methods can be described in detail because of space limitations; however, an extensive list of references at the end of the chapter provides the reader with additional sources of information about most of the well-known

0-8493-0124-6/02/$0.00+$1.50

HF techniques. In particular, four books,[3,4,17,41] two book chapters,[24,55] as well as a collection of articles[15] are good sources of additional information.

This chapter is organized as follows: first, the modern geometric optics (GO) ray technique is introduced in Section 10.2. Although this method is sufficient for some applications at optical frequencies, it does have severe deficiencies at radio frequencies (RF) and microwave frequencies. To overcome its deficiencies, the geometric theory of diffraction (GTD) was originally developed by Keller and collaborators[19,20] around 1951. The GTD and its uniform version, commonly referred to as the uniform geometric theory of diffraction (UTD), are discussed in Section 10.3. UTD solutions are presented for edge diffraction as well as for convex surfaces.

Traditionally, most of the early GTD solutions were for perfect electric conducting (PEC) surfaces, which are good models for metallic conductors. However, the effect of materials (penetrable and impenetrable) is important in many applications. Thus, Section 10.4 discusses UTD solutions valid for nonmetallic surfaces and includes a discussion on the boundary conditions that have been developed to represent nonmetallic surfaces. Note that all the previously mentioned techniques use rays to calculate the fields. However, there is an alternative way to calculate the fields using surface currents. The original technique based on this concept is physical optics (PO), where approximate HF currents are used to obtain the fields. To improve the accuracy of PO, the physical theory of diffraction (PTD) was developed in the former Soviet Union by Ufimtsev.[76-78] PTD improves the accuracy of PO in the same manner as GTD improves the accuracy of GO. The PO and PTD methods are discussed in Sections 10.5 and 10.6, respectively. Another HF technique that is frequently used is the equivalent current method (ECM). This method, discussed in Section 10.6, is used when the GTD/UTD ray techniques fail. A few examples applicable to a reflector antenna are given in Section 10.7 to illustrate the use of the techniques presented. Finally, some concluding remarks are given in Section 10.8, including a brief list of other HF techniques not considered in this chapter. This is done for completeness and to provide the reader with information about most of the well-known techniques. (*Note*: It is assumed here that the fields are time harmonic with an $e^{j\omega t}$ time dependence that is suppressed.)

10.2 Modern Geometric Optics

Geometric optics is an HF approximation of Maxwell's equations that employs rays to describe EM field propagation. According to GO, HF fields propagate along ray paths that satisfy Fermat's principle and are orthogonal to the wavefronts in an isotropic homogeneous medium. Note that Fermat's principle states that the ray trajectory is such that the optical path length is stationary. This normally implies that the path length must be a minimum, subject to some constrains. Although there are several ways to rigorously develop the GO field representation, the Luneberg–Kline series expansion[21,36] is used here. This is done to clearly show that the GO field is simply an asymptotic approximation to Maxwell's equations.

The first step is to expand the electric field $\mathbf{E}(\mathbf{r}, \omega)$ in terms of the series

$$\mathbf{E}(\mathbf{r},\omega) \sim e^{-jk\Psi(\mathbf{r})} \sum_{n=0}^{\infty} \frac{\mathbf{E}_n(\mathbf{r})}{(j\omega)^n} \tag{10.1}$$

where ω is the angular frequency, $\mathbf{r}$ is the position vector, k is the wavenumber of the homogeneous isotropic medium, and $\psi(\mathbf{r})$ is the phase function. Note that the preceding series is in inverse powers (integers) of ω. As the frequency becomes large, we only need to keep the leading ($n = 0$) term that is referred to as the GO field, namely,

$$E(\mathbf{r},\omega) \sim \mathbf{E}_o(\mathbf{r}) e^{-jk\Psi(\mathbf{r})} \qquad \omega \to \infty \tag{10.2}$$

Substituting Eq. (10.1) into the source-free Maxwell's equations yields the well-known eikonal equation

$$|\nabla\Psi|^2 = 1 \tag{10.3}$$

as well as the 0th order transport equation

$$\left(2(\nabla\Psi\cdot\nabla)+\nabla^2\Psi\right)\mathbf{E}_o = 0 \tag{10.4}$$

By letting $\hat{s} = \nabla\psi$, Eq. (10.4) can be rewritten as

$$\left(2\frac{d}{ds}+\nabla^2\Psi\right)\mathbf{E}_o = 0 \tag{10.5}$$

Note that $\hat{s}$ is perpendicular to the equiphase surface $\psi = constant$ and also turns out to be the direction of energy flow. Solving for these equations, we obtain the GO electric field

$$\mathbf{E}(s) = \mathbf{E}(0)\sqrt{\frac{\rho_1\rho_2}{(\rho_1+s)(\rho_2+s)}}\,e^{-jks} \tag{10.6}$$

and magnetic field

$$\mathbf{H}(s) = \hat{s}\times\frac{\mathbf{E}(s)}{\eta} \tag{10.7}$$

at a distance s from the reference location $s = 0$ where η is the intrinsic impedance of the medium. If the medium becomes free space, η and k become η_o and k_o, respectively. The parameters ρ_1 and ρ_2 are the principal radii of curvature of the wave front at the reference location $s = 0$ as depicted in Fig. 10.1 where a thin tube (or pencil) of rays is depicted along the ray propagation direction $\hat{s}$. This direction is straight in a homogeneous isotropic medium. The distance s is positive in the direction of wave propagation and negative in the opposite direction.

It can also be shown that

$$\hat{s}\cdot\mathbf{E} = \hat{s}\cdot\mathbf{H} = 0 \tag{10.8}$$

which implies that the GO fields do behave (locally) as plane waves.

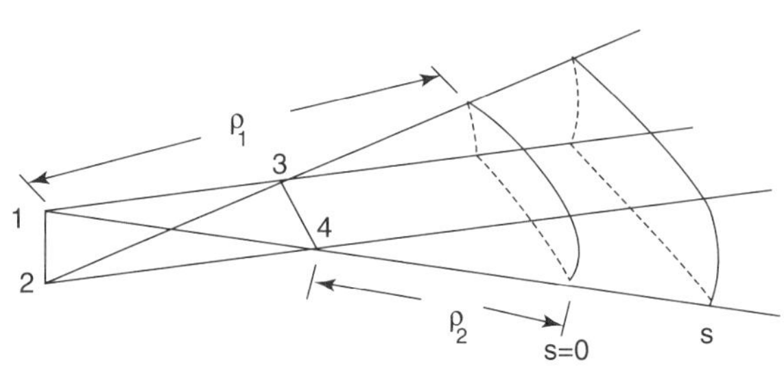

FIGURE 10.1 Astigmatic tube of rays. (After Kouyoumjian and Pathak [23], ©1974 IEEE.)

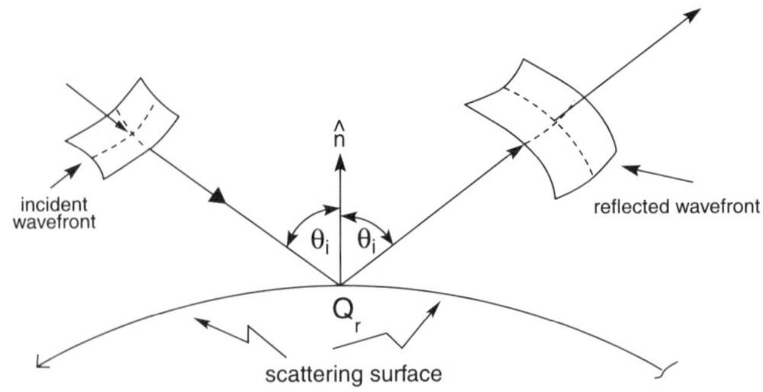

FIGURE 10.2 GO incident and reflected fields.

Figure 10.1 is useful in providing a geometric interpretation of the GO field. For example, the square root expression in Eq. (10.6) states that the cross-sectional area of the tube of rays ensures that the power within the tube is conserved in a lossless medium. It is important to note that the expression for the GO field fails at the location where $s = -\rho_1$ or $s = -\rho_2$. These locations correspond to the region where the cross-sectional area of the ray tube vanishes. Locations 1-2 and 3-4 in Fig. 10.1 correspond to the centers of curvature of the wave front (in two orthogonal directions) at the reference location $s = 0$. The ray tubes merge at these locations that are referred to as caustics or focal lines. The radii ρ_1 and ρ_2 are >0 (<0) if the locations 1-2 and 3-4 are located behind (in front of) the reference point $s = 0$.

The general expression for the astigmatic ray tube given in Eq. (10.6) reduces to well-known solutions. If both radii of curvature become infinitely large ($\rho_{1,2} \to \infty$, Eq. (10.6) becomes a plane wave. If $\rho_1 \to \infty$ while ρ_2 remains finite, Eq. (10.6) reduces to a cylindirical wave. To write Eq. (10.6) as a standard cylindrical wave, let us move the reference point $s = 0$ to the caustic line location ($s = -\rho_2$). This change of reference point location reduces Eq. (10.6) to

$$\mathbf{E}(\mathbf{r}) = \mathbf{A}^s \frac{e^{-jks}}{\sqrt{s}} \tag{10.9}$$

where $\mathbf{A}^s = \lim_{\rho 2 \to 0} \sqrt{\rho_2}\mathbf{E}(0)$ is finite and related to the strength of the line source generating the cyclindrical wave. A spherical wave is obtained by letting $\rho_1 = \rho_2 = \rho^i$. Moving the reference point to the source location (point caustic), namely, $s = -\rho^i$, yields

$$\mathbf{E}(\mathbf{r}) = \mathbf{B}^s \frac{e^{-jks}}{s} \tag{10.10}$$

where $\mathbf{B}^s = \lim_{\rho i \to 0} \rho^i \mathbf{E}(0)$ is also finite and related to the strength of the point source.

Now that we have introduced the GO field, we can consider the situation where this field is incident on an electrically large, smooth structure as depicted in Fig. 10.2. It is well known that the incident field is reflected and if the structure is penetrable, there could also be a transmitted field. This discussion initially concentrates on the reflected field. In Section 10.4 we consider the transmitted field.

By assuming that the incident GO field is given in Eq. (10.6), then the reflected GO field can be written as follows:

$$\mathbf{E}^r(s^r) = \mathbf{E}^i(Q_r) \cdot \overline{\overline{R}} \sqrt{\frac{\rho_1^r \rho_2^r}{(\rho_1^r + s^r)(\rho_2^r + s^r)}}\, e^{-jks^r} \tag{10.11}$$

where $\overline{\overline{\mathbf{R}}}$ is the dyadic (tensor, rank 2) reflection coefficient, $\mathbf{E}^i(Q_r)$ is the incident field at the point of reflection Q_r, and the reflected field is evaluated at a distance s^r from the point of reflection. The magnetic

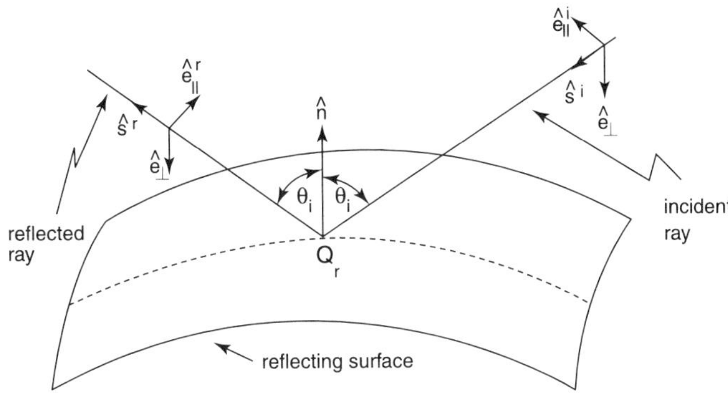

FIGURE 10.3 Ray-fixed coordinate system for GO fields.

field can be obtained from Eq. (10.7) by replacing E^i by E^r and $\hat{s}$ by $\hat{s}^r$ where $\hat{s}^r$ is the direction of propagation of the reflected field. Note that the reflected GO field has the same form as the incident GO field, except that its radii of curvature is dependent on the radii of curvature of the surface and the incident field wave front in the neighborhood of Q_r as well as the direction of incidence. Because Eq. (10.11) is similar to Eq. (10.6), the reflected GO field also fails at the caustics. The radii of curvature $\rho^r_{1,2}$ can be written as follows:

$$\frac{1}{\rho^r_{1,2}} = \frac{1}{2}\left(\frac{1}{\rho^i_1} + \frac{1}{\rho^i_2}\right) + \frac{1}{F_{1,2}} \tag{10.12}$$

where ρ^i_1 and ρ^i_s are the radii of curvature of the incident field at the point of reflection Q_r and the parameter $F_{1,2}$ is a fairly complex expression given in Reference 23.

The dyadic reflection coefficient can be significantly simplified if the fields are expressed in the proper coordinate system. This system, which applies to the incident and reflected (and transmitted if they exist) fields, is fixed to the incident and reflected rays as shown in Fig. 10.3. The unit vectors $\hat{e}^{i,r}_{||}$ are parallel to the plane of incidence (plane spanned by $\hat{n}$ and $\hat{s}^i$), while $\hat{e}_\perp$ is perpendicular to the same plane.

By using the preceding coordinate system, the reflected field can be written in terms of two components perpendicular to the direction of propagation, namely,

$$\begin{bmatrix} E^r_{||}(s) \\ E^r_{\perp}(s) \end{bmatrix} = \begin{bmatrix} R_{||\,||} & R_{||\,\perp} \\ R_{\perp\,||} & R_{\perp\,\perp} \end{bmatrix} \begin{bmatrix} E^i_{||}(Q_r) \\ E^i_{\perp}(Q_r) \end{bmatrix} \sqrt{\frac{\rho^r_1 \rho^r_2}{(\rho^r_1 + s^r)(\rho^r_2 + s^r)}}\, e^{-jks^r} \tag{10.13}$$

where the incident field is given by $\mathbf{E}^i = \hat{e}^i_{||}E^i_{||} + \hat{e}_\perp E^i_\perp$ and the reflection coefficients depend on the material properties of the reflecting surface. For the case of a PEC, it reduces to well-known simple expressions ($R_{||\,||} = 1$, $R_{||\,\perp} = 0 = R_{\perp\,||}$, $R_{\perp\,\perp} = -1$).

The two fields we defined earlier can now be used to calculate the field in the presence of an impenetrable structure (no transmitted field). Consider the convex structure with an edge as shown in Fig. 10.4. According to geometric optics, the total field at an observation point P, in the presence of the opaque structure, is given by

$$\mathbf{E}^{GO}(P) \sim \mathbf{E}^i(P)U_i + \mathbf{E}^r(P)U_r \tag{10.14}$$

where U_i and U_r are unit step functions (Heaviside). U_i is one if the incident field is not blocked by the structure; otherwise it is zero. Likewise, U_r is one if the reflected field exists at the observations point P; otherwise it is zero. In Fig. 10.4, the incident field becomes discontinuous along the incident shadow

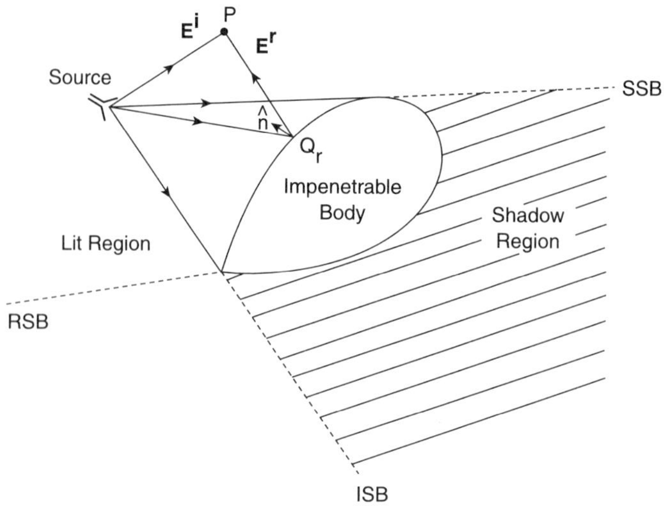

FIGURE 10.4 Convex impenetrable structure with an edge.

boundary (ISB) caused by the knife edge of the structure. As the name implies, the shadow region is the region where the incident field is zero, whereas the lit region is the region where the incident field exists. The reflection shadow boundary (RSB) also originates at the edge of the structure. In the lit side of this boundary, the reflected field exists, whereas in the shadow side there is no reflected field because of the sharp discontinuity in the surface normal at the edge. In addition to the shadow boundaries generated by the edge, there is another shadow boundary as a result of the convex nature of the surface. This shadow boundary, depicted in Fig. 10.4, is referred to as the surface shadow boundary (SSB). At the SSB, where the incident field is at grazing incidence, both the incident and reflected fields become discontinuous.

Based on the preceding description, an analysis based solely on GO fields predicts discontinuous fields along various shadow boundaries. Although this approximation provides the leading terms of an asymptotic solution to Maxwell's equations, which becomes more accurate as the frequency increases, it is not accurate enough at RF and microwave frequencies. The next section discusses the reasons for its failure and introduces diffraction concepts that greatly improve the accuracy of the GO fields.

Before we leave this section, it is important to remember three facts about GO fields. First, the GO fields behave locally as plane waves and propagate along ray paths that satisfy Fermat's principle. Second, the polarization of GO fields is perpendicular to the ray direction. Third, the energy is conserved (lossless medium) in a narrow tube (or pencil) or rays surrounding the central ray.

10.3 Geometric Theory of Diffraction

We ended the previous section by discussing some of the shortcomings of the GO fields, namely, the discontinuous fields at the shadow boundaries and the prediction of zero fields in the shadow region (assuming we have an impenetrable structure). By keeping in mind that the GO fields were obtained from the asymptotic Luneberg–Kline series [Eq. 10.1], we may be tempted to simply keep additional terms in the series expansion to remove the preceding limitations. Adding additional terms may improve the solution (although that is not necessarily guaranteed because asymptotic series are not always convergent); however, it does not remove the deficiencies of the GO field because the Luneberg–Kline series assumes that the field can be expanded in inverse integer powers of ω. In fact, it can be shown that to properly describe the fields scattered by the object of Fig. 10.4, it is necessary to expand the fields in an asymptotic series in inverse fractional powers of ω. It has been shown that the diffracted rays are proportional to $\omega^{-\alpha}$, where α is not an integer. In other words, the Luneberg–Kline series in Eq. (10.1) is not a complete representation of the fields.

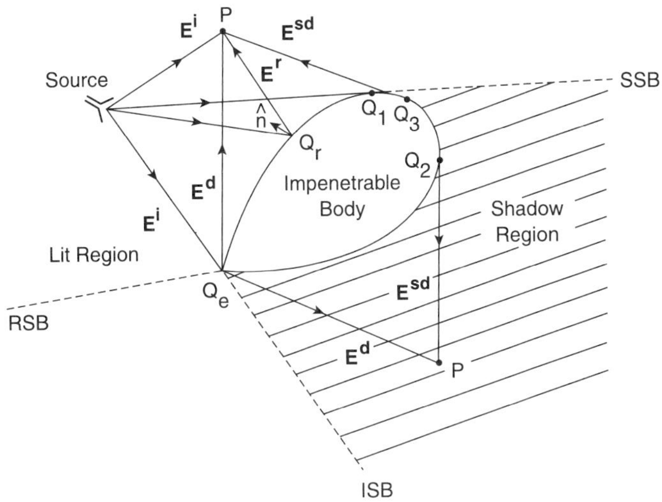

FIGURE 10.5 Ray paths for diffracted fields.

To overcome the deficiencies of the GO fields, a method referred to as GTD was introduced by Keller,[19,20] as previously mentioned. This method introduces diffracted rays that can penetrate the shadow region as depicted in Fig. 10.5.

Note that a ray hitting the edge of the structure can radiate into the shadow region as well as the lit region. Likewise, the incident field at grazing incidence on the convex surface can excite a surface diffracted field, which travels along the surface (following a geodesic path) and penetrates the shadow region.

Although solutions for the scattering of an object with an edge were obtained before Keller's work (Sommerfeld's half-plane solution[72]), the key to Keller's contribution was the interpretation in terms of rays of the various components of the solution. Keller's GTD is based on the following postulates:

1. Diffraction, like reflection and transmission, is a local phenomenon at HFs. It depends on the nature of the surface of the scatterer and the incident field in the neighborhood of the point of diffraction.
2. The diffracted field propagates along rays that are determined by a generalization of Fermat's principle that includes points on the surface of the scatterer in the ray trajectory.
3. The diffracted wave propagates along its ray so that
   a. Power is conserved in a tube of rays surrounding the central ray,
   b. The phase delay along the ray path equals the product of the wave number of the medium and the distance.

Based on the preceding postulates, we can write the general form of a variety of ray diffracted fields (edge diffraction, surface diffraction, vertex diffraction, etc.) in terms of diffraction coefficients. To determine the actual diffraction coefficients we can use postulate 1 (local nature of HF phenomenon) to replace the original complex scatterer with a canonical object whose surface properties match the original scatterer in the neighborhood of the point of diffraction. Much of the research on GTD has concentrated on solving canonical problems to obtain diffraction coefficients in closed form. If it is not possible to obtain a closed form expression for a given canonical structure, numerical techniques can be used to obtain the diffraction coefficient. Once these coefficients are found, the GTD analysis becomes very efficient because it is reduced to a ray-tracing problem where in general, only the dominant rays need to be included. Postulate 1 is very important because it breaks down a complex scattering or radiation problem into a number of simple problems associated with specific parts of the scatterer.

The total GTD field can thus be written as

$$\mathbf{E}(P) \sim \mathbf{E}^{GO}(P) + \mathbf{E}^{d}(P) \tag{10.15}$$

where $\mathbf{E}^d(P)$ is the diffracted field and $\mathbf{E}^{GO}(P)$ is the GO field defined in the previous section. The general form of the diffracted field, away from the point of diffraction, is

$$\mathbf{E}^{d}(s^{d}) = \mathbf{E}^{d}(0)\sqrt{\frac{\rho_1^d \rho_2^d}{(\rho_1^d + s^d)(\rho_2^d + s^d)}}\, e^{-jks^d} \tag{10.16}$$

where $\mathbf{E}^d(0)$ is the diffracted field at the reference point $s^d = 0$. Note that the diffracted field, away from the point of diffraction, has the same form as the GO field. To obtain this field in terms of the incident field and a diffraction coefficient, which plays the same role as the reflection coefficient, it is necessary to move the reference point to the diffraction point. The diffraction coefficient depends on the nature of the surface in the neighborhood of the diffraction point.

In its original form, the diffracted fields developed by Keller were not continuous at the shawdow boundaries. In fact, in some cases these diffracted fields become singular (unbounded) at these boundaries. These solutions are referred to as nonuniform solutions. This serious deficiency can be overcome by two methods, namely, the boundary layer method and uniform methods.[4] The uniform theories are more suitable for practical applications because they allow us to develop representations of the fields that remain valid away from and in the neighborhood of the shadow boundaries. Uniform diffraction coefficients are presented in the next section for edge and surface diffraction phenomena.

10.3.1 Uniform Geometric Theory of Diffraction

There are two well-known uniform theories of diffraction known as the uniform geometric theory of diffraction (UTD) and the uniform asymptotic theory (UAT).[26] In this section, the UTD is discussed because it is the most widely used uniform theory. The UTD was originally developed at The Ohio State University[23] for PEC surfaces. In this section, we also introduce expressions for the diffraction coefficients for a wedge with a knife edge and for surface diffraction coefficients. In Section 10.4, we present solutions for surfaces other than PEC.

10.3.1.1 Edge Diffraction

Consider a wedge with PEC walls and a knife edge as shown in Fig. 10.6. One of the caustics of the general diffracted field expression given in Eq. (10.16) is located on the edge itself. Thus, moving the reference point to the point of diffraction Q_e, the edge diffracted electric field becomes

$$\mathbf{E}^{d}(s^{d}) = \mathbf{E}^{i}(Q_e)\cdot \overline{\overline{D}}_e(\phi, \phi', \beta; k)\sqrt{\frac{\rho_e}{s^d(\rho_e + s^d)}}\, e^{-jks^d} \tag{10.17}$$

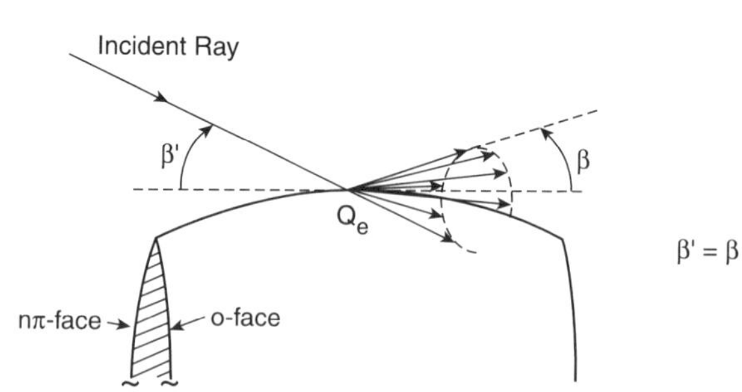

FIGURE 10.6 Wedge with PEC walls.

and the diffracted magnetic field is given by

$$\mathbf{H}^d\left(s^d\right) = \frac{\hat{s}^d \times \mathbf{E}^d\left(s^d\right)}{\eta} \tag{10.18}$$

where $\hat{s}^d$ is the direction of propagation of the diffracted field. The coefficient $\overline{\overline{\mathrm{D}}}_e$ is the edge diffraction coefficient that plays the same role as the reflection coefficient for the GO fields. It depends on the material properties of the wedge, the wedge angle, the directions of incidence and diffraction, and the wave number k. In contrast to the GO fields, away from the shadow boundaries, the diffraction coefficient is proportional to $k^{-1/2}$. Although the diffracted field can be represented in any coordinate system, the diffraction coefficient becomes much simpler when this field is represented in terms of coordinate systems fixed on the incident and diffracted rays as shown in Fig. 10.7.

For given source and observation locations, the point of diffraction Q_e on the edge can be found by means of the generalized Fermat's principle. The unit vectors $\hat{s}'$, $\hat{s}^d$ and $\hat{e}$ satisfy the law of diffraction, namely, $\hat{s}' \cdot \hat{e} = \hat{s}^d \cdot \hat{e}$. This implies that the angle of incidence β' is equal to the angle of diffraction β and also shows that the diffracted field forms a cone (Keller cone) around the unit vector $\hat{e}$. The two sets of unit vectors $(\hat{s}', \hat{\beta}', \hat{\phi}')$ and $(\hat{s}^d, \hat{\beta}, \hat{\phi})$ form an orthonormal set for the incident field and diffracted fields, respectively. The vectors $\hat{\boldsymbol{\beta}}'$ and $\hat{\boldsymbol{\beta}}$, which are parallel to the edge-fixed plane of incidence and the plane

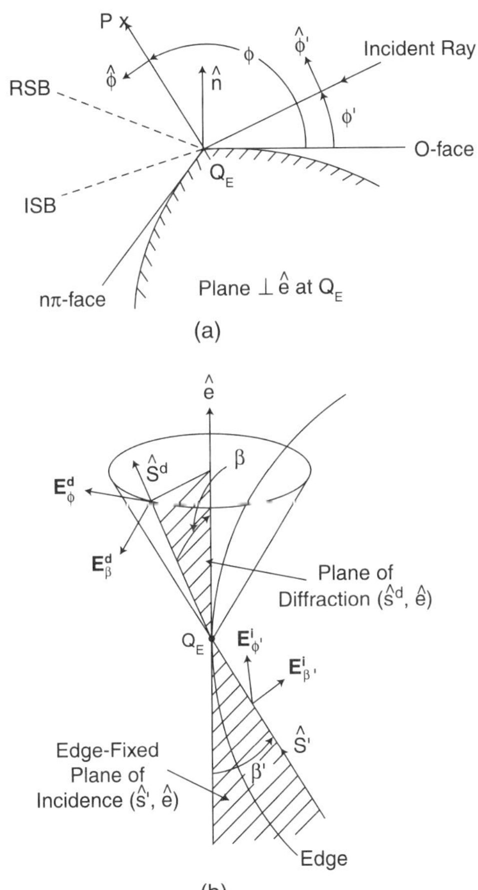

FIGURE 10.7 Ray-fixed coordinate system for edge diffracted fields: (a) unit vectors $\hat{\phi}$ and $\hat{\phi}'$, (b) angles β, and β'. (After Kouyoumjian and Pathak [23], ©1974 IEEE.)

of diffraction, respectively, are given by $\hat{\boldsymbol{\beta}}' = \hat{s}' \times \hat{\phi}'$ and $\hat{\boldsymbol{\beta}} = \hat{s}^d \times \hat{\phi}$. As depicted in Fig. 10.7(a), the unit vectors $\hat{\phi}'$ and $\hat{\phi}$ point in the direction of increasing angles ϕ' and ϕ, respectively.

The advantage of using these coordinate systems for the incident and diffracted fields is that the diffracted field can be represented in simple form, namely,

$$\begin{bmatrix} E_{\beta}^{d}(s^d) \\ E_{\phi}^{d}(s^d) \end{bmatrix} = \begin{bmatrix} -D_{es} & 0 \\ 0 & -D_{eh} \end{bmatrix} \begin{bmatrix} E_{\beta'}^{i}(Q_e) \\ E_{\phi'}^{i}(Q_e) \end{bmatrix} \sqrt{\frac{\rho_e}{s^d(\rho_e + s^d)}}\, e^{-jks^d} \tag{10.19}$$

in which the soft (D_{es}) and hard (D_{eh}) diffraction coefficients are given by

$$\begin{aligned} D_{\substack{eh\\es}}(\phi, \phi'; \beta) = \frac{e^{-j\pi/4}}{2n\sqrt{2\pi k}\sin\beta} \Bigg\{ & \cot\left(\frac{\pi + (\phi - \phi')}{2n}\right) F\left(kL^i a^+(\phi - \phi')\right) \\ & + \cot\left(\frac{\pi - (\phi - \phi')}{2n}\right) F\left(kL^i a^-(\phi - \phi')\right) \\ & \pm \left[\cot\left(\frac{\pi + (\phi + \phi')}{2n}\right) F\left(kL^{rn} a^+(\phi + \phi')\right) \right. \\ & \left. + \cot\left(\frac{\pi - (\phi + \phi')}{2n}\right) F\left(kL^{ro} a^-(\phi + \phi')\right)\right]\Bigg\} \end{aligned} \tag{10.20}$$

where $n\pi$ is the exterior wedge angle (Fig. 10.7(a)). The function F, which is known as the transition function,[23,58] makes the diffraction coefficient Eq. (10.20) uniform because its inclusion yields a solution that remains valid away from, and in the vicinity of, the shadow boundaries, including the shadow boundaries themselves. It is given by

$$\begin{gathered} F(x) = 2j\sqrt{x}\, e^{jx} \int_{\sqrt{x}}^{\infty} e^{-j\tau^2}\, d\tau \\ \frac{-\pi}{2} < Im(x) < \frac{3\pi}{2} \end{gathered} \tag{10.21}$$

Note that this function has a branch cut that runs from the origin to infinity along the negative imaginary axis in the x plane. Away from the shadow boundaries ($|x|$ large), $F(x)$ approaches one and the diffracted field becomes ray optical. This implies that the diffracted field has the same behavior as the GO fields; however, it can easily be verified from Eq. (10.20) that it varies as $k^{-1/2}$. On the other hand, when $|x|$ is small, the magnitude of $F(x)$ is proportional to $x^{1/2}$. The regions where that is true (neighborhood of shadow boundaries near a saddle point) are referred to as the transition regions. Keeping in mind that F is a function of k, the diffracted field is not a ray optical field within the transition regions because of the complicated amplitude and phase variation (in k) of the fields.

The variable $a^{\pm}(\alpha)$, where $\alpha = \phi \pm \phi'$, is a simple trigonometric function that measures the angular separation between the field point and shadow boundary, namely,

$$a^{\pm}(\alpha) = 2\cos^2\left(\frac{2n\pi N^{\pm} - \alpha}{2}\right) \tag{10.22}$$

where $N^{\pm}$ are integers that most nearly satisfy the equation

$$2\pi n N^{\pm} - \alpha = \pm\pi \tag{10.23}$$

Various values of $N^{\pm}$ at each shadow boundary are given in Table 1 of Reference 23.

The distance parameters L^i, $L^{ro,rn}$ in Eq. (10.20) contain information about the radii of curvature of the incident and reflected field wave fronts at the point of diffraction Q_e. They also contain information about the radii of curvature of both faces of the wedge in the neighborhood of Q_e. They are necessary to obtain a continuous total field at the incident and reflection shadow boundaries. These parameters can be written as follows:

$$L^i = \frac{s^d\left(\rho_e^i + s^d\right)\rho_1^i \rho_2^i \sin^2\beta'}{\rho_e^i\left(\rho_1^i + s^d\right)\left(\rho_2^i + s^d\right)} \tag{10.24}$$

$$L^{ro,rn} = \frac{s^d\left(\rho_e^{ro,rn} + s^d\right)\rho_1^{ro,rn} \rho_2^{ro,rn} \sin^2\beta'}{\rho_e^{ro,rn}\left(\rho_1^{ro,rn} + s^d\right)\left(\rho_2^{ro,rn} + s^d\right)} \tag{10.25}$$

where $\rho_{1,2}^i$ are the principal radii of curvature of the incident field wave front at Q_e and ρ_e^i is the radius of curvature of the incident field in the edge-fixed plane of incidence. Further $\rho_{1,2}^{ro}$ and $\rho_{1,2}^{rn}$ are the principal radii of curvature of the reflected field at Q_e, emanating from the 0- and n-faces of the wedges, respectively. The distance between the caustics of the diffracted field in the direction of the reflected field emanating from Q_e is $\rho_e^{ro,n}$ and can be written as follows:

$$\frac{1}{\rho_e^{ro,rn}} = \frac{1}{\rho_e^i} - \frac{2\left(\hat{n}_{o,n}\cdot\hat{n}_e\right)\left(\hat{s}'\cdot\hat{n}_{o,n}\right)}{a\sin^2\beta} \tag{10.26}$$

where $\hat{n}_{o,n}$ is normal (at Q_e) to the 0- and n-faces of the wedge, respectively; and $\hat{n}_e$ is the normal to the edge at Q_e, directed away from the center of curvature of the edge. The positive parameter a is the radius of curvature of the edge at Q_e. Finally, the distance between caustics of the diffracted field in a direction of observation $\hat{s}^d$ is ρ_e and it is given by

$$\frac{1}{\rho_e} = \frac{1}{\rho_e^i} - \frac{\hat{n}_e\cdot\left(\hat{s}' - \hat{s}^d\right)}{a\sin^2\beta} \tag{10.27}$$

Recall that the GO fields are discontinuous at the shadow boundaries. The diffracted field is also discontinuous at the same boundaries because of the presence of the transition function F. The discontinuity is such that the addition of the diffracted field to the GO fields makes the total field continuous at the shadow boundaries.

10.3.1.2 Vertex and Corner Diffraction

Two important diffracted fields for 3-D structures are the vertex- and corner-diffracted fields as shown in Fig. 10.8. As the names imply, vertex diffraction originates at the vertex of a conical structure, whereas corner diffraction originates at the corner of a plate where two edges intersect with each other. Note that corner diffraction compensates for the discontinuity of the edge diffracted fields. As the observation point moves down in Fig. 10.8b, the point of diffraction Q_e also moves down until it reaches Q_c. Beyond that point, the edge diffracted field from that particular edge becomes zero. The corner diffracted field compensates for that discontinuity.

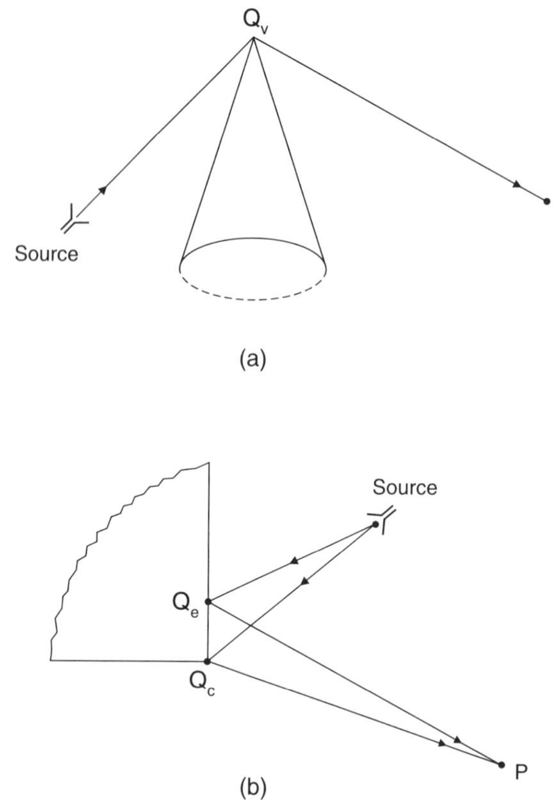

FIGURE 10.8 Vertex (a) and corner (b) diffracted fields.

The corner and vertex diffracted fields can be written as

$$\mathbf{E}^d = \mathbf{E}^i\left(Q_{c,v}\right) \cdot \overline{\overline{\mathbf{D}}}_{c,v} \frac{e^{-jks^d}}{s^d} \tag{10.28}$$

where $\overline{\overline{\mathbf{D}}}_c$ and $\overline{\overline{\mathbf{D}}}_v$ are the corner and vertex diffraction coefficients, respectively, and s^d is the distance from the point of diffraction to the observation point. Expressions for these coefficients can be found in References 12 and 13 for a cone and in Reference 70 for the corner of a flat plate. It is important to note that both diffraction coefficients vary as k^{-1} and not as $k^{-1/2}$, which is the case for the edge diffraction coefficients. The k^{-1} variation also implies that the corner and vertex-diffracted fields propagate in all directions away from the point of diffraction.

10.3.1.3 Curved Surface Diffraction

When an EM field is incident on a smooth curved surface at grazing incidence as shown in Fig. 10.5, the incident and reflected fields merge along the SSB and excite a surface ray at Q_1 that propagates on the surface to Q_2 where it leaves the surface and is able to penetrate the shadow region. According to Fermat's generalized principle, the surface ray follows a geodesic (an extremum) of the surface. As it propagates along the surface, this ray sheds energy (tangentially from the surface) continuously resulting in an exponential decay. This type of diffraction phenomenon can be separated into three distinct cases as illustrated in Fig. 10.9.

The case when the source and observation points are off the surface is referred to as the scattering problem (Fig. 10.9a). The second case, where the source is on the surface while the observation point remains off the surface, is called the radiation problem (Fig. 10.9b). Finally, when both the source and

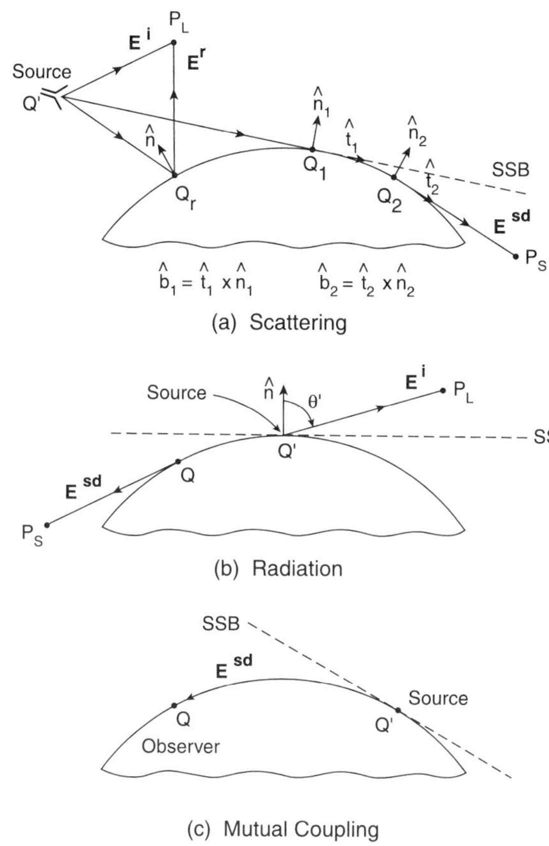

FIGURE 10.9 Three cases of surface diffraction: (a) scattering, (b) radiation, (c) mutual coupling.

observation points are on the surface, it is called the mutual coupling problem (Fig. 10.9c). Uniform asymptotic solutions have been developed for all three cases and only a brief summary of these solutions is given here. Interested readers can find more details in References 24 and 55.

Let us first consider the case when both observation and source points are off the surface and that the incident field is a ray optical field in the vicinity of the point of diffraction as shown in Fig. (10.9a). By assuming the convex surface is impenetrable, the SSB divides space in two regions, namely, the lit and shadow regions. The scattered field at the observation point $\mathbf{P}_L$ in the lit region is given by

$$\mathbf{E}(\mathbf{P}_L) \sim \mathbf{E}^i(\mathbf{P}_L) + \mathbf{E}^i(Q_r)\cdot\left(\hat{e}_\perp\hat{e}_\perp R_s + \hat{e}_\parallel^i\,\hat{e}_\parallel^r R_h\right)\sqrt{\frac{\rho_1^r\rho_2^r}{\left(\rho_1^r + s^r\right)\left(\rho_2^r + s^r\right)}}\,e^{-jks^r} \tag{10.29}$$

where Q_r is the point of reflection, whereas the unit vectors $\hat{e}_\perp$ and $\hat{e}_\parallel^{i,r}$ as well as the radii of curvature $\rho_{1,2}^r$ were already defined in Section 10.2 when the GO-reflected field was introduced.

If the observation point moves to the shadow region ($\mathbf{P}_s$), the surface diffracted field becomes

$$\mathbf{E}(\mathbf{P}_s) \sim \mathbf{E}^i(Q_1)\cdot\left(\hat{b}_1\hat{b}_2 D_s + \hat{n}_1\hat{n}_2 D_h\right)\sqrt{\frac{\rho_s}{s^d\left(\rho_s + s^d\right)}}\,d^{-jks^d} \tag{10.30}$$

where $\mathbf{E}^i$ is the incident ray optical field at Q_1 and s^d is the distance from Q_2 to the observation point. By keeping in mind that one caustic of the surface diffracted field is at Q_1, the second caustic is at a

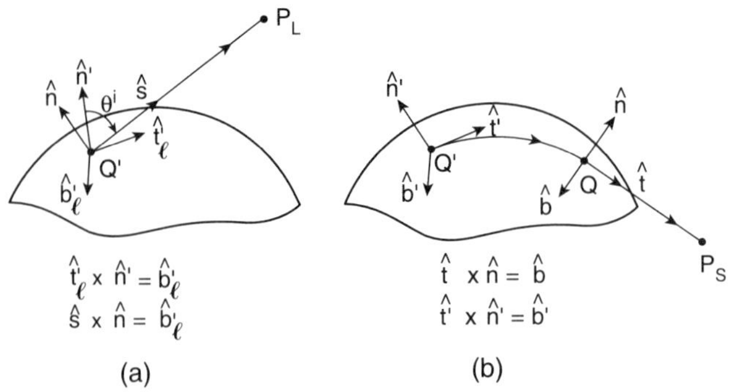

FIGURE 10.10 Ray paths and coordinates for radiation problem: (a) lit and (b) shadow regions. (After Pathak, Wang, Burnside, and Kouyoumjian, *IEEE Trans. Antennas Propagat.*, 29 (1981), 609–622, ©1981 IEEE.)

distance of ρ_s from Q_1 as shown in Fig. 10.17 of Reference 55. The unit vectors $\hat{n}_1$ and $\hat{n}_2$ are surface normals at Q_1 and Q_2, respectively. Furthermore, the unit vectors $\hat{b}_1$ and $\hat{b}_2$ are binormal vectors at Q_1 and Q_2, respectively. If $\hat{b}_1 = \hat{b}_2$, the surface ray path between Q_1 and Q_2 has zero torsion.

The coefficients $R_{s,h}$ and $D_{s,h}$ are the generalized reflection and surface diffraction coefficients, respectively. They can be written in terms of the reflection coefficients of the surface, the transition function $F(x)$ introduced in Section 10.3.1.1 and the Pekeris functions. Complete expressions for these coefficients are given in References 24 and 55. It is important to note that in the deep lit region (far away from the SSB), the reflection coefficients reduce to the usual Fresnel reflection coefficients. Likewise, in the deep shadow region, the diffraction coefficients reduce to the nonuniform coefficients introduced by Keller.[33] The uniform expressions in Eqs. (10.29) and (10.30) are valid even at the SSB. In fact, they have the same limit at the SSB, making the total field continuous.

The second case considered here is the radiation problem where the source is located on the surface of the scatterer, whereas the observation point remains away from the surface. Two types of sources are considered, namely, an aperture and a short electric monopole normal to the surface. The aperture of area da' can be modeled by a magnetic dipole moment $\mathbf{d}p_m = \mathbf{E}^a \times \hat{n}'da'$ where $\mathbf{E}^a$ is the electric field on the aperture and $\hat{n}'$ is the outer surface normal. Likewise, the monopole can be expressed in terms of an electric dipole moment $\mathbf{d}p_e = Idl'\hat{n}'$ where I is the current and dl' the length. These two types of sources can be used to model a wide variety of sources mounted on a convex surface. As in the previous case, the SSB, which in this case passes through the source location Q', divides the space into the lit and shadow regions. If the field point is in the lit region, the ray path is a straight line from the source location to the observation point (Fig. 10.10a). On the other hand, if the field point is in the shadow region, the ray travels along a surface geodesic from Q' to Q and along a straight line from Q to the field point as shown in Fig. 10.10b.

The field in the lit region has its caustic at the source location. Thus, it can be expressed as follows:

$$\mathbf{E}_{e,m}\left(\mathbf{P}_L\right) \sim \mathbf{dp}_{e,m}\left(Q'\right) \cdot \overline{\overline{\mathbf{T}}}^{\ell}_{e,m} \frac{e^{jks}}{s} \tag{10.31}$$

where the e and m subscripts refer to electric and magnetic sources, respectively, and s is the distance from Q' to $\mathbf{P}_L$. The launching coefficients $\overline{\overline{\mathbf{T}}}^{\ell}_{e,m}$ can be written as follows:

$$\overline{\overline{\mathbf{T}}}^{\ell}_{e} = -\frac{jk\eta_o}{4\pi}\left[\hat{n}'\hat{n}M + \hat{n}'\hat{b}N\right] \tag{10.32}$$

and

$$\overline{\overline{\mathbf{T}}}^{\ell}_{m} = -\frac{jk}{4\pi}\left[\hat{b}_l'\hat{n}A + \hat{t}_l'\hat{b}B + \hat{b}_l'\hat{b}C + \hat{t}_l'\hat{n}D\right] \tag{10.33}$$

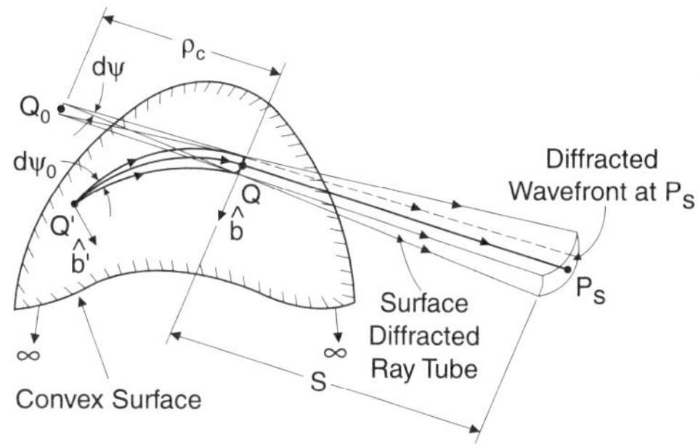

FIGURE 10.11 Ray tube behavior in shadow region where t is the arc length from the source location Q' to Q. (After Pathak, P., Techniques for High-Frequency Problems, in *Antenna Handbook,* Y.T. Lo and S.W. Lee, Eds., New York: Van Nostrand Reinhold, 1988.)

where the coefficients A, B, C, D as well as M and N are given in Table 3 of Reference 55. The unit vectors in Eqs. (10.32) and (10.33) are defined in Fig. 10.10a. Note $\hat{n}'$ is normal to the surface at Q', whereas $\hat{n}$ is normal to $\hat{s}$. All unit vectors $(\hat{n}, \hat{n}', \hat{s})$ lie in the same plane.

In the shadow region, the diffracted field has two caustics as shown in Fig. 10.11. One caustic is at the diffraction point Q, whereas the second one is at a distance of ρ_c. Therefore, the surface-diffracted field in the shadow region is given by

$$\mathbf{E}_{m,e}\left(\mathbf{P}_s\right) \sim \mathbf{dp}_{e,m}\left(Q'\right) \cdot \overline{\overline{\mathbf{T}}}_{e,m}\left(Q',Q\right) \sqrt{\frac{\rho_c}{s\left(\rho_c + s\right)}}\, e^{-jks} \tag{10.34}$$

where s is the distance from Q to $\mathbf{P}_s$ and the transfer functions from Q' to Q are given by

$$\overline{\overline{\mathbf{T}}}_e\left(Q',Q\right) = -\frac{jk\eta_o}{4\pi}\left[\hat{n}'\hat{n}T_5\left(Q'\right)H + \hat{n}'bT_6\left(Q'\right)S\right]e^{-jkt}\sqrt{\frac{d\psi_o}{\rho_c d\psi}}\left(\frac{\rho_g(Q)}{\rho_g(Q')}\right)^{1/6} \tag{10.35}$$

and

$$\begin{aligned}\overline{\overline{\mathbf{T}}}_m\left(Q',Q\right) = -\frac{jk}{4\pi}&\left[\hat{b}'\hat{n}T_1\left(Q'\right)H + \hat{t}'\hat{b}T_2\left(Q'\right)S + \hat{b}'\hat{b}T_3\left(Q'\right)S + \hat{t}'\hat{n}T_4\left(Q'\right)H\right] \\ &\cdot e^{-jkt}\sqrt{\frac{d\psi_o}{\rho_c d\psi}}\left(\frac{\rho_g(Q)}{\rho_g(Q')}\right)^{1/6}\end{aligned} \tag{10.36}$$

where the unit vectors are defined in Fig. 10.10b. Expressions for the parameters H, S, $T_1 - T_6$ as well as the radius of curvature in the direction of the ray path (ρ_g) and the caustic distance ρ_c can be found in Table 2 of Reference 55. Closed-form expressions for the ratio $\sqrt{d\psi_o/d\psi}$ can also be found for a few convex surfaces such as cylinders and spheres. However, for general convex surfaces, it has to be obtained numerically.

The mutual coupling problem, where both source and observation poins are on the surface of the convex structure, is important for antenna applications. As in the radiation case, the sources can be magnetic dipoles or electric monopoles. For a magnetic dipole, the fields are given by

$$\mathbf{H}_m\left(Q|Q'\right) \sim -\frac{jk}{4\pi\eta_o}\mathbf{dp}_m\left(Q'\right) \cdot \left\{\hat{b}'\hat{b}V_1 + \hat{t}'\hat{t}V_2 + \left(\hat{t}'\hat{b} + \hat{b}'\hat{t}\right)V_3\right\}\frac{e^{-jkt}}{t}\sqrt{\frac{td\psi_o}{\rho_c d\psi}} \tag{10.37}$$

and

$$\mathbf{E}_m\left(Q|Q'\right) \sim -\frac{jk}{4\pi}\mathbf{dp}_m\left(Q'\right)\cdot\left\{\hat{b}'\hat{n}V_4 + \hat{t}'\hat{n}V_5\right\}\frac{e^{-jkt}}{t}\sqrt{\frac{td\psi_o}{\rho_c d\psi}} \tag{10.38}$$

where t is the arc length from Q' to Q (Fig. 10.9c) and the expressions for $V_1 - V_5$ can be deduced from the equations given in References 24 and 55. The main difference between the expressions for mutual coupling and the previous two cases is that in the former, there are field components in the direction of propagation ($\hat{t}$). Similar expressions exist for an electric monopole source. They are not given here because of space limitations; however, they can be found in the literature.[24,55]

10.3.1.4 Edge Excited Surface Waves

In addition to the surface or creeping waves excited on a convex surface by a field at grazing incidence, creeping waves can also be excited by an edge on a convex surface as depicted in Fig. 10.5. It is shown there that a field incident at the edge Q_e excites a surface wave that travels counterclockwise around the curved surface until it reaches Q_3, where it leaves the surface and reaches the observation point in the lit region. It is also possible for a surface wave to reach the edge Q_e where it is diffracted to space. Approximate expressions (2D case) for these diffracted fields can be found in the literature.[10,24,45]

10.4 Nonmetallic Penetrable/Impenetrable Materials

It is well known that the scattering properties of a body are a function of its geometric (shape) as well as its material properties. All the previous solutions are applicable to PEC surfaces that are used to model metallic conductors; however, there are many applications where the surfaces are not metallic conductors. For example, to study electromagnetic wave propagation in urban environments or within buildings, which is important in the wireless communications industry, it is necessary to consider material effects. The use of composite materials is very common in the construction of aircraft and other structures where antennas are mounted. It is therefore important to study the reflection, transmission (if penetrable), and diffraction properties of structures whose surfaces are not metallic conductors and the PEC model is not applicable.

Although the GO fields scattered by material surfaces can be obtained in the same fashion as the GO fields for PEC surfaces, the Fresnel reflection coefficients are more complicated and depend on the material properties and angle of incidence. Another important difference is that for penetrable materials, there is also a transmitted field that does not exist for PEC surfaces. Furthermore, the more complex boundary conditions at the material surfaces can give rise to various types of wave phenomena not present on PEC surfaces. Thus, in addition to the reflected and transmitted fields, various other types of waves such as surface waves (slow waves), leaky waves (fast waves), and others can be excited on the surface.[14] In general, these wave phenomena can be excited two ways. Away from material and geometric discontinuities, they can be excited by the incident field depending on the type of illumination, in other words, if the incident field satisfies the proper phase matching condition on the surface. Another way they can be excited (for most types of illumination) is at material and geometric discontinuities. This excitation is based on diffraction phenomena. In general, the preceding two types of excitation can be present on a material surface. The existence of these additional wave components makes the study of scattering by material nonmetallic surfaces very challenging.

In contrast to the PEC surfaces considered in the previous sections, the calculation of the diffracted fields (edge diffraction, surface diffraction, vertex diffraction, etc.) for nonmetallic surfaces is extremely complicated. First of all, the boundary conditions (BCs) are more involved because of the presence of materials. The BCs also can couple various field components (electric and magnetic) that would normally remain uncoupled for the PEC case. The net effect is that to obtain analytic edge diffraction coefficients for structures with material surfaces, it is necessary to develop approximate BCs. These approximate BCs convert the original problem into a mixed boundary value problem that can be handled by sophisticated functional theoretical methods.

10.4.1 Approximate Boundary Conditions

Consider a finite size material body immersed in a homogeneous isotropic (usually free-space) medium. Assume that we are only interested in the fields outside the body. To uniquely solve electromagnetic scattering, radiation or propagation problems, it is necessary to apply the BCs satisfied by the fields on the material surface. Note that BCs describe the relationship between the electric and magnetic fields on the surface.

The most widely known (approximate) BC used for metallic conductors is that the tangential electric field is zero on the surface ($\mathbf{E}_{\text{tan}} = 0$). Well-known BCs also exist between two media that state the tangential electric and magnetic fields must be continuous across the boundary in the absence of surface charges or currents. The approximate BCs discussed here refer to conditions that are expressed only in terms of the fields exterior to the material body. In other words, these approximate BCs replace the body itself, making the solution of diffraction problems possible. They can be used when we are only interested in the fields outside the material body. If the fields inside the body are required, these BCs cannot be used.

BCs can be divided into two types. The one-sided BCs are applicable to situations where there is no field transmitted through the material body because of the nature of the materials. One example is the propagation of radio waves over a lossy earth. The second type are two-sided BCs where the field does propagate through the material. An example of this situation is a thin dielectric slab. The simplest one-sided boundary condition is referred to as an impedance or Leontovich boundary condition.[32,68] It can be written as follows

$$\hat{n} \times \hat{n} \times \mathbf{E} = -Z_s \hat{n} \times \mathbf{H} \tag{10.39}$$

where $\hat{n}$ is the normal to the surface and Z_s is the surface impedance that couples the electric and magnetic fields. This boundary condition is a good model for conductors and lossy materials where there is little penetration of the field. By assuming that the refractive index of a lossy dielectric material is large, $Z_s \simeq \sqrt{\mu/\varepsilon}$ where μ and ε are the permeability and permittivity of the material scatterer, respectively. Note that $Z_s = 0$ for a PEC and $Z_s = \infty$ for a perfect magnetic conductor (PMC). Although magnetic conductors do not exist, they are useful for analysis purposes. This boundary condition can also be expressed in terms of the normal (to the surface) electric or magnetic fields.[61,69] By assuming that the y-coordinate is normal to the surface and the scatterer is immersed in free space, Eq. (10.39) can be rewritten as

$$\left(\frac{\partial}{\partial y} - jk_o \delta \right) u = 0 \tag{10.40}$$

where $\delta = Z_s/\eta_o$ if $u = E_y$ and $\delta = \eta_o/Z_s$ if $u = H_y$.

Since the preceding expression contains a first-order derivative with respect to the direction normal to the surface, it is referred to as a first-order impedance BC. Although the impedance BC is a good model for conductors, lossy dielectrics, and a few other surfaces, it fails when applied to dielectric-coated conductors and other structures of interest. The accuracy of first-order BCs can be improved by obtaining BCs that contain higher order derivatives. To illustrate this method, consider a grounded dielectric slab as shown in Fig. 10.12.

Starting with the exact BCs at the PEC surface and at the dielectric free-space interface, generalized impedance boundary conditions (GIBCs) can be developed to replace the grounded slab. If properly done, these GIBCs recover the exact reflected field. The details of the derivation are given in Reference 61 and are not be repeated here. The exact GIBC for the grounded slab (at $y = 0$) can be written as follows:

$$\left\{ \frac{\partial}{\partial y} Z_e^c\left(\partial^2/\partial y^2\right) - jk_o \eta_o L_e^c\left(\partial^2/\partial y^2\right) \right\} H_y = 0 \tag{10.41}$$

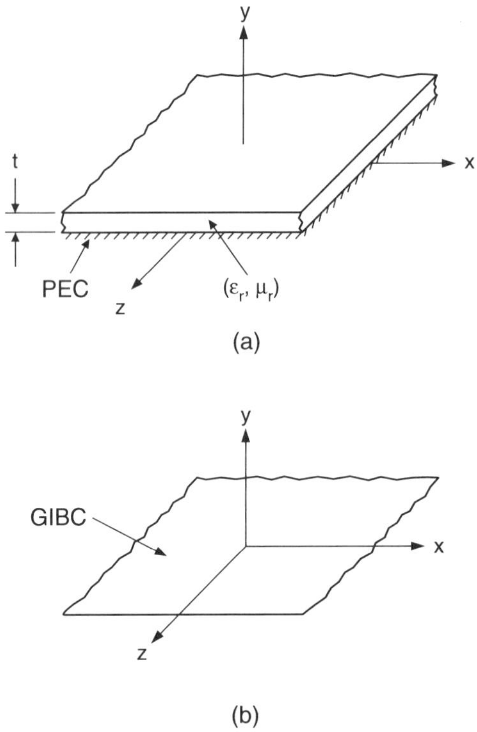

FIGURE 10.12 Grounded dielectric slab: (a) original slab, (b) GIBC at $y = 0$. (After Rojas and Al-hekail [61], ©1989 AGU.)

and

$$\left\{\frac{\partial}{\partial y} L_h^c\left(\partial^2/\partial y^2\right) - j\frac{k_o}{\eta_o} Z_h^c\left(\partial^2/\partial y^2\right)\right\} E_y = 0 \tag{10.42}$$

where $Z_{e,h}^c$ and $L_{e,h}^c$ are differential operators that can be expressed in terms of infinite series in powers of $\partial^2/\partial y^2$. Expressions for these operators can be found in the literature.[61] Obviously, the preceding exact GIBCs contain derivatives of all orders and are not suitable for the analysis of material bodies with edges. However, approximate GIBCs suitable for edge diffraction analysis can be obtained from the preceding expressions. Approximate GIBCs can be obtained in various forms, depending on the type of approximation used for the $Z_{e,h}^c$ and $L_{e,h}^c$ differential operators. For example, for a PEC surface coated with a thin dielectric layer of thickness t, BCs accurate to order $O(t^M)$ can be written as

$$\left\{\sum_{n=0}^{M} a_n(M)\frac{\partial^n}{\partial y^n}\right\} H_y = a_M(M)\prod_{n=1}^{M}\left(\frac{\partial}{\partial y} - jk_o\eta_o\lambda_n^e(M)\right) H_y = 0 \tag{10.43}$$

and

$$\left\{\sum_{n=0}^{M+1} b_n(M)\frac{\partial^n}{\partial y^n}\right\} E_y = b_{M+1}(M)\prod_{n=1}^{M+1}\left(\frac{\partial}{\partial y} - \frac{jk_o}{\eta_o}\lambda_n^h(M)\right) E_y = 0 \tag{10.44}$$

where the coefficients $a_n(M)$ and $b_n(M)$ are given in Tables 1 and 2 of Reference 61, respectively. Note that the BC for E_y has a higher order derivative than the corresponding BC for H_y. The preceding BCs

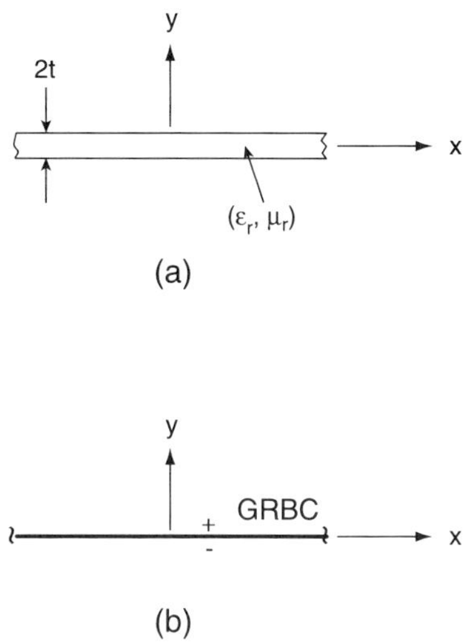

FIGURE 10.13 Dielectric layer, no ground plane: (a) original slab, (b) GRBC at $y = 0$. (After Rojas and Al-hekail [61], ©1989 AGU.)

can also be written in terms of tangential field components. For example, for $M = 1$, Eqs. (10.43) and (10.44) can be written as

$$\left(1+t\left(1-\frac{1}{\varepsilon_r}\right)\frac{\partial}{\partial y}\right)\hat{n}\times\hat{n}\times\mathbf{E} = jk_o\eta_o t\left(\frac{1}{\varepsilon_r}-\mu_r\right)\hat{n}\times\mathbf{H} \tag{10.45}$$

It is clear from Eq. (10.45) that even for a thin slab, the BC does not reduce a first-order impedance BC. That only happens if $\varepsilon_r = 1$ where the surface impedance becomes $Z_s = -jk_o t\eta_o(\mu_r - 1)$. Besides the preceding approximate GIBCs accurate to $O(t^M)$, approximate GIBCs can be obtained where the differential operators can be expanded in terms of Chebyshev polynomials, using rational functions (Padé approximation), etc. It is important to note that the development of approximate GIBCs is similar to the development of approximate absorbing BCs for numerical techniques (finite difference, finite element).

Let us now consider two-sided BCs that are used to represent material layers where the field travels through the material. Consider a planar dielectric layer without a ground plane as shown in Fig. 10.13. The exact generalized resistive boundary conditions (GRBCs) at y = 0 can replace the dielectric layer for analysis purposes. They can be written in terms of the normal components as follows:

$$Z_e^c\left(\partial^2/\partial y^2\right)\left(\frac{\partial}{\partial y}H_y^+ + \frac{\partial}{\partial y}H_y^-\right) - jk_o\eta_o L_e^c\left(\partial^2/\partial y^2\right)\left(H_y^+ - H_y^-\right) = 0 \tag{10.46}$$

$$Z_e^m\left(\partial^2/\partial y^2\right)\left(\frac{\partial}{\partial y}H_y^+ - \frac{\partial}{\partial y}H_y^-\right) - jk_o\eta_o L_e^m\left(\partial^2/\partial y^2\right)\left(H_y^+ + H_y^-\right) = 0 \tag{10.47}$$

for the H_y field and

$$L_h^c\left(\partial^2/\partial y^2\right)\left(\frac{\partial}{\partial y}E_y^+ - \frac{\partial}{\partial y}E_y^-\right) - \frac{jk_o}{\eta_o}Z_h^c\left(\partial^2/\partial y^2\right)\left(E_y^+ + E_y^-\right) = 0 \tag{10.48}$$

$$L_h^m\left(\partial^2/\partial y^2\right)\left(\frac{\partial}{\partial y}E_y^+ + \frac{\partial}{\partial y}E_y^-\right) - \frac{jk_o}{\eta_o}Z_h^m\left(\partial^2/\partial y^2\right)\left(E_y^+ - E_y^-\right) = 0 \tag{10.49}$$

for the E_y field. The fields $E_y^{\pm}$ and $H_y^{\pm}$ as well as the derivatives $\partial E_y^{\pm}/\partial y$ and $\partial H_y^{\pm}/\partial y$ are evaluated at $y = 0^{\pm}$, whereas the differential operators $Z_{e,h}^{m}$ and $L_{e,h}^{m}$ are defined in Reference 61. As in the GIBCs, approximate GRBCs can be obtained from the preceding expressions. They can be expressed in terms of the normal or tangential field components. For example, GRBCs accurate to $O(t)$ yield second-order BCs because they contain a second derivative of the normal field components. These approximate BCs can also be written in terms of the tangential components as follows:

$$-\hat{n}\times\hat{n}\times\left(\mathbf{E}^{+}-\mathbf{E}^{-}\right)+t\left(\frac{1}{\varepsilon_r}-1\right)\hat{n}\times\hat{n}\times\left(\frac{\partial}{\partial y}\mathbf{E}^{+}+\frac{\partial}{\partial y}\mathbf{E}^{-}\right)$$
$$=-jk_o\eta_o t\left(\frac{1}{\varepsilon_r}-\mu_r\right)\hat{n}\times\left(\mathbf{H}^{+}+\mathbf{H}^{-}\right) \tag{10.50}$$

and

$$-\hat{n}\times\hat{n}\times\left(\mathbf{H}^{+}-\mathbf{H}^{-}\right)+t\left(\frac{1}{\mu_r}-1\right)\hat{n}\times\hat{n}\times\left(\frac{\partial}{\partial y}\mathbf{H}^{+}+\frac{\partial}{\partial y}\mathbf{H}^{-}\right)$$
$$=j\frac{k_o}{\eta_o}t\left(\frac{1}{\mu_r}-\varepsilon_r\right)\hat{n}\times\left(\mathbf{E}^{+}+\mathbf{E}^{-}\right) \tag{10.51}$$

To conclude, another well-known BC is the resistive boundary condition. It can be used to model thin lossy materials with thicknesses smaller than the skin depth. The resistive BC states that the tangential electric field is continuous across the thin material; however, the tangential magnetic field is not. In fact, it states that the induced electric current $\mathbf{J}^{ind} = \hat{n}\times(\mathbf{H}^{+} - \mathbf{H}^{-})$ is proportional to the electric field. The resistive BC is given by

$$\hat{n}\times\left(\mathbf{E}^{+}-\mathbf{E}^{-}\right)=0 \tag{10.52}$$

and

$$\hat{n}\times\hat{n}\times\mathbf{E}=-R\hat{n}\times\left(\mathbf{H}^{+}-\mathbf{H}^{-}\right)=-R\mathbf{J}^{ind} \tag{10.53}$$

By assuming the thickness of the material is τ and σ is the conductivity, the resistivity is given by $R = 1/(\tau\sigma)$. Thin resistive sheets with resistivities ranging from 3 to 3000Ω (per square) have found many uses in antenna applications. They are now routinely used to absorb unwanted electromagnetic energy.[53] Approximate BCs for other structures such as anisotropic materials and chiral materials,[62] can also be obtained in the manner described earlier.

10.4.2 Geometric Optic Fields

The procedure to calculate the reflected and transmitted (for a penetrable structure) GO fields is similar to the one described in Section 10.2. In fact, the expression for the reflected field given in Eq. (10.13) does not change, except for the reflection coefficients. By assuming the structure is penetrable, the transmitted field exists and can be written in the same manner as the reflected field, except that the reflection coefficient is replaced by a transmission coefficient and the radii of curvature $\rho_{1,2}^{r}$ are replaced by $\rho_{1,2}^{t}$. The expressions for the reflection and transmission coefficients depend on the type of BC that is being used. For structures with edges, it is necessary to use approximate BCs to be able to obtain the diffracted fields. For structures without geometric or material discontinuities, the exact BCs can be used.

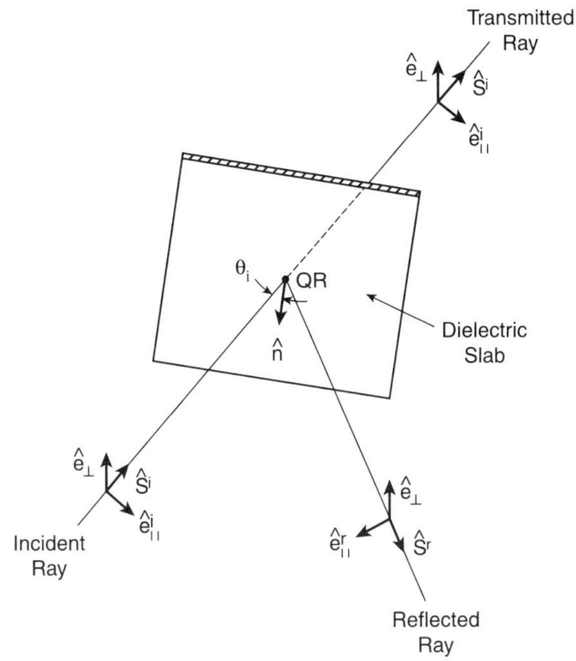

FIGURE 10.14 Planar slab showing reflected and transmitted fields.

As an example, consider a planar electric slab as shown in Fig. 10.14. The reflected field for the preceding structure is given in Eq. (10.13). The transmitted field can also be simplified if it is expressed in terms of the ray fixed coordinate system depicted in Fig. 10.14, namely,

$$\begin{bmatrix} E_{\parallel}^{t}(s) \\ E_{\perp}^{t}(s) \end{bmatrix} = \begin{bmatrix} T_{\parallel\parallel} & T_{\parallel\perp} \\ T_{\perp\parallel} & T_{\perp\perp} \end{bmatrix} \begin{bmatrix} E_{\parallel}^{i}(Q_r) \\ E_{\perp}^{i}(Q_r) \end{bmatrix} \sqrt{\frac{\rho_1^t \rho_2^t}{(\rho_1^t + s^t)(\rho_2^t + s^t)}}\, e^{-jk_o s^t} \tag{10.54}$$

where $T_{\parallel\parallel}$, $T_{\parallel\perp}$, $T_{\perp\parallel}$, and $T_{\perp\perp}$ are the transmission coefficients, referenced to the point Q_r. The radii of curvature of the transmitted field at Q_r are $\rho_{1,2}^t$ and s^t is the distance from Q_r to the observation point. In general, all the elements of the reflection and transmission matrices are not zero. However, for a homogeneous isotropic dielectric material, as shown in Fig. 10.14, the off-diagonal components become zero.

10.4.3 Diffracted Fields

As mentioned before, the calculation of diffracted fields for material structures with edges or material discontinuities is considerably more complex than for PEC surfaces. The BCs, more complex, couple the magnetic and electric fields.

Two functional theoretical methods can solve canonical structures such as the ones depicted in Fig. 10.15. The method introduced by Maliuzhinets[40] is the most general because it can handle wedge-shaped as well as planar geometries. The total field, which is expressed in cylindrical coordinates, is represented as a spectrum of planes over a two-fold Sommerfeld contour with an unknown spectral (or weight) function. By applying the BCs, radiation condition, and appropriate edge condition, functional difference equations are obtained in terms of the spectral function. The solution of these equations yields the unknown spectral function. Once the spectral function is obtained, the diffracted field can be evaluated by performing an asymptotic evaluation of the original spectral integral. Such a solution yields the GO fields as well as the edge-diffracted field and (if they exist) surface and or leaky waves.

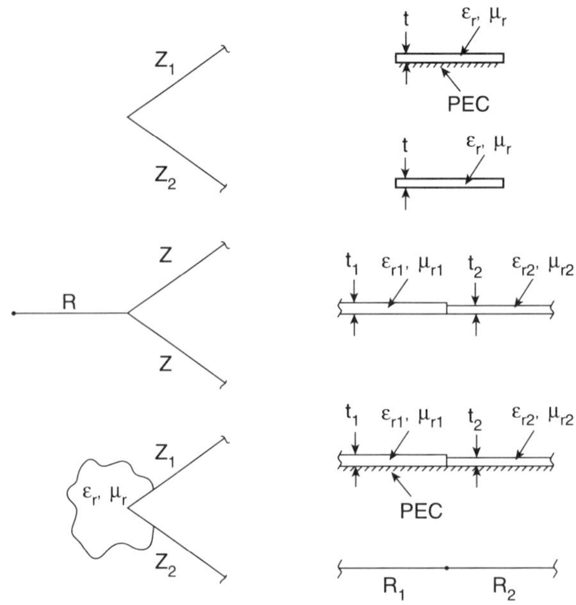

FIGURE 10.15 Canonical structures with penetrable and impedance materials.

The second analysis method, applicable to planar structures, is the Wiener–Hopf technique.[50] In this method, Fourier transforms as well as the BCs, radiation condition, and appropriate edge conditions are used to obtain the Wiener–Hopf equation. There are several ways to obtain the Wiener–Hopf equation as discussed in Reference 50. This equation is then solved, using the Wiener–Hopf procedure to obtain the spectral or weight function. Once the Wiener–Hopf equation is solved, an inverse Fourier transform is taken to obtain an integral representation of the field. The integral can be evaluated asymptotically to recover the diffracted field as well as other field components.

Although higher order BCs provide more accurate models for material surfaces, their use in the analysis of structures with edges introduces mathematical complications. By keeping in mind that the wave or Helmholtz equation is a second-order partial differential equation, the usual edge conditions are not sufficient to obtain unique solutions when BCs of second or higher order are used. This is because of the additional degrees of freedom introduced by the higher order BCs. To overcome this difficulty, generalized edge conditions must be used.[37,38,69]

Regardless of the method used, the asymptotic expression for the total electric field can be expressed as follows:

$$\mathbf{E}(P) \sim \mathbf{E}^{GO}(P) + \mathbf{E}^{d}(P) + \mathbf{E}^{sw}U_{sw} + \mathbf{E}^{lw}U_{lw} + \cdots \tag{10.55}$$

where (assuming they are excited) $\mathbf{E}^{sw}$ and $\mathbf{E}^{lw}$ are surface and leaky waves, respectively. The step functions U^{sw} and U^{lw} indicate that these two field components exist within certain regions of space and are zero elsewhere. The GO field is given by

$$\mathbf{E}^{GO}(P) \sim \mathbf{E}^{i}(P)U_{i} + \mathbf{E}^{r}(P)U_{r} + \mathbf{E}^{t}(P)U_{t} \tag{10.56}$$

where $\mathbf{E}^{i}$ and $\mathbf{E}^{r}$ were defined in Section 10.2, $\mathbf{E}^{t}$ is the transmitted field, and U_t is a unit step function. Obviously, this last field is zero for an impenetrable object. Similar expressions can be written for the magnetic field and are not repeated here.

The diffracted field for the structures depicted in Fig. 10.15 can be written as

$$\begin{bmatrix} E_\beta^d(s^d) \\ E_\phi^d(s^d) \end{bmatrix} = \begin{bmatrix} -D_{\beta\beta} & -D_{\beta\phi} \\ -D_{\phi\beta} & -D_{\phi\phi} \end{bmatrix} \begin{bmatrix} E_{\beta'}^i(Q_e) \\ E_{\phi'}^i(Q_e) \end{bmatrix} \sqrt{\frac{\rho_e}{s^d(\rho_e + s^d)}}\, e^{-jk_o s^d} \tag{10.57}$$

The preceding expression is similar to the diffracted field for PEC surfaces, except that the off diagonal components of the diffraction matrix are, in general, not zero. In other words, there is a coupling between the E_β^d and E_ϕ^d field components. Expressions for the diffracted fields as well as the GO and surface wave fields for the structures depicted in Fig. 10.15 can be found in the literature[1,2,9,11,37-40,51,52,59,60,63,69,74,79] for the case of plane wave illumination where $\rho_e \to \infty$ if the edge of the structure is assumed to be straight. They are not included here because of space limitations. Additional canonical solutions exist, such as curved impedance surfaces,[7,8] anisotropic impedance surfaces,[56,80] and others. Besides the usual plane wave excitation, solutions have been obtained for line source excitation,[75] including a Green's function for an impedance wedge.[54] Note that the references provided here are only a representative sample of the vast literature that exists on this subject.

Because of the complex analysis required to obtain diffraction coefficients, it is often necessary to use numerical techniques (MOM, FEM, and FD method) to obtain the diffraction coefficients for complex structures.[52] However, by using a combination (hybrid technique) of ray and numerical methods, it is possible to analyze EM propagation, radiation, and scattering phenomena in complex environments where material effects play an important role. An alternative approach to the rigorous methods described earlier is to use heuristically derived diffraction coefficients.[5] This approach yields reasonable results when the diffracted field is not strong. It has the advantage of having simple expression for the approximate diffraction coefficients.

10.5 Physical Optics

All the preceding asymptotic methods were in terms of fields and rays. An alternative approach is to start with the source of the fields itself. In other words, it is possible to obtain HF asymptotic expressions for the currents generating the fields. Once the currents are obtained, the fields can be evaluated using radiation integrals. Note that integral techniques are very useful in caustic regions where the UTD and other ray techniques fail.

One of the most commonly used methods for antenna pattern analysis, especially for reflector antennas, is the PO technique, also known as the surface current method. PO is an HF method that can be used to calculate the fields scattered from an electrically large object. To illustrate the method, let us consider a reflector illuminated by fields radiated by a feed as shown in Fig. 10.16. In PO, the scattered electric field, $\mathbf{E}^s(\mathbf{r})$, at a given field observation point with position vector $\mathbf{r}$, is approximately given by

$$\mathbf{E}^s(\mathbf{r}) = \frac{-jk_o\eta_0}{4\pi} \int_s \left[\mathbf{J}_s(\mathbf{r}') - \left(\hat{r}\cdot\mathbf{J}_s(\mathbf{r}')\right)\hat{r}\right] \frac{e^{-jk_o r}}{r}\, dS' \tag{10.58}$$

where $\mathbf{r}'$ is the position vector for a given point on the surface of the reflector. The induced surface current, $\mathbf{J}_s(\mathbf{r}')$, on the surface of a PEC, is given by

$$\mathbf{J}_s(\mathbf{r}') \approx 2\hat{n}(\mathbf{r}') \times \mathbf{H}^i(\mathbf{r}') \tag{10.59}$$

for the lit region (illuminated region) and

$$\mathbf{J}_s(\mathbf{r}') = 0 \tag{10.60}$$

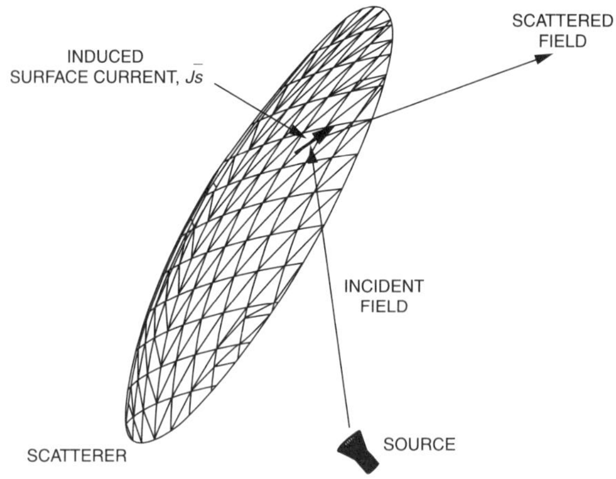

FIGURE 10.16 Physical optics method for reflector pattern analysis.

for the shadow region. In Eq. (10.59), $\hat{n}(r')$ and $\mathbf{H}^i(\mathbf{r}')$ are the unit normal vector and the incident magnetic field at the surface point $\mathbf{r}'$, respectively. A more general expressions for the PO current in Eq. (10.59) can be obtained for nonmetallic surfaces. It would be expressed in terms of the incident and reflected fields. With the approximation of the induced surface currents given in Eqs. (10.59) and (10.60), PO is only accurate in the region where specular reflection occurs. Away from this region, PO approximations are not as accurate and get worse.

In PO, the equivalent currents need to be integrated over the surface to calculate the scattered field. Obviously, the larger the surface, the more time it takes to integrate the currents. Integration can be performed in two ways. One approach is to integrate using asymptotic techniques such as the method of stationary phase.[14] This method is based on the idea that the contribution to the scattered field comes from a few isolated points on the surface as well as diffraction points along the edges. The former are called stationary points, whereas the latter are endpoints or diffraction points. As long as the stationary points are isolated and away from the edges of the surface or lit-shadow boundaries, the method yields good results. In fact, the stationary points yield the correct GO fields. The endpoint contributions along the edges contribute to the PO diffracted fields $\mathbf{E}^d_{PO}$.

A more general method to obtain the fields is to perform a numerical integration. Consequently, it is necessary to divide the surface of the object into small patches so that accurate results can be obtained. One obvious way for dividing the surface is by means of rectangular patches that are quite easy to define for a given surface with a well-defined rim shape. However, each rectangular patch has to be defined by four surface points that may not lie on the same plane. Consequently, the surface model can be in error, particularly if the surface has a large curvature. Alternatively, one can divide the surface by triangular patches where each patch is defined by three points. These three points should lie on the same plane and provide better surface modeling for the PO technique if the spacing between surface points is sufficiently small. For example, Fig. 10.16 shows a large reflector modeled by many triangular patches so that the equivalent induced surface current at each patch can be defined and integrated to obtain the field scattered by the reflector.

10.6 Physical Theory of Diffraction

As mentioned in the previous section, the PO technique is accurate near and within the specular reflection region and becomes erroneous farther away from this region. This also implies that the PO-diffracted field $\mathbf{E}^d_{PO}$ is not accurate because the PO currents are not accurate near edges. To improve the accuracy of the PO fields, it is necessary to improve the accuracy of the currents, especially in the regions where diffracted effects are important. As stated previously, PTD, originally developed by Ufimtsev,[76] is an

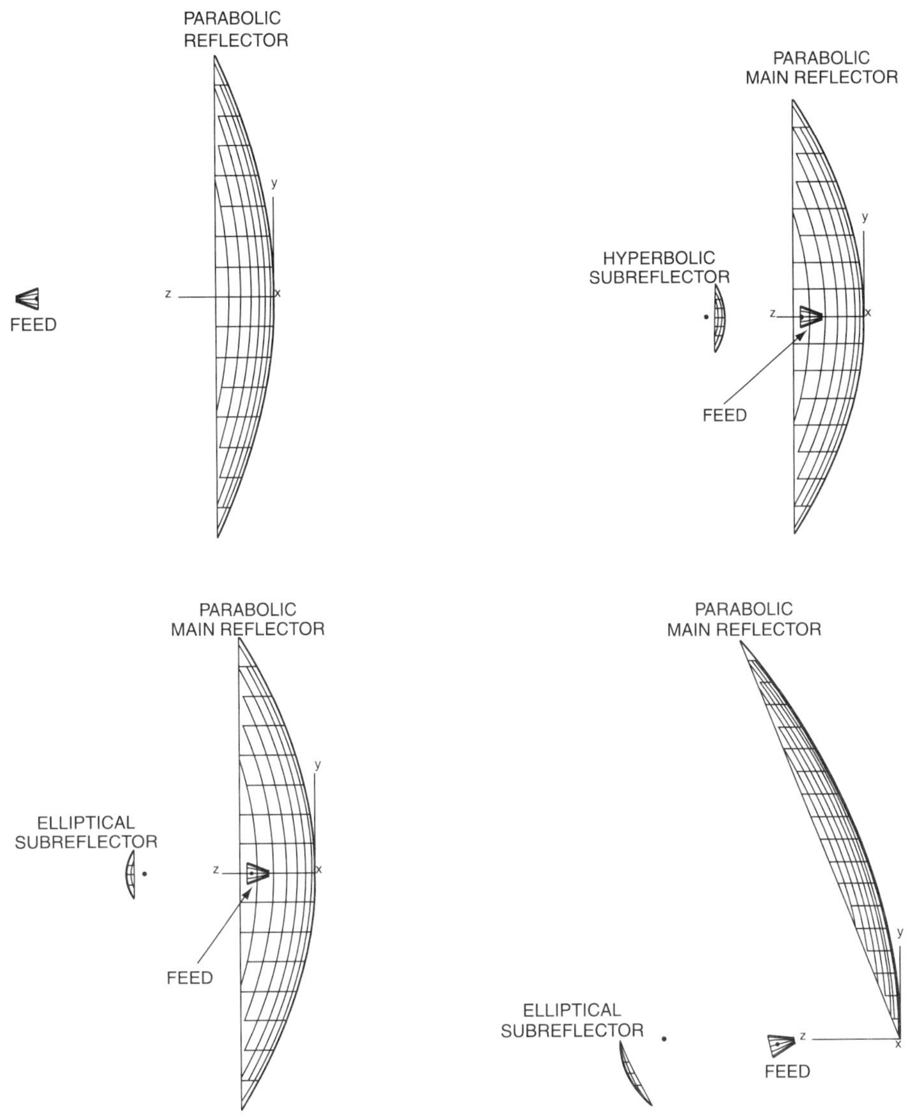

FIGURE 10.17 Configuration of commonly used reflector antennas.

extension to PO where the induced surface current given in Eqs. (10.59) and (10.60) is improved by including a correction that accounts for diffraction effects such as the discontinuity of the surface currents at the lit and shadow region boundaries and near the edges of the surface.

The PTD refines the PO solution just like UTD refines GO. A general form of the PTD, based on the concept of elementary edge waves is given in Reference 78. The PO current introduced earlier is corrected by a "nonuniform" current that accounts for diffraction effects, namely,

$$\mathbf{J}_s(\mathbf{r}') = \mathbf{J}_{PO}(\mathbf{r}') + \mathbf{J}_{PTD}(\mathbf{r}') \tag{10.61}$$

Details of how to determine $\mathbf{J}_{PTD}$ are given in References 76 and 78. The field scattered by the above surface currents are more accurate because the currents account for diffraction effects. Again, the surface integration can be performed asymptotically or numerically. The corrected scattered field resulting from the surface currents can be expressed as

$$\mathbf{E}^s(\mathbf{r}) = \mathbf{E}^s_{PO}(\mathbf{r}) + \mathbf{E}^d_{PTD}(\mathbf{r}) \tag{10.62}$$

where $\mathbf{E}^d_{PTD}(\mathbf{r})$ is the PTD correction term to the PO scattered field, $\mathbf{E}^s_{PO}(\mathbf{r})$. It is important to keep in mind that the PTD diffracted field $\mathbf{E}^d_{PTD}$ is not the same diffracted field as the GTD-diffracted field. The field $\mathbf{E}^d_{PTD}$ is radiated by the nonuniform or "fringe" currents only, whereas the GTD edge diffracted field is radiated by the total currents in the vicinity of the edge.

An approach that simplifies the evaluation of the surface integrals is referred to as the method of equivalent currents (MEC). The MEC is useful because it reduces the evaluation of surface integrals for the nonuniform or fringe surface currents to line integrals around the edges of the surface of the scatterer. This is accomplished by introducing fictitious electric and magnetic line currents that flow around the edges of the scatterer. Although the MEC was introduced heuristically several years ago,[46,67] Michaeli[43,44] showed the connection between the nonuniform surface currents from the PTD and the equivalent line currents. He also obtained more accurate expressions for these currents. It turns out that the correction term, $\mathbf{E}^d_{PTD}(\mathbf{r})$ in Eq. (10.62), can be represented by using the MEC[55] as

$$\mathbf{E}^d_{PTD}(\mathbf{r}) = \frac{jk_o}{4\pi} \oint_c \left(\eta_0 \hat{r} \times \hat{r} \times \mathbf{I}_f + \hat{s} \times \mathbf{M}_f\right) \frac{e^{-jk_o r}}{r} dl \tag{10.63}$$

where $\mathbf{I}_f$ and $\mathbf{M}_f$ are the equivalent fringe electric and magnetic line currents, respectively. Expressions for these currents can be found in the literature.[44]

10.7 Example: Analysis of Reflector Antennas

In the past 40 years, the reflector antenna has become one of the most useful and widely used antennas for communications at microwave frequencies. Traditionally, reflector antennas have been used in point-to-point communications that require antennas with high gain and a narrow beam to maintain the quality of the communication link. With the growing market in satellite communications, reflector antennas onboard satellites have also been designed to cover a given geographic region on Earth for global communication needs. Although there are other types of antennas that can be used to serve the same purposes, reflector antennas are typically much cheaper to make.

Several reflector configurations have been widely used for communications, as shown in Fig. 10.17. The prime focus-fed reflector consists of a parabolic reflector surface and a feed antenna that is located at the focus of the parabola. The Cassegrain dual-reflector system consists of a parabolic main reflector and a hyperbolic subreflector and the Gregorian dual-reflector system consists of a parabolic main reflector and an elliptical subreflector. The feed antenna for the Cassegrain and Gregorian reflectors typically is located at one of the foci associated with the subreflector, whereas the other focus matches with the focus of the parabolic main reflector. Although Fig. 10.17 shows the surfaces that are axially symmetrical, it is also very common to use just part of the surfaces so that the antennas become offset reflector systems. Most of the classical and important work related to reflector antennas has been collected in Reference 34; readers are encouraged to read this reference to understand how the reflector antenna works.

Traditionally, the PO and aperture integration (AI) techniques[71] have been widely used in reflector antenna pattern analysis. These methods have been used successfully to predict the main beam and first few sidelobes of the reflector antenna patterns. However, both methods do not predict accurate results for wide-angle sidelobes and back lobes of the reflector antenna. Consequently, alternative methods are needed to supplement these two techniques to calculate the full volumetric pattern of the reflector. In fact, the UTD[23] has been widely used to complement the PO and AI methods to obtain the complete pattern of the reflector. The PTD[55] has also been used to correct the PO method so that an all-around accurate pattern can be predicted. In this section, these commonly used techniques for analyzing reflector antennas are reviewed.

10.7.1 Physical Optics Analysis

The reflector antenna is a very good candidate to use PO because it is typically electrically large and the pattern region of interest is near and within the specular reflection region of the reflector. With the PO approach, the incident magnetic fields at the reflector surface can be found from ray tracing starting from the feed antenna. The induced surface currents can then be calculated from Eq. (10.59) and integrated by Eq. (10.58) to obtain the pattern. This approach provides good accuracy in the main beam and near sidelobes of the antenna but not for the pattern at wide angles. Because the reflector surface is electrically large, the number of patches for PO integration can become huge, resulting in a significant amount of computation time even for today's fast computers. Many techniques have been developed[47,48] to speed up the computation. More recently, a Gaussian beam (GB) technique, which is based on the asymptotic integration of the PO currents, has been developed and applied to analyze large reflector antennas. It has been found that the GB method is many times faster than conventional PO integration, especially for a reflector that is electrically large.

10.7.2 Physical Theory of Diffraction Analysis

PO can only predict the main beam and first few sidelobes of the reflector but not the wide-angle sidelobes because the termination of the surface current at the reflector rim does not provide a correct edge diffracted field. It is even worse if one attempts to simply use PO to treat a widely defocused reflector because the edges of the reflector are typically strongly illuminated. For such a situation, the edge diffracted fields from the reflector rim are significant and must be correctly included. Using the PTD, introduced in the previous section, can overcome that problem. In the PTD, the equivalent currents, $\mathbf{I}_f$ and $\mathbf{M}_f$, need to be defined around the rim of the reflector and integrated based on Eq. (10.63) to obtain the PO correction. The addition of the equivalent currents yields more accurate results for a wider region than PO alone.

10.7.3 Aperture Integration Method

The AI technique[18,27] has been used for many years to compute the main beam and the near sidelobes of the reflector antenna pattern. In general, the radiation pattern, $\mathbf{E}^s(\mathbf{r})$, of a large aperture antenna, including the reflector, can be determined by the field distribution on the aperture by

$$\mathbf{E}^s(\mathbf{r}) = \frac{-jk_o}{4\pi} r \times \iint_A \left[\hat{n} \times \mathbf{E}^a - \eta_0 \hat{r} \times \left(\hat{n} \times \mathbf{H}^a\right)\right] \frac{e^{-jk_o r}}{r} dA \tag{10.64}$$

where $\mathbf{E}^a$ and $\mathbf{H}^a$ are the electric and magnetic aperture fields, respectively. The aperture integration is carried over the entire aperture A of the antenna, and $\hat{n}$ is the unit vector normal to the aperture plane. It can be shown that for a parabolic reflector antenna with its axis coincident with the z-axis and its aperture defined in the xy plane, the radiation pattern of the antenna given in Eq. (10.64) becomes

$$\mathbf{E}^s(\mathbf{r}) = \frac{-jk_o}{2\pi} \iint_A \left[\mathbf{F}_x E_x^a + \mathbf{F}_y E_y^a\right] \frac{e^{-jk_o R}}{R} dx\, dy \tag{10.65}$$

with E_x^a and E_y^a as the x- and y components of the aperture electric field, respectively, and R is the distance from the given aperture point to the observation point $P(\mathbf{r})$. The aperture field components E_x^a and E_y^a can be obtained simply by ray tracing from the feed through the reflector surface to the aperture plane of the reflector where the size of the aperture is the projection of the reflector rim on the xy plane as shown in Fig. 10.18. The modified vector element patterns, $\mathbf{F}_x$ and $\mathbf{F}_y$, associated with two Huygen's sources[35]

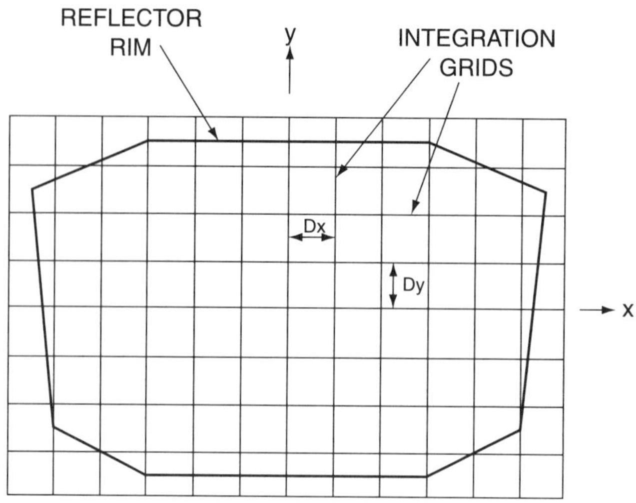

FIGURE 10.18 Geometry for aperture integration method.

(crossed electric and magnetic dipoles) each having its electric field vector parallel to the x-axis and y-axis, respectively, are given by

$$\mathbf{F}_x = \left[\hat{\theta}\cos\phi - \hat{\phi}\sin\phi\right]\cos\left(\frac{\theta}{2}\right)$$
$$\mathbf{F}_y = \left[\hat{\theta}\sin\phi - \hat{\phi}\cos\phi\right]\cos\left(\frac{\theta}{2}\right) \quad (10.66)$$

where θ and ϕ are the conventional spherical coordinate angles. Similar to PO, Eq. (10.65) is also a 2D surface integration over the aperture plane that typically requires significant computation time to obtain the radiation pattern of the antenna. However, because of the characteristics of the aperture field associated with the parabolic reflector, a fast Fourier transform (FFT) can be used to quickly determine the radiation pattern. Alternatively, one can choose a larger grid size in conjunction with the overlapping subaperture and rotated grid methods[27,30] to reduce the 2D surface integration to two 1D integrations that significantly improve the computational efficiency.

One should know that the AI technique could only be applied to a parabolic reflector because it has a well-defined aperture and the phase distribution of the aperture field is fairly smooth. On the other hand, PO can be used in any shape of reflector because the surface current is defined by the actual reflector surface. However, both methods do not predict accurate wide-angle sidelobes and back lobes.

10.7.4 Uniform Theory of Diffraction Analysis

To complement the PO and AI methods, the UTD method has been used to include the edge diffracted field from the reflector rim to predict accurate wide-angle sidelobes and back lobes of the reflector pattern.[16,42,57,65] UTD has also been used to calculate the full pattern of the hyperbolic and elliptical subreflectors for the Cassegrain and Gregorian dual-reflector systems, respectively.[29,64] In addition, if the edge of the reflector rim is not smooth (it has corners), UTD corner diffraction[6] has also been implemented in a computer code[30] to predict the sidelobe levels. Basically, the scattered field, $\mathbf{E}^s(\mathbf{r})$, from a given reflector at a given observation point, $\mathbf{r}$, as shown in Fig. 10.19, can be determined by UTD using the following expression.

$$\mathbf{E}^s(\mathbf{r}) = \mathbf{E}^r(\mathbf{r}) + \mathbf{E}^d(\mathbf{r}) \quad (10.67)$$

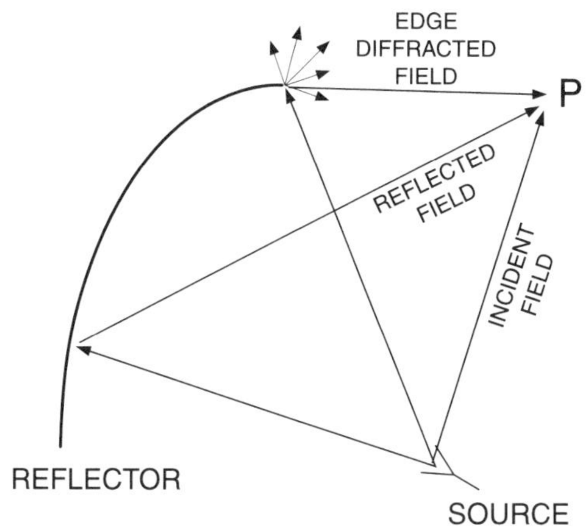

FIGURE 10.19 UTD analysis of reflector antenna.

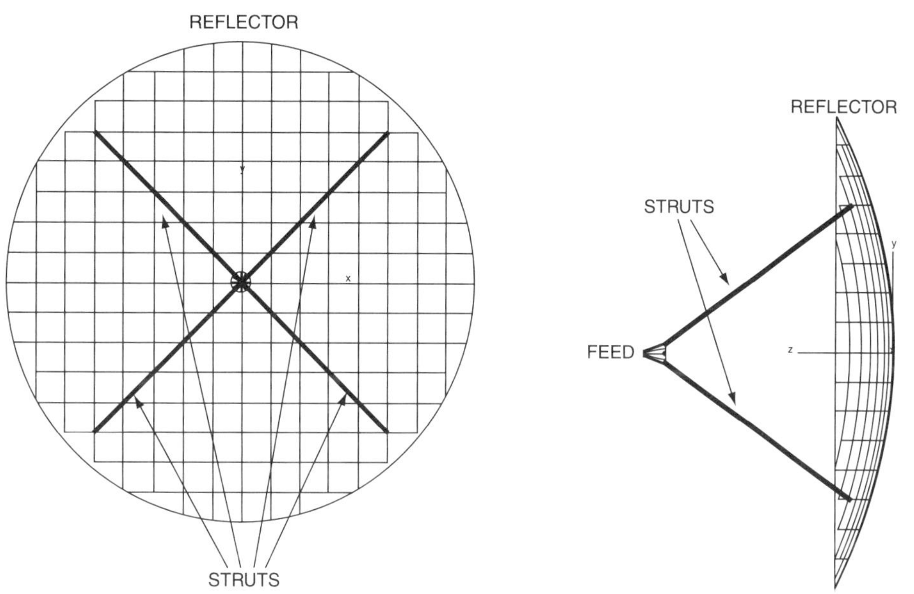

FIGURE 10.20 Feed support struts.

where $\mathbf{E}^r(\mathbf{r})$ and $\mathbf{E}^d(\mathbf{r})$ are the reflected and diffracted fields from the reflector, respectively. Note that Eq. (10.67) cannot be used for calculating the far-field pattern in the main beam region of a parabolic reflector because that is the caustic region associated with the reflected field. In this caustic region ray methods, including the UTD, fail and it is necessary to use the PTD or MEC integral techniques. The total field for the whole reflector can be obtained by adding the direct radiated field from the feed (source) to the scattered field $\mathbf{E}^s(\mathbf{r})$.

Figure 10.20 illustrates that the feed/subreflector has to be connected physically to the main reflector. Consequently, the feed/subreflector and its supporting struts affect the radiated field from the main reflector because they are located in front of the main reflector. The scattered field from the feed/subreflector and its supporting struts decrease the gain of the antenna and also raise its sidelobe level. In many applications, it is very important to take into account this scattered field. The blockage by the feed/subreflector can be treated either by blocking out the aperture field or surface current corresponding to the blocking area before one carries out the PO or AI integration. Alternatively, one can calculate the incident

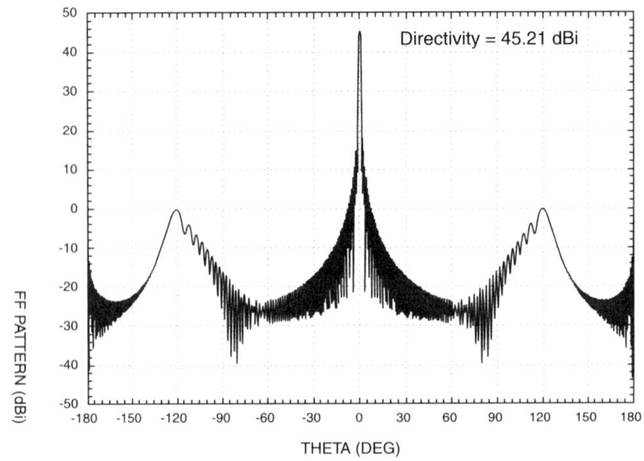

FIGURE 10.21 Calculate far-field pattern at 8 GHz of a focus-fed circular symmetrical parabolic reflector with a diameter of 96 in. and a focal length of 48 in.

field from the main reflector to the blockage and perform the PO integration over the blockage to obtain the forward scattered field from the blockage and then add that scattered field to the radiation pattern of the reflector without the presence of the blockage. For the strut-scattering calculation, one can define an equivalent strut diffraction coefficient[28] and then multiply by the incident field from the reflector to obtain the scattered field from the struts. This is feasible because the struts are normally electrically thin (less than one wavelength).

10.7.5 Numerical Results

To analyze and design reflector antennas, computer codes, such as the NEC-Reflector Antenna Code,[30] have been developed based on the techniques described in the previous sections. A center-fed single parabolic reflector with a diameter of 96 in. and a focal length of 48 in. is used here as an example. A circular symmetrical feed pattern of $\cos^{10}(\theta/2)$ located at the focus is used to illuminate the reflector. The feed antenna is supported by four circular struts and the diameter of the struts is 1 in. The angular separation between adjacent struts is 90°, similar to the one shown in Fig. 10.20. The two endpoints of the struts, in a cylindrical coordinate system, are at (ρ = 2 in., z = 48 in.) and (ρ = 42 in., z = 9 in.), respectively. A circular plate with a diameter of 4 in. located at the focus is used as the feed blockage. Fig. 10.21 shows the far-field pattern of this reflector at 8 GHz without the feed blockage and strut-scattered fields. The directivity of the antenna is about 45.21 dBi for the given feed illumination. If the feed blockage and strut-scattered fields are included, as shown in Fig. 10.22, it is found that the directivity drops to 44.93 dBi and there are significant strut-scattered fields present in the far-field pattern, especially for the pattern around $\theta \approx \pm 70°$ where the strut-scattered field is strongest because of its scattering characteristics.[66] Readers can also find another example, given in References 25 and 31, that includes both the calculated and measured patterns of two 8-ft diameter reflectors. It can be seen from the comparison between the calculated and measured patterns that the analytic techniques described here indeed provide an excellent tool for engineers designing reflector antenna systems. However, one should keep in mind that if any computer code based on these techniques is going to be used to design antennas, it is very important to first validate its accuracy.

10.8 Summary and Conclusions

A brief review of the most widely known HF techniques applicable to electromagnetic scattering and radiation phenomena has been presented in this chapter. Because of space limitations, complete expressions

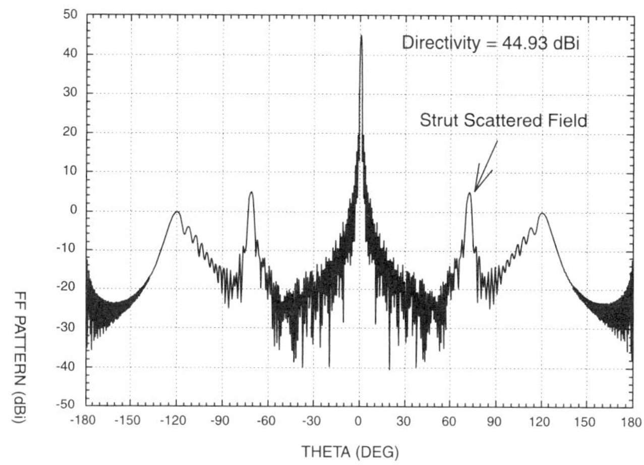

FIGURE 10.22 Calculate far-field pattern at 8 GHz of a focus-fed circular symmetrical parabolic reflector with a diameter of 96 in. and a focal length of 48 in. Feed blockage and strut-scattered fields are included.

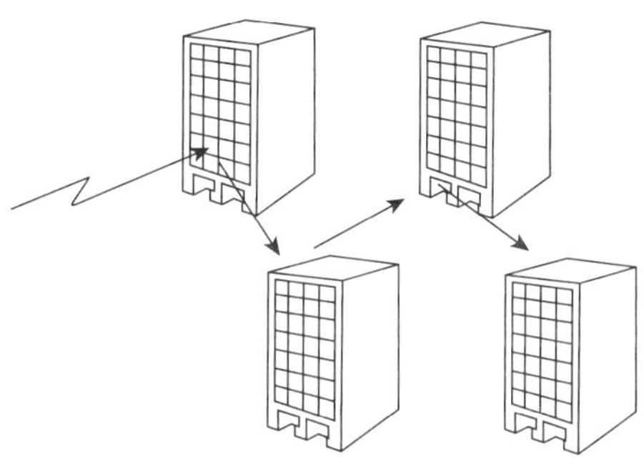

FIGURE 10.23 EM propagation in urban environment.

for various diffraction coefficients were not given, however, an extensive list of references is provided where interested readers can find additional information. Although an attempt was made to discuss, even briefly, most of the known HF techniques, several of them were not included in this chapter. Among the methods not discussed here are the uniform asymptotic theory (UAT),[26] the spectral theory of diffraction (STD),[49,73] and the concept of incremental length diffraction coefficients (ILDC).[22] The STD is useful when the incident field is not ray optical because an arbitrary incident field can, in principle, be expanded in terms of plane waves (homogeneous and inhomogeneous). This method allows us to find diffracted fields resulting from arbitrary incident fields by using the results obtained for plane wave incidence.

All the preceding techniques, in conjunction with ray-tracing algorithms, can be used to study the important problem of EM propagation in urban environments as depicted in Fig. 10.23. Appropriate ray-tracing algorithms can be used to determine the ray paths. Various ray-tracing algorithms are currently being developed for this particular application. The reflection, transmission, and diffraction coefficients can be found by the most appropriate method (analytic, numerical, hybrid) discussed in this text or found in the literature.

References

1. Bernard, J. Diffraction by a metallic wedge covered with a dielectric material, *J. Wave Motion 9* (1987), 543–561.
2. Bernard, J. On the diffraction of an electromagnetic skew incident wave by a nonperfectly conducting wedge. *Ann. Telecommun. 45* (1990), 30–39.
3. Borovikova, V., and Kinber, B. Y. *Geometrical Theory of Diffraction*. London, United Kingdom: IEE, 1994.
4. Bouche, D., Molinet, F., and Mittra, R. *Asymptotic Methods in Electromagnetics*. Berlin, Germany: Springer-Verlag, 1994.
5. Burnside, W., and Burgener, K. W. High frequency scattering by a thin lossless dielectric slab. *IEEE Trans. Antennas Propagat. 31* (January 1983), 104–110.
6. Burnside, W., Wang, N., and Pelton, E. Near-field pattern analysis of airborne antennas. *IEEE Trans. Antennas Propagat. 28* (May 1980), 318–327.
7. Büyükaksoy, A., and Uzgören, G. High-frequency scattering from the impedance discontinuity on a cylindrically curved surface. *IEEE Trans. Antennas Propagat. 35* (1987), 234–236.
8. Büyükaksoy, A., and Uzgören, G. Diffraction of high-frequency waves by a cylindrically curved surface with different face impedances. *IEEE Trans. Antennas Propagat. 36* (1988), 690–695.
9. Büyükaksoy, A., Uzgören, G., and Serbest, A. Diffraction of an obliquely incident plane wave by the discontinuity of a two part thin dielectric plane. *Int. J. Eng. Sci. 27* (1989), 701–710.
10. Chuang, C., and Liang, M. A uniform asymptotic analysis of the diffraction by an edge in a curved screen. *Radio Sci. 23* (1988), 781–790.
11. Demetrescu, C. A general solution of the second-order difference equation occurring in diffraction theory. *Radio Sci. 31* (1996), 461–467.
12. Felsen, L. Asymptotic expansion of the diffracted wave for a semi-infinite cone. *IRE Trans. Antennas Propagat. 5* (1957), 402–404.
13. Felsen, L. Plane wave scattering by small angles cones. *IRE Trans. Antennas Propagat. 5* (1957), 121–129.
14. Felsen, L., and Marcuvitz, N. *Radiation and Scattering of Waves*. Englewood Cliffs, NJ: Prentice-Hall, 1973.
15. Hansen, R., Ed. *Geometrical Theory of Diffraction*. New York, NY: IEEE Press, 1981.
16. James, G., and Kerdemelidis, V. Reflector antenna radiation pattern analysis by equivalent edge currents. *IEEE Trans. Antennas Propagat. 21* (January 1973), 19–24.
17. James, G. L. *Geometrical Theory of Diffraction for Electromagnetic Waves*, Third ed. London, United Kingdom: Peter Peregrinus, 1986.
18. Kauffman, J., Crosswell, W., and Powers, L. Analysis of radiation patterns of reflector antennas. *IEEE Trans. Antennas Propagat. 24* (January 1976), 53–65.
19. Keller, J. Geometrical theory of diffraction. In *Calculus of Variations and Applications, Ed. L. M. Graves*. New York: McGraw-Hill, 1958, pp. 27–52.
20. Keller, J. Geometrical theory of diffraction. *J. Opt. Soc. Am. 52* (February 1962), 116–130.
21. Kline, M. An asymptotic solution of Maxwell's equation. *Commun. Pure Appl. Math. 4* (1951), 225–262.
22. Knott, E. The relationship between Mitzner's ILDC and Michaeli's equivalent currents. *IEEE Trans. Antennas Propagat. 33* (1985), 112–114.
23. Kouyoumjian, R., and Pathak, P. A uniform geometrical theory of diffraction for an edge of a perfectly conducting surface. *Proc. IEEE 62* (November 1974), 1448–1461.
24. Kouyoumjian, R., Pathak, P., and Burnside, W. A uniform GTD for the diffraction by edges, vertices and convex surfaces. In *Theoretical Methods for Determining the Interaction of Electromagnetic Waves with Structures, Ed. J. K. Skwirzynski*. Netherlands, Sijthoff and Noordhoff, 1981.
25. Lambert, K., Rudduck, R., and Lee, T. A new method for obtaining antenna gain from backscatter measurements. *IEEE Trans. Antennas Propagat. 38* (June 1990), 896–902.

26. Lee, S., and Deschamps, G. A uniform asymptotic theory of EM diffraction by a curved wedge. *IEEE Trans. Antennas Propagat. 24* (January 1976), 25–34.
27. Lee, S., and Rudduck, R. Aperture Integration and GTD Techniques used in the NEC Reflector Antenna Code. *IEEE Trans. Antennas Propagat. 33* (February 1985), 189–194.
28. Lee, S., Rudduck, R., Klein, C., and Kouyoumjian, R. A GTD Analysis of the Circular Reflector Antenna Including Feed and Strut Scatter. Tech. Rep. 4381-1, Ohio State University ElectroScience Lab., Dept. of Electrical Engineering, May 1977.
29. Lee, T. Geometrical Optics and GTD Analysis of Subreflectors in Cassegrain and Gregorian Reflector Antennas. Master's thesis, Ohio State University, June, 1984.
30. Lee, T., and Rudduck, R. OSU Reflector Antenna Code — Version 3.0 (NECREF Version 3.0), User's Manual. Tech. Rep. 318021-1, Ohio State University ElectroScience Lab., Dept. of Electrical Engineering, February 1994.
31. Lee, T., Ruddock, R., and Lambert, K. Pattern Measurements of Reflector Antennas in the Compact Range and Validation with Computer Code Simulation. *IEEE Trans. Antennas Propagat. 38* (June 1990), 889–895.
32. Léontovich, M. *Investigation on radio wave propagation.* Part II, Moscow: Printing House of the Academy of Sciences, 1948, pp. 5–12.
33. Levy, B., and Keller, J. Diffraction by a smooth object. *Commun. Pure Appl. Math.* 12 (1959), 159–209.
34. Love, A. *Reflector Antennas.* New York: IEEE, 1978.
35. Ludwig, A. The definition of cross polarization *IEEE Trans. Antennas Propagat. 21* (January 1973), 116–119.
36. Luneberg, R. *Mathematical Theory of Optics.* Providence, RI: Brown Univeristy Press, 1944.
37. Ly, H., and Rojas, R. Analysis of diffraction by material discontinuities in thin material-coated planar surfaces based on maliuzhinets method. *Radio Sci. 28* (May–June 1993), 281–297.
38. Ly. H., Rojas, R., and Pathak, P. EM plane wave diffraction by a planar junction of two thin material half-planes — oblique incidence. *IEEE Trans. Antennas Propagat. 41* (April 1993), 429–441.
39. Lyalinov. M. Diffraction of a plane wave by a wedge with thin material coatings. *Radio Sci. 31* (1996), 1721–1732.
40. Maliuzhinets, G. Excitation, reflection and emission of surface waves from a wedge with given face impedances. *Sov. Phys. Dokl., Engl. Transl. 3* (1959), 752–755.
41. McNamara, D., Pistorius, C., and Malherbe, J. *Introduction to the Uniform Geometrical Theory of Diffraction.* Norwood, MA: Artech House, 1990.
42. Mentzer, C., Pathak, P., and Peters, Jr., L. Pattern analysis of an offset Fed Parabolic Reflector Antenna. Tech. Rep. 3220-2, Ohio State University ElectroScience Lab., Dept. of Electrical Engineering, June 1972.
43. Michaeli, A. Equivalent edge currents for arbitrary aspects of observation. *IEEE Trans. Antennas Propagat. 32* (March 1984), 252–258.
44. Michaeli,A. Elimination of Infinities in Equivalent Edge Currents. Part I: Fringe Current Components. *IEEE Trans. Antennas Propagat. 34* (July 1986), 912–918.
45. Michaeli, A. Transition functions for high-frequency diffraction by a curved perfectly conducting wedge; Part I: Canonical solutions for a curved sheet; Part II: A partially uniform solution for a general wedge angle; Part III: Extension to overlapping transition regions. *IEEE Trans. Antennas Propagat. 27* (1989), 1073–1092.
46. Millar, R. An approximate theory of the diffraction of an electromagnetic wave by an aperture in a plane screen. *Proc. Inst. Elec. Eng., Pt. C 103* (1956), 177–185.
47. Mittra, R., Ko, W., and Sheshadri, M.A. Transform technique for computing the radiation pattern of prime-focal and Cassegrainian reflector antennas. *IEEE Trans. Antennas Propagat. 30* (May 1982), 520–524.
48. Mittra, R., Rahmat-Samii, Y., Galindo-Israel, V., and Norman, R. An efficient technique for the computation of vector secondary patterns of offset paraboloid reflectors. *IEEE Trans. Antennas Propagat. 27* (May 1979), 294–304.

49. Mittra, R., Rahmat-Samii, Y., and Ko, W. Spectral theory of diffraction. *Appl. Phys. 10* (1976), 1–13.
50. Noble, B. *Methods Based on the Weiner-Hopf Technique.* New York: Pergamon Press, 1958.
51. Osipov, A. V. Diffraction by a wedge with higher-order boundary conditions. *Radio Sci. 31* (1996), 1705–1720.
52. Otero, M., and Rojas, R. EM scattering from a complex material cylinder in the presence of an impedance wedge. *J. Electromagn. Waves and Appl. 10* (1996), 1563–1581.
53. Otero, M., and Rojas, R. Resistive treatment to reduce edge diffraction from large wedge shaped objects and planar antennas. *Radio Sci. 32* (1997), 1745–1759.
54. Otero, M., and Rojas, R. Two dimensional Green's function for a wedge with impedance faces. *IEEE Trans. Antennas Propagat. 45* (December 1997), 1799–1809.
55. Pathak, P. Techniques for High-Frequency Problems. In *Antenna Handbook,* Y.T. Lo and S.W. Lee, Eds., New York: Van Nostrand Reinhold, 1988, ch. 4.
56. Pelosi, G., Manara, G., and Nepa, P. A UTD solution for the scattering by a wedge with anisotropic impedance faces: skew incidence case. *IEEE Trans. Antennas Propagat. 46* (April 1998), 579–588.
57. Ratnasiri, P., Kouyoumjian, R., and Pathak, P. The Wide Angle Sidelobes of Reflector Antennas. Tech. Rep. 2183-1, Ohio State University ElectroScience Lab., Dept. of Electrical Engineering, March, 1970.
58. Rojas, R. Comparison between two asymptotic methods. *IEEE Trans. Antennas Propagat. 35* (December 1987), 1489–1492.
59. Rojas, R. Electromagnetic diffraction of an obliquely incident plane wave field by a wedge with impedance faces. *IEEE Trans. Antennas Propagat. 36* (July 1988), 956–970.
60. Rojas, R. Wiener-Hopf analysis of the EM diffraction by an impedance discontinuity in a planar surface and by an impedance half-plane. *IEEE Trans. Antennas Propagat. 36* (January 1988), 71–83.
61. Rojas, R., and Al-hekail, Z. Generalized impedance/resistive boundary conditions for electromagnetic scattering problems. *Radio Sci. 24* (January–February), 1989, 1–12.
62. Rojas, R., and Chou, L. Generalized Impedance/Resistive Boundary Conditions for a Planar Chiral Slab. In *1991 North American Radio Science Meeting* (June 1991), URSI, p. 358.
63. Rojas, R., and Otero. M. Scattering by a resistive strip attached to an impedance wedge. *J. Electromagn. Waves Appl. 7* (1993), 373–402.
64. Rusch, W. Scattering from a hyperboloidal reflector in a Gassegrain feed system. *IEEE Trans. Antennas Propagat. 11* (July 1963), 414–421.
65. Rusch, W., and Sorensen, O. The geometrical theory of diffraction for axially symmetric reflectors. *IEEE Trans. Antennas Propagat. 23* (May 1975), 414–419.
66. Rusch, W., Sorensen, O., and Baars, J. Radiation cones from feed support struts of symmetric Paraboloidal antennas. *IEEE Trans. Antennas Propagat. 30* (July 1982), 786–790.
67. Ryan, C., and Peters, L.J. Evaluation of edge diffracted fields including equivalent currents for the caustic regions. *IEEE Trans. Antennas Propagat. 17* (1969), 292–299. (See also corrections, Vol. 18, pp. 275, 1979).
68. Rytov, S. Calcul du skin-effet par la méthode des perturbations. *J. Phys. USSR 2* (1940), 233–242.
69. Senior, T., and Volakis, J. *Approximate boundary conditions in electromagnetics.* Herts, United Kingdom: IEE, 1995.
70. Sikta, F., Burnside, W., Chu, T., and Peters, Jr., L. First-order equivalent current and corner-diffracton scattering from flat-plate structures. *IEEE Trans. Antennas Propagat. 31* (July 1983), 584–589.
71. Silver, S. *Microwave Antenna Theory and Design*, vol. 12 of *MIT Radiation Laboratory.* New York: McGraw Hill, 1949.
72. Sommerfeld, A. Mathematische theorie der diffraktion. *Math. Ann. 47* (1986), 314–374.
73. Tiberio, R., Manara, G., Pelosi, G., and Kouyumjian, R. High-frequency electromagnetic scattering of plane waves from double wedges. *IEEE Trans. Antennas Propagat. 37* (1989), 1172–1180.
74. Tiberio, R., Pelosi, G., and Manara, G. A uniform GTD formulation for the diffraction by a wedge with impedances faces. *IEEE Trans. Antennas Propagat. 33* (August 1985), 867–873.

75. Tiberio, R., Pelosi, G., Manara, G., and Pathak, P. High frequency scattering from a wedge with impedance faces illuminated by a line source, Part I. *IEEE Trans. Antennas Propagat. 37* (February 1989), 212–218.
76. Ufimtsev, P. Method of edge waves in the physical theory of diffraction from Russian: Medthod Krayevykh voin v fizicheskoy teorii difraktsii, 1962. Translation prepared by the U.S. Air Force Foreign Technology Division, Wright-Patterson AFB, Ohio, Sept. 1971.
77. Ufimtsev. P. Comments on the comnparison of three high-frequency diffraction techniques. *Proc. IEEE 63* (1975), 1734–1737.
78. Ufimtsev, P. Elementary edge waves and the physical theory of diffraction. *Electromagnetics 11* (April–June 1991), 125–160.
79. Vaccaro, V. The generalized reflection method in electromagnetism. *AEU 34* (1980), 493–500.
80. Yazici, A., and Serbest, A. Scattering of plane waves by an anisotropic dielectric half-plane. *IEEE Trans. Antennas Propagat. 47* (September 1999), 1476–1484.

PART C

Antenna Developments and Practical Antennas

11

Outdoor and Indoor Cellular/Personal Handy Phone System Base Station Antenna in Japan

Hiroyuki Arai
Yokohama National University

11.1 Introduction

This chapter presents outdoor and indoor base station antenna systems, which are commercially used for the Japanese cellular system and the personal handy phone system (PHS). The antenna system consists of an antenna element, a feeding network, a phase control unit, and a diplexer or a duplexer. This chapter is organized in four parts. The first part describes various components of a base station antenna system. A unique feature of the Japanese cellular system is a dual-band operation and the beam-tilting technique for a high-capacity system. The key components are dual-band antenna elements and phase shifters. The second topic covered is the configuration of a base station antenna with a coverage area bounded by a circle with the radius of 1.5 to 5 km. When the antenna gain is greater than or equal to a nominal level, the base station antenna needs an array design. In addition, every base station uses a diversity system to receive a weak uplink signal affected by the multipath fading. Two kinds of diversity antenna configurations, space and polarization diversities, are presented. The third topic covered is the antennas for a microcellular system. This system provides services inside buildings, tunnels, and subway stations. The coverage area is 20 m to 1 km, and the antenna gain of such a base station is relatively low. The last antennas described in this chapter are for the PHS.

0-8493-0124-6/02/$0.00+$1.50

11.2 Base Station Antennas for a Cellular System

A cellular system services terrestrial area by dividing it into a number of cells. The radiation pattern and the output power of the base station antenna determine the shape of the coverage area of each cell. The design parameters for the base station antenna are a vertical plane pattern and a horizontal plane pattern. The vertical pattern shape is achieved by an array antenna configuration, and the antenna element and reflector determine the horizontal pattern.

The current cellular system adopts a small size cell in radius of 1.5 km to increase the capacity of subscribers. In the frequency division multiple access (FDMA)/time division multiple access (TDMA) system, the same frequency band is used in nearby two cells, and the interference between the cells using the same frequency band is suppressed to a very low level by the radiation pattern in the vertical plane. To shape the vertical pattern, the cell is divided into several sectors in the horizontal plane to suppress the interference between mobile terminals. The FDMA/TDMA system uses different frequency bands between adjoining sectors, whereas the code division multiple access (CDMA) system uses the same frequency band in all sectors. The sector-shaped pattern is also very important for the CDMA system to increase the capacity of subscribers.

Because the CIR inside one sector degrades as the number of mobile terminals increases, the total number of subscribers should be restricted by adjusting the radiation pattern of the base station. This is usually achieved by the sector-shaped beams in the horizontal plane. An omnidirectional radiation pattern is used to expand coverage area in the suburbs. The element antenna characterizes the shape of the radiation pattern in the horizontal plane. A reflector and parasitic elements attached to a radiation element further shape the sector pattern.

The second technique to restrict the coverage area of the main beam is by tilting it downward from the horizontal axis in the vertical plane. There are two methods for this beam tilting: one is a mechanical beam tilting by leaning the antenna, and the other is an electrical beam tilting by adjusting relative phases of antenna elements.

The uplink signal at the base station antenna fluctuates as a result of fading caused by the multiple reflections at buildings and obstacles. A diversity reception is used to increase the uplink signal level at the base station and to reduce the effect of the multipath fading. Though a sector-shaped pattern in the horizontal plane is a kind of antenna pattern diversity, many base stations have another antenna to provide the space diversity. Polarization diversity is often used to decrease the number of antenna at a base station, because a reception antenna can be mounted at the same position with the common use antenna for both transmission and reception in the polarization diversity.

The main components of a base station antenna are an antenna element, a feed circuit network, and a phase shifter. The antenna gain of 10 to 20 dBi is required for the macrocellular system with the coverage area of 1 to 5 km in radius. This high gain is obtained by the array structure. A simple array structure uses the same amplitude and phase excitation for the broadside radiation; however, the phase of each element is adjusted to obtain the tilted beam pattern in the vertical plane. This section presents a basic antenna element for the cellular base station, dual-band antenna elements for the practical usage in the current system, and a triband antenna element for the introduction of IMT-2000 in Japan. The triband antenna element should have three-frequency bandwidth for 900 MHz, and 1.5 and 2 GHz. Discussions on the design and use of feed network circuit and variable phase shifters and on passive intermodulation are also included in this section.

11.2.1 Cellular Base Station Antenna Elements

A printed dipole antenna is the most popular element for the base station antennas of the cellular system, because a half wavelength dipole antenna has a wide-frequency bandwidth more than 15% for the input voltage standing wave ratio (VSWR) less than 2. The omnidirectional radiation pattern in the H-plane of the dipole antenna is formed to be a sector beam by mounting above the reflector. The feed circuit network of the printed dipoles consists of microstrip line for the array structure in the vertical plane.

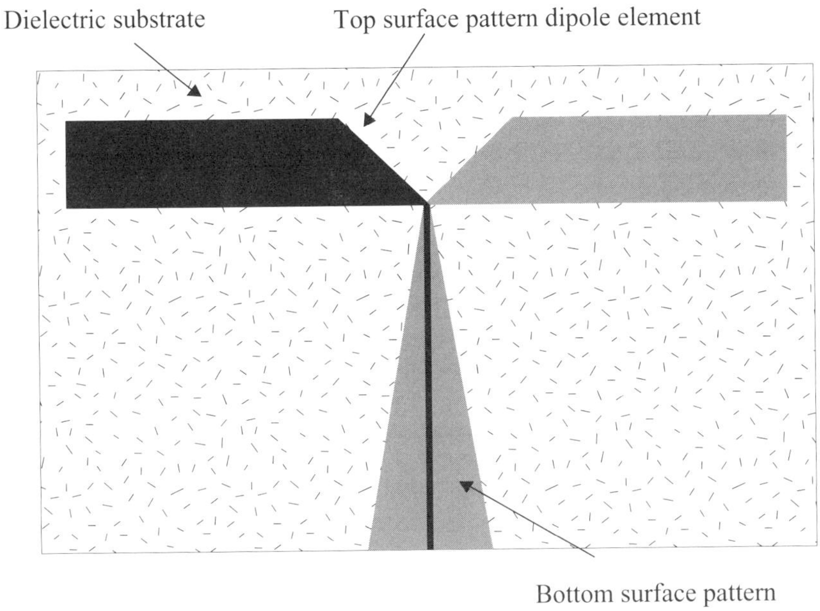

FIGURE 11.1 (a) Printed dipole antenna with taper balun.

The balanced feed of the dipole antenna is connected to a balance-to-unbalance transformer (balun) for the unbalanced feed of the microstrip line. For the mass production of the antenna element, the printed dipole with the balun circuit is used.

Figure 11.1 shows two types of printed dipole antennas [1]. The built-in balun is a taper-shaped ground plane as shown in Fig. 11.1(a) and an inverted J-shaped strip line on the wide conductor strip in Fig. 11.1(b). The input return loss characteristics of both types of feeding structures are shown in Fig. 11.2

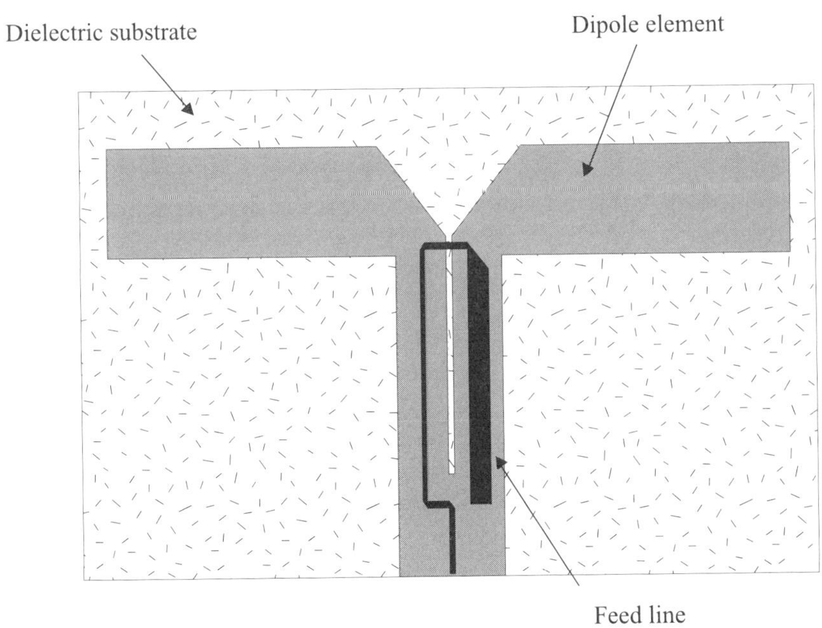

FIGURE 11.1 (b) Printed dipole antenna with built-in balun.

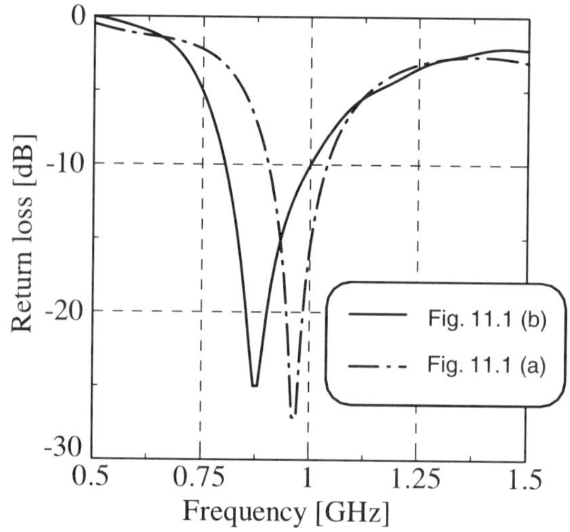

FIGURE 11.2 Input return loss characteristics of printed dipole antenna.

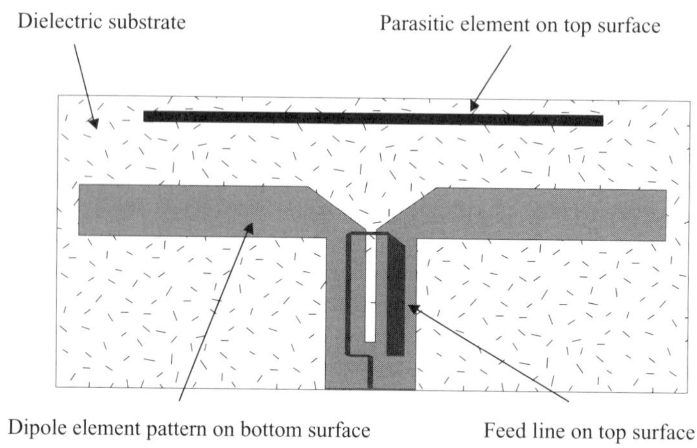

FIGURE 11.3 Printed dipole antenna with parasitic element.

by assuming that the characteristic impedance of the strip line is 50 Ω. An undesired resonance of the feed line appears in the taper-shaped ground plane; however, it does not appear in the inverted J-shaped strip line. In addition, the resonant frequency of Fig. 11.1(b) is shifted to lower frequency band by the bent current flowing on the dipole element. This effect reduces the size of the antenna. Because of their advantages, the inverted J-shaped strip line is widely used for the feed circuit of the printed dipole element.

The printed dipole antenna is mounted on a reflector to obtain a sector beam in the horizontal plane. Mainly, the reflector shape adjusts the beam width. In the Japanese personal digital cellular (PDC) system, the half power beam width in the horizontal plane is 120° or 90°. The beam width of 60° is used for IMT-2000. A wide-frequency bandwidth is necessary not only for the input impedance characteristics but also for the horizontal radiation pattern. The parasitic element is added to the printed dipole to keep the same radiation pattern in the horizontal plane as shown in Fig. 11.3 [2].

To obtain narrow beam width, a twin dipole antenna is often used as an antenna element. The half power beam width of the microstrip antenna is less than 90°. A disadvantage of the microstrip antenna element is its narrow frequency bandwidth. For an antenna made by dielectric substrate with the thickness ranging from 0.8 to 3.2 mm the bandwidth is about 2 to 3%, whereas the required frequency bandwidth

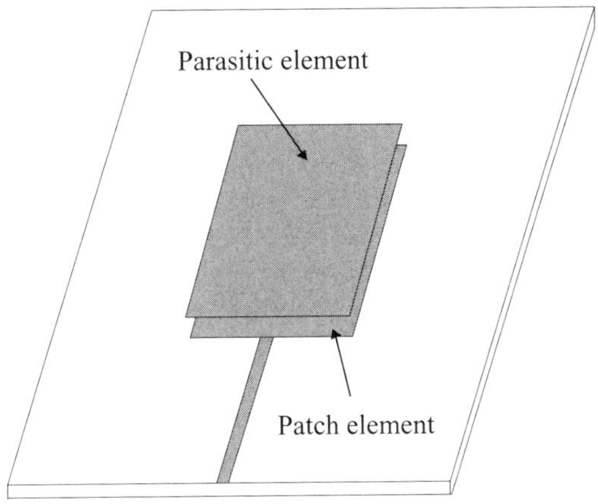

FIGURE 11.4 Wideband patch antenna with parasitic element.

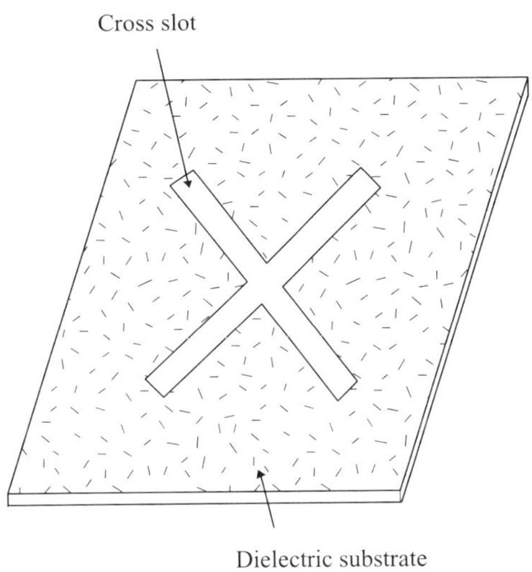

FIGURE 11.5 Cross-slot antenna for slant diversity.

for the diversity reception in the cellular system is 8 to 13%. To expand the frequency bandwidth of the microstrip antenna, a parasitic element is placed above the radiating element as shown in Fig. 11.4.

The vertical polarization is mainly used for the cellular system; however, the horizontal polarization and ± 45° slant polarization are often used for the polarization diversity system. A printed twin dipole antenna is used for both vertical and horizontal polarization diversity antennas. An element for the slant diversity is a printed cross-slot antenna as shown in Fig. 11.5.

The omnidirectional pattern antenna is installed in the suburbs with relatively low subscribers. Its antenna element is a sleeve antenna for the low-gain type and is a collinear array for high-gain antenna. For the simple structure in the array design, one wavelength slot etched on the dielectric substrate is used for the omnidirectional antenna as shown in Fig. 11.6 [3]. This element is inserted into the conducting cylinder to obtain wide-frequency characteristics. This cylinder has a function of the parasitic element.

11.2.2 Dual- and Triple-Band Antenna Elements

Figure 11.7 shows the frequency band plan of cellular telephone service and PHS in Japan. Two operators are assigned to a 900-MHz band and three operators to a 1.5-GHz band. A major operator, Nippon

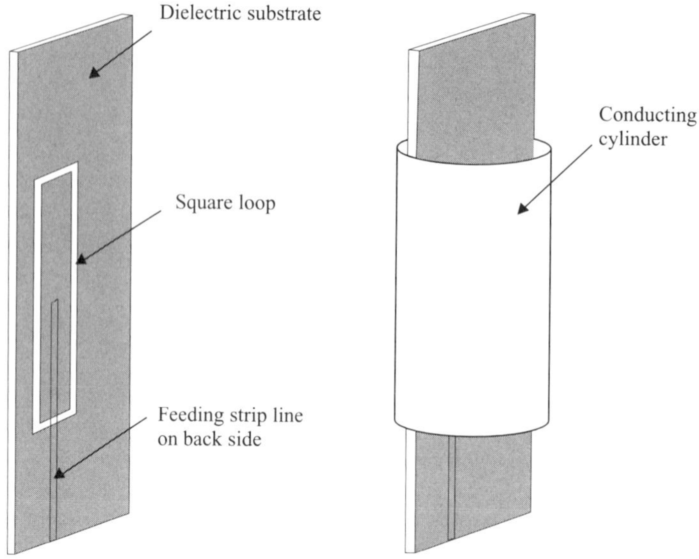

FIGURE 11.6 Wideband antenna element for omnidirectional pattern in horizontal plane.

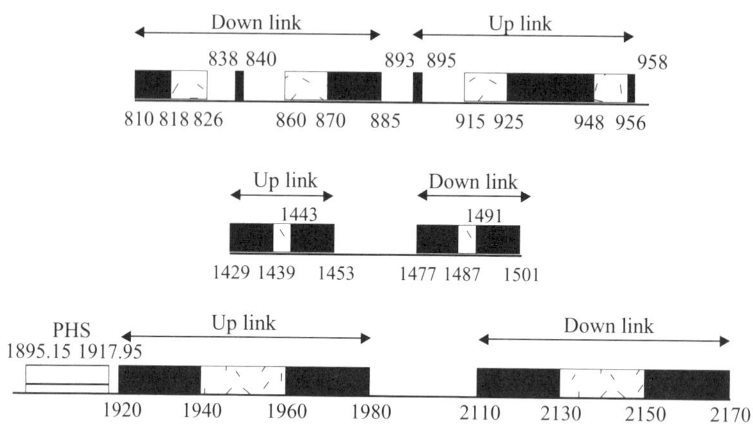

FIGURE 11.7 Frequency band plan of Japanese cellular system.

Telephone and Telegraph (NTT) DoCoMo has been assigned to both frequency bands, and uses a dual-frequency band antenna for the base station antenna [4]. The dual-band antenna has two resonant frequencies for a single-input port. Signals with two different frequency bands are fed through a combiner. A multiresonant antenna element is necessary for the dual-band antenna. A simple method for adding another resonant frequency of f_2 to the antenna element with a dominant resonant frequency of f_1 ($f_2 > f_1$) uses a parasitic element placed in the vicinity of the antenna. Figure 11.8 shows an example of a half wavelength dipole antenna for f_1 with a parasitic element for f_2. The multiresonant antenna can be designed by adding the parasitic element for another resonant frequency. In the printed antenna used for the cellular base station, the parasitic element is placed on the same substrate. It is easy to obtain a multiresonant antenna by using the parasitic element to meet input impedance characteristics. However, it is difficult to realize equally divided sector beams in the horizontal plane for every frequency. This is because the electrical length between the antenna and the reflector is different from that between the parasitic element and the reflector. The pattern of the dual-frequency antenna can be adjusted by varying the reflector shape and by adding another parasitic element [5].

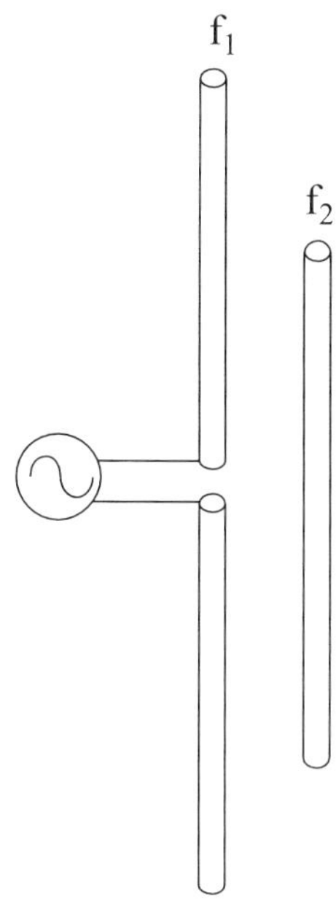

FIGURE 11.8 Dual-band dipole antenna using parasitic element.

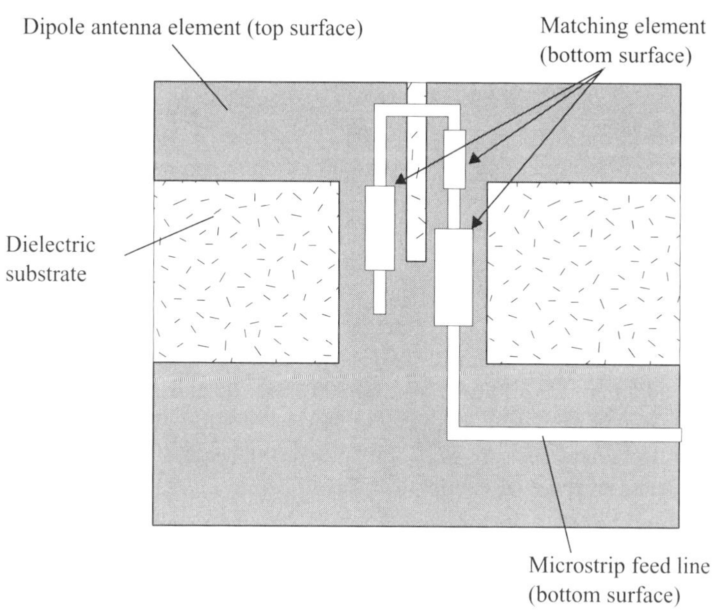

FIGURE 11.9 Dual-band antenna using matching network.

Another technique for obtaining a dual-band operation is to use a matching network circuit, while keeping the electrical spacing from the antenna to the reflector the same and exciting the antenna by the same feed line as shown in Fig. 11.9 [6]. Figure 11.10 shows an example of the dual-band antenna using a matching network. The height of the flange is adjusted to obtain the same beam width in the horizontal plane as that in the vertical plane. This dual-band antenna has a vertically mounted conductor plate (short stub) for the wideband impedance matching at 900 MHz.

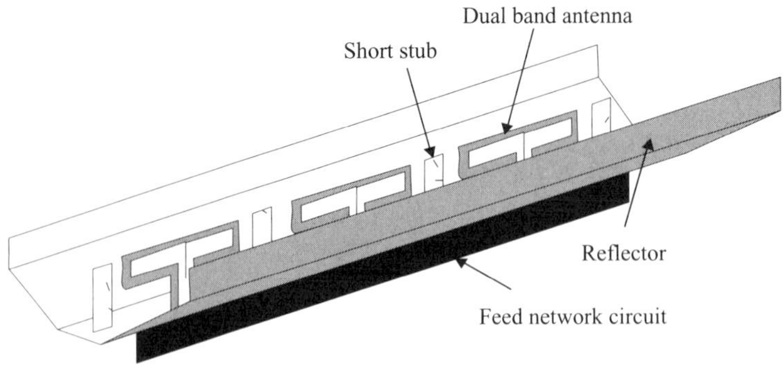

FIGURE 11.10 Dual-band base station antenna.

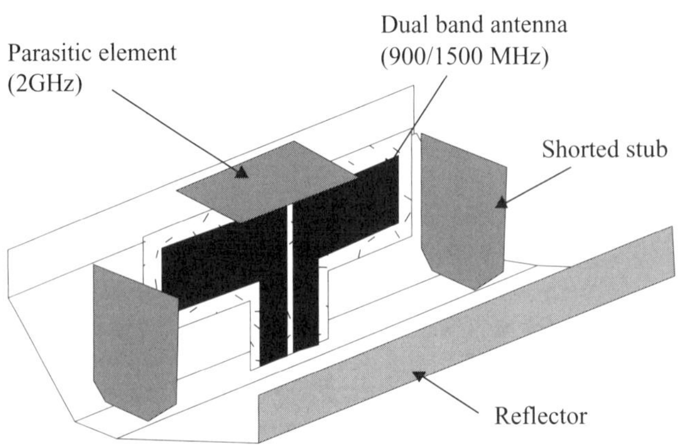

FIGURE 11.11 Triple-band antenna element.

The introduction of IMT-2000 uses an additional frequency band at 2 GHz, which requires installation of an another base station antenna or exchange of the present antenna system to a multiband one. The remarkable increase of the subscribers for the mobile communication system has left few places for new base station building. Therefore, a multiband antenna to replace the present system is required. The dual-band antenna for 900 MHz/2 GHz and for 1.5/2 GHz is obtained by the same technique described earlier. However, a triple-band antenna is also used to cover 900 MHz, 1.5 and 2 GHz. In the IMT-2000 system, a sector beam of 60° half-power beam width is also used in addition to the conventional 120° and 90° sector antenna. The triple-band antenna with the same sector beam of 120° is obtained by adding an another parasitic element of the dual-band antenna as shown in Fig. 11.11 [7]. The matching plate for 900 MHz is directed not to excite the grating lobes in 900 MHz by reducing the element spacing. This antenna element gives a triple-frequency band with almost the same half power beam width in the horizontal plane. The deviation of the beam width in three frequencies is $120 \pm 7°$.

11.2.3 Antenna Feeding Networks

Base station antennas in the Japanese cellular system consist of array structure for the high gain and beam tilting to suppress the interference with an adjoining cell. Although adjusting the phase of each antenna element tilts the beam, the antenna array is divided into several subarrays and the phases of sub-arrays are changed by using phase shifters as shown in Fig. 11.12. The array consists of seven subarrays with four antenna elements in each subarray. Phase shifters control phases of upper three and lower three subarrays using the central subarray as a reference [8].

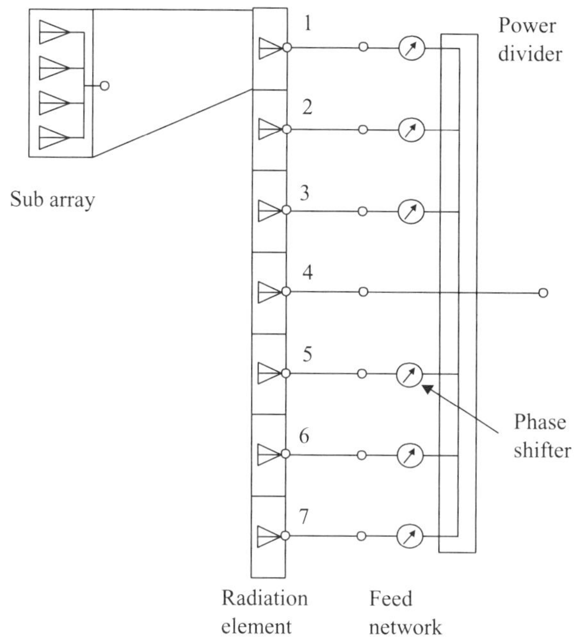

FIGURE 11.12 Feeding network and phase shifter.

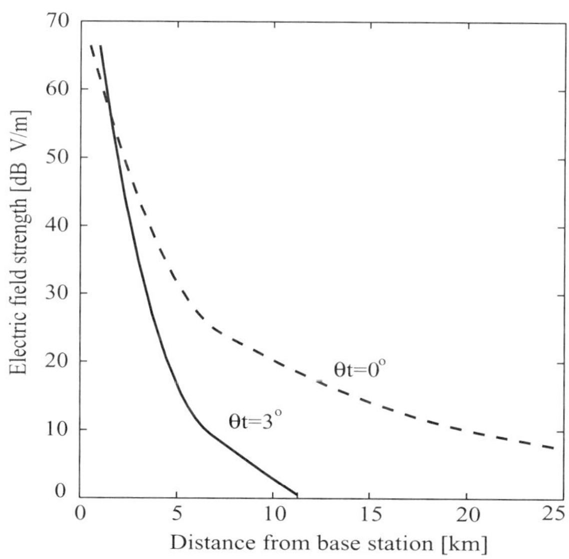

FIGURE 11.13 Propagation loss for tilt beam.

The propagation loss using a tilted beam is more than that for an untilted case as shown in Fig. 11.13. The tilt angle θ_t is given as $\theta_t = \tan^{-1}(h/a)$, where h (m) is the antenna height and a (m) is the radius of the cell. The use of tilted beams on a base leads to smaller cell size because of increased propagation loss.

The theoretical tilt angle is about 2° for $h = 50$ m and $a = 1.5$ km. An actual tilt angle should be 1° or 2° larger than this value to avoid interference with the adjoining cell. In addition to the beam tilt, the sidelobe level should be suppressed below –15 to –20 dB for $\theta < \theta_t$ to reduce the interference to the other cells [9]. Figure 11.14 shows an example of the vertical plane pattern of the base station antenna. In this

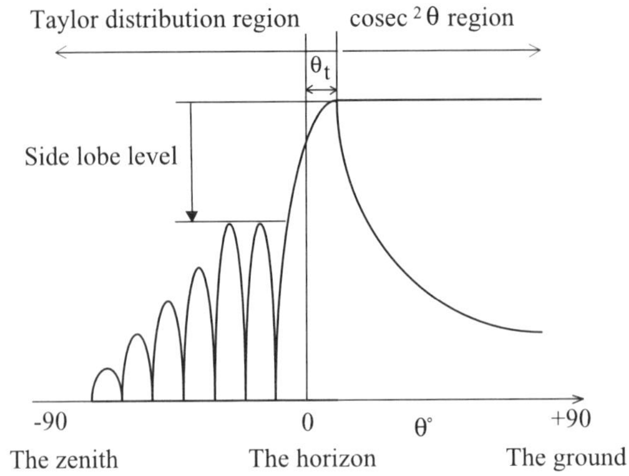

FIGURE 11.14 An example of vertical plane pattern of base station antenna.

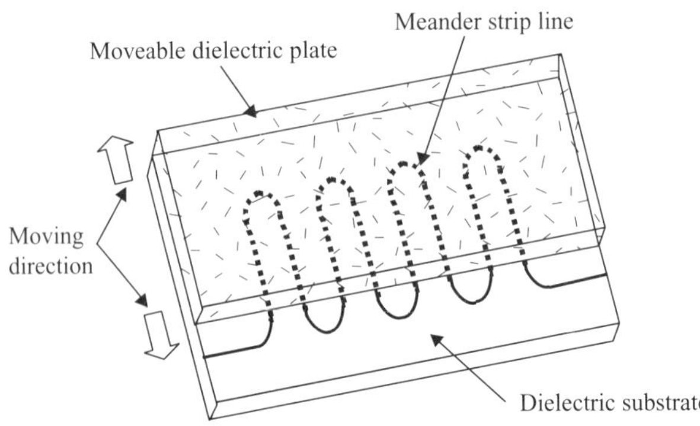

FIGURE 11.15 Variable phase shifter using meander strip line.

example, the vertical pattern is designed to be cosecant squared or $\text{cosec}^2\,\theta$ for $\theta > \theta_t$. The sidelobe level is specified by Taylor distribution for $\theta < \theta_t$. The horizontal direction is taken to be 0°. The excitation coefficient of an array element is determined by the method of pattern synthesis.

11.2.4 Variable Phase Shifters

Phase shifters with a variable phase are used in a feed network to control the phase of each subarray as shown in Fig. 11.12. The range of phase change for the variable phase shifter should be between −360° and 360°. This range could be only a few degrees if the phase of each antenna element is allowed to change. However, this configuration where the phase of each element changes is not suitable for the base station antenna because it increases the number of mechanical movable parts. The digital phase shifter using a pin diode is also not suitable because it causes passive intermodulation effects, as described in the next section. A conventional variable phase shifter consists of movable dielectric block on the meander strip line as shown in Fig. 11.15. By moving the dielectric block, the phase velocity of the strip line is changed and its amount depends on how much strip line is covered by the dielectric block. The change in the phase velocity controls the phase shift caused by the phase shifter. Thus, the location of the movable block changes the phase difference between input and output of the strip line. The insertion loss of this phase shifter is about 1 to 2 dB, which cannot be neglected.

A low loss variable phase shifter consists of triplate transmission line and movable dielectric plates sandwiching the strip line as shown in Fig. 11.16. The relative dielectric constant of the plate is 50 for

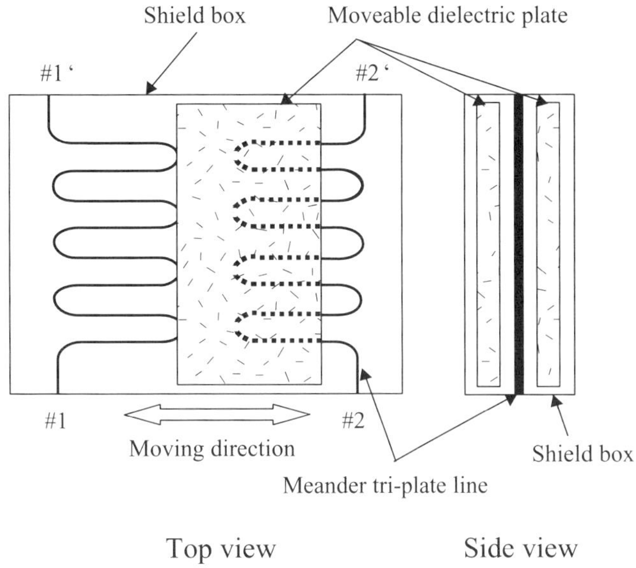

FIGURE 11.16 Variable phase shifter by sandwich dielectric plates.

the miniaturization of the phase shifter. This phase shifter has two input and two output ports designed for the base station antenna beam tilting. The phase delay difference between two output ports is $\pm\theta_d$ by offset of the dielectric plates. The phase origin is determined when the sandwich of dielectric plates is in the center. The maximum range of phase change is ±130° with the error of ±8°, and the insertion loss is about 0.2 dB at 2 GHz [10]. Figure 11.17 shows the feed circuit network for a seven-subarray antenna structure. This tournament feed network using a two-port phase shifter enables beam tilting in the vertical plane of the base station antenna.

11.2.5 Passive Intermodulation

The base station antenna is used for both transmission and reception and its transmitting power level is several watts per channel. The uplink receiving electric field strength is less than 60 dBμV, which represents the difference from the transmitting power of around 140 dB with multichannel input to base station antennas, and causes the passive intermodulation (PIM) in the passive device. The PIM is a phenomenon in which an undesired frequency is generated by combining frequency components that are inputs to a circuit with nonlinear input–output characteristics. Although the antenna is a passive device, a very weak electric discharge phenomenon occurs at the junctions of the antenna parts during high power operation. Such discharges often appear at the surface of the junctions when these are made of different kinds of metals.

Most current cellular systems are duplex, with the transmitting and receiving frequency bands allocated at a certain interval. Duplex cellular systems use the frequency bands for transmission and reception simultaneously. When the PIM occurs at an antenna, it appears on the receiving frequency band and then the receiving channel cannot be used, because of interference at the base station.

In the case of two angular frequencies input ω_1 and ω_2 ($\omega_1 < \omega_2$), the interference frequency of the PIM is given by denoting p as a positive integer: $(1 + p)\omega_2 - p\omega_1$ and $(1 + p)\omega_1 - p\omega_2$. The order of PIM is defined by $2p + 1$, which indicates that the odd-order PIM causes the most serious problems. Figure 11.18 shows the allocation of uplink and downlink frequency bands for cellular systems. The frequency interval should be allocated so that both of the frequency bands are not affected by the third- and fifth-order PIM. PIM signals are expressed as absolute values in decibels measured and relative values in decibels, normalized by the input signal.

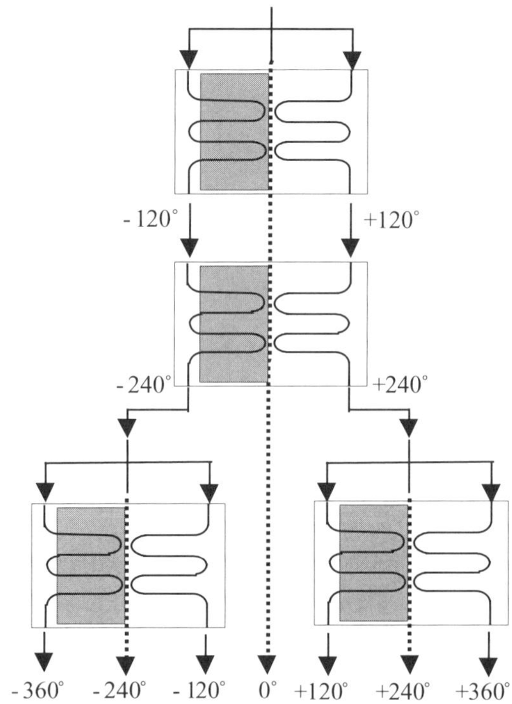

FIGURE 11.17 Feed circuit network using two-port variable phase shifter.

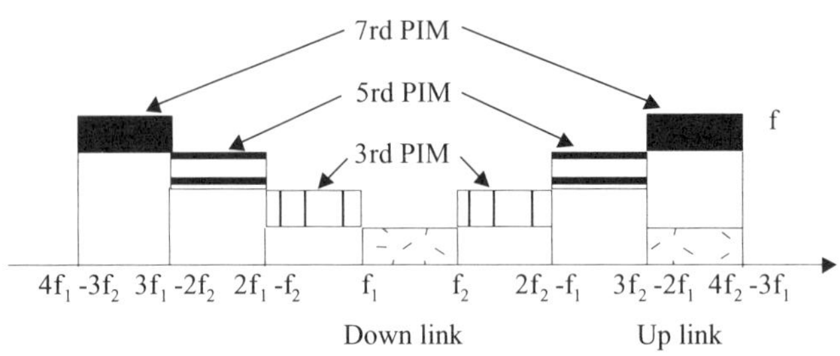

FIGURE 11.18 Frequency allocation for higher order PIM.

11.3 Diversity Antennas for a Macrocellular System

In this section, antennas for a macrocellular system are described. Uplink diversity reception is mandatory for base station antennas to decrease multipath fading and to compensate for power imbalance between uplink and downlink signals, because the uplink signal is weaker than the downlink signal. Two major diversity schemes, space diversity and polarization diversity antennas, are presented.

11.3.1 Space Diversity Antennas

For a space diversity system, antennas are installed with the interelement distance of more than 10 λ. The correlation coefficient ρ_e for this space diversity is calculated by assuming that the uplink signals originate from inside the angle ψ_m with uniform distribution as shown in Fig. 11.19. The calculated correlation is shown in Fig. 11.20. For a design value of $\rho_e < 0.5$, the distance between antennas should be more than 10 λ to provide $\psi_m > 1.5°$ as estimated from propagation measurements in the cities [11].

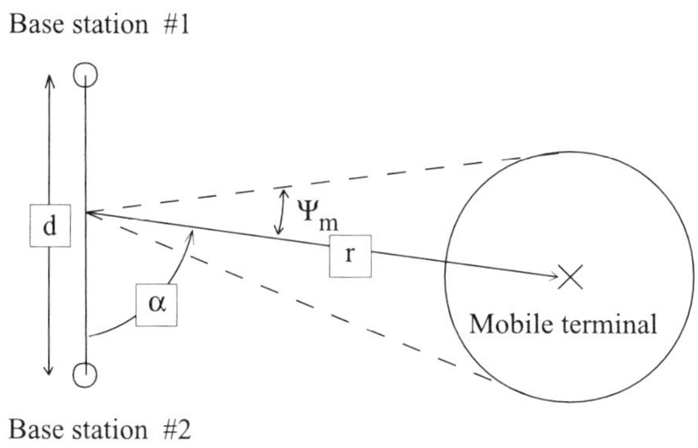

FIGURE 11.19 Location of mobile terminal station in horizontal plane.

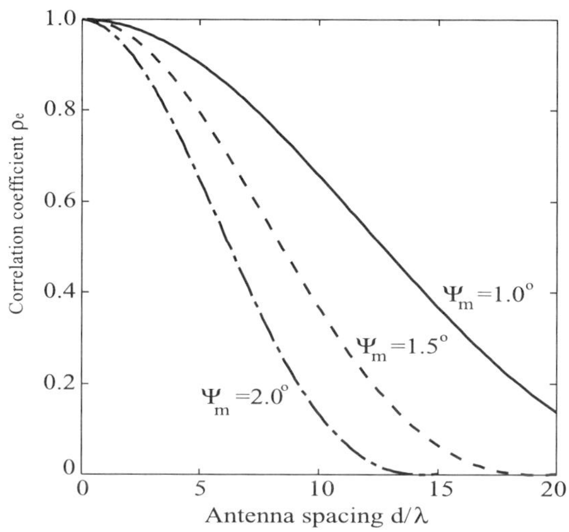

FIGURE 11.20 Correlation coefficient for antenna spacing.

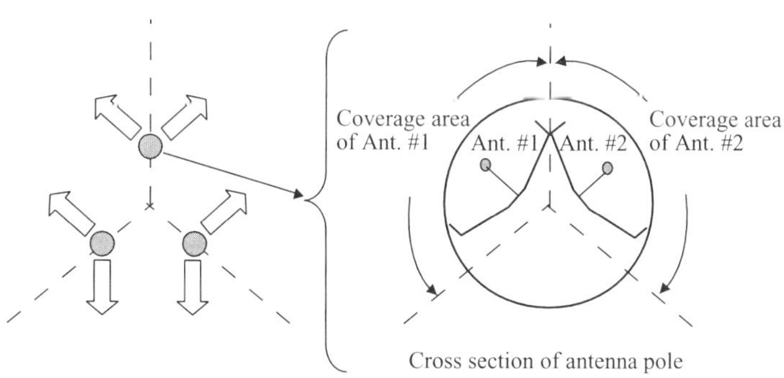

FIGURE 11.21 Three-sector space diversity antenna arrangement.

For a three-sector zone in the horizontal plane, a total of six base station antennas are required in the space diversity system if each antenna is installed independently. Because of space limitation for base station antenna placements in big cities, the location of the antenna has become very restricted. Thus, the number of antennas in one base station should be minimized. For this reason a three-sector antenna system uses a radome housing for two antennas facing different sectors as shown in Fig. 11.21 [7].

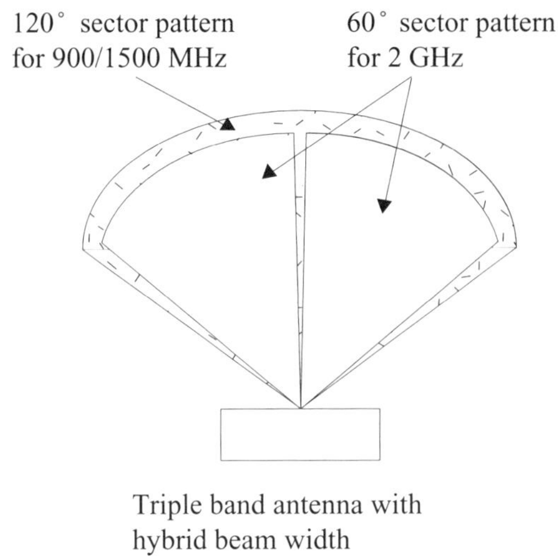

FIGURE 11.22 Hybrid horizontal pattern of triple-band antenna.

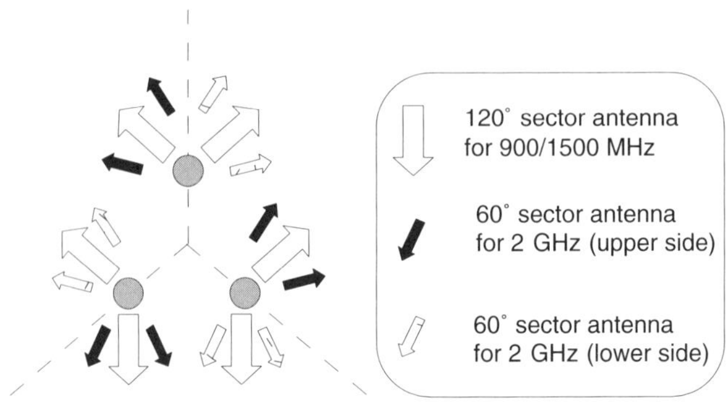

FIGURE 11.23 Antenna arrangement for hybrid pattern antenna.

A three-sector zone is mainly used for Japanese cellular systems at 900-MHz and 1.5-GHz bands. However, a six-sector zone is adopted in the IMT-2000 system in the 2-GHz band in addition to the three-sector zone. To replace present dual-band antennas with triple-band antennas, two types of base station antennas are required. One type is with 120° beam width for triple band and the second is with 120° beam width for 900 MHz/1.5 GHz, and 60° beam width for 2 GHz. The latter case needs three more 2-GHz antennas to cover the whole area. To decrease the number of base station antennas, two 60° beam width antennas facing ±30° offset in about the center of a 120° beam of lower frequency is built in the dual-band antenna, as shown in Fig. 11.22. The 2-GHz antennas are placed vertically in the form of an array. The cell allocation using this hybrid antenna is shown in Fig. 11.23. This antenna covers the present cellular system with three sectors in the 900- to 1500-MHz band and with six sectors in the 2-GHz band.

11.3.2 Polarization Diversity Antennas

Developments in electronics technology have produced very small and light portable handsets. As a result, everyone now uses a handset placed near an ear at a tilted angle. According to the statistical data presented in Reference [12], the tilt angle is about 60°. This decreases the vertical radiation component by 6 dB and the horizontal component by 5 dB. This operating condition requires increased horizontal electric field component in the uplink, making the polarization diversity an effective diversity tool.

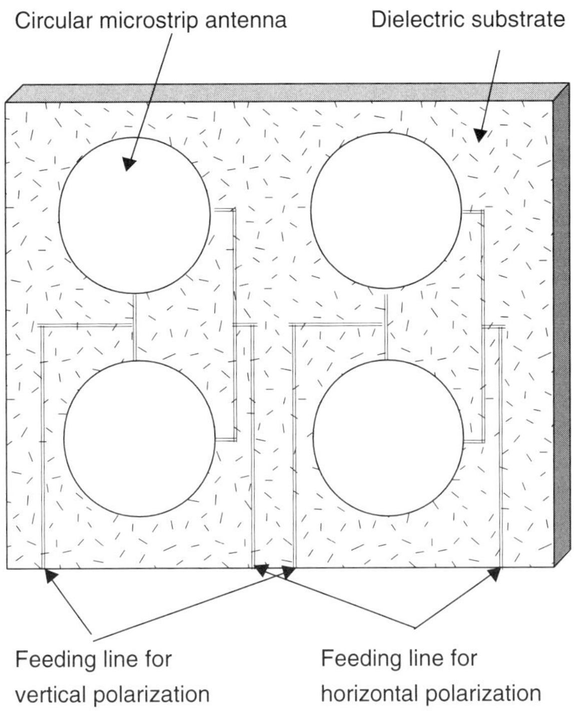

FIGURE 11.24 Antenna element for polarization diversity.

The polarization diversity gain measurement using a handset has been found about 7 dB more than the space diversity case for the line of sight (LOS) region, and 1 dB more outside the LOS region [13]. The polarization diversity system using vertical and horizontal components also has the merit of minimizing the antenna installation space, and thus has been adopted for Japanese cellular systems.

A base station antenna employing polarization diversity in both vertical and horizontal polarization uses a pair of circular patch antenna as shown in Fig. 11.24 [14]. The antenna is excited by adjusting the phase difference to obtain the same beam width in the E- and H-plane. As a 90° beam width antenna of polarization diversity, a printed twin dipole element, with parasitic element as shown in Fig. 11.25, is also used. The parasitic element expands the frequency bandwidth while keeping the beam width at the same value. This twin dipole antenna has equal beam widths in the E- and H-planes. Antennas are arranged in a vertical array as shown in Fig. 11.26. Three polarization diversity arrays are built in one radome to reduce the number of base station antennas.

Another polarization diversity system uses a slant polarization of ±45° [15]. The correlation coefficient is about 0.5, which is larger than the diversity using vertical and horizontal components. Because the equal received signal levels for both diversity ports give diversity gain, the diversity gain using this slant polarization is the same as that obtained using vertical and horizontal diversity systems described earlier. A typical antenna element of ± 45° polarization diversity is the printed cross-slot antenna as shown in Fig. 11.27.

11.4 Antennas for Microcellular and Picocellular Systems

Cellular phone service has begun to cover terrestrial areas, and the coverage area is expanding to include tunnels, subway stations, and large buildings such as shopping malls and hotels. A system covering these areas is referred to as an in-building system. This system uses a booster for the relay station between the outside base station and the new coverage area, or a microcellular system using an exclusive base station. The booster system receives a downlink signal from the outdoor base station, re-radiates its downlink

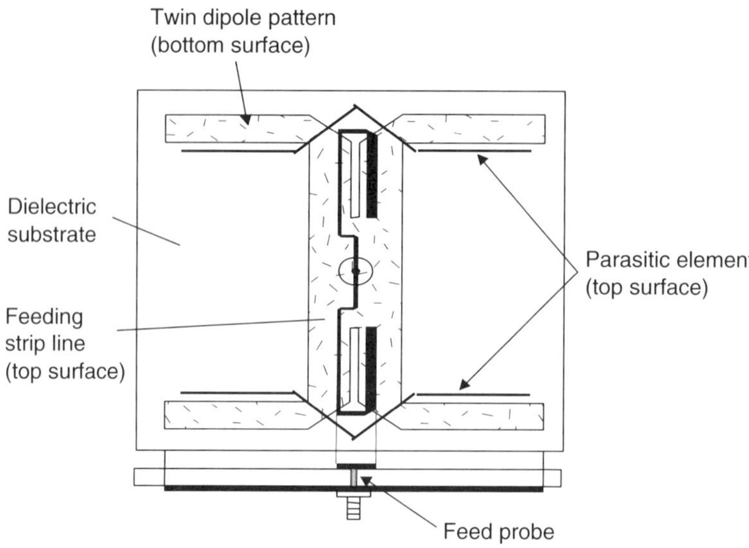

FIGURE 11.25 Printed twin dipole antenna element with parasitic element.

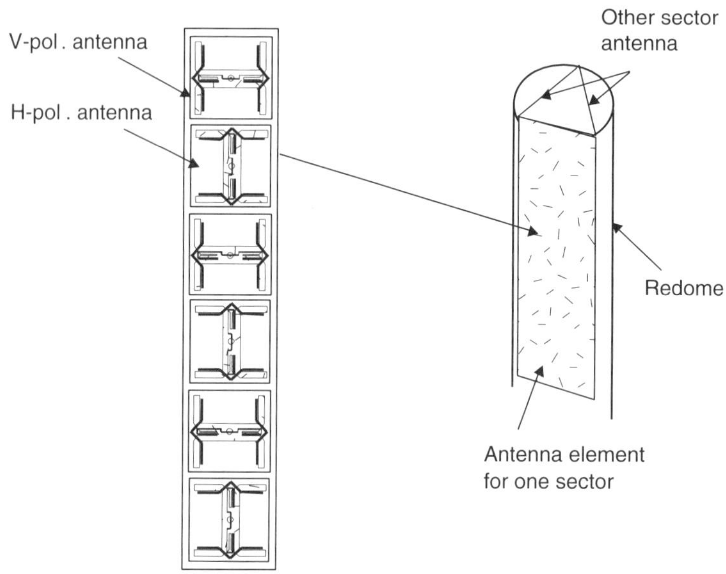

FIGURE 11.26 Polarization diversity antenna using printed twin dipole (230 mm,f @900 MHz).

signal after amplification, and vice versa for the uplink. A leaky coaxial waveguide has been widely used for a 900-MHz cellular system to expand its coverage area to the underground tunnels in the metropolitan highways in Tokyo [16]. An in-building system was installed in the newly built undersea tunnel in 1994. This novel system uses an optical fiber as a transmission line and a flat antenna as a radiating element to provide 1.5-GHz band and 900-MHz band service [17]. In this system, an optical laser diode is directly modulated by the downlink radio frequency (RF) signal from the outdoor base station, and its modulated optical signal is transmitted to inside the tunnel by an optical fiber. A very small transmission loss and a wide bandwidth enables the optical fiber to carry two frequency bands at 900 MHz and 1.5 GHz. After

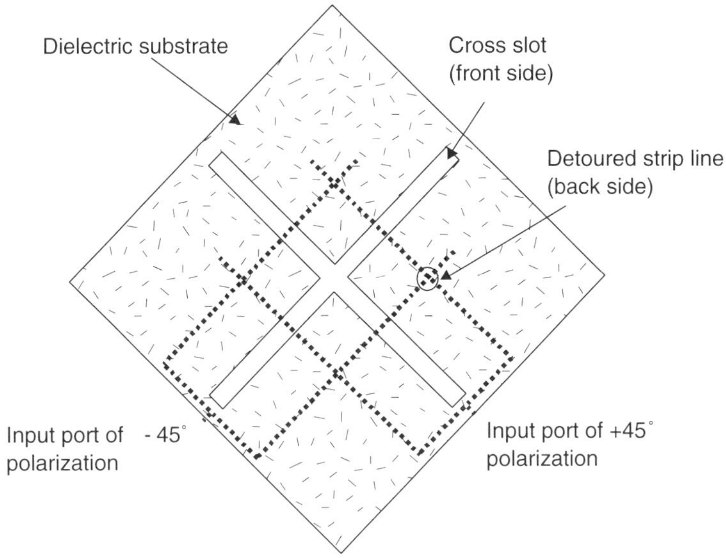

FIGURE 11.27 Polarization diversity antenna using printed cross slot.

the transmission by the optical fiber, the downlink signal is re-radiated by low-gain flat antennas installed on the sidewall at a few hundred meters apart. The reverse is done for uplink.

In addition to the highway tunnels, the in-building system has also been introduced to subway stations. This system has also been introduced in private facilities, such as huge shopping malls and big hotels.

This section describes low profile and small antennas used for present in-building systems and presents a method to expand frequency band corresponding to IMT-2000 (2 GHz). The system using leaky coaxial cable is still used for areas where the length of the cable is not very long; however, the discussion that follows only covers antennas for systems using optical fiber as a transmission line.

Small antennas for in-building systems are categorized as the sidewall-mounted antennas used inside tunnels, and ceiling-mounted antennas used in subway stations and buildings. The former antennas are required to have bidirectional radiation patterns in the horizontal plane because they are very long in longitudinal direction. The radiation patterns of ceiling-mounted antennas are omnidirectional in the horizontal plane. For this system a bidirectional antenna may also be used in a rare case when interior geometry of the building so requires. As for the tunnel antennas, ceiling mounting is the best choice from the propagation characteristics viewpoint; however, the construction to mount an antenna to the ceiling is very difficult and thus the antennas are mounted on the top of the sidewalls inside a tunnel.

11.4.1 Tunnel Booster Antennas

The required characteristics for an antenna used inside a tunnel are a bidirectional radiation pattern and a low profile structure. The height of an antenna should be less than one tenth the wavelength. The shape of the bidirectional pattern has a figure of eight in both the E- and H-planes, with a null position in front of the antenna. No uplink and downlink connection should exist just in front of the antenna. However, this kind of low-gain antenna radiates cross-polarized signals to the front side, with its level less than −20 dB of the main beam and causes multiple reflections from surroundings. Thus, in reality no dead zone exists in the bidirectional pattern.

This pattern is obtained by a two-element, half a wavelength dipole array as shown in Fig. 11.28. The elements are excited out of phase and are spaced a half wavelength apart. This array gives a figure eight pattern in the H-plane; however, installing a ground plane near the antenna element drastically changes this pattern. The tunnel booster antenna needs a mounting conductor plate because the electrical characteristics of the tunnel walls are affected by the wet or dry surface conditions of the concrete wall. The

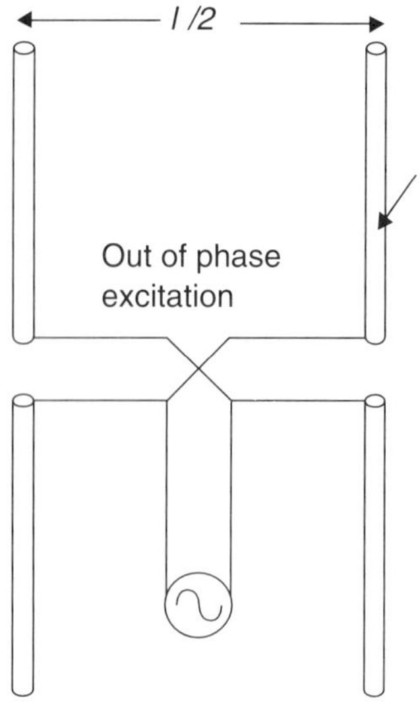

FIGURE 11.28 Bidirectional dipole antenna array.

ground plane is necessary for the tunnel antenna so that it is not affected by the surface conditions of the wall.

The frequency bandwidth of a 900-MHz cellular system is about 9% for the Japanese cellular system. This bandwidth covers all the operators in Japan. The highway tunnel is regarded as a public space, and the in-building system is for common use. An element for the highway tunnel antenna is a pair of notch antennas excited out of phase, with a wide-frequency bandwidth in spite of low profile structure [18]. The original notch antenna is cut on a large ground plane, and is regarded as a complementary structure of a monopole antenna [19]. A pair of notch antennas cut on a small ground plane are shown in Fig. 11.29. They do not have a dominant resonance of notch but instead have a resonance of feeding strip line with a length of a quarter wavelength [20]. The current flowing on the ground plane radiates the cross-polarization component, and a crank-shaped ground plane may be used to suppress the undesired

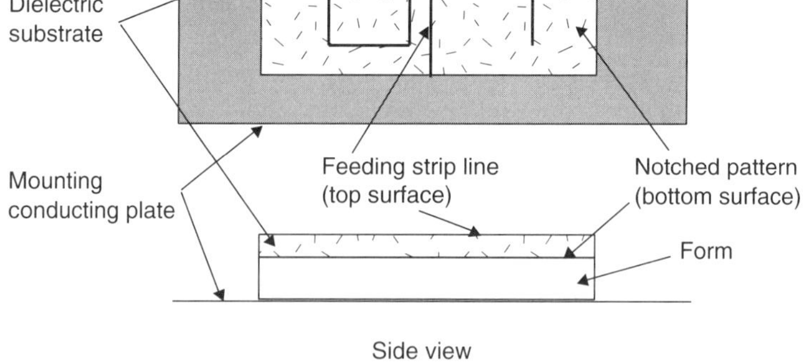

FIGURE 11.29 Bidirectional notch antenna.

radiation [21]. However, the radiation of cross-polarization is not a serious problem for low-gain antennas in the mobile communication systems, because it often increases frequency bandwidth at the input port and thus is not necessary to suppress cross-polarization to the negligible level.

A low-gain bidirectional antenna is installed at intervals of 200 m for short tunnels less than 2 km. After opening an undersea tunnel at Tokyo Bay in 1996, high-gain types have been introduced for tunnels more than 2 km long. As the antenna gain increases, the number of electrical to optical and optical to electrical converter units becomes smaller and the cost of the system becomes cheaper.

Directors attached to the bidirectional notch antenna increase the directivity gain as in the yagi-uda array. A few directors of 0.4 λ are placed in parallel to the feeding strip line on both sides, as shown in Fig. 11.30, which increases the antenna gain by 3 to 5 dB [22]. The maximum gain is restricted by the size of the ground plane. When the director is located close to the edge of the ground plane, an increase in the backward radiation limits the directivity gain. Figure 11.31 shows a photo of a tunnel booster

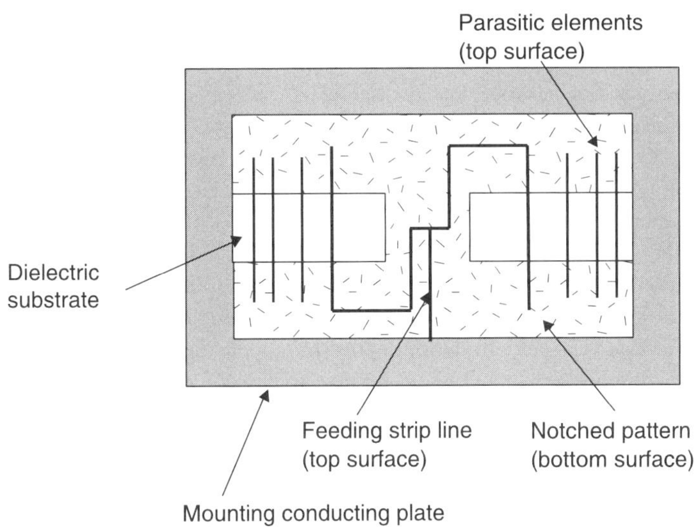

FIGURE 11.30 High-gain bidirectional notch antenna (top view).

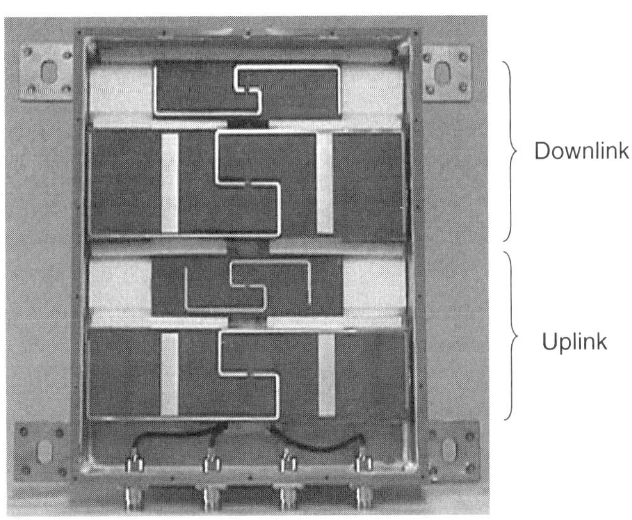

FIGURE 11.31 Photo of bidirectional notch antenna (370 × 280 × 57 mm).

antenna consisting of four elements for uplink and downlink transmission at 900 MHz and 1.5 GHz. In the commercial model, a folded ground plane for wide-frequency bandwidth and a crank-shaped ground plane for the suppression of cross-polarization are used.

For the triple-band operation of the tunnel booster antenna, one technique that may be used is to replace the dual-band notch antenna by a single-band element. The dual-band notch antenna is designed by mounting the parasitic element above the feeding strip line [23], which is effective for a low-gain antenna structure. To keep the same interval between booster antennas, the antenna gain of a 2-GHz band is adjusted to compensate for the propagation loss, which increases in proportion to the operating frequency. In the initial service of IMT-2000, a high-gain notch antenna for a 2-GHz band is introduced in addition to the current antenna.

11.4.2 In-Building Antennas

The PHS, started in 1995, uses a microcellular system in which the base station is directly connected to a public integrated services digital network (ISDN) to simplify the function of the base station and to reduce its size and weight. This compact base station uses small, low-gain antenna with the advantage of a microcellular system that is installed on the ceiling of a subway station. The cellular phone operators are required to produce the in-building system to compete with the PHS. The booster system used inside the tunnel is also applicable for this in-building cellular system; however, the base station of the microcell operates independent of the outside base station to increase the channel capacity.

The antenna shape installed on the ceiling depends on the ceiling condition. When protuberance decorations cover the ceiling, wire antennas such as sleeve and collinear arrays are easier to install than flat antennas. On the other hand, a low-profile antenna is appropriate for a flat ceiling. Low-profile antennas used for in-building systems are modified to a semi-circular-shaped monopole antenna with an ultrawide band [24], a very low profile top-loaded monopole antenna (TLMA)[25][26], and miniaturized TLMA filled with dielectric materials [27]. These antennas are shown in Fig. 11.32. The impedance matching at the feed point of the TLMA is achieved by placing shorted posts near the feed probes as shown in Fig. 11.32. A circular disk is used for disk loaded monopole antennas. As the current flowing on the disk is canceled by the image current, the disk shape does not disturb the principal radiation pattern. The TLMA is miniaturized by inserting dielectric materials under the top plate and by adding shorted posts. This antenna is used for the pager booster system in a 280-MHz band.

Figure 11.33 shows a photo of a small in-building system antenna [28]. Places to install such small antennas are on the ceilings of a department store building, a hotel, and an underground parking lot. The uplink and downlink antennas are mounted separately for both 900-MHz and 1.5-GHz bands. A

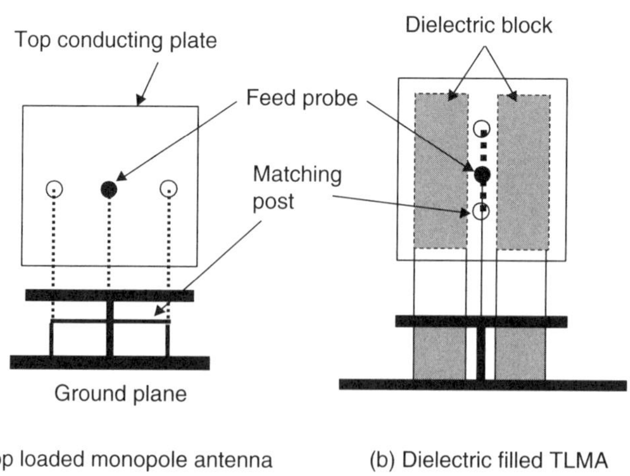

FIGURE 11.32 Top-loaded monopole antenna.

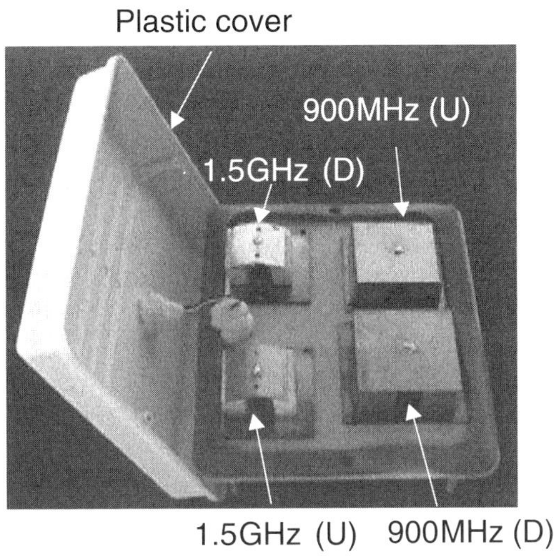

FIGURE 11.33 In-building system antenna U: uplink, D: downlink (140 × 170 × 25 mm).

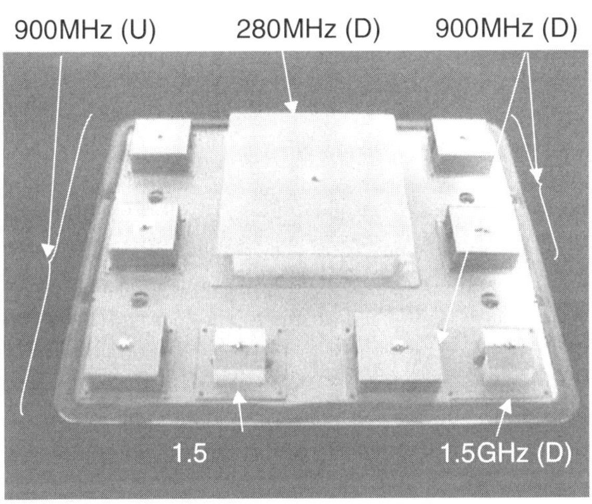

FIGURE 11.34 In-building system antenna for subway station U: uplink, D: downlink (268 × 318 × 39 mm).

rooftop-shaped TLMA with two short pins is used for the 1.5-GHz band, and a dielectric-filled TLMA is used for the 900-MHz band. The rooftop-shaped plate increases the frequency bandwidth of the TLMA. The dielectric block is effective in reducing the antenna size; however, it decreases the antenna gain up to about 2 dB because of the loss of dielectric and the radiation of the cross-polarization. The principal polarization is vertical, similar to the terrestrial system; and the cross-polarization is horizontal. This miniaturized antenna is able to achieve the minimum requirement for the antenna gain for an in-building system which is 0 dBd and is obtained by this miniaturized antenna.

Another example is the ceiling antenna for the subway station. This antenna operates in frequency bands of 280 and 900 MHz and of 1.5 GHz, covering all the frequency bands allocated for pager and cellular phone systems in Japan. As shown in Fig. 11.34, three dielectric-filled TLMAs are used to cover a split frequency band plan in 900 MHz [28]. The subway station is a public space, and in-building systems are operated by all the cellular operators. The antenna elements are arranged to minimize the distortion of radiation pattern and to obtain more than 15 dB of isolation between uplink and downlink antennas.

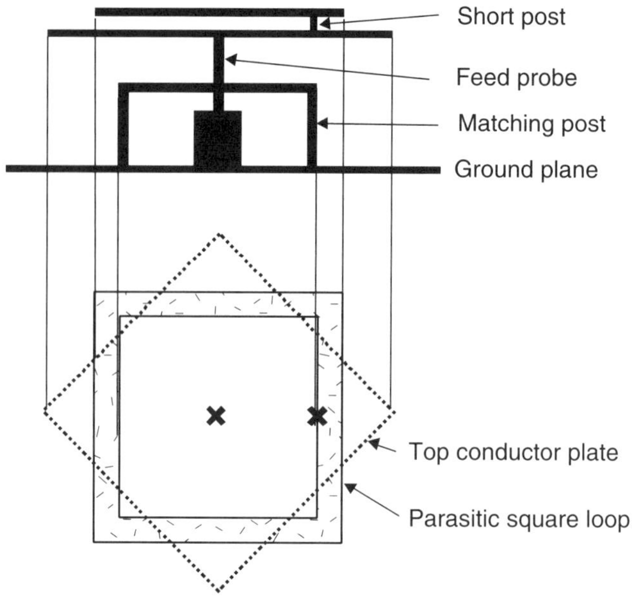

FIGURE 11.35 Dual-band top loaded monopole antenna (21 × 21 × 15 mm).

The introduction of IMT-2000 has forced operators to add a new antenna in 2-GHz band or to replace the current dual-band antenna with a new triple-band antenna. A new antenna can be easily installed by adding small uplink and downlink antennas only for a 2-GHz band. For the antenna replacement case, the current TLMA for 1.5 GHz is exchanged for dual-band TLMA operating at 1.5-GHz and 2-GHz bands. A dual-band monopole antenna is normally obtained by adding a parasitic element in the vicinity of the original monopole antenna; however, this technique cannot be applied for dual-band TLMA. For the dual-band TLMA, a square loop is attached on the top disk as shown in Fig. 11.35 [29]. The loop is electrically connected to the top plate by a short pin to make a resonance in the high-frequency band for TLMA. The original resonance of the TLMA moves down to the lower side because of the mutual coupling between the square loop and the top plate. The upper square loop is rotated by 45° so as to protrude the loop corner from the square plate. These protruding parts make additional resonance. The triple-band antenna is thus obtained by replacing the TLMA for 1.5 GHz by the dual-band TLMA with a square loop. The current flowing on the loop and plate does not radiate excessive horizontal polarization, because their image currents in the ground plane cancel each other. This composite structure produces quasi-monopole radiation pattern in the vertical plane and the omnidirectional pattern in horizontal plane.

11.5 Personal Handy Phone System Base Station Antennas

The PHS is a microcellular phone service introduced in 1995. The PHS was originally planned as a digital wireless indoor phone at 1.9 GHz next to the analog wireless phone at home using 280 to 350 MHz. The low-cost design of the PHS uses ADPCM and time division duplex (TDD)/TDMA digital system.

To expand the coverage area to stations, shopping malls, and other public places, a microcellular system with its coverage area of 100 to 500 m in radius is used. As described in the previous section, the PHS base station has a simple structure. Its antenna gain of 0 to 10 dBd is lower than the cellular base station. The radiation pattern is omnidirectional in the horizontal plane, and the main beam is tilted in the vertical plane. This section describes these base station antennas for the PHS as well as an adaptive array antenna for high-speed data transmission.

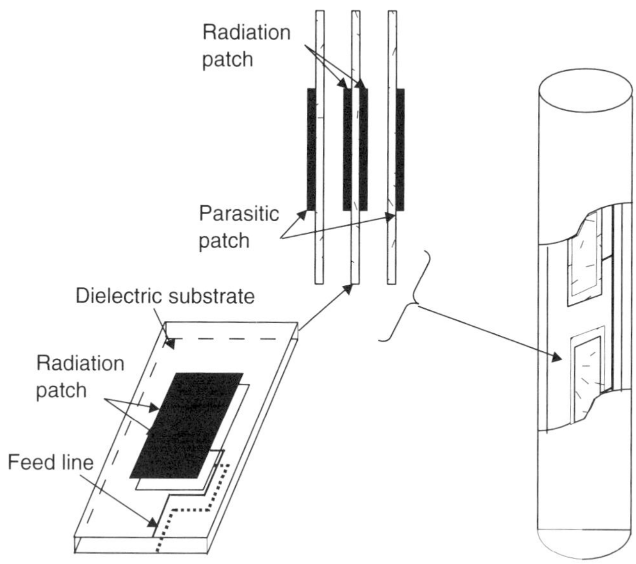

FIGURE 11.36 PHS bidirectional antenna.

11.5.1 Antenna Configuration

The PHS base station also uses diversity reception to compensate for the power imbalance between uplink and downlink antennas, and to decrease the effects of the multipath fading. Two or four antenna elements are widely adopted for the space diversity. Antenna elements used are low-gain sleeve antennas and high-gain collinear array antennas. The antenna spacing for the diversity system is 1 to 2 λ. The polarization diversity antenna is also used. This is because the mobile terminals are handy phones and are usually used at a tilted angle radiating larger horizontally polarized signal than the vertical ones when an LOS path exists between the base station and a mobile terminal. The LOS propagation environment is dominant in microcellular PHS systems. Thus, it is advantageous to use the polarization diversity system.

When the base station antenna height is lower than the surrounding buildings, the bidirectional antenna such as a tunnel booster antenna gives a good coverage area along streets. The bidirectional antenna expands the coverage area along streets by increasing antenna directivity. The antenna element used for the bidirectional antenna is a rectangular patch antenna on the narrow ground plane as shown in Fig. 11.36 [30]. This element is easy to make using an array-feeding line on a microstrip line. The combination of a rectangular patch and a slot element gives the polarization diversity for the bidirectional antenna as shown in Fig. 11.37 [31].

11.5.2 Adaptive Array System

A commercially available adaptive array antenna was introduced for the PHS system in 1998 by Kyosera and DDI Pocket [32, 33]. This adaptive array is used to increase the channel capacity in Tokyo. Figure 11.38 shows a block diagram of the adaptive array system. The RF input signal from a four-element circular array is digitized at the base band frequency after the demodulation by a sampling frequency at 1.536 MHz. This gives eight times oversampling for the data rate. The digital processing unit consists of a digital signal processor (DSP) with 66 million instructions per second (MIPS), and its calculation is performed using a fixed decimal point. Unfortunately, the algorithm used for adaptive pattern forming is not available in the open literature. The antenna spacing is about 5 λ, which excites many grating lobes and many null points. A fast rotation of such a pattern eliminates the interference waves to increase the carrier to interference (C/I) ratio. The antenna gain of array element is 10 dBi. The PHS uses the TDD system. The optimum radiation pattern for reception is also used for the transmission.

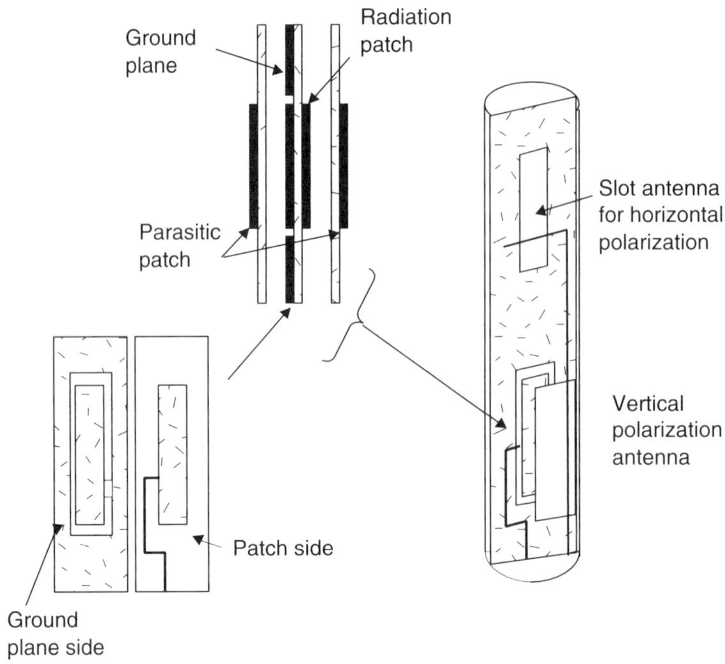

FIGURE 11.37 PHS bidirectional polarization diversity antenna.

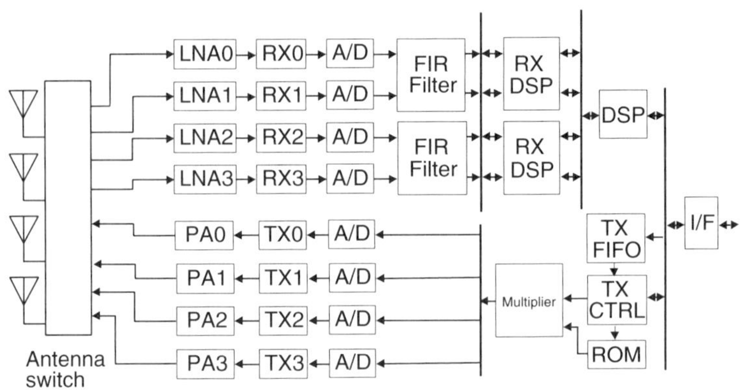

FIGURE 11.38 Block diagram of adaptive array.

References

1. Michishita N. et al., "FDTD Analysis for Printed Dipole Antenna with Balun," Asia-Pacific Microwave Conf., Sydney, pp. 739–742, Dec. 2000.
2. Karikomi M. and Ebine Y., "Characteristics of Printed Dipole Antenna with Parasitic Elements," Technical Report of IEICE Japan, AP89-2, Apr. 1989 (in Japanese).
3. Kakefuda H. et al., "A Wide Freqeuncy Bandwidth Omnidirectional Antenna for 800 MHz Band," Fall Conference, IEICE Japan, B-1-77, Sept. 1999 (in Japanese).
4. Ebine Y., "Dual Frequency Base Station Antennas for PDC Systems in Japan." IEEE AP-S, Int, Symp, and USNC/URSI National Radio Science Meeting, Orlando, FL, pp. 564–567, July 1999.
5. Ebine Y. et al., "Dual-freqeuncy Antenna with 60° Beam-Width by Cross Feeding Dipole Elements," Spring Conference, IEICE Japan, B-55, Mar. 1995 (in Japanese).

6. Kijima M., Ebine Y., and Yamada Y., "Development of a Dual-frequency Base Station Antenna for Cellular Mobile Radio," IEICE Trans. Commun., vol. 82-B, no. 4, pp. 636–644, Apr. 1999.
7. Ebine Y. "Design of Base Station Antennas for Next Generation Cellular Mobile Radios (IMT-2000)," Technical Report of IEICE Japan, AP2000-4, Apr. 2000 (in Japanese).
8. Shintaku M. and Ebine Y., "A Study of the Structure of Array Antennas for Mobile Base Stations," Fall Conference, IEICE Japan, B-1-78, Sept. 1999 (in Japanese).
9. Fujii T. et al., "Handling capacity expansion of land mobile communication system," *NTT Rev.*, vol. 2, no. 3, pp. 55–61, Mar. 1990.
10. Asaka S., Kariomi M., and Ebine Y. "The Structure and Characteristics of a Revolving Variable Phase Shifter Using Slot Coupling," *Proc. Int. Symp. Antennas Propagation,* Fukuoka, Japan, pp. 305–308, Aug. 2000.
11. Yamada Y., Ebine Y., and Tsunekawa K., "Base and mobile station antennas in land mobile radio systems," *IEICE Jpn, Trans.*, vol, E74, no. 10, pp. 3202–3209, June 1991.
12. T. Taga and K. Tsunekawa, "A Built-In Antenna for 800 MHz Band Portable Radio Units," in Proc. of ISAP'85, Kyoto, 121-1, pp.425–428, 1985.
13. Nakano M., Satoh T., and Arai H., "Up-link polarization diversity and antenna gain measurement of handheld terminal," *IEEE AP-S, Dig.*, pp. 1940–1943, July 1995.
14. Moriyasu S. et al., "Printed Dipole Pair with Modified Parasitic Elements and Reflector, Spring Conference, IEICE Japan, B-1-150, Mar. 1999 (in Japanese).
15. Lindmark B. et al., "Dual-Polarized Array for Signal-Processing Applications in Wireless Communications," *IEEE Trans. Antennas Propag.*, vol. 46, no. 6., pp. 758–763, Jan. 1998.
16. Kosono S. and Shintaku M., "Experimental Study of Propagation Characteristics Inside Rectangular Tunnel," vol. 28, no. 3, pp. 537–556, 1979-06 (in Japanese).
17. Fuke Y. and Ebine Y., "Optical Transmission Booster," *NTT DoCoMo Tech. J.*, vol. 5, no. 1, pp. 29–32, 1997-04 (in Japanese).
18. Arai H. and Goto N., "Flat antennas for indoor cellular system," Conf. Dig., 1994 IEEE AP-S, Int. Symp., Seattle, WA, pp. 344–347, June 1994.
19. Johnson W.A., "The notch aerial and some applications to aircraft radio installations," Proc. Inst. Elect. Eng., pt. B, vol. 102, pp. 211–218, 1955-05.
20. Arai H. and Kohzu K., "A Bi-directional Notch Antenna," Conf. Dig., 1996 IEEE AP-S, Int. Symp., Baltimore, MD, pp. 42–45, 1996-07.
21. Kohzu K. and Arai H. "A Bidirectional Flat Antenna" Proc., Int. Symp. Antennas and Propagation, Chiba, Japan, pp. 553–556, 1996-09.
22. Arai H., Kohzu K., Mukaiyama T., and Ebine Y. "Bi-directional Notch Antenna with Parasitic Elements for Tunnel Booster System" Conf. Dig., 1997 IEEE AP-S, Int. Symp., Montreal, Canada, pp. 2218–2221, 1997-07.
23. Kohzu K. and Arai H. "Dual Band Bidirectional Antenna," Asia-Pacific Microwave Conf., New Delhi, India, pp. 856–859, 1996-12.
24. Ihara, T. and Tsunekawa K., "Broadband Characteristics of Rounded Semi-Circular Antenna," Proc., Int. Symp. Antennas Propagation, Chiba, Japan, pp. 513–516, Sept. 1996.
25. Arai H. and Kuga N., "Small Antennas at 2.45 GHz for Spread Spectrum Applications," IEEE 2nd Int. Symp. Spread Spectrum Techniques and Applications, Yokohama, Japan, Nov. 1992, 5–16.
26. Jiang H. and Arai H. "Analysis of Patch Antenna with Short Pin by Using Non-Uniform Mesh FDTD." IEEE AP-S, Int. Symp. and USNC/URSI National Radio Science Meeting, Orlando, FL, pp. 610–613, July 1999.
27. Jiang, H. and Arai, H. "FDTD Simulation of Top Loaded Low-Profile Monopole Antenna" Asia-Pacific Microwave Conf., Singapore, pp. 150–153, pp. 1493–1496, Dec. 1999.
28. Arai H., "Base Station Antennas Inside Tunnels and Subway Stations, and Outdoor Compact Base Station Antennas for PDC Systems in Japan." IEEE AP-S, Int. Symp. and USNC/URSI National Radio Science Meeting, Orlando, FL, pp. 568–571, July 1999.

29. Rikuta Y., Arai H., and Ebine Y., "A Two-layered Patch and Loop Antenna with Dual Freqeuncy," to be presented at Asia-Pacific Microwave Conf., Sydney, pp. 1493–1496, Dec., 2000.
30. Cho K. and Hori T. "Bidirectional Rod Antenna Composed of Narrow Patches," Conf. Dig., 1994 IEEE AP-S, Int. Symp., Seattle, WA, pp. 174–177, June 1994.
31. Hori T. et al., "Dual Polarized Bidirectional Antenna for Microcell Base station," Technical Report of IEICE Japan, AP95-94, Jan. 1996 (in Japanese).
32. Yokota T. et al., "The Development of PHS Base Station with Adaptive Array Antenna," Fall Conference, IEICE Japan, B-5-74, Sept. 1999 (in Japanese).
33. Uchibori J., "The Radiation Pattern of Adaptive Array Antenna for PHS Base Station," Fall Conference, IEICE Japan, B-5-76, Sept. 1999 (in Japanese).

12

Handheld Antennas

Atef Z. Elsherbeni
University of Mississippi

Chun-Wen Paul Huang
University of Mississippi

Charles E. Smith
University of Mississippi

12.1 Introduction

12.1.1 Background

The explosive growth in the demand for wireless communication and information transfer using handsets and personal communication system (PCS) devices has created the need for major advancements of antenna design as a fundamental part of wireless systems. This move to enhance the design of antennas began with the monopole or "whip" antennas of yesteryear, but has moved in different directions because of the complex environment in which the handheld antenna radiates energy. In most cases, the handheld antenna is placed on a small plastic/conducting box that is in close proximity to biological tissue that moves within its environment — a user talking on a cell phone walking in a shopping mall. At the same time, the system must radiate low power and provide reliable communication of voice and possibly data. Added to the operational requirements, the user and service providers demand wireless units with antennas that are small and compact, cost effective for manufacturability, low profile, and easy to integrate into the wireless communication system.

The antenna designer must also consider the electrical characteristics of the antenna which include

- Return loss (input impedance)
- Bandwidth
- Gain and radiation pattern
- Operating frequencies
- Diversity

These design considerations have led antenna designers to consider a wide variety of antenna structures to meet the often conflicting needs for wireless systems. Examples of these structures that are being used fall into the following classes that have been widely studied [1]:

0-8493-0124-6/02/$0.00+$1.50

- Monopole (whip type)
- Loaded monopole
- Loop
- Helical
- Planar inverted-F antennas (PIFA)
- Patch/plate or slot-type, built-in antennas
- Multiple antennas for diversity

12.1.2 Monopole Antennas

The monopoles shown in Fig. 12.1 are probably the most widely recognized antennas used for handheld applications, and they are often used because of their classic omnidirectional radiation characteristics [2–3]. These types of dipole and the related monopole antennas have been studied extensively by King [4], as presented in the classic text, *The Theory of Linear Antennas*, and it remains the authoritive source on linear antennas. This, along with studies and investigations of many others, has provided a wide background of knowledge on this fundamental antenna element. Current work in this area provides a wide range of theoretical, numerical, and measured results to predict the operational characteristics of the center-fed cylindrical dipole and the monopole. In practice, the monopole is usually fed by a transmission line above a ground plane, in most cases, as a vertical element. The size of the ground plane and the length and radius of the monopole greatly influence the radiation characteristics. For a thin cylindrical monopole antenna above an ideal ground plane, its characteristic at resonance is

Input resistance approximately 36.5 (1/2 that of a dipole)
Radiation pattern — omnidirectional in plane predictions to the monopole (the classic donut-shaped patterns)

Variation of antenna length and shape, such as conical [5] and folded [6], can be used to increase the input impedance and/or bandwidth while maintaining the same basic radiation pattern. Another variation of the dipole/monopole is the sleeve antenna where the feed point is moved away from the ground plane in the case of the monopole. This configuration has been used for handheld wireless antennas to obtain more broadband voltage standing wave ratio response and omnidirectional type of radiation

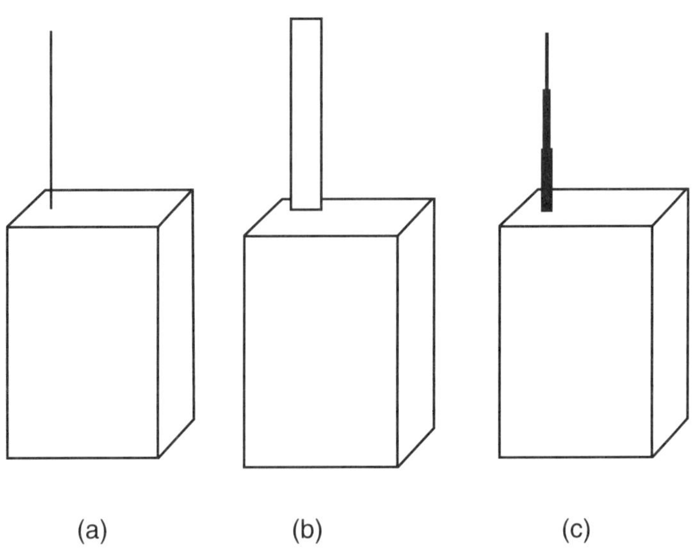

FIGURE 12.1 Monopole antenna on top of a handset. (a) Wire monopole, (b) strip monopole, (c) retractable monopole.

patterns. Wong and King obtained an octave bandwidth application for such an antenna above a reflector [7]. Also, this type of feed minimized the ground plane, or body effect, in adaptations of monopoles on handheld wireless telephones. Studies have investigated the behavior of a monopole mounted on a conducting box that is comparable to a wavelength [8, 9]. These studies indicated that the magnitude of the conductance and subsistence decreased as the monopole is moved from the center of the box to the corner. Thus, the placement of the antenna on the handheld wireless transmitter/receiver adds another variable in the complex operational environment for the handheld antenna.

However, in all cases, if the length (height) of the handheld monopole is reduced for low-profile applications, the radiation efficiency goes down; this limits communication range, if power output is not increased. Therefore, the handset interacts with the monopole, as well as other antennas in the practical case, to possibly alter the radiation pattern, the bandwidth, and the input impedance, as compared with those of the $\lambda/4$ monopole over an infinite ground plane. Even so, in many cases, the relative broad-bandwidth characteristic and simple construction of monopoles have led to their wide use in many applications. Loading (both inductive and capacitive) can also be added to a basic monopole-type antenna to obtain better operating characteristics with a shortened length for a reduced size handset, as shown in Figs. 12.2 and 12.3. The loading can be used to reduce the physical height or the effective height of the monopole by maintaining a more constant current distribution for larger field strength or the effective height. Also, the antenna impedance match can be improved with loading.

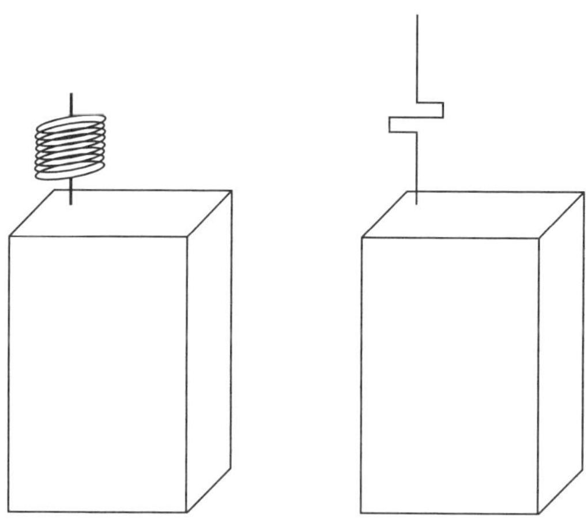

FIGURE 12.2 Different types of inductive loading of a monopole antenna on top of a handset.

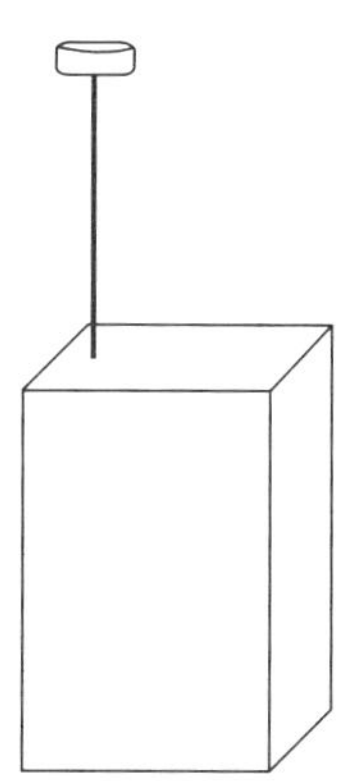

FIGURE 12.3 Capacitive loading of a monopole antenna on top of a handset.

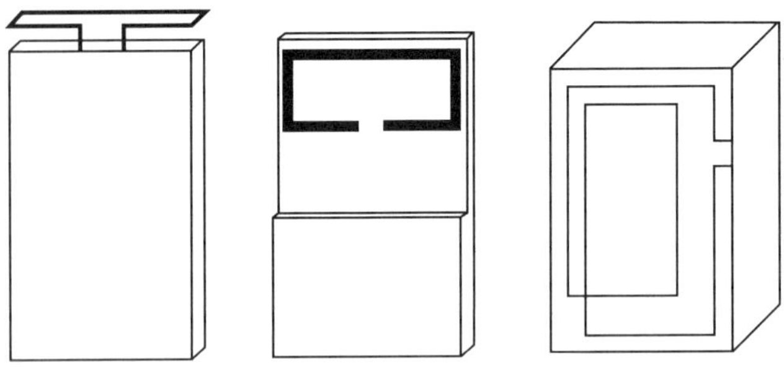

FIGURE 12.4 Loop antennas for wireless systems. (a) Top mounted, (b) case or side mounted, (c) multiple loop case mounted.

12.1.3 Loop Antennas

The loop-type structures of Fig. 12.4 also provide omnidirectional type of patterns and can be integrated into the wireless handset or PCS system having a reasonably low-profile construction. As wireless frequencies increase, the size of the required loop becomes larger relative to wavelength, and the loop can be matched much easier than the electrically small loop at lower frequencies. Although the loop has been used primarily with pagers, this higher frequency usage trend could lead to applications on high frequency wireless handsets in the future.

In the case of the loop antenna with a small total conductor length as compared with the wavelength in free space, the current distribution can be considered to be uniform and the far-field radiation can be obtained readily [10]. The radiation resistance can become very small (fractions of an ohm) for the electrically small loop antenna and extend up to over 100 Ω for antenna where area/(wavelength)^2 approaches 1. The radiation efficiency of the small single-turn loop is also consequently low, but can be increased by using multiple turns and/or a ferrite core. This low-radiation efficiency can limit the range for transmission for low-power applications, as was the case for electrically small monopole antennas.

For large loop antennas the current distribution can no longer be considered uniform and the performance of the loop can be changed significantly when the circumference is greater than 0.1 λ. The magnitude of the radiation pattern in the plane of the loop is decreased and does not resemble the figure eight pattern of the small loop. However, the resonant loop has been used widely because of its moderate input resistance and directivity, along with its symmetrical field pattern and ability to easily incorporate the loop into the handset case of the wireless phone and/or pager.

12.1.4 Helical Antennas

The helical antenna is simple wire wound around a uniform cylinder as shown in Fig. 12.5, with multiple turns like a wire-wound spring [11]. This type of antenna, which is based on the helix used in a traveling wave tube, has an axial mode of radiation and a normal mode of radiation that is perpendicular to the axis of the helix. The axial mode helix has been widely used as an endfire directional antenna; however, resonant normal-mode helical antennas are useful as short, vertically polarized radiators, similar to the monopole [12]. However, the input impedance of the normal-mode helical antennas is sensitive to changes in frequency, which results in narrower bandwidth. Even so, the short normal-mode helix has been widely used in mobile communications systems where a reduced dipole length is needed [13].

The normal-mode helical antenna of Fig. 12.5 has thus found application in portable systems, primarily because it can be shorter than the $\lambda/4$ monopole. The shorter length/low-profile of the helix and other variations have lead to the popularity of this type of antenna by manufacturers in the last few years. Several variations are in use, and multihelices have been used to optimize wireless system performance.

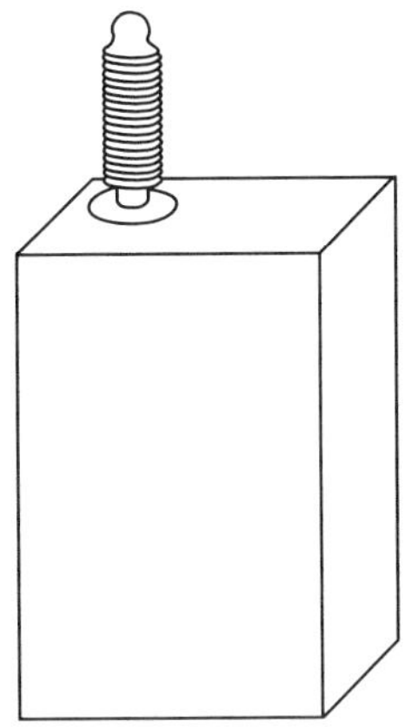

FIGURE 12.5 Helix antenna on top of a handset.

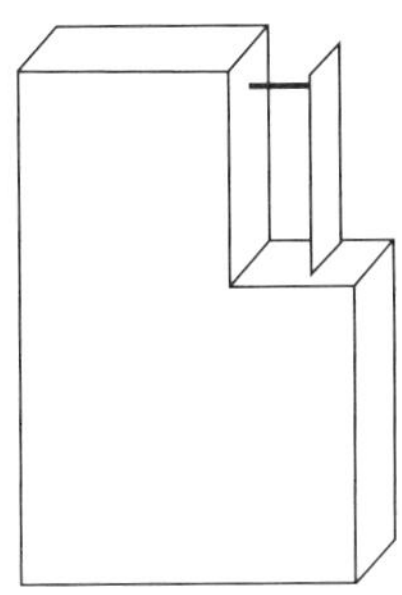

FIGURE 12.6 The basic configuration of an F antenna with a handset.

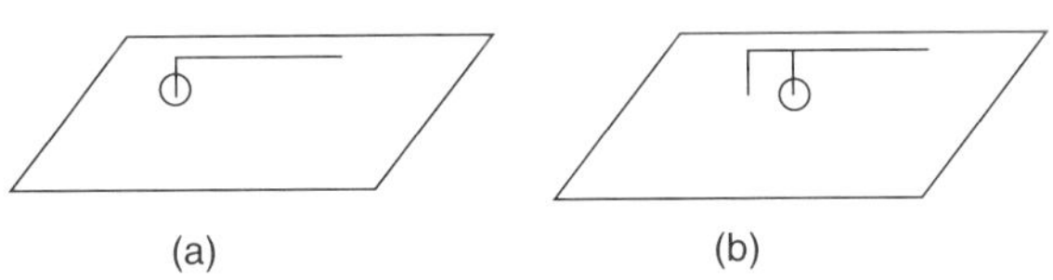

FIGURE 12.7 The evolution of the F antenna from the L antenna. (a) L antenna, (b) F antenna.

12.1.5 Planar Inverted-F Antennas

The PIFA has been studied extensively because it has a low profile, and it can easily be incorporated into wireless handsets [14–17]. The basic structure as shown in Fig. 12.6 evolved from the L antenna (shown in Fig. 12.7) of the early missile era of the 1960s, where an antenna tap was added to obtain impedance matching for both wire and strip-type configurations. This L or inverted L antenna is a variation of the monopole antenna where the monopole is bent over in an L shape with respect to a ground plane [18]. Here again, the basic monopole structure is modified to reduce the height of the antenna while obtaining a lower resonant frequency than that of a comparable electrically short monopole. The short arm of the L antenna radiates in an omnidirectional pattern in the plane perpendicular to the vertical element, and some radiation also is obtained from the long arm of the L antenna. This reduction in height reduces the radiation resistance and bandwidth, and changes the radiation pattern as a result of radiation on the long arm of the L structure. The hula hoop antenna is a variation of the L antenna where the long arm is formed into a circle to reduce the horizontal size of the antenna. A boxlike or "square loop" version could be useful in mobile communication applications for handsets.

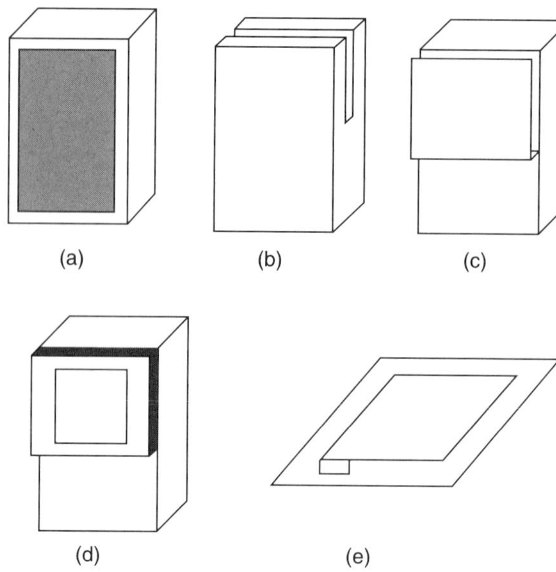

FIGURE 12.8 Patch and slot antennas for wireless applications. (a) Patch, (b) internal slot, (c) side-mounted slot, (d) combination of patch and slot, (e) top-mounted patch.

The F antenna is a variation of the L antenna that uses a "tap" on the antenna to obtain an input impedance transformation [19]. This is just another application of delta matching used for feeding a half-wave, continuous wire, dipole antenna, or "tapping" a resonant LC circuit to obtain impedance transformation. However, for the wireless antenna designer, this provides some freedom in terms of controlling the input impedance [20], which is a major advantage when designing an antenna to match a particular wireless handset configuration. This added feature gave this antenna its F image and name. Bandwidths as high as 10% have been obtained when the PIFA is located on a small ground plane or conducting handset body. Also this antenna can be easily incorporated within the wireless handset case as shown in Fig. 12.6.

12.1.6 Patch or Slot Antennas

Patch, or microstrip-type, antennas are one of the most widely used and popular low-profile antennas in use at frequencies above the very high frequency (VHF) range [21–25]. These types of structures typically consist of a rectangular or circular patch on a dielectric substrate coated ground plane as shown in Fig. 12.8. The relative dielectric constant of the substrate ranges for 2 to 4, and the thickness is in the range of 1/100 of the wavelength. The patches are usually excited with a microstrip line, a coaxial cable (coax) probe, or a slot underneath the patch placed to obtain an impedance match. By varying the feed point and effective dielectric constant, a wide range of input resistances can be obtained for impedance matching. This ability to match this antenna in the complex electromagnetic environment of the wireless handset is a significant advantage in the use of these types of structures.

For a rectangular patch, the horizontal electric field along the edges of the patch has the same direction, which results in a maximum broadside radiation pattern. Thus, the radiation pattern is similar to a small loop antenna. Higher dielectric constants result in smaller, more compact antennas, but efficiency and impedance bandwidth are sacrificed. Bandwidths are, in general, proportional to the dielectric substrate thickness, which is usually very small in terms of wavelengths. Resulting bandwidths are narrow, which can be a disadvantage when used in some mobile handsets for broadband wireless communications. Stacking of parasitic patches can be used to increase the bandwidth by a factor of 2 when using appropriate fed arrangements, which further supports the idea that bandwidth is a function of the antenna volume.

The slot antenna, which consists of a slot in a ground plane, has an electric field within the slot that can be obtained between the slot antenna and a complementary wire antenna (Babinet's principle) as established by Booker [26]. Radiation from slots is usually constrained to one side of the ground plane where one side is completely enclosed, or cavity backed, which has a significant influence on the antenna impedance. Excitation is applied across the slot at an appropriate point to obtain an impedance match similar to the PIFA impedance matching. Depending on the slot shape, the radiation characteristics can be deduced from a similar dipole (rectangular slot) or loop (circular slot) and, in general, antenna performance is similar to the comparable dipole or loop antennas. Variations of the slot can be made similar to those used with a dipole for obtaining broader bandwidth, etc. The bent slot (similar to the L antenna) in a small ground plane can provide radiation in a cross-polarized form that is useful for multipath and diversity applications for mobile communications systems.

Thus, the patch or slot type antennas presented in Fig. 12.8 are two other important classes of antennas that have emerged as viable wireless antennas because they can be almost completely integrated into the wireless handset case — a very desirable feature from the user's point of view for low profile and manufacturability. Patches and slots can be configured to obtain vertical, horizontal, or circular polarization, and they are usually resonant at higher frequencies and have relatively broad bandwidth.

12.1.7 Diversity Antennas

Signal fading in mobile communications is one of the more difficult problems to address in the complex propagation environment that exists in practice. Multiple reflections from stationary structures and moving bodies can cause serious fading in handheld reception systems within buildings and on streets in a large city with high-rise buildings. Usually, diversity is a technique used at the receiving site, which can take the form of space, field component, directivity, polarization, and/or frequency diversity for the improvement of the signal reception.

In mobile, or handheld communications, the two-antenna approach to diversity has been widely used, and this approach has been studied extensively by a large number of individuals and groups [27–32]. Two diversity signals are received from two antennas, depending on the diversity technique employed, and the two signals are combined to maximize the signal to noise (SNR) ratio after signal combing, which is a complete area of study in signal processing in its own right. However, depending on what diversity technique is employed, combinations of any of the types of antennas reviewed here can be used.

Handheld wireless communication systems quite often have to operate in such areas as inside and outside of buildings in a city environment where multipath propagation is a severe problem. Thus, the use of multiple antennas can provide some diversity in these situations, even on a small handset in areas where signal strength is weak. As a result, multiple handheld antennas have received a great deal of attention, as industry tries to optimize system performance [29, 33, 34]. Also, a recent article has investigated the diversity performance of a dual-antenna handset near the head, and the authors concluded that, though the mean effective gains were reduced by 3 to 5 dB, the tissue had little influence on the overall handset diversity [35]. Examples of approaches to this diversity problem are the combination of a monopole antenna and dual PIFAs as depicted in Fig. 12.9.

12.1.8 Summary

These example antenna classes have a wide variety of characteristics, including common traits of large physical size and/or narrow bandwidths. With this in mind, efforts have been made for improving these antenna characteristics, using similar techniques to reduce the physical size of the antenna and to increase bandwidth for a planar handled antenna that can be incorporated in the handset. One attempt to address these issues has led to a planar meander line structure, as shown in Fig. 12.10 [36]. This type of antenna has been studied because of its reduced resonant antenna length, which is desirable for the design of low-profile, compact antenna/wireless systems. This study also incorporates antenna tapering and loading to obtain larger antenna bandwidth and possible multifrequency operational modes for projected future

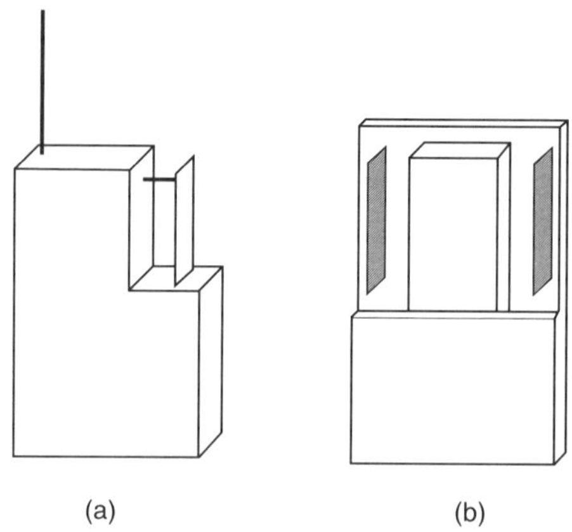

FIGURE 12.9 Handset with two antennas for diversity. (a) Monopole and PIFA, (b) dual PIFA.

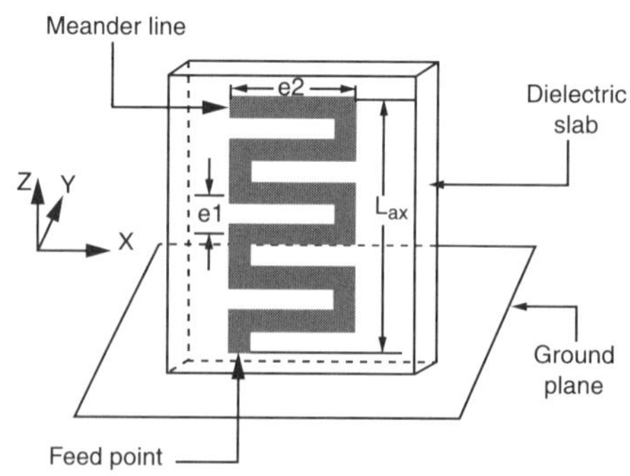

FIGURE 12.10 Meander line antenna on top of a finite size ground plane.

applications. A detailed presentation of a study on the characteristics of the planar meander line antenna for personal wireless communication systems is presented in the sections that follow.

12.2 Meander Line Antenna for Personal Wireless Communications

12.2.1 Introduction

In this chapter, the finite difference time domain (FDTD) technique is used to analyze several types of printed meander line antennas. This technique is used because it provides a successful, low-cost approach to the analysis of a very complex boundary value problem for the design and performance simulation of this type of antenna. The FDTD technique used in this study employs a modeling region consisting of a three-dimensional computational domain enclosing the antenna with boundaries terminated with Berenger's perfectly matched layers (PMLs) [37]. The details of this type of FDTD analysis are presented in Chapter 7 in this book. From this approach, empirical expressions are derived from the numerical analysis of various types of meander line antennas, which provide designers with the intrinsic properties of meander line antenna configurations. Optimum designs are obtained through the analysis process and experimental verifications are performed. The meander line antennas presented in this chapter are uniform vertical and horizontal segmental ratio meander line antennas, taper meander line antennas, top-shorted meander line antennas, back-to-back printed meander line antennas, and meander line antennas for cellular handsets. Detailed discussions are provided in each subsection.

12.2.2 Equal Vertical and Horizontal Segment Length Meander Line Antennas

In this subsection, meander line antennas having uniform vertical and horizontal segmental ratio (e1 to e2) of the antenna traces are studied, as shown in Fig. 12.10. The parameters e1 and e2 are not necessarily chosen to be the same value. The analyses of this category of meander line antennas starts with equal e1 and e2. The meander line antennas with different line widths and different e1 to e2 ratios are also investigated. Meander line monopole antennas are analyzed on a 59×25.4 mm^2 ground plane to simulate the performance of the antennas on top of a PCS handset.

12.2.2.1 Meander Line Monopoles

In this subsection, a meander line monopole antenna with line width W_l = 1 mm, equal vertical and horizontal segments e1 = e2 = 3 mm, substrate thickness t_s = 3.17 mm, and substrate width W_s = 11 mm is considered. First, the input impedance is studied as a function of various numbers of segments as shown in Fig. 12.11. Also shown in Fig. 12.11, the increase of the vertical antenna length, L_{ax}, decreases the value of the first resonance frequency and increases the resonance impedance. Additionally, when the number of segments is increased to 48 segments (73 mm for L_{ax}), the input impedance at both the first- and third-resonance frequency is approximately 25 Ω. This design shows that the antenna has resonance at both 0.95 and 2.39 GHz (dual-operational bands) with 25-Ω nominal impedance. In a previous publication, a bent wire with a 2.5-mm meander line antenna with e1 = e2 = 8 mm on a large ground plane has the impedance of 23.2 and 15.5 Ω for the first and the third resonance, respectively [38]. Therefore, additional tuner traces are needed in these types of antennas to match the traditional 50-Ω impedance of the front-end circuitry of the communication device. The empirical equations for the first-resonant frequency and impedance for various values of L_{ax} are derived from regression curve fit of the data plotted in Figs. 12.12 and 12.13, respectively. In Fig. 12.12, the regression curve is also

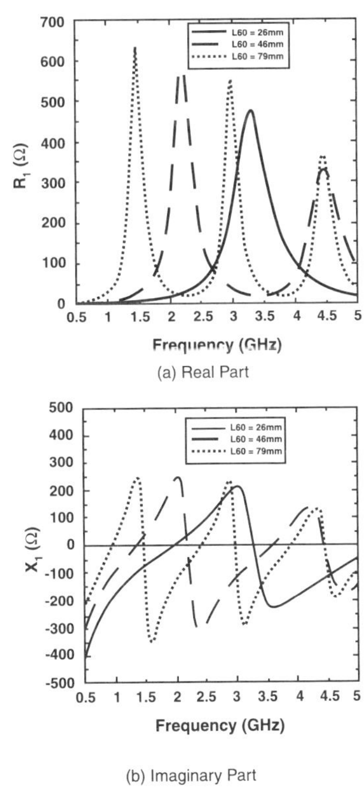

FIGURE 12.11 The input impedance of a meander line antenna vs. L_{ax}.

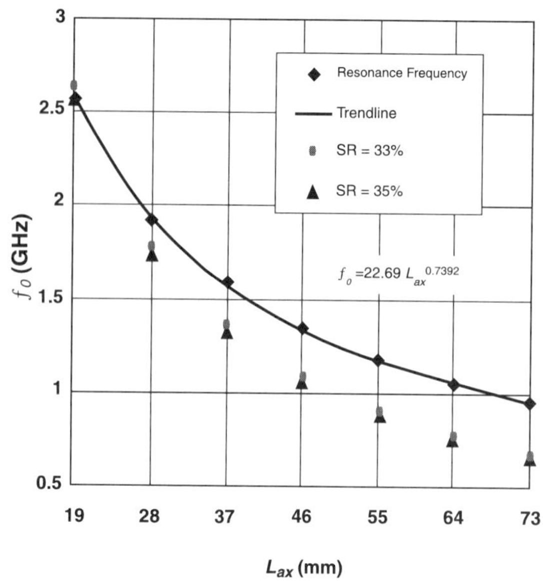

FIGURE 12.12 The first-resonant frequency of a meander line antenna vs. L_{ax}.

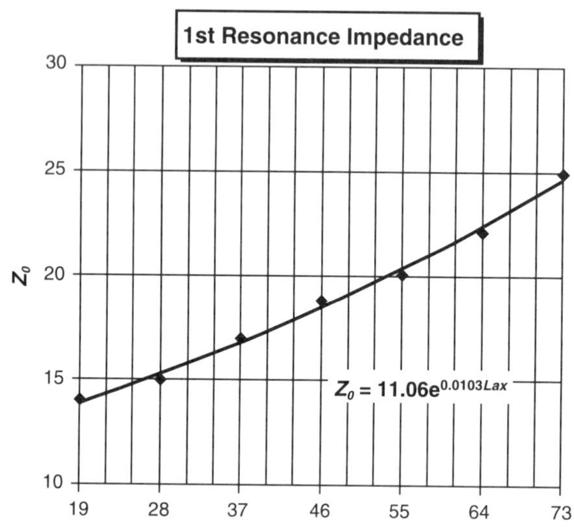

FIGURE 12.13 The first-resonant impedance of a meander line antenna vs. L_{ax}.

compared with the first-resonant frequency computed by the empirical equation presented in References [38–40], which define the length shortening ratio (SR) as

$$SR = \frac{\lambda/4 - L_{ax}}{\lambda/4} \tag{12.1}$$

Reference [40] suggests using an SR between 33 and 35% for computing the first resonant frequency of meander line antennas. From Fig. 12.12, the resonant frequency computed by Eq. (12.1) departs from the currently presented regression curve as L_{ax} increases. The empirical equations for the first resonant frequency of meander line antennas using 1 mm trace, e1 = e2 = 3mm, t_s = 3.17 mm, and W_s = 11 mm, on a 59 x 25.4 mm^2 ground plane is found to be

$$f_o = 22.69\ L_{ax}^{-0.7392} \tag{12.2}$$

and the first resonant impedance shown in Fig. 12.13, can be modeled by

$$Z_o = 11.06e^{0.0103L_{ax}} \tag{12.3}$$

These new Eqs. (12.2) and (12.3) can be used to solve for possible L_{ax} when a desired impedance is specified, and predict the first resonant frequency and impedance for a chosen vertical antenna height L_{ax}.

In addition to changing L_{ax}, different values of the substrate dielectric constant are adopted to examine its influence on both resonant frequency and impedance. From Figs. 12.14 and 12.15, both resonant

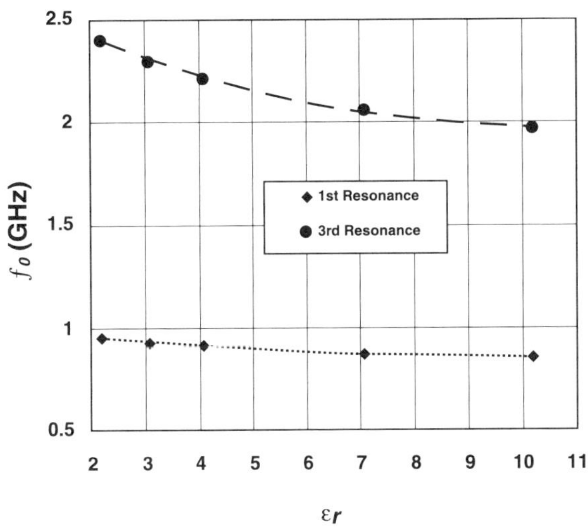

FIGURE 12.14 The first- and third-resonant frequencies of a meander line antenna vs. the substrate dielectric constant.

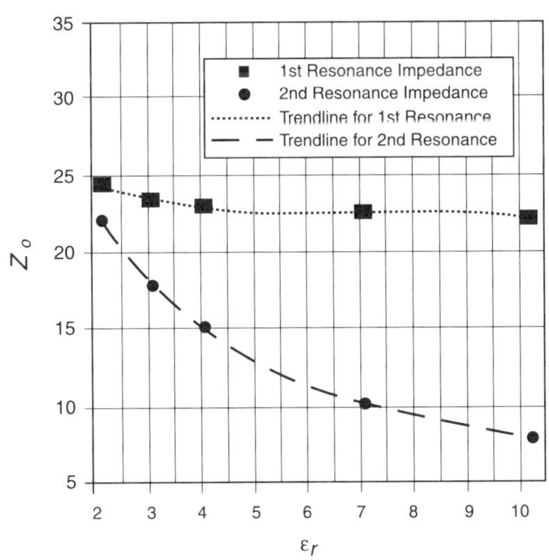

FIGURE 12.15 The first- and third-resonant impedance of a meander line antenna vs. the substrate dielectric constant.

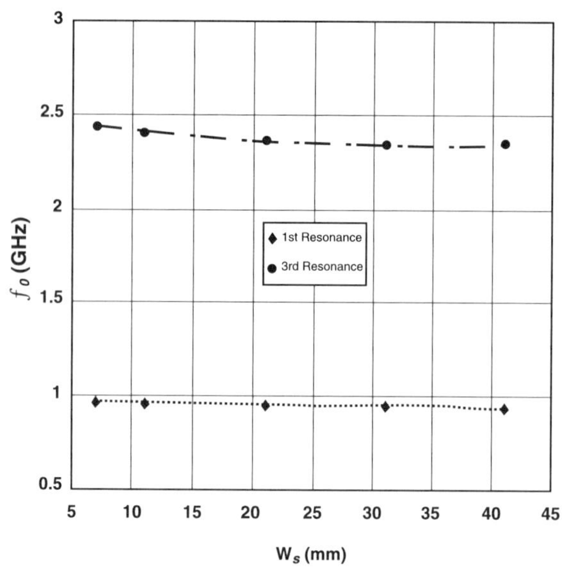

FIGURE 12.16 The first- and third-resonant frequencies of a meander line antenna vs. the substrate width.

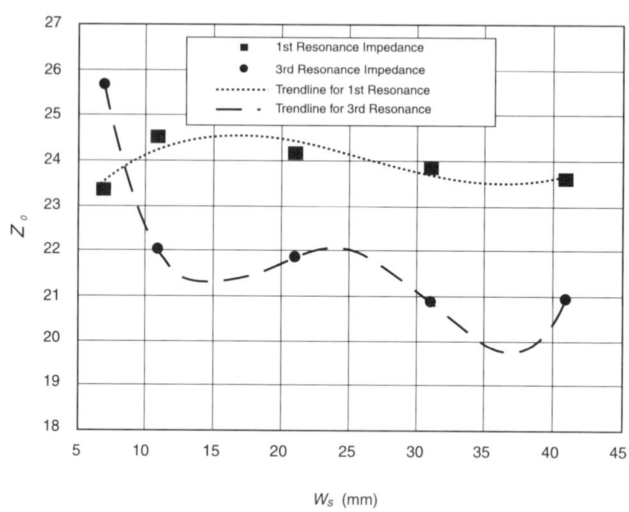

FIGURE 12.17 The first- and third-resonant impedance of a meander line antenna vs. the substrate width.

frequency and impedance decrease with the increase of the dielectric constant. The effects on resonant frequencies are not as strong as the effect on the impedance.

The effects of the width of the dielectric slab are also investigated. From Fig. 12.16, the variations of the first- and third-resonant frequencies resulting from the change of slab width are insignificant. However, the effects on the resonant impedance are clearly noticeable in Fig. 12.17. The increase of the slab width causes the reduction of the resonant impedance. Furthermore, reducing the slab width from 11 to 7 mm results in higher impedance with 23.33 and 25.65 Ω for the first and third resonance, respectively, as shown in Fig. 12.17. From the results shown in Fig. 12.17, an appropriate method for fine-tuning the resonant impedance can be achieved by simply adjusting the slab width, which is also feasible in antenna realizations.

The effect of the dielectric slab thickness is also investigated. It is found that the thickness of the dielectric slab affects the resonant frequencies within 0.02 GHz for the first resonance and 0.1 GHz for the third resonance as shown in Fig. 12.18. However, the resonant impedance of the first and third modes

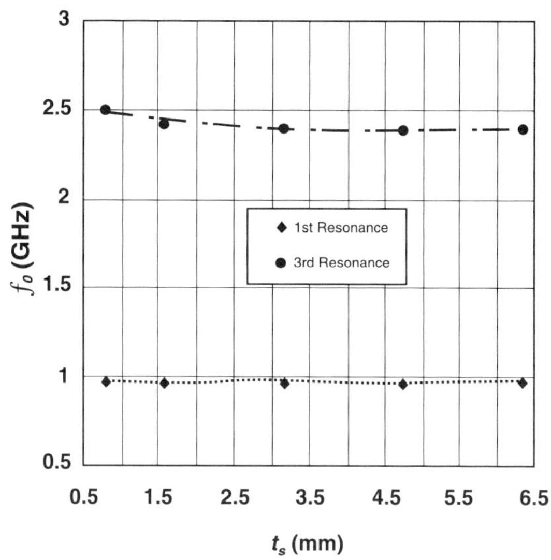

FIGURE 12.18 The first- and third-resonant frequencies vs. the substrate thickness.

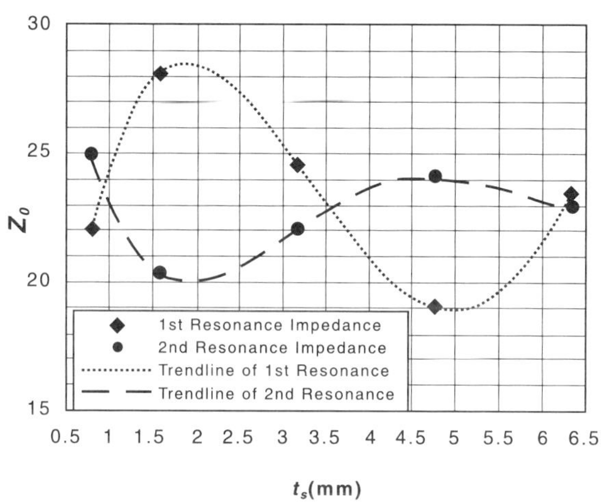

FIGURE 12.19 The first- and third-resonant impedance vs. the substrate thickness.

is changed over a range of 8 Ω, as shown in Fig. 12.19. Therefore, the variation in the slab thickness does not appear to be an effective procedure for fine-tuning this type of meander line antenna.

The analyses in this section show that meander line antennas with 1-mm line width and e1 = e2 = 3 mm on a finite small ground plane employ an odd mode resonance impedance in the range of 25 Ω. If the transceiver has an impedance transformation from a 50-Ω interior circuit to a 25-Ω antenna output port, then these antennas can be used directly on PCS applications. However, nominal impedance of most radio frequency (RF) transceivers connecting to antennas is designed to be 50 Ω. Therefore, in the subsequent two subsections, different line widths and values for e1 and e2 are used to study their effects on resonance impedance.

12.2.2.2 Meander Line Dipole Antennas

As observed in Section 12.2.2.1, the 1-mm meander line antennas with e1 = e2 = 3 mm operating on a small finite ground plane have a resonance impedance around 25 Ω when L_{ax} is about 70 mm. When L_{ax} = 73 mm, dual-operating bands are observed, which is a desired characteristic for PCS antennas. Because the input impedance of a dipole antenna is twice that of a monopole on an infinite ground

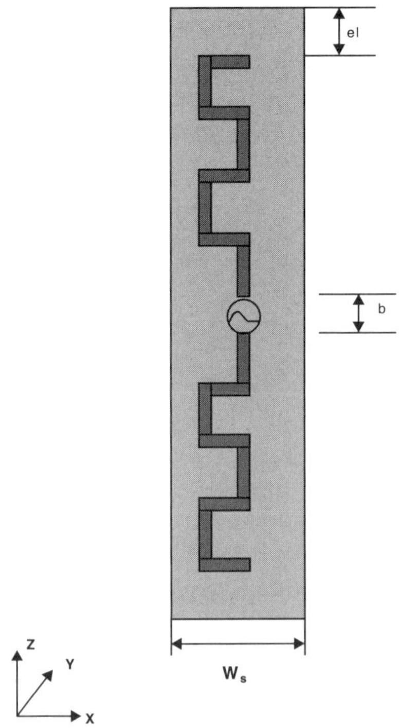

FIGURE 12.20 Meander line dipole antenna.

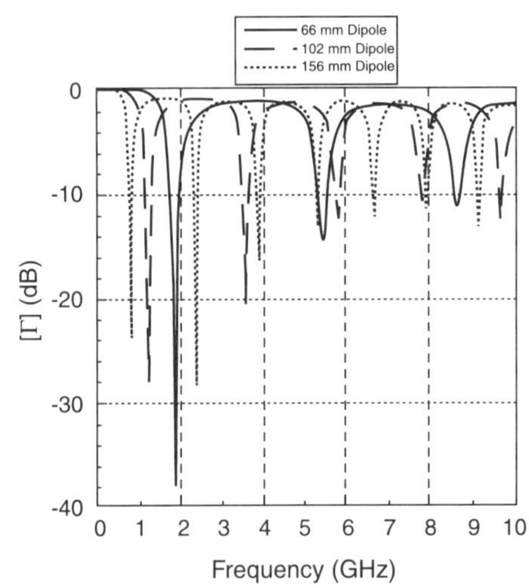

FIGURE 12.21 Meander line dipoles of L_{ax} = 28, 46, and 73 mm.

plane, it is possibe to achieve the 50 Ω input impedance while retaining the dual-operating bands by printing this meander line in a dipole format. The meander line dipoles investigated are for the cases of L_{ax} = 28, 46, and 73 mm presented in the previous section. The distance of the feed gap is modeled by using twice the length of that used for a meander line monopole as shown in Fig. 12.20.

The results of these meander line dipoles in terms of return loss normalized to a 50-Ω system are shown in Fig. 12.21. From the return loss results, the input impedance at the first and third resonance may be considered as very good 50-Ω loads, which is the same conclusion reported in Reference [39],

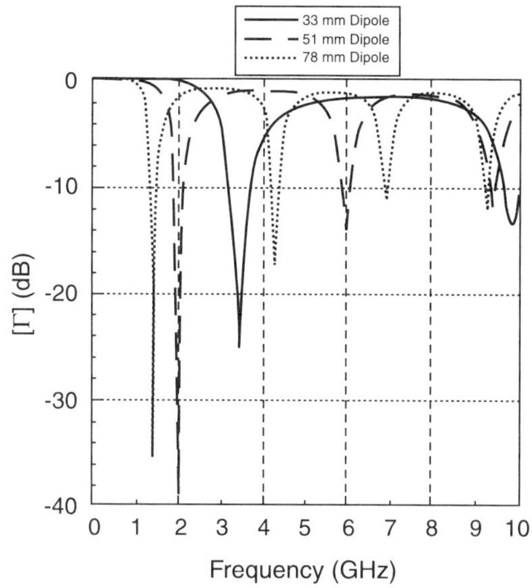

FIGURE 12.22 Meander line dipoles of L_{ax} = 14, 23, and 37 mm.

for a wire meander dipole. The first several modes of those dipoles meet the requirements for frequency and bandwidth of current and future PCS applications.

When the dipoles are made with L_{ax} = 46 and 73 mm, the total length may be longer than a cellular handset. Therefore, three additional designs are also presented based on reducing the three L_{ax}'s used in Fig. 12.21 by one half. The results of these three short dipoles are shown in Fig. 12.22. These short dipoles are also very good 50-Ω antennas, but operating at higher frequencies 3.4, 2.1, and 1.4 GHz, respectively. The lengths are also not long enough to create dual-operating frequencies in the PCS frequency range.

Based on the lengths used in the preceding two sets of meander line dipoles, a 50-Ω antenna can be designed at a given operating frequency by readjusting the trace length. However, special care is needed to prevent the induced current on the interior printed circuit traces, RF shields, and the battery pack from canceling the dipole radiated fields, when the dipole is printed on the back or on the side of a cellular or cordless handset. Dielectric slabs of high permittivity can be placed between the metal regions and the dipole to prevent the image-like canceling of radiation currents by providing an electrically large distance between the antenna and handset circuits. In this case, the metal regions may become antenna reflectors, instead of ground plane, parallel to the dipole. When the dipole is printed on the outer surface of the flip-up keypad protection cover in a cellular handset, the image-canceling problems no longer exist. The length of this plastic cover is usually designed around 60 mm, which is applicable for the three meander line dipoles presented in Fig. 12.22. The three meander line dipoles in Fig. 12.21 can be used as vehicular mobile antennas printed on auto glasses, which also minimizes the image-canceling effects by the surrounding metal parts.

The radiation patterns and directivities of this class of antennas are presented in the next section.

12.2.2.3 Effects of Ground Plane Size on Meander Line Monopoles

In Section 12.2.2.1, the characteristics of 1-mm meander line monopoles with e1 = e2 = 3 mm are analyzed. To simulate the antenna performance on a small cellular handset, a small ground plane is used. However, if the antenna is used for large mobile communication appliances, such as vehicular mobile antennas or within the presence of large metal surroundings underneath the antenna, then the effects of increasing the ground plane size on the performance of meander monopoles need to be studied. In Section 12.2.2.1, the finite, small ground plane is 59 × 25.4 mm^2, and the finite, large ground plane used in this section is 159 x 184 mm^2. Resonant frequency and impedance of meander line monopoles on infinite ground planes can also be obtained by computing the meander dipole input impedance and normalizing it by a factor of 2.

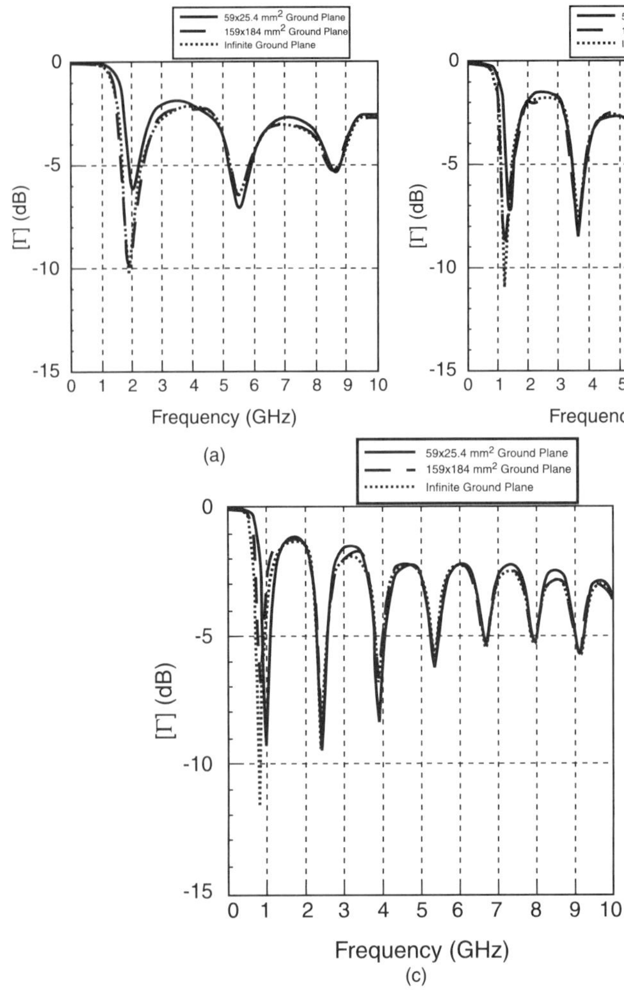

FIGURE 12.23 The effects of ground plane size on meander line antennas of (a) L_{ax} = 28 mm, (b) L_{ax} = 46 mm, and (c) L_{ax} = 73 mm.

Meander line antennas investigated in this section are those with L_{ax} = 28, 46, and 73 mm. Return losses of these three antennas on three different sizes of ground planes are shown in Fig. 12.23.

The differences in return losses of monopoles on the large finite ground plane and infinite ground planes are reduced as the frequency increases, because of the reduction in wavelength as frequency increase makes the 159 × 184 mm² ground plane perform like an electrically semi-infinite ground plane. Furthermore, the first resonant impedance of a meander line of L_{ax} = 34 mm is greatly reduced, when the ground plane size is reduced from 172 × 49.6 to 50 × 20 mm². In Reference [41], the 172-mm side of the larger ground plane is almost 1 λ at its operating frequency of 1.6 GHz, and the 50-mm side of the small ground plane is only 0.27 λ at 1.6 GHz. Therefore, a significant electrical change in resonant impedance is expected. Similarly, in the case of L_{ax} = 28 mm meander line antenna, the change of ground plane size from 59 × 25.4 to 159 × 184 mm² is also an electrically significant increase. The first resonance of this L_{ax} = 28 mm antenna is around 2 GHz, whereas the resonance impedance increases from 15 to 25 Ω. The 59-mm side length is about 0.39 λ, and the 159-mm side length is 1.06 λ at 2 GHz, respectively. However, for both antennas of L_{ax} = 46 and 73 mm, their first resonance is around 1.4 GHz and 0.9 GHz, respectively. Thus, all the sides of ground plane sizes are from less than 0.27 λ to less than 0.75 λ for

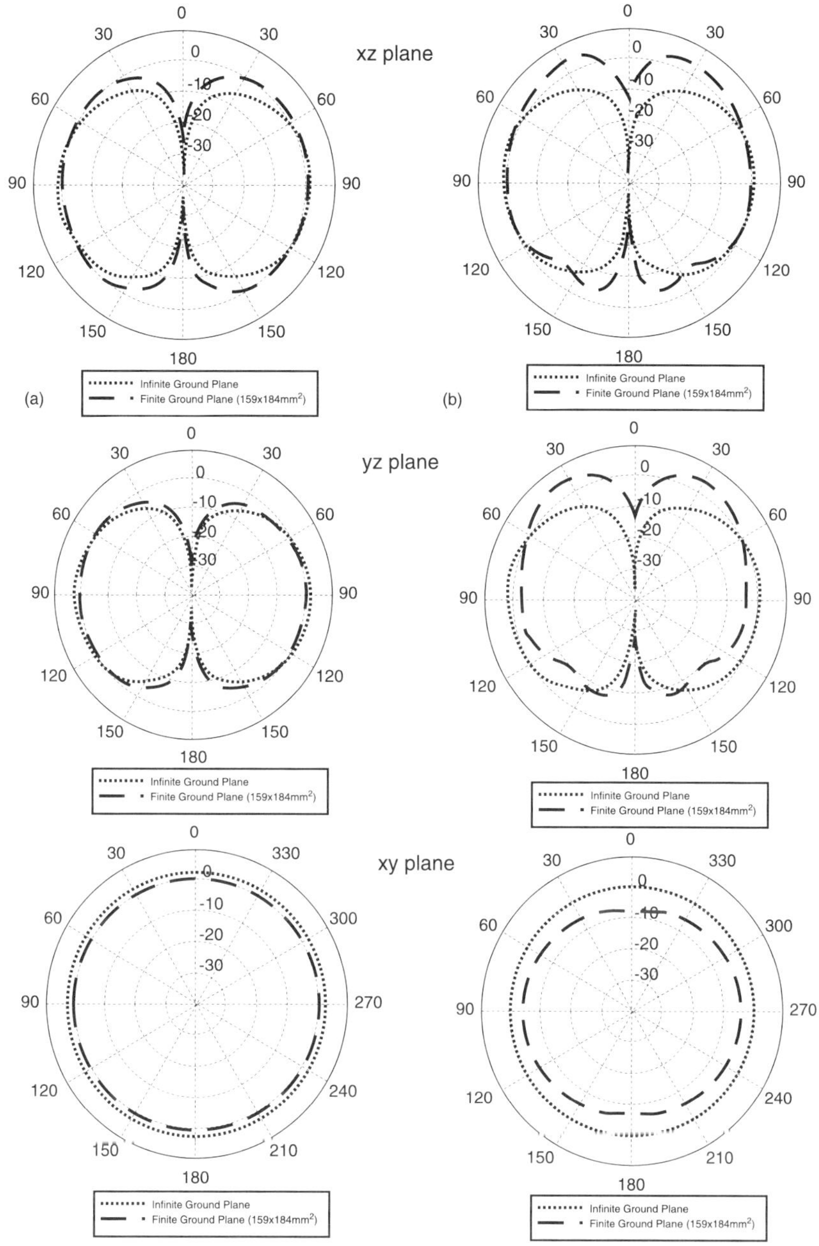

FIGURE 12.24 The radiation directivity of a meander line antenna on a large and an infinite ground plane with L_{ax} = 73 mm, (a) at 0.95 GHz, (b) at 2.39 GHz.

both cases. Therefore, the effects of ground plane size increase on these two antennas are different from those of the L_{ax} = 28 mm, which have less values of the first resonance, as shown in Fig. 12.23.

The radiation directivities of the L_{ax} = 73 mm meander line monopole are computed at 0.95 and 2.39 GHz, as shown in Fig. 12.24a and b, respectively, where the unit is dBi. The radiation directivity at 0.95 GHz is similar to the results in Reference [41] at 1.6 GHz. However, the directivity at 2.39 GHz is distorted because of the presence of the roughly 1 λ ground plane, whose distorted radiation patterns on *xz* and *yz* planes are similar to those of a quarter wavelength monopole on a 2 λ square ground plane in Reference [42]. The distorted circle-like *xy* plane directivity at 2.39 GHz is caused by the different side

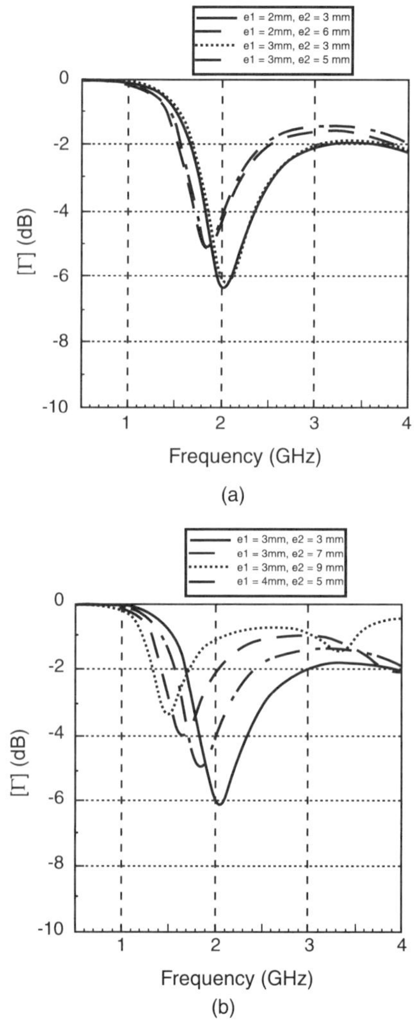

FIGURE 12.25 The return loss of 1-mm meander line antennas for various e1 and e2 values.

lengths of the rectangular ground plane, having 159 mm along the *x*-axis and 184 mm along the *y*-axis. Based on the results in Fig. 12.24a and b, the radiation directivities of the unilateral printed meander line antennas are similar to short monopoles on small ground planes except for the tilt of radiation patterns. The tilt in these patterns may be a result of the meandering current distribution on the bent trace.

12.2.2.4 Meander Line Monopoles with Different Values of e1 and e2

In Section 12.2.2.1, the characteristics of 1-mm meander line antennas with e1 = e2 = 3 mm are presented. In this subsection, the 1-mm meander line antenna resonant impedances are investigated for different values of e1 and e2.

The length of the vertical segments is e1, and the length of the horizontal segment is e2, as shown in Fig. 12.10. In Fig. 12.25, the return loss for different e1 and e2 are compared with that of e1 = e2 = 3 mm. The input impedance of meander monopole is purely real at resonance; hence, a smaller return loss indicates a closer impedance to 50 Ω. To have unbiased comparisons, all the cases shown in Fig. 12.25 have approximately the same value of the vertical heights L_{ax}. Because of the differences among e1's, the L_{ax} in the case of e1 = 2 mm is 29 mm, but the L_{ax} used for cases of e1 = 3 and 4 mm is 28 mm.

TABLE 12.1 The Trace Lengths e1 and e2 and the Resonant Frequencies of Various Values of e1 and e2

e1 and e2 (mm)	L_{ax} (mm)	L_{trace} (mm)	*SR* (%)	f_o (GHz)	f_o' (GHz)
e1 = 2, e2 = 3	29	57	25.8	1.92	1.32
e1 = 2, e2 = 5	29	85	32.7	1.74	0.88
e1 = 3, e2 = 3	28	46	26.8	1.96	1.63
e1 = 3, e2 = 5	28	64	34.5	1.755	1.17
e1 = 3, e2 = 7	28	82	41.4	1.57	0.92
e1 = 3, e2 = 9	28	100	46.6	1.43	0.75
e1 = 4, e2 = 5	28	73	34.3	1.76	1.03

When e1 is assigned a smaller value than 3 mm, the distance between two horizontal segments is smaller. From Fig. 12.25, the resonant impedance is higher when two horizontal segments are printed closer to each other. The difference of resonant frequencies of all the test cases results from the different printed trace lengths L_{trace}, which is shown in Table 12.1. The values of the *SR* defined in Eq. (12.1) are also provided in Table 12.1. One should also notice that the shortening ratio increases with e2. The f_o' is the frequency whose quarter wavelength is the same as the trace length. The frequency for the quarter wavelength equals to L_{ax} used in those cases is around 2.68 GHz. When the same L_{ax} is used, the longer trace length of an antenna results in the lower resonant frequencies. In addition, when the ratio of the L_{ax} to L_{trace} is closer to 1, the difference between f_o and f_o' is reduced. Therefore, from Table 12.1 and Fig. 12.25, to lower the resonant frequency of a meander line monopole, one should assign a longer horizontal segment length e2, whereas the increase of the resonant impedance can be achieved by reducing the vertical spacing e1.

12.2.2.5 Meander Line Monopoles with Different Line Widths

In Section 12.2.2.1, the resonant impedance of a 1-mm meander line monopole with e1 = e2 = 3 mm is determined to be between 15 and 25 Ω. To increase the resonant impedance, effects of different e1 and e2 are used, as concluded in Section 12.2.2.4. The effects of the trace widths on the resonant impedance are studied in this section to determine the relation between the line width and the resonant impedance.

The 1-mm meander line monopole with L_{ax} = 28 mm in Section 12.2.2.1 is used for comparison with the same antenna using 0.5-mm line width, as shown in Fig. 12.26. From the return loss results in Fig. 12.26, the reduction in line width results in the decrease of the first- and third-resonant impedances, which is similar to the larger thickness of straight monopole resulting in higher odd resonant impedance [42]. The L_{ax} = 28 mm meander antenna with e1 = e2 = 3 mm cannot be realized using 2-mm line width, because the spacing between the two parallel segments is occupied using 2 mm. To compare the effect of the line width change from 1 to 2 mm, the meander line is changed by increasing e1 and e2 to 6 mm and the L_{ax} and the L_{trace}, as shown in Table 12.2. From Fig. 12.26, the first and third resonant impedance of 2-mm line is the highest among those three cases. Therefore, the larger line width results in higher odd resonant impedance.

12.2.2.6 Dual-Sleeve Meander Line Monopoles

The input impedance of meander line monopole with 1-mm line width and e1 = e2 = 3 mm printed on a dielectric substrate of thickness t_s = 3.17 mm and width W_s = 11 mm has been analyzed rigorously in Section 12.2.2.1. However, these antennas are appropriate for applications with nominal impedance of 25-Ω systems. In this section, a very simple tuning method, that does not complicate the design profiles, has been adopted to readjust the input impedance and enlarge the impedance bandwidth of the antennas presented in Section 12.2.2.1, This method is referred to as dual-sleeve tuning [38]. The dual-sleeve tuning technique has been successfully applied to create wideband characteristic PCS antennas in References [38, 43, 44].

In References [38], the numerical analysis and measurement results of a bent wire meander line with a dual-sleeve tuner on large ground plane are presented. To integrate with modern printed circuit

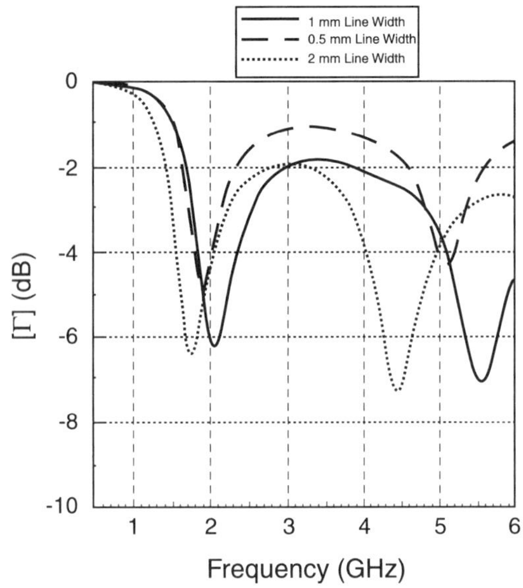

FIGURE 12.26 The return loss of meander line monopole vs. trace width.

TABLE 12.2 L_{ax}, L_{trace}, and the Resonant Frequencies of a Meander Line Monopole vs. Different Line Widths

Case	L_{ax} (mm)	L_{trace} (mm)	f_o (GHz)	*SR* (%)
W_l = 0.5 mm	28	54.5	1.96	26.8
W_l = 1.0 mm	28	46	1.83	31.7
W_l = 2.0 mm	32	50	1.67	28.7

technology applied in mobile communication, instead of using wire antennas, the meander line monopoles and sleeve tuners are printed on a dielectric substrate. The dual-sleeve meander line monopole is depicted in Fig. 12.27, which is a meander line trace printed on a dielectric slab with dual-printed sleeves (passive radiators) sitting on a finite perfectly conducting ground plane (59 x 25.4 mm^2). In this study, e1 and e2 are also chosen to be 3 mm, and the width of the printed trace is 1 mm. The distance between the edge of the dielectric slab and the edge of the ending segment of the meander line is set equal to e1. The width (W_s) and thickness (t_s) of the dielectric slab are set to the same values as those in Section 12.2.2.1, 3.17 and 11 mm, respectively. To simplify the design processes, the dual sleeves are set to have the same length *l*. In Reference [38], dual-sleeve meander wire monopoles on large finite ground planes have been successfully tuned to 50 Ω with broadband characteristics, when making the tuner length *l* one half of L_{ax}. However, the meander line antenna in this study is printed on a dielectric slab and sitting above a small ground plane. Therefore, the first step in this study of printed dual-sleeve meander line antenna is to determine the optimum length of those sleeves. From Fig. 12.28, the optimum tuning effects for a printed meander line monopole with L_{ax} = 73 mm can be observed when the sleeve length *l* is equal to one half of the antenna height L_{ax}, which agrees with the conclusion in Reference [38]. Therefore, to simplify the analysis procedures, all the following dual-sleeves meander line monopole discussed in this section have the tuning sleeve length equal to one half of the length of the antenna height.

By considering the operating frequencies as the frequencies where the return losses are local minimum, it is clearly shown in Fig. 12.29 that the existence of the dual sleeves makes the operating frequencies more appropriate for a working antenna by having a return loss well below –10 dB. The resonant frequencies for these antennas with the tuning sleeves were found to decrease with L_{ax}, as shown in Fig. 12.30a. The first resonance is found to be around 10 Ω, as shown in Fig. 12.30b. However, except for the antenna with L_{ax} = 19 mm, the first and second operating frequencies of these antennas are found

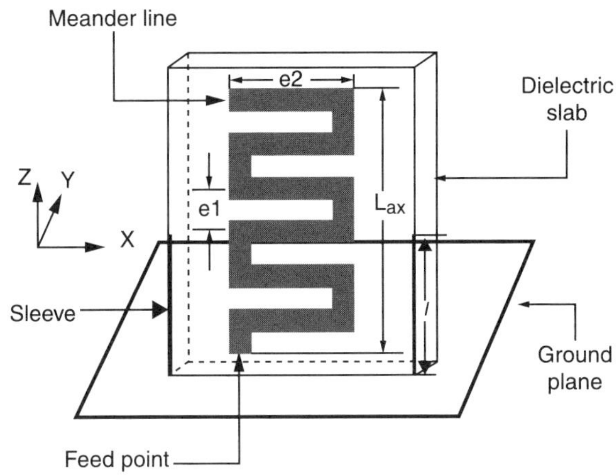

FIGURE 12.27 A dual-sleeve meander line monopole.

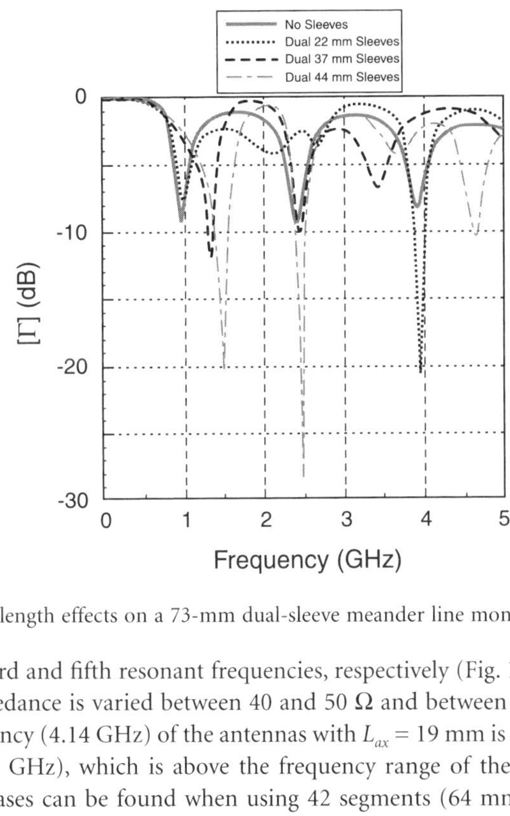

FIGURE 12.28 The sleeve length effects on a 73-mm dual-sleeve meander line monopole.

at the vicinities of the third and fifth resonant frequencies, respectively (Fig. 12.30a and c), whose corresponding resonance impedance is varied between 40 and 50 Ω and between 35 and 50 Ω (Fig. 12.30b). The first operating frequency (4.14 GHz) of the antennas with L_{ax} = 19 mm is between the first (2.5 GHz) and third resonance (7.5 GHz), which is above the frequency range of the current PCS applications. Additionally, optimum cases can be found when using 42 segments (64 mm for L_{ax}) and 48 segments (73 mm for L_{ax}), where the third and fifth resonance impedance is 42 and 40 Ω, 40.5 and 47 Ω, respectively. The third and fifth resonant frequencies for these two designs are in the frequency range of 0.9 to 3.0 GHz, which indicate that for L_{ax} between 64 and 73 mm, dual-frequency operation is achievable within the current or future PCS applications.

Furthermore, these antennas are adequately matched to the traditional 50-Ω impedance of the front-end circuitry. This 50-Ω input impedance feature should simplify the PCS system designs by removing the impedance-matching network between the feed and the antenna. From Fig. 12.30, the bandwidths of antennas with L_{ax} between 64 and 73 mm are around 12 and 4.5% for the first and second operating

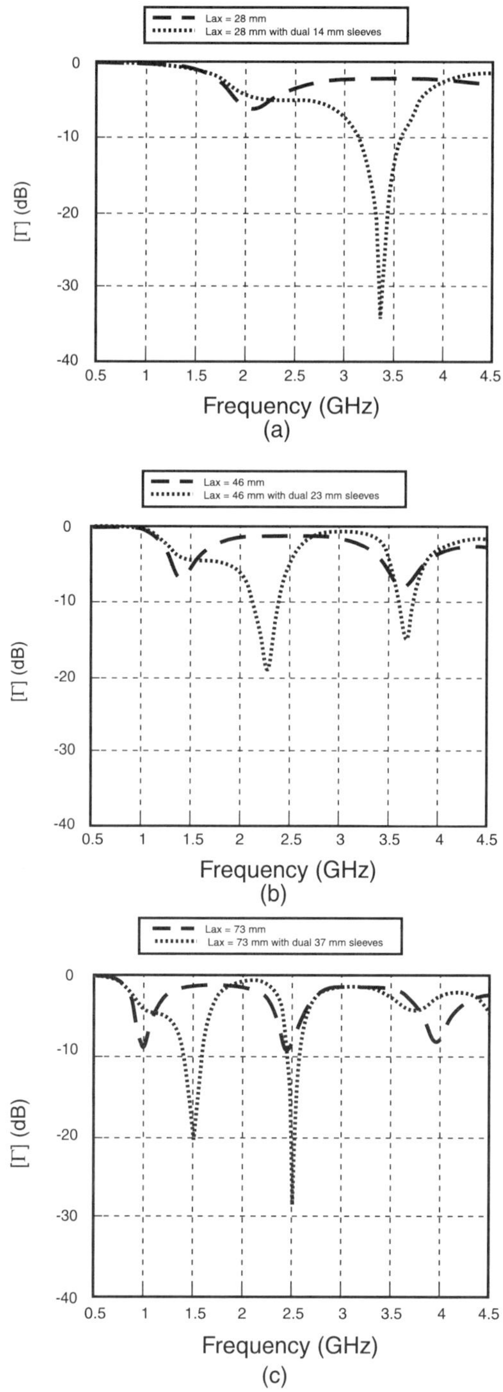

FIGURE 12.29 The return loss of a meander line antenna with dual ½ L_{ax} sleeves, (a) $L_{ax} = 28$ mm, (b) $L_{ax} = 46$ mm, (c) $L_{ax} = 73$ mm.

frequencies, respectively. The empirical equations for the first and second operating frequencies and third and fifth resonance impedance for various values of L_{ax} are derived from the regression curve fit of the data plotted in Fig. 12.30. The empirical equations for the operating frequencies are

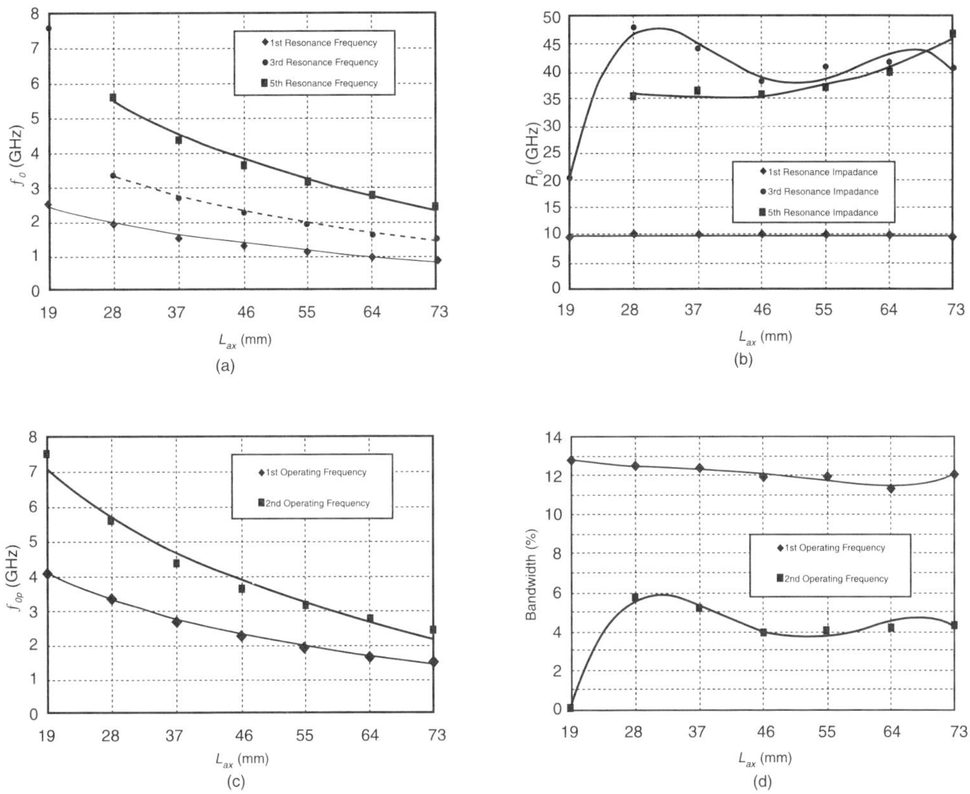

FIGURE 12.30 Radiation parameters of a meander line antenna with dual ½ L_{ax} sleeves vs. L_{ax}. (a) The resonant frequency, (b) the resonant resistance, (c) the operating frequency (d) the bandwidth.

$$f_{op1} = 10.03 - 2.0143 \ln\left(L_{ax}\right) \tag{12.4a}$$

$$f_{op2} = 18.152 - 3.7251 \ln\left(L_{ax}\right) \tag{12.4b}$$

while the equations for the third and fifth resonance impedance are given by

$$\begin{aligned} R_{o3} = {} & 9.8226\times10^{-8}\left(L_{ax}\right)^6 - 2.5210\times10^{-5}\left(L_{ax}\right)^5 + 0.0025026\left(L_{ax}\right)^4 \\ & -0.11898\left(L_{ax}\right)^3 + 2.6549\left(L_{ax}\right)^2 - 20.48\left(L_{ax}\right) \end{aligned} \tag{12.5a}$$

$$R_{o5} = 0.0096\left(L_{ax}\right)^2 - 0.7454\left(L_{ax}\right) + 49.261 \tag{12.5b}$$

These empirical equations are for meander line antennas with dual sleeves (l = ½ L_{ax}) using a 1-mm trace, e1 = e2 = 3 mm, t_s = 3.17 mm, and W_s = 11 mm; a substrate relative permittivity of 2.2; and a 59×25.4 mm² ground plane. It is worth noting that the first operating frequency for L_{ax} within 46 and

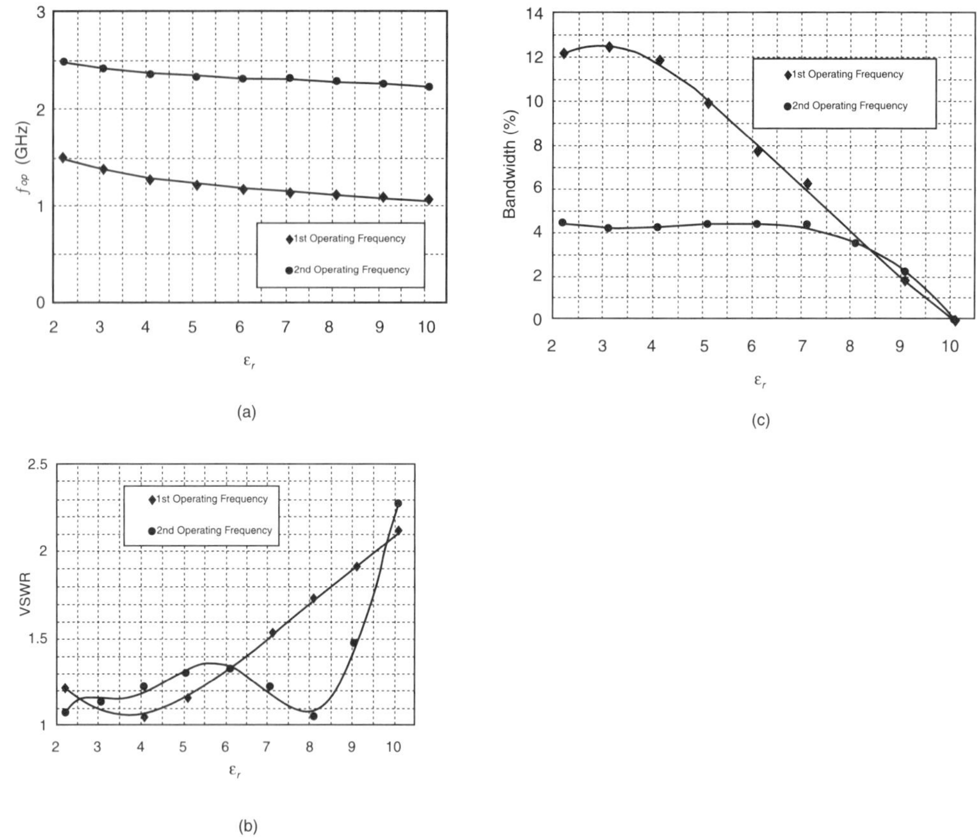

FIGURE 12.31 Radiation parameters of a meander line antennas with dual ½ L_{ax} sleeves vs. ε_r. (a) The operating frequency, (b) the VSWR, (c) the bandwidth.

64 mm features a bandwidth of 12% that may support the upcoming third generation cellular systems, operating around 2 GHz, as shown in Fig. 12.30.

In addition to changing L_{ax}, different values of the substrate dielectric constant are adopted to examine their influence on both resonant frequency and impedance in the optimum case (L_{ax} = 73 mm). From Figs. 12.31, both resonant frequency and impedance decrease with the increase of the dielectric constant. From Fig. 12.31a, when ε_r is less than 6, the change of dielectric slab can tune the operating frequencies down. When ε_r is larger than 6, both input impedance and bandwidth are becoming unacceptable for 50-Ω PCS application, as clearly shown from the corresponding voltage standing wave ratio (VSWR) values in Fig. 12.31b. Figure 12.31c shows that the bandwidth values are also acceptable for values of $\varepsilon_r < 6$.

The effect of the dielectric slab thickness t_s is also investigated. It is found that the thickness of the dielectric slab changes the resonant frequencies within 0.1 GHz for both the first and second operating frequencies as shown in Fig. 12.32a. The first and second operating frequencies are also in the vicinities of the third and fifth resonance, respectively. However, the changes of the input impedance are detectable as shown in Fig. 12.32b. When the thickness is small, the better VSWR is observed, but the bandwidth of the first band is reduced from 12 to 10%, as shown in Fig. 12.32c. The empirical equations for the first and second operating frequencies vs. t_s are

$$f_{op1} = 1.5814 - 0.064\,\ln\left(t_s\right) \tag{12.6a}$$

$$f_{op2} = 2.5353 - 0.0577\,\ln\left(t_s\right) \tag{12.6b}$$

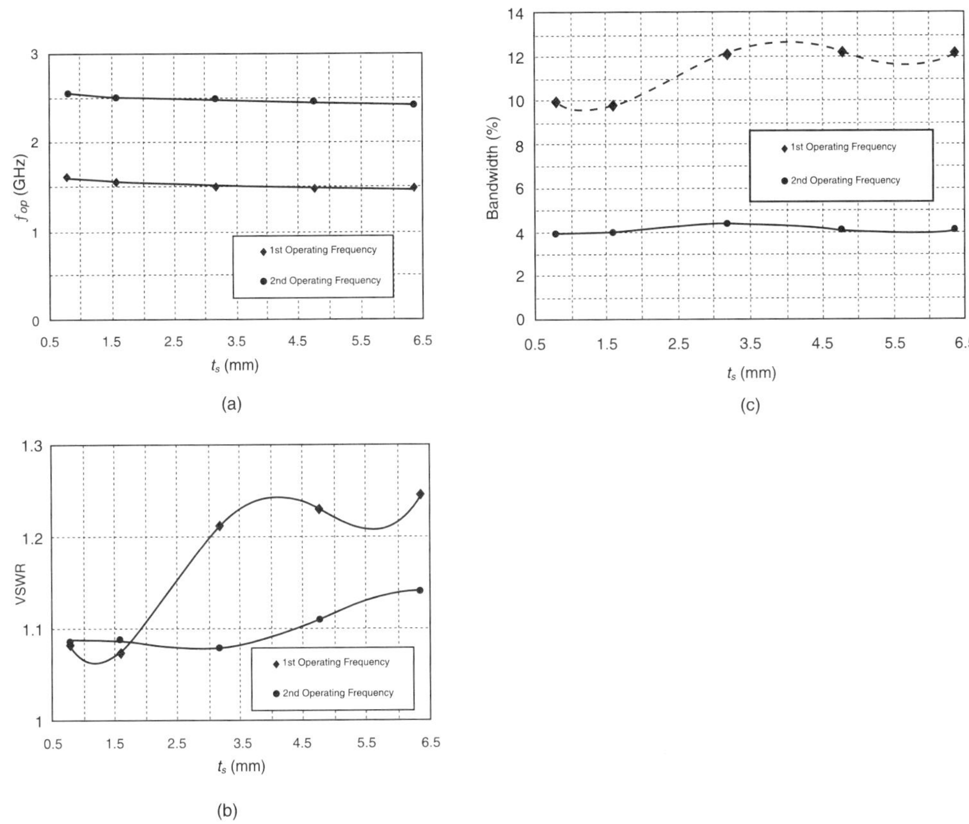

FIGURE 12.32 Radiation parameters of a meander line antennas with dual ½ L_{ax} sleeves vs. t_s. (a) The operating frequency, (b) the VSWR, (c) the bandwidth.

whereas empirical equations for the third and fifth resonant impedance vs. t_s are

$$R_{o3} = -0.012729(t_s)^6 + 0.15862(t_s)^5 - 0.51952(t_s)^4 + 0.84411(t_s)^2 + 46.647 \tag{12.7a}$$

$$R_{o5} = 0.004196(t_s)^6 - 0.0051357(t_s)^5 + 0.16667(t_s)^4 - 0.47453(t_s)^2 + 46.998 \tag{12.7b}$$

The effects of the width of the dielectric slab are also investigated for the antenna with L_{ax} = 73 mm. From Fig. 12.33, the variations of the operating frequencies and bandwidth resulting from the change of the slab width are insignificant. However, the effects on the input impedance are clearly noticeable. From Fig. 12.33b, the increase of the slab width causes the reduction of the resonant impedance and VSWR at the operating frequencies. The operating frequencies are in the vicinities of the third and fifth resonance. When the width is less than 20 mm, this antenna performs as good as a 50-Ω dual-band antenna. This feature can be applied to the antenna printed on a small circular cylinder whose curvature is comparably small to the wavelength to reduce the size of the substrate. The relation between the operating frequencies vs. W_s can be described as

$$f_{op1} = 1.5262 - 0.0193\ \ln(W_s) \tag{12.8a}$$

$$f_{op2} = 2.6080 - 0.0552\ \ln(W_s) \tag{12.8b}$$

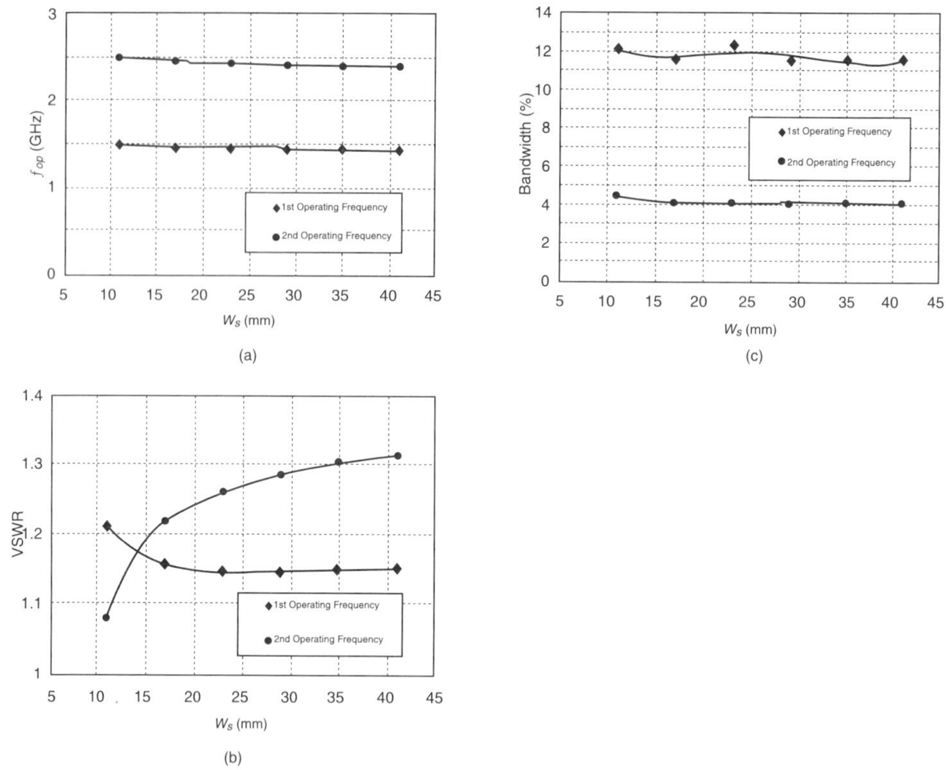

FIGURE 12.33 Radiation parameters of a meander line antennas with dual ½ L_{ax} sleeves vs. W_s. (a) The operating frequency, (b) the VSWR, (c) the bandwidth.

while relations between the third and fifth resonance impedance vs. W_s are

$$R_{o3} = 1.4673\times10^{-7}(W_s)^6 - 1.6513\times10^{-5}(W_s)^5 + 5.8583\times10^{-4}(W_s)^4 - 0.0016801(W_s)^3 - 0.33882(W_s)^2 + 7.0505(W_s) \tag{12.9a}$$

$$R_{o5} = -1.0341\times10^{-6}(W_s)^6 - 1.4712\times10^{-4}(W_s)^5 - 8.2505\times10^{-3}(W_s)^4 + 0.23041(W_s)^3 - 3.3021(W_s)^2 + 21.699(W_s) \tag{12.9b}$$

The radiation characteristics of the dual-sleeve meander line antenna are computed for the optimum case with L_{ax} = 73 mm. As shown in Figs. 12.34 and 12.35, the directivity at 1.5 GHz (see Fig. 12.34) is similar to the pattern without sleeves presented in Section 12.2.2.3. However, the directivity patterns at 2.39 GHz, as shown in Fig. 12.35, are distorted because of the existence of dual sleeves. At 1.5 GHz, the sleeve length is 0.19 λ. However, at 2.39 GHz, the sleeve length is almost 0.3 λ. Therefore, the distortion at 2.39 GHz is found at θ = 90° on *xz* and *yz* planes, because the electrical lengths of the sleeves are more than a quarter wavelength.

In this section, an optimization of the printed meander line antenna with dual sleeves for personal wireless communication has been presented. Empirical equations are also obtained and used to achieve

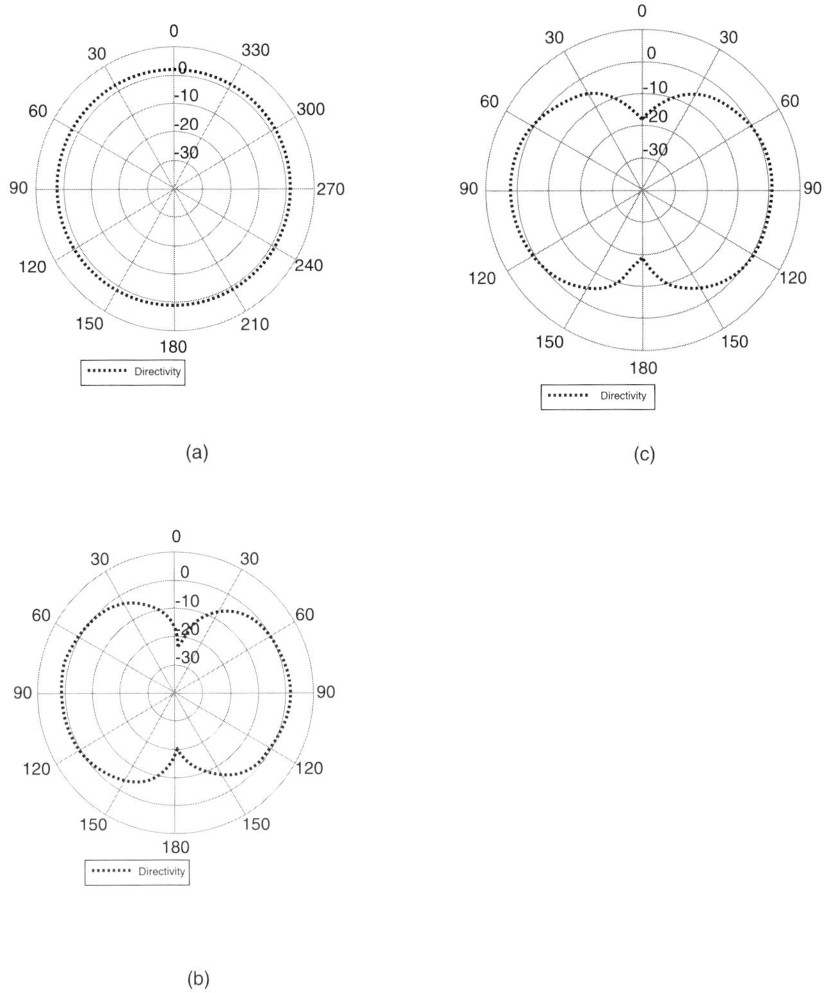

FIGURE 12.34 Directivity of a 73-mm meander line antenna with dual 37-mm sleeves at 1.5 GHz on (a) *xy* plane, (b) *xz* plane (c), *yz* plane.

a design in which an external matching network is not necessary. An additional fine-tuning for the operating frequencies and input impedance is achieved by varying dielectric slab parameters. The presented meander line with dual sleeves covers the frequency range from 1.25 to 3.0 GHz with VSWR less than 1.4 and a bandwidth varied within 120 and 340 MHz, which may effectively support the upper PCS bands and the future third-generation cellular applications. The application around 0.8 to 1 GHz may be realized by extending L_{ax} and using a higher permittivity substrate to shorten the necessary increase of L_{ax}. The radiation directivity at low-operating frequency is considered similar to that of a meander monopole without sleeves, and radiation pattern distortion is expected, when the sleeve length is longer than a quarter wavelength.

12.2.3 Taper Meander Line Antennas

In Section 12.2.2, meander line monopoles with the same e1 and e2 are analyzed. In this section, novel types of meander line monopoles are proposed as shown in Fig. 12.36a, b, and c. In Fig. 12.36a, the horizontal segment length of the taper line increases along the *z*-axis, referred as the ascendant taper line. The two other types of taper meander line, as shown in Fig. 12.36b and c, has the horizontal segment

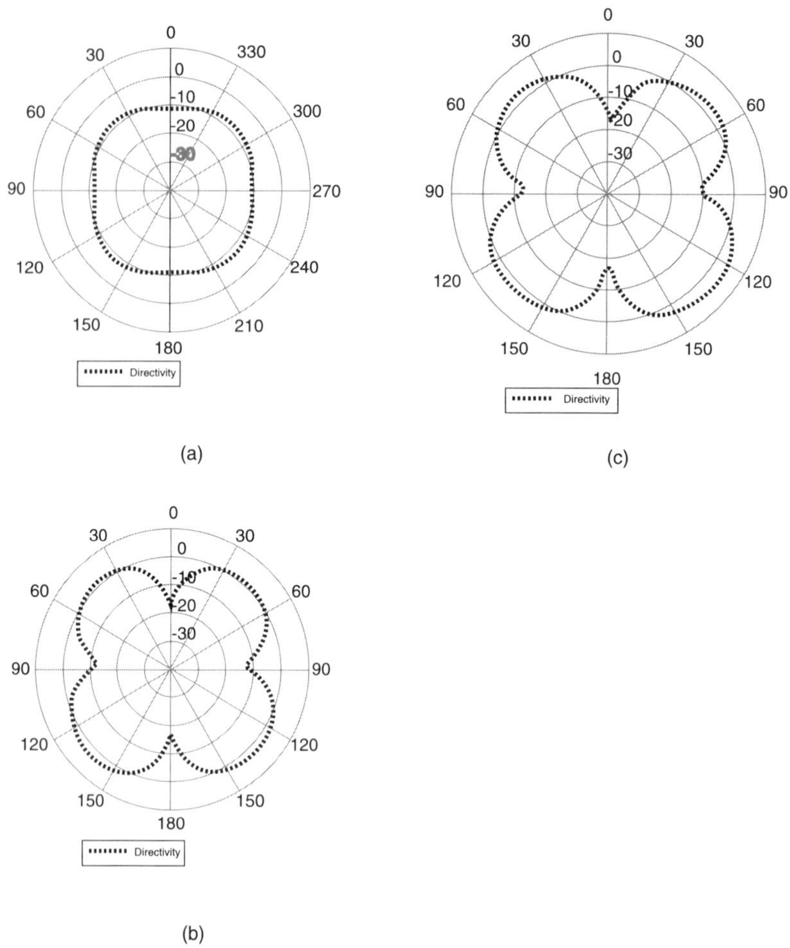

FIGURE 12.35 Directivity of a 73-mm meander line antenna with dual 37-mm sleeves at 2.39 GHz on (a) *xy* plane, (b) *xz* plane, (c) *yz* plane.

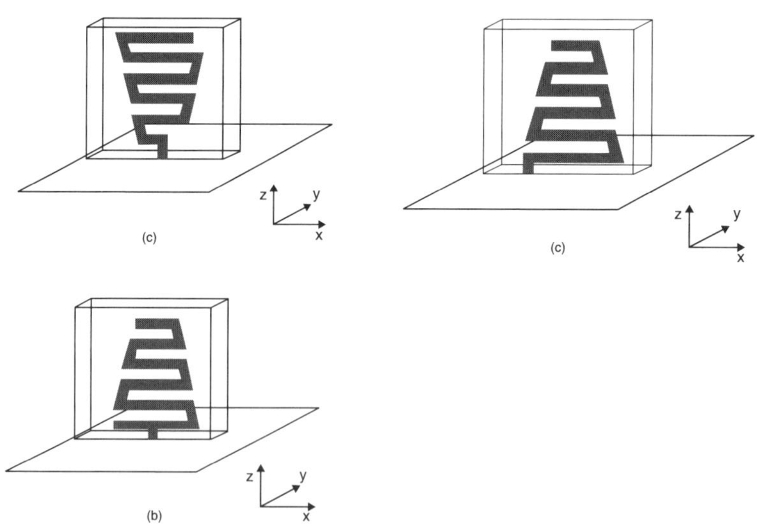

FIGURE 12.36 Taper meander line antennas. (a) Ascendant, (b) centrally fed descendant, (c) side-fed descendant.

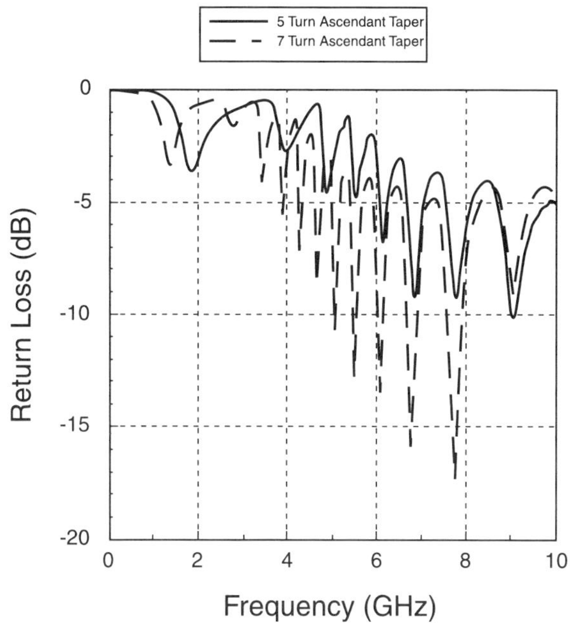

FIGURE 12.37 The return losses of five-turn and seven-turn ascendant taper monopole.

length decreasing along the *z*-axis, referred as the descendant taper line. Emphases of the analysis for taper monopoles are on their resonant frequencies and impedances.

The first test case is related to antennas designed by printing both ascendant and descendant taper lines on a 21-mm wide dielectric slab. The vertical segment length e1 = 2 mm, and the horizontal segment length e2 starting from 4 mm ending at 14 mm with a 1-mm increment for each horizontal segment, results in L_{ax} = 21 mm. The return losses of both antennas are computed and analyzed. It is found that the ascendant type of taper meander monopole has a more desirable wideband characteristic and the better return loss as the frequency increases [45]. Although the side-fed descendant meander line has a very wideband mode, the frequency is too high for PCS applications. Therefore, ascendant taper meander line antennas are analyzed in this section.

The first and third resonances of this ascendant taper line are around 1.9 and 3.98 GHz, respectively, which are too high for PCS dual-band application. In addition, the resonant impedance is also too small. Therefore, the descendant taper meander line is re-designed by adding more segments. This design example uses the same e1 but the ending e2 segment is 17 mm, which is achieved by extending two more turns from the previous example and the antenna vertical height L_{ax} = 29 mm. The return loss of this antenna is shown in Fig. 12.37 where the first and the third resonance of this 4- to 17-mm taper line is at 1.2 and 2.7 GHz, respectively, which may be tuned for a 50-Ω system.

To simplify the tuning process, the dual-sleeve method used in Section 12.2.2.6 is adopted to improve the input impedance as shown in Fig. 12.38. Because the taper line does not have uniform segments, the optimum sleeve length *l* is not necessarily one half of L_{ax}. Therefore, there is a need to study the optimum sleeve length and position. By choosing the spacing as 1-mm away from the longest horizontal segment, the effects of different sleeve length for the lower frequency end is found when the sleeve length is 24 mm, 83% of L_{ax} [45]. Because the optimum sleeve length is known, the following analysis is to determine the best location for dual sleeves. From Fig. 12.39 the optimum return loss is found when the spacing between the longest segment and a sleeve is 3 mm. The antenna is operating at 2.55 GHz with 230 MHz bandwidth, which is effective for current and future wideband mobile applications.

Next the analysis of the case with sleeve tuners not connected to the ground plane (i.e., floating printed tuner lines) is examined. To simplify the analysis procedures, the center of the tuner lines are lined up

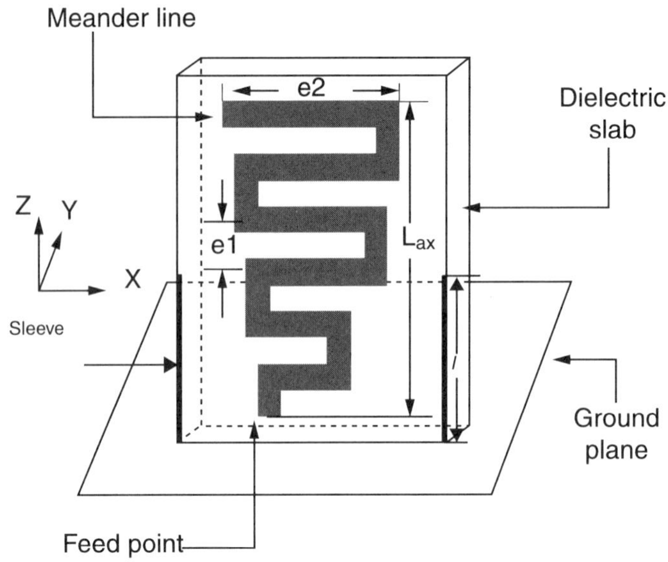

FIGURE 12.38 The dual-sleeve taper meander line monopole on a comparably small ground plane.

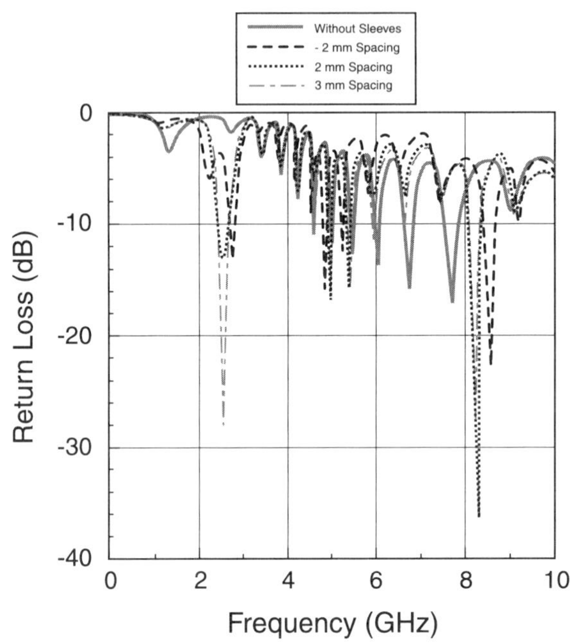

FIGURE 12.39 The return losses of seven-turn ascendant taper monopoles tuned with 24-mm dual sleeves vs. the spacing between the sleeves and the longest meander line segment.

with the center of the taper meander line trace. The effects of the lengths of the floating tuner lines are found insignificant at the frequencies below 5 GHz [45].

In this section, a set of ascendant taper meander lines is presented. The optimum case of a wideband and 50 Ω at 2.55 GHz can be obtained by using sleeve tuners, which are adequate for future mobile communication applications. The current generation of PCS applications can be achieved by using longer meander line taper trace.

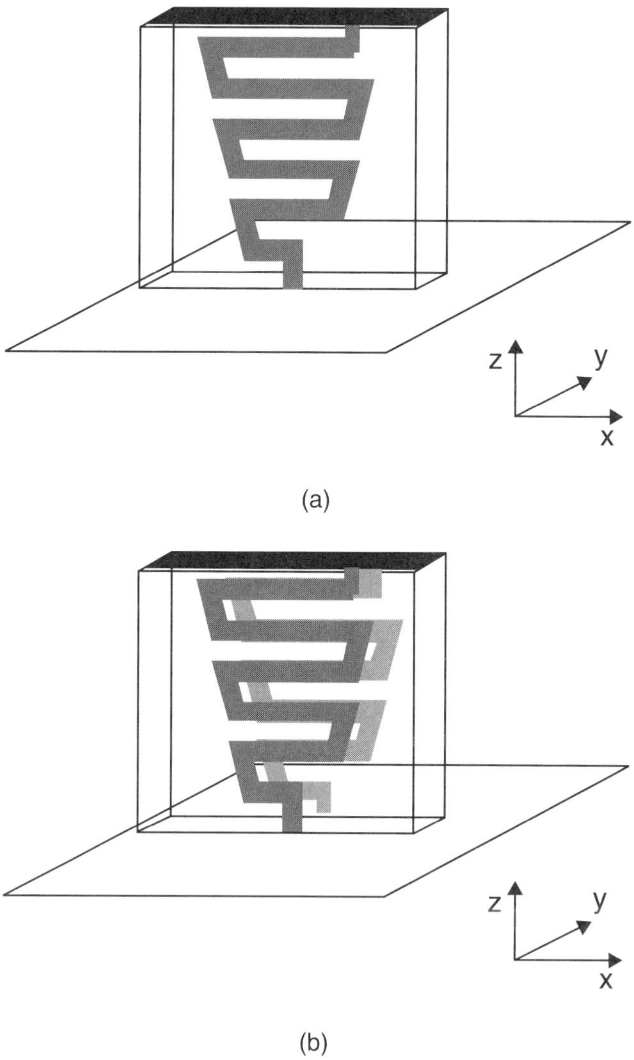

FIGURE 12.40 Top-shorted printed taper meander line antennas. (a) Unilaterally, (b) bilaterally.

12.2.4 Top-Shorted Meander Line Antennas

In Section 12.2.3, the characteristics of taper meander monopoles have been investigated. Ascendant taper meander monopoles have wider bandwidth than uniformly segment ratio meander lines at higher frequencies. In this section, top-shorted meander line antennas as shown in Figs. 12.40a and b are presented. For these antennas, the taper meander line at the front side is the same trace studied in Section 12.2.3, which has e1 = 2 mm, e2 starting from 3 to 17 mm with 1-mm increments, and L_{ax} = 29 mm (28 segment and 7 turns). The meander trace is printed on a dielectric slab of permittivity 2.2, 25 mm in width, 3.18 mm in thickness, and 31 mm in height.

The return loss of top-metalized unilaterally printed taper meander line antenna shown in Fig. 12.40a is presented in Fig. 12.41 and is contrasted with the tapered antenna without modifications. From Fig. 12.41 the top-metalized taper monopole has better return loss values at higher frequency. The first- and second-operating modes are not affected significantly. However, a very wideband is found between 4 and 7.5 GHz.

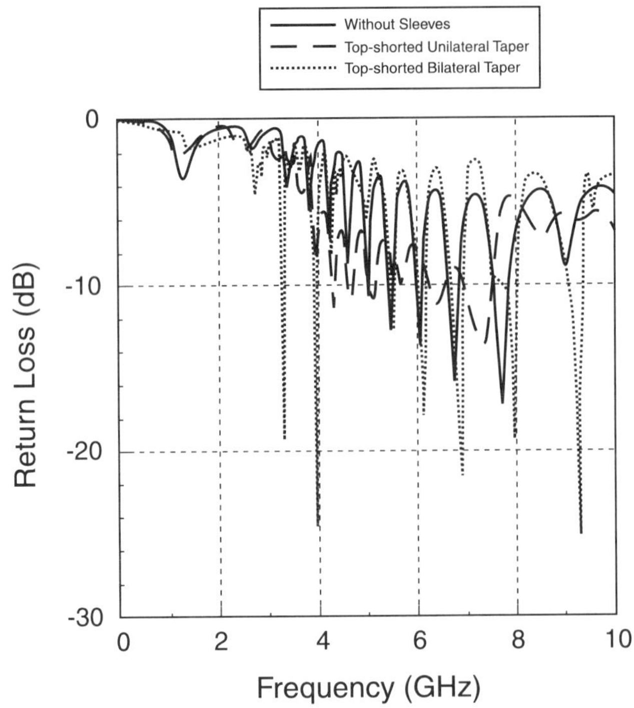

FIGURE 12.41 The return losses of unilaterally and bilaterally top-shorted taper meander line antennas.

The other type of top-shorted meander lines is an image printed of the trace on the back of the dielectric substrate and connected to the front-side trace with a metal strip as shown in Fig. 12.40b. The return loss of this antenna is shown in Fig. 12.41. The tuning effects are at frequencies higher than 2.55 GHz, which is above the PCS application range.

Therefore, by using additional tuning traces the unilateral taper meander line with top metalization may result in a useful 50-Ω antenna with wideband operation between 4 and 7.5 GHz. A longer taper trace may be used to tune this wideband characteristic to the frequency range of PCS applications if the physical space allowed for the antenna permits.

12.2.5 Back-to-Back Meander Line Antennas

In this section, bilaterally printed taper meander line antennas with and without metal strip connections on the top of the substrate are investigated, as shown in Fig. 12.42a and b. In Fig. 12.42a, the meander traces printed on the backside can be considered as the trace extension of that at the front. However, instead of connecting the image trace to the ground plane, the alternative is to image-print several turns of meander traces on the back of the dielectric slab without connecting the trace to the ground. The taper meander line at the front side is the same trace studied in Section 12.4.3, which has e1 = 2 mm, e2 starting from 3 to 17 mm with a 1-mm increment, and L_{ax} = 29 mm (28 segment and 7 turns). The meander trace is printed on a dielectric slab of permittivity 2.2, 25 mm in width, 3.18 mm in thickness, and 31 mm in height.

The traces on the back are images of the front-side traces. The return losses of these antennas with a different number of turns of imaged traces are analyzed [46]. It is found that the resonant modes are tuned toward 50 Ω as the number of turns increase. When the number of turns is increased to seven turns, between 5 and 7.5 GHz, a wideband characteristic caused by the improvement of the return losses is observed. If some tuning trace is used, then these resonant modes can become multiple broadbands.

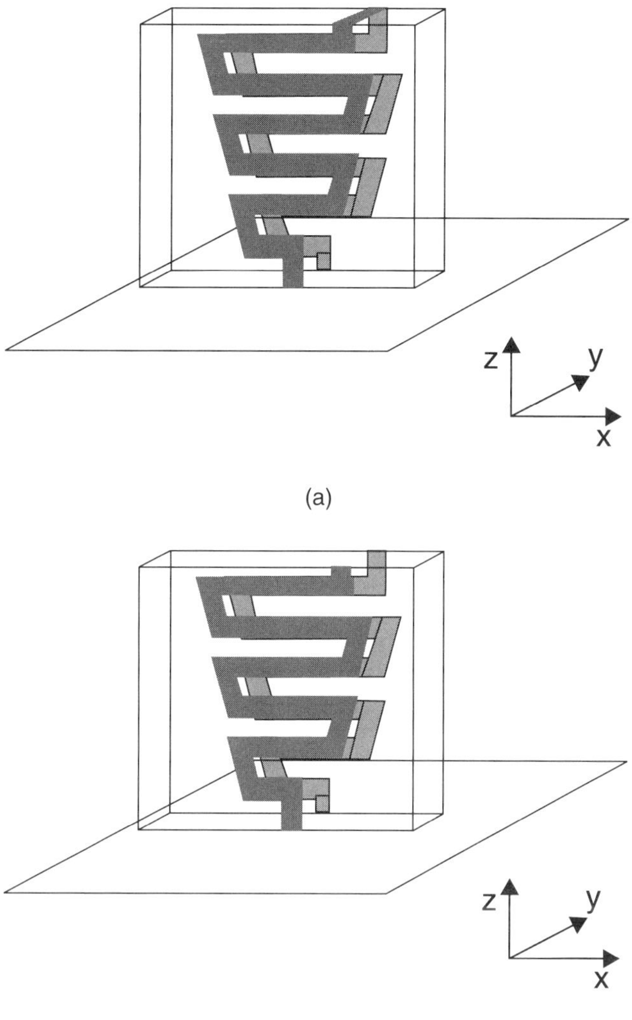

FIGURE 12.42 Bilaterally printed taper meander line antennas. (a) Top-shorted, (b) top-open.

However, the frequency range is not within the PCS frequency ranges. When the end of the seventh turn is grounded, the geometry is similar to the top-shorted meander line presented in Section 12.2.4. There are some sharp nulls within PCS frequency range, but the bandwidth is too small to be used. Wideband nulls are also at higher frequency, but they are outside the frequency range of PCS applications.

The other type of back-to-back meander line antennas is bilaterally printed taper meander traces without connection on the top as shown in Fig. 12.42b. This antenna has the same taper line on the front end with different turns of imaged traces on their backs. It is found that when the number of turns increase, more and more sharp nulls are observed in their return losses, which means the antennas at those frequencies are tuned close to 50-Ω resonant impedance. However, the sharp nulls at PCS frequencies are generally too narrow to be used. The wideband characteristics are only found above 5 GHz, when more than six turns are printed on the back.

From the preceding two study cases, these back-to-back taper meander line antennas are found useful for the applications beyond 5 GHz with wideband characteristics. A longer taper meander trace is necessary to tune the operating frequency down for PCS application.

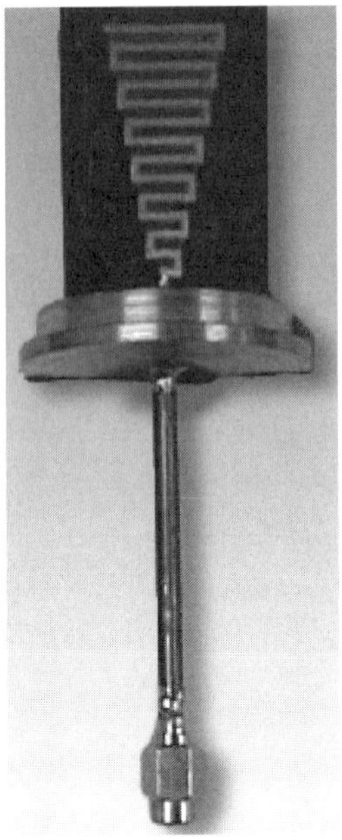

FIGURE 12.43 The prototype of taper meander line with e1 = 2 mm and e2 starting from 3 mm and ending with 17 mm on a 29-mm dielectric slab.

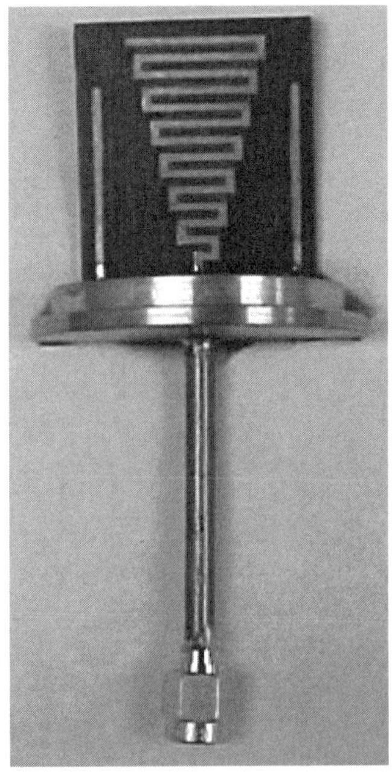

FIGURE 12.44 The prototype of dual-sleeve taper meander line with e1 = 2 mm and e2 starting from 3 mm and ending with 17 mm on a 29-mm dielectric slab.

12.2.6 Realization of Meander Line Antennas

In this section, the realization process for a meander line antenna is presented. After performing several simulations for the analysis, optimum designs may be obtained, and these optimum designs can be prototyped and tested before integration with other devices. The purpose of the tests for the designs can reassure the effectiveness of the numerical models, thus more effective analysis methods and numerical models can be achieved to shorten the design cycle. In this section, the taper meander monopole with e2 starting from 3 mm and ending with 17 mm with and without dual sleeves is fabricated, tested, and presented. The prototypes of this taper monopole without and with sleeves are shown in Figs. 12.43 and 12.44, respectively.

The taper meander is printed on a RF printed circuit board (PCB) with dielectric constant 2.2. The fabrication is realized by a computer-controlled PCB cutter. In the fabrication process, to prevent bending of the dielectric substrate, the dielectric width is increased 2 mm more in each side than the width used for the antennas in previous sections. The thickness of the PCB is one half of that used for the antennas discussed in previous sections. In Section 12.2.3, the simulation is computed on a small ground plane. However, the performance of this antenna needed to be measured on a large ground plane to avoid the influence of surrounding objects. The ground plane used in measurement is 1.5×2 m, which is effective to simulate a semi-infinite ground plane for PCS applications.

As shown in Figs. 12.43 and 12.44, a 43-mm coax is connected to the fixture of the antenna served as antenna feed. To have accurate measurement results, the effect of this coax needs to be excluded from measurement results. Therefore, a time domain measurement procedure is introduced to eliminate the phase and magnitude effects caused by this cable. After calibration, the analyzer is switched to time domain mode, and the antenna characteristics are measured in the time domain. To locate the starting plane of the antenna trace, the antenna is shorted to ground at the feed point, which results in total reflection with 180° of phase. Thus, locating the reflection with a magnitude of one and the measurement of the time difference between this reflection spike and the analyzer reference plane provided information to translate

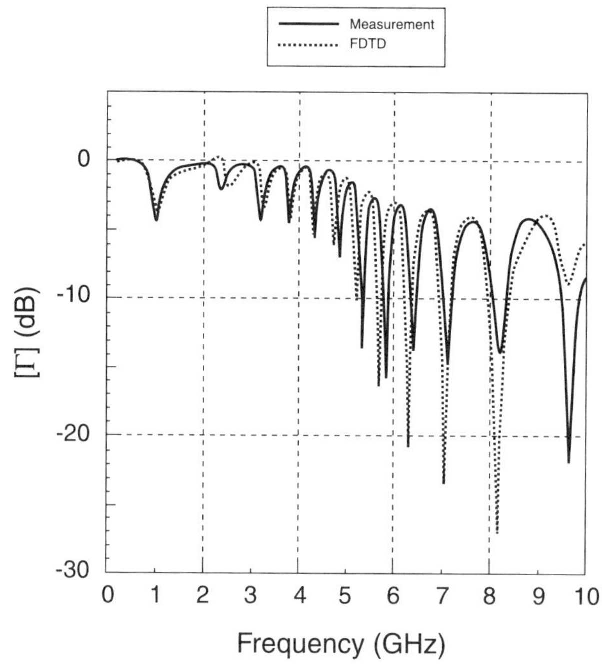

FIGURE 12.45 Comparison of measurement and simulation of a refined FDTD model for the return loss of the taper meander line antenna.

the network analyzer reference to the apparent antenna reference plane. Hence, the actual electrical delay can be determined by using one half of the time difference between the analyzer reference plane and the total reflection spike. This is because the analyzer presents the time domain reflectometry (TDR) data for twice the traveling time between source and load to detect this total reflection. The electrical length of this cable was measured to be found as 212.5 psec. Therefore, by port extending the reference plane of the analyzer by 212.5 psec, the phase delay effects of this feed cable is excluded in the measurements.

Performing the measurements using a large ground plane, the characteristics of the fabricated taper meander line dipole are obtained. The impedance for a monopole placed on an infinite ground plane can be obtained by computing half of the corresponding dipole impedance. The return loss using this dipole model is presented along with the measurement and is shown in Figs. 12.45 and 12.46. Very good agreement is obtained up to 10 GHz between numerical results and measurements. The cell size in this model is 0.5 mm, two cells for modeling the strip. The slight frequency shift at higher frequency range is mainly a result of the different feeds used in the two methods and the dielectric loss difference at higher frequencies. The coaxial and gap voltage sources result in different gap capacitance, which results in different frequency shifting at higher frequencies. In simulation, the loss only exists in the PCB with loss tangent equal to 0.01, but the prototype has both dielectric and conductor losses, which can explain the magnitude of differences in the return loss and phase change of reflection coefficient at higher frequencies. From a comparison of the results presented in Figs. 12.45 and 12.46, the effectiveness of the numerical model for the tapered monopoles and methodology are reassured, which can accurately predict the characteristics of printed antennas and serve as a robust design tool.

To validate the design of a wideband dual-sleeve taper meander line monopole presented in Section 12.2.3, the dual-sleeve taper meander line monopole is also fabricated. There is a slight difference between the prototype and the numerical model given in Section 12.2.3, which is the substrate thickness of 1.5875 mm instead of 3.175 mm. The difference is a result of the availability of the PCB. However, the effectiveness of this design can still be validated using the same test procedures for taper meander line antenna. The antenna prototype is shown in Fig. 12.44. The correction for the antenna feed is also removed using the same procedures as in the previous validation example. The measured magnitude and

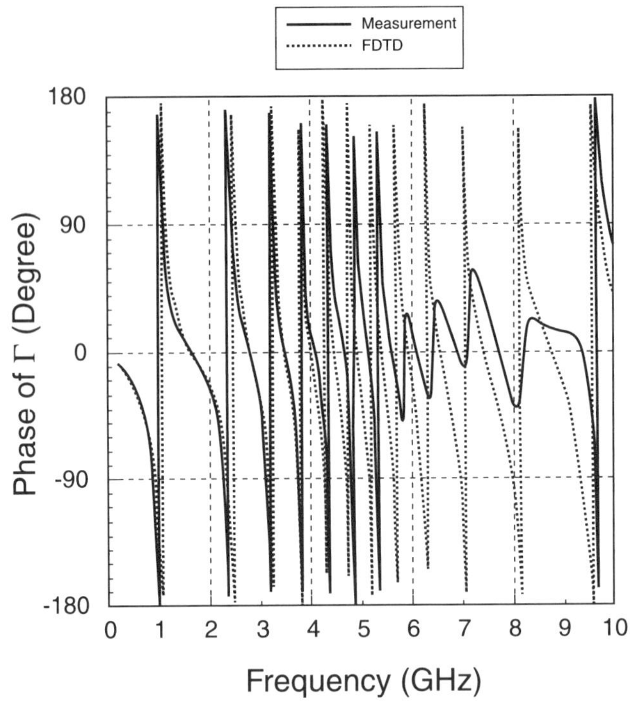

FIGURE 12.46 Comparison of measured and simulated phase change of the reflection coefficient of the taper meander line monopole.

phase of the reflection coefficient of this design are plotted in Figs. 12.47 and 12.48, respectively. From both figures, the disagreement of results from FDTD and measurement are insignificant. The slight differences may be caused by several imperfections in the prototype. The grounding of the dual sleeves is accomplished by soldering the sleeves to the fixture, as shown in Fig. 12.44, which results in the imperfect grounding at higher frequencies.

However, this imperfect grounding may be improved utilizing grounding through the antenna connectors in PCS. The dielectric and conductivity losses are also different in the prototype and the FDTD numerical model. The dual-sleeve tuning effect at 2.55 GHz is still detectable by comparing the return losses in Figs. 12.45 and 12.47. However, based on the agreement between FDTD simulation and measurements, the effectiveness of the wideband dual-sleeve taper monopole on a comparably small ground plan is reassured.

12.2.7 Meander Line Antennas on Cellular Handsets

In this section, an antenna designed on a 59×25.4 mm^2 ground plane presented in the previous section is simulated to examine its performance on a cellular handset. A wireless handset is simulated by a dielectric-coated metal box, as shown in Fig. 12.49; this box has the same width and length as those of the small ground plane, but extends 100 mm longitudinally, which is similar in size to the Ericsson GF 788e handset. The thickness of the dielectric coating is 1 mm along x- and z-directions, and 1.5875 mm along y direction. The front side of RF circuitry inside a handset is isolated by the front-panel circuitry. These front-panel circuit traces and display can be simplified by a metal shield. A wireless antenna is often placed on top of the handset and above the transceiver shield; thus, the top of a handset can be simulated as a metal shield. The back of a handset is always designed as an RF shield. The lower part of a handset on the backside has a battery pack. Therefore, a handset may be simplified as a simple dielectric coated metal box. The permittivity of the dielectric is 2.6, the same value used in Reference [47], and has a loss tangent of 0.01 at 10 GHz.

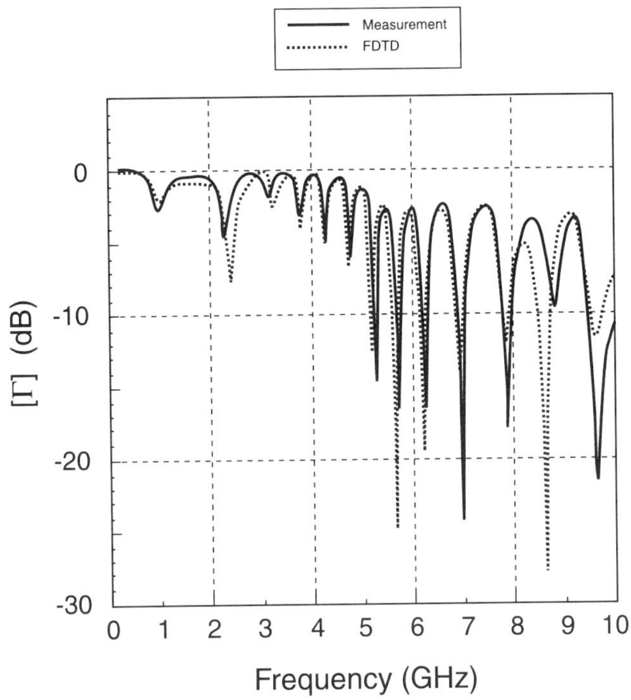

FIGURE 12.47 Comparison of measured and FDTD simulation of the magnitude of the return loss for the dual-sleeve taper meander line monopole.

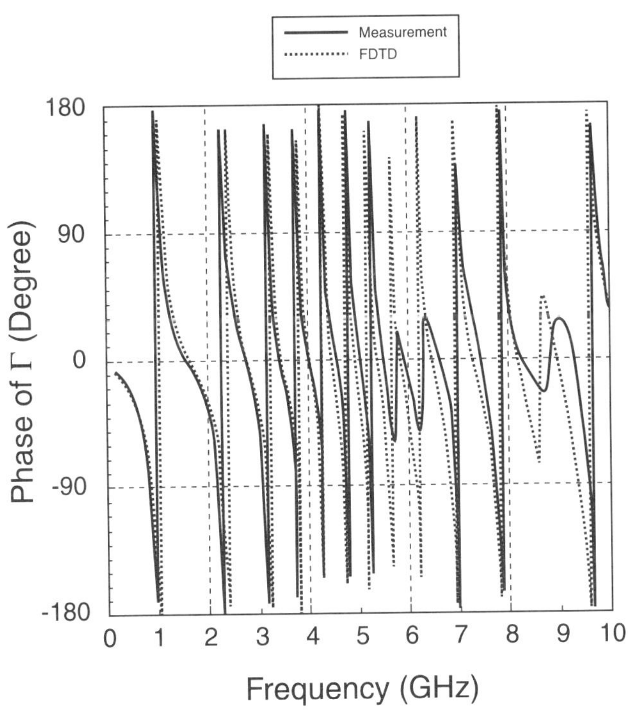

FIGURE 12.48 Comparison of measured and FDTD simulation of the phase of the reflection coefficient for the dual-sleeve taper meander line monopole.

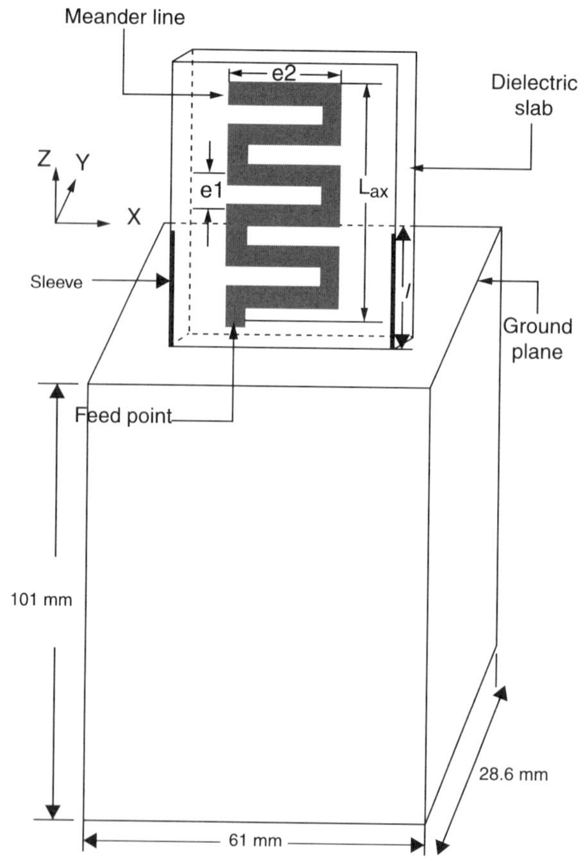

FIGURE 12.49 A dual-sleeve meander line monopole sitting on conducting box loaded with a lossy dielectric material. Outside dimensions are shown in the figure.

The antenna used in this simulation is the unilaterally printed dual-sleeve meander line of $L_{ax} = 73$ mm, that has dual-operational bands. From Fig. 12.50, the differences of operational frequencies between the antenna on the 59×25.4 mm^2 ground plane and a PCS handset are insignificant. However, the degradation of return losses found in the operational frequencies is a result of the higher resistance values for the antenna sitting on a handset. A feasible fine-tuning is made by extending the sleeve length to 49 mm, which recovers the dual-band feature with a slight frequency shift.

12.3 Conclusions

The characteristics of the planar meander line antennas for handheld wireless applications presented in this chapter are obtained using a FDTD methodology with experimental verification. This type of configuration shortens the antenna height as a result of its meander configuration, while maintaining a lower resonant frequency. Thus, the meander-type antenna can be a useful alternative for wireless handset design that provides a low-profile, or shortened length, replacement for the monopole or "whip" antenna.

Various meander configurations have been studied in this chapter; and results have been presented for ascending and descending tapers, dual-meander lines, and a meander antenna line with parasitic tuning elements. These configurations were selected with the objective to obtain desirable and optimum characteristics of good impedance match, bandwidth, and radiation pattern; and of multifrequency operation for wireless handheld systems. The ascending vertical taper, planar meander line characteristics exhibited wideband characteristics, 50-Ω input impedance, and omnidirectional radiation pattern with

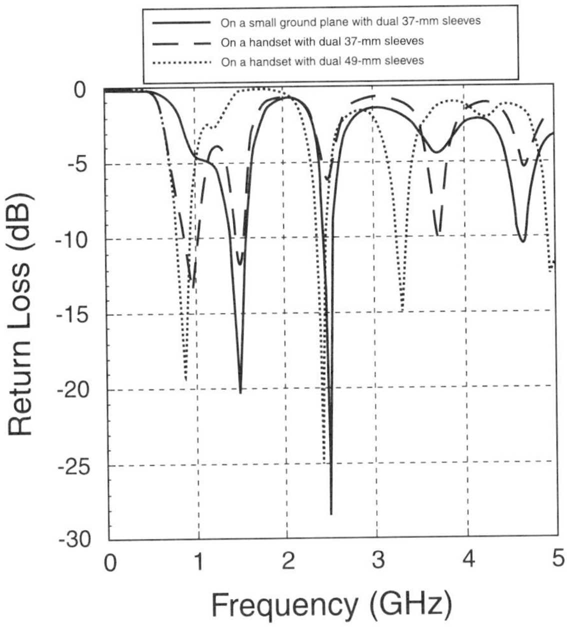

FIGURE 12.50 Comparison of the return loss of an e1 = e2 = 3- and 73-mm meander line antenna with dual 37-mm sleeves on a small ground plane and a PCS handset.

tuning capabilities using dual-sleeve parasitic elements. These impedance and bandwidth characteristics show very good agreement between the computed simulation results and the measurements that were used for verification. Empirical expressions have also been derived for the analyses of the various meander line antennas that provide a very valuable tool for the antenna designer to determine the intrinsic properties of this type antenna.

This novel design has the desirable characteristics and features of small, compact size required for modern handset design, and the meander line antenna can be used to support current and future generations of wireless cell phone and PCS.

References

1. K. Fugimoto and J. R. James, *Mobile Antenna Systems Handbook*, Artech House, Boston, 1994, p. 574.
2. R. Luebbers, L. Chen, T. Uno, and S. Adachi, "FDTD calculation of radiation patterns, impedance, and gain for a monopole antenna on a conducting box," *IEEE Trans. Antennas Propag.*, vol. 40, no. 12, pp. 1577–1583, 1992.
3. L. Chen, T. Uno, S. Adachi, and R. J. Luebbers, "FDTD analysis of a monopole antenna mounted on a conducting box covered with a layer of dielectric," *IEICE Trans. Commun.*, vol. E76-B, no. 12, pp. 1583–1586, 1993.
4. R. W. P. King, *The Theory of Linear Antennas*, Harvard University Press, Cambridge, MA, 1956.
5. S. A. Schelkunoff, *Advanced Antenna Theory*, John Wiley & Sons, New York, 1952.
6. S. Uda and Y. Mushiake, *Yagi-Uda Antenna*, Maruzen Co., Tokyo, 1954, p. 19.
7. J. L. Wong and H. E. King, "An experimental study of a balun-fed open sleeve dipole in front of a metallic reflector," *IEEE Trans. Antennas Propag.*, vol. AP-20, p. 201, March 1972.
8. S. Bhattacharya, S. A. Long, and D. R. Wilton, "The input impedance of monopole antenna mounted on a cubical conducting box," *IEEE Trans. Antennas Propag.*, vol. AP-35, pp. 756–762, July 1987.

9. A. W. C. Chu, S. A. Long, and D. R. Wilton, "The radiation pattern of a monopole antenna attached to a conducting box," *IEEE Trans. Antennas Propag.*, vol. AP-38, pp. 1907–1912, December 1990.
10. J. D. Kraus, *Antennas*, McGraw-Hill, New York, 1988, Chapter 6, pp. 238–264.
11. J. D. Kraus, "Helical beam antenna," *Electronics*, vol. 20, pp. 109–111, April 1947.
12. R. C. Johnson, *Antenna Engineering Handbook*, McGraw-Hill, New York, 1993, Chapter 13, pp. 13–18 and 19.
13. Y. Hiroi and K. Fujimoto, "Practical Usefulness of Normal Mode Helical Antennas," Digest of IEEE Int. Symp. on Antennas and Propagation, 1976, pp. 238–241.
14. T. Taga, "Analysis of Planar inverted F antennas and antenna design for portable radio equipment," in Analysis, Design, and Measurement of Small and Low Profile Antennas, K. Hirasawa and M. Haneishi, Eds., Artech House, Boston, Chapter 5, 1992.
15. M. A. Jensen and Y. Rahmat-Samii, "EM interaction of handset antennas and a human in personal communications," *Proc. IEEE,* vol. 83, no. 1, pp. 7–17, 1995.
16. Liu, Z. D., Hall, P. S., and Wake, D., "Dual frequency planar inverted F antenna," *IEEE Trans. Antennas Propagat.,* vol. 45, no. 10, pp. 1451–1458, Oct. 1997.
17. Liu, Z. D. and Hall, P. S., "Dual Band Antennas for Hand Held Portable Telephones," *Electron. Lett.,* vol. 32, no.7, pp. 609–610, March 28, 1996.
18. R. W. P. King et.al., "Transmission line antennas," *IRE Trans. Antennas Propag.*, vol. AP-8, pp. 88–90, 1960.
19. Hirasawa, K., and K. Fujimoto, "Characteristics of Wire Antennas on a Rectangular Conducting Body," *Trans. IECE Jpn.*, vol. J65-B, pp. 1113–1139, 1982.
20. Fujimoto, K., A. Henderson, K. Hirasaura, and J. R. James, *Small Antennas*, Research Studies Press, United Kingdom, 1987, p. 117–122 (distributed by John Wiley & Sons).
21. R. E. Munson, "Conformal microstrip antennas and microstrip phased arrays," *IEEE Trans. Antennas Propag.,* vol. AP-22, pp. 74–78, January 1974.
22. R. E. Munson, "Microstrip antennas," *Antenna Engineering Handbook*, McGraw-Hill, New York, Chap. 7, 1984.
23. A. Derneryd, "A Theoretical Investigation of the Rectangular Microstrip Antenna Element," *IEEE Trans. Antennas Propag.*, AP-26, pp. 532–535, July 1978.
24. K. R. Carver and J. W. Mink, "Microstrip antenna technology," *IEEE Trans. Antennas Propag.*, vol. AP-29, pp. 25–38, January 1981.
25. D. M. Pozar, "An update on microstrip antenna theory and design including some novel feeding techniques," *IEEE Antennas Propag. Soc. Newsl.*, vol. 28, pp. 5–9, October 1986.
26. H. G. Booker, "Slot aerials on their rotation to complementary wire aerials," *J. IEE (London) part III A*, vol. 93, pp. 620–626, 1946.
27. W. C. Y. Lee, *Mobile Communications Engineering*. John Wiley & Sons, New York, 1982.
28. W. C. Jakes Jr., *Microwave Mobile Communications*. John Wiley & Sons, New York, 1974.
29. R. G. Vaughan and J. B. Andersen, "Antenna diversity in mobile communications," *IEEE Trans. Veh. Technol.*, vol. VT-36, pp. 149–172, Nov. 1987.
30. R. G. Vaughan, "Polarization diversity in mobile communications," *IEEE Trans. Veh. Technol.*, vol. 39, pp. 177–186, Aug. 1990.
31. K. Ogawa and T. Uwano, "A diversity antenna for very small 800 MHz band portable telephones," *IEEE Trans. Antennas Propag.*, vol. 42, pp. 1342–1345, Sept. 1994.
32. J. S. Colburn, Y. Rahmat-Samii, M. A. Jensen, and G. J. Pottie, "Evaluations of personal communications dual antenna handset diversity performance," *IEEE Trans. Veh. Technol.*, vol. 37, pp. 737–746, Aug. 1998.
33. M. A. Jensen and Y. Rahmat-Samii, "Performance analysis of antennas for hand-held transceivers using FDTD," *IEEE Trans. Antennas Propag.*, vol. 42, no. 8, pp. 1106–1113, 1994.
34. M. A. Jensen and Y. Rahmat-Samii, "Characterization of antennas for personal wireless communication application," *Int. J. Wireless Inf. Networks*, vol. 1, no. 3, pp. 165–176, 1994.

35. B. M. Green and M. A. Jensen, "Diversity performance of dual-antenna handsets near operator tissue," *IEEE Trans. Antennas Propag.*, vol. 48, no. 7, pp. 1017–1024, July 2000.
36. C. P. Huang, A. Z. Elsherbeni, J. B. Chen, and C. E. Smith, "FDTD characterization of meander line antennas for RF and wireless communications," *J. Electromagnet. Waves and Appl.*, vol. 13, pp. 1649–1651, 1999.
37. J.P. Berenger, "A perfectly matched layer for the absorption of electromagnetic waves," *J. Comput. Phys.*, vol. 114, pp. 185–200, Oct. 1994.
38. M. Ali, S. S. Stuchly, and K. Caputa, "A wide-band dual meander-sleeve antenna," *J. Electromagnet. Waves and Appl.*, vol. 10, no. 9, pp. 1223–1236, 1996.
39. H. Nakano, H. Tagami, A. Yoshizawa, and J. Yamauchi, "Shortening ratio of modified dipole antennas," *IEEE Trans. Antennas Propag.*, vol. AP-32, no. 4, pp. 385–386, April 1984.
40. H. Y. Wang and M. J. Lancaster, "Aperture-coupled thin-film superconducting meander line antennas," *IEEE Trans. Antennas Propag.*, vol. 47, no.5, pp. 829–836, May 1999.
41. A. Z. Elsherbeni, J. B. Chen, and C. E. Smith, "FDTD analysis of meander line antennas for personal communication applications," *PIERS*, Cambridge, MA, July 1997.
42. C. A. Balanis, *Antenna Theory Analysis and Design*, 2nd Edition, John Wiley & Sons, New York, 1997.
43. M. Ali, M. Okoniewski, M. A. Stuchly, and S. S. Stuchly, "Dual frequency strip-sleeve monopole for laptop computer," *IEEE Trans. Antennas Propag.*, vol. 47, no. 2, pp. 317–328, Feb. 1999.
44. C. P. Huang, A. Z. Elsherbeni, J. B. Chen, and C. E. Smith, "FDTD characterization of meander line antennas for RF and wireless communications," *PIERS*, vol. 24, pp. 185–200, 1999.
45. C. P. Huang, A. Z. Elsherbeni, and C. E. Smith, "Analysis and design of tapered meander line antennas for mobile communications," *Appl. Comput. Electromagn. Soc. J. (ACES)*, vol. 15, no. 13, pp. 159–166, 2000.
46. C. P. Huang, Analysis and design of printed antennas for wireless communications using the finite difference time domain technique, Ph.D. Dissertation, Electrical Engineering Department, University of Mississippi, December 1999.
47. K. D. Katsibas, C. A. Balanis, P. A. Tirkas, and C. R. Birtcher, "Folded loop antenna for mobile handheld units," *IEEE Trans. Antennas Propag.*, vol. 46, no. 2, pp. 260–266, Feb. 1998.

13

Aeronautical and Maritime Antennas for Satellite Communications*

Shingo Ohmori
Communication Research Laboratory

13.1 Introduction

The International Maritime Satellite Organization (INMARSAT) provides worldwide mobile satellite communication services for maritime, land, and aeronautical users. The INMARSAT-A is the most basic system and has provided analog voice [frequency modulation (FM)] and telex services since the inauguration of the INMARSAT in 1982. The INMARSAT-C has been providing message (nonvoice) and low-speed data services since 1991; the INMARSAT-B and -M systems have been providing digital communications since the second-generation systems were introduced in 1993 and 1992, respectively. In 1990, INMARSAT introduced aeronautical satellite communications for passengers onboard and operational communications for airline companies. In 1996, INMARSAT introduced the Mini-M system. The Mini-M system is basically similar to the M system, however, its terminal size is smaller than the M system. In 2001, the F system is scheduled to be introduced to provide high-speed data services at 64 kbps. Except data speed, the mechanical and electrical characteristics are almost similar to the B system. The mobile terminals of A, B, and F systems are designed to be installed on ships and that of the Aero system

*Note: Some parts of this chapter are reprinted, with permission, from *Mobile Satellite Communications* by Shingo Ohmori, Hiromitsu Wakana, and Seiichiro Kawase. Artech House, Norwood, MA (**www.artechhouse.com**).

0-8493-0124-6/02/$0.00+$1.50

is for aircraft, so they cannot be used for land use or for portable use. On the other hand, the M, Mini-M, and C systems can be used not only by maritime users but also by land mobile users such as portable users, long-haul trucks, and trains. Table 13.1 lists the main features and characteristics of these systems.

This chapter introduces maritime and aeronautical antennas for satellite communications mainly used for the INMARSAT system. Mobile earth stations used in other systems, however, and some terminals that are still in research and development (R&D) stages are also discussed.

13.2 INMARSAT-A, -B, and -F

13.2.1 System Configuration

The characteristics of the INMARSAT-A, -B, and -F systems are almost the same, especially their antenna characteristics. Thus, in this section, the INMARSAT-A system is described as a typical example. A terminal of these systems, called a ship earth station (SES), is usually installed on a ship. As shown in Fig. 13.1, a ship earth station consists of mainly two parts: above-deck equipment (ADE), and below-deck equipment (BDE). Figure 13.2 is a photograph of the ADE covered with a radome, which is installed on a ship. Figure 13.3 shows an antenna unit inside a radome. The radome protects the antenna unit from severe environmental conditions. The ADE consists of an antenna, an antenna mount, a low-noise amplifier (LNA), a high-power amplifier (HPA), a diplexer (DIP), a stabilizer, and an antenna controller. In many cases, parabolic-type antennas have been used because of their good electrical and mechanical characteristics. The components such as LNA, HPA, DIP, and antenna controller are attached on the antenna mount to get a weight distribution suitable for good tracking performance. The technical specifications of SES for standard-A ship earth stations of the INMARSAT are given in Reference [1]. The most important requirement is that the figure of merit [gain/transmission (G/T)] be over –4 dBK. The gain and diameter of an antenna satisfying this requirement are typically about 23 dBi and 1 m.

Figure 13.4 is a photograph of the BDE installed in a ship cabin that consists of a telephone handset, a display terminal, a printer, a facsimile, a computer, and a main unit. The main unit consists of such components as a diplexer, a modulator, a demodulator, a baseband processor, terminal interfaces, and a power supply. Peripherals such as a telephone, a telex, a facsimile, and a personal computer are connected to baseband processors. Signals from the ship's gyrocompass and from several other sensors are fed to a terminal interface to give the antenna controller information about ship motions.

13.2.2 Shipborne Antenna and Tracking System

Antennas for the INMARSAT-A, -B, and -F SESs are typically parabolic antennas with gains of 20 to 23 dBi. These antennas are used for simple structure and high-aperture efficiency. Satellite tracking is an essential requirement because of ship motions and the small half-power beam width. A typical bandwidth is about 10°. The four-axis (*xy*-Az-El) stabilizer is the one most commonly used. A stable horizontal plane is obtained by controlling motion about the *x*- and *y*-axes. This stabilized antenna system installed on the *xy*. The antenna mounted on the stabilized pedestal is directed to the satellite by controlling azimuth (Az) and elevation (El) axes. This type of *xy*-axis stabilizer needs antenna pedestal control circuits with servomotors to control the axes and needs sensors such as an accelerometer, a rate sensor, and a level sensor to provide information about ship motions. Figure 13.5 shows a block diagram of an antenna pedestal control unit used for one axis (the similar units are also used for the other axes). The unit consists of a servoamplifier, an axis driver, and a central processor unit (CPU). The accuracy of the controlled angle is about 0.1°.

A flywheel stabilizer has sometimes been used to avoid the need for antenna-control sensors and electric circuit. This kind of stabilizer makes use of the inertial force generated by one or two rapidly rotating flywheels. Although these four-axis stabilizers are easy to control, and are stable and reliable, they are relatively complex, large, and heavy. A three-axis stabilizer was therefore developed, and it has

TABLE 13.1 Main Characteristics of the INMARSAT Systems

		INMARSAT-A	INMARSAT-B	INMARSAT-M	INMARSAT-Mini-M	INMARSAT-C	INMARSAT-AERO	INMARSAT-F
Main features		The first services in analog	The successor to A system in digital	Compact B system	Compact M system introduced by spot beam satellites	Store and forward communication system	For aeronautical	High-speed data system for ships
Start of service		1982	1993	1992	1996	1991	1990	Scheduled in 2001
Type of telecommunications services		Voice (FM) Telex Facsimile (4.8 kbps) Data (4.8 kbps) High-speed data (64 kbps)	Voice (16 kbps) Telex Facsimile (9.6 kbps) Data (9.6 kbps) High-speed data (64 kbps)	Voice (6.4 kbps) Facsimile (2.4 kbps) Data (2.4 kbps)	Voice (4.8 kbps) Facsimile (2.4 kbps) Data (2.4 kbps)	Telex (Facsimile: earth to ship) Data (600 bps)	Voice (9.6 kbps) Data (300 bps) High speed data (64 kbps)	Voice (4.8 kbps)
Antenna system	Antenna	Parabola (80–120 cm)	Parabola (80–120 cm)	Array	Array	Helical, cross-dipole patch	Phased array (high gain) Helical (low gain)	Parabola (80 cm)
	Gain	20–23 dBi	20–23 dBi	12–15 dBi	7–10 dBi	0–3 dBi	12 dBi (high gain) 0 dBi (low gain)	21 dBi
	G/T	–4 dBK	–4 dBK	–10 dBK (maritime) –12 dBK (land)	–17dBK	–23dBK	–13 dBK (high gain) –26 dBK (low gain)	–4 dBK
	EIRP	36 dBW	25–33 dBW	21–27 dBW (maritime) 19–25 dBW (land)	11–17 dBW	12–16 dBW	22.5–25.5 dBW (high gain) 13.5 BW (low gain)	25–33 dBW
	Weight	Approx. 120 kg	Approx. 100 kg	Approx. 25 kg	Approx. 7 kg	Approx. 10 kg	Approx. 20 kg (high gain) Approx. 3 kg (low gain)	Approx. 100 kg
Portable land use model	Terminal weight	Nonportable	20–30 kg	Approx. 13 kg	Approx. 2.3 kg	Approx. 5 kg	Nonportable	Nonportable

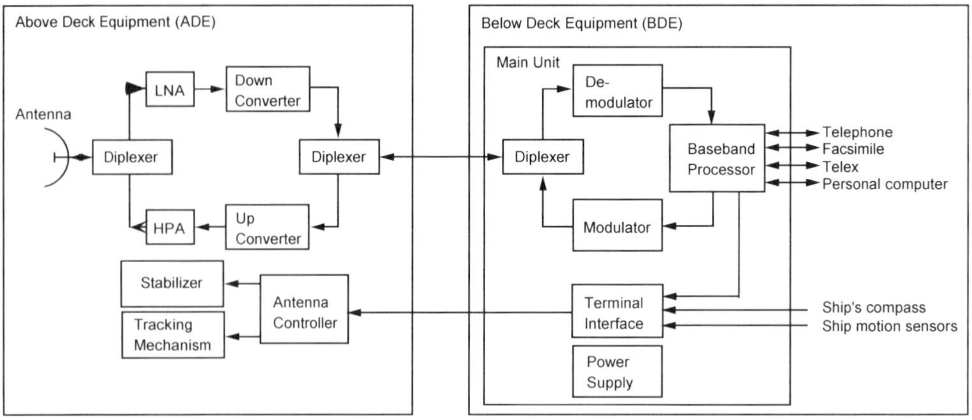

FIGURE 13.1 Block diagram of basic configuration of a ship earth station (SES).

FIGURE 13.2 Photograph of above-deck equipment (ADE) installed on a ship. An antenna unit is covered with a radome. (Courtesy of Anritsu).

been used in some recent INMARSAT-A and -B systems [2]. Figure 13.6 shows a schematic configuration of a three-axis-stabilized antenna system for the INMARSAT-A SES.

The present INMARSAT-A and -B SESs have generally used a closed-loop system for satellite tracking because of its simple configuration. In maritime satellite communications in which a ship uses a high-gain antenna with a narrow half-power beam width, the received signal is very stable. This helps to avoid sea reflections, which cause fading. The most popular open-loop tracking is a step track method. In this method an antenna is driven in elevation and azimuth directions alternatively by a step angle of 0.5° in such a way that the received signal level is kept as high as possible.

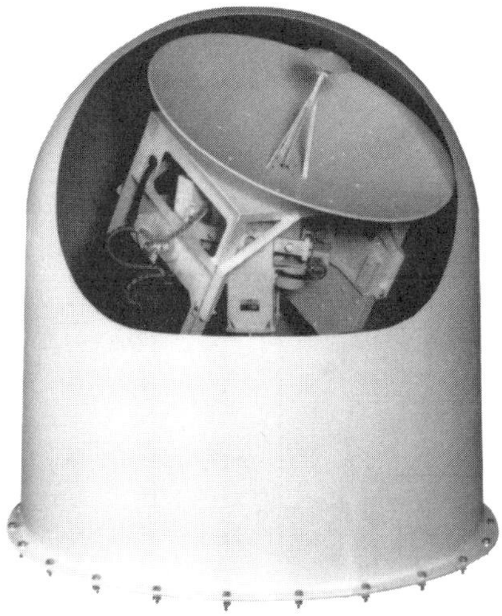

FIGURE 13.3. An antenna unit inside a radome. (Courtesy of Anritsu.)

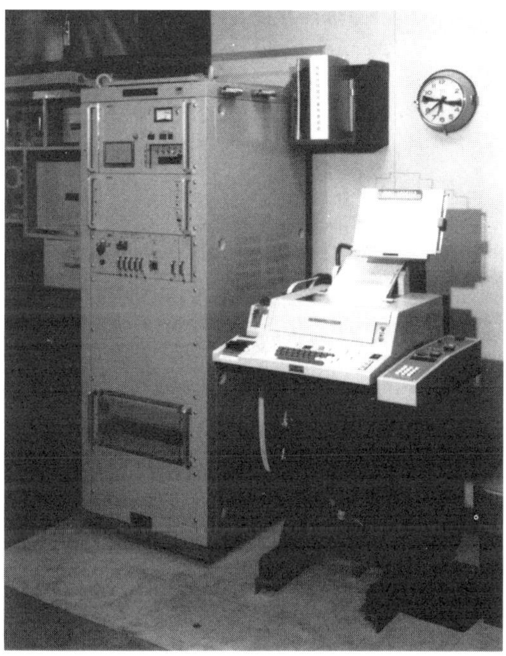

FIGURE 13.4. Photograph of below-deck equipment (BDE) installed in a ship cabin. (Courtesy of Anritsu.)

13.3 INMARSAT-C

13.3.1 System Configuration

The INMARSAT-C system was introduced in 1991 to provide data/messaging communications by terminals small enough to be hand-carried or fitted to any vessel, aircraft, or vehicle. The Aero-C was introduced several years after the introduction of the basic-C service for maritime and land mobile vehicles. The INMARSAT-C mobile earth station (MES) has a small, omnidirectional antenna, which

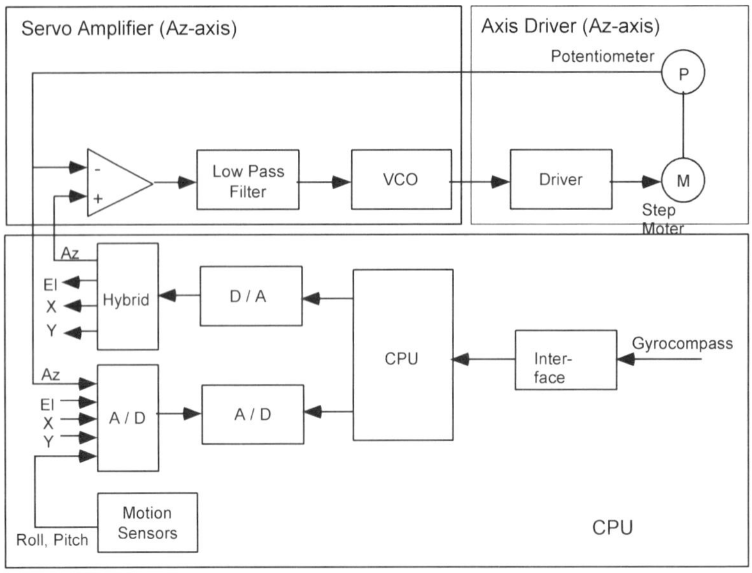

FIGURE 13.5. Block diagram of an antenna pedestal control circuit.

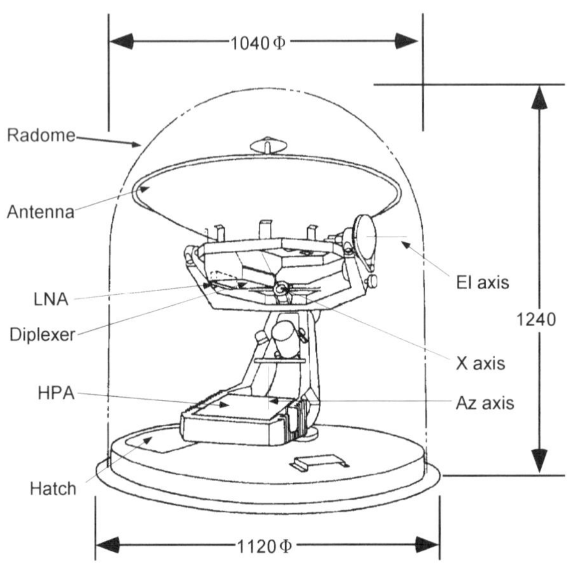

FIGURE 13.6. Three-axis-stabilized antenna system for the INMARSAT-A ship earth station.

because of its light weight and simplicity can be easily mounted on a vehicle or vessel or a handheld terminal. The main unit of a terminal is compact and weighs only 3 to 4 kg. Some terminals have message preparation and display facilities incorporated, whereas others have standard interfaces so that users can connect their own computer equipment. Handheld versions are also available. Figure 13.7 shows a photograph of the INMARSAT-C terminal for small ships. The photograph shows an externally mounted antenna unit (appears on the left), the main computer terminal (appears in the middle), and another unit.

Transmission speed between a satellite and a mobile station is 600 bps, and the signal is coded by a convolutional coding with R = 1/2 and K = 7. Bit interleaver has been adopted to disperse the burst error caused by burst fading The required G/T and equivalent istropically radiated power (EIRP) are over –23 dBK and 12 to 16 dBW, respectively.

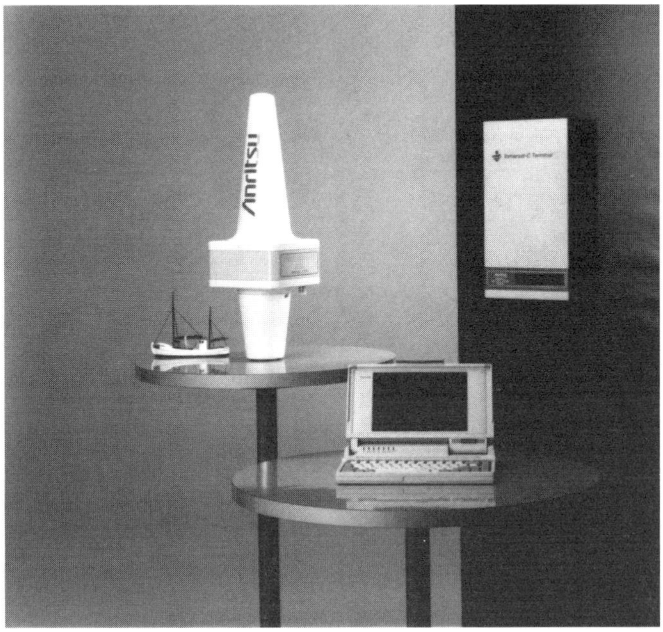

FIGURE 13.7 The INMARSAT-C terminal for small ship. (Courtesy of Anritsu.)

13.3.2 Antenna and Tracking System

The antenna has no tracking function. The antennas used for INMARSAT-C are usually omnidirectional ones such as a quadrifiler helix, a crossed-drooping dipole, and a microstrip patch. A quadrifiler antenna has been the most popular type for installation on ships because of its good performance of widebeam coverage under the condition of ship motion. A microstrip antenna is the best one to be used in handheld or briefcase terminals because of its very low profile characteristic.

13.4 INMARSAT-M

13.4.1 System Configuration

The INMARSAT-M system was introduced in 1992 to provide high-quality communications with small terminals by using digital communication technologies. Although the INMARSAT-C system cannot provide telephone services, the INMARSAT-M system can provide voice communications with very small and lightweight terminals similar to those used for the INMARSAT-C system. The M terminals may be installed in small private cruisers and cars or may be carried in a briefcase. These terminals can also be used as solar-powered emergency phones along highways as well as handheld telephones in remote areas.

The technical features of the terminal have been described in the system definition manual (SDM) of the INMARSAT [3]. Table 13.1 shows the main characteristics of an MES for the INMARSAT-M system. The required G/T values are over –10 dBK for maritime applications, and over –12 dBK for land applications. The EIRP can be set to 27 dBW in high-power and to 21 dBW in low-power modes, respectively. Figure 13.8 shows a photograph of the INMARSAT-M terminal used in a remote area.

13.4.2 Antenna and Tracking System

The kinds of antennas used for the M system depend on whether the terminals are used for maritime, land mobile, or handheld applications. In case of maritime applications, the configuration of externally mounted equipment for ships is basically the same as that of the INMARSAT-B system. A short backfire

FIGURE 13.8 Photograph of an INMARSAT-M portable-type terminal. (Courtesy of NEC.)

(SBF) antenna is one of the favorite M terminal antennas for maritime use with typical antenna gain of about 15 dBi. Although the SBF antenna is a compact shipborne antenna with a simple configuration and high efficiency, it has a narrow frequency bandwidth of about 3%, which is too narrow to cover the required frequency bandwidth of about 8% [4]. The electrical characteristics of the conventional SBF antennas are improved by changing the main reflector from a flat disk to a conical or a step plate and by adding a second small reflector. The improved SBF antenna has aperture efficiency of about 80% and the frequency bandwidth of 20% for voltage standing wave ratio (VSWR) less than 1.5 [5]. The antenna gain also improves by about 1 dB without any change in sidelobe levels [6]. Figure 13.9 shows a photograph

FIGURE 13.9 An improved short backfire (SBF) antenna installed on the deck for the ETS-V experiment. The antenna has a gain of 15 dBi, is 40 cm in diameter, and weighs about 40 kg including a stabilizer.

of an improved SBF antenna for the ETS-V experiment. The antenna has a gain of 15 dBi, is 40 cm in diameter, and weighs about 40 kg including a stabilizer.

These antennas generally have two axes (Az-El) stabilized. These are simple and compact antennas suitable for mounting in small ships. A step-track tracking method is the most commonly used, as in the INMARSAT-A and -B systems. Although aperture types of antennas such as the parabolic and SBF types have been very popular in maritime satellite communications, a phased-array antenna has sometimes been used for the M terminal. A phased array antenna for an INMARSAT-M terminal uses a two-axis (Az-El) stabilizer, and a beam can be scanned within ±30° in Az directions by electrical scanning to avoid gimbal lock [7]. In the case of the Az-El stabilizer, very high speeds and accelerations are required for accurate satellite tracking when elevation angles are near 90°. Thus, it becomes very difficult to track a satellite when elevation angles are near 90°. This condition is called gimbal lock. A step track method is also used by this antenna system.

13.5 INMARSAT-Aero

13.5.1 System Configuration

Worldwide commercial aeronautical satellite communication services have been provided by the INMARSAT since 1990. The system consists of a space segment, a ground earth station (GES), a network coordination station (NCS), and an aeronautical earth station (AES). The INMARSAT system has been described in the aeronautical SDM [8]. The AES portion of the system for commercial aviation has been described in Aeronautical Radio Incorporated (ARINC) Characteristics 741 [9].

The ARINC 741 categorizes the operational types of the AES into four classes as listed in Table 13.2. Class 1 provides only low-speed data services such as aeronautical operational control (AOC) and aeronautical administrative communication (AAC) services by using a low-gain antenna (0 dBi). In the future, this type of station will be used for air traffic control (ATC). Class 2 provides only voice services, mainly for passengers in the cabin (aeronautical passenger communication; APC), with a high-gain antenna (12 dBi). Class 3 provides both voice and high-speed data service by adding data systems to class 2. Class 4 combines the features of class 1 and class 3 and is expected to provide all kinds of aeronautical communications with high-gain and low-gain antennas. Figures 13.10 and 13.11 show block diagrams of an AES with low-gain (class 1), and high-gain antennas (class 3), respectively. ARINC 741 describes the characteristics of a satellite data unit (SDU), a radio frequency unit (RFU), a DIP/LNA, an HPA, and a high-gain antenna (HGA) with its associated beam-steering electronics. Figure 13.12 is a schematic illustration showing the location of AES equipment on a Boeing 747.

TABLE 13.2 Operational Types of AES for ARINC 741 Standard

Type	Antenna	Voice/Data	Service	User (Air)	User (Ground)
Class 1	Low-gain antenna (0 dBi)	Low-speed data	AOC AAC ATC [in the future]	Pilot Cabin crew Pilot	Airline company Airline company Control authority
Class 2	High-gain antenna (12 dBi)	Voice	APC	Passenger	Subscriber
Class 3	High-gain antenna (12 dBi)	Voice and high-speed data	AAC APC	Cabin crew Passenger	Airline company Subscriber
Class 4	Low-gain antenna (0 dBi) and high-gain antenna (12 dBi)	Voice and high- and low-speed data (Class 1 and Class 3)	AOC AAC APC ATC [in the future]	Pilot Cabin crew Passenger Pilot	Airline company Airline company Subscriber Control authority

AOC: aeronautical operational control, AAC: aeronautical administrative communication, ATC: air traffic control [in the future], and APC: aeronautical passenger communication.

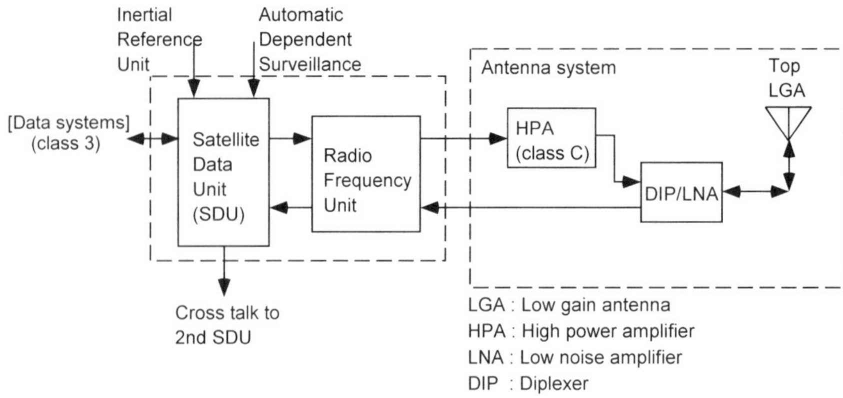

FIGURE 13.10 Block diagram of an AES with a low-gain antenna (class 1).

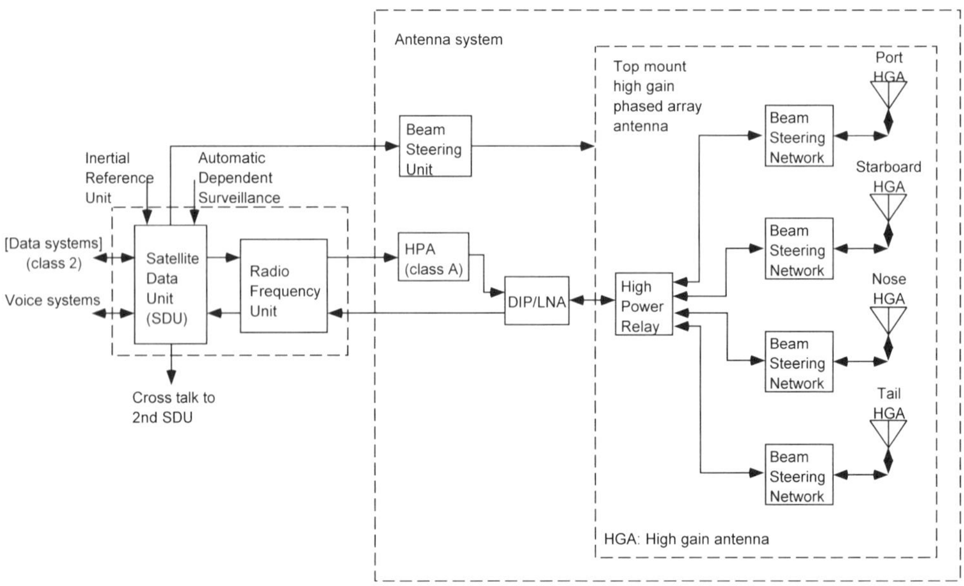

FIGURE 13.11 Block diagram of an AES with a high-gain antenna (class 2).

13.5.2 Airborne Antenna and Tracking System

As mentioned in the previous section, there are two types of airborne antennas for satellite communications: a low-gain antenna and a high-gain antenna with nominal gains of 0 and 12 dBi, respectively. New systems with these antennas are described.

13.5.2.1 Low-Gain Antenna Subsystem

A low-gain antenna system consists of an antenna element, a DIP, an LNA, and a C-class HPA. Its gain is about 0 dBi, and its radiation pattern is omnidirectional to cover over 85% of the upper hemisphere above an elevation angle of 5°. The main specifications of a low-gain and a high-gain antenna subsystems are shown in Table 13.3. Figure 13.13 shows a photograph of a low-gain antenna with an A-class HPA and a DIP/LNA.

13.5.2.2 High-Gain Antenna Subsystem

As shown in Fig. 13.12, there are two types of phased-array antennas. The first is a top-mount type, which is installed on the top of a fuselage. The second is a side-mount type, which is installed on both

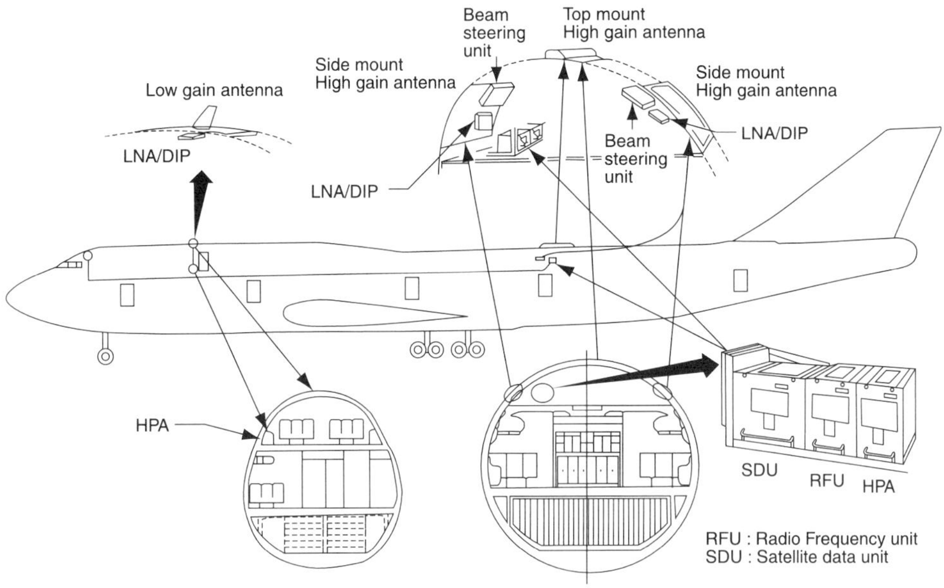

FIGURE 13.12 Location of AES equipment on the Boeing 747.

TABLE 13.3 Main Specifications of Low-Gain and High-Gain Antenna Subsystems

	Low-Gain Antenna	High-Gain Antenna
Frequency	1530.0–1559.0 MHz (receive) 1626.5–1660.5 MHz (transmit)	
Polarization	Right-hand circular	
Axial ratio	Less than 6 dB	
Figure of merit (G/T)	Over –26 dBK	Over -13 dBK
Radiation power (EIRP)	Over 13.5 dBW	Over 25.5 dBW
Antenna gain	Over 0 dBi	Over 12 dBi
Coverage for semisphericity above elevation angle 5°	Over 85%	Over 75%
Tracking	None (Omni-directional)	Program tracking

sides (port and starboard) of the fuselage. The top-mount type has the advantage of eliminating keyhole areas where the beam cannot be scanned, but it has the disadvantage of increasing air drag. The conformal type, on the other hand, has the advantage of low air drag, but has the disadvantage of keyholes. An HGA subsystem consists of a phased-array antenna, a DIP, an LNA, an A-class HPA, and a beam-steering unit (BSU), which steers the beam to track a satellite. A program-tracking method carries out the satellite tracking. A BSU steers the beam by controlling digital phase shifters of the phased-array antenna; and the information used to operate the BSU is calculated from the signals from the inertial navigation system (INS), which gives the position, heading direction, and altitude of the aircraft. To track the satellite, the HGA has to steer the beam to cover over 75% of the hemisphere above an elevation angle of 5°. When a phased-array antenna is used, these inevitably are keyhole areas in the fore and aft directions (Fig. 13.16) because it is very hard for a phased-array antenna to scan the beam to wide angular areas over 60°.

An A-class HPA is used to avoid channel intermodulation in a multicarrier operation. The main specifications of an HGA subsystem are listed in Table 13.3. Figure 13.14 is a photograph of a top-mount type of HGA with an A-class HPA and a DIP/LNA. A prototype of this antenna was developed by the Communications Research Laboratory and was installed on a Japan Air Line Boeing 747 to carry out the experiments over the transpacific flight routes between Tokyo and Anchorage in the ETS-V project

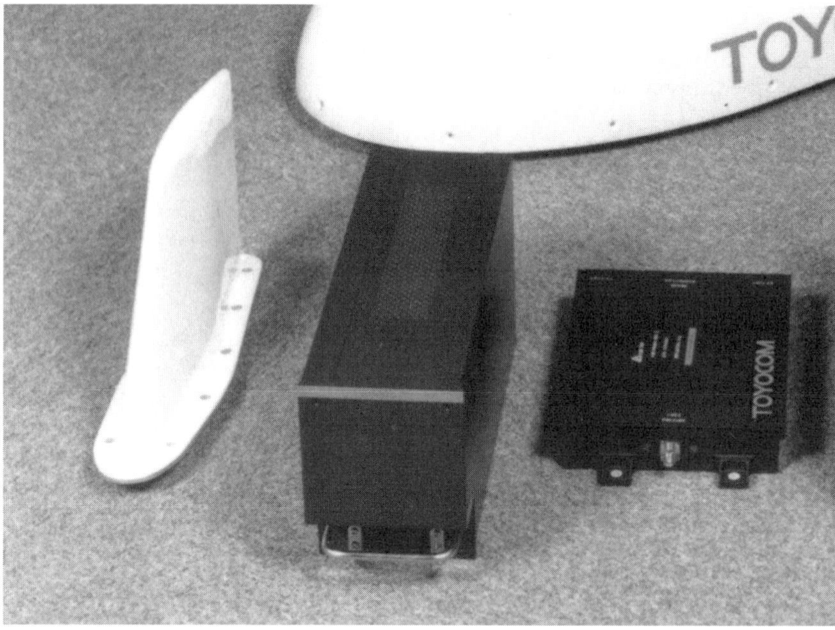

FIGURE 13.13 Photograph of a low-gain antenna with a C-class HPA and a DIP/LNA. (Courtesy of TOYOCOM.)

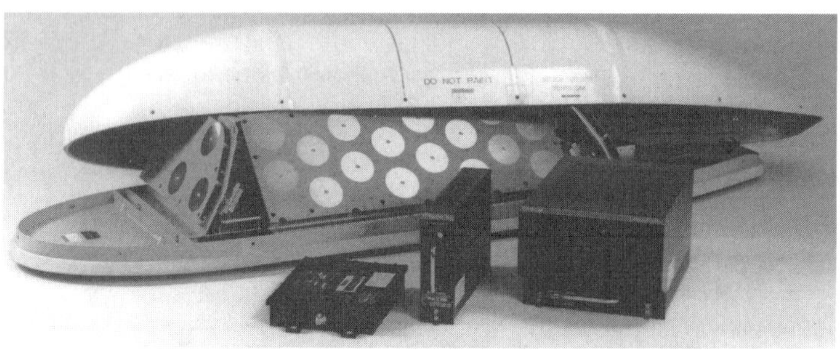

FIGURE 13.14 Photograph of a high-gain antenna with an A-class HPA, a DIP/LNA, and a BSU. (Courtesy of TOYOCOM.)

[10, 11]. As shown in Fig. 13.14, this high-gain, phased-array antenna has four phased-array units. The nose and tail array units have been used to make keyhole areas narrow enough to keep communication links over 85% in the upper hemisphere. Figure 13.15 shows a low-gain antenna and a top-mount HGA installed on a Boeing 747.

13.6 Land Mobile Earth Station

13.6.1 Overview

In the INMARSAT system, INMARSAT-M and -C earth stations have been used not only by maritime users but also by land mobile users such as trucks and trains. The INMARSAT-C earth station for land mobile users is exactly the same as that for shipboard users because its antenna is omnidirectional without tracking function. As mentioned in the previous section, a quadrifiler antenna is the most popular and suitable for the INMARSAT-C earth station. This type of earth station has been widely installed on

FIGURE 13.15 Photograph of low-gain and top-mount high-gain antennas installed on a Boeing 747.

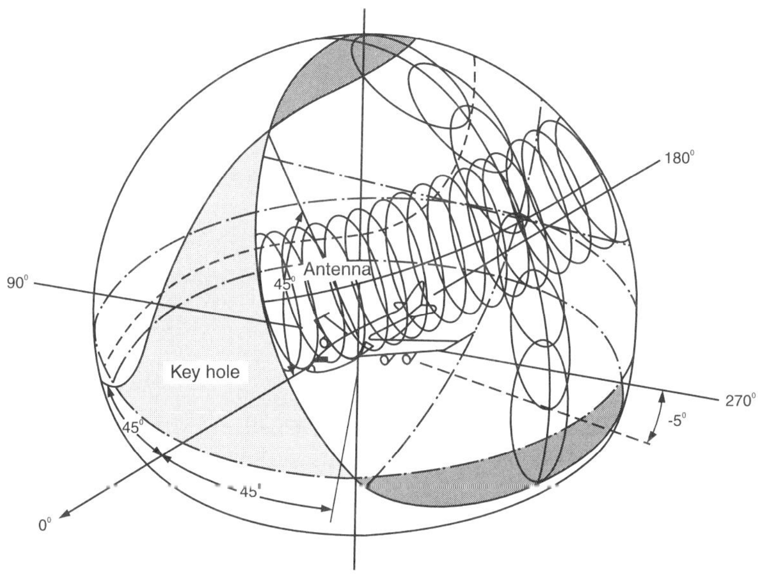

FIGURE 13.16 Beam-steering coverage and keyholes of a phased-array antenna.

relatively large trucks to provide message/data communications. Although the quadrifiler antenna is a good antenna for trucks, it is too large for installation on small cars.

The INMARSAT-M earth station for land mobile stations is basically the same as that for use on ships except that is uses different antennas and tracking methods. Antennas for land mobile use have to have a low profile and be very compact, especially the antennas used on private cars. As medium-gain antennas with gains of 12 to 15 dBi, phased-array antennas are considered the best use on small private cars because of their low profile and high-speed electrical tracking. For satellite tracking, it is very difficult to use closed-loop methods such as the step tracking, very popular in SESs because of severer propagation conditions such as fading, blocking, and shadowing. Open-loop tracking methods using new technology such as optical-fiber gyros have been developed for satellite tracking.

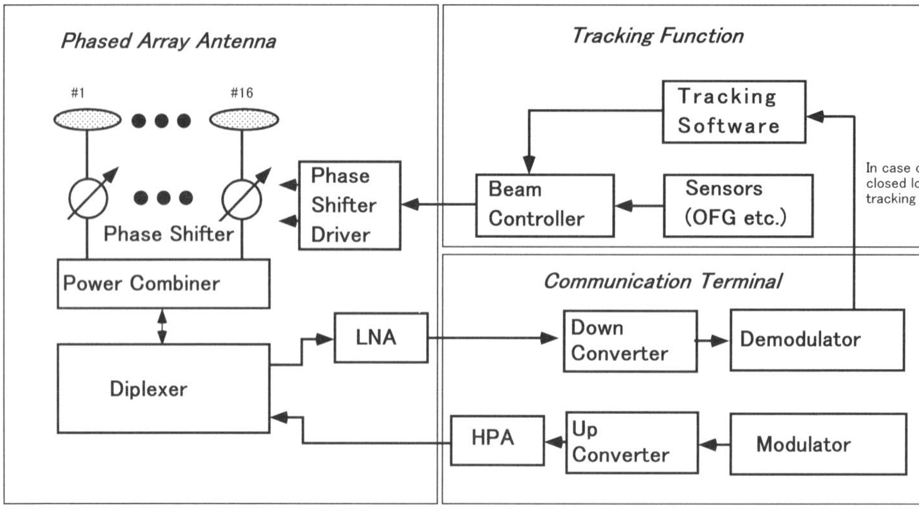

FIGURE 13.17 Block diagram of the phased-array antenna system.

13.6.2 Antenna and Tracking System

13.6.2.1 Antennas

Directional antennas have been expected to provide links for voice signals and high-speed data signals not only for long-haul trucks but also for small private cars. Cost is a very important factor to be taken into consideration when designing such antennas. Phased-array antennas are considered the best kind of antennas for vehicular use because of their low profile, high-speed tracking, and potentially low cost. Phased-array antennas with tracking functions have not yet been used in commercial land mobile satellite communication systems, but many R&D projects have been focused on phased-array antennas and on the tracking method they would use. Mainly in the United States, Canada, and Japan, several types of phased-array antennas have been developed in MSAT-X, PROSAT, and ETS-V experiments [12, 13].

Figure 13.17 shows a general block diagram of a land MES, which has a phased-array antenna with tracking function. In many cases, 19 circular patch elements are used to get system gain of about 13 dBi (G/T is about –13 dBK), and an axial ratio of about 4 dB is obtained in an elevation angle of 45°. A typical antenna is of 4-cm height, 60-cm diameter, and 5-kg weight.

Figure 13.18 is a photograph of a phased-array antenna developed by the Japanese Communications Research Laboratory (CRL) [14]. The antenna has electromagnetically coupled antenna elements printed on a very thin film (Fig. 13.18) and has frequency-dependent 3-b PIN diode phase shifters [15]. The number of PIN diodes of phase shifters used in the system is half that of the conventional phase shifters for the similar function. These characteristics help keep cost low and reduce tracking errors. A decrease in the number of PIN diodes greatly contributes to the reduction of the cost, because the cost of PIN diodes mainly dominates the total cost of phased-array antenna. Table 13.4 shows main characteristics of the phased-array antenna developed for the ETS-V experiments.

13.6.2.2 Tracking System

For land mobile satellite communications, open-loop tracking is more suitable than closed-loop tracking for the following reasons:

1. Because of shadowing and blocking effects, signal levels from and to the satellite are not always stable.
2. A closed-loop method such as a mono-pulse, a sequential lobbing, or a step track method requires otherwise unnecessary lowering of signal levels to determine the satellite direction under the condition in which a very small link margin is expected.
3. Movement of land vehicles is more complicated than that of ships and aircraft. Land vehicles change direction and speed very quickly.

FIGURE 13.18 A 19-element phased-array antenna printed on a very thin film. The G/T is about −13 dBK at an elevation angle of 45°, and the performance was evaluated using the ETS-V satellite and an open-loop tracking method.

TABLE 13.4 Main Characteristics of the Phased-Array Antenna Developed for Land Mobile Systems in the ETS-V Program

Frequency	1642.5–1650.0 MHz (Tx) 1543.0–1548.0 MHz (Rx)
Polarization	Left-handed circular polarization
Antenna element gain	19-Element phased array Circular patch antenna 12 dBi (El. = 45°)
HPBW	28°
Axial ratio	4 dB
Scanned angle	El. 30°–90°(1°step) Az. 0°–360°(1°step)
System noise	340 K
G/T	−13 dBi (El. = 45°)
EIRP	22 dBW (El. = 45°)
HPA	10 W
Tracking	Open-loop method using optical-fiber Gyro
Volume	600 mm (D), 40 mm (H)
Weight	5 kg

In land mobile satellite communication systems operated at L-band frequencies, satellite tracking is required only in the Az direction. Because the half-power beam width of the antenna is about 30°, and everywhere the land vehicle moves within a radius of few hundred kilometers, the elevation angle of the satellite is almost constant. In open-loop tracking, the characteristics of sensors are very important because the total communication performance of the MES is determined primarily by the tracking capability.

Several kinds of sensors give information of vehicle motion. In the ETS-V experiments, a geomagnetic sensor and an optical-fiber gyro were evaluated. As mentioned in Section 13.4, these two are typical kinds of sensors. A geomagnetic sensor is very attractive because of its low cost and its ability to determine absolute direction, but it is affected by environmental conditions such as buildings, bridges, and power poles. An optical-fiber gyro, on the other hand, has an excellent performance in determining relative directional changes independent of environmental condition. However, it cannot carry out the initial

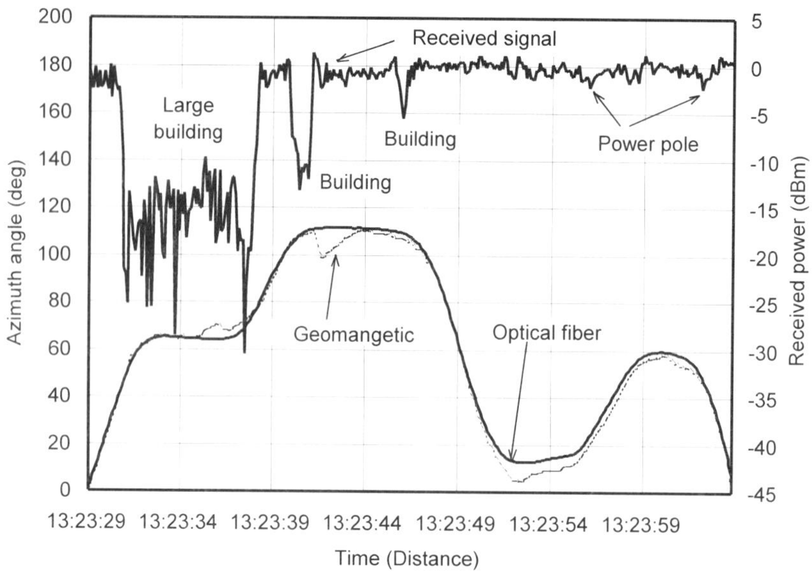

FIGURE 13.19 Tracking performance of phased-array antenna with an open-loop tracking method using an optical-fiber gyro.

acquisition of the satellite because it cannot determine the absolute direction. In open-loop tracking, an optical-fiber gyro is usually used to get the information about vehicle motion, and a geomagnetic sensor gives an absolute direction that is required to calibrate the accumulative error of the optical-fiber gyro at appropriate time intervals.

Figure 13.19 shows an example of the results obtained in an experimental evaluation of tracking performance. The result shows good tracking performance of open-loop method using an optical-fiber gyro. The phased-array antenna installed on the test van tracked the ETS-V satellite by using an optical-fiber gyro. Although the test van was changing directions, the signal level received from the satellite was found to be almost constant except for the blocking effects caused by buildings and power poles. It was demonstrated that, in short time ranges, the optical-fiber gyro has excellent performance, which is not effected by environmental conditions. The output of a geomagnetic sensor is also shown in Fig. 13.19 and the geomagnetic sensor had a tracking error (relative to that of the optical-fiber gyro) of about 10°.

13.6.2.3 Tracking Algorithm

Open-loop tracking requires two kinds of sensors. The first one is a geomagnetic sensor, which determines the absolute direction of the vehicle to carry out the initial acquisition of the satellite and to calibrate the accumulative error of the second sensor, which gives an accurate relative direction over short time ranges. The following example describes an open-loop tracking algorithm with Fig. 13.20 showing its flowchart [16, 17]:

1. Before the vehicle starts to move, an initial direction of the antenna beam is measured by using a geomagnetic compass. The accuracy of this direction depends on the performance of the geomagnetic compass. Typical accuracy of such a compass is about 1°. These data are also used to set the initial direction of the optical-fiber gyro. After this initial procedure, the antenna begins to track the satellite.
2. Data from the optical-fiber gyro and the geomagnetic sensor are sent to a computer, which processes the data to control the antenna beam directions.
3. If a standard deviation of the data from the geomagnetic compass is more than 1°, it is estimated that a magnetic disturbance has occurred or that a vehicle has changed its direction. In this case, the antenna beam is controlled by the optical-fiber gyro. Otherwise, go to step (5).

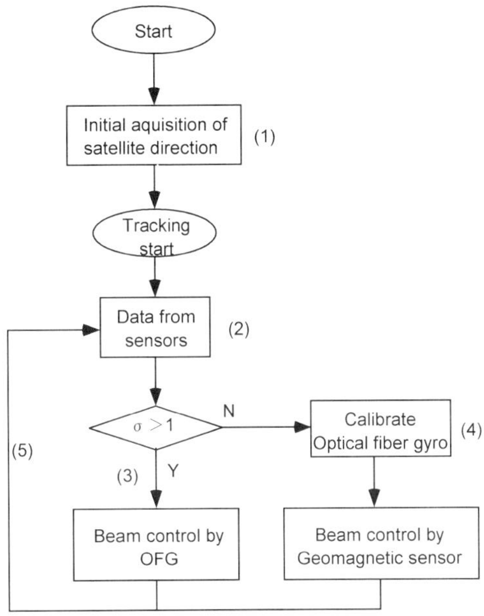

FIGURE 13.20 Flowchart of an open-loop tracking algorithm.

4. When the value of standard deviation is within 1°, the geomagnetic compass controls the antenna beam, and the optical-fiber gyro is calibrated by the geomagnetic sensor to cancel out its accumulative errors.
5. Return to step 2.

In step 4 it is assumed that the geomagnetic compass is not affected by magnetic turbulence, and is reliable for determining vehicle direction. Therefore, the angle information output from a geomagnetic compass can be used as a "true" Az angle to cancel out the accumulative errors of the optical-fiber gyro.

13.7 Handheld Terminals

Handheld terminals are used mainly as the INMARSAT -C and -Mini-M earth stations. As described before, the INMARSAT-C earth station cannot provide voice communication services and can transmit and receive only low-speed data and messages. The INMARSAT-Mini-M earth station, on the other hand, can provide digital voice and data communication services. Figure 13.21 is a photograph of a handheld INMARSAT-C earth station. Antennas and receiver are assembled in one unit (left), and data are entered by a keyboard and are displayed on a conventional personal computer screen (right). The weight of the station is about 4 kg excluding the personal computer. An antenna has an array of four patch-antenna elements, and its gain is about 8 dBi. Because of its wide beam width, the antenna can be directed to the satellite by hand. This terminal can be used outdoors for about 50 min under battery operation.

The hand-carried INMARSAT-Mini-M earth station is expected to become very popular for providing outdoor voice and data communications for relief agencies, news reporters, and oil/mining companies. Figure 13.22 shows a photograph of a handheld INMARSAT-Mini-M earth station.

Before the introduction of the INMARSAT-C system, several small terminals were developed [18, 19]. In the ETS-V program [20], for example, an experimental briefcase-size message communication terminal was developed for the experiments [21]. Figure 13.23 shows a photograph of this attaché-case-size message communication terminal for the ETS-V experiments. Its main characteristics are listed in Table 13.5. Such small terminals must have a simple configuration even though such simplification reduces basic performance such as frequency stability and transmission power. When using a low transmission power, a very narrowband transmission technique should be employed to reduce power. A

FIGURE 13.21 Hand-carried INMARSAT-C earth station. A receiver with a flat antenna weighs about 4 kg and has a compact size (320-mm weight × 210-mm height × 62-mm depth). (Courtesy of Toshiba.)

FIGURE 13.22 Hand-carried INMARSAT-M earth station. (Courtesy of NEC.)

burst-mode transmission scheme is very effective in using power efficiently. A 100-bps message terminal using only 1 W of power was developed. A key technology of this 100-bps transmission system is to achieve an initial acquisition and to keep a synchronization of a signal received by the handheld terminal. In an environment a signal suffers frequency deviation caused by frequency instability of oscillators, both in the onboard transponder and the earth station. The frequency compensation function is implemented only in the base station terminal to keep the handheld terminal as small as possible. The terminal has only one local oscillator, which is commonly used for both transmission and reception. It has no frequency

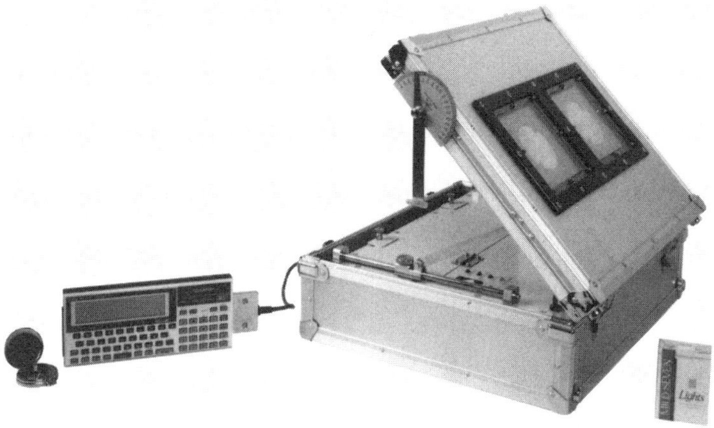

FIGURE 13.23 Photograph of the hand-carried terminal developed for the ETS-V experiments.

TABLE 13.5 Main Characteristics of the Attaché-Case-Size Message Communication Terminal Developed for the ETS-V Experiments

Frequency	1644.7–1644.8 MHz (Tx) 1543.2–1543.3 MHz (Rx)
Antennas	Two patch antennas for Rx and Tx Gain Tx: 7.5 dBi, Rx: 7.0 dBi
Polarization	Left-handed circular polarization
Tracking	Hand operation
G/T	–17 dBK
EIRP	6.5 dBW (HPA: 1 W)
Modulation	Digital FM, 100 bps, Burst mode
Frequency stability	Approximately 1×10^{-6}
Weight	13 kg

compensation function. The base station can detect the frequency deviation of the signal from the terminal by using an FFT technique [22] and control its local oscillators for reception and transmission to communicate with the hand-carried terminal.

References

1. Technical Requirements for INMARSAT Standard-A Ship Earth Station, July 1981.
2. T. Hoshikawa et al., "INMARSAT Ship Earth Station Type RSS401A," *Anritsu Tech. J.*, No. 56, pp. 41–49, September 1988.
3. INMARSAT-M System Definition Manual, November 1991.
4. H. W. Ehrenspeck, "The Backfire Antenna, a New Type of Directional Line Source," *Proc. IRE*, 48, pp. 109–110, January 1960.
5. S. Ohmori, S. Miura, K. Kameyama, and H. Yoshimura, "An Improvement in Electrical Characteristics of A Short Backfire Antenna," *IEEE Trans. Antennas Propag.*, AP-31, No. 4, pp. 644–646, July 1983.
6. T. Shiokawa and Y. Karasawa, "Compact Antenna Systems for INMARSAT Standard-B Ship Earth Stations," IEE 3rd Int. Conf. Satellite Systems for Mobile Communication and Navigation, London, June 1983.
7. K. Eguchi, "Small Antenna System for Inmarsat-M Ship Earth Station," Symposium Papers for RTCM Annual Assembly Meeting, Florida, 1992.

8. INMARSAT Aeronautical System Definition Manual, Module I, 1990.
9. Aeronautical Radio Inc., ARINC Characteristic 741: Aviation Satellite Communication Systems, 1990.
10. S. Taira, M. Tanaka, and S. Ohmori, "High Gain Airborne Antenna for Satellite Communications," *IEEE Trans. Aerospace Electron. Syst.*, Vol. 27, No. 2, pp. 354–360, March 1991.
11. S. Ohmori, Y. Hase, H. Wakana, and S. Taira, "Experiments on Aeronautical Satellite Communications Using ETS-V Satellite," *IEEE Trans. Aerospace Electron. Syst.*, Vol. 28, No. 3, pp. 788–796, July 1992.
12. K. Woo, J. Huang, V. Jamnejad, D. Bell, J. Berner, P. Estabrook, and A. Densmore, "Performance of a Family of Omni and Steered Antennas for Mobile Satellite Applications," Proc. Int. Mobile Satellite Conf., pp. 540–546, Ottawa, Canada, 1990.
13. N. Nishikawa, K. Sato, and M. Fujino, "Phased Array Antenna for Land Mobile Satellite Communications," *Trans. IEICE,* J72-B-II, No. 7, pp. 323–329, July 1989.
14. S. Ohmori, K. Tanaka, S. Matsunaga, and M. Tsuchiya, "A Phased Array Tracking Antenna for Vehicles," Proc. Int. Mobile Satellite Conf., pp. 519–522, Ottawa, Canada, June 1990.
15. S. Ohmori, S. Taira and A. Austin, "Tracking Error of Phased Array Antenna," *IEEE Trans. Antennas Propag.*, AP-39, No. 1, pp. 80–82, January 1991.
16. S. Yamamoto, K. Tanaka, H. Wakana, and S. Ohmori, "An Antenna Tracking Method for Land-Mobile Satellite Communications Systems," *Electron. Commun. Jpn.*, Part 1, Vol. 78, No. 9, pp. 91–102, 1995.
17. K. Tanaka, S. Yamamoto, H. Wakana, S. Ohmori, M. Matsunaga, and M. Tsuchiya, "Antenna and Tracking System for Land Vehicles on Satellite Communications," IEEE Proc. Veh. Technol. Conf., Denver, CO, May 1992.
18. R. J. Fang, U. Bhasker, F. Hemmati, K. M. Mackenthun, and A. Shenoy, "Design of an MSAT-X Mobile Transceiver and Related Base and Gateway Stations," *COMSAT Tech. Rev.*, Vol. 17, No. 2, pp. 421–466, 1987.
19. P. Rossiter, D. Reveler, and L. Tibbo, "L-band Briefcase Terminal Network Operation," *Proc. Int. Mobile Satellite Conf.*, pp. 279–284, Ottawa, Canada, 1990.
20. Y. Hase, S. Ohmori, and N. Kadowaki, "Experimental Mobile Satellite System (EMSS) Using ETS-V Satellite," *IEEE Denshi Tokyo,* No. 25, pp.10–12, 1986.
21. N. Kadowaki, S. Maruyama, R. Suzuki, Y. Hase, and S. Ohmori, "The Attach-Case-Size Message Communication System via Satellite Links," *IEEE Trans. Veh. Technol.*, Vol. 41, No. 4, November 1992.
22. N. Kadowaki, Y. Hase, and S. Ohmori, "ETS-V/EMSS Experiments on Message Communications with Handheld Terminal," *Proc. ICC'89,* pp. 211–215, 1989.

14

Fixed and Mobile Antennas for Satellite Communications

Marek E. Bialkowski
University of Queensland

Nemai C. Karmakar
Nanyang Technological University

Paul W. Davis
University of Queensland

Hyok J. Song
HRL Laboratories, LLC

14.1 Introduction

Since 1945, with the introduction by Arthur C. Clarke of the concept of an artificial satellite, antennas have played a significant role in the development of satellite communications.[1-3] The simplistic role of antennas in this case can be viewed as follows. The signal is beamed into space by an *uplink* antenna, electronically processed onboard the satellite, sent back to earth using a *downlink* antenna, received by the *earth station* antenna and processed by the electronic receiver. The involvement of antennas in the satellite communications system is shown in Fig. 14.1.

As seen in Fig. 14.1, antennas form input and output ports to the satellite communication system, which is divided into space and ground segments. A spacecraft with onboard communications equipment in conjunction with a *telemetry, tracking*, and *telecommand* (TT&C) earth station forms the space segment, whereas a hub or gateway station and a number of other types of earth stations form the ground segment.

The types of antennas that are used in this system depend on a number of factors. These, to a large extent, are related to the distance between the satellite and the earth. The distances involved in the present stage of satellite communications are illustrated in Fig. 14.2.[4]

As can be seen in Fig. 14.2, the orbits in which satellites can be positioned are divided into three categories geostationary earth orbit (GEO), medium-earth orbit (MEO), and low-earth orbit (LEO).

0-8493-0124-6/02/$0.00+$1.50

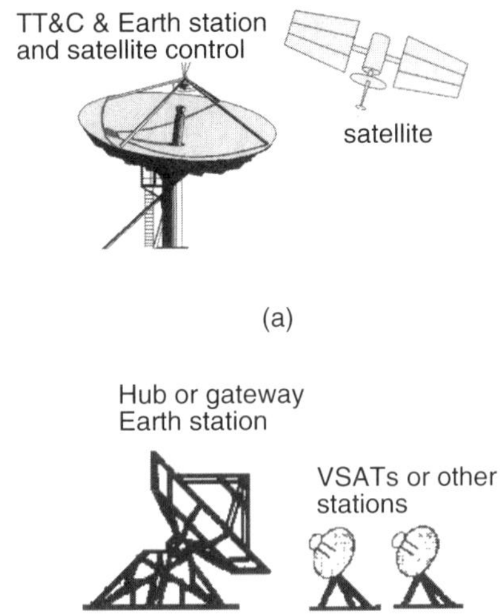

FIGURE 14.1 Basic configuration of a satellite communications system showing (a) space segment and (b) ground segment.

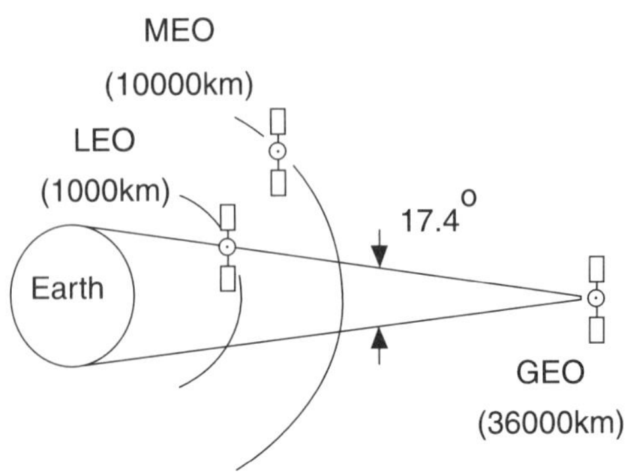

FIGURE 14.2 Location of LEO, MEO, and GEO orbits with respect to the earth.

Nearly all communication satellites are positioned in the geostationary orbit also known as the *Clarke orbit*. In this orbit, the velocity of a satellite matches the surface velocity of the earth so that the satellite appears stationary with respect to the observer on the ground. This is convenient from the point of view of the ground station antenna, because it has to only be in a *footprint* of the satellite antenna and permanently pointed toward the satellite to create the link.

One inconvenience associated with the use of the geostationary orbit is its large distance from the ground, which requires powerful rockets to raise the satellite to this position. Another problem is the delay of the communication signal that has to travel a large distance. Because of the large distance, both satellite and ground station antennas have to have a sufficiently high-directive gain to overcome space losses. This requirement translates into the demand for large aperture, pencil-beam antennas.

The requirement for high-gain antennas for the space and ground segments can be relaxed using nongeostationary MEO or LEO satellites. The use of MEO or LEO satellites reduces the distance between transmitting and receiving sites and therefore decreases the signal delay. However, as the satellites become nonstationary with respect to the ground, intricate communications links are required.

The present chapter provides an overview of fixed and mobile antennas for satellite communications. First, it commences with general principles of satellite communications, which establishes a firm ground for understanding the role of antennas. Following this introduction, practical designs of antennas for fixed and mobile satellite communications are presented.

14.2 Satellite Communications

14.2.1 Frequency Bands

To establish the satellite communication link between a ground user and a satellite only selected frequencies can be used. Generally, the frequency bands that are suitable for this purpose include the frequency range between 1 and 30 GHz. These comprise L-, S-, C-, Ku-, and Ka-bands for communication purposes; and ultrahigh frequency (UHF) and S-bands for telemetry and TT&C. The spectrum allocations are given as follows: L-band, 1.5–1.65 GHz; S-band, 2.4–2.8 GHz; C-band, 3.4–7.0 GHz; X–band, 7.9–9 GHz; Ku-band, 10.7–15 GHz; and Ka-band, 18.0–31.0 GHz. The band excluded from the present considerations is the X–band, 7.9–9.0 GHz, which is allocated for military purposes.

The frequencies that are mostly used at present time cover the range between 1.5 and 18 GHz. The L-band is least affected by rain, because rain attenuation is negligible at this frequency range. However, the ionosphere introduces a source of significant link degradation. Because of ionospheric scintillation, which results in the signal splitting into direct and refracted paths, signals combine with random phase causing signal fade. The ionosphere also rotates linear polarization. To reduce the effect of rotation, L-band links use circular polarization. S-band suffers less from ionospheric effects than L-band. However, it suffers slightly higher atmospheric attenuation as a result of rain. C-band exhibits a modest amount of fading from rain and ionospheric scintillation. It offers much larger bandwidth than L- or S-band. As a result, C-band remains the most heavily developed and used segment of the satellite spectrum. The only drawback of C-band is the requirement for relatively large size earth station antennas, with diameters between 1 and 3 m as a norm. The most significant advantage of C-band is its immunity against rain attenuation. This explains why this band is mainly used in tropical regions, which feature an abundance of heavy rains.

Ku- and Ka-bands are most affected by rain attenuation. However, because of the frequency scaling for the same gain, antennas can be of smaller size. To counter the effects of rain, Ku-band transponders have to deliver higher values of effective isotropically radiated power (EIRP).

The Ka-band seems to be the last to be exploited for commercial satellite communications. It offers 2 GHz for uplink and downlink and at present is relatively abundant. The major challenge comes from the greatest rain attenuation that it suffers in comparison with the lower frequency bands.

14.2.2 Space-Segment Antennas

Onboard the satellite, antennas are a part of the satellite communications payload module. Their location depends on the type of satellite. In the spin-stabilized type of satellite, the antenna module is positioned in the upper part of the spinning cylinder. In fact, beside the lower part on which is reserved for the apogee motor, this is the only part on which the antenna can be positioned. In the three-axis stabilized type of the satellite, more sides of the cube-shaped satellite are available for positioning of the antenna modules. In practice, up to three sides can be used for this purpose.

The satellite antennas are generally divided into wire, horn, reflector, and array antenna configurations. The TT&C antenna is the first antenna that establishes the communication link between the satellite and earth stations, which monitor the satellite launch. This antenna usually operates in the very high frequency (VHF) or S-band. The remaining antennas are used for satellite communications. The choice of a particular frequency band depends on services that the satellite has to deliver. Fixed satellite services (FSS) typically use Ku- and C-band, broadcast satellite services (BSS) mostly use Ku-band, and mobile satellite services (MSS) utilize the L- or S-band frequency spectrum.

Conical horns are used to achieve global coverage of the earth. A typical horn's beam width of 20° matches well with the earth's subtended angle of 17.4°, as observed from a geostationary orbit. To achieve beam width smaller than 10°, reflector antennas employing clusters of feeds are used. A single-feed antenna illuminates an elliptical region on the earth's surface that is within the small beam width of the antenna. This is known as a single-feed footprint. In practice, a shaped footprint is required to cover a part of a continent, a single country, or a part of this country. This task is achieved using a cluster of feeds. The use of clusters of feeds has advantages such as in-built redundancy and frequency reuse. Also, providing that the feeds are phased, the beam shape can be reconfigured. These antennas are suitable for both FSS and BSS services. They operate in conjunction with high-gain ground station antennas.

Opposite from FSS and BSS, MSS satellites are designed to provide mainly voice transmission services for users with mobile transceivers featuring low power and low antenna gain.[4,5] This means that the downlink transmit power density has to be much higher per voice channel than for a typical FSS system. Also because of the small bandwidths that are available in L- and S-band, a high degree of frequency reuse has to be applied to serve the needs of the many users of the MSS system. To achieve these aims GEO, MEO, or LEO satellite systems can be used. Present systems use GEO satellites, which transmit voice, fax, or data between a mobile user, located in a remote place, and the Public-Switched Telephone Network (PSTN). The L- or S-band is used for getting the communication link between the user and the satellite, whereas the overall satellite telecommunication traffic is handled by Ku-band (or C-band as is the case of INMARSAT). Examples of such GEO MSAT telecommunications systems include INMARSAT (global system), and Mobilesat (Australian), AMSC (North America), and N-star (Japan) (domestic satellite systems).

To obtain a footprint the size of the Australian continent that allows communication via a portable transceiver the *Optus B* transponder uses a 150-W amplifier and 48 short backfire (SBF) array of antennas. Their radiation is equivalent to 48 dBW EIRP over continental Australia. INMARSAT-M offers a similar solution, however, on a global scale, it uses a sequence of 48 dBW footprints over the continents. In these two cases, a ground terminal antenna of 10 dBic gain meets the good quality link requirements, with a margin of about 5 dB.

The AMSC satellites feature separate Tx and Rx 5.5 m in diameter reflectors that produce a sequence of footprints covering the region of Canada, the United States, Mexico, the Caribbean, and Hawaii. As in the case of Mobilesat, the Ku-band is used to provide a feeder link to PSTN.

One inconvenience of GEO MS systems is that the large distance between the satellite and the ground user causes signal delay and introduces the requirement for steerable antennas for mobile ground users. These shortcomings can be overcome using MEO or LEO satellite systems. A number of global MEO and LEO satellite systems have been proposed and some of them are under development.[2,4]

14.2.3 Earth-Segment Antennas

14.2.3.1 Gateway Antennas

One of the major requirements of an earth station is high directive gain. This is to maximize the signal strength in a given direction and to reject any undesired signals that create interference. The most demanding gain is on a gateway station because of the large traffic load it has to handle in a most reliable manner. International bodies such as INTELSAT, EUTELSAT, and INMARSAT standardize the size of gateway station antennas. According to these standards, gateway station antennas are divided into standard A, B, C, and so on, with the highest gain featured by standard A antenna. For example, standard A operating at C-band is specified as a 30-m (now reduced to 18 m because of increased satellite radiated power) diameter antenna or larger. Standard B at C-band is required to have a diameter of 11 m or larger. Center-fed dual-reflector designs are used to achieve the desired gain. To track the satellite, three main methods, monopulse, step-track, and conical scan, can be applied.

14.2.3.2 Other Ground Segment Antennas

The introduction of more powerful satellites has led to reducing size of earth stations. One of the already realized possibilities are very small aperture terminals (VSATs) for data networks. A VSAT is a complete earth station that offers special capability to introduce business television (TV) at a relatively low cost. These include teleconferencing, distance education, and telemedicine. Antennas in sizes of 2 to 3 m are sufficient to bring such services.

The other areas that are targeted by satellite communications to bring in fast revenue include direct to home (DTH) TV and mobile telephony. DTH TV instantaneously serves millions of users once the satellite is in the sky. Typically, an antenna less than 1 m in diameter is sufficient to access DTH TV. Examples of DTH TV systems include BskyB in Europe, DIRECTV in the United States, and ASIASAT in Asia.

To access mobile satellite telephone services including voice, fax, and data, many manufacturers offer suitable land or maritime equipment. All this equipment, including an antenna is portable but not handheld. The situation is changing with the introduction of the LEO MSS services, which allow for the use of handhelds.

14.3 Microstrip Antennas for Fixed and Mobile Satellite Communications

After accomplishing a brief introduction to satellite communications systems we are ready to move to the main topic of this chapter, which are antennas for fixed and mobile communications.

As it has already been mentioned, the introduction of more powerful satellite transponders as well as the emergence of new commercial applications of satellite communications has been the major driving force behind the development of small size ground antennas. In addition to small size, there has been a drive for low-profile, low-weight, and low-cost antenna designs. The main technology to achieve attributes such as low profile, low weight, and low cost is the microstrip patch antenna technology. This technology not only makes the antenna design low profile, lightweight, and low cost but also enables the direct integration of active modules and beamforming networks with the radiating layers in active and phased arrays. This goal can be achieved by using a multilayer approach to the construction of the antenna and by using electromagnetic coupling between active, beamforming, and radiating layers.

An alternative to the patch is a slot radiator. Its use, similar to that of a patch radiator, leads to low-profile and low-cost array antennas. The choice of slots becomes beneficial in large-size arrays when a low-loss feeding structure is required. The matter of obtaining low-loss feeding and signal distributing networks becomes important at upper microwave and higher frequencies at which long microstrip lines forming such networks introduce significant conduction losses.

The following sections are concerned with designs of microstrip and slot antennas for fixed and mobile satellite communications. These include a variety of antennas such as corporate-fed microstrip arrays, reflect arrays, and slot arrays for fixed communications. Following these presentations, switched-beam and phased-array antennas in microstrip technology for mobile satellite communications are described.

14.3.1 Microstrip Antenna Arrays with Circuit-Type Feeding Networks

A rectangular microstrip patch is one of the most popular elements in the class of microstrip patch antennas.[6,7] To achieve suitable radiation efficiency, polarization, and other attributes over a given frequency band, shape, height, feeding arrangements, and dielectric substrate of this planar radiator can be optimized. The most popular shapes of the patch antenna are rectangular and circular shapes, whereas the feeding methods used most often include edge-feed, coaxial probe, proximity coupling, and slot or aperture coupling.

Irrespective of the method of feeding, a single microstrip patch antenna features low gain typically in the range of 4 to 7dB, which is not enough in many applications. To obtain higher gain, the antenna aperture has to be increased. This can be achieved by co-planar arraying of many microstrip patches and by combining signals received by them. The resulting array usually follows a regular pattern with a rectangular or triangular lattice being the most popular. The spacing between adjacent elements cannot be too small to avoid adverse effects of mutual coupling. Also, it cannot be too large to avoid grating lobes in the radiation pattern. As a result, a typical spacing is chosen in the range of 0.6 to 0.8 free-space wavelength to meet optimum conditions.

To combine power from individual elements, series, parallel or series/parallel combining networks, formed by different types of waveguides (microstrip, stripline, rectangular waveguides) or by free space, can be used. These networks establish suitable amplitude and phase relations at individual elements to shape the radiated beam and to point it in a chosen direction. The usual requirement is that the assumed amplitude and phase distribution should hold over a prescribed frequency band. The resulting bandwidth depends on the type of a waveguiding medium and types of connections (series, parallel, or combinations) used in the feeding network.

Because of demands of the commercial market, in addition to technical specifications, other design aspects become important. These include minimizing the cost of substrates, and achieving design and development of prototypes in the minimum time. These additional requirements explain the preference for connection-less feeding arrangements such as proximity coupling and slot coupling. These two feeding methods avoid costly connectors used in probe-fed arrays. Thus, they minimize both the materials cost as well as an assembling time. To further minimize the design and development time, the use of modern antenna CAD and CAM techniques become very important. Equally important is the testing stage, because it is the last step to verify that the realized prototype indeed performs according to initial specifications. The design procedures should allow for small errors because of the approximate nature of CAD and manufacturing tolerances.

The following section presents the use of CAD and CAM techniques to develop a planar 16 × 16-element multilayer microstrip patch array antenna for DTH-TV reception at Ku-band with the center frequency of about 12.5 GHz and a 10% operational bandwidth. The array uses a corporate-feeding network for signal combining.

14.3.1.1 Corporate-Fed Array Design

The design commences with a single aperture-coupled microstrip patch (ACMP) followed by a 2 × 2-element subarray, which serves as an elemental module (cell) in the process of creating a corporate fed array.[8,9] To obtain a highly predictable design, a 2.5 dimensional, full-wave electromagnetic analysis software package, Ensemble®, from Ansoft is used.[10]

The ACMPA element configuration considered here is shown in Fig. 14.3. Two dielectric substrate layers and a foam layer form this antenna element.

The substrate used for the patch and feed layers is Ultralam® 2000 from Rogers. It features relative dielectric constant, ε_r, of 2.50, and thickness, h, of 0.76 mm. The foam material, chosen here as a spacer between the two substrates to enhance the antenna impedance bandwidth, is Rohacell® 51 with $\varepsilon_r = 1.07$ and $h = 1.5$ mm. The use of the low dielectric constant foam requires the aperture (slot) length to be increased.

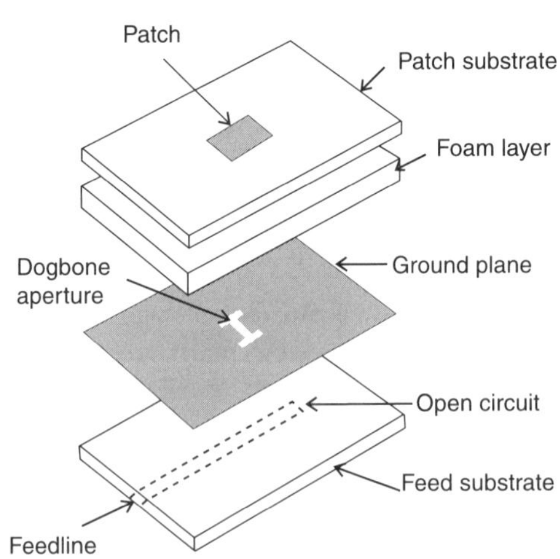

FIGURE 14.3 Configuration of aperture-coupled microstrip patch antenna (ACMPA).

This is to match the antenna to the 50-Ω characteristic impedance of the microstrip feedline. By using an end-loaded slot,[11,12] having a dog bone shape, the required impedance bandwidth of 10% can be achieved with a slot length reduction of approximately 10%.

When using Ensemble, several iterations are required to design a single antenna. The theoretical predictions and experimental results are compared and any deviations from the design specifications are removed by the next design iteration. This iteration process accounts for the approximate nature of the theoretical approach, possible lack of precise knowledge of substrate electrical parameters, and manufacturing tolerances.

After accomplishing satisfactory experimental results for the single antenna, the next step includes the design and development of the 2 × 2-element subarray. This subarray is a basic module to create large-size arrays using a corporate feeding network. The subarray element spacing in E- and H-plane directions is determined by making a compromise between the antenna gain and the sidelobe levels. Also, consideration is given to the available board size to manufacture the entire 16 × 16-element array. Here the element spacing of 0.725 and 0.875 λ_0 is chosen for E- and H-plane directions, respectively.

After chosing the optimum element spacing, the corporate feed network of the 2 × 2-element subarray is designed. This process is aided with Ensemble® using its "black box" tool. A black box is a tool that, in general, can be used in the analysis process to incorporate lumped elements such as capacitors, inductors, resistors, or even antenna subarrays. All that is required to use the black box is an S-parameter file that represents the lumped elements or antenna subarrays. In the present case, the measured values of S_{11} for a single ACMPA element were used in the black box to design the feed network for the 2 × 2-element subarray. This concept is illustrated in Fig. 14.4 and more details can be found in Song et al.[13]

Prior to proceeding with the design of a larger array, the simulation accuracy of the subarray using the black box tool is confirmed. In this case, the measured and simulated results for the return loss of the 2 × 2 subarray are compared. Once satisfactory results for the 2 × 2 subarray are obtained, the design proceeds with the (2 × 2) × 2 element subarray. Again, the black box tool of Ensemble® is employed. The measured S_{11} parameter of the manufactured 2 × 2-element subarray, over the desired frequency band (in this case from 12 to 13 GHz), provides a new black box representation. This strategy is continued to create larger subarrays until the design of the entire array is accomplished. This stage is followed by laying out the array, etching, and assembling. This is done automatically by using layouts generated in Ensemble.

The measured return loss of the manufactured 16 × 16-element array is shown in Fig. 14.5. The predicted values were obtained using the measured S_{11} of the ACMPA and its 2 × 2-element subarray and the black box utility of Ensemble. The measured values compare well with the values predicted by Ensemble simulation.

The radiation characteristics of the fabricated 16 × 16-element array, as measured in the Fresnel zone (using a 3-m spacing between the transmitting horn antenna and the antenna under test) in an anechoic chamber are shown in Fig. 14.6. Two principal plane (E- and H-plane) patterns are presented here.

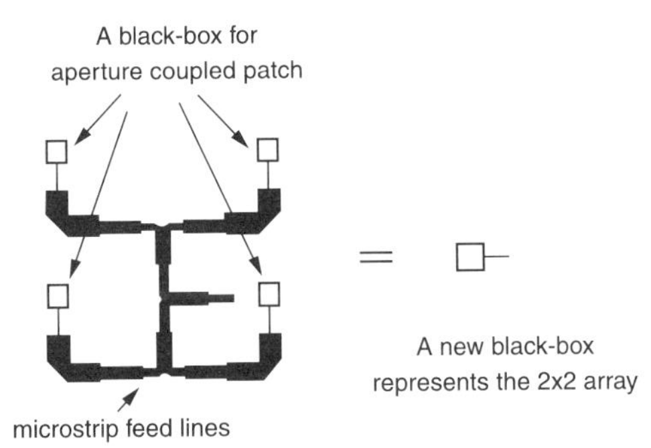

FIGURE 14.4 Illustration of the black box representation of ACMPA in the 2 × 2-element subarray feed network design.

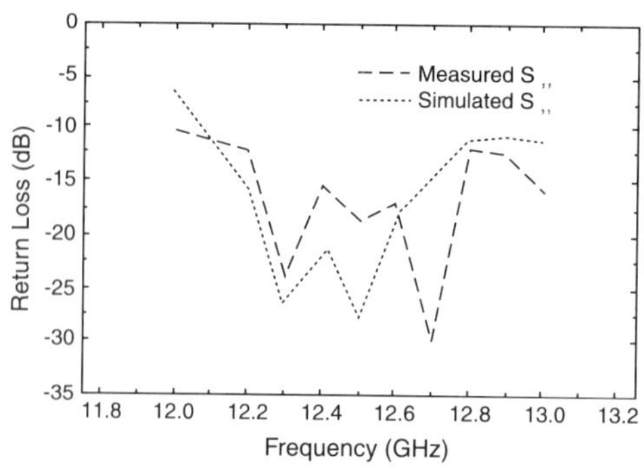

FIGURE 14.5 Comparison of measured and simulated results for return loss of the 16 × 16-element array.

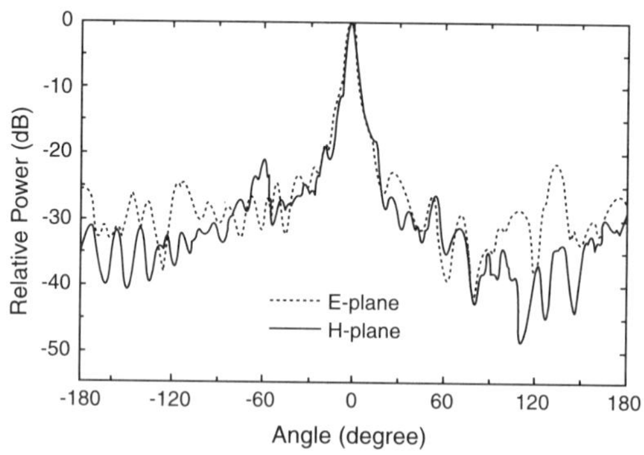

FIGURE 14.6 Radiation pattern of the 16 × 16-element ACMPA array measured in two principal (E- and H-) planes.

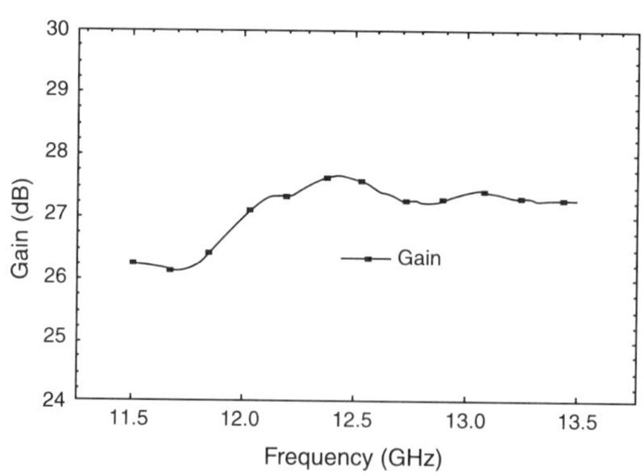

FIGURE 14.7 Measured gain of the 16 × 16-element ACMPA array.

The array produces a pencil beam in the broadside direction. This radiation pattern shows a good sidelobe level performance, which is well below –20 dB. An approximate value for the antenna directivity can be obtained from the measured 3-dB beam width in E- and H-planes. They are 4.8° and 4.5° at the center frequency of 12.5 GHz, respectively, and result in a directivity of 31.8 dB.

Figure 14.7 shows a measured gain performance of the ACMPA array with 28 dB recorded at the design frequency of 12.5 GHz. This is 3.8 dB less than the measured directivity of 31.8 dB. This gain reduction is likely a result of losses of the corporate-feed network. Also, antenna misalignment during the gain measurement could be another contributing factor to the gain reduction.

The presented results are very satisfactory, considering only a single prototyping cycle has been performed for this antenna. The obtained results show that the presented design strategy does not leave much space for developmental errors.

14.3.2 Microstrip Reflect Arrays

To complete the arraying process, signals received or transmitted by individual elements have to be suitably combined. The previous section has demonstrated the use of a microstrip corporate network to feed a Ku-band 16 × 16-element aperture-coupled microstrip patch array. This circuit-type network has a relatively large length and as a result introduces considerable conduction (and dielectric) losses that reduce the gain of the array. In the presented example of the 16 × 16-element patch array, because of these losses the expected 32-dB gain was reduced by 4 to 28dB.

The conduction and dielectric losses resulting from a circuit-type combiner aggravate further at upper microwaves and higher frequencies. This is because the conduction loss increases as a square root of frequency and the dielectric loss increases linearly with frequency. It becomes apparent that to decrease signal-combining losses, a new type of network featuring smaller conduction and dielectric losses has to be used at upper microwave frequencies.

One alternative is a metal waveguide, which exhibits smaller losses per wavelength than a microstrip line. Unfortunately, high realization costs of this nonplanar waveguiding medium offset its low-loss performance. A more viable alternative is offered by a reflect array, which entirely eliminates a circuit-type combiner replacing it by a spatial power combiner.[14,15] A basic configuration of a horizontally polarized microstrip reflect array is presented in Fig. 14.8.

As shown in Fig. 14.8, a horn antenna in conjunction with free space acts as a spatial power combiner. This combining network introduces no conduction or dielectric losses.

The role of the reflector is to convert a spherical wave front produced by a horn into a planar wave front. To perform this function, the reflector has to emulate a parabolic reflector and this is achieved by suitable phasing of individual patches.

The required phase shift ϕ_i for element i is determined by Eq. (14.1) from Huang,[15] and Targonski and Pozar[16] as

$$\phi_i = k_0\left(R_i - \tilde{\mathbf{r}}_i \cdot \tilde{\mathbf{r}}_o\right) \tag{14.1}$$

where R_i is the distance from the phase center of the feed to the element, $\tilde{\mathbf{r}}_i$ is the vector from the center of the array to the element, and $\tilde{\mathbf{r}}_o$ is the unit vector in the main beam direction.

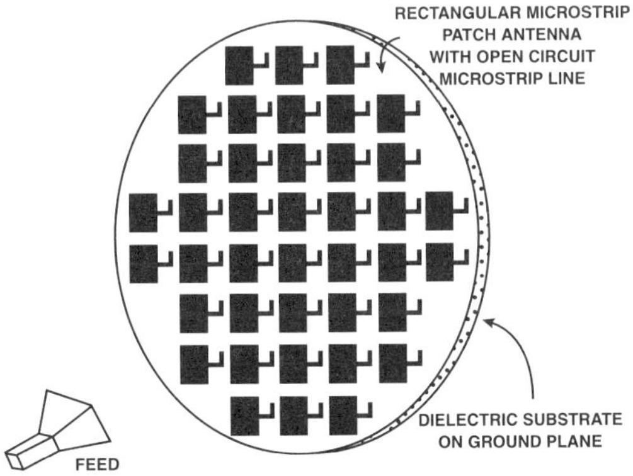

FIGURE 14.8 A reflect array formed by microstrip patch antennas with open-ended stubs operating as phasing elements for horizontally polarized signals.

Two methods frequently used to accomplish the required phasing include variable-length open-ended stubs[15] and variable-size patches.[16] Figure 14.8 demonstrates the use of open-ended stubs to accomplish the required task.

An extensive theoretical and experimental analysis of linearly polarized reflect arrays using variable size patches and different types of feeds has been performed in Pozar et al.[17] The presented design procedure assumes that the reflection from an individual patch surrounded by patches of different sizes can be approximated by the reflection from an infinite array of patches of equal size. In particular, a full-wave, moment-method solution has been applied to determine phase shifts resulting from various lengths of the patch. It has been found that for several analyzed reflect arrays a model using reflection from a single patch (disregarding mutual coupling effects) provides results in close agreement with the infinite array model.

The configuration of the linearly polarized reflect array shown in Fig. 14.8 can be extended to dual and circular polarization. Such an extension to a multiple-polarized microstrip reflect-array antenna using two variable length stubs attached to orthogonal feeding points of the square patch has been described in Javor et al.,[18] and Chang and Huang.[19] The presented design is capable of achieving horizontal and vertical polarization, circular polarization, and dual-linear or dual-circular polarization. The choice is dependent on the polarization of the feed. The cross-polarization level of the single polarized reflect array depends on isolation between the two polarizations. Special orientation of stubs has been devised to reduce the cross-polarization levels. The work presented[17-19] has demonstrated flexibility in choosing feeding methods for a reflect array showing that this antenna can employ prime-focus, offset feed, or other types of feeds already explored with parabolic reflector antennas.

One has to note that similarly as for reflectors, feed spillover, amplitude taper, and additionally dielectric and phase errors create gain reduction in reflect arrays. One of the few electrical advantages of the reflect array over a parabolic reflector is improved cross-polarization performance, which is achieved using rectangular patches.

The designs presented in the literature[17-19] feature one shortcoming, which results from the use of a single radiating and phasing layer of the reflect array. The use of a single layer in these designs creates the difficulty to simultaneously control the element phasing and bandwidth. To overcome this problem, ACMP elements as an alternative to edge feed and variable size patches to create reflect arrays can be used. Such a new approach has been proposed in Robinson et al.[20] In this case, the ACMP antenna elements offer flexibility to lower characteristic impedance of the phasing stubs, for example, to 50 Ω. Another advantage is obtaining a very high isolation between two orthogonal ports in multipolarization designs. An example of dual-polarized, aperture-coupled microstrip patch antenna featuring 50-Ω ports with very high isolation between the two orthogonal ports in the range of 35 dB over 10% bandwidth at Ku-band has been demonstrated in Bialkowski et al.[21]

Another advantage of ACMP elements is the possibility of including active elements. An active aperture-coupled microstrip reflect-array employing 137 transistor amplifiers has been presented in Robinson et al.[22] The schematic configuration of this active reflect array is shown in Fig. 14.9. As shown in Fig. 14.9, this configuration allows for the integration of amplifier stages with antenna elements, a feature not possible with the open-circuit stubs or variable-size patches used in the passive reflect arrays in the literature.[16-19] Note that to realize isolation between input and output ports of the amplifiers to avoid oscillations, the use of the orthogonal feeds results in a rotating polarization plane of the retransmitted signal. To provide the required phase shift to beam the reradiated wave in a specified direction, the length of the transmission line joining two orthogonal feeds of a dual-feed, aperture-coupled microstrip patch antenna is varied, as shown in Fig. 14.9.

Similarly, as in the earlier presented corporate 16 × 16-element microstrip array, the individual elements of the active array,[22] were designed using Ensemble, whereas the design of transistor amplifiers was accomplished by using HP-EEsof (now Agilent EEsof) Touchstone with Academy Layout and by using manufacturer's data for Mitsubishi MGF1302 transistors.

The developed prototype provided the required broadside radiation, however, with a significantly large side lobe level. This high sidelobe level could result from phase errors, which occurred because of the use of separate panels to construct the active reflect array.

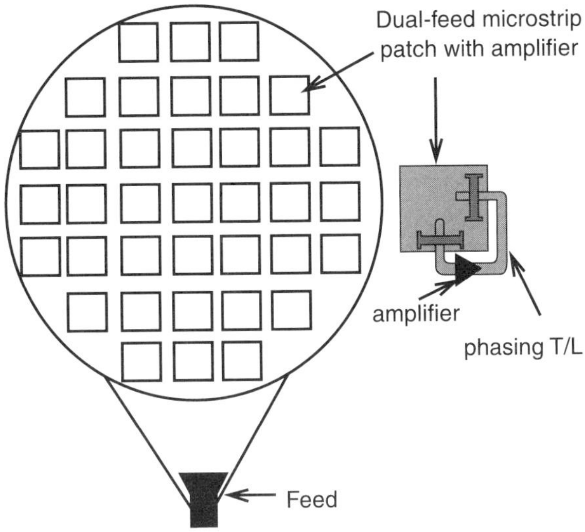

(a)

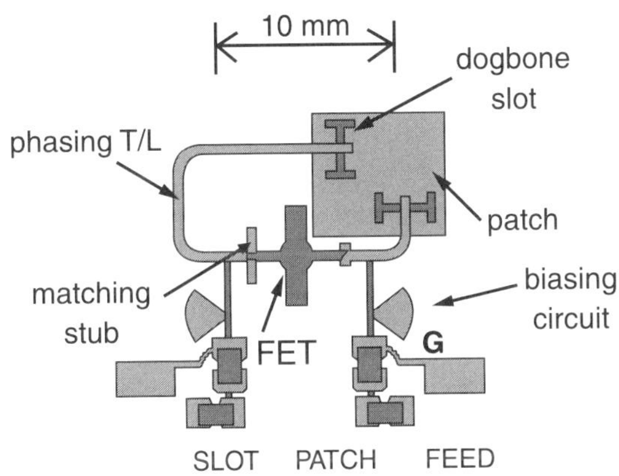

(b)

FIGURE 14.9 (a) Configuration of an active circular reflect array with dual-feed, aperture-coupled microstrip patch antenna elements showing the phasing arrangement using a variable-length microstrip transmission line; (b) layout of a unit cell amplifier of the 137-element active reflect-array. (From Robinson, A. W. et al., *Microwave Optical Technol. Lett.*, 26, 147, 2000. With permission.)

14.3.3 Radial Line Slot Arrays

Besides microstrip patch technology, slot antenna technology using narrow slots in planar conductors can be used to achieve a flat antenna design.[23] A natural way of feeding slot arrays is by using waveguides. Because of low conduction and dielectric (mainly air dielectric is used) losses attainable in such structures, the resulting gain of a slot array antenna is usually higher than that of the same size microstrip patch array. Another advantage is its high power level handling capabilities.

Similar to microstrip patch arrays, different pattern configurations and feeding/combining methods can be applied to develop slot arrays. The simplest case is a one-dimensional array formed by a rectangular waveguide with slots appearing at regular intervals in its surface. In this case, a rectangular waveguide accomplishes a series circuit-type combiner. An extra parallel-combining network is necessary to extend such a linear array to a two-dimensional array. This method of using rectangular waveguides to form slot arrays, although very frequently used in radar systems, is cumbersome and hence less attractive for commercial applications such as DTH TV.

At the end of the 1950s, Kelly[24] proposed a radial waveguide as a feeding network for a two-dimensional distribution of slots and demonstrated its use in the early 1960s.[25,26] The concept known as a radial line slot array (RLSA) antenna is very attractive because only a single waveguide, instead of its multiple, is used as a feeding network to a two-dimensional distribution of slots.

In its original construction, a circular waveguide supporting a TE_{11} mode fed a slotted radial waveguide using a suitable transition located on the side devoid of slots. The desired antenna polarization (linear, circular, or elliptical) was attained via the choice of circular-to-radial guide transition and a suitable distribution of slots. In the original design, concentric annular arrays of crossed slots were used.

In 1980, Goto and Yamamoto[27] proposed the modified concept of the slotted waveguide antenna by introducing an alternative slot arrangement that would allow for circular polarization to be obtained from a double layer radial cavity, fed by a simple probe feeding structure centrally located in the lower level of the double-layer cavity. The slot arrangement on the upper cavity surface took the form of a spiral array; each element in the array consists of two slots, spaced so as to be phased in quadrature, hence forming a unit radiator of circular polarization. This proposed antenna had removed the complexity of the feed structure, but had added a manufacturing complexity resulting from its double-layered nature; this addition was necessary for maintaining constant amplitude aperture illumination, requiring an E-bend to get the radiated field to and from the upper and lower cavities.

A simplified single-layer structure proposed by Takahashi et al.[28] is suitable for overcoming the manufacturing complexities brought about by the double-layer cavity design, but introduces the problem of an intolerably tapered aperture illumination profile resulting from the naturally decaying outward traveling radial wave in the feeding cavity.

Single- and double-layer versions of this antenna including both circular (CP) and linear polarization (LP) cases have since been introduced and investigated in the literature. Figure 14.10 shows the construction details of single- and double-layer RLSA antennas introduced in the literature.[29-31] A disk-ended coaxial probe can replace the conical probe shown in Fig. 14.10.[32]

Advantages of the single-layer RLSA antenna include a potentially high radiation efficiency; extremely low profile; ease of installation; and immunity to leaf, water, and snow buildup as a result of its flat surface. Despite this remarkable flexibility, there is, however, an inherent flaw in this RLSA antenna performance when linear polarization is required. This is the case for some European satellite systems as well as for the Australian Optus B-series satellites that employ linearly polarized transponders for their direct broadcast services.

14.3.3.1 Single-Layer RLSA Antenna Design

In its standard form the antenna consists of two plates spaced a distance d apart, with the upper plate bearing the radiating slot pattern, and the rear plate left unmodified. The radial guide formed between these plates is filled with a dielectric material of relative permittivity $\varepsilon_r > 1$ (in practice, between 1.5 and 2.5) to avoid grating lobes in the radiation pattern.

The operation of the antenna can be considered in either receive or transmit modes of operation. Both are equally as valid because of the reciprocity theorem. In the transmit mode of operation, the signal fed to the antenna via the coaxial cable is launched by the feeding mechanism into an outward traveling axially symmetrical wave inside the radial cavity. The area of radius ρ_{min} around the feed probe is left devoid of slots to allow the formation of an axially symmetrical traveling wave. In turn, this cavity mode is coupled into a radiated free-space wave via the slot pattern on the upper cavity surface. Depending

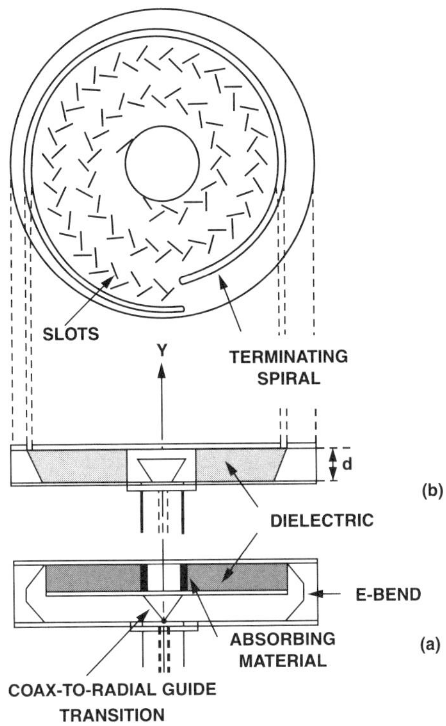

FIGURE 14.10 Radial-line slot waveguide array (a) double layered, (b) single layered.

on the slot orientation and positioning, different types of wave polarization can be radiated (linear, circular, or elliptical). Any residual power not coupled by the slot pattern is lost in the cavity termination.

To accomplish the design and development of RLSA antennas, a number of design strategies have been devised.[27,30,33-35] In Bialkowski and Davis,[36] a unified approach to the design of a single-layer RLSA antenna of arbitrary polarization, covering steps from starting performance specifications to the development of a full working prototype, has been proposed. In this approach, two types of feeding arrangements, as shown in Fig. 14.11, have been considered. The design strategies developed are for both CP and LP. The design sequence can be summarized by a flowchart, which is presented in Fig. 14.12.

The design commences with a set of specifications, which include parameters such as design frequency, polarization, and gain. While considering the feeding structure, the designer can choose between two available configurations, one of Fig. 14.11a and the other of Fig. 14.11b. The probe feed configuration of Fig. 14.11a is suitable for both CP and LP varieties of RLSA antenna. The recessed cavity structure of Fig. 14.11b is a proper choice only for a small diameter CP RLSA. The dimensions of the feed are obtained in an iterative manner using computer algorithms developed to analyze coaxial-to-radial guide transitions.[32,37,38] These algorithms are very accurate. Consequently, the designed feed does not require iterative experimental prototyping and stays the same throughout the entire process of design and development of the RLSA antenna.

The radiating surface design aims to achieve maximum gain and thus maximum aperture efficiency in the desired main beam direction. To meet such a goal, the pattern of slots has to provide uniform aperture illumination. In the present design, the slot length is the primary factor behind the control of coupling and hence aperture illumination, because the slot orientation is set by the polarization requirements. The realization of a given coupling function is achieved by using the theory of a single slot in a waveguide.[33] By including other factors, such as avoiding slot overlap, the slot location, orientation, and length are obtained in the form of a computer algorithm. The output data can be used to theoretically

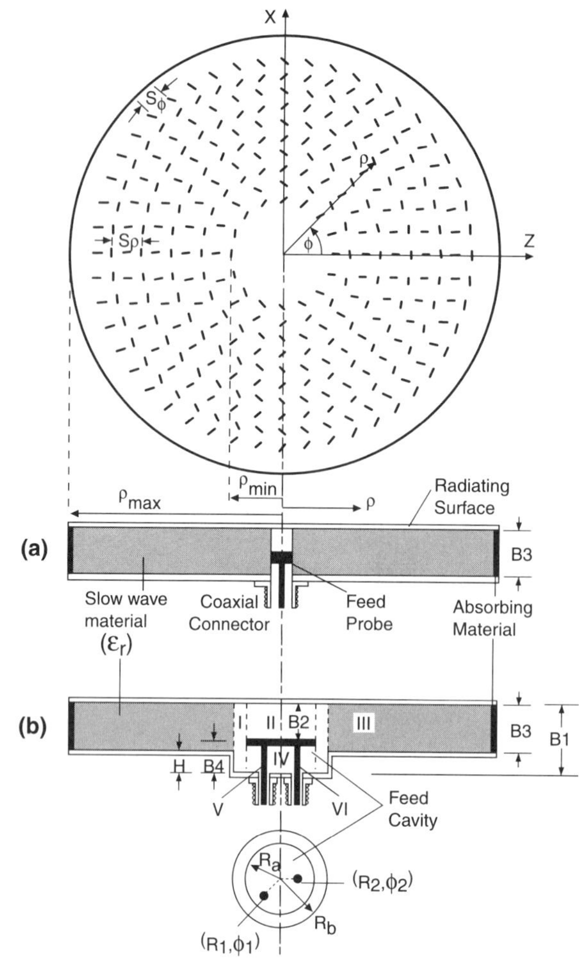

FIGURE 14.11 Single-layer RLSA with different feeds (a) probe feed, (b) recessed cavity feed.

model the radiation pattern prior to the realization of the antenna. Alternatively, the output data is used to generate a physical layout of the radiating surface, ready for computer-aided manufacture.

To overcome the inherent poor return loss performance, the LP RLSA antenna radiating surface includes additional (reflection canceling) slots. An alternative solution involves tilting of the main beam direction,[34,35] which does not require introducing additional slots. Irrespective of the method, the final result is a slot pattern, which produces high return loss at the feed point.

By assuming different beam squint angles for different sets of adjacent rings, or alternatively different sectors of the radiating surface, beam broadening or beam shaping can also be achieved.[39] The computer algorithm developed to predict the radiation pattern of a given pattern of slots aids this process.[40] In this pattern-modeling approach, each radiating slot is replaced by an equivalent magnetic dipole and the field pattern resulting from an entire set of slots in the radiating surface is obtained using the principle of superposition. The slot excitation coefficients are obtained in an approximate manner using a small coupling theory of slots. The developed computer algorithm produces plots of the radiation pattern of the RLSA antenna in an arbitrarily chosen plane.

Once a suitable arrangement of slots has been confirmed by the modeling process, the physical realization of the radiating surface is accomplished using either an aluminum foil (low-cost option) or a copper foil. Aluminum foil is used in early prototyping stages and the final designs use copper foil. The reason for selecting copper foil for the final prototype is that for a 60-cm diameter RLSA antenna,

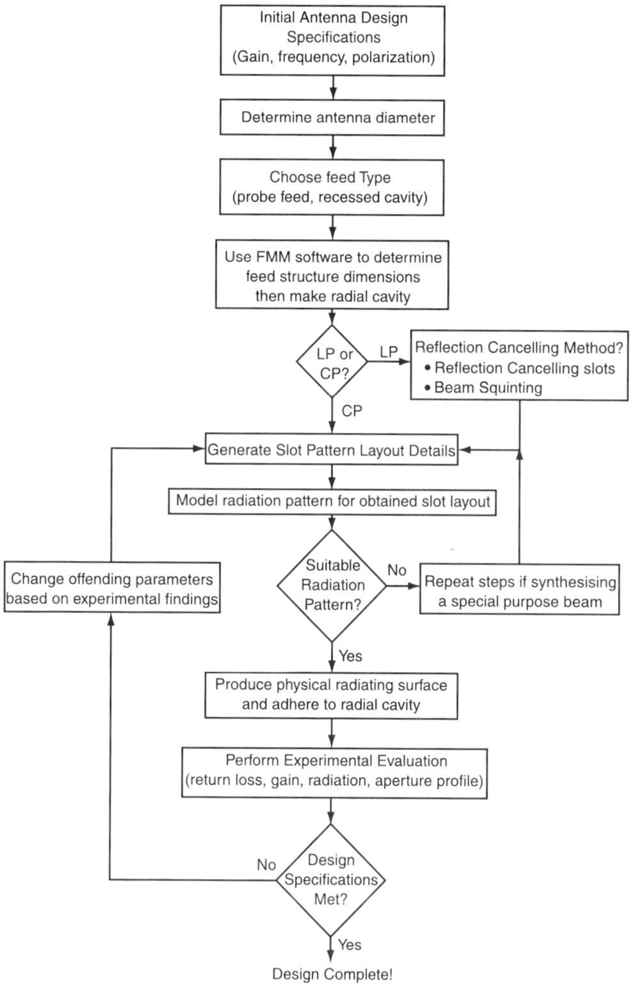

FIGURE 14.12 Flowchart representing a unified design of RLSA antennas.

the radiating surface developed in aluminum foil introduces conduction losses in the range of between 5 and 6 dB at 12 GHz. Irrespective of the conducting material used, the developed prototype antenna undergoes a number of experimental tests, including return loss and power gain, as well as near- and far-field radiation pattern tests.

The previously described unified design has been fully implemented and applied to the development of linearly and circularly polarized RLSA antennas for receiving direct-broadcast satellite service (DBS) TV programs in Ku-band (center frequency of 12.5 GHz). These developments concern antennas of approximately 60-cm diameter having gain in the range of 30 to 34 dBi and are described next.

The first example is the linearly polarized RLSA antenna with its main beam in the bore sight direction. Because this antenna in its standard form would exhibit poor return loss performance, reflection-canceling slots have been incorporated to improve the return loss figure. The resulting radiating slot pattern on a much reduced scale is shown in Fig. 14.13. The experimentally measured radiation pattern for this antenna, measured in the Fresnel zone (3-m separation from an X-band pyramidal horn) is presented in Fig. 14.14. The radiating surface of this antenna was constructed out of copper shim. The measured radiation pattern 3-dB beam width of 2.57° in both principal planes indicates a directivity of 36.9 dBi, and an aperture efficiency of 85%.

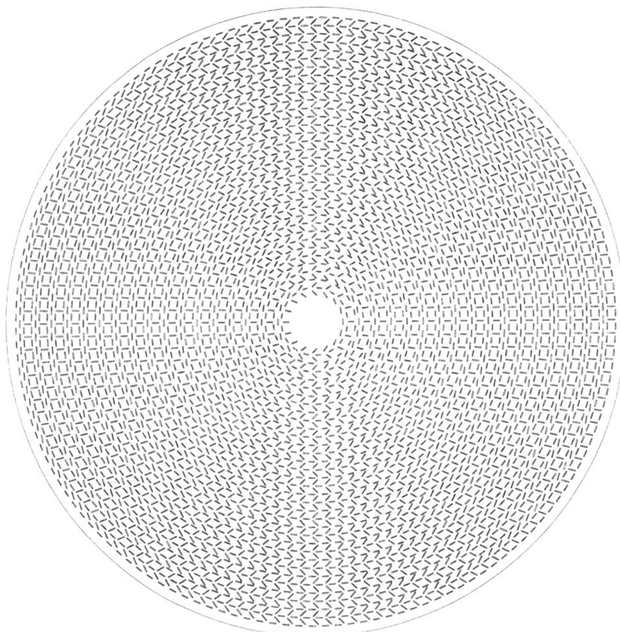

FIGURE 14.13 LP RLSA antenna slot distribution for bore sight radiation pattern.

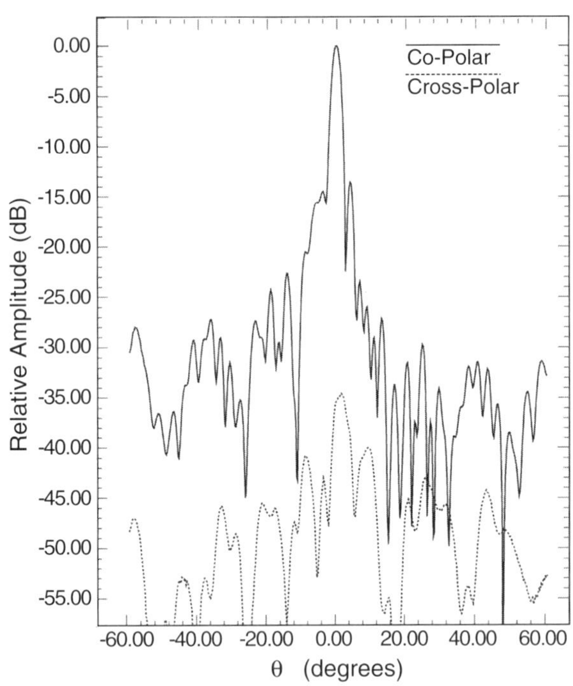

FIGURE 14.14 E-plane co- and cross-polar radiation patterns for the LP-RLSA antenna with its main beam in the bore sight direction.

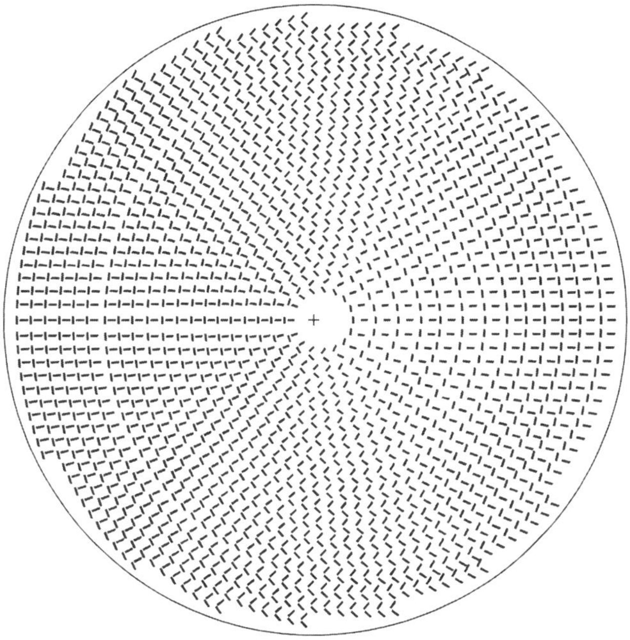

FIGURE 14.15 LP RLSA antenna slot distribution for beam width of 7° and beam squint of 20°.

The second example concerns an LP RLSA antenna with a broadened beam in the E-plane that is squinted from the bore sight direction to a beam target angle of 20°. This prototype demonstrates the concepts of both beam squinting and beam broadening, as were mentioned earlier. This prototype was developed using an aluminum foil for its radiating surface.

Figure 14.15 shows the slot distribution and Fig. 14.16, the theoretical radiation pattern compared against the experimentally measured pattern at 13 GHz for this Ku-band, 55-cm diameter prototype antenna. The beam width target of 7° represents an increase of more than double the standard 55-cm RLSA beam width of 3°.[39] The design frequency was assumed to be 13.42 GHz. A slight frequency shift of the optimum prototype performance point from the design frequency results from the change of effective dielectric constant of the radial guide by the presence of slots, which is not taken into account in theory.

The + symbol in Fig. 14.15 identifies the phase center of the antenna. It can be observed that slots arranged in oval rings form the radiating surface and that the outer ovals are incomplete because of the desired circular shape of the radiating surface. The comparison between predicted and measured radiation pattern shows good agreement in the main beam region. Of particular note in both theoretical and measured patterns is the presence of quite high sidelobe levels on the lower angular side. Because of the nature of the designed surface, the outer rings in the Fig. 14.15 pattern are those of lowest squint angle, and hence it can be deduced that the high sidelobes are a consequence of having incomplete rings in the outer few rings of the surface.

The final example is a 55-cm CP RLSA antenna with its main beam in the bore sight direction. From the slot pattern distribution shown in Fig. 14.17, the fundamental differences of the CP RLSA slot pattern to that of the LP RLSA pattern can be seen. In the CP RLSA slot pattern, all subsequent slots are spaced by one quarter guide wavelength in the radial direction from their predecessor in a spiral pattern. For this reason the CP RLSA antenna does not suffer in its standard form from the poor return loss problems plaguing the LP RLSA antenna.

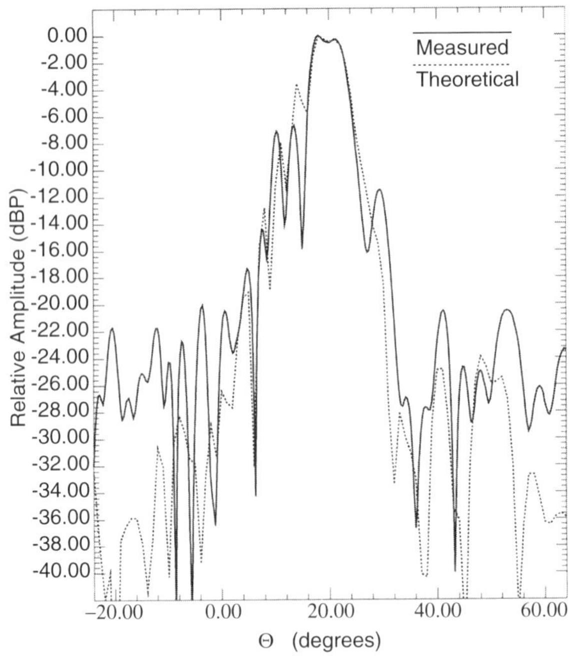

FIGURE 14.16 E-plane radiation pattern for the RLSA antenna with broadened beam.

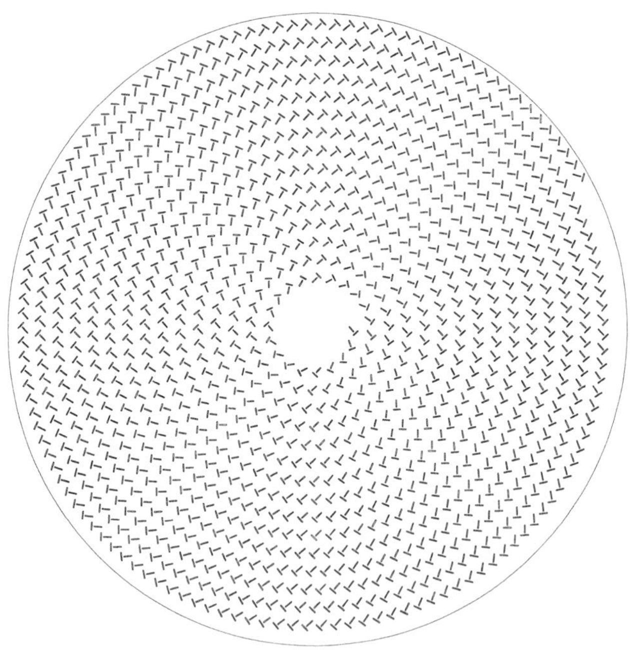

FIGURE 14.17 CP RLSA antenna slot distribution for single-layer LH CP antenna.

14.4 Mobile Antennas for Receiving Direct Broadcast Satellite Television

Microstrip patch arrays and slot arrays described in the previous sections provide the solution to receiving DBS TV at stationary positions. To receive DBS TV by mobile users, antennas have to be flat or conformal to a vehicle body. In addition, a suitable satellite tracking system is also necessary. The Toyota Central Research and Development Laboratories (CRDL) have presented a working prototype of a low-profile microstrip antenna system for operation with the DBS TV in Japan.[41] This antenna system is described as follows.

In Japan, to receive quality DBS TV pictures, an antenna featuring RHCP and gain of about 21 dBc over the frequency range from 11.7 to 12 GHz has to be used.[42] The schematic of the antenna system including a receiver and a satellite-tracking electronics, developed by CRDL researchers meeting these specifications, is shown in Fig. 14.18.

This system consists of an antenna assembly, a motor assembly, and a receiver assembly. The outdoor unit consisting of the antenna and the motor assembly is mounted onto the rooftop or rear trunk top of a car and an indoor unit is installed inside the car. The signal from the satellite in the 12-GHz band is received by a circular planar array antenna, and is converted to the 1.3-GHz band by a frequency downconverter. The converted signal is carried through a coaxial cable into the car and divided into two signals. One is carried to a receiver, and the other is carried to a signal level detector. A controller determines a rotating angle of the planar array antenna by using both the detected signal level and a gyroscope signal and tracks the satellite accordingly as the car moves.

The beam of the horizontally positioned antenna is suitably tilted toward the satellite. For most of areas in Japan, the required beam tilt is 50° from zenith with ±6° deviations. This specification is met by producing a beam pattern, which is broadened in elevation and narrowed in azimuth. The configuration of the proposed antenna is shown in Fig. 14.19. The antenna consists of a coaxial probe fed radial guide, which acts as a feeding structure to circularly polarized ring microstrip antenna (RMSA) elements arrayed in concentric circles on a dielectric substrate. The elements are closely spaced in a radial direction to suppress grating lobes. The number of the concentric circles in the radial direction is 11 and the total number of elements is 396 to satisfy the required gain. Each RMSA element has a feed pin, which is connected at the center and inserted into the radial waveguide through a small hole. The signal from the satellite is received by the RMSA element and propagates to a center feed probe through the radial waveguide. The length of the feed pin of each RMSA element is adjusted to get a uniform amplitude

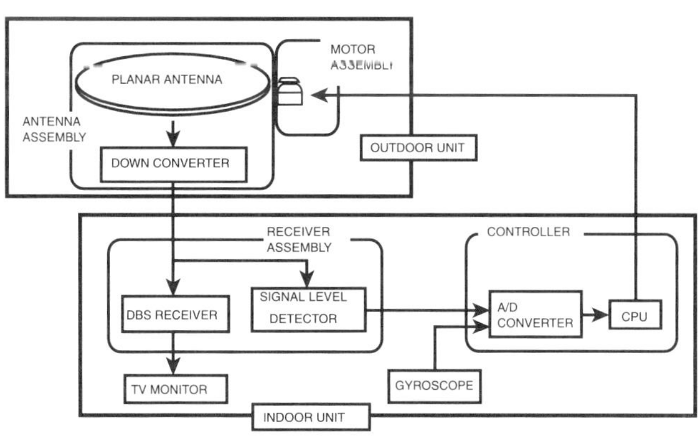

FIGURE 14.18 Block diagram of the DBS antenna system developed by Toyota CRDL. (From Ogawa, M. et al., *Proc. 1996 Int. Symp. Antennas Propag., ISAP'96*, Chiba, Japan, 1996, 1197. With permission.)

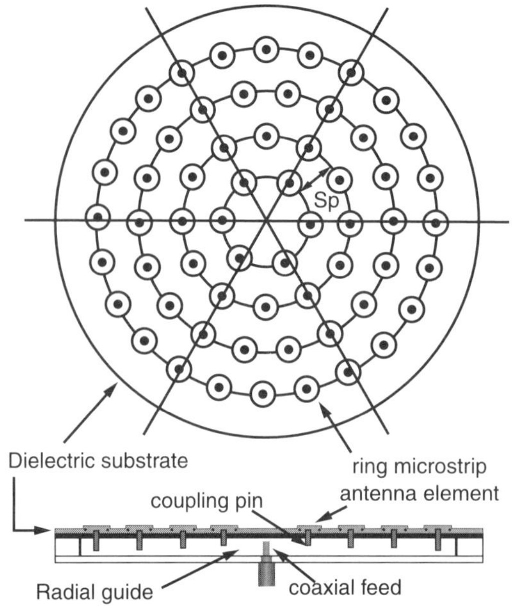

FIGURE 14.19 Configuration of a circular array antenna to receive DBS TV.

distribution over the array surface. In addition, the rotation of each RMSA element around the axis of its center feed pin, which corresponds to the variation in the excitation phase of the RMSA element, is determined so as to get a desired beam pattern. The single RMSA antenna element 3-dB axial ratio bandwidth is approximately 260 MHz, which is not enough for the current application. However, this axial ratio bandwidth is improved for the array using the sequential feeding technique.

The required beam broadening is achieved using a similar principle to that described in this chapter while discussing a beam-tilted RLSA antenna, however, with some differences. The antenna elements in the array are divided into four subarrays for which the excitation phases are selected to achieve subbeam directions of 48°, 54°, 58°, and 61°. Note that in the present antennas the required phasing is achieved while maintaining circular rings. For the RLSA antenna, beam squinting required deforming circular rings into ovals. This maneuver led to the appearance of a grating lobe. The proposed method produces beam broadening in elevation of not less than 12° across the required operational bandwidth. The diameter of the manufactured antenna is 400 mm and the height is approximately 30 mm.

The satellite tracking is required only in the azimuth. In the solution presented by the Toyota CRDL, this task is accomplished using a hybrid open/closed-loop approach employing the gyroscope (open loop) and the received signal (closed loop). The tracking algorithm proposed there is as follows. When the received signal level is higher than a predetermined threshold signal level, the controller turns the array antenna to the counter direction of turn of the car using only the gyroscope signal. The gyroscope signal has to be recalibrated at suitable time intervals because it drifts with time. Consequently, if only the gyroscope was used for satellite tracking, the beam direction would drift off the direction of the satellite and the receiving signal level would fade. To counter this situation, when the received signal level becomes lower than the threshold level, the controller intentionally generates a fluctuation of the received signal level by slightly rotating the array antenna right and left to search for the satellite, and adjusts the beam direction as the receiving signal level goes up.

A field test was conducted to confirm the tracking performance with the presented algorithm by using the BS-3 satellite of Japan. These tests confirmed that the tracking system worked very well. The deterioration of the C/N value was less than 3.5 dB even when the value of yaw rate was higher than 40°/s.

The presented concept of a mobile DBS TV antenna can be extended to other types of antennas, which are low profile and provide a suitable beam tilt directed toward a satellite. The satellite tracking for such antennas can be achieved by mechanical rotation in azimuth and by using the satellite system tracking similar to the one developed by Toyota CRDL. One such alternative solution using a slot array antenna has been presented in Hirokawa et al.[43] Similarly, Huang[44] proposed a microstrip reflect array to receive DBS TV using a mechanical satellite-tracking mechanism.

14.5 Satphone Antenna Systems

Receiving DBS TV programs using stationary or mobile systems requires antennas having gain at least on the order of 20 dB or more. In fact, typical DBS systems require antennas with a gain in the range of about 34 dBi.[2,3] This value of gain results in a very narrow beam of the ground antenna and becomes the source of its easy misalignment with the satellite. To point its beam accurately toward the satellite, a very fine positioning or tracking system is required. An alternative to a mechanical satellite tracking system is a motionless electronic tracking system. Such a system can be accomplished using a switched-beam antenna or a phased-array antenna. The most flexible solution is offered by a phased array. Unfortunately, the current stage of the phased-array technology at X- and Ku-bands at which DBS satellites operate introduces a high price tag. This is one of the reasons why this technology has been implemented primarily on commercial planes where the phased-array cost is offset by much higher costs associated with the manufacture of the plane.

A significant cost reduction can be obtained for switched-beam and phased arrays operating at low microwave frequencies such as L-band or S-band frequencies because components in these frequency bands are less expensive. These frequency bands are allocated for satellite communications concerning voice, low-rate data, and facsimile transmission. Because of the reduced bandwidth required to offer such services, a suitable communication link is achieved using medium gain antennas of about 10 dBi.

The following sections cover the design and development examples of switched-beam and phased-array antennas for voice and low-rate data mobile satellite communications. These antennas, named satphone antennas, are described here for operation with the Australian Mobilesat system. The extension of these antennas for other domestic and global satellite systems is straightforward, and hence this extension is not further elaborated. While meeting electrical performance specifications, particular attention is paid to minimizing manufacturing costs of these antenna systems. This goal is achieved by using low-cost substrates and low-cost switching diodes and by applying planar or conformal approaches to the development of these antennas. Other types of antennas for use with MSAT systems developed by other researchers in different parts of the world are also briefly covered.

14.5.1 MSAT Antenna

The complete functional block diagram of an MSAT antenna system for voice communications including a fully duplex satellite transceiving amplifier (AEU), a transceiving unit (TU), and satellite tracking electronics is shown Fig. 14.20. By replacing the handset by a new unit, data and facsimile communications can also be achieved using the equipment shown in Fig. 14.20.

The role of the antenna is to enable transmission and reception of microwave signals and to track the satellite as the vehicle moves. One possible solution is a mechanically steered array antenna similar to the one described for receiving DBS TV. Such a type of mechanically steered antenna is not discussed here. Instead considerations are directed toward electronically steered antennas including switched-beam and phased array antennas.

The functional block diagram of the electronically steered antenna system for land mobile satellite systems (LMSS) of switched-beam or phased array type is shown in Fig. 14.21. This antenna system consists of antenna elements, a beamforming network (including power dividers, couplers, switches, phase shifters, and biasing networks) and tracking electronics all conforming to given specifications. Such specifications with respect to the Australian Mobilesat are presented next. Note that very similar specifications concern the American AMSC and the global INMARSAT system.

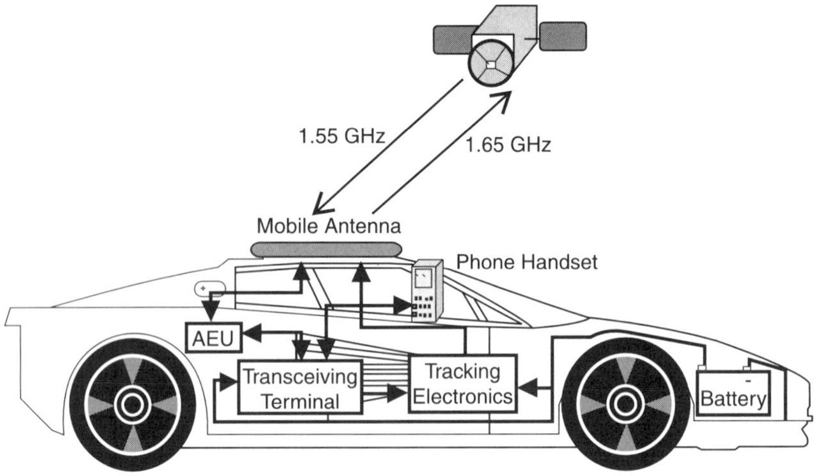

FIGURE 14.20 Functional block diagram of the MSAT terminal and an electronically steered antenna fitted to a car.

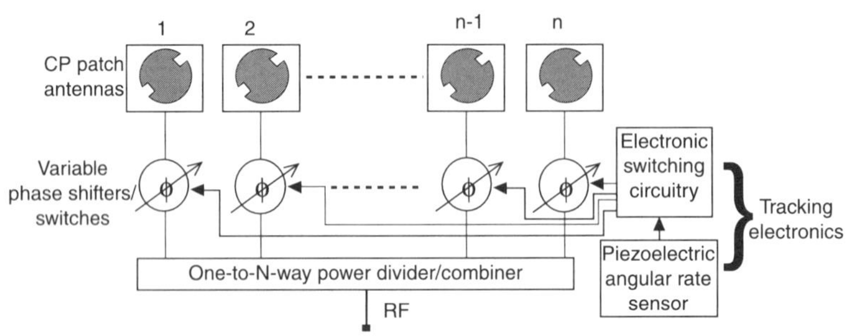

FIGURE 14.21 Functional block diagram of the electronically steered antenna system for the LMSS.

14.5.2 Antenna Specifications for Mobilesat

To be used with the Australian Mobilesat system, the vehicle antenna being a part of the mobile unit, needs to meet the following specifications:[45]

- Polarization —right-hand circular
- Frequency — 1646.5–1660.5 MHz for transmission; 1545.0–1559.0 MHz for reception
- Coverage — full 360° in azimuth and 28° to 66° in elevation
- Gain — 10 to 12 dBi desirable with a minimum of 7 dBi

From the L-band frequency requirements, it can be seen that the vehicle antenna needs a bandwidth of at least 8% to cover both receive and transmit frequencies. The gain requirement is determined by a link budget, which takes the EIRP of the satellite and the required channel quality (carrier to noise power density ratio) into account. Here, the value is obtained by assuming G/T of –18 dB/K for 20 kbit/s voice communications via a stationary satellite. The required elevation coverage of 28° to 66°, on the other hand, is determined based on knowledge of the satellite geostationary position in conjunction with the size of the Australian continent.

Low elevation angles create an extra challenge to designers, restricting them to special types of array antennas that can cater for such low-look angles. To ensure a better match with the other subsystems, such as the user's terminal, the antenna must have better than 10 dB input return loss (RL). Also while used with the Mobilesat transceiver, it has to withstand 3 W of power in the transmission mode of operation.

The desired mechanical features of the antenna should include low profile and conformal arrangement to the vehicle roof on which the antenna has to reside. Low weight to reduce payload while the vehicle is on the move is also a desirable feature. The antenna must be tough and watertight to withstand the vibrations and climatic conditions. It should operate in the extreme cold and hot weather ranging from –20 to 50°C.

14.5.3 Antenna Elements for MSAT Arrays

The specified antenna operational bandwidth of about 8% concerns both impedance and axial ratio. The impedance bandwidth is defined as a range of frequencies over which the input return loss is not smaller than a designated value, usually 10 dB, whereas the axial ratio bandwidth is defined as a range of frequencies over which the axial ratio (the ratio of major to minor axis of polarization ellipse) does not exceed a designated value, usually 3 dB. Fulfilling these requirements simultaneously can be attempted using microstrip patch antenna technology.

Getting 8% impedance bandwidth with microstrip elements has been a challenge for some time. However, this problem is now solved, resulting in patch antenna designs meeting a 40% impedance bandwidth.[46] Obtaining an equivalent ellipticity bandwidth has not been covered that well, because it is a more challenging task. It not only involves increasing the patch height above the ground plane, as is the case in impedance bandwidth enhancement for linearly polarized patches, but also requires optimization of the polarizing elements.

The most elegant configuration for bandwidth enhancement is an aperture- (or slot) coupled microstrip patch antenna.[6,7,46] This configuration not only preserves a fully planar form of the antenna system but also eliminates the forward residual radiation of the feeding structure, which is encountered, for example, with edge-fed patches.

In Karmakar and Bialkowski,[47] three configurations of aperture-coupled antennas have been investigated as possible candidates in MSAT antenna arrays. These are shown in Fig. 14.22 and include

1. Single-feed, aperture-coupled circular patch with perturbation segments
2. Single-feed, aperture-coupled stacked circular patch with perturbation segments
3. Dual-feed, aperture-coupled circular patch

The antenna elements assume Rogers Ro3003 material — with a relative dielectric constant of 3.02 and a thickness of 0.76 mm for a feed layer and a very thin and inexpensive epoxy laminate from International Laminate Ltd., with a thickness of 0.076 mm and a relative dielectric constant of 4.7 — is used for patch layers. A Pilon Plastic Ltd. foam with a relative dielectric constant of 1.07 is applied as an insert to increase the operational bandwidth.

The three configurations have been thoroughly theoretically and experimentally studied with the following results. For a single-feed, aperture-coupled patch antenna configuration the axial ratio bandwidth has been found to be only 23% of the return loss bandwidth. Because of this constraint, this antenna is unable to deliver the required ellipticity bandwidth even with very thick foam layers, which produce about 15% impedance bandwidth. The required impedance and ellipticity bandwidth can, however, be obtained by introducing a second patch and by modifying the structure to a stacked patch configuration. An alternative way to obtain the required impedance and ellipticity bandwidth is to use a dual-aperture arrangement with an external polarizer in the form of a 3-dB branch line coupler.

The last two antenna elements feature the desired impedance and axial ratio bandwidth in addition to a 9-dBi gain across the specified band. The dual-feed patch offers a better axial ratio performance. However, this performance is achieved at the expense of the extra space taken by an external polarizer in the feed network layer. This solution may be found unattractive in a phased-array design where the space for the beamforming network is at a premium. In such cases, the single-feed, aperture-coupled stacked patch can be a preferable option.

14.5.4 Compact MSAT Switched Beam Antenna

A standard MSAT switched beam antenna uses a truncated conical structure (or in more a general case, a hemispheric structure) on which individual antenna elements are located. This antenna is medium or

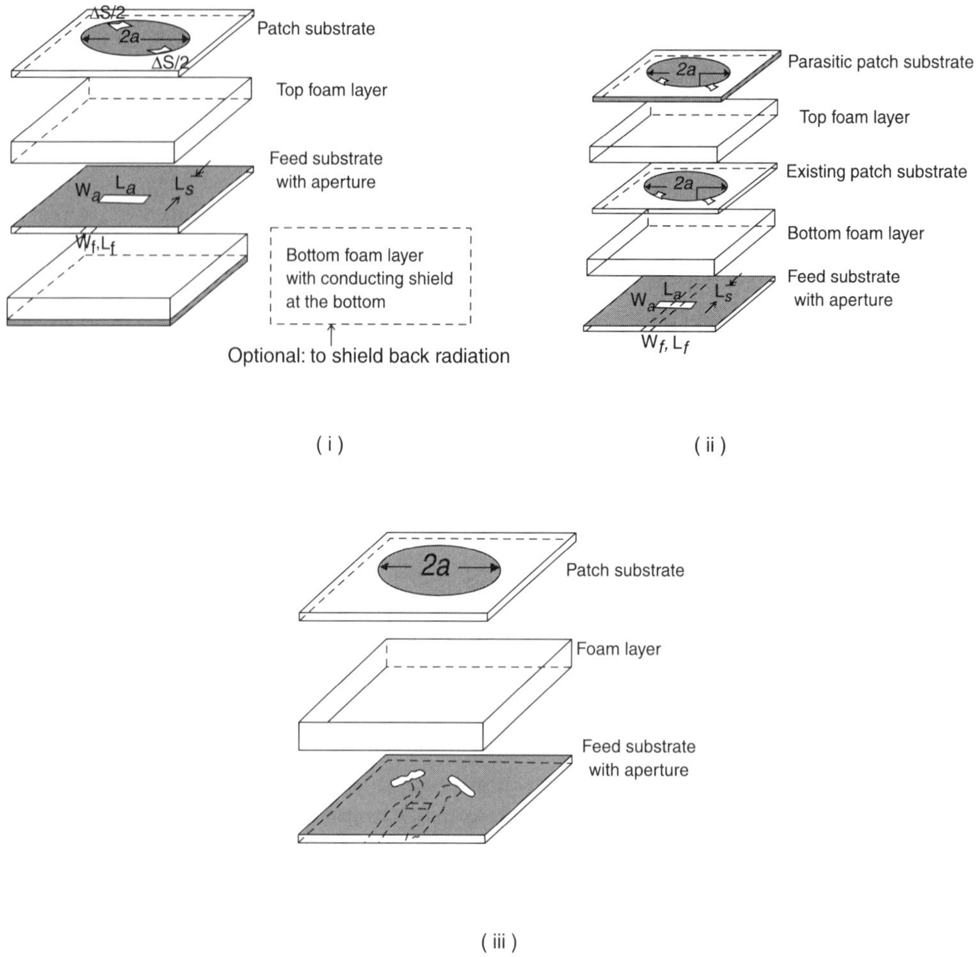

FIGURE 14.22 Configurations of three aperture-coupled microstrip patch antennas investigated for applications in MSAT arrays; (1) single patch perturbation segments, (2) stacked patch with perturbation segments, and (3) dual-feed, aperture-coupled circular patch antenna.

high profile and as such is best suited for very low satellite look angles where a planar phased-array antenna has difficulty in meeting the required gain performance. The communication link between a mobile unit and a satellite is maintained by turning on adjacent elements facing the satellite and by deactivating the remaining elements. One problem with this type of antenna is that it features considerable nulls between two adjacent beams, which can be the source of signal fading when the antenna switches over to the next beam.

The solution to this problem has been presented in Karmakar and Bialkowski,[48] where 1-b (one bit) phase shifters haven been added to a radial switching system controlling an eight-element, switched-beam array to achieve smoother transitions from one beam to the other. The configuration of this switched-beam array antenna is shown in Fig. 14.23.

As shown in Fig. 14.23, the system comprises of an array of eight radiating antenna elements and a feed network. The antenna elements are arranged in a circle on the side of a truncated cone whose inclination angle matches a *look angle* toward the satellite. The feed network comprises an eight-way, 2-elements on radial switch, and 1-b phase shifters of 50° connected to eight outputs of the switch. The function of the feed network is to provide suitable phase excitation to two adjacent elements that face the satellite and deactivate the remaining six elements. The inclusion of 1-b phase shifters in each antenna

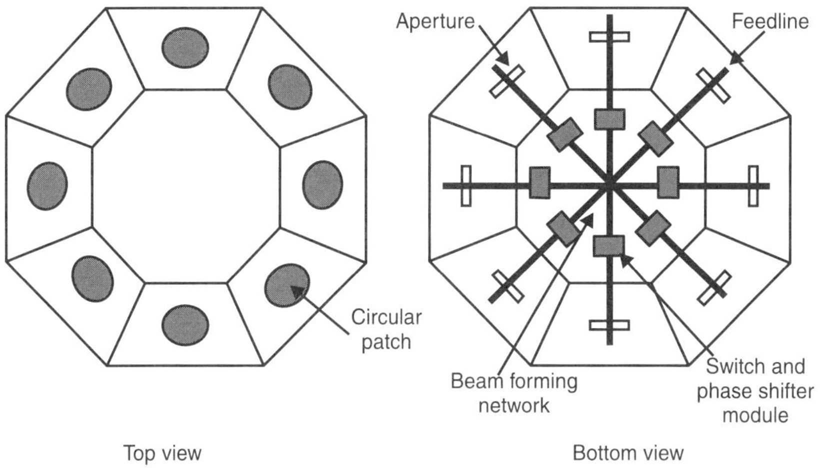

(a)

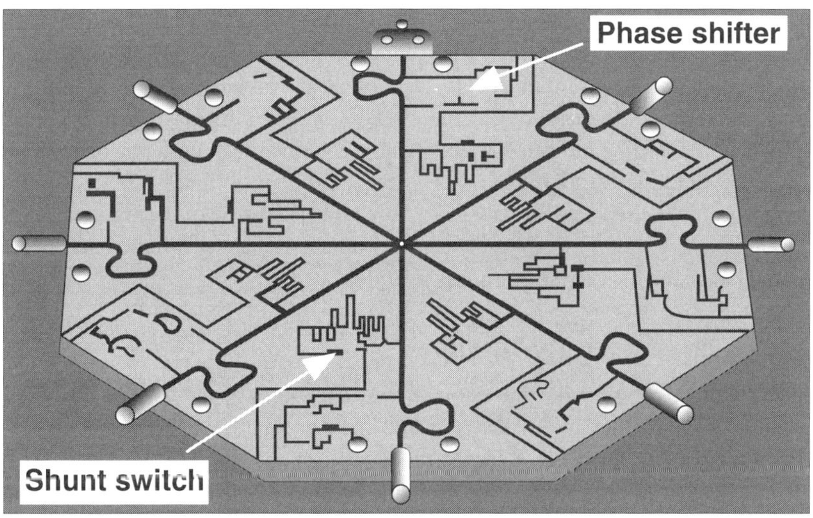

(b)

FIGURE 14.23 (a) Schematic top and bottom view of an 8-element switched-beam, phased-array antenna; (b) photograph of the integrated 8-way, 2-element, on radial switch and eight 1-b (one bit) phase shifters.

element (or equivalently, each output of the radial switch) leads to 24 independent antenna beams, which provide the required radiation pattern smoothness. Without 1-b phase shifters, the antenna is able to produce only eight beams.

The selected antenna elements are right-hand circular polarization (RHCP), dual-feed, aperture-coupled, circular patch antennas discussed earlier. The required circular polarization is achieved using an external polarizer in the form of a 3-dB branch line coupler. The two outputs of this coupler feed two orthogonal slots that in turn couple the microwave power to the circular patch. A virtual short is introduced at the isolation port of the coupler to relinquish the requirement for a plated through hole for the 50-Ω termination, making the design fully conformal.

The RF power is supplied (or received) through a coaxial line that is centrally positioned in the radial switch. The switch is activated using electronic drivers that supply suitable sequences of voltages via a 17-line ribbon cable. The completed antenna array has an overall diameter of 400 mm, a height of 95 mm, and a weight of 4.5kg.

The developed antenna has been tested indoors and outdoors. The return loss measurements, at the input coaxial port of the manufactured antenna, were performed for the case when selected two-element subarrays were active. DC +5 V (with 470-Ω current limiting resistor) and 10 mA current in the forward bias state and −15 V in the reverse bias state were used to switch the diodes in the beamforming network. In all the investigated cases, the RL was always better than 10 dB over the required Mobilesat™ frequency band.

Following satisfactory return loss results, radiation pattern and gain measurements for the array were performed in an anechoic chamber. The measurements were done at three different frequencies of 1.55, 1.6, and 1.65 GHz; and three different elevation angle settings of 30°, 50°, and 70°. The three angles correspond to the look angles for Perth and Tasmania (30°), Torres Strait (70°), and the average look angle for continental Australia (50°). The gain pattern measurements revealed a minimum of 7.8 dBc across the investigated angle and frequency range, which was a very satisfactory result.

Following indoor tests, outdoor field trials were attempted. The experiment was conducted in the southeast Queensland region in Australia where the antenna look angle was 55°. In this case, the antenna was connected to an NEC S1 transceiving terminal[49] to communicate via Mobilesat™. A testing jig was used, which allowed for the 360° rotation of the antenna in azimuth and ±40°. This setup emulated the conditions for the antenna being mounted on the roof of a vehicle turning in arbitrary direction or traveling along steep hills. In the case of an antenna horizontally positioned with two elements facing the satellite, the handset reading indicated levels between 8 and 9, meaning that a relatively strong signal was received. A level between 8 and 16 indicates a very good quality signal. Level 4 is a threshold level at which signal drop-off occurs. Levels close to 20 cause an overloading of the receiver. For the obtained setting, a telephone conversation via *Optus* satellite with an ordinary telephone user was attempted. A very good voice quality for the conversation was recorded.

14.5.5 MSAT Phased-Array Antenna

One inconvenience of the presented eight-element, switched-beam array is its height of 95 mm, which makes it a medium profile antenna. An alternative to the switched-beam array is a phased array. This antenna can be fully planar and hence it can be made very low profile. This feature is realized at some expense. The requirement of phasing antenna elements by multibit phase shifters causes the shortage of real estate in the beamforming network. To overcome real estate constraints, only a special type of antenna elements can be chosen to construct a phased array.

In the design presented in detail in Bialkowski and Karmakar,[50] single-feed stacked patch elements of Fig. 14.22 discussed earlier are selected as the best possible candidates for constructing a planar phased array. The exploded view of the proposed multilayer phased-array antenna system is shown in Fig. 14.24.

This multilayer array includes stacked CP patch antenna arrays and two foam layers to increase antenna bandwidth, a slotted ground layer for coupling, and an antenna beamforming network layer. The assembly makes an eight-layer bonded package. In addition, the antenna requires a radome as a protecting layer from the environment.

The 12 elements of this array are arranged in two rings of 8 and 4 elements. The beamforming network is comprised of two-way power dividers in two stages and a three-way power divider connected to the second stage of the two-way power divider. This is equivalent to having a 12-way corporate feed power division/combination network as required for the 12-patch array system. Twelve 3-b phase shifters (including a 180° switched line phase shifter and 45° and 90° loaded line phase shifters), located at the

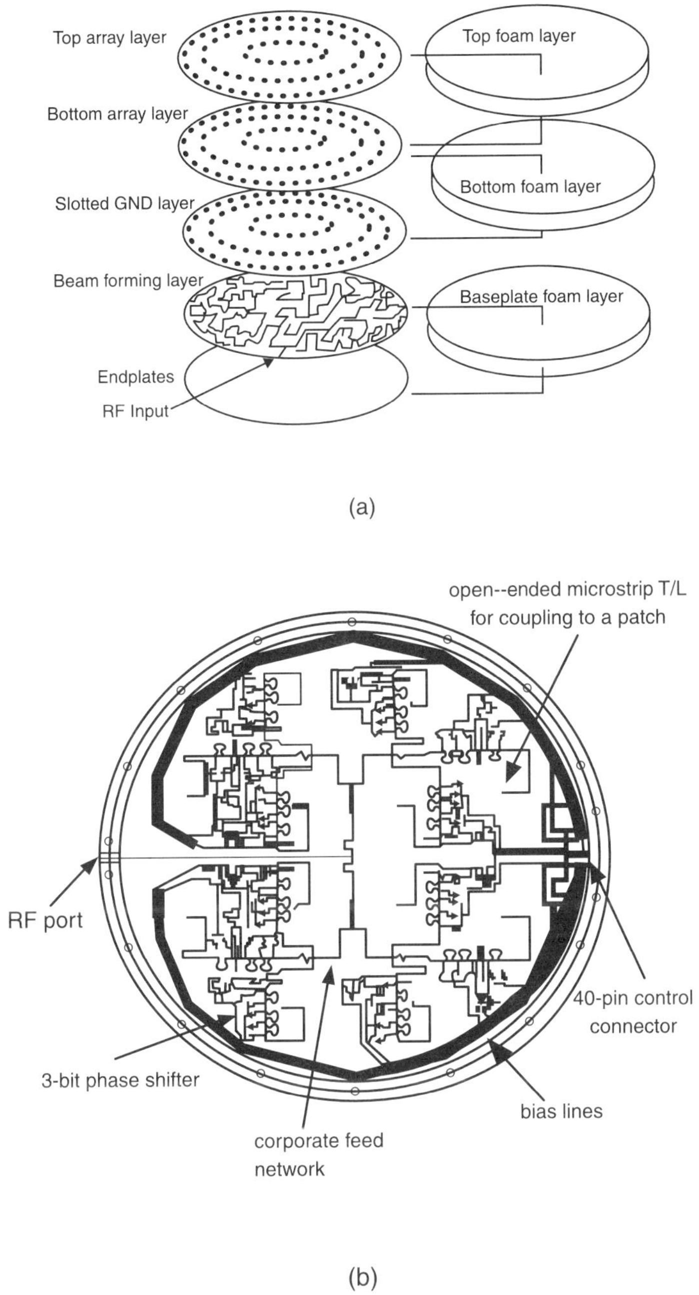

FIGURE 14.24 (a) Exploded view of the multilayer phased-array antenna and (b) details of the beamforming layer.

outputs of the 12-way divider, are coupled to individual antenna elements using an aperture-coupling mechanism.

The phase shifter modules use LC-compensated PIN diode switches.[50-53] The selected diodes are Philips BA682 surface-mounted switch diodes that normally are aimed for operation at 500 MHz. They feature low series resistance of less than 1 Ω in the on state, making them a good short circuit. The diodes exhibit a large stray capacitance in the off state, which disqualifies them as an open circuit in that state. At 1.6 GHz, the stray capacitance lowers isolation between the diode terminals to about 4 dB, making it useless for L-band applications. Applying the LC compensation technique changes the situation completely.

The compensated diode features at least 20-dB isolation over the required band. Consequently, using the LC compensated diodes phase shifters features excellent performance. Their use in conjunction with low-cost antenna elements utilizing low-cost substrates and foam materials leads to a substantial cost reduction in the manufacture of the array. The assembled antenna is 46 cm in diameter, is 4 cm in height, and weighs only 2.5 kg.

Similar to its switched-beam counterpart, the developed phased array underwent a series of indoor and outdoor trials. The RL measurements were performed at the input coaxial port of the manufactured antennas in cases when the beamforming network diodes were energized with DC +5 V and 10 mA current, in the forward bias state, and with –15 V in the reverse bias state showing the return loss being better than 10 dB over the frequency range from 1.42 to 1.75 GHz. This 10-dB RL bandwidth exceeds the Mobilesat™ frequency band.

On achieving satisfactory return loss results, radiation pattern and gain measurements for the array were performed. These measurements took place both in the far-field and near-field ranges. The measurements were performed at two different frequencies of 1.55 GHz that represented the Mobilesat receiving frequency, and 1.65 GHz that represent the Mobilesat transmitting frequency. Three different elevation angle settings of 70°, 50°, and 30° were selected. The minimum recorded gain was 8.7 dB for these investigated cases. The beam squint phenomenon (exhibiting by different beam pointing at receive and transmit frequencies) was observed at a low-elevation angle. It did not exceed 10°. This error was insignificant because for small elevation angles the main beam was relatively flat. In addition to these measurements, axial ratio measurements were taken showing less than 4 dB in extreme investigated cases.

Once satisfactory results for the gain and radiation pattern of the developed phased array were achieved, outdoor field trials were also performed. The experiment was conducted in the southeast Queensland region in Australia. At this location, the antenna look angle was 55°. The antenna was connected to an NEC S1 transceiving terminal[49] to communicate via Mobilesat™. The beamforming network was connected to the electronic driver circuit using a 40-line ribbon. Both the terminal and the driver circuit were energized from a 12-V car battery. A phone handset, which displayed the strength of the received signal, was attached to the terminal for voice communications via satellites. In the experiment, the handset reading indicated level 12. A telephone conversation via a satellite with a PSTN user was attempted. A very good voice quality during the conversation was recorded. The presented switched-beam and phased-array antennas require satellite tracking to maintain the communication link between the satellite and a mobile user.

The compact switched-beam array described here features an increased beam width in elevation and a narrow beam in azimuth. The elevation beam width is governed by a beam width of a single patch element, which is in the range of 60° to 70°, whereas the reduced beam width in azimuth results from two elements arraying. Because of similar beam features, the switched-beam array can use a one-dimensional tracking algorithm as presented for the DBS TV antenna developed by the CRDL research team described earlier. The only difference is that the switched-beam array achieves the required satellite tracking in a motionless manner using electronic switching. Fully developed and tested hybrid closed/open loop tracking systems, similar to that in Ogawa et al.,[41] for operation with switched-beam arrays have been presented in the literature.[54-56]

The methods of satellite tracking developed for switched-beam antennas are applicable to phased arrays, however, with some modifications. Assuming that the vehicle travels on flat terrain, the phased array described here would mostly require satellite tracking in azimuth. An occasional tracking in elevation, for example, resulting from the vehicle's movement along inclines, could be achieved by adding another angular sensor, which would detect the antenna rotation in elevation. This new function should not be difficult to accomplish by incorporating an extra angular sensor.

14.5.6 Other Electronically Steered MSAT Antenna Systems

The 8-element switched-beam array and the 12-element phased-array antenna described here are only examples of many MSAT antennas, which have been developed in different parts of the world. To provide a fair coverage of MSAT antennas, the authors briefly describe other antennas that have been brought to their attention.

14.5.6.1 Phased-Array Antennas

Ball Aerospace Corporation[57] and Teledyne Ryan Electronics[58] in the United States, Toyota CRDL[59] in Japan, and CSELT in Italy[60] have developed low-profile, phased-array antennas consisting of 19 elements, a 19-way power divider/combiner, and 18 digital phase shifter arrays. These antennas are capable of capturing signals in 0° to 90° elevation and 360° azimuth coverage. The basic differences in these designs are the antenna radiating elements. The only disadvantages of these designs include low gain at low elevation angles and relatively high cost because of the use of expensive PIN diodes in beamforming networks.

14.5.6.2 Switched-Beam Array Antennas

Bialkowski et al.[55] propose an antenna system of 14 elements around a truncated cone structure with 3 elements active at a time. For such an antenna, a 14-way, 3-element, on radial switch is used to energize the radiating elements. Similar switched-beam arrays were presented in References 61 and 62. The disadvantages of these antennas are a higher profile and a considerable null between adjacent beams.

14.5.6.3 Adaptive Array Antennas

In *adaptive antenna arrays*, beamforming is carried out by a digital signal processor (DSP), so that both excitations and phases can be controlled more precisely than in conventional phased arrays. Alonso et al.[63] has reported a six-element *adaptive array* for integrated global positioning system (GPS) and MSAT applications. Miura et al.[64] and Chiba[65] have reported a 4 ×4-element DBF square array. These antennas have the capability of allowing the nulls to be steered in the directions of interference signals. Their disadvantage is that they are very hardware and software intensive and, therefore, very costly.

14.5.6.4 Self-Scanned Array Antennas

Shafai[66] has proposed a new approach of beam steering called a *self-scanning* array. Two microstrip patch antennas that operate at two adjacent modes are stacked one above/over the other. The beam steering is achieved by introducing a phase shift between the stacked elements resulting in the development of a low-cost self-scanning array element. Two different types of antennas are designed. One is a single-element unit (with stacked patches operating at TM_{11} and TM_{21} modes, respectively) providing a moderate gain of about 7 dBi at 32° elevation angle, and the other is a seven-element array with a peak gain of about 14 dBi. Two different sets of phase shifters (seven 2-b or 3-b phase shifters and three 4-b phase shifters) were necessary to scan beams of the seven-element array.

14.5.7 MSAT Briefcase Antennas

One disadvantage of a vehicle-mounted antenna is that it is permanently attached to a vehicle. This can be inconvenient in situations in which vehicle access is denied. This inconvenience can be overcome using a briefcase antenna system. Such a system is cheaper and possibly more reliable because it requires no satellite tracking system.

The concept of low-cost and portable briefcase systems for satellite communications is reasonably well established. One of the first designs referred to as an INMARSAT-A terminal comprised quite a large dish antenna that unfolded like an umbrella.[67] At present, much smaller systems can be used to access INMARSAT-M, AMSC, and Mobilesat services. These systems including antennas and transceivers are relatively compact units and can be run from a vehicle battery.

A photograph of the completed two-element antenna briefcase system by Jellet and Bialkowski[68] with the NEC S1 transceiver for communications with the Australian Mobilesat is shown in Fig. 14.25a. The manufactured system has a weight of 16 kg when the internal battery is fitted, and 13.5 kg when the battery is removed. Note that 8 kg of either weight results from the transceiving terminal and the phone handset, which in this case was unavoidable.

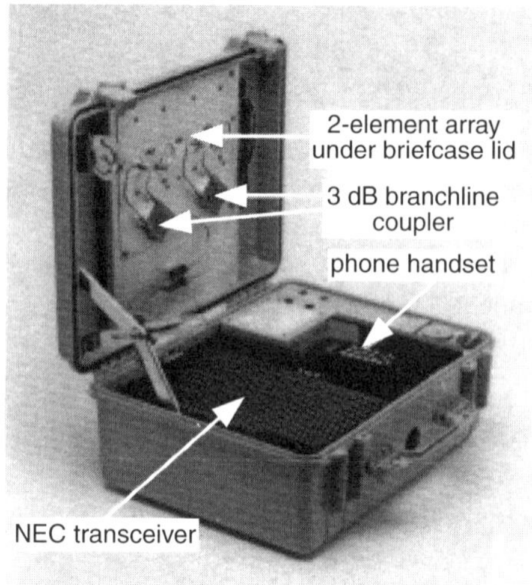

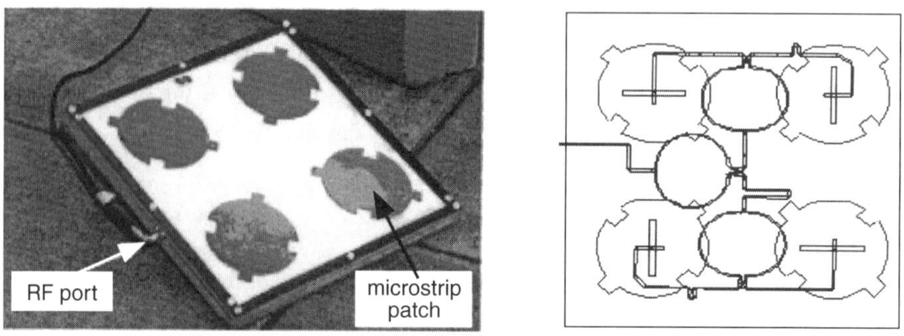

FIGURE 14.25 (a) The two-element briefcase antenna system developed for use with the Australian Mobilesat, (b) a photograph and a transparent view of the 2 × 2-element array antenna as an alternative to the two-element array.

With its lid raised as shown in Fig. 14.25a, the internal contents of the briefcase system can clearly be seen. Although the two array elements are hidden within the lid, the two 3-dB branch line couplers (used to obtain CP), the Wilkinson power divider/combiner, and the connecting semirigid cables can be observed attached to the base plate. Mounted to the lid is the lid elevation guide and adjustable arm. In the main compartment of the briefcase, the phone handset, the transceiving terminal, the power connector panel, and other components of the system are visible.

To test its performance, the briefcase system was taken to remote areas in different parts of Australia. During this testing, the setting up of the briefcase was observed to be very simple, taking approximately 1½ min each time. When calls via satellite were attempted, the result was extremely good. That is, the quality of the voice communications was very high and the indicated received power levels on the handset were in the range of 11 to 12. Such readings represent a strong signal according to the user's manual of the Mobilesat transceiving terminal.[49]

One inconvenience of the developed antenna system is its development cost and time. The elements are coaxial-fed, cavity-enclosed patches, which are labor intensive to manufacture. To reduce manufacturing time and costs of this antenna system, an alternative 2 × 2-element, aperture-coupled microstrip patch array was designed. This array, shown in Fig. 14.25b uses microstrip patch antenna elements and a sequential rotation feeding method to achieve CP.[69,70] It uses only a single RF connector to connect it to a transceiver. This antenna system has been tested indoors and outdoors showing excellent performance. When connected to the NEC S1 terminal, the handset has shown signal strength of 13 to 14 when accessing Mobilesat services in the southeast region of Queensland. A telephone conversation via a satellite with a PSTN has shown very good voice quality during the conversation.

Acknowledgments

The authors would like to acknowledge the financial support of the Australian Research Council, which has sponsored the research work on antennas for mobile satellite communications. Also acknowledged are the contributions of the former members of the University of Queensland Wireless Communications Group, Dr. S. Jellett, Dr. R. Varnes, and Dr. A. Robinson. The contribution of Toyota CRDL concerning a mobile antenna for receiving DBS TV is also appreciated.

References

1. Williamson, M. *The Communication Satellite*, Adam Hilger, Bristol, U.K., 1990.
2. Elbert, B. R. *The Satellite Communication Applications Handbook*, Artech House, Norwood, MA, 1997.
3. Breeds, J. *Satellite Book*, 2nd Edition, Swift Television Publications, Cricklade, Witts, England, 1993.
4. 1994 Asia-Pacific Microwave Conference, *Workshops Digest*, Tokyo, Japan, Dec. 9, 1994.
5. Bialkowski, M. E. Editor, Antennas and Associated Systems for Mobile Satellite Communications, Research Signpost, India, 1998.
6. James, J. R. and Hall, P. S. *Handbook of Microstrip Antennas* (IEE Electromagnetic Waves Series 28), Peter Peregrinus, London, U.K.
7. Pozar, D. M. and Schaubert, D. H. *Microstrip Antennas, The Analysis and Design of Microstrip Antennas and Arrays*, IEEE Press, Piscataway, NJ, 1995.
8. Efanov, A. E. and Tim, H. W. Corporate-fed 2 × 2 Planar Microstrip Patch Subarray for the 35GHz Band, *IEEE Antennas Propag. Mag.*, 37, 49, 1995.
9. Song, H. J. and Bialkowski, M. E. A Multilayer Microstrip Patch Antenna Subarray Design Using CAD, *Microwave J.*, 40, 22, 1997.
10. Ensemble® Version 5.1 User's Guide, Ansoft Corporation, Pittsburgh, PA, 1998.
11. Pozar, D. M. and Targonski, S. D. Improved Coupling for Aperture Coupled Microstrip Antennas, *Electron. Lett.*, 27, 1129, 1991.
12. Rathi, V., Kumar, G. and Ray, K. P. Improved Coupling for Aperture Coupled Microstrip Antennas, *IEEE Trans. Antennas Propag.*, 44, 1196, 1996.
13. Song, H.J. and Bialkowski, M.E. Ku-band 16 × 16 Array of Aperture Coupled Microstrip Patches, *IEEE Antennas Propag. Mag.*, 40, 25, 1998.
14. Barry, D. G. et al. The Reflect Array Antenna, *IEEE Trans. Antennas Propag.*, AP-11, 645, 1963.
15. Huang, J. Microstrip Reflect Array and Its Applications, *Proc. ISAP 96*, Chiba, Japan, 1996, 1177.
16. Targonski, S. D. and Pozar, D. M. Analysis and Design of a Microstrip Reflectarray Using Patches of Variable Size, *IEEE Symp. Antennas Propag. Dig.*, 3, 1820, 1994.
17. Pozar, D. M., Targonski, S. D., and Syrigos, H. D. Design of Millimeter Wave Microstrip Reflectarrays, *IEEE Trans. Antennas Propag.*, 45, 287, 1997.
18. Javor, R. D., Wu, X. D., and Chang, K. Design and Performance of a Microstrip Reflectarray Antenna, *IEEE Trans. Antennas Propag.*, 43, 932, 1995.
19. Chang, D. C. and Huang, M. C. Multiple-Polarization Microstrip Reflect Array Antenna with High Efficiency and Low Cross-Polarization, *IEEE Trans. Antennas and Propag.*, 43, 829, 1995.
20. Robinson, A. W., Bialkowski, M. E., and Song, H. J. A Passive Reflect-Array with Dual-Feed Microstrip Patch Elements, *Microwave Optical Technol. Lett.*, 23, 141, 1999.
21. Bialkowski, M. E., Robinson A. W., and Song, H. J. Spatial Power Combiner Using an Active Reflect-Array, *Proc. 30th European Microwave Conf.*, Paris, France, 2000.

22. Robinson, A. W., Bialkowski, M. E., and Song, H. J. A 137 Element Active Reflect-Array with Dual-Feed Microstrip Patch Elements, *Microwave Optical Technol. Lett.*, 26, 147, 2000.
23. Elliot, R. S. *Antenna Theory and Design*, Prentice-Hall, Englewood Cliffs, NJ, 1981.
24. Kelly, K. C. Recent Annular Slot Array Experiments, *IRE Natl. Convention Red.*, 5, 144, 1957.
25. Goebels, F. J. Jr. and Kelly, K. C. Arbitrary Polarization from Annular Slot Planar Antennas, *IRE Trans. Antennas Propag.*, AP-9, 342, 1961.
26. Kelly, K. C. and Goebels, F. J. Jr, Annular Slot Monopulse Antenna Arrays, *IEEE Trans. Antennas Propag.*, AP-12, 391, 1964.
27. Goto, N. and Yamamoto, M. Circularly Polarized Radial Line Slot Antennas, *IECE Techn. Rep.* (in Japanese), AP80-57, 43, August 1980.
28. Takahashi, M. et al. A Single-Layered Radial Line Slot Antenna, *IECE Techn. Rep.*, AP89-54, October 1989.
29. Ando, M. et al. A Radial Line Slot Antenna for 12GHz Satellite TV Reception, *IEEE Trans. Antennas Propag.*, AP-33, 1347, 1985.
30. Ando, M. et al. A Linearly Polarised Radial Line Slot Antenna, *IEEE Trans. Antennas Propag.*, AP-36, 1675, 1988.
31. Ando, M. New DBS Receiver Antennas, *Proc. 23rd European Microwave Conf.*, Madrid, Spain, 1993, 84.
32. Davis, P. W. and Bialkowski, M. E. Experimental Investigations into a Linearly Polarized Radial Slot Antenna for DBS TV in Australia, *IEEE Trans. Antennas Propag.*, AP-45, 1123, 1997.
33. Takahashi, M. et al. A Slot Design for Uniform Aperture Field Distribution in Single-Layered Radial Line Slot Antennas, *IEEE Trans. Antennas Propag.*, 39, 954, 1991.
34. Davis, P.W. and Bialkowski, M.E. Linearly Polarised Radial Slot Antennas with Improved Return Loss Performance, *IEEE Antennas Propag. Mag.*, 41, 52, 1999.
35. Takada, J., Ando, M., and Goto, N. A Beam-Tilted Linearly-Polarized Radial Line Slot Antenna, *Electron. Commun. Jpn.*, 72, 27, 1989.
36. Bialkowski, M. E. and Davis, P. W. Unified Design of a Single-Layer Radial Line Slot Array Antenna, *Proc. 2000 European Microwave Conf.*, Paris, France, 2000
37. Bialkowski, M. E., Davis, P. W., and Varnes, R. S., Analysis of a Recessed Cavity Radiating into a Radial Waveguide, *Proc. 1995 Asia-Pacific Microwave Conf. (APMC'95)*, Taejon, Korea, 1995, 800.
38. Davis, P. W. and Bialkowski, M. E. Modelling, Manufacturing and Testing Radial Line Slot Array Antennas for Wireless Communications Applications, *Proc. the Millennium Conference on Antennas and Propagation, AP2000*, Davos, Switzerland, April 9–14, 2000.
39. Davis, P. W. and Bialkowski, M. E. Beam Synthesis in Linearly Polarized Radial Line Slot Array Antennas, *Proc. 2000 IEEE Antennas Propag. Symp. (IEEE AP-S)*, Salt Lake City, UT, 2000.
40. Bialkowski, M. E. and Davis, P. W. Predicting the Radiation Pattern of a Radial Line Slot Array Antenna, *Proc. 1999 Asia Pacific Microwave Conf. (APMC'99)*, Singapore, 1999, 162.
41. Ogawa, M. et al. Mobile Antenna System for Direct Broadcasting Satellite, *Proc. 1996 Int. Symp. Antennas Propag., ISAP'96*, Chiba, Japan, 1996, 1197.
42. Ito, Y. and Yamazaki, S. A Mobile 12 GHz DBS Television Receiving System, *IEEE Trans. Broadcast.*, 35, 56, 1989.
43. Hirokawa, J., et al. Single-Layer Slotted Leaky Waveguide Arrays for Applications to Mobile Satellite Communications, *Proc. 1996 Int. Symp. Antennas Propag., ISAP'96*, Chiba, Japan, 1996, 1201.
44. Huang, J. Microstrip Reflect Array and Its Applications, *Proc. 1996 Int. Symp. Antennas Propag., ISAP'96*, Chiba, Japan, 1996, 1177.
45. Optus Communications Technical Documents, *Mobilesat™ System Description*, copyright 1992.
46. Pozar, D. M. A Review of Aperture Coupled Microstrip Patch Antennas: History, Operation, Development, and Applications, http:\\www.rfglobnet.com., pp.1–12.
47. Karmakar, N. C. and Bialkowski, M. E. Circularly Polarised Aperture Coupled Circular Microstrip Patch Antennas for L-band Applications, *IEEE Trans. Antennas Propag.*, 47, 933, 1999.

48. Karmakar, N. C. and Bialkowski, M. E. A Compact Switched-Beam Array Antenna for Mobile Satellite Communications, *Microwave Optical Technol. Lett.*, 21, 186, 1999.
49. *NEC Satellite Telephone S1 User Manual*, Document No. 53-21476-3005, copyright 1994, NEC Australia.
50. Bialkowski, M. E. and Karmakar, N. C. A Two-ring Circular Phased Array for Mobile Satellite Communications, *IEEE Antennas Propag. Mag.*, 41, 14, 1999.
51. Karmakar, N. C. and Bialkowski, M. E. RF Switches and Phase Shifters for Beam Forming Network at L-Band, in *Antennas and Associated Components for Mobile Satellite Communications*, Bialkowski, M. E., Editor, Research Signpost, India, 1998, pp. 197–208.
52. Karmakar, N. C. and Bialkowski, M. E. An L-band 90° Phase Shifter Using UHF-band PIN Diodes, *Microwave Optical Technol. Lett.*, 21, 51, 1999.
53. Bialkowski, M.E. and Karmakar, N.C. Design of Compact L-Band 180° Phase Shifters, *Microwave and Optical Technology Letters*, John Wiley & Sons, New York, 1999.
54. Varnes, R. S., Ball, J. A. R., and Bialkowski, M. E. A Tracking System for a Vehicle Mounted Mobile Satellite Telephone Antenna, in *Antennas and Associated Systems for Mobile Satellite Communications*, Bialkowski, M. E., Editor, Research Signpost, India, 1998, pp. 185–196.
55. Bialkowski, M. E., Jellett, S. T., and Varnes, R. S. Electronically Steered Antenna System for the Australian Mobilesat, *IEE Proc. Microwave Antennas Propag.*, 143, 347, 1996.
56. Ohmori, S., Yamamoto, S., and Tanaka, K. Phased Array Tracking Antenna System for Vehicles, *Proc. IREECON*, Sydney, Australia, 1991, 17.
57. Schmidt F. and Cox, B. Electronic Steered Phased Array Antenna for MSAT, *MSN*, 18, 41, 1988.
58. Schaffner, G. Low Cost Cartop Phased Array Steering, *Proc. IEEE MTT-S*, 1987, 949.
59. Nishikawa, K. et al. Phased Array Antenna for Land Vehicle Satellite Communications, *Denshi Tokyo*, 29, 87, 1990.
60. Ansbro, P., Finotto, D., and Gianola, P. A Phased Array Antenna System for Mobile Communications in L-band, *Proc. 1995 Asia Pacific Microwave Conference (APAC'95)*, Taejon, Korea, 1995, 944.
61. Craggs, J. S. et al. The Commercial Development of a Vehicle-Mounted Antenna for the L-band Land Mobile Satellite Services, *Proc. 9th National Space Engineering Symposium*, Sydney, Australia, 1994, 81.
62. Petosa, A., Strickland, P., and Wight, J. S. Microstrip Antenna Array for Mobile Satellite Communications, *Proc. IEEE APS Int. Symp.*, 1, 1992, 503.
63. Alonso, J. I. et al. Low Cost Electronically Steered Antenna and Receiver System for Mobile Satellite Communications, *Trans. IEEE MTT*, 44, 2438, 1996.
64. Miura, R. et al. A Land Mobile Satellite Experiment on a DBF Self-beam Steering Array Antenna, *Proc. Int. Symp. Antennas Propag., ISAP '96*, Chiba, Japan, 1996, 1189.
65. Chiba, I., Adaptive and Digital Beam Forming Antennas for Mobile Communications, *Proc. Asia-Pacific Microwave Conf.*, Tokyo, Japan, 1994, 831.
66. Shafai, L., MSAT Vehicular Antennas with Self Scanning Array Elements, *Proc. 2nd Int. Mobile Satellite Conf.*, Ottawa, Canada, 1990, 523.
67. Fujimoto, K. et al. *Handbook of Microstrip Antennas*, James, J. R. and Hall P. S., Editors, Peter Peregrinus, London, U.K., 1989.
68. Jellett, S. T. and Bialkowski, M. E. A Low Cost Briefcase Antenna for Accessing the Optus Mobilesat Service, *Proc. Workshop Appl. Radio Science, WARS'93*, Canberra, Australia, 1995, Session 2.
69. Karmakar, N. C. and Bialkowski, M. E. A 2×2-Element Sequentially Rotated Patch Antenna Array for a Mobilesat Briefcase Antenna System, *Microwave and Optical Technology Letters*, John Wiley & Sons, New York, 13, 7, 1996.
70. Karmakar, N. C. and Bialkowski, M. E. Low Cost Briefcase Satphone Terminal Antennas, *Microwave J.*, August 1999.

15

Shaped-Beam Antennas

Carlos A. Fernandes

Instituto Superior Técnico/Instituto de Telecomunicações

15.1 Introduction

Antennas have a decisive influence on the performance of wireless communication systems: depending on radiation characteristics, antennas differently affect cell coverage, multipath discrimination, and co-channel interference, which directly or indirectly impact on system performance.

Antenna beam shaping is a simple strategy to act on these parameters in a controlled way. No added complexity results to the system in the case of fixed shaped-beam antennas; on the contrary, with appropriate beam shaping, propagation channel impairments can be mitigated to the point that equalizer complexity can be significantly reduced. Beam agility can also be added to a shaped beam to further enhance system performance. The beam-shaping concept has not been explored extensively in present-day ultrahigh frequency (UHF) and microwave wireless systems, but beam shaping is surely a compelling strategy to enable emerging wireless broadband communications at millimeter waves, when moderate to high-terminal mobility is required.

One such system, the cellular Mobile Broadband System (MBS),[1,2] was developed in the framework of Research into Advanced Communications for Europe (RACE II) and Advanced Communications Technologies and Services (ACTS) Research and Development (R&D) programs. MBS is intended to extend to mobile users, the capacity available in fixed Broadband Integrated Services Digital Network (B-ISDN). It aims at data rates up to 155 Mbps over the air interface, with mobile velocities up to 100 km/h, with the quality of service of the fixed network. Provisional subbands were allocated by the European Radio-Communications Office for these systems at 39.5 to 40.5, 42.5 to 43.5 GHz and 62 to 63, 65 to 66 GHz. A fully functional ATM-based trial platform was built in the framework of ACTS-System for Advanced Mobile Broadband Applications (SAMBA) project for the 40-GHz band, offering full-duplex user data rates up to 27 Mbps over the air interface in complex environments, with full mobility.[3,4] Shaped-beam lens antennas were used both at the base station and the mobile terminals, to enhance system performance.[5]

Beam shaping can be also quite effective in broadband wireless local area networks (B-WLAN), contributing to significantly increase transmission capacity when compared with conventional antennas.[6]

)-8493-0124-6/02/$0.00+$1.50

Other uses for shaped-beam antennas are found in millimeter-wave telematic applications,[7] and also in millimeter-wave fixed services such as Local Multipoint Distribution Systems (LMDS).[8]

When moving into millimeter waves, it may be not just enough to scale down antennas designed for low-frequency wireless communication systems. Not only physical difficulties and limitations may occur because of the reduced wavelength, but also new antenna requirements must be considered to deal with more harsh propagation channel behavior. The fixed terminal antenna, typically located near the ceiling of a room or on a lamppost or hoist in outdoor environments (Fig. 15.1), is required to evenly distribute over the coverage area the limited power that is available from solid-state devices at millimeter waves. Fading depth may be very high at this band, and signal time dispersion is more critical then for lower bit rate systems. Antennas may also be required to discriminate reflection paths from walls at the edges of the cell, and desirably also from the ground. Shaping the beam does not necessarily entail mobility restriction even if the shaped beam on the base station is fixed, as will be seen.

More sophisticated solutions like adaptive antennas may be considered with the same objective, but these antennas tend to be less affordable at millimeter waves, and more complex. Adaptive antennas are addressed later in this book.

Two main approaches may be used for beam shaping: antenna arrays or shaped-aperture antennas. Printed arrays are compact and very flexible for beam shaping.[9,10] However, losses in the feeding circuit can be excessive at millimeter waves and must be compensated with an active circuit. Metal antennas like biconical horns, monopoles with shaped ground plane, single- and dual-reflector antennas or dielectric lenses can also be designed to produce shaped beams.[11-14] Unlike printed arrays, aperture antennas are limited to microwaves and millimeter waves because of the structure bulk at lower frequencies.

Figure 15.2 shows the basic configuration of some shaped-beam aperture antennas for wireless communications. Some of these solutions are shortly addressed in the present chapter, but the focus is on shaped dielectric lens antennas.[14] Lenses are quite an effective solution to produce different types of highly shaped radiation patterns, even allowing for the conciliation of beam shaping with beam agility, using reasonably compact assemblies. Dielectric lenses may significantly enhance system performance,[5,6] yet they are inexpensive and adequate for mass production, because the required fabrication tolerance is compatible with molding techniques. On the drawback side, internal reflections may degrade lens performance in some cases, and the required lens size may be unattractive for some applications. This chapter presents design guidelines for these antennas along with the discussion of some practical aspects, focusing on mobile applications.

15.2 General Antenna Requirements

A few arguments that are specific to wireless communications can influence the choice of antenna configuration and its design parameters. This choice becomes especially challenging for millimeter-wave systems, where some of the antenna requirements appear to be conflicting.

15.2.1 Beam Shape

The most simple beam-shaping requirements for wireless communications are found in lower frequency applications, related with the need for sectored coverage of base station antennas, or with the requirement for low radiation of the mobile terminal antenna in the direction of the user head.[15] Beam shaping in the azimuth plane seldom requires pattern synthesis.

Shaping in the elevation plane of a base station antenna is usually related with constant flux illumination requirement within the coverage region. It corresponds to the secant $(\sec)^2\,\theta$ radiation pattern, which ideally compensates the free-space loss associated with the path length d at any direction (θ, φ) within the cell (see Fig. 15.1). In the case of omnidirectional patterns for mobile terminals, some shaping in the elevation plane may be useful at millimeter waves to restrict the angular interval of radiation, and thus favor gain and reduce multipath pickup.

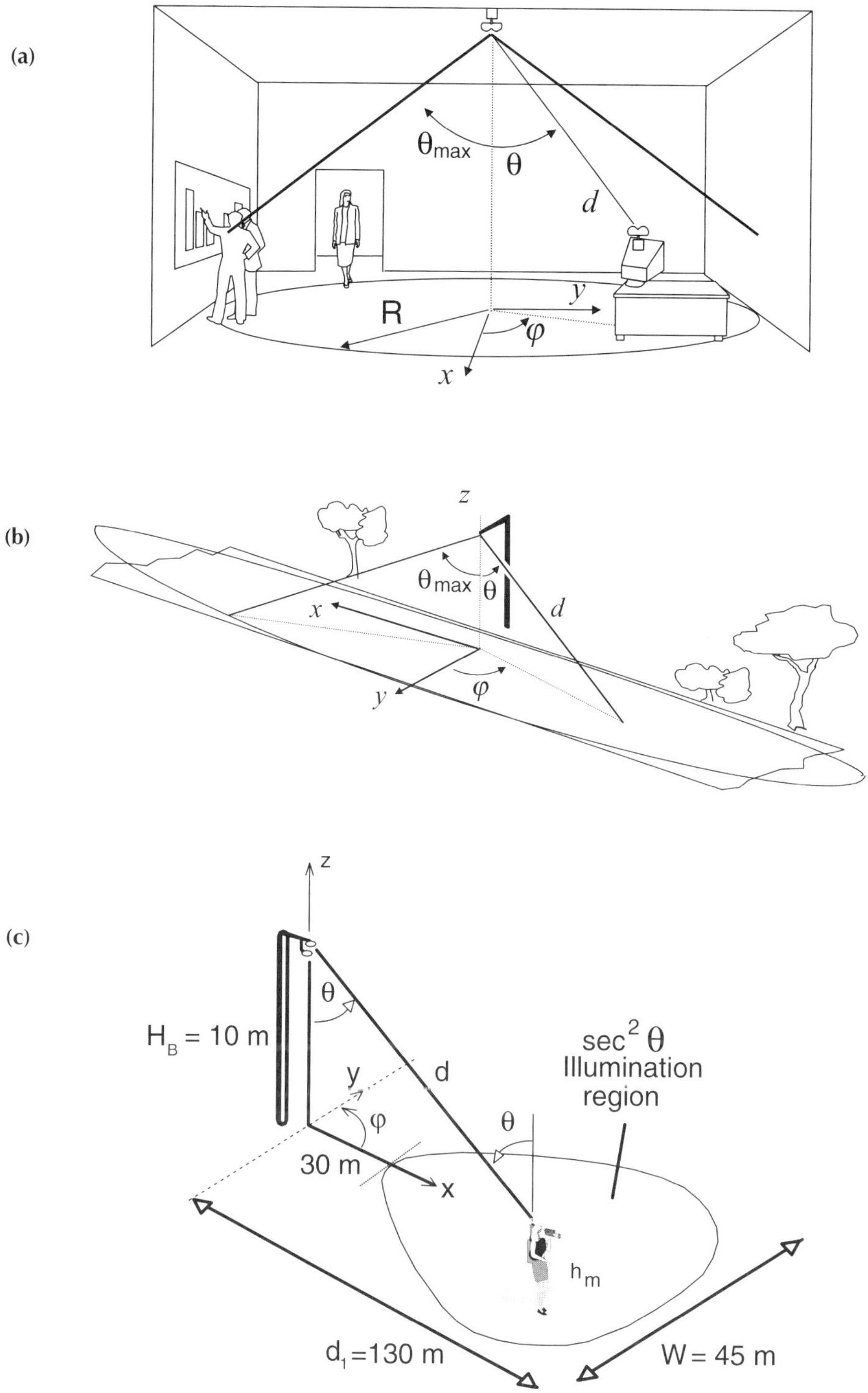

FIGURE 15.1 Scenarios for wireless millimeter-wave communications, and problem geometry. (a) Indoor circular cell; (b) elongated cell; (c) sectored cell. (From Fernandes, C.A., *IEEE Antennas Propag. Mag.*, 41, 141, 1999. With permission.)

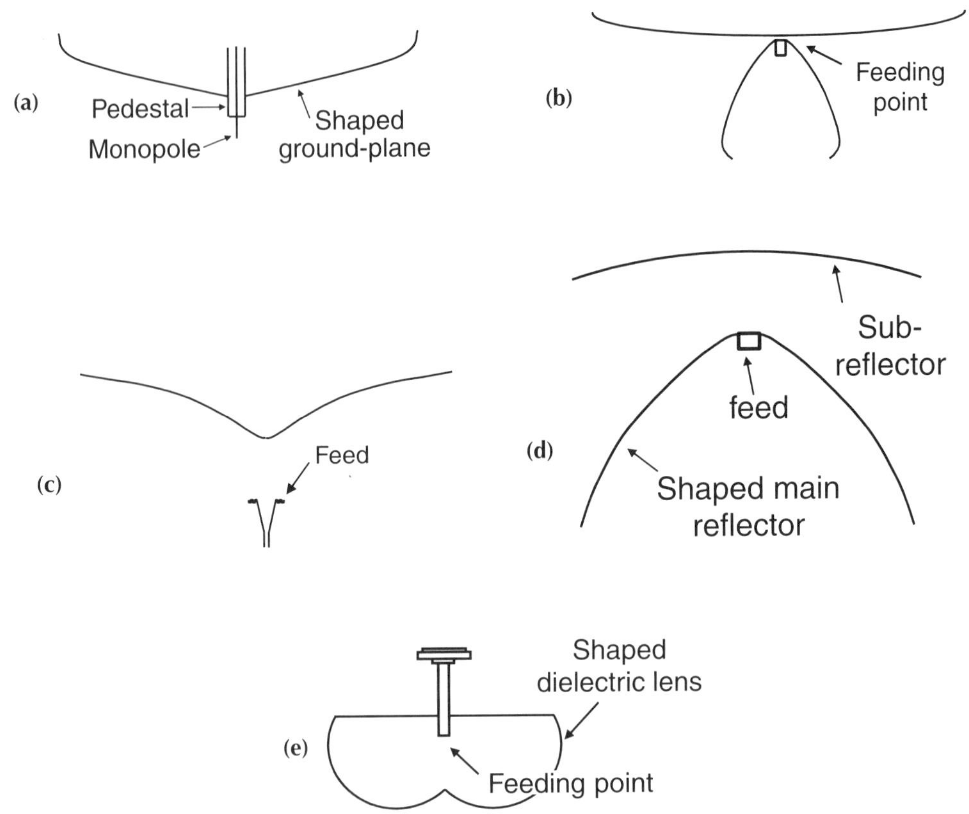

FIGURE 15.2 Different configurations of shaped-beam aperture antennas for wireless millimeter wave applications. (a) Monopole with shaped ground plane; (b) shaped biconical horn; (c) single-reflector antenna; (d) double-reflector antenna; (e) shaped dielectric lens antenna.

The following interrelated aspects must be considered when selecting the adequate antenna:

- To reduce line of sight (LOS) blockage probability, base station antennas should be located at the highest possible position, and preferably at the center of the cell. In constant flux applications this usually results in a very wide angle coverage requirement, with maximum elevation direction θ_{max} being well off-axis (typically $60° < \theta_{max} < 85°$). In some cases θ_{max} changes with azimuth φ according to the scenario (as in Fig. 15.1b), imposing a shaping condition both in the azimuth and elevation planes.[16]
- The $\sec^2 \theta$ characteristic may be provided by the base station antenna alone, or by the combined radiation patterns of the base station and mobile terminal antennas. The first case applies when a hemispheric radiation pattern is adopted at the mobile terminal to favor mobility. The second case is more appropriate for quasi-static terminals as in WLANs or for terminals with restricted movement like trains. Examples are presented later for both cases.
- Besides uniform coverage, it is most desirable that the base station antenna is able to confine the radiation to the cell limits as a goal to conserve power and to minimize co-channel interference with neighboring cells. Also the illumination of the walls at the edge of the cell should be kept to a controlled level to ensure an adequate compromise between undesirable multipath effects, and the need for alternative paths in the case of LOS blockage.
- The combined gain of base station and mobile terminal antennas must be enough to ensure appropriate received power level up to the edge of the cell. Typical values for the combined gain may range from 10 to 20 dBi. Cell size for broadband millimeter-wave systems ranges from a few

meters in indoor environments up to a few hundred meters outdoor. The range can be extended up to 1 or 2 km using highly directive antennas, for fixed point to multipoint links. A compromise must be found between antenna gain requirement for the link power budget, and terminal mobility.

Consider the scenario of Fig. 15.1a, where the base station antenna, located near the ceiling at height h_1, is required to produce a constant flux illumination within a circular region of radius R below the antenna, at height h_2. This implies that the elevation pattern $G(\theta)$ of the base station antenna ideally compensates the free-space attenuation associated with the distance d between fixed and mobile terminal antennas:

$$G(\theta) = G(\theta_{max}) \frac{\sec^2\theta}{\sec^2\theta_{max}} = G(\theta_{max}) \frac{d^2}{(h_1 - h_2)^2 + R^2}, \quad \theta \le \theta_{max}$$

$$G(\theta) < G_0, \quad \theta > \theta_{max} \tag{15.1}$$

Function $G(\theta)$ is minimum at $\theta = 0$, and increases up to its maximum value at the edge of the cell, for $\theta = \theta_{max}$. $G(\theta)$ is required to be below a certain level G_0 for $\theta > \theta_{max}$ to prevent significant illumination beyond R. Figure 15.3a shows an example of this ideal target pattern with $\theta_{max} = 76°$.

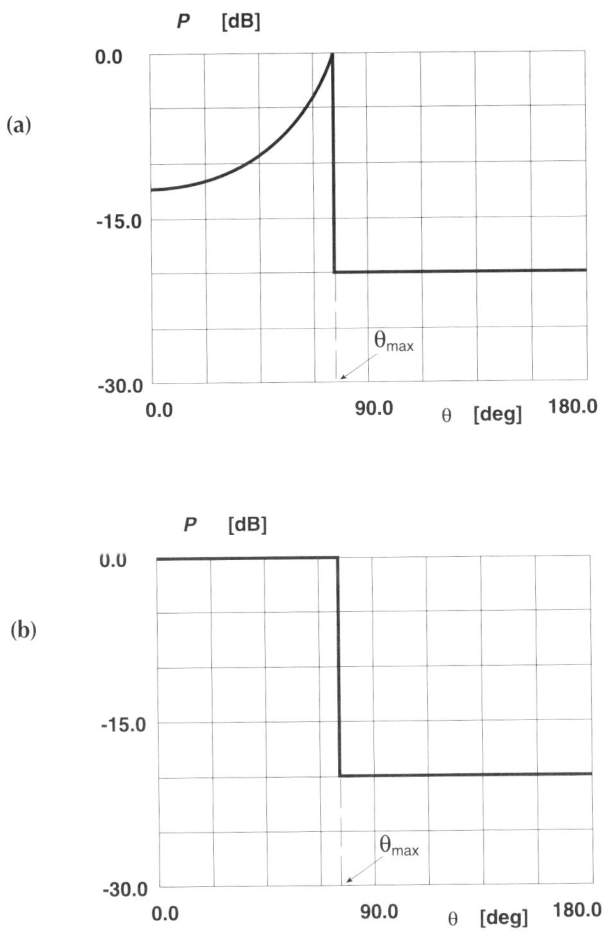

FIGURE 15.3 Target omnidirectional power patterns with shaping in the elevation plane, $\theta_{max} = 76°$. (a) Ideal $\sec^2\theta$ base station antenna pattern; (b) ideal mobile terminal flat-top pattern.

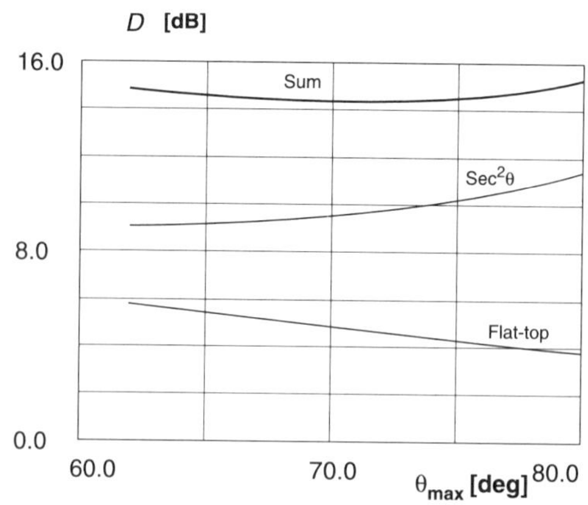

FIGURE 15.4 Maximum directivity of ideal $\sec^2\theta$ and flat-top patterns vs. θ_{max}.

It is interesting to note that once the antenna is constructed and θ_{max} is fixed, the radius of the covered region R can still be changed simply by modifying the antenna height $(h_1 - h_2)$:

$$R = \left(h_1 - h_2\right)\tan\theta_{max} \tag{15.2}$$

This is very convenient not only because a single antenna design can be used for different room sizes, but also because it allows for on-site adjustment of R and, consequently, simple control of the illumination of the walls.

At the mobile side the antenna radiation pattern must be constant for the upper hemisphere to maintain the received power level independent of the mobile terminal location and tilt angle within the cell. Although this is appropriate for terminal mobility, the trade-off is reduced gain, and increased potential for multipath pickup. A slight improvement can be obtained by restricting the mobile antenna radiation in the elevation plane to $\theta < \theta_{max}$, where θ_{max} may have the same value as the base station antenna (Fig. 15.3b).

Figure 15.4 shows the maximum possible value of $G(\theta_{max})$ in Eq. (15.1) vs. θ_{max}, calculated as the directivity of an ideal omnidirectional pattern with $\sec^2\theta$ characteristic in elevation up to $\theta = \theta_{max}$. The result of a similar calculation for a flat-top pattern with the same θ_{max} is superimposed. Figure 15.4 also shows the combined directivity of the two antennas, representing the maximum possible contribution of such base station and mobile terminal antennas for the link power budget as a function of θ_{max}. For increasing values of θ_{max}, the beams of the two antennas broaden, decreasing the directivity until a point where the sharp characteristic of the $\sec^2\theta$ near θ_{max} starts to dominate. Thereafter, the directivity increases.

The combined gain of real antennas is of course lower than the ideal directivity values presented in Fig. 15.4, depending on antenna losses and radiation pattern degradation resulting from diffraction effects. These effects tend to aggravate with increasing values of θ_{max}. If higher gain is required for the link power budget, further restriction of the antenna beam width must be adopted. This subject is detailed in the examples given Section 15.3.5.

15.2.2 Polarization

Unlike most cases at lower frequencies, at millimeter waves the link is expected to rely more on the LOS path. In this case, polarization-related aspects can cause coverage and terminal mobility restrictions, so this topic must be carefully considered when selecting the antenna configuration.

Suppose that linear polarization is chosen for the example addressed in the previous section, and E_θ is the dominant E-field component, with pattern circular symmetry. For increasing distances from the base station antenna, field direction tends to vertical. However, below the antenna the E-field direction becomes horizontal, and its contributions for all φ angles mutually cancel at $\theta = 0$ (antenna nadir) creating an underilluminated zone, which cannot be effectively filled with antenna shaping. The same type of E_θ polarization must be used at the mobile terminal as the most effective way to maintain polarization match for all positions and orientations, but this choice aggravates the blank region problem.

A linearly polarized feed with its E-plane coincident with the $x = 0$ plane can be used as an alternative. When using the same polarization at the mobile terminal, constant received power can be ensured for all positions within the cell including the region bellow the base station antenna. However, rotation of the mobile terminal in the horizontal plane produces polarization mismatch that can interrupt the link despite appropriate power density ensured everywhere within the cell. This solution was adopted for the coverage of a very elongated cell, with the base station located at half the cell length[16] (see Fig. 15.1b).

Circular polarization provides a better solution to solve the blank region problem, because it does not restrict mobility. Furthermore, circular polarization acts as a multipath filter, by reducing the first-order reflections. The challenge, however, is to obtain an antenna that is capable of maintaining a reasonable circular polarization purity within the wide angular region $\theta < \theta_{max}$ along with the $\sec^2 \theta$ characteristic. A reasonable solution can be found for circular symmetrical cells both for the base station and mobile terminal, as will be seen in Section 15.3.5, but this may not be possible for elongated cells. Polarization aspects are much easier to handle for quasi-static terminals, where antenna orientation can be manually adjusted for optimum transmission.

15.2.3 Bandwidth

A further challenge to the antenna designer is that the previously discussed characteristics and also antenna input impedance must remain acceptable within a prescribed bandwidth. If the antennas are required to accommodate both the transmitter and receiver frequency bands, this corresponds typically to a 10% bandwidth.

It is noted, however, that specifications for pattern degradation or cross-polarization for wireless mobile communication systems are less stringent than for many other services like space applications.

15.2.4 Other Requirements and Constraints

Some aspects of different nature must also be considered when selecting an antenna for wireless-millimeter wave systems:

- At millimeter waves, antenna losses can be quite high, adversely affecting noise figure and consequently receiver sensitivity. In some cases losses can be minimized by adequate design and by appropriate choice of materials, but in other cases it may be also necessary to incorporate active circuits to compensate for losses. The use of active components at millimeter waves may increase the antenna cost to some extent.
- Space diversity is an essential strategy to mitigate the multipath effects for millimeter-wave wireless systems, both at the mobile and at base station. The challenge at the mobile terminal is to find an antenna configuration that not only is compatible with the previous requirements but also allows positioning of the diversity antenna elements without significant mutual blocking for all observation directions. This may also restrict the choice of the feeding circuit. A coaxial arrangement is analyzed in Section 15.3.5.
- Especially at the mobile terminal, antenna compactness is mostly desirable. Unfortunately, aperture antennas designed for the previous beam-shaping specifications tend to be bulky and large, even at millimeter waves. Array antennas are more effective in this particular point.
- Antennas must be rugged, based on affordable technology, with viable manufacturing tolerances, and little tuning requirements in large-scale production.

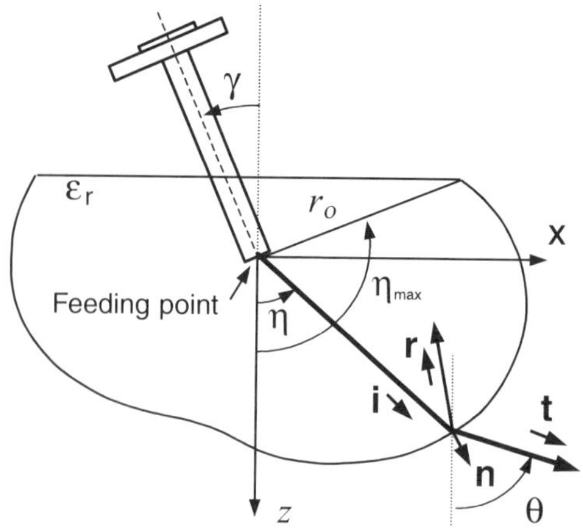

FIGURE 15.5 Geometry of shaped dielectric lens antenna.

15.3 Shaped Dielectric Lens Antennas

There is no best antenna solution to cope with all the previous requirements. The choice between the available solutions must be made according to each service specific needs. Shaped dielectric lenses are one solution that is attractive for its simplicity, flexibility, and effectiveness to comply with different specifications for wireless millimeter wave systems.

15.3.1 Lens Antenna Configuration

Figure 15.5 shows a dielectric lens antenna configuration with potential for wireless applications. Generally, it is a homogeneous dielectric body with an arbitrary three-dimensional shaped surface, which is designed to transform the feed radiation pattern into the desired output pattern. Possible feeds may be the aperture of rectangular or circular waveguides, coaxial feeds, or printed elements, either with linear or circular polarization.

Unlike conventional lens configurations, the feed is embedded in the lens material. This has important advantages over the usual configuration where the feed is located outside the lens body as follows:

- Broad-beam feeds can be used to illuminate the widest possible area of the lens surface with less spillover, favoring lens radiation for wide θ angles using moderate dielectric permittivity values.
- It is easier to promote an effective power transfer from the embedded feed into the lens. Details on this subject are given in Section 15.3.4.
- The total antenna depth is smaller when compared with alternative configurations where the feed is located at some distance away from the lens.

In this configuration only one of the lens surfaces intercepts the main beam of the feed, to shape it into a prescribed target radiation pattern: this is a single surface lens. It means that only one degree of freedom is available for pattern synthesis. Because amplitude shaping of the output beam is the only requirement for most of the wireless millimeter-wave applications addressed in this chapter, a single degree of freedom is enough for the lens design.

Although the dielectric lens may have an arbitrary three-dimensional shape, this chapter addresses only axis-symmetrical cases. Still, various useful forms of asymmetrical cell shapes can by obtained simply by appropriate off-axis positioning of the feed, inside the axial-symmetrical lens.

15.3.2 Design Formulation for Axial-Symmetrical Lenses

The design objective is to obtain the lens profile $r(\eta)$ that transforms the known radiation pattern $U(\eta)$ of the feed inside the dielectric into a prescribed target pattern $G(\theta)$. For now it will be assumed that both the feed radiation pattern and the target pattern have axial symmetry, so the feed tilt γ is set to zero (Fig. 15.5). The formulation for the general case can be found in Salema.[17]

The design procedure comprises two steps. In the first one, the lens profile is obtained from the geometric optics formulation, therefore assuming that lens dimensions are very large in terms of wavelength. Physical optics is used in the second step to obtain the corresponding output radiation pattern, taking diffraction effects into account. This pattern may be slightly different from the geometric optics prediction depending on the lens size. A slight lens profile correction may be introduced to bring the output radiation pattern closer to the target.

Consider an elementary ray tube that originates at the feed phase center, and is directed along **i** (angle η). The intersection of this ray tube with the lens surface $r(\eta)$ defines an elementary surface dS, with unit normal **n**. The power associated with the incident wave in this ray tube is reflected and transmitted at the interface. The associated directions are **r** and **t**, respectively. Power conservation along the elementary ray tube can be expressed as

$$U(\eta)T\sin\eta d\eta = K\,G(\theta)\sin\theta d\theta \tag{15.3}$$

Function T is the transmissivity, which corresponds to the ratio of the power P_t transmitted across dS to the incident power P_i.

$$T = \frac{U_{\|}\left|t_{\|}^{2}\right| + U_{\perp}\left|t_{\perp}^{2}\right|}{U}\frac{1}{\sqrt{\varepsilon_r}}\frac{\cos\beta}{\cos\alpha} \tag{15.4}$$

$$\cos\alpha = \mathbf{i}\cdot\mathbf{n} \tag{15.5}$$

$$\cos\beta = \mathbf{t}\cdot\mathbf{n} \tag{15.6}$$

where $t_{//}$ and $t_{\perp}$ represent the Fresnel transmission coefficients for parallel and perpendicular polarization, respectively. K is a normalization constant to be determined from the balance between total power inside the lens and total power outside the lens as follows:

$$K = \frac{\int_0^{\eta_{max}} U(\eta)T\sin\eta d\eta}{\int_0^{\theta_{max}} G(\theta)\sin\theta d\theta} \tag{15.7}$$

Once the transformation $\theta(\eta)$ relating **t** and **i** is found, application of Snell law for refraction then yields the surface slopes that agree with the pairs (**i**, **t**):

$$\left[\sqrt{\varepsilon_r}\,\mathbf{i} - \mathbf{t}\right]\times\mathbf{n} = 0 \tag{15.8}$$

where ε_r is the relative permittivity of the lens material, and **n** is the surface normal at the incidence point.

Equations (15.3) and (15.8) can be expanded and solved as a system of equations to provide the solution for $\theta(\eta)$ and $r(\eta)$:

$$\frac{d\theta}{d\eta} = \frac{T}{K}\frac{U(\eta)}{G(\theta)}\frac{\sin\eta}{\sin\theta} \tag{15.9}$$

$$\frac{dr}{d\eta} = \frac{r\sin(\theta - \eta)}{\sqrt{\varepsilon_r} - \cos(\theta - \eta)} \tag{15.10}$$

By starting from an initial solution ($\theta(\eta_{max})$, $r(\eta_{max})$) Eqs. (15.9) and (15.10) are integrated using second-order Runge-Kuta formulas. The choice of η_{max}, $\theta(\eta_{max})$, and $r(\eta_{max})$ must take into consideration the primary feed pattern, the required antenna coverage, and the antenna dimensions to comply with the geometric optics approximation, but it is otherwise arbitrary. The ε_r-value is also arbitrary, but a compromise must be found between moderately high values to favor high-refraction angles, and lower values to reduce internal reflections.

Function T in Eqs. (15.7) and (15.9) depends indirectly on the unknown function $r(\eta)$. An iterative process is adopted, considering a constant value for T in the first step. The obtained solution $r(\eta)$ is then used to calculate T for the next evaluation of Eqs. (15.9) and (15.10), and the process is repeated until convergence; few iterations are usually required.

Because diffraction effects are not accounted for in the geometric optics formulation, usually some discrepancies may be expected between the desired pattern $G(\theta)$ and the actual pattern yielded by the lens, which reflect its finite size. Some additional correction of the calculated lens shape may be required.[18] The actual pattern produced by the lens can be predicted using the physical optics formulation:

$$\mathbf{E}(P) = j\frac{\exp(-jkR)}{2\lambda R}\int_S \left[Z(\mathbf{n}\times\mathbf{H}(P'))\times\mathbf{R}_1 + (\mathbf{n}\times\mathbf{E}(P'))\right]\times\mathbf{R}_1 \exp(jk\rho\cdot\mathbf{R}_1)dS \tag{15.11}$$

The integral in Eq. (15.11) is taken over the lens surface. $\mathbf{E}(P')$ and $\mathbf{H}(P')$ represent the field produced by the feed over the lens surface. $\mathbf{R}_1$ is a vector directed from the origin toward the observation point P, and ρ is a vector directed from the origin toward point P' on the lens surface.

Results obtained from Eq. (15.11) compare well with measured results, so this is used not only to predict lens performance prior to fabrication but also to perform parametric studies. A dedicated software package based on the formulation of this section was developed to support lens design and analysis.

15.3.3 Dielectric Materials

The following sections show that adequate ε_r-values for the present application typically range from 2 to 3. This represents a compromise between the ability to obtain wide-angle lens refraction and the requirement for the lowest internal reflection losses. Because of large lens dimensions in terms of wavelength, dissipation losses are also an important aspect to mind. Total dissipation loss depends on lens dimensions, shape, and illumination function. Loose bounds for dissipation losses can be calculated for minimum and maximum radius r using

$$L = 27.3\sqrt{\varepsilon_r}\,\frac{r}{\lambda}\tan\delta \quad [\text{dB}] \tag{15.12}$$

Considering that typical lens radius that copes with the geometric optics approximation range from 10 to 30 λ, a value of $\tan\delta = 10^{-3}$ corresponds to dissipation losses ranging from about 0.4 to 1.3 dB.

It is not usual to find in manufacturer data sheets permittivity and loss tangent values for the 40 and 60 GHz bands. Although the values do not change much in the microwave and millimeter wave bands, extrapolation may not yield enough accuracy. For a good lens design it is important to know ε_r-values typically within 1 to 3% accuracy, depending on specific design constrains. For low-loss materials, loss tangent does not significantly affect lens shape so the required $\tan\delta$ accuracy is dictated by the user need to predict dissipation losses.

TABLE 15.1 Electric Characteristics of Materials Used for Lens Prototypes

Material	Permittivity	Loss Tangent	Frequency
PTFE	2.1	<0.0001	60-66 GHz
Polypropylene	2.20	<0.005	50–60 GHz
Polyethylene	2.35	<0.0005	50–60 GHz
Polystyrene	2.43	<0.001	50–60 GHz
Acrylic glass	2.53	<0.012	60–66 GHz

Several methods are described in the literature for the determination of the complex permittivity of dielectric samples.[19] Table 15.1 lists the results obtained for the materials used in the prototypes described in this chapter. A graphic procedure based on a waveguide method was used, which proved to be convenient and accurate for millimeter waves.[20]

Poly(tetrafluoroethylene) PTFE has the lowest loss tangent value in Table 15.1, but its mechanical rigidity is not adequate for precision milling. All the other materials on the list have good mechanical characteristics. Despite high losses, acrylic glass was used in early laboratory prototypes for lens design software validation. Like polystyrene it has good properties for milling, and appropriate glue can be used to fix struts at the back of the lens. Polyethylene has very low losses and it can be used for milling, but it requires proper sharp tools and adequate spinning speed.

15.3.4 Primary Source Radiation

The success of the lens design to obtain a prescribed radiation pattern depends partly on the adequate choice of the feed. Besides the feed direct implication on lens output pattern and polarization characteristic, the feed beam shape may affect the lens depth, the amount of internal reflection, and the quality of the approximation to the target pattern for a given lens size. Some optimization of the feed may be useful to enhance these parameters.

Feed characteristics that are relevant for the lens design are

- Pattern shape (beam width, circular symmetry)
- Polarization
- Phase center location with respect to feed aperture
- Matching at the lens input port

Three types of primary sources are briefly addressed in this chapter: waveguide feed, coaxial feed, and printed element feed. Each one has its own advantages and drawbacks, which must be carefully weighted to obtain the best solution for each case. Note that for the present antenna configuration the preceding characteristics must be known when the feed is immersed in the lens material.

Three simplifying assumptions may be introduced in the analysis of lens embedded feeds, which were confirmed by experimental evidence:

- For the typical lens size, the radiating surface may be considered to be in the far-field region of the feed.
- Feed radiation pattern inside the lens is not strongly dependent on the lens shape, and it does not change significantly even if the dielectric medium is unbounded.
- Unlike elliptical or hemispheric lenses, scarce rays internally reflected at the lens-shaped surface travel back to the feed aperture. Thus, for the lenses addressed in this chapter, the impedance seen by the embedded feed does not depend strongly on the lens shape.

Provided that only one dielectric is involved, the last assumption enables accounting for the dielectric simply by scaling the feed geometry by the $\sqrt{\varepsilon_r}$ factor. This may happen with waveguide feeds. The scaling strategy makes the available results for waveguide feeds in vacuum applicable to the dielectric embedded case.

Printed feeds involve at least two dielectrics: the substrate and the lens material. Unless both materials have the same characteristics, scaling is not possible in this case. However, it is still easier to analyze this feed when immersed in an unbounded medium, instead of being embedded in the shaped lens. Most of the feed simulation results presented in this chapter were obtained with the professional version of the electromagnetic solver WIPL-D,[21] which can arbitrarily analyze three-dimensional structures formed by wires, plates, and dielectrics. WIPL-D enables the arbitrary setting of the dielectric properties of the unbounded medium.

Measurement of the feed radiation pattern inside a solid dielectric material cannot be done directly. However if the feed is placed at the center of a hemispheric dielectric lens of a few λ radius, the tangential field components can be measured outside the dielectric, close to its surface. Noting that the tangential fields components are continuous across the dielectric-air interface, and that the surface is everywhere normal to the propagation direction of the feed originated wave, the measured tangential fields can be considered as a fair estimation of the feed radiation pattern inside the dielectric. Once the actual feed pattern inside the dielectric is obtained by indirect measurement, it may be entered into the numerical lens synthesis procedure as a lookup table.

15.3.4.1 Waveguide Feeds

Waveguide feeds are a natural solution when waveguide technology is used at the millimeter wave head. Waveguide feeds are simple and easy to integrate into the lens: in circular symmetrical structures the lens can be simply screwed into the waveguide end like a lamp bulb. The general drawback of waveguide structures is the bulk. An additional constrain in the case of circular waveguide is that it cannot be easily bent as required by the setup; a transition to rectangular waveguide must be introduced.

The radiation from the waveguide aperture into an unbounded media is a canonical problem that has received considerable attention.[22-28] These methods are well documented elsewhere, and are not addressed here. Instead, some experimental and simulated results are presented together with practical aspects that may be useful for lens design.

Circular Waveguide, TE_{11} Mode

TE_{11} is the fundamental mode of the circular waveguide, which propagates for $k_0a > 1.841/\sqrt{\varepsilon_r}$. Its excitation can be inherently wideband, with low reflection loss at the lens input-port. To ensure pattern circular symmetry the TE_{11} must be excited with circular polarization. Circular polarization is useful in the context of wireless communications for its inherent capability to filter out first-order reflections as previously mentioned.

Figure 15.6 shows an example of calculated and measured radiation patterns corresponding to waveguide radius a = 1.5 mm and wall thickness t = 0.35 mm, at 62.5 GHz. Near its open end the waveguide is filled with the same dielectric material as the surrounding medium, with permittivity ε_r = 2.53. In the simulation the surrounding dielectric medium was considered unbounded. A 6-λ radius hemispheric dielectric lens was used to obtain the experimental pattern, according to the previously described procedure. Note the good agreement between both results. Also note in this example the very good symmetry between E_θ and E_φ components for η up to 90°, a necessary condition for circular polarization.

Although narrower than for the same structure in air, the waveguide radiation pattern shows the typical shape, with its maximum at $\eta = 0$ and decreasing radiation for increasing η. This is exactly the opposite of the $\sec^2\theta$ characteristic required for the lens output pattern, with its minimum at $\theta = 0$ and maximum near θ = 90° (see Fig. 15.3a). In fact, this feed choice may be problematic when high values of θ_{max} are specified for the lens $\sec^2\theta$ target pattern. To obtain a good approximation using a reasonable lens size, the 3-dB beam width of the TE_{11} radiation pattern must be at least on the order of 60°.

Figure 15.7 shows the evolution of 3- and 10-dB beam widths vs. waveguide radius a, for wall thickness $t = \lambda_m/6$. The radius is normalized to the wavelength in the unbounded dielectric medium (λ_m). Decreasing

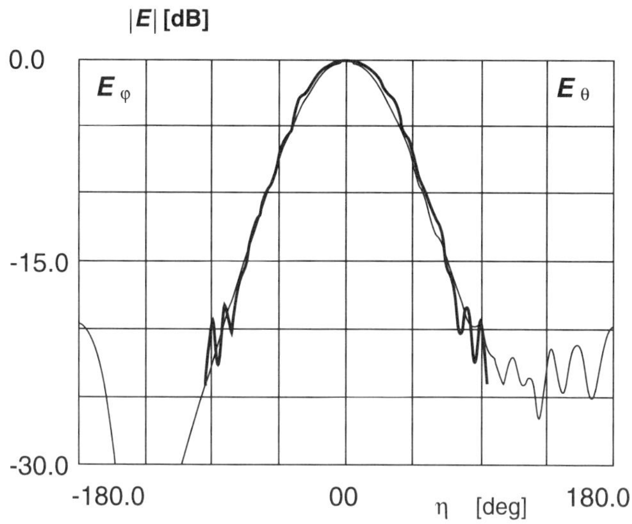

FIGURE 15.6 Radiation pattern of the TE_{11} mode of a circular waveguide immersed in unbounded dielectric medium with $\varepsilon_r = 2.53$; $f = 62.5$ GHz, $a = 1.5$ mm ($a = 0.5\ \lambda_m$), $t = 0.35$ mm. Thin line — calculated; thick line — measured.

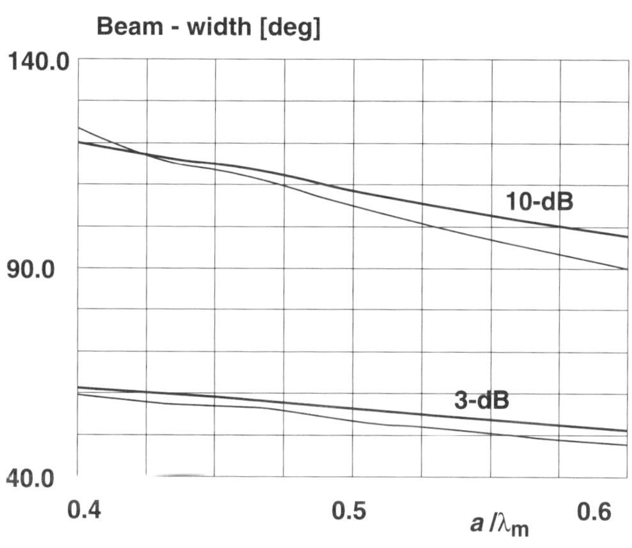

FIGURE 15.7 Beam width of TE_{11} mode radiation pattern vs. circular waveguide radius. The waveguide is immersed in unbounded dielectric medium; $t = \lambda_m/6$. Thin line — E-plane; thick line — H-plane.

the radius for larger beam width may correspond to increased spillover at the lens edge (at η_{max}), giving rise to ripple in the lens radiation pattern. However, considering that the edge of the lens can be arbitrarily set for any value of η_{max} larger than 90°, and that half 10-dB beam width is well below this value in Fig. 15.7, the spillover effect is weak for the cases of interest using this feed.

Simulations show that for the range of waveguide radius in Fig. 15.7, E-plane and H-plane beam widths are almost equal, which is useful to provide circular symmetrical patterns. This characteristic is not much sensitive to waveguide thickness in the range $\lambda_m/6 < t < \lambda_m/2$. Simulations also show that for the above range of a and for $\eta < 60°$, the distance from the E-plane or the H-plane phase centers to the waveguide aperture is less than 0.07 λ_m, whereas the distance between E-plane and H-plane phase centers is less than 0.1 λ_m.

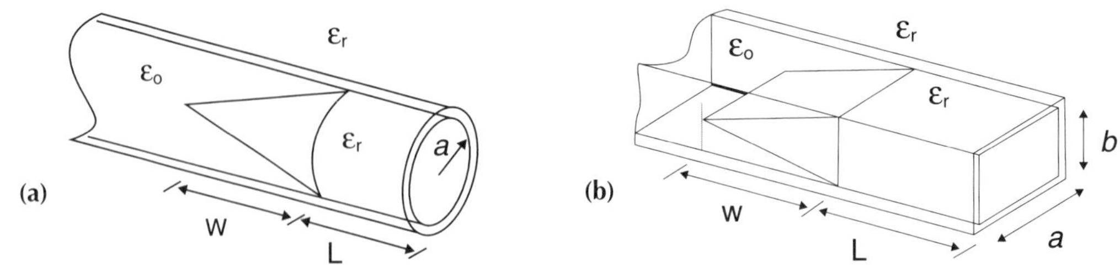

FIGURE 15.8 Waveguides with a matching dielectric load, immersed in unbounded dielectric medium. (a) Circular waveguide; (b) rectangular waveguide.

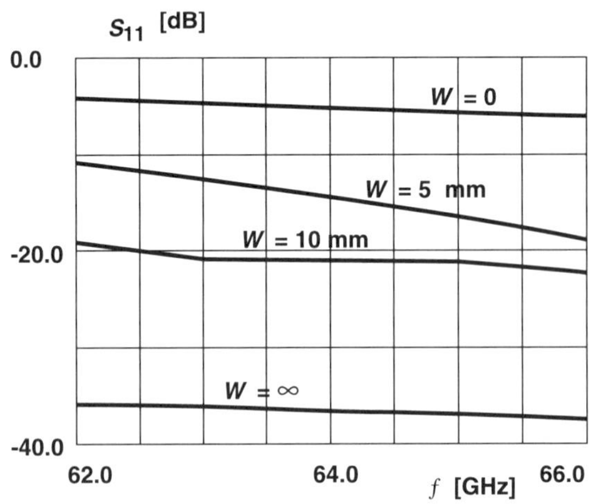

FIGURE 15.9 Effect of cone length W on TE_{11} return loss of a dielectric loaded circular waveguide, immersed in unbounded dielectric medium with $\varepsilon_r = 2.53$; $a = 1.5$ mm, $t = 0.5$ mm, $L = 5$ mm.

Another important point is to promote an adequate power transfer between waveguide and lens. A good match can be obtained by inserting a dielectric cone of the same material into the waveguide as shown in Fig. 15.8. The dependence of return loss with the cone length W is presented in Fig. 15.9, for a circular waveguide with radius $a = 1.5$ mm and wall thickness $t = 0.5$ mm, immersed in unbounded dielectric media. A good match is obtained for W slightly larger than one waveguide wavelength. For practical reasons the waveguide dielectric inset can not be machined as an integral part of the lens body; it is mounted in the waveguide as a separate piece, which comes into contact with the lens port when the waveguide is positioned inside the lens.

Circular Waveguide, TM_{01} Mode

The dominant E-field component of the TM_{01} mode aperture radiation pattern is the E_θ. This is useful as a lens feed to produce a circular symmetrical pattern with linear polarization. An example of the radiation pattern is shown in Fig. 15.10. It corresponds to a circular waveguide with radius $a = 2.1$ mm and wall thickness $t = 0.35$ mm, embedded in dielectric material with permittivity $\varepsilon_r = 2.53$. The feed radiation maximum is shifted to $\eta = 34°$, which favors lens radiation for large θ angles. However, intrinsic to linear polarization in circular symmetrical patterns, a null appears for $\eta = 0$ as discussed in Section 15.2.2. This null is also present in the lens radiation pattern, which may be a drawback in some cases.

Unlike the fundamental mode, the TM_{01} may be more difficult to excite with large bandwidth because lower order modes have to be filtered out. Figure 15.11 shows a basic setup proposed in Silver:[22] the circular waveguide is excited at the lateral wall by a rectangular waveguide carrying the TE_{10} mode. A tuning screw at the short-circuited end of the circular waveguide is adjusted to offer high impedance to

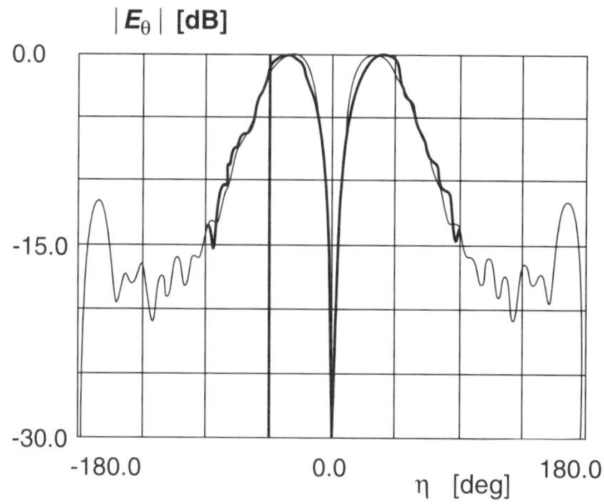

FIGURE 15.10 Radiation pattern of the TM_{01} mode of a circular waveguide immersed in unbounded dielectric medium with $\varepsilon_r = 2.53$; $a = 2.1$ mm, $t = 0.35$ mm, $f = 62.5$ GHz ($a = 0.7\ \lambda_m$). Thin line — calculated; thick line — measured.

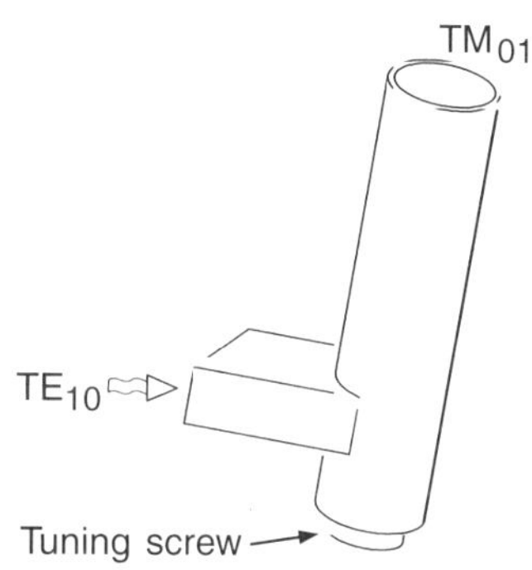

FIGURE 15.11 Set-up for the excitation of the TM_{01} mode.

its TE_{11} mode at the rectangular to circular waveguide junction. In practice this is tuned by monitoring the reduction of the TE_{11} on-axis radiation. Return loss better than –10 dB within an 8% bandwidth was obtained for a prototype tested at the 60-GHz band.

Besso[8] improved this configuration by symmetrically adding a second rectangular waveguide fed in-phase, thus inhibiting the TE_{11} mode excitation. Provided that there is no phase imbalance between the two rectangular waveguides and asymmetry in the cross-junction, the resulting inhibition of the TE_{11} is frequency independent. Naturally the power splitter between the rectangular waveguides and the cross-junction at the circular waveguide depend on the frequency, but the structure is optimized in Besso[8] to present a –20-dB return loss within 5% bandwidth.

Figure 15.12 shows calculated elevation angles that correspond to 3-dB level, 10-dB level, and maximum level, represented vs. normalized waveguide radius. In the simulations the circular waveguide was excited by a monopole mounted coaxially inside the waveguide at its short-circuited end. In a practical model this would correspond to the excitation by a coaxial line with its central conductor extended inside the waveguide. Because of the perfect circular symmetry of the resulting structure, only the TM_{01} mode is excited within a large bandwidth. Up to the direction of maximum radiation, the inner part of the lobe is reasonably independent of a; waveguide radius only affects the outer part of the lobe. Increasing the wall thickness in the range $0.35\ \lambda_m < t < 0.65\ \lambda_m$ slightly widens the blank region and increases the direction of the maximum without significantly changing the radiation level at $\eta = 90$. Compared with the TE_{11} mode, the TM_{01} mode allows less control of pattern beam width. For the above range of waveguide

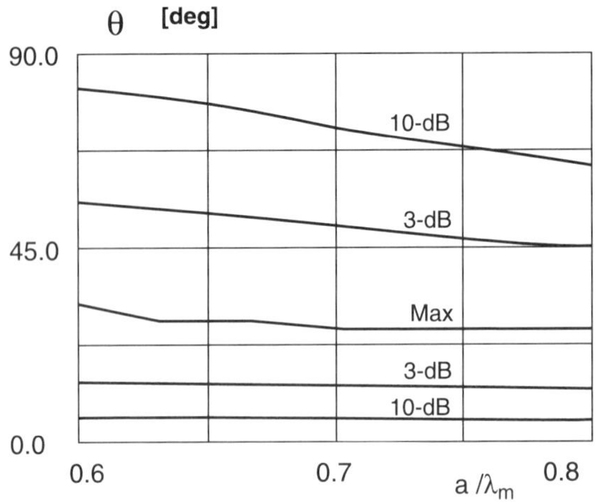

FIGURE 15.12 Elevation angle of 10-dB levels, 3-dB levels, and maximum level for TM_{01} radiation pattern vs. waveguide radius. Wall thickness: $t = \lambda_m/6$.

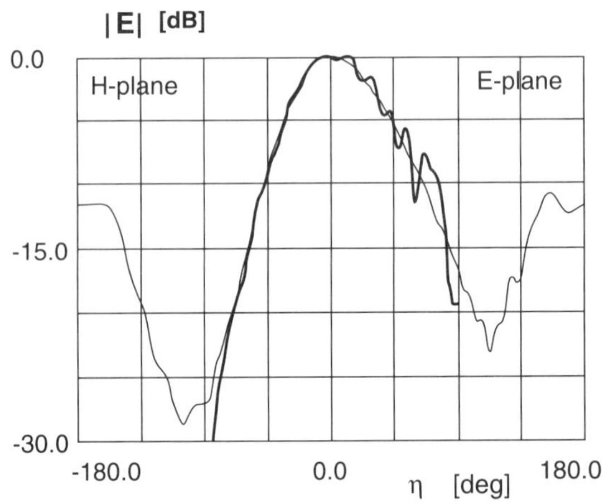

FIGURE 15.13 Radiation pattern of the TE_{10} mode of a rectangular waveguide immersed in unbounded dielectric medium with $\varepsilon_r = 2.53$; $f = 62.5$ GHz, $a = 3.8$ mm, $b = 1.9$ mm, $t = 1$ mm. Thin line — calculated; thick line — measured.

radius, phase center is well defined for $\eta < 75°$, and it is located at a distance less than 0.1 λ_m away from the waveguide aperture plane.

Rectangular Waveguide

The rectangular waveguide is usually operated in the TE_{10} mode. It is useful for linear polarization, when pattern axial symmetry is not essential. Figure 15.13 shows the calculated and measured radiation pattern of a standard WR-15 waveguide with wall thickness $t = 1$ mm ($\lambda_m/3$), when embedded in dielectric medium with permittivity $\varepsilon_r = 2.53$.

The agreement between simulation and the experimental pattern obtained with the hemispheric lens procedure is satisfactory. Note that the radiation level for $\eta = 90°$ is low in both planes. Simulation predicts high backlobe radiation. A dielectric inset with a wedge in the E-plane is positioned at the waveguide end, as sketched in Fig. 15.8b, to promote a good power transfer into the lens. Reflection loss at the waveguide port is better than –15 dB. The phase center is located at $z = -0.695$ mm (0.23 λ_m) in the E-plane for $\eta < 40°$, and $z = -0.247$ mm (0.08 λ_m) in the H-plane for $\eta < 50°$. Some control of E-plane beam width can be obtained in the usual way by flaring or tapering the waveguide in this plane.

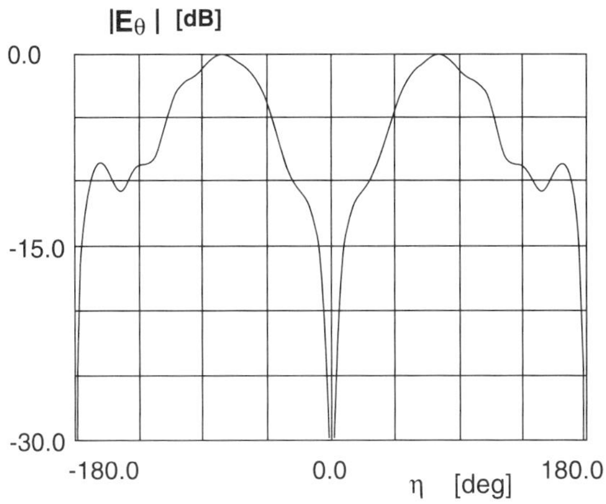

FIGURE 15.14 Calculated radiation pattern of coaxial line with extended central conductor, immersed in unbounded dielectric medium with $\varepsilon_r = 2.35$; $f = 43$ GHz.

15.3.4.2 Coaxial Feed

A monopole with a finite ground plane is an alternative to the TM_{01} mode of the circular waveguide as feed for a lens. It produces the same type of linear polarization but the blank region is wider, the maximum can be more easily shifted for larger values of η, and radiation extends beyond $\eta = 90°$. This may be useful as a lens feed to produce certain shaped fan-beam patterns, as is to be seen in the section on the flat-top fan beam.

An example of radiation pattern is shown in Fig. 15.14, calculated for 43 GHz. It corresponds to the aperture of a standard semirigid coaxial cable with a 0.29-mm central conductor diameter, and a 0.94-mm inner diameter of the outer conductor. The coaxial is inserted into a metallic cylindrical sleeve to enlarge the coaxial outer conductor diameter to 8 mm. At the coaxial aperture, the cross section of the metal sleeve acts as a finite circular ground plane. The coaxial central conductor is extended by 2.1 mm, and this feed is embedded into a dielectric with permittivity $\varepsilon_r = 2.35$.

Reducing the length of the extended part of the central conductor increases the radiation in the central region without affecting radiation beyond η of maximum radiation. This length has reduced impact on the phase center position. This is located at $z = +\ 0.9$ mm for the preceding example.

Coaxial cables are very lossy at millimeter waves, so they cannot be used to interconnect the antenna and the millimeter-wave front end, unless this is located immediately below the antenna. Either a frequency down-conversion must be made close to the lens base, or a transition to waveguide must be introduced. The latter solution was adopted in the example that is to be presented in the section on the flat-top fan beam.

15.3.4.3 Printed Feed

Printed feeds allow for future integration of transceiver millimeter-wave integrated circuits (MMIC) at the lens base. Examples are reported in the literature relative to hemispheric and elliptical lens fed by slots etched in the ground plane at the base of the lens.[29,30] For the shaped lenses addressed in this chapter, such a ground plane may be illuminated by lens internally reflected rays and originate important ripple in the overall lens radiation pattern.

A possible alternative for these shaped lenses could be a modified V-antenna[31,32] with the profile given by

$$W(\zeta) = \pm\left\{W_1 + W_2\left[1 - \cos\left(\pi\frac{\zeta}{L}\right)\right]\right\} \tag{15.13}$$

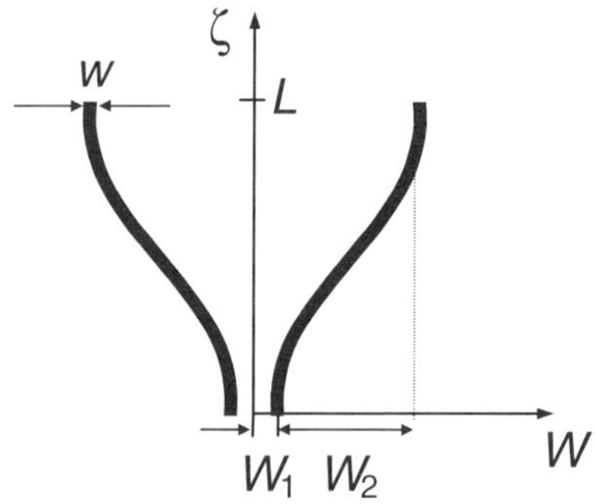

FIGURE 15.15 Geometry of printed modified V antenna.

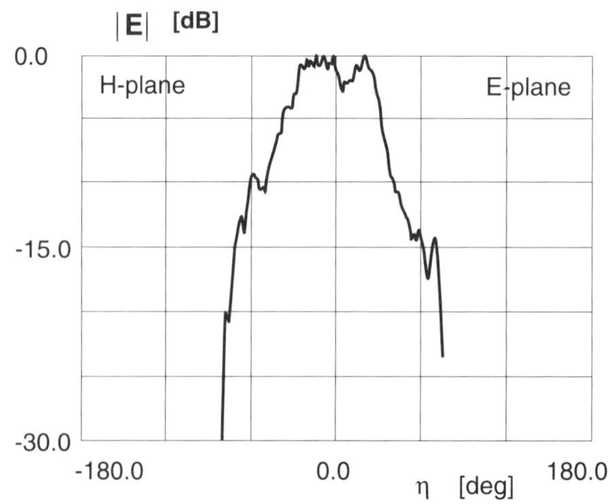

FIGURE 15.16 Measured radiation pattern of modified V antenna immersed in dielectric medium with $\varepsilon_r = 2.53$; $f = 43$ GHz, $W_1 = 0.1$ mm, $W_2 = 2$ mm, $L = 5.5$ mm, $w = 0.5$ mm.

The parameters in Eq. (15.13) are defined in Fig. 15.15. This is an end-fire radiator, which must be positioned inside the lens with the ζ-axis coincident with the lens *z*-axis. With this orientation, and because the V antenna does not have a ground plane, the metal inside the lens has little implication on internal reflections.

This feed produces a linearly polarized field, with the E-field parallel to printed feed plane. Changing the length *L* has opposite effects in E- and H-plane beam widths, making it possible to select the length that yields a symmetric power pattern.[31] Unlike the Vivaldi antenna, E- and H-plane pattern symmetry can be obtained for large beam widths. The phase center is well defined in the H-plane, but it is restricted in the E-plane to a smaller angular interval.

Figure 15.16 shows the radiation pattern of a prototype measured at 43 GHz, when immersed in a dielectric medium with permittivity $\varepsilon_r = 2.53$. Because a MMIC was not available for tests, in this prototype the modified V antenna was fed by a rectangular waveguide, through a printed tapered slot line inserted in E-plane of the waveguide, and extended outside to form the V antenna. However, the waveguide flange acts like a small ground plane that negatively affects the frequency behavior of the V antenna. The ripple in the measured pattern has to do with the setup used to embed the feed substrate inside the lens. This difficulty should disappear when a molding process is used to fabricate the lens.

15.3.5 Axial-Symmetrical Lens Examples

This section presents simulated and measured results corresponding to representative examples of circular-symmetrical lenses, which produce circular-symmetrical patterns with amplitude shaping in the elevation plane. The examples illustrate the influence of the design options on the final antenna performance in terms of radiation pattern, power efficiency, and operation bandwidth. The analysis in the following section is more extensive for the $\sec^2\theta$ lens because most of the conclusions also apply for the other examples.

15.3.5.1 Base Station Antennas

Consider the design of a shaped dielectric lens to produce a $\sec^2\theta$ coverage over a 6-m radius circular cell, for a broadband wireless B-WLAN operating at the 60-GHz band. At this point the designer is free to choose θ_{max}, feed configuration and its design parameters, lens material, η_{max}, and lens dimensions. Although these parameters become interrelated in the design process, the next sections try to point out the individual influence of some of them.

Lens Size

As an example, we choose $\theta_{max} = 76°$, which corresponds to a 1.5-m height difference between fixed and mobile terminal antennas [Eq. (15.2)]. Results are compared for two feeds: the TE_{11} mode of the circular waveguide excited with circular polarization, and the TM_{01} mode. In the first case, the waveguide radius is $a = 1.5$ mm, and in the second case, $a = 2.1$ mm. Waveguide wall thickness is $t = 0.35$ mm. Dielectric permittivity of the lens material is $\varepsilon_r = 2.53$, and it is assumed lossless for now. Dielectric losses are introduced in the continuation. The corresponding feed radiation patterns inside the dielectric material are shown in Figs. 15.6 and 15.10.

The lens profile is obtained from the integration of Eqs. (15.9) and (15.10). The initial value for r can be arbitrarily set. It does not influence the lens shape, and it acts only as a scaling factor. The larger the lens dimensions are the better the approximation to the target pattern; however, terminal compactness is highly desirable in wireless communications so a compromise must be found.

Figure 15.17 shows the profiles obtained for $r(\eta_{max}) = 66.6$ mm, which corresponds to a lens diameter of 28 λ. The lens edge is set at $\eta_{max} = 90°$, where the feed radiation is 20 dB below its maximum in the TE_{11} case, and 15 dB for the TM_{01} case. The first observation is the lens-reduced depth (depth to diameter ratio on the order of 0.3). The second observation is that the difference between the feeds is manifested mainly in the central part of the lens profile. The two profiles are almost coincident for high values of η (where the profiles approach an ellipse); this is common to $\sec^2\theta$ lenses with the same $r(\eta_{max})$ and the same θ_{max}.

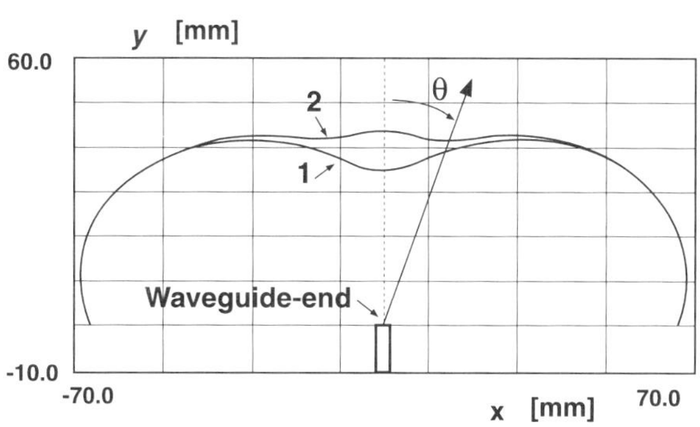

FIGURE 15.17 Profiles of two circular symmetrical dielectric lenses, fed by a circular waveguide, to produce a $\sec^2\theta$ target pattern with $\theta_{max} = 76°$; $\varepsilon_r = 2.53$, $f = 62.5$ GHz. 1 — TE_{11} mode excitation; 2 — TM_{01} mode.

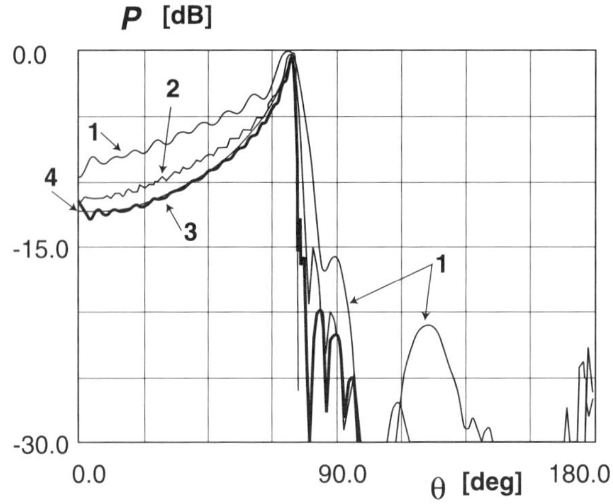

FIGURE 15.18 Power pattern of shaped lenses designed to produce $\sec^2 \theta$ pattern, with $\theta_{max} = 76°$, TE_{11} excitation, $f = 62.5$ GHz. Curves 1 to 3 — diameters ϕ =14, 28, and 56 λ; curve 4 — target pattern.

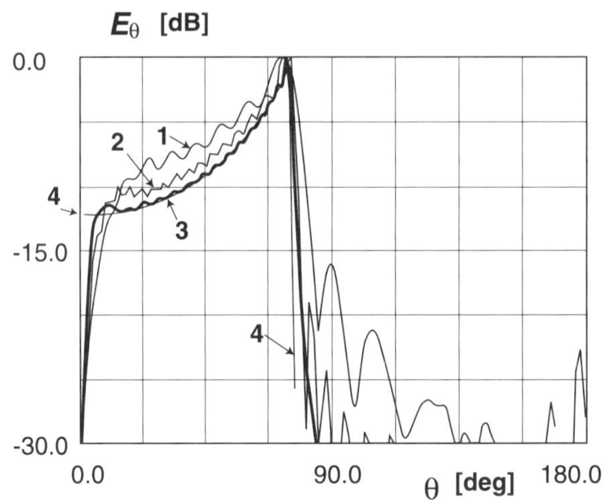

FIGURE 15.19 Power pattern of shaped lenses designed to produce $\sec^2 \theta$ pattern, with $\theta_{max} = 76°$, TM_{01} excitation, $f = 62.5$ GHz. Curves 1 to 3 — diameters ϕ =14, 28, and 56 λ; curve 4 — target pattern.

The corresponding radiation patterns calculated from physical optics [Eq. (15.11)] are shown in Figs. 15.18 and 15.19 marked with label 2. Internal reflections are not considered at this point. Superimposed on the same figures are the radiation patterns obtained for the same profiles but scaled-down (curve 1) and scaled-up (curve 3) by a factor of 2. The target $\sec^2 \theta$ pattern is marked with label 4.

As for reflector antennas, the physical optics power pattern tends to the geometric optics target pattern as the antenna aperture increases. Apart from the small ripple resulting from edge diffraction at $\eta = \eta_{max}$, the calculated pattern is almost coincident with the target for 58-λ lens. The maximum near θ_{max} becomes more peaked for increasing lens size, and the decrease of radiation beyond θ_{max} is steeper. The null near $\theta = 0$ for the TM_{01} case becomes narrower as lens dimensions increase, but it does not fill in within practical lens diameters.

A lens with 58-λ diameter is rather large, not only for terminal compactness but also because of dissipation losses. For a moderately low-loss material with $\tan \delta \sim 10^{-3}$, the resulting attenuation for the above 58-λ lens is on the order of 1.3 dB.

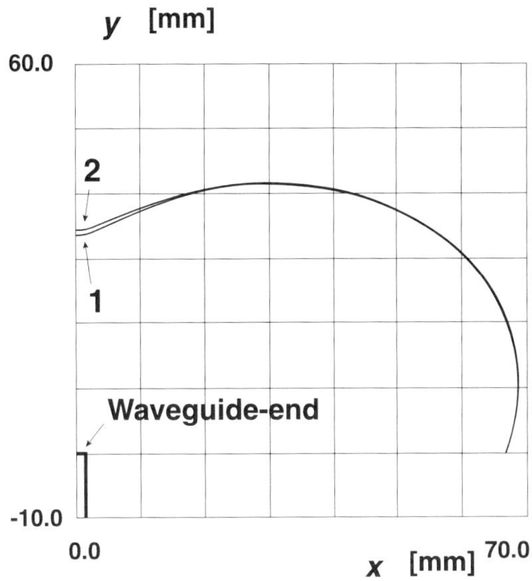

FIGURE 15.20 Circular symmetrical dielectric lens, fed by the TE_{11} mode of a circular waveguide, to produce a $\sec^2\theta$ target pattern with $\theta_{max} = 76°$; $\varepsilon_r = 2.53$, $f = 62.5$ GHz. 1 — Modified lens profile; 2 — original geometric optics profile.

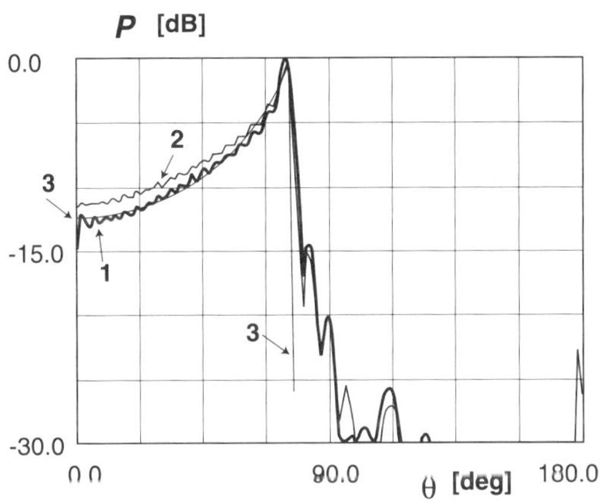

FIGURE 15.21 Power pattern of a shaped lens designed to produce $\sec^2\theta$ pattern, with $\theta_{max} = 76°$, TE_{11} excitation, $f = 62.5$ GHz, diameter $\phi = 28\ \lambda$. Curve 1 — radiation pattern of modified profile; Curve 2 — radiation pattern of original GO profile.

It must be noted that for higher loss materials like acrylic glass, lenses cannot be scaled because path attenuation varies exponentially with r, and r is not constant with η. In that case, no linear scaling relation exists between profiles obtained for different initial values $r(\eta_{max})$.

It is interesting to point out that the result obtained in the instance of the 28-λ lens can be improved without the need to increase the lens size. One alternative is to modify the previous geometrical optics profile to produce a better approximation to the target pattern[18] valid within a more restricted frequency band. One such profile is presented in Fig. 15.20 (curve 1) superimposed on the original profile (curve 2). The corresponding radiation pattern for $f = 62.5$ GHz is shown in Fig. 15.21. The coincidence with the target $\sec^2\theta$ is quite good, but the lens becomes more sensitive to frequency and the amount of internally reflected power increases, as is explained in the next section. The same procedure can be applied to

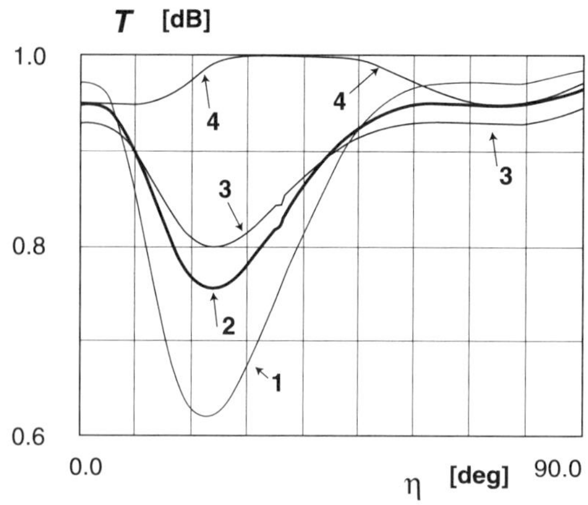

FIGURE 15.22 Transmissivity $T(\eta)$ at $\sec^2\theta$ lens surface. Curves 1 to 3 — $\varepsilon_r = 2, 2.53, 3$, TE_{11} excitation; curve 4 — $\varepsilon_r = 2.53$, TM_{01} excitation.

smaller lenses, but the larger the alteration is with respect to the geometric optics profile, the worse the implications on internal reflection and the frequency dependence.

Dielectric Permittivity and Internal Reflections

Permittivity impacts on the lens shape and on the amount of internal reflections. In the limit, high values of ε_r favor high-refraction angles, allowing smaller lenses to obtain highly shaped target patterns. However, high values of ε_r also favor internal reflections at the dielectric-air interface, which reduce lens efficiency and produce ripple over the radiation pattern and increased sidelobes.

However, because of the various interrelated design parameters, it may be that a small change on the permittivity value produces results that locally contradict the general trend stated earlier. Figure 15.22 shows the transmissivity $T(\eta)$ for a $\sec^2\theta$ lens excited by the TE_{11} mode, calculated for three permittivity values (curves 1 to 3). The same feed radiation pattern is adopted in the three cases, to isolate the results from feed dependency on ε_r. The dip in the transmissivity curves corresponds to the slope in the central region of the $\sec^2\theta$ lens surface (see Fig. 15.17), where refracted rays leave the lens close to the tangent direction. This slope decreases with increasing ε_r, and so does reflection in this region. In other parts of the lens where the incidence is closer to normal, reflection increases with ε_r as usual. The global balance between the two opposite behaviors depends on the specific design conditions. In the preceding example, the lens power efficiency is 78.4, 84.8, and 86.2% for permittivities of 2, 2.53, and 3, respectively. For the modified profile of Fig. 15.20 with $\varepsilon_r = 2.53$, the efficiency reduces to 82.8%.

Transmissivity for the TM_{01} excitation is also represented in Fig. 15.22 for $\varepsilon_r = 2.53$ (curve 4). Because the maximum of the feed radiation is shifted away from $\eta = 0$, there is no depression of the lens profile in the central region, thus favoring power transmission across the lens surface. Also recall that E_θ is the dominant E-field component that enables total transmission to occur at part of the lens surface. In this case, the lens power efficiency is 98.7%.

The power efficiency can also be improved by acting on other design parameters. Traditional strategies to match the lens surface using quarter-wave layers are not appropriate for these lenses, because the rays emerge at wide angles from the local normal; besides, matching layers are narrowband. A broadband alternative is suggested in Section 15.3.7, which can significantly reduce the internally reflected power.

Internal reflections can be reduced, but not completely eliminated. However, their effect on the radiation pattern can be minimized. Ray tracing shows that first-order internally reflected rays tend to concentrate near the $\eta = 90°$ edge of the lens, thus adding its contribution to edge diffraction effects. The ripple in the radiation pattern can be significantly reduced by extending the lens for $\eta > 90°$. Lowering the lens base has a second effect: rays reflected on this surface have an increased path length to reach the

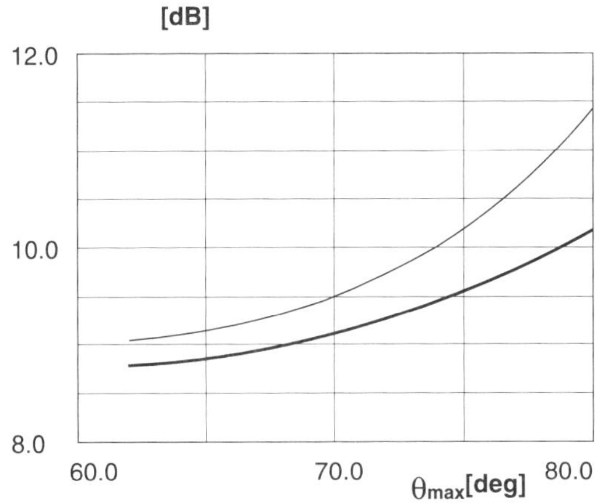

FIGURE 15.23 Ideal directivity (thin line) and gain (thick line) for $\sec^2\theta$ lens, excited by the circular waveguide TE_{11} mode. $\varepsilon_r = 2.53$, $a = 1.5$ mm, $f = 62.5$ GHz.

lens surface, and thus increased dissipation losses. Actually, this lens extension is also useful to mechanically hold the feed in place.

Choice of θ_{max}

Previous examples were calculated for $\theta_{max} = 76°$ but other values may be used, provided that $(h_1 - h_2)$ is set according to Eq. (15.2) to obtain a prescribed cell radius R. The choice of θ_{max} affects the lens performance, besides influencing the propagation channel behavior.[6]

By reducing θ_{max} the effort required on the lens to divert the front radiation of the TE_{11} is smaller, and thus smaller lens diameters can be used to effectively comply with the target $\sec^2\theta$. The lens depth to diameter ratio increases, but this favors the reduction of internal reflection in the lens central region.

Figure 15.23 compares the evolution of the ideal directivity and gain vs. θ_{max} for a $\sec^2\theta$ lens fed by the TE_{11} mode. The result refers to a lossless lens with permittivity $\varepsilon_r = 2.53$, and does not take into account pattern degradation resulting from the finite size of the lens. It is a measure of lens internal reflection loss as a function of θ_{max}.

Frequency Behavior

As previously mentioned, a 10% bandwidth specification is typical for millimeter-wave wireless communications antennas. Because the lens shape is obtained from the geometric optics formulation, it is valid for any frequency provided that the lens size is compatible with the geometric optics approximation. This assumes that the feed radiation pattern is constant, which is far from being true. In fact, the frequency behavior of the lens radiation pattern is mostly related with the frequency behavior of the feed.

Two aspects have to be evaluated: the alteration of feed radiation beam width with frequency, and the change of phase center position with respect to feed aperture. As a general rule, lowering the frequency corresponds to an increase of the feed beam width, and consequently a better approximation to the target pattern. Increasing the frequency has the opposite effect. The exact effect of beam width alteration on the lens radiation pattern has to be evaluated for each case. For a fixed lens profile, the alteration of the phase center impacts on the angle of maximum radiation of the lens θ_{max}. As a rule, a shift of the feed phase center for $z > 0$ slightly increases θ_{max}, with otherwise little degradation of the $\sec^2\theta$ characteristic; a small shift for $z < 0$ decreases θ_{max}. For most feeds, the shift in phase center position for 10% frequency variation is negligible, with little consequence to the radiation pattern. For the TE_{11} feed of the preceding example, this shift is on the order of $0.03\ \lambda_m$.

The frequency behavior of the TE_{11} feed may be estimated from Fig. 15.7. As an example, Fig. 15.24 shows the corresponding frequency dependence of the radiation pattern of previous 28-λ-lens fed by the TE_{11} mode. Within 10% frequency variation, the lens radiation pattern alteration does not exceed 2 dB.

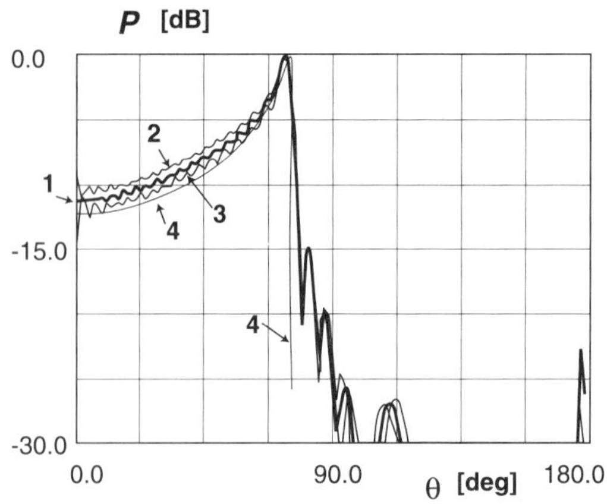

FIGURE 15.24 Effect of frequency on $\sec^2\theta$ lens radiation pattern. 28-λ lens, TE_{11} excitation; curves 1 to 3 — f = 62.5 GHz, +5%, –5%; curve 4 — target pattern.

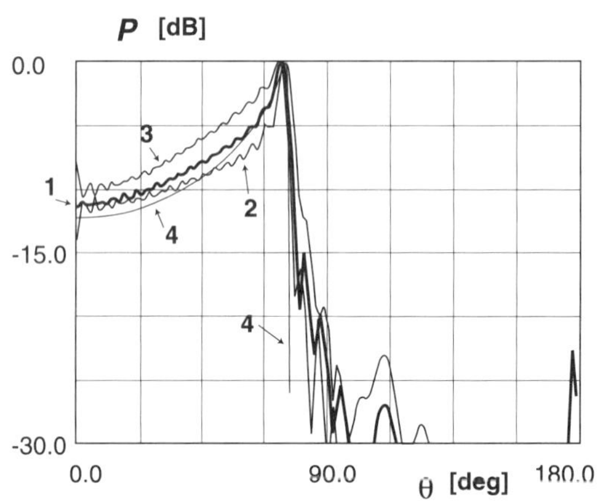

FIGURE 15.25 Effect of permittivity tolerance on $\sec^2\theta$ lens radiation pattern. 28-λ lens, TE_{11} excitation; curves 1 to 3 — ε_r = 2.53, +5%, –5%; curve 4 — target pattern.

It is also interesting to note that there is no significant shift of θ_{max}. This direction of the maximum is mostly related with the direction of the major axis associated with the previously identified elliptical part of the lens profile.

Ray tracing inside this type of lens shows that few rays are reflected back to the feed.[13] Provided that this condition is verified, the input impedance at the lens feeding point is independent of the lens shape. So its frequency dependence only reflects the behavior of the feed radiating into an unbounded dielectric.

Fabrication Tolerances

Molding is an attractive technique for large-scale production of dielectric lenses. Waveguide feeds and polarizer can be produced by electroforming. Once the lens mold is fixed, the parameter that may fluctuate from batch to batch is lens material permittivity. The ε_r tolerance depends on the specific design parameter combination. Figure 15.25 shows the influence of ±5% spread of the material permittivity, for a fixed profile corresponding to ε_r = 2.53, excitation by the TE_{11} mode, and 28-λ diameter. Figure 15.26 shows the same type of information for the TM_{01} mode excitation. It is clear that in this example tolerance is stricter for the TE_{11} excitation.

Another issue is the feed-positioning accuracy inside the lens. For rotational symmetric lenses and feeds, positioning errors are only expected along the *z* coordinate. These positioning errors in fact

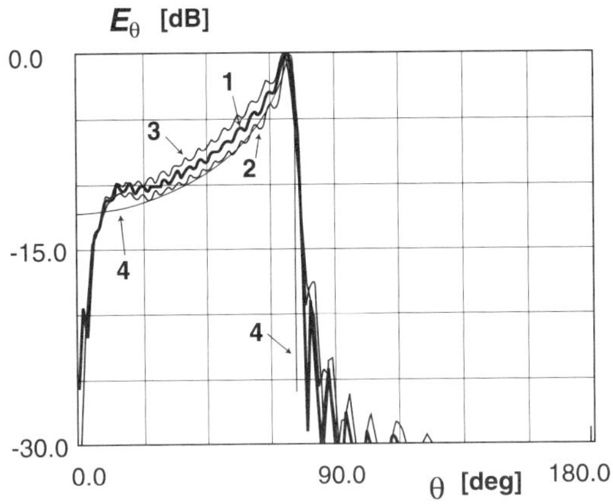

FIGURE 15.26 Effect of permittivity tolerance on $\sec^2\theta$ lens radiation pattern. 28-λ lens, TM_{01} excitation; curves 1 to 3 — $\varepsilon_r = 2.53$, +5%, –5%; curve 4 — target pattern.

correspond to phase center shift with respect to lens focus, which has been already discussed. A slight shift of the feed aperture for $z > 0$ increases θ_{max}, without degradation of the $\sec^2\theta$ characteristic, and a shift for $z < 0$ decreases θ_{max}. This fact can be used for minor adjustments of θ_{max}, without the need to modify the lens mold. For the preceding example, a shift of the circular waveguide aperture with respect to lens focus by $\pm 0.33\ \lambda_m$ changes θ_{max} by ±1.5°.

The most critical part of the $\sec^2\theta$ lens surface for the TE_{11} excitation is the depressed region at the center. The comparison between the radiation patterns of Fig. 15.21, corresponding to the original and to the modified lens profiles represented in Fig. 15.20, gives a good perception of the effect of small deviations of lens profile.

Prototype Results

Prototypes for both excitations were machined on acrylic glass. This is a readily available inexpensive material useful for laboratory prototyping when some dissipation losses can be accepted. The photograph in Fig. 15.27 corresponds to a 28-λ lens excited by the TE_{11} mode with circular polarization. At $3\ \lambda_m$ behind the feed aperture plane, the exterior side of the circular waveguide wall is enlarged and threaded, so that the lens can be screwed like a lamp bulb. In this way the waveguide also acts as the lens only strut. The lens base is 5 mm behind the waveguide aperture plane, and it is padded with a thin sheath of absorber to reduce the effect of the internally reflected rays on radiation pattern. The lens profile is slightly different from the lossless case of Fig. 15.20, because dielectric losses introduce a dependence with η that modifies the feed illumination of the lens surface.

The corresponding measured radiation pattern is shown in Fig. 15.28 (thick line), superimposed on the target $\sec^2\theta$ pattern. The agreement up to θ_{max} is satisfactory despite some small ripple resulting from lens internal reflections, and edge diffraction. The fall of radiation beyond θ_{max} is quite steep. Radiation for $\theta = 90°$ and beyond horizon is below –20 dB, which is a significant contribution to combat multipath and neighboring cell interference. Directivity calculated from the measured pattern is 9.2 dBi, which corresponds to 1-dB degradation with respect to the ideal value (see Fig. 15.23). The circular polarization axial ratio for $\theta < \theta_{max}$ is better than 2.5 dB (not shown).

The power pattern correponding to the TM_{01} fed prototype is also shown in Fig. 15.28 (thin line). Lens diameter is 18 λ in this case, but despite the reduced size, the agreement with the target is again satisfactory outside the blank region. This is the result of the better illumination of the lens surface, as previously discussed. Directivity calculated from the measured pattern is 8.7 dBi.

Although an indoor cell with a 6-m radius was used as motivation for the design of these $\sec^2\theta$ lens examples, the same lenses can be used for larger outdoor cells simply by increasing the base station antenna height. Another possible use for this lens antenna is the base station of LMDS services.

FIGURE 15.27 Prototype of 28-λ lens for $\sec^2 \theta$ coverage of circular cells, at the 60-GHz band. (From Fernandes, C.A., *IEEE Antennas Propag. Mag.*, 41, 141, 1999. With permission.)

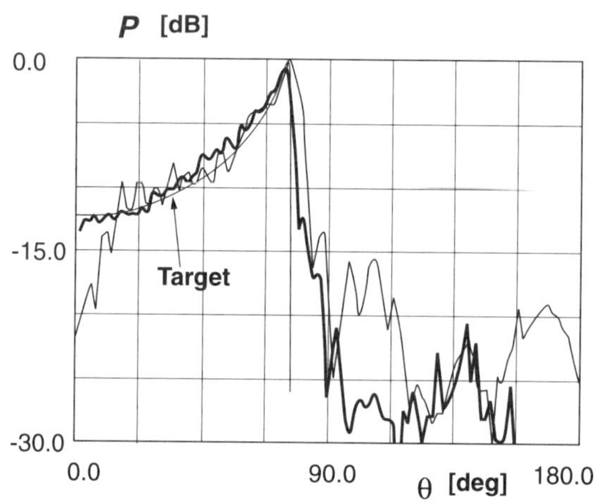

FIGURE 15.28 Measured power pattern of two acrylic lenses designed to produce $\sec^2 \theta$ pattern, with $\theta_{max} = 76°$, $f = 62.5$ GHz. Thick line — 28-λ lens, TE_{11} excitation. Thin line — 18-λ lens, TM_{01} excitation.

15.3.5.2 Mobile Terminal Antennas

Hemispheric Pattern

As previously discussed, the choice of the hemispheric pattern for the mobile terminal antenna is compatible with the $\sec^2 \theta$ characteristic of the base station, to ensure constant received power within the cell limits. This pattern enables full mobility of the mobile terminal, although with the penalty of reduced gain and potential for multipath pickup. As in previous examples, radiation may also be restricted to the interval $\theta < \theta_{max}$.

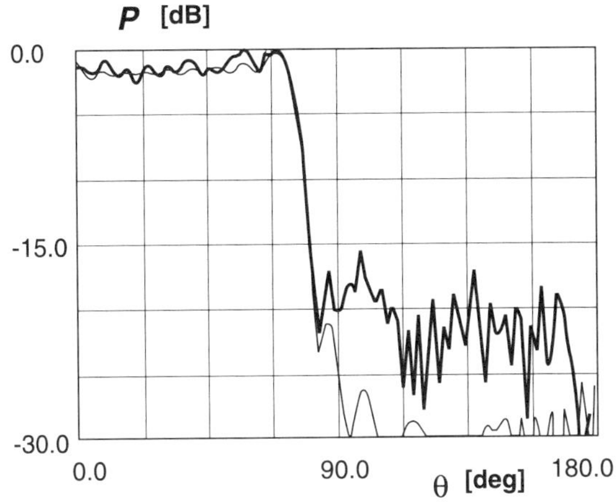

FIGURE 15.29 Power pattern of 14-λ acrylic lens designed to produce a hemispheric pattern, with $\theta_{max} = 76°$, TE_{11} excitation, $f = 62.5$ GHz. Thick line — measured; thin line — calculated.

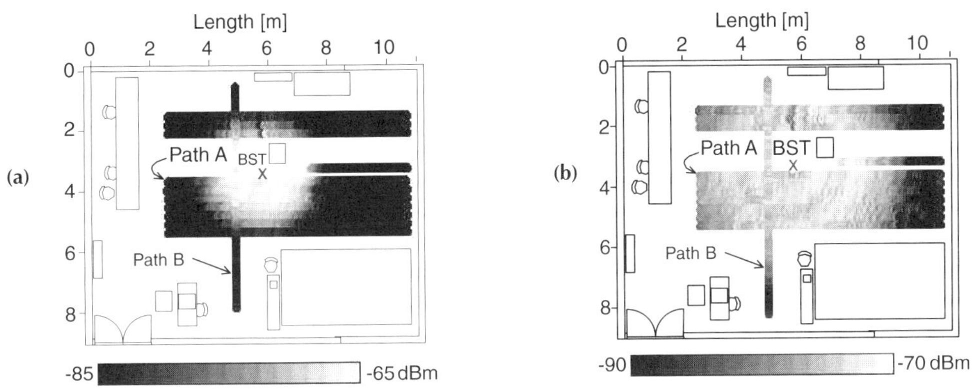

FIGURE 15.30 Received power measured along straight paths in a room illuminated by a $\sec^2 \theta$ base station lens antenna. $f = 62.5$ GHz. (a) Antenna height difference $\Delta h = 0.5$ m, (b) $\Delta h = 1$ m. (From Fernandes, C.A., *IEEE Antennas Propag. Mag.*, 41, 141, 1999. With permission.)

A lens prototype with this characteristic was designed to operate at the 60-GHz band, fed by the TE_{11} mode with circular polarization. The lens with a 14-λ diameter was machined on acrylic glass. The corresponding measured power pattern is shown in Fig. 15.29, superimposed on the calculated pattern. Despite lens-reduced size, a reasonable flat-top pattern is obtained for $\theta < \theta_{max}$, with a steep fall of radiation beyond θ_{max}. This is consistent with the fact that the target pattern is less demanding than the previous $\sec^2 \theta$, requiring less effort on the lens to divert the front radiation of the TE_{11} feed. The pattern characteristic for $\theta > \theta_{max}$ blocks the ground-reflected paths.

The combination of this mobile terminal antenna with the previous $\sec^2 \theta$ base station antenna yields almost constant received power level for $\theta < \theta_{max}$, and more than a 35-dB reduction of this level for $\theta > \theta_{max}$. The latter characteristic is responsible for very sharp cell boundaries. This is observed in Fig. 15.30, which shows the measured received power in one indoor cell for the above combination of antennas. Measurements were made at 60 GHz, and clearly show the direct relation between antenna height difference Δh and cell radius. This gives an effective way to control the illumination of walls near the edge of the cell. Furthermore, circular polarization also filters out first-order reflections. For this reason, no deep fading was measured within the cell limits, enabling very high data rate transmission.[6]

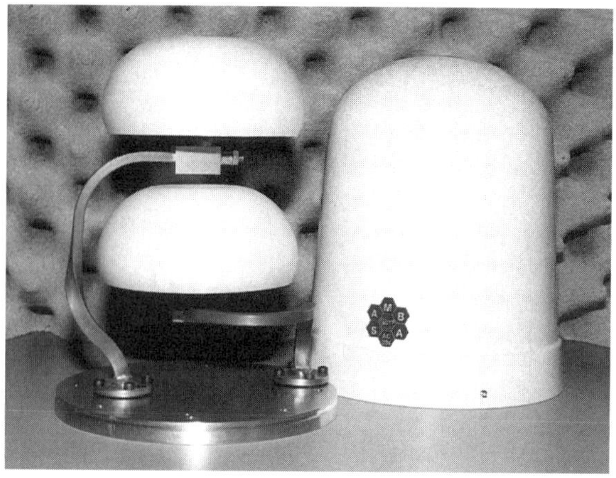

FIGURE 15.31 SAMBA mobile terminal antenna assembly, 39.5 to 43.5 GHz. (From Fernandes, C.A., *IEEE Antennas Propag. Mag.*, 41, 141, 1999. With permission.)

Flat-Top Fan Beam

A difficulty exists with antenna diversity implementation using hemispheric pattern antennas or $\sec^2 \theta$ pattern antennas that illuminate to nadir. It is not possible to position the second antenna element without blocking the first one for some direction within its field of view.

For linear polarization, the obvious choice is to mount the diversity antenna elements coaxially, taking advantage of the direction of null radiation. Vertical separation combats ground reflections, and the random axis tilt of the mobile antenna assembly may introduce some discrimination in wall-reflected paths. In Fig. 15.31, the transmitter and receiver share the top lens, whereas the lower lens is used for receiver diversity. In this way, if blocking occurs for some directions, it only affects diversity gain.

This antenna was developed for the mobile terminal of SAMBA trial platform, operating in the 40-GHz band. It was designed to produce an omnidirectional pattern with a flat-top characteristic within the elevation interval $60° < \theta < 95°$. Unlike previous examples, here θ_{max} exceeds 90°: this allows some antenna tilting while the user walks carrying the terminal as a backpack. Both lenses are equal, with a 50-mm radius (7 λ), machined on polyethylene. The lenses are fed inside its body by the aperture of a coaxial waveguide with an extended central conductor (see Section 15.3.4.2).

Figure 15.32 shows the measured radiation pattern of the top lens, when mounted on the assembly. The radiated field is linearly polarized (E_θ) and the measured gain is 4 dBi. The ripple is partly caused by feed radiation spillover at the lens edge ($\eta_{max} = 110°$), and partly to internal reflections. The lens base was padded with a thin sheet of absorbing material to reduce these effects, and to improve the isolation between the two lenses in the assembly. Note that the blank region of the lens radiation pattern already gives some contribution (although modest) to this isolation.

Because attenuation in coaxial guides is extremely high at millimeter waves, a dedicated transition from coaxial to WR-22 waveguide was introduced right below the lens feeding point. This rectangular waveguide winds for about 20 λ distance to connect to the millimeter-wave head interface. The measured return loss is shown in Fig. 15.33. It takes into account the long twisted waveguide, the transition from rectangular waveguide to coaxial feed, and the load of the lens. Return loss is better than –10 dB for both lenses, within the uplink and downlink bands, and is quite insensitive to the presence of nearby objects, even if touching the lenses.

Although a thin-wall radome was used in this prototype, the preferable solution is to completely immerse the antenna assembly in low-loss, closed-cell foam. Such a volume radome also serves as a mechanical support for the lenses, which are otherwise supported only by the feeding waveguide to avoid blocking and reflection by struts.

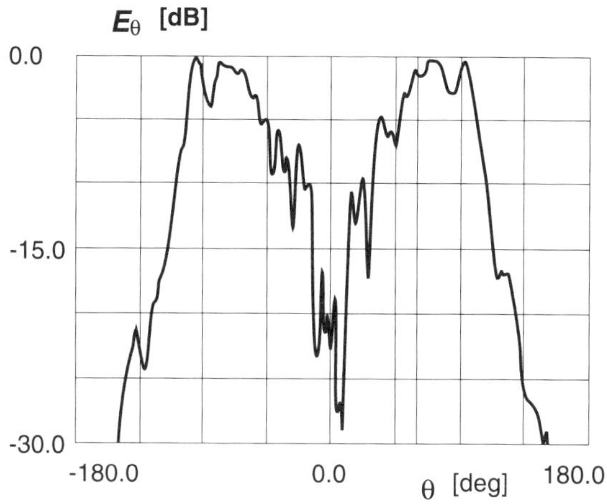

FIGURE 15.32 Measured elevation pattern of the upper lens (Tx-Rx) of SAMBA mobile terminal antenna assembly. $f = 43.5$ GHz, G = 4.0 dBi.

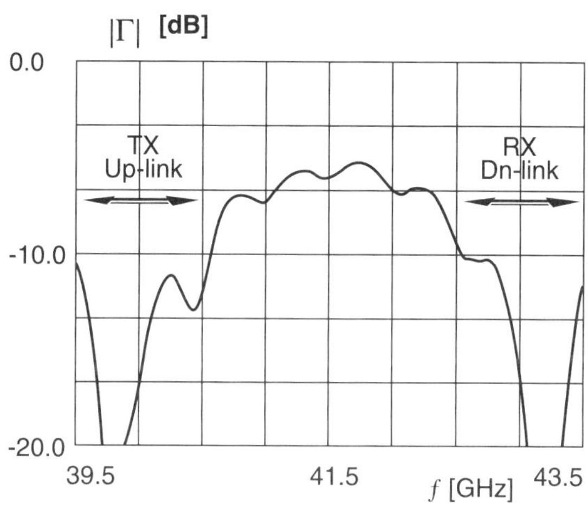

FIGURE 15.33 Measured input return loss of SAMBA mobile terminal antenna.

15.3.6 Modified Axial-Symmetrical Lenses

In many cases the $\sec^2\theta$ illumination is required for elongated cells or for sectored cells, instead of circular symmetric cells addressed in the previous sections. The resulting lens shapes are no longer axial symmetrical. Although the analytic formulation for the geometric optics calculation of an arbitrary three-dimensional lens exists,[17] the formulation of Section 15.3.2 for axial-symmetrical lenses can still be used for these cases, being enough to shift the feed to an off-axis position and tilt it if necessary.[33]

15.3.6.1 Lens for Sectored Cell Base Station

Consider the design of a base station lens antenna to produce a $\sec^2\theta$ radiation pattern within a sectored cell. The base station is located outside the cell, and the elevation coverage is restricted to the interval $71.5° < \theta < 86°$ (see Fig. 15.1c). The feed is a rectangular waveguide with the E-plane coincident with the lens *xz* plane. Appropriate waveguide tilt γ is chosen as a compromise between adequate illumination of the lens, and radiation level at the lens edge. The approximate lens design procedure comprises two steps:[33]

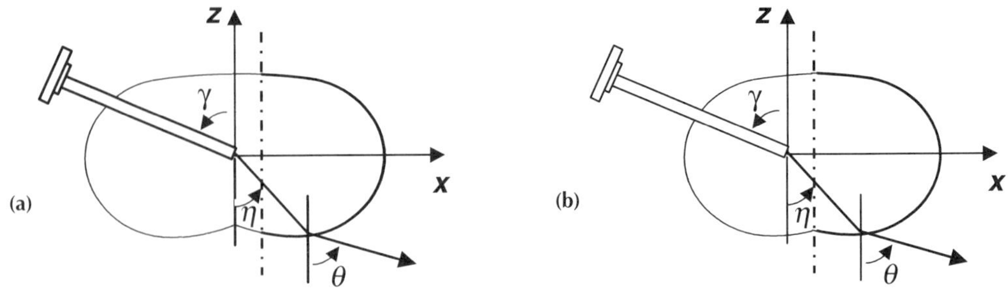

FIGURE 15.34 Geometry for axial-symmetrical lens with off-axis excitation. (a) First design step; (b) final step lens. (From Fernandes, C.A., *IEEE Antennas Propag. Mag.*, 41, 141, 1999. With permission.)

FIGURE 15.35 SAMBA base station lens antenna assembly for sectored cell coverage, 39.5 to 43.5 GHz, G = 19.5 dBi.

- First, the actual waveguide radiation pattern U is replaced by a circular symmetrical pattern obtained by repeating the tilted waveguide principal E-plane pattern inside the lens for all the lens azimuth angles. The axial-symmetrical lens design formulation of Section 15.3.2 is used to obtain the lens profile.
- Second, a new lens is generated by rotating the previous profile about a shifted vertical axis contained in the xz plane (dashed line in Fig. 15.34). The resulting lens is then fed by the actual waveguide, and the corresponding radiation pattern is calculated using physical optics. Appropriate displacement of the new rotation axis for $x > 0$ extends the principal plane $\mathrm{cosec}^2\,\theta$ elevation pattern to a given azimuth interval.

Figure 15.35 shows the lens assembly developed for the SAMBA trial platform, which is compatible with the mobile terminal antenna described in the previous section. The same antenna could be used also for sectored coverage in LMDS application; as previously mentioned, cell dimensions are scaled to the antenna height.

The two equal 130-mm diameter lenses are machined on polyethylene. One is shared by transmitter and receiver channels, and the other is used by the receiver diversity branch. The lenses are arranged in a way that no mutual blocking exists for directions contained within the cell limits.

The measured elevation pattern is shown in Fig. 15.36, superimposed on the ideal $\sec^2\theta$ pattern. The field is linearly polarized (E_θ), and measured gain is 19.5 dBi. Good agreement is obtained within the specified interval ($71.5° < \theta < 86°$), marked on the figure. The ripple for $\theta < 71.5°$ is partly explained by internal reflections and diffraction, but it corresponds to regions outside the cell.

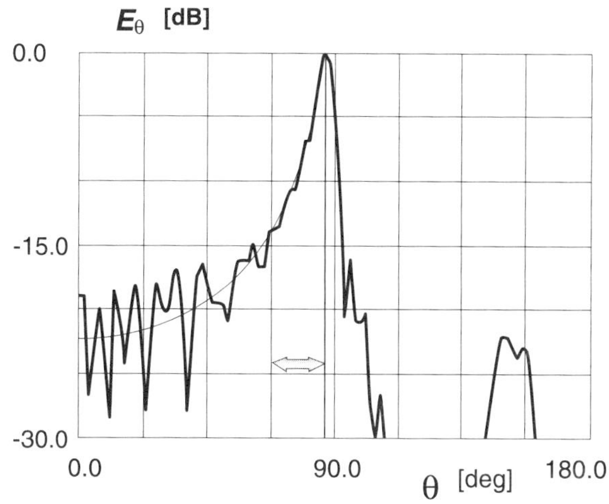

FIGURE 15.36 Measured elevation radiation pattern of SAMBA base station antenna. f = 43 GHz, ε_r = 2.35, G = 19.5 dBi. (From Fernandes, C.A., *IEEE Antennas Propag. Mag.*, 41, 141, 1999. With permission.)

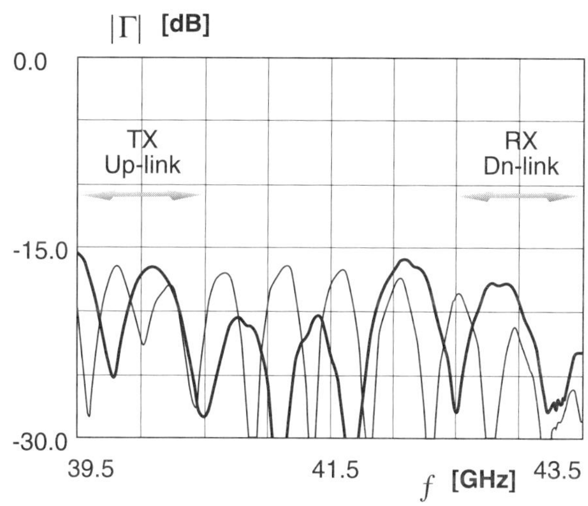

FIGURE 15.37 Input return loss of SAMBA base station lens antennas, measured at the millimeter-wave head interface port. Thin line — Tx/Rx lens; thick line — Rx. (From Fernandes, C.A., *IEEE Antennas Propag. Mag.*, 41, 141, 1999. With permission.)

The antenna assembly includes a level gauge for correct alignment in the site. Failure to level the $\sec^2 \theta$ antenna compromises the constant received power characteristic, and the control of cell extension. In the previous example, 1° error in elevation alignment extends the cell radius by 30%.

Figure 15.37 shows the return loss of the lens antennas, including the effect of the bent waveguide that connects to the millimeter-wave head interface. The input return loss is better than 15.5 dB for the upper lens (receive — Rx), and better than 16.5 dB for the lower lens (transmit–receive — Tx/Rx) within the whole frequency range from 39.5 to 43.5 GHz.

15.3.6.2 Beam Agility

The second outcome of the offset feed configuration suggested in the previous lens design is that more shaped beams can be added to the same lens, all of them exactly equal to each other, separated by a prescribed azimuth interval. Although the phase center is offset from the lens axis, the waveguide axis

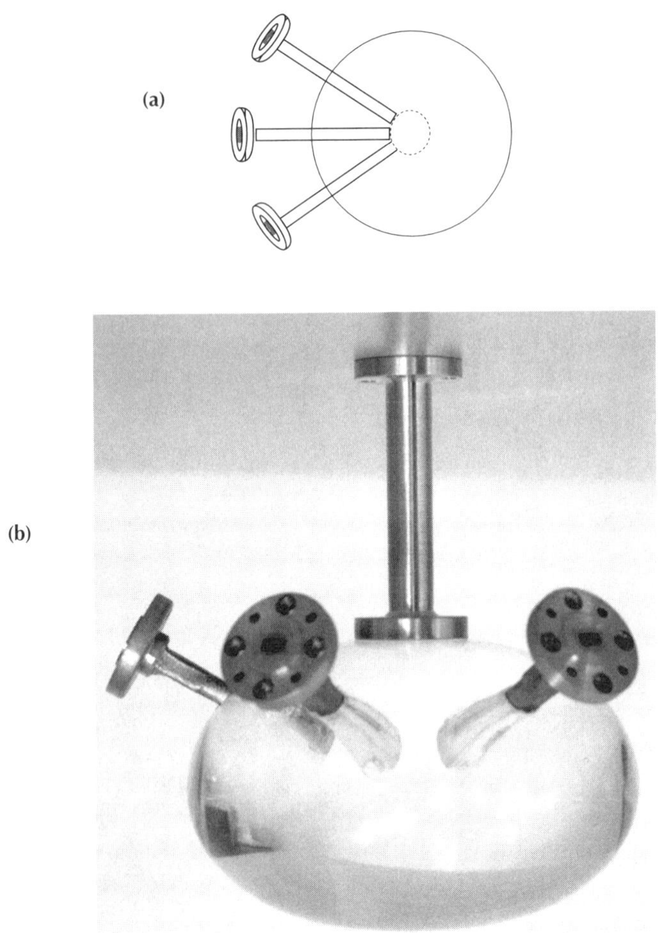

FIGURE 15.38 Three-sector lens with $\sec^2\theta$ elevation pattern. (a) Geometry, (b) prototype. (From Fernandes, C.A., *IEEE Antennas Propag. Mag.*, 41, 141, 1999. With permission.)

in the design of Section 15.3.6.1 is still contained in the lens constant φ-plane. By using this fact and the axial symmetry of the lens, other feeds can be added at other φ-planes of the same lens, with the phase centers positioned on a circle (Fig. 15.38a). Driven by a single-pole, triple-throw (SP3T) RF switch, this compact setup can produce mutually equal switchable beams with $\sec^2\theta$ shaping in the elevation plane. Switching time of 5 nsec is possible with commercially available devices for millimeter waves.

A lens prototype fed by three rectangular waveguides is shown in Fig. 15.38b. The radiation pattern of each beam has the same characteristic of the previous single beam example, and it is virtually unperturbed by the other feeds. Isolation with respect to the adjacent feed is better than 30 dB, and better than 16 dB for the farther waveguide. The same lens design was also tested with printed patch feed. Because of the smaller element size, a ring of printed patches could be used instead of the waveguide apertures to feed the lens and produce a horizontal scanning beam with $\sec^2\theta$ shape in elevation.

15.3.6.3 Coverage of Elongated Cells

Figure 15.39 shows another lens designed according to the previous approximate procedure. It was developed for the evaluation of the MBS concept.[5,16] It may be used in applications where some mobility restriction is acceptable, as in train cells. Base station and mobile terminal use equal lens antennas, each one producing a $\sec\theta$ pattern so that the combined gain corresponds to the $\sec^2\theta$ characteristic within a highly elliptical cell (see Fig. 15.1b).

FIGURE 15.39 Lens assembly for elongated cell illumination, 39.5 to 43.5 GHz, G = 14 dBi.

The lens feed is a rectangular waveguide with the E-plane coincident with lens *xz*-plane as before, but not tilted ($\gamma = 0$). The first step of the design procedure is the same as described in the previous section. In the second step, the obtained profile is rotated about a shifted horizontal axis contained in the *xz*-plane. An oblong lens is obtained, which must be oriented in the cell such that the axis of rotational symmetry is aligned with the longitudinal axis of the elongated cell, both for the base station and the mobile lenses. The lenses were machined on polystyrene, for operation in the 40-GHz band.

Figure 15.40 shows the elevation pattern for the H-plane (cell transverse plane near nadir) and for the E-plane (longitudinal plane), measured at 43 GHz. Measured gain is 14 dBi. The polarization is linear, but provided that the axis of the mobile terminal lens remains parallel to the axis of the base station lens, the polarization is matched even when the mobile crosses the *y* axis. Simulation and measured results confirmed that received power is reasonably constant within a narrow strip with length to width ratio on the order of 30, with the base station located at midlength of the cell.[5]

15.3.7 Two-Shell Lens

A few design strategies were previously discussed to reduce lens internal reflections, and to minimize its influence on the radiation pattern. Acceptable results are obtained in most cases, but further improvement may be required for stringent specifications.

It was previously referred that common methods to improve power transmission across the lens surface based on resonant-matching layers are not appropriate for these highly shaped lenses. These layers are narrowband, and incompatible with the high-refraction angles that characterize the shaped lenses addressed in the present chapter. One alternative proposed in Silveirinha[34] is to consider a double-material lens, where both dielectric interfaces are appropriately shaped using a generalization of the geometric optics formulation of Section 15.3.2. By using two interfaces a second degree of freedom is gained, opening the possibility to control another lens characteristic besides radiation pattern.

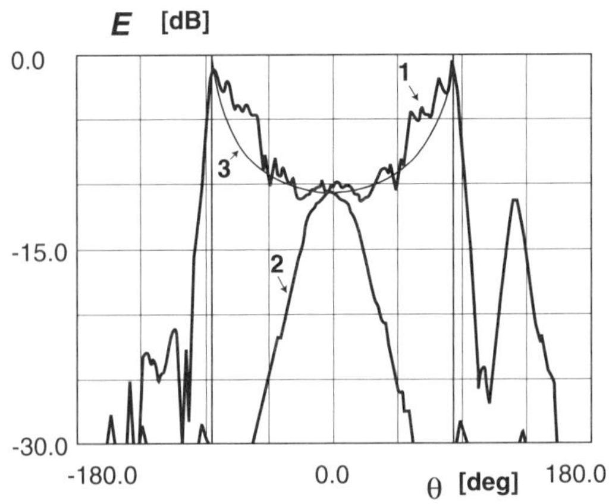

FIGURE 15.40 Measured radiation pattern of lens antenna for elongated cell, $f = 43$ GHz. Curve 1 — E-plane; curve 2 — H-plane; curve 3 — target sec θ pattern.

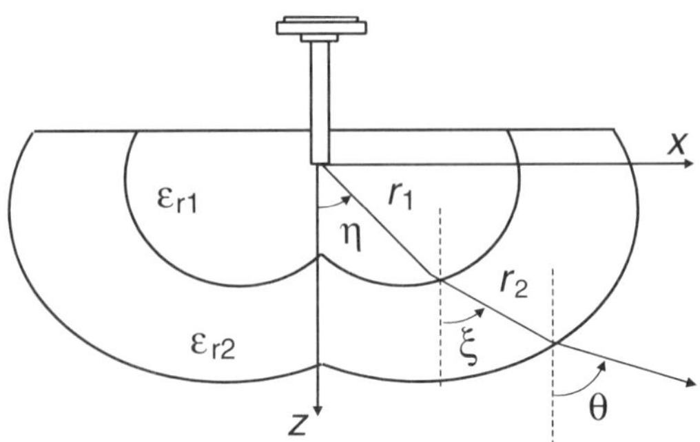

FIGURE 15.41 Geometry of a two-shell-shaped dielectric lens.

The lens geometry is shown in Fig. 15.41. It represents an axial-symmetrical lens formed by two shells, with the feed embedded in the inner lens. The permittivity of the outer shell is lower than the permittivity of the inner lens. By imposing the energy conservation condition in a ray tube and using Snell laws, the geometric optics formulation leads to a system of three coupled differential equations:[34]

$$\frac{d\theta}{d\eta} = \frac{\left[T_{\parallel} U_{\parallel}(\eta) + T_{\perp} U_{\perp}(\eta)\right] \sin\eta}{K\,G(\theta)\,\sin\theta} \tag{15.14}$$

$$\frac{dr_1}{d\eta} = r_1 \frac{\sin(\xi - \eta)}{n_{12} - \cos(\xi - \eta)} \tag{15.15}$$

$$\frac{dr_2}{d\eta} = r_2 \frac{d\xi}{d\eta} \frac{\sin(\theta - \xi)}{n_{23} - \cos(\theta - \xi)} + r_1 \frac{n_{12}\cos(\xi - \eta) - 1}{n_{12} - \cos(\xi - \eta)} \frac{\sin(\theta - \xi)}{n_{23} - \cos(\theta - \xi)} - n_{12} \frac{dr_1}{d\eta} \tag{15.16}$$

where n_{12} and n_{23} are the dielectric contrasts at the first and second interface, respectively, G represents the target radiation pattern, and U represents the feed radiation pattern. K is a normalization constant, similar to Eq. (15.7). Functions $T_{//}$ and $T_{\perp}$ are the total transmissivities for parallel and perpendicular polarization

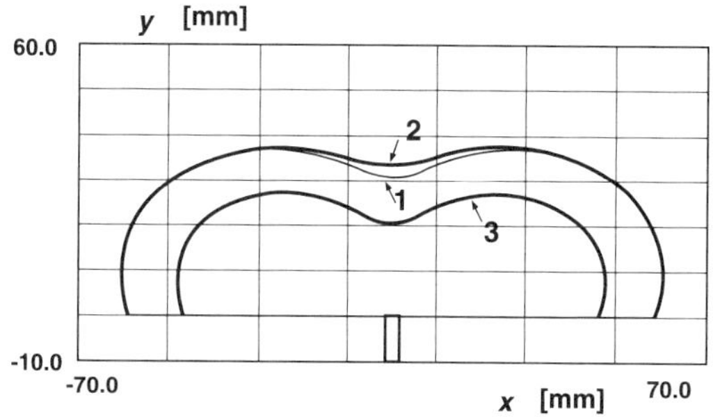

FIGURE 15.42 Profile of a double-material lens (curves 2 and 3), superimposed on the profile of an equivalent single material lens (curve 1). $\varepsilon_{r1} = 2.53$, $\varepsilon_{r2} = 1.59$. Feed is located at (0, 0).

$$T_{\|} = \left(1 - \left|\rho_{\|1}\right|^2\right)\left(1 - \left|\rho_{\|2}\right|^2\right) \tag{15.17}$$

$$T_{\perp} = \left(1 - \left|\rho_{\perp 1}\right|^2\right)\left(1 - \left|\rho_{\perp 2}\right|^2\right) \tag{15.18}$$

where $\rho_{//i}$ and $\rho_{\perp i}$ are the Fresnel reflection coefficients calculated at each of the lens interfaces.

There are four unknowns (r_1, r_2, ξ, and θ) and only three equations, so a further condition can be introduced to maximize power transmission across the lens. The optimum relation between the permittivity values of each shell remains to be defined. Not surprisingly, it can be shown that the lowest reflection loss occurs when dielectric contrast n_{ij} is equal in the two interfaces.

Figure 15.42 shows the profile of a 28-λ two-shell lens designed to produce a $\sec^2\theta$ pattern at 62.5 GHz, with $\theta_{max} = 76°$.[34] The feed is the same circular waveguide carrying the TE_{11} mode with circular polarization used in the examples of Section 15.3.5. The dielectric permittivity is 2.53 for the inner region, and 1.59 for the outer region. Figure 15.42 also shows the profile of a single-material dielectric lens with $\varepsilon_r = 2.53$, designed for the same target pattern (curve 1). The dielectric materials were considered as lossless in the calculations to isolate the effect of internal reflections.

In this example, the fraction of feed radiated power that is lost to internal reflections is reduced from 15% in the single material lens to about 7.5% for the two-shell lens. This behavior is maintained for more than a 10% bandwidth without compromising the $\sec^2\theta$ characteristic. The volume of the double-shell lens is similar to the equivalent single material lens.

Although the choice of dielectric constants is limited in practice to some available discrete values, it is still possible to find a combination that is not far from the equal contrast condition. The double-shell lens approach is preferable to the resonant layer strategy in terms of frequency behavior.

15.4 Other Beam-Shaping Antennas

As referred to in Section 15.1, there are other approaches to shape the beam, not only for millimeter-wave wireless systems but also for lower frequency applications. This section briefly points out some alternative solutions and related practical aspects.

15.4.1 Shaped Metal Antennas

The simplest metal antenna that can be used to produce an omnidirectional pattern with $\sec^2\theta$ elevation is the monopole with shaped ground plane.[35] The basic configuration of this antenna is shown in Fig. 15.2a. The monopole is mounted on a small pedestal on top of a shaped circular reflector. The design procedure comprises three steps: in the first step the reflector is taken as a conical surface, and the cone

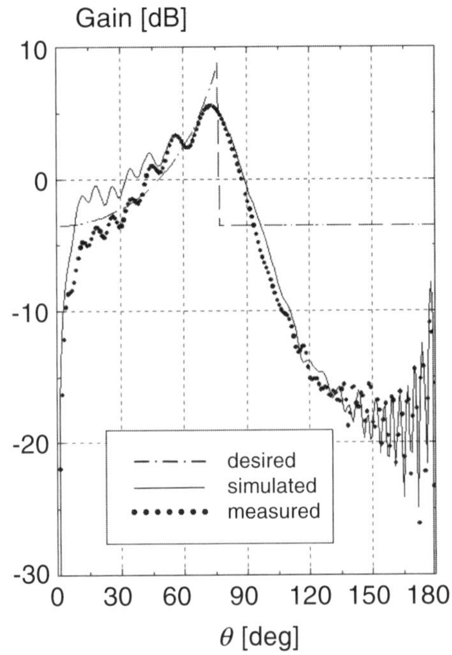

FIGURE 15.43 Radiation pattern of monopole antenna with shaped ground plane, $\theta_{max} = 76°$, $f = 60$ GHz. (Adapted from Kolundzija, B., Brankovic, V., and Zimmermann, S., *Proc. IEE Int.. Conf. Antennas Propag.*, 1997, 1460. With permission.)

angle is adjusted so that the direction of maximum radiation coincides with the desired θ_{max}; in the second step the edge of the cone is smoothed to remove the corresponding ripple from the radiation pattern; in the final step the pedestal is optimized so that sidelobes are fitted to the target $\sec^2 \theta$ curve. The latter step does not affect the previously established direction of the maximum. The design is supported on numerical optimization.

Figure 15.43 from Kolundzija[35] shows simulated and measured radiation patterns obtained for a 16-λ antenna at 60 GHz. The target elevation pattern is the $\sec^2 \theta$, with $\theta_{max} = 76°$. Except for the null in the central region that is intrinsic to linear polarization, the results show a good agreement with the target pattern. Although not very steep, there is a deep fall of the radiation beyond θ_{max}. The return loss is better than −13 dB within the 57 to 63 GHz frequency range.[11] The performance of this antenna is very sensitive to pedestal optimization, making this a critical point for fabrication.

A more elaborate metal antenna is also proposed by Kolundzija et al.[11] to produce the same type of shaped radiation pattern. It is a biconical horn with the two conical surfaces appropriately shaped to bring the radiation pattern close to the $\sec^2 \theta$ elevation characteristic. The basic configuration is shown in Fig. 15.2b. The remarkable feature of this antenna is its operation bandwidth of more then one octave,[11] although this largely exceeds the usual bandwidth requirements for wireless millimeter wave applications. For narrower bandwidths, the antenna can be made smaller than equivalent reflector antennas.

Several references exist in the literature on the use of reflector antennas to produce the type of shaped beams discussed in the previous sections.[8,36] Point to multipoint LMDS coverage at millimeter waves has been one motivation for the development of omnidirectional $\sec^2 \theta$ reflectors. Design principles similar to those used for lenses can be applied for reflectors: a condition for power conservation in a ray tube, and Snell laws for reflection.

Usually only an amplitude-shaping condition is required, so a single-reflector antenna should satisfy it. Bergmann[12] has studied such a single-reflector solution to produce an omnidirectional $\sec^2 \theta$ radiation pattern. The basic antenna configuration is represented in Fig. 15.2c. However compactness and better control of the feed spillover can be gained by introducing a subreflector.[36] A waveguide aperture or horn is placed in the focus of a parabolic reflector that illuminates the conical main reflector with a plane wave (Fig. 15.2d). The main reflector may be shaped in elevation to produce the prescribed elevation pattern. Shaping of the main reflector also in the azimuth has been reported in one case,[37] to produce square cell coverage. The TM_{01} mode is used on the feed to obtain vertical polarization, and the TE_{01}, for horizontal polarization.[38] The circular symmetry of the corresponding radiation pattern is compatible with the requirement for omnidirectional coverage. A dielectric cylinder is used to support the main reflector and subreflector. It acts as a radome, and further ensures that both reflectors are perfectly coaxial.

This antenna configuration may exhibit a wide shadow region at nadir not only because of the linear polarization but also because of the blockage by the main reflector. The conical main reflector tends to be deep, to minimize diffraction effects. The resulting total antenna depth is significantly larger than for an equivalent dielectric lens.

15.4.2 Array Antennas

Arrays of printed elements or dipoles can be designed to produce shaped beams. The advantage of these arrays over the previous solutions is the considerable reduction of volume and weight, making array antennas particularly attractive for lower frequency mobile applications.

Array synthesis methods are well known, and plenty of information on array antennas is given in subsequent chapters of this handbook. Besides analytic methods,[39-41] global optimization methods like genetic algorithms presented previously in Chapter 9 are attractive for the design of beam shaping array antennas. Figure 15.44 shows the calculated radiation pattern corresponding to a linear array of eight unequally spaced dipoles, optimized with genetic algorithms. WIPL-D is used as the electromagnetic solver for the genetic algorithm code, so coupling between the dipoles, the effect of dipole finite radius, and dipole radiation pattern are accounted for.

A total of 23 variables are used: dipole complex excitation coefficients, and distances between dipoles. The cost function specifies not only the $\sec^2\theta$ characteristic for the elevation interval $10° < \theta < 85°$, but also the frequency interval 860 MHz $< f <$ 940 MHz where the shaped characteristic must hold. Binary coding is used, with 5 b per variable. The size of the population in each generation is 48, with 33% superposition between successive generations. The resulting optimized variables are listed in Table 15.2; the array length is 8 λ. The radiation pattern is close to the target and is reasonably stable with frequency. The directivity of this omnidirectional pattern is 9.6 dBi, a value that compares well with what was previously obtained for very large apertures in terms of wavelength. Acceptable results were still obtained when fixing the excitation coefficients to a feasible distribution, and letting the genetic algorithm search only for the different spacing between dipoles.

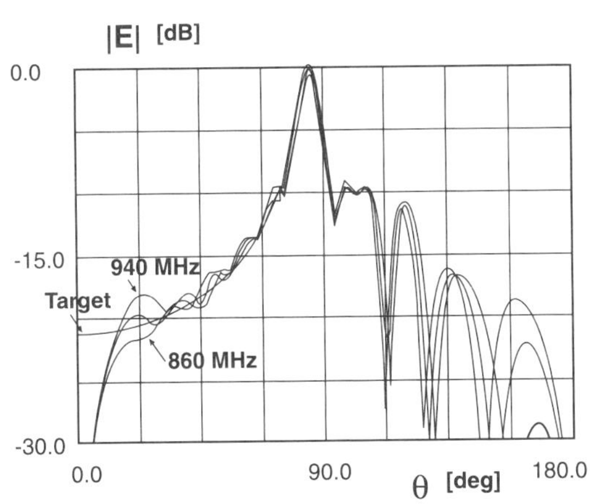

FIGURE 15.44 Calculated radiation pattern of linear array of eight dipoles optimized using genetic algorithms.

TABLE 15.2 Excitation Coefficients of Linear Array of Unequally Spaced Dipoles for $\sec^2 \theta$ Pattern at 900 MHz

Dipole No.	Re	Im	Distance [m]
1	–0.59	0.51	0.233
2	0.3	0.68	0.276
3	0.53	0.71	0.273
4	0.3	0.57	0.340
5	0.3	0.12	0.312
6	0.15	–0.35	0.180
7	0.35	–0.39	0.210
8	0.28	–0.12	—

The mast could also be included on the antenna model used for parameter optimization. However, simulation for the previous array shows that the influence of a vertical mast with 20-mm radius has negligible effect on the shaped radiation pattern provided that it is installed more than 2 λ away from the array axis.

Shaped-beam arrays are also used at millimeter waves, using printed antenna technology. At these frequencies dissipation losses and radiation in the feeding network may be high. Dissipation losses must be compensated with amplification, which may drive up the antenna cost. Besso[10] designed a 10 × 22 printed elements array to produce a cosec θ elevation pattern within a sectored region in the azimuth. The reported difficulty is the lack of space for the complex feeding circuit, resulting in coupling between radiating elements and feeding lines. Murakami et al.[9] developed a switchable four-sector printed antenna with a shaped beam in elevation, intended for base stations of wireless local area networks (WLANs) operating at the 60-GHz band. The antenna is formed by four linear arrays of four printed elements each, designed to produce a shaped elevation pattern with circular polarization. The antenna assembly is to be suspended from the ceiling. Each array is mounted on one side of a truncated pyramid, and is connected to an MMIC amplifier. Sector beam switching is obtained by on-off switching of the four amplifiers direct current (DC) bias.

Acknowledgments

I express my appreciation to all my undergraduate and graduate students who have participated in the development and evaluation of some of the lens antennas described in this chapter. Also, I thank Prof. Carlos Salema for his valuable comments and discussion, and Prof. Afonso Barbosa for the comments on the manuscript.

References

1. Fernandes, L., Developing a System Concept and Technologies for Mobile Broadband Communications, *IEEE Personal Communications Magazine*, 2, 54, 1995.
2. Dinis, M. et al., SAMBA — A Step to Bring MBS to the People, in *Proc. ACTS Mobile Communications Summit 1997*, Christiansen, C. (Ed.), Aalborg, Denmark, 1997, 495.
3. Plattner, A. et al., A Compact, Portable 40 GHz Transceiver for the Mobile Broadband System, in *Proc. ACTS Mobile Communications Summit 1998*, Rhodes, Greece, 1998, 843.
4. Dinis, M. et al., The SAMBA Trial Platform in the Field, in *Proc. ACTS Mobile Communications Summit 1999*, Sorrento, Italy, 1999, 1013.
5. Fernandes, J.G. and Fernandes, C.A., Impact of Shaped Lens Antennas on MBS Systems, in *Proc. PIMRC'98 — The 9th Int. Symp. on Personal, Indoor, and Mobile Radio Communications*, Boston, MA, 1998, 744.

6. Fernandes, C.A. and Fernandes, J.G., Performance of Lens Antennas in Wireless Indoor Millimeter-Wave Applications, *IEEE Transactions on Microwaves Theory and Techniques*, 47, 732, 1999.
7. Searle, A.D., Lewis, R.A. and Pirollo, B.P., A Millimeter Wave Reflector Antenna for Telematic Applications, in *Proc. IEE National Conference on Antennas and Propagation*, York, UK, 1999, 136.
8. Besso, P. et al., Azimuthally Omnidirectional Antenna with Shaped Elevation Pattern for LMDS Systems in the Band 40.5–42.5 GHz, in *Proc. JINA 98 — NICE International Symposium on Antennas*, Nice, France, 1998, 542.
9. Murakami, Y. et al., A Switchable Four-Sector Shaped Beam Antenna for mm-Wave Wireless LANs, in *Proc. PIMRC'98 — The 9th Int. Symp. on Personal, Indoor, and Mobile Radio Communications*, Boston, MA, 1998, 2248.
10. Besso, P. et al., Design and Characteristics of a Sector/Cosec^2 Shaped Beam Microstrip Antenna, in *Proc. IEE International Conference on Antennas and Propagation*, IEE, Eindhoven, Netherlands, 1995, 85.
11. Kolundzija, B., Brankovic, V. and Zimmermann, S., Feasibility Study of mm-Wave Metal Antennas, in *Proc. RACE Mobile Telecommunications Summit*, Granada, Spain, 1996, 295.
12. Bergmann, J.R. and Hasselmann, F.J., A Shaped Single Reflector Antenna for Omnidirectional Coverage in Point to Multipoint Services, in *Proc. Millennium Conference on Antennas and Propagation — AP2000*, Danesy, D. and Hugeete Sawaya (Eds.), Publisher, Davos, Switzerland, 2000, 530.
13. Fernandes, C.A. et al., Dielectric Lens Antennas for Wireless Broadband Communications, *Wireless Personal Communications Journal*, 10, 19, 1998.
14. Fernandes, C.A., Shaped Dielectric Lenses for Wireless Millimeter-Wave Communications, *IEEE Antennas and Propagation Magazine*, 41, 141, 1999.
15. McEwan, N.J. et al., Pattern Shaping for Handset Antennas, in *Proc. International Conference on Antennas and Propagation*, IEE, Eindhoven, Netherlands, 1995, 106.
16. Fernandes, C.A., Francês, P.O. and Barbosa, A.M., Shaped Coverage of Elongated Cells at Millimeter-waves Using a Dielectric Lens Antenna, in *Proc. 25th European Microwave Conference*, Nexus Media Ltd., Bologna, Italy, 1995, 66.
17. Salema, C., Fernandes, C. and Jha, R., *Solid Dielectric Horn Antennas*, Artech House, Boston, 1998, chap. 7.
18. Lemaire, C. et al., A Method to Overcome the Limitations of G.O. in the Design of Axis-Symmetrical Lenses, *Journal of Infrared and Millimetrewaves*, 17, 1377, 1996.
19. Von Hippel, A.R., Dielectric Materials and Applications, Artech House, Boston, 1995.
20. Pires, J., Almeida, J. and Fernandes, C.A., Determination of the Complex Permittivity of PLEXIGLAS and TEFLON at the 60 GHz Band, Report MBS/WP462/IST079.1, RACE-2067 project, IST Lisbon, 1994.
21. Kolundzija, B., Ognjanovic, J. and Sarkar, T., *WIPL-D: Electromagnetic Modeling of Composite Wire and Plates Structures — Software and User's Manual*, Artech House, Boston, 2000.
22. Silver, S. (Ed.), *Microwave Antenna Theory and Design*, McGraw-Hill Book, 1949, chap. 10 and 11.
23. Lewin, L., *Advanced Theory of Waveguides*, London, 1951, chap. 6.
24. Fradin, A. Z., *Microwave Antennas*, Pergamon Press, London, 1961, 263.
25. Olver A.D. et al., *Microwave Horns and Feeds*, IEE Press, London, 1994, chap. 7.
26. Baudrand, H., Tao, J-W. and Atechian, J., Study of Radiating Properties of Open-Ended Rectangular Waveguides, *IEEE Transactions on Antennas Propag.*, 36, 1071, 1988.
27. Wu, D. and Kanda, M., Comparison of Theoretical and Experimental Data for the Near Field of an Open-Ended Rectangular Waveguide, *IEEE Transactions on Electromagnetics Compatibility*, 31, 353, 1989.
28. Moheb, H. and Shafai, L., Numerical Computation of Radiation Characteristics of Rectangular Waveguides, in *Proc. IEEE Antennas and Propagation Society Int. Symp.*, Dallas, Texas, 1990, 1843.
29. Filipovic, D. et al., Double-Slot Antennas on Extended Hemispherical and Elliptical Silicon Dielectric Lenses, *IEEE Transactions on Microwave Theory and Technology*, 41, 1738, 1993.

30. Neto, A. et al., Mutual Coupling Between Slots Printed at the Back of Elliptical Dielectric Lenses, *IEEE Transactions on Antennas and Propagation*, 47, 1504, 1999.
31. Rodrigo, M. and Fernandes, C.A., A New Broadband Printed Antenna Element for Endfire Radiation, in *Proc. II Conf. de Telecomunicações*, Sesimbra, Portugal, 1999, 446.
32. Rodrigo, M. and Fernandes, C.A., Shaped Dielectric Lens fed by a Printed Element for WLANS at 40 GHz, in *Proc. Millennium Conference on Antennas and Propagation — AP2000*, Danesy, D. and Hugeete Sawaya (Eds.), Publisher, Davos, Switzerland, 2000, 446.
33. Fernandes, C.A., Design of Shaped Lenses for Non-Symmetric Cells in MBS, in *Proc. IEEE Antennas and Propagation International Symposium*, Orlando, FL., 1999, 2440.
34. Silveirinha, M. and Fernandes, C.A., Shaped Double-Shell Dielectric Lenses for Wireless Millimeter Wave Communications, in *IEEE Antennas and Propagation International Symposium*, Salt Lake City, UT, 2000, 1674.
35. Kolundzija, B., Brankovic, V. and Zimmermann, S., Design of Monopole Antenna Mounted on a Shaped Circular Reflector for mm-Wave Application, in *Proc. IEE Int. Conference on Antennas and Propagation*, Edinburg, 1997, 1460.
36. Norris, A.P. and Waddoup, W.D., A Millimetric Wave Omnidirectional Antenna with Prescribed Elevation Shaping, in *Proc. IEE Int. Conference on Antennas and Propagation*, IEE, Coventry, UK, 1985, 141.
37. Orefice, M., A Shaped-Omnidirectional Pattern Dual Reflector Antenna for mm-Waves, in Proc. *IEEE Antennas and Propagation Society Int. Symp.*, Atlanta, Georgia, 1998, 844.
38. Vallauri, R. et al., Azimuthally Omnidirectional Antenna with Vertical or Horizontal Polarisation for LMDS in the 40 GHz Band, in *Proc. Millennium Conference on Antennas and Propagation — AP2000*, Danesy, D. and Hugeete Sawaya (Eds), Publisher, Davos, Switzerland, 2000, 555.
39. Orchard, H.J., Elliot, R.S. and Stern, G.J., Optimizing the Synthesis of Shaped Beam Antenna Patterns, *Proc. IEE*, 132-H, 63, 1985.
40. Buckley, M., Synthesis of Shaped Beam Antenna Patterns Using Implicitly Constrained Current Elements, *IEEE Transactions on Antennas Propagation*, 44, 192, 1996.
41. Gonzalez, J.J., Sierra, M. and Montesano, A., Synthesis of Shaped Beam Antenna Patterns Using Window Currents, in *Proc. JINA 98 — NICE Int. Symp. Antennas*, Nice, France, 1998, 358.

PART D

Smart Antennas and Array Theory

16

Basic Array Theory and Pattern Synthesis Techniques

Boon Poh Ng
Nanyang Technological University

Meng Hwa Er
Nanyang Technological University

16.1 Introduction

16.1.1 Background

In signal reception system design, several possible diversity techniques, such as time, frequency, and space diversities, can be used to obtain satisfactory signal reception performance even under harsh operation environments. For the space diversity technique, diversity is achieved basically using multiple sensors, arranged in a certain pattern, which is known as an array. The use of an array has long been an attractive solution for many applications such as sonar, radar, communications, and seismology. Particularly, because of the explosion of wireless communication applications such as land mobile and wireless local loop (WLL) communications, spatial filtering is of great interest to the wireless communication system designers to achieve space division multiple access (SDMA).

In spatial filtering, the objective of using an array is to extract the desired signal copy while filtering out the unwanted interference, and environmental and system noise. This filtering in the spatial domain

0-8493-0124-6/02/$0.00+$1.50

using an array is also known as array beamforming. In digital signal processing, filtering can be broadly classified into filter synthesis and adaptive filtering. As for array beamforming, it can also be divided into two types: array synthesis and adaptive array beamforming. Generally, adaptive array is able to provide better signal reception as compared with the synthesized array. This is because the adaptive processing of an array is performed on a signal data dependent basis or as a statistically optimum process, whereas the synthesized array always has a set of fixed weights that produces a fixed array response. However, this does not suggest that the adaptive array is always preferred as compared with the synthesized array because an adaptive array system is a more costly system, in the sense that a fast speed computing machine is required to calculate and adjust the array weights according to the data collected. For a special class of array synthesis, the desired array beam pattern may even be achieved, which involves no signal weighting at all, by arranging the array in a certain geometric pattern.

Although adaptive beamforming techniques have been in existence for many years, there is still a major obstacle on the route to practical implementation. The problem is, when a physical sensor array is constructed, the array manifold associated with the array is always not completely known. Therefore, the estimation of array manifold parameters to minimize array uncertainties could pose a problem in the design of a modern array system. Signal cancellation, resulting from multipath or correlated signals, could also be a problem in some adaptive array processing algorithms. In many cases, a switched beam system is preferred as compared with a fully adaptive antenna array system (such a system is also known as *Smart Antennas*), because of its simplicity in the system itself and retrofitting to existing wireless communication technologies. In a switched beam system, a number of fixed beams are used at the base station or an antenna site. On receiving a signal, the receiver selects a beam to perform spatial filtering so as to provide the best signal enhancement while suppressing the interferences and noise. Besides its importance in switched beam system design, beam pattern synthesis is also an issue in the adaptive array system, or smart antenna system. This is because when an adaptive array is turned on, before it adjusts its weights according to the signal environment, its array weights have to be initialized while the array is adapting or being trained. One possibility is for the array to be initialized to certain presynthesized weights. In cases where a quiescent pattern is required, array synthesis techniques can be used to design such quiescent patterns.

16.1.2 Overview of Array Pattern Synthesis

As discussed in the previous section, a fixed beam can be synthesized by array weight or array geometry control. The two approaches are further discussed next.

16.1.2.1 Array Weight Synthesis

An array may be regarded as a spatial filter: to allow signal from a certain direction to pass through while rejecting or stopping all other signal sources (from other directions) impinging on the array. To achieve this filtering objective, naturally, one would like to design an array with high gain at the desired signal direction (look direction) and low gain at the nonlook directions. Theoretically, the array should be designed with a maximum directivity so as to achieve maximum signal to noise plus interference ratio (SNIR) at the output of the array beamformer. However, this is only true if the interferences (assumed equi-power) are evenly distributed, or assume certain distribution patterns over the whole spatial domain; the maximum directivity design may not be the best design. Therefore, if the designer of the array beamformer does not know the distribution of the directions of the interferences, *a priori*, an alternative design such as equi-sidelobe design may be preferred. In the case of linear array design, two important types of arrays are the equi-spaced uniformly excited array and the Chebyshev excited array. The former is an array with high resolution but with a relatively high sidelobe, whereas the latter generally has a larger beam width as compared with the former, its sidelobes are controllable and of equiheight. Dolphs[7] provides, in his classical paper, a procedure to obtain the Chebyshev linear array. For a rectangular array or circular planar array, using other modern approaches, Chebyshev patterns can also be obtained or approximated through an iterative numerical process or optimization process. In most cases, an array is synthesized with the assumption that the array elements are isotropic. However, for practical array sensors, each sensor may have its own characteristics (such as polar and gain/phase response). It is therefore undesirable to always assume that the elements are isotropic. In this respect, synthesis methods, such as those of Er et al.,[10] Ng et al.,[19] Olen and Compton,[22] and Sim and Er,[25] that do not have the preceding shortcomings are very useful.

16.1.2.2 Array Geometry Synthesis

In the case of array weight synthesis, if one is to consider the array weights as part of the array element characteristics, by adjusting the array weight vector, one is actually adjusting the performance of the array manifold. The array geometry, the parameter of the array manifold, can therefore be adjusted with all the current excitation kept uniform or at unity somehow, to achieve the objective of array synthesis. In this regard, synthesizing an array beam pattern is termed here as array geometry synthesis. One possible approach to synthesize an array geometry is the so-called minimum-redundant or nonredundant array. This approach basically gives rise to a high-resolution array, at the expense of high-sidelobe level and very long aperture length. Another area of geometry synthesis is to adjust the array element positions so as to perform null steering, to cancel unwanted interference. One such example is given by Ismail and Mahmoud.[13] Other examples, such as those illustrated by Ng et al.,[18,20,21] use the quadratic programming methods to solve for the array geometry. Nevertheless, there has not been very much research work reported so far in the literature on array geometry synthesis. This may result from the fact that optimizing the array pattern with respect to the array geometry is a highly nonlinear and difficult problem.

In most cases, linear array synthesis techniques assume the array element spacing to be a half wavelength; this is, however, not always a necessary assumption. In fact, another way of improving the beam width of an equispaced linear array while maintaining the sidelobe level is simply by extending the array spacing until just before spatial aliasing occurs. In this chapter, the expression or process to achieve the limit of the aperture length of a uniform or a Chebyshev array is given. In this case, the aperture length is obtained by setting the sidelobe level, at the two endfire directions, to be the same as the highest sidelobe level of the original half wavelength spaced array.

16.1.3 Chapter Outline

This chapter comprises five sections, including this introductory section.

Besides defining the mathematical notation used to describe an array beamformer, Section 16.2 also introduces some of the basic concepts and antenna array geometries used in array beamforming. In the next two sections, design techniques are briefly explained and illustrated with design examples.

Section 16.3 describes the various array pattern synthesis techniques — from the classical ones such as uniform, Chebyshev and Taylor array synthesis to the modern ones such as the adaptive array, quadratic programming, and genetic algorithm based methods. A brief comparison of the various approaches are also included in this section.

Section 16.4 addresses the issues and techniques for optimum array geometry synthesis. This section provides the limits of the aperture length of an equispaced uniform linear array and Chebyshev array. The corresponding expressions for the array beam widths are also given. The section also describes a method to synthesize an array with lower sidelobe and smaller beam width with the array aperture maintained more or less the same as that of a half wavelength equispaced linear array. This synthesis may also be applied to a planar array. To reduce the computation load of the synthesis method, the first-order Taylor series expansion is used to calculate the array geometry in an approximation approach.

The last section, Section 16.5, concludes the chapter with a summary on the techniques described and provides some suggestions on further reading materials, should one need to explore alternatives for system design.

16.2 Basic Theory of Antenna Array

This section introduces the mathematical notation, the array beamforming structure, and some definitions of the performance measure of an array synthesis, so as to provide a framework for discussion on the various synthesis techniques described in the next three sections.

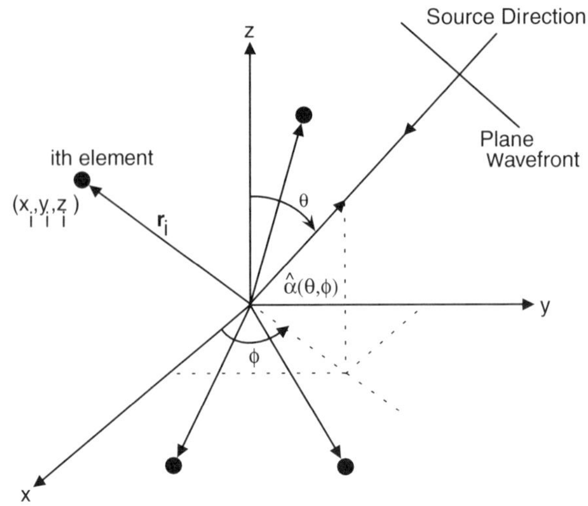

FIGURE 16.1 Reference coordinate of an arbitrary array.

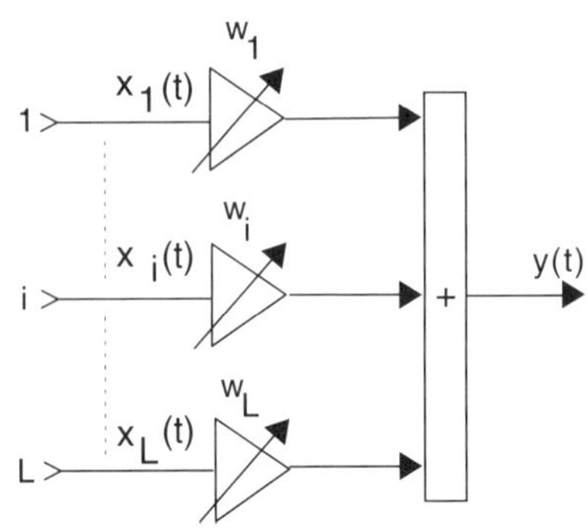

FIGURE 16.2 Narrowband antenna array beamformer with L elements.

16.2.1 Array Response and Beamforming Structure

In this chapter, it is assumed that the array is located in the far field of the point source. Therefore, as far as the array is concerned, the directional signal impinging on the array is considered as a plane wave. With the array reference coordinates, defined in Fig. 16.1, and the structure of the L-element antenna array beamformer, which basically performs the space-sampling and weigh-and-sum operation to produce the output signal $y(t)$, shown in Fig. 16.2, one can define the array response at certain operating frequency and to a plane wave front with unity amplitude arriving in direction (θ, ϕ) as follows.

The array response of the narrowband beamformer can be written as

$$H(f,\theta,\phi,\mathbf{g},\mathbf{r}) = \mathbf{w}^H \mathbf{s}(f,\theta,\phi,\mathbf{g},\mathbf{r}) \tag{16.1}$$

where the superscript H denotes Hermitian transpose and $\mathbf{w}$ is the L-dimensional complex vector of adjustable weights given by

$$\mathbf{w} = [w_1, w_2, \ldots, w_L]^T \tag{16.2}$$

where $[\bullet]^T$ denotes transpose and $\mathbf{s}(f, \theta, \phi, \mathbf{g}, \mathbf{r})$ is the L-dimensional vector, which is known as the steering vector, defined as

$$\mathbf{s}(f,\theta,\phi,\mathbf{g},\mathbf{r}) = diag\left[\mathbf{g}(f,\theta,\phi)\right]\mathbf{u}(f,\theta,\phi,\mathbf{r}) \tag{16.3}$$

where $\mathbf{u}(f, \theta, \phi, \mathbf{r})$ and $\mathbf{g}(f, \theta, \phi)$ are L-dimensional vectors defined by

$$\mathbf{u}(f,\theta,\phi,\mathbf{r}) = \left[\exp(j\mathbf{r}_1\mathbf{k}),\ \exp(j\mathbf{r}_2\mathbf{k}),\ \ldots,\ \exp(j\mathbf{r}_L\mathbf{k})\right]^T \tag{16.4}$$

$$\mathbf{g}(f,\theta,\phi) = \left[g_1(f,\theta,\phi),\ g_2(f,\theta,\phi),\ \ldots,\ g_L(f,\theta,\phi)\right]^T \tag{16.5}$$

Note that Eq. (16.4) can also be expressed as

$$\mathbf{u}(f,\theta,\phi,\mathbf{r}) = \left[\exp(j2\pi f\tau_1(\theta,\phi,\mathbf{r}_1)),\ \exp(j2\pi f\tau_2(\theta,\phi,\mathbf{r}_2)),\ \ldots,\ \exp(j2\pi f\tau_L(\theta,\phi,\mathbf{r}_L))\right]^T \tag{16.6}$$

The wave number vector $\mathbf{k}$ in Eq. (16.4) and the array geometry vector $\mathbf{r}$ are given by

$$\mathbf{k}(f,\theta,\phi) = (2\pi f/\nu)\left[\cos(\phi)\sin(\theta)\ \sin(\phi)\sin(\theta)\ \cos(\theta)\right]^T \tag{16.7}$$

$$\mathbf{r} = \left[\mathbf{r}_1^T\mathbf{r}_2^T\ldots\mathbf{r}_L^T\right]^T \tag{16.8}$$

where ν is the speed of propagation of the incident wave, $\mathbf{r}_i$ is the ith sensor location vector given by $\mathbf{r}_i = [x_i, y_i, z_i]$, and $g_i(f, \theta, \phi)$ is the complex response of the ith array sensor given by

$$g_i(f,\theta,\phi) = \left|g_i(f,\theta,\phi)\right| \exp(j\varphi_i(f,\theta,\phi)) \qquad i = 1, 2, \ldots, L \tag{16.9}$$

and τ_i is the propagation delay which can be obtained from Fig. 16.1. Mathematically, it is given by

$$\tau_i(\theta,\phi,\mathbf{r}_i) = \frac{\hat{\alpha}(\theta,\phi)\cdot\mathbf{r}_i}{\nu} \qquad i = 1, 2, \ldots L \tag{16.10}$$

where "·" is the dot product operator, $\hat{\alpha}(\theta, \phi)$ is the unit direction vector, and $\mathbf{r}_i$ is the array element position vector defined in Fig. 16.1. The expression in Eq. (16.10) can also be written as

$$\tau_i(\theta,\phi,\mathbf{r}) = \frac{1}{\nu}\left[(x_i\cos(\phi) + y_i\sin(\phi))\sin(\theta) + z_i\cos(\theta)\right] \qquad i = 1, 2, \ldots, L \tag{16.11}$$

Note that the signals existing in the signal field are assumed to have all their signal energy concentrated around $f = f_0$ for a narrowband beamforming structure.

16.2.2 Array Geometry

An array geometry is described by the array geometry matrix vector **r**, defined in Eq. (16.8), for the reference coordinate system given in Fig. 16.1. The elements of the **r** are written in terms of fraction or number of half wavelength ($\lambda/2$), where λ_0 is used to denote the wavelength corresponding to the operating frequency f_0 of the array processing system, shown as follows:

$$f_0\lambda = \nu \tag{16.12}$$

With the preceding normalized scheme, array sensor position matrices vectors can be easily obtained for the linear, circular-ring arrays as well as planar arrays such as rectangular and circular grid arrays.

16.2.3 Important Array Beamforming Concepts and Terminologies

16.2.3.1 Radiation Pattern

From the reciprocity theorem, it can be shown that the radiation pattern of an antenna remains the same whether it is used as a transmitting antenna or a receiving antenna. Similarly, the radiation pattern of a transmitting array is identical to its array response.

16.2.3.2 Array Factor and Pattern Multiplication

With the array response defined in Eq. (16.1), an array factor is the special case of array response H when **g** is equal to $[1\ 1\ \ldots\ 1]^T$. This means that when the antenna elements are all isotropic, the antenna array response is the array factor. It is also clear from Eqs. (16.1), (16.3) and (16.5) that if all g_i's are identical, an array response is the product of the array factor and the element factor (or element pattern, i.e., $g_i(f, \theta, \phi)$). This is the so-called rule of *pattern multiplication*.

16.2.3.3 Array Beam Width

The array beam width refers to the angular width that the array main beam or array main lobe occupies when measured at a particular constant power level. Commonly, the beam width is measured at the half-power point or the –3dB point. This beam width is therefore known as *half-power beam width* or *3-dB beam width*. The other alternative is to measure the beam width between the first nulls on either side of the main beam. In this case, the beam width is known as the *first-null beam width*. However, in most cases, when the term beam width is used, it usually refers the 3-dB beam width.

16.2.3.4 Directivity and Gain

In many applications, the primary objective of an antenna array is to shape a response or beam pattern such that radiation (or reception) in a certain direction is enhanced and the reception in other directions is suppressed. A useful measure of the sharpness of the array is array directivity, which is defined as the ratio of the power radiated by an array in a particular desired direction to the average of the power radiated by the array in all directions. In the context of array synthesis, as the losses in antennas and antenna circuits are not considered, array gain is frequently used interchangeably with array directivity. However, it is important to note that although array directivity and array gain are related, they are not the same.

16.2.3.5 Grating Lobes

Grating lobes are sidelobes that have an intensity equal to that of the main beam. Because the main purpose of array synthesis is sidelobe reduction for interference suppression, grating lobes are undesirable. Grating lobes occur if the array element spacing is greater than or equal to one wavelength ($1\ \lambda$) for a uniform array. As a result, an array is usually designed with an array spacing close to a half wavelength.

16.2.3.6 Array Manifold

Consider an array of L sensor elements and form the steering vector **s** corresponding to the direction (θ, ϕ). The continuum of the vector $\mathbf{s}(\theta, \phi)$ is a function of (θ, ϕ) and is a two-dimensional continuum lying in an L-dimensional space. The continuum is known as the array manifold. The array manifold can be calculated and stored for a particular array only from the knowledge of the locations and directional characteristics of all array elements. Thus, the array manifold completely characterizes any array and provides a representation of the real array in L-dimensional complex space.

16.3 Array Weight Synthesis Techniques

In sensor array weight synthesis, the main concern is to determine the excitations of a given array so as to produce a beam pattern that suitably approximates a desired pattern. The desired beam pattern can vary widely depending on the application. Among these applications, there is a major class of pattern synthesis[9] that aims to achieve a high mainlobe to sidelobe ratio or high-array resolution (i.e., narrow main beam). Array weights are calculated to achieve a certain desired array response. One typical example for the type of pattern synthesis method is the Dolph–Chebyshev method,[7] which designs the array pattern with equi-sidelobe response using an uniformly spaced linear array. From the literature, it is noted that most of these synthesis methods[4,7,23,24] are only applicable to the design of linear array but not other arrays such as planar arrays or three-dimensional arrays.

From an implementation viewpoint, it is desirable to synthesize an array with the fewest assumptions made about the array characteristics, such as isotropic elements, so that the synthesized array excitation gives the true array pattern for a set of selected array sensors with certain responses or characteristics. In the literature, most of the synthesis techniques[7,9,27,31] assume that the array elements are isotropic and the final array pattern is then obtained by pattern multiplication. Therefore, even if the synthesis is an optimum one, it is only optimum for an array of isotropic elements and may not be optimum for the actual array selected. Moreover, most of the synthesis methods are only tailored to a certain array geometry. In this respect, a conceptually easy and highly flexible synthesis method is desirable. The methods given in References 2, 8, and 22 offer some flexibility in the design using iterative process. The convergence to a solution, of course, depends on the design specifications and the numerical robustness of the algorithm itself. Another class of synthesis techniques is given in References 10, 18, 19, and 25. The basic idea of one such technique[19] is to form a quadratic program with its cost function given by the mean square error between the array response and a properly selected pattern described by a known mathematical function. This quadratic program can be a constrained or unconstrained optimization problem depending on the requirements of the desired array pattern. Because the computation power available from a personal computer has increased manyfold in the past 10 years, most of the techniques mentioned earlier are implementable on a personal computer for an antenna array with more than several hundreds of elements.

16.3.1 Uniform Array

For a uniform linear array, the array response can be simplified from Eq. (16.1) to

$$H(z) = \sum_{k=1}^{L} z^k \tag{16.13}$$

where

$$z = e^{j\psi} \tag{16.14}$$

$$\psi = \pi\beta\cos(\phi) \tag{16.15}$$

Note that the variable z in Eqs. (16.13) and (16.14) for the z plane should not be confused with the spatial coordinate z.

16.3.1.1 Broadside Linear Array

From Section 16.3.1, it can be shown that the array factor $f(\phi)$ of a broadside linear array is given by

$$f(\phi)=\frac{\sin^2\left(\frac{1}{2}\pi L\beta\cos(\phi)\right)}{L^2\sin^2\left(\frac{1}{2}\pi\beta\cos(\phi)\right)} \tag{16.16}$$

For large arrays the main beam occupies a small solid angle at broadside. For this case, the sine function in the denominator of Eq. (16.16) can be replaced by its argument, yielding

$$f(\phi)=\left[\frac{\sin\left(\frac{1}{2}\pi L\beta\cos(\phi)\right)}{\frac{1}{2}\pi L\beta\cos(\phi)}\right]^2=\left[\frac{\sin(\pi\chi)}{\pi\chi}\right]^2=\mathrm{sinc}^2(\chi) \tag{16.17}$$

where

$$\chi=\tfrac{1}{2}L\beta\cos(\phi) \tag{16.18}$$

Note that the pattern of Eq. (16.17) is a sinc function and has its half-power point occurs at

$$\pi\chi=1.39 \tag{16.19}$$

From Eqs. (16.18) and (16.19), the beam width of the array in degree is obtained as

$$\Theta=2\sin^{-1}\left(\frac{2.78}{\pi L\beta}\right) \tag{16.20}$$

where the sine function is substituted for the cosine function to shift the reference angle from the line of the array to the broadside. For large arrays, the sine function can be replaced by its argument. Therefore, from Eq. (16.20), the beam width expression in degree is simplified further to

$$\Theta=\frac{1.77}{L\beta} \tag{16.21}$$

16.3.1.2 Scanned Linear Array

For a scanned array, for a scanned angle of ϕ_0, ψ is given by

$$\psi=\pi\beta\left(\cos(\phi)-\cos(\phi_0)\right) \tag{16.22}$$

which gives rise to the array factor with the expression

$$f(\phi)=\frac{\sin^2\left(\frac{1}{2}\pi L\beta\left(\cos(\phi)-\cos(\phi_0)\right)\right)}{L^2\sin^2\left(\frac{1}{2}\pi\beta\left(\cos(\phi)-\cos(\phi_0)\right)\right)} \tag{16.23}$$

For large arrays, the scanned main beam occupies a small solid angle if it is located around the broadside. For this case, the sine function in the denominator of Eq. (16.23) can be replaced by its argument, yielding

$$f(\phi)=\left[\frac{\sin\left(\frac{1}{2}\pi L\beta\left(\cos(\phi)-\cos(\phi_0)\right)\right)}{\frac{1}{2}\pi L\beta\left(\cos(\phi)-\cos(\phi_0)\right)}\right]^2=\left[\frac{\sin(\pi\chi)}{\pi\chi}\right]^2=\mathrm{sinc}^2(\chi) \tag{16.24}$$

where

$$\chi=\tfrac{1}{2}L\beta\left(\cos(\phi)-\cos(\phi_0)\right) \tag{16.25}$$

Note that the pattern of Eq. (16.24) is a sinc function with a peak value of unity. Therefore, its two half-power points can be evaluated from the following expression:

$$\frac{\sin(\pi\chi)}{\pi\chi}=\frac{1}{\sqrt{2}} \tag{16.26}$$

which can be solved and yields

$$\pi\chi=\pm 1.392 \tag{16.27}$$

From Eqs. (16.25) and (16.27), two solutions of ϕ are obtained, that is

$$\phi_1=\cos^{-1}\left(\cos(\phi_0)+\frac{0.886}{\beta L}\right) \tag{16.28}$$

$$\phi_2=\cos^{-1}\left(\cos(\phi_0)-\frac{0.886}{\beta L}\right) \tag{16.29}$$

From Eqs. (16.28) and (16.29), the beam width of the scanned array in degree is obtained as follows:

$$\Theta=\phi_2-\phi_1=\cos^{-1}\left(\cos(\phi_0)-\frac{0.886}{\beta L}\right)-\cos^{-1}\left(\cos(\phi_0)+\frac{0.886}{\beta L}\right) \tag{16.30}$$

16.3.2 Dolph–Chebyshev Array Pattern Synthesis

In this section, the problem to be addressed is the special case where equi-sidelobe array response is to be designed on a linear equispacing linear array. It is well known that the equi-sidelobe array response can be obtained by going through a procedure that first determines the root locations of the Chebyshev polynomial.[7] This procedure is very time consuming and therefore more efficient methods have subsequently been reported in the literature. One such method can be found in Reference 9.

Let

$$\mu_0=\cosh\left[\frac{1}{L-1}\cosh^{-1}\left(10^{-sll/20}\right)\right] \tag{16.31}$$

where *sll* denotes the sidelobe level of the desired beam pattern in terms of decibels.

By using the synthesis method by Elliott,[9] for an odd number linear array with 2Γ + 1 (i.e., L) elements, the weight vector elements are given by

$$w_{\Gamma-n+1} = \sum_{p=n}^{\Gamma}(-1)^{\Gamma-p}\frac{\Gamma}{\Gamma+p}\binom{\Gamma+p}{2p}\binom{2p}{p-n}\mu_0^{2p} \qquad \text{for } n = 0, 1, 2, \ldots, \Gamma \tag{16.32}$$

and, for an array with 2Γ (i.e., L) elements, the weight vector elements are determined by

$$w_{\Gamma-n+1} = \sum_{p=n}^{\Gamma}(-1)^{\Gamma-p}\frac{2\Gamma-1}{2(\Gamma+p-1)}\binom{\Gamma+p-1}{2p-1}\binom{2p-1}{p-n}\mu_0^{2p-1} \qquad \text{for } n = 0, 1, 2, \ldots, \Gamma \tag{16.33}$$

The beam width of a Chebyshev array with half-wavelength element spacings is given by

$$\Theta_{Chebyshev} = \Theta_{uniform} \times \text{bbf} \tag{16.34}$$

where bbf is the so-called beam broadening factor that can be approximated by the following linear equation for the range of sidelobe level from 25 to 55dB:

$$bbf = 10^{-2} \times (1.723 \times sll + 62.6) \tag{16.35}$$

The directivity of the array is linearly depending on the length of the array and is also proportional to the reciprocal of the beam width. The directivity is approximated by

$$D \doteq \frac{101.5^{o}}{\Theta_{Chebyshev}} \tag{16.36}$$

Figure 16.3 gives an example of Chebyshev array pattern. The associated array weight distribution is also shown. One can see that the array weights at the two endfire ends are relatively large as compared with those of the other array elements. This may be a problem in array beamformer hardware implementation. For example, if a microstrip patch array is to be designed with a cooperate feeding network, the large ratios between the array weights may suggest a feeding network that is not realizable. For the technique to be introduced, that is, the Taylor synthesis technique, such a problem usually does not occur. Alternatively, if an array beamformer is implemented on a software radio architecture, the existence of large array weights may not be an issue at all.

16.3.3 Taylor Array Pattern Synthesis

Taylor synthesis is another commonly used technique for designing a reduced sidelobe array beam pattern. This technique differs from Chebyshev synthesis in that the relation between beam width and sidelobe level is not optimum. In this case, only a specific number of close-in sidelobes are of constant level but for the rest of the sidelobes, the heights decrease away from the broadside of the array. Though the Taylor pattern is different from the Chebyshev pattern, the beam width–sidelobe relationship of Taylor pattern is generally quite close to optimum. Moreover, the array weights for the end elements usually do not increase as much as compared with those of Chebyshev arrays. As a result, Taylor synthesis is of considerable practical interest.

Similar to that of the Chebyshev array, several approaches can synthesize the array weights of a Taylor pattern. Over here, only the aperture sampling method is introduced.

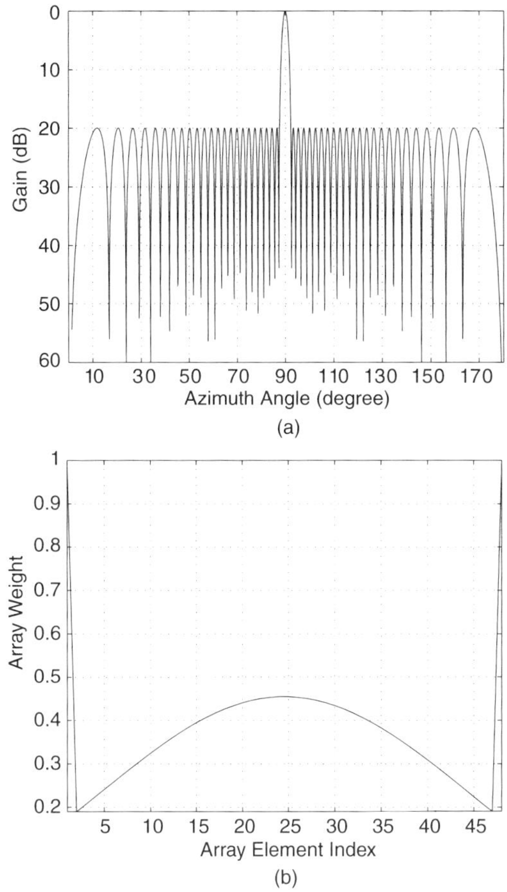

FIGURE 16.3 (a) The beam pattern and (b) the array weight distribution of the 48-element, Chebyshev linear array. The interelement spacing is a half wavelength and the sidelobe level is 20 dB.

Consider a linear array of L-element and let

$$N = L - 1 \tag{16.37}$$

then the array weights $\{w_i, i = 1, 2, \ldots, L\}$ are given by

$$w_i = \frac{1}{L}\left\{1 + 2\sum_{n=1}^{\bar{n}-1} f(n, A, \bar{n})\cos\left(\frac{2\pi n z_i}{L}\right)\right\} \tag{16.38a}$$

where the parameter $\bar{n}$ is the number used to decide the number of close-in sidelobes to be set with a constant sidelobe level. In practice, $\bar{n}$ is usually set in the range of 2 to 6. Other parts of the preceding expression are defined as

$$z_i = i - \left(\frac{N}{2} + 1\right) \tag{16.38b}$$

$$f(n,A,\bar{n}) = \begin{cases} \dfrac{[(\bar{n}-1)!]^2}{(\bar{n}-1+n)!(\bar{n}-1-n)!} \prod_{m=1}^{\bar{n}-1} \left[1-\left(\dfrac{n}{u_m}\right)^2\right] & |n| < \bar{n} \\ 0 & |n| \geq \bar{n} \end{cases} \tag{16.38c}$$

$$u_m = \begin{cases} \pm a\sqrt{A^2 + \left(m - \dfrac{1}{2}\right)^2} & 1 \leq m < \bar{n} \\ \pm m & \bar{n} \leq m < \infty \end{cases} \tag{16.38d}$$

$$a = \frac{\bar{n}}{\sqrt{A^2 + (\bar{n}-0.5)^2}} \tag{16.38e}$$

$$A = \frac{1}{\pi}\cosh^{-1} R \tag{16.38f}$$

$$R = 10^{sll/20} \tag{16.38g}$$

The beam width of the Taylor array in degree is approximated as

$$\Theta_{Taylor} \doteq 2\sin^{-1}\left\{\frac{a}{L\pi}\sqrt{\left(\cosh^{-1} R\right)^2 - \left(\cosh^{-1}\frac{R}{\sqrt{2}}\right)^2}\right\} \tag{16.39}$$

and the directivity as

$$D \doteq \frac{101.5^o}{\Theta_{Taylor}} \tag{16.40}$$

Figure 16.4a shows the Taylor array pattern with its array weight distribution shown in Fig. 16.4b. One can observe that the array weights at the two ends do not vary as much as those of the Chebyshev array.

16.3.4 Array Pattern Synthesis Using Quadratic Programming

The basic idea of this technique is to form a quadratic program with its cost function given by the mean square error between the array response and a properly selected pattern described by a known mathematical function. This quadratic program can be a constrained or an unconstrained optimization problem depending on the requirements of the desired array pattern. With this technique, one can synthesize an array of arbitrary shape to any appropriate pattern with the characteristic of the array elements taken into consideration, so long as one is able to model the array accurately. With this technique, these desirable features are also brought about by synthesizing (1) an equi-sidelobe array, either circular array or endfire linear array, without mainlobe ambiguity; (2) a circular planar array with equi-sidelobe response; (3) the polar response of a continuous antenna on an antenna array; and (4) a digital filter frequency response as the polar response of an array, which is not necessarily a linear array. To illustrate the method, design examples are given to show the array patterns obtained for different array geometries, namely, linear and single-ring circular arrays, as well as rectangular and circular planar arrays, using different design template functions that include several Chebyshev polynomials and zero-value function.

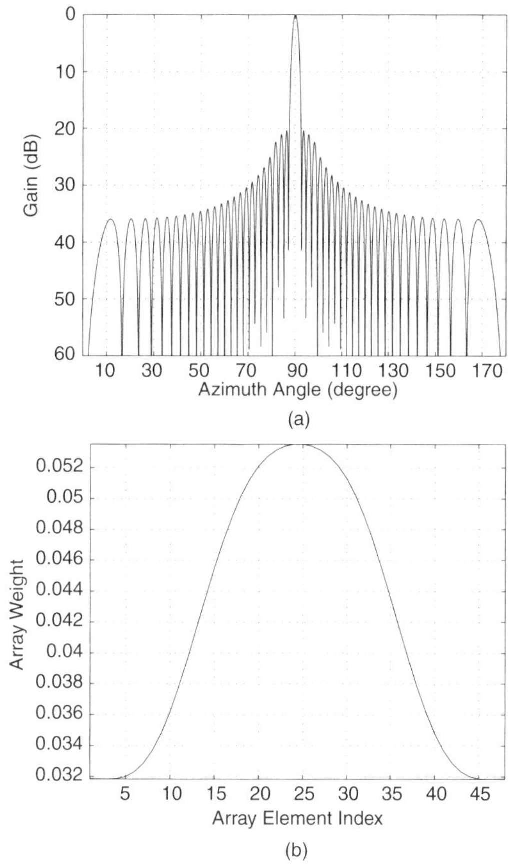

FIGURE 16.4 (a) The beam pattern and (b) the array weight distribution of the eight-element linear array using Taylor synthesis method. The interelement spacing is a half wavelength and the sidelobe level is 20 dB.

16.3.4.1 General and Flexible Formulation for Pattern Synthesis

The general and flexible formulation for array synthesis follows:

$$\underset{\mathbf{w}}{\text{minimize}} \sum_k \left(\frac{1}{\Delta\theta_k}\right) \int_{\theta_{l_k}}^{\theta_{u_k}} \sum_j \left(\frac{1}{\Delta\phi_j}\right) \int_{\theta_{l_j}}^{\theta_{u_j}} \sum_i \left(\frac{1}{\Delta f_i}\right) \int_{f_{l_i}}^{f_{u_i}} \left| H\left(f,\theta,\phi,\mathbf{g}_0,\mathbf{r}_0\right) - A\left(f,\theta,\phi\right) \right|^2 df d\phi d\theta \tag{16.41a}$$

$$\text{subject to } H\left(f_n,\theta_n,\phi_n,\mathbf{g}_0,\mathbf{r}_0\right) = H_0\left(f_n,\theta_n,\phi_n,\mathbf{g}_0,\mathbf{r}_0\right) \quad \text{for some } f_n, \theta_n \text{ and } \phi_n\text{'s} \tag{16.41b}$$

where $\Delta f_i = f_{u_i} - f_{l_i}$, $\Delta\phi_j = \phi_{u_j} - \phi_{l_j}$, $\Delta\theta_k = \theta_{uk} - \theta_{lk}$ and $H_0(f_n, \theta_n, \phi_n, \mathbf{g}_0, \mathbf{r}_0)$ is the desired constrained pattern for some f_n, θ_n, and ϕ_n; the function $A(f, \theta, \phi)$ is the template function that describes the desired array pattern. Note that the constraint equations may or may not be needed to obtain the desired response depending on the template function used. In the case where the template function fully describes the desired array pattern, then one is only required to solve an unconstrained optimization problem, namely, Eq. (16.41a); however, if additional sidelobe control is required, then constraints may be added. Examples are pattern null constraints and main beam steering.

By using the matrix notation, Eq. (16.41) can also be written as a quadratic program, shown as follows:

$$\underset{\mathbf{w}}{\text{minimize}}\, \mathbf{w}^H\mathbf{Q}\mathbf{w} - \mathbf{w}^H\mathbf{p} - \mathbf{p}^H\mathbf{w} + c \qquad (16.42a)$$

$$\text{subject to } \mathbf{w}^H\mathbf{D}(f_n, \theta_n, \phi_n, \mathbf{g}_o) = \mathbf{f}^T \qquad (16.42b)$$

where **Q**, **p**, and c are the $L \times L$ dimensional matrix, L-dimensional vector, and real scalar, respectively, given by

$$\mathbf{Q} = \sum_k \left(\frac{1}{\Delta\theta_k}\right)\int_{\theta_{l_k}}^{\theta_{u_k}} \sum_j \left(\frac{1}{\Delta\phi_j}\right)\int_{\phi_{l_j}}^{\phi_{u_j}} \sum_i \left(\frac{1}{\Delta f_i}\right)\int_{f_{l_i}}^{f_{u_i}} \mathbf{s}\mathbf{s}^H df d\phi d\theta \qquad (16.42c)$$

$$\mathbf{p} = \sum_k \left(\frac{1}{\Delta\theta_k}\right)\int_{\theta_{l_k}}^{\theta_{u_k}} \sum_j \left(\frac{1}{\Delta\phi_j}\right)\int_{\phi_{l_j}}^{\phi_{u_j}} \sum_i \left(\frac{1}{\Delta f_i}\right)\int_{f_{l_i}}^{f_{u_i}} A * \mathbf{s} df d\phi d\theta \qquad (16.42d)$$

$$c = \sum_k \left(\frac{1}{\Delta\theta_k}\right)\int_{\theta_{l_k}}^{\theta_{u_k}} \sum_j \left(\frac{1}{\Delta\phi_j}\right)\int_{\phi_{l_j}}^{\phi_{u_j}} \sum_i \left(\frac{1}{\Delta f_i}\right)\int_{f_{l_i}}^{f_{u_i}} AA * df d\phi d\theta \qquad (16.42e)$$

where **s** is defined earlier in Eq. (16.3); **D** and **f** are some matrix and vector used to achieve the constrained pattern $\mathbf{H}_0$ (f_n, θ_n, ϕ_n, $\mathbf{g}_0$, $\mathbf{r}_0$).

By using the Lagrange multiplier method,[11,15] Eq. (16.42) can be solved to yield the following solution weight vector:

$$\mathbf{w}_0 = \mathbf{Q}^{-1}\mathbf{D}\left(\mathbf{D}^H\mathbf{Q}^{-1}\mathbf{D}\right)^{-1}\left(\mathbf{f} - \mathbf{D}^H\mathbf{Q}^{-1}\mathbf{p}\right) + \mathbf{Q}^{-1}\mathbf{p} \qquad (16.43)$$

Note that the matrix **Q** is usually positive definite and Eq. (16.43) can be applied directly to obtain the array weight vector. If **Q** is only positive semidefinite, Eq. (16.43) that involves matrix inversion cannot be used directly. For this case, one can still solve a set of linear equations for the optimum array weight vector using the Lagrange necessary conditions.[11,15] Also, when the template function A is chosen properly, it is possible to obtain $\mathbf{w}_0$ such that the cost function in Eqs. (16.41a) or (15.42a) is minimized to zero.

From Eq. (16.41), it can be seen that no assumption has been made on the array elements gain/phase responses. In fact, it is always possible to include the mutual coupling effect into account when the array weights are solved for. This may be accomplished by including an impedance matrix in the optimization problem (because of the large variety of design problems to be addressed in the chapter, the inclusion of mutual coupling effect in the design of array pattern is not illustrated here).

With the formulations in Eqs. (16.41) and (16.42), a natural question arises as to what the possible choices of the template function $A(f, \theta, \phi)$ are. Though this may be a subject of study, two basic equations may be of interest to yield a variety of desired array patterns. One possibility is

$$A(f, \theta, \phi) = 0 \qquad (16.44)$$

which results in minimum average sidelobe level for a specific beam width if the sidelobe level is minimized over the desired sidelobe regions; another possibility results from the well-known Chebyshev function:[7,9]

$$A(f,\theta,\phi)=\cos\Big((\eta-1)\cos^{-1}\Big(\mu\cos\big(\beta\pi(\cos(\alpha\theta)-\cos(\alpha\theta_0))\big)\Big)\Big) \tag{16.45a}$$

for a range of azimuth angle ϕ, or

$$A(f,\theta_0,\phi)=\cos\Big((\eta-1)\cos^{-1}\Big(\mu\cos\big(\beta\pi(\cos(\alpha\phi)-\cos(\alpha\phi_0))\big)\Big)\Big) \tag{16.45b}$$

for a particular θ_0, where

$$\mu=\cosh\left[\frac{1}{\eta-1}\cosh^{-1}\left(10^{-sll/20}\right)\right] \tag{16.45c}$$

where θ_0 or ϕ_0 denotes the desired look direction; *sll* denotes sidelobe level in terms of decibels; and η, β, and α are some parameters (not necessarily positive integers) that can be manipulated to describe or approximate a desired pattern for a particular array geometry. Note also that η may not necessarily be the number of array elements used. However, if the array used is an equispaced linear array, for an optimum sidelobe–beam width relationship, then η is given by the number of array elements, β is equal to 0.5, and α is equal to 1. Examples are also shown on how these parameters may be manipulated to achieve the required array pattern. To avoid confusion, it may be important to point out that Eq. (16.45) appears to be a single variable equation because for a planar array, one would normally like to have a constant low sidelobe response for all ϕ-cuts for a particular operating frequency. It is always possible to express the desired array pattern in terms of the physical *x*- and *y* dimensions (see Reference 3 for the expression), in which case, two variables appear on the right-hand side of the template function and the optimization problems given in Eqs. (16.41) and (16.42) have to be rewritten after substitution of variables. Another point to note is that in Eqs. (16.41), (16.42), (16.44) and (16.45), f is included in the expressions because bandwidth control is possible. For a pure narrowband application, f equals f_0.

16.3.4.2 Linear and Circular Array Design Examples

In this section, several linear arrays are designed to yield different array patterns using Eqs. (16.42), (16.44), and (16.45). Figures 16.5 to 16.12 show the patterns that are obtained on the azimuth plane for a 32-element, half-wavelength-spaced linear array. Unless otherwise stated, array elements used in the pattern design are assumed to be isotropic.

Figure 16.5 shows the proposed approach in the array weight synthesis using different sidelobe-suppressed regions {[0°, 84°], [96°, 180°]} and {[0°, 80°], [100°, 180°]} for the case $A(f, \theta, \phi) = 0$. It is seen that when the beam width is increased slightly, the sidelobe level is improved on the order of tens of decibels. Note that in the subsequent figures shown in this section, all the results are obtained for a half-wavelength-spaced uniform linear array with 32 array sensors having unity response. Note also that if the array sensor response is nonunity and nondirectional, one can always design the array with the assumption that the array sensor response is unity but with the notion that the effect of sensor response is contained in the array weights [see Eqs. (16.1) and (16.3)].

To illustrate the possibility of broad null control, null constraints are introduced at directions {74.5°, 76°, 77.5°, 79°} and {45°, 48°} in the design. It is clearly shown in Fig. 16.6 that the all-desired nulls are achieved at a relatively low level using Eq. (16.42), whereas in the case of conventional array, the second set of null constraints does not produce satisfactory results. For this design example, the sidelobe suppression sectors used are {[0°, 84°], [96°, 180°]}.

Figure 16.7 shows the broadside array patterns obtained using Eqs. (16.42a) and (16.45) for the integration sectors [0°, 180°] and [85°, 95°]. Note that an integration over a small main beam region yields almost the same array pattern as for the case where integration was carried out for both sidelobe

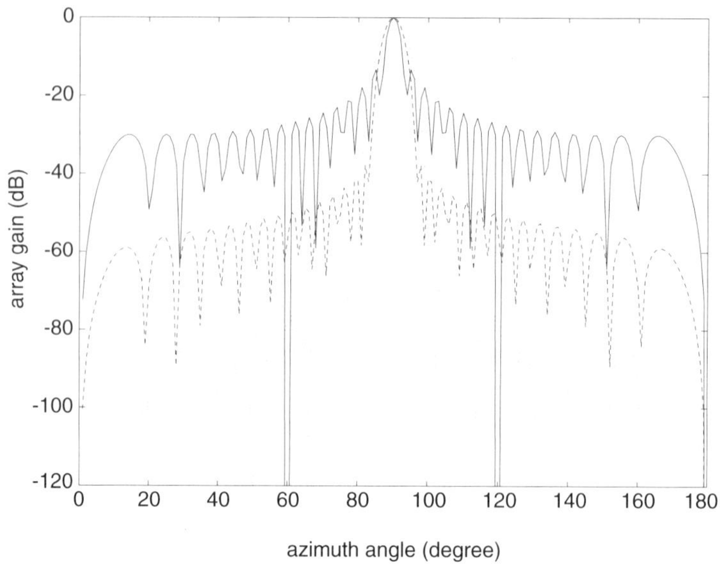

FIGURE 16.5 Array beam pattern obtained using different methods for a 32-element, uniform linear array (____ conventional array, ---- quadratic programming approach using the zero-function as a template and two sidelobe suppression sectors [0°, 84°] and [96°, 180°], ... quadratic programming approach using zero function as a template and two sidelobe suppression sectors [0°, 80°] and [100°, 180°]).

and main beam regions. Note also that when integration is carried out over the region [0°, 180°] and if the function A in Eq. (16.45) is expressed as $\mathbf{w}_C^H\mathbf{u}(\phi)$ where $\mathbf{w}_C$ is the Chebyshev weight vector[7,9] and $\mathbf{u}(\phi) = \mathbf{u}(f_0, \theta_0, \phi, \mathbf{r}_0)$ — then it is clear that the solution weight vector (trivial solution) for the unconstrained version of Eq. (16.41), that is

$$\underset{\mathbf{w}}{\text{minimize}} \frac{1}{\pi}\int_0^{\pi}\left|\left(\mathbf{w}^H - \mathbf{w}_C^H\right)\mathbf{u}(\phi)\right|^2 d\phi \tag{16.46}$$

is given by $\mathbf{w} = \mathbf{w}_C$ because the cost function in this optimization problem is nonnegative.

Figure 16.8 illustrates a multibeam design. The dotted line is the minimum norm design with two unity response constraints imposed at the two desired look directions, 70° and 140°. The solid-line pattern is obtained using two integration sectors [0°, 90°] and [90°, 180°] which correspond to two template functions, derived from Eq. (16.45), steered to 70° and 140°, respectively. The *sll*'s used for the two integration sectors are both –40 dB. It can be seen from the figure that though the sidelobe level has been degraded by a few decibels because of the inclusion of one extra main beam, the sidelobe level obtained can still be very much lower than that of the conventional minimum norm design (of course, at the expense of larger beam width).

Figure 16.9 shows the performance of the proposed method used with null control. In this example, a –40 dB *sll* with an integration sector [0°, 180°] is used and four point constraints are applied to fix a broad null at [55°, 57°, 59°, 61°]. It can be seen from the example that the equi-sidelobe performance of a Chebyshev array is well preserved.

The design given in Fig. 16.10 uses two integration sectors, [0°, 85°] and [85°, 180°], with a *sll* of –30 dB. For the left-side sector, the template function is the weighted version of Eq. (16.45), given by (for simplicity's sake, f and θ are dropped from the expression)

$$A(\phi) = \phi^4 \cos\left(31\cos^{-1}\left(\mu\cos\left(0.5\pi\left(\cos(\phi)\right)\right)\right)\right) \tag{16.47}$$

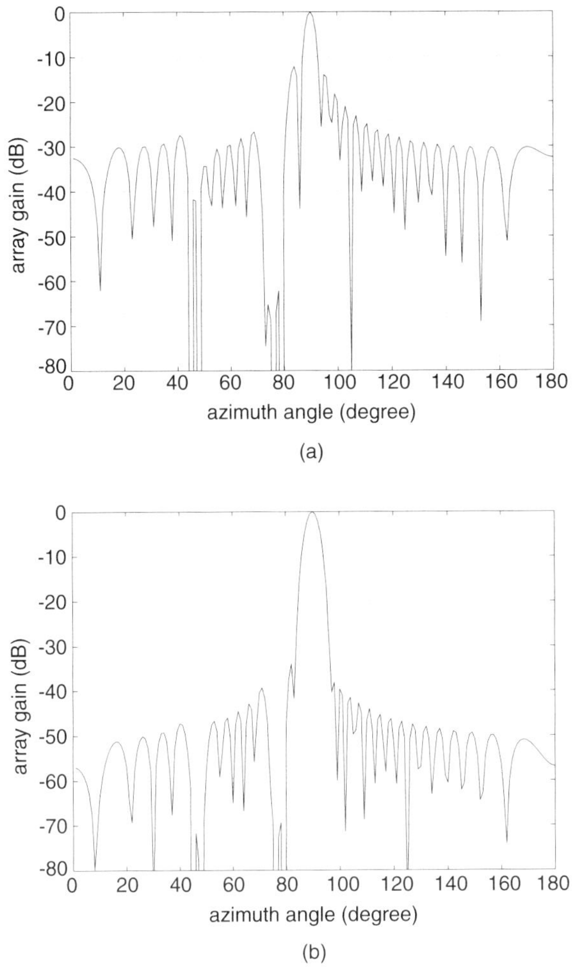

FIGURE 16.6 Array beam pattern obtained for a 32-element, uniform linear array: (a) conventional array; (b) array using the quadratic programming approach with null constraints imposed at 74.5°, 76°, 77.5°, 79°, and 45°, 48° (sidelobe suppression sectors are [0°, 84°] and [96°, 180°]).

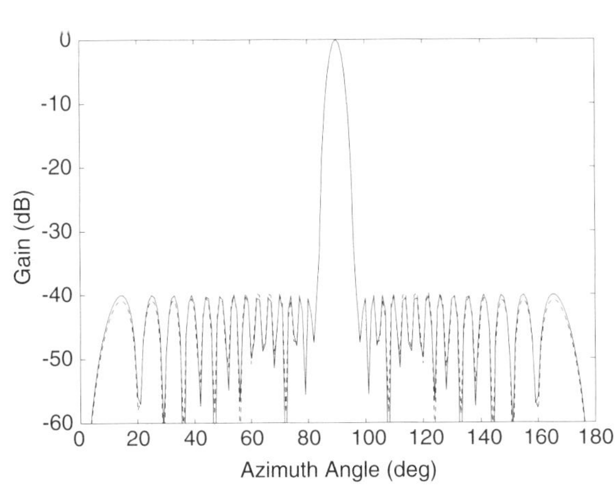

FIGURE 16.7 Approximated Chebyshev array beam patterns for a 32-element, half-wavelength-spaced linear array: ____ integration over [0°, 180°]; ---- integration over [85°, 95°].

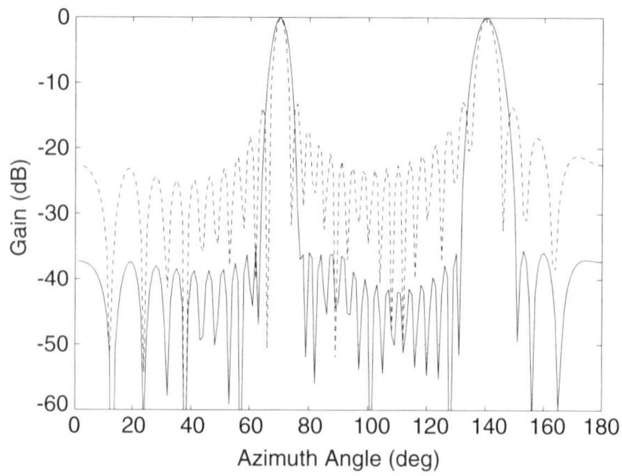

FIGURE 16.8 Multibeam array patterns for a 32-element, half wavelength-spaced linear array: ____ quadratic programming method with two integration sectors [0°, 90°], [90°, 180°] corresponding to the two mainlobes at 70° and 140°; ---- minimum power design.

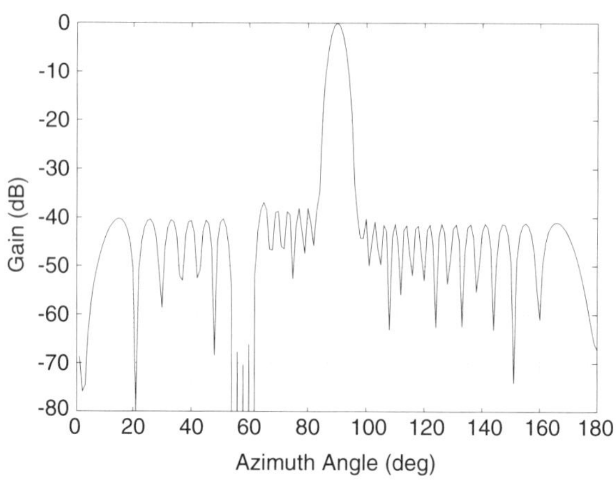

FIGURE 16.9 Approximated Chebyshev array beam pattern obtained using the quadratic programming method for a 32-element, half wavelength-spaced linear array with null constraints imposed at 55°, 57°, 59°, and 61°. Integration sector used: [0°, 180°].

Figure 16.11 shows a flat-top, beam-shape design with the design template function ($A(f_0, \theta_0, \phi)$) adopted directly from a 32-tap digital finite impulse response filter function. By doing so, the frequency response of the digital filter is mapped onto the array beam pattern from the range of normalized frequencies $[-\pi, \pi]$ onto the range of azimuth angles [0°, 180°]. The digital filter coefficients are obtained from a filter design package using Remez algorithm. This sort of pattern can be realized on a circular array with the directional ambiguity being removed.

Figure 16.12 shows the array pattern of an array of 32 short dipoles. The short dipoles are assumed to be aligned linearly with the array axis and spaced every half wavelength. In the figure, the dotted line indicates the response obtained using the Chebyshev array weights directly. The solid line is obtained using the proposed method with the polar response of the individual short dipole taken into account directly in the design. It can be seen that the sidelobe response has improved approximately by 3 dB.

Figure 16.13 shows the beam pattern on the azimuth plane for a 32-element circular array with a radius of 2.5 wavelengths. To avoid ambiguity, the α parameter in Eq. (16.45) is chosen to be 0.5. As one can verify from the template function of Eq. (16.45) that, when α is set to 1 (solid line), two mainlobes

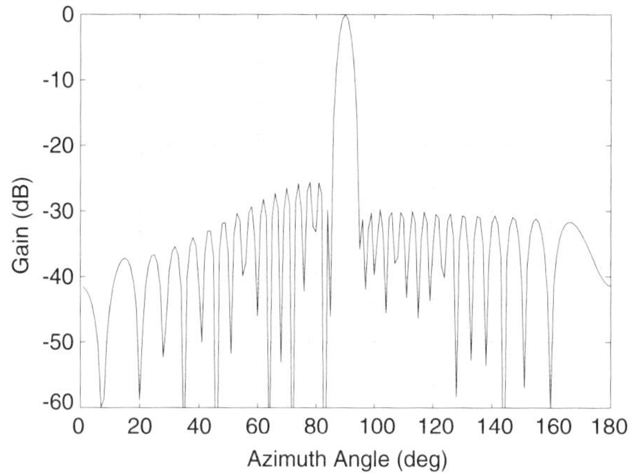

FIGURE 16.10 An asymmetrical array pattern design for a 32-element, half-wavelength-spaced linear array using the quadratic programming method. Two integration used: [0°, 85°] and [85°, 180°].

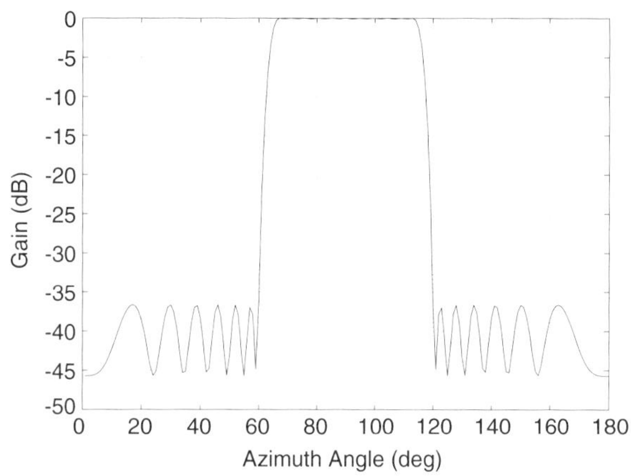

FIGURE 16.11 A flat-top, beam-shaping design for a 32-element, half-wavelength-spaced linear array using the Remez algorithm.

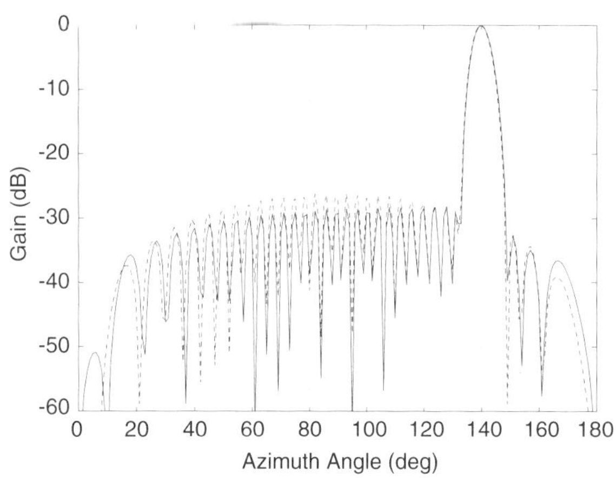

FIGURE 16.12 Array patterns for a 32-element, half-wavelength-spaced short-dipole linear array: ____ quadratic programming method with an integration sector of [0°, 180°]; ---- using Chebyshev weights for isotropic elements.

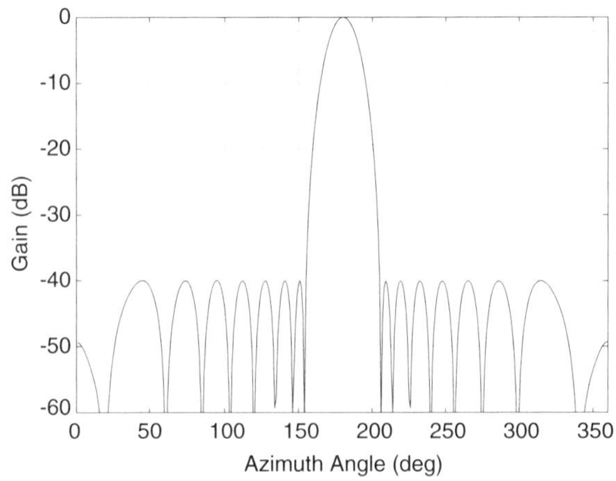

FIGURE 16.13 Approximated Chebyshev array beam pattern obtained using the quadratic programming method for a 32-element, 2.5-wavelength-radius circular array. Integration sector used: [0°, 360°].

occur in every cycle, whereas one mainlobe is only observed if α is set to 0.5. By using the same approach as in Fig. 16.11, the flat-top, beam-shape design in Fig. 16.11 can be easily extended to the circular array with the directional ambiguity removed.

16.3.4.3 Planar Array Design Examples

In this section, Eq. (16.44) and (16.45) are used to design array patterns for both rectangular and circular planar arrays. Equation (16.45) is used to obtain Figs. 16.14 and 16.15, whereas Eq. (16.44) is used for Fig. 16.16.

Array patterns shown in Fig. 16.14b correspond to the array geometry shown in Fig. 16.14a. It can be clearly seen that the Chebyshev response is closely approximated throughout all azimuth angles. For this example, η is set to 5 and the integration sectors for azimuth- and elevation planes are both [0°, 180°].

In Fig. 16.15b, Chebyshev patterns are closely approximated for the array geometry given in Fig. 16.15a for all azimuth angles. The η parameter is equal to 6.7 in this case and the integration sectors are the same as those of Fig. 16.14. It is interesting to note from here that Chebyshev-like pattern is possible on a circular planar array.

Figure 16.16 shows the minimum-average-sidelobe design using Eq. (16.44) as a template, for an integration sector of [0°, 180°] in the azimuth plane and various integration sectors for beam width control in the elevation plane. The array geometry used is the same as that given in the previous example, shown in Fig. 16.14a. Those solid-line patterns given in Fig. 16.16 are the azimuth cuts corresponding to a conventional array. To illustrate the possibility of sidelobe suppression, two sets of elevation–integration sectors that determine the beam width, [–90°, –20°], [20°, 90°] and [–90°, –30°], [30°, 90°] — which corresponds to the dashed lines and dotted lines in Fig. 16.16, respectively — are used with a unity gain response constraint imposed at $(\phi, \theta) = (0°, 0°)$. As one can see from the figures, sidelobe suppression can be improved significantly if a certain amount of beam width is sacrificed.

16.3.5 Adaptive Array-Based Method

Adaptive beamforming techniques are basically synthesis techniques that synthesize an array on the fly in a changing signal environment. With such a notion, one could also make use of adaptive array algorithms to design the array weights. In this case, the signals in the signal fields are artificial signals and they are selected according to the settings of the desired array pattern. One of the first few techniques that uses this concept is based on that of Olen and Compton.[22] Their technique is briefly explained and introduced here with an illustrated example.

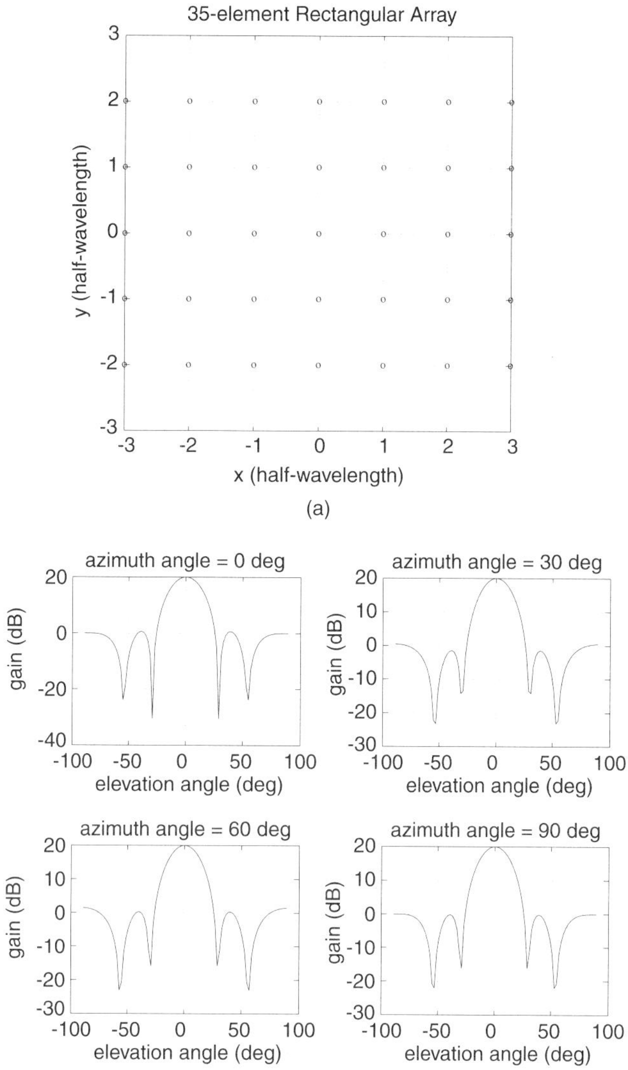

FIGURE 16.14 (a) A 35-element, rectangular planar array geometry; (b) approximated Chebyshev array patterns for various azimuth cuts using the quadratic programming method (azimuth and elevation integration sectors are both [0°, 180°], $n = 5$).

Refer to Eq. (16.1) for simplicity, where the specific frequency, elevation angle, antenna element pattern, and array geometry, letting **R** be the covariance matrix is given by

$$\mathbf{R}(k) = \sigma^2 \left[\mathbf{I} + \sum_{m=1}^{M} \xi_m(k) s(\phi_m) s^H(\phi_m) \right] \tag{16.48}$$

where M is the number of artificial interference and $\{\theta_m, m = 1, 2, \ldots, M\}$ is the set of directions of the interferences. M is usually chosen to be two to three times of the number of array elements. The term σ^2 is the artificial noise power. From the expression, it can be seen that $\xi_m(k)$ is the power of the mth artificial interference updated at the instance k.

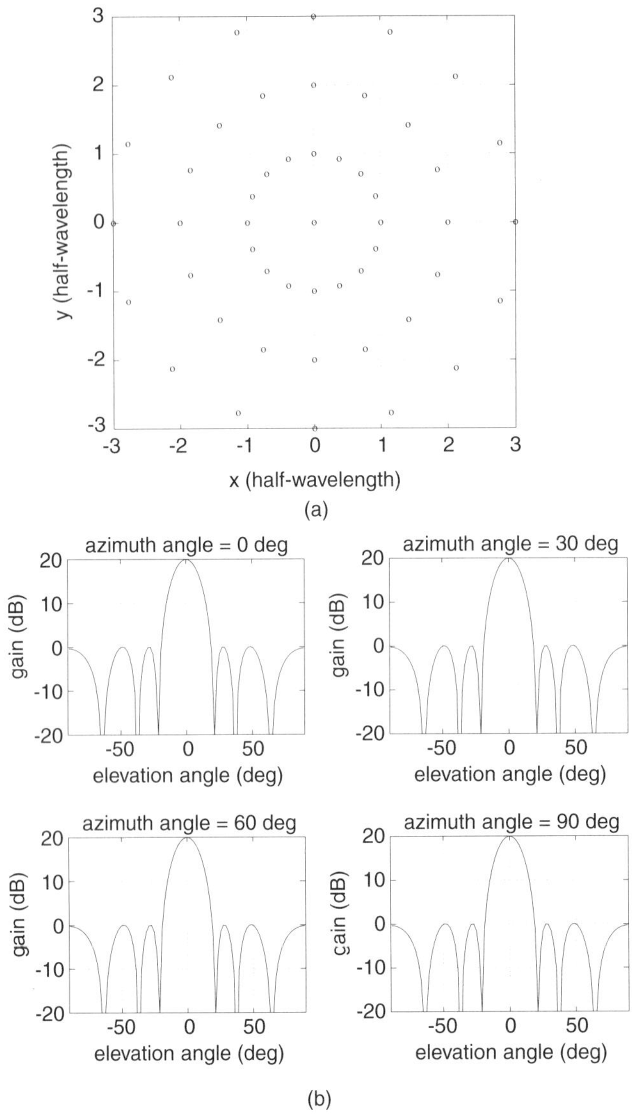

FIGURE 16.15 (a) A 49-element circular planar array geometry; (b) approximated Chebyshev array patterns for various azimuth cuts using the quadratic programming method (azimuth and elevation integration sectors are both [0°, 180°], $n = 6.7$).

$$\xi_m(k+1) = \begin{cases} 0 & \phi_m \in [\phi_L(k), \phi_R(k)] \\ \max[0, \Gamma_m(k)] & \text{otherwise} \end{cases} \tag{16.49}$$

where $[\phi_L(k), \phi_R(k)]$ defines the interference-free region and

$$\Gamma_m(k) = \xi_m(k) + K[p(\phi_m, k) - d(\phi_m, k)] \tag{16.50}$$

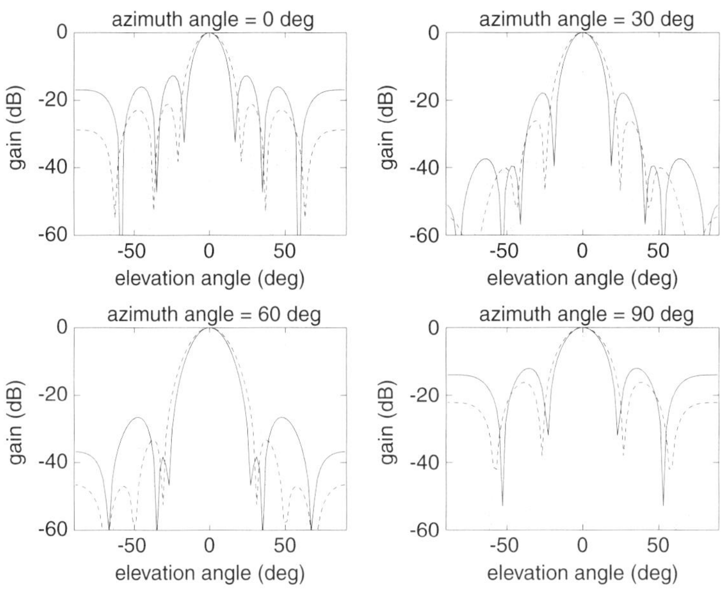

FIGURE 16.16 Array patterns for various azimuth cuts for the array geometry given in Fig. 16.14a: ____ using unity weights; ---- using the quadratic programming method for a minimum-average-sidelobe level in elevation sectors [–90°, –20°] and [20°, 90°] (azimuth integration sector [0°, 180°]); .... using the quadratic programming method for a minimum-average-sidelobe level in elevation sectors [–90°, –30°] and [30°, 90°] (azimuth integration sector [0°, 180°]).

$$d(\phi_m, k) = \frac{P(k)}{10^{sll_m/20}} \tag{16.51}$$

where K is a scalar constant called the iteration gain; sll_m is the desired sidelobe level in decibels at direction ϕ_m; $P(k)$ is the peak value of the main beam at the current instance k. With the preceding definitions, the array weight vector is then given by

$$\mathbf{w}(k+1) = \mu \mathbf{R}(k)^{-1} \mathbf{s}(\phi_d) \tag{16.52}$$

where μ is an arbitrary nonzero scalar; ϕ_d is the desired look direction, and $p(\phi_m, k)$ is the array amplitude response at the mth interference direction given by

$$p(\phi_m, k) = \left|\mathbf{w}^H(k)\mathbf{s}(\phi_m)\right| \tag{16.53}$$

The preceding algorithm may be initialized with $\mathbf{w}$ set to $\mu/\sigma^2\ \mathbf{s}(\phi_d)$. An example is given in Fig. 16.17. The array beam pattern is obtained in three iterations for a half-wavelength-spaced linear array of 16 elements.

16.3.6 Genetic Algorithm-Based Synthesis

There has been a lot of interest in the applications of genetic algorithms to antenna array synthesis problems.[1,33] This may result from the fact that a genetic algorithm is a global optimization technique that is able to find an optimum or near optimum solution in a solution space with many local minima.

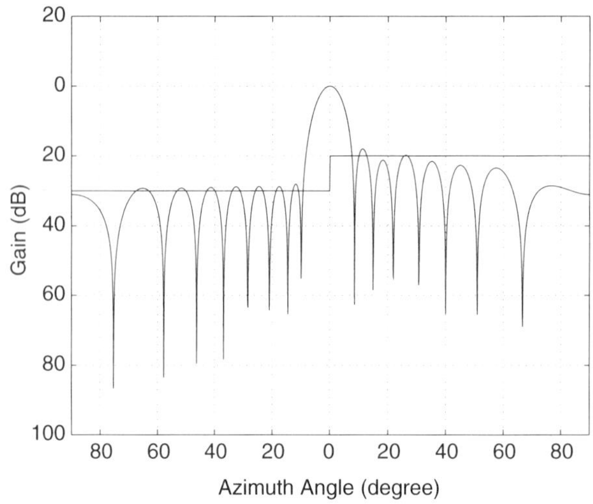

FIGURE 16.17 Array beam pattern obtained for a sidelobe envelope function: –30 and –20 dB sidelobe levels for the regions [–90°, 0°) and (0°, 90°], respectively.

Also, because the array synthesis problems are mostly highly nonlinear, genetic algorithms are therefore well suited for such problems.

16.3.6.1 Finding the Equivalence: Genetic Algorithms and Array Synthesis

Because the objective of array synthesis is to find a good array weight vector so as to produce a desirable array beam pattern, the beam patterns may be seen as the living beings. As a result, the array weights are then corresponding to the chromosomes. The basic idea of applying the genetic algorithm to synthesis is to start with some (arbitrarily) chosen array weight vectors (parents) to allow them to generate some new and better weight vectors (children) to achieve the desired array response.

16.3.6.2 Population and Initialization

A population in this case is a collection of the array weight vectors, perhaps some good ones and some bad ones or perhaps none of the members of the population are good enough to generate a satisfactory beam pattern. In the initial population, just as in any adaptive or iterative algorithm, the initialization of the algorithm plays a very important part in the convergence of the algorithm. Usually, the initial population is generated randomly. However, if the other synthesis techniques are readily available for use, one can always make use of such algorithm to generate the initial population.

16.3.6.3 Reproduction

The reproduction process is composed of three basic genetic operations, namely, mating, crossover, and mutation.

1. Mating — This is referring to picking two parent chromosomes to produce child chromosomes. The assignment of mates may be based on proportionate, tournament, genitor, or ranking selection. Generally, a ranking selection is able to produce better results. In this case, the highly fit parent chromosomes are selected for producing the child chromosomes.
2. Crossover — For array synthesis, from a linear algebraic viewpoint, crossover may simply be seen as forming linear combinations (the child chromosomes) of old array weight vectors (parent chromosomes). The coefficients of the linear combination, of course, determine the quality of the next generations and how fast the algorithm converges.
3. Mutation — This is a process to alter the genetic materials of the chromosomes (elements of the array weight vectors). This is to prevent the optimum weight searching being stuck at a local minimum.

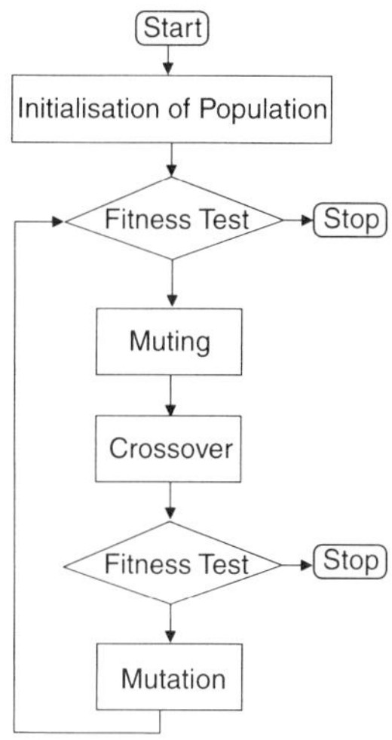

FIGURE 16.18 Genetic algorithm cycle.

16.3.6.4 Fitness Function and Survival Selection

The child chromosomes (array weight vectors) are evaluated using a fitness function. The fitness values are used to reflect how good a weight vector performs. Based on the principle of survival of the fittest, good performers survive to continue to compete in the population, and the poor ones die off. Depending on the design objective, the fitness function varies from case to case. For example, the cost function in Eq. (16.41a) can be a possible fitness function used to determine how fit a weight vector is to produce a desired beam pattern. With the fitness value, one can decide whether to accept the solution or allow the next generation to be produced. In Fig. 16.18, the flow of a genetic algorithm cycle is depicted.

16.3.7 Comparisons and Summary

As mentioned earlier, the Dolph–Chebyshev array produces the smallest beam width for a specific constant sidelobe level. The synthesis technique is basically applicable to equi-spacing linear array with isotropic elements. The Taylor array approximates the Chebyshev array quite closely. For the Taylor array, equi-sidelobes occur only around the vicinity of the main beam and the sidelobe levels decrease away from the main beam. Generally, this gives rise to smaller array weights at the two ends of the array as compared with that of the Chebyshev array. The approach in Section 16.3.4 uses a quadratic program to solve for the array weight vector. An important element of the approach is the pattern template. With a suitably chosen template, a vast variety of array patterns can be designed taking into account most of the system parameters. The next few points highlight the advantages brought about by the generalized quadratic formulations:

1. Besides synthesizing a broadside array, with the quadratic programming formulation, one is also able to synthesize an equi-sidelobe response on the azimuth plane without ambiguity (i.e., only one main beam throughout all azimuth angles).
2. The Chebyshev polynomial can be used to approximate the desired equi-sidelobe response for a circular planar array.
3. No assumptions are made about the gain/phase response or characteristics of the individual array element to synthesize the array.
4. Continuous aperture excitation can be discretized or sampled for an array using the quadratic programming formulation.
5. A digital filter design package can be used in conjunction with the quadratic programming formulation to synthesize a linear or even nonlinear array.

TABLE 16.1 A Comparison of the Various Synthesis Techniques

Method Feature	Uniform	Dolph–Chebyshev	Taylor	Quadratic Program	Adaptive Array	Genetic Algorithm
Array geometry	Linear	Linear, extendable to rectangular array	Linear, extendable to rectangular and circular arrays	Arbitrary	Arbitrary	Arbitrary
Non-isotropic elements	No	No	No	Yes	Yes	Yes
Arbitrary pattern	No	No	No	Yes	Yes	Yes
Iterative	No	No	No	No	Yes	Yes
Speed of convergence	NA	NA	NA	NA	Moderate	Slow
Closed-form solution	Yes	Yes	Yes	Yes	No	No
Weight element control	No	No	No	Easy	Easy	Difficult
Pattern hard constraint	No	No	No	Yes	Yes	No
Element failure recovery	No	No	No	Yes	Yes	Yes

In summary, the quadratic programming method is flexible. As long as one is able to find a suitable mathematical template function, one is able to synthesize an array with the desired pattern. In fact, there exist at least two useful template functions, namely, the zero function and the Chebyshev polynomial, which can be used to synthesize an array for a variety of array patterns, including the approximation of the pattern of a continuous aperture, for an arbitrary array of sensor elements that may not be isotropic or may not have identical characteristics.

The adaptive array- and the genetic algorithm-based approaches are also highly flexible. These techniques do not make assumption on the array element characteristics and the array geometry. The pattern to be designed could also be arbitrary if the solution does exist. Perhaps, the drawback of the these two classes of synthesis is, because of the iterative nature of the algorithm, no convergence (or fast convergence) can be guaranteed. Table 16.1 sums up the features offered by the methods introduced earlier.

16.4 Array Geometry Consideration for Pattern Adjustment

In array geometry synthesis, the main concern is to determine the array shape so as to produce a beam pattern that suitably approximates a desired pattern. The desired beam pattern can vary widely depending on the application, among which there is a major class of pattern synthesis methods that aims to suppress the sidelobe level while maintaining the mainlobe gain. In the case of linear array geometry synthesis, this is achieved by adjusting the spacings between the array elements (while keeping the array weights uniform) according to certain optimization criteria. Examples for such synthesis methods are the minimum redundancy and nonredundancy array methods[16,30] that design a linear array, for a given number of array elements, with maximum resolution (minimum beam width) at the expense of relatively high peak sidelobe level as compared with its uniform counterpart. Another example such as that given by Reference 13 designs an array with null control using array shape perturbation technique.

Those methods mentioned earlier[13,16,30] are only applicable to the design of linear arrays but not to other arrays such as planar arrays or three-dimensional arrays. One possible approach to synthesize a general array geometry is to apply a technique similar to that described in Section 16.3.4. The basic idea of this approach is to minimize the mean square value of the array response in certain directional regions while imposing one or several directional gain constraints at one frequency or over a frequency band of interest with respect to the array geometry or other array parameters that define the array manifold. Note that to solve for the array geometry using this approach, one has to perform multidimensional gradient search that involves heavy computation. To reduce the computational load involved, one can reformulate the preceding approach using the first-order Taylor expansion approximation. This leads to a simple quadratic program that is relatively easy to solve. However, this reformulation is only applicable to linear array geometry synthesis. This section also addresses the issues and problems with regard to the maximum array aperture length of a linear array, which is either excited uniformly or uses the Chebyshev array weights, to achieve its highest resolution.

16.4.1 Theoretical Limits of Aperture Length of Uniformly Excited Equispaced Linear Array

16.4.1.1 Broadside Array

One way to improve on the resolution of a fixed-number uniformly excited equispaced linear array is to extend the aperture length of the array, that is, to increase the interelement spacing of the array. To avoid the problem of grating lobe in the array response, it is well known that the array interelement spacing has to be less than 1 λ. Therefore, a natural question to ask would be: "To what extent can an array be stretched so as to obtain the highest array resolution?" In the literature,[9] it has been shown that, for an L-element linear array, the maximum possible array spacing is given by the following expression:

$$d_{\max} = \frac{K}{K+1} \tag{16.54}$$

where $K = L - 1$.

However, it is possible to extend the aperture length of the array such that the sidelobe levels at the two endfire directions are still smaller than or equal to any other sidelobe levels. Without loss of generality, from Eq. (16.1), if the phase center location is placed on the first array element, then Eq. (16.1) is simplified to

$$H(z) = 1 + \sum_{k=1}^{K} z^k \tag{16.55}$$

By the fundamental theorem of algebra, the amplitude of $H(z)$ can be written in the following factorized form:

$$H(z) = \prod_{k=1}^{K} (z - z_k) \tag{16.56}$$

which gives rise to

$$|H(z)| = \prod_{k=1}^{K} |z - z_k| \tag{16.57}$$

From the polynomial in Eq. (16.56), it can be seen that a point in the far field at an angle ϕ maps to a point on the unit circle in the z plane. The roots of the polynomial, which occur on the unit circle, correspond to nulls in the array beam pattern. That means the locations of these roots in the complex z plane are directly related to the beam pattern in real ϕ space. From Eq. (16.15), it can be observed that as ϕ varies in real space from 0 to π, the definitions in Eq. (16.14) and (16.15) require that ψ varies from $\psi_s = \pi\beta\cos(0) = \pi\beta$ and $\omega_f = \pi\beta\cos(\pi) = -\pi\beta$. This is illustrated in Fig. 16.19. Note that ψ_s and ψ_f denote ψ start and ψ finish, respectively.

From Eq. (16.15) and Fig. 16.19, it can be seen that for ϕ in the real space, the visible range of ψ is determined by the value of interelement spacing β. One possible solution to the determination of the maximum interelement spacing is given in Reference 9. In this case, it is suggested that

$$\psi_s = -\psi_f = 2\pi - \frac{2\pi}{K+1} = \frac{2\pi K}{K+1} \tag{16.58}$$

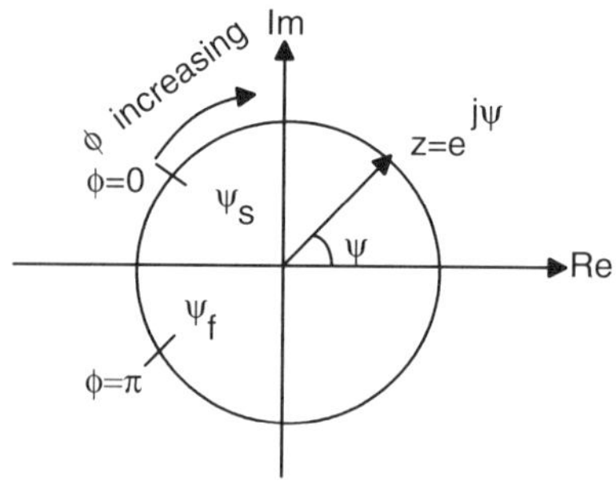

FIGURE 16.19 Schelkunoff's unit circle.

which gives rise to maximum interelement spacing

$$d_{max} = \frac{2K}{K+1} \quad \text{(half wavelength)} \tag{16.59}$$

An alternative solution is given as follows. It is observed that when the interelement spacing of the array is set at d_{max}, nulls occur at azimuth angles 0° and 180°, and if d is extended from d_{max} to 2, the array amplitude response increases from zero (null) to L (the height of the grating lobe that is equal to the peak value of the main beam). It can be seen that for $d_{max} < \beta < 2$, there exists a range of d that produces sidelobes, at 0° and 180° azimuth angles, which are lower than the first (highest) pair of sidelobes adjacent to the main beam. The upper limit of the interelement spacing can be obtained by solving the following equation:

$$\left|H(z)\right|^2 = \left|1 + \sum_{k=1}^{K} z^k\right|^2 = f(\psi_m) \tag{16.60}$$

where $f(\psi_m)$ is the array factor evaluated at $\psi = \psi_m$ and is given by

$$f(\psi_m) = \frac{\sin^2(L\psi_m/2)}{L^2 \sin^2(\psi_m/2)} \tag{16.61}$$

where

$$\psi_m = \frac{2}{N}\left[\frac{2p+1}{2}\right]\pi \tag{16.62}$$

for $p = \pm 1$, which corresponds to the first pair of the sidelobe.

By using an appropriate solution out of the K solutions of (60), $z_0 = e^{j\psi_0}$, one obtains the new upper limit of the interelement spacing:

$$\beta_{max} = \left|\frac{\psi_0}{\pi}\right| \quad \text{(half wavelength)} \tag{16.63}$$

This upper limit may also be approximated by the following formula derived from the computer studies:

$$\psi_s = -\psi_f = 2\pi - \left(\frac{\zeta - 1}{\zeta}\right)\left(\frac{2\pi}{K+1}\right) = 2\pi \times \frac{\zeta K + 1}{\zeta(K+1)} \tag{16.64}$$

where the scaling parameter ζ is set to be sufficiently large, say 5.5, or other values such as [3.5, 5.5], this yields

$$\beta_{\max} \approx 2\left[\frac{5.5K+1}{5.5(K+1)}\right] = 2\left[\frac{K}{K+1} + \left(\frac{1}{5.5}\right)\frac{1}{K+1}\right] \tag{16.65}$$

Note that as ζ increases to infinity, the scaling ratio $((\zeta - 1)/\zeta)$ approaches unity. By comparing Eq. (16.58) and (16.64), one can see that when ζ is equal to infinity, the two expressions give rise to the same results. From here, one may treat Eq. (16.58) as a special case of Eq. (16.64).

From Eqs. (16.59) and (16.65), it can also be seen that

$$\beta_{\max} > d_{\max} \tag{16.66}$$

This means that $\beta_{\max}$ always gives rise to an array with a better resolution. Only when the number of array elements is large, the two expressions give the same result, that is

$$\lim_{K\to\infty} d_{\max} = \lim_{K\to\infty} \beta_{\max} \tag{16.67}$$

Figure 16.20a illustrates the accuracy of the formula in Eq. (16.65). One can see that the sidelobe level at $\phi = 0$ and $\phi = \pi$ is always lower than or equal to the height of the first sidelobe.

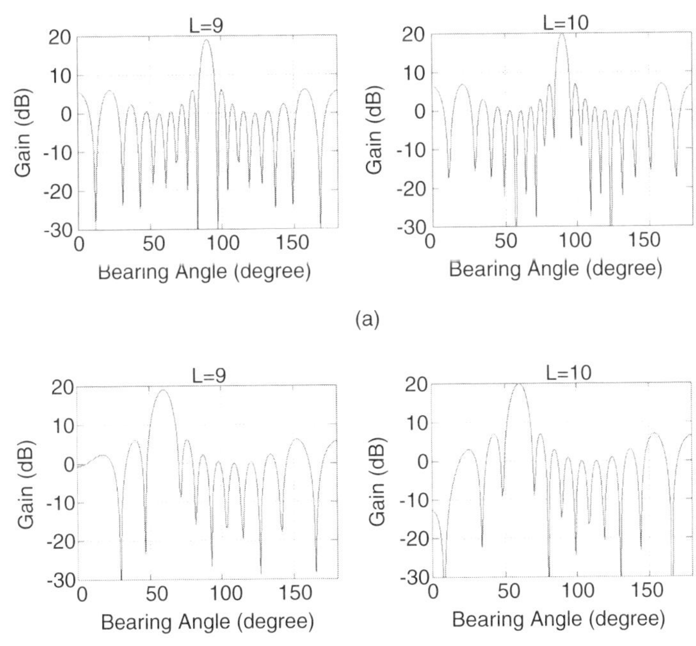

FIGURE 16.20 Array beam patterns obtained using $\beta_{\max}$ calculated from Eqs. (16.65) and (16.75) for $L = 9{,}10$ for look directions of (a) 90° and (b) 60°, respectively.

From Eq. (16.61) and after some manipulations, one obtains the beam width of the array as

$$\Theta = \frac{1.77}{L\beta} \quad (\text{radian}) \tag{16.68}$$

Based on the empirical formula given in Eq. (16.65), the beam width of the array is approximated by

$$\Theta_{\min} \approx \frac{0.885}{L + 0.818} \quad (\text{radian}) \tag{16.69}$$

16.4.1.2 Scanned Array

To obtain the expression of $\beta_{\max}$ for a scanned array, which has its beam looking (scanning) at an angle ϕ_0, one has to alter the starting and ending values of ψ as follows:

$$\psi_s = \pi\beta\left[\cos(0) - \cos(\phi_0)\right] = \pi\beta\left[1 - \cos(\phi_0)\right] \tag{16.70}$$

$$\psi_s = \pi\beta\left[\cos(\pi) - \cos(\phi_0)\right] = -\pi\beta\left[1 + \cos(\phi_0)\right] \tag{16.71}$$

where ϕ_0 is the main beam direction of the array.

By using the formula in Eq. (16.65), one obtains

$$\psi_s \geq -\left(2\pi - \left(\frac{\zeta - 1}{\zeta}\right)\left(\frac{2\pi}{K+1}\right)\right) = -2\pi \times \frac{\zeta K + 1}{\zeta(K+1)} \quad \text{if} \quad \phi_0 < \frac{\pi}{2} \tag{16.72}$$

$$\psi_s \leq 2\pi - \left(\frac{\zeta - 1}{\zeta}\right)\left(\frac{2\pi}{K+1}\right) = 2\pi \times \frac{\zeta K + 1}{\zeta(K+1)} \quad \text{if} \quad \phi_0 > \frac{\pi}{2} \tag{16.73}$$

From these two inequalities, the following condition is imposed

$$\beta\pi\left(1 + \left|\cos(\phi_0)\right|\right) \leq 2\pi \times \frac{\zeta K + 1}{\zeta(K+1)} \tag{16.74}$$

which yields the following approximation for $\zeta = 5.5$:

$$\beta_{\max} \approx \frac{2}{1 + \left|\cos(\phi_0)\right|} \times \frac{5.5K + 1}{5.5(K+1)} \tag{16.75}$$

Again, for the scanned array, the preceding expression satisfies the following conditions:[9]

$$\beta_{\max} > d_{\max} = \frac{2}{1 + \left|\cos(\phi_0)\right|} \times \frac{K}{K+1} \tag{16.76}$$

$$\lim_{K \to \infty} d_{\max} = \lim_{K \to \infty} \beta_{\max} \tag{16.77}$$

Figure 16.20b illustrates the beam patterns obtained using the formula in Eq. (16.75). For a scanned array, its beam width is given by the following expression:

$$\Theta = \cos^{-1}\left[\cos(\phi_0) - \frac{0.886}{\beta L}\right] - \cos^{-1}\left[\cos(\phi_0) + \frac{0.886}{\beta L}\right] \tag{16.78}$$

Note that this expression is valid for $0 < \phi_0 \leq \pi/2$ and the beam width is well defined in the *scan limit*, which means the second term (excluding the negative sign) is greater than or equal to zero. If ϕ_0 is falling in the range $[\pi/2, \pi)$, then the second term of the expression has to be less than or equal to π. For the case when scanning is near the broadside, the preceding expression can be simplified to the following form:

$$\Theta = \frac{1.77}{L\beta \sin(\phi_0)} \tag{16.79}$$

which yields the following expression for minimum beam width:

$$\Theta_{min} \approx \frac{0.885}{(L+0.818)\sin(\phi_0)} \quad (\text{radian}) \tag{16.80}$$

16.4.2 Maximum Aperture Length of Chebyshev Linear Array

16.4.2.1 Broadside Array

For a fixed Chebyshev array weight vector, a sidelobe starts to grow (from 0 value or the constant sidelobe level to the main beam height, depending on whether the array number is even or odd) at $\phi = 0$ and $\phi = \pi$ as the array spacing β increases from 1 to 2. For an odd or even number array, there exists at least one value of β that is greater than 1 such that the sidelobe level is equal to the sidelobe level at the directions away from the two endfire directions. Figure 16.21 shows a plot of sidelobe level at one of the endfire directions against the array spacing β.

Because the length of the array aperture determines the resolution of the array, it is desirable to have β to be as large as possible. If nulls are desired at both the endfire directions, it is suggested[9] that the maximum array spacing, which has to be less than two half wavelengths, be determined corresponding to the roots of the Chebyshev polynomial, which set the nulls at $\phi = 0$ and $\phi = \pi$. Mathematically, this means solving

$$H\left(e^{j\pi\beta\cos(\phi)}\right) = 0 \quad \text{or} \quad \left|H\left(e^{j\pi\beta\cos(\phi)}\right)\right| = 0 \tag{16.81}$$

for $\phi = 0$ or $\phi = \pi$ for the largest possible β, which is less than 2. However the condition in Eq. (16.81) that imposes nulls at $\phi = 0$ or $\phi = \pi$ can be altered if the objective of the design is to achieve maximum resolution, whereas sidelobe level is to be maintained at a constant sidelobe level at *sll* where *sll* is given in Eq. (16.45b). In this case, the maximum array spacing determined from the following expression yields a larger β as compared with that from Eq. (16.81):

$$H\left(e^{j\pi\beta\cos(\phi)}\right) = b \tag{16.82}$$

where $\phi = 0$ or $\phi = \pi$ and b = 1 or b = −1 for even number or odd number Chebyshev array, respectively.

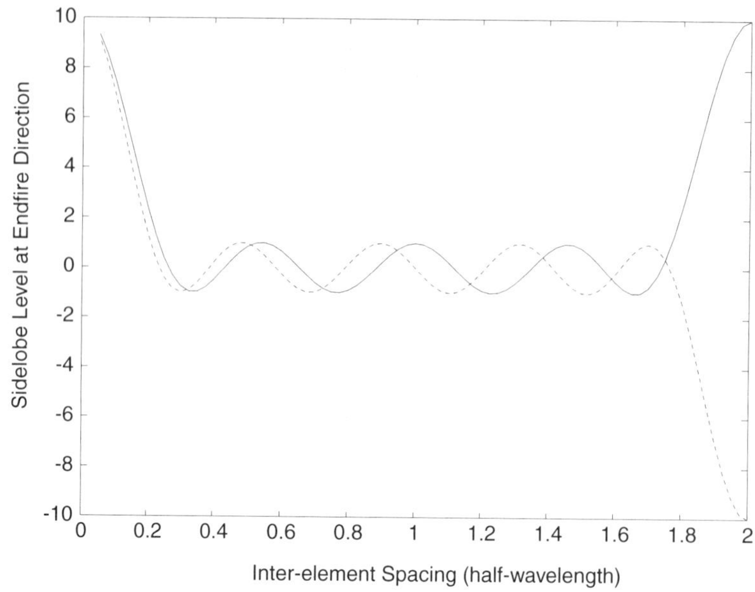

FIGURE 16.21 Sidelobe level at 0° endfire direction vs. the array interelement spacing for odd and even number linear array for a Chebyshev array with a main beam to sidelobe ratio of 20 dB. (____: 9-element, ----: 10-element)

Note that $10 \log_{10} b^2 = 0$ dB and the height of the main beam is *sll* dB above the sidelobe. Note also that for the expression in Eq. (16.82) to be a real one, the phase center location is taken to be the center of gravity of the array geometry.

Now, let $z = e^{j\pi\beta\cos(\phi)}$ for $\phi = 0$ or $\phi = \pi$, Eq. (16.82) becomes

1. For odd number array

$$\sum_{k=1}^{L} w_k z^{k-1} - bz^{0.5(L-1)} = 0 \tag{16.83}$$

2. For even number array

$$\sum_{k=1}^{L} w_k \left(z^{0.5}\right)^{2(k-1)} - b\left(z^{0.5}\right)^{0.5k+1} = 0 \tag{16.84}$$

where $\{w_i: i = 1, 2, \ldots, L\}$ are the Chebyshev weights. In fact, if the array weight vector is a symmetrical one, that is $\{w_i = w_{L-i+1}: i = 1, 2, \ldots, L\}$, the preceding equations always apply.

By solving the preceding polynomials of z and $z^{0.5}$, one gets a set of complex roots. After performing unwrapping of the phase angles of the complex roots and equating the phase angles to $\pi\beta\cos(\phi)$ for $\phi = 0$ or $\phi = \pi$, one obtains a series of values of β, from which the largest possible value of β that is smaller than 2 is β_{max}.

With the maximum array spacing available, one is able to determine the beam width of the Chebyshev array. However, there is no simple expression available for one to calculate the the beam width with respect to the array aperture, the number of array elements, and the interelement spacing, directly.

16.4.2.2 Scanned Array

In the case of the scanned Chebyshev array, the two expressions in Eqs. (16.83) and (16.84) are still valid, except that the following points must be observed:

1. When the scan angle is less than π, ϕ in z has to be set to π;
2. When the scan angle is greater than π, ϕ in z has to be set to 0.

Note that the value of β_{max} for a scanned array is always less than that of the broadside array.

16.4.3 Computer-Aided Approach to General Array Geometry Synthesis

In the previous section, it has been shown that by adjusting the array spacing of the array to β_{max}, one is able to achieve a better array resolution for a fixed main beam to sidelobe ratio. For a large array, this is achieved with an aperture length that is almost twice that of the array aperture when the interelement spacing is a half wavelength. This is sometimes undesirable because

1. The assumption of far field may not be valid anymore if the aperture length of the array is relatively large with respect to the distance of a point in the array field of view.
2. The physical space available to mount the array may be limited.

As a result, it is of interest to be able to synthesize an array geometry with beam width and sidelobe level control.

In this section, Eq. (16.41a) is used to formulate an array geometry synthesis cost function to design the shape of an array so as to achieve minimum average sidelobe for the minimum beam width or certain specified beam width. This formulation can be extended to design an array with minimum average sidelobe with null response constraint in the sidelobe region. Numerical results are shown to illustrate the performance of such a design approach.

16.4.3.1 Formulation for Array Geometry Synthesis with Minimum Average Sidelobe Level

To simplify the formulation of the minimum average sidelobe level (MASL) approach, the azimuth plane array pattern is first considered for a narrowband array with certain array sensor response (i.e., $f = f_0$, $\theta = \theta_0 = 90°$ and $\mathbf{g}$ is the L-dimensional vector given by $\mathbf{g} = \mathbf{g}_0 = [1\ 1\ \ldots\ 1]^T$). This gives rise to the following optimization problem from which the optimum array geometry, $\mathbf{r}$, is obtained

$$\underset{\mathbf{r}}{\text{minimize}} \sum_i \left(\frac{1}{\Delta\phi_i}\right) \int_{\phi_{l_i}}^{\phi_{u_i}} \left|H\left(f_0, \theta_0, \phi, \mathbf{g}_0, \mathbf{r}\right)\right|^2 d\phi \tag{16.85a}$$

$$\text{subject to } \left|H\left(f_0, \theta_0, \phi_0, \mathbf{g}_0, \mathbf{r}\right)\right|^2 = H_0^2 \tag{16.85b}$$

where H_0 is some appropriate complex constant; ϕ_0 is the desired mainlobe direction; and $\Delta\phi_0 = \phi_{ui} - \phi_{li}$. Note that $[\phi_{li}, \phi_{ui}]$'s are the spatial regions in which sidelobes are to be suppressed. By using Eq. (16.1), Eq. (16.85) can also be written as

$$\underset{\mathbf{r}}{\text{minimize}} \sum_i \left(\frac{1}{\Delta\phi_i}\right) \int_{\phi_{l_i}}^{\phi_{u_i}} \left|\mathbf{w}_0^T \mathbf{u}\left(f_0, \theta_0, \phi, \mathbf{g}_0, \mathbf{r}\right)\right|^2 d\phi \tag{16.86a}$$

$$\text{subject to } \left|\mathbf{w}_0^T \mathbf{u}\left(f_0, \theta_0, \phi, \mathbf{g}_0, \mathbf{r}\right)\right|^2 = H_0^2 \tag{16.86b}$$

where $\mathbf{w}_0$ is the L-dimensional vector given by $[1\ 1\ \ldots\ 1]^T$.

For a linear array, this optimization can be further simplified to an unconstrained optimization problem, as shown later, if the desired mainlobe direction is set to zero (i.e., broadside of the array):

$$\underset{\mathbf{r}}{\text{minimize}} \sum_i \left(\frac{1}{\Delta\phi_i}\right) \int_{\phi_{l_i}}^{\phi_{u_i}} \left|\mathbf{w}_0^T \mathbf{u}\left(f_0, \theta_0, \phi, \mathbf{g}_0, \mathbf{r}\right)\right|^2 d\phi \tag{16.87}$$

The mainlobe constraint is not required in this case because when the signal is impinging on the array from the broadside direction, the signals on the array elements are all in phase. This, in fact, inherently constrains the array response at the broadside to be maximum regardless of the arrangements of the array elements (on a straight line).

From Eq. (16.86), it can be seen that the equation involves an integration term that may not be expressed in closed form. In this case, the continuous integration may be replaced by discrete summation, shown as follows:

$$\underset{r}{\text{minimize}} \frac{1}{M_b} \sum_{i=1}^{M_b} \left| \mathbf{w}_0^T \mathbf{u}(f_0, \theta_0, \phi_i, \mathbf{g}_0, \mathbf{r}) \right|^2 \tag{16.88}$$

where M_b is the number of "bearing bins" (i.e., ϕ_i's, analogous to the terminology "frequency bins" used in discrete Fourier transform) used to cover the desired field of view. Note that the solution of the problem given in Eq. (16.87) is obtainable using a line search algorithm that is readily available in most of the mathematical texts or software packages.[11]

So far, the preceding equations are only restricted to the azimuth plane pattern design. However, one can always extend the optimization problem in Eq. (16.85) to the following more general one:

$$\underset{\mathbf{r}}{\text{minimize}} \sum_k \left(\frac{1}{\Delta\theta_k}\right) \int_{\theta_{l_k}}^{\theta_{u_k}} \sum_j \left(\frac{1}{\Delta\phi_j}\right) \int_{\phi_{l_j}}^{\phi_{u_j}} \sum_i \left(\frac{1}{\Delta f_i}\right) \int_{f_{l_i}}^{f_{u_i}} \left| H(f, \theta, \phi, \mathbf{g}_0, \mathbf{r}) \right|^2 df d\phi d\theta \tag{16.89a}$$

$$\text{subject to } \left| H(f_0, \theta_0, \phi_0 \, \mathbf{g}_0, \mathbf{r}) \right|^2 = H_0^2 \tag{16.89b}$$

where

$$\Delta f_i = f_{ui} - f_{li} \tag{16.89c}$$

$$\Delta\phi_j = \phi_{uj} - \phi_{lj} \tag{16.89d}$$

$$\Delta\theta_k = \theta_{uk} - \theta_{lk} \tag{16.89e}$$

where $[\theta_{li}, \theta_{ui}]$ and $[\phi_{li}, \phi_{ui}]$ define the field of view and $[f_{li}, f_{ui}]$ is the operating frequency range of interest.

16.4.3.2 Design Examples

In this section, design examples are shown for both linear and planar arrays. In Fig. 16.22, the array patterns are plotted for a 32-element linear array. Once again, one can see that the MASL approach, using again the sidelobe suppression sector [0°, 180°], not only reduces the sidelobe level to about 3 to 5 dB lower than that of the conventional array (in fact, at the locations of the first few sidelobes, that is, the sidelobes beside the mainlobes, the response is improved remarkably by more than 5 dB), but also reduces the beam width greatly.

In certain applications, one may like to maintain the mainlobe of the array response to be of certain width so as to ensure that the signal coming from the desired look direction is received by the array with sufficient array gain. MASL approach can be used in the design of array pattern under this situation. That is, the beam width can be preserved as that of the uniform array but with the sidelobe level being suppressed. Note that in this case the number of sidelobe suppression sectors used is two. For example, {[0°, 90° – $\Delta\phi$], [90°+$\Delta\phi$,180°]} where $\Delta\phi$ is a small angle.

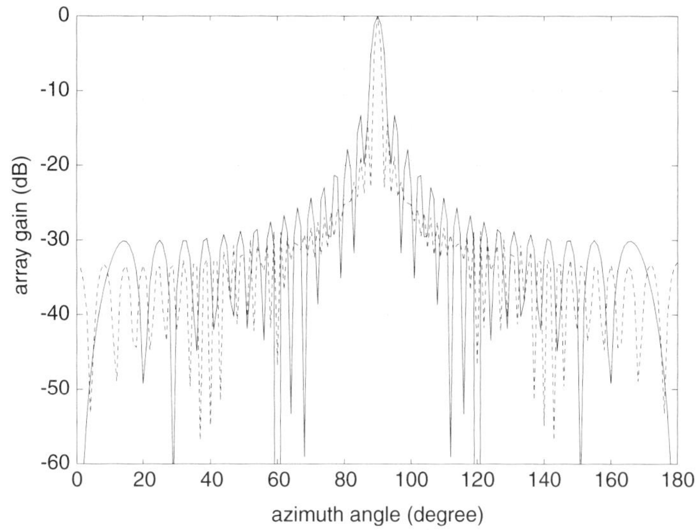

FIGURE 16.22 Array beam patterns obtained using different methods for a 32-element linear array. (____ uniform array; ---- MASL approach using one sidelobe suppression sector [0°, 180°])

To illustrate the application of the MASL approach in multidimensional array design, the result obtained for a six-element rectangular planar array is shown in Fig. 16.23. The sidelobe suppression sector used in this design example is [0°, 180°]. Once again, the property of low sidelobe and narrow beam is observed.

16.4.4 Linear Array Geometry Synthesis with Minimum Sidelobe Level and Null Control Using a First-Order Taylor Expansion Approximation Method

In this section, a first-order Taylor expansion approximation method is used to formulate a geometry synthesis cost function to design a linear array with a minimum average sidelobe for a certain specified beam width. This formulation can also be enhanced to design an array with minimum average sidelobe with null response control in the sidelobe region.

16.4.4.1 Synthesis with Minimum Average Sidelobe Level

Section 16.4.3 basically suggests the following optimization problem to obtain the array geometry:

$$\underset{\mathbf{r}}{\text{minimize}} \sum_{i} \frac{1}{\Delta\phi_i} \int_{\phi_{li}}^{\phi_{ui}} \left| H\left(f_0, \theta_0, \phi, \mathbf{g}_0, \mathbf{r}\right) \right|^2 d\phi \tag{16.90}$$

where $[\phi_{li}, \phi_{ui}]$'s are the spatial regions in which sidelobes are to be suppressed and $\Delta\phi_i = \phi_{ui} - \phi_{li}$; f_0 is the operating frequency of the array; $\theta_0 = 90°$ (i.e., array on the azimuth plane); $\mathbf{g}_0$ is the L-dimensional vector given by $[1\ 1\ \ldots\ 1]^T$).

From Eqs. (16.1) and (16.3), it can be seen that Eq. (16.90) cannot be readily solved in closed form. One possible way to obtain the solution of Eq. (16.90) is the gradient search method. However, this is sometimes not desirable because the computational load is heavy, especially when the number of array elements involved is large. One way to tackle the issue is to apply the two-term Taylor expansion to reformulate the opimization problem.

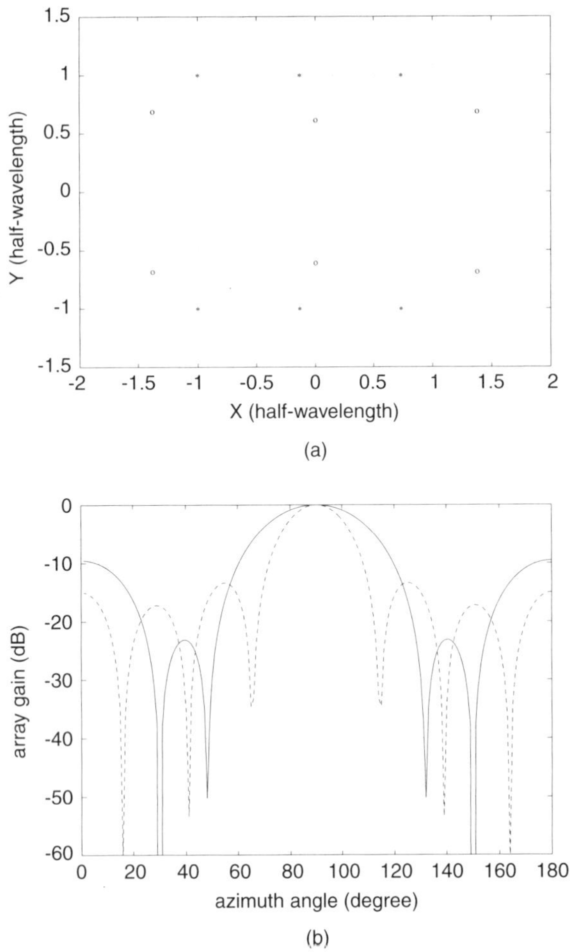

FIGURE 16.23 (a) Planar array geometries. (*, rectangular array; o, MASL approach using one sidelobe suppression sector [0°, 180°]); (b) array beam patterns obtained using different methods in (a). (____ uniform array; ---- MASL approach using one sidelobe suppression sector [0°, 180°])

Let the desired array position vector be given by

$$\mathbf{r} = \mathbf{r}_0 + \tilde{\mathbf{r}} \tag{16.91}$$

where $\mathbf{r}_0$ is the nominal array position vector and $\tilde{\mathbf{r}}$ is the array position perturbation vector.

From Eqs. (16.91) and (16.3), and assuming the vector norm of $\tilde{\mathbf{r}}$ to be sufficiently small, one may express the array gain vector in Eq. (16.3) in the following form using the two-term Taylor expansion:

$$\mathbf{s}(f_0, \theta_0, \phi, \mathbf{g}_0, \mathbf{r}) = \mathbf{s}(f_0, \theta_0, \phi, \mathbf{g}_0, \mathbf{r}_0) + j\pi\cos(\phi)\operatorname{diag}\left[\mathbf{s}(f_0, \theta_0, \phi, \mathbf{g}_0, \mathbf{r}_0)\right]\tilde{\mathbf{r}} \tag{16.92}$$

From Eqs. (16.1) and (16.92), one can easily rewrite Eq. (16.90) in the form of an unconstrained quadratic program shown as follows:

$$\underset{\tilde{\mathbf{r}}}{\text{minimize}}\, \tilde{\mathbf{r}}^T \mathbf{Q}\tilde{\mathbf{r}} - 2\tilde{\mathbf{r}}^T \mathbf{p} + \mathbf{w}_0^T \mathbf{Q}_0 \mathbf{w}_0 \tag{16.93a}$$

where $\mathbf{Q}$ and $\mathbf{Q}_0$ are the L × L dimensional matrices and $\mathbf{p}$ is the L-dimensional vector given by

$$\mathbf{Q} = \sum_i \frac{1}{\Delta\phi_i} \int_{\phi_{li}}^{\phi_{ui}} \left(\pi\cos(\phi)\right)^2 \mathbf{s}\left(f_0,\theta_0,\phi,\mathbf{g}_0,\mathbf{r}_0\right)\mathbf{s}^H\left(f_0,\theta_0,\phi,\mathbf{g}_0,\mathbf{r}_0\right)d\phi \tag{16.93b}$$

$$\mathbf{p} = \sum_i \frac{1}{\Delta\phi_i} \int_{\phi_{li}}^{\phi_{ui}} \pi\cos(\phi) Im\left[\left(\mathbf{s}^H\left(f_0,\theta_0,\phi,\mathbf{g}_0,\mathbf{r}_0\right)\right)\mathbf{w}_\phi \mathbf{s}^H\left(f_0,\theta_0,\phi,\mathbf{g}_0,\mathbf{r}_0\right)\right]d\phi \tag{16.93c}$$

$$\mathbf{Q}_0 = \sum_i \frac{1}{\Delta\phi_i} \int_{\phi_{li}}^{\phi_{ui}} \mathbf{s}\left(f_0,\theta_0,\phi,\mathbf{g}_0,\mathbf{r}_0\right)\mathbf{s}^H\left(f_0,\theta_0,\phi,\mathbf{g}_0,\mathbf{r}_0\right)d\phi \tag{16.93d}$$

where $[\cdot]^H$ denotes conjugate transpose and $Im[\cdot]$ denotes imaginary part.

Equation (16.93) yields the following solution vector:

$$\tilde{\mathbf{r}} = \mathbf{Q}^{-1}\mathbf{p} \tag{16.94}$$

By substituting Eqs. (16.94) back to Eq. (16.91), one obtains a new array geometry that minimizes the average sidelobe level with slight geometry perturbation.

16.4.4.2 Synthesis with Minimum Average Sidelobe Level and Null Control

To introduce fixed nulls into the beam pattern at the desired directions, one may solve Eq. (16.93) with additional linear null constraints given by

$$\mathbf{w}_0^T \mathbf{s}\left(f_0,\theta_0,\phi_i,\mathbf{g}_0,\mathbf{r}_0\right) = 0 \qquad \text{for } i = 1, 2, \ldots, N_r \tag{16.95}$$

for some even-number directions (i.e., N_r is even), which are symmetrical with respect to the normal of the line array, at which nulls are to be imposed.

From Eq. (16.92), (16.93) and (16.95), the following constrained optimization problem is obtained

$$\underset{\tilde{\mathbf{r}}}{\text{minimize}}\ \tilde{\mathbf{r}}^T\mathbf{Q}\tilde{\mathbf{r}} - 2\tilde{\mathbf{r}}^T\mathbf{p} + \mathbf{w}_0^T\mathbf{Q}_0\mathbf{w}_0 \tag{16.96a}$$

$$\text{subject to } j\pi\cos(\phi_i)\tilde{\mathbf{r}}^T\mathbf{s}\left(f_0,\theta_0,\phi_i,\mathbf{g}_0,\mathbf{r}_0\right) = \mathbf{w}_0^T\mathbf{s}\left(f_0,\theta_0,\phi_i,\mathbf{g}_0,\mathbf{r}_0\right) \text{ for } i = 1, 2, \ldots, N_r \tag{16.96b}$$

which yields the following solution weight vector by means of Lagrange multiplier method:[11,15]

$$\tilde{\mathbf{r}} = \mathbf{Q}^{-1}\mathbf{C}\left(\mathbf{C}^H\mathbf{Q}^{-1}\mathbf{C}\right)^{-1}\left[\mathbf{d} - \mathbf{C}^H\mathbf{Q}^{-1}\mathbf{p}\right] + \mathbf{Q}^{-1}\mathbf{p} \tag{16.97a}$$

where $\mathbf{C}$ is the $L \times N_r$ dimensional matrix and $\mathbf{d}$ is the L-dimensional vector given by

$$\mathbf{C} = \text{conj}\left[j\pi\cos(\phi_1)\mathbf{s}\left(f_0,\theta_0,\phi_1,\mathbf{g}_0,\mathbf{r}_0\right), \ldots, j\pi\cos(\phi_{Nr})\mathbf{s}\left(f_0,\theta_0,\phi_{Nr},\mathbf{g}_0,\mathbf{r}_0\right)\right] \tag{16.97b}$$

$$\mathbf{d} = -\left[\mathbf{w}_0^T\mathbf{s}\left(f_0,\theta_0,\phi_1,\mathbf{g}_0,\mathbf{r}_0\right), \ldots, \mathbf{w}_0^T\mathbf{s}\left(f_0,\theta_0,\phi_{Nr},\mathbf{g}_0,\mathbf{r}_0\right)\right]^T \tag{16.97c}$$

Note that conj[·] denotes complex conjugate.

16.4.4.3 Numerical Stability Consideration

The formulations given in Eq. (16.93) and (16.96) may not always produce a satisfactory solution. This results from the fact that the quadratic programs in Eq. (16.96) and (16.99) are derived based on the assumption given by Eq. (16.92); that is, the elements of the perturbation vector obtained by solving Eq. (16.93), and Eq. (16.96) may be so large that the approximation using the two-term Taylor expansion may not be valid anymore. As a result, the perturbation vector obtained from Eqs. (16.93) and (16.96) does not always generate an improved array beam pattern. To overcome this numerical instability problem, a further constraint may be added into the two optimization problems in Eqs. (16.96) and (16.99), which yield

$$\underset{\tilde{\mathbf{r}}}{\text{minimize}}\ \tilde{\mathbf{r}}^T\mathbf{Q}\tilde{\mathbf{r}} - 2\tilde{\mathbf{r}}^T\mathbf{p} + \mathbf{w}_0^T\mathbf{Q}_0\mathbf{w}_0 \tag{16.98a}$$

$$\text{subject to } \tilde{\mathbf{r}}^T\tilde{\mathbf{r}} \leq \varepsilon \tag{16.98b}$$

and

$$\underset{\tilde{\mathbf{r}}}{\text{minimize}}\ \tilde{\mathbf{r}}^T\mathbf{Q}\tilde{\mathbf{r}} - 2\tilde{\mathbf{r}}^T\mathbf{p} + \mathbf{w}_0^T\mathbf{Q}_0\mathbf{w}_0 \tag{16.99a}$$

$$\text{subject to } j\pi\cos(\phi_i)\tilde{\mathbf{r}}^T\mathbf{s}(f_0,\theta_0,\phi_i,\mathbf{g}_0,\mathbf{r}_0) = -\mathbf{w}_0^T\mathbf{s}(f_0,\theta_0,\phi_i,\mathbf{g}_0,\mathbf{r}_0) \quad \text{for } i = 1, 2, \ldots, N_r \tag{16.99b}$$

$$\tilde{\mathbf{r}}^T\tilde{\mathbf{r}} \leq \varepsilon \tag{16.99c}$$

where ε is a preset small real quantity to ensure that the two-term Taylor expansion is a valid one.

By solving Eqs. (16.98) and (16.99), the following solutions are obtained, respectively:

$$\tilde{\mathbf{r}} = \mathbf{Q}_a^{-1}\mathbf{p} \tag{16.100}$$

$$\tilde{\mathbf{r}} = \mathbf{Q}_a^{-1}\mathbf{C}\left(\mathbf{C}^H\mathbf{Q}_a^{-1}\mathbf{C}\right)^{-1}\left[\mathbf{d} - \mathbf{C}^H\mathbf{Q}_a^{-1}\mathbf{p}\right] + \mathbf{Q}_a^{-1}\mathbf{p} \tag{16.101}$$

where

$$\mathbf{Q}_a = \mathbf{Q} + \alpha\mathbf{I} \tag{16.102}$$

where $\alpha \geq 0$ is the Lagrange multiplier,[11,15] which can be obtained by solving the following transcendental equation if the norm bound constraints in Eq. (16.98b) and (16.99c) are active:

$$\tilde{\mathbf{r}}^T(\alpha)\tilde{\mathbf{r}}(\alpha) = \varepsilon \tag{16.103}$$

Note that the Lagrange multiplier solved from Eq. (16.103) is only optimum with respect to the error bound quantity ε and the array beam pattern is very much dependent on the error bound selected. Therefore, in the design examples shown in the next section, α is used directly as a weighting factor to control the norm of $\tilde{\mathbf{r}}$.

16.4.4.4 Design Examples

In Fig. 16.24, the beam patterns are obtained for a ten-element linear array with unity weights. One can see that the conventional array demonstrates relatively high sidelobe level. In the case of the quadratic

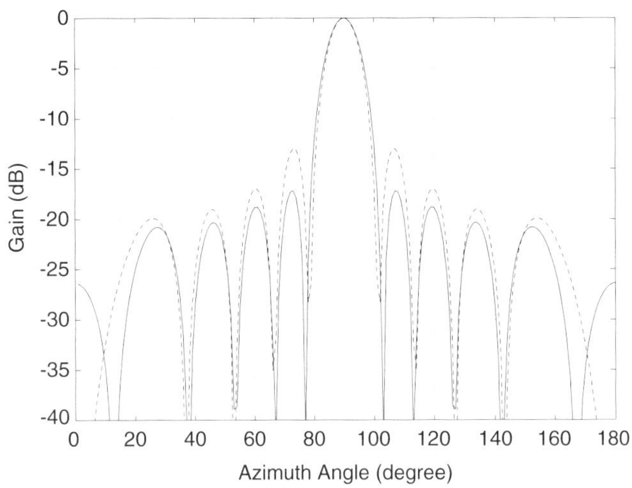

FIGURE 16.24 Array patterns for ten-element linear arrays. (---- conventional, ____ quadratic programming method, $\alpha = 0$, suppressed regions: [0°, 82°], [98°, 180°])

TABLE 16.2 The Ten-Element Array Geometry for Conventional Array and Optimum Linear Array with Minimum Average Sidelobe

Conventional	±0.50	±1.50	±2.50	±3.50	±4.50
Optimum	±0.44	±1.25	±2.20	±3.22	±4.42

programming approach, with the use of zero weighting factor α and the sidelobe suppressed regions [0°, 82°] and [98°, 180°], it offers an improvement of 1.5 to 5 dB in terms of sidelobe level suppression. This is achieved in exchange for a larger beam width. However, one can see from the same figure that the 3-dB beam width has only increased slightly, which is less than a degree. Table 16.2 shows the array geometries corresponding to the array patterns in Fig. 16.24. As one can see from Table 16.2, the aperture size obtained from the proposed approach is more or less the same as that of the conventional array. In fact, this is also observed for the rest of the design examples shown next.

Figure 16.25 shows a prescribed broad null design. For this case, α is equal to 100 and sidelobe suppressed region is [0°, 180°]. Nulls are imposed at 55°, 57.5°, 60°, 120°, 122.5°, and 125°. One can see that a broad null as deep as 50 dB is easily available. The corresponding array geometry is shown in Table 16.3.

Figure 16.26 shows a design with sidelobe suppression and null control. It is interesting to note that with a deep null (lower than –50 dB) imposed just beside the first sidelobe (at 81° and 99° directions), the first sidelobe level is still maintained at 5 to 6 dB, below that of the conventional array, whereas most of the sidelobes are still being suppressed or maintained. The sidelobe suppressed regions used are [0°, 87°] and [93°, 180°] and the weighting factor α used is 0.008. For this design example, the corresponding array geometry is shown in Table 16.4.

16.5 Conclusion and Suggested Readings

This chapter has basically covered two broad classes of array pattern synthesis techniques. The first is to design a set of array weighting coefficients to achieve the desired beam pattern; the second is to design array geometry so as to adjust the array pattern. For each channel, by adjusting the location of the array element, only the signal phase is manipulated. As such, the beam patterns achievable by using the second approach are rather limited as compared with those of the first.

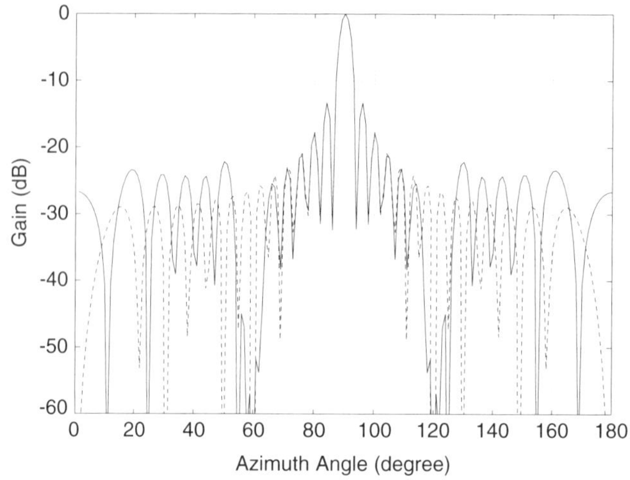

FIGURE 16.25 Array patterns for 28-element linear arrays with prescribed broad null. (---- conventional, ____ quadratic programming method, $\alpha = 100$, suppressed regions: [0°, 180°], null directions: 55°, 57.5°, 60°, 120°, 122.5°, 125°)

TABLE 16.3 The 28-Element Array Geometry for Conventional Array and Optimum Linear Array with Broad Null

Conventional	±0.50	±1.50	±2.50	±3.50	±4.50	±5.50	±6.50	±7.50	±8.50	±9.50	±10.50	±11.50	±12.50	±13.50
Optimum	±0.56	±1.54	±2.44	±3.49	±4.54	±5.49	±6.52	±7.52	±8.42	±9.51	±10.65	±11.42	±12.30	±13.67

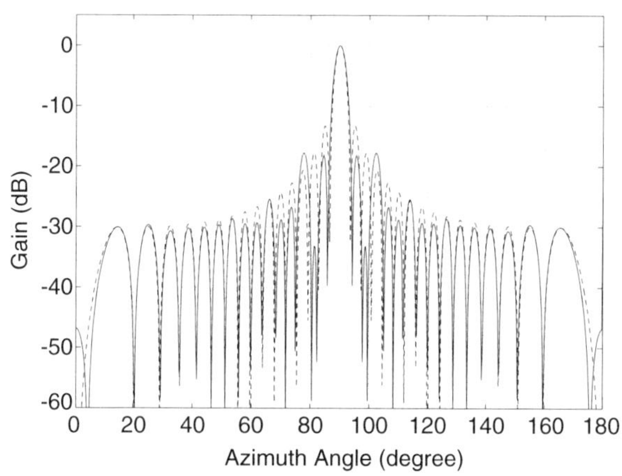

FIGURE 16.26 Array patterns for 32-element linear arrays with minimum sidelobe level and null control. (---- conventional, ____ quadratic programming method, $\alpha = 0.008$, suppressed regions: [0°, 87°], [93°, 180°], null directions: 81°, 99°)

TABLE 16.4 The 32-Element Array Geometry for Conventional Array and Optimum Linear Array with Minimum Average Sidelobe and Null Control

Conventional	±0.50	±1.50		±2.50	±3.50	±4.50	±5.50	±6.50	±7.50	±8.50
Optimum	±0.49	±1.45		±2.35	±3.19	±3.96	±4.67	±5.37	±6.11	±6.92
Conventional	±9.50	±10.50	±11.50	±12.50	±13.50	±14.50	±15.50			
Optimum	±7.85	±8.89	±10.06	±11.33	±12.65	±14.01	±15.35			

For array weight synthesis, besides covering the three classical topics on uniform, Chebyshev and Taylor arrays,[7,27] the chapter also introduces the three important computer-aided techniques using quadratic programming,[19] adaptive array theory,[22] and genetic algorithm.[1,6] Readers are referred to the book by Elliott,[9] the articles by Autrey,[3] Tseng and Cheng,[28] for Taylor pattern designs on rectangular/circular arrays; and for Chebyshev pattern designs on rectangular arrays, for both separable and nonseparable excitation coefficients.

To deal with optimum beam pattern designs using quadratic programming, further reading materials are available from articles by Carlson and Willner,[5] Er et al.[10] Some of the useful articles the readers may refer to for adaptive array based synthesis include Dufort,[8] Griffiths and Buckley,[12] Laxpati,[14] Tseng and Griffiths,[28] Zhou and Ingram.[35] In many cases, an adaptive array beamforming problem is formulated using the covariance matrix (second-order statistic), and the adaptive weight is commonly solved using a constrained quadratic program. As a result, some of the adaptive array-based approaches are quadratic programming based. One of the interesting applications not touched on in the chapter is array element failure correction. The quadratic programming and the adaptive array approaches are useful for such application.[25] Element failure correction can also be done using the genetic algorithm. One such example is given by Yeo and Lu.[34] An alternative approach to the synthesis problem is simulated annealing.[17] Examples are available in these references. For general reading, the readers may also refer to References 26 and 32.

For array beam pattern control by geometry synthesis, the chapter has provided the mathematical expressions or steps to determine the maximum aperture length and beam width for a uniform or Chebyshev linear array. Quadratic programming is also applied to compute the array element locations for sidelobe reduction, beam width suppression and/or null control. Besides quadratic programming, genetic algorithm is perhaps another useful approach for array elements positioning.

References

1. Ares-Pena, F.J., Rodriguez-Gonzalez, J.A., Villanueva-Lopez, E. and Rengarajan, S.R., Genetic algorithms in the design and optimization of antenna array patterns, *IEEE Trans. Antennas Propag.*, Vol. 47, No. 3, 506, March 1999.
2. Ares, F. and Moreno, E., New method for computing Dolph–Chebyshev arrays, and its comparison with other methods, *IEE Proc.*, Vol. 135, Pt. H, No. 2, 129, April 1988.
3. Autrey, S.W., Approximate synthesis of nonseparable design responses for rectangular arrays, *IEEE Trans. Antennas Propag.*, Vol. AP-35, No. 8, 907, August 1987.
4. Bresler, M.A. et al., A new algorithm for calculating the current distributions of Dolph–Chebyshev arrays, *IEEE Trans. Antennas Propag.*, 28, No. 6, 884, 1980.
5. Carlson, B.D. and Willner, D., Antenna pattern synthesis using weighted least squares, *IEE Proc.*, Vol. 139, Pt. H, No. 1, 11, February 1992.
6. dMarcano, D., Jimenez, M., Duran, F. and Chang, O., Synthesis of antenna arrays using genetic algorithms, *Proc. 1995 First IEEE Intl. Caracas Conf. on Devices, Circuits, and Systems,* 328, 1995.
7. Dolph, C.L., A current distribution for broadside arrays which optimizes the relationship between beamwidth and sidelobe level, *Proc. IRE*, Vol. 34, 335, 1946.
8. Dufort, E.C., Pattern synthesis based on adaptive array theory, *IEEE Trans. Antennas Propag.*, Vol. 37, No. 8, 1011, August 1989.
9. Elliott, R.S., Antenna Theory and Design, Englewood Cliffs, NJ: Prentice-Hall, 1981, Chapters 5 and 6.
10. Er, M.H., Sim, S.L. and Koh, S.N., Application of constrained optimization techniques to array pattern synthesis, *Signal Process.*, Vol. 34, 323, 1993.
11. Fletcher, R., *Practical Methods of Optimization,* 2nd ed., John Wiley & Sons, Chichester, 1987.
12. Griffiths, L.J. and Buckley, K.M., Quiescent pattern control in linearly constrained adaptively arrays, *IEEE Trans. Acoustics, Speech Signal Process.*, Vol. 35, No. 7, 917, July 1987.
13. Ismail T.H. and Mahmoud M. Dawoud, Null steering in Phased arrays by controlling the element positions, *IEEE Trans. Antennas Propag.*, Vol. 39, No. 11, 1561, November 1991.

14. Laxpati, S.R., Planar array synthesis with prescribed pattern nulls, *IEEE Trans. Antennas Propag.*, Vol. 30, No. 6, 1176, 1982.
15. Luenberger, D.G., *Linear and Nonlinear Programming,* Addison-Wesley, Reading, MA, 1984.
16. Moffet, A.T., Minimum-redundancy linear arrays, *IEEE Trans. Antennas Propag.*, Vol. 16, No. 2, 172, 1968.
17. Murino, V., Trucco, A. and Regazzoni, C.S., Synthesis of unequally spaced arrays by simulated annealing, *IEEE Trans. Signal Process.*, Vol. 44, No. 1, 119, January 1996.
18. Ng, B.P., Designing array patterns with optimum inter-element spacings and optimum weights using a computer-aided approach, *Int. J. Electron.*, Vol. 73, No. 3, 653, 1992.
19. Ng, B.P., Er, M.H. and Kot, C.A., A flexible array synthesis method using quadratic programming, *IEEE Trans. Antennas Propag.*, Vol. 41, No. 11, 1541, November 1993.
20. Ng, B.P., Er, M.H. and Kot, C.A., Linear array aperture synthesis with minimum sidelobe level and null control, *IEE Proc.-Microwave Antennas Propag.*, Vol. 141, No. 3, 162, June 1994.
21. Ng, B.P., Array synthesis using a simple computer-aided approach, *IEE Electron. Lett.*, Vol. 26, No. 5, March 1, 1990.
22. Olen, C.A. and Compton, JR. R.T., A numerical pattern synthesis algorithm for array, *IEEE Trans. Antennas Propag.*, Vol. 38, No. 10, 1666, October 1990.
23. Safaai-Jazi, A., A new formulation for the design of Chebyshev arrays, *IEEE Trans. Antennas Propag.*, Vol. 42, No. 3, 439, March 1994.
24. Shpak, T.J. and Antoniou, A., A flexible optimization method for the pattern synthesis of equispaced linear arrays with equiphase excitation, *IEEE Trans. Antennas Propag.*, Vol. 40, No. 10, 1187, October 1992.
25. Sim, S.L. and Er, M.H., Sidelobe suppression for general arrays in presence of element failures, *Electron. Lett.*, Vol. 33, No. 15, 1278, July 17, 1997.
26. Stutzman, W.L. and Thiele, G.A., *Antenna Theory and Design,* John Wiley & Sons, New York, 1981.
27. Taylor, T.T., Design of line-source antennas for narrow beamwidth and low sidelobes, *IRE Trans. Antennas Propag.*, Vol. AP-3, January 16, 1955.
28. Tseng F.I. and Cheng D.K., Optimum scannable planar arrays with an invariant sidelobe level, *Proc. IEEE*, 56, 1771, 1968.
29. Tseng C.Y. and Griffiths, L.J., A simple algorithms to achieve desired patterns for arbitrary arrays, *IEEE Trans. Signal Process.*, Vol. 40, No. 11, 2737, November 1992.
30. Vertatschitsch E. and Haykin, S., Non-redundant arrays, *Proc. IEEE*, Vol. 74, No. 1, 217, January 1986.
31. Villeneuve, A.T., Taylor pattern for discrete arrays, *IEEE Trans. Antennas Propag.*, Vol. AP-32, 1089, October 1984.
32. Wolff, E. A., *Antenna Analysis,* Artech House, Norwood, MA, 1998.
33. Yan, K.K. and Lu, Y., Sidelobe reduction in array-pattern synthesis using genetic algorithm, *IEEE Trans. Antennas Propag.*, Vol. 45, No. 7, 1117, July 1997.
34. Yeo, B.K. and Lu, Y., Array failure correction with a genetic algorithm, *IEEE Trans. Antennas Propag.*, Vol. 47, No. 5, 823, May 1999.
35. Zhou, P.Y. and Ingram, M.A., Pattern synthesis for arbitrary arrays using an adaptive array method, *IEEE Trans. Antennas Propag.*, Vol. 47, No. 5, 862, May 1999.

17

Electromagnetic Vector Sensors with Beamforming Applications*

Arye Nehorai
University of Illinois at Chicago

Kwok-Chiang Ho
Addest Technovation Pte. Ltd.

B. T. G. Tan
National University of Singapore

17.1 Introduction

Direction-of-arrival (DOA) estimation and beamforming for electromagnetic (EM) waves are two common objectives of array processing. Early work on DOA estimation and beam forming has been based on scalar sensors, each of which provides measurements of only one component of the electric or magnetic field induced. Subsequent research has investigated the use of sensors that measure two components of the electric or magnetic field, and tripole sensors that measure three complete components of the electric field. In recent years, researchers have proposed the use of EM vector sensors that measure the three complete components of the electric field and three components of the magnetic field at one point, for DOA estimation.[1,2]

*The work of A. Nehorai was supported by the Air Force Office of Scientific Research under Grants F49620-97-1-0481, F49620-99-1-0067, and F49620-00-1-0083, the National Science Foundation under Grant MIP-9615590; and the Office of Naval Research under Grant N00014-98-1-0542.

0-8493-0124-6/02/$0.00+$1.50

17.1.1 Advantages of Using Electromagnetic Vector Sensors

EM vector sensors are sensitive to both the DOA and polarization information in the incoming waves. The polarization provides a crucial criterion for distinguishing and isolating signals that may otherwise overlap in conventional scalar-sensor arrays. When a single EM vector sensor is used for DOA estimation or beamforming, it has the following advantages and capabilities:

- DOA estimation/beamforming in three dimensional (3D) while occupying very little space
- Resolution of very closely spaced (even coincident) sources based on polarization differences
- Ability to process wideband signals in the same way as narrowband signals
- Handling of sources with either single- or dual-message signals
- Isotropic response
- No need for location calibration and time synchronization among different components

Some of these advantages result from the fact that no time delays are used. In contrast, conventional scalar–sensor methods require a two-dimensional (2D) array for DOA estimation/beamforming in a 3D space, need accurate location calibration and time synchronization, and require much higher computational cost to process wideband instead of narrowband signals.

An array of spatially distributed EM vector sensors can additionally exploit time delays among the sensors. In general, arrays of EM vector sensors can achieve better performance than scalar–sensor arrays, while occupying less space. They can also be spaced farther apart o increase aperture and hence performance without introducing ambiguities.

17.1.2 Historical Development

The subject of electromagnetic vector sensors was first introduced to the signal processing community by Nehorai and Paldi.[1,2] EM vector sensors are commercially available and actively researched. EMC Braden Ltd. in Baden, Switzerland, manufactures them for a 75Hz to 30MHz frequency range, and Flam and Russell, Inc. in Horsham, Pennsylvania, for 2 to 30MHz. Lincoln Laboratories at MIT has performed preliminary localization tests with the EM vector sensors of Flam and Russell, Inc.[3] Other research on sensor development is reported by Kanda[4] and Kanda and Hill.[5]

The optimum accuracy of source parameter estimation for vector–sensor arrays is analyzed by Nehorai and Paldi[1,2] in terms of Cramér–Rao bounds (absolute limits on the accuracy of the class of unbiased parameter estimators). Quality measures are defined for estimating DOA and orientation in 3D space, including mean-square angular error and covariance of vector-angular error. Lower bounds on these measures give concrete results on expected performance.

A fast algorithm for DOA estimation using an EM vector sensor has been proposed.[1,2] Inspired by the Poynting theorem, it forms the cross product of the electric field vector with the complex conjugate of the magnetic vector and averages over time. The asymptotic performance under general conditions is shown to be close to optimum. Some cross-product algorithms for source tracking using an EM vector sensor have also been proposed.[6]

A minimum-noise-variance beamformer for interference cancellation employing a single EM vector sensor has been proposed.[7] It assumes that the DOA and polarization of the source are known. This enables suppression of uncorrelated interference, even if it comes from the same direction as the source, based on polarization as well as location differences. Analysis of the signal to interference-plus-noise ration (SINR) showed the beamformer to be very effective, particularly when the signal and interference are differently polarized.

Some high-resolution DOA estimation algorithms have been developed for EM vector-sensor arrays for various scenarios. These include those algorithms customized for scenarios with only completely polarized signals,[8,10] those for scenarios with only incompletely polarized signals,[9] and those for more general scenarios where completely and incompletely polarized signals may coexist.[11] On a separate note, some articles reported preliminary results on the application of EM vector sensors to communication problems.[12,13]

The identifiability and uniqueness issues associated with EM vector sensors have been rigorously studied in References 14 to 18. In particular, it has been established that one EM vector sensor can estimate uniquely the DOA and polarization ellipses of up to three sources.[15,16] In comparison, one would need at least four appropriately spaced scalar sensors to determine uniquely the DOA of one source.

An active and a passive models for estimating polarimetric parameters associated with EM waves have also been proposed,[19] in which EM vector sensors are used as a receiving device.

17.1.3 Contents of This Chapter

This chapter presents a minimum-noise-variance type beamformer for a single EM vector sensor[7] (which first appear in Reference 7) and also analyses the performance of the beamformer. Such a beamformer requires the knowledge of the DOA and polarization parameters of the signal, and assumes that the signal, inference, and noise are mutually uncorrelated. The beamformer minimizes the output variance while maintaining the gain in the direction of the signal. This has the effect of preserving the signal while minimizing contributions to the output due to interference and noise arriving from directions other than the direction of the signal.

The investigation of the performance of the beamformer is restricted to scenarios where there exist one signal and one interference that are uncorrelated. Two types of signals are considered: one carries a single message, and the other carries two independent messages simultaneously[1,2] The former is called a single-message (SM) signal, and the latter a dual-message (DM) signal. On the other hand, the interference under consideration takes a general form that can be completely polarized (CP) or incompletely polarized (IP). Note that SM signals are CP, while DM signals are IP.

First, explicit expressions for the SINR are obtained for both SM signals and DM signals. Then some physical implications associated with the SINR expressions are discussed. In particular, it is deduced that for the two types of signals of interest, the SINR rises with an increase in the separation between the DOAs and/or the polarizations of the signal and the interference for all DOAs and polarizations (scalar-sensor arrays and a single tripole do not have such properties). Moreover, a strategy for effectively suppressing an interference with an EM vector sensor is identified. The SINR expression for the SM signal also provides a basis for generating a DM signal in which the two message signals have minimum interference effect on one another.

The outline of this chapter is as follows. Section 17.2 presents problem formulation for beamforming and some preliminary discussions. The analyses concerning SM and DM signals are presented in Sections 17.3 and 17.4, respectively. Section 17.5 presents numerical examples. Section 17.6 analyses the characteristics of the mainlobe and sidelobe of the beampattern of an EM vector sensor, and compares them with other types of sensor arrays. Section 17.7 presents conclusions.

17.2 Beamforming Problem Formulation and Preliminary Discussion

17.2.1 Abbreviations and Notations

Abbreviations

EM	electromagnetic
SINR	signal to interference-plus-noise ratio
DOA	direction of arrival
DM	dual message
SM	single message

DOP	degree of polarization
CP	completely polarized
IP	incompletely polarized
UP	unpolarized
PD	polarization difference
ULA	uniform linear array
UCA	uniform circular array

Notations

$(\cdot)^T$, $(\cdot)^H$, *	transpose, Hermitian and complex conjugate
$\mathbf{I}_n$	$n \times n$ identity matrix
$\mathbf{y}_s(t)$, $\mathbf{y}_d(t)$	6×1 complex envelope (phasor) measurement (both electric and magnetic fields) received at an EM vector sensor at time t associated with SM signal or DM signal
$\mathbf{y}_E(t)$	3×1 complex envelope (phasor) electric field measurement
$\mathbf{y}_H(t)$	3×1 complex envelope (phasor) magnetic field measurement
$\mathbf{e}_E(t)$, $\mathbf{e}_H(t)$	3×1 complex envelope (phasor) electric and magnetic noise
$s_s(t)$	complex envelope of an SM signal
$s_{d,1}(t)$, $s_{d,2}(t)$	complex envelopes of the first and second message signals of a DM signal
$\mathbf{R}_s$, $\mathbf{R}_d$	covariance matrices of $\mathbf{y}_s(t)$, $\mathbf{y}_d(t)$
$\mathbf{R}_i$	interference covariance matrix
$\mathbf{w}$	weight vector of the minimum-noise-variance beamformer
ϕ, ψ	azimuth and elevation association with a DOA
α, β	orientation and ellipticity angles associated with a polarization of a CP signal
θ	vector denoting $[\phi, \psi, \alpha, \beta]^T$
$\mathbf{a}(\theta)$	steering vector of an EM vector sensor for a CP signal with parameter θ
$\mathbf{B}(\phi, \psi)$	steering matrix of an EM vector sensor for a UP signal with DOA (ϕ, ψ)
$\mathbf{Q}(\alpha)$	rotation matrix with angle α
$\mathbf{h}(\beta)$	2×1 unit-norm vector representing ellipticity of a polarization
$\sigma_{i,c}^2$, $\sigma_{i,u}^2$	powers of the CP and UP components of an interference
σ_s^2	power of an SM signal
$\sigma_{d,1}^2$, $\sigma_{d,2}^2$	powers of the first and second message signals of a DM signal
σ^2	power of the electric/magnetic noise
Δ_i^s	difference between the polarizations of an SM signal and interference
$\Delta_i^{d,1}$	difference between the polarizations of the first message signal (of a DM signal) and interference
$\Delta_i^{d,2}$	difference between the polarizations of the second message signal (of a DM signal) and interference
γ	angular separation between the DOAs of the signal and interference
λ	wavelength

Note that the subscripts s, d, d,1, d,2, and i are used to associate some symbols with SM signal, DM signal, the first and second messages of a DM signal, and interference, respectively. For example, the symbols β_s, $\beta_{d,1}$, $\beta_{d,2}$, and β_i denote the ellipticity angles associated with SM signals, the first and second messages of a DM signal, and interference, respectively.

17.2.2 Beamforming Problem Formulation for Single-Message Signal

This and the next subsections describe the data models as proposed by Nehorai and Paldi[1,2] for a single EM vector sensor receiving an SM signal or a DM signal. Note that an SM signal and a DM signal are two types of signals encountered in communications application, and they are results of two different methods of signal transmission. For an SM signal, one message signal is transmitted from a source using a fixed polarization, such as a linear polarization or a circular polarization. For DM signal, two independent messages are transmitted simultaneously from the same source using two different polarizations. [We shall call such transmission method dual-signal transmission (DST)]. Various DST forms exist. One DST form uses two linearly polarized signals that are spatially and temporally orthogonal with an amplitude or phase modulation (see, e.g., References 29 and 30). Another DST form uses two circularly polarized signals with opposite spins. In general, it can be stated that the advantage of the DST method is that it doubles the bandwidth of a communication system. However, the preceding DST methods may suffer from possible cross-polarization (see, e.g., Reference 29), multipath effects, and other unknown distortions from the source to the sensor.

Note that an SM signal is common in most communications applications, and transmission of a DM signal is used in some communication applications such as digital television (TV) broadcasting described in Reference 31, microwave radio system manufactured by HeliOss,[32] cellular radio network,[33] and optical communication systems.[34]

With the notation stated in Subsection 17.2.1, the complex (phasor) sensor measurement obtained by an EM vector sensor at time t, induced by an SM signal in the presence of an interference and additive noise is given by

$$\mathbf{y}_s(t) \triangleq \begin{pmatrix} \mathbf{y}_E(t) \\ \mathbf{y}_H(t) \end{pmatrix} = \mathbf{a}(\theta_s)s_s(t) + \mathbf{B}(\phi_i, \psi_i)\xi_i(t) + \mathbf{e}(t) \tag{17.1}$$

where

$$\begin{aligned}
\mathbf{a}(\theta) &= \mathbf{B}(\phi,\psi)\mathbf{Q}(\alpha)\mathbf{h}(\beta) \\
\theta &= [\phi,\psi,\alpha,\beta]^T \\
\mathbf{B}(\phi,\psi) &= \begin{pmatrix} \mathbf{v}(\phi,\psi) & \tilde{\mathbf{v}}(\phi,\psi) \\ \tilde{\mathbf{v}}(\phi,\psi) & -\mathbf{v}(\phi,\psi) \end{pmatrix} \\
\begin{pmatrix} \mathbf{v}(\phi,\psi) & \tilde{\mathbf{v}}(\phi,\psi) \end{pmatrix} &= \begin{pmatrix} -\sin\phi & -\cos\phi\sin\psi \\ \cos\phi & -\sin\phi\sin\psi \\ 0 & \cos\psi \end{pmatrix} \\
\mathbf{Q}(\alpha) &= \begin{pmatrix} \cos\alpha & \sin\alpha \\ -\sin\alpha & \cos\alpha \end{pmatrix} \\
\mathbf{h}(\beta) &= \begin{pmatrix} \cos\beta \\ j\sin\beta \end{pmatrix}
\end{aligned} \tag{17.2}$$

$\mathbf{e}(t) = [\mathbf{e}_E^T(t), \mathbf{e}_H^T(t)]^T$, $s_s(t) \in \mathbb{C}^1$, $\xi_i(t) \in \mathbb{C}^{2\times 1}$, $\mathbf{y}_E(t)$, $\mathbf{y}_H(t)$, $\mathbf{e}_E(t)$, $\mathbf{e}_H(t)$, $\in \mathbb{C}^{3\times 1}$.

The first, second, and third terms on the right-hand side of Eq. (17.1) correspond to measurements induced by the signal, interference, and noise, respectively. Physically, $\mathbf{y}_E(t)$ and $\mathbf{y}_H(t)$ are the three-component measurements of the electric and magnetic fields at the sensor at time t, respectively, and $\mathbf{e}_E(t)$ and $\mathbf{e}_H(t)$ are the noise components in these measurements. The parameters $\phi \in (-\pi, \pi]$ and $\psi \in [-\pi/2, \pi/2]$ are the azimuth and elevation of the signal, respectively; and $\alpha \in (-\pi/2, \pi/2]$ and $\beta \in [-\pi/4, \pi/4]$ are the polarization parameters referred to as the orientation angle and ellipticity, respectively. The vector $\mathbf{a}(\theta)$ is the steering vector of an EM vector sensor associated with an SM signal with parameter θ, and $\mathbf{v}(\phi, \psi)$ and $\tilde{\mathbf{v}}(\phi, \psi)$ are unit vectors that span the same plane as the electric and magnetic field vectors of the incoming signal with DOA (ϕ, ψ). The variable $s_s(t)$ is the complex envelope of the signal and $\xi_i(t)$ the complex envelopes of the interference.

The covariance of $\xi_i(t)$ determines the state of polarization of the interference. Indeed, the interference covariance matrix, $\mathbf{R}_i \triangleq E(\xi i(t)\xi_i^H(t))$, can be expressed as[19]

$$\mathbf{R}_i = \frac{\sigma_{i,u}^2}{2}\mathbf{I}_2 + \sigma_{i,c}^2\mathbf{Q}(\alpha_i)\mathbf{h}(\beta_i)\mathbf{h}^H(\beta_i)\mathbf{Q}^H(\alpha_i) \tag{17.3}$$

The first term on the right-hand side of Eq. (17.3) is the UP component with power $\sigma_{i,cu}^2$, and the second is the CP component with power $\sigma_{i,c}^2$. The degree of polarization (DOP) of the interference is defined as the ratio between the power of the CP component and the total power of the interference (i.e., $\sigma_{i,c}^2/(\sigma_{i,u}^2 + \sigma_{i,c}^2)$. The interference is said to be CP if $\sigma_{i,c}^2 \neq 0$ but $\sigma_{i,u}^2 = 0$, IP if $\sigma_{i,c}^2 \neq 0$ and $\sigma_{i,cu}^2 \neq 0$, and UP if $\sigma_{i,u}^2 \neq 0$ but $\sigma_{i,c}^2 = 0$.

The output of a beamformer in this case is

$$\hat{s}_s(t) = \mathbf{w}_s^H \mathbf{y}_s(t) \tag{17.4}$$

where $\mathbf{w}_s \in \mathbb{C}^{6\times 1}$ is a weight vector. Suppose the DOA and polarization parameters of the signal are known; then, for the minimum-noise-variance beamformer, the weight vector is obtained through the following constrained minimization:

$$\begin{aligned} \mathbf{w}_s &= \arg \min_{\mathbf{w}\in\mathbb{C}^{6\times 1}} \mathbf{w}^H\mathbf{R}_s\mathbf{w} \\ \text{subject to} \quad & \mathbf{w}^H\mathbf{a}_s = 1 \end{aligned} \tag{17.5}$$

where $\mathbf{R}_s = E(\mathbf{y}_s(t)\mathbf{y}_s^H(t))$ is the data covariance matrix, and $\mathbf{a}_s$ denotes $\mathbf{a}(\theta_s)$. The beamformer attempts to suppress all incoming interference except for the desired signal with steering vector $\mathbf{a}_s$.

17.2.3 Beamforming Problem Formulation for Dual-Message Signal

The complex (phasor) sensor measurement obtained by an EM vector sensor at time t, induced by a DM signal in the presence of an interference and additive noise is given by

$$\mathbf{y}_d(t) = \mathbf{a}(\theta_{d,1})s_{d,1}(t) + \mathbf{a}(\theta_{d,2})s_{d,2}(t) + \mathbf{B}(\phi_i,\psi_i)\xi_i(t) + \mathbf{e}(t) \tag{17.6}$$

where

$$\theta_{d,1} = (\phi_d, \psi_d, \alpha_{d,1}, \beta_{d,1}) \quad \text{and} \quad \theta_{d,2} = (\phi_d, \psi_d, \alpha_{d,2}, \beta_{d,2})$$

The first and second terms on the right-hand side of Eq. (17.6) correspond to measurements induced by the first and second message signals, respectively, associated with the DM signal, whereas the third and fourth terms correspond to the interference and noise, respectively. The variables $s_{d,k}(t)$ and $\mathbf{a}(\theta_{d,k})$, where $k = 1, 2$, are the complex envelope and steering vector of the kth message signal. Note that the two steering vectors $\mathbf{a}(\theta_{d,1})$ and $\mathbf{a}(\theta_{d,2})$ have the same DOA (ϕ_d, ψ_d) but different polarizations (i.e., $(\alpha_{d,1}, \beta_{d,1}) \neq (\alpha_{d,2}, \beta_{d,2})$). In Section 17.4, an appropriate choice of $(\alpha_{d,1}, \beta_{d,1})$ and $(\alpha_{d,2}, \beta_{d,2})$ that minimizes the interference effect on one message signal resulting from the other will be proposed.

The outputs of a beamformer for the first and second message signals are, respectively

$$\hat{s}_{d,1}(t) = \mathbf{w}_{d,1}^H \mathbf{y}_d(t)$$

and

$$\hat{s}_{d,2}(t) = \mathbf{w}_{d,2}^H \mathbf{y}_d(t)$$

where $\mathbf{w}_{d,1}, \mathbf{w}_{d,2} \in \mathbb{C}^{6\times 1}$ are the corresponding weight vectors. Note that to optimize the recovery of the message signals, a specific weight vector is used for each message signal separately. Suppose the DOA and polarization parameters of the signal are known. Then, for the minimum-noise-variance beamformer, the weight vector for the kth message signal, where $k = 1,2$ is obtained through the following constrained minimization:

$$\mathbf{w}_{d,k} = \arg \min_{\mathbf{w} \in \mathbb{C}^{6\times 1}} \mathbf{w}^H \mathbf{R}_d \mathbf{w}$$
$$\text{subject to} \quad \mathbf{w}^H \mathbf{a}_{d,k} = 1 \tag{17.7}$$

where $\mathbf{R}_d = E(\mathbf{y}_d(t)\mathbf{y}_d^H(t))$ is the data covariance matrix, and $\mathbf{a}_{d,k}$ denotes $\mathbf{a}(\theta_{d,k})$.

17.2.4 Assumptions

The analyses to be carried out are based on the following:

Assumption 1. *The DOA and polarization parameters of the signal are known.*

Assumption 2. *The complex envelopes of $s_s(t)$, $s_{d,1}(t)$ and $s_{d,2}(t)$, and of each component of* $\mathbf{e}_E(t)$ and $\mathbf{e}_H(t)$ *are all zero-mean Gaussian random variables.*

Assumption 3. *The signal is uncorrelated with the interference.*

Assumption 4. *The various components of the noise are uncorrelated among themselves, and also uncorrelated with both the signal and interference.*

Assumption 5. *The powers of the electric noise and magnetic noise are all equal to σ^2 (i.e., the noise covariance matrix is equal to $\sigma^2\mathbf{I}_6$).*

Under Assumptions 2 to 5, the data covariance matrix is

$$\mathbf{R}_s = \sigma_s^2 \mathbf{a}_s \mathbf{a}_s^H + \mathbf{B}_i \mathbf{R}_i \mathbf{B}_i^H + \sigma^2 \mathbf{I}_6$$

for the case of SM signal, where $\sigma_s^2 = E(s_s(t)s_s^*(t))$ is the power of the signal, and

$$\mathbf{R}_d = \sigma_{d,1}^2 \mathbf{a}_{d,1} \mathbf{a}_{d,1}^H + \sigma_{d,2}^2 \mathbf{a}_{d,2} \mathbf{a}_{d,2}^H + \mathbf{B}_i \mathbf{R}_i \mathbf{B}_i^H + \sigma^2 \mathbf{I}_6$$

for the case of DM signal, where $\sigma_{d,k}^2 = E(s_{d,k}(t)s_{d,k}^*(t))$ is the power of kth message signal, $k = 1, 2$.

17.2.5 Performance Measures

The ratio between the output power of the signal and output power of the interference and noise (SINR) is used to evaluate the beamformer performance. The SINR measure has been used as performance indicator for beamformers in many studies. For an SM signal, the SINR is given by

$$SINR_s \stackrel{\Delta}{=} \frac{\sigma_s^2 \mathbf{w}_s^H \mathbf{a}_s \mathbf{a}_s^H \mathbf{w}_s}{\mathbf{w}_s^H \left(\mathbf{R}_s - \sigma_s^2 \mathbf{a}_s \mathbf{a}_s^H\right) \mathbf{w}_s} \tag{17.8}$$

For a DM signal, the SINR for the kth message signal, $\hat{s}_{d,k}(t)$, is

$$SINR_{d,k} \stackrel{\Delta}{=} \frac{\sigma_{d,k}^2 \mathbf{w}_{d,k}^H \mathbf{a}_{d,k} \mathbf{a}_{d,k}^H \mathbf{w}_{d,k}}{\mathbf{w}_{d,k}^H \left(\mathbf{R}_d - \sigma_{d,k}^2 \mathbf{a}_{d,k} \mathbf{a}_{d,k}^H\right) \mathbf{w}_{d,k}} \tag{17.9}$$

where $k = 1, 2$.

In this chapter, explicit expressions for $SINR_s$, $SINR_{d,1}$, and $SINR_{d,2}$ are obtained, and their characteristics in terms of the various parameters of the signal, interference, and noise are investigated.

To interpret the SINR expressions, a parameter that provides a measure for the difference between the polarizations of two signals (using the Poincaré sphere polarization representation)[23] is introduced. First, let $(\phi_1, \psi_1, \alpha_1, \beta_1)$ and $(\phi_2, \psi_2, \alpha_2, \beta_2)$ be the DOAs/polarizations of two signals, and consider a new coordinate system where the DOAs of the two signals both lie in the *xy* plane (such a coordinate system can always be obtained with an appropriate coordinate rotation). In such a new coordinate system, the ellipticity angle β_k of the signal remain unchanged. However, the orientation angle α_k will change, which shall be denoted by α'_k. According to the Poincaré sphere representation, a polarization (α'_k, β_k) is represented by a point (referred to as Poincaré point for convenience) on a sphere whose center is at the origin and radius is 1. The position vector of that point is

$$\mathbf{P}_k = \left[\cos 2\alpha' \cos 2\beta, \sin 2\alpha' \cos 2\beta, \sin 2\beta\right]^T \tag{17.10}$$

Such a representation has two desirable properties. First, for two polarizations with the same orientation angle (with respect to the new coordinate system), the larger the difference is in their ellipticity angles, the larger the distance between two Poincaré points associated with the two polarizations. Second, for two polarizations with the same ellipticity angle (with respect to the new coordinate system), the larger the difference in their orientation angles*, the larger the distance between two Poincaré points associated with the two polarizations. Thus, it is meaningful to take the *difference between the polarizations of the two signals* to be Δ_2^1, the shorter arc length joining $\mathbf{p}_1$ and $\mathbf{p}_2$, where $\mathbf{p}_1$ and $\mathbf{p}_2$ are the representations for the polarizations (α'_1, β_1) and (α'_2, β_2) on the Poincaré sphere, respectively.

REMARKS. (1) To obtain the difference between the polarizations of two signals, there is a need to know the polarizations, as well as the DOAs of these signals. (2) It can be shown that the difference between the polarizations of two signals is independent of the coordinate system. (3) When dealing with the difference between the polarizations of two signals, only the polarizations of the CP components of the signals are of concern. (4) The range of Δ_2^1 is $[0, \pi]$. (5) The arc length Δ_2^1 is related to the orientation and ellipticity angles through Lemma 1.

LEMMA 1 (Compton[21]). *Consider polarizations (α_1, β_1) and (α_2, β_2) associated with two signals. Let (α'_1, β_1) and (α'_2, β_2) be the polarizations in a coordinate system such that the DOAs of these signals both lie in the xy plane. Then*

$$\cos^2\left(\frac{\Delta_2^1}{2}\right) = \left|\mathbf{h}^H(\beta_2)\mathbf{Q}^H(\alpha'_2)\mathbf{Q}(\alpha'_1)\mathbf{h}(\beta_1)\right|^2 \tag{17.11}$$

17.2.6 A Useful Result

Under Assumptions 1 to 5, it can be shown that the weight vectors satisfying respectively Eqs. (17.5) and (17.7) are

$$\mathbf{w}_s = \frac{\mathbf{R}_s^{-1}\mathbf{a}_s}{\mathbf{a}_s^H\mathbf{R}_s^{-1}\mathbf{a}_s} \quad \text{and} \quad \mathbf{w}_{d,k} = \frac{\mathbf{R}_d^{-1}\mathbf{a}_{d,k}}{\mathbf{a}_{d,k}^H\mathbf{R}_d^{-1}\mathbf{a}_{d,k}} \quad \text{for } k = 1, 2 \tag{17.12}$$

*The increase in difference between the orientation angles is valid within a certain (useful) range of the orientation angles.

By substituting Eq. (17.12) directly into the expressions for SINR given by Eqs. (17.8) and (17.9), the DOAs and polarizations of the signal and interference, as well as the noise power, are hidden in two matrices whose inverses need to be evaluated. For ease of interpreting the dependence of SINR on the signal, interference, and noise, the following result, which is useful for simplifying the analysis of SINR expressions, is needed.

LEMMA 2 (Cox[24]). *Let* $\mathbf{R} = \alpha_k^2 \mathbf{a}\mathbf{a}^H + \mathbf{G} \in \mathbb{C}^{6\times 6}$, *and*

$$\hat{\mathbf{w}} = \arg \min_{\mathbf{w}\in\mathbb{C}^{6\times 1}} \mathbf{w}^H \mathbf{R}\mathbf{w}$$

$$\text{subject to} \quad \mathbf{w}^H \mathbf{a} = 1$$

where $\mathbf{a} \in \mathbb{C}^{6\times 1}$ *is as defined in Eq. (17.2),* and σ_k *is a real constant. If* $\mathbf{G}$ *is nonsingular, then*

$$\frac{\sigma_k^2 \hat{\mathbf{w}}^H \mathbf{a}\mathbf{a}^H \hat{\mathbf{w}}}{\hat{\mathbf{w}}^H \mathbf{G}\hat{\mathbf{w}}} = \sigma_k^2 \mathbf{a}^H \mathbf{G}^{-1}\mathbf{a}$$

To further simplify the analysis of the SINR expression to be presented later, it is assumed hereafter that $(\phi_s, \psi_s) = (\phi_{d,1}, \psi_{d,1}) = (\phi_{d,2}, \psi_{d,2}) = (0, 0)$ and $\psi_i = 0$ (i.e., the DOA of the signal is parallel to the x-axis and that of the interference is in the xy plane)*. With such a setup, the separation between the DOAs of the signal and interference is simply ϕ_i. In addition, the difference between the polarizations of the signal and interference, Δ_i^s, satisfies $\cos^2(\Delta_i^s/2) = |\mathbf{h}^H(\beta_i)\mathbf{Q}^H(\alpha_i)\mathbf{Q}(\alpha_s)\mathbf{h}(\beta_s)|^2$.

17.3 Signal to Interference-plus-Noise Ratio for Single-Message Signal

For convenience, the angular separation between the DOAs of the signal and interference is referred to as DOA separation, and is denoted by γ. Moreover, the difference between the polarizations of the signal and interference is referred to as polarization difference (PD). Theorem 1 below expresses the $SINR_s$ explicitly in terms of the DOA separation, PD, and powers of the signal, interference, and noise.

THEOREM 1. *The expression of* $SINR_s$*, as given in Eqs. (17.8), can be expressed as*

$$SINR_s = \sigma_s^2\left(\frac{2}{\sigma^2} - \frac{(1+\cos\gamma)^2}{(\sigma^2+\sigma_{i,u}^2)}\left(\frac{\sigma_{i,u}^2}{2\sigma^2} + \frac{\sigma_{i,c}^2 \cos^2\frac{\Delta_i^s}{2}}{2\sigma_{i,c}^2 + \sigma^2 + \sigma_{i,u}^2}\right)\right) \tag{17.13}$$

PROOF. See Reference 7.

REMARK. For UP interference, the PD Δ_i^s is undefined and can take any value within $[0, \pi]$. However, $\sigma_{i,c}^2 = 0$ in this case and the last term of Eq. (17.13) is zero regardless of the value of Δ_i^s.

Clearly, $SINR_s$ increases with an increase in the signal power, σ_s^2, but decreases with an increase in the noise power, σ^2, as well as the power of the CP (i.e., $\sigma_{i,c}^2$) or UP (i.e., $\sigma_{i,u}^2$) component of the interference.

However, the dependencies of $SINR_s$ on PD and DOA separation are nontrivial, and are established in the following corollaries.

*This can be achieved with an appropriate coordinate rotation. It can be shown that with such a coordinate rotation, $SINR_s$, $SINR_{d,1}$, and $SINR_{d,2}$ as defined in Eqs. (17.8) and (17.9) remain invariant. Moreover, the separation between the DOAs and difference between the polarizations of the signal and interference remain unchanged (the latter follows from the definition of the difference between two polarizations presented in Section 17.2.5).

COROLLARY 1. *If* $\sigma_{i,c}^2 \neq 0$ *and* $\gamma \neq \pi$, *then* $SINR_s$ *is an increasing function of* Δ_i^s.

COROLLARY 2. *If* $\sigma_{i,u}^2 \neq 0$ *or* $\Delta_i^s \neq \pi$, *then* $SINR_s$ *is an increasing function of* γ.

COROLLARY 3. *If* $\sigma_{i,c}^2 = 0$, *then* $SINR_s$ *is independent of* Δ_i^s.

COROLLARY 4. $SINR_s$ *attains the maximum value,* $SINR_s^{max} = 2\sigma_s^2/\sigma^2$, *when either* $\gamma = \pi$, *or both* $\Delta_i^s = \pi$ *and* $\sigma_{i,u}^2 = 0$ *are true. Moreover,* $SINR_s^{max}$ *simply takes the value of* $SINR_s$ *in the absence of interference.*

COROLLARY 5. *For given (fixed)* σ_s^2, $\sigma_{i,u}^2 + \sigma_{i,c}^2$, σ^2, *and* $\gamma \neq \pi$, *the minimum of* $SINR_s$ *is attained when* $\Delta_i^s = 0$ and $\Delta_{i,u}^s = 0$.

PROOF. See Reference 7.

REMARKS. 1. Corollary 1 means that $SINR_s$ generally increases with an increase in the PD Δ_i^s, except for two special cases: (a) $\sigma_{i,c}^2 = 0$, or (b) $\gamma = \pi$. Note that case (a) corresponds to scenarios where the interference is UP, and case (b) to scenarios where the DOA of the signal is exactly opposite to that of the interference. For case (a), the interference has no CP component and thus the PD should not affect $SINR_s$ (see Corollary 3). On the other hand, by Corollary 4, $SINR_s$ for case (b) always attains the maximum value $SINR_s^{max}$ regardless of the other signal parameters.

2. A special case of Corollary 1 is that even if the DOAs of the signal and interference are identical, one can still increase the value of $SINR_s$ by increasing the PD Δ_i^s. This is a feature that scalar-sensor arrays lack. Indeed, for a scalar-sensor array, if the DOA of the interference is identical to that of the signal, interference suppression is impossible regardless of the PD, the number of sensors, and array aperture.

3. Corollary 2 means that $SINR_s$ generally increases with an increase in the DOA separation γ, except for the case where both $\sigma_{i,u}^2 = 0$ and $\Delta_i^s = \pi$ hold. For the case where $\sigma_{i,u}^2 = 0$ and $\Delta_i^s = \pi$, $SINR_s^{max}$ can always be attained regardless of the other parameters (see Corollary 4). Note that $\sigma_{i,u}^2 = 0$ means that the interference is CP, and $\Delta_i^s = \pi$ means that the PD is the largest possible. For a coordinate system where the DOAs of the signal and interference both lie in the *xy* plane, such a PD arises when the polarizations associated with the signal and interference satisfy $(\alpha_s, \beta_s) = (\alpha_i \pm \pi/2, -\beta_i)$. Physically, the two polarization ellipses associated with the polarizations (α_s, β_s) and (α_i, β_i) have the same shape but have their major axes orthogonal to each other, and at the same time, the directions of spin of the electric fields associated with the two polarizations are opposite.

4. By Corollary 3, if the interference is UP, then it is not possible to increase the $SINR_s$ by varying the polarization of thc signal (α_s, β_s).

5. Corollary 4 means that $SINR_s$ attains the largest possible value, $SINR_s^{max}$, when either the DOAs of the signal and interference are opposite, or when the interference is CP with largest possible polarization difference, π. In either case, $SINR_s^{max}$ obtained is equivalent to the $SINR_s$ when there is no interference regardless of the interference's power (i.e., the interference becomes completely ineffective).

6. Corollary 5 means that, for any given DOA separation, $SINR_s$ attains its lowest value when the interference is CP with polarization difference equal to 0.

The fact that $SINR_s$ increases with an increase in the DOA separation or PD for all DOAs and polarizations (see Corollaries 1 and 2) is an important feature associated with an EM vector sensor. Indeed, this feature is desirable because it is natural to expect a higher *SINR* with a larger DOA separation or PD. In contrast, for scalar–sensor arrays and a single tripole, $SINR_s$ does not necessarily increase with an increase in the separation in DOAs or polarizations (the case of a tripole is elaborated in Section 17.3.2).

The preceding corollaries are potentially useful in some applications. For example, one can exploit the fact that $SINR_s$ increases with an increase in the PD (Corollary 1) to effectively suppress an interference if the DOA and polarization of the interference are known. Indeed, for a fixed DOA separation γ, one can maximize $SINR_s$ by transmitting the signal with polarization such that the PD is the largest possible (i.e., $\Delta_i^s = \pi$). This would lead to $SINR_s = \sigma_s^2[2 - (1 + \cos\gamma)^2\sigma_{i,u}^2/2(\sigma^2 + \sigma_{i,u}^2)]/\sigma^2$. Clearly, if the interference is CP (i.e., $\sigma_{i,u}^2 = 0$), then $SINR_s$ attains $SINR_s^{max} = 2\sigma_s^2/\sigma^2$, which is the value when there is no interference, regardless of the DOA separation and the interference power.

17.3.1 Comparisons with Scalar-Sensor Arrays

Beamformers using scalar-sensor arrays have been addressed in the literature.[20] This subsection shall discuss some advantages of using an EM vector sensor as compared with scalar-sensor arrays for beamforming in 3D space. First, for a scalar-sensor array, at least three sensors are needed to perform beamforming, which means that it occupies a larger space than an EM vector sensor. Second, when the DOA of the interference is identical to that of the signal, interference suppression is impossible regardless of the number of scalar sensors and the array aperture. In contrast, a single vector sensor can suppress an interference if the difference between the polarizations of the signal and interference is nonzero (see Remark 2 of the corollaries to Theorem 1). Third, consider a signal and an interference with sufficiently large DOA separation. Then, to suppress the interference with arbitrary DOA, only one EM vector sensor is needed. However, for the case of a scalar–sensor array, at least four appropriately spaced scalar sensors are needed. Indeed, to suppress an interference, the steering vector associated with the interference must be linearly independent of that associated with the signal. In this connection, it has been shown that to ensure every two steering vectors with distinct DOAs to be linearly independent, one EM vector sensor is sufficient,[18] but at least four scalar sensors with inter-sensor spacings all less than a half wavelength are needed for the case of scalar-sensor array.[25] This is a result of the fact that an EM vector sensor searches in both the polarization and DOA domains, whereas a scalar-sensor array uses only time delay information. Fourth, the SINR for a vector sensor is isotropic, whereas for a scalar-sensor array, it very much depends on the array geometry, and does not necessarily increase with an increase in the DOA separation.

17.3.2 Comparisons with a Single Tripole

A beamformer using a single tripole has been addressed by Compton.[21] Compton investigated the performance of a single tripole in suppressing a CP interference on receiving an SM signal. From the results obtained by Compton, one can deduce that, unlike the case of an EM vector sensor, the SINR for a single tripole does not necessarily increase with an increase in the DOA separation or the PD. Two examples shall be used to illustrate this property. First, consider a signal and an interference with DOAs lying in the *xy* plane and vertically and linearly polarized. Then, the electric fields induced by the signal and interference are identical (except for a scale constant); thus, it is possible to discriminate the signal and interference regardless of their DOA separation. Thus, the SINR remains unchanged (which is the smallest possible) regardless of the DOA separation. Next, consider a signal and an interference with opposite DOAs and both lying in the *xy* plane, and suppose that both of them are circularly polarized. Then, the SINR when the signal and interference have the same spin (the PD is 0) is larger than when they have opposite spins (the PD is π). (This is because the electric fields induced at a tripole due to the signal and interference, with opposite DOAs, are identical (except for a scale constant) if their directions of spin are opposite, but are distinct if their directions of spin are identical. Consequently, the SINR does not necessarily increase with an increase in the PD.

Compton has also established that the SINR for a single tripole is the lowest (with SINR being equal to $\sigma_s^2/(\sigma^2 + \sigma_{i,c}^2)$) if one of the following three conditions holds:

1. The interference has the same DOA and polarization as the signal.
2. The DOA of the signal is opposite to that of the interference, and the polarizations of the signal and interference satisfy $\alpha_s = -\alpha_i$ and $\beta_s = -\beta_i$.
3. The signal and interference are both linearly polarized, and their electric fields are parallel to each other.

Now let us examine the preceding three conditions for an EM vector sensor (for scenarios where there exist a CP interference and an SM signal). $SINR_s$ for an EM vector sensor is lowest only if condition 1 is satisfied, and the lowest $SINR_s$ equals to $2\sigma_s^2/(\sigma^2 + 2\sigma_{i,c}^2)$, which is higher than the lowest SINR obtained with a single tripole. As for condition 2, Corollary 4 states that as long as the DOA of the signal is opposite to that of the interference, $SINR_s$ for an EM vector sensor always attains the maximum value, $SINR_s^{max}$, regardless of the polarizations of the signal and interference. On the other hand, high $SINR_s$ can be obtained for an EM vector sensor even when condition 3 is met. Indeed when condition 3 is met, the DOA separation may range from 0 to π. By Corollary 2, $SINR_s$ can be increased by increasing the DOA separation, and by Corollary 4, $SINR_s$ attains the maximum value, $SINR_s^{max}$, when the DOA of the signal is opposite to that of the interference.

Thus, a single EM vector sensor genrally outperforms a single tripole in suppressing a CP interference when receiving an SM signal.

17.4 Signal to Interference-plus-Noise Ratio for Dual-Message Signal

A DM signal consists of two SM signals (or CP signals) with the same DOA but different polarizations. The effective polarization of such a DM signal varies with time, and thus, the state of polarization of a DM signal can either be IP or UP. To transmit a DM signal (consisting of two uncorrelated message signals), it is desirable that the interference effect of one message signal on the other be minimal. Because the DOA parameters associated with the two message signals are identical, it is possible to exploit the difference only in the polarization parameters to reduce the interference effect. In this connection, Corollary 4 of Theorem 1 provides a good way for choosing the polarizations. Indeed, consider the scenario where there is no external interference and view one message signal as the desired CP signal, and the other message signal as a CP "interference". Then, by Corollary 4 of Theorem 1, both $SINR_{d,1}$ and $SINR_{d,2}$ attain their maximum values if the difference between the polarizations of the two message signals is equal to π (i.e., when extracting one message signal, there is theoretically no interference effect from the other). Therefore, it is assumed hereafter that the polarizations of the two message signals are chosen in such a way that the PD is π, meaning that the polarizations satisfy $(\alpha_{d,1}, \beta_{d,1}) = (\alpha_{d,2} \pm \pi/2, -\beta_{d,2})$ (refer to Remark 3 of the corollaries to Theorem 1 for a relevant physical meaning).

For convenience, the difference between the polarizations of the first message signal and the interference (i.e., $\Delta_i^{d,1}$) is referred to as the first PD, and the difference between the polarizations of the second message signal and the interference (i.e., $\Delta_i^{d,2}$) is referred to as the second PD. Theorem 2 that follows expresses $SINR_{d,1}$ and $SINR_{d,2}$ explicitly in terms of the DOA separation, the first and the second PDs, and the powers of the two message signals, interference and noise.

Theorem 2: If $(\alpha_{d,1}, \beta_{d,1}) = (\alpha_{d,2} \pm \pi/2, -\beta_{d,2})$, then

$$\left.\begin{aligned} SIN\,R_{d,1} &= \sigma_{d,1}^2\left(\frac{2}{\sigma^2} - \frac{(1+\cos\gamma)^2}{\sigma^4}\left(\mu + \nu\cos^2\frac{\Delta_i^{d,1}}{2} + \frac{\nu^2}{4\sigma^4\delta_1}(1+\cos\gamma)^2\sin^2\Delta_i^{d,1}\right)\right) \\ SIN\,R_{d,2} &= \sigma_{d,2}^2\left(\frac{2}{\sigma^2} - \frac{(1+\cos\gamma)^2}{\sigma^4}\left(\mu + \nu\cos^2\frac{\Delta_i^{d,2}}{2} + \frac{\nu^2}{4\sigma^4\delta_2}(1+\cos\gamma)^2\sin^2\Delta_i^{d,2}\right)\right) \end{aligned}\right\} \tag{17.14}$$

where

$$\mu = \frac{\sigma^2\sigma_{i,u}^2}{2(\sigma^2+\sigma_{i,u}^2)}, \quad \nu = \frac{\sigma^4\sigma_{i,c}^2}{(\sigma^2+\sigma_{i,u}^2)(2\sigma_{i,c}^2+\sigma^2+\sigma_{i,u}^2)} \tag{17.15}$$

$$\left.\begin{aligned} \delta_1 &= \frac{\sigma^2+2\sigma_{d,2}^2}{\sigma_{d,2}^2\sigma^2} - \frac{(1+\cos\gamma)^2}{\sigma^4}\left(\mu + \nu\sin^2\frac{\Delta_i^{d,1}}{2}\right) \\ \delta_2 &= \frac{\sigma^2+2\sigma_{d,1}^2}{\sigma_{d,1}^2\sigma^2} - \frac{(1+\cos\gamma)^2}{\sigma^4}\left(\mu + \nu\sin^2\frac{\Delta_i^{d,2}}{2}\right) \end{aligned}\right\} \tag{17.16}$$

PROOF. See Reference 7.

REMARK. Theorem 2 is derived based on the assumption that $\sigma_{d,1}^2$ and $\sigma_{d,2}^2$, the powers of the first and second message signals, respectively, are nonzero. If $\sigma_{d,1}^2$ or $\sigma_{d,2}^2$ is equal to zero, Theorem 2 reduces to the case of SM signals that have been addressed in Section 3, and the derivation of the SINR expression is somewhat different from those of $SINR_{d,1}$ and $SINR_{d,2}$.

COROLLARY 1. *If* $\sigma_{i,c}^2 \neq 0$ *and* $\gamma \neq \pi$, *then* $SINR_{d,k}$ *is an increasing function of* $\Delta_i^{d,k}$, *for* $k = 1, 2$.

COROLLARY 2. *If* $\sigma_{i,u}^2 \neq 0$ or $\Delta_i^{d,k} \neq \pi$, *then* $SINR_{d,k}$ *is an increasing function of* γ *for* $k = 1, 2$.

COROLLARY 3. *If* $\sigma_{i,c}^2 = 0$, *then* $SINR_{d,k}$ *is independent of* $\Delta_i^{d,k}$, *for* $k = 1, 2$.

COROLLARY 4. $SINR_{d,k}$ *attains the maximum value,* $SINR_{d,k}^{max} = 2\sigma_{d,k}^2/\sigma^2$, when either $\gamma = \pi$, or both $\Delta_i^{d,k} = \pi$ *and* $\sigma_{i,u}^2 = 0$ *are true, for* $k = 1, 2$. *Moreover,* $SINR_{d,k}^{max}$ *simply takes the value of* $SINR_{d,k}$ *in the absence of intereference.*

COROLLARY 5. *For given (fixed)* $\sigma_{d,1}^2$, $\sigma_{d,2}^2$, $\sigma_{i,u}^2 + \sigma_{i,c}^2$, σ^2, and $\gamma \neq \pi$, *the minimum of* $SINR_{d,k}$ *is attained when* $\Delta_i^{d,k} = 0$ and $\sigma_{i,u}^2 = 0$, for $k = 1, 2$.

PROOF. See Reference 7.

REMARKS. 1. The dependence of $SINR_{d,k}$ on $\Delta_i^{d,k}$, γ, $\sigma_{d,k}^2$, σ^2, $\sigma_{i,c}^2$, and $\sigma_{i,u}^2$ as presented in Corollaries 1 to 5 of Theorem 2 is basically identical to that of $SINR_s$ on Δ_i^s, γ, σ_s^2, σ^2, $\sigma_{i,c}^2$, and $\sigma_{i,u}^2$ as presented in Corollaries 1 to 5 of Theorem 1. Therefore, the discussion concerning Corollaries 1 to 5 of Theorem 1 in Section 17.3 is applicable to Corollaries 1 to 5 of Theorem 2.

2. Because $\mathbf{p}_{d,1}$ and $\mathbf{p}_{d,2}$ as defined in Eq. (17.10) (which correspond to the representation of $(\alpha_{d,1}, \beta_{d,1})$ and $(\alpha_{d,2}, \beta_{d,2})$, respectively, on the Poincaré sphere) are 2 antipodal points on the Poincaré sphere, it can be shown that the sum of the first PD, $\Delta_i^{d,1}$, and the second PD, $\Delta_i^{d,2}$, is equal to a constant π. Thus, an increase in the first PD leads to a decrease in the second PD, and vice versa. This has two implications. First, by Corollary 1 of Theorem 2, increasing the value of the first (or second) PD leads to an increase in $SINR_{d,1}$ (or $SINR_{d,2}$), but a decrease in $SINR_{d,2}$ (or $SINR_{d,1}$). Consequently, the values of both $SINR_{d,1}$ and $SINR_{d,2}$ cannot be increased simultaneously with a change in the polarization of the interference. Second, by Corollary 4 of Theorem 2, if the DOA separation is not equal to π, then $SINR_{d,1}$ attains its maximum value when the interference is CP and the first PD is equal to π. However, the second PD becomes zero, and thus by Corollary 5, $SINR_{d,2}$ attains its minimum value. Thus, for each DOA separation that is not equal to π, $SINR_{d,1}$ attains its maximum value if and only if $SINR_{d,2}$ attains its minimum value.

Because $SINR_{d,1}$ and $SINR_{d,2}$ are generally not identical, it is not easy to address the SINR for the DM signal as a whole. Here, let us consider the worst case and used $SINR_{min} \triangleq \min\{SINR_{d,1}\ SINR_{d,2}\}$ (which gives the smaller value between $SINR_{d,1}$ and $SINR_{d,2}$) as a measure of the effective SINR. Next, because it is reasonable to assume that both message signals are equally important, one may take the powers of the two message signals to be identical. With these considerations, one can easily verify from Corollaries 1 to 3 of Theorem 2 that

1. $SINR_{min}$ increases with an increase in DOA separation.
2. $SINR_{min}$ increases when the first (or the second) PD increases from 0 to $\pi/2$, but decreases when the first (or the second) PD increases form $\pi/2$ to π.
3. $SINR_{min}$ is independent of the first and second PDs if the interference is UP.

Comparable results are not available for scalar sensors, simply because scalar sensors cannot receive two (independent) message signals simultaneously.

17.5 Numerical Results

This section presents some numerical examples to assess the reliability of the theoretical prediction of the performance of an EM vector sensor as presented in Sections 17.3 and 17.4. First, exact data covariance matrices $\mathbf{R}_s$ and $\mathbf{R}_d$ were used in the experiments for checking the SINR expressions derived in Sections 17.3 and 17.4. Next, to make the experiments realistic, $\mathbf{R}_s$ and $\mathbf{R}_d$ generated from finite number of snapshots were used.

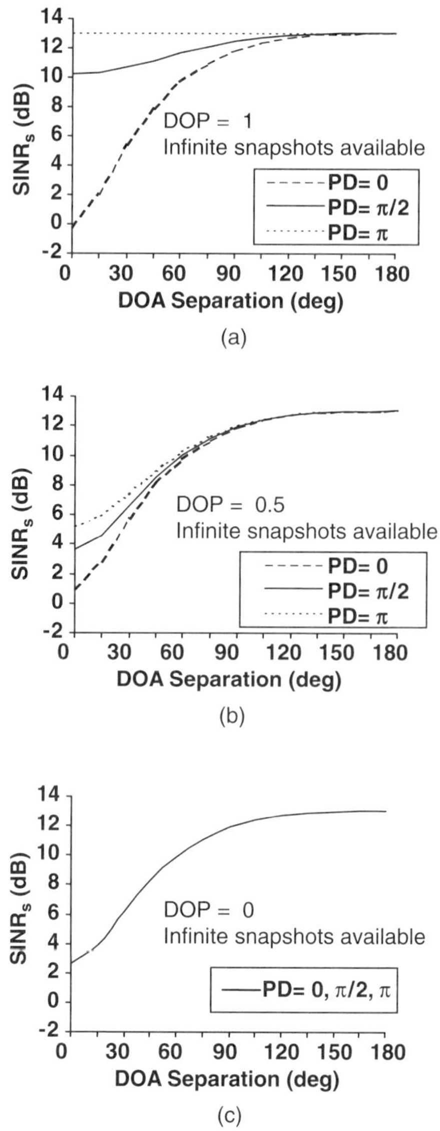

FIGURE 17.1 (a) Graphs of $SINR_s$ vs. DOA separation for one SM signal and one interference uncorrelated with $SNR = INR = 10$ dB. The three curves correspond to PD = 0, PD = $\pi/2$ and PD = π. The DOP of the interference is 1 and true covariance is used. (b) As in Fig. 17.1a, but the DOP of the interference is 0.5. (c) As in Fig. 17.1a, but the DOP of the interference is 0. ©1999 IEEE. With permission.

The case of one SM signal and one interference impinging on an EM vector sensor was investigated. The signal was circularly polarized with positive spin (i.e., $\beta_s = \pi/4$). The signal, interference, and noise were uncorrelated, and the signal-to-noise ratio (SNR) and interference-to-noise ratio (INR) were both 10 dB. The three graphs of Figure 17.1 show the result for the case where infinitely many snapshots (i.e., exact $\mathbf{R}_s$ and $\mathbf{R}_d$) were used, the $SINR_s$ was computed based on the "raw" expression (without any simplifications) given by Eq. (17.8). Figure 17.1a shows the values of $SINR_s$ as a function of DOA separation when the DOP of the interference was 1 (i.e., $\sigma_{i,u}^2 = 0$ and hence the interference was CP). The values of the PD considered were 0, $\pi/2$, and π, which correspond to interferences with polarizations that were circular with positive spin (i.e., $\beta_i = \pi/4$), linear (i.e., $\beta_i = 0$), and circular with negative spin (i.e., $\beta_i = -\pi/4$), respectively. The scenarios in Fig. 17.1b and c were identical to those of Fig. 17.1a except that the DOPs of the interferences were 0.5 (corresponding to IP interference with $\sigma_{i,c}^2 = \sigma_{i,u}^2$) and 0 (i.e., $\sigma_{i,c}^2 = 0$ and hence the interference was UP), respectively. The $SINR_s$ in Fig. 17.1a to c confirm Corollaries 1 to 4 of Theorem 1, that is, that $SINR_s$ increases with an increase in the DOA separation or the PD, $SINR_s$ is independent of the PD when the interference is UP (see Fig. 17.1c), and $SINR_s$ attains the maximum value, $2\sigma_s^2/\sigma^2$, when the DOA separation is π or when the interference is CP (DOP = 1) and the PD is π.

Also, simulations for scenarios identical to those of Fig. 17.1a to c, but with 200 snapshots, were conducted (see Reference 7 for more details). Two observations were obtained. First, the $SINR_s$ obtained using 200 snapshots differs from that obtained using infinitely many snapshots by less than 2 dB. Moreover, the dependencies of $SINR_s$ on the various parameters are similar.

Finally, simulations for scenarios similar to those of Fig. 17.1a to c, but for a DM signal, were also conducted, and the results were similar (see Reference 26 for more details).

17.6 Beampattern of an Electromagnetic Vector Sensor

This section first analyzes the beampattern of an EM vector sensor, and then makes a comparison with two other types of sensor arrays. First, consider an EM vector sensor that has been steered toward (or focused in) the direction/polarization θ_F, and assume that there is no noise and interference (an assumption adopted in some relevant studies[22,27]). Then the normalized response (or beampattern) of the EM vector sensor resulting from an incident signal with direction/polarization θ_k is given by

$$g(\theta_F, \theta_k) = \left|\mathbf{a}^H(\theta_F)\mathbf{a}(\theta_k)\right|^2 / 4 \tag{17.17}$$

(The function $g(\theta_F, \theta_k)$ reaches the maximum when $\theta_k = \theta_F$, and the maximum value attained is 1. Because the magnitude squared of $\mathbf{a}(\theta)$ is 4, a denominator is introduced on the right-hand side of Eq. (17.17) so that the magnitude of $g(\theta_F, \theta_k)$ is normalized to 1 when $\theta_k = \theta_F$). Note that, unlike scalar-sensor arrays with beampatterns that are only functions of DOA, the beampattern of an EM vector sensor is dependent on both the DOA and polarization. To facilitate the analysis of the beampattern, let us consider the coordinate system where $\phi_F = \psi_F = \phi_k = 0$. Let the separation between the DOAs (ϕ_F, ψ_F) and (ϕ_k, ψ_k) be γ_k^F. Then Eq. (17.17) can be expressed as

$$g(\theta_F, \theta_k) = \frac{\left(1 + \cos\gamma_k^F\right)^2}{4} \cos^2 \frac{\Delta_k^F}{2} \tag{17.18}$$

where Δ_k^F is the difference between the polarization toward which the EM vector sensor is steered and the polarization of the incident signal. Although Eq. (17.18) is derived using the coordinate system where $\phi_F = \psi_F = \phi_k = 0$, it holds for any (ϕ_F, ψ_F) and (ϕ_k, ψ_k). This is because Eq. (17.18) is a function of only two parameters γ_k^F and Δ_k^F, which are both independent of the actual coordinate system.

From the expression of $g(\theta_F, \theta_k)$ given by Eq. (17.18), several properties of the beampattern of an EM vector sensor can be deduced. First, the response of an EM vector sensor in the direction/polarization θ_k decreases with an increase in γ_k^F or Δ_k^F. Second, when $\gamma_k^F = \pi$ (i.e., at the direction opposite to the beam steer direction), or when $\Delta_k^F = \pi$ (i.e., if the difference in polarizations is the largest possible), an EM vector sensor does not have any response. Finally, because $g(\theta_F, \theta_k)$ attains its maximum if and only if $\theta_k = \theta_F$, the beampattern of an EM vector sensor does not contain grating lobe* (i.e., sidelobe that is as high as the mainlobe). In contrast, the beampatterns for scalar sensors with uniform linear or uniform circular array geometry contain grating lobes (this property is demonstrated in the latter part of this section), and such scalar sensors are not able to suppress interferences arriving in the directions of the grating lobes.

*Such a property is also seen in acoustic vector sensor, which measures the acoustic pressure and all three components of the acoustic particle velocity induced by acoustic signals.[28]

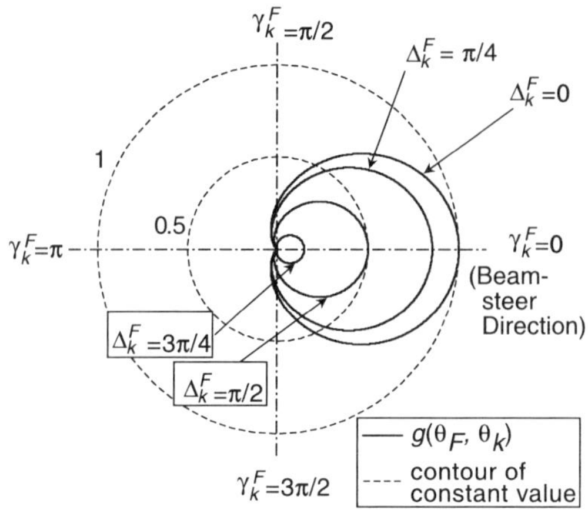

FIGURE 17.2 Polar plot of a cross section of the beampattern of an EM vector sensor. ©1999 IEEE. With permission.

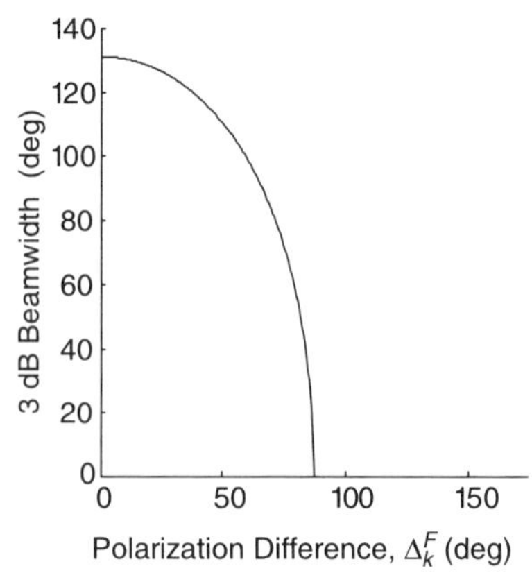

FIGURE 17.3 The 3-dB beamwidth of an EM vector sensor against the polarization difference, Δ_k^F. ©1999 IEEE. With permission.

Figure 17.2 shows the polar plot of any cross section of the beampattern that contains the beam steer direction, for $\Delta_k^F = 0, \pi/4, \pi/2, 3\pi/4$. Note that regardless of the beam steer direction/polarization, the shape of the beampattern is identical to that shown in Fig. 17.2.

Now the 3-dB (or half-power) beamwidth of the mainlobe is analyzed. Note that the beampattern of an EM vector sensor is dependent on polarization in addition to DOA, and thus is very complex. Here, the analysis of the 3-dB beamwidth shall be restricted to the case of a fixed value of Δ_k^F (as a result, the beampattern depends only on the separation in DOAs). It can be deduced from Eq. (17.18) that, for a fixed Δ_k^F, the 3-dB beamwidth is given by

$$\begin{cases} 2\cos^{-1}\left(\sqrt{2}\Big/\cos\dfrac{\Delta_k^F}{2} - 1\right) & \text{if } \Delta_k^F \in \left[0, \pi/2\right] \\ 0 & \text{if } \Delta_k^F \in \left[\pi/2, \pi\right] \end{cases}$$

Figure 17.3 plots the 3-dB beamwidth as a function of Δ_k^F. This beamwidth decreases gradually from $13\pi/18$ to 0 if $\Delta_k^F \in [0, \pi/2]$. Beyond this interval (i.e., $\Delta_k^F \in [\pi/2, \pi]$), it is identically zero. This indicates the excellent ability of the EM vector sensor to distinguish signals and interferences that have sufficiently large differences in polarizations.

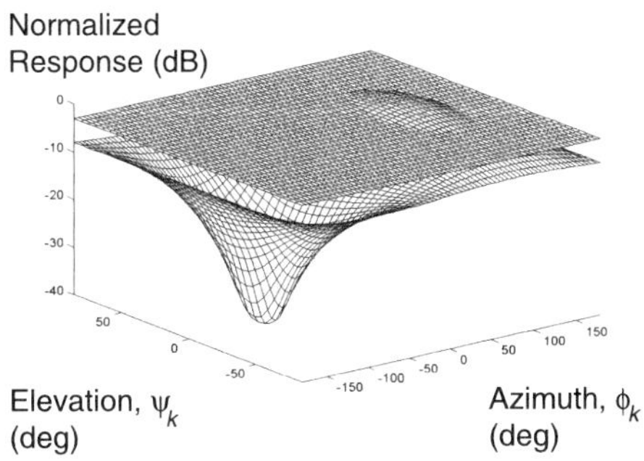

FIGURE 17.4 Beampattern of an EM vector sensor with beam steer direction $(\phi_F, \psi_F) = (\pi/2, 0)$. The value of Δ_k^F is $7\,\pi/12$. The –3 dB plane is also shown. ©1999 IEEE. With permission.

The performance of an EM vector sensor is further illustrated in Fig. 17.4. Indeed, Fig. 17.4 shows the beampattern when an EM vector sensor is steered toward the direction $(\phi_F, \psi_F) = (\pi/2, 0)$ and an arbitrary polarization, with Δ_k^F being fixed at $7\pi/12$. For ease of visualizing the beampattern variation with respect to a reference fixed at –3 dB, the horizontal plane cutting the z-axis at –3 dB is also shown. Regardless of the beam steer direction, the shape of the beampattern is identical to that shown in Fig. 17.4, except for a shift in position. (In Fig. 17.4, as well as the other figures to be presented subsequently, the value of a beampattern response is truncated to –40 dB if it is smaller than –40 dB).

Note that because the beampattern of an EM vector sensor is dependent on both DOA and polarization, it is not obvious how to define a sidelobe for this sensor. However, for a fixed Δ_k^F, the beampattern is a decreasing function of γ_k^F (with maximum value at $(\phi_k, \psi_k) = (\phi_F, \psi_F)$). Consequently, there is effectively no sidelobe for a fixed Δ_k^F.

Now let us compare the beampattern of an EM vector sensor with two types of sensors/arrays. Consider arrays of six isotropic scalar sensors that measure only one component of the electric or magnetic field induced. Two common sensor configurations are considered: a six-sensor uniform linear array (ULA) lying along the y-axis with inter-sensor spacing equal to $\lambda/2$, where λ is the wavelength of the signal of concern, and a six-sensor uniform circular array (UCA) with sensor coordinates $(\cos \pi\, \tau/3, \sin \pi\, \tau/3, 0)\lambda$, for $\tau = 0, \ldots, 5$. Note that, unlike an EM vector sensor, the shapes of the beampatterns of the ULA and UCA are dependent on the beam steer direction. Thus, to analyze the beampatterns, simulations for many different beam steer directions are conducted. An undesirable property of the beampatterns of the ULA and UCA is that they have grating lobes or sidelobes. Moreover, many grating lobes/sidelobes occur at directions that are very far from the beam steer direction. For example, Figs. 17.5 and 17.6 plot the beampatterns of, respectively, the ULA and UCA, when the arrays are steered to $(\pi/2, 0)$. In each figure,

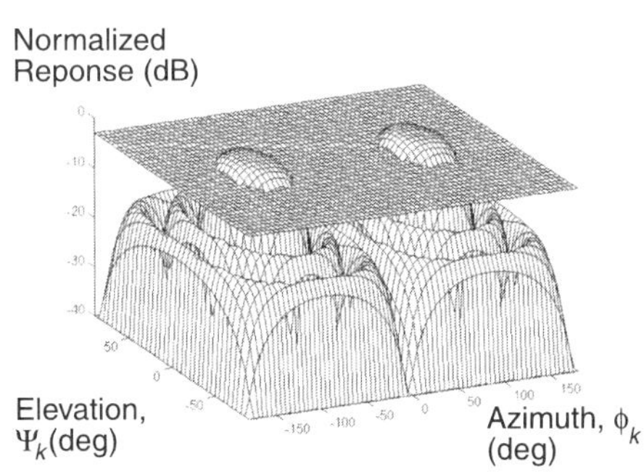

FIGURE 17.5 Beam pattern of a six-scalar sensor ULA lying along the y-axis with intersensor spacing equal to half wavelength. The beam-steer direction (ϕ_F, ψ_F) is $(\pi/2, 0)$. The –3-dB horizontal plane is also shown. ©1999 IEEE. With permission.

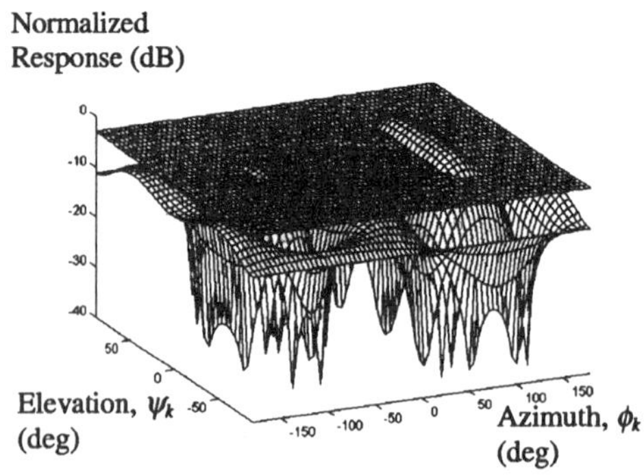

FIGURE 17.6 Same as Fig. 17.5 except that the array is a six-scalar sensor UCA with sensor coordinates ($\cos \pi\, \tau/3$, $\sin \pi\, \tau/3, 0)\lambda$, for $\tau = 0, \ldots, 5$. ©1999 IEEE. With permission.

a –3 dB plane is also plotted. For ULA, it can be seen that the beampattern contains a grating lobe. As for UCA, there is a sidelobe with strength greater than –3 dB.

17.7 Conclusions

The chapter discussed a minimum-noise-variance type of beamformer employing an EM vector sensor for one signal and one interference that are uncorrelated. Both SM and DM signals were considered, and the state of polarization of the interference under consideration ranged from completely polarized to unpolarized. To analyze the beamformer performance, an explicit expression for the SINR of an SM signal was first obtained. It was then deduced that the SINR of an SM signal increases with an increase in the separation between the DOAs and/or the polarizations, for all DOAs and polarizations (scalar-sensor arrays and a single tripole[21] do not have such properties). It was also deduced that a single EM vector sensor can suppress an (uncorrelated) interference that has the same DOA as the signal and distinct polarizations (this is impossible for scalar sensors regardless of the number of sensors and the array aperture). In addition, a strategy for effectively suppressing the interference was identified. Moreover, the SINR expression for the SM signal also provided a basis for generating DM signals of which the two message signals have the minimum interference effect on one another.

An explicit expression for the SINR of such DM signal was also derived. Subsequently, it was deduced that the previously mentioned characteristics for the SINR of an SM signal were also valid for the SINR of a DM signal. Fairly extensive computer simulations were conducted, and the results obtained were in good agreement with the analysis presented in this chapter. Finally, the chapter also analyzed the characteristics of the mainlobe and sidelobe of the beampattern of an EM vector sensor, and demonstrated the advantage of an EM vector sensor over some scalar-sensor arrays. In particular, it has been shown that the beampattern of an EM vector sensor does not contain grating lobes. In contrast, the beampatterns of a six-sensor uniform linear array and a six-sensor uniform circular array have grating lobes. Moreover, many grating lobes occur at directions that are very far from the beam steer direction.

The proposed beamformer can be extended easily to handle multiple sources with diverse polarizations using multiple vector sensors as receivers. Some possible follow-up studies are (1) investigation of the beamforming performance of an EM vector sensor for multiple signals and multiple interferences; (2) performance with multiple EM vector sensors; (3) performance for the signal and interference that are correlated; (4) performance when the powers of the electric noise and magnetic noise at EM vector sensors are not identical; (5) effects of channel depolarization on the signal; and (6) comparison of an EM vector sensor with other types of EM sensors.

Acknowledgment

We are grateful to Bert Hochwald of Lucent Technologies, United States, and Kah-Chye Tan of Addest Technovation Pte. Ltd. Singapore, for their useful suggestions.

References

1. Nehoral, A. and Paldi, E., Vector-sensor processing for electromagnetic source localization, *Proc. 25th Asilomar Conf. on Signals, Syst. and Comput.*, Pacific Grove, CA, 1991, pp. 566.
2. Nehoral, A. and Paldi, E., Vector-sensor array processing for electromagnetic source localization, *IEEE Trans. Signal Proc.*, Vol. 42, pp. 376, 1994.
3. Hatke, G. F., Performance analysis of the SuprCART antenna array, MIT Lincoln Laboratory, Lexington, MA, No. AST-22, 1992.
4. Kanda, M., An electromagnetic near-field sensor for simultaneous electric and magnetic-field measurements, *IEEE Trans. Electromag. Compat.*, Vol. 26, pp. 102, 1984.
5. Kanda, M. and Hill, D., A three-loop method for determining the radiation characteristics of an electrically small source, *IEEE Trans. Electromag. Compat.*, Vol. 34, pp. 1, 1992.
6. Nehorai, A. and Tichavsky, P., Cross-product algorithms for source tracking using an EM vector sensor, *IEEE Trans. Signal Proc.*, Vol. 47, pp. 2863, 1999.
7. Nehorai, A., Ho, K.-C., and Tan, B. T. G., Minimum-noise-variance beamformer with an electromagnetic vector sensor, *IEEE Trans. Signal Proc.*, Vol. 47, pp. 601, 1999.
8. Li, J., Direction and polarization estimation using arrays with small loops and short dipoles, *IEEE Trans. Antennas Propag.*, Vol. 41, pp. 379, 1993.
9. Ho, K.-C., Tan, K.-C., and Tan, B. T. G., Efficient method for estimating directions-of-arrival of partially polarized signals with electromagnetic vector sensors, *IEEE Trans. Signal Proc.*, Vol. 45, pp. 2485, 1997.
10. Wong, K. T. and Zoltowski, M. D., Uni-vector-sensor ESPRIT for multiscore azimuth, elevation, and polarization estimation, *IEEE Trans. Antennas Propag.*, Vol. 45, pp. 1467, 1997.
11. Ho, K.-C., Tan, K.-C., and Tan, B. T. G., Estimating directions of arrival of completely and incompletely polarized signals with electromagnetic vectgor sensors, *IEEE Trans. Signal Proc.*, Vol. 47, pp. 2845, 1999.
12. Zoltowski, M. D. and Wong, K. T., Polarization diversity and extended-aperture spatial diversity to mitigate fading-channel effects with a sparse array of electric dipoles or magnetic loops, *Proc. IEEE Int. Vehic. Technol. Conf.*, 1997, pp. 1163.
13. Chua, P.-H, See, C.-M. S., and Nehorai, A., Vector-sensor array processing for estimating angles and times of arrival of multipath communiction signals, *Proc. IEEE Int. Conf. Acoust., Speech, and Signal Process.*, Seattle, WA, 1998, pp. 3325.
14. Hatke, G. F., Conditions for unambiguous source localization using polarization diverse array, *27th Asilomar Conf. Signals, Syst. Comput.*, Los Alamitos, CA, 1993, pp. 1365.
15. Ho, K.-C., Tan, K.-C., and Ser, W., An investigation on number of signals whose directions-of-arrival are uniquely determinable with an electromagnetic vector sensor, *Signal Process.*, Vol. 47, pp. 41, 1995.
16. Hochwald, B. and Nehorai, A., Identifiability in array processing models with vector-sensor applications, *IEEE Trans. Signal Proc.*, Vol. 44, pp. 83, 1996.
17. Tan, K.-C., Ho, K.-C., and Nehorai, A., Uniqueness study of measurements obtainable with arrays of electromagnetic vector sensors, *IEEE Trans. Signal Proc.*, Vol. 44, pp. 1036, 1996.
18. Tan, K.-C., Ho, K.-C., and Nehorai, A., Linear independence of steering vectors of an electromagnetic vector sensor, *IEEE Trans. Signal Proc.*, Vol. 44, pp. 3099, 1996.
19. Hochwald, B. and Nehorai, A., Polarimetric modeling and parameter estimation with applications to remote sensing, *IEEE Trans. Signal Proc.*, Vol. 43, pp. 1923, 1995.
20. Van Veen, B. D. and Buckley, K. M., Beamforming: A versatile approaach to spatial filtering, *IEEE Acoust. Speech Signal Proc. Mag.*, Vol. 5, pp. 4, 1988.
21. Compton, R. T., Jr., The tripole antenna: An adaptive array with full polarization flexibility, *IEEE Trans. Antennas Propag.*, Vol. AP-29, pp. 944, 1981.
22. Monzingo, R. A. and Miller, T. W., *Introduction to adaptive arrays*, John Wiley & Sons, New York, 1980.

23. Deschamps, G. A., Geometrical representation of the polarization of a plane electromagnetic wave, *Proc. IEEE*, Vol. 55, pp. 2143, 1967.
24. Cox, H., Resolving power and sensitivity to mismatch of optimum array processors, *J. Acoust. Soc. Am.*, Vol. 54, pp. 771, 1973.
25. Godara, L. C. and Cantoni, A., Uniqueness and linear independence of steering vectors in array space, *J. Acoust. Soc. Am.*, Vol. 70(2), pp. 467, 1981.
26. Nehorai, A., Ho, K.-C., and Tan, B. T. G., Minimum-noise-variance beamformer with an electromagnetic vector sensor, *Report No. UIC-EECS-97-3*, Department of Electrical Engineering and Computer Science, University of Illinois at Chicago, 1997.
27. Cray, B. A. and Nuttall, A. H., A comparison of vector-sensing and scalar-sensing linear arrays, *NUWC-NPT Technical Report 10632*, Naval Undersea Warfare Center Division, Newport, RI, 1997.
28. Hawkes, M. A. and Nehorai, A., Acoustic vector-sensor beamforming and capon direction estimtion, *IEEE Trans. Signal Process.*, Vol. 46, pp. 2291, 1998.
29. Schwartz, M., Bennett, W. R., and Stein, S., *Communication systems and techniques*, McGraw-Hill, New York, 1966.
30. Keiser, B. E., *Broadband coding, modulation, and transmission enginering*, Prentice-Hall, Englewood Cliffs, NJ, 1989.
31. Monnier, R., Rault, J. B. and Couasnon, T. de, Digital television with spectral efficiency, *1992 Int. Broadcasting Conf.*, 1992, pp. 380–384.
32. http://www.helioss.com
33. Lindmark, B., A dual polarized dual band microstrip antenna for wireless communications, *IEEE Trans. Aerospace Conf.*, Vol. 3, 1998, pp. 333–338.
34. Blaikie, R. J., Taylor, D. P., and Gough, P. T., Multilevel differential polarization shift keying, *IEEE Trans. Commun.*, Vol. 45, pp. 95–102, 1997.

18

Optimum and Suboptimum Transmit Beamforming

Mats Bengtsson
Royal Institute of Technology (KTH)

Björn Ottersten
Royal Institute of Technology (KTH)

18.1 Introduction

Beamforming for transmission in a network is, in several aspects, more difficult than beamforming for reception. At first sight, the problems appear to be equivalent; because the radio channel is reciprocal, a good uplink beamformer should also work well in the downlink. However, whereas a beamformer at the receiver side only affects the signal quality for one specific user, a signal transmitted from an antenna array is received not only by the desired user but also by all other users in the surroundings. Thus, the design of downlink beamformers is less a matter of the individual links and more a matter of the system gain, because all the different beamformers in a system jointly affect the system performance.

Another fundamental difference is the channel knowledge. A receiver equipped with an antenna array can estimate the channel, or train a beamformer adaptively, using known pilot symbols in the transmitted signal. The so-called blind methods do not require any training symbols but exploit knowledge about the transmitted signal modulation or the geometry of the antenna array. With a transmitting array antenna, on the other hand, direct estimation of the channel is only possible at the mobile, which would require an additional feedback link to the beamforming base station. An alternative, which we describe in some depth, is to infer knowledge about the downlink channel from previous uplink measurements.

0-8493-0124-6/02/$0.00+$1.50

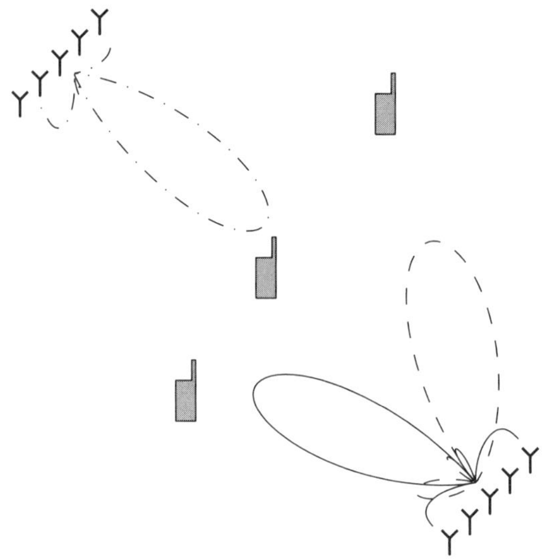

FIGURE 18.1 Downlink beamforming.

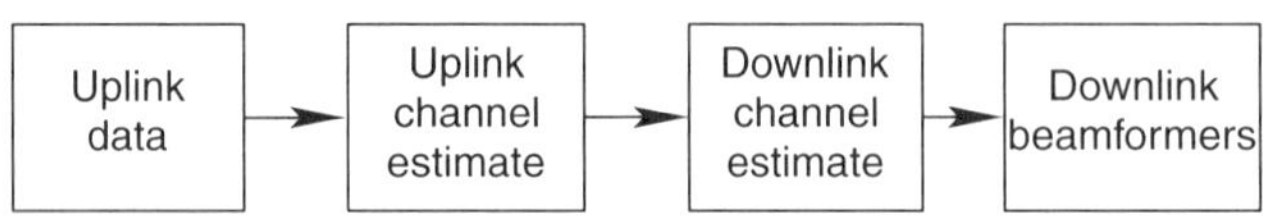

FIGURE 18.2 General strategy.

In the narrowband cellular systems of today, each operator owns a number of carrier frequencies that are allocated to the different cells according to some frequency plan. Different frequencies are used in neighboring cells to reduce the interference level from co-channel users in other cells. When antenna arrays are introduced in the system to increase the capacity, the idea is to use the directivity of the antenna to reduce the co-channel interference. This makes it possible to use a tighter frequency plan or even to use the same channel for several mobiles within the same cell. The latter alternative is sometimes called spatial division multiple access (SDMA). General presentations on the use of antenna arrays in cellular systems can be found in other chapters of this book and in several books and review articles, for example, in References 1 to 4.

18.1.1 Scope of the Chapter

We envision a system with a number of base stations, equipped with antenna arrays, and a number of single antenna mobiles. We concentrate mainly on narrowband systems, but give some indications on how the principles could be extended to wideband systems.

The main problem treated here is how to design beamformers for transmission from the antenna arrays to the mobiles. This is done taking all the co-channel users of the system into account, at least for the optimum design schemes. An example of the optimum beam patterns for a simple scenario is shown in Figure 18.1. The design algorithms use, as input, some characterization of the propagation channels from the base stations to the mobiles. The basic steps of the beamforming design are outlined in Fig. 18.2.

In addition, we touch on a number of other aspects, including

- How to get a characterization of the downlink channel from measurements on the uplink.
- The basic principles of power control. For some of the strategies described later, both the spatial characteristics and the actual transmission power of each beamformer is determined jointly; otherwise the power levels have to be determined in a separate step.

- To minimize the signaling between the base stations, it is preferable that each base station can determine the beamformers for its own users separately, based on information available locally at the base station. Unfortunately, the optimum algorithms described later require a centralized implementation with global system knowledge.
- How to handle additional constraints enforced by the physical implementation of the system.
- Robustness to miscalibration, estimation inaccuracies, or other errors.

Note that this chapter does not cover so-called space-time coding,[5-8] which has appeared as an interesting technique for efficient communication using multiple antennas. The idea of space-time coding is that the transmitter does not have any knowledge about the actual propagation conditions but introduces redundancy in the transmitted signal, both over space and time, that allows the receiver to recover the signal even in difficult propagation situations. Most work so far on transmitting diversity has concentrated on a single-user link. In contrast, we assume a multiuser system where all the channels between base stations and user terminals are known, completely or partially, and find a transmitting scheme that exploits this knowledge in an optimum way. It is actually shown in Section 18.4.2 that as long as the quality of service requirements are expressed in terms of signal to interference levels, the optimum strategy is to use a time-invariant beamformer for each signal, and a space-time code does not give better carrier to interference ratio. However, space-time codes may be better in terms of bit-error rate (BER) and mutual information. Another difference to space-time coding is that only mobiles with a single antenna are considered here.

18.2 Channel Characterizations

The basic channel model for receiving antenna arrays has been presented in previous chapters of this handbook. Here, we describe a corresponding model for the downlink and discuss how to obtain an estimate of the downlink channel from uplink measurements in different types of systems.

18.2.1 General Model

To begin with, a simple baseband model for frequency flat communication channels is presented. The underlying assumption is that all signals are narrowband and that the time dispersion of the radio link is small. Later, in Section 18.3.5, it is shown how the downlink algorithms can be extended to wideband signals and frequency-selective channels.

We consider a system with K different base stations and d mobiles, all sharing the same physical channel. The baseband signal $r_i(n)$ received at mobile i is given by

$$r_i(t) = \sum_{k=1}^{K} \mathbf{h}_{i,k}^{*} \mathbf{x}_k(t) + n_i(t) \tag{18.1}$$

Here, $\mathbf{x}_k(t)$ is the vector of signals transmitted at the antenna elements of base station k, $\mathbf{h}_{i,k}$ is the vector channel from base station k to mobile i, and $n_i(t)$ is additive noise. We assume that the noise is white with power $\sigma_i^2 = E[|n_i(t)|^2]$. Note that $\mathbf{h}_{i,k}$ includes not only the spatial characteristics of the channel but also the effects of shadow fading, fast fading, and path loss.

Because of the time and/or frequency distance between the uplink and the downlink, the fading of the uplink and downlink spatial response vectors may be uncorrelated; thus, it is in generally impossible to obtain good estimates of the instantaneous downlink channel vectors $\mathbf{h}_{i,k}$. However, as shown next, it is reasonable to have access to the channel correlation matrix.

$$\mathbf{R}_{i,k} = E\left[\mathbf{h}_{i,k}\mathbf{h}_{i,k}^{*}\right]$$

averaged over the fast fading.

We introduce the notation $\kappa(i)$ for the base station allocated to mobile i and, conversely, the notation $I(k) = \{i; \kappa(i) = k\}$ for the set of mobiles allocated to base station k. Depending on the frequency planning, zero, one, or several mobiles may be allocated to a specific base station, but each mobile is allocated to exactly one base. We no not consider soft handover situations.

The vector of signals transmitted at base station k is given by

$$\mathbf{x}_k(t) = \sum_{i \in I(k)} \mathbf{w}_i s_i(t) \tag{18.2}$$

where $\mathbf{w}_i$ is the beamforming vector and $s_i(t)$ is the modulated signal intended for mobile i. We assume that the different data signals $s_i(t)$ are uncorrelated and have normalized power $E[|s_i(t)|^2] = 1$.

To summarize, the notation is

$\mathbf{h}_{i,k}$	Channel vector from base station k to mobile i, including the channel gain
$\mathbf{R}_{i,k}$	Channel correlation matrix from base station k to mobile i, including the channel gain
$r_i(t)$	Signal received at mobile i
σ_i^2	Noise level at mobile i
$\kappa(i)$	Base station allocated to mobile i
$I(k)$	Set of mobiles allocated to base station k
$\mathbf{w}_i$	Beamforming vector for transmission from base station $\kappa(i)$ to mobile i
$s_i(t)$	Signal intended for mobile i
d	Total number of co-channel mobiles
$\mathbf{x}_k(t)$	Signal vector transmitted at base station k
K	Total number of base stations

18.2.2 Time Division Duplex Systems

In a system with time division duplex (TDD), the uplink and downlink transmission shares the same frequency using different time slots. As long as the delay between the uplink and downlink time slots is short compared with the coherence time of the channel, the reciprocity principle of electromagnetics shows that the spatial signature of the downlink is identical to that of the uplink. Thus, the instantaneous uplink channel can be estimated, using training data, and applied as a good estimate of the downlink channel $\mathbf{h}_{i,k}$ (Fig. 18.3). For beamforming algorithms using the channel correlation matrix, simply set $\mathbf{R}_{i,k} = \mathbf{h}_{i,k}\mathbf{h}_{i,k}^*$, which gives a correlation matrix of rank one. Instead of training data, it is possible to estimate the uplink channel using so-called blind methods, for example, the constant modulus algorithm[9,10] that exploits the known modulation of the transmitted signal, or using some direction-based method[11] that exploits the known array geometry.

If however, the duplex time separation is larger than the coherence time, the instantaneous downlink channel is more or less uncorrelated with the estimated uplink channel. Still, we could estimate a channel correlation matrix from uplink data collected over a sufficiently long time period to average over the fast fading but short enough that the shadow fading and main directions of the incoming waves remains constant. This correlation matrix can be used as a good estimate of $\mathbf{R}_{i,k}$ also for the downlink channel (Fig. 18.4). This means that the beamforming algorithms are not able to find the optimum solution for the instantaneous channel realizations, but they can still give the best performance averaged over the fast

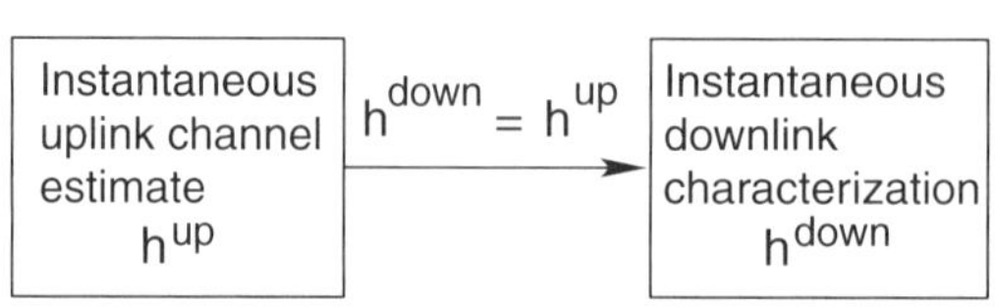

FIGURE 18.3 TDD strategy (short duplex distance).

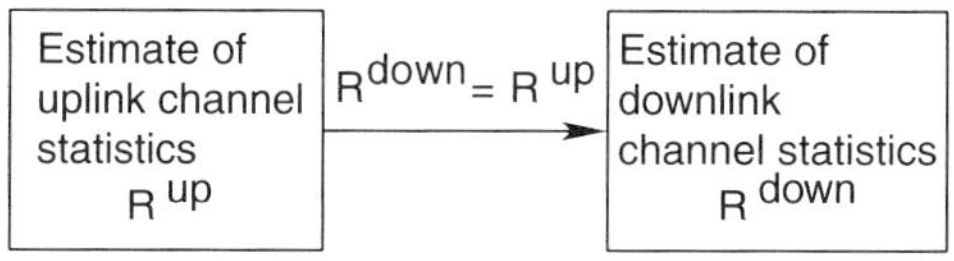

FIGURE 18.4 TDD strategy (long duplex distance).

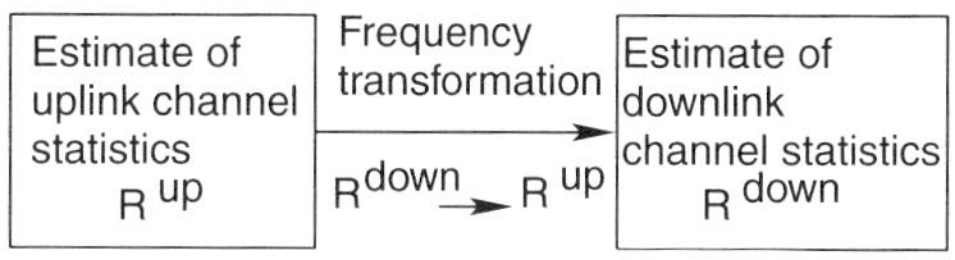

FIGURE 18.5 FDD strategy.

fading. Some empirical studies of the application of antenna arrays in TDD systems have been reported, among others, in References 12 to 14.

18.2.3 Frequency Division Duplex Systems

In frequency division duplex FDD systems, we encounter the same problem as that for TDD systems. If the duplex distance (the difference between the uplink and downlink carrier frequencies) is larger than the coherence bandwidth, the uplink channels fade independently. The solution is the same, namely, to collect a correlation matrix from the uplink data to obtain an estimate of the second-order statistics of the downlink channel. In Reference 15 it is shown that for carrier frequencies around 1800 MHz, the uplink and downlink channels are practically uncorrelated when the frequency separation is above 10 MHz. See also the measurement results in Reference 12. However, the frequency offset between uplink and downlink introduces an additional problem, because the spatial response of the antenna array changes with frequency. Therefore, some transformation is necessary to convert the uplink channel measurements into reasonable estimates of the downlink channel (Fig. 18.5).

The easiest solution is to ignore the frequency offset and use the uplink measurements as is. Several authors have analyzed the corresponding performance loss. In Reference 16 with a relative FDD frequency offset $(f^{\text{up}} - f^{\text{down}})/f^{\text{up}} = 7.6\%$ [the worst case in Global System for Mobil Communications (GSM)], the resulting loss in the signal to interference ratio is shown to be at most 1.7 dB. See also related results in References 17 and 18 and the numerical example that follows. Note that the conclusions may differ widely depending on the specific scenario and the beamforming strategy.

The remainder of this section describes briefly a number of more sophisticated methods, from the literature, for the conversion of an estimated uplink channel vector or correlation matrix to the corresponding downlink channel. The presentation concludes with a numerical comparison.

18.2.3.1 Matched arrays

A hardware-based alternative to compensate for the frequency offset, suggested in Reference 18, is to use separate arrays for the uplink and the downlink that are matched to the respective frequencies, to provide the same spatial response in uplink and downlink. Reference 19 shows how such a pair of m element arrays can be implemented with only $m + 1$ physical antenna elements. If the element separations between successive elements form a geometric series with quotient $f^{\text{up}}/f^{\text{down}}$, element 1 through m can be used for the uplink and 2 through $m + 1$ for the downlink. Even though the array calibration is not required in the actual signal processing, it is still necessary to have equivalent radio chains for the uplink and downlink.

18.2.3.2 Mobile Feedback

A more drastic solution is to let the mobile estimate the downlink channel and transmit the information back to the base station (see References 20 and 21). This requires that different training sequences are transmitted at the different antenna elements of the base station array and cannot be implemented in existing systems because it requires additional processing in the mobiles and a protocol for the feedback informtion. In principle, this is the only method that makes it possible to track the fading fluctuations of the downlink and that does not require any specific array calibration. However, with fast fading, the feedback rate may exceed the total uplink capacity and the only practical alternative is to estimate the channel correlation matrix in the mobile to keep an acceptable data rate of the feedback information.

18.2.3.3 Direction-Based Transformations

Most transformation principles require a calibrated array and are based on the idea that the uplink and downlink propagation conditions are the same. Let $\mathbf{a}(\theta; f)$ denote the array response vector of a plane wave with direction θ and carrier frequency f. Because the channel vector is the superposition of a number of reflected paths, the uplink channel vector from a mobile to a base station can be written

$$\mathbf{h}^{\mathrm{up}} = \sum_{l=1}^{L} \alpha_l^{\mathrm{up}}\, \mathbf{a}\left(\theta_l^{\mathrm{up}}; f^{\mathrm{up}}\right)$$

with correlation matrix (assuming for simplicity that the separate reflected rays have uncorrelated phase)

$$\mathbf{R}^{\mathrm{up}} = \sum_{l=1}^{L} E\left[\left|\alpha_l^{\mathrm{up}}\right|^2\right] \mathbf{a}\left(\theta_l^{\mathrm{up}}; f^{\mathrm{up}}\right) \mathbf{a}^*\left(\theta_l^{\mathrm{up}}; f^{\mathrm{up}}\right)$$

As was stated at the beginning of this section, we assume that all time-delay differences between the different paths are small and are included as phase shifts in the complex values amplitude coefficients α_l. The corresponding downlink channel is

$$\mathbf{h}^{\mathrm{down}} = \sum_{l=1}^{L} \alpha_l^{\mathrm{down}}\, \mathbf{a}\left(\theta_l^{\mathrm{down}}; f^{\mathrm{down}}\right)$$

with correlation matrix

$$\mathbf{R}^{\mathrm{down}} = \sum_{l=1}^{L} E\left[\left|\alpha_l^{\mathrm{down}}\right|^2\right] \mathbf{a}\left(\theta_l^{\mathrm{down}}; f^{\mathrm{down}}\right) \mathbf{a}^*\left(\theta_l^{\mathrm{down}}; f^{\mathrm{down}}\right) \tag{18.3}$$

Because the propagation paths are the same, we assume that $\theta_l^{\mathrm{down}} = \theta_l^{\mathrm{up}}$ and that $E[|\alpha_l^{\mathrm{down}}|^2] = E[|\alpha_l^{\mathrm{up}}|^2]$. This is the basis for all the strategies described later.

One obvious idea is to estimate the physical parameters θ_l and $E[|\alpha_l|^2]$ in the uplink and form an estimate of the downlink correlation matrix by inserting these estimates into Eq. (18.3). In most situations with multipath propagation, it is typically not possible to resolve all the paths. If the incoming rays arrive from a few clusters of directions, an alternative is to identify the main angle and angular spread of each cluster; see, for example, References 22 and 23. Note also that the different incoming rays are correlated, which means that several direction finding methods, such as multiple signal classification (MUSIC),[24] do not work. Direction estimation of correlated sources can be handled by WSF/MODE[25,26] or using spatial smoothing.[27] If each user has its training sequence embedded in the signal, this knowledge should be exploited in the direction estimation.[28,29] A general overview of different direction estimation methods can be found in Reference 11.

18.2.3.4 Spatial Spectra

Instead of estimating discrete rays or clusters of rays one could use a smooth estimate of the spatial power spectrum of each user in the uplink. One suggestion[30] is to use the Capon method (also called minimum variance distortionless response or MVDR)[31,32] to estimate the spatial spectrum based on the uplink correlation matrix

$$\hat{P}(\theta) = \frac{1}{\mathbf{a}^*\left(\theta; f^{\text{up}}\right)\left(\mathbf{R}^{\text{up}}\right)^{-1}\mathbf{a}^*\left(\theta, f^{\text{up}}\right)}$$

and estimate the corresponding downlink correlation matrix from

$$\hat{\mathbf{R}}^{\text{down}} = \int_0^{2\pi} \hat{P}(\theta)\mathbf{a}\left(\theta; f^{\text{down}}\right)\mathbf{a}^*\left(\theta; f^{\text{down}}\right)d\theta$$

A closely related method for robust uplink beamforming is presented in Reference 33, where a non-linear function of the Capon spectrum is used to weight together point source beamformers, based on a Bayesian statistical framework. In the case of diffuse multipath scattering,[34,35] the spatial power spectrum can be estimated using a truncated Fourier expansion. In a cellular system, each base station typically serves mobiles located in a sector $[-\pi/Q, \pi/Q]$. The Fourier series representation of $P(\theta)$ in this sector is

$$P(\theta) = \sum_{l=-\infty}^{\infty} c_l e^{jlQ\theta}$$

Truncation of this series gives a smooth version of the spatial spectrum

$$P(\theta) = \sum_{l=-M+1}^{M-1} c_l e^{jlQ\theta}$$

parameterized by $c_{-M+1}, \ldots, c_{M-1}$. The idea is to estimate the Fourier parameters directly from the estimated uplink correlation matrix using a least squares fit

$$\left\{\hat{c}_l\right\} = \arg\min_{\{c_l\}} \left\| \hat{\mathbf{R}}^{\text{up}} - \sum_{l=-M+1}^{M-1} c_l \Sigma_l^{\text{up}} \right\|_F^2$$

where $\|\cdot\|_F$ denotes the Frobenius norm and

$$\Sigma_l^{\text{up}} = \int_{-\pi/Q}^{\pi/Q} \mathbf{a}\left(\theta; f^{\text{up}}\right)\mathbf{a}^*\left(\theta; f^{\text{up}}\right)e^{jlQ\theta}\,d\theta$$

The downlink correlation matrix is estimated using

$$\hat{\mathbf{R}}^{\text{down}} = \sum_{l=-M+1}^{M-1} \hat{c}_l \Sigma_l^{\text{down}}$$

More details on the calculation of the Fourier parameters can be in Reference 35.

18.2.3.5 Linear Transformations

Even though the exact transformation of $\mathbf{a}(\theta; f^{up})$ to $\mathbf{a}(\theta; f^{down})$ is a nonlinear function, dependent of θ, using a linear transformation as an approximation has been suggested.[15,36] The transformation, given by a square matrix **T** should give $\mathbf{a}(\theta; f^{down}) \approx \mathbf{Ta}(\theta; f^{up})$ for all directions θ in the sector of interest, $\theta_{min} \le \theta \le \theta_{max}$. A least squares fit gives the following expression for **T**

$$\mathbf{T} = \int_{\theta_{min}}^{\theta_{max}} \mathbf{a}\left(\theta; f^{down}\right)\mathbf{a}^*\left(\theta; f^{up}\right)d\theta \left[\int_{\theta_{min}}^{\theta_{max}} \mathbf{a}\left(\theta; f^{up}\right)\mathbf{a}^*\left(\theta; f^{up}\right)d\theta\right]^{-1}$$

By using this transformation matrix, the downlink channel vectors and correlation matrices are estimated by

$$\hat{\mathbf{h}}^{down} = \mathbf{T}\hat{\mathbf{h}}^{up}$$

$$\hat{\mathbf{R}}^{down} = \mathbf{T}\hat{\mathbf{R}}^{up}\mathbf{T}^*$$

An explicit approximate formula for **T** is given for the case of uniform circular antenna arrays in Reference 28.

The same type of linear transformations is also used in direction estimation to transform measurements from an antenna array with irregular geometry into data from a virtual uniform linear array, to use root-MUSIC or any other, computationally efficient algorithm.[37]

18.2.3.6 Numerical Comparison

In Fig. 18.6 and 18.7, three of these methods are compared. Because the most critical task of downlink beamforming is to limit the interference, we have only compared the ability to suppress the signal toward a mobile, not the ability to direct the signal toward the mobile. This also makes the comparison more or less independent of the specific beamforming strategy. The idea is to use the estimate of the downlink channel correlation matrix, obtained from a transformation of the true uplink correlation matrix, and find the beamformer that transmits as little power as possible toward the single mobile, using a fixed total transmission power. To obtain this result, the beamformer is selected as the eigenvector of $\hat{\mathbf{R}}^{down}$ with the smallest eigenvalue. The figures show the power level received at the mobile when the true downlink channel is used. However, because it is difficult to distinguish between eigenvalues smaller than the background noise level, the figure shows the worst case (i.e., the largest interference level) among all

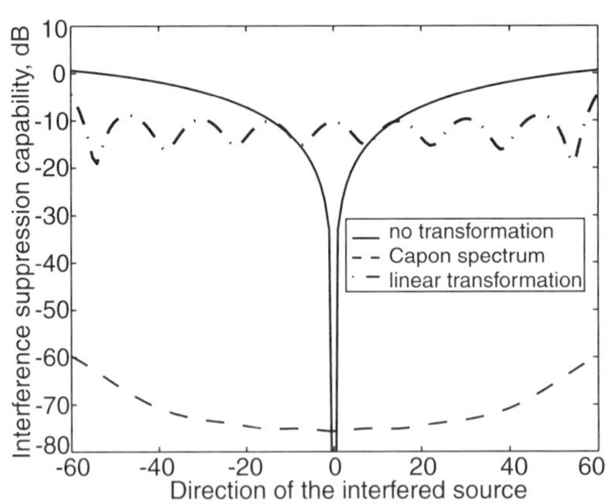

FIGURE 18.6 Interference suppressing capability for a single point source.

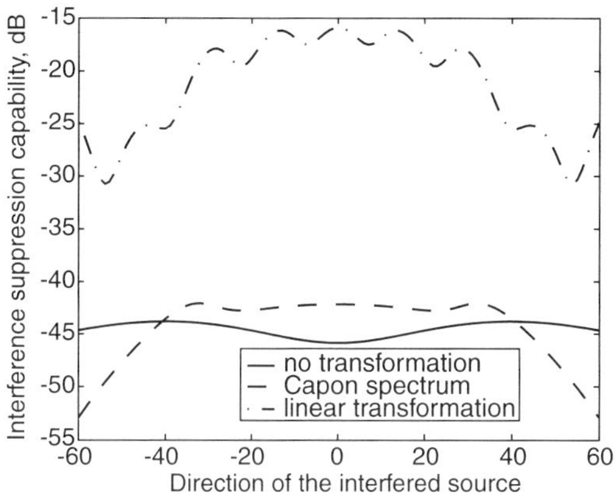

FIGURE 18.7 Interference suppressing capability for a single scattered source.

beamformers that suppress the interference to the same level as the background noise, according to $\hat{\mathbf{R}}^{\text{down}}$. In both figures, the true correlation matrix of a single source plus spatially white noise was used as the uplink estimate. The uniform linear array had eight elements separated by 0.45 wavelengths at the uplink frequency and the relative frequency separation between uplink and downlink was 7%. The noise level corresponded to a signal to noise ratio (SNR) of –40 dB (in practice the estimate is obtained using the known training sequence). The linear transformation matrix $\mathbf{T}$ was designed for the sector [–60°, 60°]. Figure 18.6 shows the performance for a point source located at different angles between –60° and 60° and Figure 18.7 shows the corresponding results for a source surrounded by local scatterers with a spread angle of 3°. Obviously, the scattered source is less sensitive to pointing errors, at least in a worst-case scenario. Note, however, that the best case behavior is more or less identical for point sources and scattered sources. During the experiments, it was observed that the transformation using the Capon spatial spectrum is sensitive to the noise level. The parametric direction-based method was not included in this evaluation because it provides perfect performance as long as the true correlation matrices are used and the parameterized model includes the true channel. A fair comparison with the parametric method would involve more realistic estimation methods of the uplink channel.

18.2.4 Channel Uncertainties

As discussed in the previous section, the available estimate of the downlink channel is at best a reasonable approximation of reality. It is well known from the literature on uplink array processing that certain beamforming strategies are very sensitive to channel estimation errors that could lead to severe signal cancellation.[38] A structured approach to handle these problems is to include the uncertainty in the data model and the algorithm design. One possible uncertainty model for a channel vector is

$$\mathbf{h} = \hat{\mathbf{h}} - \tilde{\mathbf{h}}$$

where $\mathbf{h}$ denotes the true channel and $\hat{\mathbf{h}}$ is the estimate. The estimation error $\tilde{\mathbf{h}}$ can be modeled as a deterministic vector bounded by, for example $\|\hat{\mathbf{h}}\| \le \varepsilon_h$. Alternatively, $\tilde{\mathbf{h}}$ can be seen as a stochastic vector with a given covariance matrix $E[\tilde{\mathbf{h}}\tilde{\mathbf{h}}^*]$.

The same type of uncertainty model can also be used for an estimated channel correlation matrix. Later in this chapter, we consider lower and upper bounds, $\underset{\smile}{\mathbf{R}}$ and $\widehat{\mathbf{R}}$, respectively, on the true correlation matrices

$$\underset{\smile}{\mathbf{R}}_k \preceq \mathbf{R}_k \preceq \widehat{\mathbf{R}}_k \tag{18.4}$$

Here, the notation $\mathbf{A} \preceq \mathbf{B}$ means that the matrix $\mathbf{B} - \mathbf{A}$ is positive semidefinite. One simple example is to use

$$\begin{aligned} \underset{\smile}{\mathbf{R}} &= \mathbf{R} - \varepsilon_R \mathbf{I} \\ \widehat{\mathbf{R}} &= \hat{\mathbf{R}} + \varepsilon_R \mathbf{I} \end{aligned} \tag{18.5}$$

corresponding to a bound ε_R on the eigenvectors of the estimation error.

For errors related to moving sources or incorrectly modeled multipath scattering, it is possible to artificially smooth out each source spatially, using so-called covariance matrix tapers. The idea is simply to multiply the estimated covariance matrix element by element (also called Schur Hadamard product, noted by $\odot$) with a fixed matrix

$$\mathbf{R}_{\text{robust}} = \hat{\mathbf{R}} \odot \mathbf{B} \tag{18.6}$$

For uniform linear arrays, one common choice for $\mathbf{B}$ is

$$[\mathbf{B}]_{k,l} = \text{sinc}(|k-l|\gamma)$$

where $\gamma \approx \Delta_\theta/180°$ corresponds to the angular extent Δ_θ (in degrees) of an artificially widened point source. This and other choices of the matrix tapers are discussed in References 39 to 41.

18.3 Beamforming Strategies

In this section, we describe a number of different beamforming strategies and algorithms. For optimum beamforming, we only provide an overview here and treat the algorithm details separately in Section 18.4. The other strategies are described in detail here, because they are less complex.

18.3.1 Optimum Beamforming

Optimum beamforming may be defined in many ways. Our main choice is based on the operators perspective; the system should provide an acceptable quality of service (QOS) as cheaply as possible, serving as many users as possible.

We will express the QOS requirement in the form of a lower threshold γ_i on the received signal to interference plus noise ratio (SINR) at each mobile. Under this constraint, we wish to minimize the total transmitted power at all the base stations. This results in an optimization problem, informally formulated by

$$\begin{aligned} &\min \text{ total transmitted power} \\ &\text{s.t.} \quad \frac{\text{received signal power at mobile } i}{\text{received interference} + \text{noise power at mobile } i} \geq \gamma_i \end{aligned} \tag{18.7}$$

To express this in mathematical terms, note from Eqs. (18.1) and (18.2) that the signal received at mobile i is

$$r_i(t) = \underbrace{\mathbf{h}_{i,\kappa(i)}^* \mathbf{w}_i s_i(t)}_{\text{desired signal}} + \underbrace{\sum_{n \neq i} \mathbf{h}_{i,\kappa(n)}^* \mathbf{w}_n s_n(t)}_{\text{interference}} + \underbrace{n_i(t)}_{\text{noise}}$$

It follows directly that the received SINR is

$$\mathrm{SINR}_i = \frac{\mathbf{w}_i^* \mathbf{R}_{i,\kappa(i)} \mathbf{w}_i}{\sum_{n \neq i} \mathbf{w}_n^* \mathbf{R}_{i,\kappa(n)} \mathbf{w}_n + \sigma_i^2} \tag{18.8}$$

Because the power used at base station $\kappa(i)$ to transmit the signal $s_i(t)$ is $E[\|\mathbf{w}_i s_i(t)\|^2] = \mathbf{w}_i^* \mathbf{w}_i$, the optimum downlink beamforming problem in Eq. (18.7) can be written

$$\begin{aligned} &\min \sum_{n=1}^{d} \mathbf{w}_n^* \mathbf{w}_n \\ &\text{s.t.} \frac{\mathbf{w}_i^* \mathbf{R}_{i,\kappa(i)} \mathbf{w}_i}{\sum_{n \neq i} \mathbf{w}_n^* \mathbf{R}_{i,\kappa(n)} \mathbf{w}_n + \sigma_i^2} \geq \gamma_i \\ &i = 1, 2, \ldots, d \end{aligned} \tag{18.9}$$

or equivalently

$$\begin{aligned} &\min \sum_{n=1}^{d} \mathbf{w}_n^* \mathbf{w}_n \\ &\text{s.t. } \mathbf{w}_i^* \mathbf{R}_{i,\kappa(i)} \mathbf{w}_i - \gamma_i \sum_{n \neq i} \mathbf{w}_n^* \mathbf{R}_{i,\kappa(n)} \mathbf{w}_n \geq \gamma_i \sigma_i^2 \\ &i = 1, 2, \ldots, d \end{aligned} \tag{18.10}$$

This is a quadratic optimization problem with quadratic nonconvex constraints. In general, such problems could be NP-complete, which very loosely speaking means that they cannot be solved in reasonable time. However, this specific formulation has an inherent structure that makes it possible to find the global optimum efficiently. Two different algorithms for this problem have been presented.[23,42-44] Both algorithms are described in detail in Section 18.4. Note that for the optimum $\mathbf{w}_i$, all contraints hold with equality. Assume, for example, that the first inequality is strict; then $\mathbf{w}_1$ can be scaled down to give an equality in the first constraint. This gives a smaller value of the cost function but does not break any of the other constraints, which contradicts that the solution was optimum. Note also that scaling all noise levels σ_i^2 by a common factor just scales the beamforming vectors correspondingly. In systems with mixed services, different users may require different QOS. This is easily handled using different γ_i for the different user groups.

Because all the beamformers in the system are involved in all the constraints, we have a clear indication that the solution must be calculated centrally for the whole system. Also, the channel between every combination of base station and mobile has to be known. This is one of the main objections to implementing this strategy in a real system. Therefore, the main application may be as a benchmark in the evaluation of more realistic strategies in system studies.

Several other optimally formulations have been suggested in the literature. One idea,[21] is to maximize the worst signal to interference ratio (SIR) among all the mobiles. Another idea[45] for the case of wideband channels and frequency-dependent beamformers is to maximize the worst-case (SINR) among all users. The same idea is used in Reference 46 for the narrowband case but with an additional constraint on the maximum transmitted power. However, the only known approach to find the optimum in these cases is to use general nonlinear optimization techniques or to find an approximate solution using some other suboptimal strategy. Also, trying to suppress the interference levels more than actually necessary for an acceptable QOS typically results in a more distorted beam pattern with increased sidelobe levels. This transmitted excess power increases the overall interference level in the system for all mobiles that were not specifically taken into consideration in the beamforming design and will also makes it more difficult to add more users in the system.

18.3.2 Decentralized Heuristic Solutions

The easiest beamforming strategy is to ignore all other users and to transmit as much energy as possible to the mobile of interest. This so-called conventional beamforming method maximizes the received signal power $\mathbf{w}_i^* \mathbf{R}_{i,\kappa(i)} \mathbf{w}_i$ at mobile i with an upper constraint on the transmitted power $\|\mathbf{w}_i\|^2$. If the channel vector $\mathbf{h}_{i,\kappa(i)}$ is known, the solution is given (up to a scaling) by $\mathbf{w} = \mathbf{h}_{i,\kappa(i)}$. For general channel correlation matrices, the solution is given by the eigenvector of $\mathbf{R}_{i,\kappa(i)}$ with the highest eigenvalue. This is simply spatial filtering using a matched filter. An alternative is to approximate this solution with a number of fixed beams. The advantage is that the fixed beams can be implemented in analog hardware with a Butler matrix, for example. Such fixed-beam systems have been evaluated in large-scale field trials (see, for example, Chapter 24).

Next, we look at algorithms that try to reduce the interference at other mobiles in the network. If all channel vectors are known (not only the correlation matrices) and if the total number of users d is less than the number of antenna elements m, it is possible to find a set of zero-forcing beamformers that cancel all interference in the system. The solution is given by the following linear system of equations for base station k:

$$\mathbf{h}_{i,k}^* \mathbf{w}_n = \begin{cases} 1 & i = n, i = 1, \ldots, d, n \in I(k) \\ 0 & i \neq n, i = 1, \ldots, d, n \in I(k) \end{cases}$$

Notice that the constraints involve the channel vectors of all users but only the beamformers used at the specific base station. This method is studied in Reference 47.

In general, it is impossible to reduce all the interference and the zero-forcing solution may be very sensitive to channel estimation errors. Also, forcing zeros may cause a distorted beam pattern with high sidelobes that raise the background interference level in the system. For this reason we turn our attention to schemes that reduce the interference level in a more balanced way. Maximizing the harmonic mean of the SINR at all the mobiles leads to a problem that can be solved with a decentralized algorithm. Let us first specify a constraint on the desired power level for the received signal of interest at each mobile.

$$\mathbf{w}_i^* \mathbf{R}_{i,\kappa(i)} \mathbf{w}_i = \beta_i$$

Maximizing the harmonic mean is equivalent to minimizing the summed inverse of the $SINR_i$,

$$\sum_i \mathrm{SINR}_i^{-1} = \sum_{i=1}^{d} \frac{\sum_{n \neq i} \mathbf{w}_n^* \mathbf{R}_{i,\kappa(n)} \mathbf{w}_n + \sigma_i^2}{\mathbf{w}_i^* \mathbf{R}_{i,\kappa(i)} \mathbf{w}_i} = \sum_i \sum_{n \neq i} \mathbf{w}_n^* \frac{\mathbf{R}_{i,\kappa(n)}}{\beta_i} \mathbf{w}_n + \text{const}$$

$$= \sum_n \mathbf{w}_n^* \left(\sum_{i \neq n} \frac{\mathbf{R}_{i,\kappa(n)}}{\beta_i} \right) \mathbf{w}_n + \text{const}$$

This sum can be minimized separately for each $\mathbf{w}_n$,

$$\mathbf{w}_n = \arg\min \mathbf{w}_n^{*}\left(\sum_{i\neq n}\frac{\mathbf{R}_{i,\kappa(n)}}{\beta_i}\right)\mathbf{w}_n \qquad \text{s.t. } \mathbf{w}_n^{*}\mathbf{R}_{n,\kappa(n)}\mathbf{w}_n = \beta_n \tag{18.11}$$

and the minimum is given by the eigenvector with smallest eigenvalue λ of the following generalized eigenvalue problem:

$$\left(\sum_{i\neq n}\frac{\mathbf{R}_{i,\kappa(n)}}{\beta_i}\right)\mathbf{w}_n = \lambda\mathbf{R}_{n,\kappa(n)}\mathbf{w}_n$$

properly scaled. Algorithms for generalized eigenvalue problems can be found in Reference 48 and are implemented in, for example, MATLAB. Notice that the solution only involves channel correlation matrices related to base station $\kappa(n)$ and can be solved locally at each base station. The interpretation of Eq. (18.11) is that each beamformer tries to send as little power as possible to all known interfered mobiles.

What is the proper choice of the signal levels β_i? Several authors[18,30,49] have used a constant level for all users, for example, $\beta_i = 1$. However, it has been noted[15,35] that this solution, which gives an excellent uplink beamformer, may not be the best choice in the downlink. As an example, consider a single base station transmitting to two mobiles, A and B, where the path gain to B is 10 dB higher than that to A. Then, both the desired signal for mobile B and the interfering signal intended for mobile A experience the same path gain, so there is no need to let the beamformer for A suppress the interfering signal to B more than the beamformer for B suppresses the interference to A. Thus, it makes more sense to normalize each channel correlation matrix for the interferers to give the same path gain. By using the same argument for a scenario with several base stations, it can be seen that it is reasonable to normalize each channel correlation matrix with the path gain between the interfered mobile and its own base station, not the base station transmitting the interfering signal. This corresponds to the choice $\beta_i = \mathrm{Tr}[\mathbf{R}_{i,\kappa(i)}]$.

One disadvantage is that with this choice, the algorithm requires channel knowledge collected at other base stations. However, this time only a single scalar variable has to be transferred for each mobile. An alternative that also reduces the need to estimate all channels in the system, is to only consider mobiles within the single cell (in an SDMA system) or in the directly adjacent cells. If all other unidentified mobiles are completely ignored, the resulting beamformer may cause unnecessarily high interference at some of the unidentified mobiles. One solution is to approximate the channel correlation matrices of these mobiles by a scaled identity matrix. If, for simplicity, it is assumed that we are transmitting to mobile 1, and the neighboring mobiles 2, …, d_I have been identified, all the remaining unidentified channels are approximated by

$$\sum_{i=d_I+1}^{d}\frac{\mathbf{R}_{i,\kappa(n)}}{\beta_i} \approx \alpha\mathbf{I}$$

which leads to the following algorithm

$$\mathbf{w}_1 = \arg\max \frac{\mathbf{w}_1^{*}\mathbf{R}_{1,\kappa(1)}\mathbf{w}_1}{\mathbf{w}_1^{*}\left(\sum_{i=2}^{d_I}\frac{\mathbf{R}_{i,\kappa(1)}}{Tr\left[\mathbf{R}_{i,\kappa(i)}\right]}+\alpha\mathbf{I}\right)\mathbf{w}_1} \tag{18.12}$$

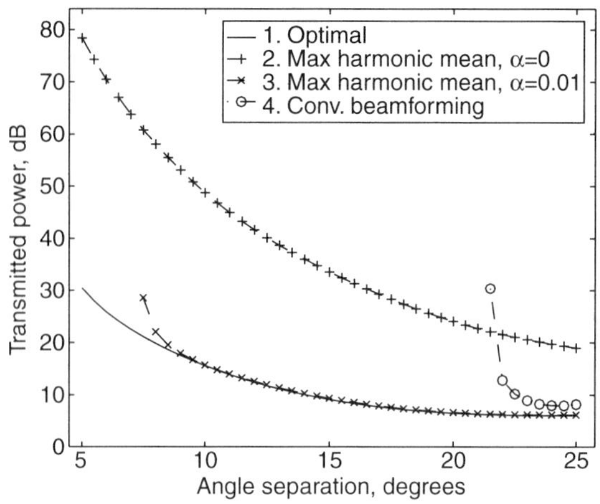

FIGURE 18.8 Total transmitted power needed to provide SINR ≥ 10 dB for all three users.

Adding such a term to the diagonal elements of the interference correlation matrix is also called diagonal loading or regularization and is a common trick in robust beamforming and for least squares methods, in general (see, for example, Reference 38). The parameter a can be determined by considering the average path loss to the unidentified mobiles.[50] This type of beamformer has also been studied in References 21, 22, 28, 50, 51.

Finally, do not forget that basically any beamformer designed for the uplink can also be used as a reasonable ad-hoc solution for the downlink.[38,41,52,53] If each carrier frequency is not used in every cell, the unidentified co-channel users are confined to certain angular sectors. The channel to all mobiles in such a sector $[\theta_1, \theta_2]$ can be approximated by

$$\sum \frac{\mathbf{R}_{i,\kappa(n)}}{\beta_i} \approx \alpha \int_{\theta_1}^{\theta_2} \mathbf{a}(\theta)\mathbf{a}^*(\theta)d\theta$$

In Fig. 18.8, we have compared the optimum and the suboptimum schemes for a simulated scenario with three mobiles served by a single base station. The base station has an eight-element uniform circular array with a radius of one wavelength. One mobile was fixed at 20° and the other two were located at directions 20° ± Δ, where the angular separation Δ was varied between 5° and 25°. Each user was surrounded by local scatterers corresponding to a spread angle of 1°. The following four strategies have been compared:

1. The optimum downlink beamformers according to Eq. (18.9) with $\gamma_i = 10$ dB
2. The decentralized algorithm Eq. (18.12) with $\alpha = 0$
3. The decentralized algorithm Eq. (18.12) with $\alpha = 0.01$
4. Conventional beamforming (i.e., using the eigenvector of $\mathbf{R}_{i,\kappa(i)}$ with largest eigenvalue)

For the three latter strategies, optimum power control was applied to provide exactly 10 dB received SINR for each mobile. For the conventional beamformer this was only possible when the sources were separated more than 21°, and for the decentralized algorithm with $\alpha = 0.01$, a separation of at least 7° was required. Note that the factor α (which can be seen as a regularization, see Section 18.5.2) is essential to reduce the excess power transmitted in the system. However, increasing α too much gives a solution more and more similar to the conventional beamformer. Thus, α determines the trade-off between low sidelobes in general and signal suppression toward the interfered mobiles known to the base station.

Figure 18.9 gives an illustration of how the choice of downlink beamforming algorithm can affect the system performance. The experiment setup was a simulated system with four base stations, each equipped

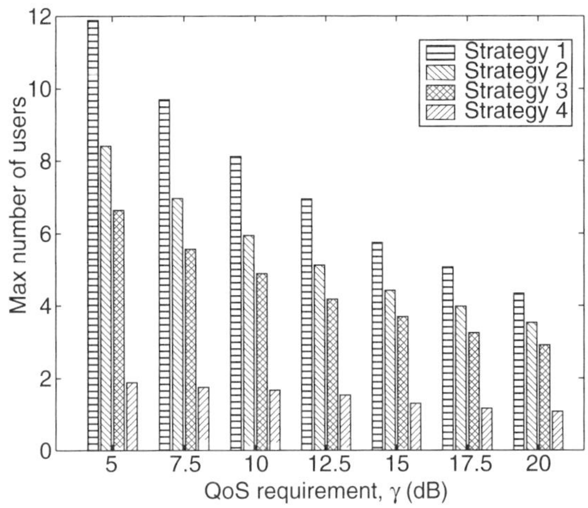

FIGURE 18.9 Max number of users for (1) optimum beamforming and allocation, (2) optimum beamforming and standard assignment, (3) decentralized beamforming, and (4) conventional beamforming.

with an eight-element uniform linear array. Mobiles were placed at random and the channel to each base station was characterized by path loss, lognormal shadow fading, and local scattering with random spread angle. For each of the following four strategies, as many mobiles as possible were admitted into the system, until it was impossible to provide sufficient SINR for all users.

1. Optimum beamforming according to Eq. (18.9) combined with the jointly optimum assignment of mobiles to base stations.
2. Optimum beamforming but no optimized assignment
3. The decentralized beamformer [Eq. 18.12]
4. Conventional beamforming as described at the beginning of this section

By default, each mobile was assigned to be base station with the highest channel gain, but obviously this is not the optimum assignment.[54] The results in Fig. 18.9 show that the maximum number of admitted users in the system averaged over 300 random scenarios.

18.3.3 Power Control

The optimum downlink beamforming strategy of Section 18.3.1 determines both the power control and the spatial filtering jointly. For the other algorithms, standard power control schemes, originally developed for single-antenna systems, can be applied to improve the system performance. The following short presentation is intended to give a quick overview of the standard power control scheme. More information can be found in References 55 through 57. The goal is to find the optimum scale factor for each beamforming vector. Use the notation $\mathbf{u}_i$ for the beamformers obtained using one of the preceding algorithms. The beamformers to be used in the system are given by $\mathbf{w}_i = \sqrt{p_i}\mathbf{u}_i$. The goal is to achieve the SINR constraints of Eq. (18.9) with equality. Introduce the vectors $\mathbf{p} = [p_1, \ldots, p_d]^T$ and $\eta = [\gamma_1\sigma_1^2, \ldots, \gamma_d\sigma_d^2]^T$, the diagonal matrix $\mathbf{D}$ with diagonal elements

$$[\mathbf{D}]_{i,i} = \mathbf{w}_i^{\star}\mathbf{R}_{i,\kappa(i)}\mathbf{w}_i$$

and the matrix $\mathbf{G}$, given by

$$[\mathbf{G}]_{i,n} = \begin{cases} 0 & i = n \\ \gamma_i \mathbf{w}_n^{\star}\mathbf{R}_{i,\kappa(n)}\mathbf{w}_n & i \neq n \end{cases}$$

By using this notation, the constraints of Eq. (18.10) will be fulfilled with equality if $(\mathbf{D} - \mathbf{G})\mathbf{p} = \eta$ [i.e., $\mathbf{p} = (\mathbf{D} - \mathbf{G})^{-1}\eta$]. However, it turns out that if the matrix inversion is calculated using an iterative scheme,

$$\mathbf{p}(t+1) = \mathbf{D}^{-1}\mathbf{G}\mathbf{p}(t) + \mathbf{D}^{-1}\eta$$

each element of $\mathbf{p}(t + 1)$ can actually be updated separately. The corresponding calculation amounts simply to estimating the SINR level at the corresponding mobile and transmitting information back to its base station on how much the transmitted power should be increased or decreased to give the desired SINR level γ_i. It can be shown that this iteration does always converge to the correct value $(\mathbf{D} - \mathbf{G})^{-1}\eta$ if the problem has a feasible solution where all the resulting power levels are nonnegative. If no such feasible power allocation exists, the iteration diverges. Interestingly enough, there is no need to perform the power updates synchronously and it is even possible to simplify the algorithm even further, just transmitting a single bit of information (increase/decrease) back from the mobile to the base station.[58]

18.3.4 Additional Constraints

To provide a practically useful solution, some additional constraints may be imposed on the beamformers. The power amplifiers for each transmitting antenna element, for example, will have a limited dynamic range that gives additional constraints to be included in the beamformer design.

Ideally, one would like to be able to put upper and lower limits on the instantaneous amplitude of the signals but such constraints are very difficult to handle. An easier problem is to handle upper and lower constraints on the average signal power at each antenna element,

$$\mu_{k,n}^{L} \le \mathrm{E}\left[\left|\left[\mathbf{x}_k\right]_n\right|^2\right] \le \mu_{k,n}^{U}$$

which corresponds to the following constraints on the beamforming vectors,

$$\mu_{k,n}^{L} \le \sum_{i \in I(k)} \left|\left[\mathbf{w}_i\right]_n\right|^2 \le \mu_{k,n}^{U} \tag{18.13}$$

A related problem is to get a balanced solution where all antenna elements of base station k transmit with the same power. In terms of the beamformer elements, this is expressed by the additional constraints

$$\sum_{i \in I(k)} \left|\left[\mathbf{w}_i\right]_n\right|^2 = \sum_{i \in I(k)} \left|\left[\mathbf{w}_i\right]_l\right|^2, \; l \neq n \tag{18.14}$$

At the end of Section 18.4.2, it is shown how this type of constraints can easily be handled in the optimization. Figure 18.10 shows the signal trajectories for the baseband signals transmitted at a four-element antenna array communicating with two co-channel mobiles. GMSK modulation is used for each of the two data signals. The left half of the figure shows the unconstrained optimum whereas the solution in the right half is constrained to give equal power on all antennas, using Eq. (18.14).

18.3.5 Frequency-Selective Channels and Beamformers

All the strategies described earlier can be extended to wideband systems. This section gives a brief sketch of the solution. The first step is to describe the sampled baseband representation of the frequency-selective channel as a tapped delay line $\mathbf{h}(t)$, where the signal contribution from a single base station k to mobile i is

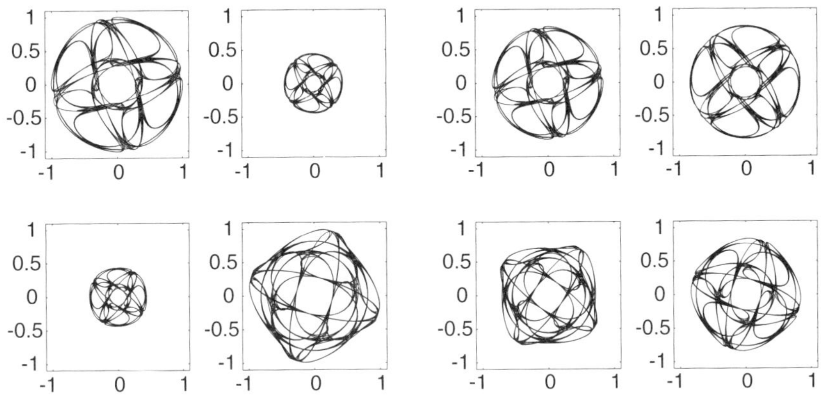

FIGURE 18.10 Signal constellation at the four antenna elements for the unconstrained optimum (left) and the constrained optimum (right).

$$\mathbf{h}_{i,k}^{\star}(t) * \mathbf{x}_k(t) = \sum_{n=0}^{L_h} \mathbf{h}_{i,k}^{\star}(n)\mathbf{x}_k(t-n)$$

resulting in a total received signal at mobile i of

$$r_i(t) = \sum_{k=1}^{K} \mathbf{h}_{i,k}^{\star}(t) * \mathbf{x}_k(t) + n_i(t)$$

For FDD systems, the second-order statistics of the channel is given by

$$\mathbf{R}_{i,k}(t_1, t_2) = E\left[\mathbf{h}_{i,k}(t_1)\mathbf{h}_{i,k}^{\star}(t_2)\right] \tag{18.15}$$

For a multitap channel, the beamformer can be used to suppress not only co-channel interference but also intersymbol interference. This can be done both with traditional single tap beamformers as well as frequency-selective multitap beamformers. With multitap beamformers, the transmitted signal is formed by convolving the data sequence $s_i(t)$ with the beamformer taps $\mathbf{w}_i(t)$.

$$\mathbf{x}_k(t) = \sum_{i \in I(k)} \mathbf{w}_i(t) * s_i(t) = \sum_{i \in I(k)} \sum_{n=0}^{L_w} \mathbf{w}_i(n) s_i(t-n)$$

The resulting impulse response from data sequence $s_j(t)$ to mobile i is given by

$$f_{i,j}(t) = \mathbf{h}_{i,\kappa(j)}^{\star}(t) * \mathbf{w}_j(t) = \sum_{n=0}^{t} \mathbf{h}_{i,\kappa(j)}^{\star}(n)\mathbf{w}_j(t-n)$$

and has length $L_f + 1 = L_h + L_w + 1$, if the channel is described by $L_h + 1$ taps and the beamformer has $L_w + 1$ taps. To obtain a problem formulation similar to the narrowband case, it is convenient to define the impulse response vector $\bar{f}_{i,j} = [f_{i,j}(0), f_{i,j}(1), \ldots, f_{i,j}(L_f)]^T$. If the beamformer taps are stacked into a tall vector $\bar{\mathbf{w}}_i = [\mathbf{w}_i^T(0), \mathbf{w}_i^T(1), \ldots, \mathbf{w}_i^T(L_w)]^T$ and the channel taps are put into a convolution matrix

$$\bar{\mathbf{H}}_{i,k} = \begin{bmatrix} \mathbf{h}_{i,k}^*(0) & 0 & \cdots & 0 \\ \vdots & \mathbf{h}_{i,k}^*(0) & \ddots & \vdots \\ \mathbf{h}_{i,k}^*(L_h) & \vdots & \ddots & 0 \\ 0 & \mathbf{h}_{i,k}^*(L_h) & \vdots & \mathbf{h}_{i,k}^*(0) \\ \vdots & \ddots & \ddots & \vdots \\ 0 & \cdots & 0 & \mathbf{h}_{i,k}^*(L_h) \end{bmatrix}$$

the impulse response vector is given by

$$\bar{f}_{i,j} = \bar{\mathbf{H}}_{i,\kappa(j)} \bar{\mathbf{w}}_j$$

If the data signals $s_j(t)$ are temporally white with normalized power 1, the total received power contribution at mobile i by signal j is

$$E\left[\sum_{n=1}^{L_f} \left|f_{i,j}(n)\right|^2\right] = \bar{\mathbf{w}}_j^* \, E\left[\bar{\mathbf{H}}_{i,\kappa(j)}^* \bar{\mathbf{H}}_{i,\kappa(j)}\right] \bar{\mathbf{w}}_j \overset{\text{def.}}{=} \bar{\mathbf{w}}_j^* \bar{\mathbf{R}}_{i,\kappa(j)} \bar{\mathbf{w}}_j$$

Normally for the desired signal, only the first tap $f_{i,i}(0)$ of the impulse response contributes to the desired signal power, and the remaining cause intersymbol interference. However, if the receiver is equipped with a RAKE receiver,[59] it will be able to combine the contribution of several taps of the impulse response. In general, assume that the first L_{des} taps contribute to the desired signal power and the remaining $L_f + 1 - L_{\text{des}}$ taps cause intersymbol interference. Denote the top L_{des} rows of $\bar{\mathrm{H}}_{i,\kappa(j)}$ by $\bar{\mathrm{H}}_{i,\kappa(j)}^{\text{des}}$ and the remaining lines by $\bar{\mathrm{H}}_{i,\kappa(j)}^{\text{ISI}}$, and define the correlation matrices

$$\bar{\mathbf{R}}_{i,\kappa(j)}^{\text{des}} = E\left[\bar{\mathbf{H}}_{i,\kappa(j)}^{\text{des}}\left(\bar{\mathbf{H}}_{i,\kappa(j)}^{\text{des}}\right)^*\right]$$

$$\bar{\mathbf{R}}_{i,\kappa(j)}^{\text{ISI}} = E\left[\bar{\mathbf{H}}_{i,\kappa(j)}^{\text{ISI}}\left(\bar{\mathbf{H}}_{i,\kappa(j)}^{\text{ISI}}\right)^*\right]$$

Note that these matrices are easily calculated from the $\mathbf{R}_{i,k}(t_1, t_2)$ defined in Eq. (18.15). By using this notation, the total received signal to noise ratio at mobile i is given by

$$\text{SINR}_i = \frac{\bar{\mathbf{w}}_i^* \bar{\mathbf{R}}_{i,\kappa(i)}^{\text{des}} \bar{\mathbf{w}}_i}{\bar{\mathbf{w}}_i^* \bar{\mathbf{R}}_{i,\kappa(i)}^{\text{ISI}} \bar{\mathbf{w}}_i + \sum_{n \neq i} \bar{\mathbf{w}}_n^* \bar{\mathbf{R}}_{i,\kappa(n)} \bar{\mathbf{w}}_n + \sigma_i^2} \qquad (18.16)$$

Except for the first term of the denominator, this expression has exactly the same form as the narrowband SINR [Eq. (18.8)]. Fortunately, both of the algorithms presented in the next section can easily be extended to solve also the more general formulation of optimum downlink multitap beamforming:

$$\min \sum_{n=1}^{d} \overline{\mathbf{w}}_n \overline{\mathbf{w}}_n$$

$$\text{s.t.} \frac{\overline{\mathbf{w}}_i^* \overline{\mathbf{R}}_{i,\kappa(i)}^{\text{des}} \overline{\mathbf{w}}_i}{\overline{\mathbf{w}}_i^* \overline{\mathbf{R}}_{i,\kappa(i)}^{\text{ISI}} \overline{\mathbf{w}}_i + \sum_{n \neq i} \overline{\mathbf{w}}_n^* \overline{\mathbf{R}}_{i,\kappa(n)} \overline{\mathbf{w}}_n + \sigma_i^2} \geq \gamma_i$$

$$i = 1, 2, \ldots, d$$

This formulation has previously been considered for the case of deterministic multitap channels and single-tap beamformers.[43]

Also, the decentralized narrowband strategies, presented earlier, can be extended in a similar way. Asté et al.[28] show that it may be reasonable to assume that

$$\mathbf{R}_{i,k}(t_1, t_2) = \delta_{t_1,t_2} \mathbf{R}_{i,k}$$

(i.e., that the channel has a flat spectrum). Based on that assumption, it is shown that a multitap beamformer derived from the max harmonic mean SINR criterion of Section 18.3.2, does not provide any improvements in terms of average SINR. In Forster et al.,[60,61] the extra degrees of freedom offered by a multitap beamformer are used to minimize the power variations caused by the fast channel fading.

For code division multiple access (CDMA) systems, the processing gain should be taken into account in the expressions for the SINR constraints of (16). A zero-forcing solution for CDMA systems, taking both co-channel interference and interchip interference into account, is described by Schubert and Boche.[62]

18.4 Algorithms for Optimum Beamforming

18.4.1 An Algorithm Based on Power Control

This algorithm has been derived independently in References 43 and 44, based on a slightly different algorithm[63] that produced a feasible but nonoptimum solution. The algorithm is based on the following result.

Lemma 1. *If* $\{\mathbf{w}_i\}$ *are the optimum beamformers for the downlink beamforming problem [Eq. (18.10)] and* $\{\mathbf{u}_i\}$ *is the optimum solution to the following problem*

$$\min_{\mathbf{u}_i, \rho_i} \sum_{n=1}^{d} \rho_n$$

$$s.t. \frac{\mathbf{u}_i^* \rho_i \mathbf{R}_{i,\kappa(i)} \mathbf{u}_i}{\mathbf{u}_i^* \left(\sum_{n \neq i} \rho_n \gamma_n \mathbf{R}_{n,\kappa(i)} + \mathbf{I} \right) \mathbf{u}_i} \geq 1 \tag{18.17}$$

$$\|\mathbf{u}_i\|^2 = 1$$

then for all the beamformers; $\mathbf{w}_i = \sqrt{p_i}\mathbf{u}$ *for some positive constants* p_i.

Proof. See the appendix.

The reformulated problem [Eq. (18.17)] has exactly the same form as an uplink problem that determines receiving beamformers $\mathbf{u}_i$ at the base stations and transmitting power levels ρ_i at the mobiles, so it is called a *virtual uplink problem*. Note that this virtual uplink problem is different from the true uplink problem for the system, even if we have a TDD system with identical uplink and downlink channels.

An extended version[64] of the power control loop has been presented that finds the optimum solution for joint optimum uplink beamforming and mobile power control by iteratively finding the maximum SINR uplink beamformers for given power levels and then doing a power control update based on these beamformers. Because Eq. (18.17) has the same form as an uplink problem, the algorithm can be applied, although not in the distributed fashion that is possible for the true uplink problem. When the beamformers $\mathbf{u}_i$ have been calculated, the power levels p_i are determined using standard power control (which can be done noniteratively because the algorithm is centralized). The resulting algorithm is summarized in the following steps (index t is used as an iteration counter, not to be confused with the time index of the channel samples):

1. Initialize $\rho_i(1) = 1$ for $i = 1, 2, \ldots, d$ (or use some other nonnegative initial value).
2. For $t = 1, 2, \ldots$ until convergence, iterate the following steps

   Beamformer update: Find

$$\mu_i = \max_{\|\mathbf{u}_i\|=1} \frac{\rho_i(t)\mathbf{u}_i^* \mathbf{R}_{i,\kappa(i)}\mathbf{u}_i}{\mathbf{u}_i^* \left(\sum_{n \neq i} \rho_n(t)\gamma_n \mathbf{R}_{n,\kappa(i)} + \mathbf{I} \right) \mathbf{u}_i} \tag{18.18}$$

   and the corresponding vector $\mathbf{u}_i$, for $i = 1, 2, \ldots, d$.

   Power control update: Let

$$\rho_i(t+1) = \frac{\gamma_i}{\mu_i}\rho_i(t)$$

3. After convergence,

$$\boldsymbol{\eta} = \left[\gamma_1\sigma_1^2, \ldots, \gamma_d\sigma_d^2\right]^T \tag{18.19}$$

$$[\mathbf{F}]_{i,n} = \begin{cases} \mathbf{u}_i^* \mathbf{R}_i \mathbf{u}_i & i = n \\ -\gamma_i \mathbf{u}_n^* \mathbf{R}_{i\kappa(n)} \mathbf{u}_n & i \neq n \end{cases} \tag{18.20}$$

$$\mathbf{p} = \mathbf{F}^{-1}\boldsymbol{\eta} \tag{18.21}$$

$$\mathbf{w}_i = \sqrt{[\mathbf{p}]_i}\,\mathbf{u}_i \tag{18.22}$$

Note that the maximum μ_i and $\mathbf{u}_i$ in Eq. (18.18) are given by the maximum eigenvalue and the corresponding eigenvector, respectively, of the generalized eigenvalue problem

$$\rho_i(t)\mathbf{R}_{i,\kappa(i)}\mathbf{u}_i = \mu_i \left(\sum_{n \neq i} \rho_n(t)\gamma_n \mathbf{R}_{n,\kappa(i)} + \mathbf{I} \right) \mathbf{u}_i \tag{18.23}$$

By using the theory of *standard interference functions*,[57] it is easy to prove that the iterations in step 2 converge to the optimum of the virtual uplink problem (see also Reference 64).

18.4.2 An Algorithm Based on Semidefinite Optimization

Assume for the moment that we allow for time-varying beamformers $\mathbf{w}_i(t)$, drawn at random from some distribution with correlation matrices $\mathbf{W}_i = \mathbf{E}[\mathbf{w}_i(t)\mathbf{w}_i^*(t)]$. This may open extra degrees of freedom in the search for optimum beamforming strategies. We show that for the optimum choice of the beamforming correlation matrices, all $\mathbf{W}_i$ are actually rank one, which means that the optimum solution is in fact to use standard fixed beamformers, even for this extended problem. The advantage of this strategy is that solving for the matrices $\mathbf{W}_i$ is in some aspects easier than solving for the vectors $\mathbf{w}_i$ directly.

From the rule $\mathrm{Tr}[\mathbf{AB}] = \mathrm{Tr}[\mathbf{BA}]$ (see, for example, Reference 32), it follows that $\mathbf{w}^*\mathbf{R}\mathbf{w} = \mathrm{Tr}[\mathbf{R}\mathbf{w}\mathbf{w}^*] = \mathrm{Tr}[\mathbf{RW}]$. Thus, the original downlink problem cost function and constraints [Eq. (18.10)] can be rewritten into

$$\min \sum_{n=1}^{d} \mathrm{Tr}\left[\mathbf{W}_n\right]$$

$$\text{s.t. } \mathrm{Tr}\left[\mathbf{R}_{i,\kappa(i)}\mathbf{W}_i\right] - \gamma_i \sum_{n\neq i} \mathrm{Tr}\left[\mathbf{R}_{i,\kappa(n)}\mathbf{W}_n\right] \geq \gamma_i \sigma_i^2$$

Because the matrices $\mathbf{W}_i$ should represent correlation matrices, we have to add constraints to guarantee that they are Hermitian and positive semidefinite. The complete optimization problem to find the $\mathbf{W}_i$ is therefore given by

$$\begin{aligned} &\min \sum_{n=1}^{d} \mathrm{Tr}\left[\mathbf{W}_n\right] \\ &\text{s.t. } \mathrm{Tr}\left[\mathbf{R}_{i,\kappa(i)}\mathbf{W}_i\right] - \gamma_i \sum_{n\neq i} \mathrm{Tr}\left[\mathbf{R}_{i,\kappa(n)}\mathbf{W}_n\right] \geq \gamma_i \sigma_i^2 \\ &\mathbf{W}_i = \mathbf{W}_i^* \\ &\mathbf{W}_i \succeq 0 \end{aligned} \tag{18.24}$$

where the notation $\mathbf{W} \succeq 0$ means that $\mathbf{W}$ is positive semidefinite. Note that this gives a cost function that is linear in the elements of $\mathbf{W}_i$. The same holds for the two first rows of constraints. The semidefinite constraints are nonlinear and even nondifferentiable but nevertheless the problem can be solved efficiently because these constraints are convex. Thus, the problem itself is convex, which means that there are no local minima and the set of optimum solutions is convex (typically a single point). In general, problems with linear cost function, linear constraints, and semidefiniteness constraints are called *semidefinite problems*. A number of so-called interior point methods or central-path methods have appeared that can solve semidefinite problems and other convex problems very efficiently. The basic principles of theory and algorithms can be found in References 65 and 66, whereas References 67 through 69 provide more in-depth coverage. Also, several program packages are freely available, for example, SeDuMi.[70] Semidefinite optimization has found several applications in control theory and signal processing.[71,72]

Going back to (Eq. (18.24)), note that with the additional constraints rank $[\mathbf{W}_i] = 1$, the optimization problem in Eq. (18.24) would be exactly equivalent to Eq. (18.10), but because those constraints are not included, Eq. (18.24) is a relaxation of Eq. (18.10), which means that we could expect to find a solution with a lower cost than Eq. (18.10) but with high rank $\mathbf{W}_i$. This is typical result when using this so-called Lagrangian relaxation method to get lower bounds on the optimum of a general quadratic optimization problem. However, for this specific formulation, the additional structure of the problem enables us to prove the following result.

LEMMA 2. *The semidefinite problem [Eq. (18.24)] does always have at least one optimum solution where all* $\mathbf{W}_i$ *have rank one.*

PROOF. *see the appendix.*

This lemma shows that Eq. (18.24) is not a strict relaxation, but actually an equivalent reformulation of Eq. (18.10). In theory, there may be situations where the solution is not unique, and in such cases, an interior point algorithm typically does not provide the rank one solution. However, from practical experience, this situation never appears in practice, except for constructed examples with exact symmetry. Nevertheless, if the algorithm does produce a solution where not every $\mathbf{W}_i$ has rank 1, the method given by Lemma 5, presented in the appendix, can be used. To summarize, the following algorithm produces optimum downlink beamformers $\mathbf{w}_i$.

1. Solve

$$\min \sum_{n=1}^{d} \mathrm{Tr}\left[\mathbf{W}_n\right]$$

$$\text{s.t. } \mathrm{Tr}\left[\mathbf{R}_{i,\kappa(i)}\mathbf{W}_i\right] - \gamma_i \sum_{n \neq i} \mathrm{Tr}\left[\mathbf{R}_{i,\kappa(n)}\mathbf{W}_n\right] \geq \gamma_i \sigma_i^2$$

$$\mathbf{W}_i = \mathbf{W}_i^*$$

$$\mathbf{W}_i \succeq 0$$

using (standard) software for semidefinite optimization.

2. Find $\mathbf{w}_i$, such that $\mathbf{w}_i\mathbf{w}_i^* = \mathbf{W}_i$, for $i = 1, 2, \ldots, d$.
3. (If some of the $\mathbf{W}_i$ do not have rank 1, find $\mathbf{u}_i \in$ span $[\mathbf{W}_i]$, for $i = 1, 2, \ldots, d$ and use Eqs. (18.20 to 18.23) to calculate the $\mathbf{w}_i$.)

This method provides large flexibility in the choice of problem formulation and additional constraints. For the additional constraints from Section 18.3.4, the key observation is that if $\mathbf{W} = \mathbf{w}\mathbf{w}^*$, then $\left|[\mathbf{w}]_n\right|^2 = [\mathbf{W}]_{nn}$, which means that constraints involving the squared magnitude of a beamformer coefficient can be expressed in terms of the diagonal elements of the beamformer correlation matrix, in the semidefinite formulation of the problem. Thus, the constraint in Eq. (18.13) can be handled by adding the following constraints to Eq. (18.24):

$$\mu_{k,n}^{L} \leq \sum_{i \in I(k)} \left[\mathbf{W}_i\right]_{nn} \leq \mu_{k,n}^{U} \tag{18.25}$$

Similarly, the constraints in Eq. (18.14) on equal power for all antenna elements are expressed by

$$\sum_{i \in I(k)} \left[\mathbf{W}_i\right]_{nn} = \sum_{i \in I(k)} \left[\mathbf{W}_i\right]_{ll}, \; l \neq n \tag{18.26}$$

As long as the additional constraints, as in these examples, are linear in the elements of $\mathbf{W}_i$, the resulting optimization problem will still be semidefinite. Note that with additional constraints, there is in general no guarantee that the resulting solution will have rank one. Such a high-rank solution can be seen as a nonobtainable lower limit on the transmitted power used by any feasible solution, which still provides useful information in system simulations, for example. However, a solution with high rank $\mathbf{W}_i$ matrices can actually be implemented in practice using time varying beamforming vectors drawn at random from some distribution with correlation matrix $\mathbf{W}_i$. With sufficient channel coding and interleaving, such a solution may provide excellent performance. A third alternative is to use some rank one approximation of the optimum $\mathbf{W}_i$ to obtain a heuristic solution with normal fixed beamformers. From practical experiments, it seems that Eq. (18.26) always provides rank one solutions. The same holds for Eq. (18.25) unless $\mu_{k,n}^{L}$ is set to force a larger power than that used by the unconstrained solution. However, these are empirical observations and are not based on any mathematical analysis of the algorithms.

18.4.3 Comparisons

Here, the two methods described in the previous section are compared. Normally, the algorithm of Section 18.4.1 is the fastest, converging in about 10 to 40 iterations. However, when the problem is close to infeasible, corresponding to a highly loaded system, the convergence can be very slow and in some situations, the algorithm may require several thousand iterations until it converges. The semidefinite formulation of Section 18.4.2, on the other hand, has been observed to converge in about 10 to 40 iterations regardless of the beamforming scenario. Theoretically, interior point methods are guaranteed to converge in polynomial time, but several authors have reported similar experience from different applications, that in practice, the algorithms converge in a few tenths of iterations. Because each iteration has higher complexity, this method is slower in normal situations but may be significantly faster than the power control-based method for almost infeasible scenarios. Another advantage of this method is the flexibility to changes in the problem formulation or addition of extra constraints.

18.4.4 Algorithms for Known Channel Vectors

For TDD systems where the estimate of the uplink channel can be used as a good estimate of the instantaneous downlink channel and for other situations where the channel correlation matrices have rank one.

$$\mathbf{R}_{i,k} = \mathbf{h}_{i,k}\mathbf{h}_{i,k}^{*} \tag{18.27}$$

more efficient algorithms can be found.

In the algorithm of Section 18.4.1, the beamformer calculation in Eq. (18.18) for each iteration is then given by

$$\mathbf{u}_i = \left(\sum_{n \neq i} \rho_n(t)\gamma_n \mathbf{R}_{n,\kappa(i)} + \mathbf{I} \right)^{-1} \mathbf{h}_{i,k} \tag{18.28}$$

because this is the only eigenvector of Eq. (18.23) with a nonzero eigenvalue. The solution can be scaled to give $\|\mathbf{u}_i\| = 1$ but this is not necessary, because the remaining algorithm works regardless of the choice of scaling of $\mathbf{u}_i$.

An alternative is to make the problem convex without introducing the matrices $\mathbf{W}_i$. By using Eq. (18.27), the basic beamforming problem in Eq. (18.10) can be written as follows:

$$\min \sum_{n=1}^{d} \mathbf{w}_n^{*}\mathbf{w}_n$$

$$\text{s.t.} \left| \mathbf{w}_i^{*}\mathbf{h}_{i,\kappa(i)} \right|^2 \geq \gamma_i \sum_{n \neq i} \mathbf{w}_n^{*}\mathbf{R}_{i,\kappa(n)}\mathbf{w}_n + \gamma_i\sigma_i^2$$

$$i = 1, 2, \ldots, d$$

This is a nonconvex problem because if, for example, $\mathbf{w}_n$ are fixed for all $n \neq i$, the constraints force $\mathbf{W}_i^{*}\mathbf{h}_{i,\kappa(i)}$ to fall into the region outside a circle in the complex plane. However, because our beamforming strategy leaves the phase of each beamformer unspecified, there is no loss of generally to add a constraint on the phase of each $\mathbf{W}_i$,

$$
\begin{aligned}
&\min \sum_{n=1}^{d} \mathbf{w}_n^{\star}\mathbf{w}_n \\
&\text{s.t.} \left(\mathbf{w}_i^{\star}\mathbf{h}_{i,\kappa(i)}\right)^2 \geq \gamma_i \sum_{n \neq i} \mathbf{w}_n^{\star}\mathbf{R}_{i,\kappa(n)}\mathbf{w}_n + \gamma_i\sigma_i^2 \\
&\operatorname{Im}\left[\mathbf{w}_i^{\star}\mathbf{h}_{i,\kappa(i)}\right] = 0 \\
&\mathbf{w}_i^{\star}\mathbf{h}_{i,\kappa(i)} \geq 0 \\
&i = 1, 2, \ldots, d
\end{aligned}
\tag{18.29}
$$

This is actually a convex problem because the constraints for each i are linear transformations of the so-called second-order convex cone.[73]

$$
\left\{(x, \mathbf{y}) x \in \mathbb{R}, \mathbf{y} \in \mathbb{C}^N, x \geq \left\|\mathbf{y}\right\|\right\}
$$

Also, for convex problems with second-order cone constraints, several efficient interior point methods are readily available.[70,73] In contrast to the semidefinite formulation in Eq. (18.24), there is no need to introduce the additional matrices $\mathbf{W}_i$ and the resulting optimization involves significantly fewer variables. Note that both Eqs. (18.28) and (18.29) only require the channel $\mathbf{h}_{i,\kappa(i)}$ to the desired mobile to have rank one; all the remaining $\mathbf{R}_{i,k}$ can have any rank.

18.5 Robust Beamforming

Section 18.2.4 presents different strategies to describe uncertainties in the channel estimates. This section illustrates how such a model can be incorporated into the beamformer design to give a more robust solution.

18.5.1 Optimum Robust Beamforming

The idea of optimum robust beamforming is that the beamformer should provide sufficient SINR to all mobiles, for all possible values of the channel correlation matrices that fall within the upper and lower bounds [Eq. (18.4)]. By using our usual cost function, minimum transmitted power, it is easy to show using a worst-case analysis that the optimum beamformers are given as the solution of

$$
\begin{aligned}
&\min \sum_{n=1}^{d} \mathbf{w}_n^{\star}\mathbf{w}_n \\
&\text{s.t. } \mathbf{w}_i^{\star}\underset{\smile}{\mathbf{R}}_{i,\kappa(i)}\mathbf{w}_i - \gamma_i \sum_{n \neq i} \mathbf{w}_n^{\star}\widehat{\mathbf{R}}_{i,\kappa(n)}\mathbf{w}_n \geq \gamma_i\sigma_i^2 \\
&i = 1, 2, \ldots, d
\end{aligned}
$$

Because the resulting problem has exactly the same form as the normal optimum beamforming problem in Eq. (18.10), the same algorithms from Section 18.4 can be applied directly. In practice, the worst case is extremely unlikely and this strategy may give an unnecessarily defensive solution.

18.5.2 Robust Decentralized Schemes

In uplink beamforming, signal cancellation is a well-known phenomenon. With a small channel mismatch, an adaptive algorithm may try to reduce the power of the desired user instead of the interference. Several approaches have been suggested to reduce this problem, including diagonal loading (regularization)[38] and different null-widening techniques.[39,41,74] Similar problems may occur in downlink beamforming, for example, for the strategies of Section 18.3.2.

When using Eq. (18.11) with $\beta_i = 1$, it may be tempting to exploit that

$$\frac{\mathbf{w}_n^\star\left(\sum_{i\neq n}\mathbf{R}_{i,\kappa(n)}\right)\mathbf{w}_n}{\mathbf{w}_n^\star\mathbf{R}_{n,\kappa(n)}\mathbf{w}_n}$$

is minimized by the same $\mathbf{w}_n$ as

$$\frac{\mathbf{w}_n^\star\left(\sum_{i=1}^{d}\mathbf{R}_{i,\kappa(n)}\right)\mathbf{w}_n}{\mathbf{w}_n^\star\mathbf{R}_{n,\kappa(n)}\mathbf{w}_n}$$

In a TDD system with low noise level, the sum in the numerator is well approximated by the covariance matrix of the received uplink signal, so it suffices to estimate the channel of the desired signal and collect the sample covariance matrix of the total received signal. However, the resulting quadratically constrained minimum variance problem is closely related to the traditional linearly constrained minimum variance beamformer for uplink processing,[52,53,75] which is very sensitive to signal cancellation.

One popular method to decrease the signal cancellation problems is diagonal loading, simply adding a constant on the diagonal of the estimated interference correlation matrices. This corresponds to increasing the parameter α in Eq. (18.12). In addition to the motivation of the parameter α given in Section 18.3.2, the regularization may also be derived from the specific form in Eq. (18.5) of the uncertainty bounds on the channel matrices, using similar arguments as in the previous section.

A deep notch in the beampattern pointing to a certain interfered mobile, gives large sensitivity to pointing errors and other channel mismatches. Therefore, several methods have been suggested to artificially widen the zeros of the beampatterns. One method described in References 39 to 41 is to modify the channel matrices using covariance matrix tapers in Eq. (18.6) to correspond to spatially smoothed versions of the interfered mobiles. Other methods involve using derivative constraints or estimating a number of discrete angles which are replaced by pairs of closely separated directions.[76]

18.6 Appendix: Proofs and Lemmas

This appendix contains the proofs of Lemma 1 and 2. First, some useful results for the proofs are presented as separate lemmas.

LEMMA 3. *If*

$$\min\sum_{n=1}^{d}\mathbf{w}_n^\star\mathbf{w}_n$$

$$\text{s.t.}\quad \frac{\mathbf{w}_i^\star\mathbf{R}_{i,\kappa(i)}\mathbf{w}_i}{\sum_{n\neq i}\mathbf{w}_n^\star\mathbf{R}_{i,\kappa(n)}\mathbf{w}_n+\sigma_i^2}\geq\gamma_i$$

$$i=1,2,\ldots,d$$

and

$$\min_{\mathbf{u}_i,\rho_i} \sum_{n=1}^{d} \gamma_n \sigma_n^2 \rho_n$$

$$\text{s.t.} \quad \frac{\mathbf{u}_i^{\star} \rho_i \mathbf{R}_{i,\kappa(i)} \mathbf{u}_i}{\mathbf{u}_i^{\star}\left(\sum_{n \neq i} \rho_n \gamma_n \mathbf{R}_{n,\kappa(i)} + \mathbf{I}\right)\mathbf{u}_i} \geq 1$$

$$\left\|\mathbf{u}_i\right\|^2 = 1$$

then $\mathbf{w}_i = \sqrt{p_i}\, \mathbf{u}_i$ *for some* p_i.

PROOF. Introduce separate variables for the power $p_i \geq 0$ and the normalized beamformers $\mathbf{u}_i$; $\|\mathbf{u}_i\|^2 = 1$ such that $\mathbf{w}_i = \sqrt{p_i}\mathbf{u}_i$. Also, introduce the vectors $\mathbf{p} = [p_i, \ldots, p_d]^T$, $\eta = [\gamma_1\sigma_1^2, \ldots, \gamma_d\sigma_d^2]^T$, $\omega = [\|\mathbf{u}_1\|^2, \ldots, \|\mathbf{u}_d\|^2]^T$, the diagonal matrix $\mathbf{D}$ with diagonal elements

$$\left[\mathbf{D}\right]_{i,i} = \mathbf{w}_i^{\star} \mathbf{R}_{i,\kappa(i)} \mathbf{w}_i$$

and the matrix $\mathbf{G}$, given by

$$\left[\mathbf{G}\right]_{i,n} = \begin{cases} 0 & i = n \\ \gamma_i \mathbf{w}_n^{\star} \mathbf{R}_{i,\kappa(n)} \mathbf{w}_n & i \neq n \end{cases}$$

Because all constraints in Eq. (18.10) are fulfilled with equality at the optimum, the problem can be written

$$\begin{aligned} &\min \mathbf{p}^T \boldsymbol{\omega} \\ &\text{s.t.} \left(\mathbf{D} - \mathbf{G}\right)\mathbf{p} = \boldsymbol{\eta} \\ &p_i > 0 \\ &\left\|\mathbf{u}_i\right\| = 1 \end{aligned} \tag{18.30}$$

Notice that $\mathbf{D}$ is invertible and that $\mathbf{D}^{-1}\,\mathbf{G}$ has only nonnegative elements and is irreducible.[77] This means that we can use the elegant Frobenius Perron theory for matrices with nonnegative elements[77] to show that Eq. (18.30) equivalently can be written

$$\begin{aligned} &\min \mathbf{p}^T \boldsymbol{\omega} \\ &\text{s.t.} \left(\mathbf{D} - \mathbf{G}\right)\mathbf{p} = \boldsymbol{\eta} \\ &\lambda_{\max}\left\{\mathbf{D}^{-1}\mathbf{G}\right\} < 1 \\ &\left\|\mathbf{u}_i\right\| = 1 \end{aligned} \tag{18.31}$$

To show this, assume that the constraints of Eq. (18.30) hold, then $(\mathbf{I} - \mathbf{D}^{-1}\,\mathbf{G})\mathbf{p} = \mathbf{D}^{-1}\,\boldsymbol{\eta}$ and looking at the ith row, we conclude that $p_i > 0$ and

$$\frac{\left[\mathbf{D}^{-1}\mathbf{G}\mathbf{p}\right]_i}{p_i} = 1 - \left[\mathbf{D}^{-1}\boldsymbol{\eta}\right]_i < 1$$

The spectral radium $\lambda_{\max}\{\mathbf{F}\}$, of a nonnegative irreducible matrix $\mathbf{F}$ is given by

$$\lambda_{\max}\left\{\mathbf{F}\right\} = \min_{x_i \geq 0} \max_{1 \leq i \leq d} \frac{\left[\mathbf{F}\mathbf{x}\right]_i}{x_i}$$

which shows that

$$\lambda_{\max}\left\{\mathbf{D}^{-1}\mathbf{G}\right\} \leq \max_{1 \leq i \leq d} \frac{\left[\mathbf{D}^{-1}\mathbf{G}\mathbf{p}\right]_i}{p_i} < 1$$

and, thus, the feasible set of Eq. (18.30) is included in the feasible set of Eq. (18.31). Conversely, if $\mathbf{p}$ and $\mathbf{u}$ are feasible according to Eq. (18.31), the spectral radius of the irreducible nonnegative matrix $\mathbf{D}^{-1}\,\mathbf{G}$ is less than 1. Then, $\mathbf{I} - \mathbf{D}^{-1}\,\mathbf{G}$ is invertible and the inverse has only nonnegative elements. Consequently, $\mathbf{p} = (\mathbf{I} - \mathbf{D}^{-1}\,\mathbf{G})^{-1}\,\mathbf{D}^{-1}\,\boldsymbol{\eta}$ is nonnegative and we have a feasible point also of Eq. (18.30). We can now solve for $\mathbf{p}$ and get the equivalent problem

$$\begin{aligned} &\min \boldsymbol{\eta}^T \left(\mathbf{D} - \mathbf{G}\right)^{-T} \boldsymbol{\omega} \\ &\lambda_{\max}\left\{\mathbf{D}^{-1}\mathbf{G}\right\} < 1 \\ &\text{s.t. } \left\|\mathbf{u}_i\right\| = 1 \end{aligned}$$

If we introduce the vector $\boldsymbol{\rho} = (\mathbf{D} - \mathbf{G})^{-T}\,\boldsymbol{\omega}$, which using the same argument will be all nonnegative, we finally arrive at

$$\begin{aligned} &\min \boldsymbol{\eta}^T \boldsymbol{\rho} \\ &\text{s.t. } \left(\mathbf{D} - \mathbf{G}^T\right)\rho = \boldsymbol{\omega} \\ &\rho_i \geq 0 \\ &\left\|\mathbf{u}_i\right\| = 1 \end{aligned}$$

or in other words,

$$\begin{aligned} &\min_{\rho_i} \sum_{i=1}^{d} \gamma_i \sigma_i^2 \rho_i \\ &\text{s.t. } \mathbf{u}_i^* \left(\mathbf{I} - \rho_i \mathbf{R}_{i\kappa(i)} + \sum_{n \neq i} \rho_n \gamma_n \mathbf{R}_{n\kappa(i)}\right) \mathbf{u}_i = 0 \\ &\left\|\mathbf{u}_i\right\| = 1 \\ &\rho_i \geq 0, \; i = 1, \ldots, d \end{aligned} \tag{18.32}$$

LEMMA 4. *The following two problems have the same unique optimal solution* $\boldsymbol{\rho} = [\rho_1, \rho_2, \ldots, \rho_d]$.

$$\begin{aligned} &\min_{\rho_i} \sum_{i=1}^{d} \alpha_i \rho_i \\ &\text{s.t. } \mathbf{u}_i^{\star}\left(\mathbf{I} - \rho_i \mathbf{R}_{i\kappa(i)} + \sum_{n \neq i} \rho_n \gamma_n \mathbf{R}_{n\kappa(i)}\right)\mathbf{u}_i = 0 \\ &\|\mathbf{u}_i\| = 1 \\ &\rho_i \geq 0, i = 1, \ldots, d \end{aligned} \tag{18.33}$$

$$\begin{aligned} &\max_{\rho_i} \sum_{i=1}^{d} \beta_i \rho_i \\ &\text{s.t. } \mathbf{I} - \rho_i \mathbf{R}_{i\kappa(i)} + \sum_{n \neq i} \rho_n \gamma_n \mathbf{R}_{n\kappa(i)} \succeq 0 \\ &i = 1, \ldots, d \end{aligned} \tag{18.34}$$

for any positive constants α_i and β_i.

PROOF. First, we show that both problems have optimum solutions in the set given by the following constraints:

$$\begin{aligned} &\mathbf{I} - \rho_i \mathbf{R}_{i\kappa(i)} + \sum_{n \neq i} \rho_n \gamma_n \mathbf{R}_{n\kappa(i)} \succeq 0 \\ &\mathbf{I} - \rho_i \mathbf{R}_{i\kappa(i)} + \sum_{n \neq i} \rho_n \gamma_n \mathbf{R}_{n\kappa(i)} \nsucc 0 \\ &\rho_i > 0 \qquad i = 1, \ldots, d \end{aligned} \tag{18.35}$$

Note that Eq. (18.35) means that all eigenvalues of $\mathbf{I} - \rho_i \mathbf{R}_{i\kappa(i)} + \Sigma_{n \neq i} \rho_n \gamma_n \mathbf{R}_{n\kappa(i)}$ are nonnegative and at least one of the eigenvalues is exactly zero.

If we, for example, assume that the factor within the parenthesis of constraint number i in Eq. (18.33) has both positive and negative eigenvalues at the optimum, then the cost can be reduced by decreasing ρ_i without violating any of the constraints, which gives a contradiction. A similar argument shows the result for Eq. (18.34). Assume now that we can find two different vectors $\boldsymbol{\rho}_1 \neq \boldsymbol{\rho}_2$ in the set given by Eq. (18.35). Because no ρ_i is zero, we can always find an i and a constant α such that $\rho_i^1 = \alpha \rho_i^2$ and $\rho_n^1 \leq \alpha \rho_n^2$ for all $n \neq i$. Assume without loss of generality that $\alpha > 1$. Then constraint i gives

$$\alpha \rho_i^2 \mathbf{R}_{i\kappa(i)} = \rho_i^1 \mathbf{R}_{i\kappa(i)} \preceq \mathbf{I} + \sum_{n \neq i} \rho_n^1 \gamma_n \mathbf{R}_{n\kappa(i)} \preceq \mathbf{I} + \sum_{n \neq i} \alpha \rho_n^2 \gamma_n \mathbf{R}_{n\kappa(i)} \prec \alpha \left(\mathbf{I} + \sum_{n \neq i} \rho_n^2 \gamma_n \mathbf{R}_{n\kappa(i)}\right)$$

which contradicts that $\boldsymbol{\rho}^2$ is in the set. Thus, Eq. (18.35) defines a single unique point that is the optimum solution to both Eqs. (18.33) and (18.34). The result is illustrated in Fig. 18.11 for a two-user example.

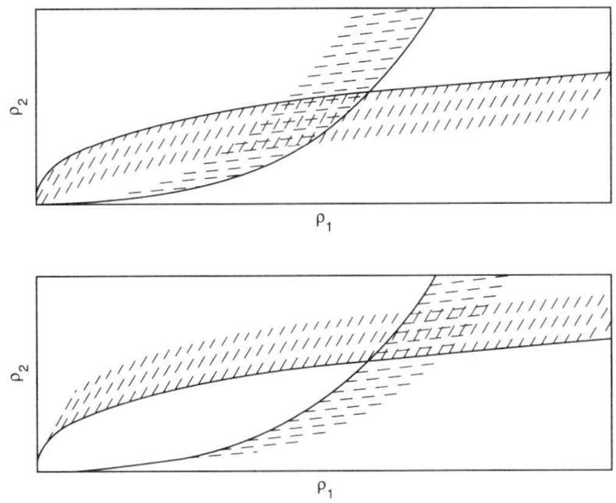

FIGURE 18.11 Example of the feasible regions of the two problems [Eq. (18.33), top, and Eq. (18.34), bottom].

LEMMA 5. *If* $\mathbf{W}_i$ *provide one optimum solution to Eq. (18.24), then the following algorithm gives a solution to Eq. (18.10) with the same cost function.*

- *Take some* $\mathbf{u}_i \in \text{span}\,[\mathbf{W}_i]$
- *Form*

$$\mathbf{y} = \left[\gamma_1\sigma_1^2, \ldots, \gamma_d\sigma_d^2\right]^T$$

$$\left[\mathbf{F}\right]_{i,n} = \begin{cases} \mathbf{u}_i^\star \mathbf{R}_i \mathbf{u}_i & i = n \\ -\gamma_i \mathbf{u}_n^\star \mathbf{R}_{i\kappa(n)} \mathbf{u}_n & i \neq n \end{cases}$$

- *Let* $\mathbf{p} = \mathbf{F}^{-1}\boldsymbol{\eta}$
- *Let* $\mathbf{w}_i = \sqrt{[\mathbf{p}]_i}\,\mathbf{u}_i$

PROOF. Any $\mathbf{u}_i$ in the null space of $\mathbf{Z}_i = \mathbf{I} - \rho_i \mathbf{R}_{i\kappa(i)} + \Sigma_{n\neq i}\rho_n\gamma_n \mathbf{R}_{n\kappa(i)}$ will be an optimum solution of Eq. (18.32). The complementary conditions[65,66] between the relaxation and its dual show that at the optimum, $\text{Tr}[\mathbf{W}_i\mathbf{Z}_i] = 0$ (i.e., $\mathbf{W}_i\mathbf{Z}_i = 0$ because both matrices are positive semidefinite). Thus, if our semidefinite problem gives a high rank solution, any $\mathbf{u}_i \in \text{span}\,[\mathbf{W}_i]$ will solve the virtual uplink problem and the corresponding scaling factors p_i are calculated from the system of linear equations.

Now we are in the position to prove Lemma 1 and 2.

PROOF OF LEMMA 1. According to Lemma 3, Eq. (18.10) is algebraicly equivalent to Eq. (18.17) if the cost function is changed to min $\Sigma\gamma_n\sigma_n^2\rho_n$, but from Lemma 4 it follows that the solution does not depend on the coefficients of p_i in the cost function.

PROOF OF LEMMA 2. The idea is to compare the optimum value of the following three problems:

1. The original problem Eq. (18.10)
2. The semidefinite relaxation Eq. (18.24)
3. The Lagrange dual of the semidefinite relaxation

We have already seen that the optimum value in problem 2 is lower than that in problem 1. A Lagrange dual does always provide a lower bound on the cost function,[65,78] so the optimum of problem 3 is lower than that in problem 2. Thus, if we can prove that problems 1 and 3 have the same optimum cost, all three problems must be equivalent.

The Lagrange dual of Eq. (18.24) is given by

$$\begin{aligned}&\max_{\rho_i}\sum_{n=1}^{d}\gamma_n\sigma_n^2\rho_n\\&\text{s.t. } \mathbf{I}=\rho_i\mathbf{R}_{i,\kappa(i)}+\sum_{n\neq i}\rho_n\gamma_n\mathbf{R}_{n,\kappa(i)}\succeq 0\end{aligned} \tag{18.36}$$

According to Lemma 4, the optimum of Eq. (18.36) is the same as the optimum of

$$\begin{aligned}&\min_{\mathbf{u}_i,\rho_i}\sum_{n=1}^{d}\gamma_n\sigma_n^2\rho_n\\&\text{s.t. } \frac{\mathbf{u}_i^*\rho_i\mathbf{R}_{i,\kappa(i)}\mathbf{u}_i}{\mathbf{u}_i^*\left(\sum_{n\neq i}\rho_n\gamma_n\mathbf{R}_{n,\kappa(i)}+\mathbf{I}\right)\mathbf{u}_i}\geq 1\\&\left\|\mathbf{u}_i\right\|^2=1\end{aligned}$$

which, according to Lemma 3, is algebraicly equivalent to the original problem in Eq. (18.10).

References

1. Godara, L.C., Applications of antenna arrays to mobile communications. I. Performance improvement, feasibility, and system considerations, *Proc. IEEE*, 85, 1031, July 1997.
2. Godara, L.C., Applications of antenna arrays to mobile communications. II. Beam-forming and direction-of-arrival considerations, *Proc. IEEE*, 85, 1195, August 1997.
3. Paulraj, A.J. and Papadias, C.B., Space-time processing for wireless communications, *IEEE SP Mag.*, 14, 49, November 1997.
4. Liberti, Jr., J.C. and Rappaport, T.S., *Smart Antennas for Wireless Communications: IS-95 and Third Generation CDMA Applications*, Prentice-Hall, Englewood Cliffs, NJ, 1999.
5. Tarokh, V., Seshadri, N., and Calderbank, A.R., Space-time codes for high data rate wireless communications: Performance criterion and code construction, *IEEE Trans. IT*, 44, 744, March 1998.
6. Raleigh, G.G. and Cioffi, J.M., Spatio-temporal coding for wireless communication, *IEEE Trans. Commun.*, 46, 357, March 1998.
7. Narula, A., Trott, M.D., and Wornell, G.W., Performance limits of coded diversity methods of transmitter antenna arrays, *IEEE Trans. IT*, 45, 2418, November 1999.
8. Foschini, G.J., layered space-time architecture for wireless communication in a fading environment when using multi-element antennas, *Bell Labs Tech. J.*, 1, 41, Autumn 1996.
9. Godard, D.N., Self-recovering equalization and carrier tracking in two dimensional data communication systems, *IEEE Trans. Commun.*, 28, 1867, November 1980.
10. Johnson, Jr., C.R., Schniter, P., Endres, T.J., Behm, J.D., Brown, D.R., and Casas, R.A., Blind equalization using the constant modulus criterion: A review, *Proc. IEEE*, 86, 1927, October 1998.
11. Krim, H. and Vibert, M., Two decades of array signal processing research: The parametric approach, *IEEE SP Mag.*, 13, 67, July 1996.
12. Jeng, S., Lin, H., Xu, G., and Vogel, W., Measurements of spatial signature of an antenna array, in *Proc. PIMRC, IEEE*, 1995, vol. 2, 669–672.

13. Wigard, J., Mogensen, P.E., Frederiksen, F., and Nørklit, O., Evaluation of optimum diversity combining in DECT, in *Proc. PIMRC, IEEE*, 1995, vol. 2, 507–511.
14. Yang, W., Kavak, A., Kim, S.Y., Xu, G., and Hansen, L., Evaluation of spatially selective receiving/transmission techniques for a smart antenna system operating at 1.8 GHz in non-stationary scenarios, in *Proc. VTC, IEEE*, May 1999, vol. 1, 862–866.
15. Zetterberg, P., Mobile cellular communications with base station antenna arrays: Spectrum efficiency, algorithms and propagation models, Ph.D. thesis, Royal Institute of Technology, Stockholm, Sweden, June 1997.
16. Farsakh, C., Pensel, K., and Nossek, J.A., An SDMA system based on spatial covariances, in *Proc. VTC, IEEE*, 1998, vol. 2, 1171–1175.
17. Giancola, D., Margherita, F., Parolari, S., Picciriello, A., and Spagnolini, U., Analysis of the spectral efficiency of a fully adaptive antenna array system in GSM/DCS networks, in *Proc. VTC, IEEE*, May 1999, vol. 1, 812–815.
18. Raleigh, G.G., Diggavi, S.N., Jones, V.K., and Paulraj, A., A blind adaptive transmit antenna algorithm for wireless communication, in *Proc. ICC, IEEE*, 1995, vol. 3, 1494–1499.
19. Hochwald, B.M. and Marzetta, T.L., Adapting a downlink array from uplink measurements, in *Proc. SPIE Aerosense Conf.*, 1999.
20. Gerlach, D. and Paulraj, A., Adaptive transmitting antenna arrays with feedback, *IEEE SP Lett.*, 1, 150, October 1994.
21. Gerlach, D. and Paulraj, A., Base station transmitting antenna arrays for multipath environments, *Signal Process.*, 54, 59, October 1996.
22. Zetterberg, P. and Ottersten, B., The spectrum efficiency of a base station antenna array for spatially selective transmission, *IEEE Trans. VT*, 44, 651, August 1995.
23. Bengtsson, M., Antenna array signal processing for high rank data models, Ph.D. thesis, Royal Institute of Technology, Stockholm, Sweden, December 1999, TRITA–S3–SB–9938.
24. Schmidt, R.O., A signal subspace approach to multiple emitter location and spectral estimation, Ph.D. thesis, Stanford University, Stanford, CA, November 1981.
25. Viberg, M. and Ottersten, B., Sensor array processing based on subspace fitting, *IEEE Trans. SP*, 39, 1110, May 1991.
26. Stoica, P. and Sharman, K.C., Maximum likelihood methods for direction-of-arrival estimation, *IEEE Trans. ASSP*, 38, 1132, July 1990.
27. Xu, G. and Liu, H., An effective transmission beamforming scheme for frequency-division-duplex digital wireless communication systems, in *Proc. IEEE ICASSP*, 1995, vol. 3, 1729–1732.
28. Asté, T., Forster, P., Féty, L., and Mayrargue, S., Downlink beamforming for cellular mobile communications (GSM system), *Ann. Télécommun.*, 53, 435, November–December 1998.
29. Li, J., Halder, B., Stoica, P., and Viberg, M., Computationally efficient angle estimation for signals with known waveforms, *IEEE Trans. SP*, 43, 2154, September 1995.
30. Hugl, K., Laurila, J., and Bonek, E., Downlink beamforming for frequency division duplex systems, in *Proc. GLOBECOM*, 1999, vol. 4, 2097–2101.
31. Capon, J., High resolution frequency wave number spectrum analysis, *Proc. IEEE*, 57, 1408, August 1969.
32. Stoica, P. and Moses, R., *Introduction to Spectral Analysis*, Prentice-Hall, Englewood Cliffs, NJ, 1997.
33. Bell, K.L., Ephraim, Y., and Trees, H.L.V., A bayesian approach to robust adaptive beamforming, *IEEE Trans. SP*, 48, 386, February 2000.
34. Goldberg, J. and Fonollosa, J.R., Downlink beamforming for cellular mobile communications, in *Proc. VTC'97*, 1997, vol. 2, 632–636.
35. Goldberg, J. and Fonollosa, J.R., Downlink beamforming for spatially distributed sources in cellular mobile communications, *Signal Process.*, 65, 181, March 1998.
36. Asté, T., Forster, P., Féty, L., and Mayrargu, S., Downlink beamforming avoiding DOA estimation for cellular mobile communications, in *Proc. IEEE ICASSP 98*, 1998, vol. 6, 3313–3316.

37. Friedlander, B., The root-MUSIC algorithm for direction finding with interpolated arrays, *Signal Process.*, 30, 15, January 1993.
38. Cox, H., Zeskind, R.M., and Owen, M.M., Robust adaptive beamforming, *IEEE Trans. ASSP*, 35, 1365, October 1987.
39. Riba, J., Goldberg, J., and Vàzquez, G., Robust beamforming for interference rejection in mobile communications, *IEEE Trans. SP*, 45, 271, January 1997.
40. Zatman, M., Comments on Theory and application of covariance matrix tapers for robust adaptive beamforming, *IEEE Trans. SP*, 48, 1796, June 2000.
41. Guerci, J.R., Theory and application of covariance matrix tapers for robust adaptive beamforming, *IEEE Trans. SP*, 47, 977, April 1999.
42. Bengtsson, M. and Ottersten, B., Optimal downlink beamforming using semidefinite optimization, in *Proc. 37th Annual Allerton Conf. Communication, Control, and Computing*, September 1999, 987–996.
43. Rashid-Farrokhi, F., Liu, K.R., and Tassiulas, L., Transmit beamforming and power control for cellular wireless systems, *IEEE J. Select. Areas Commun.*, 16, 1437, October 1998.
44. Visotsky, E. and Madhow, U., Optimum beamforming using transmit antenna arrays, in *Proc. VTC'99, IEEE*, Houston, TX, May 1999, vol. 1, 851–856.
45. Montalbano, G. and Slock, D.T., Spatio-temporal array processing for matched filter bound optimization in SDMA downlink transmission, in *URSI Int. Symp. Signals, Systems and Electronics*, 1998, 416–421.
46. Yang, W. and Xu, G., The optimal power assignment for smart antenna downlink weighting vector design, in *Proc. VTC*, May 1998, vol. 1, 485–488.
47. Mecklai, H.K. and Blum, R.S., Transmit antenna diversity for wireless communications, in *Proc. ICC 95*, 1995, vol. 3, 1500–1504.
48. Golub, G.H. and Loan, C.F.V., *Matrix Computations*, 3rd ed., Johns Hopkins University Press, London, 1996.
49. Faraskh, C. and Nossek, J.A., Spatial covariance based downlink beamforming in an SDMA mobil radio system, *IEEE Trans. Commun.*, 46, 1497, November 1998.
50. Zetterberg, P., A comparison of two systems for downlink communication with base station antenna arrays, *IEEE Trans. VT*, 48, 1356, September 1999.
51. Alexiou, A. and Yan, R.H., Downlink capacity enhancement by employing SDMA in GSM, in *Proc. IEEE Sensor Array and Multichannel Signal Process. Workshop*, 2000, 413–417.
52. Monzingo, R.A. and Miller, T.W., *Introduction to Adaptive Arrays*, John Wiley & Sons, New York, 1980.
53. Veen, B.D.V. and Buckley, K.M., Beamforming: A versatile approach to spatial filtering, *IEEE ASSP Mag.*, 4–24, April 1988.
54. Bengtsson, M., Optimal transmission using smart antennas, in *Proc. IST Mobile Commun. Summit*, Ireland, October 2000, 359–364.
55. Zander, J., Distributed cochannel interference control in cellular radio systems, *IEEE Trans. VT*, 41, 305, August 1992.
56. Foschini, G.J. and Miljanic, Z., A simple distributed autonomous power control algorithm and its convergence, *IEEE Trans. VT*, 42, 641, November 1993.
57. Yates, R.D., A framework for uplink power control in cellular radio systems, *IEEE J. Select. Areas Commun.*, 13, 1341, September 1995.
58. Herdtner, J.D. and Chong, E.K.P., Analysis of a class of distributed asynchronous power control algorithms for cellular wireless systems, *IEEE J. Select. Areas Commun.*, 18, 436, March 2000.
59. Proakis, J.G., *Digital Communications*, 3rd ed., McGraw-Hill, New York, 1995.
60. Forster, P., Féty, L., and Bot, M.L., Spatio-temporal filters for downlink processing in FDD systems, in *Proc. IEEE ICASSP*, 2000, vol. 5, 3017–3020.
61. Forster, P., Féty, L., and Bot, M.L., Downlink space-time processing for FDD systems, in *Proc. EUSIPCO, EURASIP*, 2000, vol. 3, 1425–1428.

62. Schubert, M. and Boche, H., Downlink beamforming for TD/CDMA multipath channels, in *Proc. IEEE ICASSP,* 2000, vol. 5, 2993–2996.
63. Rashid-Farrokhi, F., Liu, K.R., and Tassiulas, L., Transmit beamforming for cellular wireless communications, in *Proc. Conf. Info. Sci. Sys.*, (CISS-97), Johns Hopkins, Baltimore, MD, March 1997, 92–97.
64. Rashid-Farrokhi, F., Tassiulas, L., and Liu, K.R., Joint optimal power control and beamforming in wireless networks using antenna arrays, *IEEE Trans. Commun.*, 46, 1313, October 1998.
65. Boyd, S. and Vandenberghe, L., Introduction to convex optimization with engineering applications, Course Notes, 1999. http://www.stanford.edu/class/ee364/.
66. Vandenberghe, L. and Boyd, S., Semidefinite programming, *SIAM Rev.*, 38, 49, March, 1996.
67. Wolkowicz, H., Saigal, R., and Vandenberghe, L., editors, *Handbook of Semidefinite Programming. Theory, Algorithms and Applications*, Kluwer Academic Publishers, Dordrecht, the Netherlands, 2000.
68. Nesterov, Y. and Nemirovskii, A., *Interior-Point Polynomial Algorithms in Convex Programming*, SIAM Publications, Philadelphia, PA, 1994.
69. Rockafellar, R.T., *Convex Analysis*, Princeton University Press, Princeton, N.J., 1997, reprint of the 1970 original.
70. Sturn, J.F., *Using SeDuMi 1.0x, a MATLAB Toolbox for Optimization over Symmetric Cones*, Department of Quantitative Economics, Maastricht University, the Netherlands, http://www.unimaas.nl/~sturm/software/sedumi.html, 1999.
71. Boyd, S.P., editor, *Linear Matrix Inequalities in System and Control Theory*, SIAM, Philadelphia, PA, 1994.
72. Davidson, T.N., Luo, Z.Q., and Wong, K.M., Design of orthogonal pulse shapes for communications via semidefinite programming, *IEEE Trans. SP,* 48, 1433, May 2000.
73. Lobo, M.S., Vandenberghe, L., Boyd, S., and Lebret, H., Applications of second-order cone programming, *Linear Algebra Appl.*, 284, 193, 1998.
74. Er, M. and Cantoni, A., An alternative formulation for an optimum beamformer with robustness capability, *IEEE Proc., Pt. F,* 132, 447, October 1985.
75. Bresler, Y., Reddy, V.U., and Kailath, T., Optimum beamforming for coherent signal and interferences, *IEE Trans. ASSP,* 36, 833, June 1988.
76. Hugl, K., Laurila, J., and Bonek, E., Downlink performance of adaptive antennas with null broadening, in *Proc. VTC'99, IEEE*, Houston, TX, May 1999, vol. 1, 872–876.
77. Gantmacher, F.R., *The Theory of Matrices*, vol. II, Chelsea Publishing Company, New York, 1959.
78. Bertsekas, D.P., *Nonlinear Programming*, Athena Scientific, Belmont, MA, 1995.

19

Spatial Diversity for Wireless Communications

Ramakrishna Janaswamy
Naval Postgraduate School

19.1 Introduction

The ill-effects of multipath fading* in wireless systems can be reduced by employing spatial diversity, where a multielement antenna at the receiver is employed. When fading is present, a higher than otherwise average carrier power is needed to perform at a given bit-error rate (BER). In an array antenna, the signals received by the various elements can be weighted appropriately to result in a combined signal that fluctuates less rapidly than the individual signals. This array then requires less power to achieve a given BER than the case where only one element is used. For spatial diversity to work effectively, the signals, received by the various antenna branches must be sufficiently decorrelated so that if one of the elements is in deep fade, there is still hope of recovering the signal by receiving it at other antenna terminals. This can always be achieved in practice by choosing the element spacing appropriately. The spacing required between antenna elements to maintain certain decorrelation depends on mutual coupling and the disposition of scatterers causing the multipath transmission. For instance, in the absence of mutual coupling, spacings of about $\lambda/2$ should be sufficient at a mobile terminal that is usually surrounded uniformly by scatterers. On the other hand, spacings of the order of 10λ or more may be necessary to maintain the same decorrelation value at an elevated base station. A second condition that is required for diversity benefits is that the mean signal strengths of the diversity paths should approximately be the same. An in-depth treatment of linear combining techniques is given in Reference [13], and to a lesser extent, in Reference [6]. Some developments of antenna diversity with applications to mobile communications are treated in Reference [14].

*Much of the material found in this chapter is extracted from Chapter 7 of Reference 7.

0-8493-0124-6/02/$0.00+$1.50

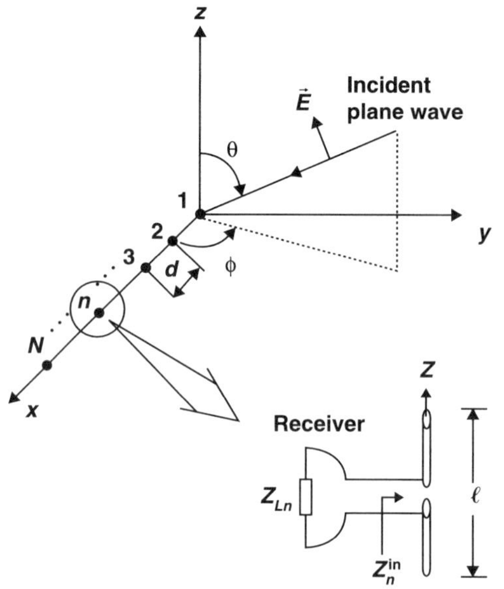

FIGURE 19.1 An array comprised of N elements along x-axis.

In this chapter, the basic principles of spatial diversity combining are treated. By assuming that the elements are spaced appropriately so that perfect decorrelation among various branches exists — a branch is assumed to mean an antenna — the improvement that can be accomplished by using diversity arrays with various combing techniques is first demonstrated. The effect of branch correlation caused by either mutual coupling between elements or angular despread of the incoming ways is shown later. The BER performance of basic modulation schemes with a diversity array is also shown. An attractive alternative to spatial diversity where space is a limiting factor is polarization diversity where two antennas with orthogonal polarization are almost colocated. In a highly scattering environment, the signal is received in almost all polarizations irrespective of the transmitter polarization. Although the decorrelation between the signals received in two orthogonally polarized states is not as good as in well-separated spatially diverse antennas, the loss in diversity improvement at the 1% probability level was shown to be less than 1 dB at 842 MHz [4]. Polarization diversity is not treated in this chapter.

19.2 General Receive Array Theory

Let us consider an array comprised of N equispaced elements placed along the x-axis as shown in Fig. 19.1. The position of the nth element is $x_n = (n - 1)d$. Although the elements of the array could be arbitrary, we consider the case where the elements are identical dipole antennas, with the dipole axes along the z-axis. A plane wave of polarization $\hat{\theta}$ and radian frequency ω is assumed incident from angles (θ, ϕ):

$$\mathbf{E}^{\text{inc}} = \hat{\theta} E_o e^{jk_o\left(x\sin\theta\cos\phi + y\sin\theta\sin\phi + z\cos\theta\right)} \tag{19.1}$$

where $k_o = \omega\sqrt{\mu_0\varepsilon_0}$ is the wave number in free space. The nth element is assumed to be terminated in a center-fed load impedance Z_{Ln}. In the absence of mutual coupling, the open-circuited voltage induced in the isolated nth dipole can be written as [3]

$$V_n^{\text{oc}} = \mathbf{h} \cdot \mathbf{E}^{\text{inc}}\left(x_n, 0, 0\right) \tag{19.2}$$

where $\mathbf{h}$ is the vector effective length of the dipole given by

$$\mathbf{h} = \frac{\mathbf{E}^{\text{rad}}}{jk_o \eta_o I_{\text{in}}} 4\pi r e^{jk_o r} \tag{19.3}$$

$\eta_o \approx 120\pi$ is the free-space intrinsic impedance, I_{in} is the input current to the antenna and $\mathbf{E}^{\text{rad}}$ is the field radiated by the antenna, both in the transmit mode, and r is the distance from the center of the dipole to the observation point. For a dipole of length ℓ carrying a sinusoidal current distribution, $I(z) = I_o \sin[k_o(\ell/2 - |z|)]$, the vector effective length can be obtained as

$$\mathbf{h} = \hat{\theta}\frac{\lambda}{\pi} \frac{F(\theta,\phi)}{\pi \sin k_o \ell} \tag{19.4}$$

where $F(\theta, \phi)$ is the radiation pattern of the dipole antenna:

$$F(\theta,\phi) = \frac{\cos\left(\frac{k_o \ell}{2}\cos\theta\right) - \cos\left(\frac{k_o \ell}{2}\right)}{\sin\theta} \tag{19.5}$$

The azimuth plane pattern ($\theta = \pi/2$) of a z-directed dipole is uniform. Inserting the preceding and Eqs. (19.1) into (19.2), we get

$$\begin{aligned} V_n^{\text{oc}} &= \frac{\lambda}{\pi} E_o \frac{F(\theta,\phi)}{\sin\frac{k_o \ell}{2}} e^{j(n-1)k_o d \sin\theta\cos\phi} \\ &= \frac{\lambda}{\pi} E_o \frac{F(\theta = \pi/2,\phi)}{\sin\frac{k_o \ell}{2}} e^{j(n-1)k_o d \cos\phi} \end{aligned} \tag{19.6}$$

for incidence in the xy-plane.

For simplicity, let us assume that the plane wave is incident from the xy-plane so that Eq. (19.6) applies. The open-circuited voltage drives a series equivalent network system composed of the load impedance Z_{Ln} and the antenna input impedance Z_n^{in}, Fig. 19.2. The voltage signal presented to the load is

$$\begin{aligned} V_n &= \frac{Z_{Ln}}{Z_{Ln} + Z_n^{\text{in}}} V_n^{\text{oc}} \\ &\triangleq V_{on} F(\pi/2,\phi) e^{j(n-1)k_o d\cos\phi} \end{aligned} \tag{19.7}$$

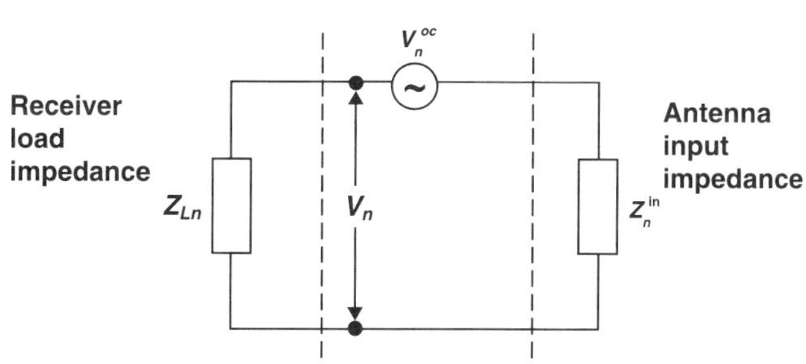

FIGURE 19.2 Antenna open-circuited voltage driving a series system composed of input impedance and receiver.

Let us now assume that the input impedance of the dipoles are all the same and equal to Z^{in}, the load impedances Z_{Ln} are all equal to Z_L, and the corresponding V_{on}'s are all equal to V_o. Denoting the vector $[V_1, V_2, \ldots, V_N]^T$ by **v**, and the vector $[1, e^{jk_o d\cos\phi}, \ldots, e^{j(N-1)k_o d\cos\phi}]^T$ by ψ, we can express the received signals compactly as

$$\mathbf{v} = V_o F(\pi/2, \phi)\psi \tag{19.8}$$

where the superscript T denoting the matrix transpose operator. The vector ψ is known as the *steering vector* of the array and is seen to depend only on the properties of the incident wave and the geometry of the array. The other factor $V_o F(\pi/2, \phi)$ contains information about the nature of the element and is known as the element factor. If the elements are changed from dipoles to loops, for example, the element factor gets changed; the steering vector of the array remains the same. The voltage vector presented to the load is different for different angles ϕ of the incident wave. Hence, the array has a directional response to the incoming plane waves. In the study of array antennas, one often assumes the elements to be isotropic (i.e., $F(\theta, \phi) = 1$) and focuses on the study of the array factor. In the case of isotropic elements,

$$\mathbf{v} = V_o \psi \tag{19.9}$$

The more general case is described by Eq. (19.7).

19.3 Combining Techniques

Basically, the three ways of combining the signals are

1. Selection combining
2. Maximal ratio combining
3. Equal-gain combining

We briefly look at each of these combining techniques and show the diversity improvement by considering the statistics of the combined signal. Although the treatment is valid for both uplink (mobile-to-base station) or downlink (base station-to-mobile), we specifically consider the case of uplink only where the mobile transmits and the base station receives.

The signals received by the elements of the receive antenna are combined linearly as shown in Fig. 19.3. The weights for combining are chosen to be $w_1^*, w_2^*, \ldots, w_N^*$, where * denotes complex conjugation. A

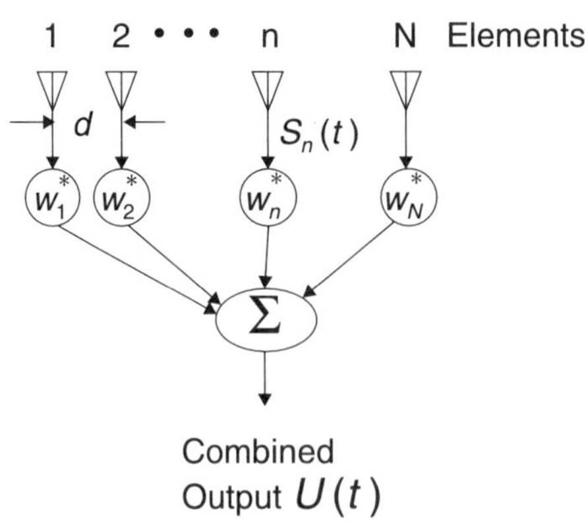

FIGURE 19.3 Diversity array consisting of N elements.

complex conjugate is included in the definition of the weights only for mathematical convenience so that the combined output appears in a convenient form as $\mathbf{w}^{\dagger}\mathbf{s}$.

It is assumed that the spacing is chosen so that signals received are sufficiently decorrelated. The signal received at one antenna element constitutes one diversity branch. Because of multipath and movement of the mobile, the signal received at each element is not a constant, but fluctuates at the fading rate. The fading rate depends on the speed of the mobile and frequency of the radio signal and is approximately equal to the maximum Doppler shift [7]. The maximum Doppler shift f_{dM} corresponding to a radio frequency of f_G GHz for a mobile moving at a speed v mi/h is

$$f_{dM} = 1.4815 f_G v \tag{19.10}$$

For a mobile moving at 100 mi/h, the maximum Doppler shift is 133.33 Hz at 900 MHz and 266.67 Hz at 1800 MHz. Coherence time of the radio channel is approximately equal to the reciprocal of the fading rate. In the preceding example, the coherence time is 7.5 ms at 900 MHz and 3.75 ms at 1800 MHz. The multipath medium between the transmitting antenna and the receiving antenna can be thought of as a linear time-varying filter and each of the N branches is characterized by an equivalent low-pass transfer function $T_n(f; t)$, $n = 1, \ldots, N$, with the t argument signifying the time variations of the radio channel responses and the f argument signifying frequency selective nature of the channel. Assuming frequency nonselective or flat fading on each diversity branch allows us to express this transfer function as $T_n(f; t) = g_n(t)$, where each $g_n(t)$ is a zero-mean complex Gaussian process. Thus, the signals received in the diversity branches may be represented in the form

$$s_n(t) \stackrel{\Delta}{=} \Re e\left[r_n(t) e^{j2\pi f_c t}\right] = \Re e\left[g_n(t) u(t) e^{j2\pi f_c t}\right] \tag{19.11}$$

where f_c is the nominal carrier frequency, $u(t)$ is the complex envelope of the transmitted signal, and $r_n(t)$ is the complex envelope of the received signal. The assumption of flat fading is valid for narrowband transmission where the delays experienced by the various multipath components are all much less than the symbol duration. It is seen that flat fading results in a multiplicative propagation factor for the transmitted signal. This is characteristic of Rayleigh faded narrowband signals. Furthermore, it is assumed that the symbol period T_s is much less than the reciprocal of the fading rate so that the fading pattern does not change over the symbol duration. For convenience we normalize the transmitted signal such that mean power is constant, namely,

$$E_T\left(\left|u(t)\right|^2\right) \stackrel{\Delta}{=} \frac{1}{T_s} \int_{-T_s/2}^{T_s/2} \left|u(t)\right|^2 dt = 1 \tag{19.12}$$

where E_T is the time-expectation or time-averaging operator in the sense of the preceding integral. The complex envelope of additive noise in the nth receiver branch is assumed to be $\nu_n(t)$ with the mean intensity over each symbol period or over any longer interval given by

$$\frac{1}{2} E_T\left(\left|\nu_n(t)\right|^2\right) = \frac{1}{2}\left\langle \left|\nu_n(t)\right|^2 \right\rangle = P_n = P_{\text{no}} \tag{19.13}$$

where the angle brackets denote statistical means and it has been assumed that all P_n are equal to P_{no}. Implicit in the preceding equation is the assumption of ergodicity of the noise process. We define the instantaneous, γ_n, and the average, Γ_n, carrier-to-noise ratios (CNRs) for the nth branch as

$$\gamma_n \stackrel{\Delta}{=} \frac{\textit{local mean} \text{ carrier power per branch}}{\text{mean noise power per branch}} \tag{19.14}$$

$$= \frac{E_T\left(\frac{1}{2}\left|r_n(t)\right|^2\right)}{P_{\text{no}}} \tag{19.15}$$

$$\approx \frac{\left|g_n(t)\right|^2}{2P_{\text{no}}} \quad \left[\text{since } g_n(t) \text{ is approximately constant over } T_s\right] \tag{19.16}$$

$$\Gamma_n \stackrel{\Delta}{=} \frac{\textit{statistical mean} \text{ carrier power per branch}}{\text{mean noise power per branch}} \tag{19.17}$$

$$= \langle \gamma_n \rangle \stackrel{\Delta}{=} \frac{P_{\text{o}}}{P_{\text{no}}} \tag{19.18}$$

where P_o is the statistical average of $|g_n|^2/2$ over the fading interval (coherence time). Note that as a result of the normalization Eq. (19.12), we have used the fact that the local average of the received signal envelope across each branch is

$$E_T\left(\left|r_n(t)\right|^2\right) = E_T\left(\left|g_n(t)\right|^2 \left|u(t)\right|^2\right) \approx \left|g_n(t)\right|^2 \tag{19.19}$$

Thus, $|(g_n t)|$ may be regarded as the local mean envelope of the received signal. By assuming Rayleigh fading, the probability density function (**pdf**) for the received signal envelope is

$$p\left(|g_n|\right) = \frac{|g_n|}{P_{\text{o}}} e^{-|g_n|^2/2P_{\text{o}}} \tag{19.20}$$

and that of γ_n is

$$p\left(\gamma_n\right) = \frac{1}{\Gamma_n} e^{-\frac{\gamma_n}{\Gamma_n}} \tag{19.21}$$

19.4 Selection Combining

The ideal selection combiner is defined as one where the branch with the highest CNR is chosen for the system output. As far as the statistics of the system output are concerned, it is immaterial where the selection is done at the intermediate frequency (IF) (predetection) or at postdetection. Because it is assumed that all the receivers have the same noise level, the highest value of γ_n also corresponds to the highest value of $|g_n|^2/2 + P_n$. Accordingly, the receiver selected is that which also has the highest total *instantaneous* power. Mathematically, we state the weight condition as

$$w_n = \begin{cases} 1, & n = M, \text{ where } M \middle| \gamma_M(t) = \max_n\left[\gamma_n(t)\right] \\ 0, & \text{otherwise} \end{cases} \tag{19.22}$$

In selection combining the unused branches at any instant do not contribute to the output signal. These unused terminals may then be left open to minimize the effects of antenna mutual coupling. In terms of implementation, selection combiner would need no more than a comparator and a fast switch. Most modern cellular systems make use of dual diversity ($N = 2$) with selection combining. A practical alternative to picking the branch with the highest CNR is to switch to the first branch that remains above a certain threshold relative to Γ_n. This is known as *switched combining*. Of course the branch picked is not necessarily the one with the highest CNR.

The probability that the CNR in one branch is less than or equal to specified value γ_s is

$$\begin{aligned}\text{Prob}\left(\gamma_n \le \gamma_s\right) &= \int_0^{\gamma_s} p\left(\gamma_n\right) d\gamma_n \\ &= 1 - e^{\frac{\gamma_s}{\Gamma_n}}\end{aligned} \tag{19.23}$$

Over the short-term fading, one can now describe all those events in which the selector output CNR γ is less than or equal to γ_s as those events in which each of the branch CNRs is simultaneously below γ_s. Because fading is assumed independent in each of the N branches, this would be equal to the probability that the CNR in *all* branches is simultaneously less than or equal to γ_s, i.e.,

$$\begin{aligned}\text{Prob}\left(\gamma \le \gamma_s\right) = \text{Prob}\left(\gamma_1, \ldots, \gamma_N \le \gamma_s\right) &= \prod_{n=1}^{N} \text{Prob}\left(\gamma_n \le \gamma_s\right) \\ &= \prod_{n=1}^{N} \left[1 - e^{-\frac{\gamma_s}{\Gamma_N}}\right] = \left[1 - e^{-\frac{\gamma_s}{\Gamma_N}}\right] \\ &\triangleq P_N^{sc}\left(\gamma_s\right)\end{aligned} \tag{19.24}$$

where it has been assumed that all the diversity branches have equal mean CNR over the short-term fading (i.e., $\Gamma_n = \Gamma$). Equation (19.24) is the cumulative distribution function (**cdf**) of the variation in γ. The subscript N in Eq. (19.24) stands for the number of elements and the superscript sc stands for selection combining. The assumption of equal mean CNRs at the branches is very reasonable in view of the fact that the array length, in practice, is very small compared with the distance from the transmitter so that all elements experience approximately the same path loss and shadowing loss.

The probability density function **pdf** for γ is obtained by differentiating the **cdf** with respect to γ_s. It is

$$p_N^{sc}\left(\gamma\right) = \frac{N}{\Gamma} e^{-\frac{\gamma}{\Gamma}} \left[1 - e^{-\frac{\gamma}{\Gamma}}\right]^{N-1} \tag{19.25}$$

The mean CNR of the selected signal is

$$\left\langle \gamma \right\rangle = \int_0^{\infty} \gamma \, p_N^{sc}\left(\gamma\right) d\gamma = \Gamma \sum_{n=1}^{N} \frac{1}{n} \sim \left(C + \ln N + \frac{1}{2N}\right)\Gamma, \text{ for large } N\left(N \ge 3\right) \tag{19.26}$$

where $\mathbf{C} = 0.577215\ldots$ is the Euler's constant. Hence the mean CNR improves logarithmically with the number of branches in the case of selection diversity. Figure 19.4 shows a plot of the **cdf**, $P_N^{sc}\ (\gamma_s)$, of selection diversity with N as a parameter. To interpret the results, let us consider the point $10 \log (\gamma_s/\Gamma) = -10$ dB on the horizontal axis. For $N = 1$ there is a 10% probability ($P_N^{sc} = 0.1$) that the instantaneous

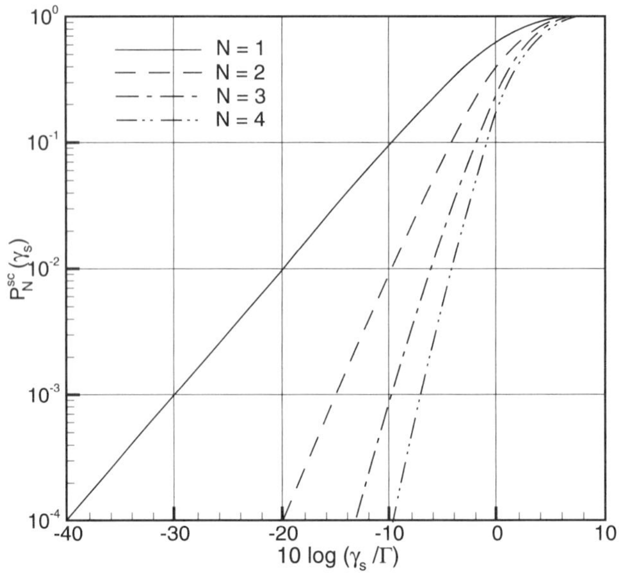

FIGURE 19.4 Cumulative distribution function of γ_s vs. γ_s/Γ for selection diversity with N as a parameter.

output CNR falls 10 dB below the mean branch CNR Γ. For $N = 4$ with selection combining, the probability is reduced to 0.01%. Thus, combining has significantly reduced the depth of fades. It is also seen that at 99% reliability (i.e., $1 - P_N^{sc}(\gamma_s) = 0.99$), selection diversity provides a 10 dB (–10 – (–20)) savings in power with two branches and 16-dB savings with four branches.

On knowing $p(\gamma)$, it is straightforward to derive the error rate performance for digital signaling. For instance, BER of ideal coherent BPSK assuming constant envelope γ is [12]

$$P_{\text{BPSK}} = \frac{1}{2}\text{erfc}\left(\sqrt{\gamma}\right) \tag{19.27}$$

where γ is the CNR per information bit. In the presence of fading, the envelope fluctuates and we view the preceding formula as a conditional probability, where the condition is that γ is fixed. To obtain error probabilities when γ is random, we must average P_{BPSK} over the **pdf** of γ as given in Eq. (19.25). One then obtains the following result:

$$\begin{aligned} P_{\text{BPSK}}^{sc} &= \int_0^\infty P_{\text{BPSK}}(\gamma)\, p_N^{sc}(\gamma)\, d\gamma \\ &= \frac{N}{2}\int_0^\infty e^{-x}\left[1-e^{-x}\right]^{N-1}\text{erfc}\left(\sqrt{\Gamma x}\right)dx \\ &= \frac{1}{2\sqrt{\pi}}\int_0^\infty \frac{e^{-\gamma}\left[1-e^{-\frac{\gamma}{\Gamma}}\right]^N}{\sqrt{\gamma}}\,d\gamma \end{aligned} \tag{19.28}$$

The preceding equation may be evaluated in a closed form for $N = 1$ and expressed as a series for $N > 1$. Experimental verification of the performance improvement of selection diversity at 920 MHz with two vehicle mounted monopole antennas is demonstrated in Reference 9.

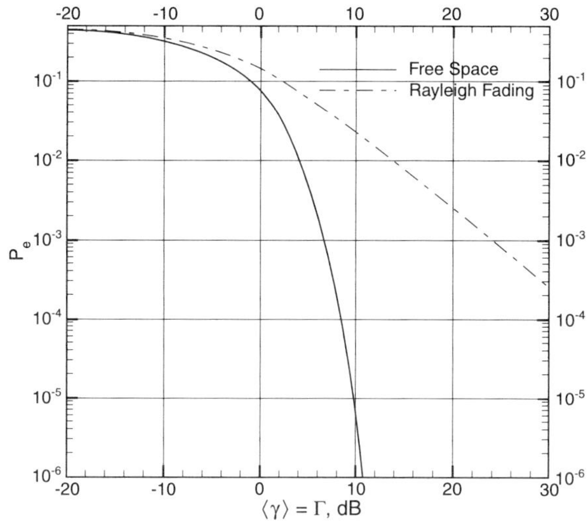

FIGURE 19.5 BER for coherent BPSK with and without Rayleigh fading.

As an aside, for $N = 1$, the substitution $\sqrt{\gamma} = t$ converts the integral into a Gaussian type of integral with the result:

$$P_{\text{BPSK}} = \frac{1}{2}\left(1 - \sqrt{\frac{\Gamma}{1+\Gamma}}\right) \tag{19.29}$$

Equation (19.29) gives the probability of bit error for BPSK in the presence of Rayleigh fading with one branch. This is in contrast to the expression in Eq. (19.27), which is valid without fading. Figure 19.5 shows a plot of the BER for BPSK with and without fading. From the plot it is seen that to maintain a BER of 10^{-3}, the required CNR at the receiver is approximately 6 and 24 dB without and with Rayleigh fading, respectively. In other words, to maintain the same BER with and without fading, approximately 18 dB more carrier power is required under Rayleigh faded conditions. Conversely, for a CNR of 10 dB, the BER without and with fading is 5×10^{-6} and 2×10^{-2}, respectively. Clearly, fading worsens BER by four orders of magnitude. It is thus seen that fading causes severe degradation in the error performance of modulation schemes.

19.5 Maximum Ratio Combining

In maximum ratio combining, the branch signals are weighted and combined so as to yield in the highest instantaneous CNR possible with any linear combining technique. By using Eq. (19.11) the total complex envelope at the nth branch in the presence of additive noise $\nu_n(t)$ can be written as

$$z_n(t) = g_n(t)u(t) + \nu_n(t) \tag{19.30}$$

If the received signals are weighted by weights w_n^* the combined output $U(t)$ of the array is

$$U(t) = u(t)\mathbf{w}^\dagger \mathbf{g} + \mathbf{w}^\dagger \nu \tag{19.31}$$

where † denotes Hermitian conjugate (transpose and complex conjugate), $\mathbf{w} = [w_1, \ldots, w_N]^T$, $\nu = [\nu_1, \ldots, \nu_N]^T$ and $\mathbf{g} = [g_1, \ldots, g_N]^T$. Assuming that the individual noise components are mutually independent, the total output noise power $P_{\text{no}}(o/p)$ is

$$P_{\text{no}}(o/p) = \frac{1}{2}\left\langle \left|\omega^{\dagger}\mathbf{v}\right|^{2} \right\rangle = \sum_{n=1}^{N} \left|\mathbf{w}_{n}^{*}\right|^{2} P_{n} \tag{19.32}$$

Therefore, the instantaneous output CNR is

$$\gamma = \frac{\frac{1}{2}E_{T}\left(\left|u(t)\omega^{\dagger}\mathbf{g}\right|^{2}\right)}{\sum_{n=1}^{N}\left|\mathbf{w}_{n}^{*}\right|^{2} P_{n}} = \frac{\frac{1}{2}\left|\sum_{n=1}^{N}\mathbf{w}_{n}^{*}g_{n}\right|^{2}}{\sum_{n=1}^{N}\left|\mathbf{w}_{n}^{*}\right|^{2} P_{n}} \tag{19.33}$$

The optimum weights are determined by the condition that the variation in γ with respect to the real and imaginary parts of w_n is zero. Alternately, the weights may be obtained by applying the Schwarz inequality to Eq. (19.33). By writing $w_n = \xi + j\eta$ and differentiating γ successively with respect to the two parameters ξ and η setting the result to zero, we get

$$w_{n}^{*} = \frac{g_{n}^{*}}{P_{n}} \tag{19.34}$$

which implies that the signals must be combined with weights made directly proportional to the complex conjugate of the branch signals and inversely proportional to the branch noise power. Thus, the branches with high signal strength are weighted more than the branches with weak signal strength. It is also noted that the weighted signals are all in phase and thus add coherently. The output CNR with the preceding weights is then

$$\gamma = \frac{\frac{1}{2}\left|\sum_{n=1}^{N}\left|g_{n}\right|^{2}/P_{n}\right|^{2}}{\sum_{n=1}^{N}\left|g_{n}^{*}\right|^{2}/P_{n}} = \frac{1}{2}\sum_{n=1}^{N}\frac{\left|g_{n}\right|^{2}}{P_{n}} = \sum_{n=1}^{N}\gamma_{n} \tag{19.35}$$

Thus, the output CNR is the sum of the CNR of the individual branches. Implementation of the maximum ratio combiner is expensive because the weights need both amplitude and phase tracking of the channel response (i.e., $g_n(t)s$). Furthermore, linear amplifiers and phase shifters over a large dynamic range of input signals are needed. In this regard, maximum ratio combining is primarily of theoretical interest. However, it serves as a benchmark against which the performance of other practical linear combining techniques can be assessed.

Because of the sum relation expressed in Eq. (19.35), the statistical distributions of the output CNR in this case can be easily derived from its characteristic function. By considering, once again, the case where all branches have equal mean CNR $\Gamma_n = \Gamma$, the **pdf** for an N-branch maximum ratio combiner can be shown to be

$$p_N^{\text{mrc}}(\gamma) = \frac{1}{(N-1)!} \frac{\gamma^{N-1}}{\Gamma^N} e^{-\gamma/\Gamma} \qquad \Gamma_n = \Gamma \tag{19.36}$$

The **pdf** is the well-known Erlang distribution [10], which is obtained by the addition of N independent and identical exponential distributions. The mean CNR at the output of the combiner is

$$\langle \gamma \rangle = \sum_{n=1}^{N} \langle \gamma_n \rangle = \sum_{n=1}^{N} \Gamma = NT \tag{19.37}$$

The corresponding **cdf** is

$$\begin{aligned} P_N^{\text{mrc}}(\gamma_s) &= \text{Prob}\,(\gamma < \gamma_s) = \int_0^{\gamma_s} p_N^{\text{mrc}}(\gamma)\,d\gamma = \frac{1}{(N-1)!} \int_0^{\gamma_s/\Gamma} x^{N-1} e^{-x} dx \\ &= 1 - e^{-\gamma_s/\Gamma} \sum_{n=0}^{N-1} \left(\frac{\gamma_s}{\Gamma}\right)^n \frac{1}{n!} \end{aligned} \tag{19.38}$$

$$= e^{-\gamma_s/\Gamma} \sum_{n=N}^{\infty} \left(\frac{\gamma_s}{\Gamma}\right)^n \frac{1}{n!} \tag{19.39}$$

$$= P_{N-1}^{\text{mrc}}(\gamma_s) - \frac{e^{-\gamma_s/\Gamma}}{(N-1)!} \left(\frac{\gamma_s}{\Gamma}\right)^{N-1} \tag{19.40}$$

Figure 19.6 shows a plot of the **cdf** $P_N^{\text{mrc}}(\gamma_s)$, of maximum ratio combining with N as a parameter. At 99% reliability maximum ratio diversity provides a 12-dB savings in power with two branches and 19-dB savings with four branches. This is in contrast to the savings of 10 and 16 dB achievable with selection diversity. Clearly, maximum ratio combining is more efficient than selection combining.

It is interesting also to look at the **pdf** in the limit as $N \rightarrow \infty$. It is very easy to show from Eqs. (19.36) and (19.37) that the **pdf** approaches a delta function for large N

$$p_\infty^{\text{mrc}}(\gamma) \stackrel{\Delta}{=} \lim_{N\rightarrow\infty} p_N^{\text{mrc}}(\gamma) = \delta(\gamma - \langle\gamma\rangle) \tag{19.41}$$

i.e., the **pdf** reduces to that of a signal received in a free-space situation with no fading.

As in the selection diversity case, the performance of any digital scheme under diversity can be evaluated by looking at the BER for a constant γ and averaging the result over the **pdf** of γ. For instance, for coherent BPSK, the BER with maximum ratio combining assuming identical noise in each branch is

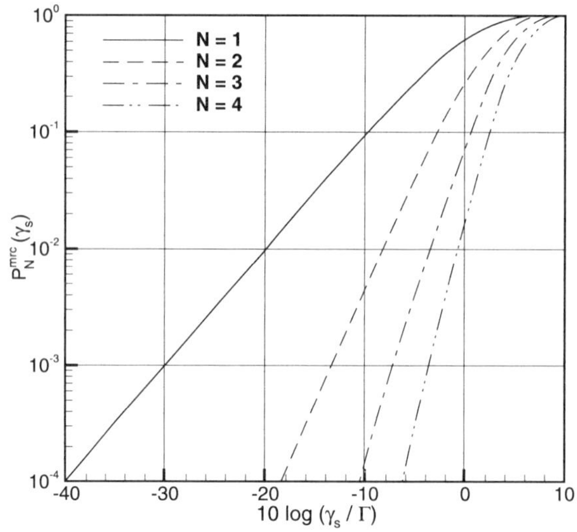

FIGURE 19.6 CDF of γ_s vs. γ_s/Γ for maximum ratio combining with N as a parameter.

$$P_{\text{BPSK}}^{\text{mrc}} = \int_0^\infty P_{\text{BPSK}}(\gamma)\, p_N^{\text{mrc}}(\gamma)\, d\gamma$$

$$= \int_0^\infty \frac{1}{2}\operatorname{erfc}\left(\sqrt{\gamma}\right) \frac{1}{(N-1)!} \frac{\gamma^{N-1}}{\Gamma^N} e^{-\gamma/\Gamma} d\gamma$$

$$= \frac{1}{2(N-1)!}\int_0^\infty \operatorname{erfc}\left(\sqrt{\Gamma x}\right) x^{N-1} e^{-x}\, dx \tag{19.42}$$

$$= \frac{1}{(N-1)!}\left(\frac{1-\mu}{2}\right)^N \sum_{n=0}^{N} \frac{(N-1+n)!}{n!}\left(\frac{1+\mu}{2}\right)^2$$

$$\sim \left(\frac{1}{4\Gamma}\right)^N \frac{(2N-1)!}{N!(N-1)!} \quad \text{for } \Gamma \gg 1 \tag{19.43}$$

where

$$\mu = \sqrt{\frac{\Gamma}{1+\Gamma}} = \sqrt{\frac{\langle\gamma\rangle}{N+\langle\gamma\rangle}} \tag{19.44}$$

It is seen that the BER decreases with Γ as $1/\Gamma^N$ for large enough N and Γ. In the limit as $N \to \infty$ the BER becomes

$$P_{\text{BPSK}}^{\text{mrc}} = \frac{1}{2}\operatorname{erfc}\left(\sqrt{\langle\gamma\rangle}\right) = P_{\text{BPSK}}^{\text{nofade}} \tag{19.45}$$

This result is to be expected because the output of the combiner would approach a steady value as N becomes large. Figure 19.7 shows a plot of the BER vs. $\langle\gamma\rangle = N\Gamma$ for BPSK with N as a parameter.

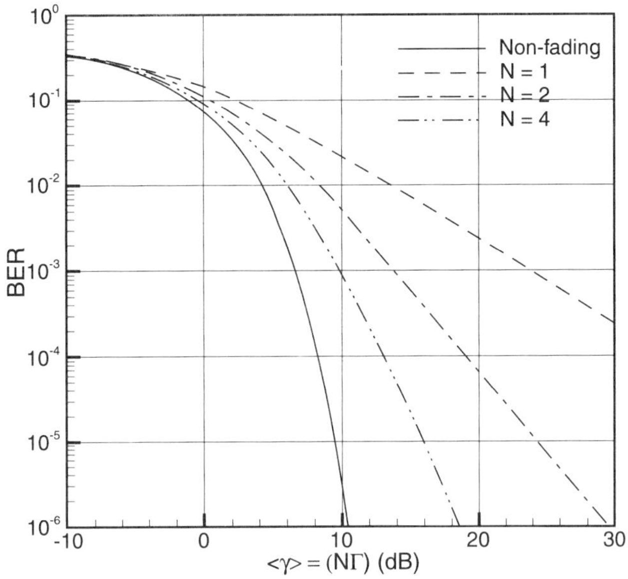

FIGURE 19.7 BER vs. average CNR per bit $\langle\gamma\rangle = N\Gamma$ for coherent BPSK under Rayleigh fading with maximum ratio diversity.

19.6 Equal Gain Combining

In equal gain combining all of the weights have the same magnitude but a phase opposite to that of the signal in the respective branch, i.e., $w_n = \exp(-j\angle g_n)$. The combined output CNR with equal gain combining is

$$\gamma = \frac{1}{2}\frac{\left[\sum_{n=1}^{N}|g_n|\right]^2}{\sum_{n=1}^{N}P_n} = \frac{r^2}{2NP_{no}} \tag{19.46}$$

assuming equal noise in the branches, and where

$$r = \sum_{n=1}^{N}|g_n| \tag{19.47}$$

is the envelope (local mean) of the combined signal. It is seen equal to the sum of N Rayleigh variables $|g_n|$. Unfortunately, no closed form expression exits for the ***pdf*** or **cdf** of the combiner output in the case of equal gain combiner. However, a closed form expression for the *mean CNR* at the output of the combiner can be obtained easily from

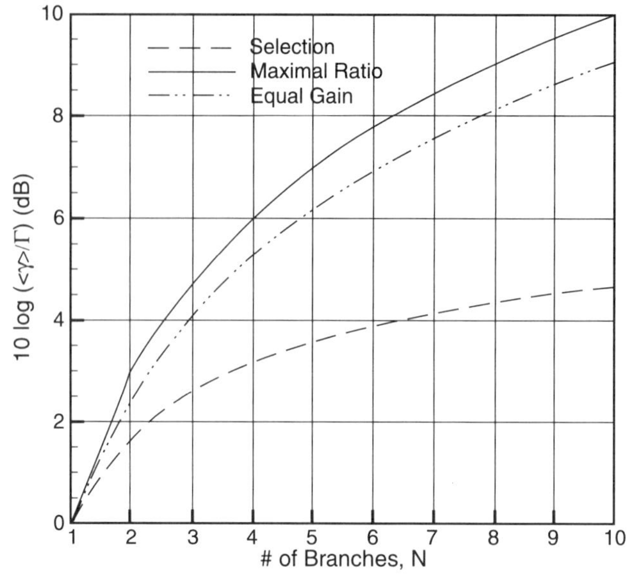

FIGURE 19.8 Improvement in mean CNR for various combining techniques.

$$\begin{aligned}\langle\gamma\rangle &= \frac{1}{2NP_{no}}\left\langle\left[\sum_{n=1}^{N}|g_n|\right]^2\right\rangle \\ &= \frac{1}{2NP_{no}}\sum_{n=1}^{N}\sum_{m=1}^{N}\langle|g_n||g_m|\rangle \\ &= \frac{1}{2NP_{no}}\left[2NP_o + N(N-1)\frac{\pi P_o}{2}\right] \\ &= \left[1+(N-1)\frac{\pi}{4}\right]\Gamma\end{aligned} \tag{19.48}$$

where we have used the fact that $\langle|g_n|^2\rangle = 2P_o$, $\langle|g_n|\rangle = \sqrt{\pi P_o/2}$, and $\langle|g_n|\,|g_m|\rangle = \langle|g_m|\rangle\,\langle|g_m|\rangle$ since the g_n are assumed uncorrelated. Similar to maximum ratio combining, equal gain diversity improves the output mean CNR in proportion to N. Reiterating once again, for large N, mean CNR improvement in selection diversity is proportional to $\ln(N)$, whereas in both maximum ratio and equal gain diversities, it increases linearly as N. Figure 19.8 shows a plot of the improvement of average CNR of a diversity combiner for selection diversity, Eq. (19.26), maximum ratio diversity, Eq. (19.37), and equal gain diversity Eq. (19.48). Both the maximum gain diversity and equal gain diversity provide superior improvement compared with the selection diversity and that the results for equal gain combining are within 1 dB of those of maximum gain combining for up to 10 branches.

19.7 Diversity Gain

Diversity gain of an N-element array is defined as the improvement in link margin for certain performance criterion. Normally, the performance criterion is taken as the BER. For example, with reference to Fig. 19.7, it is seen that to provide a 10^{-2} BER with coherent phase shift keying (PSK), an average CNR per bit of 4.3 and 13.8 dB is needed without fading and with Rayleigh fading, respectively. Clearly, an additional 9.5 dB higher average output power is needed with Rayleigh fading. Using two antennas i.e., two-branch diversity) reduces the required power to 8.4 dB and we say that the two element array provides a diversity gain of 5.4 dB (= 13.8 – 8.4). Obviously, the maximum diversity gain that one can achieve with multiple antennas under maximum ratio combining is 9.5 dB at this BER and this value will be reached asymptotically for large N as evident from Eq. (19.45).

19.8 Antenna Gain

The diversity gain must be distinguished from the antenna gain that is defined as the ratio of the output CNR of the array to the output CNR ratio of a single element for highly correlated input signals (single-incident plane wave). Recall that for a single-input plane wave, the response at various elements only differs by a phase factor $\exp(j\alpha)$, where $\alpha = k_o d \cos\phi$ depends on element spacing, the radio frequency (RF), and the angle of arrival of the plane wave with respect to the array axis. The input signal for the purpose of antenna gain then is assumed to be of the form $u(t)\sqrt{2P_o}[1, \exp(j\alpha), \exp(j2\alpha), \ldots, \exp(j[N-1]\alpha)]^T \triangleq u(t)\sqrt{2P_o}\boldsymbol{\psi}$, where P_o is the mean power at each b ranch. In the cases of maximum ratio combining and equal gain combining, the weights will be equal and proportional to $w = \sqrt{2P_o}\boldsymbol{\psi}/P_{no}$, respectively, where P_{no} is the input noise power at each branch. The combined signal plus noise voltage for a single plane wave incident is

$$\begin{aligned} U(t) &= \frac{\sqrt{2P_o}}{P_{no}}\left[\sqrt{2P_o}\,u(t)\psi^\dagger\psi + \psi^\dagger v\right] \\ &= \frac{2NP_o}{P_{no}}u(t) + \frac{\sqrt{2P_o}}{P_{no}}\psi^\dagger v \end{aligned} \tag{19.49}$$

The average carrier power at the output is $E_T(|[2NP_o u(t)/P_{no}]^2|)/2 = 2N^2(P_o/P_{no})^2$, whereas the noise power at the output is $P_o\langle\psi^\dagger v v^\dagger \psi\rangle/P_{no}^2 = 2NP_o/P_{no}$, assuming noise to be uncorrelated at various branches and using $\langle v_n^2(t)\rangle = 2P_{no}$. The output CNR is then equal to

$$\gamma = \frac{NP_o}{P_{no}} \tag{19.50}$$

from which it is clear that the antenna array gain to equal to N. Note that the *mean* CNR improvement of a maximum ratio combiner is the same whether or not the branches are correlated and the same result stated earlier would have also been obtained for the mean CNR in the absence of branch correlation. This is actually shown in the next section.

The result could also be established from standard antenna theory, which predicts that the directivity of an antenna array is directly proportional to its length. Figure 19.9 shows the directivity of an N element, uniformly excited, broadside array as a function of d/λ with N as a parameter.

For $d = m\lambda/2$ with m an integer, the directivity is equal to N. For other spacings, the directivity varies around N, reaching N asymptotically for large d/λ. For $d = m\lambda$ with m and integer, the directivity takes a sharp dip because of the onset of grating lobes.

19.9 Effect of Branch Correlation

The performance with diversity is somewhat degraded when the various branches are not perfectly decorrelated. A general account of this is given in References [13] or [6]. The branch correlation may be caused by a number of factors such as angular despreading of incoming waves or mutual coupling between

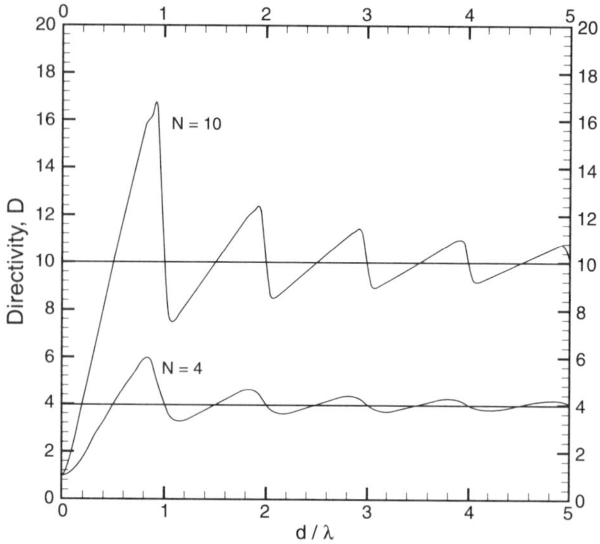

FIGURE 19.9 Directivity of an N-element, uniformly excited broadside array.

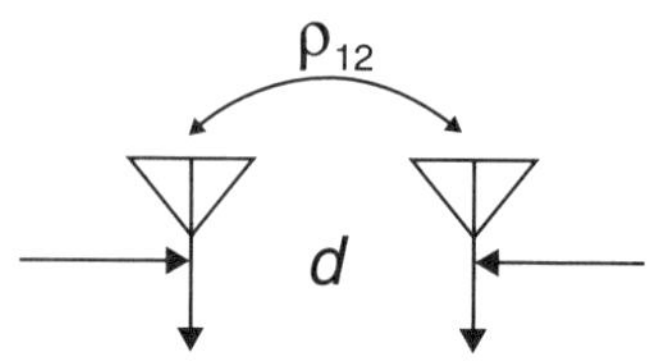

FIGURE 19.10 Two elements with correlated signals.

antenna elements. For example, with a single plane wave incident, the two branches are always correlated no matter how large the spacing is.

To illustrate the effect, consider the two-branch case with maximum ratio combining where the complex Gaussian signals at the branches are correlated with a complex correlation coefficient ρ_{12}. The correlation coefficient measures the degree of correlation between signals received at two spatial points, separated by some distance d. For waves incident from the xy plane, it is defined in Eq. (19.67) as

$$\rho_{12} = \left\langle e^{-jk_o d\cos\phi} \right\rangle \tag{19.51}$$

where the angle brackets denote statistical average with respect to the angle variable ϕ. The distance d in our case corresponds to the interelement spacing d. Figure 19.10 illustrates this case. The *envelope* correlation between the two branches is $\rho_e \approx |\rho_{12}|^2$ [7]. In the presence of branch correlations, the **cdf** of the combined signal presented in Eq. (19.39) will get modified to [6]

$$P(\gamma_s) = 1 - \frac{1}{2|\rho_{12}|}\left[(1+|\rho_{12}|) e^{-\gamma_s/(1+|\rho_{12}|)\Gamma} - 1 - |\rho_{12}| e^{-\gamma_s/(1-|\rho_{12}|)\Gamma} \right] \tag{19.52}$$

and the corresponding **pdf** changes to

$$p(\gamma) = \frac{1}{2|\rho_{12}|\Gamma}\left[e^{-\gamma/(1+|\rho_{12}|)\Gamma - e^{-\gamma/(1-|\rho_{12}|)\Gamma}} \right] \tag{19.53}$$

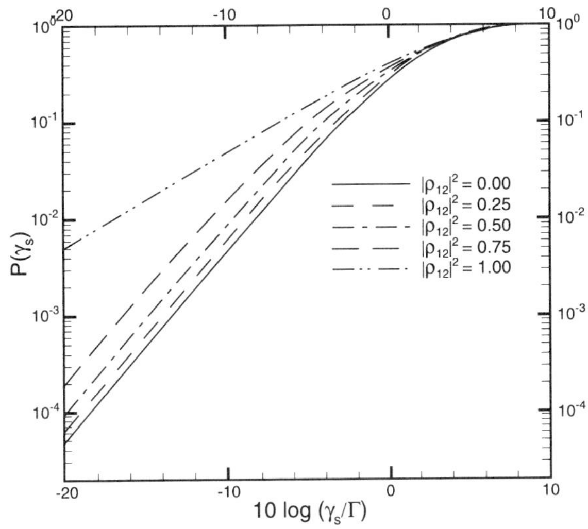

FIGURE 19.11 Effect of branch correlation on the distribution of output power in a dual-diversity, maximum ratio combiner.

The mean CNR of the combined signal remains at $\langle\gamma\rangle = 2\Gamma$ — independent of $|\rho_{12}|$ — as can be easily verified from Eq. (19.53). However, the distribution of γ depends on $|\rho_{12}|$ as seen from Fig. 19.11. The BER for a basic modulation scheme such as coherent BPSK can be carried out as in the previous section and the result is

$$P_e = \int_0^\infty \frac{1}{2}\operatorname{erfc}\left(\sqrt{\gamma}\right)p(\gamma)\,d\gamma$$

$$= \frac{1}{4|\rho_{12}|}\left[\left(1+|\rho_{12}|\right)\left(1-\frac{1}{\sqrt{1+\dfrac{1}{\left(1+|\rho_{12}|\right)\Gamma}}}\right) + -\left(1-|\rho_{12}|\right)\left(1-\frac{1}{\sqrt{1+\dfrac{1}{\left(1-|\rho_{12}|\right)\Gamma}}}\right)\right] \tag{19.54}$$

For the special case of perfect correlation (i.e., $\rho_{12} = 1$) the preceding expression reduces to Eq. (19.29) with Γ replaced by 2Γ.

The correlation coefficient depends on the distribution of incoming waves in the azimuth plane, which in turn depends on the disposition of scatterers about the transmitting and the receiving antennas. If waves arrive uniformly in the azimuth plane from all angles, as in the case of a mobile receiver situation in a highly cluttered environment, one gets the classical Clarke's two-dimensional (2D) model [6]. On the other hand, waves received at a tall base station antenna arrive from a relatively narrow angle and one of the models used to describe such a situation is the circular scattering model [11]. In the circular scattering model, scatterers are assumed to be distributed uniformly within a radius R about the mobile. Propagation from the mobile to be base station is assumed to take place via single scattering off the

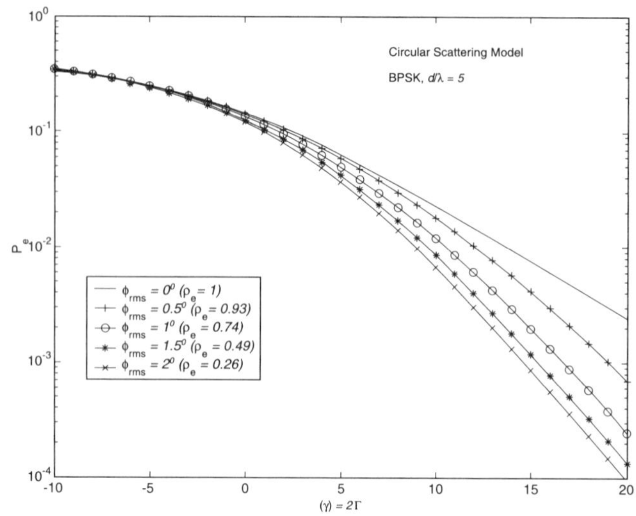

FIGURE 19.12 BER vs. $\langle\gamma\rangle$ (decibels) for dual-branch, maximum ratio combiner with correlated fadings.

scatterers. Because of the finite area of scattering, waves transmitted from the mobile arrive from a small angular region about the mean mobile direction. If the distance between the mobile and base station is D, the spatial correlation between the two elements for the circular scattering model can be derived as [7]

$$\rho_{12} = 2\frac{J_1\left(k_o dR/D\right)}{k_o dR/D} \tag{19.55}$$

where $J_n(\cdot)$ is the Bessel function of the first kind of order n. The angular spread of waves depends on the relative sizes of R and D. Using $d/\lambda = 5$, the BER performance as computed from Eq. (19.54) for various values of angular spread is shown in Fig. 19.12 as a function of $\langle\gamma\rangle$, expressed in decibels. For the model chosen, a root mean square (rms) angular spread of 1°, as seen from the base station, gives rise to an envelope correlation of 0.74. Clearly, as the angular spread increases, the branch signals become more and more decorrelated and the BER curve approaches the ideal ($\rho_{12} = 0$) 2-branch diversity curve. Comparing with Fig. 19.7 it is seen that an r.m.s. spread of 2° almost completely decorrelates the two signals. At a BER value of 10^{-2}, the diversity gain for a two-branch diversity reduces by about 5 dB (= 14 − 9 dB) when the signals change from being uncorrelated to completely correlated. In the next section, we look at another cause of correlation, namely, mutual coupling between elements, which becomes particularly important for small spacings.

19.10 Mutual Coupling

Because of electromagnetic coupling the signals received by the elements of an array will no longer be independent, but become dependent on each other. Mutual coupling will influence the cross-correlation between the received signals and will be particularly important when spacings are small. Such would be the case for antenna diversity employed at the mobile station. One way to incorporate the presence of element mutual coupling is by means of an impedance matrix. The use of impedance matrix is most convenient for the wire type of antennas. As before, we consider an array of vertical dipoles for illustration. The terminal voltages V_n and the open-circuited voltages V_n^{oc} are related through the mutual impedance Z_{nm} as

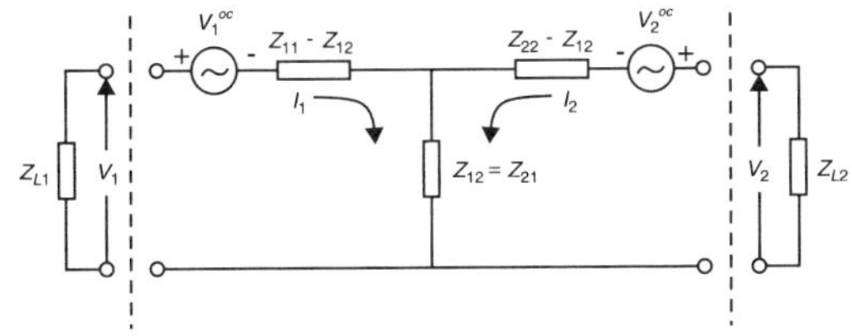

FIGURE 19.13 T-equivalent representation of antenna mutual coupling for a two-element array.

$$\begin{aligned} V_1 &= Z_{11}I_1 + Z_{12}I_2 + \cdots + Z_{1N}I_N + V_1^{oc} \\ V_2 &= Z_{21}I_1 + Z_{22}I_2 + \cdots + Z_{2N}I_N + V_2^{oc} \\ &\vdots \\ V_N &= Z_{N1}I_1 + Z_{N2}I_2 + \cdots + Z_{NN}I_N + V_N^{oc} \end{aligned} \tag{19.56}$$

The preceding equation could be written in a mateix form as

$$\mathbf{v} = \mathbf{Zi} + \mathbf{v}^{oc} \tag{19.57}$$

where $\mathbf{v}$, $\mathbf{v}^{oc}$, and $\mathbf{i}$ are the vectors of terminal voltages, open-circuited voltages, and terminal currents, respectively, and $\mathbf{Z}$ is the impedance matrix. All these quantities are in an *array environment*. Figure 19.13 illustrates the T-equivalent representation of the impedance matrix for a two-element array. The concept of impedance matrix, devised originally for lumped circuit elements, presents some difficulties when applied directly to wave propagation problems. Although we continue to use it later, the reader must at least be made aware of the difficulties. We elaborate on these by considering center-fed broadside elements in the transmit mode. According to Eq. (19.56),

$$Z_{11} = \left.\frac{V_1}{I_1}\right|_{I_2, \ldots, I_N = 0} \tag{19.58}$$

That is, Z_{11} is the driving point impedance of element 1 when the *terminal* currents at all the other elements are made zero. However, unlike in circuits, making the terminal current of an antenna to go to zero by open-circuiting its terminals does not force the current on its entire structure to go to zero. Hence, Z_{11} is not the isolated self impedance of element 1, but includes the effects of the induced currents in any and all the other elements that can flow with the terminals open. Thus, the driving point impedance Z_{nn}, $n = 1, \ldots, N$, depends on the number of elements, the interelement spacing, the element orientations, and the relative positions of the element. Even in the case of identical elements, it has a different value for different elements of the array. Similarly, Z_{mn} is slightly different because elements m and n take up different positions in the same array. Although there are means of avoiding these difficulties by resorting to full numerical approaches such as the method of moments [5], antenna engineers have continued to adopt the ideas of impedance matrix in design, with success, because the intricacies mentioned earlier tend to have a second-order effect. In a like manner, the open circuited voltage V_n^{oc} is the voltage induced by the incident plane wave across the nth terminal when all the other terminals are open circuited. It, too, is not the voltage induced in an isolated dipole, but depends on the currents flowing on the other antenna structures under open-circuit conditions. In the subsequent analysis we, however, ignore these second-order effects.

For a load impedance Z_{Ln}, the terminal voltage V_n and the current I_n (defined to flow *into* the terminals) are related by

$$V_n = -Z_{L_n} I_n \tag{19.59}$$

By using this in Eq. (19.57), we get

$$\mathbf{v} = \mathbf{Z}_L\left(\mathbf{Z}+\mathbf{Z}_L\right)^{-1}\mathbf{v}^{oc} \tag{19.60}$$

where Z_L is the diagonal matrix containing the load impedances. For convenience, we define Z^c as

$$\mathbf{Z}^c = \mathbf{Z}_L\left(\mathbf{Z}+\mathbf{Z}_L\right)^{-1} \tag{19.61}$$

and rewrite Eq. (19.60) as

$$\mathbf{v} = \mathbf{Z}^c\mathbf{v}^{oc} \tag{19.62}$$

It is clear that the presence of mutual coupling can be simply accounted for by replacing the vector $\mathbf{v}^{oc}$ with $\mathbf{Z}^c\mathbf{v}^{oc}$. For example, in the case of isotropic elements, the received signal vector (terminal voltages) in terms of the steering vector becomes

$$\mathbf{v} = V_o\mathbf{Z}^c\psi \tag{19.63}$$

Equation (19.63) is a generalization of Eq. (19.9). For a two-element array composed of vertical dipoles oriented as in Fig. 19.1, the received signals for a plane wave incident from the *xy* plane are

$$V_1 = Z_{11}^c V_1^{oc} + Z_{12}^c V_2^{oc} \tag{19.64}$$

$$V_2 = Z_{21}^c V_1^{oc} + Z_{22}^c V_2^{oc} \tag{19.65}$$

where

$$V_n^{oc} = V_o e^{j(n-1)k_o d\cos\phi} \qquad n = 1, 2 \tag{19.66}$$

Define the spatial cross-correlation between the two received signals without and with mutual coupling as

$$\rho_{12}^{oc} = \frac{\left\langle V_1^{oc} V_2^{*oc}\right\rangle}{\sqrt{\left\langle\left|V_1^{oc}\right|^2\right\rangle\left\langle\left|V_2^{oc}\right|^2\right\rangle}} = \left\langle e^{-jk_o d\cos\phi}\right\rangle \quad \text{(without mutual coupling)} \tag{19.67}$$

$$\rho_{12} = \frac{\left\langle V_1 V_2^{*}\right\rangle}{\sqrt{\left\langle\left|V_1\right|^2\right\rangle\left\langle\left|V_2\right|^2\right\rangle}} \quad \text{(with mutual coupling)} \tag{19.68}$$

where the angle brackets denote statistical average with respect to the angle variable ϕ. In the following it is assumed that the **pdf** of the arriving waves in the azimuthal plane is symmetrical about the broadside to the array axis $\phi = \pi/2$ so that ρ_{12}^{oc} is a real quantity. This is true both for the Clarke's model and the circular scattering model with the scatter region directly ahead of the array. By assuming identical elements so that $Z_{11} = Z_{22}$ and using $Z_{12} = Z_{21}$, $Z_{L1} = Z_{L2} = Z_L$, the spatial correlation expression can be simplified as

$$\rho_{12} = \frac{\rho_{12}^{oc} + \zeta}{1 + \zeta \rho_{12}^{oc}} \tag{19.69}$$

where

$$\zeta = \frac{2\Re e(\zeta_1)}{1 + |\zeta_1|^2} \tag{19.70}$$

and

$$\zeta_1 = \frac{-Z_{12}}{Z_{11} + Z_L} \tag{19.71}$$

It is clear that in the absence of mutual coupling, $Z_{12} = \Rightarrow \zeta = 0$ so that $\rho_{12} = \rho_{12}^{oc}$ as expected. For center-fed, half-wavelength dipoles, $Z_{11} \approx 73 + j42.5\ \Omega$. Mutual impedance $Z_{12} = R_{12} + jX_{12}$ between identical center-fed dipoles arranged side by side is given by Reference [8]

$$R_{12} = 30\left[\mathrm{Cin}\left[k_o\left(\sqrt{\ell^2 + d^2} + \ell\right)\right] + \mathrm{Cin}\left[k_o\left(\sqrt{\ell^2 + d^2} - \ell\right)\right] - 2\mathrm{Cin}(k_o d)\right] \tag{19.72}$$

$$X_{12} = 30\left[\mathrm{Si}\left[k_o\left(\sqrt{\ell^2 + d^2} + \ell\right)\right] + \mathrm{Si}\left[k_o\left(\sqrt{\ell^2 + d^2} - \ell\right)\right] - 2\mathrm{Si}(k_o d)\right] \tag{19.73}$$

where $\mathrm{Cin}(x)$ and $\mathrm{Si}(x)$ are cosine and sine integrals defined as [1]

$$\mathrm{Cin}(x) = \int_0^x \frac{1 - \cos t}{t} dt \tag{19.74}$$

$$\mathrm{Si}(x) = \int_0^x \frac{\sin t}{t} dt \tag{19.75}$$

The real and imaginary parts of Z_{12} are plotted in Fig. 19.14 as a function of element spacing. Note that for $d = 0$, $Z_{12} = Z_{11}$. Figures 19.15 and 19.16 show the envelope correlation $\rho_e = |\rho_{12}|^2$ for Clarke's 2D model [6] ($\rho_{12}^{oc} = J_0(k_o d)$) and the circular scattering model (azimuth angular spread of about 15°) with and without mutual coupling for load impedance $Z_L = Z_{11}^*$. As mentioned previously, the former model would correspond to array antennas employed at the mobile station, whereas the latter would correspond to array antennas employed at the base station. From the figures, the envelope cross-correlation calculated with mutual coupling differs substantially from that calculated assuming to mutual coupling. The exact effect depends on the interplay between mutual coupling and angular spread of

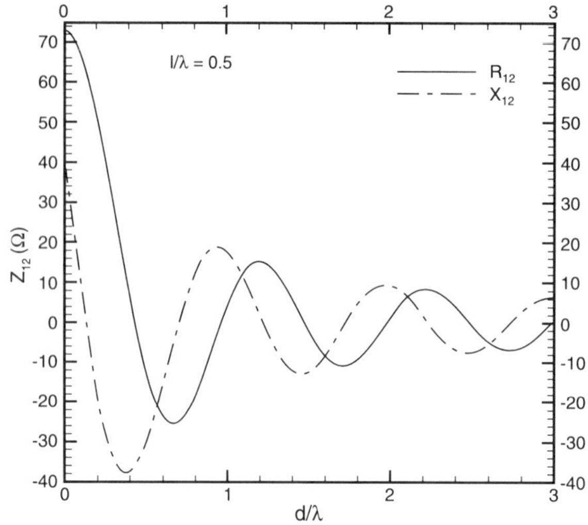

FIGURE 19.14 Mutual impedance between two side-by-side λ/2 dipoles.

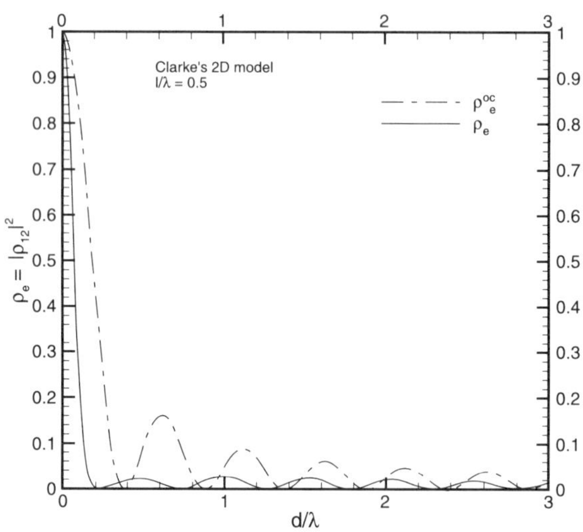

FIGURE 19.15 Spatial correlation between two elements with and without mutual coupling, Clarke's model.

arrival of the incoming waves and could lead to either increased or decreased cross-correlation between the two antennas. From Fig. 19.15, mutual coupling actually decreases the correlation between the elements when the waves arrive uniformly from all directions (Clarke's model). However, with the circular scattering model, it t ends to increase the correlation for $\rho_e < 0.6$. To provide an envelope correlation value of 0.7 with Clarke's 2D model, element separation of 0.14 and 0.05 λ is needed without and with mutual coupling. For the circular scattering model, the spacing required with and without mutual coupling is 0.18 and 0.4 λ, respectively.

Figure 19.17 shows the envelope correlation calculated using Eq. (19.69) for the Clarke's model and compared with measurements reported in References [9] and [15]. Vaughan and Scott [15] conducted experiments in the city of Wellington, New Zealand, with quarter-wave monopoles terminated in $Z_{L1} = Z_{L2} = 50\ \Omega$. Error bars shown are for the 95% confidence interval. Miki and Hata [9] conducted experiments in the city of Tokyo and provide data for envelope correlation for and element spacing of

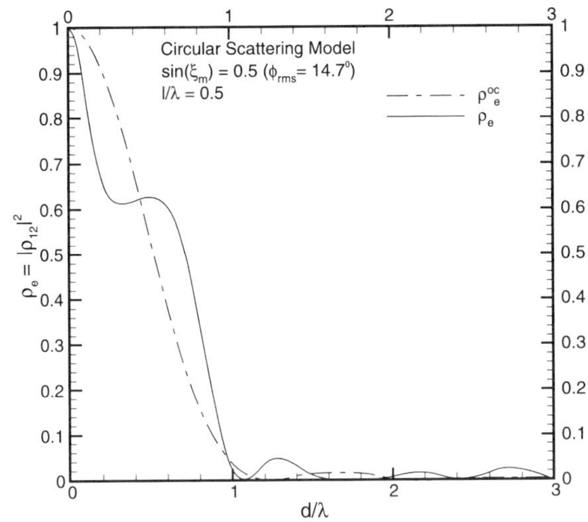

FIGURE 19.16 Spatial correlation between two elements with and without mutual coupling, circular scattering model.

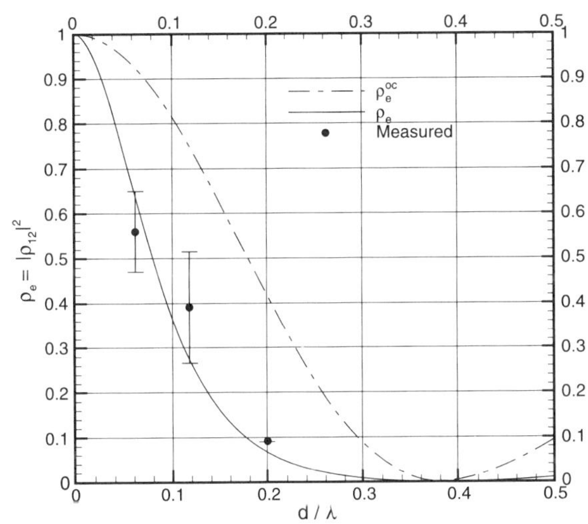

FIGURE 19.17 Spatial correlation between two elements with and without mutual coupling, Clarke's model. Measured values are from References [9] and [15].

0.2 λ. Once again, their results are for quarter-wave monopoles. Equation (19.69), derived for center-fed dipoles, can also be used for monopoles by noting that the values of Z_{11} and Z_{12} for monopoles are half of those dipoles. The values computed with mutual coupling agree quite well with the measured one.

19.11 Summary

The use of spatial diversity, whereby antenna arrays are employed at the receiver, to combat the ill-effects of fading on the reception of narrowband signals was discussed. Three techniques of combining, namely, selection combining, maximum ratio combining, and equal gain combining, together with their effects on the bit error performance of uncoded modulation schemes such as BPSK was discussed. Inclusion of angular spreading and element mutual coupling in analysis was demonstrated for arrays composed of vertical dipole elements.

References

1. M. Abramowitz and I. Stegun, *Handbook of Mathematical Functions*, New York: Dover, 1964.
2. R.E. Collin, *Antennas and Radiowave Propagation*, New York: McGraw-Hill, 1985.
3. R.E. Collin and F.J. Zucker, *Antenna Theory*, Pt. 1, Ch. 5, New York: McGraw-Hill, 1969.
4. C.B. Dietrich, Jr., et al., "Smart antennas in wireless communications: base station diversity and handset beamforming," *IEEE Antennas Propag. Mag.*, vol. 42 (5), pp. 142–151, October 2000.
5. R.F. Harrington, *Field Computation by Moment Methods*, Malabar, FL: Krieger, 1968.
6. W.C. Jakes, *Microwave Mobile Communications*, IEEE Press Classic Reissue, Piscataway, NJ: IEEE Press, 1994.
7. R. Janaswamy, *Radiowave Propagation and Smart Antennas for Wireless Communications*, Boston: Kluwer Academic Publishers, November 2000.
8. J.D. Kraus, *Antennas*, 2nd Ed., New York: McGraw-Hill, 1988.
9. T. Miki and M. Hata, "Performance of 16 kbit/s GMSK transmission with post detection selection diversity in land mobile radio," *IEEE Trans. Veh. Technol.*, vol. VT-33 (3), pp. 128–133, August 1984.
10. A. Papoulis, *Probability, Random Variables, and Stochastic Processes*, New York: McGraw-Hill, 1984.
11. P. Petrus, J.H. Reed, and T.S. Rappaport, "Geometrically based statistical channel model for macro cellular mobile environments," *Proc. IEEE Globecommun.*, pp. 1197–1201, 1996.
12. J.G. Proakis, *Digital Communications*, 2nd Ed., New York: McGraw-Hill, 1989.
13. M. Schwartz, W.R. Bennett, and S. Stein, *Communication Systems and Techniques*, IEEE Press Classic Reissue, Piscataway, NJ: IEEE Press, 1996.
14. R.G. Vaughan and J.B. Andersen, "Antenna diversity in mobile communications," *IEEE Trans. Veh. Technol.*, 36 l(4), pp. 149–171, November 1987.
15. R.G. Vaughan and N.L. Scott, "Closely spaced monopoles for mobile communications," *Radio Sci.*, 28 (6), pp. 1259–1266, November–December 1993.
16. W.L. Weeks, *Antenna Engineering*, New York: McGraw-Hill, 1968.

20

Direction-of-Arrival Estimation in Mobile Communication Environments

Mats Viberg
Chalmers University of Technology

Thomas Svantesson
Chalmers University of Technology

20.1 Introduction

Wireless communication services have experienced a tremendous growth during the last years, where speech has been the main application. With the introduction of a new generation of mobile or wireless communication systems, for example, third-generation (3G) systems, the focus is shifted toward data applications. Systems like Universal Mobile Telecommunications System (UMTS) and International Mobile Telecommunications (IMT)-2000 are designed for providing high data rates that result in an even larger demand for more efficient solutions than before. Furthermore, the general impairment of a wireless system such as fading, thermal noise, and co-channel interference still poise severe problems. Antenna array processing is considered to be an attractive technology to mitigate these problems to providing high-speed data services. By exploiting the spatial dimension offered by antenna arrays, the coverage may be improved, co-channel interference reduced, and the capacity increased. To fully achieve these goals, it is necessary to both understand and exploit the characteristics of the antenna and the wireless channel.

The underlying physical principle of all antenna array algorithms is the fact that a transmitted signal propagates along some path and then arrives at the receiving antenna. Typically, several versions of the transmitted signal impinge on the receiving antenna from different directions because of multipath. In

0-8493-0124-6/02/$0.00+$1.50

fact, the directions from which the signals arrive, the direction of arrival (DOA) is an important property when characterizing the channel as well as designing receiver algorithms. For instance, the wireless channel changes very rapidly resulting from movement of both the users as well as changes in the surrounding environment. However, the main directions of arrival do not change nearly as rapidly. Thus, characterization of the channel in terms of DOA is an interesting alternative to standard models. Another important channel property that determines the quality of the communication link is the angular spread that is closely connected to the DOA. The angular spread, among others, essentially determines the diversity gain by using an antenna array. Also, it has been proposed to employ antenna arrays to reduce the co-channel interference by transmitting energy only in the direction of a specific user and essentially no energy in the directions of other users. In these types of systems, estimating DOA forms an integral part of the system. For example, it may assist to form beams in both uplink and downlink processing — see [76, 154, 162]. Furthermore, as the mobile phone becomes more ubiquitous, the interest of employing the phone for personal locating services increases. Here, the DOA can be used for obtaining the location of the mobile phone. Thus, there are many reasons for employing DOA estimation in wireless communications.

The problem of estimating the DOA of multiple signals arriving at an array of sensors has received considerable attention for seveal decades [64, 139]. Traditionally, the most notable application has been source localization in radar and sonar. These applications have essentially driven the development of DOA algorithms since the first approaches of spatial filtering or beamforming. The conventional beamformer dates back to the Second World War, and is a mere application of Fourier-based spectal analysis to spatio-temporally sampled data. Later, adaptive beamformers and classical time-delay estimation techniques were applied to enhance one's ability to resolve closely spaced signal sources [5, 20, 42]. However, these approaches suffer from the fact that the performance directly depends on the physical size of the array (aperture). The limitations of the beamformer started a development of other algorithms that alleviated some of these limitations. Inspired by the maximum likelihood principle [60, 79, 116], a wealth of DOA estimation schemes based on parametric modeling have appeared during the last two decades. By more fully exploiting the underlying physical principles (model), a much higher performance than that of the beamformer was now possible.

The focus in this manuscript is therefore on parametric techniques for wireless communication scenarios instead of spectral beamforming techniques. Of course, the aim is not to be exhaustive. For more extended presentations on both beamforming and parametric methods, the reader is referred to other textbooks [25, 48, 50, 52, 56, 85]. Instead, a broad review of the area of DOA estimation schemes is provided that especially considers applications in wireless communications. Because DOA estimation for communication scenarios differs from the typical radar application, the wireless channel and its special characteristics is described in some detail. In particular, exploiting special structures in the transmitted signal such as training sequences are reviewed, as well as joint estimation of the time delays and the DOAs. Furthermore, because the transmitter typically is obstructed, that is, no line of sight (LOS) between the transmitter and receiver exists, the source often acts as an angularly spread source. Applying standard DOA estimation schemes in such a case can give misleading results. Therefore, several recent algorithms for spread sources that accurately account for these effects are covered. Furthermore, techniques for tracking moving sources are also presented. Although the DOA typically changes slowly in most mobile communication scenarios, the strength of the received signal may be rapidly varying resulting in sudden appearance or disappearance of scattering centers. It is therefore of interest to design tracking algorithms, because the main directions to the scattering centers changes slowly, whereas the number of sources may be highly nonstationary. Algorithms for source tracking are therefore also included along with some examples.

Although a rather large number of DOA estimation techniques and algorithms are outlined, a number of important issues have of necessity been omitted. For instance, the higher performance of the parametric DOA estimation methods is essentially achieved by exploiting the underlying data model to a larger extent. However, this performance gain is obtained at the cost of an increased sensitivity to modeling errors. In particular, calibration errors in the array response, such as gain and phase perturbations, mutual coupling errors, and sensor position errors, may have a disastrous effect on the performance. An incorrect model of the noise covariance may also be harmful. This topic has been thoroughly analyzed in the literature [71, 72, 130, 131], and the interested reader is referred to further sources and references therein. Furthermore, most of the described parametric methods assume that the number of impinging signals is known. The problem of estimating the number of emitters is often called the detection problem, and is an interesting subject of its own. A large number of detection methods have been proposed in the literature [91, 146, 147].

The chapter is organized as follows: Section 20.2 contains some physical background on both the antenna and some fundamental wave propagation. Two data models, that are used extensively in the following sections, are presented. The point sources model is the basis for most traditional array processing schemes. A more elaborate model that accounts for spread sources is also given, that may be more appropriate in wireless communications. In Section 20.3, the rich area of DOA estimation algorithms designed for point sources is surveyed, including spectral-based beamforming schemes as well as parametric methods. Section 20.4 presents techniques that are tailored to communications applications, including some recent developments for localization of spread sources. In Section 20.5, algorithms for DOA tracking in nonstationary scenarios are presented. Section 20.6 concludes the chapter.

20.2 Spatial Signal Model

To be able to estimate the DOA of an electromagnetic wave incident on an array of antenna elements, it is necessary to understand the physical properties that govern the wireless channel and the antenna. In this section, a mathematical model of the received signal is explained using fundamental electromagnetic concepts. The modeling of the received signal can essentially be divided into two separate parts, that is, antenna and channel modeling. First, a model of the antenna is presented in Section 20.2.1 that relates an incident electromagnetic wave to a measured voltage or current at the antenna. Then, a general model of the mobile communication radio channel based on fundamental propagation and scattering theory is introduced in Section 20.2.2. Some of the most widely used channel models are then reviewed and discussed.

20.2.1 Physical Modeling of the Array Antenna

The principal function of an antenna is to convert an electromagnetic wave into an induced voltage or current that is measured. If the antenna consists of several elements, a number of voltages or currents are measured. The physical principle that governs DOA estimation is that an incident wave reaches each antenna element at different time instants. A typical scenario is shown in Figure 20.1, where a wave is incident from the (θ, ϕ) direction on an array of two elements. Here, the wave reaches antenna A_1 some

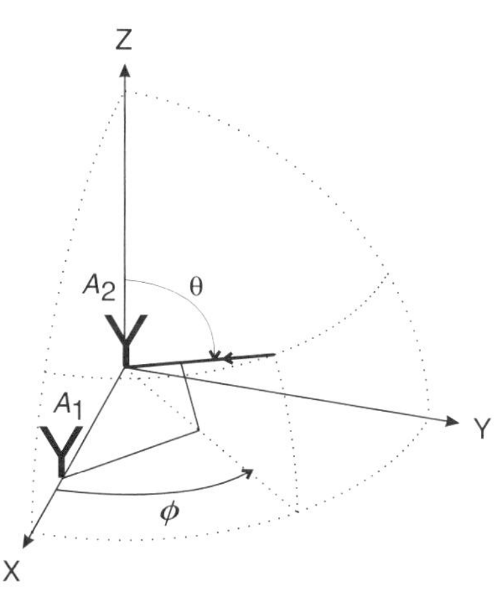

FIGURE 20.1 The geometry of the array and the relation to the coordinates.

time before antenna A_2. By exploiting a model for the measured signals at each antenna element it is possible to calculate the DOA from those signals.

Consider first a special case where the wave consists of a single sinusoidal, that is, a time-harmonic field [8]. For this case, Maxwell's equations can be written in a much simpler form and therefore most of the antenna literature uses time-harmonic fields. Of course, any signal can be written as a sum of sinusoidal signals and thus any signal can be represented using time-harmonic fields. The time-harmonic variation of the form $e^{j\omega t}$ is suppressed by using complex quantities, and the instantaneous values can be found by taking the real part of the product of the quantity and $e^{j\omega t}$. For example, the instantaneous value of a field at point r_p becomes

$$\vec{\varepsilon}\left(r_p\right) = \mathrm{Re}\left[\vec{E}\left(r_p\right)e^{j\omega t}\right] \tag{20.1}$$

If the source of the incident wave is located sufficiently far away from the array, the wave front can be considered plane. This is referred to as the *far-field* case. The field at a point r_p of an electromagnetic plane wave propagating along the **k** direction* can be written as

$$\vec{E}_{wave}\left(r_p\right) = \vec{\rho} s e^{-j\vec{k}\cdot\vec{r}_p} \tag{20.2}$$

where $\vec{\rho}$ denotes the polarization of the incident field and s the strength of the field. Note that $\vec{r}_p$ denotes the vector from origin to the point r_p, and thus the phase reference point is located at the center of the coordinate system. This type of wave is usually called a uniform plane wave (i.e., the amplitude, phase, and direction of the field is the same over a plane orthogonal to the propagation direction). For more details concerning wave propagation, see Reference [112] that treats propagation mechanisms and antenna fundamentals from a wireless communication perspective.

The action of the antenna is to convert the field into a voltage or current that is measured. How different type of antennas does this conversion is one of the fundamental topics of antenna theory. For an in-depth discussion about the properties of different antenna types, see the following classical antenna textbooks [7, 23, 33, 62, 126]. However, most antenna textbooks focus on transmitting properties and not on receiving modes. A compact summary that includes receiving properties can be found in Reference [83], and in Reference [61] equivalent circuits for antennas in receive mode are presented. In the communication literature, the impact of the antenna is usually neglected. However, in Reference [54] there is a brief introduction to antennas and propagation which can be useful.

An expression for the induced voltage can easily be derived using the concept of reciprocity and knowledge of the far-field radiation function of the receiving antenna $\vec{G}(r)$, where $\vec{r}$ is the direction of radiation. By using the results from References [61, 83], the induced voltage becomes

$$V = c\vec{G}\left(-\vec{k}\right)\cdot\vec{E}_{inc} \tag{20.3}$$

where c is a scalar defined as

$$c = -j\frac{4\pi}{k\eta I_{in}} \tag{20.4}$$

Here, k is the wave number, η the intrinsic impedance** of the propagation medium, I_{in} the input current used when calculating $\vec{G}$ and $\vec{E}_{inc}$ the incident field. Note that here $-\vec{k}$ denotes the direction from which

*The direction of the wave vector $\vec{k}$ is the direction in which the wave travels, and the magnitude $k = |\vec{k}|$ corresponds to the wave number $k = 2\pi/\lambda = \omega\sqrt{\mu\varepsilon}$, where λ is the wavelength. The symbols ε and μ denote the dielectric constant and the permeability of the medium, respectively.

**The intrinsic impedance is defined as $\eta = \sqrt{\mu/\varepsilon}$. For free space, the intrinsic impedance becomes $120\pi\,\Omega \approx 377\,\Omega$.

the wave is incident. The expression for the induced voltage is, thus, obtained once the radiation function of the antenna is obtained. For most common antenna types, the radiation function $\vec{G}$ (or approximations of it) can be found in antenna textbooks [7, 23, 33, 62, 126].

By combining Eqs. (20.2) and (20.3), the voltage can be written as

$$V = c\vec{G}\left(-\vec{k}\right)\cdot\vec{\rho}e^{-j\vec{k}\cdot\vec{r}_p}s = H\left(\theta,\phi\right)e^{-j\vec{k}\cdot\vec{r}_p}s \tag{20.5}$$

where the directional properties of the antenna and the polarization of the incident wave are combined into the scalar $H(\theta, \phi)$. Note that $H(\theta, \phi)$ corresponds to the response of the antenna to a sinusoidal signal at the carrier frequency (i.e., the frequency domain version of the antenna impulse response at the carrier frequency). Remember that the preceding voltage is written using the complex signal representation. The instantaneous value of the induced voltage is obtained in a manner similar to (20.1) as

$$V\left(t\right) = \text{Re}\left[sH\left(\theta,\,\phi\right)e^{-j\vec{k}\cdot\vec{r}_p}e^{j\omega t}\right] = \left|sH\left(\theta,\,\phi\right)\right|\cos\left(\omega t - \omega\tau + \angle sH\left(\theta,\,\phi\right)\right) \tag{20.6}$$

where $\tau = \vec{k}\cdot\vec{r}_p/\omega$ is a time delay that corresponds to the time it takes the wave to travel from the phase reference center (i.e., origin, to the point $\vec{r}_p$). Typically, the phase reference point is located at one antenna element and thus the time delay τ corresponds to the time needed by the wave to propagate over the array from the current element to the reference element.

It is important to remember that the preceding analysis was based on a single harmonic (i.e., a single frequency). In a practical application, the signal spans a band of frequencies and the field strength will be time varying. However, if the time delays τ are small compared with the reciprocal of the bandwidth of the signal, the instantaneous value of the induced voltage can be written as

$$V\left(t\right) \approx \left|s\left(t\right)H\left(\theta,\,\phi\right)\right|\cos\left(\omega t - \omega\tau + \angle s\left(t\right)H\left(\theta,\,\phi\right)\right) \tag{20.7}$$

This approximation is usually called the narrowband assumption in array signal processing [64], and essentially means that the signal $s(t)$ does not vary over the array, that is, $s(t) \approx s(t + \tau)$. To arrive at Eq. (20.7) it was also assumed that the response of the antenna element, $H(\theta, \phi)$, does not change significantly over the frequency band of interest. This is a reasonable assumption in almost all modern communication systems where the relative bandwidth usually is a few fractions of a percent. Therefore, in the light of Eq. (20.7) and the narrowband assumption, it is possible to use the complex signal representation of the voltage also for the time-varying case

$$V\left(t\right) = H\left(\theta,\,\phi\right)e^{-j\vec{k}\cdot\vec{r}_p}s\left(t\right) \tag{20.8}$$

This complex representation greatly simplifies the analysis, and is used throughout this chapter. A more detailed discussion on the preceding approximations may be found in Reference [126].

When estimating the DOA using an array of m antenna elements, m different voltages are measured. If these voltages are collected in a vector $\mathbf{x}(t)$, the resulting model becomes

$$\mathbf{x}\left(t\right) = \begin{bmatrix} x_1\left(t\right) \\ x_2\left(t\right) \\ \vdots \\ x_m\left(t\right) \end{bmatrix} = \begin{bmatrix} H_1\left(\theta,\,\phi\right)e^{-j\vec{k}\cdot\vec{r}_1} \\ H_2\left(\theta,\,\phi\right)e^{-j\vec{k}\cdot\vec{r}_2} \\ \vdots \\ H_m\left(\theta,\,\phi\right)e^{-j\vec{k}\cdot\vec{r}_m} \end{bmatrix} s\left(t\right) = \mathbf{a}\left(\theta,\,\phi\right)s\left(t\right) \tag{20.9}$$

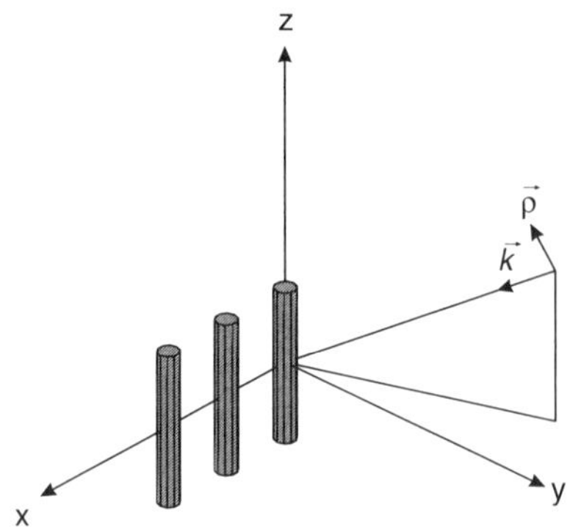

FIGURE 20.2 The geometry of the dipole array and a plane wave incident along $\vec{k}$ from the (θ, ϕ) direction.

where $H_i(\theta, \phi)$ denotes the response of antenna element i and $\vec{r}_i$ the location vector from the phase reference center to antenna element i. The $m \times 1$ vector $\mathbf{a}(\theta, \phi)$ models the spatial response of the array due to an incident plane wave from the (θ, ϕ) direction. Therefore, the vector $\mathbf{a}(\theta, \phi)$ is usually called steering vector or array response vector in the array processing literature.

Example: Uniform Linear Array of Dipoles

Consider a plane wave incident on an array of m = 3 thin dipoles of length l as illustrated in Fig. 20.2. The dipoles are oriented along the z-axis and spaced equidistantly along the x-axis with a separation distance of d. Such an arrangement is referred to as a uniform linear array (ULA). In this case, the wave reaches the element at the origin first, and after some time the other elements. The time delay for antenna element i, where $i = 0, 1, 2$, and 0 corresponds to the origin, is easily calculated as

$$\tau_i = \frac{\vec{k} \cdot \vec{r}_i}{\omega} = \frac{k}{\omega}\left(\vec{x}\sin\theta\cos\phi + \vec{y}\sin\theta\sin\phi + \vec{z}\cos\theta\right) \cdot \vec{x}di = \frac{kdi}{\omega}\sin\theta\cos\phi \tag{20.10}$$

Before the full expression for the induced voltages over the array can be obtained, the directional properties of the dipole element must be found. The far-field radiation function G of the dipole element is included in almost any antenna handbook [7, 23, 33, 62, 126] as

$$\vec{G} = \vec{\theta}\frac{j\eta I_{in}}{2\pi \sin\left(\frac{kl}{2}\right)}\left[\frac{\cos\left(\frac{kl}{2}\cos\theta\right) - \cos\frac{kl}{2}}{\sin\theta}\right] \tag{20.11}$$

Now, the antenna response $H(\theta, \phi)$ can be found using Eqs. (20.4) and (20.5) as

$$H(\theta, \phi) = c\vec{G}\left(-\vec{k}\right) \cdot \vec{\rho} = \frac{\lambda}{\pi \sin\frac{kl}{2}}\left[\frac{\cos\left(\frac{kl}{2}\cos\theta\right) - \cos\frac{kl}{2}}{\sin\theta}\right] \tag{20.12}$$

where $\vec{\rho} = -\vec{\theta}$, as indicated in Figure 20.2. The final array model is then obtained by using Eqs. (20.5) and (20.12) as

$$\mathbf{x}(t) = \frac{\lambda}{\pi \sin\frac{kl}{2}} \left[\frac{\cos\left(\frac{kl}{2}\cos\theta\right) - \cos\frac{kl}{2}}{\sin\theta} \right] \begin{bmatrix} 1 \\ e^{-jkd\sin\theta\cos\phi} \\ e^{-j2kd\sin\theta\cos\phi} \end{bmatrix} s(t) = \mathbf{a}(\theta, \phi)s(t) \tag{20.13}$$

Note that the phase shifts $\psi = kd \sin \theta \cos \phi$ (often called electrical angle) only depend on the element positions and not on the type of antenna element. The impact of the antenna element is included in the antenna response $H(\theta, \phi)$. In this example, the elements are identical and the antenna response is simply a scalar that could be absorbed into the signal $s(t)$. If the antenna response of each element is different, it should be incorporated into the steering vector $\mathbf{a}(\theta, \phi)$, as implied by Eq. (20.9). The steering vector for a dipole array was calculated in this example, but the steering vector for most antenna elements can be found in a similar manner by using formulas of the radiation function $\vec{G}$ from antenna textbooks. A more detailed derivation of the array response of a dipole array, that also includes the effects of mutual coupling, may be found in Reference [127].

The derivation of the measured voltages considered only one incident wave. When several waves are incident on the array, the superposition principle can be applied if the antenna elements and the receiver are linear. To simplify the notation in the following analysis, the waves are assumed to arrive in the *xy* plane ($\theta = 90°$) (see Fig. 20.1). The model when p uniform plane waves are incident on an array of m elements can thus be written as

$$\mathbf{x}(t) = \sum_{l=1}^{p} \mathbf{a}(\phi_l)s_l(t) = \mathbf{A}(\phi)\mathbf{s}(t) \tag{20.14}$$

where

$$\mathbf{A}(\phi) = \begin{bmatrix} \mathbf{a}(\phi_1) & \mathbf{a}(\phi_2) & \cdots & \mathbf{a}(\phi_p) \end{bmatrix} \tag{20.15}$$

$$\mathbf{s}(t) = \begin{bmatrix} s_1(t) & s_2(t) & \cdots & s_p(t) \end{bmatrix}^T \tag{20.16}$$

Note that the vector of measured voltages, at time t, $\mathbf{x}(t)$ is $m \times 1$, the steering matrix $\mathbf{A}(\phi)$ is $m \times p$, and the signal vector $\mathbf{s}(t)$ is $p \times 1$. The DOAs are contained in the $p \times 7$ parameter vector $\boldsymbol{\phi}$.

In all measurement situations, noise inevitably appears and it is typically difficult to model. Essentially, noise represents everything that does not obey the assumed model and here it is included as an additive term as

$$\mathbf{x}(t) = \mathbf{A}(\phi)\mathbf{s}(t) + \mathbf{n}(t) \tag{20.17}$$

There are many sources of noise [120], for example, environmental noise, such as cosmic noise, atmospheric absorption noise, and solar noise. It can further be man-made noise, such as jammers and power tools. The receiver also generates some noise, such as thermal noise, shot noise, and flicker noise. Often, receiver noise is the dominating noise source; and then additive white Gaussian noise is a good model. In this chapter, the noise is assumed to be both spatially and temporally white and Gaussian distributed.

$$E\{\mathbf{n}(t)\} = 0 \quad E\{\mathbf{n}(t)\mathbf{n}^H(s)\} = \sigma^2\mathbf{I}\delta_{ts} \quad E\{\mathbf{n}(t)\mathbf{n}^T(s)\} = \mathbf{0} \ \forall \ t, s \tag{20.18}$$

The motivation for this assumption is that if there are many sources of noise, the sum is Gaussian distributed according to the central limit theorem [93]. Also, the further analysis of direction finding performance is greatly simplified by assuming white Gaussian noise. Therefore, an overwhelming part of the literature of array signal processing assumes white Gaussian noise. For other noise models, such as colored noise or noise of other distributions, see Reference [51].

The transmitted signal is in a similar manner assumed to be Gaussian distributed with

$$E\{\mathbf{s}(t)\}=0 \quad E\{\mathbf{s}(t)\mathbf{s}^H(s)\}=\mathbf{P}\delta_{ts} \quad E\{\mathbf{s}(t)\mathbf{s}^T(s)\}=\mathbf{0}\ \forall\ t,s \tag{20.19}$$

Note that different source signals may be correlated, leading to a nondiagonal source covariance matrix **P**. The assumption of temporally white and Gaussian signals is not critical. It is only used when deriving the maximum likelihood estimator and the Cramér-Rao lower bound. The actual performance of the methods under study is typically independent on the source distribution, but depends only on **P** [91], unless the distribution or the signal itself is known. For other choices of signal statistics, see Reference [51]. A crucial assumption is, however, that the signal is uncorrelated with the noise.

Most DOA estimation schemes rely on the properties of the second-order moment of the measured voltages $\mathbf{x}(t)$ (i.e., the spatial covariance matrix). Therefore, some characteristics of the spatial covariance matrix of the measured voltages, $\mathbf{x}(t)$, are presented next. The latter is obtained as

$$\mathbf{R}=E\left\{\mathbf{x}(t)\mathbf{x}(t)^H\right\}=\mathbf{A}E\{\mathbf{s}(t)\mathbf{s}^H(t)\}\mathbf{A}^H+E\{\mathbf{n}(t)\mathbf{n}^H(t)\}=\mathbf{APA}^H+\sigma^2\mathbf{I} \tag{20.20}$$

Note that the cross-terms vanish because the signal and the noise are assumed to be uncorrelated and the noise is zero mean.

Many of the DOA estimation schemes presented in the literature rely on the fact that the eigendecomposition of the covariance matrix **R** can be written as a sum of two parts [64]. One part consists of eigenvectors corresponding to eigenvalues equal to the noise variance, and a second part is related to the signal

$$\mathbf{R}=\sum_{k=1}^{m}\lambda_k\mathbf{e}_k\mathbf{e}_k^H=\mathbf{E}_s\mathbf{\Lambda}_s\mathbf{E}_s^H+\sigma^2\mathbf{E}_n\mathbf{E}_n^H \tag{20.21}$$

where $\mathbf{E}_s=[\mathbf{e}_1,\ldots,\mathbf{e}_p]$ denotes the signal eigenvectors, $\mathbf{E}_n=[\mathbf{e}_{p+1},\ldots,\mathbf{e}_n]$ denotes the noise eigenvectors, and $\mathbf{\Lambda}_s=\mathrm{diag}[\lambda_1,\ldots,\lambda_p]$ denotes the signal eigenvalues. This decomposition, which sometimes is referred to as spectral factorization, is used extensively when dealing with subspace-based estimation in later sections.

Remarks

1. The model derived here for the case of an electromagnetic wave incident on an antenna array is closely connected to the sonar problem, where an acoustic wave is incident on a hydrophone array [75, 92]. Applications of similar models for medical imaging can be found in Reference [37] and also for chemical sensor arrays in Reference [88].
2. If the antenna is located close to the source of radiation, the wave front is spherical and the plane wave assumptions are no longer valid. However, if the spherical nature of the wave is correctly modeled, the range as well as the DOA can be estimated [121].
3. Only direction finding using an antenna array is considered in this chapter, but other antenna arrangements are also used. For instance, a mechanically steered antenna [120] has been used in radar for a long time. Other possibilities are to employ electromagnetic vector-sensors [87], switched parasitic elements [102], or higher order electromagnetic modes [129].
4. The linear array, which only can be used to estimate either the azimuth or the elevation, is by far the most analyzed type of array. However, two-dimensional (2D) arrays have also been analyzed.

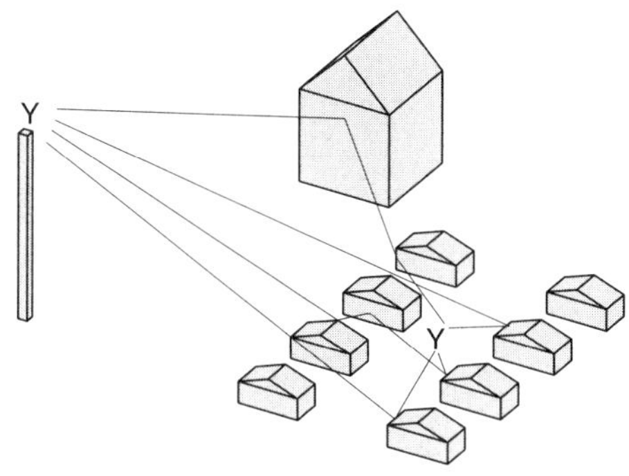

FIGURE 20.3 Typical suburban scatter environment with local scattering and a dominant scatterer.

See References [51, 52] and the references therein for results on 2D arrays, and, in particular, for circular arrays that are used to estimate both the elevation and the azimuth angles.

5. The expression for the steering vector was calculated analytically in this section. In general, these expressions do not correspond to a measured response of an antenna array. Typically, differences in temperature, aging of components, and changes in the electromagnetic environment of the array change the array response. Some types of calibration, of both the antenna system and the receiver, is therefore usually needed. Often the response of the array is measured and stored in a lookup table. For an example of the calibration procedures of a high-performance digital beam-forming antenna, see Reference [100].

20.2.2 Spatial Channel Model for Wireless Communication Systems

The propagation situation in a practical wireless communication channel is very complex. The signal that is transmitted from the mobile phone reaches the base station antenna through many different paths. At the base station antenna, it appears as if energy is arriving over an angular sector instead of a distinct direction. Therefore, the model in the previous section that assumed a plane wave incident from a distinct direction needs to be modified. Numerous textbooks on radio channel characteristics have been written with applications in cellular networks [12, 15, 94].

Consider the propagation scenario in Fig. 20.3 that contains a mobile antenna, a base station antenna, and numerous scattering objects. Several local scattering objects (houses) are located in the vicinity of the mobile and a large scattering object (large house) is located farther away. The transmitted field from the mobile experiences electromagnetic scattering, reflection, refraction, and diffraction before it reaches the base station antenna where a voltage finally is induced. For an introduction to scattering and wave propagation, see References [8, 112] and the references therein.

By exploiting fundamental physical principles, it is possible to arrive at a detailed channel model that includes most of the propagation phenomena encountered in practice. In fact, because of the recent developments in computational power, it is possible to approximately solve Maxwell's equations for the propagation scenario at hand. Through massive calculations, the channel impulse response of a practical scenario can be found if the geometry and electrical properties of the scenario are specified. Probably the most popular solution is based on a concept called ray tracing, where the propagating field is viewed as a ray. Although computationally intensive, accurate results can be obtained [22, 32, 113]. These ray-tracing schemes can in some cases replace measurement campaigns by using them to simulate data if an accurate database of the scenario is available. However, for the purpose of estimating DOA, these models are way too complex to be useful. The primary use of ray-traying schemes is in cell-planning tools.

A much simpler, yet detailed enough for the purpose of DOA estimation, channel model may be derived by revisiting Fig. 20.3. The received field at the base station antenna can be seen as a sum of many rays incident on the base station. By modeling each ray, using the antenna model from the previous subsection, a simple channel model may be obtained as

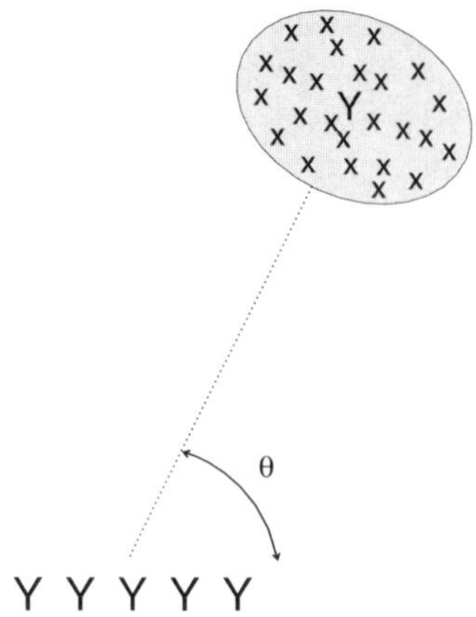

FIGURE 20.4 Local scattering model.

$$\mathbf{x}(t) = \sum_{l=1}^{p} \rho_l(t)\mathbf{a}(\phi_l(t))s(t - \tau_l(t)) \tag{20.22}$$

where p denotes the number of propagation paths; ρ_l, the complex attenuation of path l; ϕ_l, the DOA of path l; and τ_l, the time delay of path l. Note that the model is easily extended to include many mobile users, assuming that the superposition principle is applicable. Depending on the type of channel that needs to be modeled, different choices of the parameters, p, ρ, and τ can be made. In fact, a large number of different channel models that include the spatial dimension (i.e., ϕ) have been presented in the last decades. For an excellent overview of spatial channel models see Reference [35].

A common model for the propagation environment, especially in macrocell applications, is the so called local scattering model [161]. It is assumed that the transmitted signal is scattered by many scattering elements in the close vicinity of the mobile (Fig. 20.4). The direction to the center of the local cluster of scatters around the mobile is typically referred to as nominal DOA ϕ_0. At the mobile, however, the received signal appears as if coming from a distributed source because of the angular spread. Typically, it is assumed that the time delays of the different paths within the cluster can be modeled as phase shifts (i.e., a narrowband assumption is made). If, in each path, this phase shift is incorporated into a complex amplitude ρ, the channel model can be written as

$$\mathbf{x}(t) = \mathbf{v}s(t) + \mathbf{n}(t) \qquad t = 1, \ldots, N \tag{20.23}$$

As before, $s(t)$ is the signal waveform and $\mathbf{n}(t)$ is the noise vector. The vector $\mathbf{v}$ is termed the *spatial signature*, and using the preceding assumptions it can be written as

$$\mathbf{v} = \sum_{l=1}^{p} \rho_l \mathbf{a}(\phi_l) \tag{20.24}$$

where ρ_l is a complex amplitude and ϕ_l is the DOA of the lth signal path from the cluster. The number of paths in the cluser is denoted p. In the case of no multipath or negligible DOA spread, the model applies with $p = 1$ and essentially reduces to the model in Section 20.2.1.

Several possibilities for modeling the complex amplitudes ρ and the angular distribution of ϕ exist. A common choice is to use a fixed value for the amplitude of ρ, while assuming a random phase uniformly distributed over $[0, 2\pi]$ [35]. One can also argue that each ray, in fact, results from a continuum of scatterers, suggesting a Gaussian distribution for ρ (i.e., uniformly distributed phase and Rayleigh distributed amplitude). The angular distribution is accurately modeled in many types of environments as Gaussian [97]. The mean of the distribution ϕ_0, is the scatter center, whereas the standard deviation σ_ϕ depends on the type of scenario and the distance between the base station and the scatter center. In a typical urban scenario, the DOA spread σ_ϕ can be on the order of 10° [97], whereas significantly smaller clusters can be expected in a suburban or rural area. Larger delay spreads are easily included in the model by inserting multiple clusters. The latter can also be used to include several users in the model, see for instance References [35, 161]. Some more details on spread source modeling are given in Section 20.4.2.1.

20.3 Estimation for Point Sources

Estimating the DOA of a wave emanating from a point source is by far the most analyzed DOA estimation scenario. The problem of DOA estimation is significantly simplified by assuming that each source results in one plane wave arriving at the receiving antenna from a specific direction (θ, ϕ). This assumption is usually called the point source assumption, and is a reasonable assumption in radar and macrocellular communication applications where the receiver array is located relatively far from the source. For applications where the distance is smaller, such as micro- and picocell application, the source is better modeled as a spread source. In this section, DOA estimation for point sources is considered, whereas spread sources are covered in the next section.

Because the point source case is the most intensively studied case of DOA estimation, a vast literature presenting numerous DOA estimation schemes exists. Here, a few of these algorithms that are useful for DOA estimation in mobile communication environments are reviewed. Therefore, the aim is not to give a complete overview of all available methods in the literature. For a wider selection of methods, see References [43, 50, 52, 64] and further references therein.

This section first reviews the data model and the statistical assumptions in Section 20.3.1, followed by a discussion of spectral-based techniques in Section 20.3.2. Next, a few methods based on parametric estimation of the DOAs are reviewed in Section 20.3.3. The overview of algorithms is finished by presenting methods that exploit a special array structure in Section 20.3.4. Finally, the properties of the different estimators are summarized in Section 20.3.6.

20.3.1 Problem Formulation and Assumptions

The assumption that each source results in one plane wave arriving at the receiving antenna from a specific direction (i.e., point source) renders it possible to use the data model from Section 20.2.1. Thus, the model when p waves are incident on the array, can be written as

$$\mathbf{x}(t) = \mathbf{A}(\phi)\mathbf{s}(t) + \mathbf{n}(t) \tag{20.25}$$

Here, the vector of measured voltages at time t, $\mathbf{x}(t)$ is $m \times 1$, the steering $\mathbf{A}(\phi)$ is $m \times p$, the signal vector $\mathbf{s}(t)$ is $p \times 1$, and the noise vector $\mathbf{n}(t)$ is $m \times 1$. The DOAs are contained in the $p \times 1$ parameter vector ϕ. It is assumed that $m > p$ (i.e., more antennas than signals). Note that the explicit form of the steering matrix can be found using the technique outlined in Section 20.2.1. Furthermore, it is assumed that the steering matrix $\mathbf{A}$ is of full rank, and that the first and second moments of the signal and the noise are as given in Section 20.2.1 in Eqs. (20.18) and (20.19).

Because all of the DOA estimation methods in the following sections exploit the properties of the covariance matrix of the measured voltages, the fundamental properties of it are reviewed first. By using the preceding statistical assumptions, the covariance matrix of the measured voltages becomes

$$\mathbf{R} = E\left\{\mathbf{x}(t)\mathbf{x}(t)^H\right\} = \mathbf{A}E\left\{\mathbf{s}(t)\mathbf{s}(t)^H\right\}\mathbf{A}^H + E\left\{\mathbf{n}\mathbf{n}^H\right\} = \mathbf{APA}^H + \sigma^2\mathbf{I} \tag{20.26}$$

where $\mathbf{A} = \mathbf{A}(\phi)$. Note that the cross terms vanish because the signal and the noise are assumed to be uncorrelated. Assume that there is one vector $\mathbf{y}$ that is orthogonal to the steering matrix $\mathbf{A}$, or

$$\mathbf{A}^H\mathbf{y} = \mathbf{0} \tag{20.27}$$

Then $\mathbf{Ry} = \sigma^2\mathbf{y}$ (i.e., $\mathbf{y}$ is an eigenvector of $\mathbf{R}$ with corresponding eigenvalue σ^2). Because the steering matrix $\mathbf{A}$ is full rank with dimensions $m \times p$, there are $m - p$ independent vectors $\mathbf{y}$ that are perpendicular to $\mathbf{A}$. Actually, it is possible to write the eigendecomposition of the covariance matrix as a sum of two parts. One part consists of eigenvectors corresponding to eigenvalues equal to the noise variance, and a second part is related to the signal

$$\mathbf{R} = \mathbf{E}\Lambda\mathbf{E}^H = \mathbf{E}_s\Lambda_s\mathbf{E}_s^H + \sigma^2\mathbf{E}_n\mathbf{E}_n^H \tag{20.28}$$

Furthermore, by exploiting the orthonormality of the eigenvectors

$$\mathbf{E}_s\mathbf{E}_s^H + \mathbf{E}_n\mathbf{E}_n^H = \mathbf{I} \tag{20.29}$$

the covariance matrix can be written as

$$\mathbf{R} = \mathbf{E}_s\left(\Lambda_s - \sigma^2\mathbf{I}\right)\mathbf{E}_s^H + \sigma^2\mathbf{I} = \mathbf{E}_s\tilde{\Lambda}\mathbf{E}_s^H + \sigma^2\mathbf{I} \tag{20.30}$$

Comparing Eq. (20.30) with Eq. (20.26) reveals that the subspace spanned by the columns of $\mathbf{E}_s$ must be equal to the subspace spanned by the columns of $\mathbf{AP}$ or

$$\mathcal{R}\left\{\mathbf{E}_s\right\} = \mathcal{R}\left\{\mathbf{AP}\right\} \subseteq \mathcal{R}\left\{\mathbf{A}\right\} \tag{20.31}$$

with equality if the signal covariance matrix $\mathbf{P}$ is full rank. The subspace spanned by $\mathbf{A}$ (or $\mathbf{E}_s$) is usually called the signal subspace, and the corresponding space spanned by $\mathbf{E}_n$ is therefore called the noise subspace. Note that because of the orthogonality of the eigenvectors, these spaces are orthogonal. The fact that all the information about the signal is contained in a p-dimensional subset of the m-dimensional space is important in the derivation of the subspace based methods presented later.

Almost all methods discussed in the following sections exploit the properties of the covariance matrix of the measured voltages $\mathbf{R}$. However, the induced voltages $\mathbf{x}(t)$ are measured and not the covariance matrix. Usually, $\mathbf{R}$ is estimated as

$$\hat{\mathbf{R}} = \frac{1}{N}\sum_{t=1}^{N}\mathbf{x}(t)\mathbf{x}^H(t) \tag{20.32}$$

which is the unstructured maximum likelihood estimate of the covariance matrix in Eq. (20.26).

20.3.2 Spectral-Based Techniques

As the name suggests, spectral-based methods rely on calculating a spatial spectrum and finding the DOAs as the location of peaks in the spectrum. These methods were the first to be developed and are easy to apply. Probably the most widely used method of obtaining estimates of the DOAs is the beamforming method. As the name beamforming suggests, the received energy is focused to one direction (or beam) at a time. This can be expressed as

$$y(t) = \mathbf{w}^H \mathbf{x}(t) \tag{20.33}$$

where the weighing vector $\mathbf{w}$ can be seen as a spatial filter that emphasizes one particular direction. Given samples $y(1), y(2), \ldots, y(N)$, the output power is measured by

$$P(\mathbf{w}) = \frac{1}{N}\sum_{t=1}^{N} |y(t)|^2 = \frac{1}{N}\sum_{t=1}^{N} \mathbf{w}^H \mathbf{x}(t)\mathbf{x}^H(t)\mathbf{w} = \mathbf{w}^H \hat{\mathbf{R}} \mathbf{w} \tag{20.34}$$

where $\hat{\mathbf{R}}$ is defined in Eq. (20.32). Many different choices of the weighing vector $\mathbf{w}$ can be made leading to different properties of the beamforming schemes [120, 139].

20.3.2.1 Conventional Beamforming

If the weighing vector is chosen to maximize the received power in a certain direction ϕ [139], as

$$\mathbf{w}_{BF} = \frac{\mathbf{a}(\phi)}{\sqrt{\mathbf{a}^H(\phi)\mathbf{a}(\phi)}} \tag{20.35}$$

the classical spatial spectrum is obtained

$$P_{BF}(\phi) = \frac{\mathbf{a}^H(\phi)\hat{\mathbf{R}}\mathbf{a}(\phi)}{\mathbf{a}^H(\phi)\mathbf{a}(\phi)} \tag{20.36}$$

This spectrum is in array processing literature referred to as the conventional (or Bartlett) beamformer, because this actually is a natural extension of the classical Fourier-based spectral analysis with different window functions [10, 122]. In fact, if a uniform linear array of isotropic elements is used, the spatial spectrum in Eq. (20.36) is a spatial analog of the classical periodogram in time series analysis. Note that other types of arrays correspond to non-uniform sampling schemes in time-series analysis. As with the periodogram, the spatial spectrum has a resolution threshold. Waves arriving with electrical angle separation* less than $2\pi/m$ can not be resolved with this method. For example, using a five-element ULA with an element separation of $d = \lambda/2$ results in a resolution threshold of 23°. Other choices of weighting vectors $\mathbf{w}$, which result in lower resolution thresholds, were therefore investigated.

20.3.2.2 Capon's Beamformer

One of the most popular beamforming methods that to some extent alleviates the limitations of the conventional beamformer is Capon's beamformer [20, 67]. This beamformer attempts to minimize the power contributed by noise and any signals coming from other directions than the looking direction, while maintaining a fixed gain in the look direction. This type of beamformer is sometimes also referred to as minimum variance distortionless response (MVDR) filter in acoustics literature. The Capon weighting vector is

$$\mathbf{w}_{CAP} = \frac{\hat{\mathbf{R}}^{-1}\mathbf{a}(\phi)}{\mathbf{a}^H(\phi)\hat{\mathbf{R}}^{-1}\mathbf{a}(\phi)} \tag{20.37}$$

and if inserted in Eq. (20.34), the MVDR spatial spectrum becomes

*The electrical angle is defined as $kd \cos \phi$.

TABLE 20.1 Resolution Thresholds for Beamforming (BF), Capon, and MUSIC Methods

BF	Capon	MUSIC
$\Delta = \frac{2\pi}{m}$	$\Delta = 8.71\left[\frac{1}{m^5\xi}\right]^{\frac{1}{4}}$	$\Delta = \left\{\Delta: \frac{2880(m-2)}{Nm^4\Delta^4}\left[1+\sqrt{1+\frac{Nm^2\Delta^2}{60(m-1)}}\right]=\xi\right\}$

Note: Here, m denotes the number of antennas, ξ, the SNR, and N, the number of snapshots.

$$P_{BF}(\phi) = \frac{1}{\mathbf{a}^H(\phi)\hat{\mathbf{R}}^{-1}\mathbf{a}(\phi)} \tag{20.38}$$

Although more complex than the conventional beamformer, Capon's method offers significantly reduced resolution threshold. The lower resolution threshold (reduced spectral leakage) is achieved at the cost of reduced noise suppression capability [20]. A formula for calculating the resolution threshold for Capon's method assuming a perfect estimate of $\hat{\mathbf{R}}$ is given in Reference [117] (Table 20.1).

A large number of alternative methods for beamforming has been presented in the literature, see Reference [139] for an overview. An example of an application of conventional beamforming in a mobile communication scenario can be found in Reference [84] where the performance gain by exploiting conventional beamforming in a Global System for Mobile (GSM) communications network is investigated.

Still the resolution threshold for beamforming methods is quite high and that was one of the motivations for the interest in the so-called subspace methods that are described next.

20.3.2.3 Multiple Signal Classification

Subspace-based methods rely on observations concerning the eigendecomposition of the covariance matrix into a signal subspace and a noise subspace, as discussed in the previous subsection. One of the most popular subspace methods, multiple signal classification (MUSIC), was introduced in Reference [114]. The method is based on the observation in Eq. (20.27), that the noise eigenvectors are perpendicular to the steering matrix or the signal subspace. The algorithm calculates the noise subspace using an eigendecomposition of the estimated covariance matrix in Eq. (20.32). Then, the estimates of the DOAs are taken as those ϕ that give the smallest value of $\mathbf{a}^H(\phi)\,\hat{\mathbf{E}}_n$ (i.e., the values that result in a steering vector farthest away from the noise subspace). Usually this is formulated as finding the p largest peaks in the MUSIC spectrum

$$P_{MU}(\phi) = \frac{1}{\mathbf{a}^H(\phi)\hat{\mathbf{E}}_n\hat{\mathbf{E}}_n^H\mathbf{a}(\phi)} \tag{20.39}$$

Note that the eigenvectors $\hat{\mathbf{E}}_n$ are easily obtained by either an eigendecomposition of the sample covariance matrix, or a singular value decomposition (SVD) of the data matrix, and that reliable numerical routines for eigendecomposition and SVD are included in most software packages.

The main motivation for introducing the subspace methods was to reduce the resolution threshold of the beamforming methods. The resolution threshold of MUSIC depends in a complicated manner on several parameters such as number of samples, number of elements, and the signal to noise ratio (SNR). Therefore, the formula for the resolution threshold becomes more complicated than for the beamforming methods (see Table 20.1). First, the different resolution thresholds of conventional beamforming, Capon, and MUSIC are examined in a simulation example.

Example: Beamforming, Capon, and Multiple Signal Classification

The beamforming (BF), Capon, and MUSIC spectra are shown in Fig. 20.5 for the case when two waves are incident upon a ULA of five elements with half-wavelength spacing. The true directions are 85° and

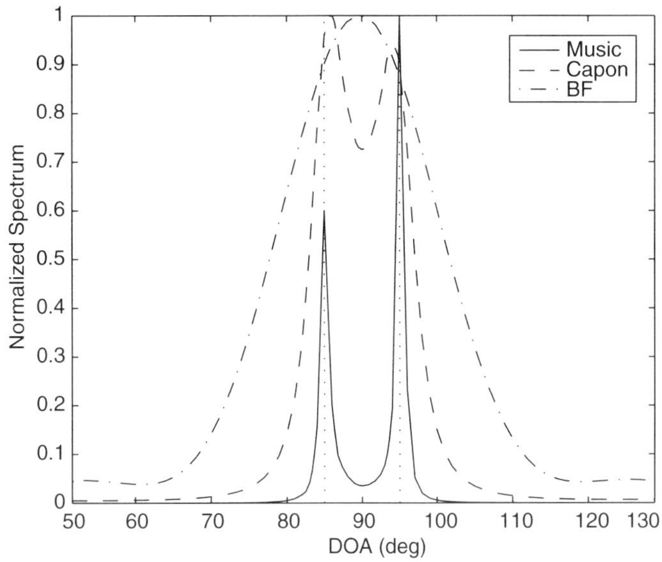

FIGURE 20.5 Normalized spectra of MUSIC (solid), Capon (dashed), and beamforming (dash-dotted) methods vs. DOA. The true DOAs are indicated by dotted vertical lines. A ULA of five elements with half-wavelength spacing is used when $N = 100$ samples of the induced voltages are taken at an SNR of 15 dB.

95° and 100 samples of the induced voltages are taken, that is, $N = 100$ in Eq. (20.32). Furthermore, the sources are assumed to be uncorrelated and of equal strength, that is, $\mathbf{P} = 10^{(15/10)}\,\mathbf{I}$ with a noise power $\sigma^2 = 1$, resulting in an SNR of 15 dB. It is clear that the conventional beamforming method [Eq. (20.36)] fails to resolve the sources, because the angular separation in this example (10°) is less then the resolution threshold of conventional beamforming (23°). Capon's method just barely resolves the sources, whereas the MUSIC method results in two well-defined peaks at the true DOAs.

20.3.2.4 Resolution Threshold

A critical property of the spectral-based methods is the ability to resolve closely spaced sources. For the conventional beamforming method a relatively simple expression for the minimum separation in electrical angle can easily be obtained as $\Delta = 2\pi/m$. A slightly more complicated expression for Capon's beamformer can be obtained for the limiting case of a perfect covariance estimate and uncorrelated sources (see Table 20.1). For correlated sources, see Reference [117]. Note that the resolution threshold of Capon's beamformer decreases with increased SNR (ξ), whereas the conventional beamformer is independent of the power. The resolution limit for the MUSIC method is more complex [58] and depends on the number of samples as well as SNR ξ and number of elements m. For more detailed descriptions on the assumptions on the resolution limits of Capon and MUSIC, see References [58, 117].

Example: Resolution Thresholds of Beamforming, Capon, and Multiple Signal Classification

The resolution threshold of the beamforming (BF), the Capon, and the MUSIC methods are shown in Fig. 20.6 for the case when two waves are incident upon an ULA with 3 to 12 elements with half-wavelength spacing. The waves are assumed to arrive symmetrically about the array normal and 100 samples of the induced voltages are taken, that is, $N = 100$ in Eq. (20.32). Furthermore, the sources are assumed to be uncorrelated and of equal strength (i.e., $\mathbf{P} = 10^{(15/10)}\,\mathbf{I}$ with a noise power $\sigma^2 = 1$ resulting in an SNR = 15 dB). It is clear that the conventional beamforming experiences the highest resolution threshold. Capon's method has a lower resolution threshold and the MUSIC method a much lower threshold. Note that if the SNR is increased, the threshold of MUSIC and Capon drop further, whereas the beamforming method is not affected. In the case of MUSIC, the threshold also drops slightly if more samples N are taken.

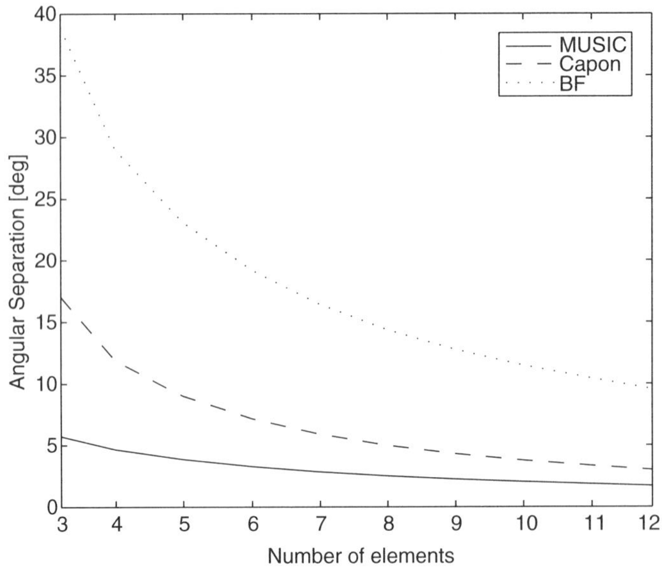

FIGURE 20.6 The resolution thresholds of MUSIC (solid), Capon (dashed), and beamforming (dash-dotted) methods vs. the number of half-wavelength-spaced antenna elements when 100 samples are taken at an SNR of 15 dB.

The high resolution of MUSIC appears to make it the preferred method over beamforming and Capon at all times. Unfortunately, the MUSIC method also suffers from a few drawbacks that prevents its application to certain problems.

1. A problem in practical applications is that it is assumed that the number of sources is known, to separate the eigenvectors into noise eigenvectors and signal eigenvectors. In theory, all noise eigenvectors should correspond to the same eigenvalue. However, in practice these differ somewhat and some type of scheme is needed to determine the number of sources. Several relatively simple and straightforward methods have been devised that solve this problem [91, 147, 149]. Still, in difficult scenarios these methods can fail and the performance of the MUSIC algorithm obviously degrades.
2. A related problem is the case when the signals are correlated. In that case, the covariance matrix is not full rank and the separation into signal and noise subspace once again becomes difficult. One typical example of correlated signals is in a multipath scenario, where several replicas of the signal arrives through different paths. However, in many scenarios there is a cluster of multipaths resulting in less correlation, thus rendering the MUSIC method still applicable. For correlations less than 0.8, MUSIC typically gives reasonable DOA estimates. If the array structure is regular, it may be possible to decorrelate the signal, thus allowing the use of MUSIC (and similar subspace-based estimators). Two popular methods are based on modifying the estimated covariance $\hat{\mathbf{R}}$ by straightforward operations. For two correlated signals the method of forward–backward averaging [101, 107] has been applied successfully. This requires essentially that the sensor placement is symmetrical with respect to the array center. If the array is a ULA, the method of spatial smoothing [41, 119] is applicable. This allows a general source correlation. Some details on these techniques are given in Section 20.3.4.
3. For scenarios with low SNR and/or a small number of samples, the resolution of the MUSIC algorithm is degraded. Several schemes for improving this situation have been designed, of which a weighted MUSIC scheme called the Min-Norm algorithm is the best known [66, 70, 109]. A comparison between MUSIC-like estimators can be found in Reference [58].

These drawbacks were the prime motivation of introducing the parametric approaches that are examined in the next section.

20.3.3 Parametric Methods

Although the spectral-based methods presented in the previous section are computationally attractive and simple to apply, they do not always yield a sufficient accuracy. This is, in particular, the case for scenarios with highly correlated signals. An alternative is to employ so-called parametric array processing methods that directly estimate the DOAs without first calculating a spectrum. These algorithms yield a higher performance in terms of accuracy and resolution by exploiting the underlying data model to a larger extent. The cost for this performance increase is a higher complexity and more computations, because typically a multi-dimensional search for the parameters is needed.

A new performance measure is needed because the DOA estimates are obtained without computing a spectrum, thus making beamwidth and resolution threshold less important. Instead, two statistical properties of the DOA estimates are usually used as a performance measure.

- *Consistency* — An estimate is consistent if it converges to the true value when the number of data tends to infinity.
- *Statistical efficiency* — An estimator is statistically efficient if it asymptotically attains the Cramér-Rao bound (CRB), which is a lower bound on the covariance matrix of any unbiased estimator (see Reference [59]).

Of these, the statistical efficiency is the most important because almost all the methods presented here are consistent. Formulas for calculating the lower bound CRB are given in Section 20.3.6.

Several parametric methods are presented in the following sections, starting with the maximum likelihood method, which is one of the main systematic techniques of statistical signal processing. An alternative method with similar performance, subspace fitting, is also introduced. Finally, methods for special array structures as well as a data preprocessing technique called beamspace processing are presented.

20.3.3.1 Maximum Likelihood

Perhaps the most well-known and frequently used model-based estimation technique in signal processing is the maximum likelihood (ML) method; see Reference [59] for a general description. The ML method assumes a model for the signals and also a statistical framework for the data generation. Here, the signal model introduced in Section 20.2.1 is used. This method is usually referred to as the stochastic maximum likelihood (SML) method, because following the assumptions in Section 20.2.1, the signal is assumed to be stochastic with a Gaussian distribution. By using the model in Eq. (20.17) and the statistical assumptions in Section 20.2.1, the negative log likelihood function to be minimized becomes (with the parameter-independent terms ignored)

$$l_{SML}\left(\phi, \mathbf{P}, \sigma^2\right) = \log\left|\mathbf{R}\right| + \mathbf{Tr}\left\{\mathbf{R}^{-1}\hat{\mathbf{R}}\right\} \tag{20.40}$$

where $\mathbf{R}$ and $\hat{\mathbf{R}}$ are defined in Eqs. (20.26) and (20.32). This expression can be put in a concentrated form [17, 53, 124]. For fixed ϕ, the minimizing values of $\mathbf{P}$ and σ^2 becomes

$$\hat{\sigma}^2\left(\phi\right) = \frac{1}{m-p}\operatorname{Tr}\left\{\mathbf{P}_A^{\perp}\hat{\mathbf{R}}\right\} \tag{20.41}$$

$$\hat{\mathbf{P}}\left(\phi\right) = \mathbf{A}^{\dagger}\left(\hat{\mathbf{R}} - \hat{\sigma}^2\left(\phi\right)\mathbf{I}\right)\mathbf{A}^{\dagger H} \tag{20.42}$$

where $\mathbf{A}^{\dagger}$ denotes the Moore-Penrose pseudoinverse of $\mathbf{A}$. By inserting these expressions into Eq. (20.40), the concentrated negative log likelihood function is obtained as

$$\hat{\phi} = \arg\min_{\phi}\left\{\log\left|\mathbf{A}\hat{\mathbf{P}}\left(\phi\right)\mathbf{A}^H + \hat{\sigma}^2\left(\phi\right)\mathbf{I}\right|\right\} \tag{20.43}$$

and the DOA estimates are taken as the minimizing arguments. One important property of the SML method is that it is efficient, because the variance of the DOA estimates attains the CRB. Here, it should be noted that the MUSIC method only attains the bound if the signals are uncorrelated, and if both the number of snapshots and the source signal powers are large. This is, of course, a major reason for choosing the SML method instead of MUSIC. Unfortunately, a numerical search is necessary because the function in Eq. (20.43) depends on the DOAs in a complicated way. This search can be computationally heavy, which is why subspace-based methods like MUSIC that are less computationally intensive, but not efficient, are attractive. The advantage of a lower computational load is thus traded for a higher variance of the estimates. An optimization method suitable for the SML minimization problem is the damped Newton method [31, 91].

Two different versions of ML have appeared in the signal processing literature [91, 64] and make different assumptions concerning the signals. Here, the vesion that assumes an SML was presented. Another possibility is to model the signal as deterministic, which leads to a nonlinear least squares (LS) fit of the model to the data [16, 44, 145]. The technique is often referred to as the deterministic maximum likelihood (DML) method. Though similar in complexity as SML, it can be shown to yield DOA estimates of somewhat higher variance. The difference in performance is, however, negligible in most scenarios of practical interest.

20.3.3.2 Subspace Fitting

Another class of methods that has attracted considerable attention is the subspace fitting methods [125, 141, 142]. With a specific weighting of the criterion, those methods are efficient and thus have properties similar to the SML method. Also, connections to spectral-based methods such as MUSIC and beam-forming can be established. The subspace fitting methods are, as all subspace methods, based on the properties of the covariance matrix in Eq. (20.21). Two versions of subspace fitting have appeared; one based on the signal subspace (SSF) and one based on noise signal subspace (NSF). The SSF version is based on the observation that the signal eigenvectors are equal to a linear combination of steering vectors. The criterion is formulated as

$$\left\{\hat{\phi},\hat{\mathbf{T}}\right\}=\arg\min_{\phi,\mathbf{T}}\left\|\hat{\mathbf{E}}_s-\mathbf{AT}\right\|_W^2 \tag{20.44}$$

where $\|\mathbf{A}\|_W^2$ denotes $\mathrm{Tr}\{\mathbf{AWA}^H\}$ and $\mathbf{W}$ is a positive definite weighting matrix. This function can be concentrated in the same manner as the SML method, and the concentrated criterion function becomes

$$\hat{\phi}_{SSF}=\arg\min_{\phi}\ \mathrm{Tr}\left\{\mathbf{P}_A^{\perp}\hat{\mathbf{E}}_s\mathbf{W}\hat{\mathbf{E}}_s^H\right\} \tag{20.45}$$

The NSF formulation of the subspace fitting criterion is based on the same observation as in MUSIC, namely, that the columns of $\mathbf{A}$ are orthogonal to the noise subspace or $\mathbf{E}_n^H\mathbf{A} = 0$. In this case, an estimate of the DOA is obtained by minimizing the following criterion

$$\hat{\Phi}_{NSF}=\arg\min_{\Phi}\left\|\hat{\mathbf{E}}_n^H\mathbf{A}\right\|_U^2 \tag{20.46}$$

where $\mathbf{U}$ is a $p \times p$ positive definite weighting matrix. Note that the criterion function is quadratic in the steering matrix, and thus estimates of parameters that enter linearly in the steering matrix can be found in closed form. For instance, calibration parameters that enter linearly, such as mutual coupling, can be included in the array response array and concentrated out from the criterion function in Eq. (20.46) [128]. Different weightings give different asymptotic properties, and by choosing a specific weighting it can be shown that the estimates calculated using Eqs. (20.45) and (20.46) are asymptotically equivalent to the SML method that is efficient [91]. Furthermore, if the weighting matrix $\mathbf{U} = \mathbf{I}$, the NSF method reduces to MUSIC. Interestingly, by choosing other types of weighting functions in Eqs. (20.45) and (20.46), many other DOA estimation schemes can be derived [141]. In a sense, the subspace fitting concept can be seen as a framework for deriving and analyzing many different DOA estimators.

In general, the weighted subspaced methods Eqs. (20.45) and (20.46) require a numerical search, and thus the computational load is higher than that of methods like MUSIC. However, the complexity is less than that of SML; see Reference [91]. One optimization method suitable for these problems is the damped Newton method [31, 91]. This method can also be used to compute the estimate when the SML method is used.

20.3.4 Exploiting Special Array Structure

The DOA estimation schemes presented so far have been derived for general array structures; however, for some special array structures, large simplifications are possible by utilizing the special structure. If the array is a ULA, it is possible to reduce the numerical searches in several algorithms to a polynomial rooting procedure. Here, only the root-MUSIC method is presented. A slightly more general case of special structure is outlined next; the ESPRIT algorithm can be applied to arrays that have a so-called shift structure. Finally, two special techniques, forward–backward (FB) averaging and spatial smoothing (SS) that are applicable to ULA structures, are presented that decorrelate highly correlated signal waveforms.

20.3.4.1 Root-MUSIC

The popular MUSIC method can be applied to any type of antenna array. However, if the antenna is a ULA, the estimation procedure can be simplified. In this case, the steering vector has elements $[\mathbf{a}(\phi)]_l = e^{-jkd(l-1)\cos\phi}$ and the expression in Eq. (20.39) can be viewed as a polynomial in $z = e^{-jkd\cos\phi}$. Therefore, the search for the DOAs in the MUSIC method described earlier can be avoided. Instead, find the $2(m-1)$ roots of the symmetrical polynomial

$$f(z) = \mathbf{a}^T(z^{-1})\hat{\mathbf{E}}_n\hat{\mathbf{E}}_n^H\mathbf{a}(z) \tag{20.47}$$

Of the $m-1$ roots inside the unit circle, pick the p that have the largest magnitude, yielding $\hat{z}_k$ for $k = 1, \ldots, p$. Then compute the DOA estimates by solving for ϕ_i in $\angle\hat{z}_i = \angle e^{-jkd\cos\phi_i}$. This method is known in the literature as the Root-MUSIC method, and was first proposed in Reference [9]. This method not only avoids the numerical search, but also improves the resolution threshold of the spectral MUSIC discussed in a previous subsection. A performance analysis in Reference [106] shows that the improvement in resolution threshold results from the fact that a small error in the radius of the root does not affect the estimation performance. Thus, only errors in the angle of the root affect the quality of the estimates, and this decreases the resolution threshold.

Several other algorithms, where the numerical search can be avoided by exploiting the additional structure provided by the ULA, can be derived. The basic idea is to parameterize the null-space of A^H directly instead of using $\mathbf{A}(\phi)$, leading to a parsimonious parameterization of $\mathbf{P}_A^\perp$. Several articles using this idea in different contexts have appeared. If the preceding parameterization is applied in combination with the DML method, it is possible to formulate an iterative algorithm based on polynomial rooting called iterative quadratic maximum likelihood (IQML) [18]. A similar extension of the SSF method was presented in Reference [125] and is sometimes called root-SSF. This algorithm is asymptotically efficient, while essentially formulated in closed form. The computations involved in all these root schemes are only eigendecompositions and polynomial rootings — no nonlinear optimization is necessary.

20.3.4.2 ESPRIT

Another subspace-based method that also relies on a special structure of the array is ESPRIT [110]. The ESPRIT method exploits a shift structure, which exists if the array can be divided into two identical arrays displaced by a known translation Δ. In this case, the DOAs are defined relative to the normal to the translation. Note that no information concerning the subarrays is needed, other than that they should be identical. Thus, problems with calibration may be avoided by using this method. The ESPRIT method relies on the steering matrix of the two subarrays to be constructed as

$$\mathbf{A}_s = \begin{bmatrix} \mathbf{A}_1 \\ \mathbf{A}_1\boldsymbol{\Phi} \end{bmatrix} \tag{20.48}$$

where $\mathbf{A}_1$ is an $l \times p$ steering matrix and $\boldsymbol{\Phi}$ is a $p \times p$ diagonal matrix with elements $\boldsymbol{\Phi}_{ii} = e^{-kj\Delta \sin \phi_i}$. Of course, not all arrays have this shift structure, but, for example, a ULA can be divided into two identical subarrays in many different ways.

The ESPRIT algorithm exploits the shift structure and the fact that the subspace spanned by the columns of $\mathbf{E}_s$ must be equal to the subspace spanned by the columns of $\mathbf{A}$, assuming that the signal covariance matrix is of full rank — Eq. (20.31). Thus, there exists a nonsingular $\mathbf{T}$ such that

$$\mathbf{E}_s = \mathbf{A}_s(\phi)\mathbf{T} \tag{20.49}$$

The basic idea is to find those ϕ:*s* that best fit this relation. Let the signal subspace matrix $\mathbf{E}_s$ be partitioned conformably with Eq. (20.48) as

$$\mathbf{E}_s = \begin{bmatrix} \mathbf{E}_{s1} \\ \mathbf{E}_{s2} \end{bmatrix} \tag{20.50}$$

Then, Eqs. (20.48 to 20.50) imply

$$\mathbf{E}_{s2} = \mathbf{E}_{s1}\boldsymbol{\Psi} \tag{20.51}$$

where $\boldsymbol{\Psi} = \mathbf{T}^{-1}\,\boldsymbol{\Phi}\mathbf{T}$. Because $\boldsymbol{\Phi}$ and $\boldsymbol{\Psi}$ are related by a similarity transformation, they share the same eigenvalues, namely, $e^{-jk\Delta \sin \phi_i}$, $i = 1, \ldots, p$.

Given an estimate $\hat{\mathbf{E}}_s$, the submatrices $\hat{\mathbf{E}}_{s2}$ and $\hat{\mathbf{E}}_{s1}$ do not exactly share the same range space. Instead, an estimate $\hat{\boldsymbol{\Psi}}$ is obtained from the approximate relation

$$\hat{\mathbf{E}}_{s2} \approx \hat{\mathbf{E}}_{s1}\boldsymbol{\Psi} \tag{20.52}$$

Solving Eq. (20.52) in an LS sense results in the LS ESPRIT method. Because there are similar errors on both sides of Eq. (20.52), a total least square (TLS) solution [45] is more natural [90, 110]. The TLS ESPRIT estimates are obtained from the eigendecomposition of the partitioned eigenvectors $\hat{\mathbf{E}}_{s1}$ and $\hat{\mathbf{E}}_{s2}$ as

$$\begin{bmatrix} \hat{\mathbf{E}}_{s1}^H \\ \hat{\mathbf{E}}_{s2}^H \end{bmatrix} \begin{bmatrix} \hat{\mathbf{E}}_{s1} & \hat{\mathbf{E}}_{s2} \end{bmatrix} = \begin{bmatrix} V_{11} & V_{12} \\ V_{21} & V_{22} \end{bmatrix} \mathbf{L} \begin{bmatrix} V_{11}^H & V_{21}^H \\ V_{12}^H & V_{22}^H \end{bmatrix} \tag{20.53}$$

where $\mathbf{L}$ is a diagonal matrix, The ith DOA estimate is then obtained as

$$\hat{\phi}_i = \sin^{-1}\left[\frac{\angle\lambda_i}{2\pi\Delta}\right] \tag{20.54}$$

where λ_i is the ith eigenvalue of the matrix

$$\hat{\Psi}_{TLS} = -V_{12}V_{22}^{-1} \tag{20.55}$$

The main advantage of ESPRIT is that it offers a high accuracy at a small computational cost, because the solution is essentially in closed form and no costly numerical search is needed. Also, the ESPRIT method avoids some problems with calibration because only the translation vector is needed in the calculations as long as the two subarrays are identical. Unfortunately, this is also the reason that the ESPRIT method experiences higher estimation errors than methods like SML and subspace fitting, which exploit the full potential of the antenna model and not only the translation. The statistical properties of the ESPRIT method have been thoroughly investigated [90, 105, 123]. Several possible strategies for forming the two subarrays are possible, and in the literature many different schemes have been proposed. In particular, if the original array is a ULA, a maximum overlapping structures is a common choice [90] (i.e., the first part consists of the first $m - 1$ elements and the second part of the last $m - 1$ elements, that is, $l = m - 1$ and $\Delta = d$). In fact, for a ULA, the statistical properties of the ESPRIT estimates can be improved to some extent by employing row-weighting [90, 123] and/or FB averaging (see later). A clever way to incorporate FB averaging, at a reduced complexity, is the unitary ESPRIT method presented in Reference [47].

The ESPRIT method has, for example, been used for directional estimation in an interference rejection scenario [76], and in a diversity investigation for a smart antenna system [55].

20.3.4.3 Forward–Backward Averaging and Spatial Smoothing

One problem with applying algorithms such as MUSIC in wireless communication scenarios is that the inherent multipath gives rise to correlated signals or even coherent signals for which MUSIC does not work. However, several techniques for mitigating these problems have been proposed for the special case of a ULA.

In the simple case of two coherent sources being present and a ULA, there is a fairly straightforward way to de-correlate the signals. The idea is to employ a FB averaging [63] as follows. Note that a ULA steering vector [Eq. (20.13)] remains invariant, up to a scaling, if its elements are reversed and complex conjugated. More precisely, let $\mathbf{J}$ be an $m \times m$ exchange matrix, whose components are zero except for ones on the antidiagonal. Then, for a ULA it holds that

$$\mathbf{J}\mathbf{a}^*(\theta) = e^{-j(m-1)\psi}\mathbf{a}(\theta) \tag{20.56}$$

where ψ corresponds to the electrical angle, [see Eq. (20.13)]. The so-called backward array covariance matrix therefore takes the form

$$\mathbf{R}_B = \mathbf{J}\mathbf{R}^*\mathbf{J} = \mathbf{A}\Psi^{-(m-1)}\mathbf{P}\Psi^{-(m-1)}\mathbf{A}^H + \sigma^2\mathbf{I} \tag{20.57}$$

where Ψ is a diagonal matrix with $e^{j\psi_k}$, $k = 1, \ldots, p$ on the diagonal, and ψ_k is the electrical angle of the kth DOA. By averaging the usual array covariance and $\mathbf{R}_B$, one obtains the FB array covariance

$$\begin{aligned} \mathbf{R}_{FB} &= \frac{1}{2}\left(\mathbf{R} + \mathbf{J}\mathbf{R}^*\mathbf{J}\right) \\ &= \mathbf{A}\tilde{\mathbf{P}}\mathbf{A}^H + \sigma^2\mathbf{I} \end{aligned} \tag{20.58}$$

where the new "source covariance matrix" $\tilde{\mathbf{P}} = (\mathbf{P} + \Psi^{-(L-1)}\mathbf{P}\Psi^{-(L-1)})/2$ generally has full rank. The FB version of any covariance-based algorithm simply consists of replacing $\hat{\mathbf{R}}$ with $\hat{\mathbf{R}}_{FB}$, defined as in Eq. (20.58).

In a more general scenario where more than two coherent sources are present, FB averaging cannot restore the rank of the signal covariance matrix on its own. A technique called "spatial smoothing" has been proposed [36, 119, 151] that is applicable to several coherent sources and uniform linear arrays. The main idea is to split the ULA into a number of overlapping subarrays. The steering vectors of the subarrays are assumed to be identical up to different scalings, and the subarray covariance matrices can therefore be averaged. Similar to Eq. (20.58), the spatial smoothing induces a phase modulation, which in turn tends to decorrelate the signals that caused the rank deficiency. A compact expression for this smoothed matrix $\tilde{\mathbf{R}}$ can be written in terms of selection matrices $\mathbf{F}_k$ as follows. Let m_s denote the number of elements in the subarrays, implying that the number of subarrays is $K = m - m_s + 1$. Then, the spatially smoothed array covariance matrix can be expressed as

$$\tilde{\mathbf{R}} = \frac{1}{K}\sum_{k=1}^{K} \mathbf{F}_k \mathbf{R} \mathbf{F}_k^T \tag{20.59}$$

with

$$\mathbf{F}_k = \left[0_{m_s \times (k-1)} \middle| \mathbf{I}_{m_s} \middle| 0_{m_s \times (m-k-m_s+1)} \right] \tag{20.60}$$

The rank of the averaged source covariance matrix $\tilde{\mathbf{P}}$ can be shown to increase by 1 with probability 1 [26] for each additional subarray in the averaging, until it reaches its maximum value p.

The drawback with spatial smoothing is that the effective aperture of the array is reduced, because the subarrays are smaller than the original array. However, despite this loss of aperture, the spatial smoothing transformation allows for the application of methods such as MUSIC that have low computational cost because a complicated multidimensional search is avoided. Note that FB and spatial smoothing essentially only are applicable to ULAs. However, some extensions are possible [36, 63, 119, 155]. A version of the ESPRIT algorithm, called Unitary ESPRIT, that employs FB and only requires real-valued computations can be found in Reference [47]. Unitary ESPRIT has been used for channel estimation in a code division multiple access (CDMA) application [14].

20.3.5 Beam Space Processing

The parametric methods presented in the previous section offer improved accuracy at the cost of increased computations. In some cases, this cost may be prohibitive. However, there are several attractive DOA estimation algorithms with significantly lower computational costs, which essentially are modifications of the schemes presented in the previous sections. If the received data are preprocessed so that only a certain spatial sector is selected to be analyzed (e.g., some prior knowledge about the broad direction of arrival of the sources may be available), one can potentially experience a performance gain. The most obvious is computational, because a smaller dimensionality of the problem usually results. It has, in addition, been shown that the bias of the estimates is decreased using this preprocessing. This preprocessing can be viewed as a spatial prefiltering of the received data, resulting in a focused beam on the selected spatial sector. This type of preprocessing is typically called beam space processing, and can be written as

$$\mathbf{z}(t) = \mathbf{T}^H \mathbf{x}(t) \tag{20.61}$$

where the columns of the matrix $\mathbf{T}$ define a bank of spatial band-pass filters. A simple choice is to let the columns of $\mathbf{T}$ be the steering vectors for a set of chosen directions (i.e., using conventional beamforming). When beam space processing is applied, the element-space steering vectors $\mathbf{a}(\phi)$ are replaced by $\mathbf{T}^H\mathbf{a}(\phi)$, and the noise covariance for the beam space data becomes $\sigma^2 \mathbf{T}^H\mathbf{T}$. For the latter reason, $\mathbf{T}$ is often orthogonalized so $\mathbf{T}^H\mathbf{T} = \mathbf{I}$, before application $\mathbf{x}(t)$.

Several algorithms have been reformulated in a beam space version, such as MUSIC [19, 158] and ESPRIT [155]. The variance of the DOA estimates is typically not improved, but a certain robustness to spatially correlated noise is achieved. The latter fact can intuitively be understood when one recalls that the spatial prefilter has a band-pass character, which clearly tends to whiten the noise. One application where beam space algorithms have been used successfully can be found [65] where a beam space version of ESPRIT is used to characterize the directional dependency in macrocell measurements made in downtown Paris.

With the exception of the beamforming-based methods, the estimation techniques discussed here require that the outputs of all elements of the sensor array be available in digital form. In many applications, the required number of high-precision receiver front ends and analog to digital (A/D) converters may be prohibitive. However, by employing a reduced dimension beam space transformation (i.e., a nonsquare matrix $\mathbf{T}$), the computational load of the digital processor can be significantly reduced. In fact, by designing the beamformers (the columns of $\mathbf{T}$) so that they focus on a relatively narrow DOA sector, the essential information in $\mathbf{x}(t)$ concerning sources in that sector can be retained in $\mathbf{z}(t)$; see, e.g., References [40, 140, 158, 164] and the references therein. With further *a priori* information on the locations of sources, the beam space transformation can in fact be performed with no loss of information [4, 34].

Note that the beam space transformation effectively changes the array propagation vectors from $\mathbf{a}(\phi)$ into $\mathbf{T}^H\mathbf{a}(\phi)$. It is possible to utilize this freedom to give the beam space array manifold a simpler form, such as that of a ULA [41]. Hence, the computationally efficient ULA techniques such as root-MUSIC are applicable in beam space. A transformation that maps a uniform circular array (UCA) into a ULA is proposed and analyzed [81], enabling computationally efficient estimation of both azimuth and elevation using root-MUSIC. This algorithm was used in a channel characterization experiment in the Netherlands [27], where accurate high-resolution DOA estimates in the city of Leidschendam were obtained.

20.3.6 Summary of Direction-of-Arrival Methods for Point Sources

In the previous subsections, many different DOA estimation algorithms with slightly different properties have been presented. This section aims to summarize the message conveyed in the algorithm descriptions. First, the main properties of DOA estimation schemes applicable to arbitrary array geometries are summarized. Next, methods that exploit some special array structure are summarized. Finally, a brief resume of algorithms employing beam space processing is included.

20.3.6.1 Arbitrary Array Structure

The DOA algorithms that are applicable to arbitrary array geometries were divided into two subclasses; spectral-based and parametric methods. In general, the spectral-based algorithms are easy to apply and of low-computational complexity. However, problems with spectral leakage resulted in difficulties to resolve closely spaced sources as well as sources of low power. One spectral-based method, MUSIC, was found to have a substantially lower resolution threshold, and is therefore one of the most popular DOA estimation methods combining low-computational demands with good performance. Unfortunately, the MUSIC method is not applicable to scenarios with coherent signals such as a multipath environment.

The more computationally demanding parametric methods, which in general obtain the DOA estimate through a numerical search, avoid the resolution problem and are able to handle coherent signals. Furthermore, the variance of the DOA estimates obtained with these methods is in general "close" to or equal to the CRB lower bound while also being consistent. Thus, these methods offer excellent performance at a higher computational cost.

The statistical properties of the different algorithms applicable to arbitrary array geometries are summarized in Table 20.2. The major computational requirements for each method are also included, assuming that the sample covariance matrix has already been acquired. Here, one-dimensional (1D) search means that the parameter estimates are computed from p 1D searches over the parameter space, whereas p-D search refers to a full p-dimensional numerical optimization.

20.3.6.2 Special Array Structure

If the array exhibits some special structure, it is possible to greatly reduce the computational cost while sometimes also increasing the estimation performance. In particular, in Table 20.3 the properties of

TABLE 20.2 Summary of Estimators Applicable to Arbitrary Array Geometries

Method	Complexity	Performance
Bartlett	1D	ML for $p = 1$, otherwise biased
Capon	1D	Low bias for high SNR
MUSIC	EVD, 1D	Consistent estimates for noncoherent signals, but resolution problem (bias) at low SNR; variance near CRB for low signal correlation
Min-Norm	EVD, 1D	Like MUSIC, but better resolution and higher variance
DML	p-D	Variance near CRB, except for very high signal correlation
SML	p-D	Efficient
WSF	EVD, p-D	Efficient

TABLE 20.3 Summary of Estimators Applicable to Uniform Linear Arrays

Method	Complexity	Performance
Root-MUSIC	EVD, polyroot	Like MUSIC, but better resolution; can work with coherent signals using FB averaging and/or spatial smoothing
ESPRIT	EVD, EVD	Like Root-MUSIC, but somewhat higher variance
IQML	Iterative LS	Like DML, but convergence problems at low SNR
Root-WSF	EVD, LS	Like WSF, but notably higher SNR threshold than using full grid search

algorithms applicable to uniform linear arrays are summarized. Note that the possibility to apply FB averaging and spatial smoothing enables DOA estimation of highly correlated, or even coherent, sources.

20.3.6.3 Beam Space Processing

Finally, by performing some preprocessing called beam space processing, it is possible to reduce the computational complexity while lowering the bias of many of the DOA estimation schemes presented previously. This preprocessing can also be implemented by employing analog hardware and thus reducing the computational load. Furthermore, by utilization *a priori* knowledge of the location of the sources, it is possible to design beam space transformations that reduce the dimensionality of the problem with no information loss.

20.4 Estimation for the Wireless Channel

In a wireless communication scenario it is possible to incorporate more information concerning the received wave than the analysis in the previous section. For instance, some information about the transmitted signal is usually available. Also, there is usually no LOS propagation, except perhaps in rural areas. The signal transmission occurs by scattering and diffraction phenomena. This means that the received signal from a single user contains contributions from several multipaths. In effect, each signal path acts as an angularly *spread* source. Applying standard DOA estimation methods in such a case can give misleading results. Furthermore, knowledge of the DOA may reveal some of the channel characteristics. However, other properties such as time of arrival (TOA) and Doppler frequency can be important in a number of cases. For instance, knowing the DOAs and the path delays can help locate mobile users, because typically the shortest path corresponds to the true angle to the user. Perhaps of greater importance is the possibility of forming beams in both space and time (space-time processing), where knowledge of the spatiotemporal properties of the channel is, of course, essential.

In this section, a number of different methods for DOA estimation that exploit some special signal structure are briefly explained, as well as methods of joint estimation of both DOA and TOA. Finally, some techniques especially designed for spread sources are outlined.

20.4.1 Exploiting Special Signal Structure

DOA estimation in wireless communication applications differs in many respects from the standard problem. Apart from the propagation conditions, a striking feature is that digitally modulated signals exhibit a rich structure, that can be exploited to improve estimation performance. In many cases, such as during channel sounding experiments, the signal waveforms can be considered known. In fact, almost all digital communications systems employ some sort of training symbols, which are also known to the receiver.

Even if the signal waveforms are not known *a priori*, the structure can be exploited to perform *blind signal separation*. During the past decade, many algorithms that are capable of separating and equalizing co-channel signals without using any information of the DOAs have been proposed. The signal properties that have been exploited include constant modulus (CM) [1, 46, 68], non-Gaussianity [21, 69, 118], cyclic correlation properties [2, 152] and finite alphabet (FA) structure [104, 132, 133].

In many cases, the blind signal estimates are of high quality, especially when relying on the FA structure. Thus, by using these techniques one can again regard the signal waveforms as perfectly known when estimating the DOAs. In wireless communication applications, a relatively small array aperture can be expected, and the available signal structure is rich (CM and/or FA). In, such a scenario, separating the signals blindly, and then finding the directions is preferable to first estimating DOAs and then applying beamforming [68, 143]. Joint DOA estimation and signal detection is generally regarded as overwhelmingly complex.

This section is focused on DOA estimation using known signals. In Section 20.4.1.1, an algorithm for estimation of point sources is presented, whereas Section 20.4.1.2 considers joint estimation of DOA and time delay.

20.4.1.1 Decoupled Direction Estimation Using Known Signals

The problem of DOA estimation using known signals has been studied, for example, in [73, 74, 143]. The method presented as follows extracts the parameters of each user separately. Provided signals corresponding to different users are uncorrelated, this simplification can be shown not to impose any appreciable performance degradation.

For simplicity, assume frequency flat fading. How to extend the results to the case of frequency-selective fading is briefly explained later. Further, the scenario is assumed stationary over the observation interval (N samples). Similar to Section 20.2.2, the signal energy is assumed to arrive in clusters. The array output is modeled as

$$\mathbf{x}(t) = \sum_{k=1}^{p} \mathbf{v}_k s_k(t) + \mathbf{n}(t) \tag{20.62}$$

$$t = 1, \ldots, N$$

As before, $s_k(t)$ is the kth waveform and $\mathbf{n}(t)$ is the noise vector. The vector $\mathbf{v}_k$ is the spatial signature of user k. The spatial signatures model the resulting spatial channel from the multipath in a cluster. Thus,

$$\mathbf{v}_k = \sum_{l=1}^{p_k} \rho_{kl}\mathbf{a}(\phi_{kl}) \tag{20.63}$$

where ρ_{kl} is a complex amplitude and ϕ_{kl} is the DOA of the lth signal path emanating from user k. The number of paths in the cluster is denoted p_k. In case of no multipath or negligible DOA spread, the model applied with $p_k = 1$.

Suppose we are given observations $\{\mathbf{x}(t)\}_{t=1}^{N}$ and a signal waveform $\{s_k(t)\}_{t=1}^{N}$ (possibly estimated), corresponding to a user of interest. By assuming the remaining signals to be uncorrelated with $s_k(t)$, the following reduced data model is postulated:

$$\mathbf{x}(t)=\mathbf{A}(\phi_k)\mathbf{b}_k s_k(t)+\mathbf{j}(t) \tag{20.64}$$

where

$$\mathbf{A}(\phi_k)=\left[\mathbf{a}(\phi_{k1}),\ldots,\mathbf{a}(\phi_{kd_k})\right] \tag{20.65}$$

$$\mathbf{b}_k=\left[\rho_{k1},\ldots,\rho_{kd_k}\right]^T \tag{20.66}$$

and where $\mathbf{j}(t)$ represents the contribution of the "interfering" signals plus noise. In deriving the ML estimate of $\phi_k = [\phi_{k1}, \ldots, \phi_{kp_k}]^T$, $\mathbf{j}(t)$ is assumed to be a white Gaussian noise. The resulting method is, of course, equally applicable to any interference, but is ML only under the assumptions made. If $\mathbf{j}(t)$ is distributed as $\mathbf{j}(t) \in \mathcal{N}(0, \mathbf{Q})$, the model in Eq. (20.64) is identical to the parameterized signals model considered in Reference [144]. Applying the result of Reference [144] yields a compact form of the ML estimator, as described next.

Introduce the following notation:

$$\hat{\mathbf{\Pi}}^{\perp}=\mathbf{I}-\hat{\mathbf{Q}}^{-1/2}\mathbf{A}\left(\mathbf{A}^H\hat{\mathbf{Q}}^{-1}\mathbf{A}\right)^{-1}\mathbf{A}^H\hat{\mathbf{Q}}^{-1/2} \tag{20.67}$$

$$\hat{\mathbf{Q}}=\hat{\mathbf{R}}_{xx}-\hat{\mathbf{r}}_{xs}\hat{r}_{ss}^{-1}\hat{\mathbf{r}}_{xs}^H \tag{20.68}$$

$$\hat{\mathbf{r}}_{xs}=\frac{1}{N}\sum_{t=1}^{N}\mathbf{x}(t)s_k^H(t) \tag{20.69}$$

where the argument of $\mathbf{A} = \mathbf{A}(\phi_k)$ has been suppressed for notational simplicity. The sample covariances $\hat{\mathbf{R}}_{xx}$ and $\hat{r}_{ss}$ are defined similarly to Eq. (20.69). The exact ML estimate of ϕ_k is obtained by solving the following p_k-dimensional optimization problem:

$$\hat{\phi}_k=\arg\min_{\phi_k}\left\|\hat{\mathbf{\Pi}}^{\perp}\hat{\mathbf{Q}}^{-1/2}\hat{\mathbf{r}}_{xs}\right\|^2 \tag{20.70}$$

Once $\hat{\phi}_k$ has been computed, the complex signal amplitudes derive from the expression

$$\hat{\mathbf{b}}_k=\left(\mathbf{A}^H\hat{\mathbf{Q}}^{-1}\mathbf{A}\right)^{-1}\mathbf{A}^H\hat{\mathbf{Q}}^{-1}\hat{\mathbf{r}}_{xs}\hat{r}_{ss}^{-1} \tag{20.71}$$

Note that the decoupled ML method requires a p_k-dimensional search for each signal waveform, as opposed to a full Σp_k-dimensional optimization that is necessary when the blind signal estimates are not exploited. In case the signals are not uncorrelated, and the full signal vector $\mathbf{s}(t) = [s_1(t), \ldots, s_p(t)]$ is available, $\hat{\mathbf{r}}_{xs}$ in Eqs. (20.70) and (20.71) should be replaced by the kth column of the matrix $\Sigma_{t=1}^N \mathbf{x}(t)\mathbf{s}^H(t)$ $[\Sigma_{t=1}^N \mathbf{s}(t)\mathbf{s}^H(t)]^{-1}$.

The decoupled method is particularly simple when $p_k = 1$. Then, Eq. (20.70) reduces to finding the highest peak in a beamforming-type criterion,

$$\hat{\phi}_k=\arg\max_{\phi}\frac{\left|\mathbf{a}^H(\phi)\hat{\mathbf{Q}}^{-1}\hat{\mathbf{r}}_{xs}\right|^2}{\mathbf{a}^H(\phi)\hat{\mathbf{Q}}^{-1}\mathbf{a}(\phi)} \tag{20.72}$$

In case the differences in path length of the terms in Eq. (20.63) are on the order of the inverse bandwidth of the signal (frequency-selective fading), the preceding approach only extracts those components that are highly correlated with the given signal at the current delay. To simultaneously estimate all DOAs corresponding to a certain user, the parameterized signals approach of Reference [144] can still be applied, but using $[s_k(t), s_k(t-1), \ldots, s_k(t-L)]$ as "basis functions," where L is the length of the channel impulse response. A slightly more advanced version of the algorithm outlined here has been presented [99], where the DOA is estimated using measured data from Düsseldorf, Germany.

20.4.1.2 Joint Direction-of-Arrival and Delay Estimation

Although knowledge of the DOA may reveal some of the channel characteristics, other properties such as TOA and Doppler frequency can be important in a number of cases. For instance, knowing the DOAs and the path delays can help locate mobile users, because typically the shortest path corresponds to the true angle to the user. Perhaps of greater importance is the possibility of exploiting both the space and time dimensions, where knowledge of the spatiotemporal properties of the channel is essential. Information of the channel characteristics can also be useful to properly evaluate and modify different existing receiver structures.

Many algorithms for joint estimation of both angle and time delay have been presented in the literature. Here, only two examples of algorithms are presented. First, an algorithm based on an ML approach is discussed, followed by an algorithm that can be used in conjunction with ML, MUSIC, or ESPRIT.

SAGE

The spacing alternating generalized expectation maximization (SAGE) algorithm [38] is essentially based on maximizing the ML function for the scenario of unknown DOA and TOA, using the technique of expectation and maximization (EM) [30]. Here, a version [96] that also includes the Doppler frequency is outlined. Consider the following signal model [differs slightly from Eq. (20.22) because of the Doppler]:

$$\mathbf{x}(t) = \sum_{k=1}^{p} \mathbf{s}(t; \theta_k) + \mathbf{n}(t) \tag{20.73}$$

where

$$\mathbf{s}(t; \theta_k) = \rho_k \mathbf{a}(\phi_k) e^{j2\pi\nu_k t} u(t - \tau_k) \tag{20.74}$$

is the received baseband equivalent signal from the kth wave when transmitting the signal $u(t)$. Each wave is characterized by its complex amplitude ρ_k, relative delay τ_k, incidence azimuth ϕ_k, and the Doppler frequency ν_k. Furthermore, the transmitted baseband signal $u(t)$ has the following form

$$u(t) = \sum_{q} g(t - qT) b_q \tag{20.75}$$

where $g(\cdot)$ is the pulse-shaping waveform, T is a single period, and b_q represents the information bits. The measured voltages $\mathbf{x}(t)$, the steering vector $\mathbf{a}(\phi)$, and the noise vector $\mathbf{n}(t)$ are defined as in Section 20.2.2. Note that the unknowns associated with each path are now $\boldsymbol{\theta}_k = [\tau_k, \phi_k, \nu_k, \rho_k]$. The unknowns of all paths may be organized in a matrix formulation as $\boldsymbol{\theta} = [\boldsymbol{\theta}_1, \boldsymbol{\theta}_2, \ldots, \boldsymbol{\theta}_p]$. These unknown parameters are estimated using ML in a manner similar to the case of DOA estimation only, except that now the delay and Doppler are also estimated. Thus, the parameters are obtained as

$$\hat{\boldsymbol{\theta}} = \arg\max_{\boldsymbol{\theta}} l_{\text{SAGE}}(\boldsymbol{\theta}) \tag{20.76}$$

Unfortunately, the maximization of the log likelihood function l_{SAGE} ($\boldsymbol{\theta}$) becomes very computationally exhaustive, because no closed form exists and the parameter matrix $\boldsymbol{\theta}$ is typically of high dimension. However, this multidimensional optimization procedure can be significantly reduced by employing the SAGE algorithm [38]. Essentially, the SAGE algorithm solves the maximization problem iteratively by solving for each wave separately, and then iterating between the solutions to reach the global optimum. The main idea is to apply a signal separation device that can estimate the contribution of each path $\mathbf{x}_k(t)$ to the full measured vector $\mathbf{x}(t)$. The maximization is then performed for each path individually, thus obtaining estimates of the parameters for path k as $\hat{\boldsymbol{\theta}}_k$. Next, these updated estimates of the parameters are used to make an updated separation, based on which the maximization is performed again. The iterative procedure is continued until a suitable convergence condition is satisfied. By using the notation introduced in Eqs. (20.73) to (20.75), the different steps in the SAGE algorithm can be written as [96]

Expectation step

$$\hat{\mathbf{x}}_k\left(t;\hat{\theta}\right) = \mathbf{x}\left(t\right) - \sum_{k'=1,k'\neq k}^{p} \mathbf{s}\left(t;\hat{\theta}_{k'}\right) \tag{20.77}$$

Maximization steps

$$\begin{aligned}
\hat{\tau}'_k &= \arg\max_{\tau_k} \left|Z_k\left(\tau_k, \hat{\phi}_k, \hat{\nu}_k\right)\right| \\
\hat{\phi}'_k &= \arg\max_{\phi_k} \left|Z_k\left(\hat{\tau}'_k, \phi_k, \hat{\nu}_k\right)\right| \\
\hat{\nu}'_k &= \arg\max_{\nu_k} \left|Z_k\left(\hat{\tau}'_k, \hat{\phi}'_k, \nu_k\right)\right| \\
\hat{\rho}'_k &= \frac{Z_k\left(\hat{\rho}'_k, \hat{\phi}'_k, \hat{\nu}'_k\right)}{\left\|\mathbf{a}\left(\hat{\phi}'_k\right)\right\|^2 E}
\end{aligned} \tag{20.78}$$

where

$$Z_k\left(\tau_k, \phi_k, \nu_k\right) = \sum_t u^*\left(t-\tau_k\right) e^{-j2\pi\nu_k t} \mathbf{a}^H\left(\phi_k\right) \hat{\mathbf{x}}_k\left(t;\hat{\theta}\right) \tag{20.79}$$

and E is the energy of the transmitted signal $u(t)$. Interestingly, the delay estimate of path k is obtained by determining the time-correlation peak between the transmitted signal $u(t)$ and the path signal $\mathbf{x}_k(t)$. Furthermore, the azimuth estimate uses a simple beamforming technique and the Doppler frequency is obtained by a Doppler correction. Thus, although the original criterion function in Eq. (20.76) is quite complicated, the actual solution consists of several simple intuitive steps. The SAGE algorithm requires an initial estimate of the parameter vector $\boldsymbol{\theta}$ and the number of impinging waves p. Typically, a sufficiently large number p is chosen manually for different scenarios, whereas the initial time-delay estimate is obtained using the MUSIC method. The azimuth, the Doppler frequency, and the complex amplitude are obtained by using the last three maximization steps in Eq. (20.78) [96]. For a detailed description of the iterative SAGE algorithm, see References [38, 39, 96]. This algorithm has successfully been used for channel characterization using measurement data in several mobile communication studies [39, 96].

JADE

The joint angle-delay estimation algorithm (JADE) [14, 137] is based on fitting a parameterized model of the channel matrix to an unstructured estimate of the channel. The unstructured estimate of the channel can easily be obtained by LS if the transmitted signal is known. If the signal is unknown, a blind channel estimation technique can be applied to obtain an estimate of the channel matrix **H**. The structured model of the channel is

$$\mathbf{H} = \sum_{k=1}^{p} \mathbf{a}(\phi_k)\rho_k \mathbf{g}^T(\tau_k) \tag{20.80}$$

where $\mathbf{a}(\phi_k)$ denotes the steering vector, ρ_k the path gain, and the $L \times 1$ vector $\mathbf{g}(\tau_k)$ denotes the sampled transmitted waveform delayed by τ_k with elements $\mathbf{g}_i(\tau_k) = g((\xi + \Delta - i)T - \tau_k)$. Here, ξ denotes the maximum integer path delay and $2\Delta T$ is the duration of the pulse-shaping waveform $g(t)$. For a more detailed description on the data model, see References [95, 137]. By defining the space-time response for a single wavefront arriving at angle ϕ_k and τ_k as $\mathbf{u}(\boldsymbol{\theta}_k, \tau_k) = \mathbf{g}(\tau_k) \otimes \mathbf{a}(\phi_k)$, the channel relation can be rewritten as

$$\mathbf{x}(t) = \mathbf{U}(\phi, \tau)\rho(t) \tag{20.81}$$

where $\mathbf{x}(t)$ is the vectorized channel matrix vec(**H**), $\boldsymbol{\rho}^T = [\rho_1, \ldots, \rho_p]$, and

$$\mathbf{U}(\phi, \tau) = \left[\mathbf{u}(\phi_1, \tau_1), \ldots, \mathbf{u}(\phi_p, \tau_p)\right] \tag{20.82}$$

By sampling at N instants, the following model for the unstructured estimate of the channel matrix results

$$\hat{\mathbf{X}} = \mathbf{U}(\phi, \tau)\mathbf{P} + \mathbf{N} \tag{20.83}$$

where $\hat{\mathbf{X}}$ is $mL \times N$, $\mathbf{P} = [\rho(1), \ldots, \rho(N)]$ is $p \times N$, and $\mathbf{N}$ ($mL \times N$) represents the error in the channel estimate $\hat{\mathbf{X}}$. Note that the length of the channel is denoted L.

The expression in Eq. (20.83) has a striking resemblance with the traditional array processing expression $\mathbf{X} = \mathbf{AS} + \mathbf{N}$. Many of the methods derived for the classical array processing expression can therefore be used to find the parameters. If the noise is assumed to be Gaussian, the ML estimate reduces to the following nonlinear LS fit: [137]

$$\left\{\hat{\phi}, \hat{\tau}\right\} = \arg\min_{\boldsymbol{\phi}, \boldsymbol{\tau}, \mathbf{P}} \left\| \hat{\mathbf{X}} - \mathbf{U}(\boldsymbol{\phi}, \boldsymbol{\tau})\mathbf{P} \right\|_F^2 \tag{20.84}$$

where $\hat{\mathbf{X}}$ corresponds to the unstructured estimate of the channel. Note that the minimization of this criterion is carried out in a similar manner as the DML method of traditional array processing, and a concentrated criterion may be obtained, where only a search for the DOAs and the TOAs is needed. Other versions of JADE, employing MUSIC and ESPRIT solutions, have appeared in the literature [137, 138], and have been analyzed mainly in simulation studies. A similar approach of fitting a parameterized model to an unstructured channel estimate has also been proposed [98], where real measurement data were analyzed.

20.4.2 Estimation for Spread Source Models

In this section, DOA estimation for the case where the sources are not accurately modeled as point sources is considered. After giving some more details on the signal model, both nonparametric (beamforming) and parametric techniques are presented.

20.4.2.1 Spread Source Modeling

The scattering environment is quite different in the uplink and downlink, respectively. The case of major concern here is transmission from the mobile to the base (i.e., the uplink). Because of the rich scattering surrounding the mobile in urban environments, DOA estimation is basically meaningless at the downlink. However, the base station is typically placed quite high, implying that most of the scattering takes place near the mobile. Because of this, it is most common to model the signal received at the base station as emanating from one or more DOA clusters. In fact, for most cases of practical interest, one cluster per source is often sufficient to accurately model the propagation [35]. For system bandwidths less than approximately 5 MHz, the various signal contributions within a cluster can be considered completely coherent. Thus, the signal received at a base station antenna array from one mobile transmitter can be modeled as

$$\mathbf{x}(t) = \mathbf{v}(t)s(t) \tag{20.85}$$

$$\mathbf{v}(t) = \sum_{l=1}^{p} \rho_l(t)\mathbf{a}(\phi_l(t)) \tag{20.86}$$

where $\{\rho_l(t)\}_{l=1}^{p}$ model the gain and phase variations of each ray, and $\{\phi_l(t)\}_{l=1}^{p}$ are the corresponding DOAs. The resulting $\mathbf{v}(t)$ is the spatial signature vector. If the terminal is moving, $\mathbf{v}(t)$ is time-varying. With a speed of 30 m/s and a wavelength of 30 cm (900 MHz), the bandwidth of $\rho_l(t)$ is on the order of 100 Hz. Thus, the propagation can be considered stationary, for example, for 1 ms. To reveal the statistical properties of the scattering may require data collected during at least 1 second (naturally, this figure depends on the system parameters).

Because of the random phase, it is reasonable to assume that $E[\rho_l(t)] = 0$, $E[\rho_k(t)] = 0$, and that $E[\rho_k(t)\rho_l^\star(t)] = \sigma_p{}^2\delta_{k,l}$. This is often referred to as a *spatially white* scattering, not to be confused with spatially white noise. Because σ_p^2 cannot be distinguished from the signal power P, it is convenient to normalize so that $\sigma_p^2 = 1/p$. The precise statistical distribution of $\rho_l(t)$ and $\phi_l(t)$ may not be of great importance for DOA estimation; only their second-order properties are crucial. It can be argued that each ray results, in fact, from a continuum of scatterers. Thus, a Gaussian model for $\rho_l(t)$ is reasonable. Further, several empirical investigations, including Reference [97], have also shown that the distribution of the DOAs are accurately modeled as Gaussian. The expected value of the distribution, ϕ_0, is the scatter center, whereas the standard deviation σ_ϕ depends on the type of scenario and the distance between the base station and the scatter center. In a typical urban scenario, the DOA spread, σ_ϕ, can be up to 10° [97], whereas significantly smaller clusters can be expected in a suburban or rural area.

To illustrate the effect of scattering on standard DOA estimation algorithms, consider the MUSIC algorithm applied to a scenario with $\phi_o = 90°$ (array broadside) and $\sigma_\phi = 5°$. The array is a ten-element ULA, and five independent data sets are generated, each with $N = 100$ time samples and different stationary ($\rho_l(t)$ and $\phi_l(t)$ constant) scattering scenarios. The number of scatters in each realization is $p = 20$. The signal subspace dimension is estimated using MDL, and the resulting MUSIC pseudospectra are shown in Fig. 20.7. As seen in the plot, the location of the highest peak can be far off from the scatter center, and one spectrum realization reveals little information on the DOA spread. A more detailed investigation concerning the effects of local scattering on standard DOA estimation schemes such as MUSIC can be found in Reference [6].

The techniques presented next do not attempt to model a single realization of $\mathbf{v}(t)$. In fact, modeling a given realization essentially reduces to DOA estimation for point sources, as described in Section 20.3. Here, the statistical properties of the scattering are sought. Of particular interest are the moments, ϕ_0 and σ_ϕ, of the DOA distribution. It is clear that claims concerning the statistical properties are meaningful only after seeing several realizations of the scattering. As previously alluded to, a data collection time longer than 1 s might be necessary in a practical scenario.

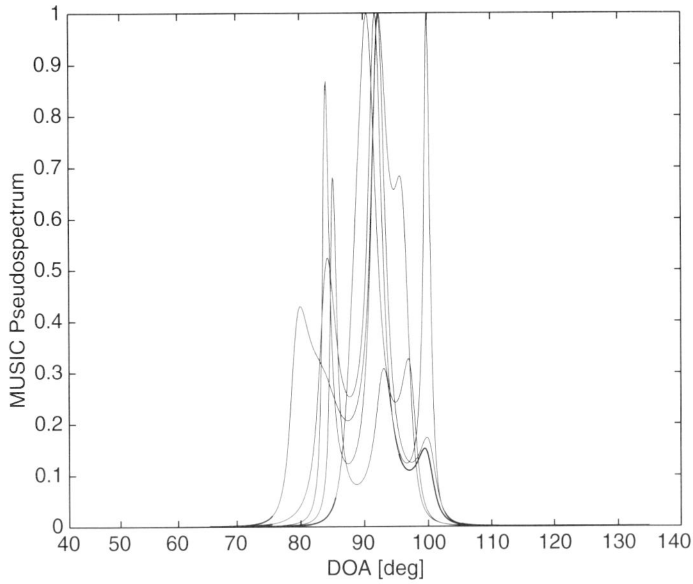

FIGURE 20.7 MUSIC pseudospectra for five different realizations of the random scattering. The cluster DOAs are distributed as $\mathcal{N}(90°, 5°)$.

20.4.2.2 Beamforming Techniques

Suppose energy is received from a single cluster. At any given time, the DOAs in the cluster are distributed according to some probability density function (**pdf**) $p(\phi)$. Spatially white scattering is assumed (i.e., the $\rho_k(t)$'s are independent and identically distributed). The signal power received at the origin, emanating from a small angular frequency band, $(\phi, \phi + d\phi)$, is then proportional to $p(\phi)$. Consequently, the angular distribution can be reconstructed from the angular power spectrum $P(\phi)$. In case the complex amplitudes $\rho_k(t)$ are not identically distributed, for example, scattering near the mobile is likely to yield larger amplitudes. $P(\phi)$ is weighted by the $E[|\rho_k(t)|^2]s$. For multiple clusters, the spectrum measures the total DOA distribution. The individual clusters cannot be identified unless they are sufficiently separated in angle. In Section 20.3.2, the conventional beamforming (spatial analog of the periodogram) and the MVDR spectra are described. The conventional beamformer is useful only for clusters wider than the Rayleigh beam width. On the other hand, the MVDR (Capon) method can yield accurate estimates even of relatively narrow spectra at high SNR. The method can be applied to the sample covariance matrix as usual, assuming a sufficiently rapid fading. If several data batches are available, each with a stationary scenario, it is reasonable to form the sample covariance from estimates of the spatial signature vector from each batch. The latter can be obtained, for example, from the dominant eigenvector of the sample covariance in each batch of stationary data. If the scattering is sufficiently time varying, the estimated sample covariance matrix yields a consistent estimate of the array covariance. From Eq. (20.85) and assuming spatially white noise, the latter takes the form

$$\mathbf{R} = E\left[\mathbf{x}(t)\mathbf{x}^H(t)\right] = E_\phi\left[E\left[\mathbf{x}(t)\mathbf{x}^H(t)\middle|\phi\right]\right] = P\int_\phi \mathbf{a}(\phi)\mathbf{a}^H(\phi)p(\phi)d\phi + \sigma^2\mathbf{I} \tag{20.87}$$

Using $\hat{\mathbf{R}} = \frac{1}{N}\sum_{t=1}^{N}\mathbf{x}(t)\mathbf{x}^H(t)$ in the MVDR spectrum gives a result nearly proportional to $p(\phi)$, provided N is large enough so that $\hat{\mathbf{R}} \approx \mathbf{R}$ and the SNR is high enough. Improvements of MVDR that potentially can work at lower SNR are discussed in Reference [135]. To illustrate the point, a simple example is provided. Suppose an eight-element standard ULA receives a signal via a single cluster. The DOAs in the cluster are distributed as $\mathcal{N}(70°, 5°)$. The sample covariance is formed from a batch of $N = 1000$ snapshots.

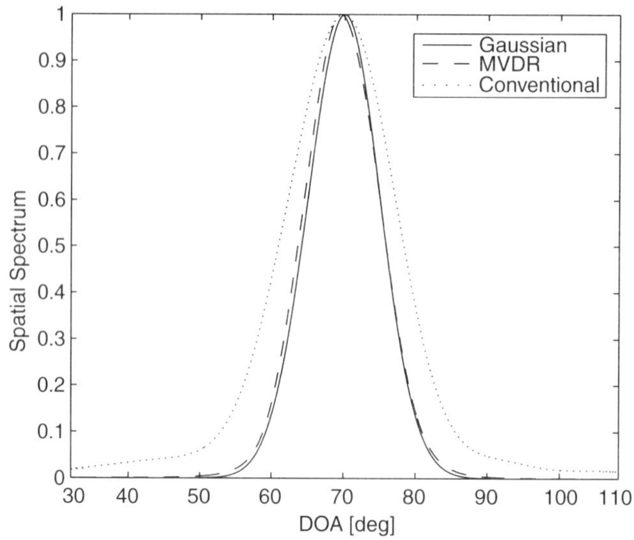

FIGURE 20.8 Conventional and MVDR spatial spectrum estimates. The true DOA distribution is $\mathcal{N}(70°, 5°)$, and is also shown.

In each data sample, 20 DOAs with random locations and reflection coefficients are drawn from the respective Gaussian distribution. The scatter parameters are independent from snapshot to snapshot. The average SNR is 30 dB in each sensor. Both the conventional and the MVDR beamformers are applied to the sample covariance. The estimated spatial spectra (normalized) are displayed in Fig. 20.8, along with the normalized Gaussian distribution. As seen in the plot, the MVDR spectrum is able to capture the essential shape of the DOA distribution, whereas the resolution of the conventional beamformer is insufficient in this scenario. An advantage, as opposed to parametric techniques, is that no model for the distribution is necessary. A drawback is that a fairly large data collection time is necessary to get a spectral estimate with reasonably low variance; see also Reference [103]. The MVDR estimates can be stabilized by adding a small multiple of the identity matrix to $\hat{\mathbf{R}}$, a technique referred to as regularization or "diagonal loading." However, this is at the expense of a decreased resolution capability.

20.4.2.3 Parametric Modeling

Similar to the case of point sources, it is of course possible to apply parameter estimation methods to find the location parameters. Essentially, the idea is to fit a spatial signal model of the form in Eq. (20.87) to the sample covariance matrix in a suitable sense. By exploiting a parametric model for the DOA distribution, meaningful estimates can be obtained using significantly less data than in the nonparametric case. Thus, suppose $p(\phi, \boldsymbol{\eta})$ is a known function of ϕ and a set of unknown parameters $\boldsymbol{\eta}$. For the Gaussian distribution, $\boldsymbol{\eta} = [\phi_0, \sigma_\phi]$. The integral form of Eq. (20.87) is not very convenient for estimation purposes. For a ULA, an approximate model has been derived. By assuming $\phi \in \mathcal{N}(\phi_0, \sigma_\phi)$, where $\sigma_\phi \ll 1$, and further that a ULA is used, the array covariance matrix can be approximated as [11, 134]

$$\mathbf{R} \approx P\left\{\mathbf{a}\left(\phi_0\right)\mathbf{a}^H\left(\phi_0\right)\right\} \odot \mathrm{B} + \sigma^2 \mathrm{I} \tag{20.88}$$

where $\odot$ denotes element wise matrix multiplication (Schur/Hadamard matrix product) [78], and the klth element of the $m \times m$ matrix $\mathbf{B}$ is given by

$$\mathbf{B}_{kl} = e^{-2\left\{\pi\Delta\left(k-l\right)\right\}^2 \sigma_\phi^2 \sin^2 \phi_0} \tag{20.89}$$

A similar approximation can be made for a uniform scatter distribution [11, 134].

Optimum Estimation Techniques

Reconsider the signal model in Eq. (20.85). Because both $\mathbf{v}(t)$ and $\mathbf{s}(t)$ are random, the distribution of $\mathbf{x}(t)$ is nontrivial. However, if $s(t)$ has constant modulus and $\mathbf{v}(t)$ is Gaussian (large number of scatters), $\mathbf{x}(t)$ is also Gaussian. Further, assuming temporally white samples, the ML parameter estimates are obtained by finding the minimizing arguments $\hat{P}$, $\hat{\sigma}^2$, $\hat{\boldsymbol{\eta}}$ of the negative log-likelihood function [134]

$$L\left(P,\sigma^2,\boldsymbol{\eta}\right)=\log\left|\mathbf{R}\right|+Tr\left\{\mathbf{R}^{-1}\hat{\mathbf{R}}\right\} \tag{20.90}$$

Similar to the standard DOA estimation case in Eq. (20.40), the preceding estimator gives highly accurate estimates even if the assumption of temporally white Gaussian observations is not met. In particular, the performance only depends on the signal power, not the specific distribution.

Unfortunately, the ML approach requires a multidimensional (four in the Gaussian case) nonlinear optimization. Because the noise-free covariance generally has full rank, no closed-form solution with respect to the signal and noise variances is known. However, because $\mathbf{R}$ is linear in P and σ^2, a covariance matching estimator (COMET) [89] results in separable minimization with respect to the powers. The COMET selects the parameters to minimize an optimally weighted LS-fit of the model covariance to the following sample version:

$$\left\{\hat{P},\hat{\sigma}^2,\hat{\boldsymbol{\eta}}\right\}=\arg\min\left\|\left(\mathbf{R}-\hat{\mathbf{R}}\right)\hat{\mathbf{R}}^{-1}\right\|_F^2 \tag{20.91}$$

After concentrating with respect to P and σ^2, the numerical search can be carried out over the DOA distribution parameters $\boldsymbol{\eta}$ only. The resulting parameter estimates can be shown to have the same large sample properties as the ML estimator [Eq. (20.90)]. The details can be found in Reference [134]. Besson and Stoica [13] have derived a COMET-based approach for the model (20.88) that only requires a numerical search over ϕ_0, thus allowing a further simplification.

Suboptimum Techniques

Although the parametric methods presented thus far have been for a single DOA cluster, the extension to several scatter centers is straightforward. The covariance model [Eq. (20.87)] then simply contains the sum of all contributions, and one set of DOA parameters is used for each cluster. However, the computational complexity of the optimum methods presented earlier then increases dramatically, because of the need for a search over a high-dimensional parameter space. This, in addition to the rather complicated form of the criteria, motivates the use of suboptimum methods of lesser complexity. Similar to the point source case, the low-rank nature of the contributions from each signal is exploited. A point source has a rank-one spatial covariance matrix. In general, a spread source model has full rank. However, for reasonably small spread angles (σ_ϕ on the order of a Rayleigh beam width), the effective rank of the kth signal contribution

$$\mathbf{R}_k\left(\boldsymbol{\eta}_k\right)=\int_\phi \mathbf{a}\left(\phi\right)\mathbf{a}^H\left(\phi\right)p_k\left(\phi,\boldsymbol{\eta}_k\right)d\phi \tag{20.92}$$

is significantly smaller than m. This observation has been used to develop MUSIC-based approaches to DOA estimation of spread sources [82] [136]. The technique by Meng et al. [82] is termed dispersed signal parametric estimation (DISPARE), and is described next. First, the eigendecomposition of the sample covariance matrix is computed, similar to Eq. (20.21),

$$\hat{\mathbf{R}}=\sum_{k=1}^{m}\hat{\lambda}_k\hat{\mathbf{e}}_k\hat{\mathbf{e}}_k^H \tag{20.93}$$

The number of "significant" eigenvalues is then identified. This can be done by thresholding, or by using a statistical test of the multiplicity of the smallest eigenvalue, as for the point source subspace-based DOA estimation schemes [147]. Let r denote the so determined signal subspace dimension. The noise subspace matrix is then estimated by

$$\hat{\mathbf{E}}_n = \left[\hat{\mathbf{e}}_{r+1}, \ldots, \hat{\mathbf{e}}_m\right] \tag{20.94}$$

If Eq. (20.92) was exactly of rank r, one would have $\mathbf{R}_k^H(\eta_k)\mathbf{E}_n = 0$ at the true cluster parameters, where $\mathbf{E}_n$ denotes the true noise subspace matrix. This motivates estimating $\boldsymbol{\eta}$ by minimizing a suitable measure of the orthogonality between $\mathbf{R}_k(\boldsymbol{\eta}_k)$ and $\hat{\mathbf{E}}_n$. Meng et al. [82] proposed using

$$\begin{aligned} &\hat{\boldsymbol{\eta}}_k \arg\min \left\| \mathbf{R}_k^H\left(\boldsymbol{\eta}_k\right)\hat{\mathbf{E}}_n \right\|_F^2 \\ &k = 1, \ldots, p \end{aligned} \tag{20.95}$$

In the case of a Gaussian DOA distribution and a ULA, the approximate model [Eqs. (20.88) to (20.89) is preferably used. In the case of multiple clusters, for example, p, one searches for the p smallest local minima of the criterion function in Eq. (20.95). If a grid search is performed, the computational complexity for the DISPARE method is approximately independent of the number of clusters.

Bengtsson and Ottersten [11] suggested using the second-order Taylor expansion

$$\mathbf{a}\left(\phi_0 + \phi\right) \approx \mathbf{a}\left(\phi_0\right) + \mathbf{d}\left(\phi_0\right)\phi + \frac{1}{2}\mathbf{d}'\left(\phi_0\right)\phi^2 + O\left(\left|\mathbf{d}''\left(\phi_0\right)\right|\phi^3\right) \tag{20.96}$$

where $\mathbf{d}(\phi) = \partial\mathbf{a}(\phi)/\partial\phi$. To be precise, Bengtsson and Ottersten [11] parameterized $\mathbf{a}(\cdot)$ using the electrical angle parameter. However, for the small ϕ where the approximation is valid (well within the mainlobe for a ULA), the difference is insignificant. Inserting Eq. (20.96) into (20.92) and performing the integration shows that $\mathbf{R}_k$ can be approximated as

$$\mathbf{R}_k \approx \mathbf{a}\left(\phi_0\right)\mathbf{a}^H\left(\phi_0\right) + \sigma_\phi^2\left[\mathbf{d}\left(\phi_0\right)\mathbf{d}^H\left(\phi_0\right) + \frac{1}{2}\left(\mathbf{a}\left(\phi_0\right)\mathbf{d}'^H\left(\phi_0\right) + \mathbf{d}'\left(\phi_0\right)\mathbf{a}^H\left(\phi_0\right)\right)\right] \tag{20.97}$$

$$\approx \frac{1}{2}\mathbf{a}\left(\phi_0 + \sigma_\phi\right)\mathbf{a}^H\left(\phi_0 + \sigma_\phi\right) + \frac{1}{2}\mathbf{a}\left(\phi_0 - \sigma_\phi\right)\mathbf{a}^H\left(\phi_0 - \sigma_\phi\right) \tag{20.98}$$

where, in the second approximation, Eq. (20.96) was invoked again. As a consequence, the data for each cluster may be regarded as generated by two point sources at $\phi_0 \pm \sigma_\phi$. Provided $p \le m/2$, a DOA estimator designed for point sources can be used to determine $2p$ estimates. The obtained DOAs are then grouped in pairs, and the corresponding cluster parameters ϕ_0 and σ_ϕ are estimated from the mean and half the angle separation, respectively. The resulting algorithm is termed *Spread-X*, where X is the employed point source method. An advantage with this approach, as compared to the weighted LS and DISPARE methods, is that the precise DOA distribution does not need to be specified. A complication is that the two-source model is only valid for very small clusters, but for DOAs that are too closely spaced the point source estimates may be of poor quality.

Example

To illustrate the ideas presented in this section, consider a scenario where an eight-element ULA receives a signal from two Gaussian clusters. The parameters are $\phi_1 = 90°$, $\sigma_{\phi 1} = 2°$, $\phi_2 = 100°$, and $\sigma_{\phi 2} = 3°$, respectively. The signals are uncorrelated, and the SNRs are 20 dB (source at 90°) and 13 dB (source at 100°). Ten independent data sets are generated, and in each the sample covariance matrix is formed from $N = 1000$ data snapshots, each using 20 scatter DOAs. Both the scattering and the noise are assumed independent between time instances. An effective rank of $r = 4$ is selected, as suggested by the MDL criterion. The DISPARE and spread-root-MUSIC methods are applied, the latter using $p = 4$ and estimating the cluster parameters as suggested by Eq. (20.98). Figure 20.9 shows a scatter plot of the parameter estimates. It can be seen that the spread-root-MUSIC estimates are more reliable in this rather difficult scenario. A contour plot of one realization of the DISPARE criterion function is shown in Fig. 20.10. By using the same data, the conventional and MVDR beamforming spectra are displayed in Fig. 20.11 along with the suitably normalized DOA distributions. The beamforming methods do not exhibit two peaks in this scenario, but the general shape of the MVDR spectrum still agrees well with the Gaussian curves.

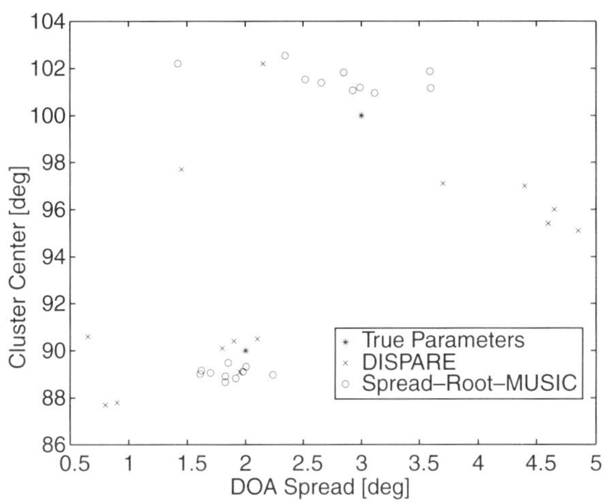

FIGURE 20.9 DISPARE and spread-root-MUSIC parameter estimates. The true DOA distributions for the two clusters are $\mathcal{N}(90°, 2°)$ and $\mathcal{N}(100°, 3°)$ respectively.

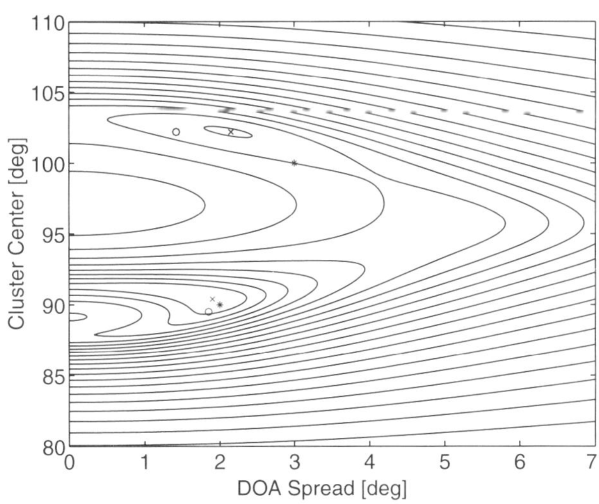

FIGURE 20.10 Contour-plot of the DISPARE criterion function for one particular data set. The minima are indicated with ×, whereas ° signifies the spread-root-MUSIC parameter estimates. The true parameters are marked with *.

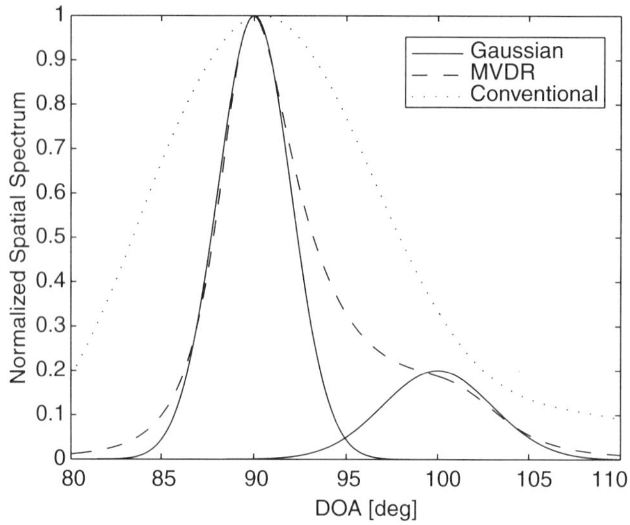

FIGURE 20.11 Conventional and MVDR spatial spectrum estimates for the same data as in Fig. 20.10. The true distributions (normalized) are also shown.

20.5 Tracking of Moving Sources

Many wireless communication systems allow a high degree of mobility. Thus, the terminals may be moving during transmission, causing a nonstationary scenario. Typically, the DOAs vary slowly, on the order of 10° per second at most. The directions to scatter centers can therefore be considered stationary for several transmitted symbols. However, the Doppler effect causes the reflection coefficients to be more rapidly time varying. This, in combination with shadowing, can cause sudden appearance or disappearance of scatter centers. Thus, the typical scenario is one where the number of DOAs is randomly changing with time, but the DOA parameters are slowly time varying. In this section, we consider tracking of point sources only. However, the extension to the spread-sources scenarios and methods described in Section 20.4.2 is in most cases straightforward.

20.5.1 Recursive Sample Covariance Updating

All DOA estimation techniques presented in this chapter exploit the data via the sample covariance matrix only. Hence, the most obvious way to create an adaptive algorithm is to use a recursive update of the sample covariance. The number of active emitters and their respective DOAs can then be computed at a suitable rate using the time-varying sample covariance. By using a fixed forgetting factor λ, $0 \ll \lambda < 1$, the latter is defined as

$$\hat{\mathbf{R}}_N = (1-\lambda)\sum_{t=1}^{N} \lambda^{N-t}\mathbf{x}(t)\mathbf{x}^H(t) \tag{20.99}$$

The exponential data-windowing causes recent data to have a greater influence on $\hat{\mathbf{R}}_N$, thus enabling tracking of time-varying signal parameters. By interpreting Eq. (20.99) as a convolution sum, we see that each component $x_k(N)x_l^*(N)$ is passed through the one-pole filter $(1-\lambda)/(1-\lambda z^{-1})$. The time constant of this low-pass filter is approximately $1/(1-\lambda)$.

From Eq. (20.99), the recursive update of the sample covariance is given by

$$\hat{\mathbf{R}}_N = \lambda\hat{\mathbf{R}}_{N-1} + (1-\lambda)\mathbf{x}(N)\mathbf{x}^H(N) \tag{20.100}$$

In general, the choice of data window is a trade-off between tracking ability (bias) and stationary behavior (variance). If the DOAs can be assumed stationary for K samples, a reasonable choice of forgetting factor is $\lambda = 1 - 1/K$. However, to avoid a large time delay in the presence of sudden changes, such as the appearance of a new source, it may be necessary to select λ much smaller. Obviously, this leads to DOA estimates with high variance. Indeed, the exponential data window implied by Eq. (20.100) has a rather poor tracking vs. variance trade-off. Other possibilities exist, such as sliding rectangular window, but Eq. (20.100) is most commonly chosen because of its computational simplicity.

A useful approach for enabling a quick number of signals detection and yet a low-estimation error variance is to postfilter the DOA estimates. Any low-pass filter can be used, for example Eq. (20.100) once again

$$\hat{\phi}_N = \lambda_\phi\hat{\phi}_{N-1} + (1-\lambda_\phi)\hat{\phi}_{prel} \tag{20.101}$$

where $\hat{\theta}_{\text{prel}}$ is the output of the DOA estimation algorithm applied to $\hat{\mathbf{R}}_N$. For "small" λ the final DOA estimation error variance is substantially reduced by selecting λ_ϕ very close to one. Such a simple low-pass smoothing approach is believed to be sufficient for the slowly time-varying applications considered here. However, if information of the source dynamics as well as the quality of the preliminary DOA estimates is available, optimum postprocessing can be achieved using the Kalman filter, as described in Section 20.5.3.

When using the MVDR spectrum, the inverse of the sample covariance matrix is needed. Because matrix inversion is expensive, on the order of m^3 floating point operations (flops), it is more efficient to update the inverse directly. To this point, the celebrated matrix inversion lemma [78] is useful:

$$(A+BCD)^{-1} = A^{-1} - A^{-1}B\left(C^{-1}+DA^{-1}B\right)^{-1}DA^{-1} \tag{20.102}$$

Here, A, B, C, and D are matrices of conformable dimensions, and all inverses are assumed to exist. Applying Eq. (20.102) to

$$\hat{\mathbf{R}}_N^{-1} = \left(\lambda\hat{\mathbf{R}}_{N-1} + (1-\lambda)\mathbf{x}(N)\mathbf{x}^H(N)\right)^{1} \tag{20.103}$$

results in

$$\hat{\mathbf{R}}_N^{-1} = \lambda^{-1}\left(\hat{\mathbf{R}}_{N-1}^{-1} - \mu_N\hat{\mathbf{R}}_{N-1}^{-1}\mathbf{x}(N)\mathbf{x}^H(N)\hat{\mathbf{R}}_{N-1}^{-1}\right) \tag{20.104}$$

where

$$\mu_N = \frac{1}{\lambda/(1-\lambda) + \mathbf{x}^H(N)\hat{\mathbf{R}}_{N-1}^{-1}\mathbf{x}(N)} \tag{20.105}$$

Thus, the "weighting vector" $\mathbf{w}_N = \hat{\mathbf{R}}_N^{-1}\mathbf{a}(\phi)$ can be updated in $O(m^2)$ flops as

$$\mathbf{w}_N = \lambda^{-1}\left(\mathbf{w}_{N-1} - \mu_N\hat{\mathbf{R}}_{N-1}^{-1}\mathbf{x}(N)\mathbf{x}^H(N)\mathbf{w}_{N-1}\right) \tag{20.106}$$

This recursion is similar in form as the recursive least squares (RLS) algorithm for adaptive filtering [49, 77]. The updating formula in Eq. (20.103) is not the best from a numerical sensitivity point of view. Similar to other RLS-based adaptive filtering algorithms, it is possible to instead update a square-root factor of $\hat{\mathbf{R}}_N$ and, if necessary, to apply regularization. See References [49, 77, 115] for details. Finally, we mention that the weight vector update in Eq. (20.106) can, of course, be replaced by a least mean squares LMS-type update, requiring only $O(m)$ flops [43, 49, 115]. However, unlike Eq. (20.106), an LMS-based algorithm does not achieve $\mathbf{w}_N = \hat{\mathbf{R}}_N^{-1}\mathbf{a}(\phi)$ in each step. In scenarios with suddenly appearing and disappearing sources, the slower convergence rate of LMS may result in a unacceptable performance loss as compared with Eq. (20.106).

20.5.2 Recursive Subspace Estimation

As seen in Section 20.3.6, several computationally efficient DOA estimation methods require computing (parts of) the eigendecomposition of the sample covariance matrix. A full eigendecomposition requires about 10 m^3 flops. By utilizing the approximate "low-rank plus $\sigma^2\mathbf{I}$" structure, the complexity can be brought down to $O(m^2p)$ [153]. However, this may also be prohibitively large in practical applications. A viable alternative is to update the signal subspace information directly, without even computing the sample covariance. Note that the sample covariance update itself costs $O(m^2)$ flops. An early survey of subspace tracking methods is given in Reference [24]. The computationally most efficient methods are the so-called spherical subspace averaging techniques [28, 29, 57] and projection approximation subspace tracking (PAST) [159, 160]. These methods require only $O(mp)$ flops for computing an (approximate) orthogonal basis for the signal subspace. Consequently, subspace tracking may require less computations than a conventional sample covariance based approach. Of the mentioned techniques, we choose to describe the PAST algorithm in some detail below. The modified version PASTd, which enables tracking of individual eigenelements, is also presented. However, first we note that a pure flops count alone is insufficient for determining whether or not a certain algorithm can operate in real time in a given implementation. For example, the subspace updating algorithm used in Reference [86] requires $O(m^2)$ flops, but its highly regular structure supports parallel implementation.

20.5.2.1 Projection Approximation Subspace Tracking and –d Algorithms

The basis of the PAST algorithm is the following criterion function:

$$J(\mathbf{W}(N)) = \sum_{t=1}^{N} \lambda^{N-t} \left\| \mathbf{x}(t) - \mathbf{W}(N)\mathbf{W}^H(N)\mathbf{x}(t) \right\|^2 \tag{20.107}$$

in which $\mathbf{W}(N)$ is an arbitrary $m \times p$-matrix, where p is the dimension of the signal subspace to be tracked. Yang [159] proved that $J(\mathbf{W}(N))$ is unimodal (i.e., the only local minimum is also the global). Furthermore, at the minimum, the columns of $\mathbf{W}(N)$ form an orthonormal basis for the space spanned by the p principal eigenvectors of $\hat{\mathbf{R}}_N$. Note, in particular, that no constraint on $\mathbf{W}(N)$ is necessary; the orthonormality is automatically fulfilled at the minimum of Eq. (20.107). Unfortunately, a closed-form solution of Eq. (20.107) does not exist in general. It is possible to apply a gradient-based search. However, a computationally more efficient and faster converging alternative is offered by Yang's projection approximation. As a result of the exponential forgetting, $J(\mathbf{W}(N))$ has a small dependence on $\mathbf{x}(t)$ for $t \le N - 1/(1-\lambda)$. By assuming a "not too rapidly" changing scenario, the approximation $\mathbf{W}^H(N)\mathbf{x}(t) \approx \mathbf{W}^H(t-1)\mathbf{x}(t)$ therefore does not change Eq. (20.107) much. This results in the modified criterion.

$$J_{\text{mod}}(\mathbf{W}(N)) = \sum_{t=1}^{N} \lambda^{N-t} \left\| \mathbf{x}(t) - \mathbf{W}(N)\mathbf{y}(t) \right\|^2 \tag{20.108}$$

$$\mathbf{y}(t) = \mathbf{W}^H(t-1)\mathbf{x}(t) \tag{20.109}$$

which is quadratic in $\mathbf{W}(N)$. The minimization of $J_{mod}(\mathbf{W})$ is therefore straightforward, and results in

$$\mathbf{W}(N) = \hat{\mathbf{R}}_{xy}(N)\hat{\mathbf{R}}_{yy}^{-1}(N) \tag{20.110}$$

$$\hat{\mathbf{R}}_{xy}(N) = (1-\lambda)\sum_{t=1}^{N} \lambda^{N-t}\mathbf{x}(t)\mathbf{y}^H(t) \tag{20.111}$$

$$\hat{\mathbf{R}}_{yy}(N) = (1-\lambda)\sum_{t=1}^{N} \lambda^{N-t}\mathbf{y}(t)\mathbf{y}^H(t) \tag{20.112}$$

Given Eq. (20.110), an RLS-type updating formula for $\mathbf{W}(N)$ is derived similar to Eqs. (20.103) to (20.106). The resulting algorithm has complexity $O(mp)$, and is summarized next.

1. Choose $\mathbf{W}(0)$ arbitrary and $\mathbf{P}(0) = C\mathbf{I}_{p\times p}$ for some large constant C ($\mathbf{P}(N)$) corresponds to $\hat{\mathbf{R}}_{yy}^{-1}(N)$; set $N = 1$
2. $\mathbf{y}(N) = \mathbf{W}^H(N-1)\mathbf{x}(N)$
3. $\mu_N = \dfrac{1}{\lambda/(1-\lambda) + \mathbf{y}^H(N)\mathbf{P}_{N-1}\mathbf{y}(N)}$
4. $\mathbf{W}(N) = \mathbf{W}(N-1) + \mu_N(\mathbf{x}(N) - \mathbf{W}(N-1)\mathbf{y}(N))\mathbf{y}^H(N)\mathbf{P}_{N-1}$
5. $\mathbf{P}_N = \lambda^{-1}(\mathbf{P}_{N-1} - \mu_N \mathbf{P}_{N-1}\mathbf{y}(N)\mathbf{y}^H(N)\mathbf{P}_{N-1})$
6. Set $N \leftarrow N + 1$ and goto 2

The resulting $\mathbf{W}(N)$ is not exactly orthonormal. However, provided there is a significant gap between the pth and $(p+1)$st eigenvalues of $\hat{\mathbf{R}}_N$, $\mathbf{W}(N)$ can be used as an orthonormal basis for the signal subspace with negligible performance loss [160]. Thus, for example, $\mathbf{W}(N)$ is used for $\hat{\mathbf{E}}_s$ in the ESPRIT algorithm, or $\mathbf{I} - \mathbf{W}(N)\mathbf{W}^H(N)$ is substituted for $\hat{\mathbf{E}}_n\hat{\mathbf{E}}_n^H$ when applying the MUSIC algorithm.

Note that PAST does not deliver the individual eigenvalues and eigenvectors of $\hat{\mathbf{R}}$ in its original version, only a basis for the signal subspace is obtained. If the eigenelements are required, they can be obtained via a deflation technique, as described in the following. When the PAST algorithm is run with a single vector $\mathbf{w}_1(N)$, it converges to the eigenvector of $\hat{\mathbf{R}}$ that corresponds to the largest eigenvalue. At convergence we have

$$\hat{r}_{yy}(N) = (1-\lambda)\sum_{t=1}^{N} \lambda^{N-t}\mathbf{w}_1^H(t-1)\mathbf{x}(t)\mathbf{x}^H(t)\mathbf{w}_1(t-1) \approx \mathbf{w}_1^H(N)\hat{\mathbf{R}}(N)\mathbf{w}_1(N) \tag{20.113}$$

Therefore, $l_1(N) = \hat{r}_{yy}(N)$ is a useful estimate of the dominant eigenvalue of $\hat{\mathbf{R}}$. Now, suppose the data vector $\mathbf{x}(N)$ is projected onto the orthogonal complement of $\mathbf{w}_1(N)$. The PAST update can then be reapplied to the projected data, using a single vector $\mathbf{w}_2(N)$. If the first step was successful, $\mathbf{w}_2(N)$ converges to the eigenvector corresponding to the second largest eigenvalue $l_2(N)$ and so on. The resulting deflation algorithm is termed PASTd [159]. A slightly modified version where r denotes the number of eigenvalue/vector pairs to be computed is summarized as follows:

1. Choosing $\mathbf{w}_1(0)$ arbitrary and $l_1(0) = 0$; set $N = 1$
2. $\mathbf{x}_1(N) = \mathbf{x}(N)$
3. For $k = 1, 2, \ldots, r$
   (a) $y_k(N) = \mathbf{w}_k^H(N-1)\,\mathbf{x}_k(N)$
   (b) $l_k(N) = \lambda l_k(N-1) + (1-\lambda)|y_k(N)|^2$
   (c) $\mathbf{w}_k(N) = \mathbf{w}_k(N-1) + \frac{1-\lambda}{l_k(N)}\,(\mathbf{x}_k(N) - \mathbf{w}_k(N-1)y_k(N))\,y_k^H(N)$

(d) $\mathbf{w}_k(N) = \mathbf{w}_k(N)/|\mathbf{w}_k(N)|$
(e) $\mathbf{X}_{k+1}(N) = \mathbf{x}_k(N) - \mathbf{w}_k(N)\, \mathbf{w}_k^H(N)\, \mathbf{X}(N)$

4. Set $N \leftarrow N + 1$ and goto 2.

As previously alluded to, $\mathbf{w}_k(N)$ and $l_k(N)$ are estimates of the kth eigenvector and eigenvalue of $\hat{\mathbf{R}}$, respectively. This offers an advantage over PAST, because the eigenvalues are useful for detecting the number of signals. Specifically, if the PASTd algorithm is applied with $r = p + 1$, where p is the anticipated number of signals, an increase in the rth eigenvalue indicates the appearance of a new source. Similarly, a discontinued source causes the $r - 1$st eigenvalue to drop to the same level as the rth. In practice, reliable detection of these events is far from trivial, in particular, for closely spaced signals, or when the noise covariance is not exactly proportional to the identity matrix. However, a detailed discussion is beyond the scope of this chapter. A computationally more expensive, but perhaps more robust, alternative is to use the estimated DOAs for detecting changes in the number of signals, for example, References [80, 150].

We conclude the discussion on subspace tracking with a remark on block processing [150]. In most digital wireless communication systems, some kind of burstwise transmission is used. In other words, the data are obtained in short sequences with a relatively long duration between each data batch. Because of the different timescales, it is then reasonable to use a variable forgetting factor. In Reference [150], it is instead suggested to treat the data as stationary in each burst, and to apply a block-based subspace tracking algorithm. In particular, a block PAST algorithm that updates a matrix square-root factor of $\hat{\mathbf{R}}_{yy}$ is outlined.

20.5.3 Source Tracking

For linear state-space models and Gaussian signals and noise, the Kalman filter [3,59] provides minimum mean square error (MMSE) estimates of the states. To apply the Kalman filter to DOA tracking, a model for the DOA evolution must be assumed. The simplest choice is the random-walk model $\phi_{N+1} = \phi_N + \mathbf{w}(N)$, where $\mathbf{w}(N)$ is a zero-mean temporally white noise vector. However, for slow time variations a smoother model of the dynamics is preferable. For example, References [108, 111] suggest a constant-acceleration model. In the current application, a constant-speed model is likely to be sufficient. By using $\phi_k(N)$ and its time derivative $\mathbf{y}_k(N) = [\phi_k(N)\ \dot{\phi}_k(N)]^T$ as state variables for the kth signal, the model is

$$\mathbf{y}_k(N+1) = \mathbf{F}\mathbf{y}_k(N) + \mathbf{w}_k(N) \tag{20.114}$$

$$\mathbf{F} = \begin{bmatrix} 1 & T \\ 0 & 1 \end{bmatrix} \tag{20.115}$$

for $k = 1, \ldots, p$. In Eq. (20.115), T is the time between two consecutive DOA updates. In the absence of process noise $\mathbf{w}_k(N)$, the DOA follows a straight line trajectory. The noise accounts for random deviations from this nominal model. Unfortunately, the measurement model in Eq. (20.17) depends nonlinearly on the DOA parameters. Thus, the Kalman filter is not directly applicable. The state-space model can, of course, be linearized locally around the current estimate, leading to the extended Kalman filter [3, 59]. However, a further complication is that the observations $\mathbf{x}(t)$ also depend on the signal waveforms $\mathbf{s}(t)$. If the signals are known (e.g., using the training sequence only), applications of the extended Kalman filter is straightforward; see, e.g., Reference [163]. In the case of unknown signals, the more pragmatic approach to use preliminary DOA estimates has been suggested in the literature; see, e.g., References [108, 111, 148].

Instead of using the array output directly as the measurement process, the (most recent) data can be used to form preliminary DOA estimates. The measurement equation complementing Eq. (20.114) is then

$$\hat{\phi}_k(N) = \mathbf{h}\mathbf{y}_k(N) + v_k(N) \tag{20.116}$$

$$\mathbf{h} = \begin{bmatrix} 1 & 0 \end{bmatrix} \tag{20.117}$$

where $\hat{\phi}_k$ represents the kth preliminary DOA estimate and $v_k(N)$ is the estimation error. In general, the different DOA estimates are correlated. However, in Reference [108] it is argued that ignoring the correlation results in an insignificant performance loss. This implies that the DOAs can be updated in parallel, using p 2-state Kalman filters. Employing the full $2p$-state filter is considerably more expensive. Define the process and measurement noise covariances as

$$\mathbf{Q}_k(N) = E\left[\mathbf{w}_k(N)\mathbf{w}_k^T(N)\right] \tag{20.118}$$

$$\sigma_{\phi,k}^2(N) = Ev_k^2(N) \tag{20.119}$$

Further, let $\hat{\mathbf{y}}_k(N+1|N)$ be the predicted state at time $N+1$ given data up to time N, whereas $\hat{\mathbf{y}}_k(N|N)$ is the filtered state estimate at time N. The respective error covariance matrices are denoted $\boldsymbol{P}_k(N+1|N)$ and $\boldsymbol{P}_k(N|N)$. The Kalman tracking algorithm for the kth signal is then given by

1. Initialization: choose $\hat{\mathbf{y}}_k(0|0)$ and $\hat{\mathbf{P}}_k(0|0)$ suitably, for example, using an initial data set; put $N = 0$
2. Prediction:

$$\hat{\mathbf{y}}_k(N+1|N) = \mathbf{F}_y\hat{\mathbf{y}}_k(N|N) \tag{20.120}$$

$$\mathbf{P}_k(N+1|N) = \mathbf{F}_y\mathbf{P}_k(N|N)\mathbf{F}_k^T + \mathbf{Q}_k(N) \tag{20.121}$$

3. Estimation: obtain a preliminary estimate $\hat{\phi}_k(N+1)$ using a suitable DOA method
4. Smoothing:

$$\mathbf{K}_k(N+1) = \frac{1}{\mathbf{h}\mathbf{P}_k(N+1|N)\mathbf{h}^T + \sigma_{\phi,k}^2(N)}\mathbf{P}_k(N+1|N)\mathbf{h}^T \tag{20.122}$$

$$\hat{\mathbf{y}}_k(N+1|N+1) = \hat{\mathbf{y}}_k(N+1|N) + \mathbf{K}_k(N+1)\left(\hat{\phi}_k(N+1) - \mathbf{h}\hat{\mathbf{y}}_k(N+1|N)\right) \tag{20.123}$$

$$\mathbf{P}_k(N+1|N+1) = \mathbf{P}_k(N+1|N) + \mathbf{K}_k(N+1)\mathbf{h}\mathbf{P}_k(N+1|N) \tag{20.124}$$

5. Update: put $N \leftarrow N+1$ and goto 2

The properties of the Kalman filter can be controlled by the noise variances $\mathbf{Q}_k$ and $\sigma_{\phi,k}^2$. For optimality, these should of course be selected to the "true values." In practice, an exact model for the source dynamics is rarely available, and the variance of the DOA estimates can only be approximated by their theoretical values at best. A more pragmatic viewpoint is to regard the noise variances as design variables, to get suitably smooth filtered estimates. Note that the Kalman gain is relatively insensitive to small variations in the noise variances, and that the filter is invariant to simultaneous scaling $\{\mathbf{Q}_k, \sigma_{\phi,k}^2\} \rightarrow \{\alpha\mathbf{Q}_k, \alpha\sigma_{\phi,k}^2\}$.

Special care should be exercised when dealing with closely spaced sources. In the measurement expression in Eq. (20.116), it is implicitly assumed that each estimated DOA can be correctly associated with a corresponding state variable. If two or more DOAs are close, this data association is a nontrivial task. At the same time, note that the utilization of state dynamics in Eq. (20.114) even allows crossing signal tracks. A simple approach to data association is to minimize the MSE between predicted DOAs and preliminary estimates. A considerably more elaborate scheme is presented in Reference [148], but for a very difficult scenario involving ambiguous estimates. In Reference [108] the association problem is avoided by collapsing nearby DOAs into clusters. This is probably the most reasonable approach for crossing sources, because any preliminary DOA estimator is bound to fail anyway at some minimum separation angle.

The Kalman smoother can, of course, be combined with any DOA estimation method. However, for slowly moving signal sources it may not be necessary to update the DOA estimate each time a new data sample is available. Each preliminary estimate can be based on a new batch of data. Alternatively, the sample covariance (or a subspace estimate) is recursively updated for each new, and a DOA estimator is applied as often as desired. For example, Reference [108] uses a batch stochastic ML method, whereas a PASTd-based MUSIC algorithm is employed in Reference [111].

Example

This section presents a simple example to illustrate the tracking methods. A mobile is assumed to travel at a constant angular speed relative the receiving eight-element standard ULA. The energy is transmitted via a Gaussian cluster, whose center is at the terminal location. The angular spread is $\sigma_\phi = 3°$ and the SNR is 10 dB. Estimates of the $r = 3$ dominant eigenvectors of the covariance matrix are updated each time a new sample arrives, using the PASTd-algorithm with a forgetting factor of $\lambda = 0.96$. Both the scattering and the signals are assumed independent between samples. The spread-ESPRIT algorithm is applied every 50 samples to produce an estimate of the scatter center. The estimate is refined by applying a Kalman filter with $\sigma^2_{\phi,1} = 1$ and $\mathbf{Q}_w = \text{diag}[10^{-9}\ 10^{-8}]$. The mobile speed is 0.001° per sample. By using a sampling rate of 1 kHz, this corresponds to $\dot{\phi} = 1°/s$ and the DOA estimates are updated at 20 Hz. The resulting PASTd-spread-ESPRIT and Kalman-filter estimates are displayed in Fig. 20.12. The corresponding PASTd eigenvalue estimates are shown in Fig. 20.13. The algorithm was applied using $r = 3$, but only the two dominant eigenvectors were used for DOA estimation.

It is clear that Kalman smoothing can improve the tracking performance considerably. The forgetting factor is deliberately chosen small for the ESPRIT estimates to be nearly unbiased. In addition, this enables starting up or discontinuing a track essentially for each new update, although this point is not explored in the shown simulation. The resulting high variance of the preliminary DOA estimates is reduced to an acceptable level by postfiltering.

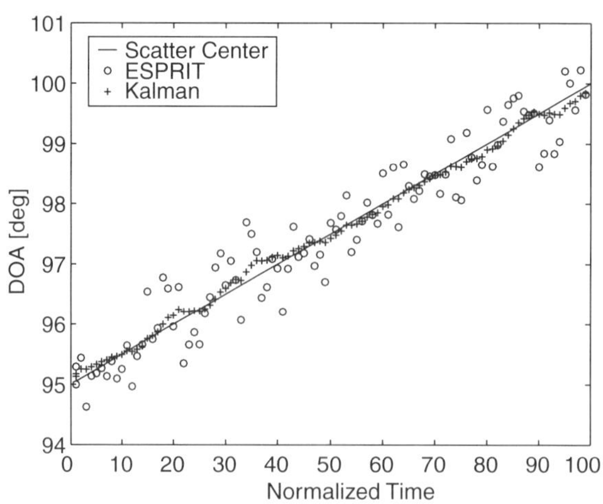

FIGURE 20.12 Tracking performance for the PASTd-spread-ESPRIT algorithm with and without smoothing by the Kalman filter. The rms errors are 0.4° for the ESPRIT estimates and 0.1° after filtering.

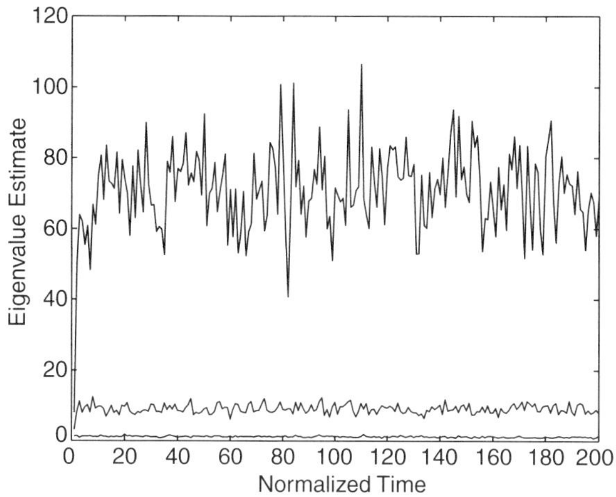

FIGURE 20.13 PASTd eigenvalue estimates for the scenario in Fig. 20.12.

20.6 Concluding Remarks

DOA estimation is a generic problem that is important in a variety of signal-processing applications. In the last few decades, a vast number of algorithms have appeared, and some of these have been presented in this chapter. The most promising techniques rely heavily on a rather simplified data model. In contrast to most signal-processing-oriented texts, the model was derived using basic electromagnetic concepts. Hopefully, this helps connecting the results presented here with other more antenna-oriented chapters. A summary of the key properties of several DOA estimation methods was given in Section 20.3.6. The subspace-based methods, in particular, when combined with beam space processing, generally offer a good balance between computational complexity and statistical performance.

Although the generic DOA estimation problem has been thoroughly studied for a long time, not many methods are designed specifically to investigate the spatial properties of the wireless channel. Some of the most promising approaches were reviewed here, but more contributions are to be expected in the future. In particular, the case of spatially spread sources deserves more attention. The problem is ill-defined in the sense that the incident wave arrives from a continuum of paths, and thus the DOA is defined only in statistical terms. Whereas the propagation conditions may be problematic in wireless communications, the rich structure of digitally modulated signals can, of course, be exploited to improve the estimation performance. Some techniques that assume the signals to be perfectly known were presented. These are applicable using, for example, the training sequence or after performing blind signal separation. The possibility of estimating the parameters of each signal at the time in a multiuser scenario offers vast advantages. Further, knowing the precise shape of the transmitted signal enables estimating the time delays between different propagation paths, which is useful for synchronization and channel characterization purposes.

Today, a great deal of the wireless communication involves moving sources. Thus, tracking time-varying DOAs is of great interest. The two main approaches to achieving this goal were presented. The first is based on subspace tracking, which enables efficient recursive implementation of subspace-based methods. In fact, combining subspace tracking with, for example, the MUSIC algorithm, requires less computations than does the conventional beamforming method. The other main tool presented is the Kalman filter. In the scenarios of interest, the DOAs vary slowly, whereas the number of significant propagation paths can be rapidly time varying. The tracking method needs to be fast to be able to track changes in the subspace dimension, but this results in DOA estimates of high variance. Applying post-filtering using the Kalman filter was shown to reduce the variability of the preliminary estimates. The combination of subspace tracking, an efficient DOA estimation method (for point or spread sources), and Kalman filtering is a powerful tool for tracking moving sources in wireless communication scenarios.

References

1. B. Agee, "Blind Separation and Capture of Communication Signals Using a Multitarget Constant Modulus Beamformer." In *Proc. IEEE MILCOM*, pages 340–346, 1989.
2. B.G. Agee, A.V. Schell, and W.A. Gardner. "Spectral Self-Coherence Restoral: A New Approach to Blind Adaptive Signal Extraction Using Antenna Arrays." *Proc. IEEE*, 78:753–767, Apr. 1990.
3. B.D.O. Anderson and J.B. Moore, *Optimal Filtering*, Prentice-Hall, Englewood Cliffs, NJ, 1979.
4. S. Andersson. "Optimal Dimension Reduction for Sensor Array Signal Processing." *Signal Processing*, 30(2), Jan. 1993.
5. S.P. Applebaum. "Adaptive Arrays (reprinted from technical report, 1966)." *IEEE Trans. AP*, AP-24:650–662, 1976.
6. D. Astély and B. Ottersten. "The Effects of Local Scattering on Direction of Arrival Estimation With MUSIC." *IEEE Trans. Signal Process.*, 47(12):3220–3234, Dec. 1999.
7. C. Balanis. *Antenna Theory: Analysis and Design*, John Wiley & Sons, New York, 1982.
8. C. Balanis. *Advanced Engineering Electromagnetics*. John Wiley & Sons, New York, 1989.
9. A.J. Barabell. "Improving the Resolution Performance of Eigenstructure Based Direction-Finding Algorithms." In *Proc. IEEE ICASSP 83*, pages 336–339, Boston, MA, 1983.
10. M.S. Bartlett. "Smoothing Periodograms from Time Series with Continuous Spectra." *Nature*, 161:686–687, 1948.
11. M. Bengtsson and B. Ottersten. "Low-Complexity Estimators for Distributed Sources." *IEEE Trans. Signal Process.*, 48(8):2185–2194, Aug. 2000.
12. H.L. Bertoni. "*Radio Propagation for Modern Wireless Applications*," Prentice-Hall, Upper Saddle River, NJ, 2000.
13. O. Besson and P. Stoica. "Decoupled Estimation of DOA and Angular Spread for a Spatially Distributed Source." *IEEE Trans. Signal Process.*, 48(7):1872–1882, July 2000.
14. J.J. Blanz, A Papathanassiou, M. Haardt, and I. Furió, and P.W. Baier. "Smart Antennas for Combined DOA and Joint Channel Estimation in Time-Slotted CDMA Mobile Radio Systems with Joint Detection." *IEEE Trans. Veh. Technol.*, 49(2):293–306, March 2000.
15. N. Blaunstein. *Radio Propagation in Cellular Networks*. Artech House Publishers, Norwood, MA, 2000.
16. J.F. Böhme. "Estimation of Source Parameters by Maximum Likelihood and Nonlinear Regression." In *Proc. ICASSP 84*, pages 7.3.1–7.3.4, 1984.
17. J.F. Böhme. "Estimation of Spectral Parameters of Correlated Signals in Wavefields." *Signal Process.*, 10:329–337, 1986.
18. Y. Bresler and A. Macovski. "Exact Maximum Likelihood Parameter Estimation of Superimposed Exponential Signals in Noise." *IEEE Trans. ASSP*, ASSP-34:1081–1089, Oct. 1986.
19. K. Buckley and X.-L. Xu. "Spatial-spectrum estimation in a location section." *IEEE Trans. ASSP*, 38:1842–1852, Nov. 1990.
20. J. Capon, "High Resolution Frequency Wave Number Spectrum Analysis." *Proc. IEEE*, 57:1408–1418, 1969.
21. J.-J. Cardoso and E. Moulines. "Asymptotic performance analysis of direction-finding algorithms based on fourth-order cumulants." *IEEE Trans. Signal Process.*, 43(1):214–224, Jan. 1995.
22. M.F. Catedra, J. Perez, F. Saez de Adana, and O. Guiterrez, "Efficient Ray-Tracing Techniques for Three-Dimensional Analyses of Propagation in Mobile Communications: Applications to Picocell and Microcell Scenarios." *IEEE Antennas Propag. Mag.*, 40(2):15–28, Apr. 1998.
23. R.E. Collin. "*Antennas and Radio Wave Propagation*." McGraw-Hill, New York, 1985.
24. P. Comon and G.H. Golub. "Tracking a Few Extreme Singular Values and Vectors in Signal Processing." *Proc. IEEE*, 78(8):1327–1343, Aug. 1990.
25. Compton, R.T., Jr., *Adaptive Antennas*. Prentice-Hall, Englewood Cliffs, NJ, 1988.
26. J.H. Cozzens and M.J. Sousa. "Source Enumeration in a Correlated Signal Environment." *IEEE Trans. Acoustics, Speech, Signal Process.*, 42(2):304–317, Feb. 1994.

27. Y.L.C. de Jong and M.H.A.J. Herben. "High Resolution Angle-of-Arrival Measurement of the Mobile Radio Channel." *IEEE Trans. Antennas Propag.*, 47(11):1677–1687, Nov, 1999.
28. R. DeGroat. "Noniterative Subspace Tracking." *IEEE Trans. Signal Process.*, 40(3):571–577, 1992.
29. R. DeGroat, E.M. Dowling, and D.A. Linebarger. "Spherical Subspace Tracking for Efficient, High-Performance Adaptive Signal Processing Applications." *Signal Process.*, 50(1–2):101–122, 1996.
30. A.P. Dempster, N.M. Laird, and D.B. Rubin, "Maximum Likelihood from the Incomplete Data via the EM Algorithm." *J. Royal Statist. Soc. Ser. B.*, 39(1):1–38, 1977.
31. J.E. Dennis and R.B. Schnabel. *Numerical Methods for Unconstrained Optimization and Nonlinear Equations*, Prentice-Hall, Englewood Cliffs, NJ, 1983.
32. M. Döttling, F. Küchen, and W. Wiesbeck. "Determination Modeling of the Street Canyon Effect in Urban Micro and Pico Cells." In *Proc. IEEE ICC'97*, pages 36–40, Montreal, Canada, 1997.
33. R.S. Elliot. *Antenna Theory and Design*. Prentice-Hall, Englewood Cliffs, NJ, 1981.
34. J. Eriksson and M. Viberg, "Adaptive Data Reduction for Signals Observed in Spatially Colored Noise." *Signal Process.*, 80(9):1823–1831, Sep. 2000.
35. R.B. Ertel, P. Cardieri, K.W. Sowerby, T.S. Rappaport, and J.H. Reed. "Overview of Spatial Channel Models for Antenna Communication Systems." *IEEE Personal Commun.*, pages 10–22, Feb. 1998.
36. J.E. Evans, J.R. Johnson, and D.F. Sun. "Application of Advanced Signal Processing Techniques to Angle of Arrival Estimation in ATC Navigation and Surveillance Systems." Technical report, MIT Lincoln, Laboratory, Jun. 1982.
37. A.J. Fenn and G.A. King, "Adaptive Nulling in the Hyperthermia Treatment of Cancer." *Lincoln Lab. J.*, 5(2):223–240, 1992.
38. J.A. Fessler and A. Hero, "Space-Alternating Generalized Expectation-Maximization Algorithm." *IEEE Trans. Signal Process.*, 42(10):2664–2677, Oct. 1994.
39. B.H. Fleury, M. Tschudin, R. Heddergott, D. Dahlhaus, and K.I. Pedersen. "Channel Parameter Estimation in Mobile Environments Using the SAGE Algorithm." *IEEE J. Selected Areas Commun.*, 17(3):434–450, Mar. 1999.
40. P. Forster and G. Vezzosi, "Application of Spheroidal Sequences to Array Processing." In *Proc. ICASSP 87*, pages 2268–2271, Dallas, TX, 1987.
41. B. Friedlander and A. Weiss, "Direction Finding Using Spatial Smoothing with Interpolated Arrays." *IEEE Trans. AES*, 28:574–587, Apr. 1992.
42. W.F. Gabriel, "Spectral Analysis and Adaptive Array Superresolution Techniques." *Proc. IEEE*, 68(6):654–666, Jun. 1980.
43. L.C. Godara, "Applications of Antenna Arrays to Mobile Communications: Part II — Beamforming and Direction-of-Arrival Considerations." *IEEE Proc.*, 85(8):1195–1245, Aug. 1997.
44. G. Golub and V. Pereyra. "The Differentiation of Pseudo-Inverse and Nonlinear Least Squares Problems Whose Variables Separate." *SIAM J. Num. Anal.*, 10:413–432, 1973.
45. G.H. Golub and C.F. VanLoan. "An Analysis of the Total Least Squares Problems." *SIAM J. Num. Anal.*, 17:883–893, 1980.
46. R.P. Gooch and J. Lundel. "The CM Array: An Adaptive Beamformer for Constant Modulus Signals," In *Proc. ICASSP 86*, pages 2523–2526, Tokyo, Japan, 1986.
47. M. Haardt and J. Nossek, "Unitary ESPRIT: How to Obtain Increased Estimation Accuracy with a Reduced Computational Burden." *IEEE Trans. Signal Process.*, 43(5):1232–1242, May 1995.
48. S. Haykin, editor, *Array Signal Processing*, Prentice-Hall, Englewood Cliffs, NJ, 1985.
49. S. Haykin. *Adaptive Filter Theory*, 2nd ed., Prentice-Hall, Englewood Cliffs, NJ, 1991.
50. S. Haykin, editor. *Advances in Spectrum Analysis and Array Processing*, Volume I–III. Prentice-Hall, Englewood Cliffs, NJ, 1991–1995.
51. S. Haykin, editor, *Advances in Spectrum Analysis and Array Processing*, Volume III. Prentice-Hall, Englewood Cliffs, NJ, 1991–1995.
52. S. Haykin and J. Litva T.J. Shepherd (eds.). *Radar Array Processing*. Springer-Verlag, Berlin, 1993.
53. A.G. Jaffer, "Maximum Likelihood Direction Finding of Stochastic Sources: A Separable Solution." In *ICASSP 88*, volume 5, pages 2893–2896, New York, Apr. 1988.

54. W. Jakes, *Microwave Mobile Communications.* Wiley-Interscience, New York, 1974.
55. S.-S. Jeng, G.T. Okamoto, G. Xu, H.-P. Lin, and W.J. Vogel. "Experimental Evaluation of Smart Antenna System Performance for Wireless Communications." *IEEE Trans. Antennas Propag.*, 46(6):749–757, Jun. 1998.
56. D.H. Johnson and D.E. Dudgeon. *Array Signal Processing — Concepts and Techniques.* Prentice-Hall, Englewood Cliffs, NJ, 1993.
57. I. Karasalo. "Estimating the Covariance Matrix by Signal Subspace Averaging." *IEEE Trans. ASSP*, ASSP-34(1):8–12, Feb. 1986.
58. M. Kaveh and A.J. Barbell. "The Statistical Performance of the MUSIC and the Minimum-Norm Algorithms in Resolving Plane Waves in Noise." *IEEE Trans. ASSP*, ASSP-34:331–341, Apr. 1986.
59. S. Kay, *Fundamentals of Statistical Signal Processing: Estimation Theory.* Prentice-Hall International Editions, Englewood Cliffs, NJ, 1993.
60. S.M. Kay, *Fundamentals of Statistical Signal Processing: Detection Theory.* Prentice-Hall, Englewood Cliffs, NJ, 1998.
61. P.-S. Kildal. "Equivalent Circuits of Receive Antennas in Signal Processing Arrays." *Microwave Optical Technol. Lett.*, 21(4):244–246, May 1999.
62. J.D. Kraus, *Antennas*, 2nd ed., McGraw-Hill, New York, 1950.
63. H. Krim and J.G. Proakis. "Smoothed Eigenspace-Based Parameter Estimation". *Automatica*, 30(1):27–38, January 1994 (Special Issue on Statistical Signal Processing and Control).
64. H. Krim and M. Viberg, "Two Decades of Array Signal Processing Research: The Parametric Approach." *IEEE Signal Process. Mag.*, 13(4):67–94, Jul. 1996.
65. A. Kuchar, J.-P. Rossi, and E. Bonek. "Directional Macro-Cell Channel Characterization from Urban Measurement." *IEEE Trans. Antennas Propag.*, 48(2):137–146, Feb. 2000.
66. R. Kumaresan and D.W. Tufts. "Estimating the Angles of Arrival of Multiple Plane Waves." *IEEE Trans. Aerosp. Electron. Syst.*, AES-19:134–139, Jan. 1983.
67. R.T. Lacross. "Data Adaptive Spectral Analysis Methods." *Geophysics*, 36:661–675, 1971.
68. A. Leshem and A.-J. Van der Veen. "Direction-of-arrival estimation for constant modulus signals." *IEEE Trans. Signal Process.*, 47(11):3125–3129, Nov. 1999.
69. A.R. Leyman and T.S. Durrani. Signal subspace techniques for DOA estimation using higher order statistic." In *Proc. ICASSP 95*, volume 3, pages 1956–1959, Detroit, MI, May 1995.
70. F. Li, J. Vaccaro, and D.W. Tufts, "Min-Norm Linear Prediction for Arbitrary Sensor Array." In *Int. Conf. Acoust. Speech, Signal Process.*, pages 2613–2616, 1989.
71. F. Li and R. Vaccaro. "Sensitivity Analysis of DOA Estimation Algorithms to Sensor Errors," *IEEE Trans. AES*, AES-28(3):708–717, Jul. 1992.
72. F. Li and R.J. Vaccaro. "Performance Degradation of DOA Estimators Due to Unknown Noise Field." *IEEE Trans. Signal Process.*, SP-40(3):686–689, Mar. 1992.
73. J. Li, B. Halder, P. Stoica, M. Viberg, and T. Kailath. "Decoupled Maximum Likelihood Angle Estimation for Signals with Known Waveforms." *IEEE Trans. Signal Process.*, 43:2154–2163, Sept. 1995.
74. J. Li and R.T. Compton, Jr. "Maximum Likelihood Angle Estimation for Signals with Known Waveforms." *IEEE Trans. Signal Process.*, 41(9):2850–2862, Sept. 1993.
75. W.S. Liggett. *Passive Sonar: Fitting Models to Multiple Time Series.* Academic Press, New York, 1973.
76. B. Lindmark, B. Ahlberg, M. Nilsson, and C. Beckman. "Performance analysis of applying up-link estimates in the down-link beamforming using a dual polarized array." In *Proc. IEEE VTC2000-Spring*, volume 2, pages 690–694, 2000.
77. L. Ljung and T. Söderström. *Theory and Practice of Recursive Identification.* M.I.T. Press, Cambridge, MA, 1983.
78. H. Lütkepohl. *Handbook of Matrices*, John Wiley & Sons, West Sussex, England, 1996.
79. V.H. MacDonald and P.M. Schulthesis. "Optimum Passive Bearing Estimation in a Spatially Incoherent Noise Environment." *J. Acoust. Soc. Am.*, 46(1):37–43, 1969.
80. C.S. MacInnes and R.J. Vaccaro. "Tracking Directions-of-Arrival with Invariant Subspace Updating." *Signal Process.*, 50(1–2):137–150, 1996.

81. C.P. Mathews and M.D. Zoltowski. "Eigenstructure techniques for 2-D angle estimation with uniform circular arrays." *IEEE Trans. Signal Process.*, 42:2395–2407, Sept. 1994.
82. Y. Meng, P. Stoica, and K.M. Wong. "Estimation of the Direction of Arrival of Spatially Dispersed Sources." *IEE Proc. Radar. Sonar Navigation,* 143(1):1–9, Feb. 1996.
83. R.P. Meys. "A Summary of the Transmitting and Receiving Properties of Antennas." *IEEE Antennas Propag. Mag.*, 42(3):49–53, Jun. 2000.
84. P. Mogensen, P. Leth-Espensen, K.I. Pedersen, F. Fredriksen, and P. Zetterberg. "Performance of Adaptive Antennas in FH-GSM Using Conventional Beamforming." *Wireless-Personal-Commun.*, 14(3):255–274, Sept. 2000.
85. R.A. Monzingo and T.W. Miller, *Introduction to Adaptive Arrays*, John Wiley & Sons, New York, 1980.
86. M. Moonen, F.J. Vanpoucke, and E.F. Deprettere. "Parallel and Adaptive High-Resolution Direction Finding." *IEEE Trans. Signal Process.*, 42(9):2439–2448, Sept. 1994.
87. A. Nehorai and E. Paldi, "Electromagnetic Vector-Sensor Array Processing." In V.K. Madisetti and D.B. Williams, editors, *The Digital Signal Processing Handbook*, CRC Press, Boca Raton, FL, 1997.
88. A. Nehorai, B. Porat, and E. Paldi. "Detection and Localization of Vapor-Emitting Sources." *IEEE Trans. Signal Process.*, 43:243–253, Jan. 1995.
89. B. Ottersten, P. Stoica, and R. Roy. "Covariance matching estimation techniques for array signal processing applications." *Digital Signal Process.*, 8(3):185–210, Jul. 1998.
90. B. Ottersten, M. Viberg, and T. Kailath. "Performance Analysis of the Total Least Squares ESPRIT Algorithm." *IEEE Trans. Signal Process.*, 39(5):1122–1135, May 1995.
91. B. Ottersten, M. Viberg, P. Stoica, and A. Nehorai. "Exact and large sample ML techniques for parameter estimation and detection in array processing." In Haykin, Litva, and Shepard, editors, *Radar Array Processing*, pages 99–51. Springer-Verlag, Berlin, 1993.
92. N.L. Owsley. Sonar array processing. In S. Haykin, editor, *Array Signal Processing*, Prentice-Hall, Englewood Cliffs, NJ. 1985.
93. A Papoulis. *Probability, Random Variables, and Stochastic Processes*, 3rd ed., McGraw-Hill, 1991.
94. D. Parsons. *The Mobile Radio Propagation Channel.* John Wiley & Sons. Chichester, England, 1996.
95. A.J. Paulraj and C.B. Papadias. "Space-Time Processing for Wireless Communications." *IEEE Trans. Signal Process. Mag.*, pages 49–83, Nov. 1997.
96. K.I. Pedersen, B.H. Fleury, and P.E. Mogensen. "High Resolution of Electromagnetic Waves in Time-Varying Radio Channels." in *IEEE Proc. Int. Symp. Pesona Indoor and Mobile Radio Communic. (PIMRC '97),* pages 650–654, Helsinki, Finland, Sept. 1997.
97. K.I. Pedersen, P.E. Mogensen, and B.H. Fleury. "A Stochastic Model for the Temporal and Azimuthal Dispersion Seen at the Base Station in Outdoor Propagation Environments." *IEEE Trans. VT,* 49:437–447, Mar. 2000.
98. P. Pelin. *Space-Time Algorithms for Mobile Communications.* PhD thesis, Chalmers University of Technology, Göteborg, Sweden, 1999.
99. P. Pelin and M. Viberg, "Direction Finding with a Dually Polarized Antenna Array Using Path-Wise Constraints with Application to Real Data." In *Proc. 32nd Asilomar Conf. Sig., Syst., Comput.,* Pacific Grove, CA, Nov. 1998.
100. L. Petterson, M. Danestig, and U. Sjöström. "An Experimental S-Band Digital Beamforming Antenna." *IEEE Aerospace Electron. Sys. Mag.*, May 1997.
101. S.U. Pillai. *Array Signal Processing*. Springer-Verlag, New York, 1993.
102. S.L. Preston, D.V. Thiel, T.A. Smith, S.G. O'Keefe, and J.W. Lu. "Base-Station Tracking in Mobile Communications Using a Switched Parasitic Antenna Array." *IEEE Trans. Antennas Propag.*, 46(6):841–844, Jun. 1998.
103. R. Raich, J. Goldberg, and H. Messer. "Bearing Estimation for a Distributed Source via the Conventional Beamformer." In *Proc. 9th Signal Process. Workshop on SSAP,* pages 5–8, Sept. 1998.
104. A. Ranheim and P. Pelin. "Decoupled Blind Symbol Estimation Using an Antenna Array." In *Proc. 8th Signal Process. Workshop on SSAP,* pages 136–139, Corfu, Greece, Jun. 1996.

105. B.D. Rao and K.V.S. Hari, "Performance analysis of ESPRIT and TAM in determining the direction of arrival of plane waves in noise. *IEEE Trans. ASSP.* ASSP-37(12): 1990–1995, Dec. 1989.
106. B.D. Rao and K.V.S. Hari, "Performance Analysis of Root-Music." *IEEE Trans. Acoustics Speech, Signal Process.*, 37(12):1939–1949, Dec. 1989.
107. B.D. Rao and K.V.S. Hari, "Weighted Subspace Methods and Spatial Smoothing: Analysis and Comparisons." *IEEE Trans. Signal Process.*, 41(2):788–803, Feb. 1993.
108. C.R. Rao, C.R. Sastry, and B. Zhou. "Tracking of the Direction of Arrival of Multiple Targets." *IEEE Trans. Signal Process.*, 42(5):1133–1144, May, 1994.
109. S.S. Reddi. "Multiple Source Location — A Digital Approach." *IEEE Trans. AES*, 15:95–105, January 1979.
110. R.H. Roy. "*ESPRIT, Estimation of Signal Parameters via Rotational Invariance Techniques.* PhD thesis, Stanford University, CA, 1987.
111. J. Sanchez-Araujo and S. Marcos. "An Efficient PASTd-Algorithm Implementation for Multiple Direction of Arrival Tracking." *IEEE Trans. Signal Process.*, 47(8):2321–2324, Aug. 1999.
112. S.R. Saunders, *Antennas and Propagation for Wireless Communication Systems*, John Wiley & Sons, New York, 1999.
113. K.R. Schaubach and N.J. Davis, IV. "Microcellular Radio-Channel Propagation Prediction." *IEEE Antennas Propag. Mag.*, 36(4):25–34, Aug. 1994.
114. R.O. Schmidt. "Multiple Emitter Location and Signal Parameter Estimation." In *Proc. RADC Spectrum Estimation Workshop*, pages 243–258, Rome, NY, 1979.
115. R. Schreiber. "Implementation of Adaptive Array Algorithms." *IEEE Trans. ASSP*, ASSP-34:1038–1045, Oct. 1986.
116. F.C. Schweppe. "Sensor Array Data Processing for Multiple Signal Sources." *IEEE Trans. IT*, IT-14:294–305, 1968.
117. G.V. Serebryakov. "Direction-of-Arrival Estimation of Correlated Sources by Adaptive Beamforming." *IEEE Trans. Signal Process.*, 43(11):2782–2787, Nov. 1995.
118. S. Shamsunder and G. Giannakis. "Signal selective localization of nonGaussian cyclostationary sources." *IEEE Trans. Signal Process.*, 42:2860–2864, Oct. 1994.
119. T.J. Shan, M. Wax, and T. Kailath. "On Spatial Smoothing for Direction-of-Arrival Estimation of Coherent Signal." *IEEE Trans. ASSP.* ASSP-33:806–811, Aug. 1985.
120. M.I. Skolnik. *Introduction to Radar System*, McGraw-Hill, New York, 1962.
121. D. Starer and A. Nehorai, "Path-Following Algorithm for Passive Localization of Near-Field Sources." *IEEE Trans. Signal Process.*, 42(3):677–680, Mar. 1994.
122. P. Stoica and R. Moses. *Introduction to Spectral Analysis.* Prentice-Hall, Upper Saddle River, NJ, 1997.
123. P. Stoica and A. Nehorai. "Performance Comparison of Subspace Rotation and MUSIC Methods for Direction Estimation." *IEEE Trans. ASSP*, Feb. 1991.
124. P. Stoica and A. Nehorai. "On the Concentrated Stochastic Likelihood Function in Array Signal Processing." *IEEE Trans. Circuit Syst.*, 14(5):669–674, 1995.
125. P. Stoica and K.C. Sharman. "Novel Eigenanalysis Method for Direction Estimation." *Proc. IEE*, F:19–26, Feb. 1990.
126. W.L. Stutzman and G.A. Thiele. *Antenna Theory and Design*, John Wiley & Sons, New York, second edition, 1998.
127. T. Svantesson. "The Effects of Mutual Coupling Using a Linear Array of Thin Dipoles of Finite Length." In *Proc. 8th IEEE Signal Process. Workshop on Statistical Signal Array Process.*, Portland, OR, September 1998.
128. T. Svantesson. "Modeling and Estimation of Mutual Coupling in a Uniform Linear Array of Dipoles." In *Proc. ICASSP 99*, pages 2961–2964, Phoenix, AZ, Mar. 1999.
129. T. Svantesson. "Multimode Based Direction Finding." In *Proc. 34th Asilomar Conf. Sig. Syst. Comput.*, Pacific Grove, CA, Oct. 2000.

130. A. Swindlehurst and T. Kailath. "A Performance Analysis of Subspace-Based Methods in the Presence of Model Errors: Part I — The MUSIC Algorithm." *IEEE Trans. Signal Process.*, SP-40(7):1758–1774, Jul. 1992.
131. A. Swindlehurst and T. Kailath. A Performance Analysis of Subspace-Based Methods in the Presence of Model Errors: Part 2 — Multidimensional Algorithms." *IEEE Trans. Signal Process.*, SP-41(7):2882–2890, Sept. 1993.
132. S. Talwar, M. Viberg, and A. Paulraj. "Blind Estimation of Multiple Co-Channel Digital Signals Arriving at an Antenna Array: Part I — Algorithms." *IEEE Trans. Signal Process.*, pages 1184–1197, May, 1996.
133. M. Torlak, L.K. Hansen, and G. Xu. "A geometric approach to blind source separation for digital wireless applications." *Signal Process.*, 73(1–2):153–167, 1999.
134. T. Trump and B. Ottersten. "Estimation of Nominal Direction of Arrival and Angular Spread Using and Array of Sensors." *Signal Process.*, 50(1–2):57–70, 1996.
135. C. Vaidyanathan and K. Buckley. "Performance Analysis of DOA Estimation Methods Based on Nonlinear Functions of Covariance Matrix." *Signal Process.*, 50(1–2):5–16, 1996.
136. S. Valaee, B. Champagne, and P. Kabal. "Parametric Localization of Distributed Sources," *IEEE Trans. Signal Process.*, 43(7):2144–2153, Sept. 1995.
137. M.C. Vanderveen, C.B. Papadias, and A. Paulraj. "Joint Angle and Delay Estimation (JADE) for Multipath Signals Arriving at an Antenna Array." *IEEE Commun. Lett.*, 1(1):12–14, Jan. 1997.
138. A. Veen, M.C. Vanderveen, and A. Paulraj. "Joint Angle and Delay Estimation Using Shift-Invariance Properties." *IEEE Signal Process. Lett.*, 4:142–145, May, 1997.
139. B.D. Van Veen and K.M. Buckley. "Beamforming: A Versatile Approach to Spatial Filtering." *Signal Proc. Mag.*, pages 4–24, Apr. 1988.
140. B.D. Van Veen and B. Williams. "Dimensionality Reduction in High Resolution Direction of Arrival Estimation." In *Proc. Asilomar Conf. Sig., Syst., Comput.*, Pacific Grove, CA, Oct. 1988.
141. M. Viberg and B. Ottersten. "Sensor Array Processing Based on Subspace Fitting." *IEEE Trans. Signal Process.*, SP-39(5):1110–1121, May 1991.
142. M. Viberg, B. Ottersten, and T. Kailah. "Detection and Estimation in Sensor Arrays Using Weighted Subspace Fitting." *IEEE Trans. Signal Process.*, SP-39(11):2436–2449, Nov. 1991.
143. M. Viberg, P. Pelin, and A. Ranheim. "Performance of Decoupled Direction Finding Based on Blind Signal Separation." In *Proc. ICASSP 97*, Munich, Germany, Apr. 1997.
144. M. Viberg, P. Stoica, and B. Ottersten. "Maximum Likelihood Array Processing in Spatially Correlated Noise Fields Using Parameterized Signals," *IEEE Trans. Signal Process.*, 45, Apr. 1997.
145. M. Wax. *Detection and Estimation of Superimposed Signals*, PhD thesis, Stanford University, Stanford CA, March 1985.
146. M. Wax. "Detection and Localization of Multiple Sources via the Stochastic Signals Model." *IEEE Trans. Signal Process.*, 39(11):2450–2456, Nov. 1991.
147. M. Wax and T. Kailath. "Detection of Signals by Information Theoretic Criteria." *IEEE Trans. ASSP*, ASSP-33(2):387–392, April 1985.
148. M. Wax and R. Tweg. "Direction of Arrival Tracking Below the Ambiguity Threshold." *IEEE Trans. AES*, 36(2):354–363, Apr. 2000.
149. M. Wax and I. Ziskind. "Detection of the Number of Coherent Signals by the MDL Principle." *IEEE Trans. ASSP*, ASSP-37(8):1190–1196, Aug. 1989.
150. R. Weber and J.A. Nossek. "Efficient DOA tracking for TDMA-based SDMA mobile communications." in *1999 IEEE 49th Vehi. Technol. Conf.*, pages 2099–2103, Houston, TX, May, 1999.
151. B. Windrow, K.M. Duvall, R.P. Gooch, and W.C. Newman, "Signal Cancellation Phenomena in Adaptive Antennas: Causes and Cures." *IEEE Trans. Antennas Propag.*, 30(5):469–478, Jul. 1982.
152. G. Xu and T. Kailath. "Direction-of-arrival estimation via exploitation of cyclostationary — a combination of temporal and spatial processing." *IEEE Trans. Signal Process.*, 40(7):1775–1786, Jul. 1992.

153. G. Xu and T. Kailath. "Fast Subspace Decomposition." *IEEE Trans. Signal Process.*, 42(3):539–551, Mar. 1994.
154. G. Xu and H. Liu, "An Effective Transmission Beamforming Scheme for Frequency-Division-Duplex Digital Wireless Communication Systems." In *Proc. ICASSP*, pages 1729–1732, Detroit, MI, 1995.
155. G. Xu, S. Silverstein, R. Roy, and T. Kailath. "Beamspace ESPRIT." *IEEE Trans. Signal Process.*, 42(2):349–356. Feb. 1995.
156. W. Xu and M. Kaveh. "Comparative Study of the Biases of MUSIC-Like Estimators." *Signal Process.*, 50(1–2):39–56, 1996.
157. X.-L. Xu and K. Buckley. "Bias Analysis of the MUSIC Location Estimator." *IEEE Trans. Signal Process.*, 40(10):2559–2569, Oct. 1992.
158. X.-L. Xu and K. Buckley. "An Analysis of Beam-Space Source Localization." *IEEE Trans. Signal Process.*, 41(1):501–504, Jan. 1993.
159. B. Yang. "Projection Approximation Subspace Tracking." *IEEE Trans. Signal Process.*, 43(1):95–107, Jan. 1995.
160. B. Yang. "Asymptotic Convergence Analysis of the Projection Approximation Subspace Tracking Algorithms." *Signal Process.*, 50(1–2):123–136, 1996.
161. P. Zetterberg. Mobile Cellular Communications with Base Station Antenna Arrays: Spectrum Efficiency, Algorithms and Propagation Models. PhD thesis, Royal Institute of Technology, Stockholm, Sweden, 1997.
162. P. Zetterberg and B. Ottersten. "The Spectrum Efficiency of a Base-Station Antenna Array System for Spatially Selective Transmission." *IEEE Trans. Veh. Technol.*, 44(3):651–660, 1995.
163. Y. Zhou, P.C. Yip, and H. Leung. "Tracking the Direction-of-Arrival of Multiple Targets by Passive Arrays: Asymptotic Performance Analysis," *IEEE Trans. Signal Process.*, 47(10):2644–2654, Oct. 1999.
164. M.D. Zoltowski, G.M. Kautz, and S.D. Silverstein. "Beamspace Root-MUSIC." *IEEE Trans. Signal Process.*, 41:344–364, Jan. 1993.

21

Blind Channel Identification and Source Separation in Space Division Multiple Access Systems

Victor Barroso
Instituto Superior Técnico

João Xavier
Instituto de Sistemas e Robótica

José M. F. Moura
Carnegie Mellon University

List of Acronyms

ACMA	analytic constant modulus algorithm
ATM	admissible transformation matrix
BSS	blind source separation
CDMA	code division multiple access
CFC_2	closed-form correlative coding
CM	constant modulus
DOA	direction of arrival
EM	expectation–maximization
EVD	eigenvalue decomposition
FDMA	frequency division multiple access
FM	frequency modulation
FSK	frequency shift keying
GDA	geodesic descent algorithm
GML	generalized maximum likelihood
GSA	Gerchberg–Saxton algorithm

0-8493-0124-6/02/$0.00+$1.50

HOS	higher order statistics
ILSE	iterative least squares with enumeration
ILSP	iterative least squares with projection
IML	iterative maximum likelihood
IOS	infinite-order statistics
ISI	intersymbol interference
LP	linear programming
LS	least squares
MAP	maximum *a posteriori*
MIMO	multiple input and multiple output
ML	maximum likelihood
OPDA	outer product decomposition algorithm
PAM	pulse amplitude modulation
PM	phase modulation
PSK	phase shift keying
QAM	quadrature amplitude modulation
QP	quadratic programming
SDMA	space division multiple access
SIMO	single input and multiple output
SML	stochastic maximum likelihood
SNR	signal to noise ratio
SOS	second-order statistics
SVD	singular value decomposition
TDMA	time division multiple access
TICC	transmitter-induced conjugate cyclostationarity

21.1 Introduction

In this chapter, we address the problem of discriminating radio sources in the context of cellular mobile wireless digital communications systems. Usually, the sources are discriminated in frequency, time, or code. In the case of frequency division multiple access (FDMA) systems, each user has assigned a different frequency band, whereas in time division multiple access (TDMA) systems the users can share the same frequency band, but transmit during disjoint time slots. Finally, code division multiple access (CDMA) systems are based in spread-spectrum techniques, where a different spreading code is assigned to each user. The spreading codes are chosen to be approximately orthogonal so that the sources present share simultaneously the same frequency bandwidth.

We also consider space division multiple access (SDMA) systems. These systems utilize the geographic diversity of the user's location in a given cell at a given time interval. Suppose that the antenna of the cell base station has a fixed multibeam beam pattern. Then, for practical purposes, users illuminated by beam i do not interfere with those illuminated by beam j, even if the users in both beams transmit at the same time, and share simultaneously the same frequency band and/or the same set of orthogonal codes. In this case, a cell with a four beam antenna has its capacity increased by a factor that is ideally four. This basic idea of sectoring a cell in several spatially disjoint smaller cells based on a single-fixed multibeam base station antenna can be translated with considerable gains into a more evolved and flexible system architecture. This relies on the concept of smart antennas and gives rise to what can be denoted smart SDMA systems, as illustrated in Fig. 21.1. A smart antenna is an antenna array with a beam pattern that is controlled electronically and/or numerically so as to illuminate the desired sources and cancel out the interferences. Usually, this is done by pointing the beam into the direction of arrival (DOA) of the desired wavefront(s), while forcing deep nulls of the beam pattern at the interference DOAs. Classically, this concept of smart antenna involves well-known DOA estimation and adaptive/blind beamforming algorithms [43, 31, 20] and [11, 12, 27, 21]. In addition, we consider a different approach, which is clarified in Subsection 21.1.1, where we set up the model of the antenna array observations data. Before doing that, we introduce notation adopted in the chapter.

Notation — $\mathbb{N}$, $\mathbb{Z}$, $\mathbb{R}$, and $\mathbb{C}$ denote the set of natural, integer, real, and complex numbers, respectively. Matrices (uppercase) and column/row vectors are in boldface type. $\mathbb{C}^{n\times m}$ and $\mathbb{C}^{n}$ denote the set of $n \times m$

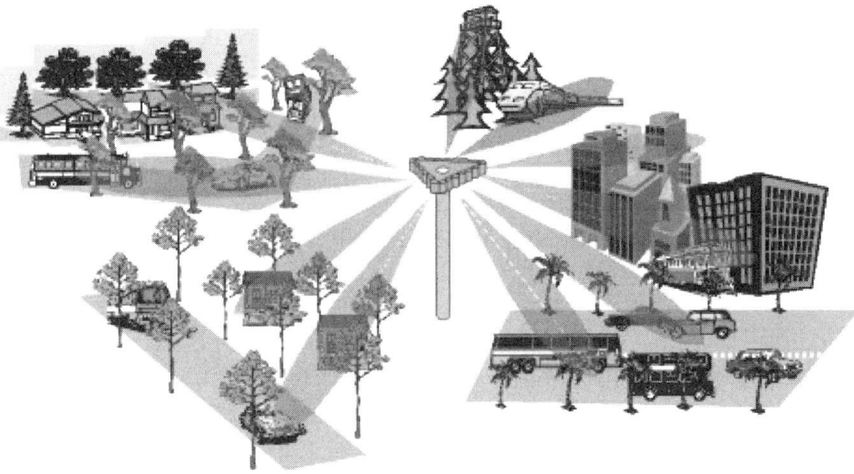

FIGURE 21.1 The concept of a smart SDMA system.

matrices and the set of n-dimensional column vectors with complex entries, respectively. The notations $(\cdot)^*$, $(\cdot)^T$, $(\cdot)^H$, $(\cdot)^\dagger$, and $\mathrm{tr}(\cdot)$ stand for the conjugate, transpose, the Hermitean, the Moore–Penrose pseudoinverse, and the trace operator, respectively; $\|\mathbf{A}\| = \sqrt{\mathrm{tr}(\mathbf{A}^H\mathbf{A})}$ denotes the Frobenius norm. The symbols $\mathbf{I}_n$, $\mathbf{0}_{n\times m}$, and $\mathbf{J}_n$ stand for the $n \times n$ identity, the $n \times m$ all-zero, and the $n \times n$ forwardshift (ones in the first lower diagonal) matrices, respectively. When the dimensions are clear from the context, the subscripts are dropped. The direct sum or diagonal concatenation of matrices is represented by diag ($\mathbf{A}_1$, $\mathbf{A}_2$, ..., $\mathbf{A}_m$); for $\mathbf{A} \in \mathbb{C}^{n\times m}$, vec ($\mathbf{A}$) $\in \mathbb{C}^{nm}$ consists of the columns of $\mathbf{A}$ stacked from left to right; and $\otimes$ represents the Kronecker product. For $\mathbf{A} \in \mathbb{C}^{n\times n}$, $\sigma(\mathbf{A}) = \{\lambda_1, \lambda_2, \ldots, \lambda_n\}$ denotes its spectrum (i.e., the set of its eigenvalues, including multiplicities). For $\mathbf{A} \in \mathbb{C}^{n\times m}$, we let row ($\mathbf{A}$) (col($\mathbf{A}$)) denote the subspace of $\mathbb{C}^m$ ($\mathbb{C}^n$) spanned by the rows (columns) of $\mathbf{A}$.

2.1.1 Array Data Model

Consider P independent digital sources, where each one generates the base band signal

$$r_p(t) = \sum_{k=-\infty}^{+\infty} s_p(k) u_p(t - kT) \tag{21.1}$$

where $u_p(t)$ is a unit energy shaping pulse, T is the baud period and $\{s_p(k)\}_{k=-\infty}^{+\infty}$ is the information sequence of independent and equally like symbols generated by the source p. These symbols are taken from a generic digital source alphabet A. We assume that the information bearing signal $r_p(t)$ modulates a radio carrier frequency ω_c. The resulting digital modulated signal is received by an antenna array of N_a omnidirectional sensors (antenna elements). At the array sensor n, the received signal has the complex envelope

$$y_{p_n}(t) = \sum_{m=1}^{M_p} \alpha_{mp} e^{-j\omega_c \tau_{mp}} r_p\left(t - \tau_{mp} - \Delta_{mp}^{(n)}\right) e^{-j\omega_c \Delta_{mp}^{(n)}} \tag{21.2}$$

where M_p is the number of propagation paths, α_{mp} and τ_{mp} are the corresponding attenuations and travel time path delay, respectively, and $\Delta_{mp}^{(n)}$ measures the intersensor propagation delay with respect to a given reference sensor. For each specific array and array/source geometry, the intersensor delays can be parameterized in terms of the DOAs.

Equation (21.2) is now rewritten into a more compact framework. We first define the overall channel impulse response between source p and sensor n, including the shaping pulse, as the time convolution

$$h_{pn}(t) = u_p(t) * h_{cha_{pn}}(t)$$

where the sensor/channel impulse responses are

$$h_{cha_{pn}}(t) = \sum_{m=1}^{M_p} \alpha_{mp} e^{-j\omega_c\left(\tau_{mp}+\Delta_{mp}^{(n)}\right)} \delta\left(t - \tau_{mp} - \Delta_{mp}^{(n)}\right)$$

Then, Eq. (21.2) is rewritten as

$$y_{p_n}(t) = \sum_{k=-\infty}^{+\infty} h_{pn}(t - kT) s_p(k) \tag{21.3}$$

This signal is time sampled with a sampling period T_s such that $T/T_s = J \geq 1$ is an integer. Then, assuming that the overall channel impulse response spans L_p baud periods, we write

$$y_{p_n}(kT + jT_s) = \sum_{l=0}^{L_p-1} h_{pn}(lT + jT_s) s_p(k-l) \qquad j = 0, 1, \ldots, J-1 \tag{21.4}$$

Here L_p determines the temporal extension of the intersymbol interferences (ISI) induced by the physical channel used by source p. The jth sample of the received signal for the kth baud period is given by Eq. (21.4) as a convolution over the variable l (i.e., over the multipath structure for that sample).

To obtain a compact expression for the received signal for an array of N_a sensors, P sources, and J samples per baud period we stack successively the samples in Eq. (21.4). We first consider a single source p. Start by stacking the received samples in Eq. (21.4) over the variable j to form the J-dimensional vector

$$\boldsymbol{y}_{p_n}(k) = \left[y_{p_n}(kT) \cdots y_{p_n}(kT + (J-1)T_s)\right]^T \qquad n = 1, \ldots, N_a$$

Then stack these N_a vectors of dimension J into the N-dimensional vector, where $N = J \times N_a$,

$$\boldsymbol{y}_p(k) = \left[\boldsymbol{y}_{p_1}(k)^T, \ldots, \boldsymbol{y}_{pN_a}(k)^T\right]^T \qquad p = 1, \ldots, P \tag{21.5}$$

Likewise, we stack the delayed replicas of the information-bearing signal for baud period k into the L_p-dimensional vector

$$\boldsymbol{s}_p(k) = \left[s_p(k)\, s_p(k-1) \cdots s_p(k - L_p + 1)\right]^T$$

and the channel impulse response for sample j from source p to sensor n into the row L_p-dimensional vector

$$\boldsymbol{h}_{pn}(j) = \left[h_{pn}(jT_s)h_{pn}(T + jT_s)\cdots h_{pn}\left((L_p - 1)T + jT_s\right)\right]$$

Define the $J \times L_p$-block matrices that collect for the J-samples these channel impulse responses from source p to sensor n

$$\mathbf{H}_{pn} = \left[\mathbf{h}_{pn}^T(0)\ \mathbf{h}_{pn}^T(1)\cdots\mathbf{h}_{pn}^T(J-1)\right]^T$$

and put together these block matrices for all N_a-sensors of the array to define the $N \times L_p$ channel matrix for source p as

$$\mathbf{H}_p = \left[\mathbf{H}_{p1}^T\ \mathbf{H}_{p2}^T\cdots\mathbf{H}_{pN}^T\right]^T$$

We can now find a concise expression for the $N = JN_a$-dimensional vector $\boldsymbol{y}_p(k)$. By using the notation just introduced, the convolution in Eq. (21.4) becomes

$$y_{p_n}(kT + jT_s) = \mathbf{h}_{pn}(j)\cdot\mathbf{s}_p(k)$$

The received signal $\boldsymbol{y}_p(k)$ for source p at all the array sensors in Eq. (21.5) is given by

$$\mathbf{y}_p(k) = \mathbf{H}_p\mathbf{s}_p(k) \tag{21.6}$$

Notice that each block matrix $\mathbf{H}_p$ determines the multipath propagation channel, including the array response, used by source p. Therefore, all the effects, such as the ISI, induced by that channel are embedded in $\mathbf{H}_p$. Naturally, sources with distinct cell locations use distinct physical channels. this means that each $\mathbf{H}_p$ acts like a source signature.

It is now easy to generalize the single-user model in Eq. (21.6) to the case of multiple users. By letting

$$\mathbf{H} = \left[\mathbf{H}_1\ \mathbf{H}_2\cdots\mathbf{H}_P\right] \in \mathbb{C}^{N\times M}$$

with $M = \Sigma_{p=1}^{P} L_p$, and the M-dimensional information signal vector

$$\mathbf{s}(k) = \left[\mathbf{s}_1^T(k)\ \mathbf{s}_2^T(k)\cdots\mathbf{s}_P^T(k)\right]^T$$

we can define the N-dimensional vector of array observations as

$$\mathbf{x}(k) = \mathbf{H}s(k) + \mathbf{n}(k),\ k = 1, 2, \ldots \tag{21.7}$$

where the N-dimensional complex vector $\mathbf{n}(k)$ represents the array observations noise at time instant (baud period) k. This is assumed to be a zero mean stationary white Gaussian vector process with covariance matrix $\sigma^2\mathbf{I}_N$.

Equation (21.7) represents the array snapshot at the kth-symbol period. If we work with K symbols, and assume that the channel matrix $\mathbf{H}$ is constant during the observation time interval of length K symbol periods, we can collect the K array snapshots in the $(N \times K)$ matrix

$$\mathbf{X} = \left[\mathbf{x}(1) \ \ \mathbf{x}(2) \ \ \cdots \ \ \mathbf{x}(K)\right]$$

Then Eq. (21.10) is compactly written as

$$\mathbf{X} = \mathbf{HS} + \mathbf{N} \tag{21.8}$$

where **S** and **N** are, respectively, the $(P \times K)$ matrix of the binary sequences of the sources and the $(N \times K)$ noise matrix,

$$\mathbf{S} = \left[\mathbf{s}(1) \ \ \mathbf{s}(2) \ \cdots \ \mathbf{s}(K)\right] \quad \text{and} \quad \mathbf{N} = \left[\mathbf{n}(1) \ \ \mathbf{n}(2) \ \cdots \ \mathbf{n}(K)\right] \tag{21.9}$$

This is the data model that we use.

Convolutive mixture model — Formally, the first term on the right-hand side of Eq. (21.7) or Eq. (21.8) represents a finite mixture of P time convolutions. This model is usually denoted as a multiple input and multiple output (MIMO) convolutive mixture model.

Instantaneous mixture model — In several year scenarios, such as in picocells, the propagation delays in Eq. (21.2) are much smaller than the baud period T. This means that the effect of ISI can be ignored. In these situations, $L_p = 1, p = 1, \ldots, P$, and the generic convolution mixture model in (21.7) or Eq. (21.8) degenerates into an instantaneous mixture model. It is clear, in any case, that we can still formally use the same representation as in Eq. (21.8)

21.1.2 Blind Source Separation

We use the same formulation for the source separation problem for the two models described in the last subsection:

Given the array data set **X**, *as defined in Eq. (21.9), where the channel mixture matrix* **H** *is unknown, find the P source information sequences* $s_p(k)$, $k = 1, \ldots, K$ *(i.e., find* **S***).*

The approach that we take to address this problem is *blind* because do not rely on any *a priori* knowledge about the channel. The solutions presented in this chapter only use the information in one or several of the following: (1) the array data model; (2) the noise statistics; and/or (3) features of the source data sequences, such as their statistics and properties of the alphabet of symbols they use.

With respect to the two mixture models considered in the last subsection — instantaneous and convolutive mixture — we can identify two distinct subproblems: blind source separation and blind equalization (ISI cancellation). With convolutive mixtures, these two subproblems need to be solved. In the simpler context of instantaneous mixtures, the ISI effect is absent and only the source separation subproblem is involved.

We present later several methods to solve these problems. In all the approaches considered, we assume that the channel matrix **H** is time invariant along the duration $T_{obs} = KT$ of K baud periods in the observation interval. For actual wireless channels, we can say that this approximation holds when T_{obs} is not very large when compared with the baud period T. This constrains the number of time samples that can be used to solve any of the subproblems referred to earlier and, as a consequence, the type of method that is used for each channel. In general, statistical methods are used to identify slow to moderately fast time-varying channels, whereas faster channels are usually identified with deterministic techniques.

21.1.3 Chapter Summary

This chapter describes several methods that can be applied to solve the blind source separation and ISI cancellation problems that we are considering. The presentation includes, in this order, deterministic or zero-order statistics (ZOS), second-order statistics (SOS), and stochastic maximum likelihood (SML) to infinite-order statistics (IOS) methods. Higher order statistics (HOS) methods are not considered because of three main reasons: (1) the large amount of data necessary to obtain efficient estimates of the HOS is compatible only with very slow varying channels, (2) the computational complexity of the resulting identification algorithms can be very high, and (3) they are very sensitive to the signal to noise ratio (SNR). Except for the stochastic maximum likelihood (SML) method where we consider only the instantaneous mixture problem, we study both the instantaneous and convolutive mixtures. We present algorithms that have been found highly relevant contributions within each class of methodologies. We include several of our own contributions.

Section 21.2 is mainly concerned with ZOS methods. We introduce several algorithms that result from a deterministic (ZOS) approach and that are suited when the channels are moderately varying. We describe the iterative least squares with projection (ILSP) and the iterative least squares with enumeration (ILSE) algorithms. These algorithms exploit the finite alphabet property of digital sources. We also present the analytical constant modulus algorithm (ACMA). This algorithm exploits a different feature of several digital modulation formats, namely, the equal energy (or constant modulus) of many signaling schemes. Like ILSP and ILSE, the ACMA requires an iterative procedure for which global convergence is not guaranteed, except when the observations are noiseless. As an alternative to iterative algorithms and to avoid their convergence problems, it is most important to have available closed-form solutions. We describe a solution that has low computational complexity. This closed-form method is based on a linear coding approach involving the sources data. We finalize the section by introducing the subspace method that solves the ISI cancellation subproblem.

Section 21.3 addresses SOS-based approaches. Essentially, the SOS are used to derive algorithms based on data prewhitening. This enables techniques that provide analytic closed-form solutions for the two subproblems that we consider (i.e., for the blind ISI cancellation and source separation). Notice that, in general, for white source sequences, the SOS-based source separation problem is ill-defined, and HOS techniques are required. The correlative coding approach, introduced in this Section, establishes a framework within which it is possible to guarantee a unique solution for the SOS-based blind source separation problem.

Section 21.4 is devoted to SML methods. The algorithms described in Section 21.2 can actually also be viewed as solutions to a maximum likelihood (ML) problem where the source signals are deterministic and the noise is white Gaussian. In Section 21.4, we solve the blind source separation problem for stochastic source signals. We restrict attention to instantaneous mixture channels. The solution that we present relies on the expectation–maximization (EM) algorithm, where the ML estimation of the channel mixing matrix is used to detect the source symbols based on the maximum *a posteriori* (MAP) criterion.

The final section of this chapter discusses new solutions to the general source separation problem, based on convex and differential geometry driven optimization techniques. These approaches exploit directly specific geometric features of the data model. The first technique falls in the class of deterministic approaches, and is restricted to high SNR scenarios and binary sources. It requires a small amount of data samples, being especially suited for fast time-varying channels. A geometric convex reformulation of the blind source separation problem is exploited to resolve the linear mixture of binary users. We also discuss an SOS-based semiblind SML technique for channel identification with white (up to second order) stationary inputs. The likelihood function for the residual unitary matrix is optimized directly over the manifold of orthogonal matrices, by a geodesic descent algorithm. This method can be applied in the case of analog or digital sources, providing semiblind and blind solutions, respectively.

21.2 Deterministic Methods

This section discusses deterministic or zero-order statistics (ZOS) methods that resolve linear mixtures of binary sources. This class of methods does not make use of the statistics or of the probability structure of the data. However, they take advantage of properties of the digital sources, for example, their finite alphabet property, or they exploit specific characteristics of the signaling/modulation used like the equal energy in signaling schemes. Although the methods described in this section can be extended to more general source alphabets, we restrict the discussion to antipodal binary alphabets $A = \{-1, +1\}$. The section addresses both cases of instantaneous and convolutive mixtures.

In Subsection 21.2.1, we present first an identifiability result and then tackle instantaneous mixtures, describing four algorithms: the ILSP algorithm; the ILSE algorithm; the ACMA; and a closed form solution based on precoding. In Subsection 21.2.2, we study convolutive mixtures and consider a subspace method for ISI cancellation.

21.2.1 Instantaneous Mixtures

In the case of instantaneous mixtures, the transmission is ISI free, and thus we take $L_p = 1, p = 1, \ldots, P$. The data model is

$$\mathbf{x}(k) = \mathbf{H}\mathbf{s}(k) + \mathbf{n}(k) \qquad k = 1, 2, \ldots \tag{21.10}$$

As described in Section 21.1, $\mathbf{x}(k)$ and $\mathbf{n}(k)$ are N-dimensional vectors denoting the array snapshot and the observation noise at time instant k, respectively, $\mathbf{s}(k)$ is the P-vector of the binary symbols generated by the P sources at time instant k; and $\mathbf{H}$ is the $(N \times P)$ channel matrix (i.e., the mixture matrix). The number P of sources is assumed known, or some estimate of it is available. Working with K symbols and grouping the K array snapshots into a single matrix, the data model is as in Eq. (21.8) repeated

$$\mathbf{X} = \mathbf{HS} + \mathbf{N} \tag{21.11}$$

The matrices in Eq. (21.11) have the following dimensions: $\mathbf{X}$ is $N \times K$; $\mathbf{H}$ is $N \times P$; $\mathbf{S}$ is $P \times K$; and $\mathbf{N}$ is $N \times K$. The noise $\mathbf{N}$ is zero mean, stationary, white Gaussian.

Problem Formulation

The problem we address is the following: given the noisy data matrix $\mathbf{X}$ in Eq. (21.11), find both $\mathbf{H}$ and $\mathbf{S}$.

Solve this problem jointly solve an estimation problem on the continuous "variable" H and a detection problem on the discrete "variable" $\mathbf{S}$. Before addressing these problems, we consider the identifiability or the uniqueness of the solution in the absence of noise.

Identifiability

Consider the noiseless, $\mathbf{N} = 0$, version of Eq. (21.11)

$$\mathbf{X} = \mathbf{HS} \tag{21.12}$$

and assume that $\mathbf{H}$ is an arbitrary full column rank matrix. The elements of $\mathbf{S}$ belong to the alphabet $A = \{-1, 1\}$. Under these conditions, any factorization

$$\mathbf{X} = \hat{\mathbf{H}}\hat{\mathbf{S}}$$

where

$$\hat{\mathbf{H}} = \mathbf{HT} \quad \text{and} \quad \hat{\mathbf{S}} = \mathbf{T}^{-1}\mathbf{S}$$

verifies Eq. (21.12), provided that $\mathbf{T}$ is one of the following $(P \times P)$ matrices: nonsingular diagonal matrix with ± 1 entries, a permutation matrix, or a product of the two. The matrix $\mathbf{T}$ causes two types of ambiguity: (1) ordering ambiguity; and (2) sign ambiguity in estimates of the signals. In any case, these ambiguities are easily removed if appropriate coding schemes are used.

Talwar et al. [34] prove an identifiability theorem that provides a sufficient condition for the existence of an admissible transform matrix (ATM) such as $\boldsymbol{T}$ cited previously. Here, we present their theorem without proof.

THEOREM 2.1 [*Identifiability*]. *Let* $\mathbf{X} = \mathbf{HS}$ *where* $\mathbf{H}$ *is an arbitrary* $(N \times P)$ *full-rank matrix with* $P < N$, *and* $\mathbf{S}$ *is a* $(P \times K)$ *full-rank matrix with* ± 1 *elements. If the columns of* $\mathbf{S}$ *include all the* 2^{P-1} *possible distinct (up to a sign) P-vectors with* ± 1 *elements, then* $\mathbf{H}$ *and* $\mathbf{S}$ *can be uniquely identified up to a* $(P \times P)$ *matrix* $\mathbf{T}$ *with exactly one nonzero element* $\{+1, -1\}$ *in each row and column.*

The probability p that, in K independent array snapshots, the columns of $\mathbf{S}$ include all the 2^{P-1} distinct (up to a sign) P-vectors with ± 1 elements is also studied in Reference [34]. This probability p is bounded

$$1 - 2^{P-1}\left(\frac{2^{P-1} - 1}{2^{P-1}}\right)^{K} \leq p \leq 1$$

Notice that p converges to 1 when K increases indefinitely. It is also clear that, for large values of P, a quite large value of K can be required to guarantee identifiability. This is only apparent. In fact, Identifiability Theorem 2.1 establishes a sufficient condition only. Most likely, smaller values of K suffice in practical situations.

At this point, we know what is the sufficient condition under which the noiseless factorization problem in Eq. (21.12) can be solved uniquely, up to an admissible transform matrix. In the following paragraphs, we discuss relevant algorithms that provide that solution.

21.2.1.1 Iterative Least Squares with Projection and Iterative Least Squares with Projection

Here, we return to the model described by Eq. (21.11), and assume that the noise is white in both the time and space dimensions, Gaussian, and zero mean, with correlation matrix $E\{\mathbf{n}(k)\mathbf{n}^H(l)\} = \sigma^2 \mathbf{I}_N \delta_{kl}$.

Under these conditions, the maximum likelihood (ML) approach is equivalent to the following separable least squares (LS) problem:

$$\left(\hat{\mathbf{H}}, \hat{\mathbf{S}}\right) = \arg\min_{\mathbf{H},\mathbf{S}} \left\| \mathbf{X} - \mathbf{HS} \right\|_F^2 \qquad (21.13)$$

where the elements of $\mathbf{S}$ are assumed deterministic and constrained to take values in $\mathscr{A}$. Recall that $\|\cdot\|_F$ is the Frobenius norm.

This minimization can be done in two steps. By noticing that $\mathbf{H}$ is an arbitrary matrix, the minimization in Eq. (21.13) with respect to $\mathbf{H}$ is unconstrained, and therefore

$$\hat{\mathbf{H}} = \mathbf{X}\mathbf{S}^{\dagger} \qquad (21.14)$$

By defining the orthogonal projection matrix

$$\mathbf{P}_{\mathbf{S}}^{\perp} = \mathbf{I}_K - \mathbf{S}^{\dagger}\mathbf{S} \qquad (21.15)$$

the substitution of Eqs. (21.14) in (21.13) yields the following constrained minimization problem:

$$\hat{\mathbf{S}} = \arg \min_{\mathbf{S} \in \mathscr{A}^{P \times K}} \left\| \mathbf{X}\mathbf{P}_{\mathbf{S}}^{\perp} \right\|_F^2 \qquad (21.16)$$

where $\mathscr{A}^{P \times K}$ denotes the set of all the $(P \times K)$ matrices with entries defined in $\mathscr{A}$. The solution to this minimization problem is achieved by enumerating over all possible choices of the binary matrix $\mathbf{S}$. It is clear that the numerical complexity of this enumeration procedure, being exponential with K and P, is prohibitive even for modest values of K and P. In the following paragraphs, we present two iterative block algorithms with lower computational complexity, the ILSP and the ILSE algorithms [33, 34].

ILSP Algorithm

For simplicity purposes, assume that the minimization problem in Eq. (21.13) is unconstrained with respect to both matrices **H** and **S**. Then, starting with an initial estimate $\hat{\mathbf{H}}_0$ of **H**, the minimization of $\|\mathbf{X} - \hat{\mathbf{H}}_0\mathbf{S}\|_F^2$ with respect to a continuous S is a simple LS problem. Each element of the solution $\bar{\mathbf{S}}_1$ is then projected back to the closest discrete value in A, producing the estimate $\hat{\mathbf{S}}_1$. The iterative algorithm runs as follows:

ILSP Algorithm

1. Given $\hat{\mathbf{H}}_0$ for $k = 0$
2. $k \leftarrow k + 1$
   a. $\bar{\mathbf{S}}_k = \hat{\mathbf{H}}_{k-1}^{\dagger}\mathbf{X}$
   b. $\hat{\mathbf{S}}_k = \text{proj}\ [\bar{\mathbf{S}}_k]$ $\quad \hat{\mathbf{S}}_k$ is the matrix in $\mathcal{A}^{P\times K}$ closest to $\hat{\mathbf{S}}_k$
   c. $\hat{\mathbf{H}}_k = \mathbf{X}\hat{\mathbf{S}}_k^{\dagger}$
3. Repeat step 2 until $(\hat{\mathbf{H}}_k, \hat{\mathbf{S}}_k) = (\hat{\mathbf{H}}_{k-1}, \hat{\mathbf{S}}_{k-1})$

In contrast with the optimum solution, which has exponential complexity in both K and P, the ILSP algorithm has polynomial complexity, more specifically, $KNP + 2P^2\ (K - P/3) + NP^2$ and $KNP + 2P^2\ (N - P/3) + KP^2$ flops per iteration to compute $\hat{\mathbf{H}}$ and $\hat{\mathbf{S}}$, respectively. Clearly, the overall complexity of the algorithms depends on its convergence rate. This complexity can be controlled if the algorithm is appropriately initialized (i.e., started with $\hat{\mathbf{H}}_0$ close to the actual **H**). In the case of low noise observations, this is likely to provide local convergency of ILSP to the optimum constrained solution in a reasonable number of iterations. However, if the mixing matrix **H** is ill-conditioned, the ILSP algorithm can diverge, essentially because of the noise enhancement produced by the least squares step to compute $\bar{\mathbf{S}}$. This important drawback of ILSP is circumvented by the ILSE algorithm.

ILSE Algorithm

The ILSE algorithm uses one property of the Frobenius norm to reduce the complexity of the enumeration procedure necessary to compute the optimum $\hat{\mathbf{S}}$. According to this property,

$$\min_{\mathbf{S}\in\mathcal{A}^{P\times K}} \left\|\mathbf{X}-\mathbf{H}\mathbf{S}\right\|_F^2 = \min_{\mathbf{s}(1)\in\mathcal{A}^{P}} \left\|\mathbf{x}(1)-\mathbf{H}\mathbf{s}(1)\right\|_F^2 + \cdots + \min_{\mathbf{s}(K)\in\mathcal{A}^{P}} \left\|\mathbf{x}(K)-\mathbf{H}\mathbf{s}(K)\right\|_F^2 \tag{21.17}$$

This means that, instead of searching among all the 2^{KP} binary matrices **S**, we can simply perform K independent enumerations, each one involving 2^P possible binary vectors $\mathbf{s}(\cdot)$,

$$\forall\, k=1,\ldots,K: \quad \hat{\mathbf{s}}(k) = \arg\min_{\mathbf{s}\in\mathcal{A}^{P}} \left\|\mathbf{x}(k)-\mathbf{H}\mathbf{s}\right\|_F^2 \tag{21.18}$$

Except for the minimization with respect to **S**, the ILSE algorithm is very similar to ILSP, each iteration being based on an alternating minimization technique as follows.

ILSE Algorithm

1. Given $\hat{\mathbf{H}}_0$ for $k = 0$
2. $k \leftarrow k + 1$
   a. With $\mathbf{H} \leftarrow \hat{\mathbf{H}}_{k-1}$ in Eq. (21.17), minimize for $\hat{\mathbf{S}}_k$ using Eq. (21.18)
   b. $\hat{\mathbf{H}}_k = \mathbf{X}\hat{\mathbf{S}}_k^{\dagger}$
3. Repeat step 2 until $(\hat{\mathbf{H}}_k, \hat{\mathbf{S}}_k) = (\hat{\mathbf{H}}_{k-1}, \hat{\mathbf{S}}_{k-1})$

The ILSE algorithm has complexity $KNP + 2P^2\ (K - P/3) + NP^2$ plus $KN{\cdot}2^P\ (P + 1)$ flops per iteration to solve $\hat{\mathbf{H}}$ and $\hat{\mathbf{S}}$, respectively. By comparing this with the complexity of ILSP, we conclude that ILSE generally requires considerably more flops per iteration to solve for $\hat{\mathbf{S}}$. Contrary to what happens with the ILSP algorithm, the ILSE algorithm has local uniform convergency and exhibits greater robustness against the observation noise, especially when **H** is ill-conditioned. These are the main advantages of ILSE over ILSP.

A detailed performance analysis of the ILSP and the ILSE algorithms is out of the scope of this chapter. The interested reader may found it in Reference [35]. Here we notice that two alternative algorithms based on successive interference cancellation concepts are introduced in Reference [22]. Like ILSE, the interference cancellation algorithms are at least monotonically convergent to a local minimum, and attain a performance similar to that of ILSE at the complexity cost of ILSP.

21.2.1.2 Analytic Constant Modulus Algorithm

While ILSP and ILSE take advantage of the finite alphabet property of digital signals, the constant modulus (CM) approach exploits the time invariance of the envelope of many communications signals such as FM and PM in the analog domain, and FSK, PSK, and 4-QAM for digital signals. The concept of modulus restoration was first introduced in the context of blind equalization problems [10, 28]. It has also been applied to solve the problem of resolving instantaneous linear mixtures of independent signals [1, 2, 14, 19, 32, 38]. Although exhibiting low computational costs, those algorithms based on gradient descent techniques have similar drawbacks. The most relevant is that there is no guarantee of convergence toward each minimum of the CM cost function, especially if the number P of sources is not known *a priori*. These very important problems are solved by the ACMA [45]. In fact, ACMA has the following properties. In the noiseless case: (1) for a number of sources $P \leq N$, $K < P^2$ array snapshots are sufficient to compute **H** and **S** exactly via an eigenvalue problem; (2) for $K > P^2$, it is possible to detect the number of CM signals present in the K array snapshots **X**; and (3) for noisy observations, robustness in finding **S** can be achieved. We introduce ACMA in this subsection. We focus on the most relevant issues that support the derivation of the algorithm.

CM factorization problem

We start from the data model in Eq. (21.12), where **H** and **S** are assumed full rank, and model the constant modulus signal property as

$$\begin{aligned} \mathbf{X} &= \mathbf{HS} \\ \left|S_{ij}\right| &= 1 \end{aligned} \tag{21.19}$$

With generality, it is assumed that only $\delta \leq P$ are CM signals. If the factorization in Eq. (21.19) is unique*, then the CM factorization problem can be formulated in an equivalent way: given the $(N \times K)$ data matrix **X** of rank P, find δ and the $(\delta \times N)$ matrix **W**, such that

$$\begin{aligned} \mathbf{WX} &= \mathbf{S}_\delta \\ \left|\left(S_\delta\right)_{ij}\right| &= 1 \end{aligned} \tag{21.20}$$

where $\delta \times K$ matrix $\mathbf{S}_\delta$ is full rank, and $\delta \leq P$ is as large as possible. Let row $(\mathbf{X})$ denote the subspace spanned by the rows of **X**, and define the set of CM signals

$$\mathcal{CM} = \left\{\mathbf{S} \,\middle|\, \left|S_{ij}\right| = 1, \forall\, i, j\right\} \tag{21.21}$$

*As in the last subsection, *uniqueness* is assumed up to an admissible transformation matrix.

To solve the CM factorization problem, we have therefore to find the rows **w** of **W** such that **wX** = **s**, with **s** ∈ row (**X**) one of the linearly independent signals in $\mathcal{CM}$. This is equivalent to finding all linearly independent signals **s** that satisfy

$$(A)\ \mathbf{s} \in \text{row}(\mathbf{X})$$

$$(B)\ \mathbf{s} \in \mathcal{CM}$$

Gerchberg–Saxton Algorithm (GSA)

The signals **s**, satisfying (A) and (B), can be found literatively using an adequate alternating projections based algorithm. Suppose that $\mathbf{y} = \mathbf{w}^{(i)}\mathbf{X}$ is a signal in the row span of **X** at iteration i. To guarantee that **y** belongs to $\mathcal{CM}$, consider the nonlinear projector onto $\mathcal{CM}$

$$\mathbf{P}_{\mathcal{CM}}(y) = \left[\frac{(\mathbf{y})_1}{|(\mathbf{y})_1|}, \ldots, \frac{(\mathbf{y})_K}{|(\mathbf{y})_K|}\right]$$

The iteration is then

$$\mathbf{w}^{(i+1)} = \left[\mathbf{P}_{\mathcal{CM}}\left(\mathbf{w}^{(i)}\mathbf{X}\right)\right]\mathbf{X}^{\dagger} \tag{21.22}$$

For each signal of interest, GSA is initialized with a different random choice of **s**.

The problem with this type of solution is that the finiteness of the data sets can preclude global convergence and may introduce spurious local minima. Reference [45] reformulates the problem in such a way that an analytic solution is provided.

Equivalent Formulation

Consider the singular value decomposition of **X** as follows:

$$\mathbf{X} = \mathbf{U\Sigma V} : \mathbf{U} \in \mathbb{C}^{N\times N},\ \Sigma \in \mathbb{R}^{N\times K},\ \mathbf{V} \in \mathbb{C}^{K\times K} \tag{21.23}$$

where **U** and **V** are unitary matrices containing the singular vectors of **X**, and Σ is a real diagonal matrix with nonnegative entries (in descending order), the singular values of **X**. If P is the number of signals, then rank (**X**) = P is the number of nonzero singular values of **X**. Thus, the first P rows of **V** form an orthonormal basis of the space row (**X**), which we collect in $\hat{\mathbf{V}} \in \mathbb{C}^{P\times K}$. Condition (A) can then be written as

$$(A'): \mathbf{s} \in \text{row}(\mathbf{X}) \Leftrightarrow \mathbf{s} = \mathbf{w}\hat{\mathbf{V}},\ \hat{\mathbf{V}} \in \mathbb{C}^{P\times K} \tag{21.24}$$

Notice that in this equivalent condition (A'), the row **w** has only P elements as a result of using the orthonormal basis in $\hat{\mathbf{V}}$ instead of the data matrix **X**.

We write the $P \times K$ matrix $\hat{V}$ as

$$\hat{\mathbf{V}} = \left[\mathbf{v}_1 \cdots \mathbf{v}_k \cdots \mathbf{v}_K\right]$$

where $\mathbf{v}_k \in \mathbb{C}^P$ is the kth column of $\hat{V}$. Define

$$\mathbf{P}_k = \mathbf{v}_k \mathbf{v}_k^H \in \mathbb{C}^{P \times P}$$
$$k = 1, \ldots, K$$

Condition (B) becomes

$$(B): \mathbf{s} = \left[(\mathbf{s})_1 \cdots (\mathbf{s})_K\right] \in \mathscr{CM} \Leftrightarrow \left[\left|(\mathbf{s})_1\right|^2 \cdots \left|(\mathbf{s})_K\right|^2\right] = [1 \cdots 1] \Leftrightarrow \begin{cases} \mathbf{w}\mathbf{v}_1\mathbf{v}_1^H \mathbf{w}^H = 1 \\ \vdots \\ \mathbf{w}\mathbf{v}_K \mathbf{v}_K^H = 1 \end{cases}$$

which is also equivalent to

$$(B'): \ \mathbf{s} \in \mathscr{CM} \Leftrightarrow \mathbf{w}\mathbf{P}_k \mathbf{w}^H = 1 \qquad (21.25)$$
$$k = 1, \ldots, K$$

From condition (B') it follows that to solve the CM factorization problem, we must find the solutions $\mathbf{w}$ to the K quadratic Eqs. (21.25). We rewrite these conditions using the Kronecker product* $\otimes$ and the following property of $\otimes$ [16]. For $\mathbf{A}_1$, $\mathbf{A}_2$, $\mathbf{B}_1$, and $\mathbf{B}_2$ matrices with appropriate dimensions

$$(\mathbf{A}_1 \cdot \mathbf{A}_2) \otimes (\mathbf{B}_1 \cdot \mathbf{B}_2) = (\mathbf{A}_1 \otimes \mathbf{B}_1) \cdot (\mathbf{A}_2 \otimes \mathbf{B}_2) \qquad (21.26)$$

Then,

$$\mathbf{w}\mathbf{P}_k \mathbf{w}^H = \mathbf{w}\mathbf{v}_k \mathbf{v}_k^H \mathbf{w}^H = \mathbf{v}_k^H \mathbf{w}^H \mathbf{w}\mathbf{v}_k = \mathbf{v}_k^H \mathbf{w}^H \mathbf{v}_k^T \mathbf{w}^T$$

These equalities follow because $\mathbf{w}$ is a $(1 \times P)$ row vector, $\mathbf{v}_k$ is a $(P \times 1)$ column vector, and $\mathbf{w}\mathbf{v}_k$ is a scalar. Because the product of two scalars commutes and equals its Kronecker product, we obtain further

$$\mathbf{w}\mathbf{P}_k \mathbf{w}^H = (\mathbf{v}_k^H \mathbf{w}^H) \otimes (\mathbf{v}_k^T \mathbf{w}^T) = (\mathbf{v}_k^H \otimes \mathbf{v}_k^T) \cdot (\mathbf{w}^H \otimes \mathbf{w}^T)$$
$$k = 1, \ldots, K$$

The last equality follows from the property in Eq. (21.26). Recall that $\mathbf{A}^T \otimes \mathbf{B}^T = (\mathbf{A} \otimes \mathbf{B})^T$ and define

$$\mathbf{y} = \mathbf{w}^H \otimes \mathbf{w}^T \in \mathbb{C}^{P^2 \times 1}$$

and

$$\mathbf{P} = \begin{bmatrix} \mathbf{p}_1 \\ \vdots \\ \mathbf{p}_K \end{bmatrix} = \begin{bmatrix} \left[\mathbf{v}_1^* \otimes \mathbf{v}_1\right]^T \\ \vdots \\ \left[\mathbf{v}_K^* \otimes \mathbf{v}_K\right]^T \end{bmatrix} \in \mathbb{C}^{K \times P^2} \qquad (21.27)$$

By using these, the K equations in conditions (B') in Eq. (21.25) are organized in matrix form

*The Kronecker product of two matrices $\mathbf{A} = [a_{ij}]$ and $\mathbf{B}$ is $\mathbf{A} \otimes \mathbf{B} = [a_{ij}\mathbf{B}]$

$$\mathbf{Py} = \begin{bmatrix} 1 \\ \vdots \\ 1 \end{bmatrix} \tag{21.28}$$

This is shown in Reference [45]. The CM factorization problem translates now into finding all linearly independent vector solutions to Eq. (21.28). Clearly, for each solution $\mathbf{w}$ of Eq. (21.28), $\mathbf{s} = \mathbf{w}\hat{\mathbf{V}}$ is the corresponding CM signal.

The solution space of Eq. (21.28) can be generally written as an affine space $\mathbf{y} = \mathbf{y}_0 + \alpha_1\mathbf{y}_1 + \cdots + \alpha_l\mathbf{y}_l$, where $\mathbf{y}_0$ is a particular solution of Eq. (21.28), and $\{\mathbf{y}_1, \ldots, \mathbf{y}_l\}$ is a basis of the kernel of $\mathbf{P}$. To work with a fully linear solution space, a linear transformation can be used. Consider a $(K \times K)$ unitary matrix $\mathbf{Q}$ such that

$$\mathbf{Q}\begin{bmatrix} 1 \\ \vdots \\ 1 \end{bmatrix} = \begin{bmatrix} K^{1/2} \\ 0 \\ \vdots \\ 0 \end{bmatrix} \tag{21.29}$$

For instance, $\mathbf{Q}$ can be chosen as the discrete Fourier transform (DFT) or a Householder transformation,

$$\begin{gathered} \mathbf{Q} = \mathbf{I}_K - 2\frac{\mathbf{q}\mathbf{q}^H}{\mathbf{q}^H\mathbf{q}} \\ \mathbf{q} = \begin{bmatrix} 1 \\ \vdots \\ 1 \end{bmatrix} - \begin{bmatrix} K^{1/2} \\ 0 \\ \vdots \\ 0 \end{bmatrix} \end{gathered} \tag{21.30}$$

which applies to $\mathbf{P}$ as follows:

$$\begin{gathered} \mathbf{QP} \doteq \begin{bmatrix} \hat{\mathbf{p}}_1 \\ \hat{\mathbf{P}} \end{bmatrix} \\ \begin{cases} \hat{\mathbf{p}}_1 \in \mathbb{C}^{1\times P^2} \\ \hat{\mathbf{P}} \in \mathbb{C}^{(\mathbb{K}-1)\times P^2} \end{cases} \end{gathered} \tag{21.31}$$

Then

$$\mathbf{Py} = \begin{bmatrix} 1 \\ \vdots \\ 1 \end{bmatrix} \Leftrightarrow \begin{cases} (\mathrm{i}) & \hat{\mathbf{p}}_1\mathbf{y} = K^{1/2} \\ (\mathrm{ii}) & \hat{\mathbf{P}}\mathbf{y} = 0 \end{cases} \tag{21.32}$$

and all the linear independent nonzero solutions $\mathbf{y}$ of Eq. (21.28) also satisfy

$$\begin{gathered} \hat{\mathbf{P}}\mathbf{y} = 0 \\ \mathbf{y} = \mathbf{w}^H \otimes \mathbf{w}^T \end{gathered} \tag{21.33}$$

Now, let $\{\mathbf{y}_1, \ldots, \mathbf{y}_{\hat{\delta}}\}$ be a basis of the kernel of $\hat{\mathbf{P}}$, where $\hat{\delta}$ is its respective dimension. Therefore, any solution $\mathbf{y}$, of $\hat{\mathbf{P}}\mathbf{y} = \mathbf{0}$ can be written as $\mathbf{y} = \alpha_1\mathbf{y}_1 + \cdots + \alpha_{\hat{\delta}}\mathbf{y}_{\hat{\delta}}$. By using another property of the Kronecker product [16], namely, that for a column vector **a** and a row vector **b**, both of the same dimension, vec $(\mathbf{a}\cdot\mathbf{b}) = \mathbf{b}^T \otimes \mathbf{a}$, the condition $\mathbf{y} = \mathbf{w}^H \otimes \mathbf{w}^T$, can also be written as $\mathbf{Y} = \mathbf{w}^T\mathbf{w}^\star$, where $\mathbf{Y} = \text{vec}^{-1}(\mathbf{y})$. Then,

$$\alpha_1\mathbf{y}_1 + \cdots + \alpha_{\hat{\delta}}\mathbf{y}_{\hat{\delta}} = \mathbf{w}^H \otimes \mathbf{w}^T \Leftrightarrow \alpha_1\mathbf{Y}_1 + \cdots + \alpha_{\hat{\delta}}\mathbf{Y}_{\hat{\delta}} = \mathbf{w}^T\mathbf{w}^\star$$

that is, the conditions in Eq. (21.33) are rewritten as a linear combination of the matrices $\{\mathbf{Y}_i\}_{i=1}^{\hat{\delta}}$, $\mathbf{Y}_i = \text{vec}^{-1}(\mathbf{y}_i)$, such that this linear combination is a rank one Hermitian matrix and hence factorizable as $\mathbf{w}^T\mathbf{w}^\star$. Linear independent solutions **w** lead to linear independent solutions **y** that in turn lead to linear independent parameter vectors $[\alpha_1 \cdots \alpha_{\hat{\delta}}]$.

The CM problem is reformulated than as follows:

Let **X** *be the data matrix from which the set of* $(P \times P)$ *matrices* $\{\mathbf{Y}_1, \cdots \mathbf{Y}_{\hat{\delta}}\}$ *are derived as described before. The CM problem is then equivalent to the determination of all independent nonzero parameter vectors* $[\alpha_1 \cdots \alpha_{\hat{\delta}}]$ *such that*

$$\alpha_1\mathbf{Y}_1 + \cdots + \alpha_{\hat{\delta}}\mathbf{Y}_{\hat{\delta}} = \mathbf{w}^T\mathbf{w}^\star \tag{21.34}$$

for each solution **w**, $\|\mathbf{w}\| = K^{1/2}$, the vector, $s = \mathbf{w}\hat{\mathbf{V}}$ *is a CM signal in* **X**.

Solution of the Noiseless CM Problem

The exact solution of the noiseless CM factorization problem is obtained in two steps: (1) computation of the number of δ of CM signals; and (2) computation of the δ row vectors **w** in Eq. (21.33).

Step 1 — Reference [45] shows that, for $K > P^2$, the dimension $\hat{\delta}$ of the kernel of $\hat{\mathbf{P}}$ equals, in general, the number δ of CM signals present in **X**. The only situations where $\hat{\delta} > \delta$ occurs is when specific phase relations exist between the signals that are present. This can be the case with BPSK and MSK signals sampled at the signal rate. These degeneracies disappear when these signals are fractionally sampled. Therefore, for almost all the cases of interest, the number of CM signals present in **X** is obtained by computing the dimension of the kernel of $\hat{\mathbf{P}}$.

Step 2 — Assume $K > P^2$ and that $\hat{\delta} = \delta$, with $\hat{\delta}$ the dimension of the kernel of $\hat{\mathbf{P}}$. Let the δ linear independent solutions to the CM problem be $\mathbf{w}_1^H \otimes \mathbf{w}_1^T, \ldots, \mathbf{w}_{\delta}^H \otimes \mathbf{w}_{\delta}^T$. They are a basis for the kernel space of $\hat{\mathbf{P}}$. It can be shown [45] that each of the matrices $\mathbf{Y}_i$ can be expressed as a linear combination of these basis vectors $\mathbf{w}_j^T\mathbf{w}_j^\star$. Writing these δ independent linear combinations for the δ matrices $\mathbf{Y}_i$ leads to the following.

Collect the $\mathbf{w}_j$ in the matrix

$$\mathbf{W} = \left[\mathbf{w}_1^T \cdots \mathbf{w}_{\delta}^T\right]^T \tag{21.35}$$

Then, the simultaneous independent linear combinations of the $\mathbf{Y}_i$ shows that the CM factorization problem is equivalent to finding the $(\delta \times P)$ matrix $\boldsymbol{W}$ with full rank δ such that

$$\begin{aligned}
\mathbf{Y}_1 &= \mathbf{W}^T\Lambda_1\mathbf{W}^\star \\
\mathbf{Y}_2 &= \mathbf{W}^T\Lambda_2\mathbf{W}^\star \\
&\cdots \\
\mathbf{Y}_{\delta} &= \mathbf{W}^T\Lambda_{\delta}\mathbf{W}^\star \\
\Lambda_1, \ldots \Lambda_{\delta} &\in \mathbb{C}^{P\times P} \text{ diagonal matrices}
\end{aligned} \tag{21.36}$$

This is a simultaneous diagonalization problem that can be solved using the *super-generalized Schur decomposition*; see Reference [45] for the details.

CM Factorization Problem with Noisy Observations
Here we consider the noisy observations model [Eq. (21.11)]

$$\mathbf{X} = \mathbf{HS} + \mathbf{N}$$

In this case, it is not possible to obtain an exact CM factorization. However, the CM factorization problem of noisy data $\mathbf{X}$ can be formulated as an optimization problem in the context of an appropriate CM metric, such as

$$\text{dist}\left(\mathbf{s}, \mathcal{CM}\right) = \sum_{k=1}^{K}\left(\left|\left(s\right)_k\right|^2 - 1\right)^2 \tag{21.37}$$

Therefore, we must find δ signals $\mathbf{s}$ that are minimizers of

$$\min\left\{\text{dist}\left(\mathbf{s}, \mathcal{CM}\right) : \mathbf{s} \in \widehat{\text{row}\left(\mathbf{X}\right)}\right\} \tag{21.38}$$

with $\widehat{\text{row}(\mathbf{X})}$ the estimated row span of $\mathbf{S}$ (i.e., the principal row span of $\mathbf{X}$ determined by an SVD); see Eq. (21.23). Letting P be the number of signals, then the principal row span of $\mathbf{X}$ is as determined before by the matrix $\hat{\mathbf{V}}$, which collects the P orthonormal P rows of $\mathbf{V}$ corresponding to the P largest singular values of $\mathbf{X}$. Like in the noiseless situation, the matrices $\mathbf{P}$ and $\hat{\mathbf{P}}$ can be constructed from $\hat{\mathbf{V}}$.

Reference [45] shows that the CM problem with noise is solved by finding all linearly independent minimizers $\mathbf{y}$ of $\|\hat{\mathbf{P}}\mathbf{y}\|^2$, subject to

$$\mathbf{y} = \mathbf{w}^H \otimes \mathbf{w}^T$$

$$\left\|\mathbf{w}\right\| = K^{1/2}$$

As in the noiseless case, the solution to this optimization problem is based on the set of d matrices $\mathbf{Y} = \text{vec}^{-1}(\mathbf{y})$. Linear combinations of these matrices should result in matrices close to rank one matrices of the form

$$\alpha_1\mathbf{Y}_1 + \cdots + \alpha_\delta\mathbf{Y}_\delta = \mathbf{Y} \simeq \mathbf{w}^T\mathbf{w}^* \tag{21.39}$$

Again, the problem of finding all the δ independent parameter vectors $[\alpha_1, \cdots, \alpha_\delta]$ can be solved based on a supergeneralized Schur decomposition. The procedure departs again from Eqs. (21.36) and starts with a QR and RQ factorizations of $\mathbf{W}^H$ and $\mathbf{W}$, respectively. Let $\mathbf{W}^T = \mathbf{Q}^H\mathbf{R}'$ and $\mathbf{W}^* = \mathbf{R}''\mathbf{Z}^H$ where $\mathbf{Q}$ and $\mathbf{Z}$ are unitary and $\mathbf{R}'$ and $\mathbf{R}''$ are upper triangular. Then premultiplying the ith equation in Eq. (21.36) on the left by the $(P \times P)$ matrix $\mathbf{Q}$ and on the right by the $(P \times P)$ matrix $\mathbf{Z}$.

$$\mathbf{QY}_i\mathbf{Z} = \mathbf{R}_i \qquad i = 1, \ldots, \delta \tag{21.40}$$

where $\{\mathbf{R} \in \mathbb{C}^{P\times P}\}_{i=1}^{\delta}$, $\mathbf{R}_i = \mathbf{R}'\Lambda_i\mathbf{R}''$, are upper triangular matrices. It is possible to show that a parameter vector $[\alpha_1 \cdots \alpha_\delta]$ satisfies the condition in Eq. (21.39) only if

$$\alpha_1\mathbf{R}_1 + \cdots + \alpha_\delta\mathbf{R}_\delta \quad \text{is rank 1}$$

Given the decomposition of the $\mathbf{R}_i$, we get the equivalent condition

$$\alpha_1\boldsymbol{\Lambda}_1 + \cdots + \alpha_\delta\boldsymbol{\Lambda}_\delta \quad \text{is rank 1}$$

Because the Λ_i are diagonal, this linear combination is diagonal. In other words, only one entry of this diagonal matrix, say entry i, is nonzero. Set this entry to one,

$$\alpha_1^i\left(\boldsymbol{\Lambda}_1\right)_{ii} + \cdots + \alpha_\delta^i\left(\boldsymbol{\Lambda}_\delta\right)_{ii} = 1 \quad i = 1, \ldots, \delta$$

By collecting these δ equations in matrix format, let $\mathbf{A}$ be a $\delta \times \delta$ matrix whose ith row in $[\alpha_1^i, \ldots, \alpha_\delta^i]$ and let $\boldsymbol{\Lambda}$ be the matrix ith row is the diagonal of $\boldsymbol{\Lambda}_i$. Then

$$\mathbf{A}\boldsymbol{\Lambda} = \mathbf{I}$$

and the rows of Λ^{-1} are the desired independent vectors $[\alpha_1 \cdots \alpha_\delta]$.

In Reference [45], it is shown that in fact one does not need to perform the factorization of the $\mathbf{R}_i$, because an equivalent result is obtained from the rows of $\mathbf{A}$:

$$\mathbf{A} = \mathbf{R}^{-1}, \ \mathbf{R} = \begin{bmatrix} \left(\mathbf{R}_1\right)_{11} & \cdots & \left(\mathbf{R}_1\right)_{\delta\delta} \\ \vdots & \vdots & \vdots \\ \left(\mathbf{R}_\delta\right)_{11} & \cdots & \left(\mathbf{R}_\delta\right)_{\delta\delta} \end{bmatrix} \tag{21.41}$$

Once these δ independent parameter vectors $[\alpha_1 \cdots \alpha_\delta]$ that verify Eq. (21.39) are found, each row vector $\mathbf{w}$ can then be estimated at the singular vector corresponding to the largest singular value of each $\mathbf{Y}$.

The simultaneous upper triangularization problem specified in Eq. (39.40) is solved using an extended **QZ** iteration described in Reference [45]. The problem with this iteration is that there is no proof of convergence, although in practice this can be achieved in a few number of iterations. We summarize the algorithm next.

Analytic Constant Modulus Algorithm

1. Estimation of row $(\mathbf{X})$:
   a. Compute an SVD of $\mathbf{X}$, Eq. (21.23)
   b. Estimate the number P = rank $(\mathbf{X})$ of signals from Σ in Eq. (21.23)
   c. Define $\hat{\mathbf{V}}$, first $\mathbf{P}$ rows of $\mathbf{V}$ in Eq. (21.23)
2. Estimation of δ = dimension of the kernel of $\hat{\mathbf{P}}$:
   a. Construct the $((K-1) \times P^2)$ matrix $\hat{\mathbf{P}}$ from $\hat{\mathbf{V}} = [\mathbf{v}_1 \cdots \mathbf{v}_K]$, using Eq. (21.27) to (21.31)
   b. Compute an SVD of $\hat{\mathbf{P}}$: $\hat{\mathbf{P}} = \mathbf{U}_P\Sigma_P\mathbf{V}_P$
   c. Estimate δ from Σ_P
   d. Define $[\mathbf{y}_1 \cdots \mathbf{y}_K]$, the last δ columns of $\mathbf{V}_P$
3. Solving of the simultaneous upper triangularization problem in Eq. (21.40):
   a. Define $\mathbf{Y}_i = \text{vec}^{-1}(\mathbf{y}_i)$, $i = 1, \ldots, \delta$
   b. Find $\mathbf{R}_i$, $i = 1, \ldots, \delta$, in Eq. (21.40)
   c. Find all vectors $[\alpha_{i1} \cdots \alpha_{i\delta}]$, $i = 1, \ldots, \delta$, from the rows of $\mathbf{A}$ in Eq. (21.41)
   d. Compute $\hat{\mathbf{Y}} = \alpha_{i1}\mathbf{Y}_1 + \cdots + \alpha_{i\delta}\mathbf{Y}_\delta$, $\hat{i} = 1, \ldots, \delta$
4. Recovery of the CM signals, for each $\hat{\mathbf{Y}}_i$
   a. Compute $\mathbf{w}_i$ such that $\hat{\mathbf{Y}}_i \simeq \mathbf{w}_i^T\mathbf{w}_i^\star$
   b. Scale: make $\|\mathbf{w}_i\| = K^{1/2}$
   c. $\mathbf{s}_i = \mathbf{w}_i\hat{\mathbf{Y}}$ with $i = 1, \ldots, \delta$, the rows of $\mathbf{S}$ that are CM signals

The ACM algorithm presents several interesting properties: (1) it is deterministic, which means that the minima of the cost function are obtained analytically; and (2) it is robust with respect to both the length of the data sequences, and to the presence of weak noise. However, because of the SVDs involved in the algorithm, it presents a high computational complexity when compared with other CM algorithms, for example, the GSA in Eq. (21.22). Even if the SVDs are computed using efficient algorithms, the complexity of the ACMA is approximately $9P^4K + 36N^2K$ flops, whereas the GSA takes $80PNK + 8N^2K$ flops. Although ACMA is an elegant analytic solution to the approximate factorization problem of the noisy observation data, the ACMA relies on an iterative procedure, the extended **QZ** iteration, which is not guaranteed to converge to the desired solution.

21.2.1.3 Closed-Form Solution Based on Linear Coding

Here, we describe a closed-form solution to the factorization problem that we have been addressing. In contradistinction with ACMA, this solution does not rely on an iterative procedure and has a much lower computational complexity. The main idea is to encode the source data using a simple linear coding scheme, which maintains the data rate and does not increase the necessary transmission bandwidth. Moreover, the resulting closed-form solution enables reconstructing the information data without first requiring the identification of the channel mixing matrix.

Consider the noiseless observations model

$$\mathbf{X} = \mathbf{H}\mathbf{Z} \tag{21.42}$$

where each row of $\mathbf{Z}$, $(z)_p$, $p = 1, \ldots, P$, represents the encoded data transmitted by user p.

$$\begin{gathered}(\mathbf{z})_p \leftarrow \text{linear encoding}\left((\mathbf{s})_p\right)\\ p = 1, \ldots, P\end{gathered} \tag{21.43}$$

and $(\mathbf{s})_p$, $p = 1, \ldots, P$, is the binary information data generated by user p. In Reference [42], the encoding scheme is determined by complex diagonal matrices $\mathbf{D}_p$, $p = 1, \ldots, P$, with symbol entries in some complex alphabet $\varepsilon \subset \mathbb{C}$, so that

$$\begin{gathered}(\mathbf{z})_p = (\mathbf{s})_p \mathbf{D}_p \in \mathbb{C}^{(1\times K)}\\ p = 1, \ldots, P\end{gathered} \tag{21.44}$$

Source Separation

Given $\mathbf{X}$ in Eq. (21.42), and assuming Eq. (21.44), the objective is to obtain all the binary signals $(\mathbf{s})_p$, $p = 1, \ldots, P$, transmitted by the sources. The algorithm starts with the SVD of the data matrix

$$\begin{aligned}\mathbf{X} &= \mathbf{U}\boldsymbol{\Sigma}\mathbf{V}^H\\ &= \left[\mathbf{U}_s \mathbf{U}_n\right]\begin{bmatrix}\Sigma_s & 0\\ 0 & 0\end{bmatrix}\begin{bmatrix}\mathbf{V}_s^H\\ \mathbf{V}_n^H\end{bmatrix}\end{aligned} \tag{21.45}$$

Assuming that $\mathbf{H}$ and $\mathbf{Z}$ are, respectively, full column and full row rank P matrices, then $\mathbf{Z}$ spans the same row space as $\mathbf{V}_s^H$. Because $\mathbf{V}_s \perp \mathbf{V}_n$ (i.e., $\mathbf{V}_s^H \mathbf{V}_n = 0$, then $\mathbf{Z}\mathbf{V}_n = 0$). By defining

$$\begin{gathered}\mathbf{V}_{n_p} = \mathbf{D}_p \mathbf{V}_n\\ p = 1, \ldots, P\end{gathered} \tag{21.46}$$

then

$$(\mathbf{s})_p \mathbf{V}_{n_p} = 0 \quad p = 1, \ldots, P \tag{21.47}$$

This constrains the encoding matrix $\mathbf{D}_p$. Taking into account that the information data is real.

$$(\mathbf{s})_p \left[\mathcal{R}\left(\mathbf{V}_{n_p}\right) \; I\left(\mathbf{V}_{n_p}\right) \right] = 0 \quad p = 1, \ldots, P \tag{21.48}$$

which doubles the constraints.

Selection of the Coding Matrices

Given the definitions in Eqs. (21.46) and (21.48), we conclude that each coding matrix $\mathbf{D}_p$, $p = 1, \ldots, P$, must be selected so as to guarantee that the left null space of $[\mathcal{R}(\mathbf{V}_{n_p}) \; I(\mathbf{V}_{n_p})]$ is 1-dimensional. As shown in Reference [42], this is achievable if and only if the mild condition $K \geq 2P - 1$ is verified.

When noise is present $\mathbf{X} = \mathbf{HZ} + \mathbf{N}$, the matrix $\mathbf{V}_n$ and all the $\mathbf{V}_{n_p}$'s [see Eq. (21.46) are noise dependent, and Eq. (21.47) is not verified exactly. This means that the selection of each coding matrix $\mathbf{D}_p$, $p = 1, \ldots, P$, must follow statistical criterion that approximates the 1-dimensional condition on the left kernel of $[\mathcal{R}(\mathbf{V}_{n_p}) \; I(\mathbf{V}_{n_p})]$, $p = 1, \ldots, P$.

The linear coding based closed-form solution performs worst than both the ACMA and the ILSP algorithms. This results from its relative simplicity, because it avoids the identification of the channel mixing matrix. The payoff is that as a closed-form algorithm, the linear coding based solution results in low computational complexity when compared with other alternatives.

21.2.2 Subspace Method for Intersymbol Interference Cancellation

Consider the linear convolutive mixture model developed in Section 21.1 for the case of P independent sources, which are transmitted through different multipath channels, all of them having the same time length L. Thus,

$$\mathbf{x}(k) = \mathbf{Hs}(k) + n(k) \quad k = 1, \ldots, K \tag{21.49}$$

where, as before, $\mathbf{x}(k)$ and $\mathbf{n}(k)$ are the N-dimensional vectors of array data and noise, respectively. Here,

$$\mathbf{s}(k) = \left[\mathbf{s}_1^T(k) \cdots \mathbf{s}_P^T(k)\right]^T$$

with

$$\mathbf{s}_p(k) = \left[s_p(k) \; s_p(k-1) \cdots s_p(k-L+1)\right]^T$$

and $\mathbf{H}$ is the (is the $N \times M$), $M = PL$, channel mixture matrix. For reasons that will become clearer shortly, we rearrange the first term on the right-hand side of Eq. (21.49) through a permutation. Let

$$\mathbf{L}_L^{PL} \tag{21.50}$$

be the permutation matrix of dimension $M = PL$ and stride L. We refer to $\mathbf{L}_L^{PL}$ as the *multipath* permutation. The multipath permutation applied on the left to the vector $\mathbf{s}(k)$ reshuffles its components by reading in the first component, then the component $L + 1$, then $2L + 1$, and so on; that is, the vector $\mathbf{s}(k)$ is rearranged into a new vector

$$\tilde{\mathbf{s}}(k) = \left[\tilde{\mathbf{s}}_0^T(k)\ \tilde{\mathbf{s}}_1^T(k) \cdots \tilde{\mathbf{s}}_{L-1}^T(k)\right]^T \tag{21.51}$$

where

$$\tilde{\mathbf{s}}_l(k) = \left[s_1(k-l)\ s_2(k-l) \cdots s_P(k-l)\right]^T \qquad l = 0, 1, \ldots, L-1 \tag{21.52}$$

It is easy to show that the inverse of the multipath permutation $\mathbf{L}_L^{PL}$ is the permutation $\mathbf{L}_L^{PL}$, referred to as the *channel multipath* permutation. Then, by inserting $\mathbf{L}_L^{PL}\ \mathbf{L}_L^{PL}$ in between $\mathbf{H}\ \mathbf{s}(k)$ in the first term of the data model in Eq. (21.49), we can write as

$$\mathbf{x}(k) = \tilde{\mathbf{H}}\tilde{\mathbf{s}}(k) + \mathbf{n}(k) \qquad k = 1, \ldots, K \tag{21.53}$$

the matrix $\tilde{\mathbf{H}} = \mathbf{H}\mathbf{L}_L^{PL}$ (i.e., it is $\mathbf{H}$ up to a column permutation). By collecting as in Section 21.1 all the K array snapshots in an $(N \times K)$ matrix,

$$\mathbf{X} = \tilde{\mathbf{H}}\tilde{\mathbf{S}} + \mathbf{N} \tag{21.54}$$

where $\tilde{\mathbf{S}}$ is a block Toeplitz matrix as can be verified by direct substitution. In fact,

$$\tilde{\mathbf{S}} = \begin{bmatrix} \tilde{\mathbf{s}}_0(1) & \ddots & \mathbf{s}_0(K-1) & \tilde{\mathbf{s}}_0(K) \\ \ddots & \ddots & \ddots & \tilde{\mathbf{s}}_1(K) \\ \tilde{\mathbf{s}}_{L-2}(1) & \tilde{\mathbf{s}}_{L-2}(2) & \ddots & \ddots \\ \tilde{\mathbf{s}}_{L-1}(1) & \tilde{\mathbf{s}}_{L-1}(2) & \ddots & \tilde{\mathbf{s}}_{L-1}(K) \end{bmatrix} = \begin{bmatrix} \mathbf{s}_1 & \ddots & \mathbf{s}_{K-1} & \mathbf{s}_K \\ \ddots & \ddots & \ddots & \mathbf{s}_{K-1} \\ \mathbf{s}_{-L+3} & \mathbf{s}_{-L+4} & \ddots & \ddots \\ \mathbf{s}_{-L+2} & \mathbf{s}_{-L+3} & \ddots & \mathbf{s}_{K-L+1} \end{bmatrix} \tag{21.55}$$

where, for $l = 0, \ldots, L-1$ and $k = 1, \ldots, K$, the (l, k)-block entries of the matrix on the right of Eq. (21.55) are $\tilde{\mathbf{s}}_l(k) = s_{k-l}$, as can be verified. By the Toeplitz condition, $\mathbf{s}_{k-l}, = \tilde{\mathbf{s}}_l(k) = \tilde{\mathbf{s}}_{l+n}(k+n) =$ for any integer n.

It is the block Toeplitz structure of the signal matrix $\tilde{\mathbf{S}}$ that is exploited by the signal subspace method. The signal subspace method essentially contributes to canceling the ISI effect for each source. Basically, the channel convolutive mixture matrix is translated into an equivalent instantaneous mixture of P-independent sources. These are then separated using one of the available algorithms, such as ILSP, ILSE, or ACMA.

In the following, we assume that $\tilde{\mathbf{H}}$ and $\tilde{\mathbf{S}}$ are matrices with full column and full row rank M, respectively. These conditions imply that the row and column spans of $\mathbf{X}$ equal the row span of $\tilde{\mathbf{S}}$ and the column span of $\tilde{\mathbf{H}}$, respectively. Formally,

$$\begin{aligned} &\tilde{\mathbf{H}} \text{ full column rank} \Rightarrow \text{row}(\mathbf{X}) = \text{row}(\tilde{\mathbf{S}}) \\ &\tilde{\mathbf{S}} \text{ full row rank} \quad\ \Rightarrow \text{col}(\mathbf{X}) = \text{col}(\tilde{\mathbf{H}}) \end{aligned} \tag{21.56}$$

The factorization $\mathbf{X} = \tilde{\mathbf{H}}\tilde{\mathbf{S}}$ can be achieved by finding either $\tilde{\mathbf{S}}$ or $\tilde{\mathbf{H}}$ with a specified row or column span, respectively; in fact, those row or column spans of $\mathbf{X}$, as expressed in the necessary conditions of Eq. (21.56). Here, we follow the first strategy, where the block Toeplitz structure of $\tilde{\mathbf{S}}$ is exploited.

Estimation of the Row Span of $\tilde{S}$

As discussed before, our assumptions guarantee that row $(\tilde{\mathbf{S}})$ can be estimated from row $(\mathbf{X})$. The SVD of $\mathbf{x}$ produces the factorization $\mathbf{X} = \mathbf{U}\Sigma\mathbf{V}^H$, where $\mathbf{U}$ and $\mathbf{V}$ are unitary matrices, and Σ is a diagonal matrix whose entries are the singular values (in descending order) of $\mathbf{X}$. In the absence of noise, $\mathbf{X}$ is rank M, and Σ has exactly M nonzero diagonal entries. In this case, we can write $\mathbf{X} = \tilde{\mathbf{U}}\tilde{\Sigma}\tilde{\mathbf{V}}^H$, where the entries of the $(M \times M)$ diagonal matrix $\tilde{\Sigma}$ are the nonzero entries of Σ, $\tilde{\mathbf{U}}$ consists of the first M columns of $\mathbf{U}$, and $\tilde{\mathbf{V}}^H$ consists of the first M rows of $\mathbf{V}^H$. Therefore,

$$\begin{aligned} \text{row}\left(\hat{\mathbf{V}}^H\right) &= \text{row}\left(\tilde{\mathbf{S}}\right) \\ \text{col}\left(\hat{\mathbf{U}}\right) &= \text{col}\left(\tilde{\mathbf{H}}\right) \end{aligned} \tag{21.57}$$

When noise is present, the rank of $\mathbf{X}$ is estimated as the number of singular values that are above the noise level. To increase the robustness against noise, this detection problem is solved based on the eigenvalues of $\mathbf{X}\mathbf{X}^H = \mathbf{U}\Lambda\mathbf{U}^H$. For white noise with covariance matrix $\sigma^2\mathbf{I}_N$, and for large enough K, the diagonal matrix Λ will have $N - M$ smallest eigenvalues $\lambda_{\text{noise}} \simeq K\sigma^2$ and M largest eigenvalues $\lambda_m \simeq (\tilde{\mathbf{S}})^2_{mm} + K\sigma^2$, $m = 1, \ldots, M$.

Forcing the Toeplitz Structure of $\tilde{S}$

Now that we have a basis $\tilde{\mathbf{V}}^H$ to span the row space of $\tilde{\mathbf{S}}$, we find a description of $\tilde{\mathbf{S}}$ that has a block Toeplitz structure with L block rows, as in Eq. (21.55). Following Reference [44], this is done using a technique denoted *row span intersections*.

The equivalence of the row spaces in Eq. (21.57) means that each row of $\tilde{\mathbf{S}} \in$ row $(\mathbf{V}^H)$. We work from this condition and reexpress it in an alternative way.

Collect the distinct block entries of $\tilde{\mathbf{S}}$ into the block row matrix

$$\mathbf{S} = \left[\underbrace{\mathbf{s}_{-L+2}\,\mathbf{s}_{-L+3} \cdots \mathbf{s}_1 \cdots \mathbf{s}_{K-L+1}}_{\text{last row block of } \tilde{\mathbf{S}}}\,\mathbf{s}_{K-L+2} \cdots \mathbf{s}_K \right] \tag{21.58}$$

in other words, $\mathbf{S}$ is the generator of the block Toeplitz matrix $\tilde{\mathbf{S}}$. For example, the last row block of $\tilde{\mathbf{S}}$ is explicitly shown in Eq. (21.58). By sliding one entry to the right under the brace we get the second to last row block of $\tilde{\mathbf{S}}$. Finally, the last K block entries of $\mathbf{S}$ are the first row of $\mathbf{S}$. Define the following shifts of the row space of $\mathbf{X}$, row $\mathbf{X}$, suitably embedded with zeros

$$\hat{\mathbf{V}}^{(l)^H} \doteq \begin{bmatrix} \mathbf{0} & \hat{\mathbf{V}}^{H} & \mathbf{0} \\ \mathbf{I}_{l-1} & \mathbf{0} & \mathbf{0} \\ \mathbf{0} & \mathbf{0} & \mathbf{I}_{L-l} \end{bmatrix} \qquad l = 1, \ldots, L \tag{21.59}$$

For example,

$$\hat{\mathbf{V}}^{(1)^H} = \begin{bmatrix} \hat{\mathbf{V}}^H & \mathbf{0} \\ \mathbf{0} & \mathbf{I}_{L-1} \end{bmatrix}$$

$$\hat{\mathbf{V}}^{(2)H} = \begin{bmatrix} \mathbf{0} & \hat{\mathbf{V}}^{H} & \mathbf{0} \\ 1 & \mathbf{0} & \mathbf{0} \\ 0 & \mathbf{0} & \mathbf{I}_{L-2} \end{bmatrix}$$

and

$$\hat{\mathbf{V}}^{(L)^H} = \begin{bmatrix} \mathbf{0} & \hat{\mathbf{V}}^{H} \\ \mathbf{I}_{L-1} & \mathbf{0} \end{bmatrix}$$

Then, it can be shown that each block row of $\tilde{\mathbf{S}}$ is in row $\mathbf{X}$ if the following L conditions are satisfied

$$\mathbf{S} \in \text{row}\left(\hat{\mathbf{V}}^{(l)^H}\right) \qquad l = 1, \ldots, L \tag{21.60}$$

In other words, $\mathbf{S}$ is in the subspace intersection of the row spaces row $(\hat{\mathbf{V}}^{(l)H})$. We interpret this condition. Consider $l = 1$.

$$\mathbf{S} = \left[\mathbf{s}_{-L+2}\,\mathbf{s}_{-L+3}\cdots\mathbf{s}_1\cdots\mathbf{s}_{K-L+1}\mathbf{s}_{K-L+2}\cdots\mathbf{s}_K\right] \in \text{row}\left(\hat{\mathbf{V}}^{(1)^H}\right)$$

This condition places no restriction on the last $(L-1)$ block entries of $\mathbf{S}$ and states that the (block) row of its first K block entries

$$\left[\mathbf{s}_{-L+2}\ \mathbf{s}_{-L+3}\cdots\mathbf{s}_1\cdots\mathbf{s}_{K-L+1}\right] \in \text{row}\left(\hat{\mathbf{V}}^{(1)^H}\right)$$

which states exactly that the last row of $\tilde{\mathbf{S}}$ is in row $(\hat{\mathbf{V}}^{(1)H})$. The remaining $(L-1)$ conditions in Eq. (21.60) are similarly interpreted.

We now consider the problem of determining the intersection subspace of the L subspaces row $(\hat{\mathbf{V}}^{(l)H})$. This intersection subspace can be computed using De Morgan's laws through the complement of the direct sum of the complements of each subspace. This direct computation of the intersection subspace is inefficient for this problem of source separation (see the discussion in the appendix of Reference [44]). An alternative is to form the matrix that stacks the orthogonal generators of each of the subspaces row $(\hat{\mathbf{V}}^{(l)H})$. One way of doing this is to compute the singular value decomposition (SVD) of the matrix that stacks the matrices $\hat{\mathbf{V}}^{(l)H}$

$$\begin{bmatrix} \hat{\mathbf{V}}^{(1)^H} \\ \vdots \\ \hat{\mathbf{V}}^{(L)^H} \end{bmatrix} \tag{21.61}$$

Because we are not interested on the left singular vectors of Eq. (21.61), it is equivalent to compute instead the singular values and the right singular vectors of

$$\mathbf{V}_{(L)} = \begin{bmatrix} \boxed{\hat{\mathbf{V}}^H} & & \mathbf{0} \\ & \boxed{\hat{\mathbf{V}}^H} & \\ \ddots & & \ddots \\ \mathbf{0} & & \boxed{\hat{\mathbf{V}}^H} \\ \boxed{\mathbf{J}_1} & & \mathbf{0} \\ \mathbf{0} & & \boxed{\mathbf{J}_2} \end{bmatrix} \tag{21.62}$$

which includes L one entry-shifted consecutive copies of $\hat{\mathbf{V}}^H$, and the matrices

$$\mathbf{J}_1 = \begin{bmatrix} \sqrt{L-1} & & & \mathbf{0} \\ & \ddots & & \\ & & \sqrt{2} & \\ \mathbf{0} & & & 1 \end{bmatrix} \tag{21.63}$$

and

$$\mathbf{J}_2 = \begin{bmatrix} 1 & & & \mathbf{0} \\ & \sqrt{2} & & \\ & & \ddots & \\ \mathbf{0} & & & \sqrt{L-1} \end{bmatrix}$$

These matrices account for the linear independent rows of the identity matrices stacked in Eq. (21.61).

The intersection of the row spans of $\hat{\mathbf{V}}^{(1)H}, \ldots, \hat{\mathbf{V}}^{(L)H}$, which determines the ISI-free P information data signals, has a basis $\mathbf{Y}$ given by the right singular vectors corresponding to the largest singular values of $\mathbf{V}_{(L)}$. When there is no noise, Reference [44] shows that $\mathbf{V}_{(L)}$ has precisely P largest singular values equal to $\sqrt{L}$, whereas the smallest are approximately equal to $\sqrt{L-1}$. Thus, we determine the interception subspace by computing the L right singular vectors of $\mathbf{V}_{(L)}$ corresponding to the largest singular value $\sqrt{L}$. Notice that for large L, this ISI filtering may be a very delicate issue.

At this point, we have a basis $\mathbf{Y}$ of the signal space where the user information data signals lie. Let $\mathbf{S}_P$ be the matrix whose P rows are these data signals. Then, we can write

$$\mathbf{Y} = \mathbf{A}\mathbf{S}_P \tag{21.64}$$

where $\mathbf{A}$ is some arbitrary matrix. Naturally, the ILSP or the ILSE algorithms can now be used to find the factorization in Eq. (21.64), subject to the elements in $\mathbf{S}_P$ being in a finite alphabet.

In this section, we have introduced the main ideas involved in the subspace method for blind ISI cancellation and source separation. For more details on the algorithm see Reference [44].

21.3 Second-Order Statistics Methods

Now we study blind multiuser signal processing techniques that exploit the information conveyed by second-order statistics (SOS) of the received data.

In Subsection 21.3.1, the SOS are used to prewhiten the observed data set. Roughly, the observed data vectors are projected on a dimension-reduced space, where the channel matrix, although still unknown, is unitary (i.e., a rotation matrix). From this algebraic property, together with other source characteristics, we derive computationally attractive source separation algorithms and/or ISI rejection techniques.

In Subsection 21.3.2, SOS are used to obtain analytic solutions for the blind channel identifiability problem. The approaches to be discussed rely on some level of preprocessing at the transmitter to attain these closed-form solutions. Basically, prefilters located at the transmitters insert a sufficiently rich structure into the correlation matrices, to assist the receiver in its blind source decoupling task.

21.3.1 Iterative Methods: Data Prewhitening

Consider the noisy linear convolutive mixture of P sources developed in Sectgion 21.1

$$\mathbf{x}(k) = \sum_{p=1}^{P} \mathbf{H}_p \mathbf{s}_p(k) + \mathbf{n}(k) \tag{21.65}$$

Hereafter, for the sake of clarity, we assume that the P channels have equal time lengths, $L_1 = L_2 = \cdots = L_P = L$ (i.e., the sources are exposed to the same degree of ISI. We rewrite, as previously done Eq. (21.65), as

$$\mathbf{x}(k) = \mathbf{H}\mathbf{s}(k) + \mathbf{n}(k) \tag{21.66}$$

where the channel matrix and the sources signal are

$$\mathbf{H} = \left[\mathbf{H}_1 \mathbf{H}_2 \cdots \mathbf{H}_P\right]$$

$$\mathbf{s}(k) = \left[\mathbf{s}_1(k)^T \ \mathbf{s}_2(k)^T \cdots \mathbf{s}_P(k)^T\right]^T$$

The assumptions on the data model in Eq. (21.66) are

ASSUMPTION 1 (A1). *The* $N \times M$ *channel matrix* $\mathbf{H}$ *is full column rank. The dimensions are* $M = PL$ and $N \geq M$;

ASSUMPTION 2 (A2). *The sources* $s_p(k)$ *denotes zero-mean, uncorrelated wide-sense stationary processes. Moreover, the sources emit uncorrelated data samples with unit power. This entails no loss of generality, because multiplicative constants are absorbed in* $\mathbf{H}$. *Thus,*

$$r_{p,q}(k,l) = \mathrm{E}\{s_p(k) s_q(l)^*\} = \delta(p-q, k-l)$$

where $\delta(n, m)$ *denotes the Kronecker delta:* $\delta(n, m) = 1$ *if* $(n, m) = (0, 0)$, $\delta(n, m) = 0$ *if* $(n, m) \neq (0, 0)$. *In matrix notation, the autocorrelation of* $\mathbf{s}(k)$, *is*

$$\mathbf{R}_{\mathbf{s}}(k) = \mathrm{E}\left\{\mathbf{s}(l)\mathbf{s}(l-k)^H\right\} = \mathbf{I}_M \delta(k)$$

ASSUMPTION 3 (A3). *For simplicity, the noise* $\mathbf{n}(k)$ *is a zero-mean spatiotemporal white Gaussian process with known variance* σ^2, *that is,*

$$\mathbf{R}_{\mathbf{n}}(k) = \mathrm{E}\left\{\mathbf{n}(l)\mathbf{n}(l-k)^H\right\} = \mathbf{I}_N \delta(k)$$

The noise $\mathbf{n}(k)$ *is independent of the source signals* $s_p(k)$.

Data prewhitening converts the unknown channel matrix $\mathbf{H}$ in Eq. (21.66) into a (still unknown) unitary matrix. The unitary structure simplifies signal processing problems such as co-channel source resolution (see Subsection 21.3.1.1), and multiuser echo suppression (see Subsection 21.3.1.2). Also, a geometric interpretation of the operation of the algorithms becomes readily available.

Let $\mathbf{R}_x(k)$ be the correlation matrix of the observations $\mathbf{x}(k)$ at lag $k \in \mathbb{Z}$, defined as $\mathbf{R}_s(k)$ and $\mathbf{R}_w(k)$,

$$\mathbf{R}_{\mathbf{x}}(k) = \mathrm{E}\left\{\mathbf{x}(l)\mathbf{x}(l-k)^H\right\}$$

Given assumptions (A2) and (A3), data prewhitening may be accomplished as follows:

$$\begin{aligned}\mathbf{R}_{\mathbf{x}}(0) &= \mathbf{H}\mathbf{R}_{\mathbf{s}}(0)\mathbf{H}^H + \mathbf{R}_{\mathbf{w}}(0)\\ &= \mathbf{H}\mathbf{H}^H + \sigma^2\mathbf{I}_N\end{aligned} \tag{21.67}$$

Denote the eigenvalue decomposition (EVD) of $\mathbf{R}_x(0)$ by

$$\mathbf{R}_{\mathbf{x}}(0) = \mathbf{U}\Sigma^2\mathbf{U}^H$$

where the $N \times N$ unitary matrix $\mathbf{U}$ and the $N \times N$ diagonal matrix Σ^2 are

$$\begin{aligned}\mathbf{U} &= [\mathbf{U}_1\,\mathbf{U}_2]\\ \Sigma^2 &= \mathrm{diag}\left(\Sigma_1^2 + \sigma^2\mathbf{I}_M, \sigma^2\mathbf{I}_{N-M}\right)\end{aligned}$$

The block $\mathbf{U}_1$ is $N \times M$, and Σ_1 is $M \times M$, diagonal, with positive diagonal entries in descending order. As seen Σ_1 is available from the EVD of $\mathbf{R}_x(0)$ by subtracting σ^2 from every diagonal entry in the upper-left $M \times M$ block Σ^2 and taking square roots of the resulting entries. From

$$\begin{aligned}\mathbf{R}_{\mathbf{x}}(0) &= \mathbf{U}_1\Sigma_1^2\mathbf{U}_1^H + \sigma^2\mathbf{I}_N\\ &= \hat{\mathbf{H}}\hat{\mathbf{H}}^H\end{aligned} \tag{21.68}$$

where we defined $\hat{\mathbf{H}} = \mathbf{U}_1\Sigma_1$. From Eqs. (21.68) and (21.67), it follows

$$\hat{\mathbf{H}}\hat{\mathbf{H}}^H = \mathbf{H}\mathbf{H}^H$$

Given assumption (A1) and standard algebra, this implies in turn that

$$\hat{\mathbf{H}} = \mathbf{H}\mathbf{Q}^H$$

for some (unknown) unitary $M \times M$ matrix $\mathbf{Q}$. Thus,

$$\hat{\mathbf{H}}^\dagger = \mathbf{Q}\mathbf{H}^\dagger$$

and whitened data samples are obtained as

$$\mathbf{y}(k) \equiv \hat{\mathbf{H}}^\dagger\mathbf{x}(k) = \mathbf{Q}\mathbf{s}(k) + \mathbf{w}(k) \tag{21.69}$$

where $\mathbf{w}(k) \equiv \hat{\mathbf{H}}^{\dagger}\mathbf{n}(k)$. Notice that $\mathbf{y}(k) \in \mathbb{C}^M$, whereas $\mathbf{x}(k) \in \mathbb{C}^N$. Thus by recalling assumption (A1), because $M \le N$, the equivalent projected data samples $\mathbf{y}(k)$ live in a dimension-reduced data space. This implies that the computational complexity of the algorithms in the remaining processing pipeline is reduced, because they typically depend on the dimensionality of the data samples.

21.3.1.1 Instantaneous Mixtures

In this subsection, we outline two iterative source separation algorithms that exploit the unitary structure of the channel matrix $\mathbf{Q}$ in the prewhiten data samples $\mathbf{y}(k)$ in Eq. (21.69): (1) the hypercube algorithm [17, 18]; and (2) the (LS) constellation derotator [46]. Here, we restrict ourselves to instantaneous mixtures, that is, $L_p = 1$ in Eq. (21.65), or equivalently, $\mathbf{Q}$ is a $P \times P$ matrix in Eq. (21.69). An SOS-based technique that converts convolutive mixtures into instantaneous ones is discussed in paragraph 3.1.2. Also, for clarity, we examine only the case of binary sources, that is, the information signal $s_p(k)$ consists of independent identically distributed (iid) data bits taken from the binary alphabet $\mathcal{A} = \{\pm 1\}$. Moreover without loss of generality, all data are assumed to be real, that is, $\mathbf{x}(k) \in \mathbb{R}^N$, $\mathbf{y}(k) \in \mathbb{R}^P$, $\mathbf{Q} \in \mathbb{R}^{P\times P}$, and so on.

Hypercube Algorithm

The hypercube algorithm is a sequential source separation technique that recursively extracts all the P transmitted data streams, one at a time. Once a user's signal is estimated, its contribution is removed from the observed mixture — the dimension of the problem is deflated by one — and the algorithm restarts searching for another signal. Suppose K is the available number of data vectors $\mathbf{y}(k)$. Then, by collecting the data vectors in a $P \times K$ data matrix $\mathbf{Y}$, we get

$$\begin{aligned} \mathbf{Y} &= \left[\mathbf{y}(1) \;\; \mathbf{y}(2) \;\cdots\; \mathbf{y}(K)\right] \\ \mathbf{Y} &= \mathbf{QS} + \mathbf{W} \end{aligned} \tag{21.70}$$

where the $P \times K$ signal and noise matrices $\mathbf{S}$ and $\mathbf{W}$ follow similar definitions as in Eq. (21.70). For noiseless samples, $\mathbf{W} = 0$; and, for a certain condition on $\mathbf{S}$, the columns of $\mathbf{Q}$ are the unique maximizers, up to a $\pm$ sign, of a certain function formulated over the observed data matrix.

Recall that for a generic vector $\mathbf{x} - [x_1 x_2 \ldots x_n]^T \in \mathbb{R}^n$ its l_p norm is given by

$$l_p(\mathbf{x}) = \left(\sum_{i=1}^{n} |x_i|^p\right)^{1/p}$$

THEOREM 3.1. *Let* $\mathbf{Y} = \mathbf{QS}$. *Suppose that* $\mathbf{S}$: $P \times 2^P$ *contains all* 2^P *combinations of* ± 1's *among its columns, and let* $\mathbf{Q}$ *be unitary. If* $\alpha \in \mathbb{R}^P$ *is a global maximizer of* $f(\alpha) = l_1(\mathbf{Y}^T\alpha)$, *subject to* $l_2(\alpha) = 1$, then $\alpha = \pm\mathbf{q}_p$, *where* $\mathbf{q}_p$ *denotes some column of* $\mathbf{Q}$.

The objective function in Theorem 3.1 is given by

$$f(\alpha) = \sum_{k=1}^{K} \left|\mathbf{y}(k)^T \alpha\right|$$

Given a global maximizer of f, say $\hat{\alpha}$, and because $\mathbf{Q}$ is unitary, then

$$\hat{\mathbf{s}}^T = \operatorname{sign}\left(\hat{\alpha}^T \mathbf{Y}\right)$$

denotes a row of $\mathbf{S}$ (i.e., some transmitted binary stream).

Maximization of the objective function $f(\alpha)$, subject to the constraint $\|\alpha\| = 1$, is achieved through a standard gradient search technique, as follows. Because for arbitrary $\mathbf{x} \in \mathbb{R}$, $|x| = \text{sign}(x)x$, then

$$f(\alpha) = \alpha^T \beta$$

where

$$\beta = \sum_{k=1}^{K} \text{sign}\left(\alpha^T \mathbf{y}(k)\right) \mathbf{y}(k)$$

Thus, assuming that $\alpha^T \mathbf{y}(k) \neq 0$, for all $k = 1, \ldots, K$

$$\nabla f(\alpha) = \beta$$

The locally convergent gradient search subroutine is given next.

1. Set $i = 0$ and choose α_0
   a. $i = i + 1$
   b. $\hat{\mathbf{s}}_i^T = \text{sign}(\alpha_{i-1}{}^T \mathbf{Y})$
   c. $\beta = \mathbf{Y}\hat{\mathbf{s}}_i / \|\mathbf{Y}\hat{\mathbf{s}}_i\|$
   d. $\alpha_i = \alpha_{i-1} + [\mathrm{I} - \alpha_{i-1}\, \alpha_{i-1}^T]\beta$
   e. $\alpha_i = \alpha_i / \|\alpha_i\|$
2. Until $\alpha_i - \alpha_{i-1} = 0$
3. $\hat{\mathbf{s}}^T = \text{sign}(\alpha^T \mathbf{Y})$

Step 1d projects the gradient β onto the tangent space of the constraint set $\|\alpha_{i-1}\| = 1$, and then moves in the direction of this projected gradient (to maximize f). Step 1e returns to the feasible set.

In practice, there are departures from the ideal conditions assumed thus far: the whitened data samples $\mathbf{Y} = \mathbf{QS} + \mathbf{W}$ are corrupted by additive colored noise $\mathbf{W}$; the signal matrix $\mathbf{S}$ possesses some correlation among its rows (i.e., the rows of $\mathbf{S}$ are not necessarily orthogonal). The hypercube algorithm takes into account these nonideal conditions: after the kth step, all the prior information is exploited to correct the estimates produced: $\boldsymbol{\alpha}_k$; $\hat{\mathbf{s}}_k^T$; and $\hat{\mathbf{q}}_k = \mathbf{Y}\hat{\mathbf{s}}_k^T / \|\mathbf{s}_k\|^2$; an estimate of the kth column of $\mathbf{Q}$. The first k columns of $\mathbf{Q}$, i.e., $\hat{\mathbf{Q}}_k = [\hat{\mathbf{q}}_1 \cdots \hat{\mathbf{q}}_k]$ are reestimated as $\hat{\mathbf{Q}}_k = \mathbf{Y}\hat{\mathbf{S}}_k^{\dagger}$, where

$$\hat{\mathbf{S}}_k = \begin{bmatrix} \hat{s}_1^T \\ \vdots \\ \hat{s}_k^T \end{bmatrix}$$

contains the curently extracted k user binary signals; the pseudoinverse handles the correlation among the rows of $\mathbf{S}$.

The estimate of α_k is improved as follows. Ideally α_k denotes the kth column of $\mathbf{Q}$, and, because it is orthogonal to the other columns of the unitary $\mathbf{Q}$, it spans the null space of

$$\tilde{\mathbf{Q}}_k^T = \left[\mathbf{q}_1 \cdots \mathbf{q}_{k-1}\, \mathbf{q}_{k+1} \cdots \mathbf{q}_P\right]^T$$

or, equivalently, the null space of

$$\tilde{\mathbf{R}}_k = \tilde{\mathbf{Q}}_k \tilde{\mathbf{Q}}_k^T$$

On the other hand, $\hat{\mathbf{R}}_k$ is the denoised correlation matrix

$$\tilde{\mathbf{R}}_k = \mathrm{E}\left\{\tilde{\mathbf{y}}(k)\tilde{\mathbf{y}}(k)^T\right\} - \mathbf{R}_w$$

with

$$\tilde{\mathbf{y}}(k) = \mathbf{y}(k) - \mathbf{q}_k s(k)$$

where the noise correlation matrix $\mathbf{R}_w$ is assumed known. Thus α_k is reestimated as the eigenvector associated with the smallest eigenvalue of

$$\frac{1}{K}\tilde{\mathbf{Y}}\tilde{\mathbf{Y}} - \mathbf{R}_w$$

where

$$\tilde{\mathbf{Y}} = \mathbf{Y} - \tilde{\mathbf{q}}_k \hat{\mathbf{s}}_k^T$$

Reference [17] offers an alternative geometric interpretation of the reestimation of α_k.

The next step deflates the dimension of the problem by one. This is achieved by applying an oblique projector to the observed data **Y**. The range of the oblique projector is set to the null space of

$$\hat{\mathbf{Q}}_k^{-1} \equiv \begin{bmatrix} \hat{\alpha}_1^T \\ \vdots \\ \hat{\alpha}_k^T \end{bmatrix}$$

and its null space to the range of $\hat{\mathbf{Q}}_k$ to reject the extracted signals and keep the unextracted ones untouched. The final step consists of reducing the dimensionality of the oblique projector. Details to implement this overall projector, say Π, can be found in Reference [5]. The complete hypercube algorithm is listed next.

1. Initialization: set $\mathbf{Y}_1 = \mathrm{Y}$, $\hat{\mathbf{Q}}_0^{-1} = \varnothing_2$ $\hat{\mathbf{S}}_0 = \varnothing$
2. for $k = 1$ to $P - 1$
   a. Call gradient search subroutine with input $\mathbf{Y}_k$, return $\mathbf{s}_k{}^T$
   b. $\hat{\mathbf{S}}_k = \begin{bmatrix} \hat{\mathbf{S}}_{k-1} \\ \hat{\mathbf{s}}_k^T \end{bmatrix}$
   c. $\hat{\mathbf{Q}}_k = \mathbf{Y}\mathbf{S}_k^{\dagger}$
   d. Reestimate α_k as the eigenvector associated with the smallest eigenvalue of $\hat{\mathbf{Y}}\hat{\mathbf{Y}}^{\mathrm{T}}/K - \mathbf{R}_w$
   e. $\hat{\mathbf{Q}}_k^{-1} = \begin{bmatrix} \hat{\mathbf{Q}}_{k-1}^{-1} \\ \alpha_P^T \end{bmatrix}$
   f. Project the observed data $\mathbf{Y}(P \times K)$: $\mathbf{Y}_{k+1} = \mathbf{\Pi Y}$ $((P - k) \times K)$
3. $\hat{\mathbf{S}}_P^T = \mathrm{sign}\ (\mathbf{Y}_P)$
4. $\hat{\mathbf{s}}_P = \begin{bmatrix} \hat{\mathbf{S}}_{P-1} \\ \hat{\mathbf{s}}_P^T \end{bmatrix}$

   $\hat{\mathbf{Q}}_P = \mathbf{Y}\ \mathbf{S}_P^{\dagger}$

Least Squares Constellation Derotator

The whitened data samples obey the model

$$\mathbf{y}(k) = \mathbf{Q}\mathbf{s}(k) + \mathbf{w}(k)$$

where the $P \times P$ matrix $\mathbf{Q}$ is orthogonal and $\mathbf{s}(k)$ belongs to the P-dimensional binary constellation

$$\mathcal{B}^P = \mathcal{B} \times \cdots \times \mathcal{B}$$

where $\mathcal{B} = \{\pm 1\}$. The cardinality of $\mathcal{B}^P$ is 2^P. Thus, geometrically, the samples $\mathbf{y}(k)$ are obtained by applying the rotation $\mathbf{Q}$ to the hypercube whose vertices are given by

$$\mathcal{H} = \mathcal{B}^P = \left\{ \left(\pm 1, \ldots, \pm 1 \right)^T \right\}$$

and adding noise $\mathbf{w}(k)$. In other words, the observations $\mathbf{y}(k)$, see Fig. 21.2, form clusters around the vertices of the rotated hypercube

$$\mathcal{H}_Q = \mathbf{Q}\mathcal{H}$$

This geometric interpretation motivates the following approach for identifying the unknown rotation $\mathbf{Q}$: find the orthogonal matrix $\hat{\mathbf{Q}}$ that "best" derotates the observed samples $\mathbf{y}(k)$, more specifically, that minimizes the LS distance of the derotated samples to the reference constellation hypercube $\mathcal{H}$. Thus, if $\text{dist}(\mathbf{x}, \mathcal{H}) = \min_{b \in H} \|x - b\|$ and $\mathcal{U} = \{P \times P \text{ orthogonal matrices}\}$, we have

$$\hat{\mathbf{Q}} = \arg\min_{\mathbf{U} \in \mathcal{U}} \sum_{k=1}^{K} \text{dist}^2 \left(\mathbf{U}^T \mathbf{y}(k), \mathcal{H} \right) \tag{21.71}$$

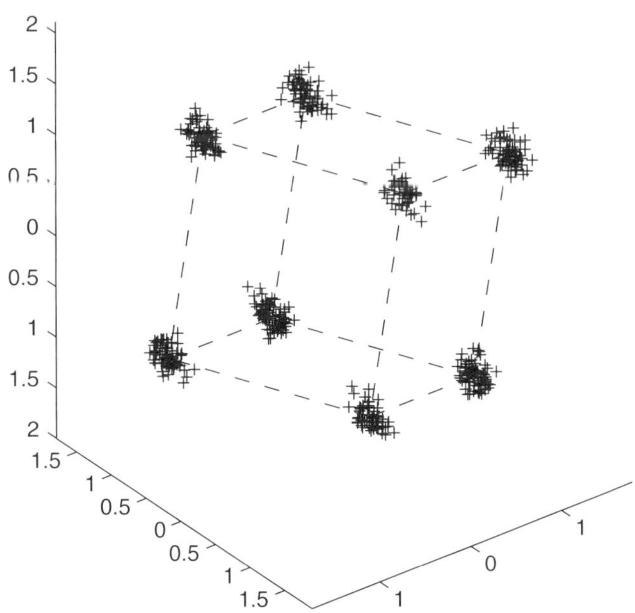

FIGURE 21.2 Geometric interpretation of $\mathbf{y}(k)$: clusters centered about the vertices of H_Q.

The function dist($\mathbf{x}$, $\mathcal{H}$) denotes the distance from the point $\mathbf{x} \in \mathbb{R}^P$ to the hypercube $\mathcal{H}$. By letting the vertex of $\mathcal{H}$ closest to $\mathbf{x}$ be

$$\mathbf{b}(\mathbf{x}) = \arg\min_{\mathbf{b}\in\mathcal{H}} \|\mathbf{x} - \mathbf{b}\|$$

and by using the fact that $\mathbf{U}$ is orthogonal, Eq. (21.71) can be rewritten as

$$\hat{\mathbf{Q}} = \arg\min_{\mathbf{U}\in\mathfrak{U}} \sum_{k=1}^{K} \left\|\mathbf{y}(k) - \mathbf{U}\mathbf{b}\left(\mathbf{U}^T\mathbf{y}(k)\right)\right\|^2 \tag{21.72}$$

The formulation in Eq. (21.72) does not admit a closed-form solution, but the alternative minimization

$$\left(\hat{\mathbf{Q}}, \hat{\mathbf{S}}\right) = \arg\min_{(\mathbf{U},\mathbf{B})\in\mathcal{U}\times\mathcal{B}} \sum_{k=1}^{K} \left\|\mathbf{y}(k) - \mathbf{U}\mathbf{b}(k)\right\|^2 \tag{21.73}$$

where $\mathbf{B} = [\mathbf{b}(1) \cdots \mathbf{b}(K)]$ and $\mathcal{B} = \{P \times K \text{ binary matrices}\}$, yields, in the first component of the ordered pair, a solution to Eq. (21.72). The minimization in Eq. (21.73) can be handled iteratively by separating the variables $\mathbf{U}$ and $\mathbf{B}$. Namely, given the nth iteration $(\mathbf{U}_n, \mathbf{B}_n)$, we let

$$\mathbf{U}_{n+1} = \arg\min_{\mathbf{U}\in\mathfrak{U}} \left\|\mathbf{Y} - \mathbf{U}\mathbf{B}_n\right\|^2 \tag{21.74}$$

$$\mathbf{B}_{n+1} = \arg\min_{\mathbf{B}\in\mathcal{B}} \left\|\mathbf{Y} - \mathbf{U}_{n+1}\mathbf{B}\right\|^2 \tag{21.75}$$

After algebraic manipulations, the subproblem in Eq. (21.74) is equivalent to

$$\mathbf{U}_{n+1} = \arg\max_{\mathbf{U}\in\mathfrak{U}} \operatorname{tr}\left(\mathbf{B}_n\mathbf{Y}^T\mathbf{U}\right)$$

whose solution is given by he transpose of the polar factor of the $P \times P$ matrix $\mathbf{B}_n\mathbf{Y}^T$, see Reference [13]; more specifically, if

$$\mathbf{B}_n\mathbf{Y}^T = \mathbf{V}_1\Sigma\mathbf{V}_2^T$$

denotes a singular value decomposition of $\mathbf{B}_n\mathbf{Y}^T$, then

$$\mathbf{U}_{n+1} = \mathbf{V}_2\mathbf{V}_1^T$$

With respect to the optimization problem expressed in Eq. (21.75), we notice that the minimization can be carried out column by column of $\mathbf{B}$, with each column found by exhaustive search in the constellation $\mathcal{B}^P$ (i.e., kth column of $\mathbf{B}_{n+1}$ is given by)

$$\mathbf{b}_{n+1}(k) = \arg\min_{\mathbf{b}\in\mathcal{B}^P} \left\|\mathbf{y}(k) - \mathbf{U}_n\mathbf{b}\right\|^2$$

21.3.1.2 Intersymbol Interference Cancellation Methods

We rework the data model of Eq. (21.69), by regrouping terms with the same delay in the multipath replicas. This is similar to what was done with Eq. (21.51) and (21.52). Recall the multipath permutation matrix $\mathbf{L}_L^M$, $M = PL$, with stride L introduced in Eq. (21.50) and its inverse the channel multipath permutation $\mathbf{L}_L^M$. Define the permuted vector $\tilde{\mathbf{s}}_l(k)$ and the matrix $\mathbf{U}$ obtained by a column permutation of the unitary matrix $\mathbf{Q}$, by applying the channel multipath permutation $\mathbf{L}_L^M$ on the right. Then Eq. (21.69) is rewritten

$$\begin{aligned}\mathbf{y}(k) &= \sum_{l=0}^{L-1} \mathbf{U}_l \tilde{\mathbf{s}}_l(k) + \mathbf{w}(k) \\ &= \mathbf{U}\tilde{\mathbf{s}}(k) + \mathbf{w}(k)\end{aligned} \tag{21.76}$$

$$\begin{aligned}\tilde{\mathbf{s}}_l(k) &= \left[\tilde{s}_1(k-l)\tilde{s}_2(k-l)\cdots\tilde{s}_P(k-l)\right]^T \in \mathbb{C}^P \\ \mathbf{U} &= \left[\mathbf{U}_0\mathbf{U}_1\cdots\mathbf{U}_{L-1}\right] \\ \tilde{\mathbf{s}}(k) &= \left[\tilde{\mathbf{s}}_0(k)^T\tilde{\mathbf{s}}_1(k)^T\cdots\tilde{\mathbf{s}}_{L-1}(k)^T\right]^T\end{aligned} \tag{21.77}$$

where we also defined the vector $\hat{\mathbf{s}}_l(k)$ that collects all source delayed replicas for symbol k. Notice that $\mathbf{U}_l$ is $M \times P$ for $l = 0, 1, \ldots, L-1$.

The data model in Eq. (21.76) describes the whitened samples $\mathbf{y}(k)$ as a linear superposition of L echos of the transmitted symbols $\{s_1(k), s_2(k), \ldots, s_P(k)\}$, that is, the replicas $\tilde{\mathbf{s}}_l(k)$ with delays $l = 0, 1, \ldots, L-1$.

The data $\mathbf{y}(k)$ is contaminated by ISI as soon as $L \geq 1$ (convolutive mixtures). The presence of ISI increases the computational burden of the blind source separation (BSS) algorithms: the echos act as additional virtual sources.

We discuss in this next section a technique that rejects the ISI in the observed samples (i.e., removes the contribution of the echos). More specifically, we outline a procedure that extracts form $\mathbf{y}(k)$, L ISI-free signals

$$\mathbf{y}_l(k) = \tilde{\mathbf{U}}_l\tilde{\mathbf{s}}_l(k) + \mathbf{w}_1(k)$$
$$l = 0, 1, \ldots, L-1$$

by exploiting statistical and algebraic properties of the data model, namely, that $\tilde{\mathbf{s}}_p(k)$ is white and $\mathbf{U}$ is unitary.

Let

$$\mathbf{J} = \begin{bmatrix} 0 & 0 & 0 & \cdots & 0 \\ 1 & 0 & \ddots & \ddots & 0 \\ 0 & 1 & \ddots & \ddots & 0 \\ \vdots & \ddots & \ddots & \ddots & \vdots \\ 0 & \cdots & 0 & 1 & 0 \end{bmatrix}$$

denote the Jordan nilpotent matrix of size $L \times L$. Then, given assumption (A2), we have

$$\mathbf{R}_{\tilde{\mathbf{s}}}(l) = \mathrm{E}\left\{\tilde{s}(k)\tilde{s}(k-l)^H\right\} = \mathbf{J}^l \otimes \mathbf{I}_P$$

for $l > 0$; here, $\otimes$ denotes the Kronecker product. By using this expression in Eq. (21.77), and by assumption (A3),

$$\mathbf{R}_y(l) = \mathbf{U}\mathbf{R}_{\tilde{s}}(l)\mathbf{U}^H = \mathbf{U}_l\mathbf{U}_0^H + \mathbf{U}_{l+1}\mathbf{U}_1^H + \cdots + \mathbf{U}_{L-1}\mathbf{U}_{L-1-l}^H \tag{21.78}$$

for $0 < l \leq L - 1$. Thus, from Eq. (21.78),

$$\operatorname{span}\{\mathbf{R}_y(l)\} = \operatorname{span}\{[\mathbf{U}_l \;\; \mathbf{U}_{l+1} \;\; \cdots \;\; \mathbf{U}_{L-1}]\}$$

This means that, through the correlation matrices $\mathbf{R}_y(1), \ldots, \mathbf{R}_y(L-1)$, the receiver has access to the orthogonal projectors $\mathbf{P}_l$ onto span $\{[\mathbf{U}_l \cdots \mathbf{U}_{L-1}]\}$, for $l = 1, \ldots, L-1$; the lth projector $\mathbf{P}_l$ is obtained from the EVD of $\mathbf{R}_y(l)$ we can write

$$\mathbf{P}_l = \sum_{i=l}^{L-1} \Pi_i$$

where $\Pi_i = \mathbf{U}_i\mathbf{U}_i^H$ if the orthogonal projector onto span $\{\mathbf{U}_i\}$. In fact, the projectors Π_i are the most interesting ones because of the orthogonality of $\mathbf{U}$,

$$\Pi_l\mathbf{U}_m = \mathbf{U}_m\delta(l-m)$$

This isolates each replica $\hat{\mathbf{s}}_l(k)$ from the observations $\mathbf{y}(k)$ in Eq. (21.76) as follows:

$$\mathbf{y}_l(k) \equiv \Pi_l\mathbf{y}(k) = \sum_{m=0}^{L-1}\left(\Pi_l\mathbf{U}_m\tilde{\mathbf{s}}_m(k)\right) + \Pi_l\mathbf{w}(k) = \mathbf{U}_l\tilde{\mathbf{s}}_l(k) + \mathbf{w}_l(k)$$

where $\mathbf{w}_l(k) = \Pi_l\mathbf{w}(k)$ denotes the component of $\mathbf{w}(k)$ in the subspace span $\{\mathbf{U}_l\}$. The projectors $\{\Pi_i, \ldots, \Pi_{L-1}\}$ can be obtained from $\{\mathbf{P}_1, \ldots, \mathbf{P}_{L-1}\}$ through the identity

$$\Pi_l = \mathbf{P}_l(\mathbf{I}_M - \mathbf{P}_{l+1})$$

Also,

$$\Pi_0 = \mathbf{I}_M - \mathbf{P}_1$$

21.3.2 Closed-Form Solutions

We discuss second-order statistics techniques that identify the unknown multiple-input, multiple-output (MIMO) channel analytically (i.e., noniteratively). These closed-form solutions rely on preprocessing at the transmitter, which judiciously shapes the statistics of the emitted signals.

Here, although we restrict ourselves to multiuser SOS techniques, we should notice that other classes of analytic solutions are available for the blind channel estimation problem. For single-user scenarios, the work by Tong et al. [37] was the first to accomplish closed-form blind identification of the communication channels based only on SOS. Another SOS-based technique is the so-called subspace method introduced by Moulines et al. [26]. Non-SOS techniques relying on redundant precoders, but with the very important property of robustness with respect to the channel spectral nulls, have been developed [9, 29, 30]. See also Reference [24] for precoding techniques in the context of undersampled multiuser communication systems.

21.3.2.1 Outer Product Method

The outer product decomposition algorithm (OPDA) was introduced in Reference [7]. It identifies the unknown coefficients of the channel matrix up to a residual unitary $P \times P$ ambiguity matrix. This residual instantaneous mixing matrix is solved in closed form by assuming prior knowledge of certain parts of the composite signal channel, for example, the pulse-shaping filters of the users [8]. Consider the signal model in Eq. (21.65), and let, for the sake of clarity, $L_1 = \cdots = L_P = L$ (i.e., all user channels have the same memory). We start once again with the model in Eq. (21.65) repeated for the sake of completeness

$$\mathbf{x}(k) = \sum_{p=1}^{P} \mathbf{H}_p \mathbf{s}_p(k) + \mathbf{n}(k) \tag{21.79}$$

and written in block notation the N-dimensional vector $\mathbf{x}(k)$

$$\mathbf{x}(k) = \mathbf{H}\mathbf{s}(k) + \mathbf{n}(k) \tag{21.80}$$

with the $N \times M$, $M = PL$, channel matrix and the PL-dimensional sources signal vector

$$\mathbf{H} = \left[\mathbf{H}_1 \mathbf{H}_2 \cdots \mathbf{H}_P\right]$$

$$\mathbf{s}(k) = \left[\mathbf{s}_1(k)^T \ \mathbf{s}_2(k)^T \cdots \mathbf{s}_P(k)^T\right]^T$$

The vectors $\mathbf{s}_p(k)$ are L-dimensional. We arrange the data with the multipath permutation $\mathbf{L}_L^{PL}$ introduced in Eq. (21.50) to get

$$\mathbf{x}(k) = \sum_{l=0}^{L-1} \tilde{\mathbf{H}}_l \tilde{\mathbf{s}}_l(k) + \mathbf{n}(k) \tag{21.81}$$

$$= \tilde{\mathbf{H}}\tilde{\mathbf{s}}(k) + \mathbf{n}(k) \tag{21.82}$$

where

$$\tilde{\mathbf{H}} = \left[\tilde{\mathbf{H}}_0 \tilde{\mathbf{H}}_1 \cdots \tilde{\mathbf{H}}_{L-1}\right] = \mathbf{H}\mathbf{L}_P^{PL} \tag{21.83}$$

$$\tilde{\mathbf{H}}_l = \left[\tilde{\mathbf{h}}_1(l) \tilde{\mathbf{h}}_2(l) \cdots \tilde{\mathbf{h}}_P(l)\right] \tag{21.84}$$

$$\tilde{\mathbf{s}}(k) = \mathbf{L}_L^{PL}\mathbf{s}(k) = \left[\tilde{\mathbf{s}}_0(k)^T \cdots \ \tilde{\mathbf{s}}_{L-1}(k)^T\right]^T \tag{21.85}$$

$$\tilde{\mathbf{s}}_l(k) = \left[s_1(k-l) s_2(k-l) \cdots \ s_P(k-1)\right]^T \tag{21.86}$$

The dimensions of these objects are: $\tilde{\boldsymbol{H}}$: $N \times PL$; $\tilde{\mathbf{H}}_l$: $N \times P$; $\tilde{\mathbf{s}}(k)$: $PL \times 1$; and $\tilde{\mathbf{s}}_l(k)$: $P \times 1$. We now stack J successive observations in the NJ-dimensional vector $\mathbf{x}_J(k)$

$$\mathbf{x}_J(k) = \left[\mathbf{x}(k)^T \ \ \mathbf{x}(k-1)^T \ \cdots \ \mathbf{x}(k-J+1)^T\right]^T \tag{21.87}$$

Likewise, we consider $P(L+J-1)$-dimensional signal vector $\tilde{\mathbf{s}}_{L+J-1}(k)$, the NJ-dimensional noise vector $\mathbf{n}_J(k)$, and the $NJ \times P(L+J-1)$ block Toeplitz channel matrix $\tilde{\mathbf{H}}$

$$\tilde{\mathbf{s}}_{L+J-1}(k) = \left[\tilde{\mathbf{s}}(k)^T \quad \tilde{\mathbf{s}}(k-1)^T \quad \cdots \quad \tilde{\mathbf{s}}(k-(L+J-2))^T\right]^T$$

$$\mathbf{n}_J(k) = \left[\mathbf{n}(k)^T \quad \mathbf{n}(k-1)^T \quad \cdots \quad \mathbf{n}(k-J+1)^T\right]^T \tag{21.88}$$

$$\tilde{\mathbf{H}} = \begin{bmatrix} \tilde{\mathbf{H}}_0 & \tilde{\mathbf{H}}_1 & \cdots & \tilde{\mathbf{H}}_{L-1} & \mathbf{0} & \cdots & \mathbf{0} \\ \mathbf{0} & \tilde{\mathbf{H}}_0 & \tilde{\mathbf{H}}_1 & \cdots & \tilde{\mathbf{H}}_{L-1} & \ddots & \vdots \\ \vdots & \ddots & \ddots & \ddots & \ddots & \ddots & \mathbf{0} \\ \mathbf{0} & \cdots & \mathbf{0} & \tilde{\mathbf{H}}_0 & \tilde{\mathbf{H}}_1 & \cdots & \tilde{\mathbf{H}}_{L-1} \end{bmatrix}$$

We have

$$\mathbf{x}_J(k) = \tilde{\mathbf{H}}\tilde{\mathbf{s}}_{L+J-1}(k) + \mathbf{n}_J(k) \tag{21.89}$$

The block Toeplitz matrix $\tilde{\mathbf{H}}$ is parameterized by the $NL \times P$ block generator matrix

$$\mathcal{H} = \begin{bmatrix} \tilde{\mathbf{H}}_0 \\ \tilde{\mathbf{H}}_1 \\ \vdots \\ \tilde{\mathbf{H}}_{L-1} \end{bmatrix} \tag{21.90}$$

which collects all the unknown channel blocks.

The OPDA estimates $\mathcal{H}$ based on the following observation. Let the $NL \times (L+J-1)P$-dimensional matrix $\tilde{\mathbf{H}}_0$ be block Hankel given by

$$\tilde{\mathbf{H}}_0 = \begin{bmatrix} \tilde{\mathbf{H}}_0 & \tilde{\mathbf{H}}_1 & \tilde{\mathbf{H}}_2 & \cdots & \tilde{\mathbf{H}}_{L-1} & \cdots & \mathbf{0} \\ \tilde{\mathbf{H}}_1 & \tilde{\mathbf{H}}_2 & \cdots & \tilde{\mathbf{H}}_{L-1} & \mathbf{0} & \cdots & \mathbf{0} \\ \tilde{\mathbf{H}}_2 & \cdots & \tilde{\mathbf{H}}_{L-1} & \mathbf{0} & \cdots & \cdots & \mathbf{0} \\ \vdots & & & & & & \vdots \\ \tilde{\mathbf{H}}_{L-1} & \mathbf{0} & \cdots & \cdots & \cdots & \cdots & \mathbf{0} \end{bmatrix} \tag{21.91}$$

It follows that the Hermitian matrix

$$\mathbf{D}_1 = \tilde{\mathbf{H}}_0\tilde{\mathbf{H}}_0^H \tag{21.92}$$

can be written as

$$\mathbf{D}_1 = \begin{bmatrix} \mathbf{D}_{11} & \mathbf{D}_{12} & \cdots & \mathbf{D}_{1L} \\ \mathbf{D}_{21} & \mathbf{D}_{22} & \cdots & \mathbf{D}_{2L} \\ \vdots & \vdots & \ddots & \vdots \\ \mathbf{D}_{L1} & \mathbf{D}_{L2} & \cdots & \mathbf{D}_{LL} \end{bmatrix}$$

with

$$\mathbf{D}_{ij} = \sum_{k=i-1}^{L-1} \tilde{\mathbf{H}}_k \tilde{\mathbf{H}}_{k+j-1}^H$$

with $\tilde{\mathbf{H}}_k = \mathbf{0}$, for $k > L - 1$. By defining the lower right block of $\mathbf{D}_1$

$$\mathbf{D}_2 = \begin{bmatrix} \mathbf{D}_{22} & \cdots & \mathbf{D}_{1L} & \mathbf{0} \\ \mathbf{D}_{32} & \cdots & \mathbf{D}_{2L} & \mathbf{0} \\ \vdots & \vdots & \vdots & \vdots \\ \mathbf{D}_{L1} & \cdots & \mathbf{D}_{LL} & \mathbf{0} \\ \mathbf{0} & \cdots & \mathbf{0} & \mathbf{0} \end{bmatrix}$$

we have the important result:

$$\Delta D = \mathbf{D}_1 - \mathbf{D}_2 = \mathcal{H}\mathcal{H}^H$$

Thus, an SVD of the outer product matrix ΔD yields $\mathcal{H}$ up to a unitary matrix, that is,

$$\hat{\mathcal{H}} = \mathcal{H}\mathbf{Q}^H$$

where the $P \times P$ matrix $\mathbf{Q}$ is unitary.

To implement this approach we need to estimate only the Hermitian matrix $\mathbf{D}_1$. We show next that this matrix is obtained directly from he correlation matrices of the system outputs, that is, from

$$\left\{\mathbf{R}_{\mathbf{x}}(l) = \mathrm{E}\left\{\mathbf{x}(k)\mathbf{x}(k-l)^H\right\} : l = 0, 1, \ldots, L-1\right\}$$

These correlations can be computed to be

$$\mathbf{R}_{\mathbf{x}}(l) = \sum_{i=0}^{L-1} \tilde{\mathbf{H}}_{i+l} \tilde{\mathbf{H}}_i^H \tag{21.93}$$

Notice also that

$$\mathbf{R}_{\mathbf{x}_J} = \mathrm{E}\left\{\mathbf{x}_J(k)\mathbf{x}_J(k)^H\right\} = \begin{bmatrix} \mathbf{R}_{\mathbf{x}}(0) & \mathbf{R}_{\mathbf{x}}(1) & \cdots & \mathbf{R}_{\mathbf{x}}(J-1) \\ \mathbf{R}_{\mathbf{x}}(1)^H & \mathbf{R}_{\mathbf{x}}(0) & \cdots & \mathbf{R}_{\mathbf{x}}(J-2) \\ \vdots & \vdots & \ddots & \vdots \\ \mathbf{R}_{\mathbf{x}}(J-1)^H & \mathbf{R}_{\mathbf{x}}(J-2)^H & \cdots & \mathbf{R}_{\mathbf{x}}(0) \end{bmatrix} \tag{21.94}$$

$$= \tilde{\mathbf{H}}\tilde{\mathbf{H}}^H + \sigma^2 \mathbf{I}_{NJ} \tag{21.95}$$

The last equation follows by direct computation from Eq. (21.87). Further, define the following block Hankel matrix

$$\mathbf{R}_0 = \begin{bmatrix} \mathbf{R}_x(0) - \sigma^2\mathbf{I}_N & \mathbf{R}_x(1) & \cdots & \mathbf{R}_x(L-1) \\ \mathbf{R}_x(1) & \mathbf{R}_x(2) & \cdots & 0 \\ \vdots & \vdots & \ddots & \vdots \\ \mathbf{R}_x(L-1) & 0 & \cdots & 0 \end{bmatrix}$$

By replacing the entries of matrix $\mathbf{R}_0$ by their definition from Eq. (21.93) for $l = 0, \ldots, L-1$, and, after some algebra, we can see that

$$\mathbf{R}_0 = \tilde{\mathbf{H}}_0 \tilde{\mathbf{H}}^H \tag{21.96}$$

where $\tilde{\mathbf{H}}$ and $\tilde{\mathbf{H}}_0$ are defined in Eqs. (21.88) and (21.91), respectively.

Because from the assumptions on the channel we can assume that $\tilde{\mathbf{H}}$ is full column rank, its pseudo-inverse $\tilde{\mathbf{H}}^\dagger = \tilde{\mathbf{H}}^H (\tilde{\mathbf{H}}\tilde{\mathbf{H}}^H)^\dagger$ is well defined and $\tilde{\mathbf{H}}^\dagger\tilde{\mathbf{H}} = \mathbf{I}$. We can then go back to the definition of $\mathbf{D}_1$ from Eq. (21.92), and obtain successively

$$\mathbf{D}_1 = \mathbf{H}_0 \tilde{\mathbf{H}}_0^H \tag{21.97}$$

$$= \tilde{\mathbf{H}}_0 \left(\tilde{\mathbf{H}}^\dagger \tilde{\mathbf{H}}\right) \tilde{\mathbf{H}}_0^H \tag{21.98}$$

$$= \tilde{\mathbf{H}}_0 \tilde{\mathbf{H}}^H \left(\tilde{\mathbf{H}}\tilde{\mathbf{H}}^H\right)^\dagger \tilde{\mathbf{H}}\tilde{\mathbf{H}}_0^H \tag{21.99}$$

$$= \mathbf{R}_0 \left(\mathbf{R}_{\mathbf{x}_J} - \sigma^2 \mathbf{I}_{NJ}\right)^\dagger \mathbf{R}_0^H \tag{21.100}$$

Equation (21.98) follows by inserting the identity matrix in between the two factors of Eq. (21.97) and by replacing it by the product $\tilde{\mathbf{H}}^\dagger\tilde{\mathbf{H}}$; Eq. (21.99) follows by replacing the pseudoinverse by its definition; Eq. (21.100) recognizes from Eq. (21.96) the first two factors and the last two factors are $\mathbf{R}_0$ and its Hermitian, and recognizes from Eq. (21.95) the middle factor as $(\mathbf{R}_{x_J} - \sigma^2\mathbf{I}_{NJ})^\dagger$.

Thus, in practice, $\mathbf{D}_1$ is estimated via Eq. (21.100), with the correlation matrices $\mathbf{R}_x(l)$ needed in $\mathbf{R}_{x_J}$ and $\mathbf{R}_0$, replaced by their consistent estimators

$$\hat{\mathbf{R}}_x(l) = \frac{1}{K-l} \sum_{k=l+1}^{K} \mathbf{x}(k)\mathbf{x}(k-l)^H$$

where K denotes the number of samples $\mathbf{x}(k)$ available.

We note that other blind SOS techniques can also convert a convolutive mixture of P users into an instantaneous one (i.e., the matrix $\mathbf{H}$ is solved up to a $P \times P$ unknown mixing matrix). We refer the interested reader to the methods in References [39, 40, 41] that achieve this goal even for infinite-impulse response (IIR) channels with minimum-phase common zeros among the subchannels. See also References [15, 25] where the MIMO channel can be recovered up to a block diagonal constant matrix; in the special case where the users are exposed to distinct channel degrees, then the convolutive mixture is completely resolved.

Solving for the residual instantaneous mixture matrix $\mathbf{Q}$ remains. As described in Reference [8], this can be accomplished analytically if part of the continuous-time composite channel $h_p(t)$ for the pth user is known *a priori* by the receiver.

Recall the derivation of the data model in Section 21.1.1, and assume, for simplicity, $N_a = 1 \Rightarrow N = J$. The composite channel $h_p(t)$ is given by the convolution of a pulse-shaping filter $u_p(t)$ with a propagation channel $c_p(t)$:

$$h_p(t) = u_p * c_p(t) \tag{21.101}$$

where $*$ denotes convolution and $c_p(t) \equiv h_{\text{cha}_{pn}} * h_{\text{rec}}(t)$ (for $n = 1$), with $h_{\text{rec}}(t)$ the receiver filter impulse response. In many communication systems, knowledge of $u_p(t)$ is available at the receiver; in this case, only the unknown part $c_p(t)$ needs to be estimated.

Assume the data are acquired with an oversampling factor of J (i.e., the sampling period is $T_s = T/J$, where T is the symbol baud period). The $J \times 1$ vectors $\tilde{\boldsymbol{h}}_p(l)$ in Eq. (21.84) are given by

$$\tilde{\mathbf{h}}_p(l) = \left[h_p((l-1)T) \;\; h_p((l-1)T + T_s) \cdots h_p((l-1)T + (J-1)T_s)\right]^T$$

that is, $\tilde{\boldsymbol{h}}_p(l)$ takes J equispaced samples of the continuous time channel $h_p(t)$ from the slot $[(l-1)T; lT)$. We now relate $\tilde{\boldsymbol{h}}_p(l)$ with the $JL \times P$ block matrix $\mathcal{H}$ given in Eq. (21.90) that collects all the channel unknowns. Reexpressing $\mathcal{H}$ in terms of its columns

$$\mathcal{H} = \left[\mathfrak{h}_1 \cdots \mathfrak{h}_p \cdots \mathfrak{h}_P\right]$$

The pth column $\mathfrak{h}_p$ of $\mathcal{H}$ is the NL-dimensional multichannel vector for the pth user

$$\mathfrak{h}_p = \left[\tilde{\mathbf{h}}_p(0)^T \;\; \tilde{\mathbf{h}}_p(1)^T \;\; \cdots \;\; \tilde{\mathbf{h}}_p(L-1)^T\right]^T$$

In other words, the pth column of $\mathcal{H}$ contains the T_s-sampled response $h_p(t)$ for source p. The discrete counterpart of Eq. (21.101) is the discrete convolution given in matrix format by

$$\mathfrak{h}_p = \mathbf{U}_p \mathbf{c}_p \tag{21.102}$$

where

$$\mathbf{U}_p = \begin{bmatrix} u_p(0) & 0 & \cdots & 0 \\ u_p(1) & u_p(0) & \ddots & \vdots \\ \vdots & \ddots & \ddots & 0 \\ u_p(L_f) & \ddots & \ddots & u_p(0) \\ 0 & u_p(L_f) & \ddots & \vdots \\ \vdots & \ddots & \ddots & \vdots \\ 0 & \cdots & 0 & u_p(L_f) \end{bmatrix}$$

$u_p(i) \equiv u_p(\mathrm{i}T_s)$ and

$$\mathbf{c}_p = \left[\mathbf{c}_p(0) \;\; \mathbf{c}_p(1) \;\; \mathbf{c}_p(2) \;\; \cdots \;\; \mathbf{c}_p(L_c)\right]^T$$

$\mathbf{c}_p(i) \equiv c_p(iT_s)$ for appropriate integers L_f, L_c. Thus, $\mathcal{H}$ can be written as

$$\mathcal{H} = \left[\mathbf{U}_1\mathbf{c}_1 \;\; \mathbf{U}_2\mathbf{c}_2 \;\; \cdots \;\; \mathbf{U}_P\mathbf{c}_P\right]$$

Denote the OPDA estimate of $\mathcal{H}$ by

$$\hat{\mathcal{H}} = \mathcal{H}Q^H$$

with the $P \times P$ matrix $Q = [q_1 \cdots q_P]$ unitary. It follows that

$$\hat{\mathcal{H}}\mathbf{q}_p = \mathbf{u}_p c_p$$

or, equivalently,

$$\underbrace{\begin{bmatrix} \mathbf{U}_p & -\hat{\mathcal{H}} \end{bmatrix}}_{\mathbf{S}_p} \begin{bmatrix} \mathbf{c}_p \\ \mathbf{q}_p \end{bmatrix} = \mathbf{0} \tag{21.103}$$

that is, the unknowns $(\mathbf{c}_p, \mathbf{q}_p)$ belong to the kernel of the matrix $\mathbf{S}_p$, which is available at the receiver. If the homogeneous system in Eq. (21.103) has a unique solution, then this yields a simple method for finding the unknowns. As shown in Reference [8], uniqueness of solutions is guaranteed provided that the z^{-1}-polynomials $H_i(z)$ $(i \neq p)$ are linearly independent of $U_p(z)L(z)$ for all $L(z)$, where $H_i(z)$ and $U_p(z)$ denote the z-transforms of the discrete time signals $h_i(\cdot)$ and $u_p(\cdot)$, respectively.

21.3.2.2 Transmitter Induced Conjugate Cyclostationarity

The transmitter induced conjugate cyclostationarity (TICC) technique for blind identification of MIMO systems was introduced in Reference [6]. It consists in inducing a cyclic frequency in each user's information sequence prior to transmission. This diversity in the cyclospectra of the emitted signals is then exploited at the receiver, to reduce the original problem of blind MIMO identification to P simpler blind single input, multiple output (SIMO) identification ones. To solve the latter problem, standard algorithms, for example, the subspace method [26], can be employed. Before giving an outline of this technique, we need some definitions.

For an $N \times M$ matrix $\mathbf{A} = [\mathbf{a}_1 \cdots \mathbf{a}_M]$ and an integer, J, we let the $N(J+1) \times (M+J+1)$ matrix $\tau_J(\mathbf{A})$ denote the block-Sylvester matrix

$$\tau_J(\mathbf{A}) - \begin{bmatrix} \mathbf{a}_1 & \cdots & \mathbf{a}_M & \mathbf{0} & \cdots & \mathbf{0} \\ \mathbf{0} & \mathbf{a}_1 & \cdots & \mathbf{a}_M & \ddots & \vdots \\ \vdots & \ddots & \ddots & \ddots & \ddots & \mathbf{0} \\ \mathbf{0} & \cdots & \mathbf{0} & \mathbf{a}_1 & \cdots & \mathbf{a}_M \end{bmatrix} \tag{21.104}$$

The conjugate cyclocorrelation sequence of a signal $z(k)$ at the frequency α, written $\mathbf{R}_z^{(\alpha)}(\boldsymbol{\tau})$, $\boldsymbol{\tau} \in \mathbb{Z}$, is given by

$$\mathbf{R}_z^{(\alpha)}(\tau) = \lim_{K\to\infty} \frac{1}{K} \sum_{k=1}^{K} \mathrm{E}\left\{ z(k+\tau) z(k)^T \right\} e^{-i2\pi\alpha k} \tag{21.105}$$

Consider the signal model in Eq. (21.65) and assume, without loss of generality, that $s_p(k)$ denote real iid symbol sequences with unit power. Cyclic frequencies are induced at the emitter by modulating each information sequence $s_p(k)$ by a complex exponential:

$$\breve{s}_p(k) = s_p(k) e^{i\pi\alpha_p k}$$

with $\alpha_p \neq \alpha_q$ for $p \neq q$, $p, q \in \{1, 2, \ldots, P\}$. Thus, we have

$$\mathbf{x}(k) = \sum_{p=1}^{P} \mathbf{H}_p \breve{s}_p(k) + \mathbf{n}(k)$$

as in Eq. (21.65). By applying the definition in Eq. (21.105) to the observations $\mathbf{x}(k)$ at the cyclic frequency β:

$$\mathbf{R}_{\mathbf{x}}^{(\beta)}(\tau) = \sum_{p=1}^{P} \mathbf{H}_p \mathbf{R}_{\mathbf{s}_p}(\tau) \breve{\mathbf{H}}_p^T \delta(\alpha_p - \beta) \tag{21.106}$$

where $\breve{\mathbf{H}}_p = [\breve{\mathbf{h}}_p(0)\ \breve{\mathbf{h}}_p(1) \cdots \breve{\mathbf{h}}_p(l_p - 1]$, $\breve{\mathbf{h}}_p(l)e^{-j2\pi\alpha_p l}$; here $\delta(\lambda) = 1$ if $\lambda = 0$ and $\delta(\lambda) = 0$ if $\lambda \neq 0$. Equation (21.106) discloses the important fact that, at the frequency $\beta = \alpha_p$, only the pth user contributes to $\mathbf{R}_{\mathbf{x}}^{(\alpha p)}(\tau)$,

$$\mathbf{R}_{\mathbf{x}}^{(\alpha_p)}(\tau) = \mathbf{H}_p \mathbf{R}_{\mathbf{s}_p}(\tau) \breve{\mathbf{H}}_p^T \tag{21.107}$$

Let $\Re_J^{(\beta)}$ denote the $N(J+1) \times N(J+1)$ block Toeplitz matrix

$$\mathcal{R}_J^{(\beta)} = \begin{bmatrix} \mathbf{R}_{\mathbf{x}}^{(\beta)}(0) & \mathbf{R}_{\mathbf{x}}^{(\beta)}(1) & \cdots & \mathbf{R}_{\mathbf{x}}^{(\beta)}(J) \\ \mathbf{R}_{\mathbf{x}}^{(\beta)}(-1) & \mathbf{R}_{\mathbf{x}}^{(\beta)}(0) & \ddots & \vdots \\ \vdots & \ddots & \ddots & \mathbf{R}_{\mathbf{x}}^{(\beta)}(1) \\ \mathbf{R}_{\mathbf{x}}^{(\beta)}(-J) & \cdots & \mathbf{R}_{\mathbf{x}}^{(\beta)}(-1) & \mathbf{R}_{\mathbf{x}}^{(\beta)}(0) \end{bmatrix} \tag{21.108}$$

From Eq. (21.107), we have the equality

$$\mathcal{R}_J^{(\alpha_p)} = \tau_J(\mathbf{H}_p)\tau_J(\breve{\mathbf{H}}_p)^T$$

If $J > L_p$ and $h_p(z) = \sum_{l=0}^{Lp-1} \mathbf{h}_p(l)z^{-1} \neq 0$ for all z, then it can be proved that both $\tau_J(\mathbf{H}_p)$ and $\tau_J(\breve{\mathbf{H}}_p)$ are full column rank. This implies that the left kernel of $\mathcal{R}_J^{(\alpha_p)}$, i.e.,

$$\mathcal{N}_l\left(\mathcal{R}_J^{(\alpha_p)}\right) = \left\{\mathbf{x} : \mathbf{x}^H \mathcal{R}_J^{(\alpha_p)} = \mathbf{0}\right\}$$

coincides with the left kernel of $\tau_J(\mathbf{H}_p)$. Let Π denote the orthogonal projector onto $\mathcal{N}_l(\mathbf{R}_J^{(\alpha_p)})$; this can be obtained from its SVD. It follows that the linear homogeneous system

$$\Pi\tau_J(\mathbf{X}) = \mathbf{0} \tag{21.109}$$

in the unknown $\mathbf{X} = [\mathbf{x}_0\mathbf{x}_1 \cdots \mathbf{x}_{Lp-1}]$ yields the solution

$$\mathbf{X} = \lambda\mathbf{H}_p$$

for some scalar λ [26]. The scalar λ can also be resolved, see Reference [6] for details. Thus, in practice, given a finite set of K data samples, the TICC technique solves P linear systems [Eq. (21.109) for $p = 1, \ldots, P$; the matrix $\Re_J^{(\beta)}$ in Eq. (21.108)] is estimated by replacing $\Re_x^{(\beta)}(\tau)$ with the consistent estimate

$$\hat{\mathbf{R}}_{\mathbf{x}}^{(\beta)}(\tau) = \frac{1}{K}\sum_{k=1}^{K-\tau} \mathbf{x}(k+\tau)\mathbf{x}(k)^T e^{-j2\pi\beta k}$$

21.3.2.3 CFC$_2$: Closed-Form Correlative Coding

The closed-form correlative coding (CFC$_2$) method for MIMO blind channel identification was formulated by Xavier and Barroso in Reference [47]. The main idea consists of prefiltering the data symbols before transmission, to induce a certain diversity in their SOS. This inserted structure enables the receiver to recover the MIMO channel by a closed-form solution. The need for coloring the emitted information sequences is a consequence of a well-known fact: in general, the MIMO channel matrix **H** — recall the data model in Eq. (21.65) — cannot be unambiguously determined from the SOS of the MIMO output, that is, from the set of correlation matrices

$$\mathcal{R}_{\mathbf{x}} = \{\mathbf{R}_{\mathbf{x}}(\tau) : \tau \in \mathbb{Z}\}$$

To make explicit the channel matrix **H**, we also refer to this set as $\mathfrak{R}_x(\mathbf{H})$.

For an illustrative example, consider the $4 \times (2 + 2)$ case $\mathbf{H} = [\mathbf{H}_1 \mathbf{H}_2]$. Then, it is easily seen that under Assumption (A2) the $s_p(k)$ are mutually independent uncorrelated symbols; thus, the following holds:

$$\mathcal{R}_{\mathbf{x}}(\mathbf{H}) = \mathcal{R}_{\mathbf{x}}(\mathbf{HQ}) \tag{21.110}$$

where the 4×4 matrix **Q** is a unitary mixing matrix given by

$$\mathbf{Q} = \frac{1}{\sqrt{2}}\begin{bmatrix} 1 & -1 \\ 1 & 1 \end{bmatrix} \otimes \mathbf{I}_2$$

Equation (21.110) says that the mapping

$$\mathbf{H} \mapsto \mathcal{R}_{\mathbf{x}}(\mathbf{H})$$

is not one-to-one in a nontrivial way: both the "true" MIMO channel matrix **H** and a mixed version of it

$$\tilde{\mathbf{H}} \equiv \mathbf{HQ}$$

induce the same SOS at the MIMO output.

We show that $\tilde{\mathbf{H}}$ is useless for source separation because the two users are still mixed. To see this, and by discarding the noise, and working with **H** we have

$$\hat{\mathbf{s}}(k) = \mathbf{H}^{\dagger}\mathbf{x}(k) = \mathbf{s}(k) = \begin{bmatrix} s_1(k) \\ s_1(k-1) \\ s_2(k) \\ s_2(k-1) \end{bmatrix}$$

that is, the sources symbols are separated, whereas, working with $\tilde{\mathbf{H}}$

$$\hat{\mathbf{s}}(k) = \tilde{\mathbf{H}}^{\dagger}\mathbf{x}(k) = \frac{1}{\sqrt{2}}\begin{bmatrix} s_1(k) + s_2(k) \\ s_1(k-1) + s_2(k-1) \\ s_2(k) + s_2(k) \\ s_2(k-1) + s_2(k-1) \end{bmatrix}$$

and the data symbols are still mixed.

In summary, the problem of identification of **H** from $\Re_x$ is not a well-posed problem, under the standard assumptions (A1) to (A3). Based on this observation, the CFC$_2$ technique replaces the original (white) symbol sequences, renamed now to $a_p(k)$, by a filtered (colored) version $s_p(k)$ of it:

$$s_p(k) = \sum_{m=0}^{M_p-1} c_p(m) a_p(k-m)$$

where $\{c_p(m): m = 0, 1, \ldots, M_p - 1\}$ is the finite-impulsive response (FIR) of the pth correlative filter. To maintain the transmitted power, we restrict ourselves to unit-power correlative filters, that is,

$$\forall p = 1, 2, \ldots, P: \quad \sum_{m=0}^{M_p-1} \left|c_p(m)\right|^2 = 1$$

The prefilters $c_p(\cdot)$ change the SOS of $s_p(k)$, and, as a consequence, of the vector process

$$\mathbf{s}_p(k) = \left[s_p(k)\ s_p(k-1) \cdots s_p(k-L_p+1)\right]^T$$

Denote by

$$\Gamma_p(\tau) = \mathbf{R}_{\mathbf{s}_p}(0)^{-1/2} \mathbf{R}_{\mathbf{s}_p}(\tau) \mathbf{R}_{\mathbf{s}_p}(0)^{-1/2} \qquad (21.111)$$

the (normalized) correlation matrix of $\mathbf{s}_p(k)$ at lag τ; this generalizes the correlation coefficient of a scalar wide-sense stationary (WSS) random process $x(k)$:

$$\gamma_x(\tau) = \frac{\mathrm{E}\left\{x(k)x(k-\tau)^*\right\}}{\sqrt{\mathrm{var}\{x(k)\}}\sqrt{\mathrm{var}\{x(k-\tau)\}}} = r_x(0)^{-1/2} r_x(\tau) r_x(0)^{-1/2}$$

Suppose that the filters $c_p(\cdot)$ are designed such that the following condition holds:

(A4) for all $p \neq q$, there is a correlation lag $\tau = \tau(p, q)$ such that

$$\sigma\left\{\Gamma_p(\tau)\right\} \cap \sigma\left\{\Gamma_q(\tau)\right\} = \emptyset \qquad (21.112)$$

where s $\{\mathbf{A}\} \subset \mathbb{C}$ denotes the set of eigenvalues of **A** (the spectrum of **A**).

Assumption (A4) requires that, for two distinct sources p and q, the condition in Eq. (21.112) must hold for *some* $\tau(p, q)$; it is not necessary, that it holds for *all* $\tau \in \mathbb{Z}$. Moreover, if (p, q) changes, $\tau(p, q)$ is allowed to change. This makes the set of correlative filters that satisfy (A4) very generic, irrespective of the length $M_p > 1$, even if we further restrict ourselves to minimum-phase filters; we refer the interested reader to Reference [50]; the minimum-phase property is attractive for suboptimal direct inversion of $c_p(\cdot)$ when recovering the input information sequence $a_p(k)$, after **H** has been identified. The CFC$_2$ method relies on the following identifiability result.

Theorem 3.2 [Identifiability]. *Consider that the signal model in Eq. (21.65) and assumptions (A1) to (A4) hold. Then, the mapping* $\mathbf{H} \rightarrow R_x(\mathbf{H})$ *is one-to-one up to a phase offset per user; that is, if* $R_x(\mathbf{H}) = R_x(\tilde{\mathbf{H}})$*, then* $\tilde{\mathbf{H}}_p = \mathbf{H}_p e^{j\theta_p}$*, for some* $\theta_p \in \mathbb{R}$, $p = 1, 2, \ldots, P$

The proof is found in Reference [50].

Also, notice that the phase ambiguity per user cannot be avoided because only SOS are considered.

In the following, we focus on the closed-form algorithm that identifies **H** from the set of correlation matrices $\Re_x$. The goal is to exploit Identifiability Theorem 2.2 and find a matrix $\tilde{\mathbf{H}}$ such that $\Re_x = R_x(\tilde{\mathbf{H}})$, that is,

$$\mathbf{R}_{\mathbf{x}}(\tau) = \tilde{\mathbf{H}}\mathbf{R}_{\mathbf{s}}(\tau)\tilde{\mathbf{H}}^H + \mathbf{R}_{\mathbf{n}}(\tau) \tag{21.113}$$

for all $\tau \in \mathbb{Z}$. If this is achieved, then Identifiability Theorem 3.2 guarantees that $\tilde{\mathbf{H}} = \mathbf{H}$ up to a phase offset per user. Let

$$\mathbf{R}(\tau) = \mathbf{R}_{\mathbf{x}}(\tau) - \mathbf{R}_{\mathbf{n}}(\tau)$$

denote the denoised correlation matrices of the MIMO channel output. Because all filters have finite memory, it suffices Eq. (21.113) holds for a finite set of τ's say $\tau = \{\tau_1, \tau_2, \ldots, \tau_I\}$. Thus, the problem is reduced to finding $\tilde{\mathbf{H}}$ such that the factorization

$$\forall_\tau \in \tau: \qquad \mathbf{R}(\tau) = \tilde{\mathbf{H}}\mathbf{R}_{\mathbf{s}}(\tau)\tilde{\mathbf{H}}^H$$

holds. The algorithm solves for $\tilde{\mathbf{H}}$ in three steps.

Step 1

Let $\mathbf{R}(0) = \mathbf{V}\mathbf{\Lambda}\mathbf{V}^H$ be an EVD of $\mathbf{R}(0)$ where the $N \times N$ matrix **V** is unitary and

$$\Sigma = \begin{bmatrix} \Lambda_1 & \mathbf{0} \\ \mathbf{0} & \mathbf{0} \end{bmatrix}$$

where $\Lambda_1 = \text{diag}(\lambda_1, \ldots, \lambda_M)$, $\lambda_1 \geq \cdots \geq \lambda_M > 0$. Define the $N \times M$ matrix $\mathbf{G}_0$ by $\mathbf{G}_0 = \mathbf{V}_1\Lambda_1^{1/2}$, where the $N \times M$ matrix $\mathbf{V}_1$ consists of the first M columns of **V**. We remark that $\mathbf{G}_0$, thus defined, satisfies the identity

$$\mathbf{G}_0 = \underbrace{\mathbf{H}\mathbf{R}_{\mathbf{s}}(0)^{1/2}}_{\mathbf{H}_0}\mathbf{Q}^H$$

where the $M \times M$ unitary matrix **Q** is unknown. This results from the fact that both $\mathbf{G}_0$ and $\mathbf{H}_0$ are full column-rank matrices verifying

$$\mathbf{R}(0) = \mathbf{G}_0\mathbf{G}_0^H = \mathbf{H}_0\mathbf{H}_0^H$$

Step 2

From this observation, it follows that

$$\mathbf{B}(\tau) \equiv \mathbf{G}_0^\dagger\mathbf{R}(\tau)\mathbf{G}_0^{\dagger H}$$

satisfies

$$\mathbf{B}(\tau) = \left[\mathbf{Q}\mathbf{R}_{\mathbf{s}}(0)^{-1/2}\mathbf{H}^\dagger\right]\left[\mathbf{H}\mathbf{R}_{\mathbf{s}}(\tau)\mathbf{H}^H\right]\left[\mathbf{H}^{\dagger H}\mathbf{R}_{\mathbf{s}}(\tau)^{-1/2}\mathbf{Q}^H\right] = \mathbf{Q}\Gamma(\tau)\mathbf{Q}^H \tag{21.114}$$

where

$$\Gamma(\tau) \equiv \mathbf{R}_{\mathbf{s}}(0)^{-1/2}\mathbf{R}_{\mathbf{s}}(\tau)\mathbf{R}_{\mathbf{s}}(0)^{-1/2}$$

Because all the $\mathbf{R}_s(\tau)$ are block diagonal,

$$\mathbf{R}_s(\tau) = \text{diag}\left(\mathbf{R}_{s_1}(\tau), \ldots, \mathbf{R}_{s_P}(\tau)\right)$$

then $\Gamma(\tau)$ is also block diagonal

$$\Gamma(\tau) = \text{diag}\left(\Gamma_1(\tau), \ldots, \Gamma_P(\tau)\right)$$

where $\Gamma_p(\tau)$ was defined in Eq. (21.111). Thus, by letting $\mathbf{Q} = [\mathbf{Q}_1 \cdots \mathbf{Q}_P]$, Eq. (21.114) reads as

$$\mathbf{B}(\tau) = \sum_{p=1}^{P} \mathbf{Q}_p \Gamma_p(\tau) \mathbf{Q}_p^H$$

Because of the orthogonality of $\mathbf{Q}$,

$$\mathbf{Q}_p^H \mathbf{Q}_q = \mathbf{I}_L \delta(p-q)$$

and $\mathbf{X} = \mathbf{Q}_p$ is a solution of the pth linear system:

$$\begin{cases} \mathbf{B}(\tau_1)\mathbf{X} - \mathbf{X}\Gamma_p(\tau_1) = 0 \\ \qquad \vdots \\ \mathbf{B}(\tau_1)\mathbf{X} - \mathbf{X}\Gamma_p(\tau_I) = 0 \end{cases} \Leftrightarrow \mathbf{B}(\tau)\mathbf{X} - \mathbf{X}\Gamma_p(\tau) = \mathbf{0},\ \tau \in \tau \tag{21.115}$$

$\mathbf{Q}_p$ is the *unique* solution (up to a scalar factor) of Eq. (21.109); see Reference [50].

Because $\mathbf{B}(\tau)$ and $\boldsymbol{\Gamma}_p(\tau)$ are available to the receiver, this provides a way to identify $\mathbf{Q}_p$. Thus, let X be a nonzero solution of Eq. (21.115) and rescale as

$$\mathbf{U}_p = \mathbf{X} \big/ \left(\|\mathbf{X}\| \sqrt{L}\right)$$

It follows that

$$\mathbf{U}_p = \mathbf{Q}_p e^{j\theta_p}$$

for some $\theta_p \in \mathbb{R}$

It is straightforward to solve the homogeneous system in Eq. (21.115) in LS sense. Let

$$\begin{aligned} f(\mathbf{X}) &= \sum_{\tau \in \tau} \left\|\mathbf{B}(\tau)\mathbf{X} - \mathbf{X}\Gamma_p(\tau)\right\|^2 \\ &= \sum_{i=1}^{I} \left\|\left(\mathbf{I}_L \otimes \mathbf{B}(\tau_i) - \Gamma_p(\tau_i)^T \otimes \mathbf{I}_M\right)\mathbf{x}\right\|^2 \\ &= \|\mathbf{T}\mathbf{x}\|^2 \\ &= \mathbf{x}^H \mathbf{T}^H \mathbf{T}\mathbf{x} \end{aligned} \tag{21.116}$$

where $\mathbf{x} = \text{vec}(\mathbf{X})$ and

$$\mathbf{T} = \left[\mathbf{T}_1^T \mathbf{T}_2^T \cdots \mathbf{T}_k^T\right]^T$$

$$\mathbf{T}_i = \mathbf{I}_L \otimes \mathbf{B}(\tau_i) - \Gamma_p(\tau_i)^T \otimes \mathbf{I}_M$$

A global minimizer of $f(\mathbf{X})$, subject to $\|\mathbf{X}\| = 1$, can be obtained by reshaping into matrix format the eigenvector associated with the minimum eigenvalue of the semidefinite positive Hermitian matrix $\boldsymbol{T}^*\boldsymbol{T}$, say $\boldsymbol{u}$, that is, $\mathbf{X} = \text{vec}^{-1}(\mathbf{u})$.

Let $\mathbf{U} = [\mathbf{U}_1 \cdots \mathbf{U}_P]$. Then, $\mathbf{U} = \mathbf{Q}$, up to a phase offset per user.

Step 3

By defining

$$\tilde{\mathbf{H}} = \mathbf{G}_0 \mathbf{U} \mathbf{R}_s(0)^{-1/2}$$

and, by what has been proved so far, we have

$$\forall_\tau \in \tau: \quad \mathbf{R}(\tau) = \tilde{\mathbf{H}} \mathbf{R}_s(\tau) \tilde{\mathbf{H}}^H$$

or, equivalently,

$$\tilde{\mathbf{H}} = \mathbf{H}$$

up to a phase offset per user.

21.4 Stochastic Maximum Likelihood Methods

Most of the maximum likelihood (ML) approaches to the source separation problem assume determin istic source signals in white Gaussian noise. In this case, the problem is reduced to an LS optimization

$$\left(\hat{\mathbf{H}}, \hat{\mathbf{S}}\right) = \arg\min_{\mathbf{H},\mathbf{S}} \left\|\mathbf{X} - \mathbf{H}\mathbf{S}\right\|_F^2$$

whose solution can be found iteratively using some alternating minimization procedure, such as, for example, the ILSP and the ILSE algorithms. Here, we use a stochastic maximum likelihood (SML) approach, in the sense that we consider a random model for the source data sequences. We discuss the approach only in the context of instantaneous mixtures. The case of convolutive mixtures is more complex and, based on our knowledge, remains unstudied.

21.4.1 Instantaneous Mixtures

We consider the same vector data model

$$\mathbf{x}(k) = \mathbf{H}\mathbf{s}(k) + \mathbf{n}(k) \qquad k = 1, \ldots, K \tag{21.117}$$

under the usual assumptions: the channel matrix $\mathbf{H}$ is assumed time invariant along the K snapshots; the noise vector $\mathbf{n}(k)$ is stationary and white in the time domain, having probability density function $p_n(n)$; the noise and signal vectors are statistically independent; in Eq. (21.117), each element $s_p(k)$ is a sample of a binary random variable that takes values in $\mathscr{A} = \{\pm 1\}$ with equal probability; the P binary data sequences generated by the P users are assumed white and statistically independent. Therefore, the signal vector takes values in the alphabet $\mathscr{A}^P$, which has cardinality $\mathscr{C}_{\mathscr{A}} = 2^P$.

To reconstruct the vector sequence $\{\mathbf{s}(k)\}_{k=1}^{K}$, we solve the multiple ($\mathscr{C}_{\mathscr{A}}$ – ary) hypothesis test problem

$$H_i: \mathbf{x}(k) = \mathbf{H}\mathbf{s}_i + \mathbf{n}(k),\ \mathbf{s}_i \in \mathscr{A}^P,\ k = 1, \ldots, K \tag{21.118}$$

Clearly, the vector $\mathbf{x}(k)$ represents the array data at snapshot k given that $\mathbf{s}_i \in \mathscr{A}^P$ was transmitted. In Eq. (21.118), the channel matrix $\mathbf{H}$ is unknown. To solve this hypothesis detection problem with unknown parameters, we follow a generalized maximum-likelihood (GML) approach: we first determine the ML estimate of $\mathbf{H}$ using the K array data snapshots, and then solve the multiple hypothesis test problem at each snapshot by replacing the unknown $\mathbf{H}$ by its ML estimate $\hat{\mathbf{H}}$.

The conditional array data probability density function at time instant k given $\mathbf{H}$ and given that $\mathbf{s}(k) = \mathbf{s}_i \in \mathscr{A}^P$ was transmitted is

$$p_{\mathbf{x}|\mathbf{H},\mathbf{s}_i}\left(x(k)\right) = p_n\left(\mathbf{x}(k) - \mathbf{H}\mathbf{s}_i\right) \tag{21.119}$$

Therefore, the conditional probability density function of the array data at time k given $\mathbf{H}$ is

$$\begin{aligned} p_{\mathbf{x}|\mathbf{H},\mathbf{s}_i}\left(\mathbf{x}(k)\right) &= \sum_{i=1}^{C_{\mathscr{A}}} p_{\mathbf{x}_i\,\mathbf{H},\mathbf{s}_i}\left(\mathbf{x}(k)\right)\Pr\left\{\mathbf{s}(k) = \mathbf{s}_i\right\} \\ &= 2^{-P}\sum_{i=1}^{C_{\mathscr{A}}} p_n\left(\mathbf{x}(k) - \mathbf{H}\mathbf{s}_i\right) \end{aligned} \tag{21.120}$$

This is a finite linear mixture model. Given the K array snapshots, we write the generalized array data likelihood function

$$\Lambda\left(\left[\mathbf{x}(1) \cdots \mathbf{x}(K)\right]\middle|\mathbf{H}\right) = 2^{-PK}\prod_{k=1}^{K}\sum_{i=1}^{C_{\mathscr{A}}} p_n\left(\mathbf{x}(k) - \mathbf{H}\mathbf{s}_i\right) \tag{21.121}$$

or the equivalent log likelihood function

$$\ln\Lambda\left(\left[\mathbf{x}(1) \cdots \mathbf{x}(K)\right]\middle|\mathbf{H}\right) = -PK\ln 2 + \sum_{k=1}^{K}\ln\left(\sum_{i=1}^{C_{\mathscr{A}}} p_n\left(\mathbf{x}(k) - \mathbf{H}\mathbf{s}_i\right)\right) \tag{21.122}$$

The ML estimate $\hat{\mathbf{H}}$ of the channel mixing matrix $\mathbf{H}$ is

$$\hat{\mathbf{H}} = \arg\max_{H}\sum_{k=1}^{K}\ln\left(\sum_{i=1}^{C_{\mathscr{A}}} p_n\left(\mathbf{x}(k) - \mathbf{H}\mathbf{s}_i\right)\right) \tag{21.123}$$

Once the estimate $\hat{\mathbf{H}}$ is obtained, we decide on the signal vector $\mathbf{s}(k) \in \mathscr{A}^P$ at each snapshot $k = 1, \ldots, K$, based on the maximum *a posteriori* (MAP) criterion:

$$\begin{aligned} \hat{\mathbf{s}}(k) &= \arg\max_{\mathbf{s}_i \in \mathscr{A}^P} \Pr\left\{\mathbf{s}(k) = \mathbf{s}_i\middle|\mathbf{x}(k), \hat{\mathbf{H}}\right\} \\ k &= 1, \ldots, K \end{aligned} \tag{21.124}$$

where the *a posteriori* conditional probabilities

$$\Pr\left\{\mathbf{s}(k)=\mathbf{s}_i \middle| \mathbf{x}(k), \hat{\mathbf{H}}\right\} = \frac{p_{\mathbf{x}|\hat{\mathbf{H}},\mathbf{s}_i}\left(\mathbf{x}(k)\right)2^{-P}}{\sum_{i=1}^{\mathscr{C}_{\mathscr{A}}} p_{\mathbf{x}|\hat{\mathbf{H}},\mathbf{s}_i}\left(\mathbf{x}(k)\right)2^{-P}}, \ \mathbf{s}_i \in \mathscr{A}^P \tag{21.125}$$

or, taking into account Eq. (21.119),

$$\hat{p}_i\left(\mathbf{x}(k), \hat{\mathbf{H}}\right) \doteq \Pr\left\{\mathbf{s}(k)=\mathbf{s}_i \middle| \mathbf{x}(k), \hat{\mathbf{H}}\right\} = \frac{p_n\left(\mathbf{x}(k)-\hat{\mathbf{H}}_{\mathbf{s}_i}\right)}{\sum_{i=1}^{\mathscr{C}_{\mathscr{A}}} p_n\left(\mathbf{x}(k)-\hat{\mathbf{H}}_{\mathbf{s}_i}\right)}, \ \mathbf{s}_i \in \mathscr{A}^P \tag{21.126}$$

Estimation of $\hat{H}$

The strong nonlinearity of the log likelihood function precludes the closed-form solution to the optimization in Eq. (21.123). In References [3, 4], this optimization problem is solved using the *expectation–maximation* (EM) algorithm [36]. When properly initialized, the EM algorithm coverages in a few number of iterations to the actual solution. For the case under discussion here, and under the Gaussian noise assumption, the solution is obtained through the iteration

$$\hat{\mathbf{H}}_{l+1} = \left(\sum_{k=1}^{K} \mathbf{x}(k) \frac{\sum_{i=1}^{\mathscr{C}_{\mathscr{A}}} p_n\left(\mathbf{x}(k)-\hat{\mathbf{H}}_l\mathbf{s}_i\right)s_i^H}{\sum_{i=1}^{\mathscr{C}_{\mathscr{A}}} p_n\left(\mathbf{x}(k)-\hat{\mathbf{H}}_l\mathbf{s}_i\right)}\right)\left(\sum_{k=1}^{K} \frac{\sum_{i=1}^{\mathscr{C}_{\mathscr{A}}} p_n\left(\mathbf{x}(k)-\hat{\mathbf{H}}_l\mathbf{s}_i\right)s_i s_i^H}{\sum_{i=1}^{\mathscr{C}_{\mathscr{A}}} p_n\left(\mathbf{x}(k)-\hat{\mathbf{H}}_l\mathbf{s}_i\right)}\right)^{-1} \tag{21.127}$$

or, using a more compact notation,

$$\hat{\mathbf{H}}_{l+1} = \left(\sum_{k=1}^{K} \mathbf{x}(k) \sum_{i=1}^{\mathscr{C}_{\mathscr{A}}} \hat{p}_i\left(\mathbf{x}(k), \hat{\mathbf{H}}_l\right)s_i^H\right)\left(\sum_{k=1}^{K}\sum_{i=1}^{\mathscr{C}_{\mathscr{A}}} \hat{p}_i\left(\mathbf{x}(k), \hat{\mathbf{H}}_l\right)s_i s_i^H\right)^{-1} \tag{21.128}$$

It is interesting to notice that Eq. (21.128) is an iterative "right inversion" of

$$\mathbf{x}(k) = \mathbf{H}\mathbf{s}(k)$$

$$k = 1, \ldots, K$$

If all the $\mathbf{s}(k)$, $k = 1, \ldots, K$ were known, then the inversion would be of the form

$$\begin{aligned} &\left(\sum_{k=1}^{K} \mathbf{x}(k)\mathbf{s}^H(k)\right)\left(\sum_{k=1}^{K} \mathbf{s}(k)\mathbf{s}^H(k)\right)^{-1} \\ &= \mathbf{H}\left(\sum_{k=1}^{K} \mathbf{s}(k)\mathbf{s}^H(k)\right)\left(\sum_{k=1}^{K} \mathbf{s}(k)\mathbf{s}^H(k)\right)^{-1} = \mathbf{H} \end{aligned} \tag{21.129}$$

assuming that the inverse exists. By looking at Eq. (21.128), we see that it can be written in a similar way as in Eq. (21.129). In fact, by defining the *a posteriori* conditional expected value

$$E_{\mathbf{x},\hat{\mathbf{H}}_l}\left\{\mathbf{s}(k)\right\} = \sum_{i=1}^{\mathscr{C}_{\mathscr{A}}} \hat{p}_i\left(\mathbf{x}(k), \hat{\mathbf{H}}_l\right)\mathbf{s}_i$$

and covariance

$$E_{\mathbf{x},\hat{\mathbf{H}}_l}\left\{\mathbf{s}(k)\mathbf{s}^H(k)\right\} = \sum_{i=1}^{\mathscr{C}_{\mathscr{A}}} \hat{p}_i\left(\mathbf{x}(k), \hat{\mathbf{H}}_l\right)\mathbf{s}_i\mathbf{s}_i^H$$

at iteration l we have from Eq. (21.128)

$$\hat{\mathbf{H}}_{l+1} = \left(\sum_{k=1}^{K} \mathbf{x}(k) E_{\mathbf{x},\hat{\mathbf{H}}_l}\left\{\mathbf{s}(k)\right\}\right)\left(\sum_{k=1}^{K} E_{\mathbf{x},\hat{\mathbf{H}}_l}\left\{\mathbf{s}(k)\mathbf{s}^H(k)\right\}\right) \tag{21.130}$$

This clarifies the operation of the EM algorithm. The right inversion in the left-hand side of Eq. (21.129) performs the maximization of the log likelihood function and corresponds to the maximization step of EM. Because the two factors involved depend on the unknown $\mathbf{s}(k)$, $k = 1, \ldots, K$, the correspond *a posteriori* conditional expected value and covariance estimates are used in Eq. (21.130). This corresponds to the E step of EM. The iteration proceeds until some stopping condition is satisfied.

Reconstruction of the Sources Signals

When convergence of the EM algorithm is achieved, then $\hat{\mathbf{H}} \simeq \hat{\mathbf{H}}_{l+1} \simeq \hat{\mathbf{H}}_l$. This means that the *a posteriori* conditional probabilities are immediately available.

$$\hat{p}_i\left(\mathbf{x}(k), \hat{\mathbf{H}}\right) \simeq \hat{p}_i\left(\mathbf{x}(k), \hat{\mathbf{H}}_l\right)$$

and the reconstruction of the data transmitted by the sources can be done using the MAP solution given in Eq. (21.124).

In Reference [4] it is shown through extensive computer simulations that, even at moderate SNRs, the EM algorithm converges to the optimum ML solution in a few number of iterations and tracks satisfactorily the channel changes. Because of its highly parallelizable capability, EM can be implemented with computational costs that compare favorably with other algorithms.

21.5 Source Separation by Model Geometric Properties

In this section, we describe new techniques to achieve source separation that exploit specific geometric features of the data model. The first technique, introduced in Subsection 21.5.1, is restricted to high SNR scenarios and binary sources, with few data samples available. The small number of observations precludes the use of SOS. A geometric convex reformulation of the blind source separation problem is exploited to resolve the linear mixture of binary users. In Subsection 21.5.2, we discuss a SOS-based semiblind ML technique for channel identification with white (up to second order) stationary inputs. The likelihood function for the residual unitary matrix is optimized directly over the manifold of orthogonal matrices, by a geodescic descent algorithm.

21.5.1 Deterministic Separation: Polyhedral Sets

Recall the data model in Eq. (21.65), and let $\mathbf{s}_p(k)$ denote binary signals, that is, $\mathbf{s}_p(k) = \pm 1$, for $p = 1, \ldots, P$ and $k = 1, \ldots, K$. For clarity, consider noiseless samples, that is, $\mathbf{n}(k) = 0$, and let all parameters in the data model be real (no complex data). Moreover, assume $\mathbf{H}$ is a $P \times P$ nonsingular matrix; this entails no loss of generality — recall the subspace method in Subsection 21.2, and Eq. (21.64). Our goal is to find equalizers $\xi \in \mathbb{R}^P$ that, when applied to the observed data $\mathbf{x}(k)$, extract a binary user from the linear mixture, that is, we seek vectors in

$$\varepsilon = \left\{ \xi \in \mathbb{R}^P : \xi^T \mathbf{x}(k) = \pm s_p(k), \text{ for some } p = 1, \ldots, P \text{ and all } k = 1, \ldots, K \right\} \tag{21.131}$$

In the sequel, we refer to ε as the set of linear equalizers. Remark that $\xi \in \varepsilon$ verifies the strict constant modulus equality

$$\text{for } k = 1, \ldots, K: \quad \left| \xi^T \mathbf{x}(k) \right| = 1 \tag{21.132}$$

Thus, $\xi \in \varepsilon$ belongs to the data-dependent polyhedron

$$\mathcal{P} = \left\{ \xi \in \mathbb{R}^P : \left| \xi^T \mathbf{x}(k) \right| \leq 1, \text{ for all } k = 1, \ldots, K \right\} \tag{21.133}$$

obtained by relaxing the equality in Eq. (21.132) to an inequality. In fact, the set ε plays a special role within $\mathcal{P}$; it is exactly the set of its vertices (extreme points), under a certain condition stated in the following theorem.

Theorem 5.1. *Consider the signal model* $\mathbf{x}(k) = \mathbf{Hs}(k)$, $k = 1, \ldots, K$, *where the* $P \times P$ *channel matrix* H *is nonsingular and suppose all distinct* 2^P *binary vectors of length* P *are represented in* $\{\mathbf{s}(k): k = 1, \ldots, K\}$. *Then, the set of extreme points of the polyhedron* $\mathcal{P}$ *in Eq. (21.133) is given by* ε *in Eq. (21.131).*

The proof can be found in Reference [48]. Thus, finding an equalizer in ε is equivalent to finding a vertex of $\mathcal{P}$. To achieve this goal, we may exploit the fact that extreme points of a compact polyhedral set S are global minima of any linear function over S. As a consequence, a $\xi \in \varepsilon$ may be found by solving a linear programming (LP) problem

$$\xi = \arg\min_{\omega \in \mathcal{P}} \mathbf{c}^T \omega$$

where $\mathbf{c}$ denotes a nonzero vector. To extract the remaining $P - 1$ equalizers, the following recursive approach may be pursued.

Suppose $1 \leq p < P$ equalizers, $\{\xi_1, \cdots \xi_p\}$ have been retrieved. We seek a new vertex ξ_{p+1} of $\mathcal{P}$, that is,

$$\xi_{p+1} \notin \mathcal{L}_p = \text{span} \left\{ \xi_1, \ldots, \xi_p \right\}$$

By recalling that the maximum of any convex function $f(\boldsymbol{\omega})$ over a compact polyhedral set S is attained at one of its extreme points, we can find ξ_{p+1} by solving a quadratic programming (QP) problem:

$$\xi_{p+1} = \arg\min_{\omega \in \mathcal{P}} \omega^T \mathbf{U}_p \Lambda \mathbf{U}_p^T \omega$$

where the $P \times (P - p)$ matrix U_p spans the orthogonal complement of $\mathcal{L}_p$ and the $(P - p) \times (P - p)$ matrix Λ denotes a positive definite diagonal matrix. The role of the convex function

$$f(\omega) = \omega^T \mathbf{U}_p \Lambda \mathbf{U}_p^T \omega$$

is to prevent convergence to the already retrieved equalizers ξ_i, $i = 1, \ldots, p$. Notice that $f(\omega) \geq 0$ with equality if and only if $\omega \in \mathcal{L}_p$. Thus $f(\xi_{p+1}) > 0$ implies that ξ_{p+1} is nonredundant (i.e., $\xi_{p+1} \notin \mathcal{L}_p$).

In the presence of noise, the conditions of the Theorem 5.1 are not met, and not all extreme points in $\mathcal{P}$ are necessarily close to equalizers in ε. This implies that convergence of the preceding optimization algorithms to spurious minima (or maxima) has to be checked, to reinitialize them with a distinct starting point. Because of this, this simple geometrically based methodology for blind separation of binary sources is only restricted to high SNR scenarios.

21.5.2 Semiblind Maximum Likelihood Separation of White Sources

We delineate a semiblind technique [49] to identify the MIMO channel matrix $\boldsymbol{H}$: recall the data model in Eq. (21.65) and assumptions (A1) to (A3). For simplicity, we also assume all data to be real. The technique is termed semiblind because we assume that, for each source $p = 1, \ldots, P$, a certain fragment

$$\mathcal{F}_p(i_p, j_p) = \{s_p(i_p), s_p(i_p+1), \ldots, s_p(j_p)\}$$

of the emitted message is known by the receiver (e.g., a portion of the frames' header). Notice that it is not required that $i_p = i_q$ (or, $j_p = j_q$) for two distinct $p \neq q$ (i.e., no synchronization among the sources is assumed, which would be difficult to satisfy in practice, but only between each source and the base station receiver).

The assumption of certain symbols being known is needed because as seen before and as it is well known, even in the noiseless case, the factorization $\mathbf{Hs}(k)$ is not unique: $\mathbf{H}$ can only be solved up to a residual $P \times P$ instantaneous mixing matrix [15, 44]. Notice further that, here, no finite-alphabet assumption is made on the sources, which is also a possible way to resolve the mixing ambiguity as seen in the preceding sections.

The technique works as follows. We assume that the data samples have been prewhitened as explained in Section 21.3.1. Thus, we find ourselves in the data model of Eq. (21.69), reproduced here for the convenience of the reader:

$$\begin{gathered}\mathbf{y}(k) = \mathbf{Qs}(k) + \mathbf{w}(k) \\ k = 1, 2, \ldots, K\end{gathered} \tag{21.134}$$

where the corresponding (unknown) $PL \times PL$ channel matrix $\mathbf{Q}$ is unitary and has fewer parameters to estimate than the associated $N \times PL$ matrix $\mathbf{H}$. For the pth source, collect all the data samples that play a role in Eq. (21.134) in the vector $\mathbf{s}_p$, that is,

$$\mathbf{s}_p = \left(s_p(2-L), s_p(3-L), \ldots, s_p(K-1), s_p(K)\right)^T$$

and stack these in the overall data vector

$$\mathbf{s} = \left[\mathbf{s}_1^T\ \mathbf{s}_2^T \cdots \mathbf{s}_P^T\right]^T$$

For further reference, we expres in matrix terms the knowledge of the data fragments as

$$\mathbf{E}_p^T \mathbf{s}_p = \eta_p$$

where the $(K - L + 1) \times (j_p - i_p + 1)$ matrix $\mathbf{E}_p$ has a single 1 in each of its columns, with all the remaining entries being 0, that selects the *a priori* known entries in s_p, and

$$\eta_p = \left(s_p(i_p), s_p(i_p+1), \ldots, s_p(j_p)\right)^T$$

contains the known data fragment. Thus, the overall knowledge for all the sources is expressed as

$$\mathbf{E}^T\mathbf{s} = \eta$$
$$\mathbf{E} = \text{diag}\left(\mathbf{E}_1, \ldots, \mathbf{E}_P\right)$$
$$\eta = \left[\eta_1^T \cdots \eta_P^T\right]^T$$

Because the statistical description of the sources is not known beyond the second order, we can say loosely that the prewhitening step has exhausted all the stochastic information about them. Thus, in the sequel, **s** is treated as a deterministic vector of parameters; the entries of **Q** are also deterministic. The joint maximum likelihood estimates of (**Q**, **s**), subject to the known constraints, is given by

$$\widehat{\left(\mathbf{Q},\mathbf{s}\right)} = \arg \max_{\mathbf{Q}\in\mathbb{O},\mathbf{E}^T\mathbf{s}=\eta} \Lambda\left(\mathbf{y}(1), \ldots, \mathbf{y}(K)\middle|\mathbf{Q},\mathbf{s}\right) \tag{21.135}$$

where

$$\mathbb{O} = \left\{\mathbf{U}\in\mathbb{R}^{PL\times PL} : \mathbf{U}^T\mathbf{U} = \mathbf{I}_{PL}\right\}$$

denotes the set of $PL \times PL$ orthogonal matrices, and $\Lambda(\cdot|\mathbf{Q}, \mathbf{s})$ is the likelihood of the prewhitened data samples $\mathbf{y}(k)$ conditioned on the pair (**Q**, **s**). Equation (21.135) is equivalent to

$$\widehat{\left(\mathbf{Q},\mathbf{s}\right)} = \arg \max_{\mathbf{Q}\in\mathbb{O},\mathbf{E}^T\mathbf{s}=\eta} \phi\left(\mathbf{Q},\mathbf{s}\right) \tag{21.136}$$

where

$$\phi\left(\mathbf{Q},\mathbf{s}\right) = \frac{1}{K}\left\|\mathbf{Y} - \mathbf{Q}\tau\left(\mathbf{s}\right)\right\|_{C^{-1}}^2 \tag{21.137}$$

Here, $\mathbf{Y} = [\mathbf{y}(1)\mathbf{y}(2) \cdots \mathbf{y}(K)]$ contains the whitened data and

$$\tau\left(s\right) = \begin{bmatrix} \tau\left(\mathbf{s}_1\right) \\ \tau\left(\mathbf{s}_2\right) \\ \vdots \\ \tau\left(\mathbf{s}_P\right) \end{bmatrix}$$

is a stack of P Toeplitz matrices, each given by

$$\tau\left(\mathbf{s}_p\right) = \begin{bmatrix} s_p(1) & s_p(2) & s_p(3) & \cdots & s_p(K) \\ s_p(0) & s_p(1) & \ddots & \ddots & s_p(K-1) \\ \vdots & \ddots & \ddots & \ddots & \vdots \\ s_p(2-L) & s_p(3-L) & \cdots & \cdots & s_p(K-L+1) \end{bmatrix}$$

TABLE 21.1 IML Algorithm

Let $\mathbf{Q}^{(0)} \in O$ for $n = 1, 2, \ldots$
i) $\mathbf{s}^{(n)} = \arg\min_{ETs=\eta} \phi\,(\mathbf{Q}^{(n-1)}\mathbf{s})$
ii) $\mathbf{Q}^{(n)} = \arg\min_{\mathbf{Q}\in O} \phi\,(\mathbf{Q}, \mathbf{s}^{(n)})$
until $\mathbf{Q}^{n} - \mathbf{Q}^{(n-1)} = 0$

TABLE 21.2 Geodesic Descent Algorithm

1. Choose $\mathbf{Q}^{(0)} \in O(\mathrm{LP})$
2. for $m = 1, 2, \ldots$
   a. Let $\mathbf{D}$ denote the projection of $-\nabla\varphi\,(\mathbf{Q}^{(m-1)})$ onto the tangent space of O at $\mathbf{Q}^{(m-1)}$
   b. Let $\mathbf{Q}(\mathrm{t}) \in O$, $t \geq 0$, denote the geodesic emanating from $\mathbf{Q}(0) = \mathbf{Q}^{(m-1)}$ in the direction $\mathbf{Q}(0) = \mathbf{D}$
   c. Minimize $\varphi\,(\mathbf{Q}(t)$ with respect to $t \geq 0$, to obtain $t_{\min}$ (a global minimizer); set $\mathbf{Q}^{(m)} = \mathbf{Q}(t_{\min})$
3. Repeat Step 2 until $\mathbf{Q}^{(m)} - \mathbf{Q}^{(m-1)} = 0$

Moreover, in Eq. (21.137), $\mathbf{C}$ denotes the covariance matrix of the noise $\mathbf{w}(k)$, and, as usual,

$$\left\|\mathbf{Z}\right\|_{C^{-1}}^{2} = \mathrm{tr}\left\{\mathbf{Z}^{T}\,\mathbf{C}^{-1}\,\mathbf{Z}\right\}$$

The optimization problem expressed in Eq. (21.137) has no apparent closed-form solution. However, an iterative cyclic coordinate descent approach may be employed. This is only locally convergent. The resulting algorithm, termed here as iterative maximum likelihood (IML), is given in Table 21.1

Solving for $\mathbf{s}^{(n)}$ in substep (i) of the IML algorithm can be formulated as a standard LS minimization problem with linear constraints, whose solution is well known; see Reference [49] for details. Substep (ii) is more involved and requires more attention. After some algebra, it can be seen that substep (ii) boils down to the minimization of

$$\varphi(\mathbf{Q}) = \mathrm{tr}\left\{\mathbf{Q}^{T}\mathbf{C}^{-1}\mathbf{Q}\hat{\mathbf{R}}_{s}\right\} - 2\,\mathrm{tr}\,\left\{\mathbf{Q}^{T}\mathbf{C}^{-1}\hat{\mathbf{R}}_{ys}\right\} \tag{21.138}$$

where

$$\hat{\mathbf{R}}_{s} = \frac{1}{K}\tau(\mathbf{s})\tau(\mathbf{s})^{T} \text{ and } \hat{\mathbf{R}}_{ys} = \frac{1}{K}\mathbf{Y}\tau(\mathbf{s})^{T}$$

subject to $\mathbf{Q} \in o$, the group of $PL \times PL$ orthogonal matrices. To exploit the curvature of this surface constraint a geodesic descent algorithm, which is a generalization of the traditional steepest gradient method in flat spaces [23] to curved manifolds, may be employed. This technique is described in Table 21.2. Thus, the geodesic descent algorithm acts as classical line search methods, except for the fact that the lines are replaced by their counterparts in the constraint surfaces: the geodesics. Derivation of all the details of the geodesic descent algorithm (GDA) is out of our scope here; see Reference [49]. A possible initialization for the GDA is given by

$$\mathbf{Q}^{(0)} = \Pi_{o}\left\{\mathbf{C}^{-1}\,\hat{\mathbf{R}}y_{s}\right\} \tag{21.139}$$

where $\Pi_{o}\,(\mathbf{Z})$ is the (nonlinear) projection of the matrix $\mathbf{Z} \in \mathrm{R}^{LP\times LP}$ onto the orthogonal group o. It may be obtained as follows by letting:

$$\mathbf{Z} = \mathbf{U}\Sigma\mathbf{V}^{T}$$

denote a SVD of **Z**; then,

$$\Pi_{\mathcal{O}}(\mathbf{Z}) = \mathbf{U}\mathbf{V}^T$$

The idea behind this initialization is that, near the global minimum,

$$\hat{\mathbf{R}}_s = \frac{1}{K}\sum_{k=1}^{K} \mathbf{s}(k)\mathbf{s}(k)^T \simeq \mathbf{R}_s = \mathrm{E}\left\{s(k)s(k)^T\right\} = \mathbf{I}_{LP}$$

Thus, the first term in Eq. (21.138) reduces to a constant, and

$$\varphi(\mathbf{Q}) \simeq -2\ \mathrm{tr}\left\{\mathbf{Q}^T\mathbf{C}^{-1}\hat{\mathbf{R}}_{ys}\right\}$$

which is precisely minimized by Eq. (21.139) [13].

Future Research Directions

The two geometrically inspired techniques for source separation presented in this section may be further developed. Future work on the polyhedral characterization of equalizers may include the extension of the proposed technique to higher cardinality digital modulation alphabets such as pulse amplitude modulation (PAM) and quadrature amplitude modulation (QAM). Also, robustness to noise and a reformulation of the concepts that we presented to avoid *a priori* channel length estimation are important topics for further study. With respect to the semiblind ML approach, more efficient optimization techniques like Newton, or conjugate gradient, matched to the specific structure of the constraint manifolds at hand are to be investigated, to speed up the convergence of the proposed methods.

21.6 Conclusions

This chapter considers the blind channel identification and source separation problems in the context of SDMA wireless communication systems. We exploit array processing techniques for smart antennas technology. The approaches described did not rely on any type of spatial structure modeling, such as array and wave front geometries. These features are assumed unknown and embedded in the MIMO channel model. For both memoryless and ISI channels, the channel model is represented by an unknown channel matrix, referred to as instantaneous and convolutive mixing matrix, respectively. For memoryless channels, we provide solutions to the blind source separation problem. For MIMO-ISI channels, we study also the blind ISI cancellation (equalization) subproblem. We organize the chapter according to the different approaches that we consider: deterministic methods, SOS-based techniques, and SML. Many of the known solutions to blind channel identification an source separation are implemented with iterative algorithms. This is the case with the ILSP and ILSE algorithms in Section 21.2, the hypercube and the LS constellation derotator methods based on SOS described in Section 21.3, and the SML–EM algorithm of Section 21.4. We have also focused our attention on analytic and/or closed-form solutions. This is a very important topic, because it avoids efficiently the global convergency problems of the iterative solutions. The ACMA and the linear coding methods for blind source separation, and the subspace method for ISI cancellation presented in Section 21.2 are in the class of analytic or closed-form deterministic solutions. Similarly, the OPDA, the TICC, and the CFC_2 approaches in Section 21.3 provide closed-form SOS solutions. With all these closed-form solution algorithms, global convergence to the optimum solution comes with a price, namely, some increase in algorithmic computational complexity. We revisit this trade-off between increased computational complexity for closed-form solutions and global convergence of iterative methods in Section 21.5 where we develop source separation algorithms that exploit distinct and specific geometric features of the data model. We present a deterministic algorithm in Subsection 21.5.1 that is based on a geometric convex formulation of the blind source separation problem. In Subsection 21.5.2, we describe a SML approach to solve the semiblind separation of white sources based on geodesic descent algorithms in specific manifolds related to the geometry of the data model. These two algorithms are the result of preliminary research on the application of convex and differential geometry driven optimization techniques to the blind source separation problem. We believe that this opens promising guidelines for future research.

References

1. B.G. Agee, "Fast Adaptive Polarization Control Using the Least-Squares Constant Modulus Algorithm," in *Proc. 20th IEEE Asilomar Conference on Signals, Systems and Computers*, pp. 590–595, 1987.
2. B.G. Agee, "Blind Separation and Capture of Communication Signals Using a Multitarget Constant's Modulus Beamforming," in *Proc. IEEE MILCOM*, vol. 2, pp. 340–346, Boston, MA, 1989.
3. V.A.N. Barroso and J.M.F. Moura, "Beamforming with correlated Arrivals in Mobile Communications," in Vijay Madisetti and Douglas Williams, editors, *Digital Signal Processing Handbook*, chapter on Array Signal Processing, edited by Mos Kaveh, CRC Press, Boca Raton, FL, 1997.
4. V.A.N. Barroso, J.M.F. Moura, and J. Xavier, "Blind Array Channel Division Multiple Access (AChDMA) for Mobile Communications," *IEEE Trans. Signal Process.*, vol. 46, pp. 737–752, March 1998.
5. R.T. Behrens and L.L. Scharf, "Signal Processing Applications of Oblique Projection Operators," *IEEE Trans. Signal Process.*, vol. 42, pp. 1413–1424, June 1994.
6. A. Chevreuil and P. Loubaton, "MIMO Blind Second-Order Equalization Method and Conjugate Cyclostationarity," *IEEE Trans. Signal Process.*, vol. 47, no. 2, pp. 572–578, February 1999.
7. Z. Ding, "Matrix Outer-Product Decomposition Method for Blind Multiple Channel Identification," in *IEEE Trans. Signal Process.*, vol. 45, no. 12, pp. 3053–3061, December 1997.
8. Z. Ding, "Blind Wiener Filter Estimation for Multi-Channel Systems Based on Partial Information," in *Proc. IEEE Int. Conf. Acoustics, Speech and Signal Process.* — ICASSP, pp. 3609–3612, Munich, Germany, 1997.
9. G.B. Giannakis, "Filterbanks for Blind Channel Identification and Equalization." *IEEE Signal Process. Lett.*, vol. 4, pp. 184–187, June, 1997.
10. D.N. Godard, "Self-Recovering Equalization and Carrier Tracking in Two Dimensional Data Communication Systems," in *IEEE Trans. Communic.*, vol. COM-28, pp. 1867–1875, November 1980.
11. L.C. Godara, "Applications of Antenna Arrays to Mobile Communications, Part I: Performance Improvement, Feasibility, and System Considerations," in *Proc. IEEE*, vol. 85, no. 7, pp. 1031–1060, July 1997.
12. L.C. Godara, "Applications of Antenna Arrays to Mobile Communications, Part II: Beamforming Direction-of-Arrival Considerations," in *Proc. IEEE*, vol. 85, no. 8, pp. 1195–1245, August 1997.
13. G. Golub and C.F. Van Loan, *Matrix Computations*, John Hopkins University Press, Baltimore and London, 1989.
14. R. Gooch and J. Lundell, "The CM Array: An adaptive Beamformer for Constant Modulus Signals," in *Proc. IEEE Int. Conf. Acoustics, Speech and Signal Process.* — ICASSP, pp. 2523–2526, Tokyo, Japan, 1986.
15. A. Gorokhov and P. Loubaton, "Subspace-Based Techniques for Blind Separation of Convolutive Mixtures with Temporally Correlated Sources," *IEEE Trans. Circuits and Syst. — I: Fundamental Theory Appl.*, vol. 44, no. 9, pp. 813–820, September 1997.
16. A. Graham, *Kronecker Products and Matrix Calculus with Applications*, Ellis Horwood Series, Mathematics and Its Applications, England, 1981.
17. L. Hansen and G. Xu, "A Fast Sequential Source Separation Algorithm for Digital Cochannel Signals," *IEEE Signal Process. Lett.*, vol. 4, no. 2, pp. 58–61, February 1997.
18. L. Hansen and G. Xu, "A Hyperplane-Based Algorithm for the Digital Co-channel Communications Problem," *IEEE Trans. Inf. Theory*, vol. 43, no. 5, pp. 1536–1548, September 1997.

19. A.V. Keerthi, A. Mathur, and J.J. Shync, "Direction Finding Performance of the Multistage CMA Array," in *Proc. 28th IEEE Asilomar Conf. Signals, Systems and Computers,* vol. 2, pp. 847–852, 1994.
20. H. Krim and M. Viberg, "Two decades of Array Signal Processing Research," *IEEE Signal Process. Mag.*, pp. 67–94, July 1996.
21. J.C. Liberti, Jr. and T.S. Rappaport, *Smart Antennas for Wireless Communications: IS-95 and Third Generation CDMA Applications*, Prentice-Hall Communications Engineering and Emerging Technologies Series, Theodore S. Rappaport, Ed., 1999.
22. T. Li and N.D. Sidiropoulos, "Blind Separation of Linear Mixtures of Digital Signals Using Successive Interference Cancellation Iterative Least Squares," in *Proc. IEEE Int. Conf. Acoustics, Speech and Signal Process. — ICASSP,* vol. 5, pp. 2703–2706, Phoenix, AZ, March 1999.
23. D. Luenberger, *Linear and Nonlinear Programming*, 2nd ed., Addison-Wesley, Reading, MA, 1989.
24. H. Lui and X.-G. Xia, "Precoding Techniques for Undersampled Multireceiver Communication Systems," *IEEE Trans. Signal Process.*, vol. 48, no. 7, pp. 1853–1864, July 2000.
25. K.A. Meraim, P. Loubaton, and E. Moulines, "A Subspace Algorithm for Certain Blind Identification Problems," *IEEE Trans. Inf. Theory,* vol. 43, no. 2, pp. 499–511, March 1997.
26. E. Moulines, P. Duhamel, J. Cardoso, and S. Mayrargue, "Subspace Methods for the Blind Identification of Multichannel FIR Filters," *IEEE Trans. Signal Process.*, vol. 43, no. 2, pp. 516–525, February 1995.
27. A.J. Paulraj and C. Papadias, "Space–Time Processing for Wireless Communications," in *IEEE Trans. Personal Commun.*, vol. 14, no. 5, pp. 49–83, November 1997.
28. Y. Sato, "A Method for Self-Recovering Equalization for Multilevel Amplitude–Modulation Problems," in *IEEE Trans. Commun.*, vol. COM-23, pp. 679–682, June 1975.
29. A. Scaglione, G.B. Giannakis, and S. Barbarossa, "Redundant Filterbank Precorders and Equalizers Part I: Unification and Optimal Designs," *IEEE Trans. Signal Process.*, vol. 47, no. 7, pp. 1988–2006, July 1999.
30. A. Scaglione, G.B. Giannakis, and S. Barbarossa, "Redundant Filterbank Precorders and Equalizers Part II: Blind Channel Estimation, Synchronization, and Direct Equalization," *IEEE Trans. Signal Process.*, vol. 47, no. 7, pp. 2007–2022, July 1999.
31. R.O. Schmidt, "Multiple Emitter Location and Signal Parameter Estimation," in *IEEE Trans. Antennas Propag.*, vol. AP-34, no. 3, pp. 276–280, March 86.
32. J.J. Shync and R.P. Gooch, "Convergence Properties of the Multistage CMA Beamformer," in *Proc. 27th IEEE Asilomar Conf. Signals, Systems and Computers*, vol. 1, pp. 622–626, 1993.
33. S. Talwar, M. Viberg, and A. Paulraj, "Blind Estimation of Multiple Co-Channel Digital Signals Using an Antenna Array," in *IEEE Signal Process. Lett.*, vol. 2, no. 1, pp. 29–31, February 1994.
34. S. Talwar, M. Viberg, and A. Paulraj, "Blind Estimation of Multiple Co-Channel Digital Signals Using an Antenna Array — Part I: Algorithms," in *IEEE Signal Process. Lett.*, vol. 44, no. 5, pp. 1184–1197, May 1996.
35. S. Talwar, M. Viberg, and A. Paulraj, "Blind Estimation of Multiple Co-Channel Digital Signals Using an Antenna Array — Part II: Performance Analysis," in *IEEE Trans. Signal Process.*, vol. 45, no. 3, pp. 706–718, March 1997.
36. D.M. Titterington, A.F.M. Smith, and U.E. Markov, *Statistical Analysis of Finite Mixture Distributions*. John Wiley & Sons, New York, 1985.
37. L. Tong, G. Xu, and T. Kailath, "Blind Identification and Equalization Based on Second-Order Statistics: A Time Domain Approach," *IEEE Trans. Inf. Theory,* vol. 40, no. 2, pp. 340–349, March 1994.
38. J.R. Treichler and M.G. Larimore, "New Processing Techniques Based on Constant Modulus Adaptive Algorithm," in *IEEE Trans. Acoustics, Speech and Signal Process.*, vol. ASSP-33, pp. 420–431, April 1985.
39. J.K. Tugnait, "On Linear Predictors for MIMO Channels and Related Blind Identification and Equalization," *IEEE Signal Process. Lett.*, vol. 5, pp. 289–291, November 1998.

40. J.K. Tugnait and B. Huang, "Blind Channel Estimation and Equalization of Multiple-Input Multiple-Output channels," in *Proc IEEE Int. Conf. Acoustics, Speech and Signal Process. (ICASSP'99)*, vol. 5, pp. 2707–2710, 1999.
41. J.K. Tugnait and B. Huang, "Multi-Step Linear Predictors-Based Blind Equalization of Multiple-Input Multiple-Output Channels" in *Proc. IEEE Int. Conf. Acoustics, Speech and Signal Process. (ICASSP'99)*, vol. 5, pp. 2949–2952, 1999.
42. P. Vandaele, G. Leus, and M. Moonen, "A Non-Iterative Blind Binary Signal Separation Algorithm Based on Leinear Coding," in *Proc. IEEE Int. Workshop Signal Processing Advances for Wireless Commun. (SPAWC'99)*, pp. 98–101, Annapolis, MD, May 1999.
43. B.D. Van Veen and K.M. Buckley, "Beam-Forming: A Versatile Approach to Spatial Filtering," in *IEEE Mag. Signal Process.*, vol. 5, pp. 4–24, April 1988.
44. A.J. Van der Veen, S. Talwar, and A. Paulraj, "A Subspace Approach to Blind Space–Time Signal Processing for Wireless Communications Systems," in *IEEE Trans. Signal Process.*, vol. 45, no. pp. 173–190, January 1997.
45. A.J. Van der Veen and A. Paulraj, "An Analytical Constant Modulus Algorithms," in *IEEE Trans. Signal Process.*, vol. 44, no. 5, pp. 1136–1155, May 1996.
46. J. Xavier and V. Barroso, "Blind Source Separation, ISI Cancellation and Carrier Phase Recovery in SDMA Systems for Mobile Communications," *Wireless Personal Communications*, special issue on *Wireless Broadband Communications*, vol. 10, pp. 36–76, Kluwer Academic Publishers, June 1999.
47. J. Xavier and V. Barroso, "Correlative Coding Approach for Blind Identification of MIMO Systems," in *Proc. 2nd IEEE Signal Processing Workshop on Signal Processing Advances in Wireless Commun. (SPAWC'99)*, pp. 263–266, Annapolis, MD, May 1999.
48. J. Xavier and V. Barroso, "Polyhedral Methods for Blind Deterministic Separation of Binary Co-Channel Signals," in *Proc. Global Telecommunications Conf. Globecom '99*, pp. 2269–2272, December 1999.
49. J. Xavier and V. Barroso, "Semi-Blind Maximum Likelihood Separation of Linear Convolutive Mixtures," in *Proc. 10th IEEE Workshop on Statistical Signal and Array Processing (SSAP'2000)*, pp. 329–333, August 2000.
50. J. Xavier, V. Barroso, and J.M.F. Moura, "Closed-Form Blind Identification of MIMO Channels by Correlative Coding: An Isometry Fitting to Second Order Statistics," *IEEE Trans. Signal Process.*, in press.

PART E

Implementation of Smart Antenna Systems

22

Smart Antenna System Architecture and Hardware Implementation

T. Bao Vu
City University of Hong Kong

22.1 Introduction

Mobile communications represents one of the greatest commercial breakthroughs in the last decade of the previous millennium. Its success undoubtedly arises from its ability to provide a low-cost answer to the demand from both private and business communities for practically instant contact with anyone at any time, from almost anywhere. Although mobile communications is not a new phenomenon, its success stemmed from the cellular concept, introduced in the 1980s, whereby the frequency spectrum can be reused. As a result, a large number of users can be accommodated by using a rather limited spectral bandwidth, but without causing undue interference with one another.

Since its inception, the cellular mobile communications industry has continued to experience a spectacular growth, with the number of users worldwide expected to pass the billion level well before the end of the first decade of this new millennium. Although this is good news for the industry, it also heralds possible troubles ahead, unless something is done to overcome its effect. The main problem is that an increase in the number of users also means more capacity will be required. Thus, to serve this huge market, the industry must develop new technology to significantly increase its capacity, and to reduce the drop-call rate, which affects the quality of service of its networks, especially in new applications requiring very high-data rates. This is certainly not a simple task, considering all the constraints imposed on designers of mobile communication systems.

At first glance, the simplest way to increase the capacity of a cellular mobile network is to increase the spectral bandwidth, but this is not always the most feasible or economical solution, because of strict governments regulations and the high cost of acquiring new spectral bandwidths. One must also consider the higher path loss associated with higher frequency bands. Another possible solution is to further

0-8493-0124-6/02/$0.00+$1.50

subdivide the cells, but a reduction in cell size is also accompanied by an increase in infrastructure cost, which may not be justifiable. Alternatively, one can optimize the overall receiver design by using more advanced signal processing techniques, including more complex adaptive time-domain equalizers, but it appears that only limited improvement may be expected in the near future. Other solutions include, among other things, adopting lower rate codec. However, a very low rate codec may not function well in the harsh mobile environment, with relatively high bit-error rates, especially when high mobility is also required. Thus, the search for the optimum solution goes on.

In general, a cellular mobile network is a complex technological system, embracing several disciplines of engineering, but it basically is also a wireless system, which employs free space as its transmission medium. Consequently, it is highly vulnerable to interference coming from multiple reflection and scattering of desired and unwanted signals by a multitude of obstacles. This multipath propagation phenomenon is responsible for delay-spread and multipath-fading problems, the main causes for high drop-call rates commonly encountered in many metropolitan areas. The problem is further aggravated by the mobile cellular structure, which allows frequencies to be reused in different cells. Consequently, interference from same-frequency channels or co-channels can be quite significant under certain operating conditions. Thus, it has long been recognized that the capacity and the quality of service of a mobile cellular system are limited by three major factors, namely, multipath fading, delay spread, and adjacent plus co-channel interference. The upshot is that the future of the mobile communications industry depends critically on its ability to overcome these problems.

Up to now, the industry has relied mainly on coding and channel equalization techniques to mitigate the effects of multipath propagation and co-channel interference. This was considered a more rational approach, because highly sophisticated coding techniques can be implemented through the use of modern digital signal processors, which are getting more powerful and less costly each year. Thus, although the performance of any wireless system depends critically on the antenna, it has continued to play a secondary role.

In first- and second-generation mobile communication systems, antennas used at base stations have been anything but smart, with a fixed broad beam determined by the required radiation coverage, but with no mechanism for making any intelligent decision. Consequently, these antennas are highly susceptible to the effect of multipath propagation, and their performance becomes increasingly inadequate as the number of users increases, even with the help of various diversity schemes. Thus, new and smarter antennas are required. The basic concept is to develop suitable algorithms, which shape the antenna radiation pattern so that the main beam points to the direction of the desired signal, and directions of the nulls are steered toward interfering sources. In most cases, a narrowband structure is assumed, which essentially performs its spatial filtering function by varying the phases and amplitudes of individual antenna elements of the array. The fact that this narrowband structure is ineffective in solving the multipath delay problem is considered unimportant, because it is assumed that any remaining problems could be successfully handled by the channel equalization and decoding stages in the mobile receiver.

However, a better solution would be to integrate the traditional spatial-antenna architecture with the temporal channel equalizer into a new smart antenna structure. The rational is that the adaptive optimization process would be more effective because it can *simultaneously* modify parameters of both space-domain beamforming and time-domain equalization sections. To be effective, a rather complex smart antenna architecture using powerful and hence costly processors would be required. Up to now, this high-cost feature of smart antennas has prevented their widespread acceptance by the industry, but significant changes are on the way. The main reason is the mounting pressure from the upcoming demand for a large increase in capacity created by the continued exponential growth in user population. The general agreement is that smart antennas indeed represent the most promising tools against interference and the detrimental effects of multipath fading on the desired user's signal. Moreover, a larger user population would also mean that operators have more resources for implementing more advanced albeit more expensive tools. Smart antennas have therefore been included as a standard in third-generation mobile systems.

22.1.1 What Is a Smart Antenna?

Despite their wide publicity in recent years,[1-3] and although the general consensus is that smart antennas would increase a mobile system capacity, there is no general agreement on the precise meaning of the term *smart antenna*. Generally speaking, the term may be used to cover a wide range of antennas, from a rather simple array with fixed orthogonal beams and limited intelligence, to sophisticated adaptive arrays with complex structures and control algorithms. What is certain, however, is that a smart antenna usually means much more than just the basic array of identical antennas. In fact, even a simple multibeam array must have a decision-making mechanism for selecting the beam that has the strongest desired signal component. More specifically, a smart antenna is a product of a successful marriage between communications and computer engineering technologies. In general, the architecture of a smart antenna consists of five basic subsystems, namely, the antenna array itself, the radio frequency (RF) front end, the downconverter section, the data acquisition system, and the digital signal processing (DSP) module. Moreover, because the latter module provides the brain power required for minimizing the effect of multipath propagation, it would be quite feasible to generalize the smart-antenna architecture to include the time-domain adaptive delay equalizer as well. As mentioned earlier, the rational is that it would be more effective to optimize the performance of the two adaptive systems as a whole, instead of considering each individual system in isolation.

This chapter, however, does not deal with the development of different adaptive algorithms, but instead focuses on the practical design and implementation of a smart antenna system. Nevertheless, as will be clear in later discussions, the hardware design principle described here is applicable to various system architectures with varying degrees of complexity. The only major modification required would be in the DSP module (e.g., with faster processors and more complex control software to implement more sophisticated adaptive algorithms). The rest of the hardware remains more or less the same, irrespective of any specific architecture chosen by the designer. The material presented in this chapter is based mainly on the research carried out by the author and his research group when he was with the University of New South Wales in Sydney, Australia. Although brief descriptions of the work have been published over the years,[4-8] this is the first time that complete systems are discussed in the open literature. Section 22.2 of this chapter first outlines the basic architecture of a smart antenna system, and compares conventional microwave-based adaptive arrays with modern systems that are dominated by digital technology, where software will play an increasingly more important role. It then describes the general method of implementing such a system. In particular, it outlines the basic structure of the hardware platform, and the role of the software. This chapter ends with Section 22.3, which highlights important design issues, with emphasis on antenna design and use of parallel processing in the DSP module.

22.2 Overview of System Architecture and Implementation

As mentioned in the introduction, a smart antenna is a system that integrates the function of many components and subsystems to achieve its main objective of minimizing the effect of interference and multipath delay. Thus, in this smart system, the antenna array itself is but one of the subsystems, albeit a very important one, because its radiation characteristics determine the ultimate quality of the signal obtainable at the output of the mobile receiver. It should be noted here that we specifically consider smart antenna systems for mobile communications. However, the architecture described in this chapter is also applicable to other related fields, such as radar and sonar, where the channel is usually an isotropic, homogenous transmission medium. A smart antenna is quite different from conventional antennas in that it is a product of many different technologies. The upshot is that a smart antenna system engineer must be familiar not only with antenna techniques but also with analog microwave/RF technology, digital receiver technology, and computer technology. Thus, in the following discussions of the smart antenna system, we assume that the reader already has a basic understanding of the subject matter, and is also familiar with cellular systems and concept. Finally, because the basic principle is the same, whether the antenna is transmitting or receiving, we concentrate mainly on the receiving case. It is also noted that smart antenna implies the use of an array of antenna elements. Consequently, the principle described in this chapter is equally applicable to base station antennas and mobile terminal antennas, provided that the mobile terminal is large enough to accommodate an antenna array, even if it is a small one.

22.2.1 Basic Smart-Antenna System Architecture

A smart antenna is often referred to as an adaptive antenna, and in that sense it has a long history. Generally speaking, the architecture of a smart antenna system is governed by the following three main tasks it must carry out to achieve its objectives:

- It must first acquire the necessary information or signals via its antenna array.
- It must then transform these signals into a format that is suitable for processing. In modern systems using digital technology, this means that the original analogue signals must be converted to digital form through the use of analog to digital converters (ADC).
- It must also modify the signals from individual antenna elements and optimally combine them to strengthen the desired signal and minimize the interference. This is the task of the adaptive processor.

In principle, we can implement adaptive antenna algorithms by using either conventional microwave-based methods or modern DSP-based techniques. In addition, adaptive antennas can be designed for both broadband and narrowband applications. In most discussions on adaptive antennas, however, narrowband signals are normally assumed so that a time delay can be represented purely by a shift in the phase of the carrier signal. Consequently, the nulling of a given interference, and the steering of the main beam to a known direction can be achieved simply by changing the phases and amplitudes of the signals in individual elements of the array. Thus, the narrowband assumption greatly simplifies the system architecture, because no delay lines are required. In the following sections, we discuss the system architecture of these narrowband systems in more detail.

22.2.1.1 Conventional Microwave-Based Architecture

Traditionally, large adaptive antennas used in military radar systems are predominantly microwave-based systems, using computer-controlled microwave digital phase shifters and attenuators to achieve the desired adaptive nulling and beamforming. Because these phase shifters and attenuators employ different numbers of switching diodes to achieve different phases and attenuation, there are inherent problems, such as unwanted and unequal insertion losses in different antenna-element channels as a result of the losses in active switching diodes. In addition, there are latent errors in adaptive weights because, to contain the cost, phases and amplitudes can only be changed in rather large discrete steps. Consequently, the antenna performance is not quite optimum. In addition, a significant weakness of conventional array is its inflexible analog-based architecture, which cannot be easily modified to take advantage of the rest of the predominantly digital receiver. A summary of these disadvantages follows:

- The system is susceptible to many types of errors, including quantization errors in phase shifters and attenuators.
- Because of the cost factor, relatively large discrete steps in phase and amplitude increments are normally adopted, so that the array cannot exactly steer beams and nulls to the desired directions. Consequently, a loss of gain and a less effective suppression of interference are expected.
- As different numbers of diodes are switched into the transmission path for different interference configurations, unequal losses of gain are experienced. Hence, the system performance is not uniform as different interference scenarios are encountered.
- There is inflexibility in reconfiguring the system architecture (e.g., once the hardware is designed and built, it is not easy to reconfigure the system to suit the current interference scenario).
- It is costly to achieve multibeams for multichannel receivers.

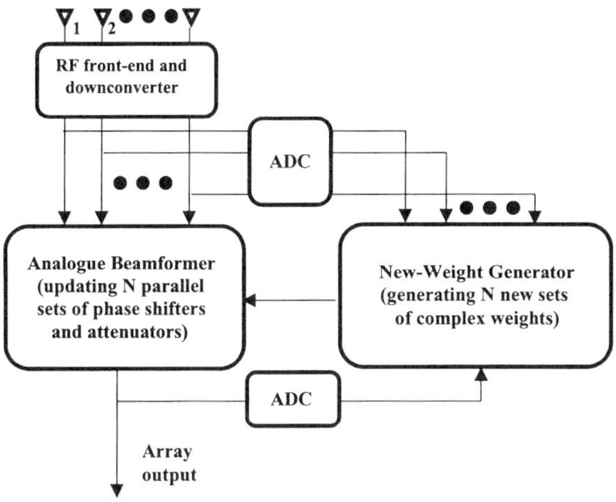

FIGURE 22.1 Block diagram of conventional analog-based adaptive antenna.

Figure 22.1 shows a block diagram of a conventional adaptive array using microwave digital phase shifters and attenuators. Here, the processor relies on the new input to the array and a known reference to calculate the required adaptive weights to maximize the signal to noise plus interference ratio (SNIR). These new adaptive weights are then converted into required phases and attenuation in the corresponding elements of the array to achieve the desired array output.

22.2.1.2 Modern Digital Signal Processor-Based Architecture

A modern smart antenna, on the other hand, is basically a digital system, where software is becoming more and more predominant. The signal from each antenna element is independently digitized before being processed in the DSP module together with digital signals from other antenna elements. Thus, smart antennas have all the flexibility and cost advantage of a DSP-based system; and consequently their cost should decrease significantly with time, whereas their power will improve with more advanced DSP developments. More important, because all the main actions of a smart antenna take place in the digital DSP module, it is an easy matter to combine the traditional antenna function with others, such as the time-domain equalization. This provides an opportunity to optimize the receiver performance *as a whole*, in contrast to the traditional approach of optimizing *individual* subsystems separately, despite the fact that concatenating a series of optimum subsystems will not necessarily produce the optimum whole. The advantage of optimizing a combined system may be exemplified by the robustness of a combination of adaptive antenna and delay equalizer against interference. The combined antenna–equalizer system would successfully handle a given large number of interferences, whereas a system consisting of an adaptive array in concatenation with an adaptive equalizer would not. The main reason is that an adaptive antenna acting alone would fail when there are more interfering sources than antenna elements in its array, which is often the case in a mobile environment.

In contrast to microwave-based systems, modern smart antennas implement adaptive weight control purely through digital signal processing. The main advantages include the elimination of the previously mentioned adaptive weight errors and insertion losses experienced in conventional systems using microwave digital phase shifters and attenuators. In addition, the DSP-based smart antenna architecture has more flexibility (e.g., it can be easily expanded to integrate the time-domain delay equalizer and the conventional space-domain adaptive antenna). Finally, DSP-based array is also capable of carrying out self-calibration to eliminate deterministic system errors caused by imperfect component design and fabrication.

Figure 22.2 shows the corresponding smart antenna block diagram. The most important information it conveys is that a smart antenna requires two main signal-processing tasks. The first task is to implement

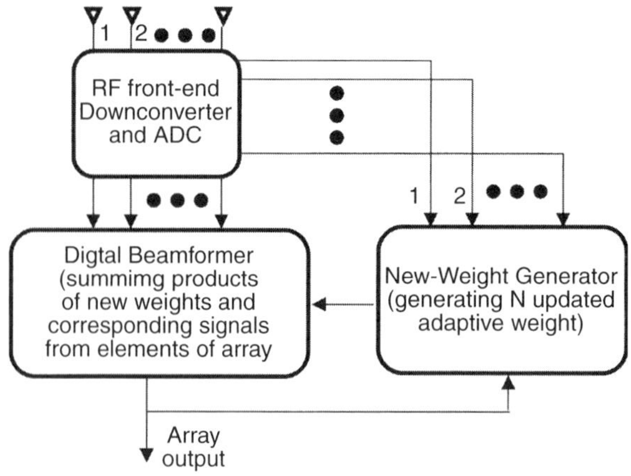

FIGURE 22.2 Block diagram of a DSP-based smart antenna.

a suitable algorithm to analyze and process new input data, and calculate the new optimum weights to achieve a given objective, for example, to maximize the signal to interference ratio (SIR). On the other hand, the second task is much more straightforward, and involves the multiplication of the optimum weights with incoming digital signals from the array elements, followed by the summing of these products to give the desired array output. The former task is referred to as weight-updating function (WUF), which is normally quite complex and time consuming, and the degree of complexity depends on the system architecture as well as the chosen algorithm. The latter, on the other hand, is referred to as beamforming function (BFF), and is quite straightforward because it simply involves the standard multiplication and accumulation (MAC) process commonly encountered in digital filter operations. In the case of the smart antenna, this may result in steering nulls to directions of interference and pointing the main beam of the array to the desired signal direction.

With most commercially available DSP, the MAC task of the BFF module can often be completed in microseconds for each new snapshot, or set of data samples received by the DSP. More importantly, the cellular mobile environment tends to remain stable over a short period of a time (e.g., a few milliseconds) under most operating conditions. Thus, there is no need to generate new adaptive weights for every new snapshot. In other words, the beamforming section of the DSP module normally reuses the old adaptive weights until a new set of optimum weights arrives from the WUF section. From this discussion, it is clear that a more effective data management can be obtained if two separate DSP are used to carry out in parallel the two main tasks of the DSP module, especially in more complex smart antenna architectures.

An extended smart antenna architecture, which combines traditional array with time-domain delay equalizer can be easily implemented by software in the DSP module, with each element of the antenna array constituting a separate channel. Figure 22.3 shows a single channel, which generates the in-phase (I) and quadrature (Q) components by using two orthogonal signals at the frequency of the second local oscillator (LO 2). An analog downconversion scheme is adopted, but it is important to point out that

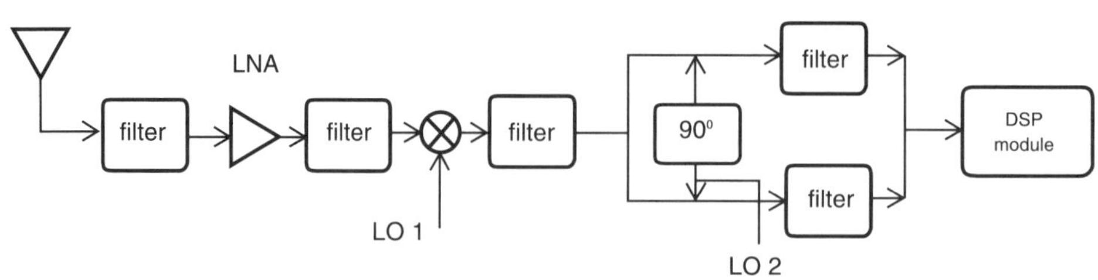

FIGURE 22.3 A smart antenna channel with quadrature component generated by hardware.

digital downconversion becomes possible. In fact, the digital scheme is technically more attractive, because it provides more flexibility in the design. With digital downconversion, undersampling is adopted, so that the sampling frequency is twice the signal bandwidth, instead of twice the highest frequency of the mobile bandwidth. It is likely that the second analog stage of the downconversion segment will be replaced by its digital counterpart in the near future. A glimpse of what to come is described by Slifkin et al.[9]

22.2.2 System Implementation

22.2.2.1 Hardware/Software Overview

Like all DSP-based systems, a smart antenna system consists of two main sections, namely, the hardware platform and the software, which controls the system hardware, and processes signals provided by the hardware to produce the desired results. In fact, the software plays the most crucial role. First, there is the main controlling software, which may be designed to examine the current mobile operating conditions, and then to select the most appropriate adaptive nulling and beamforming algorithm. Second, a monitoring software would be required to monitor the performance of individual components by analyzing the signals in each data channel. If a channel does not perform as required, the monitoring software initiates a special course of action to minimize the effect of the malfunction on the overall system performance. The monitor may also extract raw data from the DSP module, and emulates the function of the DSP to visually display the dynamic operation of the smart antenna system. Finally, a special software must also be developed for carrying out routine functions such as input–output synchronization, and direct memory access (DMA) control for efficient transfer of data. The software must also carry out other tasks like system self-calibration, maintaining optimum system performance via automatic gain control and possibly polarization switching, etc.

In general, to achieve high efficiency and fast system response, software code should be written in assembly language. On the other hand, to simplify the task of implementing the complex adaptive control algorithm, especially during the development stage, a higher level language should be adopted. With more advanced DSP, facilities are available for a highly efficient translation from a higher language to assembler code, but with older DSP it may be more effective to use a hybrid approach to software design. The software must also be designed to fully exploit the facilities provided by the DSP system (e.g., using DMA to more efficiently transfer large blocks of data to and from memory blocks). It should also be designed to minimize the number of instruction cycles required to complete a task by exploiting special features like parallel instructions. In addition, if multiprocessor architecture is adopted, the division of tasks among processors must be carefully executed to avoid an unnecessary bottleneck in the data flow. Nevertheless, because the complexity of the software is dependent on the particular DSP adopted for the system, a full discussion of the software implementation is not given here. However, samples of software developed by my previous smart antenna research group at the University of New South Wales may be found in relevant Ph.D. and Master theses.[10-13] In what follows, a brief outline of the basic hardware of a smart antenna system is given, whereas a more detailed discussion of important design issues is given in the next section.

In simple terms, except for the fact that the low-noise amplifier (LNA) and downconverter are replaced by the power amplifier and upconverter, respectively, the transmitter and receiver modules consist of similar components but in reverse order. For this reason, we only discuss the receiving case. In addition, we restrict ourselves to smart base station antennas, although the basic principle applies to mobile terminals as well.

Generally speaking, the hardware platform of a smart receiving base station antenna consists of the following main sections:

- Antenna array and associated feeding network
- RF front end, which includes LNAs and filters
- Downconverter section, which translates the frequency of the signal from each antenna element to baseband for digitization before being processed in the DSP module

- Data acquisition module (DAM), which provides the interface between the analog section and the DSP module
- DSP module, which processes raw data from all the input channels, with the ultimate aim to achieve interference-free output signals; the smart antenna output then further processed in the receiver to recover the original information

22.2.2.2 Antenna Array

In the receiving case, signals from desired and unwanted mobile terminals as well as from other interference sources are received by an array of identical antennas that are equally spaced. If discrimination against unwanted interference in only one plane (e.g., in the horizontal plane) is required, then a linear antenna array would be sufficient. On the other hand, if it is desirable to focus the array beam into a narrow shaft, then an $N \times M$ element rectangular array may be used, but such an array would result in a very heavy workload in the DSP module. However, there is no need to employ a filled planar array, with N elements in the x direction and M elements in the y direction to achieve a narrow pencil beam in the receiving case. The reason is that its radiation pattern is equal to the product of the patterns of two linear arrays, one with N elements in the x direction and the other with M elements in the y direction. Therefore, two orthogonal linear arrays can be used to achieve a pencil beam in the receiving case, because the multiplication of the two patterns can be efficiently carried out in a DSP-based smart antenna system. This approach has long been adopted by radio astronomers, and reduces the number of channels to be processed by the DSP module from $N \times M$ to $(M + N)$ (i.e., a great reduction in DSP workload especially if N and M are large). Figure 22.4 shows the pattern obtained by multiplying that of two crossed linear arrays, each with eight elements spaced $\lambda/2$ apart. Apart from the fact that the resultant main beam is a pencil beam, Fig. 22.4 clearly shows that the sidelobes are very low everywhere, but in the principal directions. Thus, even without using adaptive nulling-algorithm, the interference from unwanted sources can generally be kept very low, because it is more likely that the interferences lie outside the principal directions.

In conclusion, a properly designed antenna array is essential, because the overall performance of the smart antenna system depends critically on the strength of the desired signal just after it is received by the antenna array. However, in addition to technical issues, the shape and size of the array must also be

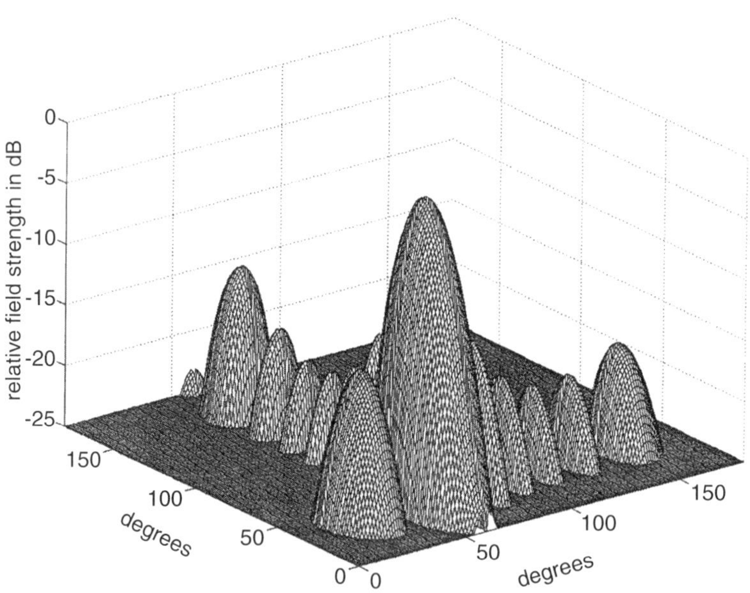

FIGURE 22.4 Product pattern of two crossed linear arrays, each with eight elements spaced $\lambda/2$ apart.

aesthetically designed to appeal to users and the public at large. A more detailed discussion of array design issues is given in Section 22.3.1.

22.2.2.3 Radio Frequency Front End

The signals received by the antenna array must be suitably amplified by an LNA and filtered before being downconverted to baseband for digitization. The main aim here is to amplify the signal to the desired level, and to reject all out-of-band signals without significantly affecting the receiver noise figure.

In principle, this means that the LNA should have very high gain and very low noise figure, so that noise contribution from subsequent devices does not significantly affect the receiver noise figure. In other words, GaAs MESFET and its derivatives (i.e., HEMT and PHEMT) should represent a better alternative to silicon devices, because they can provide a lower noise figure (i.e., less than 0.5 dB). In practice, however, one must consider contributions to the receiver noise figure from upstream and downstream devices. These include losses in diplexer or transmitter/receiver switch, on the one hand, and the effect of a high LNA output on the performance of downstream devices such as the mixer, on the other hand. If active mixers with a high noise figure are used in the downconverter section, their effect on the receiver noise must also be taken into account in choosing the LNA.

Thus, to design a suitable RF front end, a clear understanding of the behavior of individual components and the interaction between receiver sensitivity and dynamic range is essential. In addition, with the widespread availability of ultrahigh frequency (UHF) integrated circuits (ICs), the designer must be able to decide on their relative merit if an attempt is made to construct the RF front end from an assembly of commercially available ICs. Information on the design of components of the RF front end is widely available in the literature.[14-20]

22.2.2.4 Downconverter

The downconverter plays a particularly important role in a smart antenna system, because a poorly designed downconverter can introduce a great amount of spurious signals that would seriously affect the receiver performance. The downconversion process is achieved by using a mixer. In most low-cost receivers, a simple single-diode-mixer circuit is used, but more complex mixer circuits that enhance the overall mixer performance should be employed in practical smart antenna systems.

In a single-diode mixer, the RF and the local-oscillator (LO) signals are connected to a single diode through suitable filter networks. Because of the nonlinear characteristics of the diode, its output consists of signals at (RF + LO), (RF – LO) as well as other frequencies. A filter in the intermediate frequency (IF) port selects one sideband, for example, (RF – LO), and rejects the remaining products. A Schottky barrier diode is normally used because it has a very small junction capacitance, thus making the Schottky barrier diode more effective as mixer at frequencies greater than 1.0 GHz.

In smart antenna design, the mixer conversion loss is an important parameter, so it must be carefully analyzed. Because in most practical cases, the amplitude of the LO signal is much larger than the RF and IF signals, terms involving higher order powers of RF and IF signals can be neglected in the analysis of the conversion loss. For a simple single-diode mixer, this loss is usually greater than or equal to 6 dB. However, it should be noted that active mixers can have conversion gain (i.e., the power of the IF signal is larger than that of the RF signal).

Another important design parameter is the level of intermodulation distortion. When the RF signal is weak, the IF power increases linearly with RF input power; however, as the RF power increases above a certain level, the input/output curve becomes increasingly nonlinear, and results in intermodulation distortion. Moreover, if the RF input consists of two or more signals with closely spaced frequencies, as in the cellular mobile case, the IF output consists of many unwanted frequency components that fall within the IF passband. These so-called intermodulation products cannot be filtered out by the IF filter, and the third-order product tends to be more disturbing. Thus, a very high gain LNA may produce serious distortion, and a trade-off between sensitivity and dynamic range is necessary. In practice, an attenuator may be inserted between the LNA and the mixer to ensure that the RF input power to the mixer is kept small enough to avoid intermodulation distortion.

Finally, one must consider the mixer noise figure in the design. As noise in both the upper and lower sidebands are converted to noise in the IF band, this must be taken into account in designing the RF front-end filters.

The single-diode mixer, however, has many drawbacks, because it provides no isolation between the RF and LO ports. There is also poor isolation between the IF port and the RF and LO ports. In most conventional RF circuits, the IF frequency is much lower than the RF frequency. Hence, the RF and LO frequencies are very close; consequently, it is not easy to isolate the RF port from the LO port by using external filtering networks. However, good isolation between the IF port and the RF and LO ports can be achieved by placing a series LC circuit across the RF and LO ports. Here, L and C are chosen so that the circuit resonates at the IF frequency, and therefore provides a very low impedance path across the RF and LO ports. At the same time L presents a very high load across the RF and LO ports at the high RF frequency.

In a balanced mixer, two diodes with matched electrical characteristics are used to eliminate the LO signal at the IF port. Noise from the LO circuit does not appear at the IF output. To provide good isolation between the RF and LO ports, a double-balanced mixer may be used. It has the added advantage of low intermodulation distortion, because even harmonics of RF and LO signals are suppressed.

Finally, mixers can also be designed to reject unwanted signals in the image frequency band. More detailed discussion on the theory of mixers are readily available in textbooks and other references.[19-22]

22.2.2.5 Frequency Synthesizer

The LO signal required by the downconverter may be obtained by direct digital synthesis, whereby a digital approximation of a sine wave is generated. Low-pass filtering of the digital-to-analog converter output produces the required sine wave output. A dielectric resonator oscillator may also be designed, but a more common method of generating the LO signal is by indirect technique, which employs high-speed dividers and a phase-lock loop.

In a typical frequency synthesizer circuit, the output of a voltage-control oscillator (VCO) is controlled in a phase-lock loop (PLL) so that its frequency remains locked to a multiple of the frequency of a known reference. The basic principle of the PLL frequency synthesizer is quite straightforward. In the feedback path of the loop, a signal whose frequency is a suitable submultiple of the VCO frequency is generated and compared with the reference signal via a phase comparator. If their frequencies are different, the error voltage from the phase comparator forces the VCO frequency to change in such a way that this frequency difference eventually is reduced to zero. The phase comparator may be a mixer, and the reference signal may be from a crystal oscillator. With this scheme, the output frequency is a multiple of the reference frequency, and the noise performance depends on the loop configuration, the phase detector technology, and the noise characteristics of the reference signal. In particular, the noise from the reference is multiplied in the output by the ratio of the VCO frequency over the reference frequency. This is important, because in a smart antenna a very low phase noise would be required. The most extensively used PLL synthesizer architecture is the single-loop PLL. Because these devices are simple, have low cost, and are easily implemented in integrated circuits, PLL frequency synthesizers are attractive. Description of the theory and design of PLL frequency synthesizers may be found in some of the references cited earlier.[16-17]

22.2.2.6 Data Acquisition and Digital Signal Processor Module

In principle, the DAM and the DSP may reside on the same board, but in practice, they may be produced by different manufacturers. In the two models to be discussed in Section 22.3, the DAM is designed and built in-house, but the DSP card is purchased from commercial sources. Figure 22.5 is an expanded view of a typical DAM. It clearly shows how the DAM interfaces with the DSP module and the RF front end, where the signal from each antenna element is split into its I and Q channels by a quadrature demodulator. At the command of the DSP via the S/H signal, all the channels (i.e., I and Q signals from all elements of the array) are sampled at the same time to preserve their phase relationship. The sample from each channel is then selected by the DSP sequentially for digitization via control signals applied to the

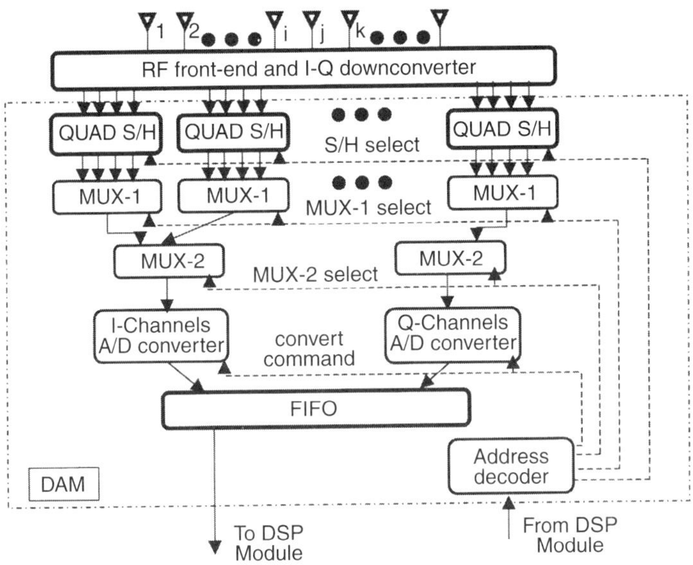

FIGURE 22.5 Expanded diagram of the data acquisition module (DAM).

multiplexers (MUXs) and analog to digital (A/D) converters. The latter are driven by the convert signal, which is derived from the DSP clock, the strobe, and R/W signals. The DSP can then read the digitized signals from the first-in, first-out (FIFO) buffer, and processes the signals by combining them in a special way so that the desired signal can be more correctly recovered from the interference plus noise background.

This architecture of the DAM, which employs a single fast analog-to-digital converter to digitize a large number of channels, is not unique, but it does reduce the number of chips required. An alternative approach may also be adopted, where one analog-to-digital converter is used for each I or Q channel. In this case, to preserve the relative phases among the N channels, all the A/D converters are synchronized to convert simultaneously. Their outputs are then sequentially selected by MUXs under the control of the DSP. However, there is no real difference between the two approaches in terms of system performance or complexity. It is also important to note that the quadrature signals may also be generated by the Hilbert transform. Therefore, instead of using the preceding hardware approach, software may be used to produce the I and Q channels. This is the approach adopted in the two models described in Section 22.3, where the Hilbert transform is approximated by a linear-phase, finite-impulsive response (FIR) filter. Thus, the savings in quadrature demodulator hardware is achieved at the expense of an increased load on the DSP module, especially if the required bandwidth is wide since a longer filter length must be used. Generally speaking, to remove the interference and noise background from the desired signal, one must first multiply the complex data sample ($I + jQ$) from a given array element by a suitable complex weight. All these products are then summed to provide the desired output. Therefore, in principle, the number of the Hilbert transform required per snapshot is equal to the number of elements in the array. Thus, the number of instruction cycles required to perform all these Hilbert transform operations is very high, especially for large arrays. However, there is a solution to this problem, because if the weights are approximated as time invariant, then the total number of Hilbert operations is reduced to two, irrespective of the number of elements in the array.[12] This approximation has been used quite successfully in the software developed for the two models described in Section 22.3.

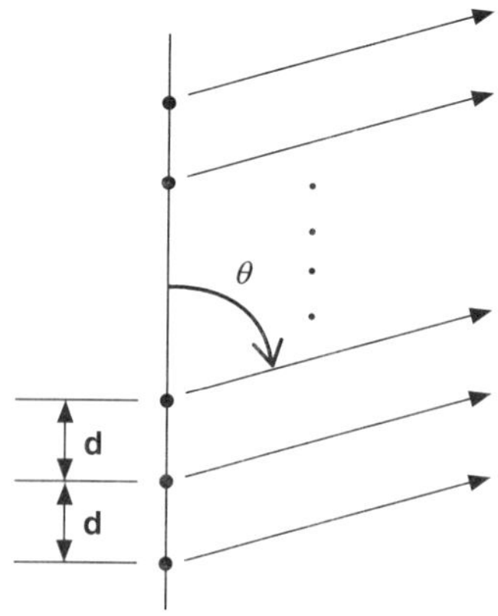

FIGURE 22.6 Geometry of a linear array with interelement spacing equal to d.

In a smart antenna system, the DSP module plays the most crucial role, because it is the source of the system intelligence. By carrying out the appropriate commands in the system software, the DSP module can strengthen the desired signal, and at the same time minimize the effect of interference and propagation delay. If a suitable adaptive algorithm is implemented, the DSP module can produce, in real time, an almost interference-free desired signal at the smart antenna output. As mentioned previously, the BFF section generates the desired smart antenna output by using and reusing the latest available weights, whereas the WUF section creates new set of optimum weights from the latest input samples it receives. In practical terms, the BFF operation is equivalent to pointing the main beam toward the desired signal, and simultaneously steering the pattern nulls toward the interference directions. The smart antenna output can then be further processed in the mobile receiver to recover the original information.

22.3 Important Design Issues

22.3.1 Antenna Design Issues

Because the overall performance of the smart antenna system depends on relative strength of the desired signal after being received by the antenna array at the front end, the array must be designed with a lot of care. First, there is the question of grating lobes, which cause ambiguity and interference if they are not kept small. Second, the radiation pattern of individual elements of the array (i.e., the unit pattern) can also have a detrimental effect, especially when the beam is scanned to the edge of the sector in a cellular mobile communication system. We briefly illustrate these effects by considering the radiation pattern of a uniform linear array. As is well known in the theory of antenna arrays, the radiation pattern of the array is the product of the unit pattern and the array-factor pattern. The latter depends only on the number of elements in the array and the direction of the main beam, with the absolute value of the array-factor (AF) pattern of a uniform linear array given by the following formula:

$$|\mathrm{AF}| = \left| \frac{\sin(0.5\,N\,\psi)}{\sin(0.5\,\psi)} \right| \tag{22.1}$$

where, referring to Fig. 22.6, N = total number of elements in the array and

$$\psi = \frac{2\pi d}{\lambda}\left(\sin\phi - \sin\phi_M\right) \tag{22.2}$$

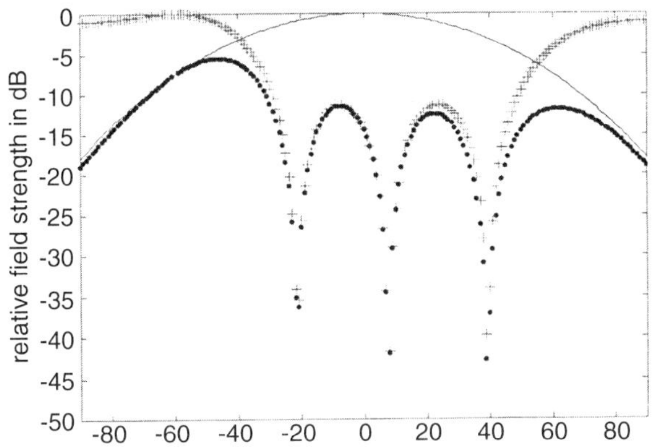

FIGURE 22.7 Effect of unit-pattern shape and beam-scan angle on grating lobe.

Additionally, ϕ = direction measure from broadside = $\theta - \pi/2$, ϕ_M = direction of the main beam of the array, λ =wavelength, d = interelement spacing. The unit pattern, however, is the radiation pattern of one single element, and is therefore different for different antennas chosen as elements of the array.

In most textbooks, it is usually stated that an interelement spacing of $\lambda/2$ should be used to avoid the grating lobe, which causes ambiguity and interference. In this section, we demonstrate that this is only true for a full-size grating lobe. However, a near full-size grating lobe (referred to hereafter as quasi-grating lobe) can still occur for $d = \lambda/2$ if the beam is scanned close to the endfire direction. Therefore, a question arises as to the maximum value d can have that still ensures the quasigrating lobe is no larger then other sidelobes. The answer to this question is given in Section 22.3.1.1. Next we simply illustrate the effect of scanning on the size of the quasi grating lobe, and the modulating effect of the unit pattern on the radiation pattern of the array.

In Fig. 22.7, the radiation pattern of an eight-element uniform linear array (circular dots) is shown together with the array-factor pattern (shown as a series of + signs) and the unit pattern (solid curve). Here, the interelement spacing is $\lambda/2$, and the main beam is steered to the direction of the desired signal at –60° from broadside (i.e., to the edge of a 120° sector in a cellular mobile communication system). The unit pattern is idealized, but its shape is similar to that of the E-plane pattern of a microstrip antenna, so that the array radiation pattern given here is similar to that of a practical array of microstrip antennas.

Two important features deserve some attention. First, if we neglect the effect of the unit pattern by assuming an array of omnidirectional elements as in many textbooks on adaptive antennas, then the radiation pattern of the array is the array-factor pattern. In that case, when the beam is scanned to the edge of the sector cell, as shown in Fig. 22.7, a large quasi-grating lobe appears in the direction of the neighboring sector. Consequently, this quasi-grating lobe can pick up strong interference from the base station and mobile terminals in that sector, even though the interelement spacing is $\lambda/2$. Thus, a smaller spacing should be used in this case. The situation would be even worse if the beam were scanned closer to the endfire direction.

Second, in many practical cases, the unit pattern is not omnidirectional, and therefore has a very strong influence on the performance of the array. On the one hand, the tapered unit pattern (as in Fig. 22.7) provides the desired effect by minimizing the quasi-grating lobe. In our example, the quasi-grating lobe has been reduced to that of the other two sidelobes as shown by the pattern of circular dots. On the other hand, because the main beam is modulated by the tapered unit pattern, its strength in the desired-signal direction (i.e., –60° from broadside) is only about 5 dB higher than the sidelobe level. The upshot is that the array would not provide a good discrimination against interference. Moreover, the peak of the beam is not where it is pointed to, but is closer to -45°. In other words, interference in that direction receives more gain than the desired signal at –60°. Thus, a sharply tapered unit pattern must be avoided. It is clear from this discussion that the designer must carefully select elements of the array to ensure that the beam remains high compared with the sidelobes, even when the beam is steered to the edge of the sector. In addition, to reduce the effect of the quasi-grating lobe, the spacing between elements of the array should be less than $\lambda/2$, where λ is the wavelength at the operating frequency. More detailed discussions of methods of achieving these objectives are given later.

22.3.1.1 Grating Lobe Consideration

An important issue in phased-array design is the correct choice of the interelement spacing so that the effect of the grating lobe can be avoided.[23] It is well known among practioners of linear arrays that this is achieved by using an interelement spacing (d) of half the wavelength. However, this rule is somewhat arbitrary because in the case of a broadside array, a full-size grating lobe only occurs if d is equal to λ. On the other hand, an almost full-size, quasi-grating lobe may occur when the beam is scanned to endfire direction, even if d is slightly less than $\lambda/2$. It is therefore important to be able to determine the maximum allowable value of d that would not make quasi-grating lobes grow above an acceptable threshold. This is particularly true, because the gain of the antenna array is directly related to its length and hence to d.

Next we present a method for calculating this maximum allowable spacing for any specific scanning range of the antenna beam. The method makes use of the unit circle in the complex plane, where each point on the unit circle corresponds to a given physical direction. With reference to the array geometry shown in Fig. 22.6, the directions $\theta = 0$ and $\theta = \pi$ correspond to the starting point (ψ_s) and the endpoint (ψ_e) of the array locus on the unit circle, respectively. It can be shown that for a given interelement spacing, the actual height of the quasi-grating lobe depends on how close ψ_s or ψ_e is to $2k\pi$, where k is a positive or negative integer. As shown later, this in turn depends on the scanning range of the antenna beam.

Effect of Scan Angle on Grating Lobe

To simplify the explanation, a uniform linear array of N isotropic elements is assumed. Because the elements are uniformly spaced along a straight line, and assuming that a plane wave is incident on the array, there is a uniform increment in delay as one moves from one element to the next. One may therefore express the array factor as follows:

$$AF = 1 + z + z^2 + \cdots + z^{(N-1)} \tag{22.3}$$

where $z = \exp(j\psi)$.

As $|z| = 1.0$, z moves on the circumference of the unit circle when θ changes. From Eq. (22.2), the starting point of the locus of z (ψ_s) and the end point (ψ_e) are given by

$$\psi_S = \frac{2\pi d}{\lambda}\left(1 - \cos\theta_M\right) = \frac{4\pi d}{\lambda}\sin^2\left(\frac{\theta_M}{2}\right)$$

and

$$-\psi_e = \frac{2\pi d}{\lambda}\left(1 + \cos\theta_M\right) = \frac{4\pi d}{\lambda}\cos^2\left(\frac{\theta_M}{2}\right) \tag{22.4}$$

Thus, ψ_s is always positive, whereas ψ_e is always negative. With reference to the unit circle in Fig. 22.8, Eq. (22.3) shows that point X where $\psi = 2k\pi$, corresponds to the maximum value of the array factor, where k is zero or an integer. In addition, the θ direction that corresponds to $k = 0$ is that of the main beam, whereas the directions correspond to $k = \pm 2, \pm 3$, etc. represent directions of full-size grating lobes. Moreover, the arc A X B of the locus corresponds to the main beam region or grating-lobe region. The array-factor pattern, however, corresponds to the visible region, which is determined by ψ_s and ψ_e. Thus,

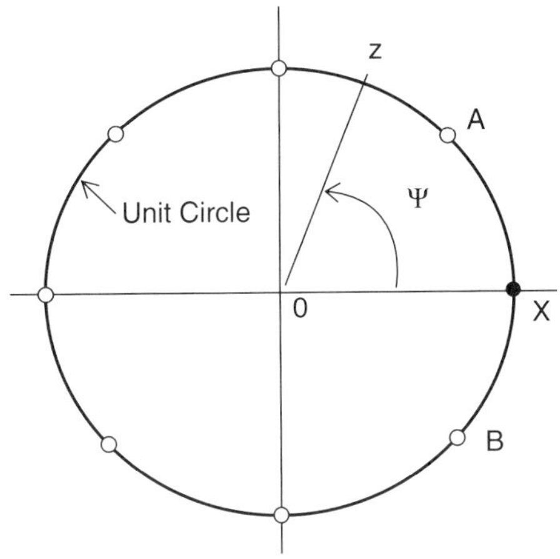

FIGURE 22.8 Distribution of zeros of array factor on unit circle when $N = 8$.

a full-size grating lobe occurs only if $\psi_s \geq 2\pi$ or $\psi_e \leq -2\pi$ (i.e., if the interelement spacing satisfies the following condition):

$$d \geq \frac{\lambda}{2} \operatorname{cosec}^2(\theta_M/2) \qquad 0 \leq \theta_M \leq \frac{\pi}{2} \tag{22.5}$$

or

$$d \geq \frac{\lambda}{2} \sec^2(\theta_M/2) \qquad \frac{\pi}{2} \leq \theta_M \leq \pi \tag{22.6}$$

Equations (22.5) and (22.6) show that whether a full-size grating lobe will occur depends on the beam direction. In other words, a broadside array (i.e., $\theta_M = \pi/2$) does not have a full-size grating lobe if d is less than λ. Similarly, an endfire array must have d smaller than $\lambda/2$ to avoid a full-size grating lobe. However, a quasi-grating lobe will exist if ψ_s or ψ_e lie within the arc A X B of the unit circle as shown in Fig. 22.8), that is, if one of the following inequalities is satisfied:

$$2\pi(1-1/N) < \psi_S < 2\pi \tag{22.7}$$

$$-2\pi(1-1/N > \psi_e > -2\pi) \tag{22.8}$$

Minimization of Quasi-Grating Lobe

The size of the quasi-grating lobe is greater if ψ_s is closer to 2π (or if ψ_e is closer to -2π). On the other hand, the quasi-grating lobe is completely absent, that is, reduced to zero if we have $\psi_s = 2\pi(1 - 1/N)$, or if $\psi_e = -2\pi(1 - 1/N)$.

Equation (22.7) or (22.8) can be used in conjunction with Eq. (22.4) to determine the allowable value of d for no quasi-grating lobe if the antenna beam is to be scanned between $\pm(\pi/2 - \theta_1)$ from the broadside direction, that is,

$$d_m = (\lambda/2)(1-1/N)\operatorname{cosec}^2(\theta_1/2) \tag{22.9}$$

However, if we want to maximize the length of the array, a slightly larger value of d may be used, because a quasi-grating lobe can still be tolerated if it is not higher than the highest sidelobe. If we assume that the maximum allowable level for the quasi-grating lobe is 'L', then it can be shown[23] that the maximum allowable d is given by:

$$d_M = (\lambda/2)\left(1 - 1/N + \psi_1/(2\pi)\right) \operatorname{cosec}^2(\theta_1/2) \tag{22.10}$$

where

$$\psi_1 = -2\pi(1 - 1/N) + L/S \tag{22.11}$$

Here, S is the slope of the tangent to the array-factor pattern in the ψ-space at the point where $\psi = -2\pi(1 - 1/N)$, and may be calculated for any desired current distribution.

22.3.1.2 Unit Pattern Consideration

Conventional base station antennas are usually designed to provide a broad main beam that covers the whole sector angle (e.g., a 120° coverage in the azimuth plane). Thus, an acceptable solution is to use a single antenna element, specially designed to have a –6-dB radiation taper at ±60° from the broadside direction, together with a sharp cutoff beyond the edge of the sector. In the receiving case, such an antenna would minimize interference from adjacent sectors, and cause a gradual degradation of the signal strength as the mobile moves away from the broadside direction. In the vertical direction, however, the required shaped beam in the elevation plane means that an array of identical elements must be used. Thus, conventional fixed beam antennas for base stations consist of a vertical column of antennas. A smart antenna, on the other hand, is required to have a narrow beam in the azimuth plane, so an array of elements must be used in the horizontal plane as well. However, as previously shown in Fig. 22.7, a –6-dB taper at the edge of the radiation coverage would not be desirable, because when the beam is steered to the edge of coverage, its strength is not much higher than that of sidelobes. These design constraints mean that one must choose the elements of the smart antenna array very carefully. For applications where the sector angle is less than or equal to 120°, rectangular or circular microstrip antennas may be used as elements of the array. However, if the advantage of the array is to be fully exploited, rectangular or circular microstrip antennas may not provide the best solution, because even their E-plane pattern would incur high losses at ±60° from the broadside direction. By using them in a smart antenna array, signals from desired mobile terminals at the edge of the sector would be seriously affected by strong interference in the directions of high sidelobes. Moreover, if a 180° sector is deemed desirable, then microstrip antennas should not be used at all, because their E- and H-plane patterns drop sharply in directions equal to ±90° from the broadside. For such applications, the radiation pattern of a quadrifilar helical antenna would be much more desirable.

In addition to the preceding technical problems, one must also consider other properties, including the aesthetic nature of the product. In this respect, microstrip patch antennas have highly desirable characteristics including very low profile and light weight, in addition to good polarization properties. They can be switched from one linear polarization to another simply by changing the feed point, and can be designed to give circular polarization as well. The quadrifilar helical antenna on the other hand, does not have a very low profile and cannot produce linear-polarized signals. In the following sections, we discuss these two main types of antennas.

Microstrip Antenna As Element of Array

Although the original concept of microstrip antennas was proposed in the early 1950s, these antennas did not receive widespread attention until the late 1970s. One of the most popular is the rectangular microstrip antenna, which consists of a length (L) of microstrip line about $\lambda_g/2$, where λ_g is the guide-wavelength in the dielectric material. However, the width of the line is much wider than that of a normal microstrip transmission line. For a nonspecialist, the antenna is but a wide patch of metal foil deposited

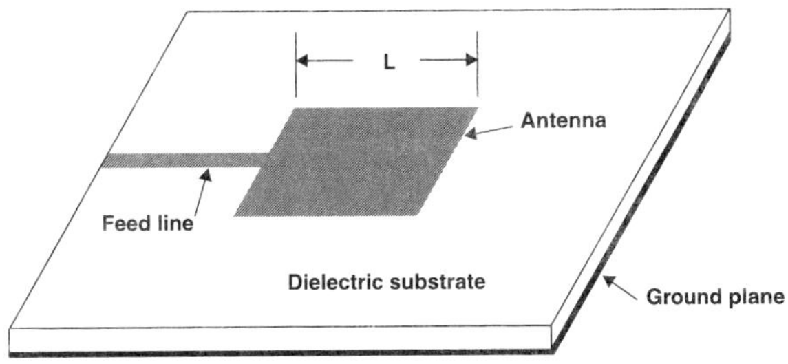

FIGURE 22.9 Rectangular microstrip antenna fed by a microstrip transmission line.

on top of a thin dielectric sheet, which in turn is backed by a metallic ground plane. Figure 22.9 shows a typical rectangular microstrip antenna structure, which is fed by a microstrip transmission line. For most applications, the antenna is designed to radiate in its dominant mode, with the strongest radiation in the direction normal to the plane of the patch. The polarization of the signal is linear, with the electric field parallel to the feed line, and consequently the E-plane pattern is the radiation pattern in the plane parallel to the microstrip feed line and normal to the ground plane. The H-plane pattern is orthogonal to both the E-plane and the ground plane.

As expected, other patch shapes and sizes can also be employed, but for most applications, circular or rectangular patch antennas are normally used. Because a microstrip patch antenna may be visualized as a cavity, which can support many modes, the radiation characteristics of a given antenna depend on the size of the patch. For most applications, however, the size of the patch is chosen so that only the dominant mode exists. This explains why the patch size is on the order of $\lambda_g/2$, with the exact size dependent on the substrate thickness. Microstrip antennas may be directly fed by a microstrip line, but having the feed line and the patch on the same level may degrade the antenna performance. A better feeding method is by a coaxial probe or through an aperture in the ground plane. Both methods allow a good isolation between the antenna on the one hand, and the feeding network and associate electronics below the ground plane, on the other hand.

The easiest way to understand the radiation mechanism of a rectangular patch is to use the transmission line model, which views the antenna as an array of two slot antennas, one at each end of the patch. Because the substrate thickness is normally very small compared with the wavelength, the shape of the radiation pattern of each slot is hemispheric in the plane normal to the slot axis, assuming an infinite ground plane. Because the length of each slot is equal to the width of the patch and because their spacing is equal to the length of the patch, the radiation pattern of the patch antenna can be easily found from this model.

The circular microstrip antenna can also be studied using either analytic or numerical techniques (e.g., the cavity model and the method of moments). Of these methods, the cavity model is simpler, yet able to produce reasonably accurate radiation patterns. Thus, the circular microstrip antenna may be considered as a leaky cavity, defined by the metallic circular patch and corresponding ground plane surface, and a cylindrical magnetic wall around the edge of the patch. Like the transmission line model, the basic cavity model does not take into consideration the effects of feeding the cavity. In addition, it is assumed that all electrical conductors are lossless; the magnetic conductor exists; the dielectric constant of the substrate filling the cavity is lossless; and there is no fringing of electric field at the edge of the antenna. By applying known theories relating to fields that exist in the cavity, and by using techniques outlined by Balanis,[24] the far-field radiation patterns can be calculated. In both the rectangular and circular cases, to take into account the feeding structure, numerical methods must be used to calculate the exact radiation pattern and the input impedance of the antenna. Fortunately, excellent textbooks[24-27] on the subject, as well as good software packages for numerical calculations, are available.

As far as applications in smart antennas are concerned, microstrip antennas have many attractive features, including their low profile, compact size, light weight, and reasonably good radiation characteristics. In addition, the antenna ground plane provides a convenient shield, which allows the mounting

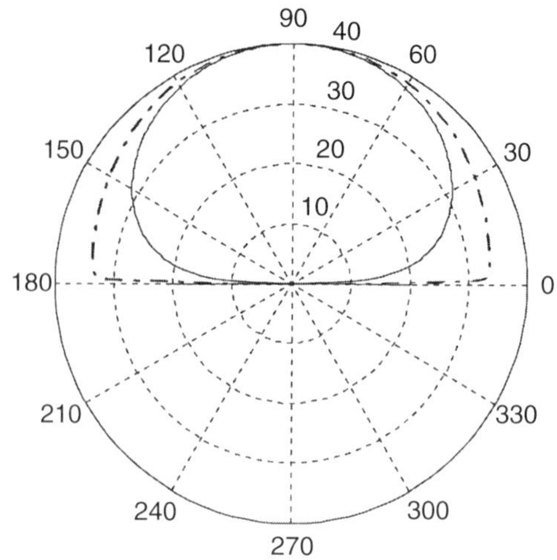

FIGURE 22.10 Typical theoretical radiation patterns of rectangular patch antenna:. E-plane pattern (dot–dash curve); H-plane pattern (solid curve).

of the associated electronic subsystems right behind the antenna structure. Consequently, a smart antenna system based on microstrip patches can be made very thin in profile. Such a smart antenna would be more pleasing to the eye as well as less threatening, especially if the radome that covers the microstrip array is painted over by non-metallic-based paint. Note, however, that microstrip antennas using a very thin substrate layer of high dielectric constant also have narrow bandwidth. They may also be seriously affected by surface wave coupling. These problems can be ameliorated by using substrates with very low dielectric constant (e.g., porous foam). In particular, microstrip antennas can be made wideband by increasing the spacing between the patch and the ground plane.[27-28]

As far as the requirement to have a broad beam width is concerned, both circular and rectangular patches with a finite ground plane may be used with some limitations. The theoretical E-plane and H-plane patterns of a rectangular microstrip patch are shown in Fig. 22.10. The patterns were calculated by using the software provided with the textbook on computer-aided design of microstrip antennas by Sainati.[25]

It is clear from Fig. 22.10 that the E-plane pattern is quite broad, with a 3-dB beam width close to 100°. The H-plane pattern, however, is much narrower. Because an infinite ground plane is assumed, there is no radiation for angles between 180° and 360°. In practice, one must take into account the effect of the size of the ground plane. In general, a small ground plane may cause fluctuation in field strength over the beam width of the pattern. In the H-plane, the infinite ground plane has a more sharply tapered pattern than the finite-size case. On the other hand, in the E-plane, the infinite ground plane gives the widest beam width.[27] The sharp cutoff of the H-plane pattern at ±60° from broadside makes a rectangular microstrip antenna suitable as fixed-beam sector antenna in cellular mobile communications, because it minimizes interference in adjacent sectors. The broad beam width in the E-plane also means that a vertical array of these patches can be used to generate the required shaped beam in the elevation plane.

However, both circular and rectangular microstrip antennas, with their relatively narrow beam width, especially in the H-plane, are not ideal as elements of the array in a smart antenna that is designed for a 120° or wider sector. As pointed out previously, when the main beam is scanned to the edge of the radiation coverage (i.e., to the edge of the sector), the tapered unit pattern would cause a serious loss in signal strength. More important, the relative strength of the main beam would be much weaker in relation to the sidelobes that fall near the broadside direction. Consequently, the ability of the smart antenna to discriminate against interference in the directions of these sidelobes would be seriously impaired. On the plus side, however, the microstrip antenna can operate in either linear or circular polarization mode. Moreover, the low profile of microstrip antenna arrays would make them less obstrusive, and hence more

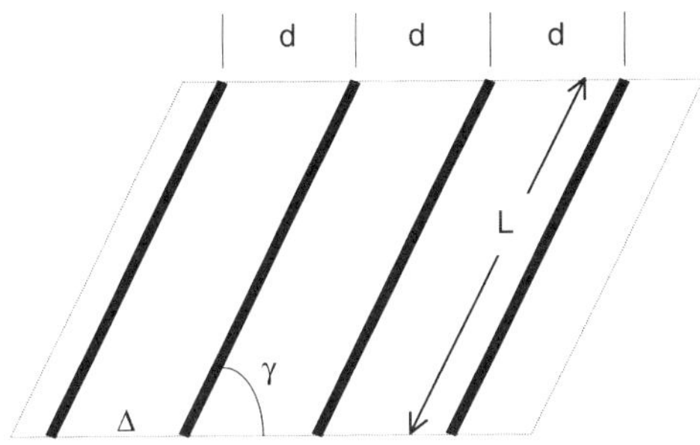

FIGURE 22.11 Construction of a quadrifilar helical antenna.

pleasing to the eye. Thus, for aesthetic reasons, microstrip antennas would be preferred to other types of antennas, including the quadrifilar helical antenna.

Quadrifilar Helical Antenna as Element of Array

A quadrifilar helical antenna (QHA) consists of four equally spaced conducting helical wires or tapes wound around a common cylindrical core. In addition, these helical wires are fed so that there is a 90° phase progression as one moves from one wire to the next around the circle. The quadrifilar helical structure can be easily constructed by first etching on a flat piece of malleable substrate four parallel straight lines that are equally spaced and at the correct pitch angle (γ). The next step is to wrap the substrate around a circular cylindrical core, the circumference of which is equal to four times the interspacing (d). With reference to Fig. 22.11, the axis of the cylinder is normal to the baseline (Δ).

QHAs have many desirable properties, which make them highly suitable as elements of an antenna array, especially when the main beam of the array must be scanned over a large angular range. Although the exact inventor of the quadrifilar antenna is not known, C. C. Kilgus was the first to demonstrate that a QHA can be designed to give near-hemispheric radiation coverage. Briefly, a QHA may be viewed as two bifilar helical antennas that are mutually orthogonal in space and fed in phase quadrature. Kilgus[29-31] showed that a half-turn bifilar helical antenna may be studied by analyzing a system consisting of a circular loop and a dipole antenna. When the two orthogonal bifilar helices are combined to give a QHA, the total electric field radiated by the system has both θ and ϕ components that have the same amplitude, but are in phase quadrature. Moreover, with a correct choice of the helical parameters, the shape of the radiation pattern can be made equal to a cardioid with maximum radiation in one direction of the helical axis and zero radiation in the opposite direction. In other words, a shape of the main beam is close to a hemisphere. As the θ and ϕ components of the electric field are in phase quadrature, the resultant electric field radiated by the QHA is circularly polarized. Figure 22.12 shows a polar plot of the theoretical radiation pattern in a plane passing through the axis of the QHA, where the radial lines give the relative strength of the radiation power in decibels.

In a QHA structure, each of the helices is fed at one end, whereas the other end can either be shorted or left open ended. The latter structure is the more practical one, especially in the development stage, because it allows easy trimming of the helical length (L) to optimize the antenna input impedance and other characteristics. Because the current amplitudes must be the same in all helices, and the phase distribution must have equal increments of 90° (i.e., the relative phase distribution must be 0°, 90°, 180°, and 270°) the feeding method can be quite complex.

Feeding by Using Balun. One compact feeding structure incorporates a balance-to-unbalance transformer (balun) to achieve the required amplitude and phase distribution. The main advantage is the relative simplicity of the design and the compactness and ruggedness of the resulting QHA. Bricker and Rickert[32] employed an infinite balun, which results in a very compact QHA structure. Considering the first bifilar of the QHA, the outer conductor of the coaxial feed is used to form its first monofilar, whereas the inner conductor is connected to its second monofilar. The second bifilar, which must be fed in phase quadrature with the first one, is constructed using the self-phasing technique. Briefly, the two bifilars are

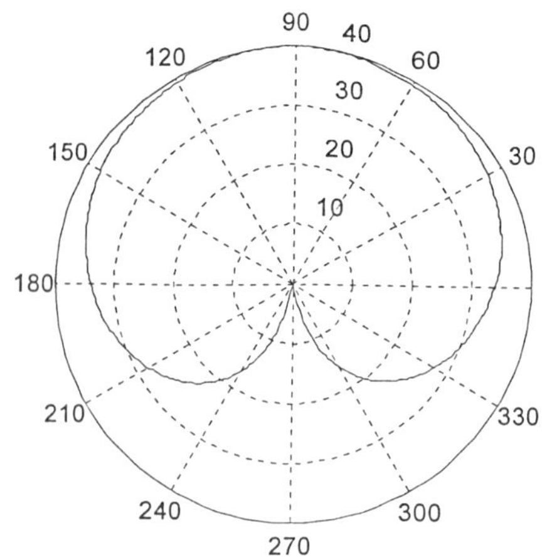

FIGURE 22.12 Theoretical pattern of a quadrifilar helical antenna (QHA).

fed in parallel, with the longer one made inductive by having its length longer than the resonant length. The shorter one, on the other hand, is made capacitive by having its length shorter than the resonant length. By adjusting the length of the large bifilar, one can make its normalized impedance equal to $z_L = (1 + j)$. Similarly, the length of the small bifilar can be designed to make its normalized impedance equal to $z_S = (1 - j)$. In other words, their normalized admittances are given by

$$y_L = \frac{1}{1+j}$$

$$y_S = \frac{1}{1-j}$$

As they are fed in parallel, their combined normalized admittance is the sum of y_L and y_S (i.e., equal to unity). The QHA therefore presents a matched load to the input coaxial line. The resulting antenna has a very broad beam, and its feeding structure is very compact. However, because one bifilar is larger than the other, the radiation pattern is not symmetrical about the axis of the helix. More importantly, as the first helix of the large bifilar is an integral part of the input coaxial cable, its length is fixed once the antenna is fabricated. As a result, this design provides no means for fine tuning the antenna performance by trimming its length.

Feeding Structure Using Power Divider. Another method of achieving the required current amplitude and phase distributions is to use a power divider. In this case, the feed may consist of a hybrid ring coupler and two 90° branchline couplers. The hybrid ring provides two outputs with equal amplitudes and with relative phases of 0° and 180°, respectively. Each of these two outputs is then connected to the input of one branchline coupler to provide two outputs that are 90° apart but with the same amplitude. To achieve a compact structure, the hybrid ring coupler may be stacked on top of the branchline couplers. The power-divider method allows the use of open-ended, equal length helices and therefore is capable of producing a radiation pattern that is symmetrical about the helix axis. It also allows the designer to fine-tune the antenna performance by trimming the open end of each helix. The measured radiation pattern of a typical open-ended QHA fed by a power divider is shown in Fig. 22.13, where the radial axis is the relative strength of the radiation power in decibels. There is a slight distortion of the pattern caused by small reflection from the feeding mechanism, but the pattern is almost hemispheric in the half plane above the horizon as required. The polarization purity is also good over the whole hemispheric coverage. As far as broad radiation coverage is concerned, the performance is certainly better than that achievable

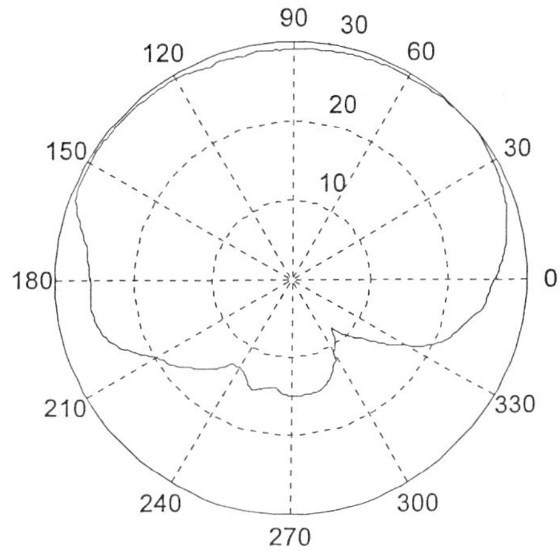

FIGURE 22.13 Measured radiation pattern of a typical QHA with open-ended helices.

with a microstrip patch antenna. In addition, the antenna input-impedance can be made very close to 50 Ω (i.e., with a voltage standing wave ratio (VSWR) equal to 1.1 or less).

In conclusion, for applications where a broad angular coverage is required, the QHA would give better performance than a microstrip antenna, except if linear polarization is the only option. The QHA radiates circular polarized waves, and would therefore lose 3 dB in signal strength if used to receive a linear polarized signal. Microstrip antennas, however, can be designed for linear or circular polarization. In addition, in the linear case, polarization diversity can be achieved by operating the antenna in either of two orthogonal polarized modes simply by switching the feed location. Thus, the choice depends on the dominant design requirements.

22.3.2 Radio Frequency Front-End Design Issues

The RF front end of a smart antenna is quite complicated, as it must carry out the necessary amplification and filtering as well as up- or downconversion to the required IF frequency. Because the main building blocks in the transmit and receive paths are similar, albeit in the reverse order, we concentrate our discussion on the receiving case. The main aim is to maximize the receiver sensitivity, and to minimize the effect of noise and out-of-band interference. Thus, the LNA must generally have low noise figure and high gain to ensure that losses in subsequent stages do not significantly affect the receiver noise figure. The filters, on the other hand, must be designed to reject out-of-band interference as well as to minimize the effect of signals in the image band. A good design also requires impedance matches at the input and output ports. However, odd-order intermodulation products and, in particular, the third-order ones that fall within the band occupied by the desired signal cannot be eliminated by filtering. These intermodulation products are caused by the mixing of strong signals on adjacent channels from nearby base stations, and can effectively mask a weak desired signal. To minimize the effect of intermodulation distortion, the gain of the LNA must not be too high. It is therefore clear that the emphasis on high gain and low noise in the LNA cannot be taken too far. This is particularly true because the equivalent antenna noise temperature is much higher than the ambient temperature in a typical urban environment. In spite of its complexity, the architecture of the RF front end of each channel of a smart antenna is very much the same as that of a normal mobile receiver or transmitter. Consequently, the task of the designer of the RF subsystem of a smart antenna is greatly simplified as a result of the widespread availability of low-cost RF components as well as integrated subassemblies. Thus, the whole receiver can be implemented by using commercially available radio frequency integrated circuits (RFICs), which would also simplify the task of preserving the relative amplitudes and phases of signals in different channels. Alternatively, a smart antenna model for laboratory demonstration purposes can be designed by assembling individual components that have been designed and optimized independently of one another. With such an approach, special care must be taken to ensure that all the channels are closely matched. In any case, a good understanding of the theory of RF engineering is essential for a successful design. In this respect, excellent textbooks on the subject are widely available.[18-19] Some of the more significant design issues are discussed in the following sections.

22.3.2.1 Low-Noise Amplifier and Filter Design Consideration

The smart antenna in a mobile communication system receives analog signals from the desired mobile handset as well as unwanted interference from other users. It must therefore process the signal in its RF front end to remove out-of-band interference without introducing extra noise and distortion before digitizing the signals for final processing in the DSP module. In general, to achieve a good performance, the RF front end should have a double-conversion superheterodyne architecture, but direct conversion to baseband may also be used if low cost is of primary concern, and lower performance is acceptable. To have a good reception under all operating conditions, the receiver must meet some minimum requirements, including a reasonably wide dynamic range. In other words, it must cater for very weak signals when the wanted mobile terminal is near the edge of the cell, as well as strong signals when the base station and the mobile terminal are close together. Because one may be dealing with multichannel systems, with strong interference from adjacent channels, the receiver architecture must be carefully designed to minimize the effect of the interference. In this respect, it is worth noting that by using a very low noise amplifier with very high gain, one can theoretically ensure a very low receiver noise floor. Nevertheless, one must also take into account the effect of intermodulation products that arises from the nonlinear input–output characteristics of the mixer when it is driven too hard. More specifically, when the signals on adjacent channels are very strong, a high gain LNA can result in unacceptable intermodulation product distortion in the mixer stage. Thus, there must be a trade-off between high sensitivities (by using LNA with a very high gain) and wide dynamic range of the receiver.

Another important issue is the location of the RF bandpass filter near the LNA. In all RF front-end subsystems, noise and unwanted signals having their frequencies fallen within the image-frequency band must be removed by filtering. Although in some receivers, the RF filter is placed ahead of the LNA, such a configuration would not protect the receiver from the LNA noise in the image-frequency band. Therefore, a better solution would be to place the filter between the LNA and the mixer, because the high gain of the LNA would ensure that losses associated with the filter would have less impact on the receiver noise. In practice, however, the problem is more complicated if commercial IC components are used. The reason is that it may be more convenient for manufacturers to integrate the LNA and the mixer into one single IC. On the other hand, to reduce cost, they may not provide a means for the insertion of the external RF filter.

Finally, note that with a careful choice of RF components or packages, it is possible to add desirable features to the smart antenna without incurring much higher costs. In fact, some RF front-end ICs also include switches, which may be used to implement system self-calibration. This may be done by connecting each LNA to a known test signal, so that the designer can measure the deterministic system errors for subsequent corrections in system calibration. Alternatively, the switches may be used for space diversity if deemed desirable. In this case, each LNA is switched to one of a pair of antennas that are located far enough apart to have significantly different performance. Similarly, polarization diversity can be achieved by using the switch to connect the LNA to the microstrip-antenna feed point that corresponds to the desired polarization.

22.3.2.2 Mixer Design Issues

In smart antenna systems, the mixer is one of the most crucial components, mainly because frequency translation is used in both the transmitter and receiver circuits. The main reason for using frequency translation in the transmitting case is to achieve a more efficient transmission of signals, and to reduce the antenna size by using a high-carrier frequency. Consequently, in the receiving case, the signal must be returned to its original low frequency. In the case of smart antennas, the DSP technology also dictates how low the frequency must be reduced before digitization and processing in the DSP module.

Excellent references on the mixer design and practical design tips are available in the literature,[22,33-35] but highlights of important issues are given here. As pointed out in Section 22.2, mixers can be designed to provide different degrees of performance. If low cost is the main design constraint, then one may adopt the single-diode unbalanced mixer, but this simple design requires external bandpass filters at both the RF and LO ports. In addition, because both RF and LO frequencies appear at the IF port, a filter is also required at the IF output to reduce unwanted noise or mixing products. On the other hand, if high performance is the deciding factor, a more effective solution is the double-balanced mixer. This requires a complex circuitry, but in return, it provides good isolation between the IF port and both the LO and RF ports as well as between the LO and RF ports. This mixer also rejects all even harmonics of both input signals.

It is also important to note that in cellular mobile communications, the RF port of the mixer may consist of signals from adjacent channels, which can result in unwanted products in the desired frequency band at the IF port. Because such image-frequency interference cannot be removed by IF filtering, a special mixer called image-rejection mixer may be used. It employs two separate balanced mixers, and by exploiting the phase difference between the desired signal and the unwanted product at the output of the mixer, it can achieve an image rejection greater than 20 dB. Other design issues are related to the noise introduced by the mixer, and nonlinear characteristics of its input–output curve, when it is driven close to saturation.

Conversion Loss

The conversion loss gives a measure of the efficiency of the mixing process, and is normally quoted in decibels in data sheets provided by manufacturers. In diode mixers, the signal is affected by many factors, which include impedance mismatch at RF and IF ports, losses associated with resistance and capacitance of the diode junction, and loss associated with the voltage drop across the diode junction. On the other hand, active mixers, which are widely available commercially, can be designed to give conversion gain as high as 20 dB. A popular active mixer is the Gilbert-cell mixer, but its high noise figure can significantly increase the receiver noise floor, even if an LNA with low noise figure is employed. As discussed previously, high RF gain may not be the solution, because it would degrade the receiver dynamic range.

Noise Figure

In principle, low noise figure is one of the most important aims in receiver design, because the lower the noise floor a receiver has, the better it is in receiving weak signals. Therefore, the noise figure of mixers is an important parameter in smart antenna design, and it is related to the conversion loss mentioned earlier. It gives a measure of the degradation the signal to noise ratio (SNR) caused by the mixing process. As a rule, the smart-antenna designer must aim for a low mixer noise, because a mixer with a poor noise figure can significantly affect the overall performance of the receiver. Manufacturers usually specify the mixer noise figure based on assumed values of the diode noise temperature and the IF noise figure. It is therefore important to ascertain the exact values they assume for these figures.

Intermodulation Products

As mentioned previously, in a cellular mobile system, two or more signals from adjacent channels may be picked up by the antenna and amplified by the LNA before being applied to the mixer RF port. The nonlinear characteristics of these devices may result in a large number of radio components called intermodulation products, the frequencies of which are combinations of harmonics of the input signals. In most practical cases, however, the third-order intermodulation product is more likely to fall within the desired signal frequency band, and affects the receiver performance unless a special measure is used to suppress it. One obvious method is to ensure that the intermodulation product stays below the noise floor by operating the devices within the linear range of the input–output curve. However, the problem is complicated by the fact that most receivers are required to operate over a wide range of signal strengths. Thus, if strong signals are present at their input, these devices may operate in the nonlinear part of their input–output curves, and third intermodulation products may grow quickly with increase in input signal strength.

In conclusion, due consideration should be given to the effect of mixer characteristics on the performance of the RF receiver. Manufacturers normally specify the 1-dB compression point (above which nonlinear effect becomes serious) and the third-order intercept point (above which intermodulation products become problematic).

22.3.3 Digital Signal Processor Module Design

In a smart antenna system, the DSP module plays a critical role, because a satisfactory system performance can only be achieved if a DSP module with efficient hardware and software can be designed. This is especially true if a complex adaptive algorithm must be implemented. Because our main targets are researchers at universities, and those who are interested in the development of a smart antenna platform for evaluating a variety of adaptive algorithms, the description given here puts more emphasis on the design principle. In particular, we emphasize the importance of parallel processing to achieve a fast response time. Thus, we view the two main adaptive functions, namely, forming the output beam and updating the weights as two independent tasks to be carried out in parallel. Moreover, because the data rate is very fast, the computations associated with the BFF need be carried out at high speed. Because the operation must be repeated for every new snapshot of data samples, it is essential to have the BFF software efficiently implemented in assembler code, even in the development stage. On the other hand, the rate of the weight update is much slower, mainly because the mobile environment remains practically unchanged within a millisecond or two. In addition, the WUF operation is highly complicated even if the chosen algorithm is relatively simple; it is therefore much more desirable to write the WUF software in a higher level language. In the development stage, this also makes it easier to implement a large variety of adaptive algorithms, and hence to compare their relative merit. The use of a higher level language also provides a more flexible way of monitoring the system performance, because visual display of the dynamic response of the system can be simply achieved.

It is therefore natural to divide the DSP module hardware into two separate sections. First, the BFF section should be implemented by using a high-speed DSP device that is designed to efficiently multiply individual channel signals with their corresponding weights, and to accumulate these products. Because these computations involve complex signals, with both I and Q components, a means must be used to generate them from signals derived from the antenna elements. Although a quadrature demodulator in the second downconversion stage may be used for this purpose, we only consider the generation of the Q signals by software in the discussions given later. Second, in the development stage, the WUF operation should be carried out in a higher level language by using a PC with a high-speed Pentium. In fact, the speed of current generation of Pentium is fast enough to achieve close to real-time operations. The main attraction of this approach is flexibility, not just in implementing a particular algorithm, but also in comparing the relative performance of different algorithms. Moreover, once the smart antenna system has been satisfactorily tested in the laboratory, the personal computer (PC) can simply be replaced by a suitable DSP device. The translation of the software from higher level language to the assembler code of the chosen DSP should pose no problems, because highly efficient translators are available from their manufacturers.

Because of limited space available, discussions given here are necessarily short, and concentrate mainly on important issues. However, these discussions are based mainly on smart antenna research by the author and his students at the University of New South Wales in Sydney, Australia. More detailed information may therefore be obtained from Master and Ph.D. theses given in the list of references.[10-13]

Briefly, two main designs are described in some detail. The first is called a concept demonstration model, a rather simple model, which implements the BFF and WUF operations in parallel in a straight-forward manner as outline earlier. The second model represents a more serious approach to smart antenna design, which is used to demonstrate the exploitation of parallel processing to improve the system throughput when more advanced DSPs are available.

22.3.3.1 Design of a Concept Demonstration Model

For a simple model designed to demonstrate the basic smart antenna concept, the cost is of prime consideration, and high-performance DSP does not need to be contemplated. In addition, the RF front end should be designed by using low-cost but reliable components that are freely available commercially.

From the architectural point of view, the conventional superheterodyne receiver with a two-stage downconverter may be replaced by a simpler receiver using direct downconversion. In other words, the receiver requires only one local oscillator, the signal from which is fed in-phase to all the mixers in individual RF channels via a power divider network. The signals from all RF channels are then sampled, digitized, and processed in the DSP module to produce the desired output. Moreover, the quadrature demodulator depicted in Fig. 22.3 does not need to be used, because the DSP module can generate the required quadrature component by software via the Hilbert transform. By using this model, the principle of target tracking and interference nulling can also be easily demonstrated using analog signals such as amplitude modulation (AM) or frequency modulation (FM) speech signals, without having to downconvert the antenna signals to baseband before digitization. This approach is particularly attractive to researchers at universities, because signal generators with facilities for analog modulation are more widely available.

As mentioned previously, the adaptive processor must carry out two distinct functions. The first, called WUF, is to process newly arrived data by implementing a given algorithm to generate a new set of weights that maintain a maximum SIR as the mobile environment changes. The second, called BFF, is to produce the desired output by multiplying the desired weights with corresponding complex signals from individual channels, and combining these products. These two functions can be carried out either by one fast DSP, or by two slower DSP working in parallel.

For the simple model under discussion, the WUF operation may be carried out at a much slower rate than in real systems. The parallel architecture is therefore adopted, but instead of a second DSP, a desktop PC is used to carry out the complex task of calculating the new weights using a higher level language. This approach makes it much simpler to compare the performance of different algorithms. Another advantage is that the PC, with the aid of sophisticated software packages like MATLAB®*, also provides efficient facilities for displaying the results. Consequently, other tasks, such as system calibration, monitoring the performance of individual channels, etc. can be easily performed during the system development phase.

The discussion given here is based on one of the earliest smart antenna models developed in the early 1990s at the University of New South Wales. The brain of this model is a Texas Instruments TMS320C30 DSP, which was one of the most versatile DSPs at that time. Next is a brief discussion of both the system hardware and software. A more detailed description of this model can be found in an M.E. thesis by Jonas.[12]

Overall System Hardware Description

The architecture of this model has three distinct sections. First is the microwave antenna array, which consists of a linear array of eight L-band antennas designed to provide a full 180° scan coverage in azimuth. The second section is composed of the eight RF heterodyne receivers, including a Wilkinson microwave power divider, LNA stages, mixers, LO, and assorted filters. The final section is the DSP module, where the data acquisition board interfaces the RF receivers with the DSP. For compactness, the eight heterodyne receivers from eight array elements are etched on the same dielectric substrate, which is mounted directly behind the ground plane of the antenna array. Moreover, the ADC multiplexers and control logic circuitry are all combined into a single DAM.

In the RF section, to achieve good frequency stability in the microwave LO, a microwave PLL frequency synthesizer is also used. These synthesizers must be correctly designed to minimize their phase noise. For simplicity of design, a single-loop PLL frequency synthesizer is adopted. A monolithic double-balanced microwave mixer from Mini-Circuits is chosen for the downconverter. The LO signal is distributed to individual mixers via an eight-way Wilkinson power divider. The substrate on which the power divider is etched is also used to mount the RF and IF components. The LNA stages also use Mini-Circuits monolithic amplifiers.

The chief component of the data acquisition module is the CLC922 ADC. That is, a single high-speed converter is selected for all eight channels, to reduce cost, to simplify the design, and to achieve compactness. Sample-and-hold amplifiers (S/H) are inserted between array inputs and the CLC922 ADC, which

*Registered trademark of the MATH WORKS Inc., Natick, MA.

has 12-b accuracy, 10 million samples per second (Msps) sampling speed, with large dynamic range and wide processing bandwidth. At the command of the DSP, the S/Hs take a snapshot across all the eight channels by holding the signals simultaneously to preserve their relative phase information. Multiplexing devices then sequentially select each channel in turn for digitization. The S/Hs selected are the AD684 quad S/Hs from Analog Devices, which have a 4-MHz small-signal bandwidth, with 12-b accuracy.

DSP Module Design

A PC houses the DSP module hardware, which comprises two boards, the Texas Instruments XDS500 hardware Emulator and the TMS32OC30 Target board. The Emulator provides a platform for high-level software and hardware programming and debugging, and performs full speed emulation of the TMS32OC30.

Hardware Design Issues. As mentioned before, the heart of the DSP module is the Texas Instruments TMS32OC30 DSP (hereafter referred to as C30), which is the Texas Instruments third-generation member of the TMS320 family of digital signal processors. The C30 used in the model is a CMOS 32-b floating-point device with a 60-nsec cycle time. Its ability to perform MAC operations in parallel is particularly attractive for the purpose of the BFF. Other features include two 1 K × 32-b single-cycle, dual-access random access memory (RAM) blocks; on-chip DMA controller for concurrent I/O and central processing unit (CPU) operation; 64 word instruction cache; circular address mode; and two internal programmable timers.[36]

The DMA controller is a very useful feature in the C30, because it allows reading from or writing to the memory or data bus independently of CPU operation. In other words, while new data are being transferred by the DMA controller, the C30 continues to process previous data, and is not slowed down by I/O operations. However, this does impose extra overheads, because the DMA source, target, and countervariables must be initialized before any background data transfer can take place.

The TMS32OC30 controls the data acquisition circuitry via its expansion bus, with external hardware logically seen as memory locations. Thus, the data transfer, channel selection, sampling, and digitization are all controlled by the C30 via the DAM. The DAM also has two DAC. If the input is analog FM signal, the DAC can be used to convert the smart antenna output to analog form for demodulation to verify the quality of the signal after being processed by the smart antenna. The DAC used is the AD767 from Analog Devices, and is a 12-b DAC with a small signal bandwidth of 1.0 MHz.

Software Design Issues. In a smart antenna system, the software plays a critical role, and therefore must receive careful attention. In principle, for fast and efficient operations, the control algorithm must be implemented in assembler code. However, for systems using complex algorithms, the development of the necessary assembler code would be very time consuming. In addition, it is more difficult to see the direct effect of varying a design parameter on the system performance. Thus, a more flexible design is to adopt an assembler code for the straightforward tasks required by the BFF, whereas the more complex adaptive control algorithm required by the WUF may be implemented in the PC by using MATLAB. This is the approach adopted in developing the concept-demonstration model discussed in this section. In any case, optimization of the C30 code for the BFF module is very important, considering its relatively slow speed compared with more advanced devices currently available. In fact, the maximum sampling rate is directly related to the code efficiency. Thus, serious consideration should be given to the use of DMA for data transfer from the DAM to the C30, and the judicious use of DSP arithmetic and address registers to minimize the need for branch operations.

Dual-Port RAM (DPRAM). Although in principle, the tasks required by both the BFF and WUF modules can be performed on the C30, this effectively halves the bandwidth of the BFF output. It is therefore better to have the BFF and the WUF carried out as two separate functions, with the C30 exclusively performing the BFF operation in real time. On the other hand, the slower WUF operation is performed on the PC in near real time.

Thus, in this simple model, the C30 acquires the signal through the data acquisition module, and performs the BFF to achieve an almost interference-free output signal. On the other hand, the PC periodically uses the input data to recalculate the weights required to maximize the SIR. Consequently, a mechanism must be available for the PC to access the input data from the C30, and to transfer the newly calculated weights back to the C30. This mechanism is provided by the DPRAM.

With our model, the DPRAM is physically located on the C30 application board, and is mapped onto the upper conventional RAM area of the PC memory as well as onto the C30 memory map. In other words, the DPRAM is accessible by both the C30 and the PC. The C30 periodically extracts the updated complex weights placed in the DPRAM by the PC. It also transfers a new block of samples to the DPRAM for the PC to collect and to generate new complex weights. The transfer of data between the two processors, via the DPRAM requires special handshaking procedures to avoid errors when both devices attempt to read or write to the same memory location at the same time. In addition, a special synchronization scheme is required, because the data transfer between the DPRAM and the PC is generally much slower than between the C30 and the DPRAM. Synchronization was accomplished via a semaphore register, so that while one processor is reading from, or writing to the DPRAM, the other processor is barred from any access.

After each set of complex weights has been calculated by the PC, their real and imaginary components are transferred to the DPRAM for the C30 to read and use in the BFF. Thus, with an eight-element array, 32 memory bytes must be reserved for them. New data blocks required by the PC to update the weights are also stored here, as well as other parameters transferred from the PC. One is the user-defined weight-update rate, which is stored in timer format for the C30 to determine how often to send a new sample block to the DPRAM, and to read new weights calculated by the PC. Next are the parameters required for the Hilbert transform (i.e., Hilbert coefficients and the Hilbert length chosen by the user). The C30 uses this information to generate quadrature components of the signal by software via Hilbert transform.

BFF Software Development. In the BFF mode of operation, two timers are used to control the sampling rate and the weight update rate. The weight updates are required to operate at between 20 and 200 Hz, whereas the sampling frequency is around 100 to 200 kHz. The sampling interrupt initiates the BFF cycle on new input data but using the same old set of weights, except when the PC updates the weights. Thus, the sampling-interrupt loop can operate at more than 1000 times the speed of the WUF loop. To track a fast-moving target, a fast weight-update rate must be used. This may be achieved by optimizing the software, but ultimately, it is limited by the complexity of the tracking algorithm as well as the speed of both the C30 and the PC. For the simple example, where the smart antenna only has to lock its beam onto the desired moving target, a high-speed Pentium can quickly update the weights, so that the target can be tracked almost in real time. This is particularly true if the scanning resolution is reduced, or if the antenna makes use of the fact that the new position of the target is close to its previous location.

The speed of the BFF operation in the C30 depends on the required length of the assembler code, which may be determined from the critical tasks involved. First, the C30 must command the S/H to simultaneously sample and then hold data in all eight channels until they are digitized and read by the C30. It then must set the DMA source, destination, and counter registers for data transfer, and start the DMA transfer of samples to internal C30 RAM. Finally, it must carry out the MAC operations and Hilbert transformation to provide the desired smart antenna output. The total code length therefore depends on the number of elements in the antenna array as well as the length of the Hilbert transform. By knowing the clock speed of the DSP (C30 in this case), the maximum sampling rate can then be determined.

WUF Software Development. For this task, the C30 is required to transfer to the DPRAM a number of consecutive samples from each of the eight data channels, with the exact data length determined by the weight-update algorithm. For example, if the task required is simply to scan the beam in azimuth to determine the direction of the strongest signal, then one sample from each of the eight channels would be sufficient. On the other hand, if an adaptive tracking algorithm is performed to minimize the interference, then a larger block of samples from each data channel would be required. It is noted here that the result may be highly sensitivity to direct current (DC) offset errors in the receiver output, because they result in a nonsymmetrical distribution of power in the positive and negative cycles of the carrier.

The data therefore vary according to where the waveforms are sampled. The important issues here are transfer synchronization, transfer speed, and data block size.

Software Development Using MATLAB. As mentioned earlier, the PC is responsible for the WUF operation, and the actual tools used for the calculation are provided by the MATLAB software package. It is noted in passing that, Turbo C may be used to develop the control algorithm; however, this approach is not ideal if extensive use of graphic display is required as in the development stage of a smart antenna project. On the other hand, MATLAB, with its extensive graphic tools, provides the ideal solution, because the user is now free to concentrate on the critical task of developing signal processing software, without having to worry about detailed graphic manipulation.

MATLAB/C30 Interface. Because both the C30 and the PC have access to the Dual-Port RAM, the MATLAB Executable (MEX) file called by MATLAB must be able to read and write to the Dual-Port RAM correctly. To achieve a satisfactory handshaking between the C30 and PC, two alternative methods are available.

The first is to define two Dual-Port RAM locations as control bytes, one exclusively for the C30 to write to, and the other for the PC alone. Simultaneous writes to the same location can therefore be avoided. These two control bytes can be allocated in the DPRAM memory map as PC interrupt and C30 interrupt, respectively. Each time the C30 finishes a task or is ready to accept a transfer of data, it can interrupt the PC, and vice versa.

The second method is to use the semaphore registers on the C30 application board. These registers are mapped to specified I/O locations, which can be set in one of three states: controlled by the PC, controlled by the C30, or free state. Before accessing a DPRAM memory location, the C30, for example, must try to gain control by writing a 0 to the semaphore register. If it subsequently reads back a 0, then this means that it has succeeded in gaining control. The other device (i.e., the PC) will then be denied access to the DPRAM until the C30 successfully releases its control. It does this by writing a '1' to the semaphore register. This process ensures that only one device will be able to access the Dual-Port RAM at any one time, and guarantees an error-free data transfer between the C30 and the PC.

22.3.3.2 Parallel Processing Using More Advanced Digital Signal Processor

In this section, we show that parallel processing can be exploited to improve the speed of system response. The basic issue is that the system throughput must be fast to implement in real time the appropriate adaptive algorithm. Consequently, the weights used in producing the smart antenna output are optimum for the mobile environment that the system is currently facing. Although the easiest solution is to use the fastest available DSP, in practice, the cost factor tends to play a dominant role. Therefore, the most cost-effective solution requires an intelligent design approach, which exploits the characteristics of the DSP to improve the system performance. We therefore discuss a new model based on a more advanced DSP, which provides a higher degree of parallel processing. Thus, the main design issues involve the DSP module, but the generation of the I and Q channels is briefly examined.

Generation of I and Q Channels

The RF front end for practical smart antennas should adopt a two-stage downconversion to achieve good performance. Most of the remarks given in the previous section on LNA, filters, and mixers also apply here. A more important issue, however, is whether to use the Hilbert transform in the DSP module, or to adopt quadrature demodulation in the RF section to generate the quadrature component. Each method has its own advantages as well as disadvantages in terms of hardware and software complexity.

In principle, to generate the I and Q components of the input signal, the quadrature demodulator requires two subcarriers of the same frequency but 90° out of phase. These I and Q components must then be digitized before being processed in the DSP module. Thus, apart from a more complex RF section, the use of quadrature demodulation also results in doubling the amount of data acquisition hardware, as well as twice the number of channels that must be processed by the DSP. In other words, there are now 2*N* (where *N* is the number of elements in the array) analog channels, instead of *N* channels if the Hilbert transform is performed by software.

On the other hand, Hilbert transforming the digital input signals from the DAM by software inevitably slows down the BFF in the DSP module. One may argue in favor of the software approach that by approximating the array weights as non-time-varying quantities, the extra load is minimum because only two Hilbert transform operations need be used for each snapshot of array output.[12] However, the increased workload in the DSP module per sample only represents part of the picture. The reason is that the amplitude response of the Hilbert filter is good only at a frequency close to $f_s/4$, where f_s is the sampling frequency, especially if the filter length is short. In fact, computer simulation shows that a minimum of five samples per period may be needed for a good Hilbert transformation using a 15-tap FIR filter.[13] In other words, a much higher sampling rate would be required if the software approach to the Hilbert transform were adopted. By contrast, if quadrature demodulation is used, the normal Nyquist rate of two samples per period is adequate. Nevertheless, in the following discussions, we assume that the I and Q channels are generated by software via the use of the Hilbert transform.

DSP Module Design Issues

The new model is based on a more advanced DSP, the Texas Instruments TMS320C40 (hereafter referred to as C40). The most important improvement comes from the reorganization of the DSP module. As in the previous model, the ADC is driven by the convert signal, which is derived from the C40 clock, the strobe, and R/W signals. The main signal processing functions are again the BFF and WUF. As a rule, the mobile environment does not change significantly within a few sampling periods, so a new set of weights is not required for each new snapshot of input data. On the other hand, the speed of the BFF operation is critical in determining the maximum sampling rate, and hence the maximum bandwidth of the system. For this reason, we concentrate on exploiting the higher degree of parallel processing provided by the C40 to improve the throughput of the BFF module. For convenience, it is again be assumed that the algorithm-dependent WUF is performed by a PC. To minimize delay caused by the slow data transfer across the PC buses, an extra DSP is used as a buffer between the PC and the C40 allocated to the BFF. In a real system, the PC would be replaced by another DSP dedicated to the WUF operation.

Development Platform. Although more powerful DSP devices are now available on the market, the C40 is chosen mainly because it is versatile enough to illustrate the principle of parallel processing. This section mainly highlights important issues involving BFF design, but more detailed descriptions of this smart antenna model can be found in a Master's thesis by Lam[13] and a Ph.D. thesis by Chen.[10]

Briefly, the C40 is a floating-point processor, designed for parallel processing. There are six communication ports for high-speed interprocessor communication; six-channel DMA coprocessor for concurrent I/O operations and CPU; two identical external data and address buses; and separate internal program, data, and DMA coprocessor buses. In addition, the CPU is capable of 176 million or more operations per second (MOPS) and 204.8 Mbytes/s throughput, and memory access performance is increased thanks to on-chip cache and dual-access, single-cycle RAM.[37]

For our system, the C40 processors reside on the TMS320C4x Parallel Processing Development System (PPDS) from Texas Instruments. The PPDS is a development platform, with four interconnected C40 processors, an expansion bus connector for external interface to the shared global memory bus, and eight external communication connectors for connecting off-board C40s and external peripherals to the PPDS C40s.

Internal communication between the four onboard processors is achieved through interconnected communication ports, with each processor directly connected to the other three. All arbitration and handshaking is handled internally by the built-in port arbitration unit (PAU). To avoid conflicts in accessing a shared memory, accesses by the processors are coordinated by the global bus controller (GBC).

Parallel Processing Architecture. The system described here is not meant to optimize the utilization of all the available C40 processors, but simply to illustrate the benefit of adopting a higher level of parallel processing in the BFF section. Therefore, communications between the PPDS and the PC is maintained via a C40 on another board, the HEPC2-M, which is a PC-AT plug-in card from Traquair Data Systems. The upshot is that the DSP module now consists of the PPDS, the HEPC2-M, and the PC, with three different languages used. In other words, the C40s on the PPDS and HEPC2-M, are programmed in assembly language, whereas the PC/HEPC2-M interface is in another language such as C. As in the C30-based model, the WUF algorithm is implemented in MATLAB.

In principle, parallel processing is most suitable when a number of tasks can be mostly processed independently of one another. As far as the BFF of this model is concerned, there are three independent subtasks, which may be processed in parallel. First, input data must be transferred from the ADC to the DSP module, with all timing and control signals obtained through the expansion bus on the PPDS. Second, the data from all the channels must be combined with the desired weights through the MAC operations. Finally, the Hilbert transform by software is required because, as mentioned previously, quadrature demodulation is not implemented. It is also noted here that the weights are assumed time invariant, so the Hilbert transform operation need be carried out only twice for each snapshot of samples. Consequently, the MAC together with a data-transfer operation require roughly the same number of instruction cycles to complete as the Hilbert transform operation. They can therefore be efficiently carried out in parallel by two of the four resident C40s.

Thus, the parallel-processing scheme is implemented by having one of the four C40s on the PPDS responsible for both data acquisition and MAC calculations (referred to hereafter as the data processor). It controls the sampling operation and reads in digitized samples through the expansion bus. It also gets complex array weights from the PC via the HEPC2-M board, and generates MAC results of the array samples. A second C40 on the PPDS (referred to hereafter as the Hilbert transform (HT) Processor) then performs the Hilbert transform on the MAC results received from the data acquirer to generate the corresponding beam output.

The transfer of digitized antenna samples from the DAM to the data processor internal memory is handled by the built-in DMA coprocessor in the background. Two separate RAM blocks are allocated for the DMA coprocessor to store data, and for the CPU to read data from, respectively. This allows the DMA coprocessor and the CPU to alternate their use of the two RAM blocks after each sampling period.

The upshot is that the two C40 on the PPDS are utilized at their maximum speed, without having to wait for the each other. This together with the careful use of parallel execution of some instruction pairs, more than doubles the speed of the BFF operation compared with that achievable without using parallel processing.

Finally, in the receiving case, two orthogonal linear arrays may be used to achieve a two-dimensional array with a pencil beam. In such a case, the system described earlier can be very efficiently utilized, because the remaining two C40 processors on the PPDS may now be dedicated to the second array. The C40 on the HEPC2-M is also more fully utilized in this case.

In conclusion, it is mainly the design of the DSP module that is influenced by the choice of new signal processors. In fact, we have also investigated smart antennas based on Texas Instruments TMS320C80 processors, where parallel processing is carried out to an even higher degree. Nevertheless, the basic principle of hardware design is the same. A brief description of the software developed for the TMS320C80-based antenna system may be found in a Ph.D. thesis by B. Xu, a member of my old research group.[11]

References

1. Rappaport, T.S., *Smart Antennas: Adaptive Antennas, Algorithms, and Wireless Position Location: Selected Readings*, IEEE Press, 1998.
2. Liberti, Jr., J.C., and Rappaport, T.S., *Smart Antennas for Wireless Communications: IS-95 and Third Generation CDMA Applications*, Prentice-Hall PTR, 1999.
3. Okamoto, G.T., *Smart Antenna Systems and Wireless LANs*, Kluwer Academic Publishers, Dordrecht, the Netherlands, 1999.

4. Vu, T.B., and Jonas, G., A DSP-Based Direction Finder, *International Journal of Electronics*, 73, 1133, 1992.
5. Vu, T.B., and Jonas, G., A DSP-Based Direction Finder, *Proc. Int. Symp. Antennas and Propagation*, Sapporo, Japan, 1992, 1093.
6. Vu, T.B., Jonas, G., and Millet, T., An Experimental DSP-Based Tracking Array, *Proc. Progress in Electromagnetics Research Symposium*, Pasadena, CA, 1993, 594.
7. Vu, T.B., et al., A Tracking Antenna Based on a Parallel DSP Architecture, *Proc. Int. Symp. Personal, Indoor and Mobile Radio Communications,* Taipei, Taiwan, 1996.
8. Chen, H., et al., A DSP-Based Adaptive Array for Mobile Communications, *Proc. Asia-Pacific Conference on Communications* (*APCC '97*), Sydney, Australia, 1997, 1111.
9. Slifkin, M., Mayan, Z., and Globinsky, Z., A DSP VHF FM push-button receiver, *Electronics World*, p. 802, 1999.
10. Chen, H., *Interference cancellation using smart antennas in TDMA mobile communication systems*, Ph.D. thesis, School of Electrical Engineering, University of New South Wales, Sydney, Australia, 1999.
11. Xu, B., *Smart antenna interference suppression techniques for mobile communications*, Ph.D. thesis, School of Electrical Engineering, University of New South Wales, Sydney, Australia, 1999.
12. Jonas, G., *High speed digital signal processing application for a multi-target tracking antenna*, Master of Engineering thesis, School of Electrical Engineering, University of New South Wales, Sydney, Australia, 1995.
13. Lam, N., *Smart antennas for GSM base stations using a parallel DSP architecture,* Master of Engineering thesis, School of Electrical Engineering, University of New South Wales, Sydney, Australia, 1998.
14. White, I., UHF Technology for the Cordless Revolution, *Electronics World + Wireless World*, 99, 542, 1993.
15. White, I., UHF Technology for the Cordless Revolution, Pt. 2, Systems on a Chip, *Electronics World + Wireless World*, 99, 657, 1993.
16. Smith, J.R., *Modern Communication Circuits*, WCB/McGraw-Hill, New York, 1998.
17. Hagen, J.B., *Radio-Frequency Electronics*, Cambridge University Press, 1996.
18. Ha, T.T., *Solid-state microwave amplifier design*, John Wiley & Sons, New York, 1981.
19. Collin, R.E., *Foundations for Microwave Engineering*, McGraw-Hill, New York, 1992.
20. Reynolds, D.L., LNA/Mixer ICs ease wireless receiver design, *Microwaves & RF*, 32, 99, Nov. 1993.
21. Maas, S.A., *Microwave Mixers*, Artech House, Norwood, MA, 1986.
22. Hosking, M., New Wave Microwaves, Pt 7: Mixers and Signal Conversion, *Electronics World + Wireless World*, 100, 829, 1994.
23. Vu, T.B. On Maximum Inter-Element Spacing in Linear Array, *Proc. Progress In Electromagnetics Research Symposium* (*PIERS*), Noordwijk, the Netherlands, 1994, 435.
24. Balanis, C.A., *Antenna Theory, Analysis and Design*, John Wiley & Sons, New York, 1997.
25. Sainati, R.A., *CAD of Microstrip Antennas for Wireless Applications* (*software included*), Artech House, Norwood, MA, 1996.
26. James, J.R., and P.S. Hall, eds., *Handbook of Microstrip Antennas, Vols. 1 and 2*, Peter Peregrinus, IEE, London, 1989.
27. Zurcher, J.F., and F.E. Gardiol, *Broadband Patch Antennas*, Artech House, Norwood, MA, 1995.
28. Pham, M.S., *Analysis and design of wideband microstrip antennas for mobile communications*, Master of Engineering thesis, School of Electrical Engineering, University of New South Wales, Sydney, Australia, 1998.
29. Kilgus, C.C., Multi-element, fractional turn helices, *Transactions Antennas and Propagation*, 16, 499, 1986.
30. Kilgus, C.C., Resonant quadrifilar helix, *Transactions Antennas and Propagation*, 17, 349, 1969.
31. Kilgus, C.C., Resonant quadrifilar helix design, *Microwaves Journal*, 13, 49, Dec. 1970.

32. Bricker, R.W., and H.H. Rickert, "S-band resonant quadrifilar antenna for satellite communications," *RCA Engineer*, p. 70, Feb./Mar. 1975.
33. Carr, J., RF receiver mixers, *Electronics World*, p. 220, 1999.
34. Carr, J., RF mixing, *Electronics World*, p. 306, 1999.
35 Carr, J., RF mixers, *Electronics World*, p. 379, 1999.
36. *TMS320C3x User's Guide*, Texas Instruments, January 1997.
37. *TMS320C4x User's Guide*, Texas Instruments, June 1992.
38. *TMS320C4x Parallel Processing Development System*, Texas Instruments, June 1992.

23

Phased Array Technology for Wireless Systems

Hiroyo Ogawa
Ministry of Posts and Telecommunications

Hiroyuki Tsuji
Ministry of Posts and Telecommunications

Ami Kanazawa
Ministry of Posts and Telecommunications

Ryu Miura
Ministry of Posts and Telecommunications

Masato Tanaka
Ministry of Posts and Telecommunications

23.1 Introduction

There have been increasing demands for phased-array antenna systems that can improve system performance as well as expand its functions. A lot of architectures for phased-array antennas have been investigated for use of radar and communication systems because of the attractiveness of beam steering without mechanical elements. In this chapter, the phased-array antenna systems for mobile, stratospheric, and satellite communication systems are described.

23.2 Phased-Array Antennas for Land Mobile Communication Systems

The use of array antennas has played an important role in mobile wireless communications as the popularity of personal communication services such as cellular phone has increased and the demand for personal and multimedia services (e.g., mobile Internet access) has also increased. There are several reasons that array antennas are introduced into wireless communication systems. The first is to increase

0-8493-0124-6/02/$0.00+$1.50

the capacity of the channel while maintaining its quality. As the number of terminals increases, techniques for increasing the number of available users in the same cell by reducing the deterioration in signal quality are required. The second reason is to respond to the growing interest in using wireless access systems to extend broadband network service to wireless terminals. The demand for multimedia services, fast access to large databases, and fast access to the Internet is increasing more than ever. Techniques that can provide high-data-rate wireless communication between terminals are therefore needed. In developing high-grade multimedia services, it is hard to solve these problems by using conventional techniques, such as equalizer and diversity techniques. Another reason is to reduce the power transmitted to reduce the drain on the terminal battery. Although mobile terminals tend to be smaller, the life of their batteries should be longer. Reducing the power transmitted can also reduce the signal to interference and noise ratio (SINR) and thereby make it possible to reduce the bit-error rate (BER) to an acceptable level.

The development of the digital signal processor (DSP) is also one of the reasons. The performance of the DSP has been vastly improved in recent years, so many algorithms needed for array signal processing can be implemented.

The application of the code division multiple access (CDMA) system, which has been used in standards like the IMT-2000, has received much attention. Also, the use of an array antenna at the base station in a CDMA system increases the reverse link budget and improves system performance. CDMA cellular systems have difficulty with the reverse link for several reasons. One of the reasons is that the power transmitted from each mobile terminal must be controlled to compensate for the near–far effect. If the power control is not performed correctly, multiple access interference increases. Reducing the levels of multiple access interference therefore increases the capacity of a CDMA system.

Array antenna systems already have been used in communication systems. For example, an adaptive array antenna is used at the base station of the personal handy phone system (PHS) in Japan. The PHS uses the π/4-shifted QPSK modulation method and the time division duplex (TDD) multiple access system. The purpose of the array antenna in the PHS is to form antenna beams in the direction of the desired signals and to form antenna beam nulls in the direction of the interfering signals to improve the quality of the communications signals. An actual array antenna system, developed by Kyocera Corporation, is introduced here. The antenna elements are located circularly at a distance of seven times the wavelength of the carrier frequency. As shown in Fig. 23.1, the array antenna system consists of several units. Because there are four time slots in one frame and one DSP is allocated for each time slot, four DSPs are used to control the antenna beams. Figure 23.2 shows an antenna beam pattern obtained

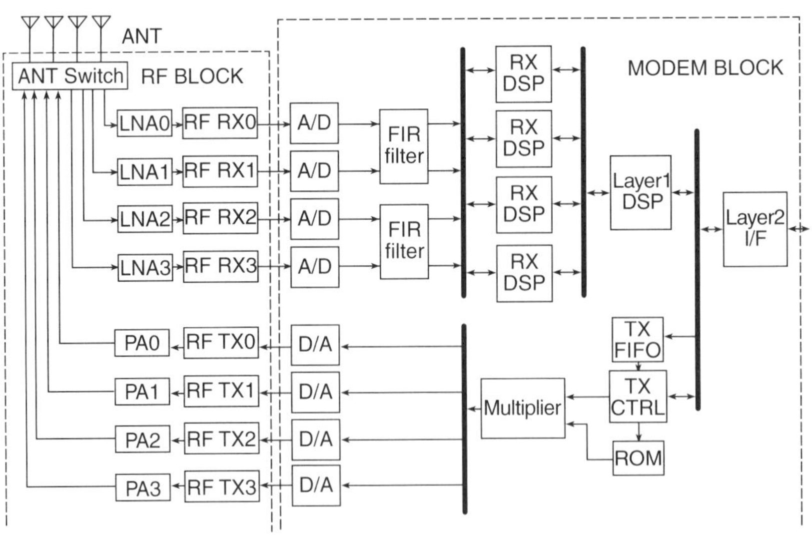

FIGURE 23.1 Structure of the PHS base station antenna.

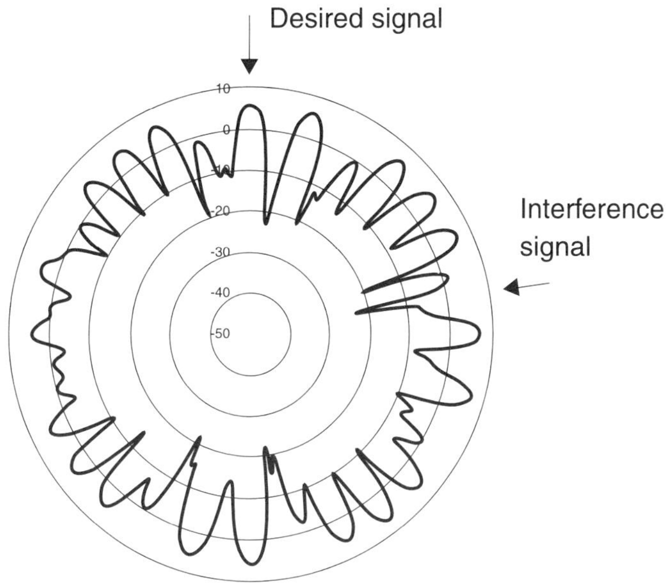

FIGURE 23.2 Beam pattern of the four-element array.

experimentally when a desired signal and an undesired signal impinged on the array. It is reported that the beamforming of the array antenna can reduce the interference power by about 16 dB and can halve the number of the channel changes or handovers in the system.[1]

The purpose of this section is to show the practical application of array antennas and their efficiency in wireless communications. We focus the applications of the array antennas in macrocell and microcell cellular systems. We first discuss some problems in mobile wireless communications and some compensatory techniques. We then discuss techniques for improving the channel quality in macrocell and microcell systems. After explaining equipment developed for array antennas in a macrocell system, we show simulation results demonstrating the increased channel capacity. Results obtained with experimental equipment are also shown. In Section 23.2.2, we discuss an array antenna for microcell systems. A new antenna beam-tracking algorithm enables the effect of multipath channel to be reduced by forming antenna beams. We also show the experimental results obtained using that antenna.

23.2.1 Development of an Array Antenna for Macrocell Control

Adaptive array antenna controls are expected to increase the channel capacity of mobile communication systems by reducing cochannel interference and multipath fading, and here we discuss the effect of using such antennas at the base stations (BS) in macrocell systems. The cell radius in a macrocell system is usually more than 1 km. Figure 23.3 shows the conventional cell arrangements used in macrocell systems. A BS covers an omni area or a sectored area. In actual systems, however, the amount of traffic varies from BS to BS because the distribution of users varies geographically and temporally. Therefore, a more effective zone configuration would therefore be expected to improve the efficiency with which the available frequency spectrum is utilized. Figure 23.4 shows the ideal zone configuration for a macrocell system. Such a configuration is obtained by using a dynamic zone configuration technique in which each BS adaptively controls the shape of its zone according to the distribution of users.

The dynamic zone configuration technique can be used advantageously with all the temporal access schemes for wireless communication systems, such as frequency division multiple access (FDMA), time division multiple access (TDMA), and code division multiple access (CDMA). It reduces cochannel interference and multipath fading because it directs the antenna beam pattern toward the desired area,

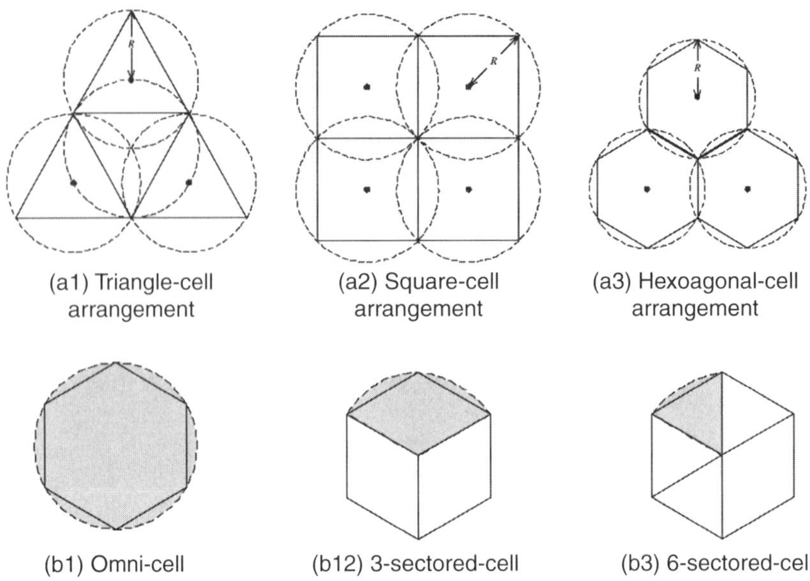

FIGURE 23.3 Conventional cell arrangements for macrocell systems. (a) Types of cell arrangement and (b) omnicell and sectorized cell.

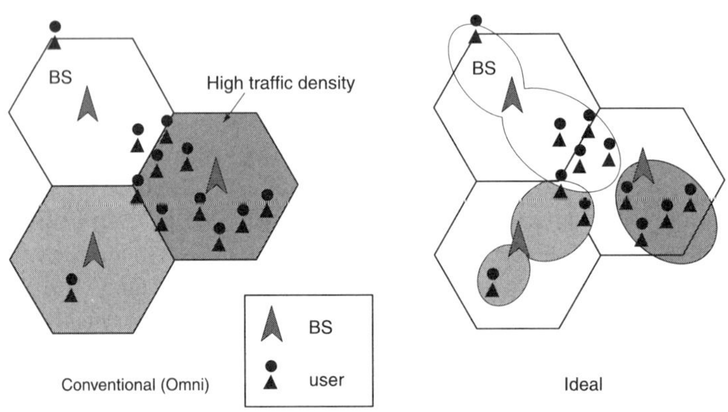

FIGURE 23.4 Conventional and ideal zone configurations in a macrocell system.

thereby reducing the outage probability and increasing both the capacity and the quality of service (QOS).[2]

In this subsection we describe two experimental prototypes for the dynamic zone configuration developed by Communication Research Laboratory, ministry of posts and telecommunications, Japan (CRL).[3]

23.2.1.1 Overview of the Prototypes Developed for Macrocell Systems

The configuration of the array is an important factor determining the characteristics of the beam pattern, and popular configurations are linear, planar, circular, etc. The developed prototypes were used circular array antenna because a BS in macrocell systems needs to communicate with surrounding users. One consists of the omnidirectional elements, and the other consists of the directional elements.

Figure 23.5a shows the former prototype, which consists of a 2-GHz band radio frequency generator, a directivity controller to control the amplitude and phase components of the array antenna, and a

(a) photograph

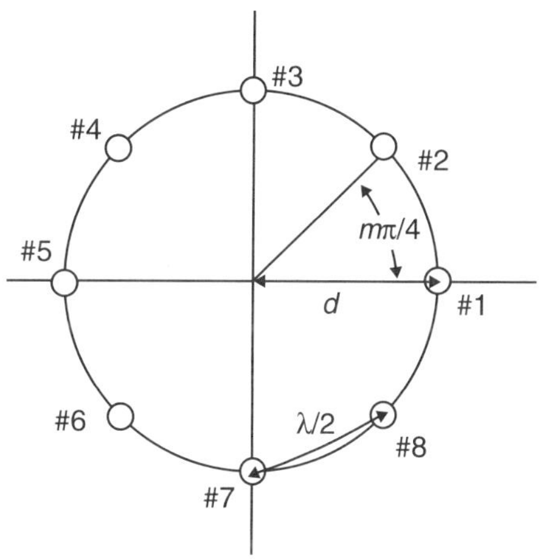

(b) horizontal arrangement of array elements.

FIGURE 23.5 Overview of adaptive zone configuration system prototype with omnidirectional elements.

circular 8-element array antenna in which the antenna elements are arranged at equal-angle intervals. Each element is a dipole antenna, and the horizontal distance between elements is λ/2 as shown in Fig. 23.5b. The amplitudes and phases of each element are controlled in analog at the RF part to form the desired antenna beam pattern. Since the prototype has a high-power amplifier, the maximum transmission power is 10 W at vertical polarization.

The later prototype, on the other hand, uses the same radio-frequency band, but has a 12-element circular array antenna. Each element is a corner reflector antenna and the distance from the center of the array to each element is 1.4 λ. This value minimizes the mutual coupling of the elements. The phase of each element can be controlled every 11.25 degrees, and the amplitude of each element can be controlled between 0 and 15 dB every 1 dB. They are controlled in digital at baseband part. Since this prototype aims high-speeed control, it can memorize 12 patterns of weight vectors, and change the weights rapidly (within 10 msec) to follow the movement of users. Figure 23.6 shows a photograph and a diagram of the arrangement of its elements.

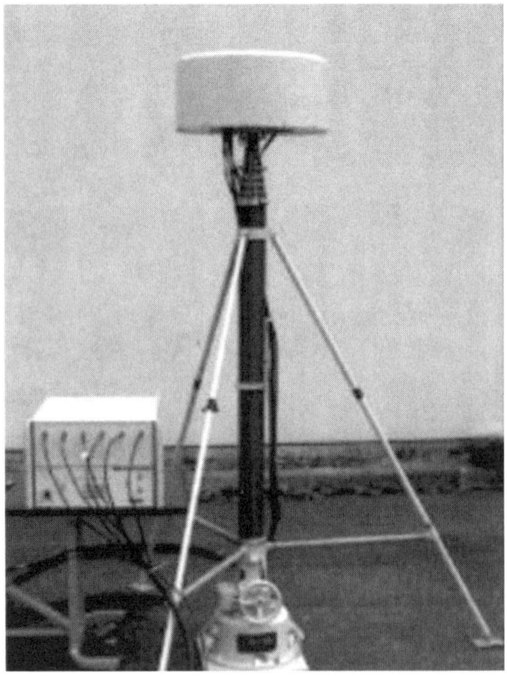

(a) photograph

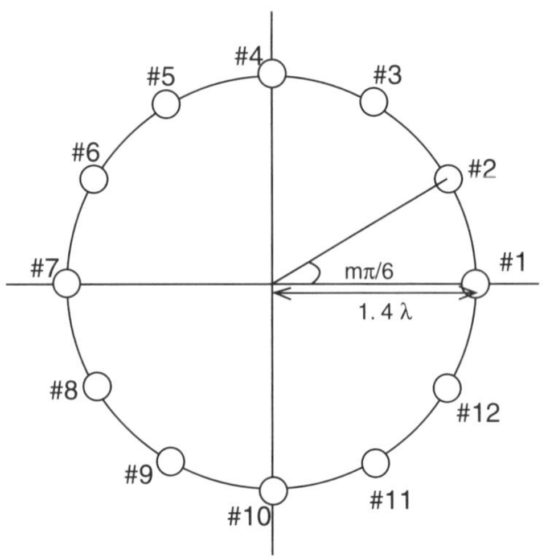

(b) array arrangement

FIGURE 23.6 Overview of adaptive zone configuration system prototype with directional elements.

Here we show the experimental zone configuration results obtained by former prototype in a residential area. Figure 23.7a shows the zone shape calculated by a computer simulation using the measured array response of each element, and Fig. 23.7b shows the measured zone shape.

23.2.1.2 Improvement of Channel Capacity

To evaluate the efficiency of the channel improvement obtained using the array antennas discussed above, we show the results calculated by computer simulation.

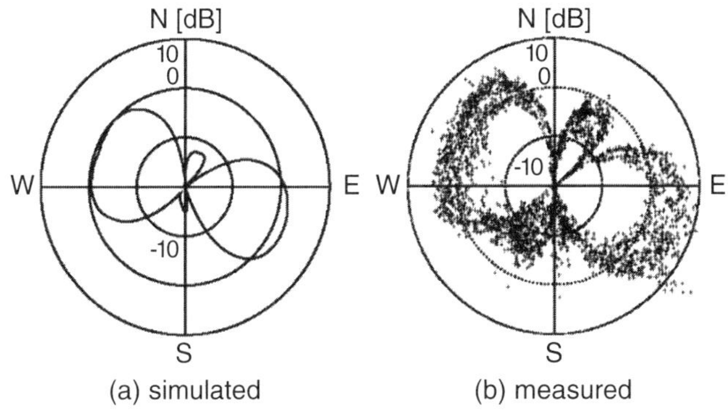

FIGURE 23.7 Simulation results and results measured using our prototypes.

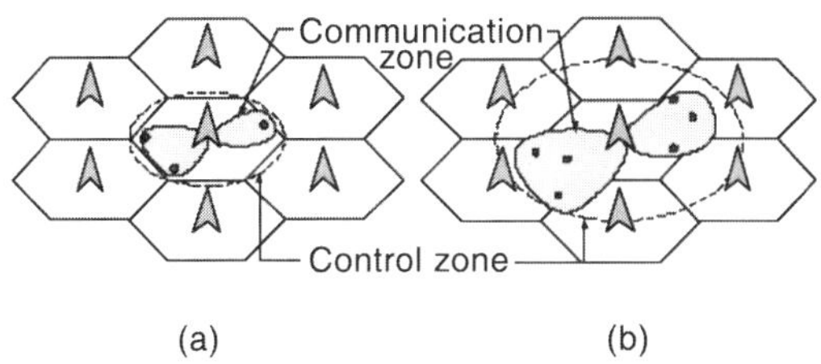

FIGURE 23.8 Communication zone configuration when (a) $z = 1$ and (b) $z = 2$.

Simulation Model

In our computer simulation model, each BS uses two zones: one is for control channels used to determine the positions of users, and the other is for communication channels used to carry information between the BS and the users. These zones are called the "control zone" and the "communication zone" and are shown in Fig. 23.8. The control zone has an omni-directional shape and its size is the same as or larger than the size of each cellular zone. Its radius is r times larger than each cellular zone. The BS uses the control channel to search for the active users requesting a new call, requesting handover from an adjacent zone, or terminating a call. Then, by using the information about the users' positions and our dynamic zone configuration technique, the BS configures the required communication zone. We compared two cases: one in which the BS antenna had 8 omni-directional elements, the other in which the BS antenna had 8 directional elements. In this simulation we assumed a narrow-band FDMA-FDD (frequency division duplex) mobile communication system. We also assumed that BS knew the users' positions in advance.

Determination for Computer Simulation Model

(a) The simulated service area was two-dimensional, there were 196 (14×14) cells, and the system performance was investigated in the 64 (8×8) cells around the center of the service area (see Fig. 23.9a).

(b) To simulate our proposed system not only in a uniformly distributed traffic environment but also in a non-uniformly distributed traffic environment, we used 49 cells with non-uniformly distributed traffic. They were arranged as shown in Fig. 23.9a.

(c) The traffic load in the cells with uniformly distributed traffic was assumed to be 6 erl/cell. In the cells with nonuniformly distributed traffic, the traffic load was u erl/cell in one sixth of the cell and was 5 erl/cell in the other five sixths. (When $u = 1$, the traffic is uniform.) The total traffic load in a cell with non-uniformly distributed traffic was thus $(u + 5)$ erl/cell, and in each cell, the high-traffic area with u erl/cell was randomly located in a sector like that shown in Fig. 23.9b.

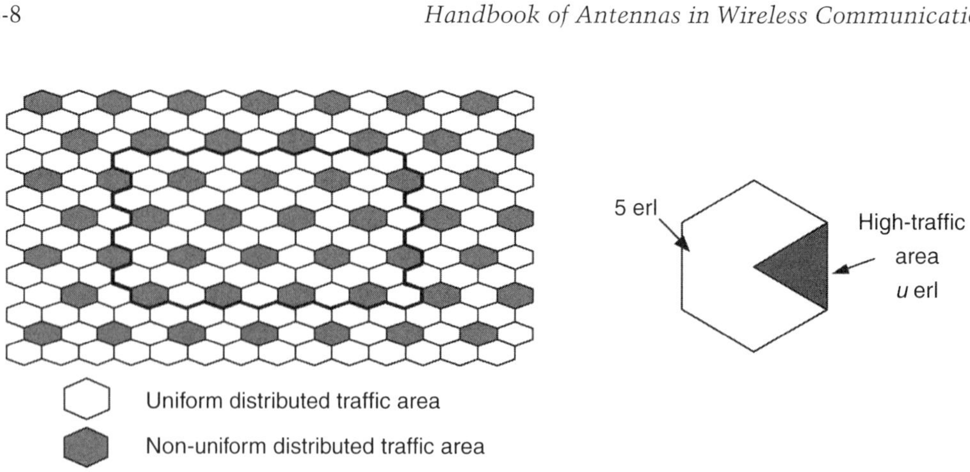

(a) Arrangement of the cells with non-uniformly distributed traffic and the cells with uniformly distributed traffic

(b) Traffic distribution in the cell with non-uniformly distributed traffic

FIGURE 23.9 Service area model.

(d) Call attempts from users were generated as a Poisson process in time. Call duration was assumed to have an exponential distribution with a mean of 120 seconds.
(e) The total number of FDMA channels was 70 for the uplink and 70 for the down-link, and these frequency channels were reused every 7 cells.
(f) The propagation factor γ was 3.4, the standard deviation of the short-term median value variation caused by fading σ was 6.5 dB, and the required power M[dBm] at the user farthest from the BS was 0.
(g) Users were at a standstill (i.e., speed = 0 km/h).
(h) CIR_{call} of 15 dB and a CIR_{over} of 10 dB. The CIR_{over} is the threshold CIR value for the channel changing. If the CIR of either the up-link or the down-link is less than CIR_{over}, the user must try to get a new channel.
(i) The radius of the omnicell z_o was 1 km.
(j) r was 1 (see Fig. 23.8).
(k) The BS changed the zone shape when a user tried to make a new call, terminated a call, or requested a handover.
(l) The weight of array antenna was calculated according to the users' distribution.[3,4]

The simulation was carried out as follows (see Fig. 23.10): when a user tried to make a call or request a handover, the user contacted the nearest BS. The BS looked for an available channel. If a channel was available, the BS checked the position of all users communicating with the BS and then created a communication zone by using our proposed method and algorithm. If no channel was available and the user was in a control zone covered by one or more neighbor BSs, the BS told the user to contact the neighbor BSs in descending order of the received power of their control channels. Each BS contacted repeated the above steps. If an available BS was not located, the call attempt was blocked.

Numerical Results

Examples of communication zone configurations generated in our simulation are shown in Fig. 23.11, and the relation between the blocking rate (BR) and the offered load over the high-traffic area (u) is shown in Fig. 23.12. The dynamic zone configuration technique had the lowest BR. Figure 23.13 shows the transmission power ratio needed to configure the communication zone. The computer simulations also showed that these results could be improved by combining the dynamic channel assignment (DCA) technique.[3,5]

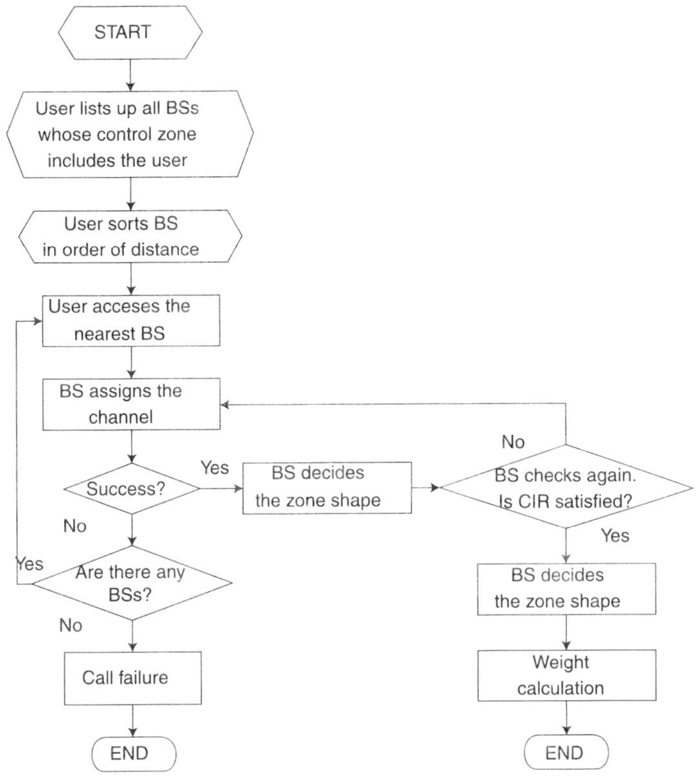

FIGURE 23.10 Flowchart for channel allocation.

23.2.1.3 Discussion

The computer simulations showed that the adaptive beam control for the multiple users in a macrocell system could reduce the blocking rate and the transmission power. Here we compare the performances that result from the difference between the characteristics of the elements. A BS with an array antenna, having omnidirectional elements, can control $(m-1)$ nulls and the width of beams freely when the antenna has m elements. However, the influence of mutual coupling becomes large because of the characteristics of the elements. Moreover, it is difficult to suppress the antenna gain for wide range. As a result, the reliable high antenna gain for the undesirable directions causes the interference. On the other hand, a BS with an antenna having directional elements can suppress the sidelobe level for undesired wide area effectively. From the viewpoint of effective frequency reuse in a macrocell system, it is expected to suppress the antenna gain for unnecessary area efficiently.

Moreover, in order to put the dynamic zone configuration technique into practice, we need to develop a technique to detect the users' positions, moving directions, and effective access schemes taking account of this useful information.

23.2.2 Development of an Array Antenna for Microcell Control

The growing demands for wireless high-rate transmission services have been increased. One way to cope with this situation is by using a microcell system in which the radius of a cell is 500 m or less. And an effective way to improve the performance of a microcell system is to use the adaptive array antenna controls based on the information about the direction of arrival (DOA) of signals from a user who is communicating or causing interference. This information will be useful for improving the QOS, simplifying the handoff process, and utilizing the available channels more effectively. Because a transmitter or

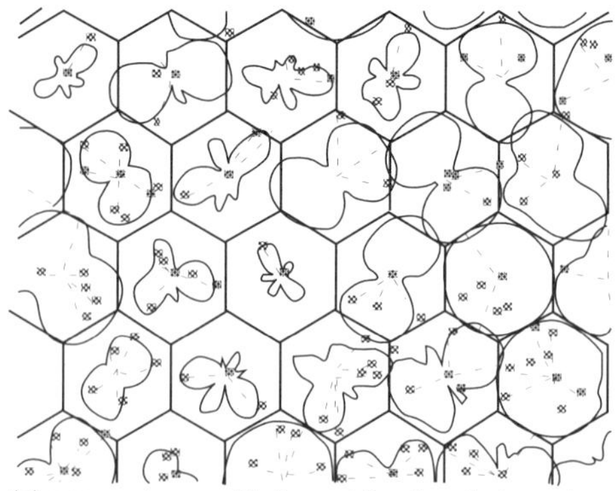

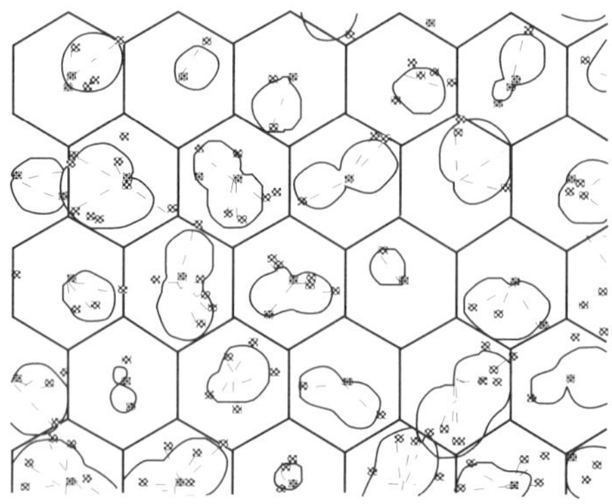

FIGURE 23.11 Examples of adaptive zone configurations ($u = 1, z = 2$), (a) array antenna with eight omnidirectional elements, (b) array antenna with eight directional elements.

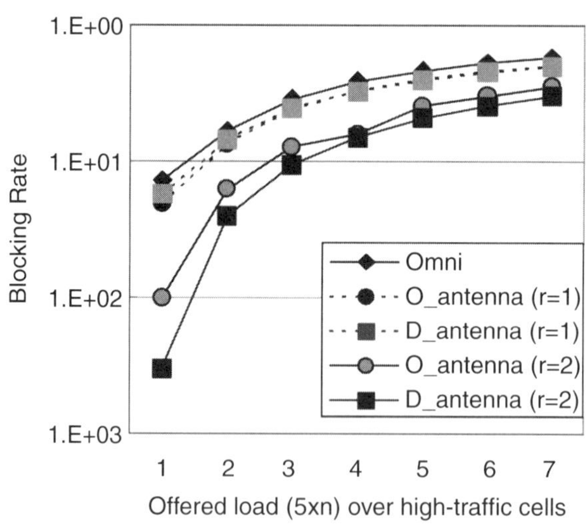

FIGURE 23.12 Blocking rate.

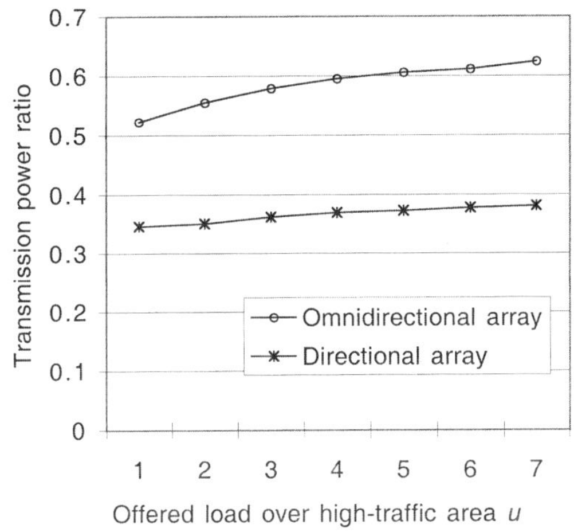

FIGURE 23.13 The transmission power ratio compared with the omnisystem.

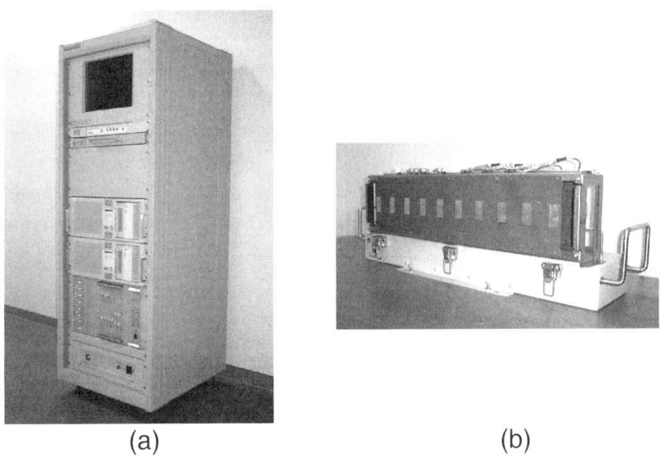

FIGURE 23.14 Components of the adaptive array antenna system, (a) control part, (b) eight-element linear array antenna.

receiver using an array antenna can increase or decrease the antenna gain in intended directions by using information about the locations of desired users and of sources of interference. Many algorithms for estimating the DOA at an array antenna have therefore been proposed.[6-12]

In this subsection, we describe the experimental array antenna prototype for DOA estimation and beamforming control, developed by CRL.

23.2.2.1 Overview of the Prototypes Developed for DOA Estimation

Figure 23.14 shows our experimental adaptive array system, and Fig. 23.15 is a block diagram of the system. It consists of an array antenna block, a frequency conversion block, and a digital processing block. The array antenna has eight equi-spaced elements with microstrip patch antennas. The distance between adjacent antennas is half a wavelength. A dummy element is placed at each end of the array in order to equalize the characteristics of the elements. The frequency of the RF signals received at the array elements is in the 2-GHz band, and these signals are converted to intermediate frequency (IF) signals (450 kHz) in the frequency conversion block. The IF signals are sent to the digital processing block, a DOA estimation block, and a beamforming block. The IF signals are digitized by an analog to digital (A/D) converter, which operates at a 1.8-MHz sampling clock rate, and are then converted to the I and Q baseband signals.

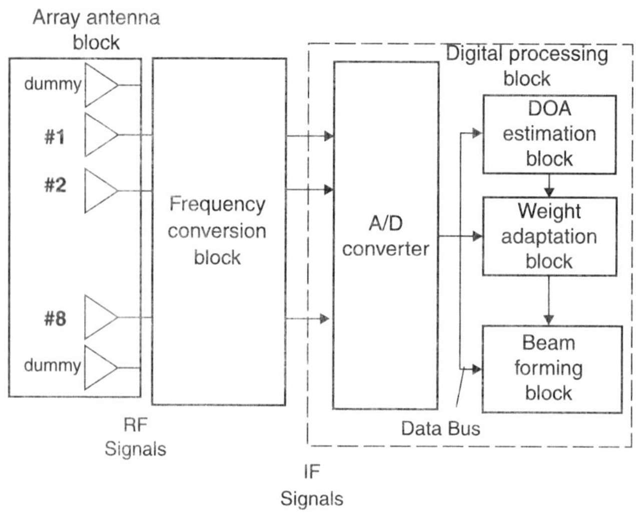

FIGURE 23.15 Configuration of the experimental system.

The data bus transfers the I/Q channel data to the DOA estimation, weight adaptation, and beamforming blocks. The DOA estimation block applies three types of DOA estimation algorithms, such as the multiple signal classification[8] with spatial smoothing processing[9] SS-MUSIC and the estimation signal parameters via rotational in variance techniques[10] with spatial smoothing processing (SS-ESPRIT) or a fast-DOA-tracking algorithm[11,12] to the received data. The weight adaptation block calculates the optimum weight vector based on the estimated DOA, and the beamforming block synthesizes the data by using the calculated weight vector.

23.2.2.2 A Fast-DOA-Tracking Algorithm[11,12]

There have been many studies of DOA estimation, but most conventional methods for estimating the DOA are not suitable for us in a microcell system because they are not intended for real-time computations and they are not accurate. Especially, the speed of DOA estimation is important because the mobile terminals in a microcell are close to the base station and they occasionally move at the speed of a car. A new tracking system has therefore been developed, one in which a new algorithm for tracking mobile terminals is implemented in a real-time processor. The points of this method are to use a model of the trajectory of the DOA and to estimate the DOA and the angular velocities of the mobile terminals simultaneously.

The basic idea of this algorithm is illustrated in Fig. 23.16. Each DOA trajectory $\hat{\theta}_n$ is approximated as a straight line:

$$\hat{\theta}_n\left(k-l|k\right)=\hat{\theta}_n\left(k\right)-\hat{\alpha}_n\left(k\right)lT \tag{23.1}$$

where $\alpha_n(k)$ is the angular velocity of the nth target at time k and T denotes the sampling period of the received data. This model is different from conventional models in that the DOA $\theta_n(k)$ and the angular velocity $\alpha_n(k)$ are estimated directly by using the past received data in the observation time. The DOAs and angular velocities at time k are estimated by minimizing a cost function $J(k)$ as follows:

$$J\left(k\right)=\sum_{l=0}^{k-1}\beta_{k-l,k}\left\|y\left(k-l\right)-\hat{\mathbf{x}}\left(k-l\,|\,k\right)\right\|^2 \tag{23.2}$$

where $\mathbf{y}(k-l)$ is an observation data vector and $\hat{\mathbf{x}}(k-l|k)$ is an estimated received data vector, which is obtained by using the parameters $\theta_n(k)$ and $\alpha_n(k)$.

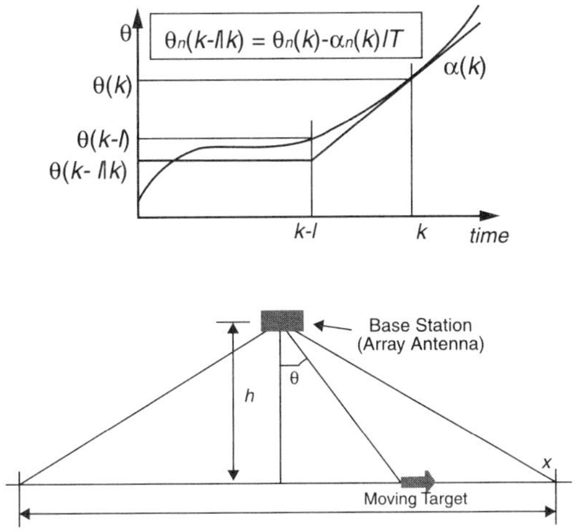

FIGURE 23.16 Model for the proposed tracking mobile system.

The angular velocity $\alpha_n(k)$ is a linear parameter in $J(k)$ and the DOA $\theta_n(k)$ is a nonlinear parameter in $J(k)$, so $\alpha_n(k)$ can be estimated analytically by using the least squares method and $\theta_n(k)$ can be calculated by using a *Newton*-type recursive algorithm.

23.2.2.3 Experimental Examples

Figure 23.17 shows the configuration of the outdoor experiments (line-of-sight [LOS] environment). Two transmitters, T × 1 (in a vehicle) and T × 2, were placed in front of the array antenna. T × 1

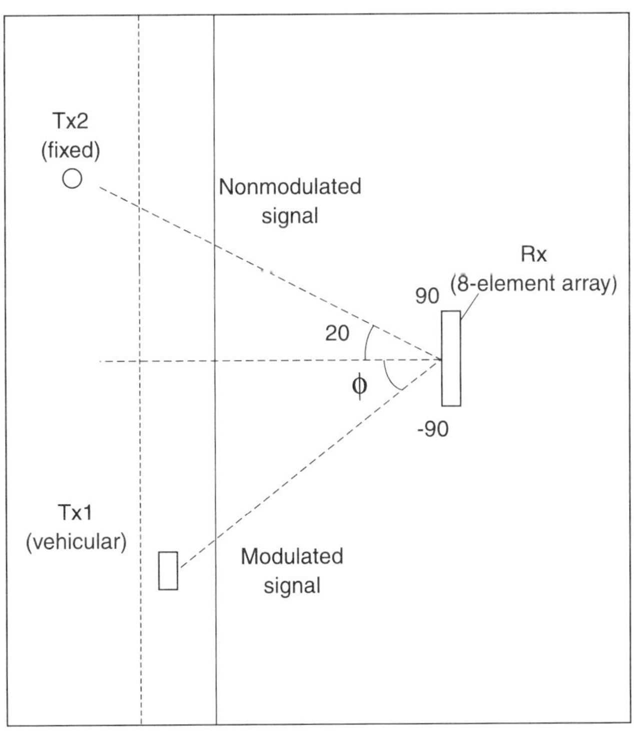

FIGURE 23.17 Configuration of outdoor experiments I.

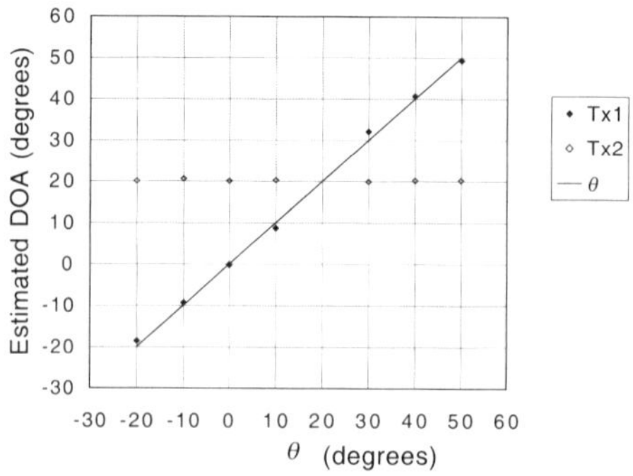

FIGURE 23.18 DOA estimated using ESPRIT (for two fixed transmitters).

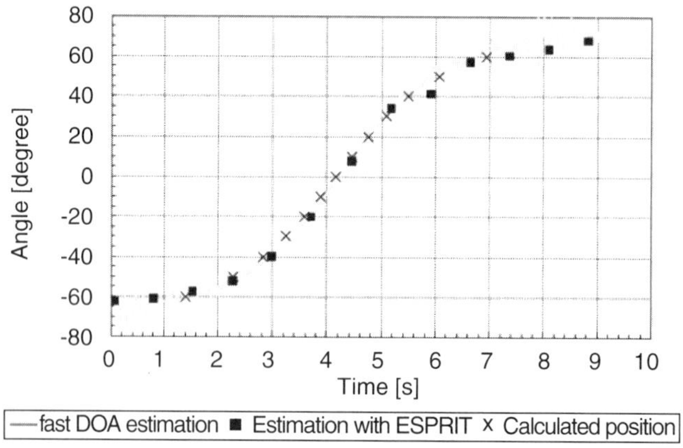

FIGURE 23.19 DOA estimated with a fast DOA-tracking algorithm.

transmitted a modulated signal as a desired signal and T × 2 transmitted a non-modulated (carrier only) signal as an interference signal. The T × 1 was positioned at several points in order to change the DOA of its signal. The DOA estimation block used 1024 samples from each element for the calculation of the MUSIC and ESPRIT. Figure 23.18 shows the results of DOA estimation. In front of the array antenna the estimated DOA was within 1 degree of the actual DOA.

Figure 23.19 shows the results of DOA estimation for a moving transmitter. For this experiment, the DOA estimation block used ESPRIT and a fast-DOA-tracking algorithm calculation together because a fast-DOA-tracking algorithm needs 1/10 calculation time needed by MUSIC or ESPRIT. The T × 1 moved down the street at 60 km/h (max. 47.7 degree/sec). In Figure 23.19, the square shows the estimated DOA calculated by ESPRIT, and the solid line shows the angle of T × 1 calculated by the fast DOA algorithm. The crosses show the actual positions calculated from the time and velocity. This figure indicates that it is possible to accurately estimate the DOA of signals from a rapidly moving user.

Figure 23.21 shows the comparison of BER performance under two different conditions:

Case 1 — The BS received the signal with one element.
Case 2 — The BS received the signal with eight elements, and the phase of each element was controlled to point the direction of T × 1 [beamforming].

The grand plan is shown in Fig. 23.20. Throughout the experiment the transmission level of T × 2 was changed, while the transmission level of T × 1 was fixed. In case 2, the CIR improved about 15 [dB] from case 1 to achieve BER = 10^{-2}.

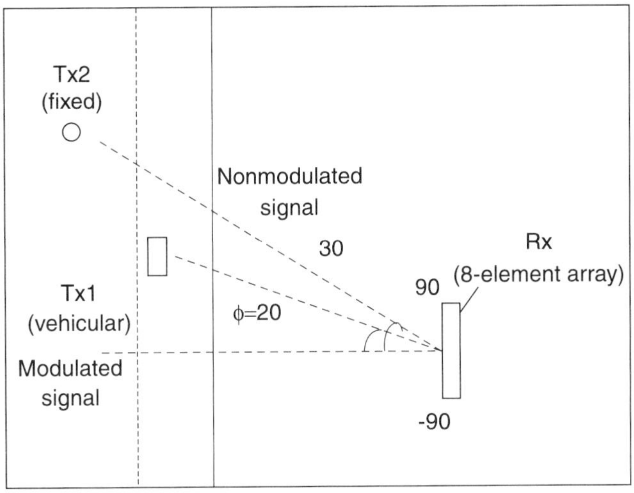

FIGURE 23.20 Configuration of outdoor experiments II.

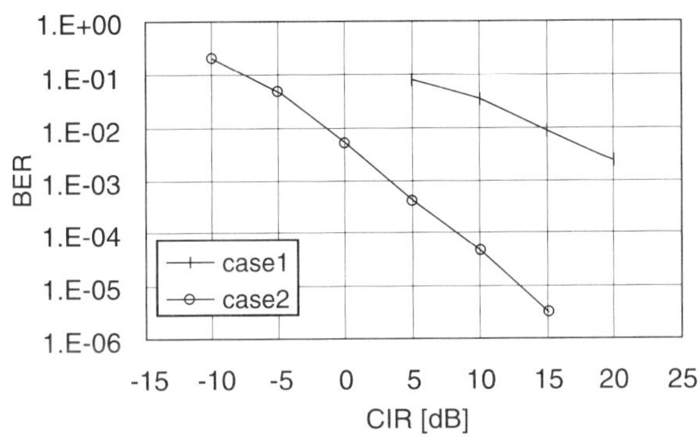

FIGURE 23.21 BER performance.

23.2.2.4 Discussion

Experimental trials confirmed that the array antenna estimated the fixed or moving transmitters accurately in LOS environment. Based on the DOA estimation results, the beam steering for the target direction improved the BER effectively.

To implement such a technique in actual microcell system, however, we need to study for DOA estimation in more detail because even a microcell contains several different microenvironments, such as the non-line-of-sight environment where the multipath wave between a user and a BS is dominant. We need to develop effective methods for DOA estimation and beamforming control that obtains reliable performance in various situations in order to construct a high speed data transmission system.

23.2.3 Conclusion

The goal of this section is to show the practical application of array antennas for land mobile communication systems. As shown, an array antenna provides a powerful technique for combating the deterioration in signal and developing high-grade multimedia communications and has now reached the practical stage. As the demand for higher communication speed increases, higher frequency bands and smaller cell sizes tend to be used. Therefore, the usage of array antennas plays an important role. Although there have been many studies of, and schemes for, array antennas, we have to evaluate suitable array antenna systems and schemes for adaptive array processing because the link budget of the intended system, the definition of interference, and so on differ from one system to another.

23.3 Phased-Array Antennas for Stratospheric Communication Systems

A wireless access network using large airships, located within the stratosphere at an altitude of about 20 km, is attracting interest worldwide.[14-19] Such airships are referred to here as stratospheric platforms (SPFs). A high-speed, high-density wireless access network using high frequencies may be an especially suitable application for an SPF system because of the high LOS availability provided by the high elevation angles in such a system.[15] Small-aperture ground terminals are available for wireless access at more than 10 Mbps. A wide service area can thus be covered by multiple platforms deployed at certain intervals and connected to each other by wireless interplatform links.

The antenna onboard an SPF for such an application would have to be light, and should have a low profile and a large capacity with multiple beams. Phased-array technology provides a solution that meets these requirements. For a beamforming network in a phased array on a commercial version of SPF, which has more than 100 beams, a digital beamforming (DBF) scheme could be superior to conventional analog beamforming schemes in terms of manufacturing cost, weight, and performance. Because the effect of radiation in the stratosphere is weaker than at satellite orbits, the use of digital devices for SPF onboard equipment should be more practical than for equipment on a satellite.

This next section describes a test-flight model of a DBF antenna under development for technical evaluation onboard a prototype SPF.

23.3.1 Communication Systems Using Stratospheric Platforms

23.3.1.1 Stratospheric Platform

A large helium-filled airship, as shown in Fig. 23.22, can be used as the SPF.[14] Power is provided by solar cells in the day and residual power charges fuel cells for operation at night. The commercial model of the airship will be 150 to 250 m long. The airship has propellers driven by electric motors for station-keeping against winds with a speed exceeding 30 m/s in the stratosphere, and for propulsion.[20]

The concept of using helium-filled balloons as a repeater platform for wireless communication has existed for some time. However, it has been difficult for the balloons to maintain a stable position in the stratosphere mainly because they require a large envelope to obtain sufficient buoyancy under atmospheric pressure that is only 5% of that on the ground and a large power is required to control the

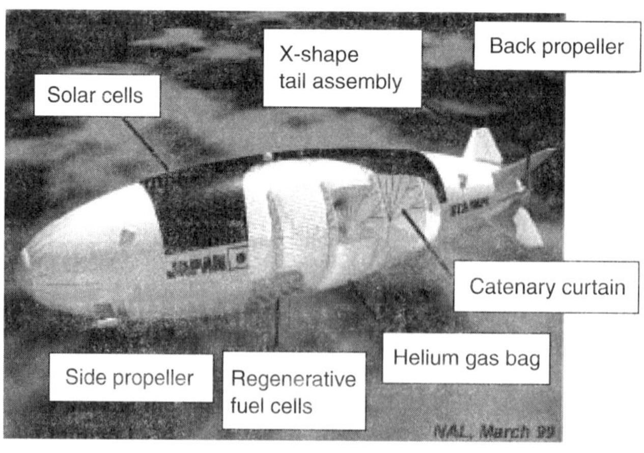

FIGURE 23.22 Conceptual design of an SPF. (Courtesy of National Aerospace Laboratory.)

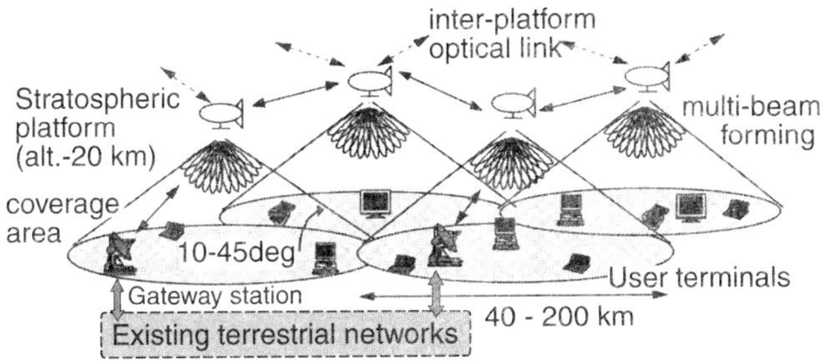

FIGURE 23.23 SPF network for high-speed wireless access.

platform against the wind. Progress in technologies including solar cells and fuel cells, however, is making that operation feasible. Platforms using those technologies is expected to carry a mission payload of up to a 1000 kg. It should also be an environmentally clean system.

23.3.1.2 Wireless Access Network Using Stratospheric Platforms

The SPF acts as a high-altitude tower, at an altitude of about 20 km, for wireless communication systems. An SPF network thus provides high LOS availability with a high-elevation angle, and with much shorter propagation distances than with satellite systems. Also, because of the large payload capacity of the SPF, much more access capacity is attainable with multiple platforms than can be achieved with geostationary-orbit (GSO) or nongeostationary-orbit (NGSO) satellites. A service using NGSO satellites cannot be started until all satellites in the system are deployed globally, whereas an SPF-based service can be started with just one SPF covering a metropolitan area and the number of SPFs can be gradually increased. Therefore, the initial cost of starting an SPF access service is much less than that for starting an NGSO access service. Moreover, maintenance is easier than it is for satellite systems because SPFs can be brought from the sky to the ground for repair, upgrades, and so on.

Mobile communication and broadcasting are also potential applications for an SPF network. When such applications are based on terrestrial networks, they require a huge number of base stations or relay stations to cover a wide area. If we use SPFs as the base/relay stations, the number of stations needed to cover the same area, and their construction cost are significantly reduced.

An SPF network for high-speed wireless access is shown in Fig. 23.23. Multiple platforms are deployed at intervals of about 40 to 200 km. Each platform provides service coverage for access links to user terminals on the ground with a minimum elevation angle of about 10° to 45°. The typical bit rate for an access link would be more than 20 Mbps per carrier and the maximum bit rate could be more than 100 Mbps to support high-speed Internet or high quality digital video services. The platforms need multibeam antennas to obtain high equivalent isotropic radiation power (EIRP) and high antenna gain to system noise temperature (G/T) ratio for high-quality broadband transmission, and to enable efficient use of frequency bands to increase the traffic capacity. Interplatform links enable users separated by a long distance to be connected by an all-wireless link, including the access links and trunk links, without using other networks such as satellite or terrestrial networks. Optical links can be used for interplatform connection to support large-capacity trunk connections at as much as 10 Gbps with fewer interference problems. An SPF network thus represents a third form of infrastructure that is an alternative to terrestrial and satellite networks.

23.3.1.3 Frequency Band

A station onboard an SPF is defined as a "high altitude platform station" in the International Telecommunications Union (ITU) Radio Regulations.[21] Frequency bands for fixed wireless systems using an SPF were designated at World Radiocommunication Conferences (WRC) in 1997 and 2000. The bands are the 47- to 48-GHz band (47.2 to 47.5 and 47.9 to 48.2 GHz) on a global basis and the 31- to 28-GHz band (31.0 to 31.3 GHz for uplink and 27.5 to 28.35 GHz for downlink) on a national basis.[22] Our test flight model of the DBF antenna uses the 31- to 28-GHz band. This band is subject to be shared with various services such as terrestrial-based fixed service, mobile service, and fixed satellite service; and must be compatible with passive sensors onboard earth exploration satellites and radio astronomy radio telescopes, both of which are allocated in an adjacent band. The DBF antenna is expected to significantly improve the sharing and compatibility situations with its high-performance beamforming capabilities.

23.3.2 Test Flight Model of a Digital Beamforming Antenna Onboard a Stratospheric Platform

23.3.2.1 Requirements for an Onboard Antenna

The link quality of an SPF network using a band above 20 GHz usually is limited by the available power margin in the link budget. As in satellite networks, this is much smaller than in terrestrial mobile networks using frequency bands below 2 GHz because of power loss resulting from long propagation distances, the RF circuits, and the need to share the frequency band with other wireless services. Angle range at the onboard antenna to cover a necessary area on the ground can be as much as 160° when the minimum service elevation angle at the ground station is 10°. Because of these conditions, the antennas onboard the platform must meet the following requirements:

1. Directional antennas must have a high EIRP and G/T.
2. The multibeam antenna must enable a large traffic capacity through efficient frequency use. Because the minimum service elevation angle may be 10° to 45° and the number of spot beams for one platform may be about 60 to 400, a broad scanning angle range is needed with little degradation in gain or the axial ratio.
3. The spot beams should not be uniform in their beam width and transmission power. Independent automatic transmission power control to compensate regional rain attenuation is necessary for each spot beam to keep high-link availability and to prevent the SPF downlink signal interfering with the uplink of satellite systems that use the same frequency channels as the SPF system.
4. Because the platform position fluctuates or drifts as a result of variation in the wind and pressure conditions, the antenna beams should be controlled to compensate for any motion of the platform and to fix the position and size of the ground footprint.
5. A broad bandwidth is needed.
6. Durability against the stratospheric environment is needed.

Two types of multibeam antenna is under development for the test flight model to evaluate how well each meets these requirements: the multibeam horn (MBH) antenna and the digital beamforming (DBF) antenna. Each has advantages and disadvantages. The MBH antenna provides fixed multiple beams with multiple horn antennas directed in different directions. This antenna gives a low-cost solution for broadband transmission (e.g., >100 MHz). The antenna is mounted on a gimbal platform that is controlled mechanically to compensate for the motion of the SPF. The rest of this section describes the DBF antenna.

23.3.2.2 Digital Beamforming Antenna

This antenna generates adaptive beams by combining an active array antenna and spatial digital signal processing in the baseband. This type of antenna is also called a smart antenna or a software antenna.[2,23] In the receiving antenna, automatic acquisition, tracking, and isolation from interference are provided through spatial signal processing. In the transmitting antenna, retro-directive spatial power combining is provided through beamforming using parameters given by the receiving antenna. DBF antennas have been studied since the 1980s, mainly for military radar applications.[24] However, DBF antennas with many antenna elements using frequencies of 10 GHz or more have not been developed for commercial use, mainly because of its high cost, large volume, and large power consumption for RF and digital devices. Insufficient signal processing speed in digital devices has also been a barrier.

Future commercial models of the onboard antenna may need to have more than 100 antenna elements to provide more than 100 multiple beams, for example, for use at frequencies above 20 GHz. Analog beamforming with so many beams could result in a very complex and lossy RF beamforming network.

In contrast, because the beamforming network of the DBF antenna is based on digital devices, the complexity and power loss, compared with that of an analog circuit in RF and IF, is significantly reduced. A DBF antenna generally needs amplifiers, mixers, and an A/D or D/A converter in each of the antenna branches, but an analog beamforming antenna also needs those devices, in quantities equal to the number of beams, when used as a multibeam antenna.

The advantages of using a DBF antenna on the SPF are as follows.

1. Flexible beam steering for each user terminal is made possible by using adaptive multiple beams. Fixed spot-beam cells are no longer necessary. This should lead to increased traffic capacity. The platform motion or drift is automatically compensated.
2. Consequently, handover in the coverage of one SPF is not needed except when some user terminals using the same channel get too close to each other. High-link quality is expected with the maximum antenna gain provided to each of the user terminals on their demand.
3. Spatial signal processing can reduce interference from undesired signals and interference with other systems such as satellite systems. The DOA of communication or illegal radio signals can be detected.
4. Partial defection of antenna elements does not seriously affect the total performance of the array antenna.
5. High-speed array calibration is enabled by signal processing in the transmitting and receiving array antennas.

The disadvantage of the DBF antenna is that the bandwidth of the antenna depends strongly on the processing speed of the digital devices. The target transmission data rate for the test flight model of the DBF antenna is about 4 Mbps at most. The achievement of a much higher rate is expected in the near future, though, because of the rapid progress in digital device technologies. Thus, the development target for the DBF antenna is high-performance beamforming instead of high-speed transmission in this prototype, whereas the MBH antenna provides the high-speed transmission.

Figure 23.24 shows a block diagram of the DBF antenna test flight model. The antenna consists of a 16-element (4×4) array with patch antennas; a meander-line polarizer; an RF section containing amplifiers, mixers, and filters; an IF section containing amplifiers, filters, and an A/D or D/A converter; and digital beamformers containing digital signal processing circuits. The transmitting and receiving bands are the 28-GHz band and 31-GHz band, respectively. The transmitting and receiving antennas are separate. This DBF antenna is used to obtain technical data for the development of a large-scale DBF antenna with more than 100 elements. Figures 23.25a and b shows an expanded view and a photograph of the DBF antenna test flight model. In the photograph, general-purpose power supplies are used, but they are replaced by compact ones for the onboard test. Table 23.1 lists its major specifications.

Two types of digital signal processing devices are used: a field programmable gate array (FPGA) and a high-speed microprocessing unit (MPU). The FPGA is suitable for high-speed parallel processing and the MPU allows easy programming.

23.3.3 Beamforming Algorithm

Onboard multibeam antennas allow space division multiple access (SDMA) in the access link of an SPF network. The same channel, the same time slot, and the same modulation scheme can be shared among different user terminals located in different directions from the platform, thereby enabling efficient frequency reuse by a large number of users under the coverage of a platform. SDMA is to be applied in conjunction with FDMA, TDMA, or CDMA. A DBF antenna should be suitable for SDMA.

To achieve SDMA with a DBF antenna in the 28- to 31-GHz band, we must overcome some algorithmic hurdles. Because the number of antenna elements in the array and the required transmission rate may increase in future commercial models, the amount of computation should be minimized. Also the beamforming performance should be robust under low carrier to noise power ratio (CNR) conditions because the receiving CNR for one antenna element could be as small as –10 dB when the number of

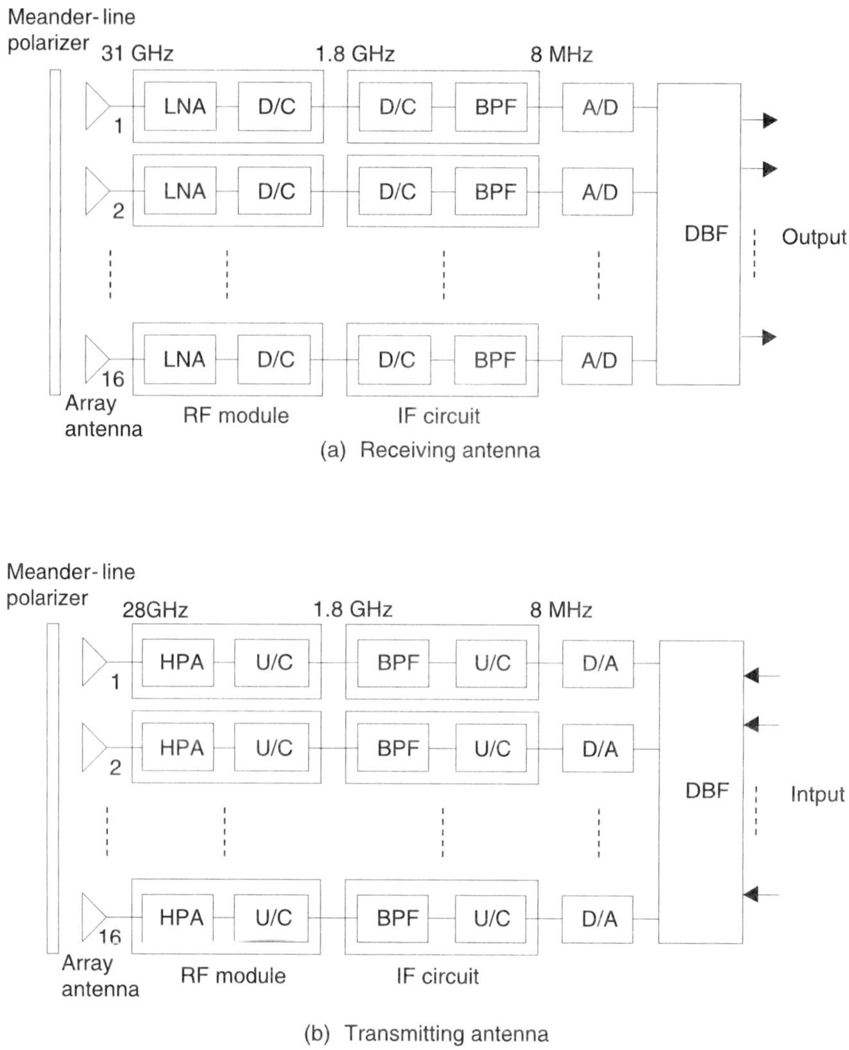

FIGURE 23.24 Block diagram of the DBF antenna test flight model.

antenna element exceeds a hundred. Fast acquisition of a desired signal out of many undesired signals is also required to handle burst-type transmission in packet networks or to reduce overhead reference sequences for beamforming.

The maximum-ratio-combining (MRC) beamformer in beam space assisted by a reference sequence is a beamforming algorithm that should meet the preceding requirements.[25] This beamformer is expressed as

$$b_n(N) = \sum_{k=0}^{m-1} c_k x_k(N) \exp\left(-j\frac{2\pi}{m}kn\right) \tag{23.3}$$

$$w_n(N) = \overline{b_n^*(N) r_i(N)} \tag{23.4}$$

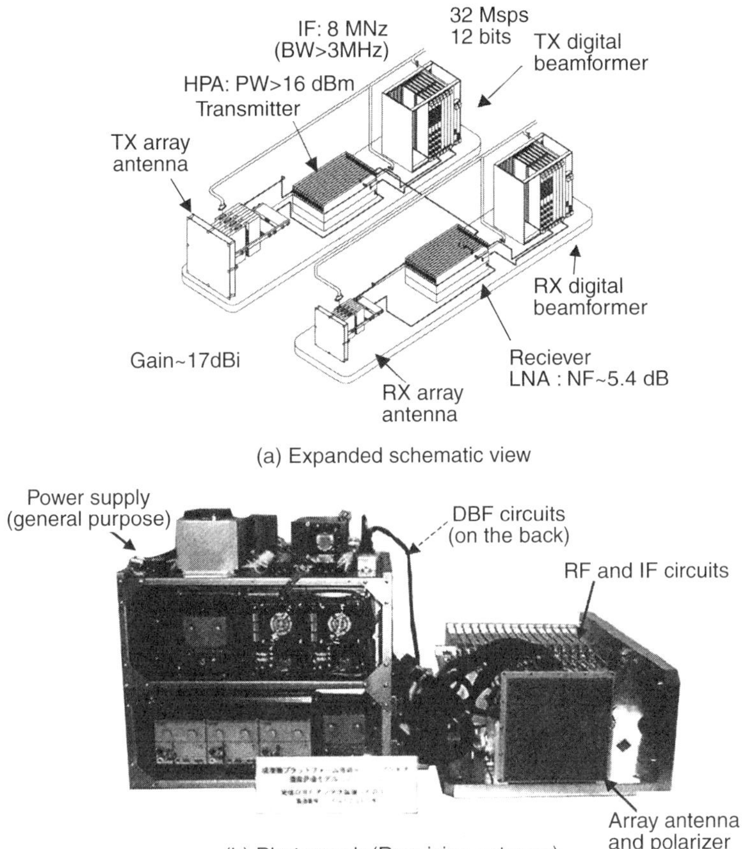

(a) Expanded schematic view

(b) Photograph (Receiving antenna)

FIGURE 23.25 Test flight model of the DBF antenna.

TABLE 23.1 Specifications (Test Flight Model)

	DBF Antenna
Frequency band	TX: 27.5–28.35 GHz (LHCP)
	RX: 31.0–31.3 GHz (RHCP)
Number of antenna elements	16 (4 × 4)
Type of antenna element	Microstrip antenna
Half-power beam width	10–13°
Antenna gain	17 dBi
Number of multiple beams	Fixed beams: 9
	Adaptive beams: 3
EIRP	11 ~ 15 dBW
G/T	–13 ~ 17 dB/K
Bandwidth	>3 MHz
Sampling rate	32 MHz
A/D quantization	12 bit
Power consumption (TX and RX)	<1.6 kW
Weight (TX and RX)	<120 kg

$$y_i(N) = \sum_{n \in CSEL} w_n(N) b_n(N) = \sum_{n \in CSEL} \overline{b_n^*(N) r_i(N)} \cdot b_n(N) \tag{23.5}$$

where $x_k(N)$ is the input signal (complex value) from the kth antenna element at the Nth sample iteration; c_k, fixed amplitude weighting (real value) for antenna branches followed by a distribution for low side lobes; m, total number of antenna elements; $b_n(N)$, nth fixed multibeam signal given by a spatial DFT of input signals $x_k(N)$; $r_i(N)$, replica of a known reference sequence contained in the ith desired signal; $w_n(N)$, beamforming weight factor for the nth fixed multibeam signal $b_n(N)$; $y_i(N)$, beamformer output signal for the ith desired signal; *, complex conjugate; $\bar{\;}$, time averaging or low-pass filtering.

Equation (23.3) gives a discrete Fourier transform (DFT) in the space domain for spatially sampled signals at the antenna elements $x_k(N)$ weighted by fixed amplitude factors c_k. This processing provides the transformation of the spatially sampled signals from element space to beam space and gives fixed multiple beams. Equation (23.4) gives the calculation of the weight factors for MRC in beam space, which enables the self-beam steering capability. The weight factor used in this algorithm directly uses the complex correlation between each of the multibeam signals and a replica of a known reference sequence. Equation (23.5) gives the beamformer output by combining the weighted multibeam signals. The condition, $n \in CSEL$, means the selection of multibeam signals through the weight factors as follows:

$$\text{If } \left|w_n(N)\right|^2 < \text{threshold, then } w_n(N) = 0 \tag{23.6}$$

Therefore, the condition in Eq. (23.6) means beam selection is done according to the correlation criterion. By this nonlinear process, the isolation between desired and undesired signals is improved and can be controlled through the fixed amplitude weighting for antenna branches c_k. This fixed amplitude weighting is effective when applied to an array antenna with many elements. The simple computation of weight factors without feedback loops or matrix operations in the preceding algorithm should enable easy implementation into high-speed digital devices with parallel and pipelined architectures.

The minimum mean square error (MMSE) scheme is a well-known group of algorithms that gives optimum beamformers.[6] Though the MRC-based beamformer does not give an optimum solution under interference conditions, we expect it to be superior to the MMSE beamformer in response speed, robustness, and computation, while maintaining sufficient isolation between multiple access signals, particularly when used with a large-scale array antenna in power-limited channels. Figure 23.26 shows the simulated performance of this beamformer when used for a 16-element linear array antenna. The response speed, stability, and amount of computation for this beamformer (MRC-T) are better than for a beamformer using the recursive least squares (RLS) algorithm, which is one of the fastest MMSE

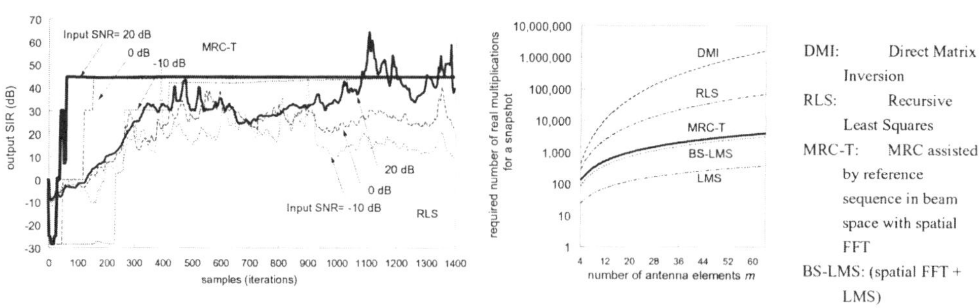

(a) Initial beamforming response
(One desired signal and one undesired signal arrive with the same power.)

(b) Amount of computation

FIGURE 23.26 Performance of the MRC beamformer assisted by a reference sequence (MRC-T).

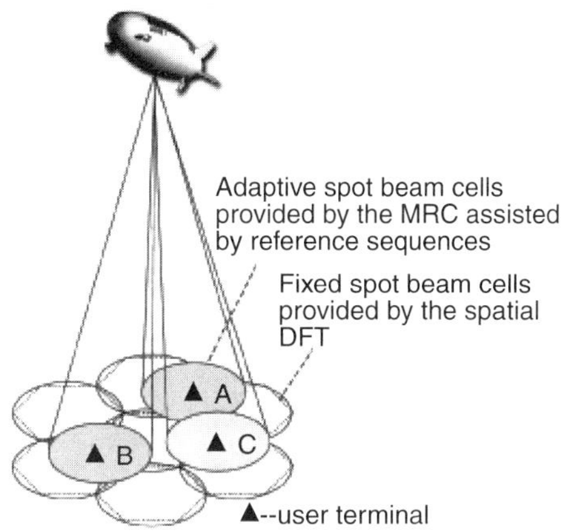

FIGURE 23.27 Footprint example with a DBF antenna.

algorithms. SDMA in a multiuser environment is achieved by using multiple beamformers with different reference sequences to separate user signals.

Figure 23.27 shows the SDMA that is to be attempted in an experiment using the test flight model of the DBF antenna. Three separate spot beams are to be provided to three user terminals on the ground. In this example, the angle separation between terminals A and B is large enough, compared with the beam width, to ensure that the signals from these terminals are spatially separated in the onboard DBF antenna and that they can use the same channel, the same time slot, and the same spreading code in their uplink. On the other hand, the angle separation between terminals A and C is not large enough to allow spatial isolation by the beamforming; thus the assigned channel, time slot, or spreading code for terminal C must differ from that of terminal A. The fixed spot beam cells provided by the spatial DFT can be used for control channels.

23.3.4 Conclusion

An SPF has various types of antennas depending on required applications and available frequencies. A DBF antenna is expected to be one of the best candidates for advanced wireless access systems that provide services to fixed and mobile user terminals on the ground. Despite its excellent potential performance, many problems must still be overcome to increase the array size; for example, the RF feeding network must be integrated into the active array antenna, the processing speed and distributed architecture of digital devices must be improved, the bandwidth of A/D and D/A converters must be increased, and the number of RF and IF cables must be reduced. The beamforming algorithm can also be combined with dynamic access control schemes and adaptive modulation/coding schemes. The application of the DBF antenna on SPFs provides an incentive to accelerate the development and commercialization of a DBF antenna with high frequencies and a large number of elements.

23.4 Phased-Array Antennas for Satellite Communication Systems

Satellite communications often require phased-array antennas. By using the satellite-borne, airborne, and car-borne phased arrays as examples, this section describes the technologies used in phased arrays.

23.4.1 Satellite-Borne Phased Array

Figure 23.28 shows a 19-element multibeam phased array antenna[26] installed in the Japanese *Engineering Test Satellite VI (ETS-VI)* launched in 1994. It is for intersatellite communication, and it operates at 2.1

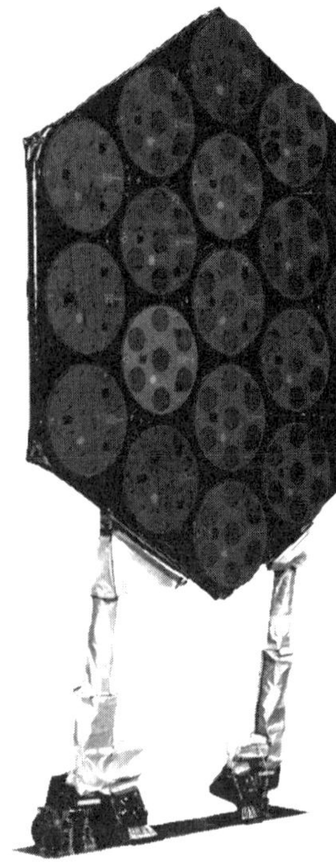

FIGURE 23.28 A 19-element satellite-borne phased array.

and 2.3 GHz. Figure 23.29 shows a block diagram of the receiving system of the phased array. The phased array employs an active array system in which each antenna element has a low-noise amplifier (LNA) and a high-power amplifier (HPA). The active array has the following advantages.

1. The insertion loss of the phase shifter does not degrade the G/T or EIRP.
2. It is robust because the failure of a single LNA and HPA does not result in a feral functional loss.
3. Small, lightweight, low-power phase shifters can be used because the transmitting phase shifter is in the stage before the HPA.
4. The overall EIRP can be large even if the HPA output power per unit is small.

On the other hand, an active array needs many phase shifters. The number of phase shifters is (the number of antenna elements) × (the number of beams). The phase shifters were reduced in size by bending the transmission lines for high-density packaging and by using a substrate with a high dielectric constant ($\varepsilon_r = 21$). Because the minimum spacing between the transmission lines is on the order of the substrate thickness, the coupling between the transmission lines affects the phase-shift error. This coupling is therefore taken into account in the design process. Figure 23.30 shows a 4-bit PIN diode phase shifter.

As shown in Fig. 23.31, the subarray of the phased array consists of seven circular microstrip patches. A microstrip antenna is suitable for use in a satellite-borne phased array because it is lightweight and resistant to the vibration occurring during the launch, but its bandwidth is generally small. The patches in the microstrip subarray are therefore printed on a Nomex honeycomb substrate of 10-mm thickness and low dielectric constant ($\varepsilon_r = 1.2$) for broadband operations. Each patch is excited at two points with 90°phase shift by the rear feeding circuit. The microstrip antenna is arranged so that each patch has small notches that cancel the elliptically polarized components generated as a result of the asymmetrical feed structure. The measured gain of this subarray is more than 16 dBi at both the 2.1-GHz transmitting frequency and the 2.3-GHz receiving frequency. The measured radiation patterns are shown in Fig. 23.32.

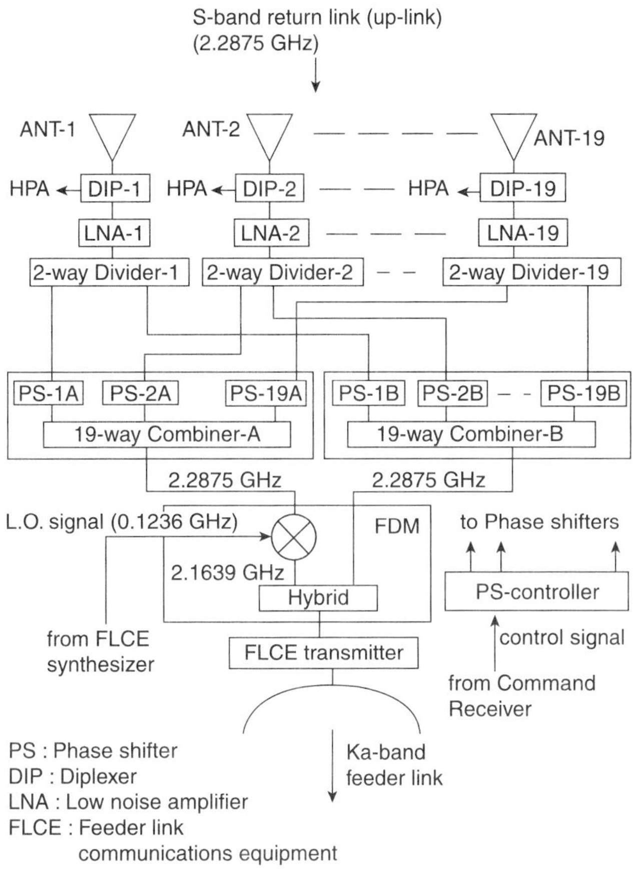

FIGURE 23.29 Block diagram of receiving system.

FIGURE 23.30 A 4-bit PIN diode phase shifter.

23.4.2 In-Orbit Measurement of Satellite-Borne Phased Arrays

It is important that the excitation amplitude and phase of each antenna element of a satellite-borne phased array be measured in orbit. Although the phase is not easy to measure there, both the phase and

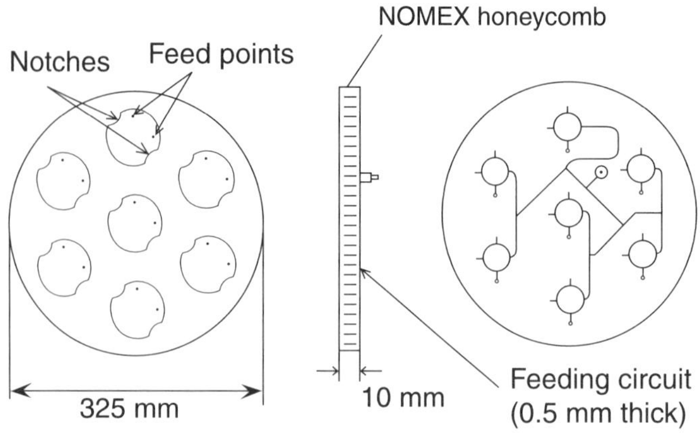

FIGURE 23.31 Subarray of 19-element satellite-borne phased array.

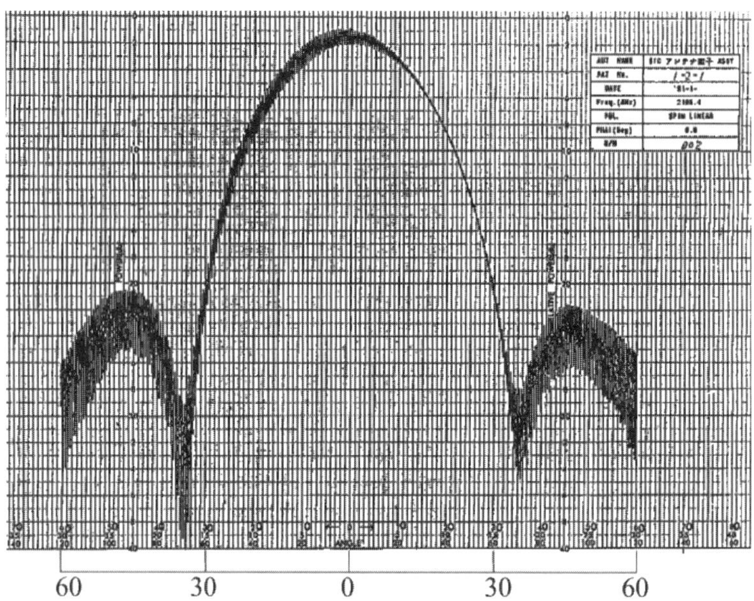

FIGURE 23.32 Radiation pattern of subarray.

the amplitude of each element can be determined by using the rotating element electric field vector method (REV method).[27] All that needs to be done is measure the amplitude change of the composite electric field of the phased array when using a phase shifter to vary the phase of an element.

The electric field vector in a given direction is represented by combining the electric field vectors of the elements, as shown in Fig. 23.33(1). When the phase of an element is varied, the electric field vector of that element rotates and the composite electric field vector changes as shown in Fig. 23.33(1). The amplitude and phase of that element can be determined by measuring the amplitude change of the composite electric field vector and processing the results with a computer.

In Fig. 23.33(1) the amplitude and the phase of the composite electric field vector $\dot{\mathbf{E}}_0$ at the initial state are denoted $\mathbf{E}_0$ and ϕ_0, respectively. Let the amplitude and the phase of the electric field vector $\dot{\mathbf{E}}_n$ of the considered nth element at the initial state be E_n and ϕ_n, and suppose that the phase of the nth element is varied by Δ. Let the amplitude of the composite electric field be E and let the composite power be Q. Q changes with the phase in sinusoidal form as shown in Fig. 23.33(2), where Δ_0 is the phase that

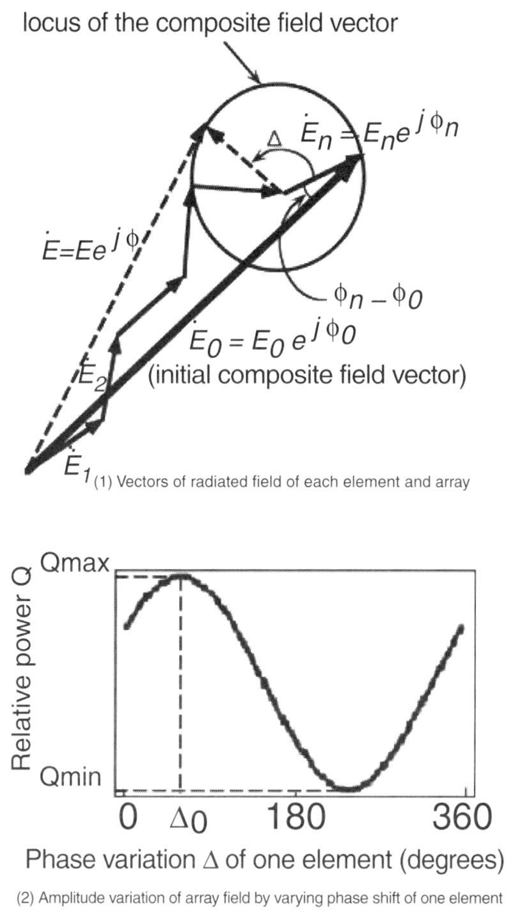

FIGURE 23.33 Principle of the rotating element electric field vector method, (1) Vectors of radiated field of each element and array, (2) Amplitude variation of array field by varying phase shift of one element.

maximizes Q. Let γ be the absolute value of the square root of the ratio of the maximum value of Q to the minimum value of Q. That is,

$$\gamma = \left| \left(\frac{Q_{max}}{Q_{min}} \right)^{1/2} \right| \tag{23.7}$$

When $\dot{E}_o$ and $\dot{E}_n$ are initially set so that

$$\left| \dot{E}_o - \dot{E}_n \right| > \left| \dot{E}_n \right| \tag{23.8}$$

The relative amplitude and the relative phase of the nth element are calculated from γ and Δ_0 by using the following expressions:

$$\frac{E_n}{E_0} = \frac{\Gamma}{\sqrt{1 + 2\Gamma \cos \Delta_0 \Gamma^2}} \tag{23.9}$$

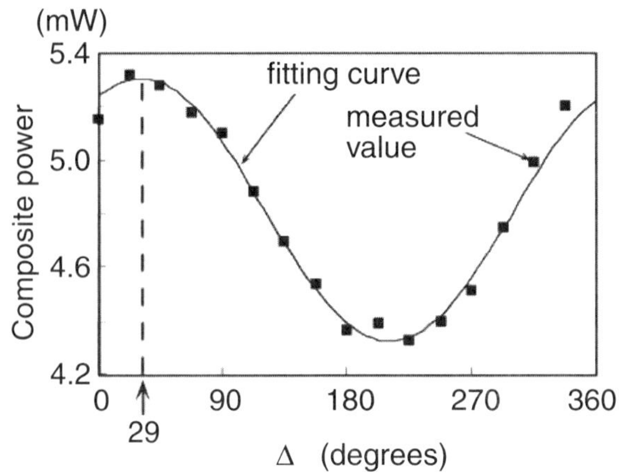

FIGURE 23.34 Example of amplitude variation of the array composite field.

and

$$\phi_n - \phi_0 = -\tan^{-1}\left(\frac{\sin\Delta_0}{\cos\Delta_0 + \Gamma}\right) \tag{23.10}$$

where

$$\Gamma = \frac{\gamma - 1}{\gamma + 1} \tag{23.11}$$

Thus, the behavior of the composite power Q of the array is measured by varying the phase of the nth element. The amplitude and the relative phase of the element are obtained from Eqs. (23.9) and (23.10) by determining the maximum and the minimum values of Q as well as the phase Δ_0 that maximizes Q. It should be noted that the relative amplitude and the phase determined by Eqs. (23.9) and (23.10) are those in the initial state.

The REV method was used to evaluate the phased-array antenna in the Japanese ETS-VI, as mentioned in previous section. The setting of the phase shifter was controlled by commands from Earth, and Fig. 23.36 shows an example of the change of the array composite power measured when the peak of the mainlobe was directed toward the earth station and the phase of an element was varied. The curve for the absolute power (mW) is shown so that the sinusoidal change of the composite power can be seen clearly. The black squares in the figure are measured values, and the curve is the sinusoidal curve obtained by least squares fitting. Because the measured value fluctuated as a result of measurement error, there can be errors in γ and Δ_0. Their actual values are therefore best determined from the obtained sinusoidal curve by using least squares approximation. The results in Fig. 23.34 were obtained on the peak of the mainlobe as the initial state. Thus, the composite electric field should be the maximum for $\Delta = 0°$, but the field is actually the maximum for $\Delta = 29°$, which indicates that there is an error in the excitation phase.

Figure 23.35 is an example of the contour representation of the amplitude and the phase errors on the aperture plane of the phased array in the ETS-VI.[28] The outer hexagon indicates the radiation plane. Each grid point of the triangle corresponds to the position of the element. The number attached to the grid point is the element number. The phased array is designed as the uniform distribution for the amplitude, and ideally the relative amplitude of the element should be equal. The error of the magnitude of 3 $dB_{p\text{-}p}$ is produced overall. As to the phase, a calibration is already made on the ground before

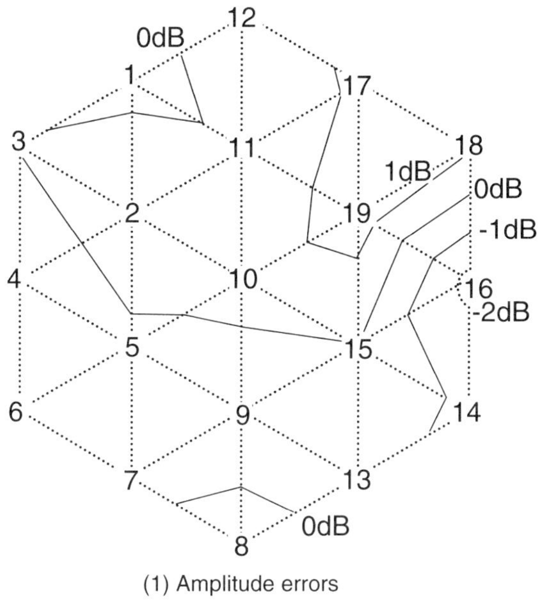

FIGURE 23.35 Contour expression of amplitude errors and phase errors on the antenna plane, (1) Amplitude errors

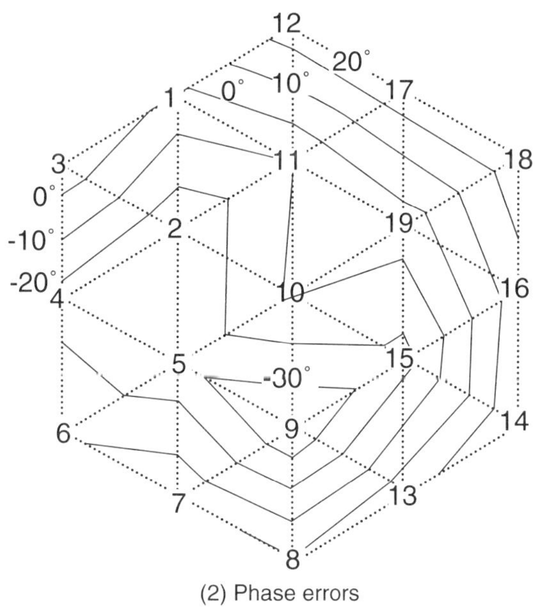

FIGURE 23.35 Contour expression of amplitude errors and phase errors on the antenna plane, (2) Phase errors.

launching, using the same REV method. Ideally, what remains as the only indeterministic component should be the quantization error of the 4-bit digital phase shifter; that is, the phase error within ± 11.25°. However, from Fig. 23.35 we see that a phase error of about $70°_{p\text{-}p}$ can be produced. Because the phase of the phased array was already calibrated by the test on the ground, the phase error in Fig. 23.35 is the deviation from the result of the test on the ground. Such a phase error seems to result from the mechanical error as well as the quantization error in the digital phase shifter.

FIGURE 23.36 Airborne phased array.

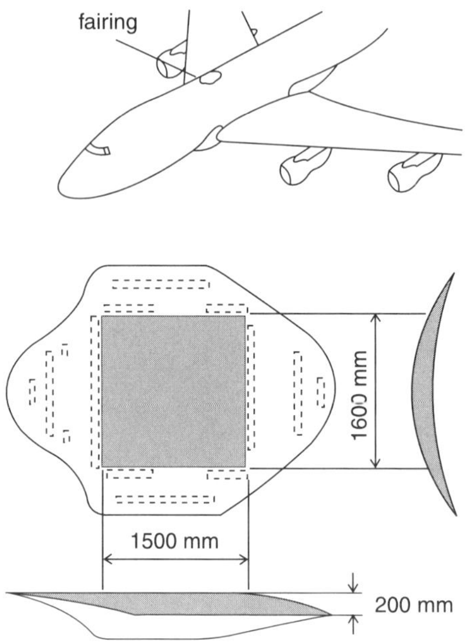

FIGURE 23.37 Fairing of a Boeing 747 Jumbo Jet.

23.4.3 Airborne Phased Array

Japanese experimental domestic mobile satellite communications, using the *ETS-V* launched in 1987, provides high-quality links for ships and aircraft at L-band. The phased array shown in Fig. 23.36 has been developed for aircraft–satellite communication.[29] It is mounted in the fairing of a Boeing 747 Jumbo Jet as shown in Fig. 23.37. From a limited space of only 20 cm in height of the fairing and the requirement for high gain over ±60° in azimuth, the phased array becomes a roof-shape structure. The phased array has two array panels to provide the required wide coverage and only one of the array panels facing the satellite is used for communication.

The phased array consists of 2 × 8 microstrip patches, and the beam can be scanned only in azimuth. Thus, each of eight 4-bit digital phase shifters is connected to two vertical elements, as shown in Fig. 23.38. By controlling the phase shifters, the antenna beam can be scanned in 4° steps. The required coverage

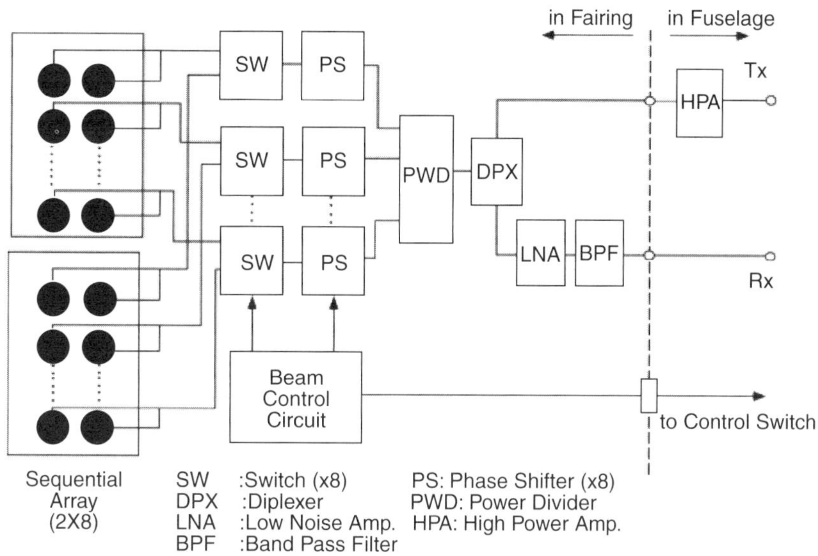

FIGURE 23.38 Configuration of airborne phased array.

TABLE 23.2 Characteristics of Airborne Phased Array

Frequency	Tx: 1545–1548 MHz
	Rx: 1647–1650 MHz
Polarization	LHCP
Gain	Tx: 14.7 dBi (nonscan)
	Rx: 13.5 dBi (nonscan)
G/T	–10.8 dBK (nonscan)
Element	Circular patch
Array	2 × 8 Sequential array
Substrate	Glass Teflon ($\varepsilon_r = 2.6$)
Axial ratio	Less than 2.0 dB (nonscan)
VSWR	Less than 1.4 (nonscan)
Tracking	Step track
Beam step	4°
Phase shifter	4-bit digital
Weight	18 kg
Volume	760 (L) × 320 (W) × 180 (H) mm

angle in elevation is as narrow as ±20°, so that the beams are not steered in elevation directions but the beam covers by its wide beam width in elevation. A microstrip antenna is suited to airborne antenna, because of its very low profile, light weight, and mechanical strength. However, one disadvantage is a very narrow frequency bandwidth, usually 1%, which does not satisfy the required value of 7%. A two-frequency resonant element with a one-point feed was used to cover both the transmitting band and receiving band. This type of element has a very poor axial ratio, but that problem was overcome by using the sequential array technique described in the next section. The characteristics of the array are listed in Table 23.2.

23.4.4 Sequential Array Technique

In a sequential array, perfect circular polarization is obtained at a boresight direction independently of the polarization of elements.[30] For the rectangular sequential array it has high cross-polarization discrimination over the wide angle in the two principle planes, which is useful for reducing fading resulting from reflection from the sea surface.

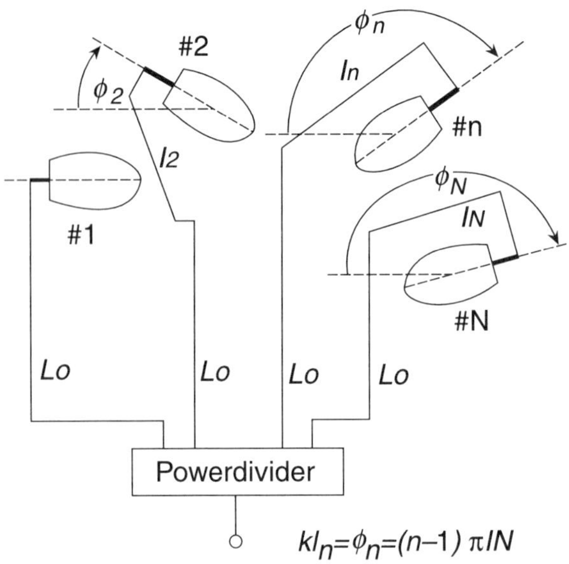

FIGURE 23.39 *N*-element sequential array.

In a sequential array, the incremental angular orientation and the excitation phase difference are provided sequentially to each element. The configuration of an *N*-element sequential array is illustrated in Fig. 23.39. The *n*th element is located at an arbitrary position on a plane, but with an orientation angle of

$$\phi_n = P(n-1)\pi/N \tag{23.12}$$

where P is an integer and $1 \le P \le N - 1$. The *n*th element is also fed with a differential phase shift of ϕ_n radians.

It is assumed that the polarization of the field radiated from the first element is elliptical in the boresight direction and is expressed by

$$\mathbf{E}_1 = a\mathbf{U}_1 + jb\mathbf{V}_1 \tag{23.13}$$

where $\mathbf{U}_1$ and $\mathbf{V}_1$ are orthogonal unit vectors, respectively, corresponding to the major and the minor axes of the polarization ellipse, and a and b are the amplitudes of both components. The total field radiated from the sequential array in the boresight direction can be derived as

$$\mathbf{E} = \sum_{n=1}^{N} \mathbf{E}_n = \frac{(a+b)}{2} N(\mathbf{U}_1 + j\mathbf{V}_1) \tag{23.14}$$

This means that the sequential array radiates perfect circularly polarized waves in the boresight direction regardless of the polarization of the elements.

If the elements of the rectangular sequential array shown in Fig. 23.40 are phase shifted to scan the beam to (θ_0, ϕ_0), the field radiated in the beam direction is

$$\mathbf{E}(\theta_0,\phi_0) = (1+\alpha)MN\exp(j\phi_0)\{E_p(\theta)\mathbf{e}_\theta + jH_p(\theta)\mathbf{e}_\phi\}/2 \tag{23.15}$$

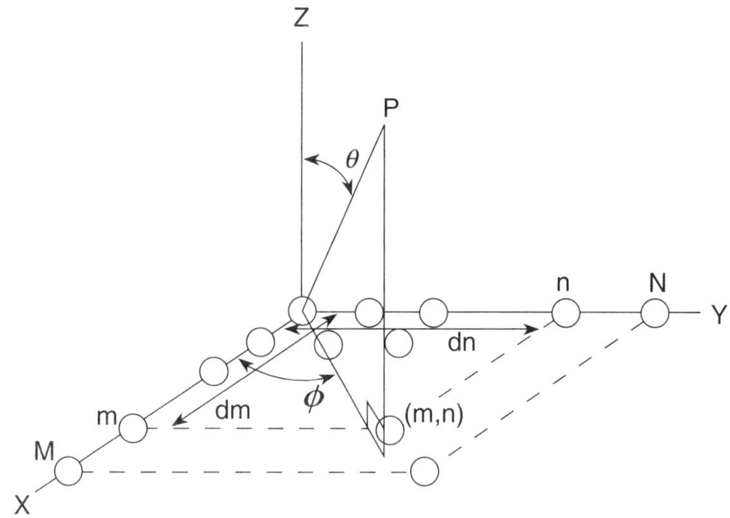

FIGURE 23.40 Geometry of a planar sequential array.

where α is a complex excitation factor of each element (α = 1 means perfect polarization), $E_p(\theta)$ and $H_p(\theta)$ are E-plane and H-plane patterns, $\mathbf{e}_\theta$ and $\mathbf{e}_\phi$ are unit vectors.[31] From the preceding equation, it can be seen that within the angular region where $E_p(\theta) = H_p(\theta)$ holds, the scanned beam has an excellent circular polarization regardless of the polarization of the elements.

23.4.5 Tracking Error of Phased Array

When the same phase shifters are used at both transmitting and receiving frequencies, the beam scanning errors are observed between both frequencies. Figure 23.41 shows the radiation patterns of the airborne phased array. The dots are the data measured at the transmitting frequency (f_T; 1.6 GHz), and the circles are the data measured at the receiving frequency (f_R; 1.5 GHz). There is an angular difference of about 8° between f_R and f_T frequencies.

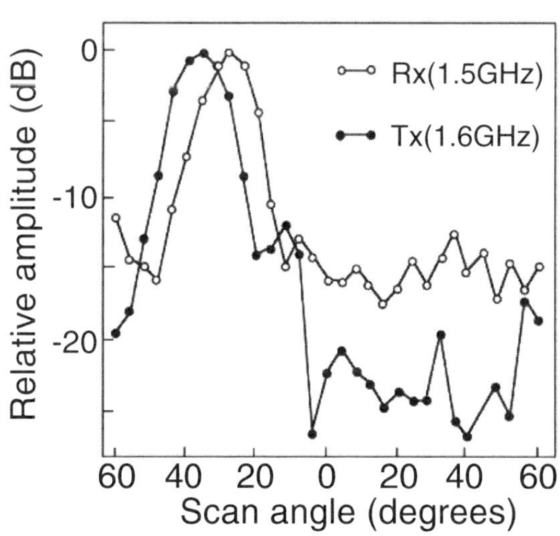

FIGURE 23.41 Radiation patterns measured in transmitting and receiving frequency.

Phased arrays direct the main beam maximum by controlling a variable phase shift between the array elements. Consider an array of N elements located along the x-axis. As is the case of the array antenna under consideration, if an array is linearly spaced at equal distances d and uniformly excited with a variable phase shift ϕ between the adjacent elements, a normalized array factor is given by the following equation:

$$\left|f(\theta)\right| = \frac{\sin\left\{(N/2)\left[\phi + (2\pi d/c) f \sin\theta\right]\right\}}{N \sin\left\{\left[\phi + (2\pi d/c) f \sin\theta\right]/2\right\}} \tag{23.16}$$

where θ, c, and f denote the beam scan angle with respect to the y-axis, the speed of light, and frequency, respectively.

The direction of the beam maximum is where the argument of the sine function in Eq. (23.16) is equal to zero. Therefore, the interelement phase difference required to scan the beam to an angle θ from the array normal is given by

$$\phi = -(2\pi d/c) f \sin\theta \tag{23.17}$$

When the frequency is changed by Δf, the beam scanning and phase deviation angles $\Delta\theta$ and $\Delta\phi$ are related by

$$(\phi + \Delta\phi) = -(2\pi d/c)(f + \Delta f)\sin(\theta + \Delta\theta) \tag{23.18}$$

When the value of phase is assumed constant or is independent of frequency, $\Delta\phi$ is equal to zero, and then the right sides of Eqs. (23.17) and (23.18) can be set equal as follows:

$$f \sin\theta = (f + \Delta f)\sin(\theta + \Delta\theta) \tag{23.19}$$

From Eq. (23.19) the value of $\Delta\theta$ is given by

$$\Delta\theta = -\theta + \sin^{-1}\left[\sin\theta / (1 + \Delta f / f)\right] \tag{23.20}$$

To eliminate the scanning errors, the following equation must be satisfied, where ϕ_R and ϕ_T denote the value of phase shifte at f_R and f_T, respectively.

$$(\phi_R / f_R) = (\phi_T / f_T) \tag{23.21}$$

From Eq. (23.21), it is found that the adoption of frequency-dependent phase shifters can potentially reduce the beam tracking errors between transmitting and receiving frequencies.[32]

23.4.6 Car-Borne Phased Array

Advanced experiments for mobile satellite communication in the Ka-band were carried out using the Japanese Communications and Broadcasting Engineering Test Satellite (COMETS), which was launched in 1998. A Ka-band active phased array was the most technically challenging component in the COMETS system development. It consists of 168 microstrip antennas with microwave monolithic integrated circuit (MMIC) phase shifters and MMIC LNAs, and at 21 GHz it provides a minimum G/T of –6.8 dB/K at an elevation angle of over 42°.[33] Figure 23.42 is a photograph of the 110 × 120-mm active phased array.

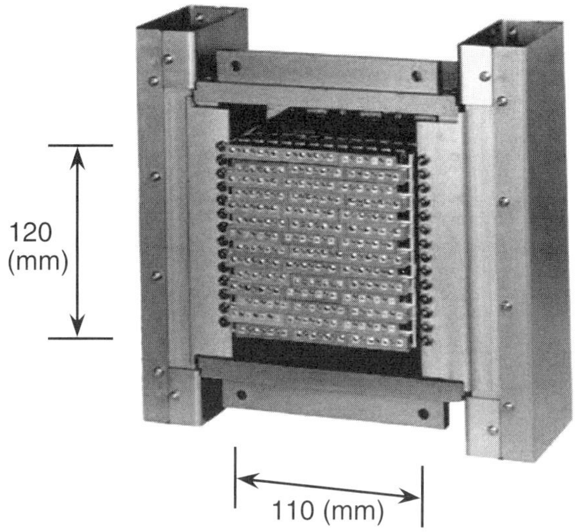

FIGURE 23.42 Ka-band car-borne active phased array.

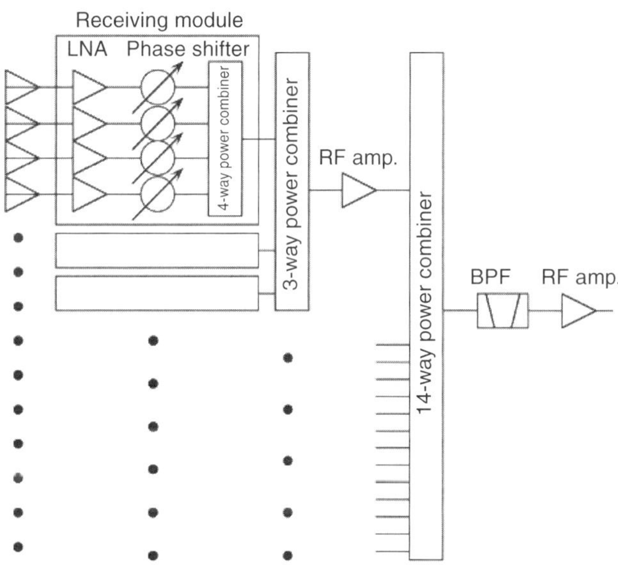

FIGURE 23.43 Block diagram of the Ka-band active phased array.

A block diagram of the array is shown in Fig. 23.43, and the array characteristics are listed in Table 23.3. It provides a receive-only function with a bandwidth of 92 MHz (less than 0.5%). A single four-way power combiner and four series of MMIC LNAs and MMIC phase shifters are assembled onto a single receiving module as shown in Fig. 23.44. The module also has control circuits to drive and control the LNAs and the phase shifters.

The beam-pointing accuracy is 1° root mean square (rms), which is one tenth of the beam width. Angle data giving the satellite direction can be input to the beam controller every 10 ms from a mobile attitude sensor in a car.

The active phased array and a phase-shifter driver unit are installed inside a platform on a roof rack. Each radiating element of the active phased array is a rectangular microstrip patch with two notched corners. These elements can radiate circular polarization by one-point feeding. Because microstrip

TABLE 23.3 Characteristics of Ka-band Active Phased Array

Frequency	21.028 GHz ± 46 MHz (only receiving)
G/T	> –6.8 dB/K
Beam scan range	El: >42°, Az: 0° to 360°
Pointing accuracy	1° rms
Polarization	LHCP
Axial ratio	<6 dB
Element number	168
Element allocation	Triangular form
Element spacing	0.6 λ
Radiating element	Rectangular microstrip antenna
Phase shifter	4-bit MMIC
LNA	3-Stage MMIC
Array size	110 × 120 mm
System dimension	L 390 × W 410 × H 230 mm
Weight	<30 kg
Temperature range	–5° to 50°

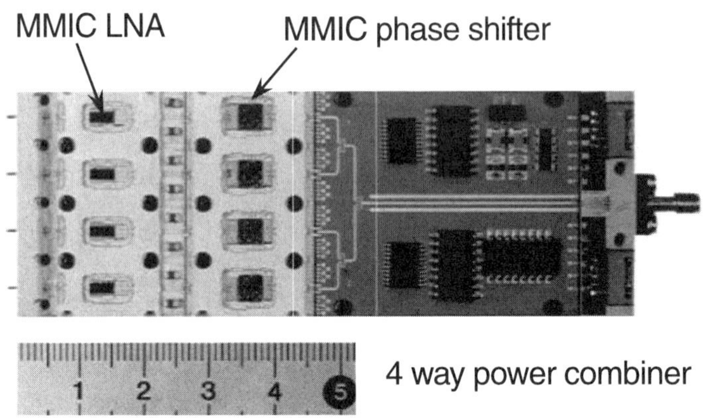

FIGURE 23.44 Receiving module of Ka-band active phased array.

patches have a very narrow bandwidth of less than 1%, the frequency band of the antenna may move out of the required frequency range when the dielectric constant of the substrate changes as a result of a change in temperature. A Teflon™ substrate with a temperature-stable dielectric constant is used to keep the resonance frequency of the microstrip patch from changing. Within the operating temperature range of the array, from –5° to +50°, the dielectric constant ranges only from 2.950 to 2.947. The resonant frequency of the rectangular microstrip patch is given by

$$f = \frac{c}{2\left(d + t/2\right)\sqrt{\varepsilon_r}} \tag{23.22}$$

where c is the velocity of light, d is the length of the patch, t is the thickness of the substrate, and ε_r is the dielectric constant. For the microstrip patch that we used, d is 3.89 mm and t is 0.254 mm. The resonant frequency change resulting from the change in the dielectric constant over the entire range of operating temperature is only 0.05%, which is less than the bandwidth of the microstrip patch.

To make a compact 21-GHz, 4-bit MMIC phase shifter, self-switched filter circuit configurations are adopted for the 180°, 90°, and 45° sections. The circuit topologies of the self-switched filter are changed between a low-pass filter and a high-pass filter by switching the state of the FETs. The layouts and insertion losses of the self-switched filter circuits are smaller than those of circuits using single-pole double throw

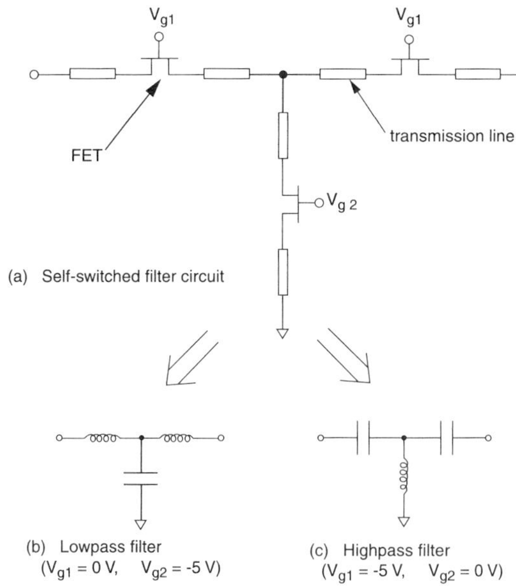

FIGURE 23.45 Circuit topology of the phase shifter using a self-switched filter circuit.

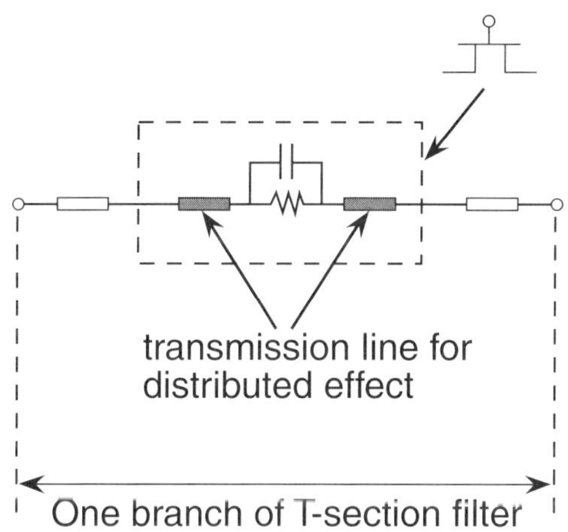

FIGURE 23.46 Equivalent circuit model of FET.

(SPDT) switches to switch between the low-pass and high-pass filter sections. Figure 23.45 shows a self-switched filter circuit with a T-section filter. The FET is inductive when the gate voltage of an FET is set to 0 V, and the FET is capacitive when it is set to –5 V. Thus, depending on the gate voltages of three FETs, the T-section filter alternates between acting as a low-pass filter and as a high-pass filter.

Figure 23.46 shows each branch of the T-section filter using the equivalent circuit model of a GaAs FET. At higher microwave frequencies, such as 21 GHz, the distributed effect of the electrode structures of an FET cannot be disregarded to the electric lengths of the transmission lines for the inductances. Therefore, a filter has been designed taking into account the distributed effects of the FET.

23.4.7 Conclusion

This chapter discusses the technologies used in phased arrays for satellite communications based on the actual examples of satellite-borne, airborne, and car-borne phased arrays. In particular, the method to measure the satellite-borne phased arrays in orbit, the sequential array technique, and the method to reduce tracking error have been provided in detail. The more advanced the satellite communications, the more the new phased arrays are used in satellite communications. If these techniques described here are useful to the research and development of the new phased arrays, the author will be pleased.

23.5 Conclusion

The phased array antennas play an important role in communication systems that require beam steering, nulling, tracking, or capturing. These functions are utilized in communication systems to increase system transmission speed and capacity. In the near future, the phased-array antenna techniques will be installed not only at the base (satellite) control station but also at the mobile terminals.

References

1. Kimura, S., Fujiwara, Y., Taniguchi, T., and Yamamoto, K., Development of adaptive array antenna for mitigating lack of frequency in high-traffic areas, *EMC*, 124–125, 37–42, 1998.
2. Litva, J. and Lo, T. K., *Digital Beamforming in Wireless Communications*, Artech House, Norwood, MA, 1996.
3. Kanazawa, A. et al., The feasibility study of the dynamic zone configuration technique with a developed circular array antenna, *IEICE Trans. Fundam.*, E82-A, 7, 1210, 1999.
4. Kanazawa, A. et al., Performance evaluation of adaptive zone configuration system based on developed array antenna and sector antenna prototypes, in *Proc. MTT-S European Wireless*, 1998, 325.
5. Ichikawa, N. et al., Performance evaluation of adaptive zone configuration system, in *Proc. PIMRC*, 1998.
6. Godara, L. C., Application of antenna arrays to mobile communications, part II: beam-forming and direction-of-arrival considerations, in *Proc. IEEE*, 85, 8, 1195, 1997.
7. Haykin, S., *Adaptive Filter Theory*, Prentice-Hall, Englewood Cliffs, NJ, 1996.
8. Schmidt, R. O., Multiple emitter location and signal parameter estimation, *IEEE Trans. Antennas Propag.*, 34, 3, 276, 1986.
9. Shan, T. J., Wax, M., and Kailath, T., On spatial smoothing for direction-of-arrival estimation of coherent signals, *IEEE Trans. Acoust., Speech, Signal Process.*, ASSP-33, 4, 806, 1985.
10. Roy, R. et al., ESPRIT — Estimation of signal parameters via rotational invariance techniques, *IEEE Trans. Acoust., Speech, Signal Process.*, 37, 984, 1989.
11. Kagiwada, H., Ohmori, H., and Sano, A., A recursive algorithm for tracking DOAs of moving targets by using linear approximations, *IEICE Trans. Fundam.*, E-81A, 1998.
12. Okada, Y. et al., Millimeter wave broadband wireless access system with tracking technology of moving targets, in *Proc. IEEE VTC'98*, Ottawa, Canada, 1998, 2057.
13. Kanazawa, A. et al., Experimental analysis of DOA estimation and BER performance using adaptive array antenna system, in *Proc. IEEE ISPAST'2000,* 549, 2000.
14. Proc. 1st Stratospheric Platform Workshop (SPSW'99), Yokosuka, Japan, May 1999.
15. Hase, Y., Miura, R., and Ohmori, S., A novel broadband all-wireless access network using stratospheric platforms, in *Proc. VTC'98*, Ottawa, Canada, 1998, 1191.
16. Djuknic, G. M., Freidenfelds, J., and Okunev, Y., Establishing wireless communications services via high-altitude aeronautical platforms: A concept whose time has come?, *IEEE Commun. Mag.* Sept., 128, 1997.
17. http://www.skystation.com/
18. Lee, Y. and Ye, H., Sky station stratospheric telecommunications system, a high speed low latency switched wireless network, in *Proc. 17th AIAA ICSSC*, Yokohama, Japan, 1998, 25.
19. Rothblatt, M., Frohbieter, J., and Ye, H., Stratospheric altitude: the key to delivering broadband wireless local loop service to consumers worldwide, *ITU News*, Aug., 37, 1997.

20. *Aerological data of Japan — 30 year period averages (1961–1990),* Japan Meteorological Agency, 1991.
21. *Radio Regulations*, 1998 Edition, 13, ITU, 1998.
22. *Provisional Final Acts and Working Documents*, WRC-2000, Istanbul, Turkey, 2000.
23. Karasawa, Y., The software antenna: A new concept of kaleidoscopic antenna in multimedia radio and mobile computing era, *IEICE Trans. Commun.*, E80-B, 8, 1214. 1997.
24. Farina, A., *Antenna-Based Signal Processing Techniques for Radar Systems*, Artech House, Norwood, MA, 1992.
25. Miura, R., Oodo, M., Kanazawa, A., and Koyama, Y., Maximal-Ratio-Combining Array Beamformer by a Training Sequence for Space Division Multiple Access in Power-Limited Channels, *IEICE Trans. Commun.*, E83-B, 2, 394, 2000.
26. Tanaka, M., Kimura, S., Teshirogi, T., Matsumoto, Y., Ito, T., Akaishi, A., Mizutamari, H., and Okubo, S., Development of multibeam phased array antenna on ETS-VI for S-band intersatellite communications, *Electron. Commun. Jpn.*, Part 1, 77, 1, 117, 1994.
27. Mano, S. and Katagi, T., A method for measuring amplitude and phase of each radiating element of a phased array antenna, *Electron. Commun. Jpn.*, Part 1, 65-B, 5, 58, 1982.
28. Tanaka, M., Matsumoto, Y., Kozono, S., Suzuki, K., Yamamoto, S., and Yoshimura, N., On-orbit measurement of phased arrays in satellites by rotating element electric field vector method, *Electron. Commun. Jpn.*, Part 1, 81, 1, 1, 1998.
29. Taira, S., Tanaka, M., and Ohmori, S., High gain airborne antenna for satellite communications, *IEEE Trans. Aerospace Electron. Syst.*, 27, 2, 354, 1991.
30. Teshirogi, T., Tanaka, M., and Chujo, W., Wideband circularly polarised antenna with sequential rotations and phase shifts of elements, in *Proc. Int. Symp. Antennas Propag.*, 117, 1985.
31. Teshirogi, T., Tanaka, M., and Ohmori, S., Airborne phased array antenna for mobile satellite communications, in *IEEE AP-S Int. Symp. Antennas Propag. Dig.*, 735, 1986.
32. Ohmori, S., Tracking error of phased array antenna, *IEEE Trans. Antennas Propag.*, 39, 1, 80, 1991.
33. Tanaka, M., Yamamoto, S., Obara, N., Saito, H., and Miura, R., Ka-band mobile-vehicular active phased array antenna system for mobile satellite communications, in *Proc. 17th Int. Commun. Satellite Syst. Conf.*, AIAA-98-1306, 431, 1998.

24

Adaptive Antennas for Global System for Mobile Communications and Time Division Multiple Access (Interim Standard-136) Systems*

Sören Andersson
Ericsson Radio Systems

Bo Hagerman
Ericsson Radio Systems

Magnus Berg
Ericsson Radio Systems

Henrik Dam
Ericsson LMD

Ulf Forssén
Ericsson Radio Systems

Jonas Karlsson
Ericsson Radio Systems

Fredric Kronestedt
Ericsson Radio Systems

Sara Mazur
Ericsson Radio Systems

Karl J. Molnar
Ericsson Incorporated

24.1 Introduction

The deployment of cellular systems based on Global System for Mobile Communications (GSM) and Time Division Multiple Access (TDMA) or Interim Standard[136] is a worldwide success with the standards

*A version of this chapter appears in *IEEE Personal Commun.*, vol. 6, no. 3, June 1999, pages 74–86.

0-8493-0124-6/02/$0.00+$1.50

in operation in, for example, Europe, China, the United States, Australia, South Africa, and large parts of Asia as well as the current buildup of networks for Personal Communications Services (PCS) services in the United States and most of the South American countries. Today, more than 100 countries use digital GSM and TDMA (IS-136) technology.

As the number of users increase, more capacity is needed. This is, in particular, so for the 800- to 900-MHz band where the bandwidth is limited. The available bandwidth at 1800 to 1900 MHz is larger, but, on the other hand, the path loss is higher. There are several ways to increase the capacity in a cellular network, such as frequency hopping, power control, introduction of microcell technology in hot spot traffic areas, and introduction of a half-rate code/decode (codec). The application of adaptive antennas may also turn out to be an important step in the system evolution process of current radio network technology, and of interest not only for third-generation wireless systems. Adaptive antennas have generated great interest in recent years, and at present several manufacturers and operators are separately or jointly performing field tests to gain more detailed knowledge of the potential of the technology; see References [1], [2], [3], [4], and [5] for good overviews and extensive reference listings of both algorithms and concepts. This chapter summarizes many of the experiences that have been collected since 1996 during field trials carried out for both GSM and TDMA (IS-136) projects by Ericsson in cooperation with two major network operators.

The chapter is organized as follows. This section gives an overview of the most basic characteristics of the air interface parts of the GSM and TDMA (IS-136) wireless systems. Then, Section 24.2 outlines some of the most important issues to consider when applying adaptive antenna techniques in already existing cellular systems; next, Section 24.3 discusses some possible system architectures suitable for implementation. Algorithm aspects are treated in Section 24.4, and then one can compare link-level simulations with link-level measurement results obtained from actual field trials in Sections 24.5 and 24.6. Section 24.5 also gives an overview of system-level simulation results. An overview of test scenarios and some results from the extensive field measurement campaigns carried out are given in Section 24.6. Packet data systems, which is the evolutionary path defined for GSM and TDMA (IS-136), are in Section 24.7 evaluated with respect to the performance improvements possible to achieve when employing adaptive antennas. Finally, we end in Section 24.8 with some concluding remarks and issues where further work is necessary.

24.1.1 Brief Overview of the Global System for Mobile Communications Air Interface

The multiple access scheme in GSM is a combination of frequency division multiple access (FDMA) and TDMA. The spectrum is divided into 200-kHz-wide *frequency slots*, where the center frequencies are the carrier frequencies. Each frequency slot is divided into eight *time slots*, which means that the TDMA factor (i.e., the maximum number of users that can share one frequency slot) is eight. The length of the time slot is 0.577 ms, or more exactly 15/26 ms. Each user is assigned one of these time-frequency resources, as depicted in Fig. 24.1.

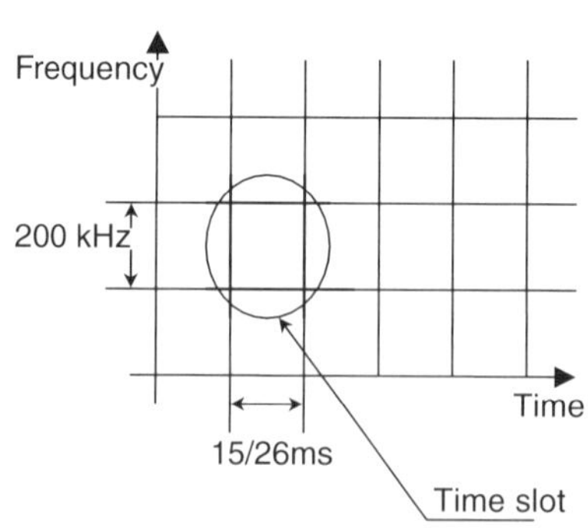

FIGURE 24.1 The combination of FDMA and TDMA used in GSM shown in a time-frequency diagram.

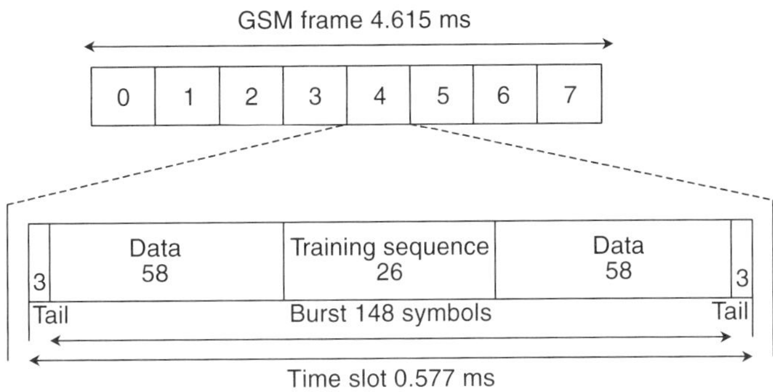

FIGURE 24.2 The structure of a normal burst in GSM. A user sends a burst every eight time slots. A set of eight time slots is called a frame.

24.1.1.1 Frame and Burst Structure for Global System for Mobile Communications

The set of eight time slots on one frequency slot is called a *frame*. A frame has in GSM a duration of 60/13 ms (= 4.615 ms). Within a time slot a *burst* of data is sent. The frame and the structure of a normal burst (that is used during a regular speech session) are shown in Fig. 24.2. Also bursts exist that have other structures, but they are mainly used for initial synchronization and/or signaling between the base station and the mobile terminal and are not further discussed here. A normal burst consists of two 58-*symbol* long packets of data surrounding a training sequence containing 26 symbols (a symbol can consist of several bits, but in GSM a symbol consists of 1 bit only). The training sequence has a known symbol pattern that is used for the on-line synchronization as well as for estimating the channel during the demodulation process. Three predetermined tail symbols are added on each side to provide the demodulation process with known final states. The total length of a normal burst is thus 148 symbols.

24.1.1.2 Frequency Hopping for Global System for Mobile Communications

Frequency hopping means that the carrier frequency is changed among a certain number of allowed frequencies at regular intervals. In GSM, this occurs for every burst (i.e., every 4.615 ms, resulting in 217 shifts per second). The advantages with frequency hopping are two. First, it combats multipath fading; a slowly moving mobile terminal may stay in a fading dip for a long time, but with frequency hopping activated the fading pattern changes, thus making only part of the transmission or reception lost. This phenomenon is commonly referred to as *frequency diversity*. Second, without frequency hopping, all strong interfering signals continuously degrade the reception of the desired signal. Frequency hopping activated in this case causes different signals to interfere with the desired signal at different times and from a system level perspective this results in an averaging effect because the interference is distributed more evenly between users; this is a phenomenon commonly referred to as *interference diversity*.

Frequency hopping can be performed either cyclically or (pseudo) randomly. *Cyclic hopping* means that all cells use the same hopping sequence, which implies that the same signals interfere with each other at all times. Thus, with cyclic hopping the interference diversity may be limited. In *random hopping*, the cells use uncorrelated pseudorandom hopping sequences, making the interference vary randomly for each burst transmitted.

24.1.2 Brief Overview of the Time Division Multiple Access (Interim Standard-136) Air Interface

In the multiple access scheme in TDMA (IS-136), no frequency hopping option is available. The spectrum is divided into 30-kHz-wide frequency slots, where the center frequencies are the carrier frequencies. Each frequency slot is divided into six time slots. Two of these are used for each full-rate speech channel, which means that three traffic channels are available per frequency carrier. The length of a time slot is 6.67 ms, or more exactly 20/3 ms.

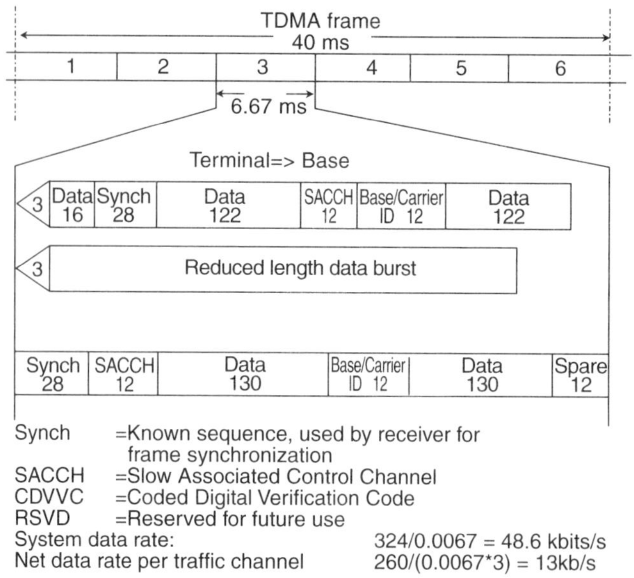

FIGURE 24.3 The structure of a normal burst in TDMA (IS-136). A set of six time slots is called a frame. A full-rate speech channel is composed of two time slots per frame.

24.1.2.1 Frame and Burst Structure for Time Division Multiple Access or Interim Standard-136

The set of six time slots on one frequency slot is called a frame. A frame has in TDMA (IS-136) a duration of 40 ms. The frame and burst structures are shown in Fig. 24.3. Each time slot carries an information sequence, *system signaling* slow associate control channel (SACCH), and an identification (ID) code giving information about the base station/operator and carrier. A synchronization sequence is also needed, which is also used as training sequence for the channel equalizer. Each time slot contains 260 traffic bits; because 25 frames per second are transmitted and two time slots per frame are used for a full-rate channel, the gross data rate for a traffic channel is 13 kb/s ($260 \times 25 \times 2$). As in the case of GSM, TDMA (IS-136) also has bursts that have other structures, mainly used for initial synchronization and/or signaling between the base station and the mobile terminal.

24.1.3 Basic Structure of Signal Processing for Global System for Mobile Communications and Time Division Multiple Access or Interim Standard-136

The block structure of a wireless system from a signal processing perspective is shown in Fig. 24.4. The same basic blocks are required for both GSM and TDMA (IS-136), but the actual implementation, of course, differs because of the difference in parameter settings for the two systems. For example, the much longer time slots in TDMA (IS-136) impose the requirement on the channel equalization to continuously adapt to changes in the impulse reponse of the propagation channel during a data burst. This renders the channel equalization more complex compared with GSM, which can use the same channel estimate over the whole burst because of the fact that the channel can be regarded as quasi stationary during that time period. Table 24.1 summarizes the most relevant parameters for an initial comparison of the air interfaces.

24.2 System-Level Aspects

During a call, the terminals and the base station need to exchange large amounts of information, especially to prepare for the next handover to another cell. Because of the presence of large amounts of interference,

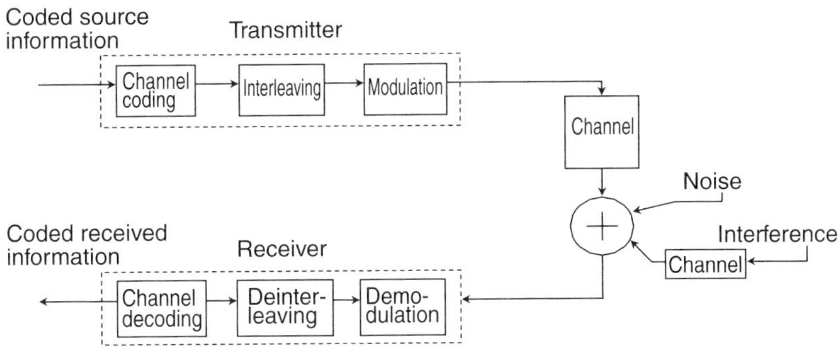

FIGURE 24.4 The signal processing block structure of a transmitter and a receiver in a GSM or TDMA (IS-136) cellular system.

TABLE 24.1 Comparison of the GSM and TDMA (IS-136) Systems

	GSM	TDMA (IS-136)
Channel spacing	200 kHz	30 kHz
Modulation	GMSK	QAM (π/4-DQPSK)
System data rate	271 kb/s	48.6 kb/s
Frame length	4.6 ms	40 ms
Time slot length	8 × 0.57 ms	6 × 6.67 ms
Bit rate full-rate channel (net)	13 kb/s	7.95 kb/s
Bit rate (with channel coding, gross)	22 kb/s	13 kb/s
Interleaving depth (full-rate codec)	8	2
Frequency hopping	Possible	No

it is also sometimes necessary to switch to a different radio channel in the same cell (i.e., perform an *intracell* handover) to be able to continue a connection with good enough quality. All the required signaling for this information exchange is specified in the standards for both GSM and TDMA (IS-136) in terms of protocol procedures for how all signaling must be conducted. Typically, there is also a number of parameters (maximum number of retransmissions, counters, timers, and so on) associated with the protocols, so that some flexibility is available. Therefore, the *main issue* for any new feature (in software and/or hardware, such as adaptive antennas) under consideration for introduction in already existing cellular systems, is that of *backward compatibility.*

Because there are very complex implementation dependent interactions between many of the system-level aspects in a cellular network, an engineer basically needs to sort out the following issues to be able to devise a conceptual solution for adaptive antennas in GSM or TDMA (IS-136) cellular systems. First is the issue of broadcast information; that is, the network-related information that must continuously be received by all mobile stations. Typically, this is information from the network concerning cell identity, which frequencies are used in a specific cell, which frequency hopping sequence is to be used in a certain cell, the maximum power levels allowed, and so on. Because this information always reaches all mobiles, the implication is that the information must be transmitted simultaneously all over the sector's coverage area (i.e., on an antenna that defines the cell in terms of coverage).

Second are the following protocol issues:

- How are the protocols defined for call setup (initial access), handover, paging, random access, and so on? This is relevant for the development of algorithms robust enough for direction estimation in both noise- and interference-limited environments.
- Are there any impacts on the interfaces between the base station and the (possibly distributed) functionality in nodes on a higher level in the network? That is, will — or must — a base station equipped with an array antenna be treated differently than a conventional base station? This may be highly relevant when it comes to implementation complexity and development cost.

Third and finally is the issue of system-level gain vs. array antenna characteristics such as width and height: the antenna size is maybe the most important dimensioning factor of the capacity and/or range extension improvements that can be achieved when applying adaptive antenna technology. In principle, the width (for a fixed height) determines the minimum beam width that can be realized by the adaptive antenna, and this width determines the amount of interference reduction that can be achieved on a system level. The height (for a fixed width) determines the amount of antenna gain increase that can be achieved and thereby also the amount of range extension improvement that can be expected. Thus, this issue ends up in a trade-off between implementation complexity/cost with respect to a selection of the number of antenna elements to implement, the desired system-level performance gain, and antenna (in)visibility requirements. Some antenna design issues are discussed in Reference [19].

24.3 Antenna System Architectures

In adaptive antenna solutions aimed at increasing system capacity, the conventional sector antennas are replaced by one or several antenna arrays. Instead of transmitting information in the entire sector, the basic principle is to direct narrow beams from the base station toward the mobile station. Several downlink transmission strategies can be used, all with different implications on the complexity and required architecture of the adaptive antenna system. The strategies may be grouped within two main classes: either when a beam can be steered directly toward the mobile station, or when a beam can be selected from a set of beams with fixed directions. The beam suitable for downlink transmission is selected or steered based on information derived from the uplink, the direction of arrival (DOA) of the received information. The DOA can be an estimate of the direction to the mobile station or simply an identification of the best uplink beam. The algorithms selecting the most suitable beam for downlink transmission may also include other elements of information, such as feedback information concerning the downlink transmission conditions.

Several approaches can be used to direct the radiated power from an antenna array in a narrow beam. The phase front on the antenna elements corresponding to a beam can be generated at baseband using digital beamforming or at radio frequency (RF) using a passive network or phase shifters. Baseband beamforming techniques always require phase coherency all the way to the antenna elements. One main advantage in terms of implementation complexity of using one beam from a passive beamforming network is that it does not require phase coherency between the radio transmitter and the beamformer. However, if multiple beams are transmitting simultaneously using the same carrier frequency, phase coherency between the beams is essential. One passive network is the Butler matrix, which generates a set of orthogonal beams, thus minimizing the beamforming loss. Orthogonal beams, however, have a high-gain loss at the beam crossover points.

Figure 24.5 depicts three schematic drawings of adaptive antenna system architectures representing different levels of implementation complexity (e.g., different requirements on uplink and downlink phase coherency between radio branches). In Fig. 24.5a a sketch of a multibeam or switched-beam adaptive antenna architecture with a passive beamforming network is shown. The multibeam approach is the most basic and straightforward solution, requiring a minimum complexity. The DOA can be an identification of the best uplink beam, and therefore no phase coherency is needed in either up- or downlink. Figure 24.5b depicts an adaptive antenna system architecture comprising a solution for switched interleaved beams in the downlink. The number of downlink beams can be increased by using a solution with switched interleaved beams. In this more complex and advanced architecture, different beamforming is used in the uplink and downlink. In the uplink, the number of beams is limited by the number of receiver branches. A DOA estimated from the uplink information is used to select one beam from a larger set of downlink beams. Several beamforming networks are present in parallel in the downlink. After beamforming the signals to the antenna elements are combined. This solution reduces the phase coherency requirements in the downlink from the steered beam approach described next. An accurate DOA estimate

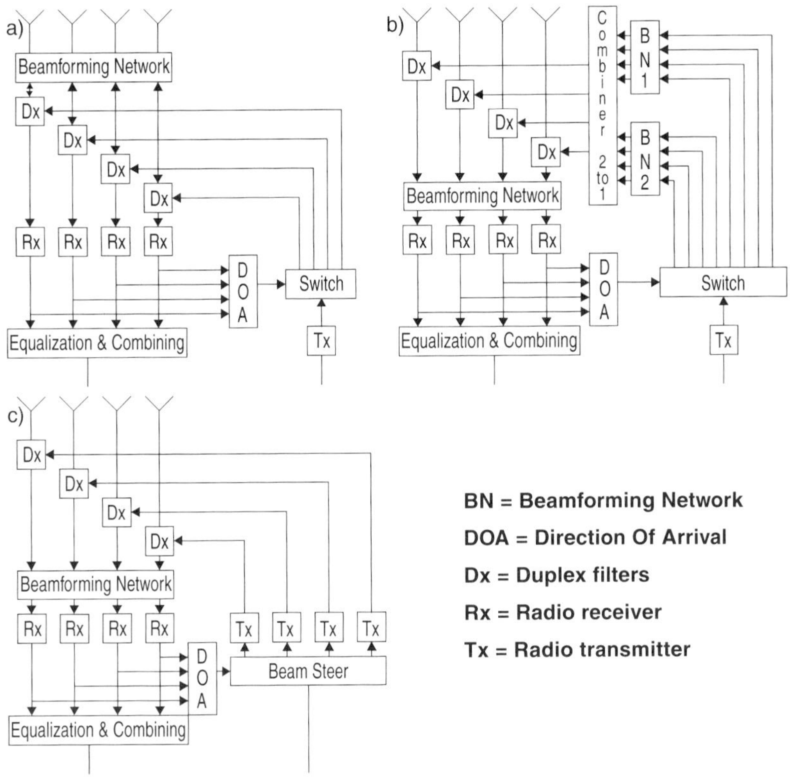

FIGURE 24.5 Schematic drawings of antenna system architectures: (a) multibeam or switched-beam adaptive antenna solution; (b) fixed interleaved beams in the downlink; (c) steered beam solution.

may, however, require coherent receiver branches and calibration of phase and/or amplitude in the uplink. Figure 24.5c outlines a more advanced adaptive antenna solution which utilizes fully steerable beams. This more complex solution requires one individual transmitter for each antenna element as well as phase coherency of the branches on both receiving and transmitting sides. The advantage is that downlink beamforming is not limited to a fixed set of beams or beam shapes. Furthermore, if the direction to the mobile stations and interferers can be estimated with sufficient accuracy, more advanced features such as downlink interference nulling may be introduced.

Antenna patterns corresponding to the possibilities with the different antenna system architectures described previously are shown in Fig. 24.6, together with a sector antenna pattern (Fig. 24.6a). The basic antenna used for the depicted example is a four-column array with a horizontal element spacing of 0.5 λ. Figure 24.6b depicts the patterns from the multibeam or switched-beam solution with four orthogonal beams, whereas Fig. 24.6c shows two interleaved sets of four orthogonal beams corresponding to the more advanced switched-beam solution. The steered beam solution is represented in Fig. 24.6d by patterns from 16 downlink beam directions formed by four interleaved sets of four orthogonal beams. For all configurations, the beam widths are around 26°, resulting from the basic antenna used. Such a four-column antenna array with an antenna gain around 18 dBi has an area slightly below 1 m^2 at the cellular band (~850 MHz) and an area about 0.25 m^2 at the PCS band (~1900 MHz).

As alternatives to the examples described in this section, antenna system architectures with separate antenna systems for uplink and downlink can, of course, also be considered. Antenna system configurations as those described earlier can be extended with multiple sets of antennas for spatial/polarization diversity arrangements if so desired.

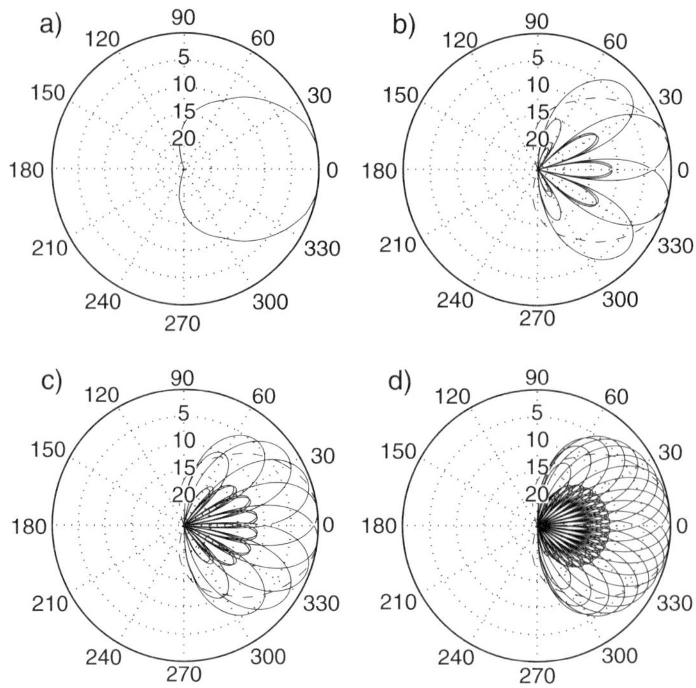

FIGURE 24.6 Antenna patterns for (a) sector antenna; (b) one, (c) two, and (d) four sets of four orthogonal beams from a four-column antenna array with a 0.5 λ horizontal element spacing.

24.4 Signal Processing Algorithms

24.4.1 Uplink Algorithms

First, we describe an interference rejection combining algorithm (IRC) for the GSM system. This general implementation can easily be transformed into the maximum ratio combining algorithm (MRC). Because the major difference between an implementation for GSM and one for TDMA (IS-136) is that of channel tracking, we end the section with a brief description of an IRC implementation for TDMA systems. More details about advantages and limitations concerning the algorithms can be found in References [6], [7], and [8]. In Reference [8], for example, improved IRC algorithms for GSM are developed that take into account that interference can also arise outside the training sequence of the received GSM burst (in which normally all channel estimation computations are made in GSM). The application then becomes suitable for nonsynchronized networks, where it is impossible to have knowledge about the time instances when interference will occur.

24.4.1.1 Multidimensional Maximum-Likelihood Sequence Estimation Equalizer

Denote the transmitted signal by $y(t)$, and model the radio channel from the mobile antenna to the first base station array antenna element as a k-tap finite impulse response (FIR) filter,

$$h_1 = \left[h_{11}, h_{12}, \ldots, h_{1k}\right] \tag{24.1}$$

Let the antenna array be composed of m elements; model the radio channel to the remaining antenna elements in the same way as to the first element; and collect the impulse responses in the channel matrix H

$$H = \left[h_1^T, h_2^T, \ldots, h_m^T\right] \tag{24.2}$$

where T denotes matrix transpose. We model the noise received by the antenna elements by

$$n(t) = \left[n_1(t), n_2(t), \ldots, n_m(t)\right] \tag{24.3}$$

and we assume that the noise process has expected value zero and the following second-order moment:

$$E\left[n(t)n^H(t)\right] = Q \tag{24.4}$$

where superscript H denotes the Hermitean matrix transpose operator (complex conjugate transpose). Finally, we construct a vector $y_v(t)$ with time-delayed copies of the transmitted signal,

$$y_v(t) = \left[y(t), y(t-1), \ldots, y(t-k+1)\right]^T \tag{24.5}$$

The signal received by the array antenna can then compactly be written according to,

$$x(t) = \left[x_1(t), x_2(t), \ldots, x_m(t)\right]^T = Hy_v(t) + n(t) \tag{24.6}$$

Given an observation $x(t)$, the maximum likelihood estimate of the transmitted sequence $y_v(t)$ is

$$\hat{y}_v = \arg\max_{\bar{y}_v} \left\{P\left(\bar{y}_v \middle| x\right)\right\} = \arg\max_{\bar{y}_v} \left[\frac{P\left(x \middle| \bar{y}_v\right) P\left(\bar{y}_v\right)}{P(x)}\right] \tag{24.7}$$

By assuming a Gaussian noise process (which must also be assumed to be temporally white), the criteria for optimization simplifies to

$$\hat{y}_v = \arg\min_{\bar{y}_v} \sum_t \left(x(t) - H\bar{y}_v(t)\right)^H Q^{-1}\left(x(t) - H\bar{y}_v(t)\right) \tag{24.8}$$

To realize this in practice, it is necessary to find an estimate of the channel matrix and of the noise covariance matrix Q. The channel matrix can be estimated using the least squares method, and for the noise covariance matrix we can use the estimate

$$\hat{Q} = \frac{1}{N} \sum_{t=1}^{N} \left(x(t) - \hat{H}\hat{y}_v(t)\right)\left(x(t) - \hat{H}\hat{y}_v(t)\right)^T \tag{24.9}$$

where N is the number of samples used for the estimation (with $N = 26$ for GSM training sequences). This way of estimating the Q matrix is well suited for GSM systems because of the quasi-stationary nature of the radio channel during one time slot (in terms of fading of the *desired* signal). The drawback with estimating the Q matrix during the training sequence only is, obviously, that it becomes impossible to adapt to sudden changes in the interference environment inside one time slot (but outside of the training sequence of the burst). More aspects of, and possible solutions to, this problem can be found in Reference [8].

To summarize, then; the MLSE-solution receiver is found to be

$$\hat{y}_v = \arg\min_{\bar{y}_v} \sum_t \left(x(t) - \hat{H}\bar{y}_v(t)\right)^H \hat{Q}^{-1}\left(x(t) - \hat{H}\bar{y}_v(t)\right) \tag{24.10}$$

The receiver is called the IRC receiver because of its capability to suppress interference. The IRC receiver can be implemented very efficiently with the Viterbi algorithm.

24.4.1.2 The Maximum Ratio Combining Algorithm

If we replace Q with a diagonal matrix D, which have the same diagonal elements as Q but zeros elsewhere, we obtain the ordinary MRC-receiver. This receiver is the result of the assumption that the noise process, in addition to our assumptions made this far, is also assumed to be *spatially* white. In MRC each signal is weighted with its signal-to-noise ratio to achieve optimum performance in white noise environments. IRC, on the other hand, takes advantage of the correlation properties of co-channel interferers by means of suppressing them, and is therefore able to achieve much better performance than MRC in co-channel interference-limited environments.

24.4.1.3 Algorithm for Interference Rejection Combining Algorithm for Time Division Multiple Access (Interim Standard-136)

The received signal at the multiple antenna elements is modeled as

$$x(t) = C_t^H y_v(t) + n(t) \tag{24.11}$$

where

$$y_v(t) = \left[y(t)\, y(t-1),\; \ldots,\; y(t-k+1)\right]^T \tag{24.12}$$

is the vector of transmitted symbols assuming k channel coefficients per antenna element. The (time dependent) channel matrix is represented by

$$C_t = \left[c_{1,t}\, c_{2,t}, \ldots, c_{m,t}\right] \tag{24.13}$$

where

$$c_{l,t} = \left[c_{l,t}(0)\, c_{l,t}(1), \ldots, c_{l,t}(k-1)\right]^T \tag{24.14}$$

is the channel response for antenna element l at time t.

The multichannel receiver performs MLSE equalization and incorporates interference rejection combining by using the following branch metric within the Viterbi algorithm:

$$M_t^{hyp} = \left(x(t) - \hat{x}(t)\right)^H \hat{Q}_t^{-1}\left(x(t) - \hat{x}(t)\right) \tag{24.15}$$

where

$$\hat{x}(t) = \hat{C}_t^H y_v^{hyp}(t) \tag{24.16}$$

is the estimated received signal based on the hypothesized symbol sequence $y_v^{hyp}(t)$ and $\hat{Q}_t$ is the (time dependent) estimated second-order moment of the noise process (or the *impairment correlation matrix*). Channel coefficients and impairment correlation matrix elements are initially estimated over the known synchronization field contained in the TDMA data burst. Then, these are tracked over the unknown data fields using LMS-based tracking algorithms [11]. Similar to the GSM implementation as described earlier in this section, by zeroing off the nondiagonal elements in the impairment correlation matrix we have instead arrived at a possible implementation of the MRC algorithm.

24.4.2 A Simple Downlink Algorithm

Here we shall only briefly describe some ideas for a robust downlink algorithm that combines the benefits of two simple and very well-known DOA algorithms: the beamforming method and a method using knowledge of the training sequence transmitted from the mobile station to the base station. Figure 24.7 depicts the case when one desired signal and an interfering signal impinge on an array antenna. The power spectrum for each method is shown. More details about the two algorithms and the computations involved can be found in Reference [9], which is also a very good tutorial for many available beamforming-type methods.

The main goal is to devise a robust algorithm that does not need to estimate the direction to any interfering signal(s), only estimate one single direction, namely, that toward the desired signal. To reach this objective, we combine the spatial power spectrum resulting from the well-known beamforming method and that of the so-called reference-signal-based approach (which is the Wiener-filter solution when the objective is to minimize the mean square error between a weighted received signal and a known *reference* signal). The very basic idea is to combine the two spectra only to estimate the location of one single peak, corresponding to the direction toward the desired signal, instead of having to first estimate the directions to all potential signals of interest and then also to be forced to detect which signal belongs to which estimated direction. The main problem arises in cases where the desired and interfering signals are fairly close to each other, because then there is indeed a rather high risk of ending up with an estimate of the direction toward the interferer instead of finding our desired signal direction.. This is especially critical if the interfering signal has high power.

More details concerning some interesting characteristics related to practical implementation aspects of such a downlink algorithm can be found in the Appendix. The main conclusion is, in short, that to have this algorithm perform well, it is necessary to find a good trade-off for the selection of which

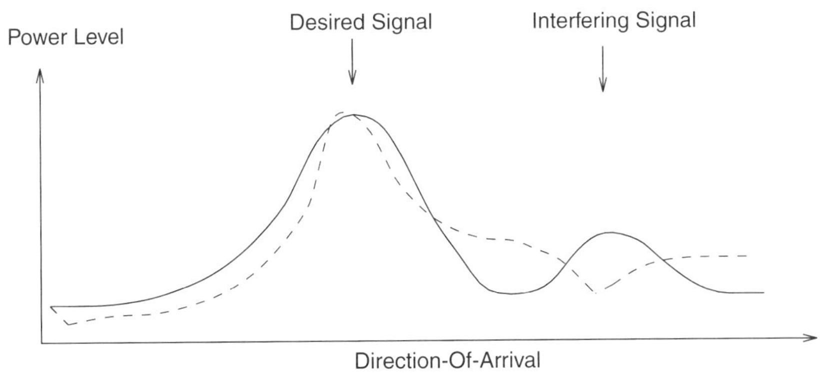

FIGURE 24.7 Beamforming method spectrum: solid; reference signal method spectrum: dashed.

weighting factors to use when combining the two different spectra. Also, certain requirements must be imposed on the sidelobe levels of the antenna diagram.

24.5 Simulation Results

This section describes simulation results obtained for the algorithms that were implemented in the field trials. Some comparisons can therefore be made with the results obtained during field measurements (see Figs. 24.17 and 24.19 later in Sections 24.6.1 and 24.6.2 for GSM and TDMA, respectively) to support the conclusion that the expected results are achieved in the field trials.

24.5.1 Link-Level Results

24.5.1.1 Global System for Mobile Communications

The MRC and IRC algorithms for uplink combining are studied in this section by means of link-level simulations. The algorithms select a subset of eight fixed beams defined in the uplink by the antenna model, and use these as the input to the detection part; see Fig. 24.5 for a comparable antenna system architecture. The channel model is, in the time domain, specified in the GSM standard. A normal distributed angular spreading around the mobile station is introduced in the spatial domain, with standard deviations of 0°, 3° and 10°. In the simulations, the desired mobile station is located in the 0° direction (broadside), whereas a (single) interfering mobile location is varied between –32° and 0°. The necessary carrier to interference (C/I) ratio to obtain a raw bit-error rate (BER) of 8% is used as performance measure for the individual combinations of algorithms and scenarios (Fig. 24.8).

The importance of the angular spread toward the necessary C/I is examined in the simulations. Figure 24.8 (left) depicts the results for the MRC algorithm. The thick line is for an angular spread of 0°, where the nulls of the antenna pattern are maintained. When an angular spread of 3° is used, the nulls in the antenna pattern almost disappear; and when the angular spread increases to 10°, the nulls are no longer visible in the graph. Another interesting property is that the angular spread can be exploited by the uplink combining algorithm even when the interference is located in the same direction as the desired mobile. This is seen in Fig. 24.8, when the interference is located at 0°. Furthermore, it is clearly indicated in Fig. 24.8 that nulling of interferers requires a wider null than the one in the antenna pattern if the angular spread is larger than a few degrees (3 to 5°). Figure 24.8 (right) depicts the same simulations, but for the IRC algorithm. The performance is better, because the IRC algorithm cancels the interference by means of making more efficient use of the degrees of freedom made available by the particular selection of antenna system architecture.

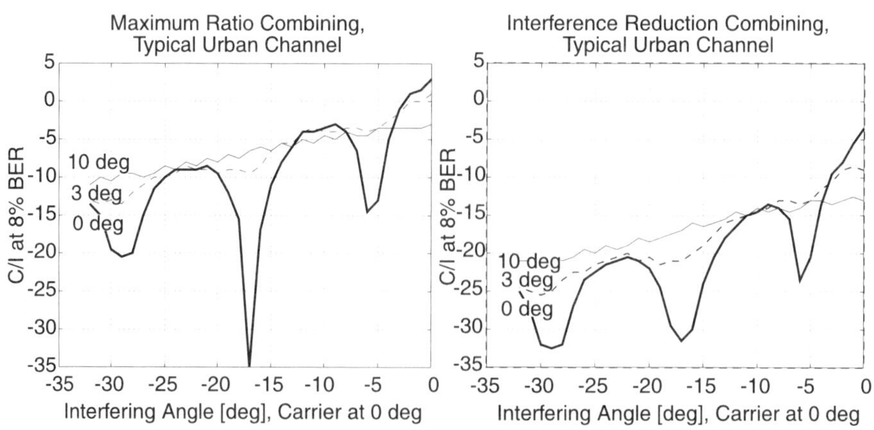

FIGURE 24.8 Simulated performance of the MRC (left) and IRC (right) algorithms when utilized in a GSM system.

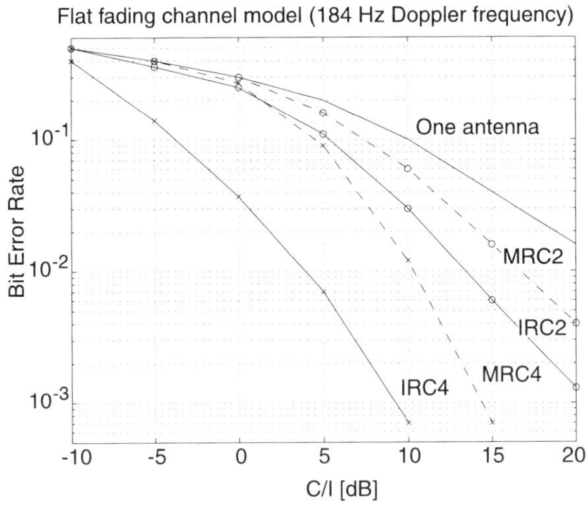

FIGURE 24.9 Simulated performance of the MRC and IRC algorithms with a TDMA flat-fading channel model and a Doppler frequency of 184 Hz corresponding to a speed of ~100 km/h when transmitting at PCS band (1900 MHz).

24.5.1.2 Time Division Multiple Access

Because the major difference between an implementation for GSM and one for TDMA (IS-136) is that of channel tracking, TDMA systems show a similar performance behavior with an equivalent antenna system configuration as does the GSM simulation mentioned earlier. However, differences in absolute performance and gain figures are found because of the different modulation schemes and expected channel characteristics. Performance results for the uplink MRC and IRC algorithms are presented in Fig. 24.9 from a TDMA link-level simulation study of a combined space–polarization diversity antenna system configuration model. The combining algorithms receive signals from either two or four antenna branches, where a conventional space-diversity sector antenna installation is assumed in the two-branch case, whereas a combination of space–polarization diversity sector antenna configuration is assumed in the four-branch case. The interference mitigation performance for the four-branch case mentioned earlier is also equivalent to the performance for an antenna system architecture consisting of a four-column antenna array with four receiver branches as the input to the detection part (see Fig. 24.5 for a comparable antenna system architecture). For the simulation results presented in Fig. 24.9, the desired mobile station and a (single) interfering mobile station are exposed to a flat-fading propagation model at vehicle speeds of 100 km/h corresponding to a Doppler frequency of 184 Hz at PCS band (1900 MHz) transmission [20]. With the selected antenna system architecture, the performance is independent of the interfering mobile azimuth angle separation from the desired mobile station (later see Fig. 24.19, which provides validation of — and comparison with — the results obtained during field measurements). Additional simulation results for some alternative antenna system configurations can be found in Reference [24].

24.5.2 System Level Results

The use of adaptive antennas mainly leads to decreased interference in a network. The narrow beams reduce the received interference in the uplink and the distribution of interference in the downlink. The result is improved network quality, for example, that can be translated to an improved C/I, which can be traded for increased network capacity.

24.5.2.1 Increasing Global System for Mobile Communications Capacity

In GSM, it is possible to increase the capacity in at least two ways. The most straightforward way is to use the C/I gain from adaptive antennas to implement a tighter frequency reuse [12] than what is used for conventional antennas in a network. For example, this could result in going from an average reuse of nine to a reuse of four when introducing adaptive antennas. Typically, the capacity improvement is about 100 to 120% for a 5 to 6 dB C/I gain. Another way to boost the network capacity is to utilize fractional loading. Fractional loading networks are characterized by a very tight frequency reuse (typically 1/1 or 1/3) that is applied in combination with frequency hopping and radio network features such as power control and discontinuous transmission (DTX). It is possible to maintain the network quality if only a fraction of the frequencies are used simultaneously (i.e., fractional loading). This is very beneficial for the frequency hopping feature. A larger gain from frequency hopping appears because a lot of frequencies are included in the hopping sequences and because frequency utilization is low [13]. This solution has received a lot of attention in the GSM community. Using adaptive antennas based on such a radio network solution generally implies that the fractional load can be increased, and more traffic can be carried for a certain reuse.

Applying adaptive antennas based on these two methods implies different consequences. With a tighter frequency reuse, the interferers come closer and thereby the interference variation becomes larger than in a scenario with sparser reuse. This means that a part of the adaptive antenna C/I gain is "consumed" because of the increased interference variance, and the capacity gain is somewhat reduced. With fractional loading, the interferers are at the same position and only the probability of interference is increased when the traffic load increases. This implies that the entire C/I gain can be used to boost capacity. Adaptive antennas applied for fractional loading networks are therefore expected to be the more efficient approach.

To estimate the capacity gain with adaptive antennas in fractional loading networks, system level simulations including multiple cells have been performed [14]. The simulated adaptive antenna system provides a 5 to 6 dB C/I gain compared with a conventional sector antenna system in terms of interference reduction. Two frequency reuse cases were investigated: 1/1 and 1/3 reuse. Figure 24.10 shows the downlink performance for different loading factors and frequency utilizations in a network planned for 1/1 frequency reuse including power control and DTX. The downlink performance is measured as the probability of having a frame erasure rate (FER) below 2%, which is assumed to reflect good speech quality. Furthermore, acceptable network performance is assumed to be achieved when the speech quality requirement is fulfilled with 90% probability. By using this measure, it can be concluded from Fig. 24.10 that with adaptive antennas it is possible to have ~72% frequency utilization, but only 19% utilization

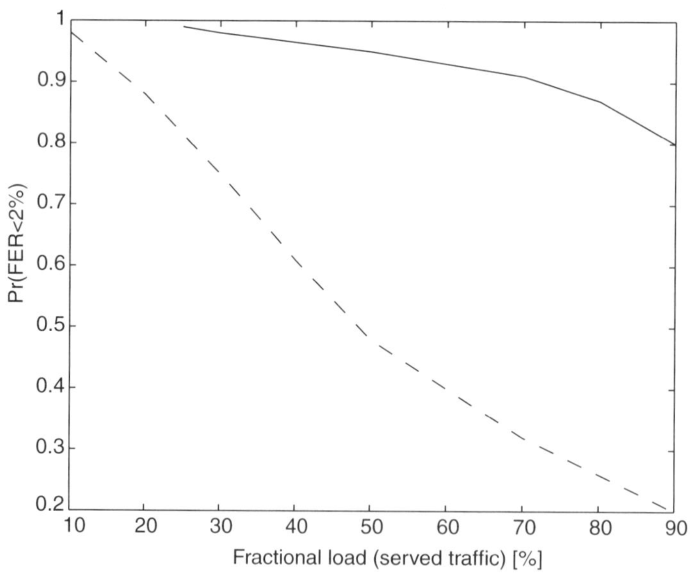

FIGURE 24.10 The probability of having a speech segment with FER below 2% in the downlink vs. channel utilization in the two GSM networks planned for 1/1 frequency reuse. Random frequency hopping, power control, and DTX were applied. Solid line — adaptive antenna system; dashed line — sector antenna system.

is possible for sector antennas (power control and DTX used in this case). The capacity increase is hence approximately 280% in a 1/1 fractional loading network. The simulation study also showed that a network planned in a 1/3 reuse pattern can cope with a fractional load of ~100% if adaptive antennas are employed, but only some 25% for sector antennas (no power control or DTX used in this case). The network capacity is limited by hard blocking in this case (i.e., the lack of traffic channels). In terms of capacity, 1/1 reuse is more efficient, because three times more frequencies are available in each cell than in the 1/3 reuse case. The 1/1 reuse is more than two times (= 72·3/100) more spectrum efficient than the 1/3 reuse case. Thus, for adaptive antennas 1/1 is the optimum frequency reuse in fractional loading networks, and this can be compared with conventional sector antennas where the optimum frequency reuse is around 1/3 [13], [15]. Consequently, an adaptive antenna system gives a solution that fits particularly well into the evolutionary path for fractional loading GSM networks, and it should not be seen as a competing technique.

24.5.2.2 Increasing Time Division Multiple Access (Interim Standard-136) Capacity

The previous discussion of basic ideas on increasing capacity for GSM networks is in general also valid for TDMA networks. However, there are some differences between the two standards concerning protocols and features such as power control on time slot basis and frequency hopping, which only is available in GSM. The TDMA IS-136 Revision A specification states that the base station output power on each carrier frequency shall be kept at a constant level for the full duration of the frame, as soon as any time slot is occupied. This obstructs the introduction of downlink beamforming and beam switching individually for each time slot. One way to approach this problem is to instead perform downlink beamforming on a carrier basis. With this approach, performance can be improved by introducing beam packing (i.e., allocating mobile stations with similar directions from the base station time slots on the same carrier frequency) [22]. For TDMA IS-136 Revision B, a specification change is proposed to allow beamforming and power control to be performed individually for each time slot. Capacity estimations by means of system-level simulations for both beamforming on a carrier basis and on a time slot basis are next presented, corresponding to IS-136 Revision A and potential Revision B mobile stations, respectively.

The system-level simulations were performed in an interference-limited, homogeneous, three-sector site, macrocellular network environment. The system model assumed time-synchronized base stations on a time slot basis. Traffic was generated with equal probability over the simulated area, and no uplink or downlink power control was applied. In the simulations the mobile transmitter was considered a point source and the angle between the base station and mobile station was assumed to be known (i.e., the estimation of the DOA was assumed perfect).

Downlink performance is measured as the C/I value at the 10% level of the downlink cumulative distribution function (CDF). In IS-136, acceptable quality is usually defined by C/I at the 10% CDF level higher than 17 dB. Served traffic is defined as the average number of users per cell divided by the number of channels in the system. Hence, a system with a 7/21 frequency reuse pattern can serve a traffic of 1/21 user/cell/channel at full load. Capacity is defined as the maximum amount of traffic possible to serve with compliance of the quality criterion. No trunking gains are accounted for in this study. For reference purposes, systems with conventional sector antennas were simulated at full load. In Table 24.2 the served traffic and the downlink C/I values at the 10% CDF level are presented for different frequency

TABLE 24.2 Served Traffic and C/I at 10% CDF Level for 7/21, 4/12, and 3/9 Systems with Sector Antennas Operating at Full Load

Frequency Reuse	Served Traffic [User/Cell/Channel]	Downlink C/I at 10% CDF Level
7/21	0.048	17.3
4/12	0.083	13.5
3/9	0.111	11.4

TABLE 24.3 Beam Crossover Loss and Downlink C/I at 10% CDF Level for a 4/12 System Operating on a Time Slot Basis at Full Load

Number of Beams	Beam Crossover [dB]	Downlink C/I at 10% CDF Level [dB]
4	–3.4	17.9
8	–0.8	18.9
16	–0.2	19.1

TABLE 24.4 Beam Widths, Beam Crossover Loss and Downlink C/I at the 10% CDF Level for Configurations with Four Beams from a Four-Column Antenna Array with 40, 30, and 20 dB Tapering

Tapering [dB]	Beam Width [degrees]	Beam Crossover [dB]	Beamforming Loss [dB]	Downlink C/I at 10% CDF Level [dB]
40	31	–2.2	2.2	17.7
30	29	–2.5	1.8	17.8
20	28	–2.9	1.2	17.9

reuse patterns. Evidently, both 4/12 and 3/9 conventional systems are far below the acceptable quality level at full load.

In Table 24.3, simulation results are summarized for 4/12 reuse systems operated at full load with beamforming performed individually for each time slot, using the different beam pattern configurations shown in Fig. 24.6. The gain loss at beam crossover is reduced from more than 3 dB to less than 1 dB when several interleaved sets of orthogonal beams are used. When a 4/12 frequency reuse pattern is employed at full load with a conventional sector antenna, downlink C/I is reduced by almost 4 dB compared with a 7/21 reuse reference system. By introducing beamforming in four orthogonal downlink beams, more than 4 dB is gained in downlink C/I. With eight interleaved beams, an additional gain of 1 dB is obtained, because of the reduced beam crossover loss. The steered beam approach, modeled with 16 downlink beam directions, yields only a minor performance improvement compared with the configuration with eight interleaved beams.

The impact on the estimated downlink C/I performance is studied by changing the antenna characteristics and utilizing either tapered beams or reduced beam width. The simulated setup is identical to the case presented in Table 24.3. For the same antenna size used for the results presented in Table 24.3, lower beam crossover losses can be obtained by introducing tapering in the beamforming. Tapering also increases the beam widths and reduces the size of the sidelobes. However, tapering also introduces an additional beamforming loss compared with utilizing orthogonal beams.

In Table 24.4, simulation results are presented for configurations with four downlink beams and different levels of tapering. From Table 24.4 it is evident that downlink performance is not improved by tapering the beamforming. On the contrary, the positive effect from the lower beam crossover losses is canceled by the increased level of interference spread by the wider beams.

Reduced beam widths can be obtained with a wider antenna array. Simulation result for an array comprising six columns with 0.4-λ element spacing are presented in Table 24.5. One, two, or four sets of the middle four orthogonal beams have been selected to give 4, 8, or 16 downlink beams. The results show that using beams with lower beam widths from a wider antenna is only beneficiary in a configuration with eight or more beams. With only four beams, the positive effect of transmitting in narrower beams is canceled by degraded sector coverage. Even with 16 downlink beams, the performance improvement is moderate, around 0.5 dB. Also, narrower beams require a more accurate DOA estimate in addition to increasing the width of the antenna.

In Fig. 24.11, the cumulative distribution functions of downlink C/I are shown for the 7/21 reuse reference system with a conventional sector antenna, and 4/12 reuse systems using beamforming configurations with

TABLE 24.5 Beam Widths, Beam Crossover Loss and Downlink C/I at the 10% CDF Level for Configurations with 4, 8, and 16 Beams from a Six-Column Antenna Array

Number of Beams	Beam Width [degrees]	Beam Crossover [dB]	Downlink C/I at 10% CDF Level [dB]
4	21	–3.6	17.8
8	21	–0.8	19.3
16	21	–0.2	19.7

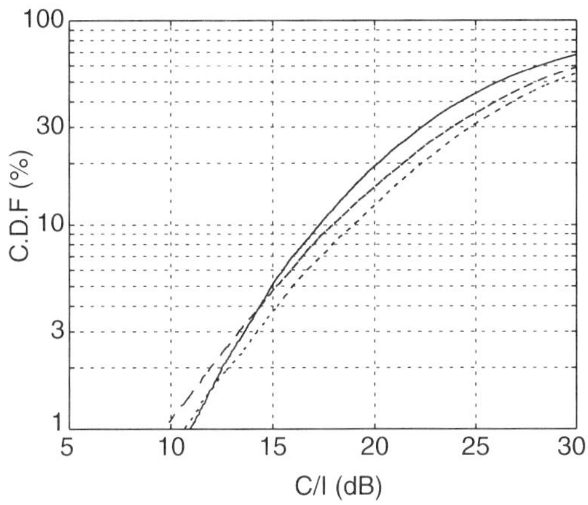

FIGURE 24.11 Cumulative distribution functions of downlink C/I for a 7/21 system with a conventional sector antenna (solid), a 4/12 system with four orthogonal downlink beams (dashed), and a 4/12 system with eight interleaved downlink beams (dotted).

TABLE 24.6 Parameters Characterizing the C/I Distributions Shown in Fig. 24.13

Reuse	Number of Beams	10% CDF Level	Mean [dB]	Median [dB]	σ [dB]	Min [dB]	Max [dB]
7/21	1, sector	17.3	27.3	26.1	9.1	–0.0	94.6
4/12	4	17.9	29.0	28.0	9.6	–3.2	87.8
4/12	8	18.9	30.0	28.9	9.7	–4.2	92.1

four orthogonal beams or eight interleaved beams. All systems are operated at full load and beamforming is performed on an individual time slot basis. Some parameters characterizing the three C/I distributions are summarized in Table 24.6. From Fig. 24.11 and Table 24.6 it is evident that the C/I distribution function adopts a different shape when transmission in narrow beams is introduced. The systems with beamforming exhibit substantially higher mean and median C/I values. However, the width of the distribution function also increases and the minimum C/I values actually decrease. Compared with the reference system, the configuration with four downlink beams improves downlink C/I for around 96% of the distribution at an increase in served traffic by a factor 1.75. With eight interleaved downlink beams, 98.5% of the distribution experience an improvement in downlink C/I at the higher served traffic. The shape of the distribution function is, of course, also influenced by other factors, such as the implementation of power control and/or adaptive channel allocation (ACA).

To estimate the increase in system capacity, the downlink C/I at the 10% CDF level is investigated as a function of served traffic. In Fig. 24.12 (top picture), a simulation results are shown for the reference 7/21 system together with results for 4/12 and 3/9 reuse systems with downlink beamforming on a time

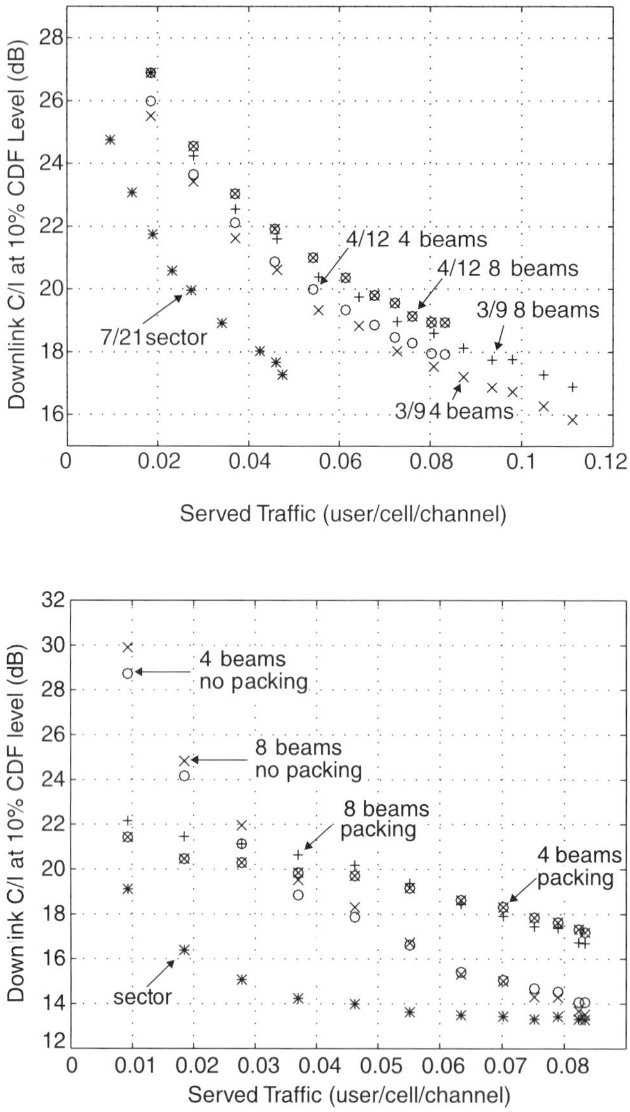

FIGURE 24.12 Downlink C/I at 10 percent CDF level as a function of served traffic. Top picture: for a 7/21 reuse system with a conventional sector antenna (*), 4/12 system with four downlink beams (o), 4/12 system with eight downlink beams (⊗), 3/9 system with four downlink beams (x), and 3/9 system with eight downlink beams (+). Bottom picture: for a 4/12 reuse system with six carriers per sector utilizing beamforming on a carrier basis with a conventional sector antenna (*), four beams without beam packing (o), eight beams without beam packing (x), four beams including beam packing (⊗), and eight beams including beam packing (+).

slot basis. Two beamforming configurations from the four-column array are included: four orthogonal and eight interleaved downlink beams. With a 4/12 frequency reuse plan, the system may be operated at full load with both beamforming configurations, corresponding to a capacity increase of 1.75 compared with the 7/21 reference system. The improvement in quality is, however, significantly larger with eight interleaved beams in the downlink, where the percentile of the distribution with poor quality is reduced to approximately 6%. With a 3/9 frequency reuse plan, the capacity can be increased by a factor of 1.9 with four orthogonal beams in the downlink, or a factor of 2.3 using eight interleaved beams. In these situations, the gain in downlink C/I is explored to give maximum capacity at maintained system quality. These capacity estimates do not consider the effect of higher trunking efficiency with a tighter frequency reuse. This effect enhances capacity as a direct consequence of the increased number of available channels per cell resulting from the lowered reuse plan.

In Fig. 24.12, the bottom picture shows simulation results for the same beam configurations as in the top picture, but introduces the IS-136 requirement of a constant power level for the full duration of the frame as soon as any time slot is occupied. Thus, a carrier may be transmitted in multiple beams simultaneously, which increases the downlink interference level and reduces the beamforming gain. Simulation results with beamforming on a carrier basis are shown both without beam packing and when beam packing is introduced. With the downlink power constant for each carrier, the downlink performance with a sector-covering antenna is degraded faster at low load. The C/I drops to the level corresponding to full load already around 50% channel utilization and is then constant with served traffic. The beamforming gain in downlink C/I without beam packing is high at low channel utilization, but the gain is reduced with increasing load, and at full load totally canceled. The performance of both beamforming configurations is virtually the same. However, the behavior is not the same when beam packing is introduced in the systems. Beam packing is achieved by assigning time slots on the same carrier frequency to mobile stations using the same beam. At low load, the C/I performance is degraded with a simple beam packing routine, because it acts to place each mobile on a separate carrier in a situation where no mobiles use the same beam.

There is, however, a substantial improvement at higher channel utilization. Compared with a system with a sector-covering antenna transmitting at constant power, beamforming gains above 4 dB are obtained for both beam configurations at realistic loads. With four beams, downlink C/I is again above 17 dB at full load. One problem with beam packing is that the mobility of the mobile stations continuously alters the preferred channel allocation. This effect is not present in these results. Either the mobility causes an increased number of handoffs in the system, or the mobiles stay on the same carrier but instead causes that carrier to be transmitted in multiple beams that increase the level of interference. Furthermore, it is reasonable to expect that a quality-based channel allocation (e.g., interference driven channel selection) in combination with beam packing improves the C/I performance at low-channel utilization.

In Fig. 24.12, the top picture shows that a configuration with eight interleaved beams improves C/I at full load by 1 dB, compared with a configuration with four orthogonal beams in a system with beamforming on a time slot basis. With beamforming on a carrier basis, the situation becomes the opposite in a system with six carriers per sector. In Fig. 24.12 (bottom picture), the configuration with four beams actually gives a better performance than the eight-beam configuration.This results from the fact that more beams cause an increased number of carriers to be transmitted in multiple beams simultaneously, which increases the level of downlink interference. Figure 24.13 shows the total gain in downlink C/I at the 10% CDF level at full load obtained with beamforming on a carrier basis in combination with beam packing as a function of number of carriers per sector. The gain is relative to a system with a conventional sector-covering antenna and the same number of carriers at full channel utilization. With five or more carriers per sector, the downlink beamforming gain obtained with a configuration with four orthogonal beams is sufficient for the use of a 4/12 frequency reuse at full load, also with beamforming performed on a carrier basis. For comparable performance with the eight-beam configuration, a minimum of seven carriers per sector is required. Evidently, the more complex solution with eight interleaved beams does not improve the beamforming gain at full load over that of the configuration with four orthogonal beams in systems with fewer than ten carriers per sector. It should, however, be noted that these results are obtained without power control. Power control on a time slot basis can be assumed to be more effective than power control on a carrier basis. Also, the performance of power control on a carrier basis can be improved by allocating mobile stations with a similar downlink power need to time slots on the same carrier (power packing).

24.5.2.3 Migration Strategies

The adaptive antenna technique has often been criticized for leading to increased costs for the network operators, particularly when introducing the technique in an existing network with conventional antennas. For instance, it has been claimed that all base stations must be employed with adaptive antennas to

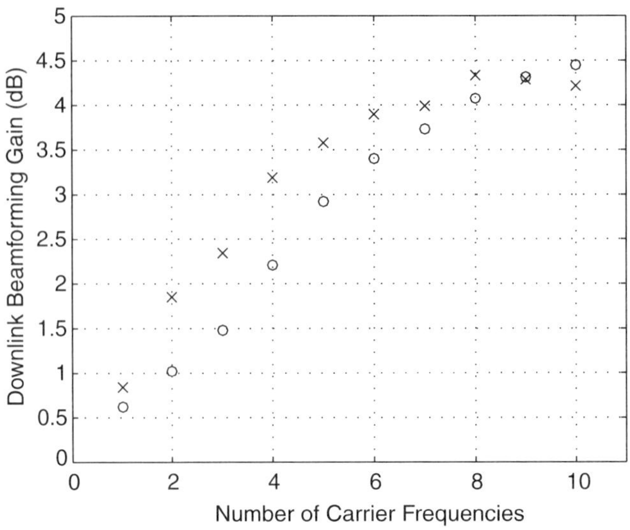

FIGURE 24.13 Downlink beamforming gain with beamforming on a carrier basis together with beam packing, as a function of number of carrier frequencies per sector. Results are shown for a 4/12 reuse system operated at full load with four beams (x) and eight interleaved beams (o).

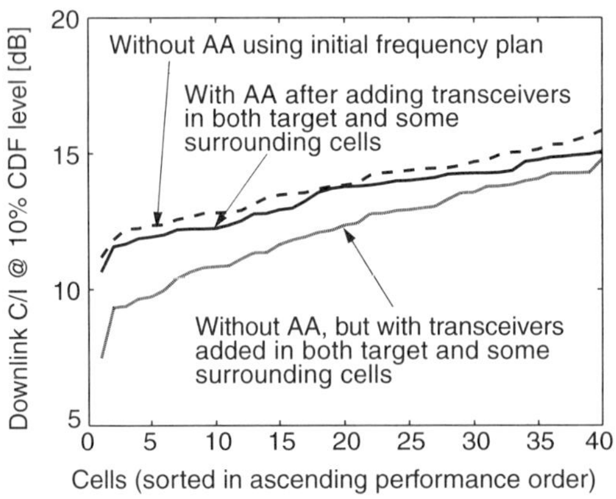

FIGURE 24.14 Downlink performance on cell level before and after introduction of adaptive antennas and addition of frequencies in the target cell and some surrounding cells. Note that the performance before and after the adaptive antenna (AA) introduction is more or less the same. The step-by-step migration has increased the capacity. The plot shows the 40 worst cells in the GSM network model used.

improve performance. This strategy is, of course, less attractive, because it implies very high initial costs for a service provider. The preferred migration strategy would be to gradually, in a step-by-step manner, install adaptive antennas in a network based on the need for more capacity. The advantage is obvious: lower initial deployment cost for the service provider.

The step-by-step migration strategy was studied for GSM in Reference [16], but one may note that the conclusions are generally also valid for TDMA (IS-136) networks. The basic idea behind this strategy is to use adaptive antennas in "hot-spot" cells, employing adaptive antennas at the base stations that serve a lot of traffic; those base stations severely disturb a large number of other cells. An example is large macrocells with many transceivers using antennas mounted on high masts. Adaptive antennas at these base stations mitigate the interference situation considerably and make it possible to locally increase the frequency reuse or fractional load. An example of this operation is illustrated in Fig. 24.14, which shows the quality (C/I) by cell level. System-level simulations were performed for a realistic city network including off-grid cells using different antenna heights and an uneven traffic distribution. A single adaptive antenna was installed in a hot-spot cell in the town center. The interference reduction obtained made it possible to double the number of transceivers (from five to ten) in the hot-spot cell. In addition, 17 new transceivers could be added to six surrounding base stations without adaptive antennas. Altogether, the capacity improvement was approximately 10% over the entire area. This could not be done if the base station covering the hot-spot cell used a conventional sector antenna; a severe quality degradation appeared in that case (see Fig. 24.14). Accordingly, the step-by-step migration strategy appears to indeed be feasible.

24.6 Summary of Performed Field Trials

This section gives a condensed summary of the field trials carried out by Ericsson. First, some pictures of the hardware employed for the tests can be found in Figs. 24.15 and 24.16.

FIGURE 24.15 Equipment used for GSM 1800-MHz field tests shown above to the left (one carrier) and a GSM 900-MHz prototype base station antenna installation (six carriers) deployed in commercial traffic is shown to the right.

FIGURE 24.16 Prototype adaptive antenna base station equipment (DMAD = DAMPS minimal array downlink) for TDMA (IS-136) 1900-MHz PCS-band deployed in commercial traffic. The size of the physical array antenna enclosure is 66 × 44 cm (height × width).

24.6.1 Link-Level Measurement Results for Global System for Mobile Communications

This section reviews the results from the GSM measurement campaign carried out in Düsseldorf, Germany, and its surroundings. This campaign started in the second half of 1996 and ended in the beginning of 1997, in the first joint research project with Mannesmann Mobilfunk. The test equipment (left in Fig. 24.15) was installed at base station sites located in a total of five different propagation environments, and link-level performance was evaluated for the MRC and IRC uplink algorithms (described in Section 24.4.1), as well as for two downlink algorithms. To the right in Fig. 24.15, an array antenna site installation is shown from one of the follow-up projects that took place during 2000, in this case with Vodafone Airtouch in London, United Kingdom. Similar equipment was used for field trials carried out under commercial traffic conditions in Giessen, Germany, during the autumn of 1998 in the second joint research project with Mannesmann. The selected base station architecture for the follow-up projects was based on the results from the first Ericsson/Mannesmann field trial project.

24.6.1.1 Brief Review of Algorithms Evaluated

The first downlink algorithm, termed fixed beam, selects the best beam for downlink transmission based on weighted and averaged-over-time data received during uplink transmission. The principle is then to direct the downlink transmission in the beam on which the most energy and/or the best quality is received in the uplink. The second algorithm, steered beam, estimates the angular direction to the desired signal based on averaged-over-time spatial power spectra computed from uplink data using a beamforming type of algorithm and steers the beam in the estimated direction during downlink transmission. The main goal was to compare the performance of these two algorithms. If no major differences can be found, the fixed beam approach is the preferred, because it does not require coherency of receive or transmit branches; that is, no calibration of the system would be necessary. The tested uplink algorithms, MRC and IRC, were described earlier in Section 24.4.1.

A more detailed description of the GSM adaptive antenna link-level field trials can be found in Reference [10]. We note here only that the test equipment consisted of eight coherent receive and transmit branches.

Figure 24.17 shows the results from all measurements performed in interference-limited scenarios at all five site locations. No line of sight (LOS) conditions were present for any of the measurement cases. Each one of the five pairs of curves depicts the results for the two different uplink algorithms, as well as for the two different downlink algorithms. We have already noted that the measured results are well in

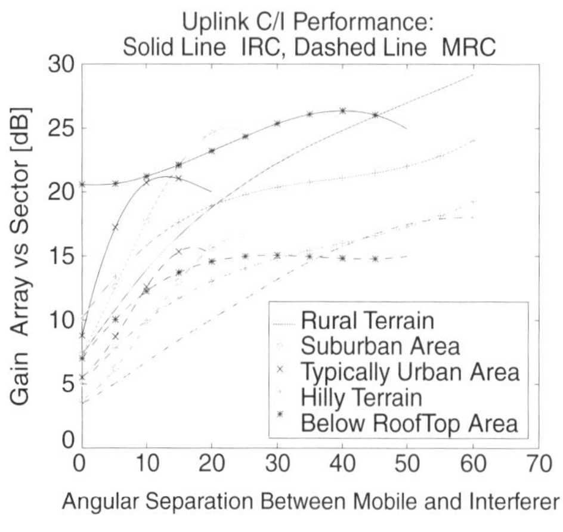

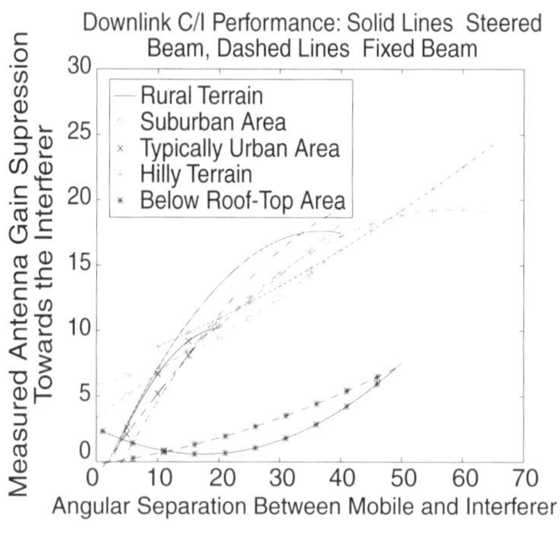

FIGURE 24.17 Summarizing results from the Ericsson/Mannesmann measurement campaign conducted in a total of five different propagation environments in Düsseldorf, Germany. (Left picture: IRC, solid; MRC, dashed. Right picture: steered beam, solid; fixed beam, dashed.)

line with simulation results. To the left in Fig. 24.17, we find the uplink C/I-improvement relative a conventional two-branch polarization diversity sector antenna utilizing the MRC algorithm. The improvements are measured at the 8% BER level for various angular separations between the desired and the interfering mobiles. Note the significant improvement achieved in the most demanding propagation environment (denoted *below rooftop*). This results from the fact that the uplink array antenna signal processing algorithms here can take full advantage of the abundance of multipath propagation present. To the right in Fig. 24.17, we see the interference suppression as a function of the angular separation between the desired and the interfering mobiles, obtained by means of base station transmission in a narrow beam. Note that all propagation environments, except for the below rooftop, shows very good agreement with the ideal narrow beam antenna pattern expected in LOS conditions. The antenna was installed well below the rooftops of the surrounding buildings for the below rooftop scenario, which very well explains the poor performance in this case. Finally, we observe that the performance of the fixed beam algorithm is similar to that of the steered beam algorithm. Thus, an array antenna system employing downlink transmission using fixed beams is a very attractive solution in terms of implementation complexity (no calibration is then required). This is also the architecture implemented in the second Ericsson/Mannesmann commercial field trial project.

24.6.2 Link-Level Measurement Results for Time Division Multiple Access (Interim Standard-136)

For the TDMA (IS-136) standard, uplink performance gains were quantified during the second half of 1996 until the beginning of 1997 for a prototype system that implemented a combination of array processing and adaptive equalization techniques. The tested antenna system configuration is identical to the architecture described in Section 24.5.1.2. The PCS-band (1900 MHz) prototype equipment was installed at a base station site located in Dallas, Texas, in a suburban propagation environment; and link-level performance was evaluated for the MRC and IRC uplink algorithms described in Section 24.4.1.

The antenna system was mounted at rooftop level, at approximately 35 m height, overlooking a segment of a highway as well as numerous local roads in the nearby residential and business areas. The base station received signals from two co-channel mobiles, one desired and one interfering. The sector coverage area was divided into test areas. Most test areas were defined as squares with a side up to 0.5 km, so that the angle of arrival to the base station varied ~5° across the square (Fig. 24.18). One test area was along the highway, providing fast fading. For each measurement scenario, the desired and interfering mobiles were driven around in one of the test areas (which may or may not be the same area) following the local traffic flow. In general, aside from the highway test area, LOS propagation was not present within the test areas. Blockage from houses and small buildings as well as local shadowing by trees was typical for the residential environment. In the residential test areas, vehicle speeds reached up to 50 km/h, whereas in the fast-fading highway test area, a 100 km/h vehicle speed was used.

Figure 24.19 shows the interference mitigation capability of the prototype configurations in interference-limited scenarios vs. the azimuth angle separation between the desired and the co-channel interfering mobile test areas. The presented results include desired and interfering mobiles located in different residential test areas averaged over pairs of test areas for a given angle. The measured performance is presented together with simulated results for vehicle speeds of 100 km/h and illustrates the independence of the azimuth angle separation between mobiles and IRC performance. The propagation model used for the simulation is the dispersive two-ray fading model in which the signal at each receive antenna consists of discrete, independent Rayleigh fading rays that are correlated in time. A second ray of each signal is delayed a quarter of a symbol period relative to the main ray. See Section 24.5.1.2 for comparison with results obtained during simulations with a different channel model. The measured results from the field test agree with predicted simulated performance, accounting for implementation losses and different propagation environments. Performance results show that the approach is robust for both low- and high-speed mobiles in the true radio environment, and illustrate the independence of IRC performance on the interfering mobile azimuth angle separation from the desired mobile with the selected antenna system

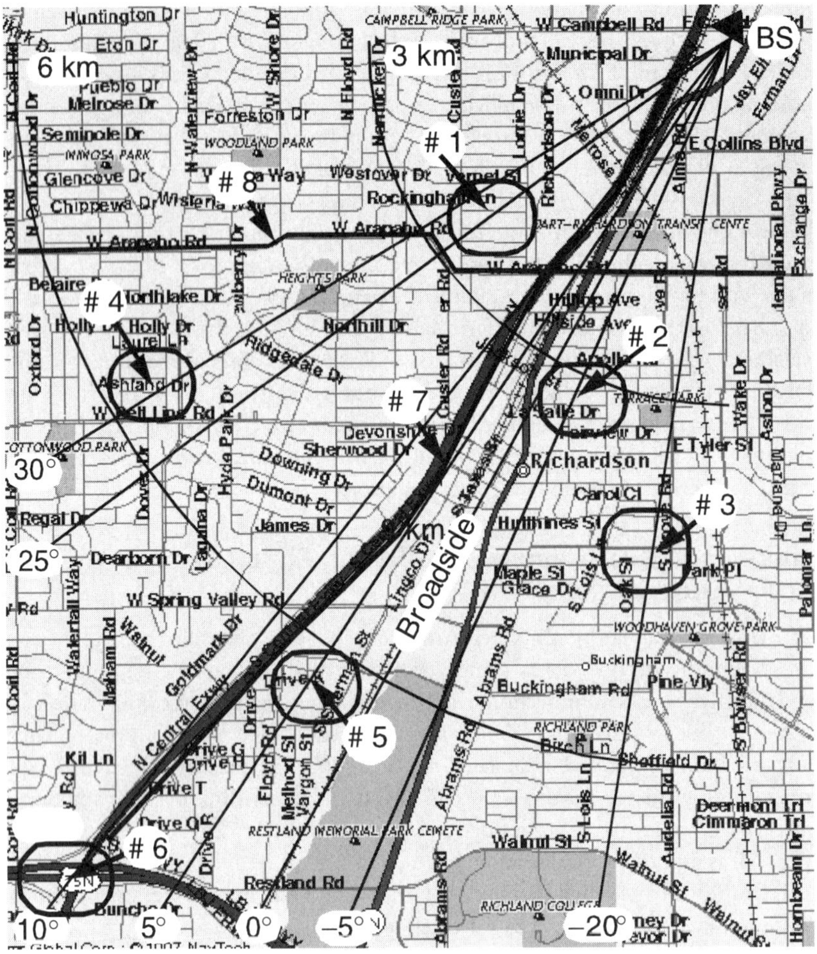

FIGURE 24.18 A map of the coverage area from TDMA link-level performance tests at a base station site located in north Dallas, Texas. The confined test areas are marked on the map.

architecture. A more detailed description of the TDMA (IS-136) link-level field trials can be found in Reference [21].

24.6.3 System-Level Measurements

24.6.3.1 Global System for Mobile Communications

With the results from the link-level field trial as prerequisites, three adaptive test base stations, compliant with the GSM standard, were developed to run a field trial with commercial traffic. The base stations were built on standard GSM equipment, such as modified Ericsson RBS2000 transceivers, standard cabinets, and combiners. The used test configuration contained one adaptive transceiver, one sector transceiver, and one transceiver used for broadcasting the BCCH.

During all parts of the tests the base stations were connected to the GSM infrastructure in the commercial Mannesmann D2 network, that is, a base station controller (BSC) and a mobile service switching center (MSC). As shown in Fig. 24.20, the three adaptive antenna base stations covered the city center of Giessen, a medium-sized German city. The interference was increased by gradually deploying tighter reuse in the three-cell system. To be able to smoothly increase the interference, every transceiver utilized random frequency hopping over three frequencies, thus providing a hardware load of 33%. When using a total of nine frequencies, an orthogonal frequency plan with no internal interference was deployed,

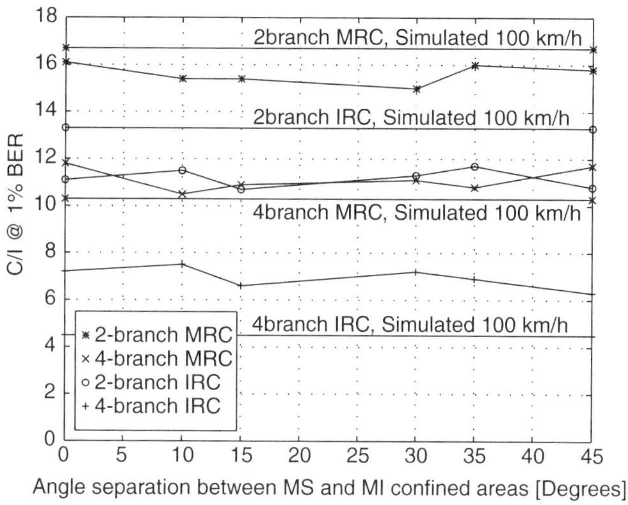

FIGURE 24.19 Summarizing the interference mitigation capability of the prototype configurations in interference limited scenarios from a measurement campaign conducted in a suburban propagation environment in Dallas, Texas.

and when using a total of only three frequencies, the same frequencies were deployed in all cells (1/1 reuse). The interference originating from outside the test area was of minimum impact, in comparison with the high 1/1 reuse interference introduced within the test area. The performance test was designed to be a comparison test between the sector antenna and the array antenna configured systems (i.e., all frequency plans were evaluated for both sector and array configurations). The quality of the different frequency plans and configurations were then compared.

When using commercial traffic in a field trial, it is important to use a quality measure that indicates a quality change before the actual degradation is noticed by the customers. The number of intracell handovers, ICHO, was identified to be such an indicator. When increasing the interference, the probability of inducing an ICHO increases. Thus, the number of ICHOs at constant traffic increase with higher interference. Furthermore, the limiting link in an interference-limited GSM system is usually the downlink and therefore the monitored quality parameter was the number of ICHOs caused by bad downlink quality. Only the traffic transceivers were used for the evaluation. The number of frequencies deployed in the system is a measure of the spectrum efficiency, with fewer deployed frequencies causing higher interference and higher spectrum efficiency. Figure 24.21 shows that the quality gain of adaptive antennas increases with increasing interference. The orthogonal frequency plan, with no internal interference, provides similar quality for both the sector and array configurationions. When the interference is introduced, the performance of the two configurations starts to differ. By monitoring the systems on an acceptable quality limit, as shown in Fig. 24.21, it can be seen that the sector system needs 120% more frequencies to provide the same quality as the array system at the selected (by the operator Mannesmann) quality level.

The results from functional field trials are described in more detail in Reference [17] and some additional results from the performance field trials have been reported in Reference [18].

24.6.3.2 Time Division Multiple Access (Interim Standard-136)

In a cooperative project between AT&T Wireless Services and Ericsson, an adaptive antenna multisite field trial was performed at the AT&T Wireless Services Phoenix PCS market. The field trial was performed during the period from October 1998 to March 1999. The trial system concept was based on simultaneous use of three modified base stations with an integrated DAMPS minimal array downlink (DMAD) system providing adaptive antenna functionality. Some pictures of the radio base station hardware employed for the tests can be found in Fig. 24.16. Some basic characteristics of the DMAD prototype system are as follows: switched multibeam downlink solution with individual user power control using four beams from a passive beamforming network; four-branch uplink interference rejection combination resulting

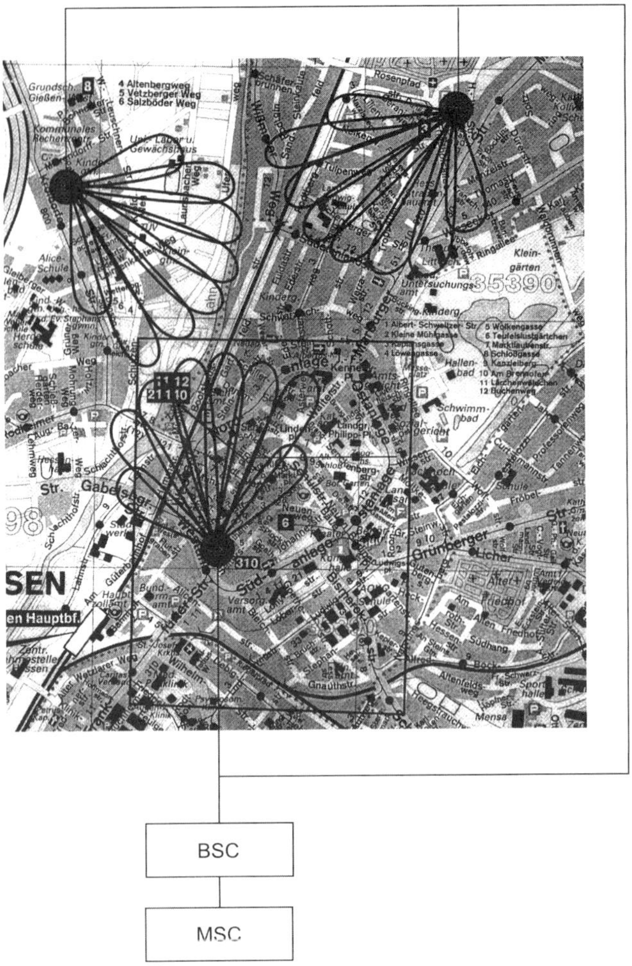

FIGURE 24.20 The commercial GSM test system setup with three adaptive antenna base stations connected to the GSM network infrastructure.

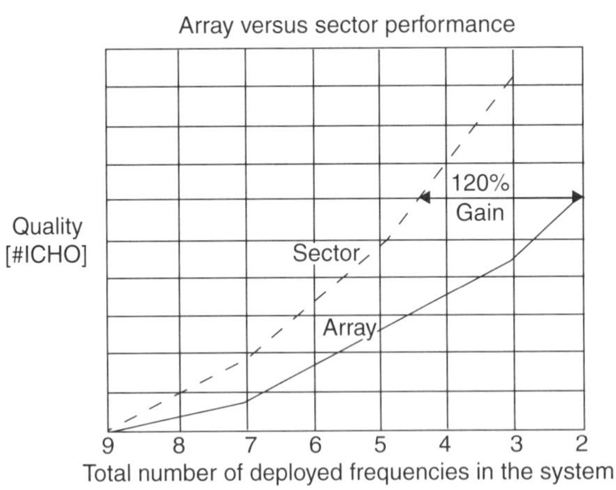

FIGURE 24.21 The quality for different frequency plans when the commercial GSM test system is configured in the sector- and the array setup, respectively. High values on the *y*-axis represent poor quality.

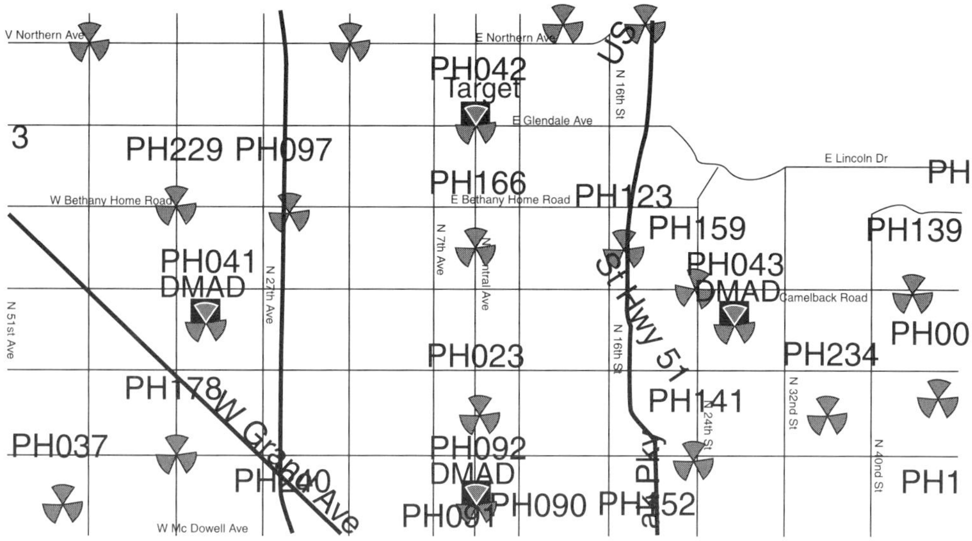

FIGURE 24.22 One system setup scenario with three prototype TDMA (IS-136) adaptive antenna base stations (DMAD = DAMPS minimal array downlink) integrated in the AT&T Wireless Services commercial Phoenix PCS-band TDMA network.

in full sector coverage with high-interference resistance; and integrated network solution where base stations distribute performance indicators and measurements per beam.

The three adaptive antenna prototype base stations were installed in a test scenario as depicted in Fig. 24.22. The base station antennas were all integrated to the north-pointing sector of the sites, with a selected (system-preferred) sector radii of ~1.5 km.The test scenario was forming an equivalent 3/9 reuse plan. During all parts of the tests, the base stations were fully integrated into the TDMA infrastructure (i.e., connected to an MSC) in the commercial network supporting call setup and intracell handoff directly into narrow beams. The downlink beam selection together with the integrated power control was handled locally by the base stations on a slot-by-slot basis. The network integrated DMAD system performance test was conducted with commercial traffic and the results show that the adaptive antenna system provides quality and capacity improvements in accordance with expectations. In the measured scenario, the downlink C/I improvement was more than 6 dB at the 10% CDF level compared to the existing deployed sector system, as expected. This can be seen in Fig. 24.23, which shows the CDF of the downlink C/I at full load, both for the adaptive antenna deployment as well as for the conventional existing 65° sector antenna deployment. For the same system configurations at the full load traffic scenario, the resulting distributions of the BER classes (BERC, defined in IS-136) are shown in Fig. 24.24. From Fig. 24.24, it is evident that the quality target is supported in the 3/9 reuse system by using adaptive antennas where the quality target is defined as minimum 95% of the samples that should have BERC better than 5 (equivalent to better than 3%, BER). The performance figures for the 7/21 reuse conventional existing 65° sector antenna reference system and for the 3/9 reuse cases dependency on traffic load are summarized in Table 24.7.

Table 24.7 shows that with the cost-efficient adaptive antenna stay-on-site solution, in addition to supporting an equivalent capacity improvement of 130% (a factor of 2.3 more traffic), the quality is increased. More details concerning the Ericsson–AT&T Wireless Services joint TDMA commercial traffic trial can be found in Reference [25].

24.7 Adaptive Antennas for Packet Data Systems

The evolution that has been outlined for GSM and TDMA (IS-136) implies that these two system standards converge toward a single packet-data-based standard [23]. In the following section, we give a

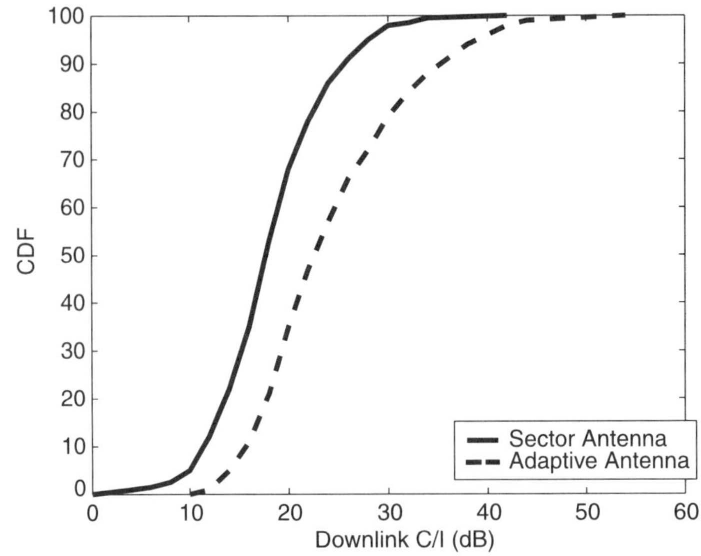

FIGURE 24.23 The cumulative distribution function (CDF) of the downlink carrier to interference (C/I) ratio for the adaptive antenna deployment and for the existing 65° sector antenna deployment. The results are for full traffic load in the system scenario depicted in Fig. 24.22.

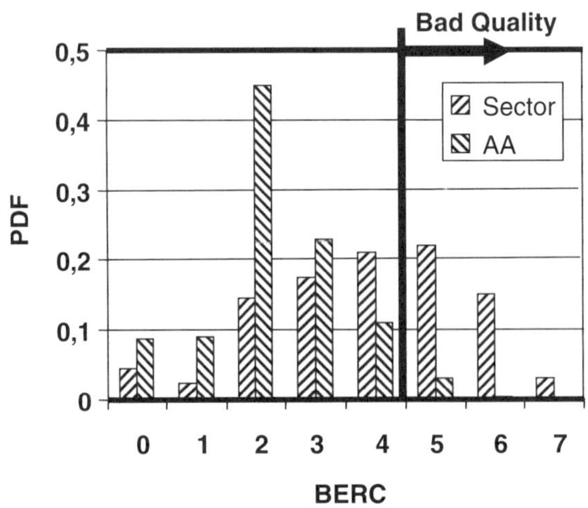

FIGURE 24.24 The BERC (i.e., the quality distribution) for the adaptive antenna system and for the existing 65° sector antenna. The results are for full traffic load in the system scenario depicted in Fig. 24.22.

short overview of the performance improvement possibilities available when employing adaptive antennas in packet-data systems.

The packet-data modes for GSM are called general packet radio service (GPRS) and enhanced GPRS (EGPRS). The major difference compared with circuit switched services is that the application layer always sees an open communication link, whereas the air interface autonomously allocates resources to the mobile terminal when data need to be transferred. This behavior is very efficient for bursty data traffic with a relatively short transfer session in conjunction with relatively long application sessions. This is the case, for example, for www traffic (i.e., Web-browsing). The major difference on the physical layer is that the number of bits transmitted in a burst is dependent on the current link quality. In GPRS, the same modulation technique as in circuit-switched GSM (GMSK) in conjunction with four different

TABLE 24.7 Summary of the Performance Dependency on Traffic Load

	Downlink C/I [dB]	Downlink Quality % BERC > 4	Load/Carrier/Sector Erlang
7/21 Reuse			
Reference System	17.3	4	≥1
4/12 Reuse			
Sector Antenna System	11.5	40	≥1
4/12 Reuse			
Adaptive Antenna System	17.7	3.4	3
4/12 Reuse			
Adaptive Antenna System	18.0	0.4	2
4/12 Reuse			
Adaptive Antenna System	19.2	0.6	1

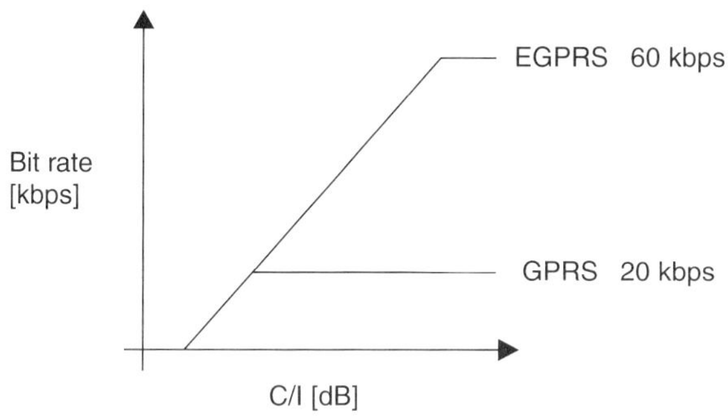

FIGURE 24.25 A schematic picture of the link quality control of GPRS and EGPRS.

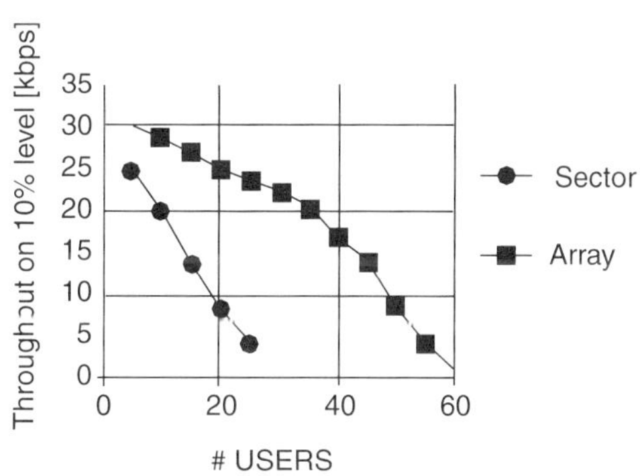

FIGURE 24.26 The throughput for different traffic loads for a sector- and an array antenna configured EGPRS system, respectively.

coding schemes (CS1-4) are used. In EGPRS two modulation techniques, GMSK and 8PSK, are used; in conjunction with different coding schemes, nine different modulation and coding schemes are provided (MCS1-9). The link layer adaptively selects the MCS that provides the highest instant throughput. This feature is called link quality control (LQC) and is schematically depicted in Fig. 24.25. The LQC feature interacts very well with adaptive antennas, because the interference reduction introduced by adaptive antennas is directly transformed into a throughput increase due to the possibility of selecting an MCS that allows for a higher bit rate.

In Fig. 24.26, the specific case of 1/3 reuse and seven time slots dedicated for EGPRS use is simulated. It can be seen that the gain of adaptive antennas in a packet data network is composed of two parts.

During peak traffic hours it is possible to serve more users and at lower traffic hours it is possible to provide higher bit rates for the same number of users.

24.8 Concluding Remarks and Future Directions

The main conclusion that can be drawn from the extensive work that has been conducted, involving link- and system-level simulations and link- and system-level field trials, is that there are major performance gains achievable for adaptive antenna base station architectures of rather limited complexity. This is mainly a result of the nature of the interference scenarios in GSM and TDMA networks, which both present a limited number of well-defined directions from which the main interference originates.

Some remaining work in the GSM and TDMA (IS-136) areas include closer studies of frequency planning techniques, and continued refinements on packet-data protocol and implementation issues for further enhancing the performance of adaptive antenna systems. Furthermore, for CDMA systems link- and system-level simulation studies similar to those presented here, as well as planning for field trial evaluations, are underway and will provide a thorough understanding concerning the most suitable application scenarios of adaptive antennas for CDMA.

24.9 Appendix: A Simplistic, Robust Downlink Algorithm

We make some evaluations of the characteristics of a simple algorithm for downlink transmission. The main goal is to devise a robust algorithm that does not need to estimate the direction to any interfering signal(s), and only estimates one single direction, namely, the one toward the desired signal. To reach this objective, we shall combine the spatial power spectra resulting from the well-known beamforming method and that of the so-called reference signal-based approach (which is the Wiener-filter solution when the objective is to minimize the mean square error between a weighted received signal and a known signal). More details about the algorithms can be found in Reference [9].

24.9.1 Basic Notation Used

Let $a(\theta) \in C^{mx1}$ denote the array response vector, that is, the response by the array from a plane wave impinging from the direction θ on an array with m elements. Furthermore, let $r(t)$ and $i(t)$, denote the desired and the interfering signal, respectively, and let finally θ_0 and θ_1 be the angular directions toward the desired and interfering signals. The general data model describing the output of the array, $x(t) \in C^{mx1}$, can then be written as

$$x(t) = \sum_{k=1}^{d} a(\theta_\kappa) \cdot s_k(t) + n(t) \qquad (24.17)$$

where d is the number of general signals, $s_k(t)$, impinging on the array and $n(t) \in C^{mx1}$ is the noise. We assume for simplicity that the noise is spatially and temporally white.

24.9.2 Beamforming Method Spectrum

Because we are addressing the case with one desired and one interfering signal, we have the array output equal to

$$x(t) = a(\theta_0) \cdot r(t) + a(\theta_1) \cdot i(t) + n(t) \qquad (24.18)$$

By assuming orthogonality between desired and interfering signals, we obtain the beamforming spectrum,

$$P_{BF}(\theta)=\frac{1}{N}\sum_{t=1}^{N}\left|a^H(\theta)\cdot x(t)\right|^2 \rightarrow \sigma_r^2\cdot\left|a^H(\theta)a(\theta_0)\right|^2+\sigma_i^2\cdot\left|a^H(\theta)a(\theta_1)\right|^2+\sigma_n^2 \tag{24.19}$$

as $N \rightarrow \infty$. Here, σ_r^2, σ_i^2, and σ_n^2 are the expected power levels of the corresponding signals and we have also made the assumption that $|a(\theta)| = 1$. Furthermore, superscript H denotes the Hermitean transpose operator (complex conjugate transpose).

24.9.3 Reference Signal Method Spectrum

In this case, the spectrum is defined by

$$P_{RS}(\theta)=\frac{1}{N}\sum_{t=1}^{N}\left|w^H(\theta)\cdot x(t)\right|^2 \tag{24.20}$$

where $w(\theta) = R_{xx}^{-1}\cdot r_{xd}(\theta)$, R_{xx} is the array covariance matrix, and

$$r_{xd}(\theta)=\frac{1}{N}\sum_{t=1}^{N}x(t)\cdot r^H(t) \tag{24.21}$$

is the vector describing the correlation between the model of the array output signals and the known reference signal. Assuming also here that $N \rightarrow \infty$, we find the final spectrum to have the form

$$P_{RS}(\theta)=\sigma_r^4\cdot a^H(\theta)R_{xx}^{-1}a(\theta) \tag{24.22}$$

If we now assume that the interferer is strong compared with the desired signal, that is, $\sigma_i \gg \sigma_r$ (which is the most interesting and difficult case), we find from simple matrix inversion rules that we can write

$$R_{xx}^{-1}=I_{mxm}-\frac{a(\theta_1)a^H(\theta_1)\cdot\dfrac{\sigma_i^2}{\sigma_n^2}}{1+a^H(\theta_1)a(\theta_1)\cdot\dfrac{\sigma_i^2}{\sigma_n^2}} \tag{24.23}$$

where I_{mxm} is the $m \times m$ identity matrix resulting from the spatial/temporal white noise assumptions. Some more straightforward computations lead us to the final expression for the combined spectrum obtained from the beamforming and the reference signal methods; the following expression results:

$$P(\theta)=\frac{1}{\sigma_n^2}\cdot\left(1-\left|a^H(\theta)a(\theta_1)\right|^2\right)\left(1+a^H(\theta)a(\theta_1)\cdot\frac{\sigma_i^2}{\sigma_n^2}+a^H(\theta)a(\theta_0)\cdot\frac{\sigma_r^2}{\sigma_n^2}\right) \tag{24.24}$$

24.9.4 Remarks on Some Characteristics of $P(\theta)$

We are mainly interested in examining the behavior of the location of the peak of the spectrum under the assumption that the interfering signal is notably stronger than the desired one (i.e., $\sigma_i > \sigma_r$), because this peak location corresponds to the estimated direction

$$\hat{\theta}_0=\arg\max_{\theta\in\Theta}P(\theta) \tag{24.25}$$

toward our desired signal $r(t)$.

Case 1. θ_1 close to θ_0: In this case we have the factor $1 - |a^H(\theta)a(\theta_1)|^2$ "small" around the true position, and we see that σ^2_i dominates the beamforming part of the spectrum. Thus, the beamforming part, which is the power part, should be "weighted up" to make the estimated direction be closer to the true direction of the desired signal.

Case 2. θ_1 very distant from θ_0: In this case we have the factor $1 - |a^H(1)a(\theta_1)|^2 \approx 1$ around the true position θ_0, and we find that the dominating term in the beamforming part of the spectrum around θ_0 is $a^H(\theta)a(\theta)_0 \cdot \sigma_r^2/\sigma_n^2$ provided that $\sigma_r^2 > \sigma_i^2 \cdot |a^H(\theta_0)a(\theta_1)|$.

Note that this last requirement is on the sidelobe level of the antenna diagram.

Hence, the conclusion of this exercise is that to have good performance with this type of direction-finding algorithm, it is necessary to find a good trade-off for the selection of which weighting factor to use for the beamforming part of the spectrum in $P(\theta)$.

References

1. Krim H. and Viberg, M., Two decades of array signal processing research — the parametric approach, in *IEEE Signal Process. Mag.*, vol. 13, no. 4, pp. 67–94, July 1996.
2. Paulraj, A. and Papadias, C., Space-time processing for wireless communications, in *IEEE Signal Process. Mag.*, vol. 14, no. 6, pp. 49–83, November, 1997.
3. Godara, L., Applications of antenna arrays to mobile communications, Part I: Performance improvement, feasibility and system considerations, in *Proc. IEEE*, vol. 85, no. 7, July 1997.
4. Godara, L., Applications of antenna arrays to mobile communications, Part II: Beam-forming and direction-of-arrival considerations, in *Proc. IEEE*, vol. 85, no. 8, August 1997.
5. TSUNAMI II, Final Workshop on adaptive antennas, September 17, 1998, Copenhagen, Denmark.
6. Bottomley, G. and Jamal, K., Adaptive arrays and MLSE equalization, in *Proceedings of VTC Conference*, pp. 50–54, Chicago, 1995.
7. Karlsson, J. and Heinegärd, J., Interference rejection combining for GSM, in *Proceedings of ICUPC Conference*, Cambridge, MA, 1996.
8. Karlsson, J., Adaptive antennas in GSM systems with non-synchronized base stations, *Licentiate's thesis*, Department of Signals, Sensors and Systems, Royal Institute of Technology, Sweden, 1997. ISSN 1103-8039.
9. Van Veen, B.D. and Buckley, K., Beamforming: a versatile approach to spatial filtering, in *IEEE ASSP Mag.*, vol. 5, no. 2, pp. 4–24, April 1988.
10. Andersson, S. et al., Ericsson/Mannesmann GSM field-trials with adaptive antennas, in *Proceedings of EPMCC Conference*, Bonn, Germany, October 1997.
11. Larsson, G., Gudmundson, B. and Raith, K., Receiver performance for the north American digital cellular system, in *Proceedings of VTC Conference*, St. Louis, MO, May 1991.
12. Engström, S. et al., Multiple reuse patterns for frequency planning in GSM networks, in *Proceedings of VTC Conference*, Ottawa, Canada, May 1998.
13. Olofsson, H. et al., Interference diversity as means for increased capacity in GSM, in *Proceedings of EPMCC Conference*, Bologna, Italy, October 1995.
14. Kronestedt, F. and Andersson, S., Adaptive antennas in frequency hopping GSM, in *Proceedings of ICUPC Conference*, Florence, Italy, October 1998.
15. Wigard, J. et al., Capacity of a GSM Network with Fractional Loading and Random Frequency Hopping, in *Proceedings of PIMRC Conference*, Taipei, Taiwan, ROC, 1996.
16. Kronestedt, F. and Andersson, S., Migration of adaptive antennas into existing networks, in *Proceedings of VTC Conference*, Ottawa, Canada, May 1998.
17. Dam, H. et al., Functional tests of adaptive antenna base stations for GSM, in *Proceedings of EPMCC Conference*, Paris France, March 9–11, 1999.
18. Dam, H. et al., Performance Evaluation of Adaptive Antenna Base Stations in a Commercial GSM Network, in *Proceedings of VTC Conference*, Amsterdam, the Netherlands, September 1999.

19. Johannisson B., Adaptive base station antennas for mobile communication systems, in *Proceedings of the IEEE Conference on Antennas and Propagation for Wireless Communications* (*APWC'98*), November 2–4, 1998, Waltham, MA.
20. Molnar, K. and Bottomley, G., D-AMPS performance in PCS bands with array processing, in *Proceedings of VTC Conference*, Atlanta, GA, May 1996.
21. Hagerman, B. et al., Field test performance results for D-AMPS in PCS bands with array processing, in *Proceedings of VTC Conference*, Phoenix, AZ, 1997.
22. Hagerman, B. and Mazur, S., Adaptive antennas in IS-136 systems, in *Proceedings of VTC Conference*, Ottawa, Canada, May 1998.
23. The Evolution of TDMA to 3G, *IEEE Personal Commun. Mag.*, vol. 6, no. 3, June 1999.
24. Hagerman, B., Molnar, K., and Molnar B., Evaluation of novel multi-beam antenna configurations for TDMA (IS-136), in *Proceedings of VTC Conference*, Houston, TX, May 1999.
25. Hagerman, B. et al., Ericsson — AT&T Wireless Services joint adaptive antenna multi-site field trial for TDMA (IS-136) systems, in *Proceedings of Sixth Smart Antenna Workshop*, Stanford University, Stanford, CA, July 1999.

PART F

Electromagnetic Radiation and the Human Body

25

Electromagnetic Interactions of Handheld Wireless Communication Antennas with the Human Body

Magdy F. Iskander
University of Utah

Zhengqing Yun
University of Utah

25.1 Introduction

With the explosive growth of the wireless communication technology, there have been concerns about the safety aspects of these devices and the potential hazardous effects associated with electromagnetic (EM) radiation interaction with human tissue. The general public is very much aware that people cook food and heat meat in microwave ovens and having handheld devices near the human head is certainly discomforting and a cause of serious concern. The mobility and convenience advantages that these wireless devices provide (connect to anyone, any time, anywhere), however, cannot be dismissed; hence, there is a need to carefully study this issue, accurately quantify these interactions, and determine their compliance with set safety standards.

0-8493-0124-6/02/$0.00+$1.50

From the antenna engineering point of view, this area of human body interactions with antennas is intriguing and represents divergence from standard procedures for designing antennas for wireless communications. It represents a near-field interaction problem whereby the human head and parts or all of the body need to be taken into account in determining the antenna characteristics. These effects are expected not only to impact the gain and radiation pattern of antennas on the handheld devices but also to affect the input impedance characteristics that are essential in optimizing the low-power consumption of these devices. Antenna engineering research has actually progressed further, and more advanced antenna designs that minimize the EM radiation interaction with the human body have been developed. This provides a solution for the often unpredictable influence of the human body on the radiation and impedance characteristics of the antenna, and certainly alleviates the concerns about the potential hazardous effects of these devices.

In this chapter, we start by briefly reviewing the safety standard for EM radiation with some focus placed on the near-field aspects as well as the set standards for handheld devices. A review of the various types of antennas usually used in handheld devices are then presented, followed by sections that describe numerical techniques and experimental methods used to quantify and characterize the interactions of radiated fields with humans. Results showing the EM power deposition patterns in the human head are then presented, and more recent data illustrating the human body effects on these calculations are discussed. The chapter concludes with a section that provides illustrative examples showing the effect of these interactions on the radiation and input impedance characteristics of antennas in handheld devices.

25.2 Exposure Standards for Radio Frequency Fields

Determination of exposure standards and safety levels of radio frequency (RF) radiation has been a research area of significant interest for over three decades and, in particular, since 1966 with the publication of a safety standard by the United States of America Standards Association, currently known as the American National Standards Institute (ANSI). Initial efforts included quantification of plane wave interactions with simplified human models [1], but calculation techniques and measurement procedures continued to progress to accurately model realistic models of humans, to address additional areas of interest such as near-field interactions [2, 3], and to identify locations and levels of hot spots in the specific absorption rate (SAR) distribution in the human body [4–6]. Among the significant observations identified in these studies are the importance of accounting for the polarization effects [7] and the need for carefully considering the location and levels of hot spots [8]. The polarization effects and the resonance absorption in the frequency range from about 30 to 100 MHz resulted in reducing the ANSI standard from 10 mW/cm^2 of incident plane wave power density to 1 mW/cm^2 in the modified standard published in 1982 [1]. To help accommodate needs to address near-field exposure issues, safety standards are also expressed in terms of electric and magnetic field intensities instead of power density values [9].

With the growing concern about the potential health hazards associated with RF radiation from handheld devices, updated ANSI/Institute of Electrical and Electronics Engineers (IEEE) RF safety guidelines were published in 1992. These ANSI/IEEE C95.1-1992 safety guidelines were given in terms of the maximum possible exposure of electric and magnetic fields, or of RF power density [10]. These guidelines, however, were not easy to use whenever highly nonuniform fields, such as those in the near-field regions of handheld devices, were encountered. In these cases, the following alternative safety guidelines, which are based on the specific absorption rate (SAR) instead of the incident RF fields, were suggested.

An exposure condition can be considered to be acceptable if it can be shown that it produces SARs "below 0.08 W/kg as averaged over the whole body, and spatial peak SAR values not exceeding 1.6 W/kg as averaged over any 1 g of tissue." As in the case of the originally published safety standards of the ANSI, there have been other standards published by U.S. and other international organizations [1]. These standards are often based on other considerations in addition to average SAR values and their distributions in a human body. For example, for cellular phones, a spatial peak SAR value of 8 W/kg as averaged over any 1 g of tissue has been issued by the Telecommunications Technology Council of Japan [11]. Yet another safety standard is set in Europe for the use of mobile telecommunication equipment by the public [12]. The European standard is set at 2 W/kg averaged over a volume equivalent to 10 g and a period of 6 min. Both the Japanese and European standards are clearly less restrictive than the ANSI/IEEE one enforced in the United States. These alternative safety guidelines and their reliance on absorbed power instead of incident fields generated a flurry of new studies focused on quantifying the absorption characteristics of RF radiation from handheld devices in human heads [13–18]. Existing cellular phones were examined and systems manufacturers initiated focused design efforts to help minimize these effects as well as to optimize the antenna performance in typical operating conditions. In the following sections, types of antennas used in wireless handheld devices are described, and the numerical and experimental procedures used to quantify their SAR characteristics in the human body are discussed.

25.3 Antenna Types for Handheld Wireless Devices

Antenna designers are often concerned with developing new antennas that meet gain, radiation pattern, and bandwidth requirements in a given frequency range; and incorporating advanced features such as beam steering capabilities in their designs. With the advent of wireless handheld devices, additional requirements emerged and became equally important factors in these designs. These include physical dimensions, appearance, cost, and minimum change in performance in a typical use in wireless communication environment. With the growing concerns about the biological effects and potential health hazards associated with these devices, new designs that minimize the RF radiation in the human head region are also being developed. Furthermore, more modern handheld wireless communication devices use multiple antennas to provide diversity advantages and overcome signal loss and multipath fading effects. In other words, there has been renewed interest in designing simple antenna elements with significant communication and health-related advantages in a typical terrestrial wireless communication environment.

Figure 25.1 shows four types of antennas commonly used in handheld transceiver units. These include the monopole antenna, side-mounted dual planar inverted-F antenna (PIFA), the top-mounted bent inverted-F antenna (BIFA), and the back-mounted PIFA. The variety of PIFA antenna mountings is known to be efficiently and conveniently integrated with handheld transceivers [14]. Alternative designs incorporate electrically short helical antennas operating in the normal mode condition, with maximum radiation perpendicular to the orientation of the antenna [19] and use of ferrite sheet attachments to reduce radiation toward the head from monopole antennas [20]. Examples of these ferrite attachment designs are shown in Fig. 25.2.

With the continued advances in improving quality of service and increasing capacity of modern wireless communication systems, there has been significant effort in incorporating signal processing algorithms to help develop diversity techniques to minimize signal losses and overcome fading resulting from

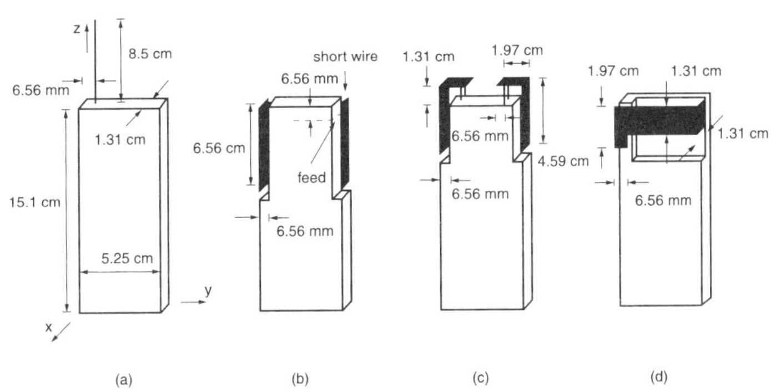

FIGURE 25.1 Antenna geometries and dimensions for handheld transceiver units: (a) monopole; (b) side-mounted dual PIFA; (c) top-mounted dual BIFA; (d) back-mounted PIFA. The chassis dimensions (105 cm^3) shown in (a) apply to all four figures. (From Jensen, M.A. and Rahmat-Samii, Y., *Proc. IEEE*, 83(1), 7, Jan. 1995. With permission.)

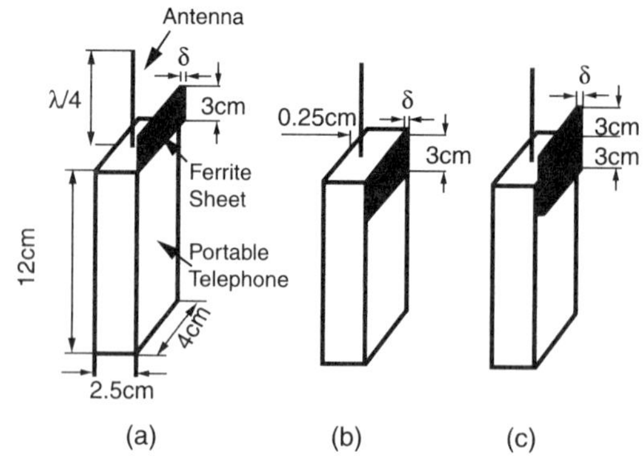

FIGURE 25.2 Portable telephones with ferrite sheets. (a) Type A. (b) Type B. (c) Type C. (From Wang, J. and Fujiwara, O., *IEICE Trans. Commun.*, E80-B(12), 1810, Dec. 1977. With permission.)

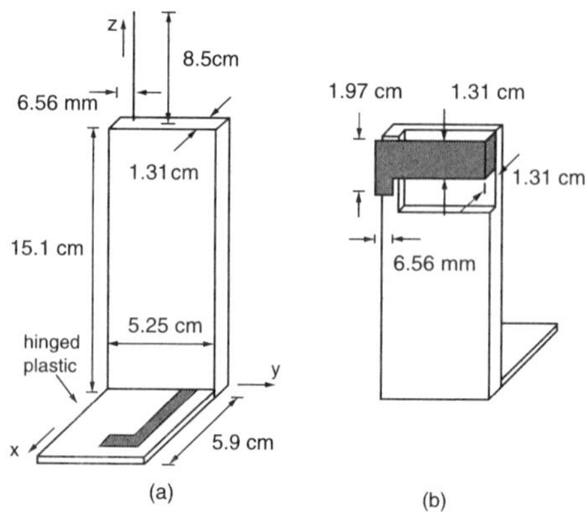

FIGURE 25.3 Two dual-antenna diversity handset configurations. (a) Monopole/mouthpiece monopole. (b) PIFA/mouthpiece monopole. (From Green, B.M. and Jensen, M.A., *IEEE Trans. Antennas Propag.*, 48(7), 1017, July 2000. With permission.)

multipath interference. Implementation of these techniques often requires the use of multiple antennas at the base station as well as in the handheld transceiver. An example of the use of multiple antennas in a mobile transceiver is shown in Fig. 25.3 [21]. This figure shows two possible configurations consisting of either two quarter-wavelength monopole antennas or a monopole and a PIFA antenna integrated on the mouthpiece of the transceiver.

Another type of antenna used in this application is the Yagi array shown in Fig. 25.4 [22]. In this case, a single two-element array consisting of a radiator and reflector or a pair of two-element arrays were used as shown in Fig. 25.5 [22]. In this case, the two arrays were intended for separate transmitting and receiving antennas or for multiband operation of the transceiver. Radiation characteristic results of these antennas as well as results from numerical and experimental evaluation of SAR values in a human head are presented in the results section (Section 25.7).

25.4 Numerical Computation of Specific Absorption Rate Distribution in the Human Head

In this section, we briefly outline the various numerical techniques used in the calculation of the SAR and its distribution in a human head. This section also includes a description of the head model often used in these calculations.

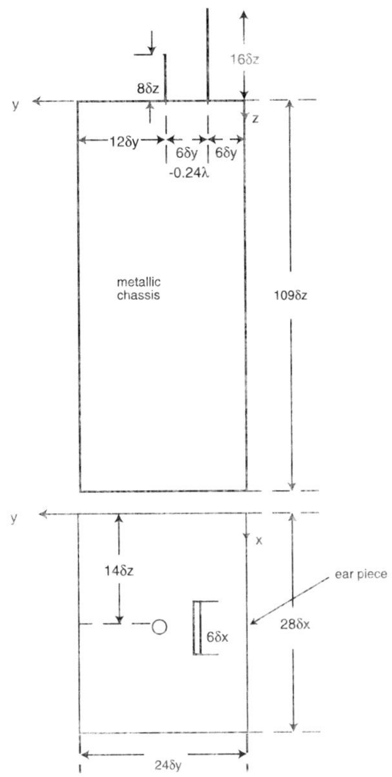

FIGURE 25.4 Geometric arrangement of the centrally placed antenna 1. The dimensions are shown in terms of the cell lengths δ_x, δ_y, and δ_z in the *x*-, *y*-, and *z* directions, respectively. For the FDTD calculations, $\delta_x = \delta = 1.974$ mm, $\delta_z = 1.5$ mm. (From Gandhi, O.P. and Chen. J.-Y., *IEEE Trans. Electromagnetic Compatibility,* 7(4), 547, Nov. 1995. With permission.)

To begin with, it should be acknowledged that advances in computational EM made it possible not only to accurately and efficiently calculate SAR in realistic models of the human head, but also, through these advances, to simulate anatomically based models of the human head with resolution as small as 1 to 2 mm. In many cases, the dimensions, materials properties, and orientation of the handheld device were also taken into account in these simulations. In general, computation techniques used in these calculations include the finite difference time domain (FDTD) method [13-17, 23], transmission line matrix (TLM) method [24], modified method of moments (MOM) [25], and a hybrid MOM/FDTD technique [26]. FDTD has been by far the method of choice in many of these simulations. In a more recent effort, however, and in an attempt to examine the effect of the human body on the SAR values in the human head, it was necessary to use the multigrid FDTD code recently developed by our group at the University of Utah [27, 28].

The calculation procedure based on the FDTD method has amply been described in literature and hence needs no further discussion. In applying the TLM procedure, however, it was necessary to modify the method to account for the frequency-dependent dielectric properties of tissue [24]. Specifically, a technique that handles second-order Debye dielectric dispersion was developed, and the SAR distribution in a relatively rough model of the human head ($7 \times 7 \times 7$ mm^3) was calculated. The obtained results are encouraging but additional refinements to increase the resolution are desirable. Other simulation efforts to better model the antenna structure on handsets [25] or to incorporate improved antenna modeling using the MOM while still calculating SAR in the head model using FDTD [26] were also reported.

As mentioned earlier, a multigrid FDTD code [28] was used to calculate SAR distribution in a human head, taking into account the effect of the human body [29]. In this case, two FDTD models were

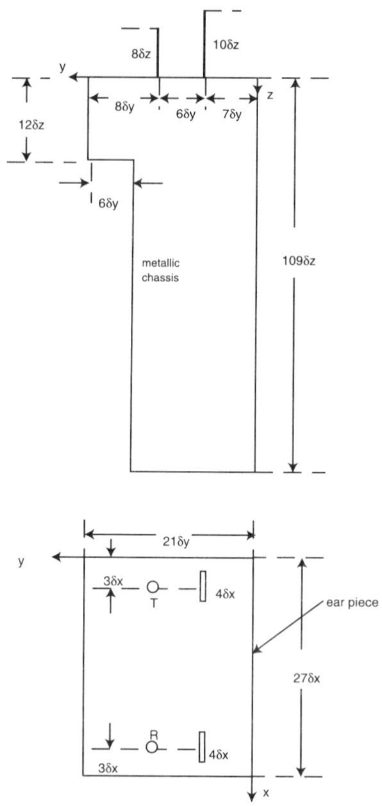

FIGURE 25.5 Geometric arrangement for antenna 2. Note that individual Yagi antennas T and R are used for transmitting and receiving antennas, respectively. (From Gandhi, O.P. and Chen. J.-Y., *IEEE Trans. Electromagnetic Compatibility,* 7(4), 547, Nov. 1995. With permission.)

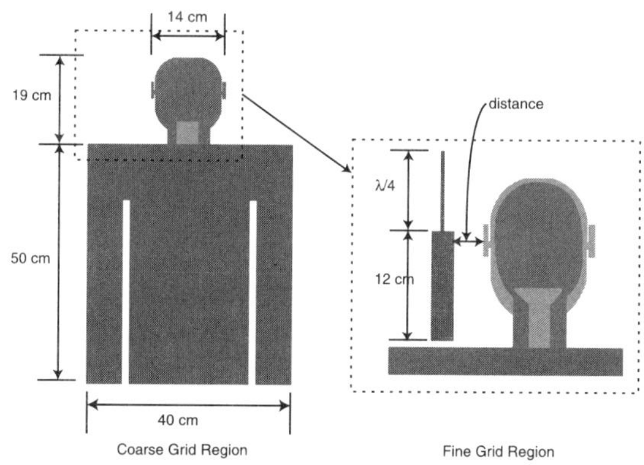

FIGURE 25.6 The FDTD regions in the multigrid FDTD simulations. (From Iskander, M.F., Yun, Z., and Quintero-Illera, R., *IEEE Trans. Microwave Theory Tech.,* vol. 48, no. 11. pp, 1979–1987, 2000. With permission.)

utilized — a fine grid for the head with a 2.45-mm cell size and a coarse grid for the human body with an 8.9-mm cell size (four times larger than the fine grid cell). Furthermore, an anatomically accurate model for the head was used, while the dielectric constant of the body was assumed homogeneous and represented by a complex permittivity value equal to 2/3 that of muscle tissue. Figure 25.6 illustrates the fine and coarse grid regions in the multigrid FDTD calculations. Further discussion of the various head models used in these and other calculations are described in the following section.

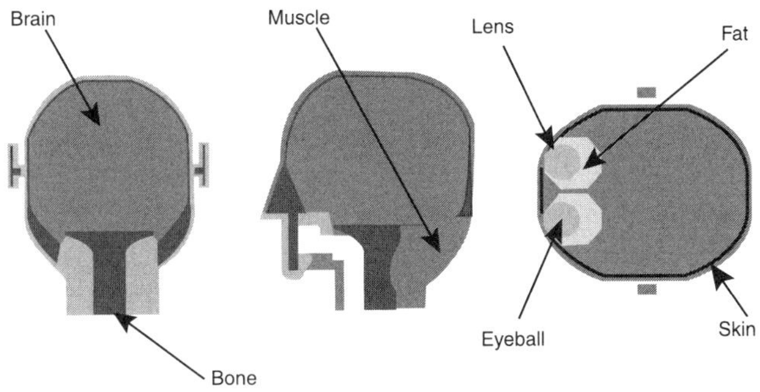

FIGURE 25.7 Human head model. A fine grid computation domain was used for modeling the head. (From Iskander, M.F., Yun, Z., and Quintero-Illera, R., *IEEE Trans. Microwave Theory Tech.*, vol. 48, no. 11, pp. 1979–1987, 2000. With permission.)

25.5 Anatomically Based Head Model for Specific Absorption Rate Calculations

A variety of models was used in the SAR calculations of a human head. Simple homogeneous spherical models were used when focus was placed on evaluating and optimizing the performance of more complex antenna designs [25, 26]. A layered eccentric sphere structure was also used to help provide a concise analytic formulation and an exact solution [30]. Other simplified canonical models were also developed as part of the European COST 244 WG3 project [31]. Perhaps the most commonly used model in these calculations is the anatomically based model of the human head. These models were based on magnetic resonance imaging (MRI) [32]. The basic parts constituting the human head are shown in Fig. 25.7. The dielectric and conductivity of the tissues in the human head model are given in Tables 25.1 and 25.2 for 900 and 1900 MHz, respectively. A more detailed list of the dielectric properties and specific gravities of various tissues was also compiled in References [13, 15]. The data presented in Reference [13] is also given in Table 25.3 for completeness. Other dielectric properties for the head models were obtained from an available dielectric database [33]. In any event, images from computerized tomography (CT) scans and MRI images made it possible to model the human head to within mm-size resolution. The use of CT scan images in EM simulations, however, goes back to the late 1970s [34], but the accuracy and the resolution of the more recent images [32] have certainly played a critical role in improving SAR predictions.

TABLE 25.1 Properties of the Tissues in the FDTD Model: 900 MHz

	Bone	Brain	Muscle	Eyeball	Fat	Skin	Lens
Dielectric constant	9.67	52.7	59.1	80.0	4.67	59.1	59.1
Conductivity [S/m]	0.0508	1.05	1.26	1.90	0.0583	1.26	1.26

From Iskander, M.F., Yun, Z., and Quintero-Illera, R., *IEEE Trans. Microwave Theory Tech.*, vol. 48, no. 11, pp. 1979–1987, 2000. With permission.

TABLE 25.2 Properties of the Tissues in the FDTD Model: 1900 MHz

	Bone	Brain	Muscle	Eyeball	Fat	Skin	Lens
Dielectric constant	7.75	46.0	55.3	80.0	9.70	59.1	55.3
Conductivity [S/m]	0.105	1.65	2.0	1.90	0.270	1.26	2.0

From Iskander, M.F., Yun, Z., and Quintero-Illera, R., *IEEE Trans. Microwave Theory Tech.*, vol. 48, no. 11, pp. 1979–1987, 2000. With permission.

TABLE 25.3 Dielectric Properties and Specific Gravities of the Various Tissues Assumed at the Midband Mobile Telephone Frequencies of 835 and 1900 MHz

Tissue	Spec. Gravity 10^3 kg/m^3	835 MHz ε_r	835 MHz σ S/m	1900 MHz ε_r	1900 MHz σ S/m
Muscle	1.04	51.76	1.11	49.41	1.64
Fat	0.92	9.99	0.17	9.38	0.26
Bone (skull)	1.81	17.40	0.25	16.40	0.45
Cartilage	1.10	40.69	0.82	38.10	1.28
Skin	1.01	35.40	0.63	37.21	1.25
Nerve	1.04	33.40	0.60	32.05	0.90
Blood	1.06	55.50	1.86	54.20	2.27
Parotid gland	1.05	45.25	0.92	43.22	1.29
CSF	1.01	78.10	1.97	77.30	2.55
Eye humor	1.01	67.90	1.68	67.15	2.14
Sclera	1.17	54.90	1.17	52.56	1.73
Lens	1.10	36.59	0.51	42.02	1.15
Pineal gland	1.05	45.26	0.92	43.22	1.29
Pituitary gland	1.07	45.26	0.92	43.22	1.29
Brain	1.04	45.26	0.92	43.22	1.29

From Gandhi, O.P., Lazzi, G., and Furse, C.M., *IEEE Trans. Microwave Theory Tech.*, 44(10), 1884, Oct. 1996. With permission.

25.6 Experimental Measurement of Specific Absorption Rate in Phantoms of the Human Head

Average SAR values and SAR distribution measurements in phantoms have long existed and provided valuable information since the early 1970s [1, 6, 8]. Phantom materials for both low and high dielectric property tissues were developed and several techniques for measuring average SAR values and SAR distribution were described [1]. Procedures for preparing saline solutions with specific conductivity values are available, and techniques for measuring SAR values in solid phantoms (saline solutions with jelling agent) have long been described [8]. As may be expected, much of the effort in evaluating the potential health effects from handheld transceivers was focused on the determination of the SAR distribution in the head. To this end, more accurate phantom preparation is required to provide accurate representation of the various tissue types in this rather complex and highly sensitive part of the human body. Gandhi and Chen [22] used ingredients in available recipes for preparing phantom materials and, through systematic variations in the percentage of contribution from each ingredient and measurement of the resulting dielectric properties, developed formulas for the preparation of a wide variety of tissue properties that can be used in head models. Table 25.4 lists compositions used to simulate the dielectric properties of tissues used in an experimental model of the head [22]. The empirical formulas used to determine the percentage of water and salt (NaCl) contents in a typical phantom mixture were calculated based on the following factors [22]:

$$\left(K_{\sigma}\right)_{H_2O} = \frac{\Delta\sigma}{\Delta\, H_2O(\%)} \cong 0.05 \tag{25.1}$$

$$\left(K_{\varepsilon_r}\right)_{H_2O} = \frac{\Delta\varepsilon_r}{\Delta\, H_2O(\%)} \cong 0.80 \tag{25.2}$$

TABLE 25.4 Compositions Used to Simulate the Dielectric Properties (ε_r, σ) of the Soft Tissues for the Experimental Model of the Head

	Desired		Test Composition (Percentage)				Measured	
Tissue	ε_r	σ S/m	H_2O	S.S.	P.E.P	NaCl	ε_s	σ S/m
Brain	30.0	5.3	70.0	7.0	22.17	0.83	30.6	5.3
Muscle	40.0	4.9	80.0	13.0	7.0	0	35.5	5.1
Eyes	30.0	3.8[a]	92.0	3.2	3.0	1.8	46.0	6.8
Vitreous humor	63.0	7.2						
Ear	23.0	2.6	70.0	7.0	23.0	0	25.0	3.5

[a] Average of the properties desired for the eyes and the vitreous humor are used to develop the tissue-simulant material for this organ.

From Gandhi, O.P. and Chen, J.-Y., *IEEE Trans. Electromagnetic Compatibility*, 37(4), 547, Nov. 1995. With permission.

TABLE 25.5 Compositions Used to Simulate Skull or Bone at the Experimental Frequency of 6 GHz

	Test Composition (Percentage)				Measured	
Mixture	Epoxy	Hardener	KCl or NaCl Solution*	Slab length (mm)	ε_r	σ S/m
1	35.7	35.7	28.6 (KCl)	15.8	6.27	0.34
				22.0	5.78	0.32
2	35.7	35.7	28.6 (NaCl)	16.2	5.76	0.31

Note: Desired $\varepsilon_r = 6.0$, $\sigma = 0.3$ S/m; the short-circuited waveguide method using different slab lengths of the material was used for the measurements.

[a] The composition used for KCl or NaCl solution was 130 g of salt mixed with 950 g of water.

From Gandhi, O.P. and Chen, J.-Y., *IEEE Trans. Electromagnetic Compatibility*, 37(4), 547, Nov. 1995. With permission.

$$\left(K_\sigma\right)_{\text{NaCl}} = \frac{\Delta\sigma}{\Delta\,\text{NaCl}(\%)} = 1.00 \tag{25.3}$$

These formulas were obtained from the slopes of measured ε_r and σ vs. H_2O% and NaCl% curves. As for the skull or bone tissue types, a composition of 36.4% epoxy, 36.4% hardener, and the rest a solution of potassium chloride (KCl) or sodium chloride in which 88% by weight is water (130 g of salt mixed with 950 g of water) is used [22]. Table 25.5 provides a list of the composition and the measured dielectric properties of the skull (bone) type of phantom material [22]. In the work by Gandhi and Chen [22], the NaCl-based composition was used to build a hollow human skull model of the head. This skull mold was then filled with various types of soft tissues to accurately represent a human head model.

As for the SAR distribution measurement techniques, one of the following three approaches is implemented [1]:

1. Use of nonperturbing temperature probes to measure SAR values at specified locations of interest — Nonperturbing temperature probes may include fiber optics probes [35] or the F probe available commercially [36].
2. Use of nonperturbing miniature electric-field probes [37] — Some of these implantable E-field probes are available commercially (Narda model 26089/BRH-15) and require calibration to determine conversion (scale) factor relating the actual measured E-field values to the actual ones inside

the phantom model. The actual field values are usually obtained based on analytic expressions in phantom models of simple geometries (i.e., spheres or slabs of dielectrics).

3. Use of split phantom and thermographic camera measurements — This technique is based on developing two symmetrical parts of the head phantom model and exposing the surface at the symmetry plane to a thermographic camera for temperature distribution measurements [8]. It should also be noted that new nondegradable dry phantom materials have been developed for this application [38–40].

As in any experimental procedure, calibration procedures need to be discussed and limitations need to be pointed out. As mentioned earlier, the implantable E-field probe needs to be calibrated so as to relate the measured output voltage (electric field) from these probes to the actual electric fields at the measured locations. Although SAR values may be directly calculated from the measured and normalized E-field values using the formula SAR = $\sigma|E|^2/\rho$ where σ is the electrical conductivity of the phantom (tissue) and ρ is the density, precautions must be exercised to minimize errors resulting from probe–phantom interfaces, potential airgaps at measurement locations, and interference caused by the resistive leads in the measuring probes. In the split phantom measurements, on the other hand, caution should be used concerning thermal diffusion and temperature distribution errors as a function of time after irradiating the phantom and in the process of opening the two halves to expose the symmetry surface to the thermographic camera. Calibration curves vs. time for these experiments need to be established; and extrapolation, if possible to the $t = 0$ time (right after terminating the phantom exposure to RF radiation), needs to be used for accurate implementation of the formula SAR = cT/t where c is the specific heat, T is the temperature rise, and t is the exposure duration time [38].

25.7 Specific Absorption Rate Distribution and Radiation Characteristics of Antenna Results

In this section, we summarize the SAR distribution results either calculated numerically or measured experimentally, and also discuss the radiation characteristics of some of the used antennas when operated near the human head. Some of the more recent results illustrating the effect of the human body on the SAR distribution in the human head are also discussed.

25.7.1 Specific Absorption Rate Distribution in the Human Head Models

Several types of SAR distribution in the human head were reported in the literature. Some were intended to illustrate the effect of rotating the operator's head while using the handheld transceiver [13, 41]; another was to illustrate differences in absorption when different types of antennas are used [14, 18, 19], whereas others were intended to examine a variety of effects such as placement of the hand [13, 18], separation distance between the handset and the head [18], and the effect of the operating frequency [13, 30, 42].

Figure 25.8 shows three configurations of the head model used to examine the effect of tilting the operator's head when using the cellular phone [41]. The obtained peak 1-g SAR for the head, for the brain, and the average SARs in some of the selected tissues are given in Tables 25.6 and 25.7. As may be noted, results were given at both 835 MHz and 1.9 GHz. It may also be worth noting that the maximum peak 1-g SAR for both the head and the brain are maximum for the case of vertical head and minimum for the case of a 30° tilt with an additional 9° rotation in the horizontal plane [41]. A similar observation may be made even for the average SAR values in different tissue types.

Another important consideration in evaluating potential hazards and health effects is the operating frequency. Tables 25.8 and 25.9 compare results for transceivers operating at 835 and 1900 MHz, respectively [13]. As it may be noted, higher frequencies (1900 MHz) cause less than 50% of the peak 1-g SAR values for the head and brain at the lower frequency. At higher frequencies, much of the power absorption occurs at the surface and, hence, the relatively higher values in the hand at these higher frequencies. A

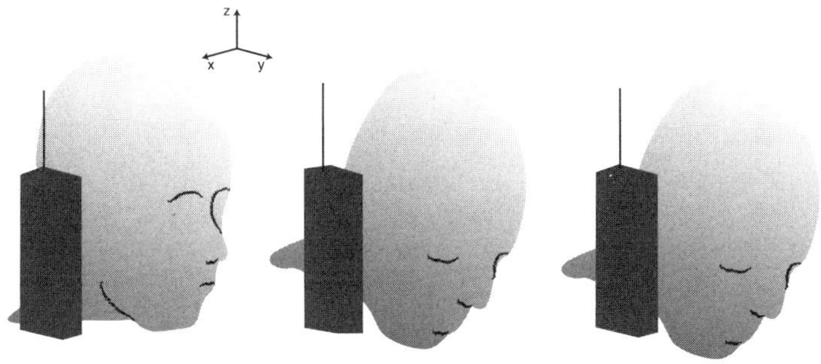

FIGURE 25.8 Visualization of the various segmented head models with the telephone: Vertical, tilted 30° with a further rotation of 9° of the face toward the telephone. (From Lazzi, G. and Gandhi, O.P., *IEEE Trans. Electromagnetic Compatibility,* 39(1), 55, 1997. With permission.)

TABLE 25.6 Peak 1-g SARs for the Head and the Brain in W/kg at 835 and 1900 MHz for the Three Head Models Considered

Frequency (MHz)		Vertical Head Model	Tilted 30° Head Model	Tiled 30° Head Model, with Further Rotation of 9°
835	Peak 1-g SAR for head	2.93 (1.01 g)	2.44 (1.03 g)	2.31 (1.10 g)
	Peak 1-g SAR for brain	1.13 (1.09 g)	0.93 (1.02 g)	0.66 (1.00 g)
1900	Peak 1-g SAR for head	1.11 (1.03 g)	1.08 (1.03 g)	1.20 (1.01 g)
	Peak 1-g SAR for brain	0.19 (1.00 g)	0.20 (1.04 g)	0.16 (1.02 g)

Note: Given in parentheses are the actual weights of the subvolumes considered for the peak 1-g SARs. The telephone is assumed to be a plastic-covered metal box with a $\lambda/4$ monopole antenna mounted above it. The radiated power is assumed to be 600 mW at 835 MHz and 125 mW at 1900 MHz.

From Lazzi, G. and Gandhi, O.P., *IEEE Trans. Electromagnetic Compatibility,* 39(1), 55, 1997. With permission.

TABLE 25.7 Average SARs in Selected Tissues at 835 and 1900 MHz for the Three Head Models Considered

Frequency (MHz)	Average SARs (mW/kg)	Vertical Head Model	Tilted 30° Head Model	Tilted 30° Head Model, with Further Rotation of 9°
835	Brain	72.3	60.6	49.1
	CSF	72.7	66.4	53.7
	Eye humor	31.8	20.6	20.7
1900	Brain	7.6	7.0	7.2
	CSF	7.9	6.7	7.9
	Eye humor	3.2	1.2	1.7

From Lazzi, G. and Gandhi, O.P., *IEEE Trans. Electromagnetic Compatibility,* 39(1), 55, 1997. With permission.

more systematic study of the effect of frequency was reported in Reference [30] for the more simplified six-layer model consisting of eccentric spheres.

There is no question that the type and location of antennas in the handset also play an important role in the amount and distribution of SAR values in the human head. Jensen and Rahmat-Samii presented results illustrating differences in SAR distributions for different antennas [14]. Figure 25.9 shows the geometric arrangement of the handset with respect to the head, and Fig. 25.10 shows results of the computed SAR distribution values [14]. As it may be seen, different antennas, even when placed at the same location from the head, would produce different SAR distributions.

TABLE 25.8 Comparison of the Powers Absorbed and Peak SARs for the λ/4 and 3λ/8 Antennas at 835 MHz

Antenna Length	Tilt	Peak 1-g SAR for Head	Peak 1-g SAR for Brain	% Power Absorbed by "Hand"	% Power Absorbed by Head and Neck
λ/4	0°	2.93 (1.00 g)	1.13 (1.09 g)	9.2	45.0
λ/4	30°	2.42 (1.03 g)	0.93 (1.02 g)	12.4	39.8
3λ/8	0°	1.60 (1.00 g)	0.65 (1.05 g)	5.6	33.7

Note: Time-averaged radiated power = 600 mW.
From Gandhi, O.P., Lazzi, G., and Furse, C.M., *IEEE Trans. Microwave Theory Tech.*, 44(10), 1884, October 1996. With permission.

TABLE 25.9 Comparison of the Powers Absorbed and Peak SARs for the λ/4 and 3λ/8 Antennas at 1900 MHz

Antenna Length	Tilt	Peak 1-g SAR for Head	Peak 1-g SAR for Brain	% Power Absorbed by "Hand"	% Power Absorbed by Head and Neck
λ/4	0°	1.11 (1.03 g)	0.20 (1.00 g)	13.8	35.6
λ/4	30°	1.08 (1.03 g)	0.20 (1.04 g)	13.9	35.5
3λ/8	0°	0.69 (1.06 g)	0.16 (1.00 g)	7.0	29.4

Note: Time-averaged radiated power = 125 mW.

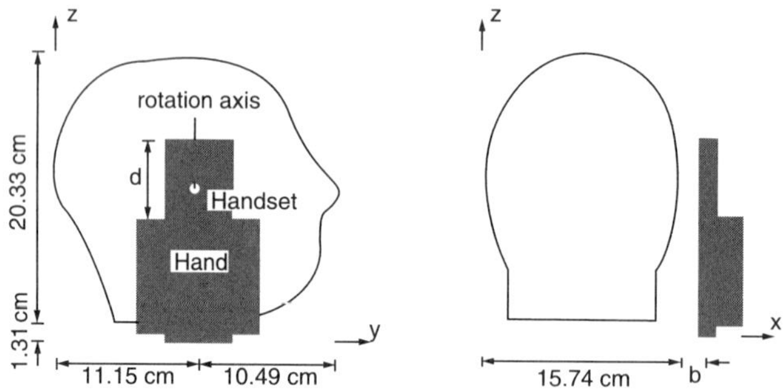

FIGURE 25.9 Side and rear views of the FDTD head/hand/handset mode showing dimensions. (From Jensen, M.A. and Rahmat-Samii, Y., *Proc. IEEE*, 83(1), 7, Jan. 1995. With permission.)

25.7.2 Effect of the Human Body on Specific Absorption Rate Values and Distributions

As may be noted from the bibliography, several research groups have examined SAR calculations in the human head when exposed to EM radiation from handheld devices. In all cases, however, the isolated human head was modeled and included in the simulation. It is of interest to examine the effect of the human body on these SAR values and their distribution in the human head. The multigrid FDTD code was used to address this issue [28]. In this approach, a fine grid region was placed around the head and the transceiver while a coarse grid was used to model the rest of the human body. Specifically, a fine grid with a 2.45-mm spatial resolution was used for the human head, resulting in a grid with 100 × 100 × 100 fine grid cells in this region. For the body model, on the other hand, a coarse 8.9-mm grid was used, and the dielectric constant of the body tissue was assumed to be homogeneous and ε_r^* the value for muscle tissue. Figure 25.11 shows the obtained SAR distribution results at 900 MHz with and without taking the human body into account [29]. From these results, it may be seen that for distances between the antenna and human head, $d \geq 4$ cm, some differences that may be significant may be observed in

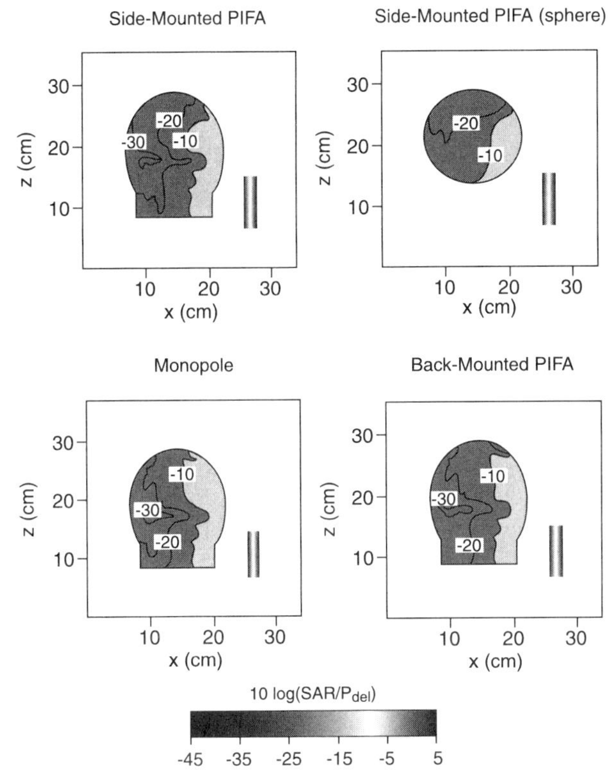

FIGURE 25.10 Computed normalized SAR distribution in a plane through the center of the head with $d = 7.21$ cm, $b = 1.97$ cm, and the plastic casing present at 915 MHz. The configurations are the side-mounted PIFA, the side-mounted PIFA with the spherical head, the monopole, and the back-mounted PIFA. SAR values were calculated based on an average over 1g of tissue and normalized to the power delivered to the antenna using 10 log (SAR/P_{del}). (From Jensen, M.A. and Rahmat-Samii, Y., *Proc. IEEE,* 83(1), 7, Jan. 1995. With permission.)

SAR values and their distributions in the head. Results at 1900 MHz for the vertical monopole case are shown in Fig. 25.12. Tables 25.10 to 25.12 summarize some of the obtained results for the case of vertically polarized monopole antenna. From these results, it should be noted that taking the effect of the human body into account is certainly justifiable and, depending on the operating frequency and the distance between the head and the antenna, differences by as much as 53% may be observed [29]. Similar results for the PIFA antenna are shown in Fig. 25.13, and from Table 25.13 it may be noted that similar differences were observed and confirm the need for taking the effect of the human body into account when assessing the health effects associated with the use of handheld devices.

25.8 Effect of the Human Body on the Radiation Characteristics of Antennas

Thus far in our discussion, we focused on the safety aspects of handheld transceivers; and, to this end, issues such as safety standards and SAR distributions in the head were described. Another important aspect in studying these issues is to learn more about the radiation characteristics of the various antennas when operating near the human body. This information can be used to optimize the antenna performance but, equally important, it can be used to explore the development of new and innovative designs that can minimize sensitivity to operating conditions (head tilt and rotation, distance from the head, etc.) and also possibly minimize radiation toward the head. In this section, we summarize some of the reported results on the radiation characteristics of antennas in handheld devices.

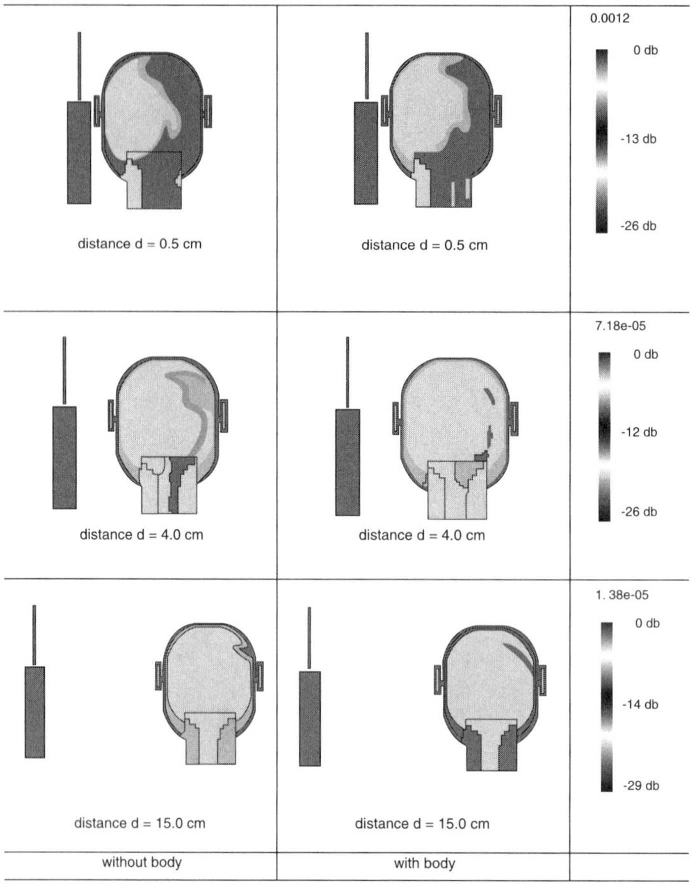

FIGURE 25.11 Vertical cross sections of the human body showing SARs distributions for vertical monopole antenna at 900 MHz with and without body. (From Iskander, M.F., Yun, Z., and Quintero-Illera, R., *IEEE Trans. Microwave Theory Tech.,* vol. 48, no. 11, pp. 1979–1987, 2000. With permission.)

Okoniewski and Stuchly [15] examined the radiation characteristics of monopole antennas and their variations as a function of distance from the human head. Figure 25.14 shows the studied geometry and some of the obtained results. As it may be seen from the θ-plane radiation patterns, there may be reduction in gain and change in the level of sidelobes as the antenna gets closer to the head. The effect of the hand placement on the receiver box was also examined and the obtained results are shown in Fig. 25.15. The effect of the hand on the input impedance of the antenna was also significant as described in Reference [14]. Figure 25.16 shows $|S_{11}|$ of a side-mounted PIFA antenna for cases with and without hand and as a function of the distance d between the hand and the antenna [14]. From these results, it may be seen that not only the presence of the hand but also its relative location on the handset causes changes in the reflection coefficient and, more importantly, may result in a shift in the resonance frequency and hence detune the antenna. The results shown in Fig. 25.17 show a significant difference in the effect of the hand between the side-mounted and the top-mounted PIFA antennas [14]. Clearly, the effect of the hand is less severe for the case of the top-mounted PIFA antenna. As a matter of fact, placing the hand at a distance of 2.62 cm for the case of the side-mounted antenna almost detuned the antenna, whereas similar effect for the top-mounted PIFA antenna would result at distance $d = 0.66$ cm from the head [14]. Other studies on the radiation characteristics of antennas in the presence of a human head may also be found in the literature [43, 44].

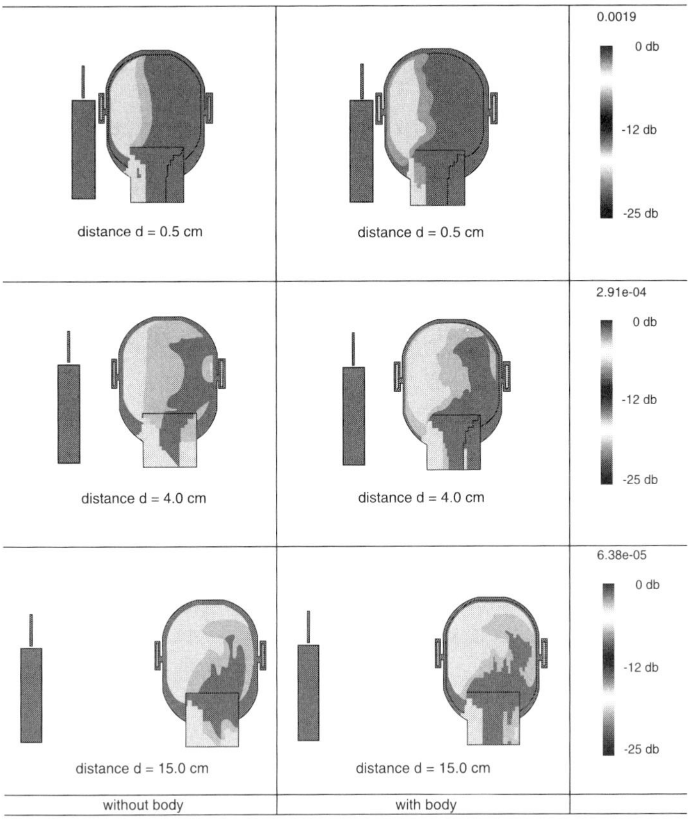

FIGURE 25.12 Vertical cross sections of the human head illustrating SARs distributions for vertical monopole antenna at 1900 MHz with and without body. (From Iskander, M.F., Yun, Z., and Quintero-Illera, R., *IEEE Trans. Microwave Theory Tech.*, vol. 48, no. 11, pp. 1979–1987, 2000. With permission.)

TABLE 25.10 Total SARs in the Human Head: Vertical Polarization

	Frequency					
	900 MHz			1900 MHz		
Distance (cm)	Without	With	Increase (%)	Without	With	Increase (%)
0.5	1.04	1.29	24	1.03	1.12	8.7
4.0	0.30	0.46	53	0.45	0.48	6.7
7.0	0.17	0.24	41	0.28	0.30	7.1
10.0	0.13	0.16	23	0.18	0.21	17
15.0	0.099	0.12	21	0.11	0.136	24

From Iskander, M.F., Yun, Z., and Quintero-Illera, R., *IEEE Trans. Microwave Theory Tech.*, vol. 48, no. 11, pp. 1979–1987, 2000. With permission.

25.9 Conclusions

In this chapter, we attempt to review ongoing research in the area of EM interactions of handheld antennas with the human body. This area of research was very active in the 1970s but received less attention after the publication of safety standards, including ANSI [9], in the 1980s. In most cases, focus was placed on interaction of plane waves with the human body [1]. Various computational techniques and experimental

TABLE 25.11 Total SARs in the Human Brain Area: Vertical Polarization Case

	Frequency					
	900 MHz			1900 MHz		
Distance (cm)	Without	With	Increase (%)	Without	With	Increase (%)
0.5	0.65	0.84	29	0.53	0.597	13
4.0	0.15	0.21	40	0.20	0.23	15
7.0	0.08	0.094	18	0.14	0.16	14
10.0	0.059	0.051	–14	0.089	0.11	24
15.0	0.046	0.032	–30	0.058	0.072	24

From Iskander, M.F., Yun, Z., and Quintero-Illera, R., *IEEE Trans. Microwave Theory Tech.*, vol. 48, no. 11, pp. 1979–1987, 2000. With permission.

TABLE 25.12 Maximum SARs in the Human Brain: Vertical Polarization

	Frequency					
	900 MHz			1900 MHz		
Distance (cm)	Without	With	Increase (%)	Without	With	Increase (%)
0.5	1.35×10^{-4}	1.75×10^{-4}	29	3.1×10^{-4}	2.9×10^{-4}	–7
4.0	4.58×10^{-5}	3.55×10^{-5}	–22	9.7×10^{-5}	8.4×10^{-5}	–13
7.0	2.36×10^{-5}	1.83×10^{-5}	–22	5.9×10^{-5}	6.0×10^{-5}	2
10.0	1.57×10^{-5}	1.39×10^{-5}	–11	3.2×10^{-5}	4.1×10^{-5}	28
15.0	1.05×10^{-5}	1.15×10^{-5}	10	1.8×10^{-5}	2.4×10^{-5}	33

From Iskander, M.F., Yun, Z., and Quintero-Illera, R., *IEEE Trans. Microwave Theory Tech.*, vol. 48, no. 11, pp. 1979–1987, 2000. With permission.

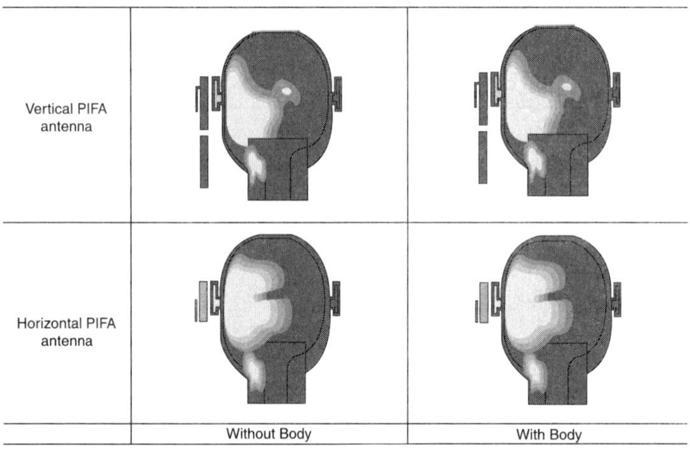

FIGURE 25.13 Vertical cross sections of the human head showing SARs distributions for vertically and horizontally oriented PIFA antenna at 900 MHz with and without the effect of the human body. (From Iskander, M.F., Yun, Z., and Quintero-Illera, R., *IEEE Trans. Microwave Theory Tech.*, 2000. With permission.)

methods were developed to quantify the average SAR and its distribution (hot spots) in the human body. As mentioned earlier, these activities slowed down after the publication of safety standards.

The continued growth and widespread use of personal communications systems renewed interest in this area of research with focus on interaction of EM radiation from handheld devices with the human

TABLE 25.13 SARs in the Human Head: f = 900 MHz; PIFA Antenna

	Polarization					
	Horizontal			Vertical		
Distance (cm)	Without	With	Increase (%)	Without	With	Increase (%)
0.5	0.018	0.019	6	0.017	0.010	–41

From Iskander, M.F., Yun, Z., and Quintero-Illera, R., *IEEE Trans. Microwave Theory Tech.*, vol. 48, no. 11, pp. 1979–1987, 2000. With permission.

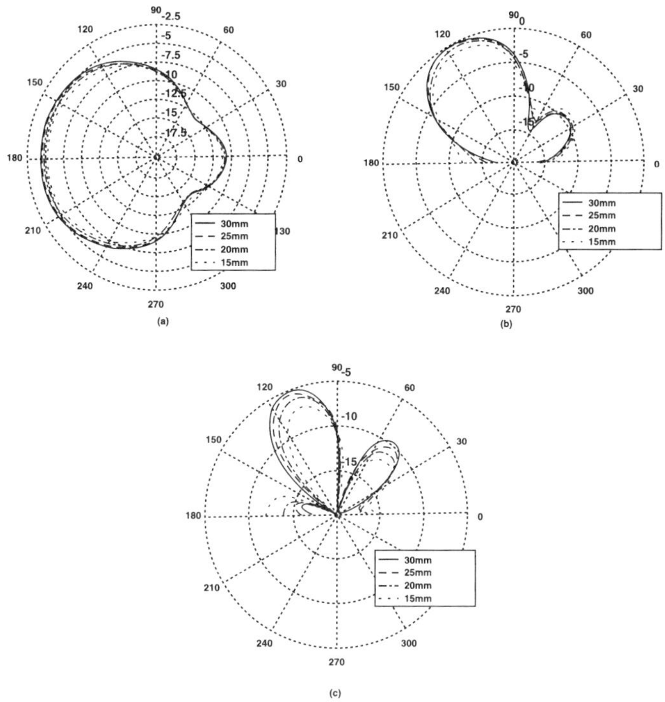

FIGURE 25.14 Radiation pattern for a monopole on a dielectric covered metal box at 915 MHz as a function of distance from the head model (no hand). (a) ϕ plane with $\theta = 90°$, (b) θ plane with $\phi = 180°$ (away from the head), and (c) θ plane with $\phi = 0°$ (toward the head). (From Okoniewski, M. and Stuchly, M.A., *IEEE Trans. Microwave Theory Tech.*, 44(10), 1855, Oct. 1996. With permission.)

head. Remarkably accurate models of the human head with resolution on the order of 1 mm were developed using MRI images and this, together with recent advances in computational EMs, helped in achieving accurate characterization and quantification of these interactions. SAR distributions in the head were calculated and compliance with safety standards for a wide variety of antennas was examined. Detailed effects such as distance between the antenna and the human head, placement of the hand on the transceiver, and position and rotation of the head while using these devices were examined and accurately quantified. In addition to these studies related to health effects and safety standards, antenna engineers took up the challenge and used these available computational and experimental methods to examine the impact of such interactions on the radiation and input impedance characteristics of antennas. Changes in radiation patterns, sidelobe levels, and gains were quantified; and the effect of the head and hand placement on the transceivers on the input impedance characteristics were studied. Even the diversity performance of multiple antennas in handheld devices was characterized, and this important

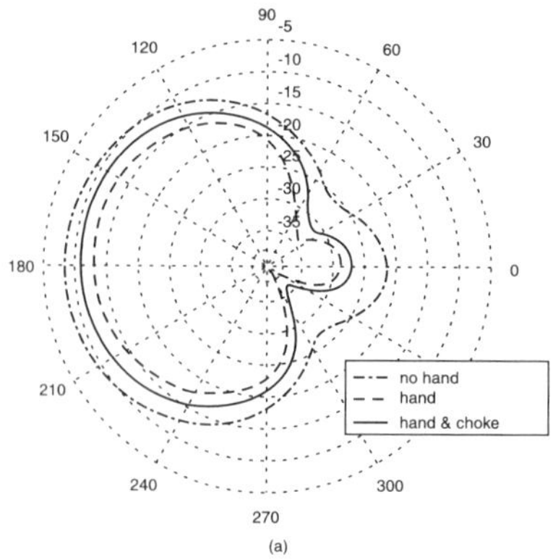

FIGURE 25.15 Radiation from a monopole on a dielectric covered metal box of 915 MHz at 2 cm away from the head model, without the hand, with the hand, and with the hand and a choke. (From Okoniewski, M. and Stuchly, M.A., *IEEE Trans. Microwave Theory Tech.,* 44(10), 1855, Oct. 1996. With permission.)

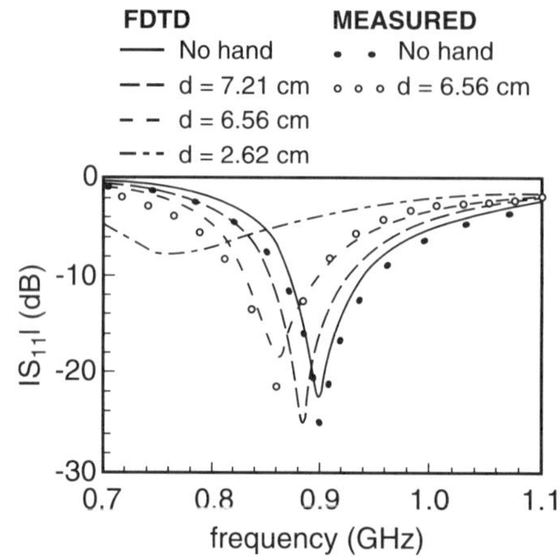

FIGURE 25.16 Computed value of $|S_{11}|$ for the side-mounted PIFA on the handset without the hand and with the hand for three different values of d. Measured values appear for the configurations with no hand and with the hand at $d = 6.56$ cm. (From Jensen, M.A. and Rahmat-Samii, Y., *Proc. IEEE,* 83(1), 7, Jan. 1995. With permission.)

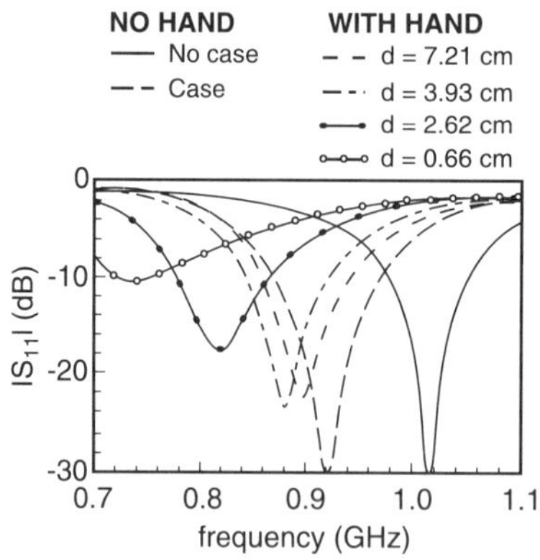

FIGURE 25.17 Computed value of $|S_{11}|$ for the top-mounted PIFA on the handset without the plastic casing, with the casing, and with the casing and hand at four different locations. (From Jensen, M.A. and Rahmat-Samii, Y., *Proc. IEEE,* 83(1), 7, Jan. 1995. With permission.)

development is expected to continue to help improve quality of services, increase channel capacity, and minimize adverse propagation effects such as fading and multipath interference. Progress has been significant, and it is hoped that the development of an antenna system that achieves optimum communication advantages while causing minimum or no effect on the human body will be a reality soon.

References

1. C. H. Durney, H. Massoudi, and M. F. Iskander, Radiofrequency Radiation Dosimetry Handbook, 4th Edition, Electrical Engineering Department, University of Utah, Prepared for USAF School of Aerospace Medicine, Oct. 1986.
2. M. F. Iskander, P. Barber, C. H. Durney, and H. Massoudi, "Irradiation of proloate spheroidal models of humans in the near-field of a short electric dipole," *IEEE Transactions on Microwave Theory and Techniques,* vol. 28, no. 7, pp. 801–807, 1980.
3. M. F. Iskander, H. Massoudi, C. H. Durney, and S. J. Allen, "Measurement of the RF power absorption in spheroidal human and animal phantoms exposed to the near field of a dipole source," *IEEE Transactions on Biomedical Engineering,* vol. 28, pp. 258–264, March 1981.
4. M. J. Hagmann, O. P. Gandhi, and C. H. Durney, "Numerical calculation of electromagnetic energy deposition for a realistic model of man," *IEEE Transactions on Microwave Theory and Techniques,* vol. 27, no. 9, pp. 804–809, 1979.
5. C. T. Tsai, H. Massoudi, C. H. Durney, and M. F. Iskander, "A procedure for calculating fields inside arbitrarily shaped, inhomogeneous dielectric bodies using linear basis functions with the moment method," *IEEE Transactions on Microwave Theory and Techniques,* vol. 34, pp. 1131–1139, 1986.
6. A. E. Guy, "Analysis of electromagnetic fields induced in biological tissues by thermographic studies on equivalent phantom models," *IEEE Transactions on Microwave Theory and Techniques,* vol. 19, pp. 205–214, 1971.
7. O. P. Gandhi, "Conditions of strongest electromagnetic power deposition in man and animals," *IEEE Transactions on Microwave Theory and Techniques,* vol. 23, no. 12, pp. 1021–1029, 1975.
8. A. E. Guy, M. D. Webb, and C. C. Sorensen, "Determination of power absorption in man exposed to high frequency electromagnetic fields by thermographic measurements on scale models," *IEEE Transactions on Biomedical Engineering,* vol. 23, pp. 361–371, 1976.
9. ANSI C95.1-1982, "Safety levels with respect to human exposure to radio frequency electromagnetic fields, 300 kHz to 100 GHz," American National Standards Institute, New York, Sept. 1, 1982.
10. ANSI/IEEE C95.1-1992, "American National Standard — Safety levels with respect to human exposure to radio frequency electromagnetic fields, 300 kHz to 100 GHz," New York: IEEE, 1992
11. "Radio frequency protection guidelines," Report of Telecommunication Technology Council for the Ministry of Posts and Telecommunications, Deliberation, No. 38, Tokyo, 1990.
12. CENELEC CLC/SC111B, European Prestandard prENV 50166-2. Human Exposure to Electromagnetic Fields High-Frequency: 10 kHz — 300 GHz, CENELEC, Brussels, Belgium, Jan. 1995.
13. O. P. Gandhi, G. Lazzi, and C. M. Furse, "Electromagnetic absorption in the human head and neck for mobile telephones at 835 and 1900 MHz," *IEEE Transactions on Microwave Theory and Techniques,* vol. 44, no. 10, pp. 1884–1897, Oct. 1996.
14. M. A. Jensen and Y. Rahmat-Samii, "EM interaction of handset antennas and a human in personal communications," *Proceedings of the IEEE,* vol. 83, no. 1, pp. 7–17, Jan. 1995.
15. M. Okoniewski and M. A. Stuchly, "A study of the handset antenna and human body interaction," *IEEE Transactions on Microwave Theory and Techniques,* vol. 44, no. 10, pp. 1855–1864, Oct. 1996.
16. V. Hombach, K. Meier, M. Burkhardt, E. Kuhn, and N. Kuster, "The dependence of EM energy absorption upon human-head modeling at 900 MHz," *IEEE Transactions on Microwave Theory and Techniques,* vol. 44, no. 10, pp. 1865–1873, Oct. 1996.
17. P. Bernardi, M. Cavagnaro, and S. Pisa, "Evaluation of the SAR distribution in the human head for cellular phones used in a partially closed environment," *IEEE Transactions on Electromagnetic Compatibility,* vol. 38, no. 3, pp. 357–366, Aug. 1996.

18. S. Watanabe, M. Taki, T. Nojima, and O. Fujiwara, "Characteristics of the SAR distributions in a head exposed to electromagnetic fields radiated by a handheld portable radio," *IEEE Transactions on Microwave Theory and Techniques,* vol. 44, no. 10, pp. 1874–1883, Oct. 1996.
19. G. Lazzi and O. P. Gandhi, "On modeling and personal dosimetry of cellular telephone helical antennas with the FDTD code," *IEEE Transactions on Antennas and Propagation,* vol. 46, no. 4, pp. 525–530, Apr. 1998.
20. J. Wang and O. Fujiwara, "Reduction of electromagnetic absorption in the human head for portable telephones by a ferrite sheet attachment," *IEICE Transactions on Communications,* vol. E80-B, no. 12, pp. 1810–1815, Dec. 1997.
21. B. M. Green and M. A. Jensen, "Diversity performance of dual-antenna handsets near operator tissue," *IEEE Transactions on Antennas and Propagation,* vol. 48, no. 7, pp. 1017–1024, July 2000.
22. O. P. Gandhi and J.-Y. Chen, "Electromagnetic absorption in the human head from experimental 6-GHz handheld transceivers," *IEEE Transactions on Electromagnetic Compatibility,* vol. 37, no. 4, pp. 547–558, Nov. 1995.
23. Hsing-Yi Chen and Hou-Hwa Wang, "Current and SAR induced in a human head model by the electromagnetic fields irradiated from a cellular phone," *IEEE Transactions on Microwave Theory and Techniques,* vol. 42, no. 12, pp. 2249–2254, Dec. 1994.
24. I. El-Babli, A. Sebak, and N. Simons, "Application of the TML method to the interaction of EM fields with dispersive dielectric bodies," *IEE Proceedings — Microwave Antennas and Propagation,* vol. 147, no. 3, pp. 211–217, Jun. 2000.
25. H. O. Ruoss, U. Jakobus, and F. M. Landstorfer, "Efficient EM analysis of handheld mobile telephones close to human head using modified method of moments," *Electronics Letters,* vol. 31, no. 12, pp. 947–948, June 1995.
26. M. A. Mangoud, R. A. Abd-Alhameed, N. J. McEwan, and P. S. Excell, "SAR reduction for handset with two-element phased array antenna computed using hybrid MoM/FDTD technique," *Electronics Letters,* vol. 35, no. 20, pp. 1693–1694, Sept. 1999.
27. M. J White and M. F. Iskander, "Development of a multigrid FDTD code for three-dimensional applications," *IEEE Transactions on Antennas and Propagation,* vol. 45, no. 10, pp. 1512–1517, October 1997.
28. M. J White, Z. Yun, and M. F. Iskander, "A new 3D FDTD multigrid technique with dielectric traverse capabilities," *IEEE Transactions on Microwave Theory and Techniques,* vol. 49, no. 3, pp. 422–430, March, 2001.
29. M. F. Iskander, Z. Yun, and R. Quintero-Illera, "Polarization and human body effects on the microwave absorption in a human head exposed to radiation from handheld devices," *IEEE Transactions on Microwave Theory and Techniques,* vol. 48, no. 11, pp. 1979–1987, November, 2000.
30. N. C. Skaropoulos, M. P. Ioannidou, and D. P. Chrissoulidis, "Induced EM field in a layered eccentric sphere model of the head: plane-wave and localized source exposure," *IEEE Transactions on Microwave Theory and Techniques,* vol. 44, no. 10, pp. 1963–1973, Oct. 1996.
31. L. Matens, "Canonical problems proposed within the frameword of the European COST 244 project for comparison of numerical methods used for electromagnetic calculations for mobile communications," in *Proc. PIERS,* Seattle, WA, July 24–28, p. 691, 1995.
32. I. G. Zubal, C. R. Harrell, E. O. Smith, Z. Rattner, G. R. Gindi, and P. H. Hoffer, "Computerized three-dimensional segmented human anatomy," *Medical Physics and Biology,* vol. 21, pp. 299–302, 1994.
33. Microwave Consultants, "Dielectric Database," Microwave Consultants, London, pp. 1–5, 1994.
34. M. F. Iskander, R. Maini, C. H. Durney, and D. G. Bragg, "A microwave method for measuring changes in lung water content: numerical simulation," *IEEE Transactions Biomedical Engineering,* vol. 28, no. 12, pp. 797–804, Dec. 1981. See also M. F. Iskander, and C. H. Durney, "A microwave method for estimating absolute value of average lung water," *Radio Science,* vol. 17, no. 5S, pp. 111S–117S, Sept.–Oct. 1982.

35. D. A. Christensen, "A new nonperturbing temperature probe using semiconductor band edge shift," *Journal of Bioengineering,* vol. 1, pp. 541–545, 1977.
36. A. Christensen, "A review of current optical techniques for biomedical physical measurements," in *Physical Sensors of Biomedical Applications,* Boca Raton, FL: CRC Press, 1980.
37. H. Bassen, M. Swicord, and J. Abita, "A miniature broadband electric field probe," *Annals of the New York Academy of Science,* vol. 247, pp. 481–493, 1974.
38. T. Nojima, S. Nishiki, and T. Kobayashi, "An experimental SAR estimation of human head exposure to UHF near fields using dry-phantom models and thermograph," *IEICE Transactions on Communications,* vol. E77-B. no. 6, pp. 708–713, Jun. 1994.
39. Y. Nidawa, M. Chino, and K. Kikuchi, "Soft and dry phantom modeling material using silicone rubber with carbon fiber," *IEEE Transactions on Microwave Theory and Techniques,* vol. 44, no. 10, pp. 1949–1953, Oct. 1996.
40. T. Kobayashi, T. Nojima, K. Yamada, and S. Uebayashi, "Dry phantom composed of ceramics and its application to SAR extimation," *IEEE Transactions on Microwave Theory and Techniques,* vol. 41, no. 1, pp. 136–140, Jan. 1993.
41. G. Lazzi and O.P. Gandhi, "Realistically tilted and truncated anatomically based models of the human head for dosimetry of mobile telephones," *IEEE Transactions on Electromagnetic Compatibility,* vol. 39, no. 1, pp. 55–61, 1997.
42. A. D. Tinniswood, C. M. Furse, and O. P. Gandhi, "Computations of SAR distributions for tow anatomically based models of the human head using CAD files of commercial telephones and the parallelized FDTD code," *IEEE Transactions on Microwave Theory and Techniques,* vol. 46, no. 6, pp. 829–833, June 1998.
43. K. W. Kim and Y. Rahmat-Samii, "Handset antennas and humans at Ka-band: the importance of directional antennas," *IEEE Transactions on Antennas and Propagation,* vol. 46, no. 6, pp. 949–950, Jun. 1998.
44. H. Arai, N. Igi, and H. Hanaoka, "Antenna-gain measurement of handheld terminals at 900 MHz," *IEEE Transactions on Vehicular Technology,* vol. 46, no. 3, pp. 537–543, Aug. 1997.

26

Safety Aspects of Radio Frequency Effects in Humans from Communication Devices

Alan W. Preece
Medical Physics University Research Centre

26.1 Introduction

The spectacular increase in personal communication systems could hardly have been predicted. In excess of 50 million mobile phones is a number that will continue to increase if the pattern in Finland (currently the world leader in proportion of the population with a mobile phone at about 72% in 2000) is an example. It is only the ingenuity of the manufacturers and network providers in providing cheap units, effective use of radio spectrum, and wide coverage that will determine the pace of increase. The public seems to have taken to a "keeping in touch" approach. However, accompanying this new communication increase is a media-driven anxiety about possible adverse health effects in the areas of cancer, cerebrovascular accidents (stroke), eye and memory damage from handset use, along with similar worries about base stations. In 2000, there were some 22,000 masts in the United Kingdom alone, many sited on schools and in areas of high population density. Additional safety concerns are associated with phone use while

0-8493-0124-6/02/$0.00+$1.50

driving, electromagnetic compatibility (EMC) problems with prostheses such as pacemakers, and interference with safety critical medical and control equipment.

This chapter considers how guidelines for human exposure to radio frequency (RF) are derived, known interactions with human tissue and their measurement, and the evidence for the existence of health effects.

26.2 Radio Frequency Effects in Biological Tissues

26.2.1 General Tissue Properties

The velocity of propagation of electromagnetic waves through tissue is decreased compared with that of free space and this can be regarded as a result of increased capacitance. Additionally, the impedance of tissue compared with that of free space is low compared with the 377 Ω of free space. This has three effects on a wave meeting a tissue. First, there is an impedance mismatch and some energy is reflected as it would in a change of coaxial line impedance; second, the wavelength of the field in the tissue is decreased; and third, the rate of attenuation is increased. The parameters that define these changes are the real and imaginary components of the relative dielectric constant ε. At microwave frequencies the values are determined largely by the water content; and because much of the content of human tissues consists of water (up to 85%), the properties of different tissues can largely be described by their water content. However, there are major differences in dielectric constant so that the value of ε is highest in blood, liver, brain, and muscle, and lower in bone, skin, and fat (in decreasing order). There are also, for example, differences between white (nerve axons) and gray (nerve cells) matter of brain, where ε is about 39 and 56, respectively, at about 1 GHz, reflecting the different fat content associated with myelin sheaths over the nerves. There is little evidence of resonances occurring in biological tissues (unlike pure water or ice) below about 100 GHz, but relaxational effects can be observed. These are degenerated resonances arising because of the sluggish nature or viscosity of the water on which the structural proteins of cells reside. These have been described by Schwan[1] as three zones labeled α, β, and γ. Figure 26.1 shows how the complex permittivity (or dielectric constant) relates to frequency for a typical tissue (muscle) and decreases in a nonlinear fashion because of the degraded resonances.

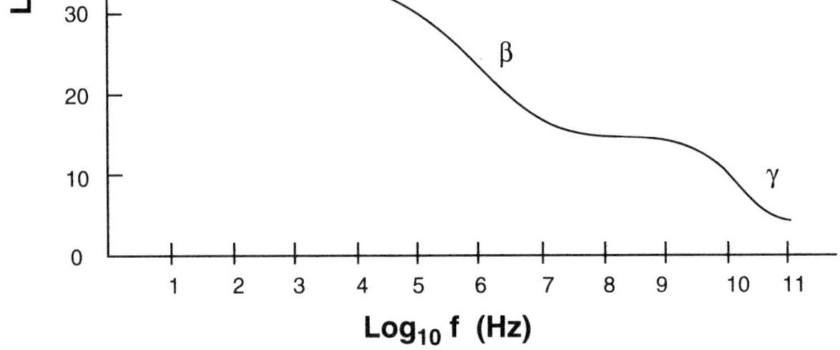

FIGURE 26.1 Complex dispersion of a typical biological tissue with frequency illustrating the very large frequency dependence.

In interpreting these relationships it becomes possible to relate the particular zones to the physical structure or characteristic of the tissue, and to some extent at low frequencies this reflects the physiological properties of the tissue. As stated previously, these zones are characterized by three regions, described as follows:

1. α region — This is dominated by counterion relaxation and electrophoretic relaxation. This characteristic is largely of live cells with intact membranes able to maintain a potential difference resulting from selective secretion of ions across the cell membrane.
2. β region — This results from inhomogeneous structures (Maxwell–Wagner effect, or interfacial polarization where an inhomogeneous structure shows frequency-dependent dielectric and conductive properties that differ from those of the constituents of the mixture). These properties characterize living or dead tissue that *has not undergone significant autolysis* (i.e., there is still structure present).
3. γ region — This is defined here by the behavior of free water and extends from a few MHz to about 20 GHz. There is a contribution from the rotational motion of amino acids, and in the region up to 2 GHz the effects of the presence of larger proteins increase the dielectric constant.

Large differences in tissue dielectric values range from ε = 5-15 for fat, ε = 49 for muscle, to ε = 56 for brain at about 1 GHz. The most important effects are (1) the contraction of wavelength in the tissue by $\sqrt{\varepsilon}$, so that, for example, the length of a resonant antenna in muscle is $n \cdot \sqrt{49}$ of the free-space size, and (2) the losses determined by conductivity σ at lower frequencies, or defined by dielectric relaxational losses at higher frequencies. Conductivity increases with increasing frequency, which results in more limited penetration with increasing frequency (the rate of energy deposition increases).

26.2.2 Limitations of Animal Models

At 1 GHz the penetration in muscle is about 3.0 cm for plane wave state (far field), and less for nonplane wave (near field). This leads to the argument that experimental animal research at a particular frequency cannot mimic human exposure. A good example is the behavioral study in mice or rats where animals are exposed to 900-MHz, cell-phone-type irradiation. In a human, this irradiates perhaps 1 to 2 cm of the brain cortex and at 10W/kg would be dispersed by the blood flow into the large heat sink of the body, whereas in the rodent this would result in whole body thermal stress. Scaling frequency to the animal is beset with other problems because the interaction with tissue also changes with frequency as described earlier. Therefore, animal research relating to RF exposure needs to be treated with caution, particularly where this involves complex physiological interactions such as for cognitive studies. Also the question of thermal stress is very different in man and animals.

26.2.3 Measurement Techniques

Measurement in plane wave (far-field) conditions are relatively straightforward even in conditions of high attenuation rate as occurs in tissue, but if the source is close to the target volume (e.g., cell phone close to the head) there are additional problems. Not only are the wave fronts curved and varying spatially along the source, but near-field measurement is difficult because of the arbitrary phase relationship of the three components of each of the fields. It is usual (in air) to infer equivalent plane wave power density D from

$$D = \frac{\left|E\right|^2}{377} \tag{26.1}$$

or

$$D = 377\left|H\right|^2$$

In high dielectric media such as water (ε = 80) or muscle (ε = 49), the apparent wavelength contraction of by $\sqrt{\varepsilon}$ as described in Section 26.2.1 suggests that the effective distance for near-field conditions of 0.5 to 1 λ usually enables plane wave conditions to be assumed.

26.2.4 Measurement Probes for Human Exposure Assessment

1. Radiation hazard meters are available for different frequency ranges, and for most applications an isotropic probe is preferable to cope with unknown polarization for near-field measurements, and where there may be multipath reflections. An H-field probe is needed for near-field conditions, in particular. Below 100 MHz, induced currents and contact currents are important and need to be considered in addition to spatially averaged E- and H-field measurements. However, at high RF and microwave frequencies it is convenient and usually accurate to measure either the E- or H-field component and to relate these on the basis of the medium in which measurements are to be made, whether this is air or tissue.
2. Probes should be specific to the field parameter (e.g., E-field probes should not respond to H-fields). This can be achieved by ensuring that there are no conducting paths between the E-field plates and that the electronics are not sensitive to induced pickup; or conversely, in B-field probes there are no open-circuit components susceptible to electric fields.
3. Probes need to be small in dimension compared with the wavelength in the medium at the highest frequency of interest. At ELF this is no problem, but if the probe is large compared with the wavelength, it becomes very sensitive to both position and orientation and underestimates the field.
4. Probes should be isotropic — this can be achieved with three orthogonal dipoles or loops. An alternative arrangement is to use a monopole set at 270° that can be rotated in the field to detect radiation in *x*-, *y*-, and *z*-axes.[2]
5. Probes need to be nonperturbing — this is attainable by use of resistive leads of small physical cross section, such as carbon-loaded plastic, or carbon monofilament of similar conductivity to the tissue or medium.
6. Probes should also be reasonably accurate — an error of <3 dB is adequate for protection purposes; although for calibration, 10% would probably be as good as could be reasonably achieved. Calibration must be done in the medium in which the probe is to be used; and although this can pose a problem, there are techniques using a loaded waveguide with a window dividing the air section from that loaded with medium in which the transition point can be used to compare in-air and in-medium sensitivities.

E-field probes have been reviewed by Chou et al.[3] and Stuchly[4] and these fall into the following different groups:

1. Simple diode-based probes are effective up to 1 Wcm^{-2} over the range 400 kHz to 12 GHz but have a complex power/output relationship because of the square law behavior. At lower levels the output voltage is proportional to $[E]^2$ or $[H]^2$, but at higher levels is proportional to E and H directly. Diodes also have a high-temperature coefficient (0.05 dB $°C^{-1}$). The use of resistive, nonperturbing leads typically with static impedance of about 100 KΩ/m to 1 MΩ/m gives a high time constant leading to difficulty in measurement of short pulses, particularly where these are high with respect to the average power. Diode probes need to be calibrated at the frequency and intensity of interest, but are relatively resistant to damage by overload.
2. Thermocouples can be linked to form a linear resistive dipole and these offer good linearity with a square law characteristic but again are slow in response.
3. Electro-optical sensors with internal or external modulation offer fast response and recording of both phase and amplitude and have been reviewed by Stohr et al.[5]
4. Multiple arrays of all these devices are possible, but particularly of diodes, to give specific absorption rate (SAR) surface mapping on real or simulated human bodies (phantoms). A typical example is an array made by Szentpali et al.[6] who used a probe fabricated by thick film technology on a 125-μm polyester substrate on which were GaAs planar doped zero bias diodes connected to silver-printed diodes. A similar construction of Schottky barrier diodes was used by Kaatee and Van Rhoon[7] to carry out their study of temperature rise and SAR induced by mobile phones.

The SAR is the rate of absorption or dissipation of energy (W) in unit mass (M):

$$SAR = \frac{d}{dt}\left(\frac{dW}{dm}\right) = \frac{d}{dt}\left(\frac{dW}{\rho dV}\right) \tag{26.2}$$

where ρ is specific density.

Because the absorption or dissipation of energy results in heat, it is also possible to measure SAR as a temperature change by knowledge of the rate of temperature rise and the thermal capacity (specific heat) of the material or tissue.

26.2.5 Practical Measurements of Radio Frequency Absorption

These methods are always invasive if they are not to be restricted to surface exposure only and are only applicable to simulation techniques or phantoms. Currently, there is no easy method of measuring RF energy disposition in real bodies, although the use of functional magnetic resonance imaging (MRI) system at 2- to 3-tesla (T) field strength makes a measurement of a relatively large SAR feasible, based on the relative shift in T_1 with temperature. Different material (e.g., fat and water) decay amplitudes have different resonant frequencies and therefore different temperature coefficients, and the combination of these can result in highly variable results. This is because the coefficients for fat and water have not only different amplitudes, but also different signs. The change observed with temperature is therefore not defined for different tissues, but MRI can be used for comparative measurements of single tissue entities.

Unfortunately, MRI is the only practical noninvasive measurement method and therefore has its uses in safety research. In practice, the very lowest temperature rise that can claim to be detected is about 0.3°C (personal study and as demonstrated by Yablonskiy et al. on a 1.5 T seimens (S) MRI),[8] which is a similar temperature rise as that resulting from prolonged exposure at the current guideline.

In practice, it is necessary to use a phantom that is constructed from tissue equivalent materials (TEM) to simulate the real (ε′) and imaginary (ε″) dielectric components at the frequency of interest. In addition, the phantom can be constructed to be complex to represent the tissue layers, or simply canonical in design. A simple phantom may consist of a plastic cube or sphere containing 1.5-kg saline (0.9% NaCl in water), or may be improved by use of a sucrose/saline or ethandiol/water brain TEM. This can be further improved by a two layer skin simulation, or by including skin, fat, muscle, bone, eye, etc. to give an anatomically correct phantom that can be imaged either for direct temperature change, or to provide a basis for modeling. Dielectric values for phantom construction can be obtained from References 9 and 10.

26.2.6 European Committee for Electrotechnical Standardization Phantom

One problem with the previously described approach is that no TEM has ever been developed that can simulate electrical properties of tissue and that is stable. Solid dielectrics can be high loss or high permittivity, but not both, in the way human tissue presents itself. Even when constructed as a gel, such TEMs contain mobile liquid that diffuses from the layers. The use of nonpermeable layers to constrain liquid movement is not possible because this introduces more interfaces. This poses a problem for international standards because no two measurement phantoms would be the same if constructed from unstable materials with a short life. The new European Committee for Electrotechnical Standardization (CENELEC) phantom uses a different approach and has designed a specific anthropomorphic mannequin (SAM) for SAR measurements consisting of a thin low-loss, low-permittivity shell ($\varepsilon = <5$, i.e., loss tangent <0.05) containing a uniform liquid. This is dielectrically a compromise with values lowered (e.g., $\varepsilon = 42$, $\sigma = 0.99$ S/m at 900 MHz) to take account of the skin, fat, and bone between the source and the brain. Because this is liquid, then scanning of the probe is possible allowing the SAR in 1 or 10 g of tissue to be calculated for different regions as is required in the guidelines described later in Section 26.5. This is likely to be the standard for all areas except perhaps the United States.

26.3 Health Effects of Radio Frequency Radiation

26.3.1 Thermal and Nonthermal

As stated in Section 26.5, most countries have set guidelines that limit the significant thermal effects of RF exposure. However, public concern has been expressed over the possibility that RF fields can cause nonthermal biological and health effects at levels below those causing thermal effects. The three terms applied to biological effects of RF exposure are *thermal*, *a-thermal*, and *nonthermal*. These are all relative terms and therefore it is not possible to define the zone at which these cross over. They can be interpreted as follows:

1. Thermal effects occur when sufficient RF energy is deposited to cause a measurable increase in the temperature of the sample, e.g., 0.1°C.
2. A-thermal effects are those that occur when sufficient energy is deposited to cause an increase in the temperature of the sample but no changes of any significance in temperature actually occur because of temperature control in the body.
3. Nonthermal effects are those defined when the energy deposited in the sample is less than those associated with normal bodily functions and is otherwise undetectable.

26.3.2 *In vitro* Research

All depositions of RF energy in the megahertz and gigahertz region will result in some degree of temperature elevation, but below 0.1°C these could be considered physiologically and biologically insignificant. There have been a number of studies of nonthermal effects and the most appropriate to these are in cellular cultures that are amenable to detailed control. Cleary[11] has carried out numerous studies of cell proliferation and the cell cycle kinetics under continuous wave RF exposure at 2450 and 27 MHz. They reported increased proliferation of cell growth at 1, 3, and 5 days following a single 2-h RF exposure at either of these frequencies. The measure of increased proliferation was the uptake of radio-labeled nucleic acids in DNA synthesis, and this was seen at specific SARs of 5 and 50 $W \cdot kg^{-1}$. Potentially, this is in the thermal region, but the design of the experiments maintained the temperature to less than 0.1°C change even in the presence of strong RF fields. They reported very similar effects on human peripheral lymphocytes exposed under the same conditions.

Stagg et al.[12] exposed a number of brain cell lines to RF signals very similar to mobile phone signals for 24 h but at considerably lower exposure levels than those of Cleary. At an SAR of 5.9 $mW \cdot kg^{-1}$ increases in nucleic acid uptake in DNA synthesis were observed but not at a lower or higher SAR. The actual growth curve of the cells was not altered by any of the systems of exposure. A counter experiment by Kwee and Raskmark[13] looked at transformed human epithelial amnion cells exposed to 960 MHz at a range of low SARs up to 2.1 $mW \cdot kg^{-1}$. A decrease in cell growth was seen for exposures of more than 30 min. These changes in DNA proliferation are suggestive of alterations in growth even though the evidence is somewhat confusing because some exposure conditions seem to cause an increase in growth whereas other ones, a decrease.

26.3.3 *In vivo* Research

The *in vitro* observation of growth gene expression (if confirmed) leads to the possibility that stimulation of growth could have carcinogenic potential, and a number of studies have sought to look at this situation and thereby assess the possible implications to human exposure. A number of studies in animal systems could be classed as *in vivo* exposure experiments. One of these[14] looked at the spontaneous tumors and carcinogen-induced central nervous system tumors in rats and showed that there was in fact a reduction of growth in tumors after exposure to 837 MHz 1 to 1.2$W \cdot kg^{-1}$ for 2 h per day, 4 days per week for exposures of 2 years. Another study by Imaida et al.[15] looked at 930 MHz 1 to 2$W \cdot kg^{-1}$ in rat liver cancer and showed no promotional effect on the development of tumors. A study by Salford et al. in 1994[16] showed no increase in implanted brain tumor cells in Fisher rats exposed to 0.4 $W \cdot kg^{-1}$ of 915-MHz RF radiation for 2/3 weeks. This showed no increase in growth of such tumors. The one outstanding piece of information that goes contrary to the preceding studies was that by Repacholi et al.[17] who exposed transgenic mice with a genetic alteration that made them prone to lymphomas. The exposure was Global System for Mobile Communications (GSM) pulses at 900 MHz of about 0.13 to 1.4 $W \cdot kg^{-1}$ for an hour per day for 18 months and in this case showed a doubling of the incidence of tumors. This particular experiment is being repeated.

Therefore, it seems difficult to demonstrate with any consistency either direct effects on cells or on whole animal systems with particularly designed tumor models. At a molecular level the evidence is equally confusing. There have been suggestions that specific genes involved with the growth processes or with control of carcinogenic development might be affected. The most researched examples are the P53 gene thought to be responsible for clearance of damaged cells by the process of apoptosis, and C-fos and C-jun known to be growth genes. These are recognized by protein markers associated with the gene functioning and many hours of research have been concentrated on these markers. They involve much new, long, and difficult technology in molecular biology and can therefore go wrong or be easily upset. Goswami et al.[18] showed no increase in one stress protein marker (*jun*) but a change in another (*fos*) at 0.6 $W \cdot kg^{-1}$ in fibroblast cells, whereas another study by Ivaschuk et al.[19] showed changes in *jun* but not in *fos* at 46 $mW \cdot kg^{-1}$. Phillips et al.[20] showed an increase in DNA damage and subsequent repair, whereas Stagg et al.[21] showed no change at about 0.1 to 59 $mW \cdot kg^{-1}$, in the same parameters as the previous study. Another study by Fritze et al.[22] to GSM-type transmissions up to 1.5 $W \cdot kg^{-1}$ showed no changes in *fos* and *jun*. On the basis of these conflicting data, it has to be assumed that the effects are either subtle or too ephemeral for detection by conventional laboratory methods.

26.3.4 Effects on DNA

The greatest controversy surrounds the occurrence of DNA strand breaks that have been demonstrated by Lai and Singh[23] on rat brain cells at 2450 MHz both continuous wave (CW) and pulsed at between 0.6 and 1.2 $W \cdot kg^{-1}$. This is demonstrated by a technique called the covariance matching estimator (COMET) assay, which looks for the production of single strand or damaged DNA by a technique of electrophoresis that separates the damaged DNA from the native source DNA. However, Malyapa et al.[24] studying the same model (i.e., rat brain cells) also at 2450 MHz at 1.2 $W \cdot kg^{-1}$ did not show this effect and claimed the previous results were an artifact of the method. Until this controversy as to whether there is DNA damage from nonthermal levels of microwave exposure is resolved the arguments about safety are not going to be concluded.

26.3.5 Epidemiology

Only two epidemiological studies have examined the possibility that mobile phone exposure may be associated with malignancy. The first was by Rothman et al.[25] that looked at 250,000 phone users and followed them for 1 year. There was no evidence that the death rate was any different in users of handheld phones compared with the suitcase-based mobile phones. Indeed the heaviest users of handheld phones had a lower malignancy risk than the other group. If there had been any significant association between RF exposure and malignancy it would have been expected to show up in such a large cohort of users. What this study does not take into account is the possible long induction period between initiation and development of a cancer, and the question of exposure/dose was not adequately addressed.

The second study was a case control study in Sweden of patients with brain tumors carried out by Hardell et al.[26] In this case the risk of brain tumors was not elevated in subjects using mobile phones, either for analog or digital models. There was, however, an association between the reported side of use of the phone and the side of occurrence of the tumors. A number of problems are associated with this study, namely, the failure to identify all the possible subjects that might have been expected and also the impossibility of eliminating recall bias when subjects are questioned about the side that they normally use the phone. As a result of concern about this possible association and the lack of robust data, through IARC (International Agency for Research on Cancer) the European Community has funded a study that is intending to recruit 2000 brain tumor subjects. This will examine the possible association of phone use with brain tumors, acoustic neuromas, and salivary gland tumors as well as examining the confounders that could be associated with phone use. One of the reasons for these concerns is a study in Poland conducted on military personnel with occupational exposure to RF. In this study Szmigielski[27] suggested a sixfold elevation in the risk of lymphatic and blood cancers. However, the methodology of this study is considered to be unsatisfactory and the various methods used in the research are not adequately described. Other studies have also been reviewed[28,29] involving groups of people exposed to RF radiation through their work or hobbies (e.g., radio amateurs). The most common reported diseases are lymphoma and leukemia and brain cancer. A number of cases of clustering of cancer incidence in the neighborhood of television and radio transmitter antennas have been reported. One example of these was studied by Dolk et al.[30,31] and showed an increase of leukemia within 2 km of a television and radio transmitter in Sutton Coldfield, United Kingdom. However, when this study was extended to other medium wave and very high frequency (VHF) transmitting antennas, no similar clusters were found in other parts of the country.

26.3.6 Cognitive Effects

26.3.6.1 In Animals

Safety aspects of RF exposure can be grouped into three broad headings, namely, neurological, driving, and electromagnetic compatibility with safety critical devices, such as prosthetic implants, or with hospital-based devices. A survey carried out in Scandinavia by Mild et al.[32] compared the effect of GSM and analog handset sources on the reported subjective symptoms in the users. The prevalent symptoms reported in this study were losses of concentration and memory loss but particularly fatigue, headache, warmth on the ear, burning of the skin, and tingling and tightness of the skin. The symptoms were more prevalent in analog than in GSM phones, which may be associated with the greater power consumption of analog phones and greater RF output. The study, however, does indicate that there may be increased levels of stress with the use of mobile phones, but why this should be greater for analog systems is a mystery. The possibility that RF exposure is associated with stress has been studied in animals. At thermal levels that involve elevation in body temperature of greater than 1° there are indications that this disturbs behavior and is probably associated with an avoidance reaction and attempts to get to areas of lower discomfort. An example is a study of the effects of $10\ \mathrm{W \cdot kg^{-1}}$ at 600 MHz for 20 min in rats,[33] which produced a dose-dependent deficit in object recognition. There seems no doubt that this is going to be a hyperthermic response to RF as a stressor. A study on spatial working memory function in rats[34] showed that exposed animals at $0.6\ \mathrm{W \cdot kg^{-1}}$ of 2.45 GHz made more errors than unexposed animals. This was stated to be related to changes the neurotransmitter activity of the brain, in particular, the intrinsic opioid levels that are associated with the "flight and fight" response and enable the organism to deal with stressful situations, and the alteration of cholinergic activities that are associated with memory and learning. The reported interference therefore could also be considered a stress response to the action of microwaves. A follow-up study by Sienkiewicz et al.[35] in mice in a similar radial-arm maze task using pulsed 900-MHz radiation at $0.05\ \mathrm{W \cdot kg^{-1}}$ showed that the exposed animals performed just as well as controls. However, there were other effects such as an increased fecal production during exposure. This again appears to be a possible stress effect. Philosophically, if $0.6\ \mathrm{W \cdot kg^{-1}}$ could be considered nonthermal, then the last experiment at tenfold less is probably of little value in negating the existence of effects.

26.3.6.2 In Humans

Identification of effects in humans has been conducted by a number of researchers. Usually evoked responses are used as indicators of response. Three studies of evoked responses to GSM-like signals have been conducted. Urban et al.[36] looked for visual sensory responses to checkerboard reversal and found that these were not affected. However, two other studies[37,38] reported positive effects. The other types of studies have involved analysis of sleep responses, and a study by Borbely et al.[39] indicated that sleep to a certain extent was affected in that the electroencephalogram (EEG) was somewhat more active, which could be interpreted as a degree of arousal occurring during RF exposure. That study involved exposure during sleep. Another group has found a similar effect, but from exposure before sleep.[40] The indication is that exposure effects last for up to 30 min after the end of a call. The obvious importance of this kind of study is in trying to analyze the possible effects of cell phone antennas or other antennas placed on buildings where the subjects are sleeping. Alteration in sleep patterns lead to other psychological effects such as depression and anxiety. Direct provocation studies on humans during cognitive testing have been conducted by Preece et al.[41] and Koivisto et al.[42] Both these studies indicated that RF exposure to the brain resulted in enhanced cognitive performance related to changes in reaction time. The sum total of the sleep, cognitive, and electrophysiological studies suggests indeed that there is an effect of RF on the brain, whether it is analog or pulsed. The mechanism for these is important. If indeed it is simply heating, then this is probably of very little physiological significance, and very likely it is unrelated to any health effects and therefore really has few implications for safety. It has been pointed out, however, that most of the exposures to handset antennas result in a very minimal degree of temperature elevation, probably not more than 0.1 of a degree, as clearly shown by both calculation and experiment by Van Leeuwen et al.[43]

Therefore, the mechanism may be somewhat more complex and there are two other possible candidates. One of these is the direct effect on the synapses in the brain, which is in the reaction time determining processes, a glutamate response. This requires the release of a neurotransmitter in response to an electrical signal. The electrical signal in the neuron that leads to the neurotransmitter release area is held at a particular potential until triggered. The supposition is that the superposition of a RF alternating signal on this potential means that the threshold for the triggering can be achieved more easily, thus leading to a speeding up of response. This is exactly the kind of reaction seen both in the Preece[41] (15 ms speeding up) and the Koivisto[42] study (19 ms speeding up) in reaction time. The third possibility, which may have more serious implications because of the possible long-term effects, is that these changes could be mediated by protein alterations. One candidate is the change in heat shock protein (in particular, HSP70) that has been demonstrated to be altered in some simple animal models. A study by de Pomerai et al.[44] shows that nonthermal levels of RF lead to the alteration in the production of HSP series of compounds. This effect has been seen in other *in vitro* systems and in cellular systems, also at electromagnetic frequencies from 50 Hz to 2.5 GHz.

There is obviously much more research needed but the existence of these effects has led to the suggestion that the intervention levels should be lowered for the general public (see the Stewart report in Section 26.6).

26.4 Other Safety Issues

26.4.1 Interaction with Safety Critical Equipment

The levels of radiation likely to cause problems for electronic devices are probably lower than for people. The EN60601-1-2, the EMC standard for medical devices, specifies an immunity of at least 3 V/m to radiation frequencies from 26 MHz to 1 GHz. This is in contrast to the recommendation for personal exposure of 50/70 V/m at these mobile phone frequencies. The Food and Drug Administration (FDA) of the United States, which licenses medical devices and reports adverse incidents, has recorded over 100 incidents in a 15-year period, some of which involve devices other than mobile phones. In particular, some medical equipment can cause problems because emission of RF energy is part of its normal function. Examples are electro-surgery and short-wave diathermy devices, both of which can produce fields of 30 to 40 V/m at short distances and are well-known sources of interference. Devices particularly at risk are ventilators, infusion pumps, hearing aids, and powered wheelchairs. There have also been reported cases of interference with physiological monitors of the kind used in intensive care units. The question of interference with pacemakers is a serious one but fortunately the risk to an implanted pacemaker is very small indeed. Many of these devices do show interference when on the bench but once implanted the conductivity of the tissue that surrounds them is a highly protective device. The field strength in air at any distance d from an antenna is approximately

$$E = \sqrt{(30PG)}\Big/d \ \ \mathrm{V \cdot m^{-1}} \tag{26.3}$$

where P is power in Watts, and G the antenna gain.

For a typical device (mobile phone) with a dipole antenna configuration giving a gain of 1.6

$$E = \frac{7\sqrt{P}}{d} \ \ \mathrm{V \cdot m^{-1}} \tag{26.4}$$

A survey was conducted by the Medical Devices Agency (MDA) in the United Kingdom that has shown it is necessary to have a safety clearance area between safety critical devices and the use of mobile phones or handheld communicators. This is fully reported in a bulletin[45] that suggests a minimum clearance of 1.5 m is necessary to protect some devices. In practice, most hospitals now have a clearance zone of 3 m that take into account the higher powered mobile communicators occasionally used by security, fire, and ambulance staff (Fig. 26.2).

There is a particular problem with antennas mounted on mobile vehicles that may carry safety critical equipment (e.g., an ambulance). For the purposes of providing adequate light to attend to patients,

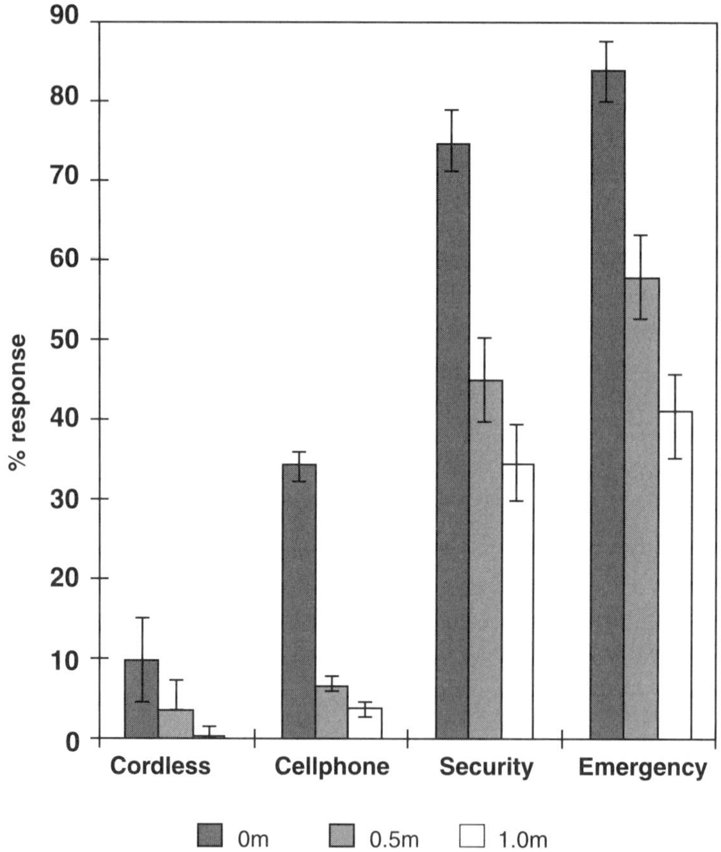

FIGURE 26.2 Probability of interaction of typical hospital critical care equipment with different communication devices at different distances.

ambulances are frequently fitted with fiberglass translucent roofs that are also transparent to RF. If, as occasionally happens, there may be a 25-W sourced antenna mounted on the front part of the roof, then equipment close to it is subjected to considerable fields.

26.4.2 Driving

It would seem obvious that the use of a handheld device inside a vehicle while driving would be detrimental to driving performance. However, it seems that the situation is more complex than this and that the mere act of conversation is not the whole problem. A number of studies in driving simulators (e.g., McKnight and McKnight,[46] Alm and Nilsson,[47,48] Lamble et al.[49]) all show deficits in driving performance by as much as a second in terms of reaction time. It seems likely that none of this is related to the RF exposure coming from the antenna, but is related to the dual-task effect and the possible effect of a complex conversation where the person driving is placing himself or herself mentally in the area of the person with whom they are holding the conversation.

26.5 Exposure Guidelines for Radio Frequency Radiation

26.5.1 NRPB

An approach to risk management is to set up defined exposure guidelines to define the level below which health effects should not occur. In the United Kingdom, the NRPB have reviewed the epidemiological evidence (NRPB GS11)[50] and concluded that the only basis for control is a thermal one. The evidence at the time, in 1988 and 1993, gave no indication that a detectable health hazard from nonthermal levels existed, and that protection from thermal overload was the correct objective. Thermal loads that can be tolerated are, in humans, about a 4°C rise, which causes distress, headaches, nausea, and disorientation. Cell killing starts to become significant over about 41 to 43°C so that a limit of 1°C has been suggested. At 1°C elevation of temperature a human simply sweats and may feel vaguely uncomfortable. Normally the body generates its own heat as a result of its basal metabolism. This is on the order of 40 kcal/m^2/h, so for a standard 70-kg man of 1.7 m^2 surface area this represents about 1.13 W/kg. Experimental studies on animals and man indicate that a thermal load from inside or outside the body in addition to the basal metabolism results in a 1°C body temperature rise, after which the thermoregulatory mechanisms control and lower temperature. This is easily achieved by internal energy expenditure occasioned by gentle jogging. The external thermal load to create this rise over about 15 min is about 4 Wkg^{-1}. RF energy is no different, so using a suggested safety factor of ten results in a whole body exposure limit of 0.4 W · kg^{-1}. However, for partial body exposures, the presence of a blood circulation means that the thermal load can be "shared" with the rest of the body and, in particular, the limbs and head, which are the most likely to receive higher exposure, are allowed 10 W · kg^{-1}. Unfortunately, body SAR is a very difficult parameter to measure and has to be inferred from the E or H components.

For use in the near-field frequency range of 10 to 100 MHz, International Nonionizing Radiation Committee (INIRC) the following calculation is recommended to give the equivalent plane wave power.

$$P_{eq} = \frac{5}{6}\left(E^2/120\pi\right) + \frac{1}{6}\left(120\pi H^2\right) \tag{26.5}$$

This can be applied to calculate the following derived exposure limits for workers and the general public. Some of these are shown in Tables 26.1 and 26.2.

These recommendations were then updated in 1993 (documents of the NRPB, vol. 4, no. 5)[51] to give the values used in the United Kingdom to date. The field strengths listed in Tables 26.1 and 26.2 define the environmental exposure limits that can be measured by physical instruments. To define the absorbed energy, as has been suggested earlier, translating this into SARs is difficult. Nevertheless, this parameter is the critical one that defines the effect on the body. Table 26.3 lists a summary of the main endpoints

TABLE 26.1 INIRC Recommended Derived Occupational Exposure Limits to Radio Frequency Electromagnetic Fields

Frequency (MHz)	Unperturbed rms Field Strength: Electric E ($V \cdot m^{-1}$)	Magnetic H (Aa/m^{-1})	Equivalent Plane Wave Power Density P_{eq} ($W \cdot m^{-2}$)
0.1–1	614	1.6/f	—
1–10	614/f	1.6/f	—
10–400	61	0.16	10
400–2,000	$3\sqrt{f}$	$0.008\sqrt{f}$	f/40
2,000–300,000	137	0.36	50

TABLE 26.2 INIRC Recommended Derived General Public Exposure Limits to Radio Frequency Electromagnetic Fields[10]

Frequency (MHz)	Unperturbed rms Field Strength: Electric E ($V \cdot m^{-1}$)	Magnetic H (Aa/m^{-1})	Equivalent Plane Wave Power Density P_{eq} ($W \cdot m^{-2}$)
1–10	87	$0.23/\sqrt{f}$	—
10–400	$87/\sqrt{f}$	$0.23/\sqrt{f}$	—
400–2,000	27.5	0.073	2
2,000–300,000	$1.375\sqrt{f}$	$0.0037\sqrt{f}$	f/200
	61	0.16	10

TABLE 26.3 Summary of the Basic Restriction and Investigation Levels for Exposure of Whole or Part Body to Electromagnetic Radiation, Based on the Specific Absorption Rate

f	SAR Av Over Body ($W \cdot kg^{-1}$)	SAR Limbs ($W \cdot kg^{-1}$)	SAR Head ($W \cdot kg^{-1}$)	Environmental Power Density ($W \cdot m^{-2}$)
100 kHz–10 MHz	0.4	20	10	Mag. flux = $23/f^2 \mu T$
10 MHz–10 GHz	0.4	20	10	Variable
10–300 GHz	—	—	—	100 Wm^{-2}

that avoid taking into account factors such as shape, resonant size, and the degree of localization of the exposure. This is the U.K. guideline that is of value in assessing whether any device is causing exposure that might be of concern.

In addition to the thermal effects, there are indirect effects that can arise, for example, those involving interaction with an object at different electrical potential from the body. One of these effects is the direct simulation of peripheral nerves and muscles. However, this is not likely to be a particular problem at RFs above 10 to 40 kHz because stimulation of nerves, either peripheral or central, normally ceases at the maximum frequency of an action potential. Nevertheless, there is always the possibility of causing burns by body parts making contact with conductive objects and producing tissue heating from the passage of current through the tissue. Guidelines to these possible health effects are normally in two levels of protection depending on whether the subjects are the general public and not controlled by health checks, protective clothing, or procedures, or subjects are workers in controlled environments or controlled working practices as explained in item 4 that follows. This is supplemented by introducing the concepts of *basic restriction* and *reference levels* and are interrelated by the following considerations:

1. Basic restrictions are based on established health effects.
2. Reference levels are derived from measurement or computed predictions. These are used to provide practical exposure assessment to determine whether the basic restrictions are likely to be exceeded.
3. If a reference level is exceeded, it does not necessarily follow that the basic restriction is exceeded. However, the compliance with the basic restriction has to be tested to comply with regulations.
4. Guidelines thus often differentiate between occupational exposure and general public exposure. These can be defined as occupational/controlled exposure that is appropriate when persons who are fully aware of the potential for exposure and are therefore able to exercise control over the exposure system or the levels in their environment. These levels would also apply if on a location that might involve temporary occupancy providing there is adequate warning in the form of signs and other indicators. For the general population one uses the term *uncontrolled* exposure, which applies when the persons who are exposed may not be aware either of the levels of fields that exist or of the possible consequences of their exposure. The NRPB guidelines (NRPB 1993)[51] consist of basic restrictions and reference levels according to Table 26.3 but this does not distinguish between members of the public and occupationally exposed workers. In a number of summaries since the production of the 1993 guidelines, NRPB has questioned the scientific justification of a blanket approach and of the wisdom of further reductions in exposure. Peak levels experienced from pulsed RF field conditions are only restricted insofar as required to avoid the effect of microwave auditory effects in people with normal hearing, and all other limits are based on average levels.

26.5.2 ICNIRP

ICNIRP[52] guidelines are set out in two levels of protection (for the general public and for the healthy worker) under the following conditions:

1. All SAR values are averaged over any 6-min period and localized SAR values are averaged over 10 g of tissue in a single mass.
2. For pulses of duration T, the equivalent frequency is determined by 0.5 T.
3. For pulsed exposures in the range 0.3 to 10 GHz and for localized exposure of the head, the specific absorption should not exceed 10 mJ · kg^{-1} averaged over 10 g.

The ICNIRP guidelines have been arrived at by general consensus by a number of European committees and do not yet have the force of law in the United Kingdom.

26.5.3 United States, Canada, and Australasia

The Federal Communications Commission (FCC)[53] sets out the following:

1. Two levels of exposure limits are based on occupational/controlled and general population/uncontrolled exposures.
2. Maximum permissible exposure limits are based on recommendations of the NCRP[54] and the IEEE.[55]

Other standards apply in Australia, New Zealand, and Canada. In 1998, Australian and New Zealand standards were merged as an interim standard (AS/NZS2772.1 [INT]; 1998). These adopted a standard allowable general public exposure limit for the frequency used by mobile phone companies, for example of 0.2 mW·cm^{-2}. This is a large factor lower than the FCC, ANSI/IEEE, ICNIRP and NCRP standards. The Canadian standard is laid out in “Health Canada: Limits of Exposure to Radio Frequency Fields at Frequencies from 10 KHz to 300 GHz, Safety Code 6, Canada Communication Group, Ottawa, Canada (1993). This appears to be very similar to the FCC standard.

The following bodies have set up guidelines that have been generally adopted starting with NCRP (1986), ANSI/IEEE (1992), NRPB (1993), CENELEC (1995), FCC,[56] ICNIRP.[57] There is as yet no consensus on which of these standards are adopted in any countries except in the United States, which has consistently adopted the FCC limits based on the original ANSI/IEEE (1992) standards. An example of these standards are those for mobile phone exposure — an example of the public exposure to radiation from antennas — as:

1. The 1.6 $W \cdot kg^{-1}$ limit by laboratory test or by computation
2. At 900 MHz, 6 W/m^{-2} exposure of the general public or 30 W/m^{-2} for occupational exposure of persons who are aware of the situation
3. Above 1500 MHz (normally 1800 MHz in the United Kingdom, 1900 MHz in the United States), 10 W/m^{-2} for general public and 50 $W \cdot m^2$ for occupational exposure
4. No exclusion clauses
5. The FCC deferring consideration of the effects of modulation as premature based on current scientific knowledge

The testing of mobile phones and the third-generation systems will be covered by new standards. In Europe this will be the CENELEC prEN50316, due to be published in April 2001; and in the United States, the new IEEE Std. 1528-200X, due to be published in March 2001. Both these refer to the design of phantoms and to the method of measuring SAR, and will almost certainly create a greater degree of harmony in setting standards, which is highly desirable in view of the international nature of communication technology.

26.6 Recent Reviews

A number of bodies have produced detailed reports on reviews of the potential health risks of RF fields, particularly based on wireless telecommunication devices. The two most recent of these are a report by the Royal Society of Canada[58] (also on Web site)[59] and by an independent expert group on mobile phones under the chairmanship of Sir William Stewart (Mobile Phones and Health 2000).[60] The Royal Society of Canada produced its report for Health Canada and looked at all peer-reviewed literature on RF fields and health, examining both epidemiological data, studies on mechanisms, and laboratory cellular and animal work. Their conclusion, not surprisingly, is that there is no good evidence to date that the few observed nonthermal effects have any links to adverse health effects. However, the review identified desirable areas of research and a set of specific questions. In particular, effects on the eye, on melatonin secretion, and on the various neurotransmitter systems need to be studied; also a better understanding is needed of the biophysical detection mechanisms that must exist. It was estimated that a 5- to 10-year program of research would be desirable. It was felt, however, that the current guidelines were adequate to protect against thermal hazards. The study is a good resource of literature review.

The Stewart report, more correctly the report of the "Independent Expert Group on Mobile Phones" (IEGMP), was produced in response to the advice of a Parliamentary Select Committee on Science and Technology,[61] which asked NRPB to set up an independent review. This examined the preexisting evidence for both thermal and nonthermal effects, but reexamined much of the evidence critically to decide what weight to apply to the published data. The conclusion of this group (available on a Web site)[62] also was that there was evidence of a biological response to low levels of RF, but again no evidence that these responses were either harmful or even potentially so in the long term. Nevertheless, as such responses were occurring below the existing guidelines and as a precautionary approach, it would be desirable to adopt the ICNIRP guidelines that took into account the difference between controlled occupational and uncontrolled public exposure. In addition, the question of public perception of risk was addressed for the first time. The recommendations were that the exemption from planning regulations for small (less than 15 m high) masts bearing antennas should be stopped, the public should be consulted through normal planning processes, siting masts on schools should be avoided, the ALARA (as low as reasonably achievable) principle should be used for public exposure, and the phones should be identified with the SAR achieved under standard test conditions. The committee also recommended that more research (a substantial research program) should be sponsored and partly financed by the industry, but initiated and monitored by an independent body. In particular, the committee recommended that a register of occupationally exposed workers be established and that cancer risks and mortality be examined to determine whether there are any harmful effects. If any adverse effects of exposure to RF radiation are identified, then the health and safety executive should establish a system of health surveillance.

Following this report the Institute of Electrical Engineers (IEE) issued a position statement that can be found on a Web site.[63] In this, with respect to RF exposure, the possibility of the existence of health effects is dismissed as unlikely, and even those studies the IGEMP had considered significant require replication before being considered to indicate a response. The overall consensus (of the IEE) was that there is no need for a precautionary approach and "that there is no scientific basis to the measures recommended" (*by the Stewart Committee*). Unfortunately, this has not resolved the public risk perception problems.

26.7 Conclusions

Thermal effects and their consequence are well established, and national guidelines, even though at variance with each other, have been set out to control these.

Nonthermal effects have been seen to occur at low levels and probably represent no more than minor physiological responses.

There is no confirmed evidence of health effects (defined as alteration of well-being) whether minor or more serious.

The Stewart Committee Report[61,63] has acknowledged that the major hazard to society is the use of communication equipment while driving.

The other hazards to health are interference with medical equipment and aircraft guidance systems and need to be controlled.[45]

It would seem sensible that the United Kingdom should align more with Europe and the United States and all should adopt common standards of exposure.

References

1. Schwan, H.P. EM-field induced force effects. in *Interactions between Electromagnetic Fields and Cells* (A. Chiabrera, C. Nicolini and H.P. Schwan, Eds.). New York, Plenum Press, 1985.
2. Taylor, H.C., Burl, M. and Hand, J.W. Design and calibration of electric field probes in the range 10–120 MHz. *Phys. Med. Biol.*, 42, 1387, 1997.
3. Chou, C-K. et al. Radio frequency electromagnetic exposure: tutorial review on experimental dosimetry. *Bioelectromagnetics*, 17, 195, 1996.
4. Stuchly, M.A. Biomedical concerns in wireless communications. *Crit. Rev. Biomed. Eng.*, 26, 117, 1998.
5. Stohr, A. et al. Novel analog optical RF E-field sensor system. *Proc. SPIE*, 3463, 158, 1998.
6. Szentpali, B., Van Tuyen, B. and Thuroczy, G. E-field probe for measuring the exposure occurred by mobile phones in phantoms. *Proc. Int. Semiconductor Conf. 20th Ed., CAS '97, Sinaia, Romania* (IEEE), 2, 373, 1997.
7. Kaatee, R.S.J.P. and Van Rhoon G.C. An electric field measurement system using a two-dimensional array of diodes. *Int. J. Hyperthermia*, 15, 41, 1999.
8. Yablonskiy, D.A., Ackerman J.J.H. and Raichle M.E. Coupling between change in human brain temperature and oxidative metabolism during prolonged visual stimulation. *Proc. Natl. Acad. Sci.*, 97, 7603, 2000.
9. http://www.brooks.af.mil/AFRL/HED/hedr/reports/dielectric/home.html
10. http://www.fcc.gov/fcc-bin/dielec.sh
11. Cleary, S.F. Cellular effects of radio-frequency electromagnetic fields. In *Biological Effects and Medical Applications of Electromagnetic Energy* (O. P. Gandhi, Ed). Englewood Cliffs, NJ, Prentice-Hall, p. 339, 1990.

12. Stagg, R.B. et al. DNA synthesis and cell proliferation in C6 glioma and primary glial cells exposed to a 836.55 MHz modulated radio-frequency field. *Bioelectromagnetics*, 18, 230, 1997.
13. Kwee, S. and Raskmark P. Changes in cell proliferation due to environmental non-ionising radiation 2. Microwave radiation. *Bioelectrochem. Bioenerg.*, 44, 251, 1998.
14. Adey, W.R. et al. Spontaneous and nitrosourea-induced primary tumours in Fischer 344 rats chronically exposed to 836 MHz modulated microwaves. *Radiat. Res.*, 152, 293,1999.
15. Imaida, K. et al. Lack of promoting effects of the electromagnetic near-field used for cellular phones (929.2 MHz) on rat liver carcinogenesis in a medium-term liver bioassay. *Carcinogenesis*, 19, 311, 1998.
16. Salford, L.G. et al. Permeability of the blood-brain barrier induced by 915 MHz electromagnetic radiation, continuous wave and modulated at 8, 16, 50, and 200 Hz. *Microsc. Res. Tech.*, 15;27(6), 535, 1994.
17. Repacholi, M.H. et al. Lymphomas in Eμ-*Pim1* transgenic mice exposed to pulsed 900 MHz electromagnetic fields. *Radiat. Res.*, 147, 631, 1994.
18. Goswami, P.C. et al. Proto-oncogene mRNA levels and activities of multiple transcription factors in C3H10T½n murine embryonic fibroblasts exposed to 835.62 and 847.74 MHz cellular phone communication frequency radiation. *Radiat. Res.*, 151, 300, 1999.
19. Ivaschuk, O.I. et al. Exposure of nerve growth factor-treated PC-12 rat pheochromocytoma cells to a modulated radio-frequency field at 836.55 MHz, effects on c-jun and c-fos expression. *Bioelectromagnetics*, 18, 223, 1997.
20. Phillips, J.L. et al. DNA damage Molt-4 T-lymphoblastoid cells exposed to cellular telephone radio-frequency fields in vitro. *Bioelectrochem. Bioenerget.*, 45, 103,1998.
21. Stagg, R.B. et al. DNA synthesis and cell proliferation in C6 glioma and primary glial cells exposed to a 386.55 MHz modulated radio-frequency field. *Bioelectromagnetics*, 18, 230, 1997.
22. Fritz, K. et al. Effect of global system for mobile communication (GSM) microwave exposure on blood-brain permeability in rat. *Acta Neuropathol.*, 94, 465, 1997.
23. Lai, H., and Singh, N.P. Acute low-intensity microwave exposure increases DNA single-strand breaks in rat brain cells. *Bioelectromagnetics*, 16, 207, 1995.
24. Malyapa, R.S. et al. DNA damage in rat brain cells after *in vivo* exposure to 2450 MHz electromagnetic radiation and various methods of euthanasia. *Radiat. Res.*, 149, 637, 1998.
25. Rothman, K.J. et al. Overall mortality of cellular telephone customers. *Epidemiology*, 7, 303, 1996.
26. Hardell, L. et al. Use of cellular telephones and the risk for brain tumours: a case-control study. *Int. J. Oncol.*, 15, 113,1999.
27. Szmigielski, S. Cancer mortality in subjects occupationally exposed to high-frequency (radio-frequency and microwaves) electromagnetic radiation. *Sci. Total Environ.*, 180, 9, 1996.
28. Moulder, J.E. et al. Cell phones and cancer: what is the evidence for a connection? *Radiat. Res.*, 151, 513, 1999.
29. Royal Society of Canada Expert Panel Report A review of the potential health risks of radio-frequency fields from wireless telecommunication devices. An Expert Panel Report prepared at the request of the Royal Society of Canada for Health Canada, Ottawa, RSC.EPR 99-1, 1999.
30. Dolk, H. et al. Cancer incidence near radio and television transmitters in Great Britain. 1: Sutton Coldfield transmitter. *Am. J. Epidemiol.*, 145, 1, 1997a.
31. Dolk, H. et al. Cancer incidence near radio and television transmitters in Great Britain: All high power transmitters. *Am. J. Epidemiol.*, 145, 10, 1997b.
32. Mild, K.H. et al. Comparisons of symptoms experienced by users of analogue and digital mobile phones: a Swedish-Norwegian epidemiological study. Report for the National Institute for Working Life. 1998.
33. Mickley, G.A. et al. Disruption of a putative working memory task and selective expression of brain c-fos following microwave-induced hyperthermia. *Physiol. Behav.*, 55, 1029, 1994.
34. Lai, H., Horitaa, A. and Guy, A.W. Microwave irradiation affects radial-arm maze performance in the rat. *Bioelectromagnetics*, 15, 95, 1994.

35. Sienkiewicz, Z.J. et al. Low-level exposure to pulsed 900 MHz microwave radiation does not cause deficits in the performance of spatial memory task in mice. *Bioelectromagnetics*, 21, 151, 2000.
36. Urban, P., Lukas, E. and Roth, Z. Does acute exposure to the electromagnetic field emitted by a mobile phone influence visual evoked potentials? A pilot study. *Centr. Eur. J. Public Health*, 6, 288, 1998.
37. Eulitz, C. et al. Mobile phones modulate response pattenrs of human brain activity. *NeuroReport*, 9, 3229, 1998.
38. Freude, G. et al. Effects of microwaves emitted by cellular phones on humans slow brain potentials. *Bioelectromagnetics*, 19, 384, 1998.
39. Borbely, A.A. et al. Pulsed high-frequency electromagnetic field affects human sleep and sleep electroencephalogram. *Neurosci. Lett.*, 275, 207, 1999.
40. Huber, R. et al. Exposure to pulsed high-frequency electromagnetic field during waking affects human sleep EEG. *NeuroReport*, 11 (15), 3321, 2000.
41. Preece, A.W. et al. Effect of a 915-MHz simulated mobile phone signal on cognitive function in man. *Int. J. Radiat. Biol.*, 75, 447, 1999.
42. Koivisto, M. et al. Effects of 902 MHz electromagnetic field emitted by cellular phones on response times in humans. *NeuroReport*, 11, 413, 2000.
43. Van Leeuwen, G.M.J. et al. Calculation of brain temperatures due to exposure to a mobile phone. *Phys. Med. Biol.*, 44, 2367, 1999.
44. de Pomerai, D., Daniells C. and David H. Non-thermal heat-shock response to microwaves. *Nature* 405(6785), 417, 2000.
45. MDA (Medical Devices Agency) Electromagnetic Compatibility of Medical Devices with Mobile Communications. London, MDA DB 9702, 1997.
46. McKnight, A. J. and McKnight, A.S. The effect of cellular phones use upon driver attention. *Accid. Anal. Prev.*, 25, 259, 1993.
47. Alm, H. and Nilsson, L. Changes in driver behaviour as a function of handsfree mobile phones — a simulator study. *Accid. Anal. Prev.*, 26, 441, 1994.
48. Alm, H. and Nilsson, L. The effects of a mobile telephone task on driver behaviour in a car following situation. *Accid. Anal. Prev.*, 27, 707, 1995.
49. Lamble, D. et al. Cognitive load and detection in thresholds in car following situations: safety implications for using mobile (cellular) telephones while driving. *Accid. Anal. Prev.*, 31, 617, 1999.
50. NRPB GS11 Guidance as to restrictions on exposure to time varying electromagnetic fields and the 1988 recommendations of the international non-ionising radiation committee. HMSO ISBN 0 85951 314 9, 1988.
51. Statement by NRPB Restrictions on human exposure to static and time-varying electromagnetic fields and radiation. Doc NRPB 4(5), 1993.
52. ICNIRP Guidelines for limiting exposure to time-varying electric, magnetic and electromagnetic fields (up to 300GHz) *Health Phys.*, 74(4) 494, 1998.
53. FCC (Federal Communications Commission) Guidelines for Evaluating the Environmental Effects of Radio-Frequency Radiation. Report and Order, ET Docket 93-62, FCC 96-326, 61 *Fed. Reg.*, 41006, 1996.
54. NCRP (National Council on Radiation Protection and Measurements). Biological effects and exposure criteria for radio-frequency electromagnetic fields. NCRP Report No. 86, NCRP, Bethesda, 1986.
55. IEEE (Institute of Electrical and Electronics Engineers) C95.3-1991 IEEE recommended practice for the measurement of potentially hazardous electromagnetic fields-RF and Microwave, Institute of Electrical and Electronics Engineers, New York, 1997.
56. FCC (Federal Communications Commission). Evaluating compliance with FCC guidelines for human exposure radio-frequency electromagnetic fields, FCC OET Bulletin 65 (edition 97-01), 1997.
57. ICNIRP (International Commission on Non-ionising Radiation Protection) Guidelines for limiting exposure to time-varying electric, magnetic and electromagnetic fields (up to 300 GHz). *Health Phys.*, 74, 494–522, 1998.

58. Royal Society of Canada Expert Panel Report. A review of the potential health risks of radiofrequency fields from wireless telecommunication devices. Ottawa, RSC.EPR 99-1, 1999.
59. http://www.rsc.ca/english/RFreport.html
60. The Stewart Report (IEGMP — Independent Expert Group on Mobile Phones), in *Mobile Phones and Health*, NPL, 2000.
61. Science and Technology Committee. Third report to scientific advisory system, in *Mobile Phones and Health*, Volume II. London: The Stationery Office, 1999.
62. http://www.iegmp.org.uk
63. http://www.iee.org.uk/PAB/Bio_effects/bio_sum2.htm

Index

A

C

D

E

F

G

H

I

J

K

L

M

N

O

P

T

U

V

W

X

Y

Z